VITAMINS AND MINERALS IN NEUROLOGICAL DISORDERS

VITAMINS AND MINERALS IN NEUROLOGICAL DISORDERS

Edited by

COLIN R. MARTIN

Professor of Clinical Psychobiology and Applied Psychoneuroimmunology & Clinical Director: Institute for Health and Wellbeing, University of Suffolk, Ipswich, United Kingdom

VINOOD B. PATEL

Reader in Clinical Biochemistry, University of Westminster, London, United Kingdom

VICTOR R. PREEDY

Emeritus Professor of Nutritional Biochemistry, King's College London, London, United Kingdom
Professor of Clinical Biochemistry and Pathology (Hon), King's College Hospital, London, United Kingdom

ELSEVIER

ACADEMIC PRESS
An imprint of Elsevier

Academic Press is an imprint of Elsevier
125 London Wall, London EC2Y 5AS, United Kingdom
525 B Street, Suite 1650, San Diego, CA 92101, United States
50 Hampshire Street, 5th Floor, Cambridge, MA 02139, United States
The Boulevard, Langford Lane, Kidlington, Oxford OX5 1GB, United Kingdom

ISBN 978-0-323-89835-5

For information on all Academic Press publications
visit our website at https://www.elsevier.com/books-and-journals

Publisher: Nikki P. Levy
Acquisitions Editor: Natalie Farra
Editorial Project Manager: Timothy J. Bennett
Production Project Manager: Sreejith Viswanathan
Cover Designer: Miles Hitchen

Typeset by STRAIVE, India

Printed in United States of America

Last digit is the print number 9 8 7 6 5 4 3 2 1

Dedication

This book is dedicated to my wonderful daughter, Dr. Caragh Brien, a caring, compassionate, and evidence-based junior doctor, of whom I am so incredibly proud.

Colin R. Martin

Contents

PART X Parkinson's disease

Contributors

Paul A. Adlard
Synaptic Neurobiology Laboratory, The Florey Institute of Neuroscience and Mental Health, The University of Melbourne, Melbourne, VIC, Australia

Hassan Imran Afridi
National Centre of Excellence in Analytical Chemistry, University of Sindh, Jamshoro, Pakistan

Renu Agarwal
Department of Pharmacology and Therapeutics, School of Medicine, International Medical University, Kuala Lumpur, Malaysia

Shiek Fareeth Ahmed
Faculty of Allied Health Sciences, Chettinad Hospital and Research Institute [CHRI], Chettinad Academy of Research and Education [CARE], Kelambakkam, Tamilnadu, India

Ansab Akhtar
Division of Pharmacology, University Institute of Pharmaceutical Sciences, Panjab University, Chandigarh, India

Sulaiman Al Mohaish
Department of Pharmacotherapy & Outcomes Science, Virginia Commonwealth University School of Pharmacy, Richmond, VA, United States

Gerald Harvey Anderson
Department of Nutritional Sciences; Department of Physiology, Faculty of Medicine, University of Toronto, Toronto, ON, Canada

Angela
Department of Neurology, Faculty of Medicine, Duta Wacana Christian University, Yogyakarta, Indonesia

Marcos Avila-Rodriguez
Department of Clinical Science, University of Tolima, Tolima, Colombia

Nehad M. Ayoub
Department of Clinical Pharmacy, Faculty of Pharmacy, Jordan University of Science and Technology, Irbid, Jordan

Md. Abdul Aziz
Department of Pharmacy, Noakhali Science and Technology University, Noakhali, Bangladesh

Christos Bakirtzis
Multiple Sclerosis Center, B' Department of Neurology, AHEPA University Hospital, Aristotle University of Thessaloniki, Thessaloniki, Greece

Desirée Bartolini
Department of Pharmaceutical Sciences; Department of Medicine and Surgery, Unit of Human, Clinical and Forensic Anatomy, University of Perugia, Perugia, Italy

Maria Angela Bellomo-Brandão
Department of Pediatrics, School of Medical Sciences, University of Campinas, Campinas, São Paulo, Brazil

Alpha A. (Berry) Fowler, III
Division of Pulmonary Disease and Critical Care Medicine, Virginia Commonwealth University School of Medicine, Richmond, VA, United States

Michele Betti
Department of Biomolecular Science, Urbino University, Urbino, Italy

Sunita Bijarnia-Mahay
Department of Clinical & Metabolic Genetics, Institute of Medical Genetics & Genomics, Sir Ganga Ram Hospital, New Delhi, India

Julita Maria Pereira Borges
Department of Science and Technology, State University of Southwest of Bahia, Jequié, Bahia, Brazil

Jamilah Borjac
Department of Biological Sciences, Faculty of Science, Beirut Arab University, Debbieh, Lebanon

Gretchen M. Brophy
Department of Pharmacotherapy & Outcomes Science, Virginia Commonwealth University School of Pharmacy, Richmond, VA, United States

Carlos-Alberto Calderón-Ospina
Center for Research in Genetics and Genomics (CIGGUR), GENIUROS Research Group, Pharmacology Unit, Department of Biomedical Sciences, School of Medicine and Health Sciences, Universidad del Rosario, Bogotá, Colombia

Iana Bantim Felicio Calou
Coordenação de Nutrição, Rua Cícero Duarte, Federal University of Piauí (UFPi), Picos, Piauí, Brazil

Larissa Marcely Gomes Cassiano
Graduate Program of Biochemistry and Immunology, Federal University of Minas Gerais; Neurogenomics/Immunopatology, René Rachou Institute, Oswaldo Cruz Foundation (FIOCRUZ), Belo Horizonte, Brazil

Lyubov Chernova
The Laboratory of Molecular Dietetics, The Center for Bioelementology and Human Ecology, Sechenov University, Moscow, Russian Federation

Aakanksha Choudhary
Institute of Home Economics, University of Delhi, New Delhi, India

Ajay Choudhary
Department of Neurosurgery, Atal Bihari Vajpayi Institute of Medical Sciences, Dr. R.M. L. Hospital, New Delhi, India

McCoy Clementson
Biomedical Sciences, College of Graduate Studies; College of Osteopathic Medicine, Midwestern University, Glendale, AZ, United States

Roney Santos Coimbra
Neurogenomics/Immunopatology, René Rachou Institute, Oswaldo Cruz Foundation (FIOCRUZ), Belo Horizonte, Brazil

David C. Consoli
Division of Diabetes, Endocrinology, and Metabolism, Vanderbilt University Medical Center, Nashville, TN, United States

Maria Camila Cortes-Albornoz
Neuroscience Research Group (NEUROS) – Centro Neurovitae, Neuroscience Unit, School of Medicine and Health Sciences, Universidad del Rosario, Bogotá, Colombia

Marina da Silva Oliveira
Neurogenomics/Immunopatology, René Rachou Institute, Oswaldo Cruz Foundation (FIOCRUZ), Belo Horizonte, Brazil

Maria Daglas
Synaptic Neurobiology Laboratory, The Florey Institute of Neuroscience and Mental Health, The University of Melbourne, Melbourne, VIC, Australia

Jan Damoiseaux
Central Diagnostic Laboratory, Maastricht University Medical Center, Maastricht, The Netherlands

Efthimios Dardiotis
Department of Neurology, Faculty of Medicine, School of Health Sciences, University of Thessaly, Larissa, Greece

Antônio-Carlos G. de Almeida
Departamento de Engenharia de Biossistemas, Laboratório de Neurociências Experimental e Computacional, Universidade Federal de São João del-Rei, São João del-Rei, Minas Gerais, Brazil

Glauce Socorro de Barros Viana
Departamento de Fisiologia e Farmacologia, Rua Coronel Nunes de Melo, Federal University of Ceará (UFC), Fortaleza, Brazil

Daniel T. Dibaba
Cancer Research And Biostatistics (CRAB), Seattle, WA, United States

Namrata Dudha
Department of Biotechnology and Microbiology, School of Sciences, Noida International University-NIU, Gautam Budh Nagar, Uttar Pradesh, India

Débora Estadella
Departamento de Biociência, Rua Silva Jardim, Federal University of São Palo (UNIFESP), Santos, São Paulo, Brazil

Gorka Fernandez
Department of Neurology, Neurometabolism Unit, AP-HP Pitié Salpétrière Hospital, Paris, France

Eduardo Galembeck
Institute of Biology, Department of Biochemistry, State University of Campinas-São Paulo, Campinas, São Paulo, Brazil

Francesco Galli
Department of Pharmaceutical Sciences, University of Perugia, Perugia, Italy

Parisa Gazerani
Department of Life Sciences and Health, Faculty of Health Sciences, Oslo Metropolitan University, Oslo, Norway; Department of Health Science and Technology, Faculty of Medicine, Aalborg University, Aalborg, Denmark

Valeria C. Gonçalves
Departamento de Neurologia e Neurocirurgia, Escola Paulista de Medicina, Universidade Federal de São Paulo, São Paulo, Brazil

Rola Hammoud
Lunenfeld-Tanenbaum Research Institute, Dr. Daniel J. Drucker Lab, Mount Sinai Hospital, Toronto, ON, Canada

Fiona E. Harrison
Division of Diabetes, Endocrinology, and Metabolism, Department of Medicine, Vanderbilt University Medical Center, Nashville, TN, United States

Yosuke Horikoshi
Division of Biochemistry, Department of Pathophysiological and Therapeutic Sciences, School of Medicine, Faculty of Medicine, Tottori University, Yonago, Japan

Igor Iezhitsa
Department of Pharmacology and Therapeutics, School of Medicine, International Medical University, Kuala Lumpur, Malaysia; Department of Pharmacology and Bioinformatics, Volgograd State Medical University, Volgograd, Russia

M. Indira
Department of Biochemistry, University of Kerala, Thiruvananthapuram, India

Mohammad Safiqul Islam
Department of Pharmacy, Noakhali Science and Technology University, Noakhali, Bangladesh

Nafisa M. Jadavji
Biomedical Sciences, College of Graduate Studies; College of Veterinary Medicine, Midwestern University, Glendale, AZ, United States; Department of Neuroscience, Carleton University, Ottawa, ON, Canada

V. Janakiraman
Faculty of Allied Health Sciences, Chettinad Hospital and Research Institute [CHRI], Chettinad Academy of Research and Education [CARE], Kelambakkam, Tamilnadu, India

Jesse Jauhal
Biomedical Sciences, College of Graduate Studies, Midwestern University, Glendale, AZ, United States

Ahmad Jayedi
Food Safety Research Center (Salt), Semnan University of Medical Sciences, Semnan, Iran

Sydney M.A. Juan
Synaptic Neurobiology Laboratory, The Florey Institute of Neuroscience and Mental Health, The University of Melbourne, Melbourne, VIC, Australia

Amal Kaddoumi
Department of Drug Discovery and Development, Harrison College of Pharmacy, Auburn University, Auburn, AL, United States

Ka Kahe
Department of Obstetrics and Gynecology and Department of Epidemiology, Columbia University, New York, NY, United States

Santanu Kaity
Formulation Development Service, Mylan Laboratories Limited, Hyderabad, India

Markos Kashiouris
Division of Pulmonary Disease and Critical Care Medicine, Virginia Commonwealth University School of Medicine, Richmond, VA, United States

Kaviraj Kaushik
Department of Neurosurgery, Atal Bihari Vajpayi Institute of Medical Sciences, Dr. R.M. L. Hospital, New Delhi, India

Raida W. Khalil
Department of Clinical Nutrition and Dietetics, Faculty of Science, University of Philadelphia, Amman, Jordan

Woon-Puay Koh
Healthy Longevity Translational Research Programme, Yong Loo Lin School of Medicine, National University of Singapore, Singapore, Singapore

Ruslan Kubant
Department of Nutritional Sciences, Faculty of Medicine, University of Toronto, Toronto, ON, Canada

Anoop Kumar
Department of Pharmacology, Delhi Pharmaceutical Sciences & Research University (DPSRU), New Delhi, India

Ashok Kumar
Department of Neurosurgery, Atal Bihari Vajpayi Institute of Medical Sciences, Dr. R.M. L. Hospital, New Delhi, India

Stefan W. Leichtle
Division of Acute Care Surgical Services, Department of Surgery, Virginia Commonwealth University School of Medicine, Richmond, VA, United States

Jimmy Li
Department of Neurology, University of Sherbrooke, Sherbrooke, QC, Canada

Ioannis Liampas
Department of Neurology, Faculty of Medicine, School of Health Sciences, University of Thessaly, Larissa, Greece

Gianfranco Liguri
Department of Biochemistry, Faculty of Medicine and Surgery, University of Florence, Florence, Italy

Marcela Almeida Linden
Department of Pediatrics, School of Medical Sciences, University of Campinas, Campinas, São Paulo, Brazil

Harjit Singh Mahay
Department of Critical Care, Fortis Hospital, New Delhi, India

S. Mahima
Faculty of Allied Health Sciences, Chettinad Hospital and Research Institute [CHRI], Chettinad Academy of Research and Education [CARE], Kelambakkam, Tamilnadu, India

Giada Marcantonini
Department of Pharmaceutical Sciences, University of Perugia, Perugia, Italy

Rita Marinelli
Department of Pharmaceutical Sciences, University of Perugia, Perugia, Italy

Amanda L. Marino
Division of Diabetes, Endocrinology, and Metabolism, Vanderbilt University Medical Center, Nashville, TN, United States

Angela Marino
Department of Chemical, Biological, Pharmaceutical and Environmental Sciences, University of Messina, Messina, Italy

Alaa Matar
Department of Biological Sciences, Faculty of Science, Beirut Arab University, Debbieh, Lebanon

Tatsuya Matsura
Department of Nutritional Sciences, Faculty of Human Ecology, Yasuda Women's University, Hiroshima, Japan

Anna Migni
Department of Pharmaceutical Sciences, University of Perugia, Perugia, Italy

Haitham S. Mohammed
Biophysics Department, Faculty of Science, Cairo University, Giza, Egypt

Yann Nadjar
Department of Neurology, Neurometabolism Unit, AP-HP Pitié Salpétrière Hospital, Paris, France

Saritha S. Nair
R&D Division, Sipwise Beverages Pvt. Ltd., Chennai, India

Kazuhiro Nakaso
Division of Biochemistry, Department of Pathophysiological and Therapeutic Sciences, School of Medicine, Faculty of Medicine, Tottori University, Yonago, Japan

Mauricio Orlando Nava-Mesa
Neuroscience Research Group (NEUROS) – Centro Neurovitae, Neuroscience Unit, School of Medicine and Health Sciences, Universidad del Rosario, Bogotá, Colombia

Adejoke Y. Onaolapo
Department of Anatomy, Behavioural Neuroscience Unit, Neurobiology Subdivision, Ladoke Akintola University of Technology, Ogbomoso, Oyo, Nigeria

Olakunle J. Onaolapo
Department of Pharmacology, Behavioural Neuroscience Unit, Neuropharmacology Subdivision, Ladoke Akintola University of Technology, Ogbomoso, Oyo, Nigeria

Ana María Paez-Hurtado
Neuroscience Research Group (NEUROS) – Centro Neurovitae, Neuroscience Unit, School of Medicine and Health Sciences, Universidad del Rosario, Bogotá, Colombia

Emanuela Pannia
Department of Molecular Genetics, Dr. Jim Dowling's Research Group, Faculty of Medicine, University of Toronto; Program in Genetics and Genome Biology, Dr. Jim Dowling's Research Group, The Hospital for Sick Children, Toronto, ON, Canada

Vinood B. Patel
School of Life Sciences, University of Westminster, London, United Kingdom

Rizaldy Taslim Pinzon
Department of Neurology, Faculty of Medicine, Duta Wacana Christian University, Yogyakarta, Indonesia

Andryawan Wahyu Pradana
Department of Neurology, Faculty of Medicine, Duta Wacana Christian University, Yogyakarta, Indonesia

Victor R. Preedy
Department of Nutritional Sciences, Faculty of Life Science and Medicine, School of Life Course and Population Sciences, King's College London, London, United Kingdom

S.H. Priyanka
Department of Biochemistry, University of Kerala, Thiruvananthapuram, India

Rajkumar Rajendram
College of Medicine, King Saud bin Abdulaziz University for Health Sciences; Department of Medicine, King Abdulaziz Medical City, King Abdullah International Medical Research Center, Ministry of National Guard Health Affairs, Riyadh, Saudi Arabia

Yugandhar P. Reddy
Department of Zoology, Government Degree College, Kurnool, Andhra Pradesh, India

Mario Rende
Department of Medicine and Surgery, Unit of Human, Clinical and Forensic Anatomy, University of Perugia, Perugia, Italy

Daniel Araki Ribeiro
Departamento de Biociência, Rua Silva Jardim, Federal University of São Palo (UNIFESP), Santos, São Paulo, Brazil

Dennis J. Rivet, II
Harold F. Young Neurosurgical Center, Department of Neurosurgery, Virginia Commonwealth University School of Medicine, Richmond, VA, United States

Gianluca Rizzo
Independent Researcher, Messina, Italy

Diego Alejandro Rodríguez-Gomez
Neuroscience Research Group (NEUROS) – Centro Neurovitae, Neuroscience Unit, School of Medicine and Health Sciences, Universidad del Rosario, Bogotá, Colombia

Mauro Ciro Antonio Rongioletti
Department of Laboratory Medicine, Research and Development Division, San Giovanni Calibita Fatebenefratelli Hospital, Rome, Italy

Jency Roshni
Faculty of Allied Health Sciences, Chettinad Hospital and Research Institute [CHRI], Chettinad Academy of Research and Education [CARE], Kelambakkam, Tamilnadu, India

Érika Pereira Rubio
Department of Science and Technology, State University of Southwest of Bahia, Jequié, Bahia, Brazil

Payam Saadat
Mobility Impairment Research Center, Health Research Institute, Babol University of Medical Sciences, Babol, Iran

Sangeeta Pilkhwal Sah
Division of Pharmacology, University Institute of Pharmaceutical Sciences, Panjab University, Chandigarh, India

Arushi Gahlot Saini
Department of Pediatrics, Postgraduate Institute of Medical Education and Research (PGIMER), Chandigarh, India

Kiarash Saleki
Student Research Committee; USERN Office, Babol University of Medical Sciences, Babol, Iran

Melissa Sandler
Department of Physical Medicine and Rehabilitation, Virginia Commonwealth University School of Medicine, Richmond, VA, United States

Gesivaldo Santos
Department of Biological Sciences, State University of Southwest of Bahia, Jequié, Bahia, Brazil

Carla A. Scorza
Departamento de Neurologia e Neurocirurgia, Escola Paulista de Medicina, Universidade Federal de São Paulo, São Paulo, Brazil

Fulvio A. Scorza
Departamento de Neurologia e Neurocirurgia, Escola Paulista de Medicina, Universidade Federal de São Paulo, São Paulo, Brazil

Sakineh Shab-Bidar
Department of Community Nutrition, School of Nutritional Sciences and Dietetics, Tehran University of Medical Sciences, Tehran, Iran

Varun Kumar Sharma
Department of Biotechnology and Microbiology, School of Sciences, Noida International University-NIU, Gautam Budh Nagar, Uttar Pradesh, India

Vasileios Siokas
Department of Neurology, Faculty of Medicine, School of Health Sciences, University of Thessaly, Larissa, Greece

Anastasia Skalnaya
Laboratory of Invasive Neurointerfaces of Research Institute "Technobiomed", A.I. Yevdokimov Moscow State University of Medicine and Dentistry, Moscow, Russian Federation

Anatoly Skalny
The Center for Bioelementology and Human Ecology, Sechenov University, Moscow, Russian Federation

Joost Smolders
Departments of Neurology and Immunology, MS Center ErasMS, Erasmus Medical Center, Rotterdam; Neuroimmunology Research Group, Netherlands Institute for Neuroscience, Amsterdam, The Netherlands

J. David Spence
Neurology & Clinical Pharmacology, Western University; Stroke Prevention & Atherosclerosis Research Centre, Robarts Research Institute, London, ON, Canada

Rosanna Squitti
Molecular Markers Laboratory, IRCCS Istituto Centro San Giovanni di Dio Fatebenefratelli, Brescia; Department of Laboratory Medicine, Research and Development Division, San Giovanni Calibita Fatebenefratelli Hospital, Rome, Italy

Febrina Eva Susanto
Department of Neurology, Faculty of Medicine, Duta Wacana Christian University, Yogyakarta, Indonesia

Dènahin Hinnoutondji Toffa
Department of Neurosciences, University of Montreal Hospital Research Centre, Montreal, QC, Canada

Lomas Kumar Tomar
Department of Biotechnology, VSPG(PG) College, CCS University, Meerut, Uttar Pradesh, India

Phan H. Truong
Synaptic Neurobiology Laboratory, The Florey Institute of Neuroscience and Mental Health, The University of Melbourne, Melbourne, VIC, Australia

Charu Tyagi
Department of Biotechnology and Microbiology, School of Sciences, Noida International University-NIU, Gautam Budh Nagar, Uttar Pradesh, India

Vanessa Veronica
Department of Neurology, Faculty of Medicine, Duta Wacana Christian University, Yogyakarta, Indonesia

Milena de Barros Viana
Departamento de Biociência, Rua Silva Jardim, Federal University of São Palo (UNIFESP), Santos, São Paulo, Brazil

Vishal Yajnik
Division of Anesthesia Critical Care, Department of Anesthesiology, Virginia Commonwealth University School of Medicine, Richmond, VA, United States

Gaku Yamanaka
Department of Pediatrics and Adolescent Medicine, Tokyo Medical University, Tokyo, Japan

Neil Victor Yang
Department of Nutritional Sciences and Toxicology, Comparative Biochemistry Graduate Group, Dr. Ronald Krauss's Research Group, University of California, Berkeley, CA, United States

Ariel F. Ying
Health Services and Systems Research Programme, Duke-National University of Singapore Medical School, Singapore, Singapore

Linda Zatini
Department of Pharmaceutical Sciences, University of Perugia, Perugia, Italy

Preface

Neurological disorders have an enormous impact on the individual, family unit, community, and society at large. For example, the World Health Organization reports that on a global basis, 6 million people die from stroke each year, 50 million people have epilepsy, and a similar number have dementia of which the most common is Alzheimer's disease. Half of the world's adult population had a headache at least once in the last year.

Many of these neurological conditions have a central feature: poor or good diet and nutrition may either worsen or lessen symptoms. Some neurological conditions arise directly as a consequence of poor nutrition, or the ingestion of toxic dietary components such as alcohol.

Much of the research on diet and nutrition is focused around single neurological diseases. However, consideration must be taken of the fact that the information relating to one neurological condition may well be relevant to other conditions.

For example, the Mediterranean diet has been investigated in amyotrophic lateral sclerosis, but not in relation to headache and migraine. On the other hand, the Dietary Approaches to Stop Hypertension diet has been studied in relation to headache but not amyotrophic lateral sclerosis.

It is therefore possible that information relating to one neurological disease may very well have relevance to understanding or treating other neurological diseases. Furthermore, studies showing the beneficial effects of plant or natural extracts provide the foundation for further rigorous studies in clinical trials. Crossing this trans-speciality divide is somewhat difficult as many textbooks are focused on single neurological conditions. There is also a trans-intellectual divide that makes some texts extremely difficult for the novice or student to comprehend and understand.

The editors have sought to overcome these limitations and advance the understanding of neurosciences and nutrition with a series of three books collectively called "Nutrition and Neurological Disorders".

The three books are as follows:

Book 1: Diet and Nutrition in Neurological Disorders

Book 2: Vitamins and Minerals in Neurological Disorders

Book 3: Treatments, Nutraceuticals, Supplements, and Herbal Medicine in Neurological Disorders.

Each chapter has the following sections that we consider unique:

- *Abstract (published online)*
- *Applications to other neurological conditions*
- *Other components of interest*

- *Key facts*
- *Mini-dictionary of terms*
- *Summary points*

The section *Applications to other neurological conditions* is particularly important as it highlights the translational aspect of nutritional research. The section *Other components of interest* emphasizes the fact that diet and nutrition are complex areas and that there may be other components that can confer health benefits or even be detrimental. The section *Key facts* gives important information about individual components in each chapter. The section *Mini-dictionary of terms* is suited for both nonexperts and those working in other fields or areas. The section *Summary points* encapsulates the entire chapter in brief sets of simple sentences.

Vitamins and Minerals in Neurological Disorders is divided into 13 parts as follows:

- Part I: Alzheimer's disease and dementias
- Part II: Amyotrophic lateral sclerosis
- Part III: Brain injury
- Part IV: Cerebral palsy
- Part V: Dietary neurotoxins
- Part VI: Epilepsy
- Part VII: Headaches and migraines
- Part VIII: Multiple sclerosis
- Part IX: Neuroinflammation
- Part X: Parkinson's disease
- Part XI: Peripheral neuropathy
- Part XII: Prenatal effects and neurodevelopment
- Part XIII: Stroke

This book is designed for research and teaching purposes. It is suitable for nutritionists, dietitians, neurologists, physicians, health scientists, healthcare workers, pharmacologists, and researchers. The audience also includes federal and state program directors. This book is valuable as a personal reference book and is also suitable for academic libraries that covers the domains of nutrition, neurology, or health sciences. Contributions are from leading national and international experts, including those from world-renowned institutions. The book is suitable for undergraduates, postgraduate, lecturers, and academic professors.

Colin R. Martin
Vinood B. Patel
Victor R. Preedy

CHAPTER 1

Vitamins and minerals in the context of reference values and requirements

Rajkumar Rajendram[a,b], Vinood B. Patel[c], and Victor R. Preedy[d]
[a]College of Medicine, King Saud bin Abdulaziz University for Health Sciences, Riyadh, Saudi Arabia
[b]Department of Medicine, King Abdulaziz Medical City, King Abdullah International Medical Research Center, Ministry of National Guard Health Affairs, Riyadh, Saudi Arabia
[c]School of Life Sciences, University of Westminster, London, United Kingdom
[d]Department of Nutritional Sciences, Faculty of Life Science and Medicine, School of Life Course and Population Sciences, King's College London, London, United Kingdom

Abbreviations

AI adequate intake
RDA recommended dietary allowance
UL tolerable upper intake

Dietary reference values

Guidelines for the intake of micro- and macronutrients or different foods are used by various countries, organizations, and statuary bodies. Essentially, these guidelines allude to targets that populations should achieve for good health. The guidelines may be food-based (food pyramids, for example) or nutrient-based (specific identified nutrients) or combination of both. Some guidelines also include precautionary notes, relating to the effects of ingesting excessive amounts. In the United Kingdom, the guidelines are encompassed in the Dietary Reference Values For Food, Energy And Nutrients (Department of Health, 1991). These national and international guidelines are occasionally simplified or amalgamated. One example includes the Dietary Guidelines for Americans Departments of Agriculture and Health and Human Services (2020). Tables 1 and 2 relate to minerals, and Tables 3 and 4 relate to vitamins. They show variations in terms of age and gender. For example, there is a marked transition in the requirements for iron in females between 19 and 30 and 51 + years representing the impact of menopause and cessation of menstruation (Table 2).

The tables are not exhaustive as for example, neither copper, manganese, nor selenium are mentioned. All guidelines have material relating to life stages such as pregnancy and lactation and the requirements of newborns, infants, and the elderly. However, different countries have guidelines that are distinct. For example, in France, the guidance is entitled Recommandations Alimentaires Du Programme National Nutrition Santé (The French National Nutrition And Health Program's Dietary Guidelines). The Food and

Vitamins and Minerals in Neurological Disorders
https://doi.org/10.1016/B978-0-323-89835-5.00044-2

Table 1 Daily nutritional goals for minerals in the younger population.

	Source of goal	Child 2–3 y	Female 4–8 y	Female 9–13 y	Female 14–18 y	Male 4–8 y	Male 9–13 y	Male 14–18 y
Calcium (mg)	RDA	700	1000	1300	1300	1000	1300	1300
Iron (mg)	RDA	7	10	8	15	10	8	11
Magnesium (mg)	RDA	80	130	240	360	130	240	410
Phosphorus (mg)	RDA	460	500	1250	1250	500	1250	1250
Potassium (mg)	AI	2000	2300	2300	2300	2300	2500	3000
Sodium (mg)	CDRR	1200	1500	2300	2300	1500	1800	2300
Zinc (mg)	RDA	5	5	8	9	5	8	11

These values are based on dietary reference intakes and dietary guidelines recommendations. *AI*, adequate intake; *CDRR*, chronic disease risk reduction level; *RDA*, recommended dietary allowance; *y*, years of age.
Data derived from U.S. Departments of Agriculture and Health and Human Services, (2020). *U.S. Department of Agriculture and U.S. Department of Health and Human Services Dietary guidelines for Americans, 2020–2025* (9th ed.). December 2020. Available at https://www.dietaryguidelines.gov/.

Table 2 Daily nutritional goals for minerals in the older population.

	Source of goal	Female 19–30 y	Female 31–50 y	Female 51+y	Male 19–30 y	Male 31–50 y	Male 51+y
Calcium (mg)	RDA	1000	1000	1200	1000	1000	1000 [*]
Iron (mg)	RDA	18	18	8	8	8	8
Magnesium (mg)	RDA	310	320	320	400	420	420
Phosphorus (mg)	RDA	700	700	700	700	700	700
Potassium (mg)	AI	2600	2600	2600	3400	3400	3400
Sodium (mg)	CDRR	2300	2300	2300	2300	2300	2300
Zinc (mg)	RDA	8	8	8	11	11	11

These values are based on dietary reference intakes and dietary guidelines recommendations. *AI*, adequate intake; *CDRR*, chronic disease risk reduction level; *RDA*, recommended dietary allowance; *y*, years of age. [*] Calcium RDA for males 71+y is 1200 mg.
Data derived from U.S. Departments of Agriculture and Health and Human Services, (2020). *U.S. Department of Agriculture and U.S. Department of Health and Human Services Dietary guidelines for Americans, 2020–2025* (9th ed.). December 2020. Available at https://www.dietaryguidelines.gov/.

Agricultural Organization (FAO) has tabulated food-based guidelines for approaching 100 countries (FAO, 2022).

Guideline values can be used in four situations: (1) assessing adequacy of micronutrient intake in individuals; (2) assessing at-risk subgroups; (3) prescribing diets; and (4), food labeling. Of relevance to neurological disorders are conditions where either vitamins (e.g., Linden & Bellomo-Branda, 2023; Marinelli, 2023; Kaddoumi, 2023; Leichtle et al., 2023; Nadjar & Fernandez, 2023), minerals (e.g., Matar & Borjac, 2023; Choudhary, 2023; Hinnoutondji Toffa & Li, 2023; Mohammed, 2023; Santos, 2023; Squitti, Rongioletti, & Liguri, 2023), or both (e.g., Afridi, 2023; Islam, 2023; Riizzo & Marino, 2023) are investigated to ultimately understand the disease process and/or enhance patient well-being.

Table 3 Daily nutritional goals for vitamins in the younger population.

	Source of goal	Child 2–3 y	Female 4–8 y	Female 9–13 y	Female 14–18 y	Male 4–8 y	Male 9–13 y	Male 14–18 y
Vitamin A (mcg RAE)	RDA	300	400	600	700	400	600	900
Vitamin E (mg AT)	RDA	6	7	11	15	7	11	15
Vitamin D (IU)	RDA	600	600	600	600	600	600	600
Vitamin C (mg)	RDA	15	25	45	65	25	45	75
Thiamin (mg)	RDA	0.5	0.6	0.9	1.0	0.6	0.9	1.2
Riboflavin (mg)	RDA	0.5	0.6	0.9	1.0	0.6	0.9	1.3
Niacin (mg)	RDA	6	8	12	14	8	12	16
Vitamin B-6 (mg)	RDA	0.5	0.6	1.0	1.2	0.6	1.0	1.3
Vitamin B-12 (mcg)	RDA	0.9	1.2	1.8	2.4	1.2	1.8	2.4
Choline (mg)	AI	200	250	375	400	250	375	550
Vitamin K (mcg)	AI	30	55	60	75	55	60	75
Folate (mcg DFE)	RDA	150	200	300	400	200	300	400

These values are based on dietary reference intakes and dietary guidelines recommendations. *AI*, adequate intake; *AT*, alpha-tocopherol; *DFE*, dietary folate equivalent; *IU*, international units; *RAE*, retinol activity equivalents; *RDA*, recommended dietary allowance; *y*, years of age.
Data derived from U.S. Departments of Agriculture and Health and Human Services, (2020). *U.S. Department of Agriculture and U.S. Department of Health and Human Services Dietary guidelines for Americans, 2020–2025* (9th ed.). December 2020. Available at https://www.dietaryguidelines.gov/.

Caveats of the dietary reference values

It is important to note that dietary reference values are designed for populations, rather than individuals. They are for healthy people, i.e., those who are not ill or suffer from either communicable or noncommunicable conditions, or injuries. It is assumed that requirements follow a normal distribution, except for iron (Department of Health, 1991).

Another important caveat is that the reference value of one micro- or macronutrient presupposes that the intake of all other micro- and macronutrients is met (Department of Health, 1991). Also, it is assumed that reference values relate to nutrients ingested with food, rather than via artificial support regimens such as enteral or parental nutrition. There are, of course, exceptions.

Reference values are used as guidelines to provide nutrients in enteral feeding regimens (Iacone et al., 2016). Another example of exceptions relate to the life-long provision of micronutrients to patients who have undergone bariatric (weight loss) surgery. After bariatric surgery, patients are prescribed commercially available mineral and vitamin supplements, which themselves are based on dietary reference values

Table 4 Daily nutritional goals for vitamins in the older population.

	Source of goal	Female 19–30y	Female 31–50y	Female 51+y	Male 19–30y	Male 31–50y	Male 51+y
Vitamin A (mcg RAE)	RDA	700	700	700	900	900	900
Vitamin E (mg AT)	RDA	15	15	15	15	15	15
Vitamin D (IU)	RDA	600	600	600 [~]	600	600	600 [~]
Vitamin C (mg)	RDA	75	75	75	90	90	90
Thiamin (mg)	RDA	1.1	1.1	1.1	1.2	1.2	1.2
Riboflavin (mg)	RDA	1.1	1.1	1.1	1.3	1.3	1.3
Niacin (mg)	RDA	14	14	14	16	16	16
Vitamin B-6 (mg)	RDA	1.3	1.3	1.5	1.3	1.3	1.7
Vitamin B-12 (mcg)	RDA	2.4	2.4	2.4	2.4	2.4	2.4
Choline (mg)	AI	425	425	425	550	550	550
Vitamin K (mcg)	AI	90	90	90	120	120	120
Folate (mcg DFE)	RDA	400	400	400	400	400	400

These values are based on dietary reference intakes and *dietary guidelines* recommendations. *AI*, adequate intake; *AT*, alpha-tocopherol; *DFE*, dietary folate equivalent; *IU*, international units; *RAE*, retinol activity equivalents; *RDA*, recommended dietary allowance; *y*, years of age. [~] Vitamin D RDA for males and females aged 71+years is 800 IU.
Data derived from U.S. Departments of Agriculture and Health and Human Services, (2020). *U.S. Department of Agriculture and U.S. Department of Health and Human Services Dietary guidelines for Americans, 2020–2025* (9th ed.). December 2020. Available at https://www.dietaryguidelines.gov/.

The dietary reference values for micronutrients are in constant stages of scientific scrutiny and advancement (Cashman & Kiely, 2013; Cashman, Ritz, Carlin, & Kennedy, 2022; Dumas et al., 2020; Martin, 2015; Powers, 2021).

A state of micronutrient equilibrium is attained when intake matches requirements (Fig. 1). It is sometimes assumed that vitamin and mineral deficiencies occur when intake alone is reduced. However, in pathological conditions, such as those pertaining to neurological disorders, at-risk situations arise when if any of the following situations occurs: intake is reduced and requirements are unchanged, intake is reduced and requirements are increased, intake is unchanged and requirements are increased, and intake is increased and requirements are increased to a much greater extent (Fig. 1).

The problem is that in any pathological condition, the requirements of minerals and vitamins are difficult to assess. There are also scenarios when there may be toxic effects or pro-oxidant activity, though adaptive mechanisms may come into play. For example, intake of exceedingly high levels of vitamin C may result in adaption. However, an abrupt transition to a lower intake may result in pseudo-vitamin C deficiency (Department of Health, 1991).

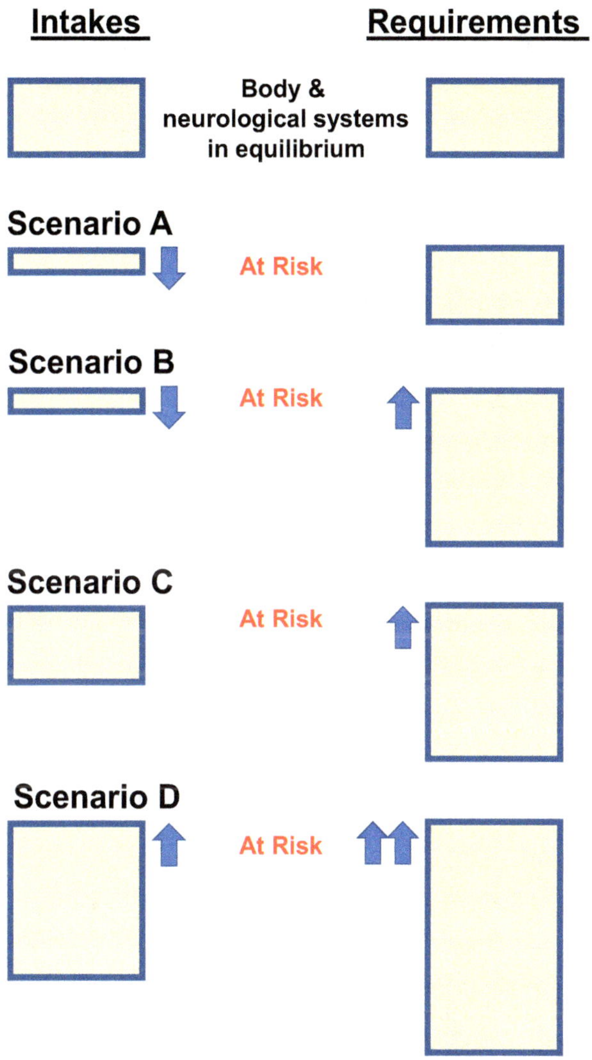

Fig. 1 Conceptual framework of different scenarios where the patient may be at risk of vitamins or minerals deficiency. In this framework of different scenarios, there is a metabolic equilibrium whereby intake meets requirements (upper component of figure). However, at-risk situations will occur if any of the following scenarios occurs: intake is reduced and requirements are unchanged (Scenario A), intake is reduced and requirements are increased (Scenario B), intake is unchanged and requirements are increased (Scenario C), and intake is increased and requirements are increased to a greater extent (Scenario D). *(From the authors.)*

In some circumstances, vitamins and minerals may act as pro-oxidants, for example, vitamin C (Mir & Khanday, 2022; Timoshnikov, Kobzeva, Polyakov, & Kontoghiorghes, 2020), alpha tocopherol (one of the 8 isomers of vitamin E) (Amel Kanane et al., 2019; Winterbone, Sampson, Saha, et al., 2007), and vitamin A/beta carotene (Palozza, 2003; Phan-Thi, Durand, Prost, Prost, & Waché, 2016). Pro-oxidant effects of minerals, alone or in complexes, may be exploited for therapeutic properties such as the work on copper (Rieber, 2020; Simunkova et al., 2021; Veiga, Alvarez, & Castellano, 2021). A comprehensive review of the pro-oxidant properties of micronutrients is found elsewhere (Sotler et al., 2019). From a physiological point, many adaptive mechanisms can counteract the high levels of minerals or vitamins that may arise. However, there is some imprecision in elucidating the levels at which adaptive mechanisms occur and toxicity and pathology arise.

Adaptions or toxic effects do not just occur when intake alone is increased (Fig. 2). They can also occur when intake and requirements are reduced, but requirements are reduced to a greater extent (Fig. 2). They may also arise when intake is unchanged but requirements are reduced or with increased intake and a lesser increase in requirements (Fig. 2). The ideal situation occurs when intake and requirements are matched (Fig. 3) or where the scenarios in Fig. 2 occur but adaptations are beneficial.

The dilemma for the dietetic and medical professions

There is a dilemma in interlinking the relationship between micronutrient status and neurological disorders, especially when there may be a reduced intake or a high metabolic demand. It raises the question as to how the dietitian, or medical doctor, prescribes vitamins and minerals to patients suffering from neurological disorders.

Many neurological disorders certainly increase the risk of malnutrition. Yet malnutrition in this cohort is frequently underrecognized and undertreated by healthcare professionals (Lamb, Parr, Lamb, & Warren, 2009). This is at least partly because the assessment of nutritional status in, and treatment of, patients with neurological disorders, such as stroke, is challenging (Sabbouh & Torbey, 2018). There is no universally accepted definition of malnutrition nor does the assessment of nutritional status have a gold standard method (Sabbouh & Torbey, 2018), though instruments such as "Malnutrition Universal Screening Tool" (MUST) and the Mini Nutritional Assessment Short-Form (MNA-SF) are of benefit in representing a starting point upon which all other screening studies can be compared. Screening for malnutrition in general across all medical conditions and cohorts is, however, less than desirable (Marino & Beattie, 2018).

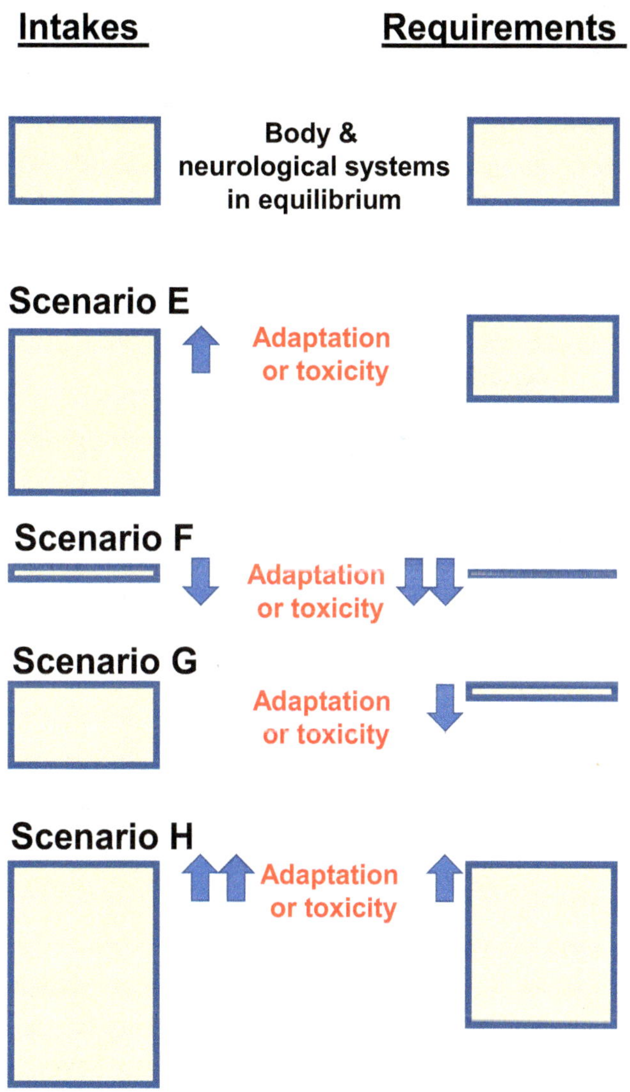

Fig. 2 Conceptual framework of different scenarios where there may be excessive vitamins or minerals. In this framework of different scenarios, there is a metabolic equilibrium whereby intake meets requirements (upper component of figure). However, there will be an imbalance resulting in excessive mineral or vitamin status if any of the following situations occur: intake is increased but requirements are unaltered (Scenario E), intake is reduced and requirements are also reduced but to a greater extent (Scenario F), intake is unaltered and requirements are reduced (Scenario G), and intake is increased to a greater extent than when requirements are also increased (Scenario H). *(From the authors.)*

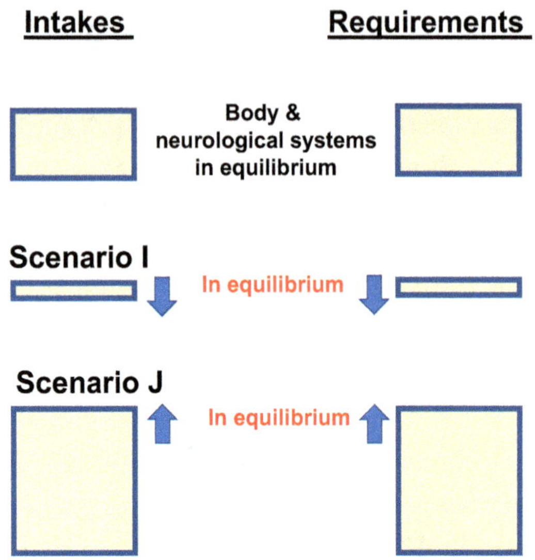

Fig. 3 Conceptual framework of different scenarios where the patient may be in equilibrium in terms of vitamin and mineral status. In this framework of different scenarios, there is a metabolic equilibrium whereby intake meets requirements (upper component of figure). Metabolic equilibrium also occurs when then is a concomitant and equal reduction in both intake and requirements (Scenario I) and when there is a concomitant and equal increase in both intake and requirements (Scenario J). *(From the authors.)*

Guidelines, such as those from the European Society of Parenteral and Enteral Nutrition (i.e., Burgos, Bretón, Cereda, et al., 2018), are available to enhance clinical practice and patient well-being. Regardless, the treatment of neurological disorders caused by malnutrition and nonneurological disorders caused by malnutrition secondary to a neurological disorder is the most straightforward.

Beriberi and Wernicke's encephalopathy, for example, are treated with thiamine (Pacei, Tesone, Laudi, et al., 2020). While serum thiamine status can and should be measured (e.g., indirectly via the activity of erythrocyte transketolase), treatment must be initiated based on clinical suspicion before the results are available. This is because delaying this nutritional intervention risks severe and irreversible damage or death (Royal College of Physicians, 2001). Yet, there are no randomized controlled trials to guide the route of administration, dose, or duration of therapy (Pacei et al., 2020). Thus, management guidelines are based on expert opinion and individual physician's preference. Regardless, patients' responses to therapy must be monitored, and their treatment should be adjusted as and when required.

The nutritional supplementation of patients with neurological disorder whose measured micronutrient levels or status are below the reference ranges is also relatively easy to supplement. The management of patients with neurological disorders that increase the risk of malnutrition but whose micronutrient levels are within their respective reference ranges is more challenging. More data are required to determine whether nutritional supplementation improves outcomes in these patients (Burgos et al., 2018). The gold standard is randomized, double-blind, placebo-controlled trials. Although simple in concept, consideration must be given to the diverse range and symptomology of neurological disorders: in other words, one treatment may not be applicable for all.

References

Afridi, H. I. (2023). Micronutrient levels in Parkinson's disease. In C. R. Martin, V. B. Patel, & V. R. Preedy (Eds.), *Vitamins and minerals in neurological disorders* (pp. 579–602). Elsevier.

Amel Kanane, A., Rouaki, F., Errahmani, M.-B., Laraba, A., Mesbah, H., & Ammouche, A. (2019). Pro-oxidant effects of a high α-tocopherol dose on kidney antioxidant biomarkers and histopathological aspects. *Journal of Biological Macromolecules, 205*, 154–168.

Burgos, R., Bretón, I., Cereda, E., et al. (2018). ESPEN guideline clinical nutrition in neurology. *Clinical Nutrition, 37*, 354–396.

Cashman, K. D., & Kiely, M. (2013). EURRECA-estimating vitamin D requirements for deriving dietary reference values. *Critical Reviews in Food Science and Nutrition, 53*(10), 1097–1109.

Cashman, K. D., Ritz, C., Carlin, A., & Kennedy, M. (2022). Vitamin D biomarkers for dietary reference intake development in children: A systematic review and meta-analysis. *American Journal of Clinical Nutrition, 115*(2), 544–558.

Choudhary, A. (2023). Vitamin D level and the outcome after traumatic brain injury. In C. R. Martin, V. B. Patel, & V. R. Preedy (Eds.), *Vitamins and minerals in neurological disorders* (pp. 157–170). Elsevier.

Department of Health. (1991). Dietary reference values for food, energy and nutrients in the United Kingdom. In *Report on health and social subjects, no. 41*. London. HMSO.

Dumas, C., Fabiani, L., Ioannidou, S., Pentieva, K., Turck, D., & Neuhauser-Berthold, M. (2020). Update by the European food safety authority of the dietary reference values for vitamins set for adults in the EU. In *Proceedings of the nutrition society, 79(OCE2)*.

FAO. (2022). *Food-based dietary guidelines*. Food and Agriculture Organization. https://www.fao.org/nutrition/nutrition-education/food-dietary-guidelines/en/. (Accessed 12 May 2002).

Hinnoutondji Toffa, D., & Li, J. (2023). Magnesium and Alzheimer's disease. In C. R. Martin, V. B. Patel, & V. R. Preedy (Eds.), *Vitamins and minerals in neurological disorders* (pp. 47–63). Elsevier.

Iacone, R., Scanzano, C., Santarpia, L., D'Isanto, A., Contaldo, F., & Pasanisi, F. (2016). Micronutrient content in enteral nutrition formulas: Comparison with the dietary reference values for healthy populations. *Nutrition Journal, 15*(1).

Islam, M. S. (2023). Correlation of reduced vitamin C and minerals with epilepsy. In C. R. Martin, V. B. Patel, & V. R. Preedy (Eds.), *Vitamins and minerals in neurological disorders* (pp. 263–274). Elsevier.

Kaddoumi, A. (2023). α-Tocopherol for Alzheimer's disease. In C. R. Martin, V. B. Patel, & V. R. Preedy (Eds.), *Vitamins and minerals in neurological disorders* (pp. 13–28). Elsevier.

Lamb, C. A., Parr, J., Lamb, E. I., & Warren, M. D. (2009). Adult malnutrition screening, prevalence and management in a United Kingdom hospital: Cross-sectional study. *British Journal of Nutrition, 102*, 571–575.

Leichtle, S. W., Al Mohaish, S., Sandler, N., Yajnik, V., Rivet, D., Kashiouris, M., et al. (2023). Vitamin C (L-ascorbic acid) in brain injury. In C. R. Martin, V. B. Patel, & V. R. Preedy (Eds.), *Vitamins and minerals in neurological disorders* (pp. 121–135). Elsevier.

Linden M. A., & Bellomo-Branda, M. A. (2023). Cerebral palsy: Evaluating vitamin D levels. In C. R. Martin, V. B. Patel, & V. R. Preedy (Eds.), *Vitamins and minerals in neurological disorders* (pp. 195–205). Elsevier.

Marinelli, R., Bartolini, D., Betti, M., Migni, A., Zatini, L., Marcantonini, G., et al., (2023). Vitamin E and neuroprotection in epilepsy. In C. R. Martin, V. B. Patel, & V. R. Preedy (Eds.), *Vitamins and minerals in neurological disorders* (pp. 229–243). Elsevier.

Marino, L. V., & Beattie, R. M. (2018). Compliance with nutrition screening in a children's hospital. *Archives of Disease in Childhood, 103*(8), 798–800.

Martin. (2015). Introduction to dietary reference values (DRVs). *Annals of Nutrition and Metabolism, 67*-(Suppl. 1), 22–23.

Matar, A., & Borjac, J. (2023). Iron, zinc and multiple sclerosis patients. In C. R. Martin, V. B. Patel, & V. R. Preedy (Eds.), *Vitamins and minerals in neurological disorders* (pp. 341–357). Elsevier.

Mir, H. A., & Khanday, F. A. (2022). Prooxidant vitamin C mechanistically exploits p66Shc/Rac1 GTPase pathway in inducing cytotoxicity. *International Journal of Biological Macromolecules, 205*, 154–168.

Mohammed, H. S. (2023). Selenium use in epilepsy. In C. R. Martin, V. B. Patel, & V. R. Preedy (Eds.), *Vitamins and minerals in neurological disorders* (pp. 245–261). Elsevier.

Nadjar, Y., & Fernandez, G. (2023). Riboflavin and lower motor neuron diseases. In C. R. Martin, V. B. Patel, & V. R. Preedy (Eds.), *Vitamins and minerals in neurological disorders* (pp. 107–118). Elsevier.

Pacei, F., Tesone, A., Laudi, N., et al. (2020). The relevance of thiamine evaluation in a practical setting. *Nutrients, 12*, 2810.

Palozza, P. (2003). Prooxidant effects of beta-carotene in cultured cells. *Molecular Aspects of Medicine, 24*, 353–362.

Phan-Thi, H., Durand, P., Prost, M., Prost, E., & Waché, Y. (2016). Food chemistry, 2016 effect of heat-processing on the antioxidant and prooxidant activities of β-carotene from natural and synthetic origins on red blood cells. *Food Chemistry, 1*(190), 1137–1144.

Powers, H. J. (2021). Approaches to setting dietary reference values for micronutrients, and translation into recommendations. *Proceedings of the Nutrition Society, 80*(3), 365–372.

Rieber, M. (2020). Cancer prooxidant therapy through copper redox cycling: Repurposing disulfiram and tetrathiomolybdate. *Current Pharmaceutical Design, 26*, 4461–4466.

Riizzo, G., & Marino, A. (2023). Cognitive impairment and micronutrients: Vitamin B12, folate, and homocysteine and implications for dementia. In C. R. Martin, V. B. Patel, & V. R. Preedy (Eds.), *Vitamins and minerals in neurological disorders* (pp. 29–46). Elsevier.

Royal College of Physicians. (2001). *Alcohol—Can the NHS afford it? Recommendations for a coherent alcohol strategy for hospitals*. London: Royal College of Physicians.

Sabbouh, T., & Torbey, M. T. (2018). Malnutrition in stroke patients: Risk factors, assessment, and management. *Neurocritical Care, 29*, 374–384.

Santos, G., Avila-Rodriguez, M., Rubio E′. P., Galembeck, E., & Borges, J. M. P. (2023). Linking copper and neurotoxic activities to the CNS. In C. R. Martin, V. B. Patel, & V. R. Preedy (Eds.), *Vitamins and minerals in neurological disorders* (pp. 209–225). Elsevier.

Simunkova, M., Barbierikova, Z., Jomova, K., Hudecova, L., Lauro, P., et al. (2021). Antioxidant vs. prooxidant properties of the flavonoid, kaempferol, in the presence of Cu(ii) ions: A ros-scavenging activity, Fenton reaction and DNA damage study. *International Journal of Molecular Sciences, 22*, 1619.

Sotler, R., Poljšak, B., Dahmane, R., Jukić, T., Pavan Jukić, D., Rotim, C., et al. (2019). Prooxidant activities of antioxidants and their impact on health. *Acta Clinica Croatica, 58*, 726–736.

Squitti, R., Rongioletti, M., & Liguri, G. (2023). Copper, oxidative stress, Alzheimer's disease and dementia. In C. R. Martin, V. B. Patel, & V. R. Preedy (Eds.), *Vitamins and minerals in neurological disorders* (pp. 65–85). Elsevier.

Timoshnikov, V. A., Kobzeva, T. V., Polyakov, N. E., & Kontoghiorghes, G. J. (2020). Redox interactions of vitamin c and iron: Inhibition of the prooxidant activity by deferiprone. *International Journal of Molecular Sciences, 21*, 3967.

U.S. Departments of Agriculture and Health and Human Services. (2020). *Dietary guidelines for Americans, 2020–2025* (9th ed.). U.S. Department of Agriculture and U.S. Department of Health and Human Services. December 2020. Available at https://www.dietaryguidelines.gov/.

Veiga, N., Alvarez, N., & Castellano, E. E. (2021). Comparative study of antioxidant and prooxidant properties of homoleptic and heteroleptic copper complexes with amino acids, dipeptides and 1,10-phenanthroline: The quest for antitumor compounds. *Molecules, 26*, 6520.

Winterbone, M. S., Sampson, M. J., Saha, S., et al. (2007). Pro-oxidant effect of α-tocopherol in patients with type 2 diabetes after an oral glucose tolerance test—A randomised controlled trial. *Cardiovascular Diabetology, 6*, 8. https://doi.org/10.1186/1475-2840-6-8.

PART I

Alzheimer's disease and dementias

CHAPTER 2

α-Tocopherol for Alzheimer's disease

Nehad M. Ayoub[a], Raida W. Khalil[b], and Amal Kaddoumi[c]
[a]Department of Clinical Pharmacy, Faculty of Pharmacy, Jordan University of Science and Technology, Irbid, Jordan
[b]Department of Clinical Nutrition and Dietetics, Faculty of Science, University of Philadelphia, Amman, Jordan
[c]Department of Drug Discovery and Development, Harrison College of Pharmacy, Auburn University, Auburn, AL, United States

Abbreviations

α-Tph	α-tocopherol
Aβ	amyloid-β
AD	Alzheimer's disease
APP	Amyloid-β precursor protein
APPsw	amyloid-β precursor protein with Swedish mutation
Bax	bcl-2-like protein 4
BBB	blood-brain barrier
Bcl-2	B-cell lymphoma 2
CSF	cerebrospinal fluid
IL-1β	interleukin-1β
IL-4	interleukin-4
IL-6	interleukin-6
MCI	mild cognitive impairment
MDD	major depressive disorder
NMDA	N-methyl-D-aspartic acid
NO	nitric oxide
PLTP	phospholipid transfer protein
PLTP-KO	PLTP-deficient mice
TTP	α-tocopherol transport protein
Ttpa$^{-/-}$	α-tocopherol transport protein knockout in mouse

Introduction

Alzheimer's disease

Alzheimer's disease (AD) is a complex neurodegenerative disorder characterized by memory loss and progressive dementia. AD is the most common cause of dementia affecting the elderly (Lage, 2006). Worldwide, AD cases account for 60%–80% of dementia cases (Prince et al., 2013). While the cause of AD is not clear, several factors play a role solely or in a combination such as genetic factors (apolipoprotein ε4 allele), germline mutations (amyloid-β precursor protein (APP) gene, presenilin-1 gene, and presenilin-2 gene), environmental and lifestyle-related factors, as well as health

Vitamins and Minerals in Neurological Disorders
https://doi.org/10.1016/B978-0-323-89835-5.00015-6

conditions such as diabetes, brain injury, and cardiovascular disease (Kamat et al., 2008). To date, no effective treatments exist, and available drugs, including acetylcholine esterase inhibitors and *N*-methyl-D-aspartic acid (NMDA) receptor antagonists, treat the symptoms rather than the underlying causes of the disease. AD is characterized by brain deposition and accumulation of amyloid-β (Aβ), neurofibrillary tangles, cerebral amyloid angiopathy, neuroinflammation, synaptic dysfunction, and neuronal loss (Selkoe, 2001). In addition, oxidative stress represents one of the main factors involved in the neuropathology of AD, attributing to both AD initiation and progression (Gugliandolo, Bramanti, & Mazzon, 2017; Kamat et al., 2008). Accordingly, antioxidants have been strongly proposed, at least in theory, to prevent initiation or slow progression of neurodegenerative conditions, including AD (Kamat et al., 2008). In this regard, several clinical and preclinical studies have investigated the beneficial effect of α-tocopherol (α-Tph) and its role in preclinical models of AD as well as in AD patients.

α-Tocopherol

Vitamin E is an essential micronutrient for humans. Vitamin E refers to a group of naturally existing lipid-soluble compounds that are divided into tocopherols and tocotrienols (Gugliandolo et al., 2017; Jiang, 2017; Lee & Han, 2018; Lloret, Esteve, Monllor, Cervera-Ferri, & Lloret, 2019). Each class of vitamin E has four naturally occurring members, which are distinguished as α-, β-, γ-, and δ-tocopherols, and as α-, β-, γ-, and δ-tocotrienols. Structurally, tocopherols have a saturated, while tocotrienols have three double bonds on the side chain (Gugliandolo et al., 2017; Jiang, 2017; Lee & Han, 2018; Lloret et al., 2019). All vitamin E forms are lipophilic antioxidants that scavenge peroxyl radicals and terminate the oxidation of polyunsaturated fatty acids essential for maintaining the integrity of cell membranes (Gugliandolo et al., 2017; Lee & Han, 2018). Currently, the recommended dietary allowance of vitamin E is 15 mg for both adult males and females (Higgins, Izadi, & Kaviani, 2020). While the main role of vitamin E is its antioxidant activity, vitamin E also has antiinflammatory and neuroprotective properties (Lloret et al., 2019). Among the eight members of naturally occurring vitamin E, α-tocopherol (α-Tph) is the most abundant form of vitamin E in tissues and blood (Jiang, 2017). The binding of α-Tph to its transport protein (TTP) maintains the high bioavailability of α-Tph in tissues and prevents its catabolism (Jiang, 2017; Lim & Traber, 2007).

α-Tocopherol in AD

Association studies of altered levels of α-Tph in AD have been reported. Several studies found an inverse relationship between α-Tph blood levels and mild cognitive impairment (MCI) and AD, suggesting that reduced α-Tph levels are associated with an increased risk of AD. For example, compared to cognitively normal individuals, AD and MCI patients demonstrated lower plasma levels of total tocopherols, including α-Tph (Mangialasche et al., 2012). Consistent with these findings, other studies reported significantly lower levels of α-Tph in the serum and cerebrospinal fluid (CSF) of AD patients (Jimenez-Jimenez et al., 1997; Tohgi et al., 1994). Jimenez-Jimenez and colleagues suggested that reduced levels of α-Tph in AD patients could be related to the dietary deficiency of α-Tph, which could increase the risk of AD (Jimenez-Jimenez et al., 1997). More recent studies have demonstrated that reduced levels of plasma α-Tph were associated with an increased risk of MCI and AD (Casati et al., 2020; Mangialasche et al., 2010). Besides its antioxidant effect, De Leeuw and colleagues suggested that the inverse relationship between α-Tph and risk of AD could also be explained, at least in part, by its ability to attenuate microglia activation as determined in brain tissues obtained from deceased individuals (de Leeuw, Schneider, Agrawal, Leurgans, & Morris, 2020).

On the other hand, another study that evaluated α-Tph levels in ventricular CSF, obtained at autopsy from brain tissues for AD diagnoses, demonstrated a weak association of α-Tph levels with low AD neuropathology, specifically neuritic plaques (Hensley et al., 2011). While the mixed results point to the need for more and larger studies, the contribution of other forms of tocopherols and tocotrienols to AD risk could be significant as they also add to overall antioxidant activity (Hensley et al., 2011). Collectively, as several studies suggested the potential neuroprotective effect of α-Tph against AD, the beneficial effect of α-Tph supplementation as a therapeutic approach to reduce the risk and/or treat AD has been studied. Below, we present available findings from preclinical and clinical studies on the therapeutic impact of α-Tph in AD.

Preclinical studies

In vitro models

Early evidence revealed that the exposure of cultured neurons to $A\beta_{42}$ increased levels of ceramides, cholesterol, and oxidative stress, which were prevented by α-Tph pretreatment, suggesting a neuroprotective activity of α-Tph (Cutler et al., 2004). Furthermore, Dai and colleagues reported the protective effect of α-Tph against the cytotoxic effect of $A\beta_{40}$-Cu(II) complex, an oligomeric complex formed by copper binding to $A\beta_{40}$ that generates neurotoxic hydrogen peroxide (Dai, Sun, & Jiang, 2007). The neurotoxicity of $A\beta_{40}$-Cu(II) complex was determined by reduced viability and increased lactate dehydrogenase release from neuronal cells in culture, which was successfully blunted by α-Tph treatment as a free radical antioxidant (Dai et al., 2007). Similarly, when tested for its effect against Aβ-induced neuronal apoptosis in differentiated SK-N-BE neurons, α-Tph effectively prevented cell apoptosis as determined by the suppression of mitochondrial proteins B-cell lymphoma 2 (Bcl-2) and bcl-2-like protein 4 (Bax), the release of cytochrome *c* in the cytosol, poly-ADP ribose polymerase cleavage, and activation of caspase 3, suggesting the powerful neuroprotective activity of α-Tph (Tamagno et al., 2003). These findings were further supported by a recent study by Pahrudin Arrozi and colleagues who reported the antioxidant effect of α-Tph in SH-SY5Y cells stably transfected with wild-type and mutant forms of the APP gene (Pahrudin Arrozi et al., 2020). Interestingly, however, using the same model of SH-SY5Y cells overexpressing APP, Grimm et al. revealed an opposite finding of the ability of all tocopherols, including α-Tph, to rather increase Aβ level by enhancing the production and decreasing the degradation of Aβ in the neuronal cell line (Grimm et al., 2015). In this context, tocopherols have been shown to enhance the expression of β- and γ-secretases, the enzymes responsible for Aβ production from APP. Indeed, additional studies are essential to explain the discrepancy; nevertheless, compared to other tocopherols, α-Tph demonstrated a minor impact on Aβ production (Grimm et al., 2015). Consistent with α-Tph-positive effect, Eckert and colleagues showed powerful antioxidant and neuroprotective effects of α-Tph against oxidative stress-induced cell death in PC12 cells stably transfected with human APPSw (Eckert et al., 2001). The neuronal protective potency of α-Tph was greater than that observed with caspase inhibitors (Eckert et al., 2001).

Furthermore, a transcriptomic analysis for the effect of α-Tph in retinoic acid differentiated SH-SY5Y neuroblastoma cells exposed to $A\beta_{42}$ revealed the ability of α-Tph to upregulate genes involved in the nonamyloidogenic processing of APP. At the same time, it downregulated the amyloidogenic pathway, thus reducing Aβ production (Gugliandolo, Chiricosta, Silvestro, Bramanti, & Mazzon, 2019). Besides its antioxidant activity, α-Tph modulated the expression of genes involved in autophagy, an important process for Aβ clearance and known to be altered in AD, suggesting a nonantioxidant activity of α-Tph (Gugliandolo et al., 2019). Additional in vitro studies compared the neuroprotective effects of both natural (+) and synthetic (±) forms of α-Tph to the female hormone 17-β estradiol, known for its natural neuroprotective effect, using mouse clonal hippocampal HT22 cells and rat cerebellar granule neurons (Behl, 2000). Both forms of α-Tph induced a greater neuroprotective antioxidant activity than estradiol and were more effective in preventing oxidative neuronal cell death induced by Aβ, hydrogen peroxide, and glutamate. Furthermore, both forms of α-Tph induced the activity of the redox-sensitive transcription factor NF-κB involved in neuronal cell survival (Behl, 2000).

In vivo models

Nishida and colleagues showed that depletion of α-Tph impaired Aβ clearance from the brain of a mouse model of AD harboring human APPSw, which promoted the accumulation of Aβ in the mouse brain (Nishida et al., 2009). In addition, the knockout of α-Tph transfer protein ($Ttpa^{-/-}$) in APPSw mice induced severe cognitive dysfunction, which was associated with increased Aβ deposits in the brain, an effect that was ameliorated with α-Tph supplementation (Nishida et al., 2006). In another study, Ishihara and colleagues assessed the effect of α-Tph on memory function in a triple transgenic mouse model of AD, and similarly, they reported that α-Tph-supplemented diet improved cognitive function and relieved oxidative stress (Ishihara et al., 2013). Moreover, the antioxidant and neuroprotective effects of α-Tph were tested in a mouse model deficient from plasma phospholipid transfer protein (PLTP), which is responsible for the transfer of α-Tph in the brain (Desrumaux et al., 2013). PLTP-deficient (PLTP-KO) mice demonstrated increased cerebral oxidative stress, increased brain $A\beta_{42}$ levels, and reduced expression of the synaptic marker synaptophysin, as compared with wild-type mice. Exposure to $A\beta_{25-35}$ in the PLTP-KO model induced memory impairment and increased oxidative stress, which were restored by chronic dietary supplementation of α-Tph (Desrumaux et al., 2013).

To evaluate the preventive versus treatment effect of α-Tph, Sung and colleagues investigated the impact of early versus late supplementation of α-Tph administered to a transgenic mouse model of AD, namely, Tg2576 (Sung et al., 2004). Early and late supplementation refers to treatment at a young age and before pathology begins, started at five months and continued until 13 months, and older age (with pathology) started at 14 months of age until 20 months. While α-Tph supplementation effectively suppressed brain lipid peroxidation at both time points, α-Tph significantly reduced Aβ levels and amyloid deposition only when administered early to young Tg2576 mice (Sung et al., 2004), suggesting α-Tph may function better as a prophylactic rather than a treatment.

Besides its antioxidant effect and reduction of Aβ brain levels by increasing Aβ clearance and reducing Aβ production, α-Tph enhanced the blood-brain barrier (BBB) integrity and function (Elfakhri, Abdallah, Brannen, & Kaddoumi, 2019; Elfakhri et al., 2018). Several studies have reported a dysfunctional BBB in AD, an effect that is correlated with dysfunctional clearance of Aβ across the BBB and formation of Aβ plaques (Sagare, Bell, & Zlokovic, 2012), neuroinflammation, oxidative stress, and neuronal damage (Bowman et al., 2007). When tested using in vitro and in vivo experimental models of AD, α-Tph protected and restored the function of Aβ-induced BBB disruption (Elfakhri et al., 2018). Consistent with these results, findings from in vitro and in vivo studies evaluated the effect of α-Tph in combination with the nonsteroidal antiinflammatory drug etodolac on Aβ-related pathology, including BBB dysfunction, Aβ brain levels, neuroinflammation, and oxidative stress, demonstrated that α-Tph alone or in combination with etodolac restored the BBB function and reduced neuroinflammation and oxidative stress (Elfakhri et al., 2019). Further, α-Tph alone or in combination decreased total Aβ levels in the brains of 5XFAD mice, a mouse model of AD (Elfakhri et al., 2019). Collectively, findings from the preclinical studies revealed the neuroprotective effects of α-Tph along with its antioxidant and nonantioxidant activities, supporting its therapeutic potential for AD.

Clinical studies of α-Tph supplementation

Unlike the overall positive findings from the preclinical studies, the evidence of α-Tph effectiveness in humans with MCI or AD is unsatisfactory and requires additional studies. A summary of available findings is presented in Table 1. Few studies evaluated α-Tph supplementation in combination with AD drugs. Sano and colleagues evaluated the effect of α-Tph (1000 IU, twice daily for 2 years) alone or in combination with selegiline for its effect on cognitive function in moderate-to-severe AD. No improvement in cognitive test scores with either treatment was observed (Sano et al., 1997). In another study, the effect of α-Tph (2000 IU) with and without memantine, an NMDA receptor antagonist used to treat the symptoms of AD, in mild-to-moderate AD patients was investigated for 4 years (Dysken et al., 2014). Findings showed that α-Tph supplementation significantly delayed the functional decline in AD patients who were taking acetylcholine esterase inhibitors compared to placebo. However, this effect was not observed in patients who received memantine alone or memantine with α-Tph (Dysken et al., 2014), suggesting α-Tph combination with acetylcholine esterase inhibitors could benefit mild-to-moderate AD patients. However, when evaluated individually or in combination with the acetylcholine esterase inhibitor donepezil in MCI subjects, α-Tph alone or combined with donepezil neither provided a therapeutic benefit nor altered the rate of AD progression compared to placebo groups (Petersen et al., 2005). Negative results were also reported in mild-to-moderate AD patients who received α-Tph (400 IU/day) with vitamin C (1000 mg/day) besides their acetylcholine esterase inhibitor therapy for one year (Arlt, Muller-Thomsen, Beisiegel, & Kontush, 2012).

Few studies, however, reported a potentially positive impact of α-Tph, at least in part, when supplemented in formulation. For example, Chan and colleagues tested α-Tph (30 IU) prepared in a formulation containing folic acid (400 μg), vitamin B12 (6 μg), S-adenosylmethionine (400 mg), *N*-acetyl cysteine (600 mg), and acetyl-L-carnitine (500 mg) in 14 patients with early-stage AD supplemented for one year. The formulation enhanced neuropsychiatric performance for more than 12 months and delayed disease progression by approximately 7 months (Chan, Paskavitz, Remington, Rasmussen, & Shea, 2008). Based on these promising results, a Phase II study in 106 individuals with AD was conducted. Consistent with the earlier study, the formulation maintained or improved cognitive performance compared to placebo (Remington, Bechtel, et al., 2015). The studies were further extended to evaluate the formulation in MCI subjects, and similarly, the formulation maintained or improved cognitive performance (Remington, Lortie, et al., 2015). While these results are promising, the findings suggest that a combinatorial approach could provide a greater neuroprotection than individual α-Tph.

Epidemiological follow-up studies to evaluate the association of dietary α-Tph intake with risk of AD were also reported (Table 2); findings suggested that vitamin E from food is beneficial, whereas its supplementation is not effective (Morris et al., 2002, 2015).

Table 1 A summary of clinical trials on the therapeutic effect of α-Tph in AD.

Reference	Treatment	AD stage	Number of participants	Outcomes
Sano et al. (1997)	α-Tph and/or selegiline	Moderate to severe	341	Slowed disease progression
Petersen et al. (2005)	α-Tph and/or donepezil	MCI	769	No benefit
Chan et al. (2008)	Nutraceutical formulation folic acid, vitamin B12, S-adenosylmethionine, *N*-acetyl cysteine, and acetyl-L-carnitine	Early AD	14	Improved neuropsychiatric performance
Lloret et al. (2009)	Vitamin E	AD	33	Mixed results, responders and nonresponders
Arlt et al. (2012)	α-Tph and vitamin C	AD	33	No benefits
Galasko et al. (2012)	α-Tph, vitamin C, and α-lipoic acid	Mild to moderate	78	Accelerated decline in Mini-Mental State Examination
Dysken et al. (2014)	α-Tph and/or memantine	Mild to moderate	561	No significant differences between those who received memantine alone or memantine plus α-Tph. α-Tph slowed functional decline and decreased caregiver burden
Remington, Bechtel, et al. (2015)	Nutraceutical formulation folic acid, vitamin B12, S-adenosylmethionine, *N*-acetyl cysteine, and acetyl-L-carnitine	AD	106	Improved cognitive performance
Remington, Lortie, et al. (2015)	Nutraceutical formulation folic acid, vitamin B12, S-adenosylmethionine, *N*-acetyl cysteine, and acetyl-L-carnitine	MCI	34	Improved cognitive performance
Kryscio et al. (2017)	α-Tph and/or selenium	Disease-free	7540	No benefits

Table 2 Epidemiological studies on α-Tph supplementation and the risk of AD.

Reference	Source	Number of participants	Outcomes
Morris et al. (1998)	Vitamin E supplement	633	Participants who received vitamin E did not develop AD.
Masaki et al. (2000)	Vitamin E supplement	3385	Vitamin E and vitamin C supplements in combination may protect against vascular dementia, but not AD. Vitamin E use was associated with better cognitive performance at a 10-year follow-up.
Engelhart et al. (2002)	Vitamin E dietary intake	5395	A high intake of vitamin E was associated with a lower risk of AD.
Morris et al. (2002)	Vitamin E dietary intake or supplements	815	High vitamin E dietary intake was associated with a decreased risk of AD in ApoE4 negative individuals. Vitamin E supplementation effect was inconclusive.
Luchsinger et al. (2003)	Vitamin E dietary intake or supplements	980	Neither dietary, supplemental, nor total intake was associated with a reduced risk of AD.
Zandi et al. (2004)	Vitamin E supplement	4740	Vitamin E and vitamin C supplements in combination reduced the incidence of AD.
Morris et al. (2005)	Vitamin E or α-Tph dietary intake	1041	Combined intake of the tocopherol forms may be more important than α-Tph alone to protect against AD.
Gray et al. (2008)	Vitamin E supplement	2969	Vitamin E supplement did not reduce the risk of AD.
Devore et al. (2010)	Vitamin E dietary intake	5395	A high dietary intake of vitamin E may modestly reduce the long-term risk of AD.
Basambombo, Carmichael, Cote, and Laurin (2017)	Vitamin E supplement	5269	Vitamin E with and without vitamin C supplements was associated with a reduced risk of cognitive decline.

Moreover, collective tocopherols rather than α-Tph alone could provide the protective effect against AD (Morris et al., 2005). While other studies suggested the use of a high dose of α-Tph supplements may reduce AD risk (Engelhart et al., 2002; Morris et al., 1998), others concluded that neither dietary nor supplemental intake of vitamin E reduced the risk of AD (Gray et al., 2008; Luchsinger, Tang, Shea, & Mayeux, 2003). Additional studies that evaluated the association of dietary α-Tph intake (in the form of vitamin E) with the risk of AD are summarized in Table 2.

Collectively, evidence supporting the clinical benefits of α-Tph against AD remains inconclusive. Differences in the design and methodology of the studies could be attributed to data discrepancy. Several factors could influence the response variability within and between reported studies such as the small number of participants, the variable dosage of α-Tph used without monitoring serum level of α-Tph at baseline and end of supplementation as well as the length of supplementation. Besides, additional factors that were not considered in previous studies and could be highly contributed to the discrepancy are related to the characteristics of study participants. The data were not stratified or analyzed by age, gender, race, genetics, the presence of comorbidities such as cardiovascular diseases, and diabetes mellitus linked to polypharmacy, and potential drug interactions. Accordingly, to better predict disease risk and who would benefit from α-Tph supplementation, well-designed and controlled clinical trials should be performed.

Applications to other neurological conditions

There are few reports on the association of α-Tph with the non-AD type of dementia. As an antioxidant, α-Tph could be beneficial for other neurological conditions. For example, the therapeutic effect and potential association of α-Tph with stroke and Huntington's disease were reported. In an epidemiological study that assessed the dietary intake of antioxidant vitamins in Japanese men and women, the authors concluded that among the evaluated vitamins, only vitamin C intake was inversely associated with the incidence of total stroke and ischemic stroke among nonsmokers, while α-Tph was not effective (Uesugi et al., 2017). Interestingly, in another study, Le and colleagues reported an association between high consumption of α-Tph and hypervitaminosis E in patients with intracerebral hemorrhagic stroke. The authors explained that such an effect could be explained by the known effects of α-Tph as a vitamin K antagonist, its antiplatelet activity, hypo-prothrombinemic effect, and its involvement in cytochrome *p*-450 interactions (Le, Kesayan, Chang, & Rose, 2020). The authors concluded that a high intake of α-Tph is not recommended as it could increase the risk of intracerebral hemorrhagic stroke. When evaluated as a potential therapy for Huntington's disease patients, α-Tph was not an effective treatment as determined by the lack of changes in neurologic and neuropsychiatric symptoms; however, α-Tph slowed the rate of motor decline when used for early intervention (Peyser et al., 1995). Moreover, pending additional evidence, preliminary findings on the role of α-Tph in major depressive disorder (MDD) demonstrated a relationship between low α-Tph serum levels and MDD symptoms (Manosso, Camargo, Dafre, & Rodrigues, 2020). MDD is characterized by inflammation and oxidative and nitrosative stress. Again, similar to AD, data on the effectiveness of α-Tph in individuals with other neurological diseases are mostly derived from preclinical studies with insufficient evidence of its usefulness generated from human studies.

Other components of interest

In this chapter, while we describe the role of α-Tph in AD and its potential therapeutic use to reduce the risk and/or treat AD, studies have also shown α-Tph quinone, the oxidation product α-Tph, could provide a beneficial effect in AD. Yang and colleagues demonstrated the beneficial effects of α-Tph quinone in a preclinical model of AD (Yang et al., 2010). α-Tph quinone suppressed $A\beta_{42}$-induced neurotoxicity in SH-SY5Y neuroblastoma cells and inhibited the formation of fibril $A\beta_{42}$ in a dose-dependent fashion, suggesting an additional mechanism for α-Tph by inhibiting Aβ aggregation via the formation of α-Tph quinone. In addition to its ability to reduce the formation of reactive oxygen species and nitric oxide (NO), α-Tph quinone produced an antiinflammatory effect by reducing tumor necrosis factor-α and interleukin-1β (IL-1β) in microglia, while it increased the formation of IL-4 (Yang et al., 2010). Subsequently, Wang et al. showed that oral administration of α-Tph quinine in APPSw/PS1dE9 transgenic mice induced several favorable activities, including improved memory function, decreased oxidative stress and inflammatory mediators such as IL-6 and IL-1β, and prevented the production of inducible NO synthase in the mouse brains (Wang et al., 2016).

Furthermore, α-Tph quinine administration reduced the levels of Aβ oligomers and inhibited microglial activation by inhibiting the NF-κB signaling pathway in the mouse brains (Wang et al., 2016). While the translational impact of these preclinical findings with α-Tph quinone is yet to be investigated, an interesting study that monitored the levels of α-Tph and α-Tph quinone in CSF from AD patients compared to healthy subjects, the AD patients had a significantly reduced level of α-Tph with unchanged α-Tph quinone. Alternatively, when tested in patients with vascular dementia of the Binswanger type, CSF α-Tph levels were unaltered compared to healthy subjects; however, α-Tph quinone levels were significantly higher, an effect that was correlated with cognitive impairment (Tohgi et al., 1994).

The antioxidant and neuroprotective effects of α-Tph were further evaluated with concurrent administration with other dietary components. Short-term oral administration (two weeks) of α-Tph and folic acid in combination has been shown to protect against Aβ-induced neurotoxicity in an animal model of AD by reducing cognitive decline, synaptic loss, and neuronal death (Figueiredo et al., 2011). The chronic treatment was ineffective in lowering Aβ-induced glial cell activation, suggesting that the combined effect is antiinflammation independent; however, the chronic treatment of α-Tph and folic acid generated an antioxidant activity as indicated by a decline in NO production and prevented $A\beta_{40}$-induced increase in the activity of mitochondrial complexes I and IV (Figueiredo et al., 2011). In another study that evaluated the combination of α-Tph (at low and high doses) with vitamin C in aged APP/PS1 mice for 4 months starting at the age of 6 months, the authors found that either combination did not alter brain Aβ deposition and that high-dose α-Tph with vitamin C was less effective at reducing oxidative stress than the combination with low-dose α-Tph. Interestingly, the high-dose combination impaired memory function (Harrison et al., 2009).

These findings are interesting because they support the importance of performing dose-dependent studies, especially in humans, to select an effective dose range that could benefit patients with MCI and AD.

Mini-dictionary of terms

- **(+) α-Tph:** Natural form of α-Tph.
- **(±) α-Tph:** Synthetic form of α-Tph.
- **α-Tph quinone:** A derivative of α-Tph produced at a slow rate during the antioxidant action of α-Tph.
- **Cerebral amyloid angiopathy:** A disease condition presents in about 70%–80% of Alzheimer's disease patients characterized by the deposition of Aβ in cerebrovasculature.
- **Neurofibrillary tangles:** Hyperphosphorylated tau protein aggregation in brains of Alzheimer's disease patients.
- **$A\beta_{40}$ and $A\beta_{42}$:** Different forms of amyloid-β peptides, the numbers refer to the number of amino acids of each peptide.
- **MCI:** The stage between the expected cognitive decline of normal aging and the more serious decline of dementia. Characterized by a slight but noticeable and measurable decline in cognitive abilities, including memory and thinking skills.
- **Binswanger's disease:** Also known as subcortical vascular dementia, is a type of dementia caused by widespread, microscopic areas of damage to the deep layers of white matter in the brain. The damage results from the thickening and narrowing of arteries that feed the subcortical areas of the brain.

Key facts of Alzheimer's disease

Alzheimer's disease is characterized by a decline in memory, thinking, behavior, and ability to perform everyday activities.

Worldwide, it is estimated that about 50 million people are living with Alzheimer's disease or related dementia.

Sporadic Alzheimer's disease represents about 95% of the cases, while familial cases due to inherited genetic mutations represent the remaining 5%.

Alzheimer's disease affects elderly people; however, it is not a normal part of aging.

Currently, there is no treatment or cure for Alzheimer's disease.

Summary points

- While available evidence from preclinical studies supports α-Tph supplementation as a potential therapy for AD, the evidence from clinical studies for its beneficial effect in AD patients is mixed and insufficient.

- Whether low blood levels of α-Tph increases the risk of AD or not remains contradictory.
- Additional studies are required to investigate the causal relationship of serum and CSF α-Tph levels with risk of AD and changes as the disease progress.
- Large controlled clinical trials are required to evaluate the therapeutic potential of α-Tph to prevent, slow the progression, and/or treat AD.
- Dose-dependent studies are essential to finding the effective dose range that could benefit AD patients.
- Whether the status of APOE variants alters AD patients' response to α-Tph or not is yet to be evaluated.
- Besides its antioxidant activity, evidence from preclinical studies in AD mouse models demonstrated that α-Tph has an antiinflammatory effect, reduces brain Aβ accumulation, reduces synaptic loss, and restores BBB function.
- Pending further evaluation in AD patients, current evidence from clinical trials suggests that α-Tph is not beneficial for AD.
- α-Tph demonstrated mixed results in MCI, AD, non-AD type dementia, and other neurological diseases, suggesting the evidence is insufficient for its recommendation against neurological disease.
- Supplementing with vitamin E could be more effective than α-Tph as a single isoform by providing synergistic interactions.

References

Arlt, S., Muller-Thomsen, T., Beisiegel, U., & Kontush, A. (2012). Effect of one-year vitamin C- and E-supplementation on cerebrospinal fluid oxidation parameters and clinical course in Alzheimer's disease. *Neurochemical Research*, *37*(12), 2706–2714. https://doi.org/10.1007/s11064-012-0860-8.

Basambombo, L. L., Carmichael, P. H., Cote, S., & Laurin, D. (2017). Use of vitamin E and C supplements for the prevention of cognitive decline. *The Annals of Pharmacotherapy*, *51*(2), 118–124. https://doi.org/10.1177/1060028016673072.

Behl, C. (2000). Vitamin E protects neurons against oxidative cell death in vitro more effectively than 17-beta estradiol and induces the activity of the transcription factor NF-kappaB. *Journal of Neural Transmission (Vienna)*, *107*(4), 393–407. https://doi.org/10.1007/s007020070082.

Bowman, G. L., Kaye, J. A., Moore, M., Waichunas, D., Carlson, N. E., & Quinn, J. F. (2007). Blood-brain barrier impairment in Alzheimer disease: Stability and functional significance. *Neurology*, *68*(21), 1809–1814. https://doi.org/10.1212/01.wnl.0000262031.18018.1a.

Casati, M., Boccardi, V., Ferri, E., Bertagnoli, L., Bastiani, P., Ciccone, S., et al. (2020). Vitamin E and Alzheimer's disease: The mediating role of cellular aging. *Aging Clinical and Experimental Research*, *32*(3), 459–464. https://doi.org/10.1007/s40520-019-01209-3.

Chan, A., Paskavitz, J., Remington, R., Rasmussen, S., & Shea, T. B. (2008). Efficacy of a vitamin/nutriceutical formulation for early-stage Alzheimer's disease: A 1-year, open-label pilot study with an 16-month caregiver extension. *American Journal of Alzheimer's Disease and Other Dementias*, *23*(6), 571–585. https://doi.org/10.1177/1533317508325093.

Cutler, R. G., Kelly, J., Storie, K., Pedersen, W. A., Tammara, A., Hatanpaa, K., et al. (2004). Involvement of oxidative stress-induced abnormalities in ceramide and cholesterol metabolism in brain aging and Alzheimer's disease. *Proceedings of the National Academy of Sciences of the United States of America*, *101*(7), 2070–2075. https://doi.org/10.1073/pnas.0305799101.

Dai, X., Sun, Y., & Jiang, Z. (2007). Protective effects of vitamin E against oxidative damage induced by Abeta1-40Cu(II) complexes. *Acta Biochimica et Biophysica Sinica Shanghai, 39*(2), 123–130. https://doi.org/10.1111/j.1745-7270.2007.00261.x.

de Leeuw, F. A., Schneider, J. A., Agrawal, S., Leurgans, S. E., & Morris, M. C. (2020). Brain tocopherol levels are associated with lower activated microglia density in elderly human cortex. *Alzheimers Dement, 6*(1), e12021. https://doi.org/10.1002/trc2.12021.

Desrumaux, C., Pisoni, A., Meunier, J., Deckert, V., Athias, A., Perrier, V., et al. (2013). Increased amyloid-beta peptide-induced memory deficits in phospholipid transfer protein (PLTP) gene knockout mice. *Neuropsychopharmacology, 38*(5), 817–825. https://doi.org/10.1038/npp.2012.247.

Devore, E. E., Grodstein, F., van Rooij, F. J., Hofman, A., Stampfer, M. J., Witteman, J. C., et al. (2010). Dietary antioxidants and long-term risk of dementia. *Archives of Neurology, 67*(7), 819–825. https://doi.org/10.1001/archneurol.2010.144.

Dysken, M. W., Sano, M., Asthana, S., Vertrees, J. E., Pallaki, M., Llorente, M., et al. (2014). Effect of vitamin E and memantine on functional decline in Alzheimer disease: The TEAM-AD VA cooperative randomized trial. *JAMA, 311*(1), 33–44. https://doi.org/10.1001/jama.2013.282834.

Eckert, A., Steiner, B., Marques, C., Leutz, S., Romig, H., Haass, C., et al. (2001). Elevated vulnerability to oxidative stress-induced cell death and activation of caspase-3 by the Swedish amyloid precursor protein mutation. *Journal of Neuroscience Research, 64*(2), 183–192. https://doi.org/10.1002/jnr.1064.

Elfakhri, K. H., Abdallah, I. M., Brannen, A. D., & Kaddoumi, A. (2019). Multi-faceted therapeutic strategy for treatment of Alzheimer's disease by concurrent administration of etodolac and alpha-tocopherol. *Neurobiology of Disease, 125*, 123–134. https://doi.org/10.1016/j.nbd.2019.01.020.

Elfakhri, K. H., Duong, Q. V., Langley, C., Depaula, A., Mousa, Y. M., Lebeouf, T., et al. (2018). Characterization of hit compounds identified from high-throughput screening for their effect on blood-brain barrier integrity and amyloid-beta clearance: In vitro and in vivo studies. *Neuroscience, 379*, 269–280. https://doi.org/10.1016/j.neuroscience.2018.03.028.

Engelhart, M. J., Geerlings, M. I., Ruitenberg, A., van Swieten, J. C., Hofman, A., Witteman, J. C., et al. (2002). Dietary intake of antioxidants and risk of Alzheimer disease. *JAMA, 287*(24), 3223–3229. https://doi.org/10.1001/jama.287.24.3223.

Figueiredo, C. P., Bicca, M. A., Latini, A., Prediger, R. D., Medeiros, R., & Calixto, J. B. (2011). Folic acid plus alpha-tocopherol mitigates amyloid-beta-induced neurotoxicity through modulation of mitochondrial complexes activity. *Journal of Alzheimer's Disease, 24*(1), 61–75. https://doi.org/10.3233/JAD-2010-101320.

Galasko, D. R., Peskind, E., Clark, C. M., Quinn, J. F., Ringman, J. M., Jicha, G. A., et al. (2012). Antioxidants for Alzheimer disease: A randomized clinical trial with cerebrospinal fluid biomarker measures. *Archives of Neurology, 69*(7), 836–841. https://doi.org/10.1001/archneurol.2012.85.

Gray, S. L., Anderson, M. L., Crane, P. K., Breitner, J. C., McCormick, W., Bowen, J. D., et al. (2008). Antioxidant vitamin supplement use and risk of dementia or Alzheimer's disease in older adults. *Journal of the American Geriatrics Society, 56*(2), 291–295. https://doi.org/10.1111/j.1532-5415.2007.01531.x.

Grimm, M. O., Stahlmann, C. P., Mett, J., Haupenthal, V. J., Zimmer, V. C., Lehmann, J., et al. (2015). Vitamin E: Curse or benefit in Alzheimer's disease? A systematic investigation of the impact of alpha-, gamma- and delta-tocopherol on ass generation and degradation in neuroblastoma cells. *The Journal of Nutrition, Health & Aging, 19*(6), 646–656. https://doi.org/10.1007/s12603-015-0506-z.

Gugliandolo, A., Bramanti, P., & Mazzon, E. (2017). Role of vitamin E in the treatment of Alzheimer's disease: Evidence from animal models. *International Journal of Molecular Sciences, 18*(12). https://doi.org/10.3390/ijms18122504.

Gugliandolo, A., Chiricosta, L., Silvestro, S., Bramanti, P., & Mazzon, E. (2019). Alpha-tocopherol modulates non-amyloidogenic pathway and autophagy in an in vitro model of Alzheimer's disease: A transcriptional study. *Brain Sciences, 9*(8). https://doi.org/10.3390/brainsci9080196.

Harrison, F. E., Allard, J., Bixler, R., Usoh, C., Li, L., May, J. M., et al. (2009). Antioxidants and cognitive training interact to affect oxidative stress and memory in APP/PSEN1 mice. *Nutritional Neuroscience, 12*(5), 203–218. https://doi.org/10.1179/147683009X423364.

Hensley, K., Barnes, L. L., Christov, A., Tangney, C., Honer, W. G., Schneider, J. A., et al. (2011). Analysis of postmortem ventricular cerebrospinal fluid from patients with and without dementia indicates association of vitamin E with neuritic plaques and specific measures of cognitive performance. *Journal of Alzheimer's Disease, 24*(4), 767–774. https://doi.org/10.3233/JAD-2011-101995.

Higgins, M. R., Izadi, A., & Kaviani, M. (2020). Antioxidants and exercise performance: With a focus on vitamin E and C supplementation. *International Journal of Environmental Research and Public Health*, *17*(22). https://doi.org/10.3390/ijerph17228452.

Ishihara, Y., Itoh, K., Mitsuda, Y., Shimada, T., Kubota, T., Kato, C., et al. (2013). Involvement of brain oxidation in the cognitive impairment in a triple transgenic mouse model of Alzheimer's disease: Non-invasive measurement of the brain redox state by magnetic resonance imaging. *Free Radical Research*, *47*(9), 731–739. https://doi.org/10.3109/10715762.2013.818218.

Jiang, Q. (2017). Natural forms of vitamin E as effective agents for cancer prevention and therapy. *Advances in Nutrition*, *8*(6), 850–867. https://doi.org/10.3945/an.117.016329.

Jimenez-Jimenez, F. J., de Bustos, F., Molina, J. A., Benito-Leon, J., Tallon-Barranco, A., Gasalla, T., et al. (1997). Cerebrospinal fluid levels of alpha-tocopherol (vitamin E) in Alzheimer's disease. *Journal of Neural Transmission (Vienna)*, *104*(6–7), 703–710. https://doi.org/10.1007/BF01291887.

Kamat, C. D., Gadal, S., Mhatre, M., Williamson, K. S., Pye, Q. N., & Hensley, K. (2008). Antioxidants in central nervous system diseases: Preclinical promise and translational challenges. *Journal of Alzheimer's Disease*, *15*(3), 473–493. https://doi.org/10.3233/jad-2008-15314.

Kryscio, R. J., Abner, E. L., Caban-Holt, A., Lovell, M., Goodman, P., Darke, A. K., et al. (2017). Association of Antioxidant Supplement use and Dementia in the prevention of Alzheimer's disease by vitamin E and selenium trial (PREADViSE). *JAMA Neurology*, *74*(5), 567–573. https://doi.org/10.1001/jamaneurol.2016.5778.

Lage, J. M. (2006). 100 years of Alzheimer's disease (1906-2006). *Journal of Alzheimer's Disease*, *9*(3 Suppl), 15–26. https://doi.org/10.3233/jad-2006-9s303.

Le, N. K., Kesayan, T., Chang, J. Y., & Rose, D. Z. (2020). Cryptogenic intracranial hemorrhagic strokes associated with hypervitaminosis E and acutely elevated alpha-tocopherol levels. *Journal of Stroke and Cerebrovascular Diseases*, *29*(5), 104747. https://doi.org/10.1016/j.jstrokecerebrovasdis.2020.104747.

Lee, G. Y., & Han, S. N. (2018). The role of vitamin E in immunity. *Nutrients*, *10*(11). https://doi.org/10.3390/nu10111614.

Lim, Y., & Traber, M. G. (2007). Alpha-tocopherol transfer protein (alpha-TTP): Insights from alpha-tocopherol transfer protein knockout mice. *Nutrition Research and Practice*, *1*(4), 247–253. https://doi.org/10.4162/nrp.2007.1.4.247.

Lloret, A., Badia, M. C., Mora, N. J., Pallardo, F. V., Alonso, M. D., & Vina, J. (2009). Vitamin E paradox in Alzheimer's disease: It does not prevent loss of cognition and may even be detrimental. *Journal of Alzheimer's Disease*, *17*(1), 143–149. https://doi.org/10.3233/JAD-2009-1033.

Lloret, A., Esteve, D., Monllor, P., Cervera-Ferri, A., & Lloret, A. (2019). The effectiveness of vitamin E treatment in Alzheimer's disease. *International Journal of Molecular Sciences*, *20*(4). https://doi.org/10.3390/ijms20040879.

Luchsinger, J. A., Tang, M. X., Shea, S., & Mayeux, R. (2003). Antioxidant vitamin intake and risk of Alzheimer disease. *Archives of Neurology*, *60*(2), 203–208. https://doi.org/10.1001/archneur.60.2.203.

Mangialasche, F., Kivipelto, M., Mecocci, P., Rizzuto, D., Palmer, K., Winblad, B., et al. (2010). High plasma levels of vitamin E forms and reduced Alzheimer's disease risk in advanced age. *Journal of Alzheimer's Disease*, *20*(4), 1029–1037. https://doi.org/10.3233/JAD-2010-091450.

Mangialasche, F., Xu, W., Kivipelto, M., Costanzi, E., Ercolani, S., Pigliautile, M., et al. (2012). Tocopherols and tocotrienols plasma levels are associated with cognitive impairment. *Neurobiology of Aging*, *33*(10), 2282–2290. https://doi.org/10.1016/j.neurobiolaging.2011.11.019.

Manosso, L. M., Camargo, A., Dafre, A. L., & Rodrigues, A. L. S. (2020). Vitamin E for the management of major depressive disorder: Possible role of the anti-inflammatory and antioxidant systems. *Nutritional Neuroscience*, 1–15. https://doi.org/10.1080/1028415X.2020.1853417.

Masaki, K. H., Losonczy, K. G., Izmirlian, G., Foley, D. J., Ross, G. W., Petrovitch, H., et al. (2000). Association of vitamin E and C supplement use with cognitive function and dementia in elderly men. *Neurology*, *54*(6), 1265–1272. https://doi.org/10.1212/wnl.54.6.1265.

Morris, M. C., Beckett, L. A., Scherr, P. A., Hebert, L. E., Bennett, D. A., Field, T. S., et al. (1998). Vitamin E and vitamin C supplement use and risk of incident Alzheimer disease. *Alzheimer Disease and Associated Disorders*, *12*(3), 121–126. https://doi.org/10.1097/00002093-199809000-00001.

Morris, M. C., Evans, D. A., Bienias, J. L., Tangney, C. C., Bennett, D. A., Aggarwal, N., et al. (2002). Dietary intake of antioxidant nutrients and the risk of incident Alzheimer disease in a biracial community study. *JAMA*, *287*(24), 3230–3237. https://doi.org/10.1001/jama.287.24.3230.

Morris, M. C., Evans, D. A., Tangney, C. C., Bienias, J. L., Wilson, R. S., Aggarwal, N. T., et al. (2005). Relation of the tocopherol forms to incident Alzheimer disease and to cognitive change. *The American Journal of Clinical Nutrition*, *81*(2), 508–514. https://doi.org/10.1093/ajcn.81.2.508.

Morris, M. C., Schneider, J. A., Li, H., Tangney, C. C., Nag, S., Bennett, D. A., et al. (2015). Brain tocopherols related to Alzheimer's disease neuropathology in humans. *Alzheimers Dement*, *11*(1), 32–39. https://doi.org/10.1016/j.jalz.2013.12.015.

Nishida, Y., Ito, S., Ohtsuki, S., Yamamoto, N., Takahashi, T., Iwata, N., et al. (2009). Depletion of vitamin E increases amyloid beta accumulation by decreasing its clearances from brain and blood in a mouse model of Alzheimer disease. *The Journal of Biological Chemistry*, *284*(48), 33400–33408. https://doi.org/10.1074/jbc.M109.054056.

Nishida, Y., Yokota, T., Takahashi, T., Uchihara, T., Jishage, K., & Mizusawa, H. (2006). Deletion of vitamin E enhances phenotype of Alzheimer disease model mouse. *Biochemical and Biophysical Research Communications*, *350*(3), 530–536. https://doi.org/10.1016/j.bbrc.2006.09.083.

Pahrudin Arrozi, A., Shukri, S. N. S., Wan Ngah, W. Z., Mohd Yusof, Y. A., Ahmad Damanhuri, M. H., Jaafar, F., et al. (2020). Comparative effects of alpha- and gamma-tocopherol on mitochondrial functions in Alzheimer's disease in vitro model. *Scientific Reports*, *10*(1), 8962. https://doi.org/10.1038/s41598-020-65570-4.

Petersen, R. C., Thomas, R. G., Grundman, M., Bennett, D., Doody, R., Ferris, S., et al. (2005). Vitamin E and donepezil for the treatment of mild cognitive impairment. *The New England Journal of Medicine*, *352*(23), 2379–2388. https://doi.org/10.1056/NEJMoa050151.

Peyser, C. E., Folstein, M., Chase, G. A., Starkstein, S., Brandt, J., Cockrell, J. R., et al. (1995). Trial of d-alpha-tocopherol in Huntington's disease. *The American Journal of Psychiatry*, *152*(12), 1771–1775. https://doi.org/10.1176/ajp.152.12.1771.

Prince, M., Bryce, R., Albanese, E., Wimo, A., Ribeiro, W., & Ferri, C. P. (2013). The global prevalence of dementia: A systematic review and metaanalysis. *Alzheimers Dement*, *9*(1), 63–75. e62 https://doi.org/10.1016/j.jalz.2012.11.007.

Remington, R., Bechtel, C., Larsen, D., Samar, A., Doshanjh, L., Fishman, P., et al. (2015). A phase II randomized clinical trial of a nutritional formulation for cognition and mood in Alzheimer's disease. *Journal of Alzheimer's Disease*, *45*(2), 395–405. https://doi.org/10.3233/JAD-142499.

Remington, R., Lortie, J. J., Hoffmann, H., Page, R., Morrell, C., & Shea, T. B. (2015). A nutritional formulation for cognitive performance in mild cognitive impairment: A placebo-controlled trial with an open-label extension. *Journal of Alzheimer's Disease*, *48*(3), 591–595. https://doi.org/10.3233/JAD-150057.

Sagare, A. P., Bell, R. D., & Zlokovic, B. V. (2012). Neurovascular dysfunction and faulty amyloid beta-peptide clearance in Alzheimer disease. *Cold Spring Harbor Perspectives in Medicine*, *2*(10). https://doi.org/10.1101/cshperspect.a011452.

Sano, M., Ernesto, C., Thomas, R. G., Klauber, M. R., Schafer, K., Grundman, M., et al. (1997). A controlled trial of selegiline, alpha-tocopherol, or both as treatment for Alzheimer's disease. The Alzheimer's disease cooperative study. *The New England Journal of Medicine*, *336*(17), 1216–1222. https://doi.org/10.1056/NEJM199704243361704.

Selkoe, D. J. (2001). Alzheimer's disease: Genes, proteins, and therapy. *Physiological Reviews*, *81*(2), 741–766. https://doi.org/10.1152/physrev.2001.81.2.741.

Sung, S., Yao, Y., Uryu, K., Yang, H., Lee, V. M., Trojanowski, J. Q., et al. (2004). Early vitamin E supplementation in young but not aged mice reduces Abeta levels and amyloid deposition in a transgenic model of Alzheimer's disease. *The FASEB Journal*, *18*(2), 323–325. https://doi.org/10.1096/fj.03-0961fje.

Tamagno, E., Parola, M., Guglielmotto, M., Santoro, G., Bardini, P., Marra, L., et al. (2003). Multiple signaling events in amyloid beta-induced, oxidative stress-dependent neuronal apoptosis. *Free Radical Biology & Medicine*, *35*(1), 45–58. https://doi.org/10.1016/s0891-5849(03)00244-2.

Tohgi, H., Abe, T., Nakanishi, M., Hamato, F., Sasaki, K., & Takahashi, S. (1994). Concentrations of alpha-tocopherol and its quinone derivative in cerebrospinal fluid from patients with vascular dementia of the

Binswanger type and Alzheimer type dementia. *Neuroscience Letters*, *174*(1), 73–76. https://doi.org/10.1016/0304-3940(94)90122-8.
Uesugi, S., Ishihara, J., Iso, H., Sawada, N., Takachi, R., Inoue, M., et al. (2017). Dietary intake of anti-oxidant vitamins and risk of stroke: The Japan public health center-based prospective study. *European Journal of Clinical Nutrition*, *71*(10), 1179–1185. https://doi.org/10.1038/ejcn.2017.71.
Wang, S. W., Yang, S. G., Liu, W., Zhang, Y. X., Xu, P. X., Wang, T., et al. (2016). Alpha-tocopherol quinine ameliorates spatial memory deficits by reducing beta-amyloid oligomers, neuroinflammation and oxidative stress in transgenic mice with Alzheimer's disease. *Behavioural Brain Research*, *296*, 109–117. https://doi.org/10.1016/j.bbr.2015.09.003.
Yang, S. G., Wang, W. Y., Ling, T. J., Feng, Y., Du, X. T., Zhang, X., et al. (2010). Alpha-tocopherol quinone inhibits beta-amyloid aggregation and cytotoxicity, disaggregates preformed fibrils and decreases the production of reactive oxygen species, NO and inflammatory cytokines. *Neurochemistry International*, *57*(8), 914–922. https://doi.org/10.1016/j.neuint.2010.09.011.
Zandi, P. P., Anthony, J. C., Khachaturian, A. S., Stone, S. V., Gustafson, D., Tschanz, J. T., et al. (2004). Reduced risk of Alzheimer disease in users of antioxidant vitamin supplements: The Cache County study. *Archives of Neurology*, *61*(1), 82–88. https://doi.org/10.1001/archneur.61.1.82.

CHAPTER 3

Cognitive impairment and micronutrients: Vitamin B12, folate, and homocysteine and implications for dementia

Gianluca Rizzo[a] **and Angela Marino**[b]
[a]Independent Researcher, Messina, Italy
[b]Department of Chemical, Biological, Pharmaceutical and Environmental Sciences, University of Messina, Messina, Italy

Abbreviations

AD	Alzheimer's disease
ADAS-Cog	Alzheimer's disease assessment scale-cognition
ApoE	apolipoprotein E
CAMCOG	Cambridge cognition examination
Cbl	cobalamin
CSF	cerebrospinal fluid
DLB	dementia with Lewy bodies
FTD	Frontotemporal Dementia
MAT	methionine adenosyltransferase
MCI	mild cognitive impairment
MMSE	mini-mental state examination
MRI	magnetic resonance imaging
MTHFR	methylene tetrahydrofolate reductase
NMDA	*N*-methyl-D-aspartate receptor
PD-D	Parkinson's disease dementia
PRE-MCI	preclinical mild cognitive impairment
P-tau	phosphorylated tau protein
PUFAs	polyunsaturated fatty acids
RBANS	repeatable battery for the assessment of neuropsychological status
SAHH	S-adenosylhomocysteine hydrolase
SAM	S-adenosyl Methionine
tHcy	total homocysteine
TICS	telephone interview for cognitive status
VaD	vascular dementia
VCIND	vascular cognitive impairment, no dementia

Vitamins and Minerals in Neurological Disorders
https://doi.org/10.1016/B978-0-323-89835-5.00004-1

Introduction

Dementia is one of the main causes of disability and healthcare expenditure in Western countries and afflicts people in almost every country (Kalaria et al., 2008). With the increase in average life span, an increasing number of elderly individuals could be affected by cognitive problems. In fact, the number of persons with dementia in the world doubles every 20 years, with projections indicating 115 million people affected by 2050, compared to 35.6 million estimated in 2010 (Prince et al., 2013). Even if the causes are not well known, dementia is a multifactorial condition that includes an etiopathogenesis based on predisposition factors and environmental/behavioral factors. Among the non-modifiable factors, there are age, genetic, sex, and family history. Modifiable factors include hypertension, low physical activity, smoking, type 2 diabetes, depression, low education, high cholesterol, stroke, and obesity (Kalaria et al., 2008). There is currently no cure for the various types of dementia, and drug treatments are aimed at reducing the disease progression. Therefore, intervention on modifiable factors has greatly stimulated the research in this field. Among the eligible modifiable factors, nutrition and lifestyle could represent cost-effective approaches conferring health benefits even for other pathologies. Total homocysteine (tHcy) levels have been identified among the possible modifiable risk factors for dementia and Alzheimer's disease (AD) and, more generally, for cognitive decline in elderly people (Smith et al., 2018). Furthermore, tHcy levels are closely correlated with B vitamins, and interventions in old individuals with impaired cognition have shown a reduction in tHcy levels with a possible effect on slowing cognitive decline (Smith et al., 2018). In addition, tHcy levels can have an indirect influence through interaction with cardiovascular diseases, which in turn influence the risk of dementia. Conversely, other causative factors such as pathologies, environmental aspects, and pharmacological treatments could act on tHcy levels, which in turn affect cognitive decline. Despite clues from the literature on the possible role of B vitamins in cognitive decline, there are still conflicting data about efficacy with tHcy-lowering treatments. This chapter aims to clarify the most recent evidence in the literature, analyzing the main available observational and intervention data. Furthermore, the possible mechanisms underlying the interaction between B vitamins and dementia are here being proposed. In particular, understanding the metabolism of B vitamins is a useful step for clarifying their influence on the physiology of cognition.

Dementia and cognitive impairment: Clinical aspects

Dementia represents a heterogeneous set of pathologies affecting the central nervous system with repercussions on various cognitive domains, mainly memory. There are different types of dementia, and among the most common, AD and vascular dementia (VaD) make up 70% and 15% of all dementias, respectively (Jiang, Yu, Tian, & Tan, 2013).

Together with these, mixed dementia, Parkinson's disease with dementia (PD-D), dementia with Lewy bodies (DLB), and frontotemporal dementia (FTD) are considered. Furthermore, there are mixed conditions and overlaps, which make the diagnosis difficult. Dementia involves some variants occurring on specific cognitive domains such as language, visual, frontal, and apractic. It is commonly accepted that dementia is preceded by a prodromal phase with less clear-cut signs not involving loss of self-sufficiency, called mild cognitive impairment (MCI), and representing a stage with a greater cognitive decline than what seen in normal aging. It is estimated that 19% of individuals over 65 years of age are affected by this condition and approximately 46% of people with MCI will develop dementia in the first 3 years (Tschanz et al., 2006). As with dementia, different types of MCI have been classified over time: amnestic MCI with episodic memory impairment, MCI due to AD that it is likely to develop AD dementia, and MCI due to vascular disease or "vascular cognitive impairment, no dementia" (VCIND). The diagnosis of MCI is even more complicated by the absence of a universally accepted definition. More and more often, the concept of a preclinical stage has been proposed, considering that decades of incubation may be likely before symptoms, defined as PRE MCI (Sperling et al., 2011). Several tests are used in clinics to check cognitive functioning with various domains. The most used are the Mini-Mental State Examination (MMSE), Short Portable Mental Status Questionnaire (SPMSQ), Alzheimer's Disease Assessment Scale-Cognition (ADAS-Cog), the Telephone Interview for Cognitive Status (TICS), Cambridge Cognition Examination (CAMCOG), Montreal Cognitive Assessment (MoCA), and Repeatable Battery for the Assessment of

Table 1 Common tests for cognitive functioning.

Test	Description
ADAS-Cog	A test for cognitive status in AD, consisting of 11 tasks, including memory, spatial and temporal orientation, language, attention, and praxis
CAMCOG	It consists of 67 items regarding orientation, memory, attention, calculation, language, praxis, perception, and abstraction domains. It includes MMSE items
MMSE	A 30-point questionnaire investigating different domains such as language, orientation, attention, recall, and calculation
MoCA	Developed for mild cognitive deterioration screening. It consists of different cognitive domains such as executive, memory, language, visuospatial, abstraction, orientation, and mathematical skills
RBANS	A 5-domains test, including language, attention, visuospatial, immediate and delayed memory. It does not include motor or executive functions
SPMSQ	It consists of 10 items exploring different domains: memory, spatial orientation, and concentration
TICS	It includes 11 items pertaining to different domains: concentration, memory, language orientation, mathematical skills, and praxis. It can be administrated by telephone or face-to-face

Neuropsychological Status (RBANS) (Table 1). However, some of these are population specific and often sensitive to the patient's educational level. The heterogeneity of these tests generates difficulties in comparing different studies. Also, AD is a disease with essentially histopathological diagnosis, and only 70% of clinical diagnoses show the AD postmortem neuropathological signs (McCaddon, 2006). Brain imaging such as magnetic resonance imaging (MRI) scanning could give a more robust cognitive marker in respect to cognitive tests.

Evidences from observational studies

Prospective studies have shown a correlation between hyperhomocysteinemia and cognitive decline. Various meta-analyses of observational studies have highlighted the correlation between increased tHcy and dementia, including AD (Table 2). However, very different pooled risk values could be partly explained by the use of various cutoffs for hyperhomocysteinemia assessment, as well as the different characteristics of the cohorts. The association between high tHcy levels and AD cannot be explained by the polymorphisms of the methylene tetrahydrofolate reductase (MTHFR) gene, as

Table 2 Meta-analyses of observational studies about the link between B vitamins/homocysteine and cognition.

Reference	Studies and participants	Main findings
Beydoun et al. (2014)	5 Studies with 4388 subjects	High tHcy levels were associated with incident AD (RR = 1.93; 95% CI = 1.50–2.49; Q = 2.64)
Cao et al. (2016)	11 Studies	B vitamins was related to the decrease of dementia (RR = 0.72; 95% CI = 0.54–0.96; $P = 0.026$; $I^2 = 41.6\%$)
Hu et al. (2016)	34 Studies with 9397 subjects	Association between tHcy levels and the risk for AD (OR = 3.37; 95% CI = 1.90–5.95; $P = 0.00029$)
Nie et al. (2014)	14 Studies with 15,908 participants	Significant correlation between high tHcy concentration and risk of cognitive decline (RR = 1.53; 95% CI = 1.23–1.91; $P = 0.0002$; $I^2 = 71\%$)
Shen and Ji (2015)	9 Studies with 4830 participants for tHcy and 6 studies with 2070 subjects for folate	Increased risk of AD was associated with high tHcy levels (RR = 1.77; 95%CI = 1.37–2.16; $I^2 = 0\%$), low folate (RR = 2.11; 95% CI = 1.51–2.71; $I^2 = 0\%$) but not with Cbl (RR = 0.79; 95% CI = 0.36–1.23; $I^2 = 44.2\%$)

Table 2 Meta-analyses of observational studies about the link between B vitamins/homocysteine and cognition—cont'd

Reference	Studies and participants	Main findings
Wald et al. (2011)	8 Studies with 8669 participants	5 μmol/L increase in tHcy was associated with the risk of dementia (OR = 1.35; 95%CI = 1.02–1.79; I^2 = 0%)
Zhang, Luo, Yuan, and Ding (2020)	21 Studies	Higher B vitamins levels were protective for cognition in cross-sectional studies (Cbl: OR = 0.77; 95%CI = 0.61–0.97; I^2 = 76.9%; folate: OR = 0.68; 95%CI; 0.51–0.90; I^2 = 56.0%) but not among prospective studies
Zhou and Chen (2019)	28 Studies	Significant correlation between high tHcy and the risk of all-cause dementia (RR = 1.58; 95% CI = 1.33 1.87; I^2 = 15.4%, n = 10). Significant correlation between high tHcy and the risk of AD dementia (RR = 1.74; 95% CI = 1.32–2.29; I^2 = 58.5%; n = 8). Significant linear correlation dose-response analysis between AD dementia and 5 μmol/L increase in tHcy levels (RR = 1.15; 95% CI = 1.04–1.26; I^2 = 56.6%, n = 5)

shown in the case of Mendelian randomization for the C677T polymorphism, normally associated with an increase in tHcy (Hu, Teng, Li, Hao, & Wang, 2016). Many of the prospective studies show a gradient relationship between cognition and tHcy levels in plasma, starting from specific thresholds. For example, in normal old people, the incidence of dementia is reduced by 5 times among those individuals with baseline values of tHcy <10.1 μmol/L compared with values >15 μmol/L (Ravaglia et al., 2005). A better cognitive performance was evidenced in the case of higher folate concentrations, while the folate deficiency in the blood showed a more severe cognitive impairment with concomitant hyperhomocysteinemia (Baroni et al., 2019). Similarly, the association between medial temporal lobe atrophy and tHcy levels was observed in patients with AD, and with a higher rate in those patients with tHcy >11 μmol/L comparted with values <11 μmol/L (Clarke et al., 1998). The influence on cognitive functioning may not depend on the mere absolute value of tHcy but on its variation over time. In fact, in a prospective study on 1670 old individuals, after 6 years there was a better performance in memory tests in individuals who did not show changes in tHcy levels, compared to those showing an increase up to 8 μmol/L (Nurk et al., 2005). Consistently, individuals exhibiting a reduction up to 6 μmol/L of tHcy levels showed a better mnemonic performance after 6 years than those who had not experienced significant changes over time. Although there is a convincing association between cognitive impairment and an increase in tHcy in prospective studies, the observational nature of the studies does not allow to draw causality conclusions.

Evidences from interventional studies

A possible inverse causality between cognitive impairment and related behavioral aspects leading to vitamin deficiencies has emerged several times, though remaining not yet fully clarified. Specifically, nutrient malabsorption, feeding difficulties, and drug treatments could increase tHcy levels. In the Folic Acid and Carotid Intima-media Thickness (FACIT) clinical trial, 818 normal old persons with high baseline tHcy were treated with 0.8 mg folic acid, daily for 3 years (Durga et al., 2007). A reduction of tHcy was observed in the treated group compared with the placebo. The intervention group showed higher scores in the speed processing information domain, with better outcomes among participants with higher tHcy at baseline. In the AD Cooperative Study Trial, 340 patients with probable AD were included (Aisen et al., 2008). The treatment consisted of 5 mg folic acid, 25 mg vitamin B6, and 1 mg vitamin B12 (Cbl), daily for 18 months. Even if there wasn't any significant difference compared with placebo, an interaction between MMSE score at baseline and treatment effect was highlighted in the subgroup analysis. Benefits from B vitamins treatment were significant in participants with higher MMSE score at baseline, i.e., in patients with a moderate stage of cognitive impairment. As emerged from the VITACOG trial using Bayesian analysis, the reduction of tHcy levels with B vitamins treatments slowed brain atrophy and cognitive decline (Douaud et al., 2013). In a meta-analysis of 11 trials out of 22,000 participants, tHcy-lowering intervention with B vitamins did not show significant effects on global outcomes or specific domains of cognitive function and cognitive aging (Clarke et al., 2014). However, most of the participants did not have a baseline cognitive assessment. Similarly, another meta-analysis of trials on the effect of B vitamins on cognition did not show a definitive effect due to short duration, low sample size, and inclusion of healthy individuals not at risk of cognitive decline or, vice versa, with severely impaired and probably irreversible cognitive status (Wald, Kasturiratne, & Simmonds, 2011) (Table 3).

It is important that in clinical trials, the control exhibits a cognitive decline, to clearly detect the effect of the intervention. This, however, does not have to be high, due to the risk of irreversibility. There is a wide heterogeneity of clinical trials involving old people: with or without signs of cognitive decline; with mild cognitive decline or diagnosed dementia; with normal or high baseline tHcy values. In this regard, an international consensus statement about the link between tHcy and dementia established the need for some decisive criteria for the design of a clinical trial that can reveal effects on cognition with

Table 3 Meta-analyses of interventional studies using B vitamins for cognitive improvement.

Reference	Trials and participants	Main findings
Behrens, Graessel, Pendergrass, and Donath (2020)	8 Trials with 7552 participants	Not significant effect on global cognitive function with B vitamins supplementation (SMD = 0.03; 95%CI = −0.004, 0.09; I^2 = 41%)
Forbes, Holroyd-Leduc, Poulin, and Hogan (2015)	3 Trials with 717 participants	B vitamins supplementation did not improve MMSE scores (MD = 0.02; 95% CI = −0.22, 0.25; P = 0.90; I^2 = 0%).
Ford and Almeida (2012)	19 Trials with 5398 participants	B vitamins supplementation did not improve cognitive functions (SMD = 0.10; 95% CI = −0.08, 0.28).
Ford and Almeida (2019)	31 Trials with 17,029	Not significant effect on MMSE scores in participants with (SMD = 0.16; 95% CI = −0.18, 0.51; I^2 = 0%) or without cognitive impairment (SMD = 0.04; 95% CI − −0.10, 0.18; I^2 = 0%)
Li et al. (2014)	5 Trials	Small effect of B vitamins supplementation on memory in MCI (SMD = 0.60; 95% CI = 0.20–1.00) but no effect on general cognition or other domains. No effect on cognition among AD patients
McCleery et al. (2018)	5 Trials with 664 individuals	B vitamins intervention had no effect on overall cognitive function and cognitive domains.
Rutjes et al. (2018)	14 Trials	B vitamins intervention had no effect on global cognitive functioning (baseline or final measurement; with different years of supplementation and for different cognitive domains in subanalyses).
Wald, Kasturiratne, and Simmonds (2010)	9 Trials with 2835 participants	Not significant protective effect on cognitive decline with B vitamins supplementation (SMD = 0.01; 95%CI = −0.08, 0.10). No protective effect for intervention in subanalyses for specific cognitive domains
Zhang, Ye, Mu, and Cui (2017)	4 Trials with 679 participants	Reduction of tHcy with B vitamins (SMD = −3.625; 95%CI = −5.642, −1.608; P < 0.001; I^2 = 87.65%) but without any improvement in MMSE scores (SMD = 0.027; 95%CI = −0.518, 0.573; P = 0.921; I^2 = 0%)

tHcy-lowering interventions (Smith et al., 2018). However, clinical trials may not be suitable for the evaluation of causal effects in phenomena with long latency.

B vitamins metabolism

B vitamins participate in energy metabolism, catalyzing the pathways of one-carbon metabolism. Cobalamin (Cbl) acts as a cofactor for the methionine synthase, which converts homocysteine into methionine. In this step, Cbl, in the form of methylcobalamin, provides the methyl group for the reaction. The methyl group of Cbl is then restored by folate in the 5-methyl tetrahydrofolate form. The tetrahydrofolate, following the loss of the methyl group, can be converted into methylenetetrahydrofolate through the involvement of serine, and then reconverted to 5-methyl tetrahydrofolate by MTHFR, or it can be directed to the synthesis of nucleic acids. Methionine acts as a substrate for the formation of S-adenosyl methionine (SAM) through the catalysis of the methionine adenosyltransferase (MAT). SAM plays the role of methyl donor in various body districts, participating in the methylation of RNA, chromatin, and histones (involved in gene expression) and in the biosynthesis of phosphatidylcholine and neurotransmitters. After releasing the methyl group, SAM transforms into S-adenosyl homocysteine, a precursor of Hcy, which re-enters the cycle to be converted back into methionine. The deficiency of Cbl and folate leads to a slowing of this metabolic pathway. Therefore, Hcy can represent a functional marker of the deficiency of the vitamin cofactors involved in the pathway. Furthermore, Hcy can be converted into cysteine through a trans-sulfuration pathway, which involves vitamin B6 as a cofactor in two consecutive steps. Besides, riboflavin (vitamin B2) acts as a cofactor for the MTHFR enzyme, and niacin (vitamin B3) acts as a cofactor for S-adenosylhomocysteine hydrolase (SAHH) and dihydrofolate reductase. In part, therefore, the increase in Hcy may also depend on the lack of other B vitamins, although in the literature the evidence is greater in the case of folate and Cbl (Fig. 1). There are clues that niacin deficiency is linked to Parkinson's disease (Wakade, Chong, Bradley, & Morgan, 2015). Moreover, also thiamine (vitamin B1) was associated with cognitive impairment (Gibson et al., 2016). The absorption of Cbl is very articulated and involves various carriers, from the oral cavity to the intestinal absorption site (Rizzo & Laganà, 2020a). Proper absorption needs some prerequisites, such as sufficient gastric acidity and the secretion of intrinsic factor, without which absorption from food can be greatly compromised. Since gastric atrophy is often related to aging, this phenomenon may be partly responsible for Cbl, which in turn could promote cognitive decline. Therefore, the assessment of B vitamins in a framework based on multiple markers can provide a more effective assessment of nutritional status (Rizzo et al., 2016).

Molecular mechanisms of interaction with central nervous system

The correlation between B vitamins deficiency and anemia is easily identifiable due to the role in the one-carbon metabolism that leads to the synthesis of purines and pyrimidines

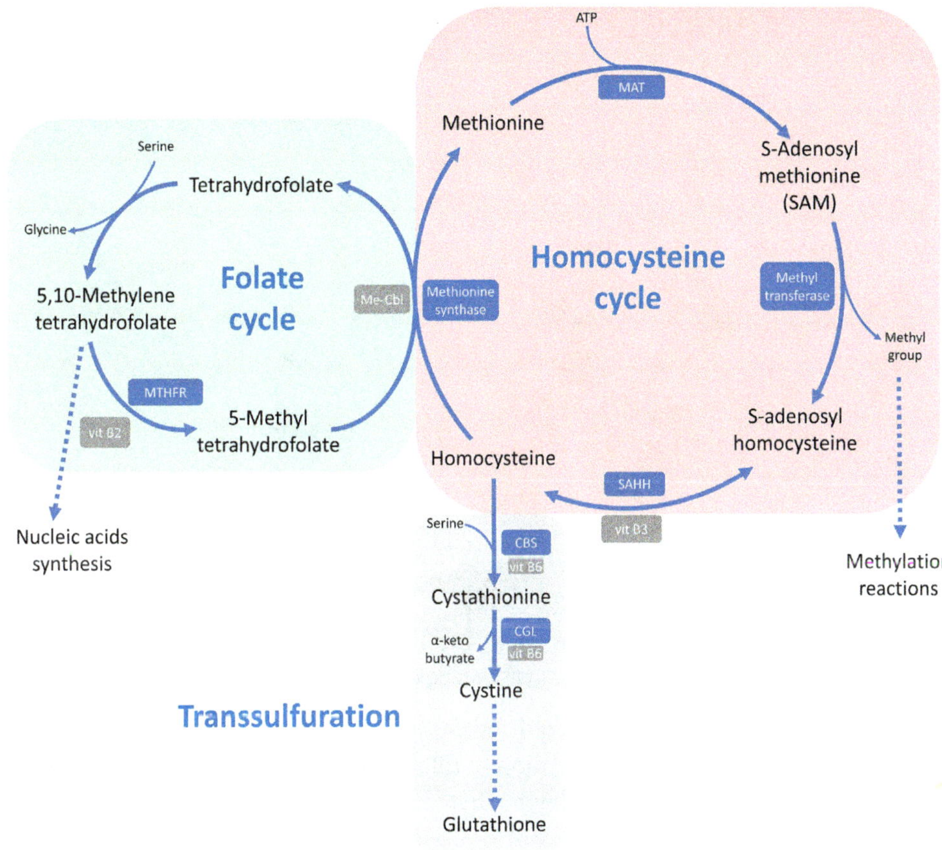

Fig. 1 B vitamins in one-carbon metabolism.

occurring in blood cells that show a high turnover within the hematopoietic process. This mechanism, however, cannot explain the influence on the brain tissue, whose cells do not have the same rapid turnover. It is much more likely that B vitamins are decisive in the fitness of neurological tissues through their role in methylation processes. In fact, both folate and Cbl are necessary for the synthesis of SAM, which acts as methyl donor in the nervous system. In turn, SAM is associated with the metabolism of proteins, membrane phospholipids, myelin, and neurotransmitters. The plausibility of a correlation between B vitamins deficiency, hyperhomocysteinemia, and cognition derives from the fact that Hcy is involved in vascular mechanisms, in epigenetic mechanisms, in the formation of the neurofibrillary tangles, amyloid plaques, and more generally in neuronal death and regional brain atrophy (Obeid & Herrmann, 2006) (Fig. 2). Deposition of insoluble proteins such as phosphorylated tau protein (P-tau) in neurofibrillary tangles and amyloid-β protein in plaques is responsible for cognitive impairment in dementia, including neuronal death. For example, folate deficiency seems to increase tau protein

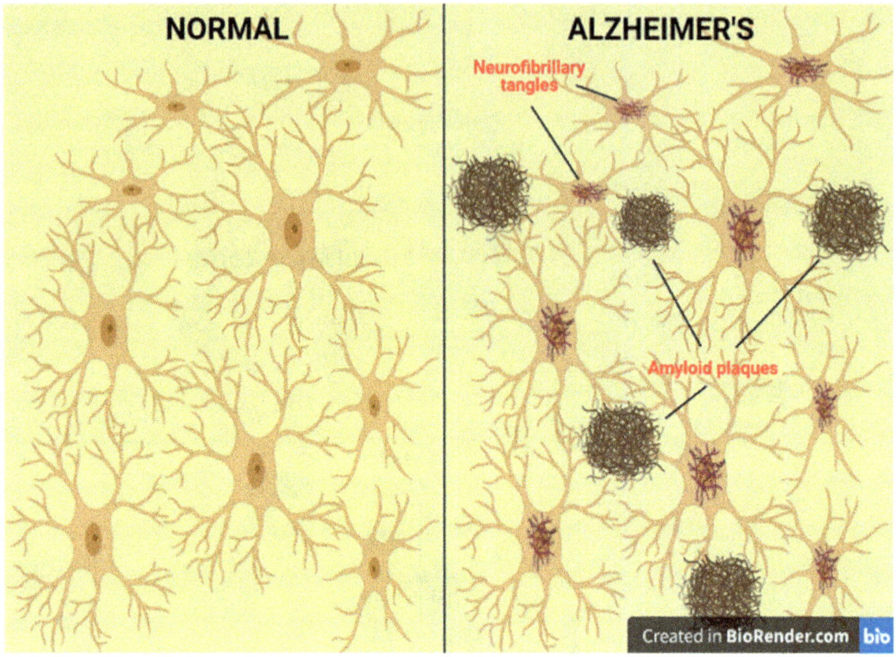

Fig. 2 Graphical representation of neurofibrillary tangles and amyloid plaques.

phosphorylation through the modulation of tHcy levels, which leads to the inhibition of phosphatase activity following enzyme hypomethylation (Chan, Alsaraby, & Shea, 2008). In fact, the first observations of the association between high tHcy levels and medial temporal lobe atrophy in patients with dementia confirm the mechanism (Clarke et al., 1998). Similarly, the association between regional brain atrophy and hyperhomocysteinemia was confirmed in the VITACOG trial (Douaud et al., 2013). The increase in tHcy levels was related to the increase in SAM in the cerebrospinal fluid (CSF), which in turn is associated with an increase in P-tau (Obeid et al., 2007). Hcy could have a toxic effect on the nervous system, exerted through neuronal excitation via the *N*-methyl-D-aspartate (NMDA) receptor, which can culminate in cell death. Hcy, in its oxidized form of homocysteic acid, is an agonist of NMDA receptor (Do, Herrling, Streit, & Cuénod, 1988). Activation of NMDA channel would lead to calcium-dependent stimulation of kinase activity (Chan et al., 2008). This receptor has been found in senile plaques and neurofibrillary tangles (Hoyer & Nitsch, 1989). The neurotoxic phenomenon of the N-homocysteinylation of lysine residues in proteins through the homocysteine thiolactone is known (Sharma, Kumar, Dar, & Singh, 2015). Noteworthy, the trans-sulfuration pathway from homocysteine to cystathionine and then to cysteine, leads to the formation of glutathione, an endogenous antioxidant. It is therefore not surprising that oxidative stress is related to homocysteine metabolism and that it acts by reducing the formation of methionine through the inhibition of

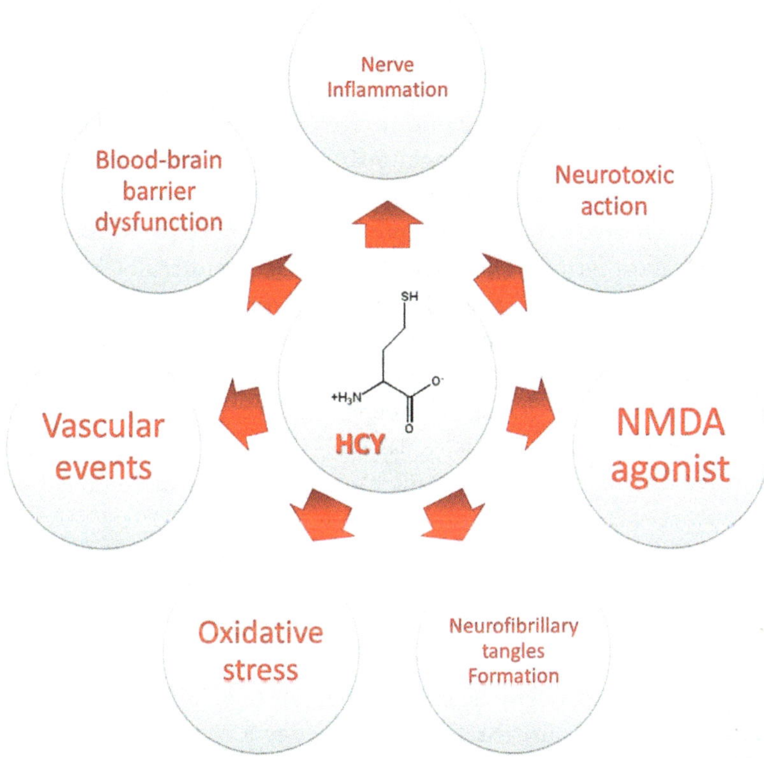

Fig. 3 Possible mechanisms of interaction between homocysteine and central nervous system.

methionine synthase and betaine-homocysteine methyltransferase, stimulating at the same time the cystathionine beta-synthase and thus favoring the formation of glutathione (Fig. 3). Among the pathophysiological signs of cognitive decline in dementia, oxidative stress and inflammation are central to the neurodegeneration process. Folic acid appears to play an advantageous role in reducing inflammation markers in AD patients treated with donepezil (Chen et al., 2016). Furthermore, the toxicity of homocysteine in the brain seems to be mediated by a mechanism of activation of reactive oxygen species and consequent oxidative stress (Gröber, Kisters, & Schmidt, 2013). Finally, homocysteine could damage the integrity of the blood-brain barrier by increasing its permeability (Beard Jr, Reynolds, & Bearden, 2011).

Application to other neurological conditions

Cbl deficiency, in addition to the classic hematological manifestations, has been related to neuronal demyelination of peripheral nerves and the central nervous system. One of the most accredited hypotheses for this phenomenon is the destruction of odd-chain fatty acids metabolism, required for the incorporation into membrane phospholipids. Cbl, in fact, in addition to participating in the conversion of homocysteine, acts as a cofactor in the form of adenosylcobalamin in the conversion of methyl malonyl CoA into succinyl CoA catalyzed by the enzyme methyl malonyl CoA mutase. High tHcy levels have been associated with central nervous system disorders, including epilepsy (Elliott, Jacobson, & Haneef, 2007) and stroke (Lehotský et al., 2016). In a retrospective study of 594,338 patients aged 18 to 70 years with no history of peripheral neuropathy, the risk of peripheral neuropathy was related to folate levels, but not to Cbl, in younger patients, while not in those over 40 years of age (Taverner et al., 2019). Furthermore, Cbl, but not folate and Hcy, appears to play a role in the development of depression (Moorthy et al., 2012). Although a clinical trial has shown positive results regarding B vitamins supplementation on depression, further confirmatory studies are needed. As in the case of cognitive decline, the link between B vitamins and depression is a correlation proposed several times without a definitive mechanistic explanation (Alpert & Fava, 1997). The possible mechanism could involve the antidepressant action of SAM (Mischoulon & Fava, 2002).

Other components of interest

The increase in Hcy levels, and therefore also B vitamins deficiency, is not only associated with dementia but also with other clinical aspects. Stroke, heart disease, and macular degeneration are aspects related to the increase in total Hcy. Furthermore, some of these clinical conditions represent the risk factors for cognitive decline. Mechanistically, it is not surprising that a vascular insufficiency in the brain caused by cerebral vasculature dysfunction can contribute to dementia (Smith & Refsum, 2016). Conversely, the potential interventions on modifiable factors concern not only B vitamins but also other nutrients possibly involved in the slowing and prevention of cognitive decline. It is important to understand the effects of such molecules that could have a parallel or adjuvant effect on B vitamins. For example, in a post hoc analysis of results from the VITACOG trial, it was shown that only participants with a better status of omega 3 polyunsaturated fatty acids (PUFAs) at baseline exhibited a beneficial effect of B vitamins treatments, slowing clinical and cognitive decline as well as brain atrophy (Oulhaj, Jernerén, Refsum, Smith, & de Jager, 2016). Polyunsaturated fatty acids appear to interact with one-carbon metabolism, stimulating various enzymes that take part in the Hcy cycle such as methylenetetrahydrofolate reductase, cystathionine-γ lyase, methionine adenosyltransferase, and other enzymes indirectly related to the degradation of Hcy via betaine, such as choline-phosphate cytidylyltransferase and phosphatidylethanolamine *N*-methyltransferase (Rizzo & Laganà, 2020b). Vitamin D deficiency seems to be related to numerous neurodegenerative processes perhaps for its antioxidant properties (Dursun & Gezen-Ak, 2019). Furthermore, low levels of vitamin D are correlated with advancing age. Vitamin D appears to act on cystathionine β-synthase levels, thus indirectly influencing Hcy metabolism (Kriebitzsch et al., 2011). The ApoE genotype represents a non-modifiable risk factor for cognitive decline. Individuals with ε4 allele of the ApoE gene show a greater correlation between low levels of Cbl and cognitive impairment (Vogiatzoglou et al., 2013). These data suggest an interaction between genome and vitamin status that deserves further investigation.

Mini-dictionary of terms

- **Dementia:** Chronic degenerative condition that includes different types of diseases. It acts with progressive deterioration of brain functions and cognitive faculties with the consequent loss of autonomy and self-sufficiency.
- **MMSE:** The most used test in medicine for assessing cognitive impairment. It consists of a 30-point developed questionnaire that includes several cognitive domains.
- **Hcy:** A sulfur-containing amino acid that does not code for proteins. It represents an independent marker for cardiovascular diseases. Its increase is related to B vitamins deficiency.
- **MTHFR:** The enzyme that converts 5,10 methylenetetrahydrofolate into 5 methyltetrahydrofolate. This is a step that can be limited due to the presence of polymorphisms in the MTHFR gene with reduction of enzymatic conversion and accumulation of homocysteine.
- **SAM:** A coenzyme that acts as a donor of methyl groups in many metabolic pathways such as methylation of nucleic acids, proteins, and lipids with regulatory purposes.
- **Amyloid-β proteins:** Peptides with a length of 36–42 amino acids that can be found in the amyloid plaques of individuals with AD. They originate from misfolding phenomena and show neurotoxicity.
- **P-tau:** Tau proteins are soluble molecules associated with the functionality of the microtubules of neuronal axons. In dementias, there is hyperphosphorylation of tau proteins (P-tau) with a reduction in solubility and formation of neurofibrillary tangles with consequent neurotoxicity.
- **NMDA:** A membrane receptor of nerve cells with the function of an ion channel. An excess of its stimulation leads to excitotoxicity and consequent neurodegeneration.

Key facts of B vitamins

- Vitamins B are water-soluble vitamins that participate in cell metabolism as enzymatic cofactors for energetic pathways, hemopoiesis, in the nervous tissue, and the biosynthesis of molecules such as nucleic acids.
- Several B vitamins participate in the one-carbon metabolism, and their deficiency is related to the increase in homocysteine levels and the disturbance of the regulatory functions that involve methylation.

- Vitamin B12 can be found only in foods of animal origin. A strictly plant-based diet that does not include supplements or fortified foods leads to vitamin B12 deficiency.
- Folate is a highly thermolabile vitamin. In circumstances where the demands are greater, such as pregnancy, the diet may fail to guarantee adequate intake.
- Folic acid is a synthetic precursor of tetrahydrofolate with greater stability than dietary folate. Being a provitamin, it must be converted to be used by the cell but the lower thermolability makes it easier to use for food fortification or supplement formulation.
- Folate or vitamin B12 deficiency shows a similar clinical picture with hematological and neurological symptoms.

Key facts of dementia

- Dementia can be classified into subsets with peculiar histological characteristics; however, there may be mixed states that present characteristics of multiple types of dementia.
- There are conditions of dementia secondary to other diseases such as AIDS, infections, liver and kidney diseases, some of which are considered reversible.
- The types of dementia can also be divided according to topodiagnostic criteria: cortical and subcortical.
- Since dementia is often a degenerative and irreversible condition, the current research effort is focusing on the prodromal stages, through the definition of recognition criteria and possible interventions.
- The main symptoms of dementia are memory loss, speech problems, disorientation, motor problems, personality disorders, and cognitive deficits of various kinds.
- The pharmacological treatments currently available are aimed at slowing down cognitive impairment but there is currently no cure for primary dementia.

Summary points

- The incidence of dementia is rising worldwide, with an estimation of 115 million people affected by 2050.
- Dementia is a multifactorial disease with non-modifiable factors such as genetics and age and modifiable factors such as lifestyle, hypertension, and homocysteine levels.
- The reduction of homocysteine levels through intervention with B vitamins is a candidate intervention to deal with cognitive decline.
- Interventions with B vitamins have shown efficacy in reducing homocysteine levels but the results on cognition gave mixed results.
- There may be a correlation between the effectiveness of treatments with B vitamins to reduce homocysteine levels and levels of PUFA or vitamin D.
- It will be necessary to conduct further clinical trials on populations at risk or with prodromes of cognitive disorders and with confirmed signs of B vitamins deficiency. Trials need an adequate time of conduction to verify the incidence of cognitive issues.

References

Aisen, P. S., Schneider, L. S., Sano, M., Diaz-Arrastia, R., van Dyck, C. H., Weiner, M. F., et al. (2008). High-dose B vitamin supplementation and cognitive decline in Alzheimer disease: A randomized controlled trial. *JAMA*, *300*, 1774–1783. https://doi.org/10.1001/jama.300.15.1774.

Alpert, J. E., & Fava, M. (1997). Nutrition and depression: The role of folate. *Nutrition Reviews*, *55*, 145–149. https://doi.org/10.1111/j.1753-4887.1997.tb06468.x.

Baroni, L., Bonetto, C., Rizzo, G., Bertola, C., Caberlotto, L., & Bazzerla, G. (2019). Association between cognitive impairment and vitamin B12, folate, and homocysteine status in elderly adults: A retrospective study. *Journal of Alzheimer's Disease: JAD*, *70*, 443–453. https://doi.org/10.3233/JAD-190249.

Beard, R. S., Jr., Reynolds, J. J., & Bearden, S. E. (2011). Hyperhomocysteinemia increases permeability of the blood-brain barrier by NMDA receptor-dependent regulation of adherens and tight junctions. *Blood*, *118*, 2007–2014. https://doi.org/10.1182/blood-2011-02-338269.

Behrens, A., Graessel, E., Pendergrass, A., & Donath, C. (2020). Vitamin B-can it prevent cognitive decline? A systematic review and meta-analysis. *Systematic Reviews*, *9*, 111. https://doi.org/10.1186/s13643-020-01378-7.

Beydoun, M. A., Beydoun, H. A., Gamaldo, A. A., Teel, A., Zonderman, A. B., & Wang, Y. (2014). Epidemiologic studies of modifiable factors associated with cognition and dementia: Systematic review and meta-analysis. *BMC Public Health*, *14*, 643. https://doi.org/10.1186/1471-2458-14-643.

Cao, L., Tan, L., Wang, H. F., Jiang, T., Zhu, X. C., Lu, H., et al. (2016). Dietary patterns and risk of dementia: A systematic review and meta-analysis of cohort studies. *Molecular Neurobiology*, *53*, 6144–6154. https://doi.org/10.1007/s12035-015-9516-4.

Chan, A. Y., Alsaraby, A., & Shea, T. B. (2008). Folate deprivation increases tau phosphorylation by homocysteine-induced calcium influx and by inhibition of phosphatase activity: Alleviation by S-adenosyl methionine. *Brain Research*, *1199*, 133–137. https://doi.org/10.1016/j.brainres.2008.01.008.

Chen, H., Liu, S., Ji, L., Wu, T., Ji, Y., Zhou, Y., et al. (2016). Folic acid supplementation mitigates Alzheimer's disease by reducing inflammation: A randomized controlled trial. *Mediators of Inflammation*, *2016*, 5912146. https://doi.org/10.1155/2016/5912146.

Clarke, R., Bennett, D., Parish, S., Lewington, S., Skeaff, M., Eussen, S. J. P. M., et al. (2014). Effects of homocysteine lowering with B vitamins on cognitive aging: Meta-analysis of 11 trials with cognitive data on 22,000 individuals. *The American Journal of Clinical Nutrition*, *100*, 657–666. https://doi.org/10.3945/ajcn.113.076349.

Clarke, R., Smith, A. D., Jobst, K. A., Refsum, H., Sutton, L., & Ueland, P. M. (1998). Folate, vitamin B12, and serum total homocysteine levels in confirmed Alzheimer disease. *Archives of Neurology*, *55*, 1449–1455. https://doi.org/10.1001/archneur.55.11.1449.

Do, K. Q., Herrling, P. L., Streit, P., & Cuénod, M. (1988). Release of neuroactive substances: Homocysteic acid as an endogenous agonist of the NMDA receptor. *Journal of Neural Transmission*, *72*, 185–190. https://doi.org/10.1007/BF01243418.

Douaud, G., Refsum, H., de Jager, C. A., Jacoby, R., Nichols, T. E., Smith, S. M., et al. (2013). Preventing Alzheimer's disease-related gray matter atrophy by B-vitamin treatment. *Proceedings of the National Academy of Sciences of the United States of America*, *110*, 9523–9528. https://doi.org/10.1073/pnas.1301816110.

Durga, J., van Boxtel, M. P. J., Schouten, E. G., Kok, F. J., Jolles, J., Katan, M. B., et al. (2007). Effect of 3-year folic acid supplementation on cognitive function in older adults in the FACIT trial: A randomised, double blind, controlled trial. *Lancet (London, England)*, *369*, 208–216. https://doi.org/10.1016/S0140-6736(07)60109-3.

Dursun, E., & Gezen-Ak, D. (2019). Vitamin D basis of Alzheimer's disease: From genetics to biomarkers. *Hormones (Athens, Greece)*, *18*, 7–15. https://doi.org/10.1007/s42000-018-0086-5.

Elliott, J. O., Jacobson, M. P., & Haneef, Z. (2007). Cardiovascular risk factors and homocysteine in epilepsy. *Epilepsy Research*, *76*, 113–123. https://doi.org/10.1016/j.eplepsyres.2007.07.005.

Forbes, S. C., Holroyd-Leduc, J. M., Poulin, M. J., & Hogan, D. B. (2015). Effect of nutrients, dietary supplements and vitamins on cognition: A systematic review and meta-analysis of randomized controlled trials. *Canadian Geriatrics Journal: CGJ*, *18*, 231–245. https://doi.org/10.5770/cgj.18.189.

Ford, A. H., & Almeida, O. P. (2012). Effect of homocysteine lowering treatment on cognitive function: A systematic review and meta-analysis of randomized controlled trials. *Journal of Alzheimer's Disease: JAD*, *29*, 133–149. https://doi.org/10.3233/JAD-2012-111739.

Ford, A. H., & Almeida, O. P. (2019). Effect of vitamin B supplementation on cognitive function in the elderly: A systematic review and meta-analysis. *Drugs & Aging*, *36*, 419–434. https://doi.org/10.1007/s40266-019-00649-w.

Gibson, G. E., Hirsch, J. A., Fonzetti, P., Jordan, B. D., Cirio, R. T., & Elder, J. (2016). Vitamin B1 (thiamine) and dementia. *Annals of the New York Academy of Sciences*, *1367*, 21–30. https://doi.org/10.1111/nyas.13031.

Gröber, U., Kisters, K., & Schmidt, J. (2013). Neuroenhancement with vitamin B12-underestimated neurological significance. *Nutrients*, *5*, 5031–5045. https://doi.org/10.3390/nu5125031.

Hoyer, S., & Nitsch, R. (1989). Cerebral excess release of neurotransmitter amino acids subsequent to reduced cerebral glucose metabolism in early-onset dementia of Alzheimer type. *Journal of Neural Transmission*, *75*, 227–232. https://doi.org/10.1007/BF01258634.

Hu, Q., Teng, W., Li, J., Hao, F., & Wang, N. (2016). Homocysteine and Alzheimer's disease: Evidence for a causal link from Mendelian randomization. *Journal of Alzheimer's Disease: JAD*, *52*, 747–756. https://doi.org/10.3233/JAD-150977.

Jiang, T., Yu, J. T., Tian, Y., & Tan, L. (2013). Epidemiology and etiology of Alzheimer's disease: From genetic to non-genetic factors. *Current Alzheimer Research*, *10*, 852–867. https://doi.org/10.2174/15672050113109990155.

Kalaria, R. N., Maestre, G. E., Arizaga, R., Friedland, R. P., Galasko, D., Hall, K., et al. (2008). Alzheimer's disease and vascular dementia in developing countries: Prevalence, management, and risk factors. *The Lancet. Neurology*, 7, 812–826. https://doi.org/10.1016/S1474-4422(08)70169-8.

Kriebitzsch, C., Verlinden, L., Eelen, G., van Schoor, N. M., Swart, K., Lips, P., et al. (2011). 1,25-dihydroxyvitamin D3 influences cellular homocysteine levels in murine preosteoblastic MC3T3-E1 cells by direct regulation of cystathionine β-synthase. *Journal of Bone and Mineral Research: the Official Journal of the American Society for Bone and Mineral Research*, *26*, 2991–3000. https://doi.org/10.1002/jbmr.493.

Lehotský, J., Tothová, B., Kovalská, M., Dobrota, D., Beňová, A., Kalenská, D., et al. (2016). Role of homocysteine in the ischemic stroke and development of ischemic tolerance. *Frontiers in Neuroscience*, *10*, 538. https://doi.org/10.3389/fnins.2016.00538.

Li, M. M., Yu, J. T., Wang, H. F., Jiang, T., Wang, J., Meng, X. F., et al. (2014). Efficacy of vitamins B supplementation on mild cognitive impairment and Alzheimer's disease: A systematic review and meta-analysis. *Current Alzheimer Research*, *11*, 844–852.

McCaddon, A. (2006). Homocysteine and cognition—A historical perspective. *Journal of Alzheimer's Disease: JAD*, *9*, 361–380. https://doi.org/10.3233/jad-2006-9402.

McCleery, J., Abraham, R. P., Denton, D. A., Rutjes, A. W., Chong, L. Y., Al-Assaf, A. S., et al. (2018). Vitamin and mineral supplementation for preventing dementia or delaying cognitive decline in people with mild cognitive impairment. *The Cochrane Database of Systematic Reviews*, *11*, CD011905. https://doi.org/10.1002/14651858.CD011905.pub2.

Mischoulon, D., & Fava, M. (2002). Role of S-adenosyl-L-methionine in the treatment of depression: A review of the evidence. *The American Journal of Clinical Nutrition*, *76*, 1158S–1161S. https://doi.org/10.1093/ajcn/76/5.1158S.

Moorthy, D., Peter, I., Scott, T. M., Parnell, L. D., Lai, C. Q., Crott, J. W., et al. (2012). Status of vitamins B-12 and B-6 but not of folate, homocysteine, and the methylenetetrahydrofolate reductase C677T polymorphism are associated with impaired cognition and depression in adults. *The Journal of Nutrition*, *142*, 1554–1560. https://doi.org/10.3945/jn.112.161828.

Nie, T., Lu, T., Xie, L., Huang, P., Lu, Y., & Jiang, M. (2014). Hyperhomocysteinemia and risk of cognitive decline: A meta-analysis of prospective cohort studies. *European Neurology*, *72*, 241–248. https://doi.org/10.1159/000363054.

Nurk, E., Refsum, H., Tell, G. S., Engedal, K., Vollset, S. E., Ueland, P. M., et al. (2005). Plasma total homocysteine and memory in the elderly: The Hordaland homocysteine study. *Annals of Neurology*, *58*, 847–857. https://doi.org/10.1002/ana.20645.

Obeid, R., & Herrmann, W. (2006). Mechanisms of homocysteine neurotoxicity in neurodegenerative diseases with special reference to dementia. *FEBS Letters*, *580*, 2994–3005. https://doi.org/10.1016/j.febslet.2006.04.088.

Obeid, R., Kasoha, M., Knapp, J. P., Kostopoulos, P., Becker, G., Fassbender, K., et al. (2007). Folate and methylation status in relation to phosphorylated tau protein(181P) and beta-amyloid(1-42) in cerebrospinal fluid. *Clinical Chemistry*, *53*, 1129–1136. https://doi.org/10.1373/clinchem.2006.085241.

Oulhaj, A., Jernerén, F., Refsum, H., Smith, A. D., & de Jager, C. A. (2016). Omega-3 fatty acid status enhances the prevention of cognitive decline by B vitamins in mild cognitive impairment. *Journal of Alzheimer's Disease: JAD*, *50*, 547–557. https://doi.org/10.3233/JAD-150777.

Prince, M., Bryce, R., Albanese, E., Wimo, A., Ribeiro, W., & Ferri, C. P. (2013). The global prevalence of dementia: A systematic review and metaanalysis. *Alzheimer's & Dementia: The Journal of the Alzheimer's Association*, *9*, 63–75.e2. https://doi.org/10.1016/j.jalz.2012.11.007.

Ravaglia, G., Forti, P., Maioli, F., Martelli, M., Servadei, L., Brunetti, N., et al. (2005). Homocysteine and folate as risk factors for dementia and Alzheimer disease. *The American Journal of Clinical Nutrition*, *82*, 636–643. https://doi.org/10.1093/ajcn.82.3.636.

Rizzo, G., & Laganà, A. S. (2020a). Chapter 6—A review of vitamin B12 (A c. Di). In V. B. Patel (Ed.), *Molecular nutrition* (pp. 105–129). Academic Press. https://doi.org/10.1016/B978-0-12-811907-5.00005-1.

Rizzo, G., & Laganà, A. S. (2020b). The link between homocysteine and omega-3 polyunsaturated fatty acid: Critical appraisal and future directions. *Biomolecules*, *10*. https://doi.org/10.3390/biom10020219.

Rizzo, G., Laganà, A. S., Rapisarda, A. M. C., La Ferrera, G. M. G., Buscema, M., Rossetti, P., et al. (2016). Vitamin B12 among vegetarians: Status, assessment and supplementation. *Nutrients*, *8*. https://doi.org/10.3390/nu8120767.

Rutjes, A. W., Denton, D. A., Di Nisio, M., Chong, L. Y., Abraham, R. P., Al-Assaf, A. S., et al. (2018). Vitamin and mineral supplementation for maintaining cognitive function in cognitively healthy people in mid and late life. *The Cochrane Database of Systematic Reviews*, *12*, CD011906. https://doi.org/10.1002/14651858.CD011906.pub2.

Sharma, G. S., Kumar, T., Dar, T. A., & Singh, L. R. (2015). Protein N-homocysteinylation: From cellular toxicity to neurodegeneration. *Biochimica et Biophysica Acta*, *1850*, 2239–2245. https://doi.org/10.1016/j.bbagen.2015.08.013.

Shen, L., & Ji, H. F. (2015). Associations between homocysteine, folic acid, vitamin B12 and Alzheimer's disease: Insights from meta-analyses. *Journal of Alzheimer's Disease: JAD*, *46*, 777–790. https://doi.org/10.3233/JAD-150140.

Smith, A. D., & Refsum, H. (2016). Homocysteine, B vitamins, and cognitive impairment. *Annual Review of Nutrition*, *36*, 211–239. https://doi.org/10.1146/annurev-nutr-071715-050947.

Smith, A. D., Refsum, H., Bottiglieri, T., Fenech, M., Hooshmand, B., McCaddon, A., et al. (2018). Homocysteine and dementia: An international consensus statement. *Journal of Alzheimer's Disease: JAD*, *62*, 561–570. https://doi.org/10.3233/JAD-171042.

Sperling, R. A., Aisen, P. S., Beckett, L. A., Bennett, D. A., Craft, S., Fagan, A. M., et al. (2011). Toward defining the preclinical stages of Alzheimer's disease: Recommendations from the National Institute on Aging-Alzheimer's association workgroups on diagnostic guidelines for Alzheimer's disease. *Alzheimer's & Dementia: The Journal of the Alzheimer's Association*, 7, 280–292. https://doi.org/10.1016/j.jalz.2011.03.003.

Taverner, T., Crowe, F. L., Thomas, G. N., Gokhale, K., Thayakaran, R., Nirantharakumar, K., et al. (2019). Circulating folate concentrations and risk of peripheral neuropathy and mortality: A retrospective cohort study in the U.K. *Nutrients*, *11*. https://doi.org/10.3390/nu11102443.

Tschanz, J. T., Welsh-Bohmer, K. A., Lyketsos, C. G., Corcoran, C., Green, R. C., Hayden, K., et al. (2006). Conversion to dementia from mild cognitive disorder: The cache county study. *Neurology*, *67*, 229–234. https://doi.org/10.1212/01.wnl.0000224748.48011.84.

Vogiatzoglou, A., Smith, A. D., Nurk, E., Drevon, C. A., Ueland, P. M., Vollset, S. E., et al. (2013). Cognitive function in an elderly population: Interaction between vitamin B12 status, depression, and apolipoprotein E ε4: The Hordaland homocysteine study. *Psychosomatic Medicine*, *75*, 20–29. https://doi.org/10.1097/PSY.0b013e3182761b6c.

Wakade, C., Chong, R., Bradley, E., & Morgan, J. C. (2015). Low-dose niacin supplementation modulates GPR109A, niacin index and ameliorates Parkinson's disease symptoms without side effects. *Clinical Case Reports*, *3*, 635–637. https://doi.org/10.1002/ccr3.232.

Wald, D. S., Kasturiratne, A., & Simmonds, M. (2010). Effect of folic acid, with or without other B vitamins, on cognitive decline: Meta-analysis of randomized trials. *The American Journal of Medicine*, *123*, 522–527. e2. https://doi.org/10.1016/j.amjmed.2010.01.017.
Wald, D. S., Kasturiratne, A., & Simmonds, M. (2011). Serum homocysteine and dementia: Meta-analysis of eight cohort studies including 8669 participants. *Alzheimer's & Dementia: The Journal of the Alzheimer's Association*, 7, 412–417. https://doi.org/10.1016/j.jalz.2010.08.234.
Zhang, C., Luo, J., Yuan, C., & Ding, D. (2020). Vitamin B12, B6, or folate and cognitive function in community-dwelling older adults: A systematic review and meta-analysis. *Journal of Alzheimer's Disease: JAD*, 77, 781–794. https://doi.org/10.3233/JAD-200534.
Zhang, D. M., Ye, J. X., Mu, J. S., & Cui, X. P. (2017). Efficacy of vitamin B supplementation on cognition in elderly patients with cognitive-related diseases. *Journal of Geriatric Psychiatry and Neurology*, *30*, 50–59. https://doi.org/10.1177/0891988716673466.
Zhou, F., & Chen, S. (2019). Hyperhomocysteinemia and risk of incident cognitive outcomes: An updated dose-response meta-analysis of prospective cohort studies. *Ageing Research Reviews*, *51*, 55–66. https://doi.org/10.1016/j.arr.2019.02.006.

CHAPTER 4

Magnesium and Alzheimer's disease

Dènahin Hinnoutondji Toffa[a] **and Jimmy Li**[b]
[a]Department of Neurosciences, University of Montreal Hospital Research Centre, Montreal, QC, Canada
[b]Department of Neurology, University of Sherbrooke, Sherbrooke, QC, Canada

Abbreviations

APP	amyloid precursor protein
BBB	blood-brain barrier
CSF	cerebrospinal fluid
ERK1/2	extracellular signal-regulated protein kinases 1 and 2
GLUT1	glucose transporter 1
GSK-3β	glycogen synthase kinase 3-beta
IFN-γ	gamma-interferon
IL-1	interleukin 1-beta
LAM	leukocyte adhesion molecule
LRP	lipoprotein receptor-related protein
LRP1	lipoprotein receptor-related protein 1
mGluR5	metabotropic glutamate receptor 5
MMP-9	matrix metallopeptidase 9
MPP	1-methyl-4-phenylpyridinium
NFTs	neurofibrillary tangles
NF-κB	nuclear factor-kappa B
NMDA	N-methyl-D-aspartate
NMDARs	NMDA receptors
PI3-K/Akt	phosphoinositide-3-kinase/protein kinase B
PICALM	phosphatidylinositol-binding clathrin assembly protein
PPARγ	peroxisome proliferator-activated receptor gamma
PrP^{C}	cellular prion protein
RAGE	receptor for advanced glycation end products
TLR4	toll-like receptors 4
TNF-α	tumor-necrosis factor alpha
VEGF	vascular endothelial growth factor

Introduction

Magnesium (Mg) is a divalent cation that plays a major role in more than 300 cytosolic enzymatic reactions related to nucleic acid metabolism regulation, protein synthesis and spatial configuration, mitochondrial energy production, and survival signaling (Allgrove, 2015). Due to these properties, the availability and metabolism of Mg have the potential to significantly alter cell survival and excitability in various situations. Several studies have

Vitamins and Minerals in Neurological Disorders
https://doi.org/10.1016/B978-0-323-89835-5.00032-6

demonstrated these properties at the cell and tissue levels (Trapani, Mastrototaro, & Wolf, 2011). However, there is still a lack of clinical data on the subject, probably due to the multifactorial nature of the diseases associated with Mg dysmetabolism. Despite the controversies, the association between Mg metabolism and several types of dementia, such as Alzheimer's disease (AD), is well established and has been garnering increasing levels of attention from the research community.

In this chapter, we will discuss the close association between Mg metabolism and key brain mechanisms involved in excitotoxicity, neuroinflammation, and other processes affecting neuronal survival and excitability. All these mechanisms can result in disorders of premature cerebral aging such as AD. We will also briefly discuss what is currently known about the therapeutic potentials of Mg in AD.

Metabolism of Mg: Generalities on physiology and neurodegeneration pathophysiology

Mg is finely regulated at both intracellular and extracellular levels. Its plasmatic concentration is normally maintained within the range of 0.75–1.4 mmol/L and is balanced via variations in intestinal absorption as well as changes in urinary excretion (Allgrove, 2015). About 47.5%–50% of plasmatic Mg is in a free ionized form, whereas 42%–46% is bound to proteins (Allgrove, 2015). The remaining extracellular Mg forms complexes with molecules such as phosphates and citrates. Regarding the intracellular pool of Mg, less than 10% is present in a free ionized form (Allgrove, 2015).

The impact of Mg on neuronal membrane permeability and various intracellular processes confers Mg a critical role in neuronal excitability, neuroglial communication, protein synthesis, and neuroglial viability (Kirkland, Sarlo, & Holton, 2018). Notably, protein synthesis, cell survival, and cell proliferation are regulated by a combined action of intracellular free Mg and Mg-ATP complexes (Allgrove, 2015). Regarding neurotransmission, Mg modulates the activity of *N*-methyl-D-aspartate receptors (NMDARs) at both synaptic and extrasynaptic levels. The main synaptic role of Mg consists in the regulation of postsynaptic excitatory responses by modulating the sensitivity of NMDARs to glutamate. On the extrasynaptic level, Mg modulates the probability of presynaptic glutamate release. Mg also modulates the release and reuptake of glutamate by astrocytes. In addition to these roles, intracellular Mg promotes cell survival or conversely induces cell death, depending on the cellular level (Trapani et al., 2011).

There are works suggesting that low levels of cellular Mg are the dominant type of dysmetabolism leading to brain degeneration (Kirkland et al., 2018; Trapani et al., 2011). This central neurodegeneration may be secondary to NMDA-related excitotoxicity or result from the primary impact of poor Mg availability. Regarding the first hypothesis, hypomagnesemia-induced seizures are well described in severe cases of congenital hypomagnesemia (Chen, Prasad, Kobrzynski, Campbell, & Filler, 2016;Franken et al., 2021;

Ndiaye et al., 2013). It has also been reported that patients with congenital hypomagnesemia develop additional progressive severe cognitive impairments and brain atrophy if left untreated (Franken et al., 2021; Ndiaye et al., 2013). On the other hand, there are studies suggesting that low Mg levels may lead to seizure-free neurodegeneration. For example, Kieboom et al. (2017) analyzed the incidence of dementia in 9569 patients with normal cognition at baseline (median follow-up of 7.8 years). Their work showed that both low serum Mg levels ($\leq$0.79 mmol/L) and high serum Mg levels ($\geq$0.90 mmol/L) increase the risk of dementia. Another study that included 101 CE patients reported a close relationship between Mg ion (but not total Mg) and AD occurrence (Barbagallo, Belvedere, Di Bella, & Dominguez, 2011). These authors found significantly lower Mg ion in the AD group as compared to an age-matched control group free of AD (0.50 ± 0.01 mmol/L vs 0.53 ± 0.01 mmol/L; $p < 0.01$) (Barbagallo et al., 2011). One of the most salient publications highlighting the pathogenic impact of low Mg levels consists of a systematic review that collected 13 studies (including 559 patients with AD, 381 healthy controls, and 126 patients with other medical illnesses) and that showed that there was a significantly lower level of Mg in the cerebrospinal fluid (CSF) of AD patients. Interestingly, such a difference was not observed for serum Mg levels (Veronese et al., 2016).

Besides its direct neuronal impact, it seems that Mg dysmetabolism may alter blood-brain barrier (BBB) function and thus potentially promote neurodegeneration. Indeed, Wolf et al. found that low extracellular Mg levels increase the oxidative stress generated in endothelial cells by hydrogen peroxide exposure, thereby potentially impairing brain microperfusion, with subsequent degenerative consequences (Wolf, Trapani, Simonacci, Ferré, & Maier, 2008).

Alzheimer's disease: Generalities

Currently, AD is considered a condition resulting from a combination of excitotoxicity and nonexcitotoxic processes, which lead to premature neuronal aging. AD is the most frequent (60%–70% of dementia cases) and well-studied neurodegenerative disease (Lane, Hardy, & Schott, 2018). While most AD cases appear sporadic, rare (<0.5%) familial forms exist and are associated with mutations in three genes: amyloid precursor protein (*APP*), presenilin 1 (*PSEN1*), and presenilin 2 (*PSEN2*) (Lane et al., 2018). Familial AD seems to develop earlier, usually between 30 and 50 years of age. AD is characterized by an early progressive dementia associated with diffuse brain atrophy (predominantly the temporal, parietal, cingulate, and anterior lateral frontal regions), extracellular plaques of aggregated beta-amyloid (A_β), and intraneuronal neurofibrillary tangles (NFTs) (Lane et al., 2018). In addition to these cardinal features, dystrophic neurites, neuropil threads, astrogliosis, microglial activation, and cerebral amyloid angiopathy are frequently observed (Lane et al., 2018).

Plaques are extracellular insoluble deposits resulting from the accumulation of abnormally folded $A_β$ polymers. These $A_β$ polymers represent various combinations of small peptides that span 39–43 amino acids (mostly $A_β$42) and that are synthesized from the fragmentation of the full-length amyloid precursor protein (APP) (Chen et al., 2017). APP is a transmembrane protein composed of a large extracellular domain, a short transmembrane domain, and a cytoplasmic tail. This protein is crucial for neuronal plasticity and survival. The processing of APP is regulated by three intracellular proteases, namely, beta- and gamma-secretase (β-secretase and γ-secretase), which are within endocytic compartments, and alpha-secretase (α-secretase), which is anchored in the cell membrane (Fig. 1) (Chen et al., 2017). During APP proteolysis, the reaction leading to the formation of $A_β$ is catalyzed by the β-secretase followed by the γ-secretase. It is important to note that while plaques of $A_β$ are the main neuropathologic hallmark of AD, soluble aggregates of $A_β$ are considered to be involved earlier in the pathogenesis of AD and are likely more determinant to the clinical severity of the disease (Cline, Bicca, Viola, & Klein, 2018; Lesné et al., 2013). These soluble aggregates are oligomers of $A_β$ ($oA_β$), of which the most studied types are amyloid-β dimers and $A_β$*56. When purified from AD brains and applied to neurons in vitro, these $A_β$ oligomers alter synaptic transmission and long-term potentiation (which is critical for memory), hyperphosphorylate tau proteins at AD-relevant epitopes, damage dendritic spines, cause neuritic dystrophy, and induce neuronal apoptosis (Cline et al., 2018).

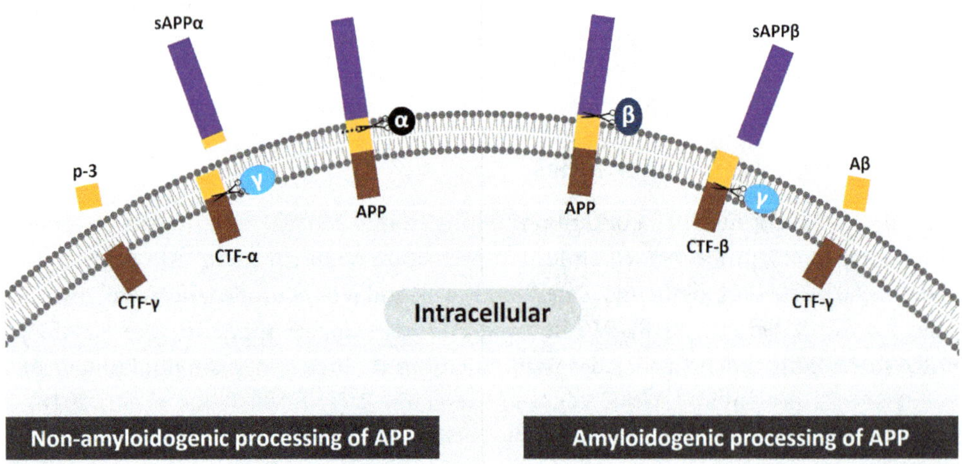

Fig. 1 Amyloidogenic versus nonamyloidogenic processing of APP. α-Secretase cleaves APP within the $A_β$ protein domain and releases soluble APP-α (sAPPα), whereas β-secretase releases $sAPP_β$. In both cases, residual chains (C-terminal fragments = CTF) remain bound to the membrane: CTF-α are residual chains from α-processing, and CTF-β are residual chains from β-processing. Then, γ-secretase releases extracellular p-3 from $CTF_α$ and $A_β$ protein from CTF-β.

Unlike A_β plaques, which are extracellular accumulations, NFTs result from the abnormal intracellular aggregation of tau proteins (Lane et al., 2018). In physiological conditions, phosphorylated tau stabilizes the microtubules and thus is crucial for cellular transport, axonal outgrowth, and cellular architecture. In the pathogenesis of AD, a hyperphosphorylation of tau proteins (mainly due to oA_β's toxicity) results in cross-pairing and accumulation of NFTs (Amar et al., 2017; Lesné et al., 2013). These disorders in tau proteins progressively alter cellular metabolism and connectivity, extending throughout the final stages of the disease long after the metabolic A_β abnormalities have reached a plateau phase (Amar et al., 2017).

Neuroglial processes underlying the potential relationship between AD neurodegeneration and abnormalities of Mg metabolism

Several pro-neurodegenerative mechanisms can be triggered or amplified by disorders of Mg metabolism. In this section, we will review these disorders and discuss the molecular bases of their potential to induce or promote AD.

Mg and abnormalities in the metabolism of tau proteins and APP

From the earliest preclinical stages of AD, impaired synthesis and storage of APP metabolites and tau protein complexes play a critical role in its pathogenesis. Even if cause-and-effect relationships are not clearly established between disorders of Mg metabolism and the development of AD, there still exists significant evidence that Mg availability potentially impacts the course of neurodegeneration in AD. Interestingly, several studies have reported that there is a significant impact of Mg concentrations on the regulation of the metabolism of APP and tau proteins (Barbagallo et al., 2011; Kieboom et al., 2017; Kirkland et al., 2018; Veronese et al., 2016).

The intracellular physiological roles of Mg make it an ion that can modulate the natural course of A_β oligomers accumulation via concentration-dependent switching between relevant pathways. In fact, the cellular concentrations of Mg regulate the activity of the enzymes involved in the main reactions which produce these AD-promoting products. Accordingly, an experimental study showed that low concentrations (0–0.4 mM) of extracellular Mg promote the β-secretase pathway and lead to an increase in A_β synthesis and an impairment in cell survival (Yu et al., 2010). Conversely, high Mg levels (1.2–4.0 mM) promote the α-secretase cleavage pathway for APP processing. Of note, the α-secretase cleavage produces CTF-α and $sAPP_\alpha$, which precludes the formation of A_β, whereas the β- and γ-cleavages lead to A_β accumulation (Zhang, Thompson, Zhang, & Xu, 2011). While A_β is neurotoxic, $sAPP_\alpha$ is neuroprotective and supports synaptic maintenance, rendering the balance of APP pathways critical for neuroglial excitability and survival. The β- to α-secretase cleavage switching at high Mg level probably results from intramembrane retention of APP (Yu et al., 2010). In contrast, low rates of intramembrane APP are observed for low Mg concentrations (Yu et al., 2010). To any avail, there is strong evidence that high rates of sAPP-α inhibit oxidative stress and activate neurotrophic pathways (Baratchi, Evans, Tate, Abraham, & Connor, 2012; Demars, Hollands, Zhao, & Lazarov, 2013; Ryan et al., 2013). The neurotrophic pathways seem to integrate multilevel transcriptional paths, including chromatin remodeling, upregulation of early gene transcription factors (AP-1, Egr1), and activation of the constitutive transcription factors CREB and NF-κB (Ryan et al., 2013).

Unlike for amyloid disorders, however, little is known regarding the potential impact of Mg metabolism on tau accumulation. An experimental study showed an increase of the inhibitory phosphorylation of glycogen synthase kinase 3-beta (GSK-3β) at serine 9 subsequent to Mg sulfate supplementation (Xu et al., 2014). This in turn decreases the rate of tau hyperphosphorylation, increases the activity of Akt at Ser473 and PI3K at Tyr458/199, and improves insulin sensitivity (Xu et al., 2014). Although these findings may hint at possible mechanisms linking Mg high levels to partial reversal of tau dysmetabolism, further experimental studies are required on the topic.

Mg metabolism, pro-neurodegenerative neurotransmission, and energetic metabolism

The impacts of Mg on synaptic transmission are various. On the one hand, Mg can directly modulate the response of NMDARs by modifying the sensitivity of this receptor to the presence of glutamate (Nikolaev, Magazanik, & Tikhonov, 2012). On the other hand, Mg contributes to the regulation of chemical reactions and neurotrophic cell signaling that modulate synapse architecture as well as the efficiency of synaptic transmission (Trapani et al., 2011; Yu et al., 2010). These properties of Mg become all the more interesting when taking in count that, in AD, synaptic alterations play a critical pathophysiological role, especially during the early stages of the disease.

Current data suggest that the earliest pathological synaptic changes in AD consist in a reduction in dendritic extensions and synaptic contacts between cortical neurons (Pozueta, Lefort, & Shelanski, 2013). Accordingly, several studies have shown that the pathogenic metabolic pathways involved in APP fragmentation significantly alter the architecture and function of brain excitatory synapses. It is precisely these pathways that are hyperactive during the early stages of AD and when Mg levels in the CSF are low (Kieboom et al., 2017; Kirkland et al., 2018; Zhang et al., 2011). Unfortunately, many of the available studies on the subject matter are based on animal models, and extrapolation to the natural history of the disease in humans is difficult to do with certainty.

During the preclinical stages of AD, even before A_β plaques' constitution, the pathologic increase in oligomers of amyloid-β (oA_β) synthesis plays the most impactful role on glutamatergic synapses (Cline et al., 2018). The pathogenic role of oA_β was first described on brain slices exposed to synthetic oA_β (without fibril). In brain slices, despite the retention of evoked action potentials, synthetic oA_β altered hippocampal long-term potentiation without lag and before neurodegeneration (Lambert et al., 1998). Further ex vivo and in vivo works confirmed the deleterious impact of $oA_β$ on synaptic function and neuroglial survival (Cline et al., 2018). Although parts of the mechanisms involved in the toxic effect of $oA_β$ in AD remain elusive, current opinion points to the role of cell signaling through cellular prion proteins (PrP^C) and metabotropic glutamate receptor 5 (mGluR5) (Brody & Strittmatter, 2018), of tau hyperphosphorylation, and of the inactivation of insulin receptor substrates (Cline et al., 2018). Considering the pathologic impact of $oA_β$, it is interesting to recall that Mg has demonstrated properties in inhibiting the synthesis of $A_β$ derivatives (Li et al., 2014; Yu et al., 2018), reducing the toxic effect of an accumulation of these metabolites (Xu et al., 2014; Yu et al., 2018), and preventing synaptic losses (Li et al., 2014) in various AD models.

The physiology of NMDARs provides insight into the disease-inducing/promoting processes described above and how Mg may impact these processes. NMDARs are excitatory glutamate-dependent channels that are permeable to calcium and sodium ions. These channels have the particularity of being blocked by Mg at membrane voltages near rest. In other words, the blockade exerted by Mg depends on the local potential and thus can be released when the surrounding electrical activity increases the local membrane potential (Hansen et al., 2018). The NMDARs are tetrameric complexes formed by a variable combination of seven types of subunits (GluN1, GluN2A to D, and GluN3A and B) (Hansen et al., 2018). These receptors are distributed between synapses (synaptic NMDARs) and the extrasynaptic cell membrane (extrasynaptic NMDARs) on spine necks, dendritic shafts, and somas (Hansen et al., 2018). A minority of extrasynaptic NMDARs are activated by ambient glutamate to maintain tonic extrasynaptic currents (Hansen et al., 2018; Le Meur, Galante, Angulo, & Audinat, 2007), enhance synaptic transmission (Hansen et al., 2018; Harris & Pettit, 2008; Zhou et al., 2015), and promote neuronal network development, mainly via massive calcium influx signaling (Hansen et al., 2018; Le Meur et al., 2007; Volianskis et al., 2015). However, most of the extrasynaptic NMDARs require higher concentrations of neurotransmitter, for example, in the case of extrasynaptic leakage of glutamate following excessive synaptic activity (Hardingham & Bading, 2010; Zhou et al., 2015). Cerebral insults and brain trauma may also result in massive ectopic release of glutamate and subsequent deleterious activation of extrasynaptic NMDARs (Hardingham & Bading, 2010; Zhou et al., 2015).

In physiological conditions, the main neuroplastic process involving NMDARs is synaptic potentiation, which is a transient or long-term enhancement of local neurotransmission. This process consists of an initial short-term potentiation (STP) leading to a long-term potentiation (LTP), which is a stable stage facilitating interneuronal communication. STP requires a fast activation of GluN2A and GluN2B subunits and a slower activation of GluN2B and GluN2D subunits (Hansen et al., 2018). LTP results afterward from an activation of primarily triheteromeric NMDARs on GluN2A and GluN2B subunits (Hansen et al., 2018).

The subunits involved in synaptic potentiation also play various roles in the metabolic processes of APP fragmentation. Thus, synaptic activation of NMDAR GluN2A subunits is associated with nonamyloidogenic processing of APP by alpha-secretase (Hoey, Williams, & Perkinton, 2009), whereas GluN2B subunits promote β-secretase pathway and A_β accumulation (Hardingham & Bading, 2010). Unlike synaptic NMDAR complexes, extrasynaptic NMDARs (containing GluN2B subunits) mostly play a role in long-term depression (LTD), neurotoxic processes, and neurodegenerative signaling related to acute brain injuries or progressive neurodegeneration (Hardingham & Bading, 2010). Interestingly, it has been shown that triheteromeric GluN1-GluN2B-GluN2D NMDARs have less affinity for external Mg(2+) than GluN1-GluN2B receptors where GluN2B are more expressed (Huang & Gibb, 2014). Moreover, for the GluN1-GluN2B NMDARs, although the stabilization of Mg blockade depends on asparagine residues provided by both GluN1 and GluN2B, it appears that the GluN2B subunits play the primary role in the efficiency of NMDAR blockade by Mg (Fedele et al., 2018). As such, a decrease in extracellular Mg availability has a greater propensity to promote the pro-amyloid metabolic activity dependent on GluN2B-NMDARs. Namely, experimental works have demonstrated that GluN2B dysfunction plays a critical role in mediating the synaptotoxic effect of oA_β (Zhou, 2014). Supporting the complex interrelation between the activity of NMDARs and APP fragmentation, there is strong evidence that abnormal metabolic pathways involved in AD genesis lead to overexpression/overstimulation of extrasynaptic NMDARs as well as synaptic excitotoxic activity of these receptors (Pozueta et al., 2013; Wang & Reddy, 2017). In fact, excessive A_β synthesis promotes the release of hydrogen peroxide and hydroxyl radicals, thereby leading to oxidative impairment of cellular membrane permeability and hyperexcitability (Zhang, Li, Feng, & Wu, 2016). Nonetheless, the clinical relevance of these properties is not clear with respect to the natural history of AD in humans. There is a lack of clinical studies clearly establishing the pathological sequence that chronologically involves the dysfunction of GluN2B-NMDARs and other physiological alterations in AD. In any case, Mg, due to its affinity and impact on the function of NMDARs and APP metabolism, potentially plays an essential role in these pathological phenomena. Like toxic A_β forms, dendritic NFTs induce local overstimulation of extrasynaptic NMDARs and lead to more tau hyperphosphorylation in a vicious circle (Zhang et al., 2016). Mg may interfere with these pathological processes by binding to the NMDAR's deep site—this is the same site used by memantine, a common drug for the treatment of moderate to severe clinical forms of AD. Similar to memantine, Mg has the potential to reduce NMDAR-mediated excitotoxicity and the β-processing of APP (Li et al., 2014; Nikolaev et al., 2012; Yu et al., 2018, 2010). Nevertheless, further work is needed to determine whether the relatively low levels of Mg found in the CSF of AD patients are initiators or rather promoters of the vicious circles described above.

Among the enzymatic functions that depend closely on the availability of magnesium, energy metabolism is also essential in the pathophysiology of Alzheimer's disease. In effect, Mg is crucial for most of the chemical reactions of cellular respiration, especially in extreme situations of glutamatergic hyperactivity such as in AD and in certain other chronic neurodegenerative or acute brain pathologies (Trapani et al., 2011). Currently, little is known regarding how CSF and intracellular levels of Mg affect energetic metabolism during the natural course of AD. However, published experimental data show adding Mg to previously physiological concentrations of this ion improves the spare respiratory capacity of neurons when exposed to excitotoxic levels of glutamate (Clerc et al., 2013). It is important to recall that mitochondrial accumulation of A_β alters the respiratory chain in the manner of a vicious circle: a dysfunction of complexes III and IV of the respiratory chain occurs, thereby increasing the intracellular concentrations of calcium, which is then followed by oxidative stress and ATP down-production, which thereafter leads to additional accumulation of calcium and worsening oxidative stress (Moreira, Carvalho, Zhu, Smith, & Perry, 2010).

Mg dysmetabolism and prodegenerative BBB alterations

The integrity of the BBB is crucial for the homeostasis of the neuronal environment. There is now solid evidence that alterations of this barrier occur quite early in the course of AD even before the appearance of NFTs and A_β plaques (Sweeney, Zhao, Montagne, Nelson, & Zlokovic, 2019; Yamazaki & Kanekiyo, 2017). These alterations are both causes and consequences of the key metabolic disorders of the disease, namely, hyperactivation of the β-secretase pathway and hyperphosphorylation of tau proteins (Steinman, Sun, & Feng, 2021). Indeed, A_β accumulation (both oA_β and plaques) is associated with chronic neuroinflammation, which disrupts BBB (Yamazaki & Kanekiyo, 2017). This inflammation is coupled with an excessive angiogenic signaling mainly mediated by thrombin and vascular endothelial growth factor (VEGF) (Jefferies et al., 2013). In addition, both thrombin and VEGF are able to promote endothelial A_β hypersecretion, thus leading to prooxidative dysmetabolism and further endothelial/neuroglial damaging (Jefferies et al., 2013).

To date, the pathological processes linking inflammation, BBB alteration, and abnormal accumulation of pro-amyloid metabolites of APP are relatively well described. During the earliest stages of AD, abnormal paracellular/transcellular transfers through the BBB result from progressive alterations in the BBB's layer components (e.g., perivascular astrocytic end-feet, pericytes, basement membrane, and endothelial cells) (Jefferies et al., 2013; Sweeney et al., 2019; Yamazaki & Kanekiyo, 2017). Several transporters of the BBB have been shown to be downregulated in AD, such as glucose transporter 1 (GLUT1), P-glycoprotein (P-gp), and low-density lipoprotein receptor-related protein 1 (LRP1) (Jefferies et al., 2013; Sweeney et al., 2019; Yamazaki & Kanekiyo, 2017).

Other BBB constituents like the receptor for advanced glycation end products (RAGE), the matrix metallopeptidase 9 (MMP-9), and the leukocyte adhesion molecule (LAM) are overexpressed (Jefferies et al., 2013; Sweeney et al., 2019; Yamazaki & Kanekiyo, 2017). In addition to these changes in BBB transporters, there is a progressive abnormal increase in the paracellular permeability of the BBB resulting from a combination of disruptions in tight junctions and the basement membrane, endothelial cell dysmorphism, and astrocyte end-foot retraction (Jefferies et al., 2013; Sweeney et al., 2019; Toffa, Magnerou, Kassab, Hassane Djibo, & Sow, 2019; Yamazaki & Kanekiyo, 2017). Destruction of the BBB is mainly due to capillary deposition of Aβ polymers (Jefferies et al., 2013; Yamazaki & Kanekiyo, 2017). Subsequent pathologic leakage through BBB leads to the accumulation of neurotoxic and pro-inflammatory deposits of blood-derived substances like fibrin, thrombin, albumin, immunoglobulin G, and hemosiderin (Jefferies et al., 2013; Sweeney et al., 2019; Yamazaki & Kanekiyo, 2017). These deposits induce chronic neuroglial membrane peroxidation and apoptosis (Jefferies et al., 2013; Sweeney et al., 2019). In summary, abnormal accumulation of A_β polymers with their subsequent pro-inflammatory cascades (both neuroglial and endothelial), oxidative stress in endothelial cells, and toxic leakage through the BBB are fundamental in AD pathogenesis (Toffa et al., 2019).

Besides these mechanisms, it has also been postulated that a primary alteration of cerebral blood flow (CBF) due to arteriolar disorders may lead to pathologic microangiogenesis involving VEGF and thrombin (Jefferies et al., 2013; Steinman et al., 2021). Reduced CBF likely increases amyloid pathology through failure of A_β trans-BBB drainage from neuroglial environment (Weller, Subash, Preston, Mazanti, & Carare, 2008) or increased rate of tau phosphorylation (Koike, Green, Blurton-Jones, & Laferla, 2010). Furthermore, an excessive NMDAR-mediated increase of the BBB's permeability has been reported and is likely to play a critical role in mid-to-late stages of AD (Xhima, Weber-Adrian, & Silburt, 2016).

The role of Mg availability in these pathologic processes is still under investigation. However, it seems that the ability of Mg to regulate APP processing, its antiinflammatory properties, as well as its anti-NMDAR effects, is key for the preservation of microvascular function. It remains unclear whether low rates of CSF Mg observed in AD patients act as promoting factors or initiating factors for vascular disruption in the natural course of the disease. Further work is needed since currently available publications usually report experimental supplementation of Mg in in vitro and in vivo animal models.

In addition to the metabolic and antiinflammatory properties of Mg, normal levels of Mg have been suggested to prevent microvascular alterations of AD by improving CBF via vessel smooth muscle relaxation, especially in elderly. Moreover, there exists evidence that Mg may facilitate A_β transcytosis through the BBB, a process that involves low-density lipoprotein receptor-related protein (LRP) as well as phosphatidylinositol-binding clathrin assembly protein (PICALM) (Zhu, Su, Fu, & Xu, 2018). Mg may also decrease the influx of $A_β$ from blood to brain through the activation of RAGE and caveolae (Zhu et al., 2018).

Mg and neuroinflammation-based neurodegeneration

One of the most recognized properties of Mg is its role in the inhibition of inflammation. This is of particular interest because, as described earlier, inflammation plays a crucial role in the mechanisms leading to neuronal loss and impairment of neuroglial functions in most of neurodegenerative diseases. In AD, the pathologic accumulation of amyloid metabolites and the toxic leakage through the BBB result in an excessive release of pro-inflammatory cytokines, namely, gamma-interferons (IFN-γ), interleukin 1-beta (IL-1β), and tumor-necrosis factor alpha (TNF-α) by astrocytes and microglial cells. The inflammatory sequence involves both IFN-γ released by reactive astrocytes and excessive $A_β$ polymers (from neurons), which activate microglial cells via IFN-γ receptors and toll-like receptors 4 (TLR4), respectively (Park et al., 2018). Microglia then increase the expression of inducible nitric oxide synthase (iNOS) and nuclear factor-kappa B (NF-κB) (Leng & Edison, 2021). Microglia also release TNF-α and cytotoxic NO, resulting in neuronal degeneration (Leng & Edison, 2021; Park et al., 2018).

There is strong evidence of an association existing between Mg levels and neuroinflammation intensity. Reported observations show that low Mg levels are associated with an increased inflammatory response, while magnesium supplementation induces the opposite effect (Nielsen, 2018). The mechanism behind this finding has been suggested to be based on a calcium-dependent inhibition of the inflammatory response by astrocytes and microglia (Nielsen, 2018). It has also been suggested that phosphoinositide-3-kinase/protein kinase B (PI3-K/Akt) and NF-κB signaling are crucial for the neuroprotective properties of Mg (Nielsen, 2018; Yu et al., 2018). Moreover, Mg inhibits neuronal gamma-secretase and microglial TNF-α, thus blocking the complex interaction between pro-amyloid processes and inflammation (Yu et al., 2018). The increase in the production of $A_β$ polymers leads to the neuronal accumulation of this substrate and subsequent overexpression of TNF-α in microglial cells (Leng & Edison, 2021). In addition, Mg suppresses $IL\text{-}1_β$ via the stimulation of extracellular signal-regulated protein kinases 1 and 2 (ERK1/2) and peroxisome proliferator-activated receptor gamma (PPARγ) (Wang et al., 2017).

Should Mg be used in the treatment of AD?

Although there are numerous well-designed studies supporting the neuroprotective potential of Mg against the principal processes leading to neurodegeneration in AD, there is a lack of clinical trials to conclude whether Mg supplementation can provide satisfactory results within its therapeutic window. It is also important to recall that animal models do not exactly reproduce the natural course of AD. As such, extrapolating the cognitive impact and adverse effects of Mg-based treatment in humans from animal studies is difficult and may prove to be imprecise (Toffa et al., 2019).

Considering that CSF Mg levels—but not serum levels—have been shown to be low in AD patients, it is likely that Mg supplementation protocols will need to address this observation to be effective. To any avail, the preventive administration of Mg can potentially slow the progression of the AD, and perhaps Mg may play a curative role as well in the future, though probably as an adjunct therapy.

Applications to other neurological conditions

Although the real impact of Mg on the natural evolution of AD is still under investigation, the currently available data have helped shed light on the pathophysiology of neurodegeneration. Considering the similarity in most critical processes involved in neurodegenerative diseases, Mg may play a comparable role in these other diseases. Uncovering the role of Mg dysmetabolism in other neurodegenerative conditions thus seems paramount to our collective understanding of neurodegeneration as a whole. Other dementias, Parkinson's disease, and amyotrophic lateral sclerosis are all pathologies of interest that may benefit from a careful investigation of Mg; in fact, NMDARs, which are tightly regulated by Mg availability, play a central role in excitotoxicity, which is critical in these aforementioned diseases. For example, it is interesting to note that lower CSF Mg levels (but not serum levels) were also reported in Parkinson's disease (PD) patients versus healthy controls (Oyanagi & Hashimoto, 2011). In experimental studies, continuous low Mg intake for two generations has been associated with selective loss of dopaminergic neurons in rat models (Oyanagi, 2005; Oyanagi et al., 2006). These findings confirm that there remains our understanding of Mg in brain excitability and neuroglial survival is far from perfect, and there certainly exists room for research.

Other components of interest

It is important to note that the role of Mg is dependent on calcium release, as shown in numerous neuroglial cascades. Thus, calcium metabolism is tightly linked to the role played by Mg in AD. This association has multiple implications. Notably, it highlights the need for studies that evaluate both Mg and calcium levels in humans and in animal models. Such works are likely to lead to more salient findings regarding the intricate mechanisms of Mg availability in neurodegeneration and neuroprotection. Moreover, currently, little is known about the relevance of adjunctive drugs targeting calcium metabolism (either calcium itself or vitamin D) in order to potentiate the neuroprotective effects of Mg. There are data suggesting that unlike Mg, excessive calcium availability could lead to competitive inhibition of the neuroprotective effects of Mg in addition to calcium's own deleterious effects on cellular hyperactivity and apoptosis (Toffa et al., 2018; Wang & Reddy, 2017). Therefore, in evaluating the therapeutic use of Mg in AD, one should consider the option of conjointly evaluating the calcium metabolism.

Mini-dictionary of terms

- **Excitotoxicity:** This term refers to severe cell alterations or cell death subsequent to acute or chronic excessive action of excitatory neurotransmitters.
- **Parkinson's disease:** Parkinson's disease (or Parkinson's) is a chronic degenerative disorder of the central nervous system that mainly affects the motor system.
- **Neurodegeneration:** This term refers to the progressive loss of structure or function of neurons, which may ultimately lead to cell death.
- **Glutamate:** Glutamate is a nonessential α-amino acid and the most abundant excitatory neurotransmitter in the vertebrate's central nervous system.
- **Blood-brain barrier:** The BBB is a highly selective semipermeable compilation of cell layers that filters blood substances crossing into the extracellular fluid of the central nervous system. This barrier allows the passive diffusion of some small molecules, as well as the active transport of specific substances.

Key facts of Alzheimer's disease (AD)

- AD results from a loss of neurons combined with the deterioration of the connections between the remaining neurons.
- There are three key processes leading to brain degeneration in AD: the accumulation of toxic forms of beta-amyloid proteins, the alteration of cellular transport and connections due to the hyperphosphorylation of tau proteins, and the hyperactivity of some specific excitatory receptors

Key facts of the role of magnesium (Mg) in AD

- Mg levels are significantly lower in the perineuronal fluid of AD patients.
- Human and animal studies revealed that a reduction in Mg availability is associated with massive neuronal impairment and loss.
- Experimental data show that Mg supplementation has the potential to block or reverse neurodegenerative processes.
- The role of Mg levels in the natural history of AD is still subject to some gray areas.
- Further work is needed, notably clinical trials, to better explore the therapeutic potential of Mg in AD patients.

Summary points

- Low levels of Mg are associated with neurodegeneration in both animal models and humans.
- Mg promotes the processing of amyloid precursor protein (APP) by alpha-secretase.
- The increase of sAPPα synthesis by Mg action enhances neuron survival as well as dendrite arborization.
- Mg plays a strong role in regulating NMDARs, thereby reducing the neurotoxic activity mediated by these receptors.
- Mg has the potential to inhibit most of the mechanisms leading to BBB disruption in AD's natural course.
- Mg blocks neuroinflammation by reducing both its causes and consequences.

References

Allgrove, J. (2015). Physiology of calcium, phosphate, magnesium and vitamin D. *Endocrine Development, 28*, 7–32. https://doi.org/10.1159/000380990.

Amar, F., Sherman, M. A., Rush, T., Larson, M., Boyle, G., Chang, L., et al. (2017). The amyloid-β oligomer Aβ*56 induces specific alterations in neuronal signaling that lead to tau phosphorylation and aggregation. *Science Signaling, 10*. https://doi.org/10.1126/scisignal.aal2021.

Baratchi, S., Evans, J., Tate, W. P., Abraham, W. C., & Connor, B. (2012). Secreted amyloid precursor proteins promote proliferation and glial differentiation of adult hippocampal neural progenitor cells. *Hippocampus, 22*, 1517–1527. https://doi.org/10.1002/hipo.20988.

Barbagallo, M., Belvedere, M., Di Bella, G., & Dominguez, L. J. (2011). Altered ionized magnesium levels in mild-to-moderate Alzheimer's disease. *Magnesium Research, 24*, S115–S121. https://doi.org/10.1684/mrh.2011.0287.

Brody, A. H., & Strittmatter, S. M. (2018). Synaptotoxic signaling by amyloid Beta oligomers in Alzheimer's disease through prion protein and mGluR5. *Advances in Pharmacology, 82*, 293–323. https://doi.org/10.1016/bs.apha.2017.09.007.

Chen, B. B., Prasad, C., Kobrzynski, M., Campbell, C., & Filler, G. (2016). Seizures related to hypomagnesemia: A case series and review of the literature. *Child Neurology Open, 3*. https://doi.org/10.1177/2329048X16674834.

Chen, G., Xu, T., Yan, Y., Zhou, Y., Jiang, Y., Melcher, K., et al. (2017). Amyloid beta: Structure, biology and structure-based therapeutic development. *Acta Pharmacologica Sinica, 38*, 1205–1235. https://doi.org/10.1038/aps.2017.28.

Clerc, P., Young, C. A., Bordt, E. A., Grigore, A. M., Fiskum, G., & Polster, B. M. (2013). Magnesium sulfate protects against the bioenergetic consequences of chronic glutamate receptor stimulation. *PLoS One, 8*, e79982. https://doi.org/10.1371/journal.pone.0079982.

Cline, E. N., Bicca, M. A., Viola, K. L., & Klein, W. L. (2018). The amyloid-β oligomer hypothesis: Beginning of the third decade. *Journal of Alzheimer's Disease, 64*, S567–S610. https://doi.org/10.3233/JAD-179941.

Demars, M. P., Hollands, C., Zhao, K. D. T., & Lazarov, O. (2013). Soluble amyloid precursor protein-α rescues age-linked decline in neural progenitor cell proliferation. *Neurobiology of Aging, 34*, 2431–2440. https://doi.org/10.1016/j.neurobiolaging.2013.04.016.

Fedele, L., Newcombe, J., Topf, M., Gibb, A., Harvey, R. J., & Smart, T. G. (2018). Disease-associated missense mutations in GluN2B subunit alter NMDA receptor ligand binding and ion channel properties. *Nature Communications, 9*, 957. https://doi.org/10.1038/s41467-018-02927-4.

Franken, G. A. C., Müller, D., Mignot, C., Keren, B., Lévy, J., Tabet, A.-C., et al. (2021). The phenotypic and genetic spectrum of patients with heterozygous mutations in cyclin M2 (CNNM2). *Human Mutation, 42*, 473–486. https://doi.org/10.1002/humu.24182.

Hansen, K. B., Yi, F., Perszyk, R. E., Furukawa, H., Wollmuth, L. P., Gibb, A. J., et al. (2018). Structure, function, and allosteric modulation of NMDA receptors. *The Journal of General Physiology, 150*, 1081–1105. https://doi.org/10.1085/jgp.201812032.

Hardingham, G. E., & Bading, H. (2010). Synaptic versus extrasynaptic NMDA receptor signalling: Implications for neurodegenerative disorders. *Nature Reviews. Neuroscience, 11*, 682–696. https://doi.org/10.1038/nrn2911.

Harris, A. Z., & Pettit, D. L. (2008). Recruiting extrasynaptic NMDA receptors augments synaptic signaling. *Journal of Neurophysiology, 99*, 524–533. https://doi.org/10.1152/jn.01169.2007.

Hoey, S. E., Williams, R. J., & Perkinton, M. S. (2009). Synaptic NMDA receptor activation stimulates alpha-secretase amyloid precursor protein processing and inhibits amyloid-beta production. *The Journal of Neuroscience, 29*, 4442–4460. https://doi.org/10.1523/JNEUROSCI.6017-08.2009.

Huang, Z., & Gibb, A. J. (2014). Mg2+ block properties of triheteromeric GluN1-GluN2B-GluN2D NMDA receptors on neonatal rat substantia nigra pars compacta dopaminergic neurones. *The Journal of Physiology, 592*, 2059–2078. https://doi.org/10.1113/jphysiol.2013.267864.

Jefferies, W. A., Price, K. A., Biron, K. E., Fenninger, F., Pfeifer, C. G., & Dickstein, D. L. (2013). Adjusting the compass: New insights into the role of angiogenesis in Alzheimer's disease. *Alzheimer's Research & Therapy, 5*, 64. https://doi.org/10.1186/alzrt230.

Kieboom, B. C. T., Licher, S., Wolters, F. J., Ikram, M. K., Hoorn, E. J., Zietse, R., et al. (2017). Serum magnesium is associated with the risk of dementia. *Neurology, 89*, 1716–1722. https://doi.org/10.1212/WNL.0000000000004517.

Kirkland, A. E., Sarlo, G. L., & Holton, K. F. (2018). The role of magnesium in neurological disorders. *Nutrients, 10*, 730. https://doi.org/10.3390/nu10060730.

Koike, M. A., Green, K. N., Blurton-Jones, M., & Laferla, F. M. (2010). Oligemic hypoperfusion differentially affects tau and amyloid-{beta}. *The American Journal of Pathology, 177*, 300–310. https://doi.org/10.2353/ajpath.2010.090750.

Lambert, M. P., Barlow, A. K., Chromy, B. A., Edwards, C., Freed, R., Liosatos, M., et al. (1998). Diffusible, nonfibrillar ligands derived from Abeta1-42 are potent central nervous system neurotoxins. *Proceedings of the National Academy of Sciences of the United States of America, 95*, 6448–6453. https://doi.org/10.1073/pnas.95.11.6448.

Lane, C. A., Hardy, J., & Schott, J. M. (2018). Alzheimer's disease. *European Journal of Neurology, 25*, 59–70. https://doi.org/10.1111/ene.13439.

Le Meur, K., Galante, M., Angulo, M. C., & Audinat, E. (2007). Tonic activation of NMDA receptors by ambient glutamate of non-synaptic origin in the rat hippocampus. *The Journal of Physiology, 580*, 373–383. https://doi.org/10.1113/jphysiol.2006.123570.

Leng, F., & Edison, P. (2021). Neuroinflammation and microglial activation in Alzheimer disease: Where do we go from here? *Nature Reviews. Neurology, 17*, 157–172. https://doi.org/10.1038/s41582-020-00435-y.

Lesné, S. E., Sherman, M. A., Grant, M., Kuskowski, M., Schneider, J. A., Bennett, D. A., et al. (2013). Brain amyloid-β oligomers in ageing and Alzheimer's disease. *Brain, 136*, 1383–1398. https://doi.org/10.1093/brain/awt062.

Li, W., Yu, J., Liu, Y., Huang, X., Abumaria, N., Zhu, Y., et al. (2014). Elevation of brain magnesium prevents synaptic loss and reverses cognitive deficits in Alzheimer's disease mouse model. *Molecular Brain, 7*, 65. https://doi.org/10.1186/s13041-014-0065-y.

Moreira, P. I., Carvalho, C., Zhu, X., Smith, M. A., & Perry, G. (2010). Mitochondrial dysfunction is a trigger of Alzheimer's disease pathophysiology. *Biochimica et Biophysica Acta, 1802*, 2–10. https://doi.org/10.1016/j.bbadis.2009.10.006.

Ndiaye, M., Toffa, D. H., Sow, A.-D., Sène, M.-S., Basse, A.-M., Fall, A.-L., et al. (2013). Familial congenital hypomagnesemia revealed by neonatal convulsions. *Archives de Pédiatrie, 20*, 1212–1218. https://doi.org/10.1016/j.arcped.2013.08.019.

Nielsen, F. H. (2018). Magnesium deficiency and increased inflammation: Current perspectives. *Journal of Inflammation Research*, *11*, 25–34. https://doi.org/10.2147/JIR.S136742.

Nikolaev, M. V., Magazanik, L. G., & Tikhonov, D. B. (2012). Influence of external magnesium ions on the NMDA receptor channel block by different types of organic cations. *Neuropharmacology*, *62*, 2078–2085. https://doi.org/10.1016/j.neuropharm.2011.12.029.

Oyanagi, K. (2005). The nature of the parkinsonism-dementia complex and amyotrophic lateral sclerosis of Guam and magnesium deficiency. *Parkinsonism & Related Disorders*, *11*(Suppl 1), S17–S23. https://doi.org/10.1016/j.parkreldis.2005.02.010.

Oyanagi, K., & Hashimoto, T. (2011). Magnesium in Parkinson's disease: An update in clinical and basic aspects. In R. Vink, & M. Nechifor (Eds.), *Magnesium in the central nervous system [Internet]*. Adelaide, AU: University of Adelaide Press. Available from: https://www.ncbi.nlm.nih.gov/books/NBK507269/.

Oyanagi, K., Kawakami, E., Kikuchi-Horie, K., Ohara, K., Ogata, K., Takahama, S., et al. (2006). Magnesium deficiency over generations in rats with special references to the pathogenesis of the Parkinsonism-dementia complex and amyotrophic lateral sclerosis of Guam. *Neuropathology*, *26*, 115–128. https://doi.org/10.1111/j.1440-1789.2006.00672.x.

Park, J., Wetzel, I., Marriott, I., Dréau, D., D'Avanzo, C., Kim, D. Y., et al. (2018). A 3D human triculture system modeling neurodegeneration and neuroinflammation in Alzheimer's disease. *Nature Neuroscience*, *21*, 941–951. https://doi.org/10.1038/s41593-018-0175-4.

Pozueta, J., Lefort, R., & Shelanski, M. L. (2013). Synaptic changes in Alzheimer's disease and its models. *Neuroscience*, *251*, 51–65. https://doi.org/10.1016/j.neuroscience.2012.05.050.

Ryan, M. M., Morris, G. P., Mockett, B. G., Bourne, K., Abraham, W. C., Tate, W. P., et al. (2013). Time-dependent changes in gene expression induced by secreted amyloid precursor protein-alpha in the rat hippocampus. *BMC Genomics*, *14*, 376. https://doi.org/10.1186/1471-2164-14-376.

Steinman, J., Sun, H.-S., & Feng, Z.-P. (2021). Microvascular alterations in Alzheimer's disease. *Frontiers in Cellular Neuroscience*, *14*, 618986. https://doi.org/10.3389/fncel.2020.618986.

Sweeney, M. D., Zhao, Z., Montagne, A., Nelson, A. R., & Zlokovic, B. V. (2019). Blood-brain barrier: From physiology to disease and back. *Physiological Reviews*, *99*, 21–78. https://doi.org/10.1152/physrev.00050.2017.

Toffa, D. H., Kpadonou, C., Gams Massi, D., Ouedraogo, M., Sow, A. D., Ndiaye, M., et al. (2018). Le magnésium et le calcium réduisent la sévérité des troubles de la mémoire spatiale pour le modèle kaïnique d'épilepsie mésiale temporale chez la souris. *Canadian Journal of Physiology and Pharmacology*, *96*, 1132–1144. https://doi.org/10.1139/cjpp-2018-0020.

Toffa, D. H., Magnerou, M. A., Kassab, A., Hassane Djibo, F., & Sow, A. D. (2019). Can magnesium reduce central neurodegeneration in Alzheimer's disease? Basic evidences and research needs. *Neurochemistry International*, *126*, 195–202. https://doi.org/10.1016/j.neuint.2019.03.014.

Trapani, V., Mastrototaro, L., & Wolf, F. I. (2011). Magnesium and the Yin-Yang interplay in apoptosis. In R. Vink, & M. Nechifor (Eds.), *Magnesium in the central nervous system [Internet]*. Adelaide, AU: University of Adelaide Press.

Veronese, N., Zurlo, A., Solmi, M., Luchini, C., Trevisan, C., Bano, G., et al. (2016). Magnesium status in Alzheimer's disease: A systematic review. *American Journal of Alzheimer's Disease and Other Dementias*, *31*, 208–213. https://doi.org/10.1177/1533317515602674.

Volianskis, A., France, G., Jensen, M. S., Bortolotto, Z. A., Jane, D. E., & Collingridge, G. L. (2015). Long-term potentiation and the role of N-methyl-D-aspartate receptors. *Brain Research*, *1621*, 5–16. https://doi.org/10.1016/j.brainres.2015.01.016.

Wang, R., & Reddy, P. H. (2017). Role of glutamate and NMDA receptors in Alzheimer's disease. *Journal of Alzheimer's Disease*, *57*, 1041–1048. https://doi.org/10.3233/JAD-160763.

Wang, P., Yu, X., Guan, P.-P., Guo, J.-W., Wang, Y., Zhang, Y., et al. (2017). Magnesium ion influx reduces neuroinflammation in Aβ precursor protein/Presenilin 1 transgenic mice by suppressing the expression of interleukin-1β. *Cellular & Molecular Immunology*, *14*, 451–464. https://doi.org/10.1038/cmi.2015.93.

Weller, R. O., Subash, M., Preston, S. D., Mazanti, I., & Carare, R. O. (2008). Perivascular drainage of amyloid-beta peptides from the brain and its failure in cerebral amyloid angiopathy and Alzheimer's disease. *Brain Pathology*, *18*, 253–266. https://doi.org/10.1111/j.1750-3639.2008.00133.x.

Wolf, F. I., Trapani, V., Simonacci, M., Ferré, S., & Maier, J. A. M. (2008). Magnesium deficiency and endothelial dysfunction: Is oxidative stress involved? *Magnesium Research*, *21*, 58–64.

Xhima, K., Weber-Adrian, D., & Silburt, J. (2016). Glutamate induces blood-brain barrier permeability through activation of N-methyl-D-aspartate receptors. *The Journal of Neuroscience*, *36*, 12296–12298. https://doi.org/10.1523/JNEUROSCI.2962-16.2016.

Xu, Z.-P., Li, L., Bao, J., Wang, Z.-H., Zeng, J., Liu, E.-J., et al. (2014). Magnesium protects cognitive functions and synaptic plasticity in streptozotocin-induced sporadic Alzheimer's model. *PLoS One*, *9*, e108645. https://doi.org/10.1371/journal.pone.0108645.

Yamazaki, Y., & Kanekiyo, T. (2017). Blood-brain barrier dysfunction and the pathogenesis of Alzheimer's disease. *International Journal of Molecular Sciences*, *18*, 1965. https://doi.org/10.3390/ijms18091965.

Yu, X., Guan, P.-P., Zhu, D., Liang, Y.-Y., Wang, T., Wang, Z.-Y., et al. (2018). Magnesium ions inhibit the expression of tumor necrosis factor α and the activity of γ-secretase in a β-amyloid protein-dependent mechanism in APP/PS1 transgenic mice. *Frontiers in Molecular Neuroscience*, *11*, 172. https://doi.org/10.3389/fnmol.2018.00172.

Yu, J., Sun, M., Chen, Z., Lu, J., Liu, Y., Zhou, L., et al. (2010). Magnesium modulates amyloid-beta protein precursor trafficking and processing. *Journal of Alzheimer's Disease*, *20*, 1091–1106. https://doi.org/10.3233/JAD-2010-091444.

Zhang, Y., Li, P., Feng, J., & Wu, M. (2016). Dysfunction of NMDA receptors in Alzheimer's disease. *Neurological Sciences: Official Journal of the Italian Neurological Society and of the Italian Society of Clinical Neurophysiology*, *37*, 1039–1047. https://doi.org/10.1007/s10072-016-2546-5.

Zhang, Y., Thompson, R., Zhang, H., & Xu, H. (2011). APP processing in Alzheimer's disease. *Molecular Brain*, *4*, 3. https://doi.org/10.1186/1756-6606-4-3.

Zhou, Q. (2014). GluN2B-NMDA receptors in Alzheimer's disease: Beyond synapse loss and cell death. *Neural Regeneration Research*, *9*, 1878–1879. https://doi.org/10.4103/1673-5374.145346.

Zhou, X., Chen, Z., Yun, W., Ren, J., Li, C., & Wang, H. (2015). Extrasynaptic NMDA receptor in excitotoxicity: Function revisited. *The Neuroscientist*, *21*, 337–344. https://doi.org/10.1177/1073858414548724.

Zhu, D., Su, Y., Fu, B., & Xu, H. (2018). Magnesium reduces blood-brain barrier permeability and regulates amyloid-β transcytosis. *Molecular Neurobiology*, *55*, 7118–7131. https://doi.org/10.1007/s12035-018-0896-0.

CHAPTER 5

Copper, oxidative stress, Alzheimer's disease, and dementia

Rosanna Squitti[a,b], Mauro Ciro Antonio Rongioletti[b], and Gianfranco Liguri[c]
[a]Molecular Markers Laboratory, IRCCS Istituto Centro San Giovanni di Dio Fatebenefratelli, Brescia, Italy
[b]Department of Laboratory Medicine, Research and Development Division, San Giovanni Calibita Fatebenefratelli Hospital, Rome, Italy
[c]Department of Biochemistry, Faculty of Medicine and Surgery, University of Florence, Florence, Italy

Abbreviations

AD	Alzheimer's disease
AIs	adequate intakes
ATPase7A	copper-transporting P-type ATPase 7A
ATPase7B	copper-transporting P-type ATPase 7B
ATSDR	Agency of Toxic Substances and Disease Registry
Aβ	beta amyloid
BBB	blood-brain barrier
BSE	bovine spongiform encephalopathy
COX	cytochrome oxidase
Cp	ceruloplasmin
CSF	cerebrospinal fluid
CTR-1	Cu transporter 1
Cu	copper
Cu/Zn SOD	copper/zinc superoxide dismutase
CuAD	copper subtype of Alzheimer's disease
EEG	electroencephalogram
Fe	iron
GSH	glutathione
IOM	Institute of Medicine
IWG	International Working Group
K_d^c	Cu dissociation constant
MCI	mild cognitive impairment
MTs	metallothioneins
Non-Cp Cu	non-ceruloplasmin copper
PD	Parkinson's disease
PET	positron emission tomography
PrP	prion protein
PrP^C	normal prion protein
PrP^{Res}	protease-resistant isoforms designated
PrP^{Sc}	prior protein scrapie
RDA	recommended dietary allowance
ROS	reactive oxygen species
ULs	upper levels of tolerable intake

Vitamins and Minerals in Neurological Disorders
https://doi.org/10.1016/B978-0-323-89835-5.00030-2

WD	Wilson disease
WHO	World Health Organization
Zn	zinc

Introduction

Alzheimer's disease (AD) is the main cause of dementia in the elderly. In 2019, the number of individuals affected by AD has been estimated on 50 million and is expected to reach more than 152 million by 2050 (Alzheimer's Disease-International, 2019).

Alzheimer's disease can be referred to as early- or late-onset depending on the disease onset (<65 or >65 years old, respectively) and is normally distinguished in familial (<5%), and sporadic, with a multifactorial complex disease etiology. The early-onset familial AD is identified on the presence of mutations in *APP*, *Presenilin1* or *Presenilin2* genes. Old age and a family history of Alzheimer's and genetics (*APOE4* gene) are the greatest risk factors for late-onset AD (Alzheimer's Disease-International, 2019). World Health Organization (2019) published recommendations to reduce the risk of cognitive decline and dementia by improving physical activity, quitting smoking, and managing cardiovascular risk factors (especially diabetes, obesity, smoking, and hypertension). In this line, previous expert committees (Barnard et al., 2014; IOM, 2015) recommend the implementation of a healthy diet, lifelong learning, and cognitive training that are associated with reduced risk of cognitive decline.

Why Cu has much to do with Alzheimer's disease

In this chapter, we will provide a short and focused overview of the role of copper (Cu) in AD and its relation to aging, with the attempt to circumstantiate the merit of evidence-based medicine to guide decision-making in AD prevention with regard to Cu imbalance. The available evidence demonstrating that a body's Cu dysfunction contributes to increase the risk of AD is substantial and dietary changes may be justified to reduce this risk (Barnard et al., 2014). In fact, not only in cases in which available evidence demonstrates risk beyond any doubt but also in cases in which evidence of risk is substantial that changes may be justified (Barnard et al., 2014; Squitti, Siotto, & Polimanti, 2014).

The rational at the basis for some practical recommendations for an adequate dietary regimen in individuals susceptible to Cu abnormalities is discussed herein.

Alzheimer's disease

Alzheimer's disease: General description

Alzheimer's disease (Alzheimer's Disease-International, 2019) is an irreversible neurodegenerative disorder, characterized by a gradual loss of recent memory and cognition eventually leading to dementia and death. As per the 2018 National Institute on

Aging—Alzheimer's Association (NIA-AA) Research Framework (Jack Jr. et al., 2018), the cognitive continuum is represented as a syndromic cognitive staging scheme divided into three categories as follows: cognitive unimpaired, mild cognitive impairment (MCI), and dementia. Dementia can be further divided into mild, moderate, and sever stage (Table 1).

Alzheimer's disease diagnosis

The International Working Group (IWG) and the US National Institute on Aging-Alzheimer's Association in 2014 defined new criteria for the diagnosis of AD (Dubois et al., 2014). The IWG-2 criteria simplified the diagnosis of AD, requiring the presence

Table 1 Symptom progression in Alzheimer's disease.

Cognitive function	Stage I (mild)	Stage II (moderate)	Stage III (severe)
Years of symptoms	2–4 years	2–10 years	1–3 years
MMSE[a]	30–21	20–11	10–0
Memory	Mild amnesia (recent memory)	Amnesia (recent and remote)	Untestable
Language	Mild impairment, anomia	Sensory aphasia	Severe impairment
Attention	Reduced	Impaired	Untestable
Perceptual skills	Mild impairment	Moderate to severe impairment	Untestable
Constructive abilities	Reduced	Impaired	Untestable
Problem solving	Mild impairment	Impaired	Untestable
Functional abilities	Almost normal, some mistakes	Unreliable	Untestable
Orientation	Mild impairment	Impaired	Untestable
Behavior	Mild changes, signs of withdrawal	Absent	Agitated, delusional
Gait	Normal	Normal	Impaired
Posture	Normal	Normal or flexed	Bedridden
Brief description of symptoms	Minor memory loss and difficulties in learning new information while long-term memory is intact; mood changes	Confusion, difficulties in self-care, daily tasks, judgment, communication, and behavior (anger, anxiety, frustration, restlessness)	Not speaking and not recognizing family members; loss of bodily function; episodes of violence and aggression

[a]MMSE, minimental state examination (Folstein, Folstein, & McHugh, 1975).

of typical or atypical clinical AD phenotype and presence of Alzheimer's pathology as revealed by pathophysiological biomarkers including decreased beta-amyloid (Aβ)1–42 together with increased T-tau or P-tau in cerebrospinal fluid (CSF); increased tracer retention on amyloid positron emission tomography (PET); AD autosomal *dominant PSEN1, PSEN2,* or *APP* mutation (Dubois et al., 2014). Atypical forms of AD represent 6%–14% of cases and vary from the typical amnestic form (Dubois et al., 2014).

The hallmark pathologies of the AD brain are the amyloid plaques, consisting in the accumulation of the protein fragment Aβ outside neurons and twisted strands of the protein tau (neurofibrillary tangles) inside neurons. Amyloid plaques are clusters of protein fibrils made of an amyloidal core, abnormal neurites, and glial cells. The amyloidal core is mainly an aggregation of Aβ, which is produced by the proteolytic cleavage of the amyloid precursor protein (APP).

Cu intake and regulation in physiology

Cu intake in human diet and drinking water

Cu is an essential micronutrient indispensable for brain development and physiology: While severe Cu deficiencies are associated with cardiac, bone, immune, and central nervous system conditions, Cu chronic excess can be mostly associated with liver damage. Cu can receive and donate electrons and, as a redox catalyst, is necessary for many enzymes' catalytic activities. Cells use Cu for respiration, blood cell line maturation, immune responses, wound heling, myelin sheath formation, and neurotransmitter synthesis.

Cu balance is determined by the equilibrium between the rates of dietary absorption from food, supplements and drinking water, and excretion through stools and bile, and it is tightly regulated (Fig. 1).

The recommended dietary allowance (RDA) for Cu, which represents the intake level sufficient to meet the nutrient requirements for 97%–98%, is 0.9–1.3 mg/day (Table 2), according to the United States (US) Institute of Medicine (IOM) (IOM, 2001), whereas the average person consumes ~2 mg/day (WHO, 1996). A Cu intake of 2–3 mg/day is safe and adequately prevents Cu deficiency. However, ingesting >5 mg/day is considered toxic (de Romana, Olivares, Uauy, & Araya, 2011).

In 2000, the Report on Copper Drinking Water (Council, 2000) stated that gastrointestinal symptoms arise from drinking water with Cu at approximately 3 mg/L.

The Agency of Toxic Substances and Disease Registry (ATSDR) in 2004 reported the median intake of Cu from food sources as 0.93–1.3 mg/day for adults in the United States (ATSDR, 2004). Dietary intake values have been provided also by the US IOM: The mean dietary intake from food is estimated at 1.0–1.7 mg Cu/day, whereas the mean

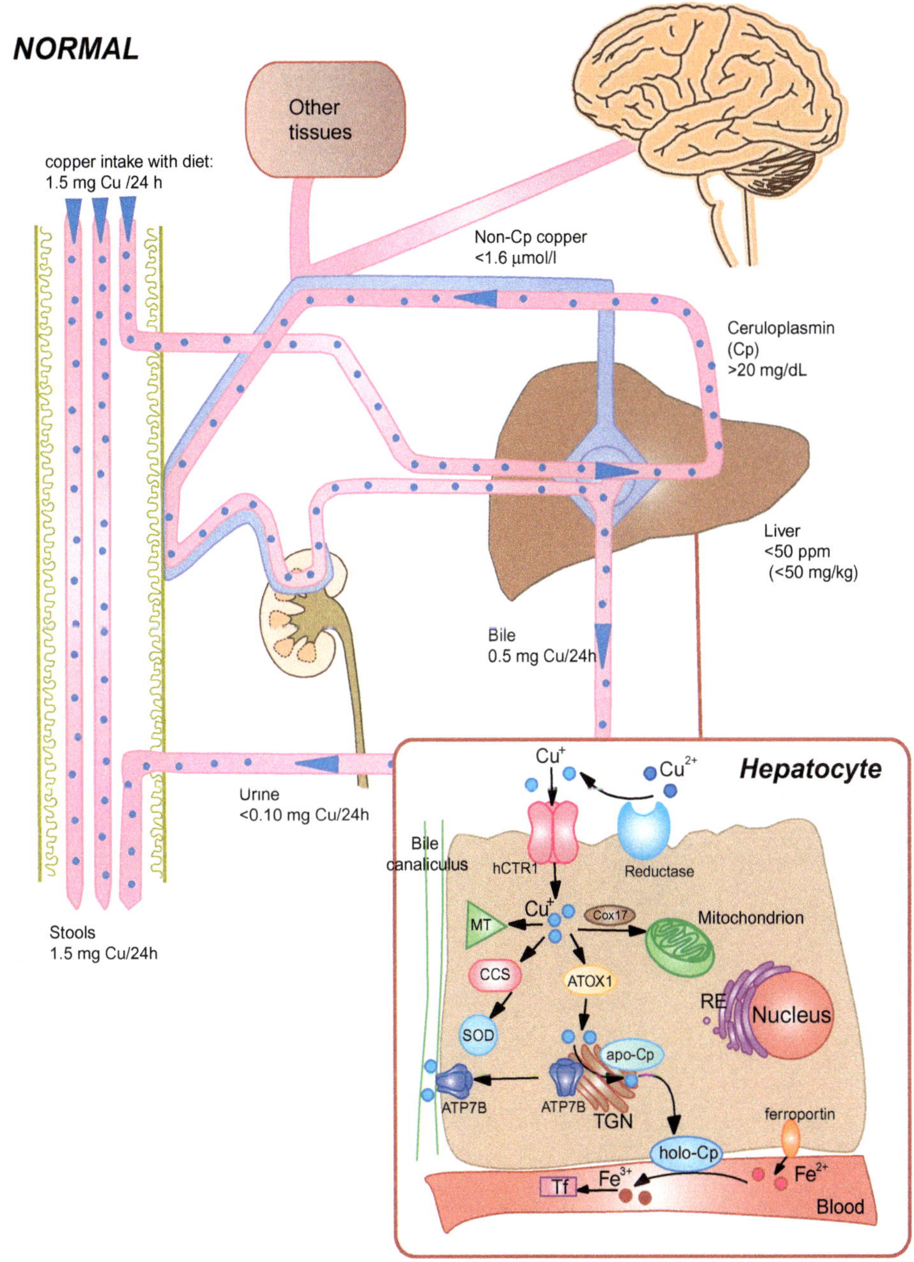

Fig. 1 The normal pathway for copper absorption. Copper dietary intake in 24h is 1.5mg; 0.5mg/day of this copper is absorbed from duodenum and transported to the liver through the portal system. In the liver, the copper is structurally incorporated into ceruloplasmin (Cp) by the copper pump ATPAse7B (ATP7B). Normal values of Cp are above 20mg/dL; the concentrations of Non-Cp copper and urinary copper after an overnight fast are below 1.6μmol/L. *Red box*: **the hepatocyte**. A still unknown reductase protein reduces Cu^{2+} to Cu^{+}, which is then imported into the cell by hCTR1. Once in the cell, copper is carried by cytochrome C oxidase assembly homolog (COX17) to the mitochondria; by the copper chaperone for superoxide dismutase (CCS) to Cu/Zn SOD and by the antioxidant protein 1 homolog (ATOX1) to the ATP7B. In the *trans* Golgi network (TGN), the ATP7B protein loads copper into nascent apo-ceruloplasmin to form active holo-ceruloplasmin (Cp). *(Reproduced with permission from Squitti, R., Siotto, M., Polimanti, R. (2014). Low-copper diet as a preventive strategy for Alzheimer's disease.* Neurobiology of Aging, 35 Suppl. 2, *S40–50. https://doi.org/10.1016/j.neurobiolaging.2014.02.031.)*

Table 2 Dietary reference intake of copper defined by the Food and Nutrition Board of the Institute of Medicine at the National Academies of Sciences.

Age/status group	RDA (μg/day)	AI (μg/day)	UL (μg/day)
Infants			
0–6 mo	–	200	ND
7–12 mo	–	220	ND
Children			
1–3 y	340	–	1000
4–6 y	440	–	1000
Males			
9–13 y	700	–	5000
14–18 y	890	–	8000
19–30 y	900	–	10,000
31–50 y	900	–	10,000
50–70 y	900	–	10,000
>70 y	900	–	10,000
Females			
9–13 y	700	–	5000
14–18 y	890	–	8000
19–30 y	900	–	10,000
31–50 y	900	–	10,000
50–70 y	900	–	10,000
>70 y	900	–	10,000
Pregnancy			
≤18 y	1000	–	8000
19–30 y	1000	–	10,000
31–50 y	1000	–	10,000
Lactation			
≤18 y	1300	–	8000
19–30 y	1300	–	10,000
31–50 y	1300	–	10,000

RDAs, recommended dietary allowances; AIs, adequate intakes, and ULs, upper levels of tolerable intake are reported. ND, not determinable.

Mo: months; y: years.

The European Food Safety Authority (EFSA), Parma, Italy (2015) proposed AIs of 1.6 mg/day for men and 1.3 mg/day for women; 0.7 mg/day for children aged 1 to <3 years, 1 mg/day for children aged 3 to <10 years, 1.3 and 1.1 mg/day for boys and girls aged 10 to <18 years, 0.4 mg/day for infants aged 7–11 months, and 1.5 mg/day for pregnant and lactating women.

Reproduced with permission from Squitti, R., Siotto, M., Polimanti, R. (2014). Low-copper diet as a preventive strategy for Alzheimer's disease. *Neurobiology of Aging, 35 Suppl. 2*, S40–50. https://doi.org/10.1016/j.neurobiolaging.2014.02.031.

dietary plus supplemental Cu intake is 1.3–2.2 mg Cu/day (IOM, 2001). Interestingly, in US as per data collected in 1986, up to 15% of US adults consumed dietary supplements containing an average of 1.3–2.2 mg Cu/day (IOM, 2001). Some authors estimated that considering the average intake values of Cu from water, food, and supplements, the average of Cu taken in by adult individuals in US approximated a total of 1.85 mg Cu/day (Taylor et al., 2020).

Cu regulation in physiology

Cu taken in through the diet is readily absorbed from the stomach, intestines (duodenum and ileum; 0.5 mg/day) (Hoogenraad, 2001), and transported to the liver (Fig. 1): It generally corresponds to 30% of the Cu consumed with the diet, supplements and drinking water (food 75%, supplements and drinking water 25%). The rate of Cu absorption decreases as dietary Cu intake increases: It can vary from 56% of the dietary Cu intake during a low-Cu diet to 12% under a high-Cu diet (Turnlund, Keyes, Kim, & Domek, 2005).

In the cells of the intestinal lining, Cu is absorbed as a pool of low molecular-weight-soluble complexes of Cu is then pumped out of the enterocyte's basolateral membrane by the copper-transporting P-type ATPase (Cu-ATPase) 7A (ATPase7A). This Cu pool is mostly bound to amino acids, peptides, micronutrients, and albumin and is transported into the serum. From the gut, the pool of low-molecular-weight Cu, known as "free," or non-ceruloplasmin (non-Cp) Cu, reaches the liver through the portal vein. Hepatocytes absorb most of these Cu complexes and limit non-Cp Cu concentration in the blood to 0.008–1.6 μmol/L (equivalent to 0.05–1 mg/dL). This value represents the normal reference range of non-Cp Cu in serum after an overnight fast (Hoogenraad, 2001) (Fig. 1).

Human Cu transporter 1 (CTR-1) facilities Cu(I) to enter the hepatocyte and Cu chaperons/transporters then accompany the metal to the sites of utilization within the cell. Here, Cu is loaded into Cu proteins and enzymes, including cytochrome oxidase (COX) for mitochondrial respiration, copper/zinc superoxide dismutase (Cu/Zn SOD) for antioxidant defenses, ceruloplasmin for regulating the iron (Fe) oxidative state, and distribution to tissues and organs (review in Squitti, Siotto, et al., 2014). The key copper-transporting P-type ATPase (Cu-ATPase) 7B (ATPase7B), similarly to the ATPase7A, is a Cu pump located in the trans-Golgi network that loads Cu into Cu-dependent enzymes, and specifically into nascent ceruloplasmin in the hepatocyte (review in Squitti, Siotto, et al., 2014). In the hepatocyte, the Cu that exceeds the needs of the cell is excreted by ATPase7B into the bile (Hoogenraad, 2001) (Figs. 1 and 2).

The choroid plexus produces and secretes Cu in the CSF, the biological fluid that surrounds the brain (range between 0.5 and 2.5 μmol/L (Squitti, Simonelli, et al., 2014)). Non-Cp Cu in rat *choroid plexus* is the main species uptaken from the blood into

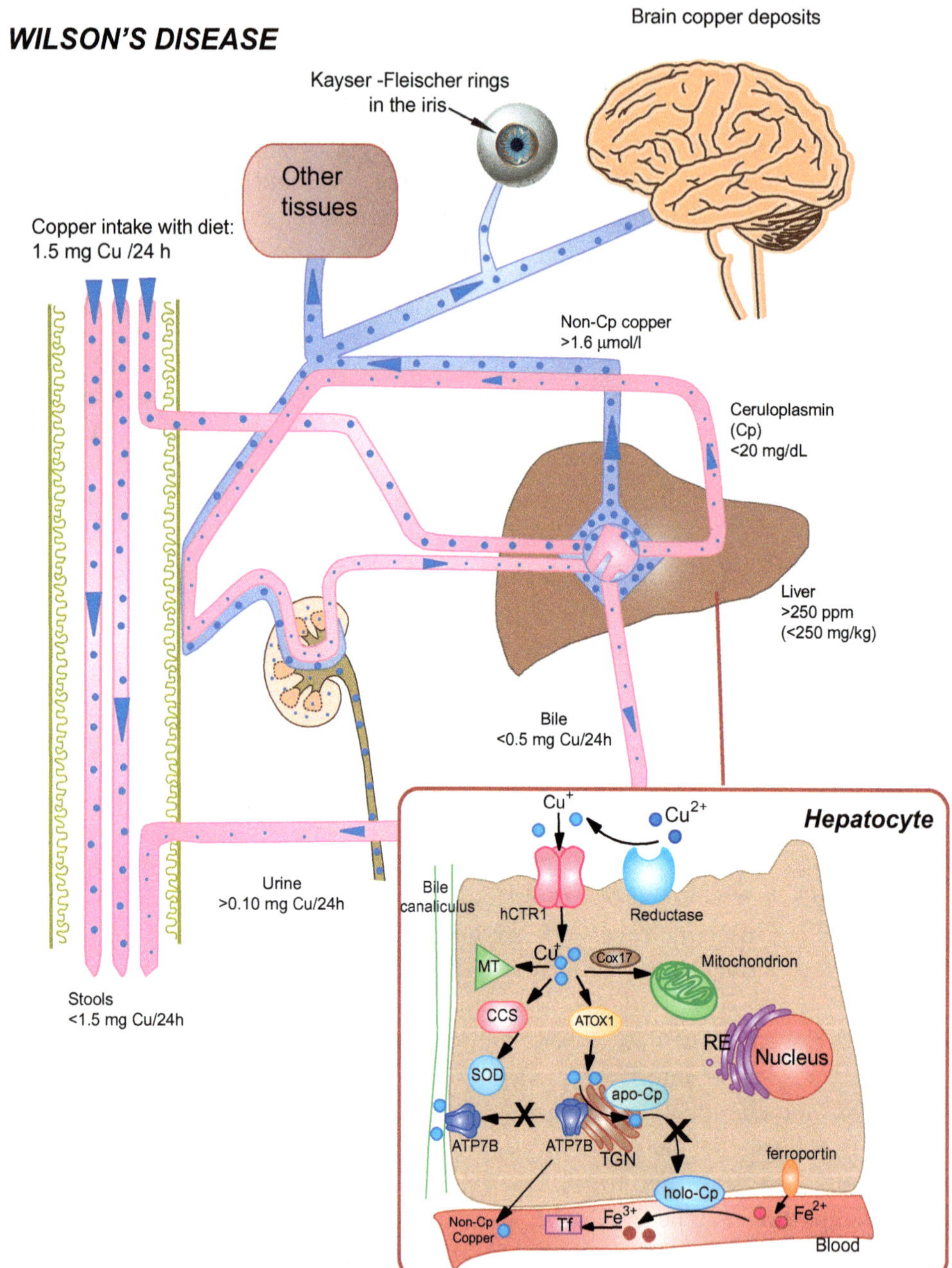

Fig. 2 Copper absorption in patients with Wilson's disease. Absorption is normal, but there is a failure in its incorporation into Cp. *Red box*: **the hepatocyte**. *ATP7B* functionality is impaired (*black* cross). This causes a decrease in copper excretion through the bile and an overflow of non-Cp copper in general circulation and urine. The overload of non-Cp copper in systemic circulation causes brain copper deposits and, in a certain percentage of patients, Kaiser-Fleisher rings in the iris of eyes. Values of Cp are below 20 mg/dL; the concentrations of non-Cp copper and urinary copper are above 1.6 μmol/L. *(Reproduced with permission from Squitti, R., Siotto, M., Polimanti, R. (2014). Low-copper diet as a preventive strategy for Alzheimer's disease.* Neurobiology of Aging, 35 Suppl. 2, *S40–50. https://doi.org/10.1016/j.neurobiolaging.2014.02.031.)*

the brain (Choi & Zheng, 2009). This is also exemplified in Wilson disease (WD) in which non-Cp Cu easily crosses the blood-brain barrier (BBB), the filter that separates the blood from the brain (Hoogenraad, 2001). Evidence in this line has been collected also for AD (review in Kepp & Squitti, 2019).

The process of Cu excretion occurs mainly through exfoliation of mucosal cells, when enterocytes are shed into the intestinal lumen and released with feces: Cells of the intestinal lining contain MTs that tightly bind Cu; their exfoliation allows for Cu removal (Hoogenraad, 2001). Zn therapy (review in Squitti et al., 2020) can modulate MTs expression in the enterocytes increasing by 25-fold the expression of these natural Cu-trapping proteins that put in place a mucosal block (Hoogenraad, 2001).

Cu dysregulation in Wilson disease and in Menkes disease

In the paradigmatic disease of non-Cp Cu toxicosis and accumulation, WD, mutations in the *ATP7B* gene encoding for the Cu pump ATPase7B, alter the structure function of the protein (Fig. 3). Defective ATPase7B fails to transfer Cu to ceruloplasmin or to excrete the metal into the *canalicula biliaris*. Cu is, therefore, abnormally released as non-Cp Cu into the bloodstream. People affected by WD show decreased blood levels of holo-ceruloplasmin (<200 mg/L), have an expansion of the non-Cp Cu pool in plasma (>1.6 μmol/L), Cu accumulation in the liver (>250 μg/g tissue), increased Cu levels the urine (>250 μg/mL), and Cu deposits in the cornea where they can be observed as Keyser-Fleisher rings (Hoogenraad, 2001).

Menkes disease is a rare disease and the paradigmatic disease for copper insufficiency and death in which mutations in gene *ATP7A* encoding for the ATPase7A lead to Cu deficiency and promote abnormalities, including those affecting the central nervous system. *ATP7A* mutations prevent Cu absorption at the intestinal level, causing a Cu deficiency. An early diagnosis of Menkes disease and Cu intravenous infusion in the very early stages of the disease increase the likelihood of the survival (review in Squitti, Siotto, et al., 2014).

Cu imbalance in Alzheimer's disease

A bulk of evidence shows that the amyloid hypothesis is strongly intertwined with Cu imbalance because APP/Aβ are Cu/Zn binding proteins with a potential role as natural Cu/Zn-buffering proteins and due to toxic functions of Cu-Aβ as summarized in Kepp and Squitti (2019). Amyloid plaques contain relevant amounts of essential trace metals such as Fe, Cu, and Zn, and it is established that Aβ is a metallopeptide and that Cu^{2+} (and Zn^{2+}) binding considerably changes the structure of the Aβ peptide, its aggregation propensity, and, mostly, its toxicity (Faller, Hureau, & La Penna, 2014). In analogy with WD, a subset of AD patients is typified by the expansion of the non-Cp Cu pool in plasma (>1.6 μmol/L) (Squitti, Simonelli, et al., 2014).

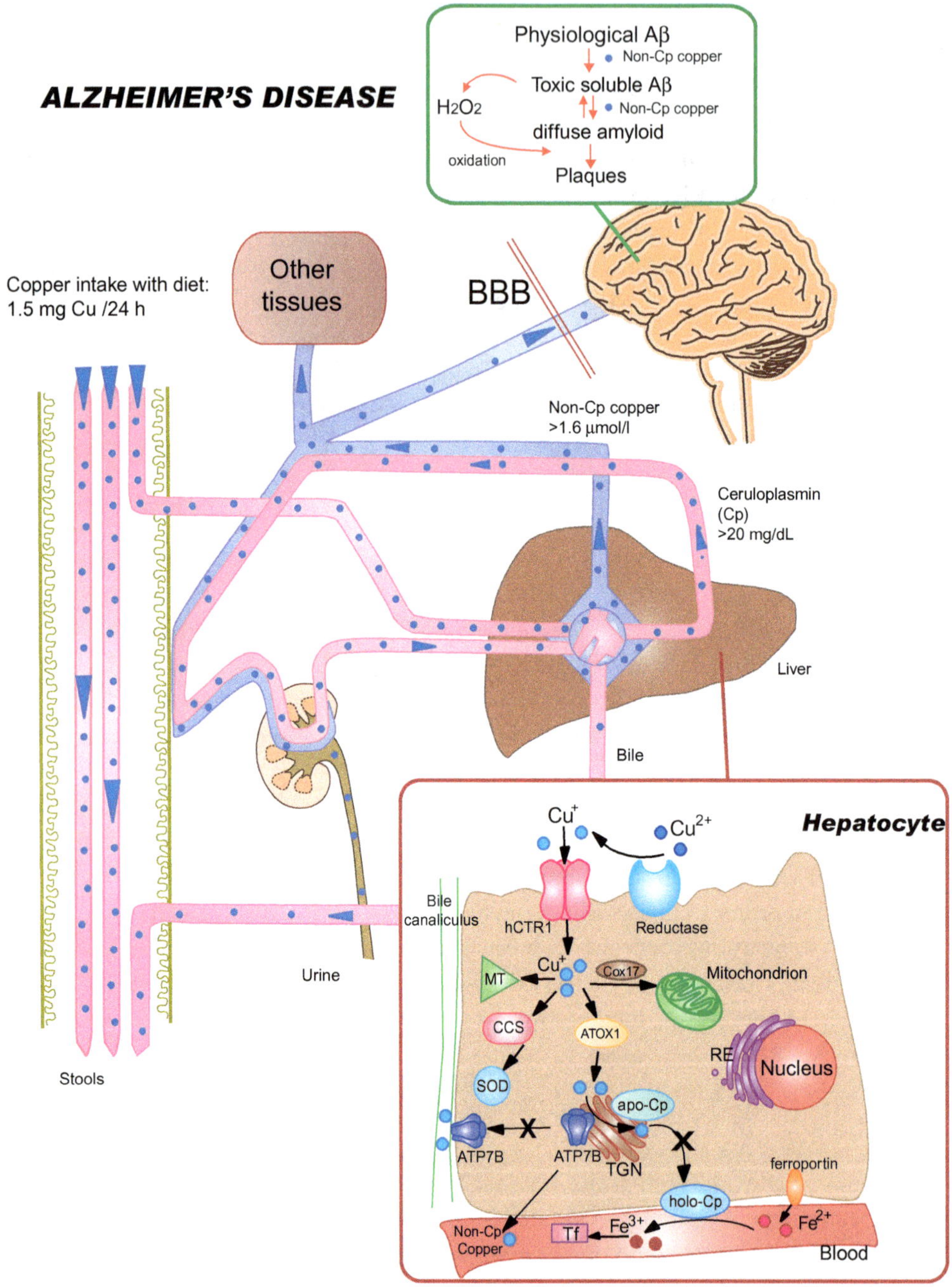

Fig. 3 Model of copper absorption in Alzheimer's disease patients. Absorption is normal, but there is a failure in its incorporation into Cp. *Red box*: **the hepatocyte**. ATP7B functionality is impaired (*black* cross), even though less severely with respect to Wilson's disease. This causes an overflow of non-Cp copper sufficiently to be detected in general circulation. *Green box*: **the brain**. When values of non-Cp copper exceed 1.6 µmol/L, it crosses the brain-blood barrier (BBB) and enters Fenton and Haber-Weiss reactions with Aβ, producing H_2O_2 and Aβ plaques deposition. Values of Cp are above 20 mg/dL, the concentrations of non-Cp copper are above 1.6 µmol/L. *(Reproduced with permission from Squitti, R., Siotto, M., Polimanti, R. (2014). Low-copper diet as a preventive strategy for Alzheimer's disease.* Neurobiology of Aging, 35 Suppl. 2, *S40–50. https://doi.org/10.1016/j.neurobiolaging.2014.02.031.)*

The model of Cu displacement and dysregulation in Alzheimer's disease

In a previous study, we have proposed a new model to explain how Cu misplacement and dysregulation might accelerate the onset and progression of AD (Kepp & Squitti, 2019) at least in a percentage of AD individuals. Summarizing the model, the Cu imbalance in AD can be described by a critical, location-dependent control variable named *Cu dissociation constant*, K_d^c (Kepp & Squitti, 2019). The model estimates that (i) the reduction of the pool of divalent Cu(II) with $K_d < K_d^c$ implies the loss of functional Cu from protein-bound pools: This misplacement of Cu is key to decrease energy production and impair oxidative stress control mainly within the brain; (ii) the increase of the pool with Cu with $K_d > K_d^c$ implies the gain of redox-toxic functions by these loosely bound Cu-complexes. In (i), the loss of functional Cu from protein-bound pools in neuronal cells impairs function and structure of crucial Cu proteins such as prion protein (PrP), α-synuclein, COX, Cu/Zn SOD, ceruloplasmin, ATPase7B, MTs, dopamine beta-hydroxylase (Kepp & Squitti, 2019), and can drive the intracellular accumulation of phosphorylated tau due to the presence of Cu-binding sites on the protein (Kepp & Squitti, 2019). In (ii), the shift toward the build-up of an increased pool of loosely bound Cu can facilitate oxidative stress via Fenton-like reactions and the activation of Cu-driven bioenergetic abnormalities in mitochondria, impair glucose utilization, inflammation, calcium dyshomeostasis, and interfere with the balance of other brain metals like Fe and Zn (Kepp & Squitti, 2019). Cu imbalance also has relevant synaptic effects by interfering with critical proteins that are involved in Cu buffering and functioning of glutamatergic synapses (Kepp & Squitti, 2019). As expressed above, this complex of processes might facilitate the aggregation of Aβ oligomers in insoluble plaques and favor the formation of amyloid plaques short-circuiting the neuronal networks to which those synapses belong.

The K_d^c model of Cu imbalance in AD find support in results from meta-analysis (Squitti, White, Granzotto, & Kepp, 2021). These studies reveal an extensive Cu imbalance in AD featured by decreased values of Cu in the brain, coexisting with increased values of non-Cp Cu in the blood, a phenomenon that can explain the reported increased serum values of total Cu in AD in comparison with healthy controls. On average, data from literature indicated a decrement of Cu in the AD brain by about 24% and a rise of Cu in the blood by about 7%, while non-Cp Cu concentration doubled, pinpointing to an overall Cu imbalance in AD (Squitti et al., 2021).

The scientific basis for considering CuAD as a subtype of Alzheimer's disease

Non-Cp Cu excess has been found in patients with mild cognitive impairment (MCI) (review in Mathys & White, 2017 and Kepp & Squitti, 2019). It is associated with a faster rate of conversion to AD: In a 6-year survey period, 50% of MCI patients with high levels of non-Cp Cu were found to convert to AD (review in Mathys & White, 2017 and Kepp & Squitti, 2019), with a hazard rate of conversion to AD three times higher than MCI with normal non-Cp Cu values. Excess non-Cp Cu provides also information about a worse prognosis of AD patients (review in Mathys & White, 2017 and Kepp & Squitti, 2019).

Meta-analysis, epidemiological, and large population studies further support for the role of Cu in the modulation of cognition (review in Kepp & Squitti, 2019). On this basis, Cu abnormalities might serve as a stratification biomarker for the disease in a subset of patients (review in Kepp & Squitti, 2019). Direct and indirect methods to detect the toxic non-Cp Cu associated with AD have been developed, and it has been demonstrated that Cu dysregulation affects approximately 50%–60% of patients with MCI or AD (review in Kepp & Squitti, 2019). The phenomenon appears specific of a subpopulation of AD who are carriers of selective *ATP7B* gene variants and have been called CuAD subtype (review in Kepp & Squitti, 2019). Further characterization of this Cu-related AD subtype revealed that this subset showed distinct electroencephalogram (EEG)-derived eLORETA cortical rhythms alterations and peculiar neuroimaging features (review in Kepp & Squitti, 2019). Wilson disease and AD have two diverse etiologies but a Cu-dependent pathogenic overlap (review in Kepp & Squitti, 2019): It is conceivable that, in the AD, brain of a selected subset of patients with non-Cp Cu excess, a displacement of Cu might trigger mechanisms of neuronal toxicity as depicted above (Kepp & Squitti, 2019).

Implementation of a Cu diet to decrease the CuAD risk

There is no public health concern about Cu deficiency and excess in humans. In WD, Indian childhood cirrhosis and idiopathic Cu toxicosis is evident the interplay between genetic predisposition and environmental Cu exposure in determining the onset of the disease (review in Squitti, Siotto, et al., 2014).

Cu diet in Alzheimer's disease

The report on Drinking Water (2000) (Council, 2000) described that about 1% of general population is a heterozygote carrier of *ATP7B* mutations for WD and that these individuals also have abnormal Cu regulation and build-up of Cu in the body. As depicted in this chapter (see "Why Cu has much to do with Alzheimer's disease" section), individual carriers of the G allele in rs1061472 and the A allele in rs732774, two functional SNPs of the *ATP7B* gene, have increased levels of non-Cp Cu of in general circulation and other Cu abnormalities linked to an increased risk of AD. This population that is sensitive to excess Cu might wish to investigate their Cu status (as e.g., to measure levels of non-Cp Cu), and on this basis decide, if using multiple vitamin supplements, to choose those without Cu as per expert committees' suggestions (Barnard et al., 2014) or implement a low-Cu diet to regaining Cu balance in their body in the attempt to decrease their chance to have AD. The recommendation is to undergo a clinical testing before initiating any diet changes (Barnard et al., 2014; Squitti, Siotto, et al., 2014).

Reference values for dietary intake of Cu have been provided by the Food and Nutrition Board of the Institute of Medicine of the National Academies of Sciences (Council, 2001) (Table 2).

Foods enriched in Cu include seafood (especially shellfish (oysters up to 8 mg Cu/100 of serving size) and (calamari 2.1 mg Cu/100 g), organ meats (animal liver 15 mg Cu/100 g), grains (sesame seeds 4.1 mg Cu/100 g) nuts (cashew 2.2 mg Cu/100 g) and cocoa (3.8 mg Cu/100 g)). Other minor food sources of Cu include cereals, potatoes, red meat, and mushrooms.

On the other side, foods with a low amount of Cu include rice, chicken, and tea that can be consumed in greater amounts and grant a safe source of Cu. Epidemiological (Morris et al., 2006) as well as preclinical models of AD (Arnal, Morel, de Alaniz, Castillo, & Marra, 2013; Sparks & Schreurs, 2003) pinpoint a combined effect of Cu and cholesterol in rising the likelihood of AD. In line with this evidence, individuals with a susceptibility to Cu dyshomeostasis should reduce foods not only rich in Cu but also high in fat, and accordingly, perform the analysis of serum non-Cp Cu coupled with the analysis of fat level in the blood (e.g., total cholesterol, high-density lipoproteins, low-density lipoprotein, and triglycerides).

Zn reduces the body's capacity to absorb Cu by potentiating the mucosal block by increasing the levels of MTs that trap Cu in the intestine (see "Cu regulation in physiology" section) but only at a Zn dosage of 100/150 mg day, as it used in WD treatment.

Requirement for a rapid test for non-Cp Cu assessment

Non-Cp Cu might serve as prognostic/stratification biomarker of the CuAD subset. An innovative approach, not including a separation step that might distort non-Cp Cu quantitation, is represented by sensor technology based on anodic stripping voltammetry. Recently, a patent employing such technology has been described (Anastasiadou, Jannakoudakis, & Girousi, 2010). Although research is still ongoing, this technology is very promising for the development of a non-Cp Cu reliable test that would be a cost-effective and fast analytical method for population-level screening.

Application to other neurological conditions

A recent study by Boyle et al. (2018) shows the relative contributions of specific neuropathologies to cognitive loss at the individual level, indicating that many neurodegenerative processes synergistically take place in the brain of persons affected by dementia, and suggests the urgent need for precision medicine approaches that embrace the heterogeneity of disease to target specific biological subsets rather than broad clinical entities.

Alzheimer's disease, Parkinson's disease (PD), and Amyotrophic lateral sclerosis share key underlying pathophysiological processes, primarily oxidative stress, protein misfolding/oligomerization, and inflammation that are linked to Cu imbalance and associated with brain aging.

Cu affects Aβ (see "Cu imbalance in Alzheimer's disease" section), but it is also involved in prion-like mechanisms of disease progression, based on prion protein (PrP) abnormalities. The modified protein PrP^{Sc} is an aggregated form of PrP and the agent of prion disease (review in Kepp & Squitti, 2019). The PrP is a glycoprotein located at synapses that can bind up to five Cu atoms. The binding of Cu to PrP and the formation of the Cu-PrP complex facilitate PrP internalization within the cell as well as PrP^{C} conversion to PrP scrapie (PrP^{Sc}). At the glutamatergic synapse, Cu-PrP modulates neurotransmission (Brown et al., 1997). The Cu-PrP complex also binds to Aβ (review in Kepp & Squitti, 2019). Interestingly, Cu can affect α-synuclein functioning, a key protein involved in PD pathology, that abnormally accumulates and aggregates in the form of Lewy's bodies. In correct physiology, α-synuclein protein is located at the synapse, but Cu^{2+} misplacement, via oxidative stress, can contribute to its oligomerization and fibrillation as well as the formation of Lewy's bodies (Miotto et al., 2015).

The processes of neurodegeneration depicted above (see "The model of Cu displacement and dysregulation in Alzheimer's disease" section) and associated to Cu imbalance in AD might take place also in these neurodegenerative diseases.

Other component of interest: Oxidative stress and Alzheimer's disease

Oxidative stress has been implicated in the pathogenesis of major chronic diseases, including neurodegenerative disorders (Jenner, 1994). To get protected against damage from excess of free radicals and reactive oxygen species (ROS), humans and other living organisms are equipped with powerful and complex antioxidant systems comprising specific enzyme systems (Cu/Zn SOD, Glutathione peroxidase, Catalase) and nonenzymatic antioxidants such as tocopherol, retinol, reduced glutathione, ascorbic, and uric acids, able to inactivate radical species. These compounds, provided with reducing power, are either produced in the body or derived from the diet (Sies, 2007). Antioxidants don't function as isolated entities but as part of complex systems with relevant interdependence and synergistic effects.

Aβ peptide and oxidative stress are at the basis of the pathophysiology of the many neurodegenerative diseases and specifically AD (Markesbery, 1997). The relationship between Aβ and free radical production is a vicious circle: Not only oxidative processes can convert nonaggregated Aβ into aggregated Aβ in vitro, Aβ itself is also a source of free radicals (Dyrks, Dyrks, Hartmann, Masters, & Beyreuther, 1992). In fact, several studies have shown that Aβ stimulates ROS production in microglia and triggers microglial release of nitrogen dioxide (NO_2) radicals in rodents (McDonald, Brunden, &

Landreth, 1997). It has been suggested that several individual antioxidants and free radical scavengers, such as ascorbate, vitamin E, and protein sulfhydryl groups, or a combination of these, can be neuroprotective and decrease the risk of AD or slow its progression (Jomova, Vondrakova, Lawson, & Valko, 2010). One of the main endogenous free radical scavenger systems in the brain is associated with glutathione (GSH). GSH is the most abundant intracellular nonprotein thiol and is able to detoxify various oxidants by direct scavenging of free radicals or by acting as a co-substrate in GSH-peroxidase-catalyzed reactions (Pinnen et al., 2011).

GSH depletion is observed in the aging brain and is accelerated in many neurodegenerative diseases including AD. Importantly, GSH loss in the brain is associated with impairments in cognitive function (Currais & Maher, 2013). The role of GSH depletion in AD appears to be pivotal considering the well-established causative relationship between oxidative stress and apoptosis, a fundamental mechanism accounting for tissue degeneration in AD (Redza-Dutordoir & Averill-Bates, 2016).

Furthermore, GSH deficiency in brain might indirectly affect AD pathogenesis since GSH has chelating properties and can reduce Cu ion excess as well as additional toxic elements in tissues (Zeevalk, Bernard, & Guilford, 2010).

GSH is not bioavailable when administered orally as it undergoes digestion by enteric peptidases. However, novel modes of delivery such as liposomal and prodrug preparations are emerging (Cacciatore, Baldassarre, Fornasari, Mollica, & Pinnen, 2012). It may be also administered intravenously, by transdermal route and via nebulizer. Recently, several glutathione derivatives have been synthesized: They have the potential to be more bioavailable than GSH when administered orally.

Cu and GSH interplay

Glutathione depletion is one of the main causes of oxidative stress-induced programmed cell death. A reduction in GSH concentrations is seen in the aging brain. Fe and Cu also potentiate cell death induced by other GSH depletes but not by compounds that induce oxidative stress via other pathways (Maher & Conti, 2018).

Applications to other areas: Cu in type 2 diabetes

Implementation of a low-Cu diet can be also considered as a potential preventive strategy in type 2 diabetes mellitus. Type 2 diabetes is another complex disease where Cu imbalance can contribute on the disease onset and progression (Cooper, 2012). The processes at the basis of Cu excess in type 2 diabetes have not been elucidated so far. However, it is known that Cu levels are higher in obesity and are associated with insulin resistance as well as with the reduction of high-density lipoprotein-cholesterol levels. Moreover, the activity of the ATPase7B Cu pump is modulated by insulin and glucagon

(Cooper, 2012). On this basis, it has been suggested that the processes that underlying type 2 diabetes may share similarities with processes of Cu dyshomeostasis.

More intriguing is the putative connection of Cu misbalance in type 2 diabetes and the cognitive deterioration associated with this disease. Indeed, type 2 diabetes is an established risk factor for AD (WHO, 2019) even though the link between the two is still elusive. The demonstration that Cu imbalance associates with an increased rate of cognitive worsening in type 2 diabetes patients and would bring almost immediate benefits for the community.

Key facts of the WHO on numbers of dementia and prevention:

As per WHO, the top seven countries in the number of deaths caused by AD are China (563.5), United States (259.5), India (140.9), United Kingdom (82.5), Indonesia (54.7), France (49.6), and Germany (48.8) thousand deaths in 2016 (WHO, 2020).

WHO in 2019 (WHO, 2019) published recommendations to reduce risk of cognitive decline and dementia.

Key facts of the WHO on Cu dietary intake:

The recommended dietary allowance (RDA) for Cu is 0.9–1.3 mg/day (IOM, 2001).

As per WHO, the average consume of Cu per person is ~2 mg/day (WHO, 1996).

Key fats of the WHO on Cu drinking water:

As per the (WHO, 2018), 104 countries have a regulation or guideline for Cu concentrations in drinking water, with a median value of 1.5 mg/L and a range of 0.05–3 mg/L.

Key facts of Cu imbalance in AD:

The gradual loss of Cu from key Cu proteins in the brain impairs mitochondrial energy production and decreases oxidative stress defenses while the increase of labile Cu pool favors oxidative stress via Fenton-type reactions, thereby participating in two major pathologies of AD that strongly associate with the aging process, the main risk factor for the disease.

Mini-dictionary of terms

Amyloid β (Aβ): peptides derived from the proteolytic and amyloidogenic cleavage of the amyloid precursor protein (APP). Aβ peptides are the primary constituent of amyloid plaques whose distribution and density typified the AD brain.

α-synuclein: key protein that abnormally accumulates and aggregates in neuronal inclusions: Lewy's bodies and Lewy neurites within neurons. These inclusions typify Dementia with Lewy Body, PD, and PD dementia that can be included in Lewy body diseases. Lewy bodies affect neuronal functioning and lead to problems with thinking, movement, behavior, and mood.

Ceruloplasmin (Cp): positive acute phase reactant in inflammation states; it is a 151-kDa protein with ferroxidase activity that controls iron oxidation rate required for the release of the metal from the storage sites. Ceruloplasmin binds 6–8 Cu atoms structurally that are required for its catalytic activity. On this basis, it accounts for 70%–85% of Cu in general circulation.

Menkes' disease: paradigmatic disease of Cu deficiency caused by mutations of the gene encoding the Cu-transporter ATPase7A. X-linked recessive disease typified by decreased levels of serum Cu and ceruloplasmin. It is a severe pediatric disease, eventually leading to death. Main symptoms are intellectual disability and developmental delay, seizures, lack of muscle tone, floppiness, and kinky hair.

Metallothioneins (MTs): cysteine-enriched low-molecular-weight proteins that bind metals in cell and tissues.

Non-ceruloplasmin copper (non-Cp Cu): Non-Cp Cu has been historically defined as "free" Cu. Non-Cp Cu terminology derives from the algorithm employed for its estimation by indirect methods, based on the subtraction of the Cu bound to ceruloplasmin from the total concentration of circulating Cu.

Prion protein (PrP): glycoprotein located at synapses that can bind up to five Cu atoms. It can exist in diverse forms: the normal PrP^{C} and protease-resistant isoforms designated PrP^{Res} such as the disease-causing $PrP^{Sc(scrapie)}$ and an isoform located in mitochondria. In particular circumstances, it can form aggregates that are involved in PrP^{C} conversion to PrP^{Sc}. The PrP^{Sc} is a misfolded isoform associated with a variety of neurodegenerative disorders including ovine scrapie, bovine spongiform encephalopathy (BSE, mad cow disease), and a variant Creutzfeldt-Jakob disease.

Reactive oxygen species (ROS): unstable molecules also known as free radicals that contain oxygen and easily reacts with other molecules in cells and tissues producing oxidative stress. Oxidative stress can cause damage to DNA, RNA, and proteins and may cause cell death. ROS can also have a crucial role as signaling molecules.

Tau proteins: proteins involved in microtubule stabilization within the neuronal axons in the brain. When hyperphosphorylated, tau proteins are prone to aggregation producing neurofibrillary tangles within the neuronal cells and promote neurodegeneration.

Wilson disease (WD): paradigmatic disease of Cu excess caused by autosomal recessive mutations in the *ATP7B* gene. WD is typified by decreased levels of ceruloplasmin, decreased level serum Cu, increased levels of non-Cp Cu. Main symptoms are fatigue, jaundice, dystonic rigidity, dysarthria, dysphagia, and tremor. The disease can have a hepatic, neurologic, or psychiatric presentation. The hepatic presentation is generally associated with a pediatric onset, while the neurological and psychiatric forms have generally an adult onset.

Summary points

Section 1. Introduction: Excess Cu is toxic and increase the risk of AD.

Section 2. Alzheimer's disease: AD is the most common form of dementia with a defined pathology at the basis of the new IWG-2 criteria.

Section 3. Cu intake and regulation in physiology: Cu is an essential trace metal tightly regulated to grant correct physiology. Cu deficiency as Cu excess is toxic. RDA for Cu is 0.9–1.3 mg/day. Defective function of ATPase7A and ATPase7B upon gene mutations is at the basis of the rare monogenetic recessive disease Menkes disease and WD.

Section 4. Cu imbalance in Alzheimer's disease: Cu displacement is at the basis of Cu dysregulation associated with AD. A *Cu dissociation constant*, K_d^c, well describes the reduction of functional Cu bound to proteins and the shift of the Cu to the non-Cp Cu pool, prone to oxidative stress. Non-Cp Cu excess in AD can easily cross the BBB, as exemplified in WD, likely promoting the aggregation of Aβ peptides, and oxidative stress. Chemical, clinical, genetic, and meta-analysis data point at Cu imbalance in AD and suggest a specific role of the metal in a subset of AD/MCI patients, referred to as CuAD.

Section 5. Implementation of a Cu diet to decrease the CuAD risk: non-Cp Cu might serve as a stratification/prognostic biomarker to identify individuals susceptible to Cu dysregulation and cognitive decline—a CuAD subset—who exhibit an increased risk for developing cognitive decline. This people might implement prevention programs based on Cu dietary restriction in the attempt to decrease their risk of cognitive worsening.

Section 6. Application to other neurological conditions: Cu can affect PrP and promote the formation of the Cu-PrP complex facilitates PrP internalization within the cell as well as PrP conversion to PrP^{Sc}, promoting prion-like mechanisms of disease progression. Cu can also affect α-synuclein functioning, a key protein involved in PD pathology, that abnormally accumulates and aggregates in the form of Lewy's bodies.

Section 7. Other component of interest: Oxidative stress and Alzheimer's disease: GSH loss is among the AD causes since it is at base of oxidative stress. Cu can potentiate either GSH loss or cell death; in turn, GSH has Cu chelating properties.

Section 8. Applications to other areas: Cu in type 2 diabetes: Diabetes type 2 is typified by Cu abnormalities associated with ATPase7B Cu pump dysfunction.

References

Alzheimer's Disease-International. (2019). *World Alzheimer report 2019: Attitudes to dementia*. London: Alzheimer's Dis. Int.

Anastasiadou, Z. D., Jannakoudakis, P. D., & Girousi, S. T. (2010). Square wave anodic stripping voltammetry determination of eco-toxic metals in samples of biological and environmental importance. *Chemical Science*, *8*(5), 999–1008.

Arnal, N., Morel, G. R., de Alaniz, M. J., Castillo, O., & Marra, C. A. (2013). Role of copper and cholesterol association in the neurodegenerative process. *International Journal of Alzheimer's Disease*, *2013*, 414817. https://doi.org/10.1155/2013/414817.

ATSDR. (2004). *Toxicological profile for copper*. Atlanta, GA: ATSDR. https://www.atsdr.cdc.gov/toxprofiles/tp132.pdf.

Barnard, N. D., Bush, A. I., Ceccarelli, A., Cooper, J., de Jager, C. A., Erickson, K. I., et al. (2014). Dietary and lifestyle guidelines for the prevention of Alzheimer's disease. *Neurobiology of Aging*, *35*(Suppl. 2), S74–S78. https://doi.org/10.1016/j.neurobiolaging.2014.03.033.

Boyle, P. A., Yu, L., Wilson, R. S., Leurgans, S. E., Schneider, J. A., & Bennett, D. A. (2018). Person-specific contribution of neuropathologies to cognitive loss in old age. *Annals of Neurology*, *83*(1), 74–83. https://doi.org/10.1002/ana.25123.

Brown, D. R., Qin, K., Herms, J. W., Madlung, A., Manson, J., Strome, R., et al. (1997). The cellular prion protein binds copper in vivo. *Nature*, *390*(6661), 684–687. https://doi.org/10.1038/37783.

Cacciatore, I., Baldassarre, L., Fornasari, E., Mollica, A., & Pinnen, F. (2012). Recent advances in the treatment of neurodegenerative diseases based on GSH delivery systems. *Oxidative Medicine and Cellular Longevity*, *2012*, 240146. https://doi.org/10.1155/2012/240146.

Choi, B. S., & Zheng, W. (2009). Copper transport to the brain by the blood-brain barrier and blood-CSF barrier. *Brain Research*, *1248*, 14–21. https://doi.org/10.1016/j.brainres.2008.10.056.

Cooper, G. J. (2012). Selective divalent copper chelation for the treatment of diabetes mellitus. *Current Medicinal Chemistry*, *19*(17), 2828–2860. Retrieved from http://www.ncbi.nlm.nih.gov/pubmed/22455587.

Council, N. R. (2000). *Copper in drinking water*. Washington, DC: The National Academies Press.

Council, N. R. (2001). *Dietary reference intakes for vitamin A, vitamin K, arsenic, boron, chromium, copper, iodine, iron, manganese, molybdenum, nickel, silicon, vanadium, and zinc*. Washington, DC: The National Academies Press.

Currais, A., & Maher, P. (2013). Functional consequences of age-dependent changes in glutathione status in the brain. *Antioxidants & Redox Signaling*, *19*(8), 813–822. https://doi.org/10.1089/ars.2012.4996.

de Romana, D. L., Olivares, M., Uauy, R., & Araya, M. (2011). Risks and benefits of copper in light of new insights of copper homeostasis. *Journal of Trace Elements in Medicine and Biology*, *25*(1), 3–13. https://doi.org/10.1016/j.jtemb.2010.11.004.

Dubois, B., Feldman, H. H., Jacova, C., Hampel, H., Molinuevo, J. L., Blennow, K., et al. (2014). Advancing research diagnostic criteria for Alzheimer's disease: The IWG-2 criteria. *Lancet Neurology*, *13*(6), 614–629. https://doi.org/10.1016/S1474-4422(14)70090-0.

Dyrks, T., Dyrks, E., Hartmann, T., Masters, C., & Beyreuther, K. (1992). Amyloidogenicity of beta A4 and beta A4-bearing amyloid protein precursor fragments by metal-catalyzed oxidation. *The Journal of Biological Chemistry*, *267*(25), 18210–18217. Retrieved from https://www.ncbi.nlm.nih.gov/pubmed/1517249.

Faller, P., Hureau, C., & La Penna, G. (2014). Metal ions and intrinsically disordered proteins and peptides: From Cu/Zn amyloid-beta to general principles. *Accounts of Chemical Research*, *47*(8), 2252–2259. https://doi.org/10.1021/ar400293h.

Folstein, M. F., Folstein, S. E., & McHugh, P. R. (1975). "Mini-mental state". A practical method for grading the cognitive state of patients for the clinician. *Journal of Psychiatric Research*, *12*(3), 189–198.

Hoogenraad, T. (2001). *Wilson disease*. Amsterdam/Rotterdam: Intermed Medical Publishers.

IOM. (2001). *Dietary reference intakes for vitamin A, vitamin K, arsenic, boron, chromium, copper, iodine, iron, manganese, molybdenum, nickel, silicon, vanadium, and zinc*. Washington, DC: National Academies Press (US). Panel on Micronutrients Institute of Medicine US. Available from: https://www.ncbi.nlm.nih.gov/books/NBK222310/.

IOM. (2015). *Cognitive aging: Progress in understanding and opportunity for action*. Washington, DC: The National Academies Press. Institute of Medicine.

Jack, C. R., Jr., Bennett, D. A., Blennow, K., Carrillo, M. C., Dunn, B., Haeberlein, S. B., et al. (2018). NIA-AA research framework: Toward a biological definition of Alzheimer's disease. *Alzheimers Dement*, *14*(4), 535–562. https://doi.org/10.1016/j.jalz.2018.02.018.

Jenner, P. (1994). Oxidative damage in neurodegenerative disease. *Lancet*, *344*(8925), 796–798. https://doi.org/10.1016/s0140-6736(94)92347-7.

Jomova, K., Vondrakova, D., Lawson, M., & Valko, M. (2010). Metals, oxidative stress and neurodegenerative disorders. *Molecular and Cellular Biochemistry*, *345*(1–2), 91–104. https://doi.org/10.1007/s11010-010-0563-x.

Kepp, K. P., & Squitti, R. (2019). Copper imbalance in Alzheimer's disease: Convergence of the chemistry and the clinic. *Coordination Chemistry Reviews*, *397*, 168–187. https://doi.org/10.1016/j.ccr.2019.06.018.

Maher, P., & Conti, B. (2018). Deciphering the pathways that protect from IL-13-mediated potentiation of oxidative stress-induced dopaminergic nerve cell death. *Cytokine*, *103*, 114–120. https://doi.org/10.1016/j.cyto.2017.09.018.

Markesbery, W. R. (1997). Oxidative stress hypothesis in Alzheimer's disease. *Free Radical Biology & Medicine*, *23*(1), 134–147. https://doi.org/10.1016/s0891-5849(96)00629-6.

Mathys, Z. K., & White, A. R. (2017). Copper and Alzheimer's disease. *Advances in Neurobiology*, *18*, 199–216. https://doi.org/10.1007/978-3-319-60189-2_10.

McDonald, D. R., Brunden, K. R., & Landreth, G. E. (1997). Amyloid fibrils activate tyrosine kinase-dependent signaling and superoxide production in microglia. *The Journal of Neuroscience*, *17*(7), 2284–2294. Retrieved from https://www.ncbi.nlm.nih.gov/pubmed/9065490.

Miotto, M. C., Valiente-Gabioud, A. A., Rossetti, G., Zweckstetter, M., Carloni, P., Selenko, P., et al. (2015). Copper binding to the N-terminally acetylated, naturally occurring form of alpha-synuclein induces local helical folding. *Journal of the American Chemical Society*, *137*(20), 6444–6447. https://doi.org/10.1021/jacs.5b01911.

Morris, M. C., Evans, D. A., Tangney, C. C., Bienias, J. L., Schneider, J. A., Wilson, R. S., et al. (2006). Dietary copper and high saturated and trans fat intakes associated with cognitive decline. *Archives of Neurology*, *63*(8), 1085–1088. https://doi.org/10.1001/archneur.63.8.1085.

Pinnen, F., Sozio, P., Cacciatore, I., Cornacchia, C., Mollica, A., Iannitelli, A., et al. (2011). Ibuprofen and glutathione conjugate as a potential therapeutic agent for treating Alzheimer's disease. *Archiv der Pharmazie*, *344*(3), 139–148. https://doi.org/10.1002/ardp.201000209.

Redza-Dutordoir, M., & Averill-Bates, D. A. (2016). Activation of apoptosis signalling pathways by reactive oxygen species. *Biochimica et Biophysica Acta*, *1863*(12), 2977–2992. https://doi.org/10.1016/j.bbamcr.2016.09.012.

Sies, H. (2007). Total antioxidant capacity: Appraisal of a concept. *The Journal of Nutrition*, *137*(6), 1493–1495. https://doi.org/10.1093/jn/137.6.1493.

Sparks, D. L., & Schreurs, B. G. (2003). Trace amounts of copper in water induce beta-amyloid plaques and learning deficits in a rabbit model of Alzheimer's disease. *Proceedings of the National Academy of Sciences of the United States of America*, *100*(19), 11065–11069. https://doi.org/10.1073/pnas.1832769100.

Squitti, R., Pal, A., Picozza, M., Avan, A., Ventriglia, M., Rongioletti, M. C., et al. (2020). Zinc therapy in early Alzheimer's disease: Safety and potential therapeutic efficacy. *Biomolecules*, *10*(8). https://doi.org/10.3390/biom10081164.

Squitti, R., Simonelli, I., Ventriglia, M., Siotto, M., Pasqualetti, P., Rembach, A., et al. (2014). Meta-analysis of serum non-ceruloplasmin copper in Alzheimer's disease. *Journal of Alzheimer's Disease*, *38*(4), 809–822. https://doi.org/10.3233/JAD-131247.

Squitti, R., Siotto, M., & Polimanti, R. (2014). Low-copper diet as a preventive strategy for Alzheimer's disease. *Neurobiology of Aging*, *35*(Suppl. 2), S40–S50. https://doi.org/10.1016/j.neurobiolaging.2014.02.031.

Squitti, R., White, A. R., Granzotto, A., & Kepp, K. P. (2021). Clinical, biochemical and genetic aspects of copper imbalance in Alzheimer's disease: Toward an etiology. *Journal of Alzheimer's Disease*, *3*(1), 23–41. https://doi.org/10.3233/JAD-201556.

Taylor, A. A., Tsuji, J. S., Garry, M. R., McArdle, M. E., Goodfellow, W. L., Jr., Adams, W. J., et al. (2020). Critical review of exposure and effects: Implications for setting regulatory health criteria for ingested copper. *Environmental Management*, *65*(1), 131–159. https://doi.org/10.1007/s00267-019-01234-y.

Turnlund, J. R., Keyes, W. R., Kim, S. K., & Domek, J. M. (2005). Long-term high copper intake: Effects on copper absorption, retention, and homeostasis in men. *The American Journal of Clinical Nutrition*, *81*(4), 822–828. https://doi.org/10.1093/ajcn/81.4.822.

WHO. (1996). Copper. In *Trace elements in human nutrition and health* (pp. 123–143). Geneva, Switzerland: World Health Organization.

WHO. (2018). *A global overview of national regulations and standards for drinking-water quality*. Geneva, Switzerland: World Health Organization.

WHO. (2019). *Risk reduction of cognitive decline and dementia: WHO guidelines*. https://www.who.int/mental_health/neurology/dementia/guidelines_risk_reduction/en/. (Accessed 13 September 2019).
WHO. (2020). *Disease burden and mortality estimates*. Available online: http://origin.who.int/healthinfo/global_burden_disease/estimates/en/. (Accessed 6 May 2020).
Zeevalk, G. D., Bernard, L. P., & Guilford, F. T. (2010). Liposomal-glutathione provides maintenance of intracellular glutathione and neuroprotection in mesencephalic neuronal cells. *Neurochemical Research*, *35*(10), 1575–1587. https://doi.org/10.1007/s11064-010-0217-0.

CHAPTER 6

Chromium and Alzheimer's disease

Ansab Akhtar* **and Sangeeta Pilkhwal Sah**
Division of Pharmacology, University Institute of Pharmaceutical Sciences, Panjab University, Chandigarh, India

Abbreviations

AD	Alzheimer's disease
AKT	protein kinase B
APOE	apolipoprotein E
APP	amyloid precursor protein
ATP	adenosine triphosphate
BACE 1	beta-site APP cleaving enzyme 1.
BBB	blood-brain barrier
BMI	body mass index
CA	Cornu Ammonis
CDK-5	cyclin-dependent kinase-5.
c-JNK	c-Jun N-terminal kinase
CRP	C-reactive protein
DM	diabetes mellitus
FAD	familial Alzheimer's disease
GLUT	glucose transporter
GSK	glycogen synthase kinase
HD	Huntington's disease
HO-1	heme oxygenase-1
5-HT	5-hydroxytryptamine
IDE	insulin-degrading enzyme
ICV	intracerebroventricular
IFN-γ	interferon-γ
IL	interleukin
IR	insulin receptor
IRS	insulin receptor substrate
mTOR	Mammalian target of rapamycin
NF-κB	nuclear factor-kappa chain of B cells
Nrf2	nuclear factor erythroid 2-related factor 2
PCO	protein carbonylation
PD	Parkinson's disease
PDK1	phosphoinositide-dependent kinase 1
PI3K	phosphoinositide 3-kinase
ROS	reactive oxygen species
STZ	streptozotocin
TBARS	thiobarbituric acid reactive substance.

* Current affiliation: School of Health Sciences and Technology, University of Petroleum and Energy Studies, Bidholi, Dehradun, India.

Vitamins and Minerals in Neurological Disorders
https://doi.org/10.1016/B978-0-323-89835-5.00036-3

TGF tumor growth factor
TNF-α tumor necrosis factor-α

Introduction

Alzheimer's disease (AD) is a progressive neurodegenerative disease occurring in elderly people above 65 years and is termed late-onset AD. On the other hand, mutations in the genes of presenilin 1, presenilin 2, and amyloid precursor protein (APP) lead to early-onset familial AD. AD comprises a wide range of signs and symptoms, including but not limited to dementia, depression, anxiety, anorexia, sleeplessness, speaking difficulty, psychosis, etc. However, cognitive impairment remains the primary one (Lanctôt et al., 2017). Compared to other chronic medical conditions (heart diseases, cancer, and stroke), AD is the fourth leading cause of death (Selkoe, 2012). According to the World Alzheimer Report 2015 led by King's College London, 46.8 million people are suffering from dementia around the globe and numbers are likely to increase every 20 years becoming 131.5 million by 2050.

Brain regions of the prefrontal cortex, entorhinal cortex, hippocampus, hypothalamus, olfactory bulb, amygdala, cerebellum, etc., are presumed to be important for learning, memory, recognition, perception, interaction, and communication (Deibel, Ehmann, & Markesbery, 1996). Further, several hypotheses are the foundations of AD pathology. These incorporate mainly amyloid and tau pathology, cholinergic dysfunction, mitochondrial dysfunction, oxidative stress, neuroinflammation, and insulin resistance. The major triggers in AD could be impairment in glucose and energy metabolism. This dysregulation could arise from numerous factors including poor dietary habits where micronutrients or trace elements play a pivotal role in brain health, implicated in the learning and memory process directly or indirectly. Chromium is one of the most important micronutrients generally taken as food or supplements, required in micrograms daily. The nutrients can pass through the blood-brain barrier (BBB) to exert their direct action or produce peripheral influence, thereby modulating brain functions (Bourre, 2006).

Alzheimer's disease

Alzheimer's disease is the most common neurodegenerative disease and is generally a multifactorial disorder, majorly constituting amyloid and tau pathology (Fig. 1). There are two main types of AD.

Sporadic Alzheimer's disease (SAD): SAD generally occurs after 65 years of age. Hence, it is termed late-onset Alzheimer's disease. It is the more prevalent type of AD and accounts for almost 90–95% of AD cases. SAD has not usually been associated with genetic alterations or manipulation. However, the APOE ὲ allele of the apolipoprotein E gene has been found to play a role here. Besides, it has been reported to arise from impaired glucose and energy metabolism. Late-onset AD has been revealed to depict

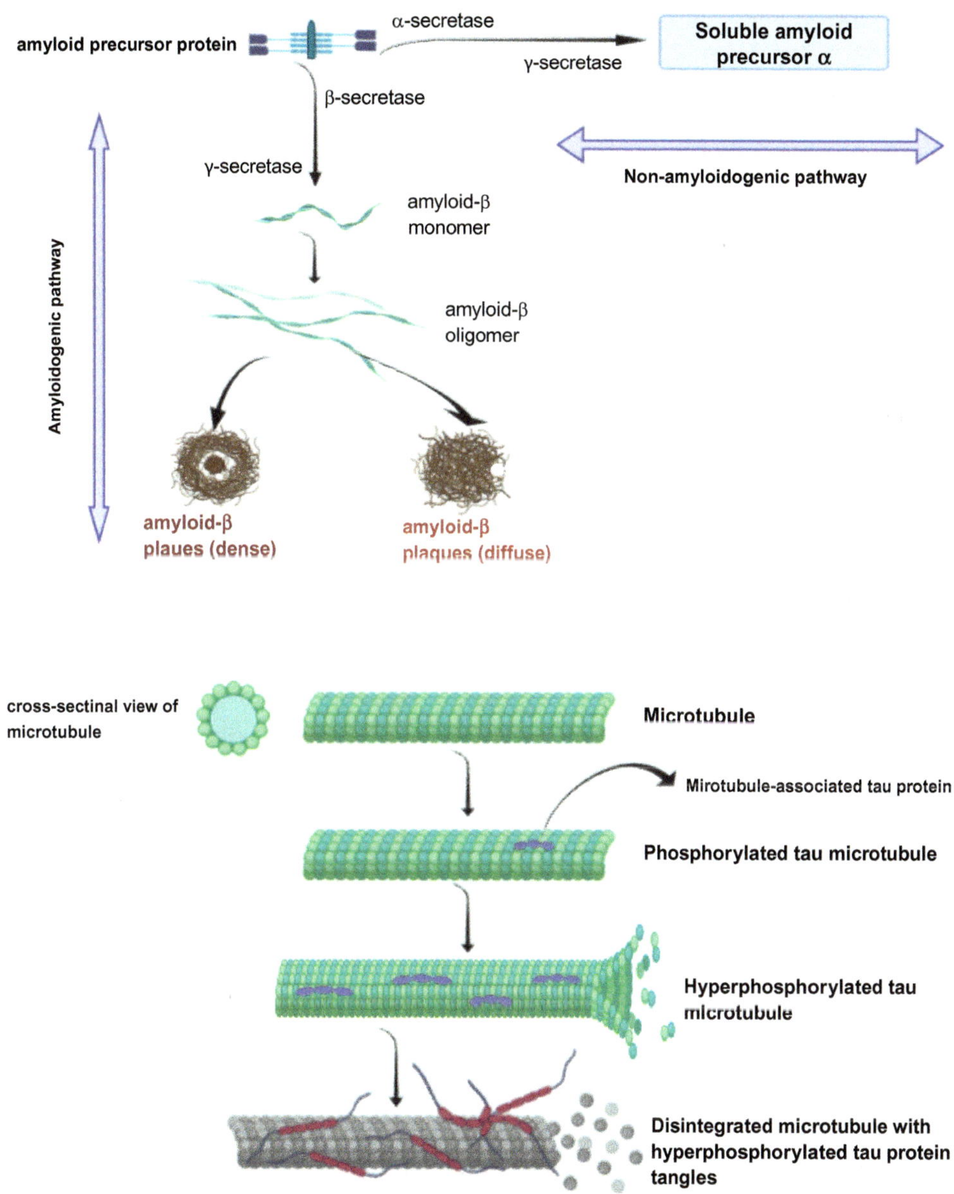

Fig. 1 Amyloid and tau pathology in Alzheimer's disease. Non-amyloidogenic and amyloidogenic pathways leading to amyloid plaques, and hyperphosphorylation of tau protein, leading to neurofibrillary tangles and destabilization of microtubules. Figures were created on Biorender.com.

a more significant amount of symptom severity. Furthermore, the symptoms progress more rapidly, and neurodegeneration is spread over a greater brain area than early-onset AD. The proteolytic activity of beta-site APP cleaving enzyme 1 (BACE 1) on amyloid precursor protein (APP) has been widely recognized to be involved in the pathogenesis of SAD (Li et al., 2004; Piaceri, Nacmias, & Sorbi, 2013).

Familial Alzheimer's disease: This is an early-onset Alzheimer's disease that arises from genetic mutations in some specific genes and has been associated with family history. Individuals below 65 years commonly suffer from this type of AD. It is less prevalent and occurs in up to 5% of people total suffering from AD. It is a type of an autosomal-dominant form of mutation. The mutations in the genes of presenilin 1, presenilin 2, APP, and p-tau have been reported to generate familial AD models (Piaceri et al., 2013; Wu et al., 2012). These mutations are responsible for the formation of amyloid peptides and tau pathology. Pathological symptoms and progression of the disease are mild to moderate in FAD.

Characteristic features of AD in terms of brain insulin resistance

Several reports suggest that AD is a degenerative metabolic disease, also labeled as "type 3 diabetes" associated with brain insulin resistance as insulin signaling is hugely involved in glucose metabolism, energy production, neuronal survival, and synaptic transmission (Ferreira, Clarke, Bomfim, & De Felice, 2014). The insulin produced by the islet of Langerhans of pancreatic glands can cross the blood-brain barrier from the peripheral circulation to the brain by insulin receptor-mediated transport. The insulin present in the brain impacts several aspects such as the neuromodulator, neuroprotective effects, and also has a role in cognition and memory. Insulin can also modulate neurotransmitter release at the synaptic areas required for the preservation of synaptic plasticity. Therefore, these findings suggest that insulin in the brain is involved in several physiological functions (Mahoney, Azpurua, & Eaton, 2016).

Brain insulin resistance can be defined as the failure of brain cells to respond to insulin as they normally would, resulting in impairments in synaptic, metabolic, and immune response functions (Spinelli et al., 2017). Upregulating insulin sensitivity or modulating insulin signaling downstream molecules through nutritional approaches could be major contributors to protect, reverse, or attenuate AD-related pathology and symptoms. In this regard, some of the critical molecules involved in the insulin cascade are IR, IRS-1, IRS-2, PI3K, AKT, GSK-3β, PDK1, mTOR, IDE, etc. (Akhtar & Sah, 2020). Moreover, insulin in the brain has recently been demonstrated to be involved in the metabolism of several nutrients including glucose which is transported to the neuronal cells through glucose transporters (GLUT). The major GLUTs present in brain regions are GLUT-1, 3, and 4 which can uptake or transport glucose primarily in the hypothalamic and hippocampal regions of the brain.

Remarkably, insulin up to a normal level in the brain promotes neuronal cell growth and survival without inflicting neuronal cytotoxicity. However, beyond a certain level, hyperinsulinemia or insulin resistance is induced. The condition of hyperinsulinemia might result in downregulated gene expression of a few insulin signaling molecules like IR, IRS, PI3K, and AKT and upregulated gene expression of certain kinases like GSK-3β, JNK, and CDK-5, resulting in tau hyperphosphorylation (Zaulkffali et al., 2019). Moreover, insulin receptors (IRs) are ubiquitously present in different brain regions.

The brain areas of the cerebral cortex, CA1, dentate gyrus regions of the hippocampus, generally attributed to cognitive functions like learning and memory, and the hypothalamus controlling the appetite, have dense concentrations of IRs. Insulin binds to α-subunit and then β-subunit of IR which is phosphorylated at the tyrosine residue which then results in a downstream cascade of phosphorylation, with the insulin receptor substrates (IRSs) being the first. IRSs are the ligand proteins, which play a vital role in insulin signaling mechanisms and act as signaling adaptor proteins (Boucher, Kleinridders, & Kahn, 2014). Tyrosine phosphorylation and upregulated gene expression of IRS-1 of the insulin-signaling cascade activates the PI3K/AKT pathway of insulin which inhibits glycogen synthase 3-β (GSK3-β). Activation of GSK3-β is highly linked to tau hyperphosphorylation. Insulin resistance leads to the downregulation of this pathway and associated phosphorylation of tau increases. However, the activation of insulin signaling results in the reversal of AD-associated pathological features (Candeias et al., 2012).

Other triggers associated with AD and brain insulin resistance

Further, AD is associated with the neuroinflammatory pathway. Neuroinflammation causes synaptic disruption and eventually, neurodegeneration progresses. Additionally, upregulated insulin signaling promotes the attenuation of amyloid-mediated neuroinflammatory pathways. Besides, insulin-signaling molecule GSK-3β has also been reported to augment inflammatory cytokine production causing insulin resistance (Najem, Bamji-Mirza, Chang, Liu, & Zhang, 2014). Activation of microglia, astrocytes, and macrophages has a profound role in neuroinflammation by releasing pro-inflammatory cytokines like IL-6, IL-1β, TGF-β, TNF-α, etc. The release of these cytokines can also be triggered by the binding of aggregated proteins like amyloid to the neurons, microglia, or astroglia. Likewise, neuroinflammation at chronic levels is the causative factor of neuronal death (Heneka et al., 2015). In this context, foodstuffs causing both peripheral and central inflammatory process need to get reduced, so that the prevalence of neurodegenerative diseases like AD can be minimized. At the same time, spices, supplements, and natural products possessing anti-inflammatory properties should be included in the diet.

Notably, another detrimental key factor in the neurodegenerative process related to AD is oxidative stress. Reactive oxygen species (ROS) have been proven to be the triggering agents for the inflammatory pathway of NF-κB and c-JNK. The activated inflammatory pathway, in turn, causes downregulation of the insulin-signaling pathway by enhancing serine phosphorylation and inhibiting tyrosine phosphorylation activation of IR and IRS. Hence, antioxidant food items could potentially boost up the scavenging process of ROS (Evans, Maddux, & Goldfine, 2005). Conversely, insulin resistance-induced impairment in insulin signaling can offer a role in oxidative stress. Irregularities in the IRS and AKT molecules of insulin signaling lead to alterations in oxidative stress biomarkers like decreased catalase expression and increased protein carbonylation (PCO) formation, signifying oxidative imbalance (Dong et al., 2006). Further, there is extensive evidence of intertwined actions of oxidative stress and mitochondrial dysfunction. ATP is generated in the inner mitochondrial membrane by the transfer of electrons through the mitochondrial enzyme complexes. However, a little part of the electron interacts with oxygen to generate reactive oxygen-free radicals in the form of superoxides and peroxides. Hence, any imbalance in mitochondrial functioning results in ROS overproduction. Again, both oxidative and nitrosative radicals might contribute to mitochondrial dysfunction, contributing a major part to neurodegeneration and AD (Bhat et al., 2015). Therefore, proper regulation of mitochondrial potential and oxidative phosphorylation through a balanced amount of diet can maintain neuronal energy homeostasis. Besides, insulin resistance could also give rise to mitochondrial dysfunction or vice versa. The mechanism involved is dysfunctional mitochondria arising from suppressed IRS-1 through the c-JNK pathway. This phenomenon also increases serine phosphorylation of IRS and reduces tyrosine phosphorylation, leading to insulin-signaling impairment. Hence, mitochondria being responsible for glucose and energy metabolism find a role in insulin sensitivity for glucose utilization. Also, impaired glucose and ATP generation in mitochondrial dysfunction make the insulin less reliable, making it appear as an insulin-resistant state (Pospisilik et al., 2007).

Chromium

The first element of group six of the periodic table is chromium with the symbol Cr and atomic number 24. Chromium is a widely recognized essential trace element or micronutrient and is generally supplemented with food or water. The name micronutrient or trace or essential element is due to its requirement by the human body in micrograms. This particular mineral is also an adjunctive of vitamin and mineral products available in the market (Keunen, Van Elburg, Van Bel, & Benders, 2015). Some widely used chromium salts include chromium picolinate, chromium histidinate, chromium chloride, chromium hydroxide, chromium sulfate, chromium nitrate, etc.

Chemical and physical properties of chromium: Chromium is a steel-like, hard, and shining element. This is one of the transition metals that means it can form chemical bonds utilizing its valence electron. It exists in oxidation states from −4 to +6. The possible shapes it can form are tetrahedral, octahedral, and planar. Trivalent chromium (Cr^{3+}) found in many forms is most widely used in nutritional and pharmacological strategies. Chromium is absorbed less than 3% from the intestine. However, the most frequently applicable organic salt, chromium picolinate (CrPic) (Fig. 2), is highly absorbed (Anderson, Bryden, Polansky, & Gautschi, 1996). The molecular weight of CrPic is 418.33 g/mol and is available in a solid amorphous state with a reddish-pink color.

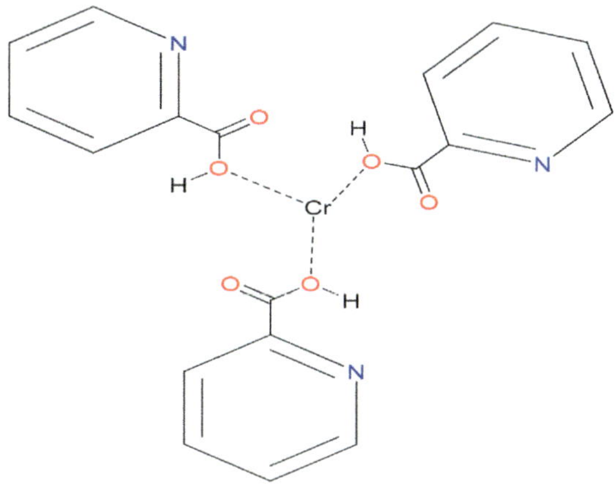

Chronium(III) picolinate

Fig. 2 Structure of a chromium salt. Structure of chromium picolinate, synthesized through trough trivalent chromium and picolinic acid, available as a marketed supplement. The structure was created with the help of ChemDraw.

Moreover, it is a coordination complex compound comprised of chromium chelated with three molecules of picolinic acid. The IUPAC name of CrPic is chromium (3 +), pyridine-2-carboxylate. Cr^{+3} binds to plasma protein transferrin and is transferred from the pool of blood to tissues. Hence, the solubility and hydrophilicity of chromium are enhanced by the salting-in process.

Sources of chromium: The natural sources of chromium include potatoes, tomatoes, oysters, broccoli, green beans, and whole grains. The fruit derivatives of chromium are apples, bananas, oranges, and grapes. Animal sources are seafood, beef, poultry, etc. Milk and dairy products also contain chromium. However, drinking water contains only a tiny fraction of chromium. Furthermore, artificial chromium is extracted through refining the ores like chromite or by refraction processes. However, soil, air, industrial wastes, rocks, etc., have minor quantities of chromium. *Leptospermum scoparium,* a flowering plant native to southeast Australia and New Zealand belonging to the family *Myrtaceae,* also possesses a little amount of chromium (Saha, Nandi, & Saha, 2011).

Concerning its nutritional and neurological perspectives given below, chromium could potentially prevent, protect, or slow down the AD-related pathological outcomes through the following pathways (Fig. 3).

Role of chromium against insulin resistance: Chromium binds to insulin receptors increasing tyrosine kinase activity, and hence, the downstream molecules such as AKT/PI3K are activated, and GSK-3β is subsequently inactivated (Hua, Clark, Ren, & Sreejayan, 2012). Importantly, it has been studied to improve insulin sensitivity in diabetes mellitus (DM) by targeting the IRS-1/PI3K/AKT/GSK-3β pathway (Hua et al., 2012; Martin et al.,

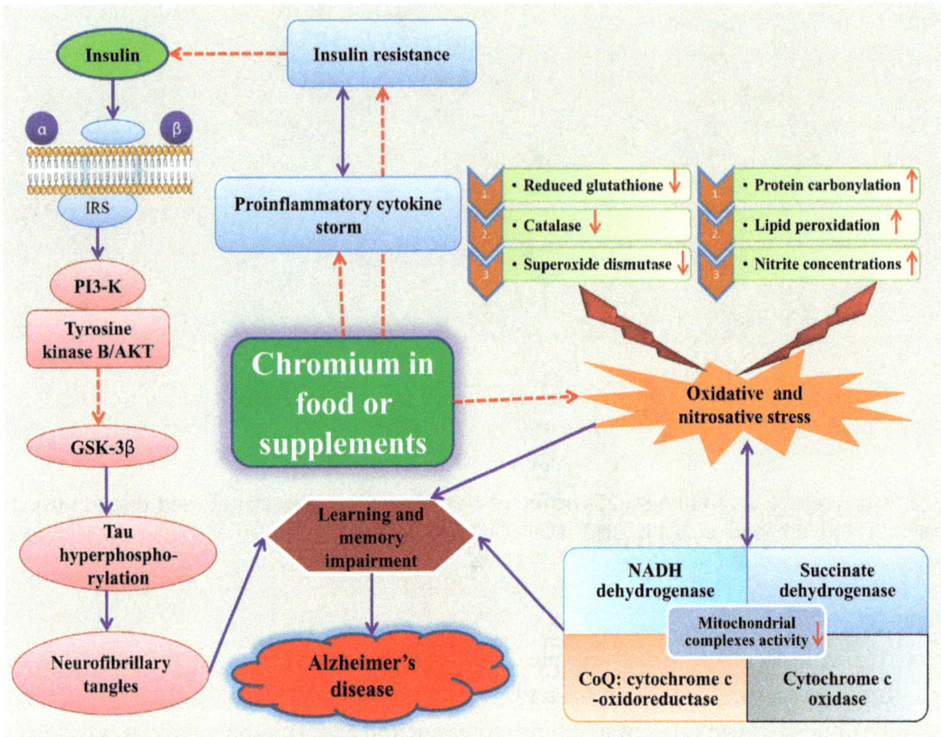

Fig. 3 Role of chromium in Alzheimer's disease protection. Inhibition of insulin resistance, neuroinflammation, oxidative stress, and mitochondrial dysfunction through nutritional food enriched with chromium.

2006). In DM, chromium augments glucose, protein, and lipid metabolism by reducing insulin resistance. With insulin utilization, it helps in the uptake of glucose into cells, thus controlling hyperglycemia. Improved insulin sensitivity by this micronutrient can also be through an increased number of insulin receptors. Chromodulin, a substance that binds to chromium, is demonstrated to restore insulin sensitivity through the use of chromium. It stimulates protein tyrosine kinase activity eight times, in the presence of insulin (Davis & Vincent, 1997). Polycystic ovarian syndrome associated with insulin resistance has also been found to be improved with chromium treatment. AD is hugely related to brain insulin resistance or insulin-signaling impairment. In this context, chromium has been studied as a potential element to improve insulin signaling, thereby attenuating insulin resistance (Akhtar et al., 2020). Chromium picolinate and propionate supplementation can restore tissue sensitivity to insulin, thereby controlling the growth and survival of the organs like brain (Spears et al., 2012).

Considering the potential of chromium in improving glucose utilization in insulin-resistant and diabetic patients, a placebo-controlled double-blind trial was conducted in 26 old adults. Chromium picolinate (CrPic) treatment for 12 weeks reduced semantic interference in learning, recall, and other tasks related to recognition memory. Functional magnetic resonance imaging demonstrated activation of cerebral regions. The study depicted that CrPic enhances cerebral function and cognitive function in older at risk of neurodegeneration (Krikorian, Eliassen, Boespflug, Nash, & Shidler, 2010). A clinical trial is being conducted to assess the effect of chromium chloride on insulin resistance in AD patients (https://clinicaltrials.gov/ct2/show/NCT03038282).

Role of chromium against oxidative stress: Chromium salt has been proved to ameliorate diabetic symptoms by increasing antioxidant enzyme levels (Anderson et al., 2001). Rats receiving daily supplementation of 1 mg chromium revealed decreased TBARS, indicating reduced oxidative stress (Lai, 2008). Chromium supplements also decreased serum malondialdehyde (MDA) levels in a study reported on hens (Onderci, Sahin, Sahin, & Kilic, 2003). This indicated that chromium diminishes lipid peroxidation. Lipids being important constituents of neuronal cells can be shielded through chromium usage. Further, in alloxan-induced diabetic rats, it attenuates ROS levels, simultaneously ameliorating the diabetic complications associated with oxidative stress (Doddigarla, Parwez, Abidi, & Ahmad, 2016). CrPic might depict their antioxidant property through the Nrf2/HO-1 pathway as well by increasing their activity (Sahin et al., 2012). So, chromium could possess a strong role to scavenge AD-associated ROS and oxidative damage. Antioxidant enzymes were reported to get balanced in chromium-treated materials. The study predicted to decrease oxidative stress (Ali et al., 2013). Hence, taking the perspective of AD being aggravated by oxidative stress, chromium becomes an important household material for its inclusion in dietary habits.

Role of chromium against mitochondrial dysfunction: The brain is the most energy-consuming organ in the body. That is why the powerhouse of energy is found abundantly in brain regions comprising neuronal cells. These mitochondria continuously generate ATP through the utilization of glucose consumed in food. Further, oxidative stress can cause mitochondrial dysfunction and vice versa, leading to neurodegeneration and AD. Hence, the property of chromium to decrease oxidative stress can indirectly benefit the stabilization and survival of mitochondria. This phenomenon can protect neurons from death, enhancing their survival capacity. Chromium also normalizes mitochondrial potential, correcting the disturbed mitochondrial functioning (Rains & Jain, 2011). However, an excessive amount of chromium can damage mitochondria causing neuronal toxicity. So, regular exercise and restricted food consumption should be taken care of.

Role of chromium against neuroinflammation: The anti-inflammatory activity of chromium can be mediated by downregulation of the NFκB-p65 pathway. Another critical salt of chromium, namely chromium histidinate, has been reported to reverse inflammation through the NF-κB pathway in a high-fat diet and STZ-induced rat model of DM (Sahin et al., 2012; Selcuk et al., 2012). It has also been reported to reverse depressive symptoms by targeting the neuroinflammatory pathway. Chromium salts can also decrease the release and functioning of pro-inflammatory cytokines like TNF-α, IL-1β, IL-6, IF-γ, etc. (Jain, Rains, & Croad, 2007). This phenomenon can significantly halt the process of peripheral inflammation and neuroinflammation. Chromium-mediated regulation of astrocytes and microglia normalizes the neuroinflammatory process (Sahin et al., 2013). Therefore, the neuroinflammation-derived pathology of AD has the chance to be combated by chromium.

Optimum requirements of chromium by humans: The daily requirement of chromium for men is 30–35 μg, and for women 20–25 μg, simultaneously, the need could be much higher, up to 1 mg per day for diseased conditions like DM, AD, or any insulin resistance-related ailment. Foods only provide less than 2 μg per diet. CrPic is a widely marketed supplement and is the second-most nutritional supplement, followed by calcium. It is available as tablets in doses ranging from 200 to 500 μg.

Toxicity and harmful effects associated with chromium: Chromium has been observed to be nontoxic in rats weighing 200–250 g even after 3 mg consumption daily for 20 weeks (Franklin & Odontiadis, 2003). Despite this, the hexavalent form of chromium (Cr^{6+}) generally released from industry-generated pollution has been reported to be carcinogenic and neurotoxic in long-term use. On the other hand, a trivalent form is prevalent in food items and is far less toxic than hexavalent chromium as well as Cu, Mn, Se, Zn, etc. (Pechova & Pavlata, 2007). Rarely, renal dysfunction has also been reported with chromium supplements (Mozaffari, Patel, Ballas, & Schaffer, 2005). Further, respiratory obstruction, pneumonia, nasal itching, etc., have been reported in the case of inhalational chromium. Besides, occasional chromosomal breakdown may happen due to chromium (Stearns, Wise Sr, Patierno, & Wetterhahn, 1995). Mitochondrial damage can occur at higher concentrations.

Precautions in handling or consuming chromium: Considering the toxicities of chromium, its hexavalent form should be avoided. Industrial wastes should properly be disposed of. Nevertheless, trivalent chromium can more frequently be used and its salts can be made as supplements. However, its consumption should be confined to micrograms and occasionally milligrams and strictly restricting its provision in grams. Furthermore, while handling chromium or its salts, touching by hands should be avoided, and gloves and masks are worn to prevent its inhalation. Drinking water and air quality in surrounding places of industries should be tried to purify as exposure to water and air pollution can incur damage to certain body organs. Thus, excessive intake of chromium-rich foods and contaminated water should be avoided.

Nutrition and neuroprotection

Neuronal proliferation, myelination, development, maturation, growth, and survival are hugely dependent on the way we eat. That means an appropriate amount of each nutrient plays a critical role in neuroprotection and the well-being of mental health. In this case, minerals are ought to pay heed as the food contains several constituents and goes through the metabolic process after ingestion and passes through the BBB to make a building block in the brain. Hence, deficiency of a few crucial nutrients can be detrimental to some specific brain regions.

Several neurodegenerative disorders can be the outcome of poor or unhealthy diets (Null, Pennesi, & Feldman, 2017). Macronutrients like protein, carbohydrate, and fat are the ones that have to be consumed in larger amounts. However, micronutrients including vitamins and minerals are required in very minor quantities. Mineral micronutrients are chromium, cobalt, iron, zinc, etc. Micronutrients are essential for the metabolic process and proper tissue functioning. Further, micronutrients provide the opportunities to manufacture precursors and biomolecules for the synthesis of neurotransmitters. These neurotransmitters are important for nerve impulse transmission, synapse formation, synaptic plasticity, and ultimately neuronal survival (Null et al., 2017). Interestingly, oxidative damage and mitochondrial breakage in neuronal cells have been reported to outburst when micronutrients are deficient (Evans & Halliwell, 2001).

Neuroscientific aspects

A clinical study has revealed the substantial role of chromium in older individuals suffering from a cognitive decline. The rationale behind chromium-mediated learning and memory improvement could be due to regulated glucose and insulin metabolism in the cerebral region (Krikorian et al., 2010). Further, chromium can modulate the serotonergic receptor. This can mimic the effect of the mood hormone serotonin (Krikorian et al., 2010). AD-associated dysregulation of the serotonergic system can be upregulated by the use of chromium compounds. Furthermore, regulation of neuronal survival and growth is crucial for cognitive development. This is dependent on neuronal somatic cell and dendrite structure, proper axonal transport, microtubule stability, adequate neurotransmitter synthesis and release, maintenance of pre- and postsynaptic density, smooth synaptic transmission, synaptic plasticity, and long-term potentiation. These phenomena are accountable for learning acquisition and memory consolidation, essential in the protection against AD (McKay, Placzek, & Dani, 2007).

Nutritional aspects

Dietary chromium with increasing doses has been reported to build up muscular strength by improving protein metabolism or halting its breakdown. The different doses of chromium used were 100, 200, and 500 μg/day (Bernao, Meseguer, Aguilar, Para, & Muñoz, 2004). Chromium picolinate and chromium niacinate supplements positively alter the levels of cholesterol, triglyceride, and glycated hemoglobin, leading to reduced chances of a diabetic condition. This could further help in improved overall health (Jain et al., 2007). Carbohydrate metabolism is regulated by chromium and prevents the body from glucose intolerance. Lipid metabolism gets regulated through proper chromium contents taken through food or supplements (Cefalu, Wang, Zhang, Baldor, & Russell, 2002). This action controls BMI and weight maintenance, which further affects brain health, as there is a strong relationship between body fat and brain health. Conversely, a chromium-aided diet improves appetite and food consumption. This might lead to slight healthy weight gain (Sahin et al., 2010).

Applications to other neurological conditions

In this chapter, we reviewed the role of chromium through various pathways like reversing oxidative stress, mitochondrial dysfunction, neuroinflammation, and insulin resistance in AD. Similar kinds of disturbances have widely been reported in other neurological diseases like Parkinson's disease (PD), Huntington's disease (HD), amyotrophic lateral sclerosis, and ischemic stroke (Athauda & Foltynie, 2016; Deng, Liu, Wang, Li, & Cai, 2017; Yan, Wang, & Zhu, 2013). Hence, the micronutrient chromium, possessing the potential to alleviate these abnormalities, could be beneficial to attenuate the symptoms of other neurodegenerative diseases. A few studies have reported AD-like dementia associated with PD and HD, thereby making chromium a relevant component. Additionally, the salt chromium picolinate supplementation too contributes to improve cerebral function and better glucose utilization in a diabetic brain (Krikorian et al., 2010). Moreover, chromium histidinate alleviates brain damage induced by a hyperglycemic condition (Sahin et al., 2013). Stroke-related cerebral ischemia was also found to get relieved upon supplying chromium add-on (Chen et al., 2016). In a randomized placebo-controlled clinical trial, chromium alleviated the depressive symptoms. Here, chromium acted by downregulating the 5HT2A receptor (Davidson, Abraham, Connor, & McLeod, 2003). Further, a preclinical study has depicted the beneficial role of chromium picolinate supplied in drinking water in an animal model of anxiety and depression (Dubey, Ansari, Vohora, & Khanam, 2015). The thymus and pineal gland are also found to be under the influence of chromium-derived hypothalamic regulation. This can be due to the insulin-sensitizing action of chromium (McCarty, 1994).

Other components of interest

In this chapter, we described the role of chromium in Alzheimer's disease. There are other trace elements in the same category which play vital roles in AD. Of these are cobalt, zinc, iodine, iron, copper, selenium, manganese, and molybdenum (Table 1).

Copper is an important cofactor in many metabolic and catalytic processes. Besides, the dysregulated copper content is accountable for the enhanced aggregation of phosphorylated tau protein. Copper also acts as a catalyst to produce ROS. Hence, the implication of copper in neurodegeneration is quite clear (Eskici & Axelsen, 2012; Hung, Bush, & Cherny, 2010).

Table 1 Essential elements and their sources and functions.

S. No.	Essential elements	Symbol and atomic number	Food sources	Major functions
1	Copper	29 Cu	Green vegetables, dark chocolates, nuts, seeds, lobster	Helps to strengthen the nervous and immune system
2	Zinc	30 Zn	Oyster, red meat, beans, grains, dairy	Neuronal growth and plasticity are assisted
3	Cobalt	27 Co	Green vegetables, nuts, fishes	Prevents demyelination of nerves and improves synaptic transmission
4	Iodine	53 I	Iodized salt, seaweed, shrimp	Nurtures growth hormone influencing brain development
5	Manganese	25 Mn	Green vegetables, nuts, legumes, rice	Implicates neural activity but accumulation can lead to neurotoxicity
6	Selenium	34 Se	Dairy products, nuts, seafoods	Selenium-dependent enzymes modulate brain functions
7	Molybdenum	42 Mo	Legumes, milk, grains	Acts as a cofactor in various brain enzymatic functions
8	Vanadium	23 V	Mushrooms, grains, wine	Vanadium compounds improve metabolic brain condition
9	Nickel	28 Ni	Chocolates, grains, nuts	Nickel is found in nucleic acids and protects the molecular structure of cells
10	Iron	26 Fe	Beans, spinach, lentils	Helps in myelin and neurotransmitter synthesis
11	Silicon	14 Si	Cereals, grains, oats	Strengthens epithelial and connective tissues of the brain

Zinc plays a role in the release of neurotransmitters and synaptic transmission, thereby implicating memory formation (Xie, Wang, Zhang, Ren, & Yang, 2012). Conversely, amyloid plaques in AD are constituted by copper and zinc. So, the chelating agents directed to these trace elements can be applied to dissolve the plaques and reverse the disease symptoms (Cherny et al., 2001).

The toxicity of hexavalent chromium in the brain can be mitigated with the moderate use of another trace element, namely selenium. Selenium has the potential to balance the oxidative system to normal. This has probably happened through the formation of a complex between chromium and selenium (Hao et al., 2017). AD-associated tau pathology was also corrected by cobalt-based complexes. Cobalt effectively decreases tau aggregation and subsequent accumulation in the brain (Gorantla et al., 2019).

Further, altered levels of almost all the trace elements were revealed in patients with epileptic seizures (Wojciak, Mojs, Stanislawska-Kubiak, & Samborski, 2013). This signifies the importance and relevance of trace elements in brain diseases.

Mini-dictionary of terms

- *Trace elements*. Required by the human body daily in very trace amounts like micrograms in the diet.
- *Micronutrients*. Nutritional components required by the human body in micrograms or milligrams along with regular food.
- *Supplements*. Food items that are marketed and provided in a predetermined quantity added to packaged food items.
- *Nutrition*. Necessary food for the proper growth and development of body and mind.
- *Neurodegeneration*. Loss or death of neuronal cells either due to damage to the cell body, axon, or synapses, resulting in atrophy of the brain.
- *Neuroinflammation*. Inflammation in the nervous tissues or neuronal cells of the brain is aggravated by injury, trauma, and metabolism, resulting in microglial activation or cytokine release.
- *Cognitive impairment*. Learning and memory impairment, unable to recognize, difficulty in speech, loss of track, distorted thoughts, etc.
- *Sporadic*. Occurring at a certain period or age, like late-onset or above 60 years.
- *Dementia*. Loss of memory, forgetfulness, cognitive decline, and impaired thinking.

Key facts of Alzheimer's disease

- German psychiatrist Alois Alzheimer was the first person to discover AD pathology in a patient namely Auguste Deter, suffering from dementia.
- In the United States, Alzheimer's Association was founded in 1984 for the upliftment and welfare of the AD healthcare system.

- Acetylcholinesterase inhibitors like rivastigmine, donepezil, galantamine, and NMDA receptor antagonist memantine are the FDA-approved drugs for AD.
- The most recent and current drugs tested for AD are gantenerumab and solanezumab, intended to clear excessive amyloid-β, and are expected to come by the end of 2021.
- The approximate mortality rate in AD is 37 per 1 lakh people.

Key facts of chromium

- French chemist Nicolas-Louis Vauquelin discovered chromium in 1797 from a mineral namely crocoite.
- Chromium is the hardest metal, does not rust, reflects 70% of visible light, and is used in stainless steel.
- According to nutritionists, the average amount of chromium present in a balanced diet of 2000 kcal is 27 μg.
- On average, men consume 54 μg/day and women consume 29 μg/day of chromium.
- One of the marketed chromium supplements is chromium picolinate made by Thorne Research, one capsule of which contains 500 μg of chromium.

Summary points

- Alzheimer's disease is a progressive neurodegenerative disease occurring either due to impaired glucose and energy metabolism or genetic mutations in brain regions of the cerebral cortex and hippocampus.
- Insulin resistance, neuroinflammation, oxidative stress, and mitochondrial dysfunction could be possible triggers of AD.
- Neurological disorders like AD could be dependent on the diet content of trace elements or micronutrients like chromium.
- Chromium requirement is fulfilled through fruits, vegetables, and supplements which are helpful in brain health and neuroprotection.
- Chromium principally improves insulin sensitivity or attenuates insulin resistance associated with AD.
- Chromium also has a strong potential to reverse oxidative stress, mitochondrial dysfunction, and neuroinflammation pertaining to AD.
- Other neurodegenerative diseases protected by chromium can be cerebral ischemia, amyotrophic lateral sclerosis, parkinsonism, and Huntington's disease.
- Other trace elements effective in the halting of AD progression include zinc, copper, cobalt, molybdenum, etc.

References

Akhtar, A., Dhaliwal, J., Saroj, P., Uniyal, A., Bishnoi, M., & Sah, S. P. (2020). Chromium picolinate attenuates cognitive deficit in ICV-STZ rat paradigm of sporadic Alzheimer's-like dementia via targeting neuroinflammatory and IRS-1/PI3K/AKT/GSK-3β pathway. *Inflammopharmacology*, *28*(2), 385–400.

Akhtar, A., & Sah, S. P. (2020). Insulin signaling pathway and related molecules: Role in neurodegeneration and Alzheimer's disease. *Neurochemistry International*, *135*, 104707.

Ali, S., Farooq, M., Jahangir, M., Abbas, F., Bharwana, S., & Zhang, G. (2013). Effect of chromium and nitrogen form on photosynthesis and anti-oxidative system in barley. *Biologia Plantarum*, *57*(4), 758–763.

Anderson, R. A., Bryden, N. A., Polansky, M. M., & Gautschi, K. (1996). Dietary chromium effects on tissue chromium concentrations and chromium absorption in rats. *The Journal of Trace Elements in Experimental Medicine*, *9*(1), 11–25.

Anderson, R. A., Roussel, A.-M., Zouari, N., Mahjoub, S., Matheau, J.-M., & Kerkeni, A. (2001). Potential antioxidant effects of zinc and chromium supplementation in people with type 2 diabetes mellitus. *Journal of the American College of Nutrition*, *20*(3), 212–218.

Athauda, D., & Foltynie, T. (2016). Insulin resistance and Parkinson's disease: A new target for disease modification? *Progress in Neurobiology*, *145*, 98–120.

Bernao, A., Meseguer, I., Aguilar, M. V., Para, M. C. M., & Muñoz, M. J. G. (2004). Effect of different doses of chromium picolinate on protein metabolism in infant rats. *Journal of Trace Elements in Medicine and Biology*, *18*(1), 33–39.

Bhat, A. H., Dar, K. B., Anees, S., Zargar, M. A., Masood, A., Sofi, M. A., et al. (2015). Oxidative stress, mitochondrial dysfunction and neurodegenerative diseases; a mechanistic insight. *Biomedicine & Pharmacotherapy*, *74*, 101–110.

Boucher, J., Kleinridders, A., & Kahn, C. R. (2014). Insulin receptor signaling in normal and insulin-resistant states. *Cold Spring Harbor Perspectives in Biology*, *6*(1), a009191.

Bourre, J.-M. (2006). Effects of nutrients (in food) on the structure and function of the nervous system: Update on dietary requirements for brain. Part 1: Micronutrients. *Journal of Nutrition Health and Aging*, *10*(5), 377–385.

Candeias, E., Duarte, A. I., Carvalho, C., Correia, S. C., Cardoso, S., Santos, R. X., et al. (2012). The impairment of insulin signaling in Alzheimer's disease. *IUBMB Life*, *64*(12), 951–957.

Cefalu, W. T., Wang, Z. Q., Zhang, X. H., Baldor, L. C., & Russell, J. C. (2002). Oral chromium picolinate improves carbohydrate and lipid metabolism and enhances skeletal muscle Glut-4 translocation in obese, hyperinsulinemic (JCR-LA corpulent) rats. *The Journal of Nutrition*, *132*(6), 1107–1114.

Chen, W.-Y., Mao, F. C., Liu, C.-H., Kuan, Y.-H., Lai, N.-W., Wu, C.-C., et al. (2016). Chromium supplementation improved post-stroke brain infarction and hyperglycemia. *Metabolic Brain Disease*, *31*(2), 289–297.

Cherny, R. A., Atwood, C. S., Xilinas, M. E., Gray, D. N., Jones, W. D., McLean, C. A., et al. (2001). Treatment with a copper-zinc chelator markedly and rapidly inhibits β-amyloid accumulation in Alzheimer's disease transgenic mice. *Neuron*, *30*(3), 665–676.

Davidson, J. R., Abraham, K., Connor, K. M., & McLeod, M. N. (2003). Effectiveness of chromium in atypical depression: A placebo-controlled trial. *Biological Psychiatry*, *53*(3), 261–264.

Davis, C. M., & Vincent, J. B. (1997). Chromium oligopeptide activates insulin receptor tyrosine kinase activity. *Biochemistry*, *36*(15), 4382–4385.

Deibel, M., Ehmann, W., & Markesbery, W. (1996). Copper, iron, and zinc imbalances in severely degenerated brain regions in Alzheimer's disease: Possible relation to oxidative stress. *Journal of the Neurological Sciences*, *143*(1–2), 137–142.

Deng, X.-L., Liu, Z., Wang, C., Li, Y., & Cai, Z. (2017). Insulin resistance in ischemic stroke. *Metabolic Brain Disease*, *32*(5), 1323–1334.

Doddigarla, Z., Parwez, I., Abidi, S., & Ahmad, J. (2016). Effect of chromium picolinate and melatonin either in single or in a combination in alloxan induced male Wistar rats. *Journal of Biomedical Science*, *42*(1), 106–114.

Dong, F., Fang, C., Yang, X., Zhang, X., Lopez, F., & Ren, J. (2006). Cardiac overexpression of catalase rescues cardiac contractile dysfunction induced by insulin resistance: Role of oxidative stress, protein carbonyl formation and insulin sensitivity. *Diabetologia*, *49*(6), 1421–1433.

Dubey, V. K., Ansari, F., Vohora, D., & Khanam, R. (2015). Possible involvement of corticosterone and serotonin in antidepressant and antianxiety effects of chromium picolinate in chronic unpredictable mild stress induced depression and anxiety in rats. *Journal of Trace Elements in Medicine and Biology*, *29*, 222–226.

Eskici, G., & Axelsen, P. H. (2012). Copper and oxidative stress in the pathogenesis of Alzheimer's disease. *Biochemistry*, *51*(32), 6289–6311.

Evans, P., & Halliwell, B. (2001). Micronutrients: Oxidant/antioxidant status. *British Journal of Nutrition*, *85*(S2), S67–S74.

Evans, J. L., Maddux, B. A., & Goldfine, I. D. (2005). The molecular basis for oxidative stress-induced insulin resistance. *Antioxidants & Redox Signaling*, 7(7–8), 1040–1052.

Ferreira, S. T., Clarke, J. R., Bomfim, T. R., & De Felice, F. G. (2014). Inflammation, defective insulin signaling, and neuronal dysfunction in Alzheimer's disease. *Alzheimer's & Dementia*, *10*, S76–S83.

Franklin, M., & Odontiadis, J. (2003). Effects of treatment with chromium picolinate on peripheral amino acid availability and brain monoamine function in the rat. *Pharmacopsychiatry*, *36*(05), 176–180.

Gorantla, N. V., Landge, V. G., Nagaraju, P. G., Priyadarshini CG, P., Balaraman, E., & Chinnathambi, S. (2019). Molecular cobalt (II) complexes for tau polymerization in Alzheimer's disease. *ACS Omega*, *4*(16), 16702–16714.

Hao, P., Zhu, Y., Wang, S., Wan, H., Chen, P., Wang, Y., et al. (2017). Selenium administration alleviates toxicity of chromium (VI) in the chicken brain. *Biological Trace Element Research*, *178*(1), 127–135.

Heneka, M. T., Carson, M. J., El Khoury, J., Landreth, G. E., Brosseron, F., Feinstein, D. L., et al. (2015). Neuroinflammation in Alzheimer's disease. *The Lancet Neurology*, *14*(4), 388–405.

Hua, Y., Clark, S., Ren, J., & Sreejayan, N. (2012). Molecular mechanisms of chromium in alleviating insulin resistance. *The Journal of Nutritional Biochemistry*, *23*(4), 313–319.

Hung, Y. H., Bush, A. I., & Cherny, R. A. (2010). Copper in the brain and Alzheimer's disease. *JBIC, Journal of Biological Inorganic Chemistry*, *15*(1), 61–76.

Jain, S. K., Rains, J. L., & Croad, J. L. (2007). Effect of chromium niacinate and chromium picolinate supplementation on lipid peroxidation, TNF-α, IL-6, CRP, glycated hemoglobin, triglycerides, and cholesterol levels in blood of streptozotocin-treated diabetic rats. *Free Radical Biology and Medicine*, *43*(8), 1124–1131.

Keunen, K., Van Elburg, R. M., Van Bel, F., & Benders, M. J. (2015). Impact of nutrition on brain development and its neuroprotective implications following preterm birth. *Pediatric Research*, 77(1–2), 148–155.

Krikorian, R., Eliassen, J. C., Boespflug, E. L., Nash, T. A., & Shidler, M. D. (2010). Improved cognitive-cerebral function in older adults with chromium supplementation. *Nutritional Neuroscience*, *13*(3), 116–122.

Lai, M.-H. (2008). Antioxidant effects and insulin resistance improvement of chromium combined with vitamin C and E supplementation for type 2 diabetes mellitus. *Journal of Clinical Biochemistry and Nutrition*, *43*(3), 191–198.

Lanctôt, K. L., Amatniek, J., Ancoli-Israel, S., Arnold, S. E., Ballard, C., Cohen-Mansfield, J., et al. (2017). Neuropsychiatric signs and symptoms of Alzheimer's disease: New treatment paradigms. *Alzheimer's & Dementia*, *3*(3), 440–449.

Li, R., Lindholm, K., Yang, L.-B., Yue, X., Citron, M., Yan, R., et al. (2004). Amyloid β peptide load is correlated with increased β-secretase activity in sporadic Alzheimer's disease patients. *Proceedings of the National Academy of Sciences of the United States of America*, *101*(10), 3632–3637.

Mahoney, R. E., Azpurua, J., & Eaton, B. A. (2016). Insulin signaling controls neurotransmission via the 4eBP-dependent modification of the exocytotic machinery. *eLife*, *5*, e16807.

Martin, J., Wang, Z. Q., Zhang, X. H., Wachtel, D., Volaufova, J., Matthews, D. E., et al. (2006). Chromium picolinate supplementation attenuates body weight gain and increases insulin sensitivity in subjects with type 2 diabetes. *Diabetes Care*, *29*(8), 1826–1832.

McCarty, M. (1994). Longevity effect of chromium picolinate—'rejuvenation'of hypothalamic function? *Medical Hypotheses*, *43*(4), 253–265.

McKay, B. E., Placzek, A. N., & Dani, J. A. (2007). Regulation of synaptic transmission and plasticity by neuronal nicotinic acetylcholine receptors. *Biochemical Pharmacology*, *74*(8), 1120–1133.

Mozaffari, M. S., Patel, C., Ballas, C., & Schaffer, S. W. (2005). Effects of chronic chromium picolinate treatment in uninephrectomized rat. *Metabolism*, *54*(9), 1243–1249.

Najem, D., Bamji-Mirza, M., Chang, N., Liu, Q. Y., & Zhang, W. (2014). Insulin resistance, neuroinflammation, and Alzheimer's disease. *Reviews in the Neurosciences*, *25*(4), 509–525.

Null, G., Pennesi, L., & Feldman, M. (2017). Nutrition and lifestyle intervention on mood and neurological disorders. *Journal of Evidence-Based Complementary and Alternative Medicine*, *22*(1), 68–74.

Onderci, M., Sahin, N., Sahin, K., & Kilic, N. (2003). Antioxidant properties of chromium and zinc. *Biological Trace Element Research*, *92*(2), 139–149.

Pechova, A., & Pavlata, L. (2007). Chromium as an essential nutrient: A review. *Veterinarni Medicina*, *52*(1), 1–18.

Piaceri, I., Nacmias, B., & Sorbi, S. (2013). Genetics of familial and sporadic Alzheimer's disease. *Frontiers in Bioscience (Elite Edition)*, *5*(1), 167–177.

Pospisilik, J. A., Knauf, C., Joza, N., Benit, P., Orthofer, M., Cani, P. D., et al. (2007). Targeted deletion of AIF decreases mitochondrial oxidative phosphorylation and protects from obesity and diabetes. *Cell*, *131*(3), 476–491.

Rains, J. L., & Jain, S. K. (2011). Hyperketonemia decreases mitochondrial membrane potential and its normalization with chromium (III) supplementation in monocytes. *Molecular and Cellular Biochemistry*, *349*(1), 77–82.

Saha, R., Nandi, R., & Saha, B. (2011). Sources and toxicity of hexavalent chromium. *Journal of Coordination Chemistry*, *64*(10), 1782–1806.

Sahin, N., Akdemir, F., Tuzcu, M., Hayirli, A., Smith, M., & Sahin, K. (2010). Effects of supplemental chromium sources and levels on performance, lipid peroxidation and proinflammatory markers in heat-stressed quails. *Animal Feed Science and Technology*, *159*(3–4), 143–149.

Sahin, K., Tuzcu, M., Orhan, C., Ali, S., Sahin, N., Gencoglu, H., et al. (2013). Chromium modulates expressions of neuronal plasticity markers and glial fibrillary acidic proteins in hypoglycemia-induced brain injury. *Life Sciences*, *93*(25–26), 1039–1048.

Sahin, K., Tuzcu, M., Orhan, C., Gencoglu, H., Ulas, M., Atalay, M., et al. (2012). The effects of chromium picolinate and chromium histidinate administration on NF-κB and Nrf2/HO-1 pathway in the brain of diabetic rats. *Biological Trace Element Research*, *150*(1), 291–296.

Selcuk, M. Y., Aygen, B., Dogukan, A., Tuzcu, Z., Akdemir, F., Komorowski, J. R., et al. (2012). Chromium picolinate and chromium histidinate protects against renal dysfunction by modulation of NF-κB pathway in high-fat diet fed and Streptozotocin-induced diabetic rats. *Nutrition & Metabolism*, *9*(1), 30.

Selkoe, D. J. (2012). Preventing Alzheimer's disease. *Science*, *337*(6101), 1488–1492.

Spears, J., Whisnant, C., Huntington, G., Lloyd, K., Fry, R., Krafka, K., et al. (2012). Chromium propionate enhances insulin sensitivity in growing cattle. *Journal of Dairy Science*, *95*(4), 2037–2045.

Spinelli, M., Fusco, S., Mainardi, M., Scala, F., Natale, F., Lapenta, R., et al. (2017). Brain insulin resistance impairs hippocampal synaptic plasticity and memory by increasing GluA1 palmitoylation through FoxO3a. *Nature Communications*, *8*(1), 1–14.

Stearns, D. M., Wise, J. P., Sr., Patierno, S. R., & Wetterhahn, K. E. (1995). Chromium (III) picolinate produces chromosome damage in Chinese hamster ovary cells 1. *The FASEB Journal*, *9*(15), 1643–1649.

Wojciak, R. W., Mojs, E., Stanislawska-Kubiak, M., & Samborski, W. (2013). The serum zinc, copper, iron, and chromium concentrations in epileptic children. *Epilepsy Research*, *104*(1–2), 40–44.

Wu, L., Rosa-Neto, P., Hsiung, G.-Y. R., Sadovnick, A. D., Masellis, M., Black, S. E., et al. (2012). Early-onset familial Alzheimer's disease (EOFAD). *Canadian Journal of Neurological Sciences*, *39*(4), 436–445.

Xie, Y., Wang, Y., Zhang, T., Ren, G., & Yang, Z. (2012). Effects of nanoparticle zinc oxide on spatial cognition and synaptic plasticity in mice with depressive-like behaviors. *Journal of Biomedical Science*, *19*(1), 1–11.

Yan, M. H., Wang, X., & Zhu, X. (2013). Mitochondrial defects and oxidative stress in Alzheimer disease and Parkinson disease. *Free Radical Biology and Medicine*, *62*, 90–101.

Zaulkffali, A. S., Md Razip, N. N., Syed Alwi, S. S., Abd Jalil, A., Abd Mutalib, M. S., Gopalsamy, B., et al. (2019). Vitamins D and E stimulate the PI3K-AKT signalling pathway in insulin-resistant SK-N-SH neuronal cells. *Nutrients*, *11*(10), 2525.

PART II

Amyotrophic lateral sclerosis

CHAPTER 7

Riboflavin and lower motor neuron diseases

Yann Nadjar and Gorka Fernandez
Department of Neurology, Neurometabolism Unit, AP-HP Pitié Salpêtrière Hospital, Paris, France

Abbreviations

***ACAD9* gene**	acyl-coA dehydrogenase 9
***AIFM1* gene**	apoptosis inducing factor mitochondria associated 1
***ETFA* gene**	electron transfer flavoprotein alpha
***ETFB* gene**	electron transfer flavoprotein beta
***ETFDH* gene**	electron transfer flavoprotein dehydrogenase
FAD	flavin adenine dinucleotide
FADD	FAD synthase
***FLAD1* gene**	flavin adenine dinucleotide synthetase 1
FMN	flavin mononucleotide
LMN	lower motor neuron
MADD	multiple acyl-CoA dehydrogenase deficiency
MFT	mitochondrial FAD transporter
RREI	riboflavin responsive exercise intolerance
RFVT1	riboflavin transporter 1
RFVT2	riboflavin transporter 2
RFVT3	riboflavin transporter 3
SLC52A2 gene	solute carrier family 52 member 2
SLC52A3 gene	solute carrier family 52 member 3
UMN	upper motor neuron

Introduction

Riboflavin (also named vitamin B2) is a water-soluble vitamin that is present in large quantities in food (dairy products, vegetables, meat, fruits, fatty fish, etc.) and can be synthesized by bacteria including human microbiota (Lienhart, Gudipati, & MacHeroux, 2013). As for all vitamins, humans need to have exogenous supply as they do not have the molecular machinery to synthesize it. Riboflavin is mainly used as a substrate to the synthesis of its two derived co-factors, FMN (flavin mononucleotide) and FAD (flavin adenine dinucleotide). The enzyme riboflavin kinase leads to FMN from riboflavin, whereas FAD synthase (FADS) leads to FAD from FMN (see Fig. 1). Both FMN and FAD act as essential cofactors for numerous and various enzymes, involved in several metabolic pathways including mitochondrial beta oxidation, mitochondrial respiratory

https://doi.org/10.1016/B978-0-323-89835-5.00014-4

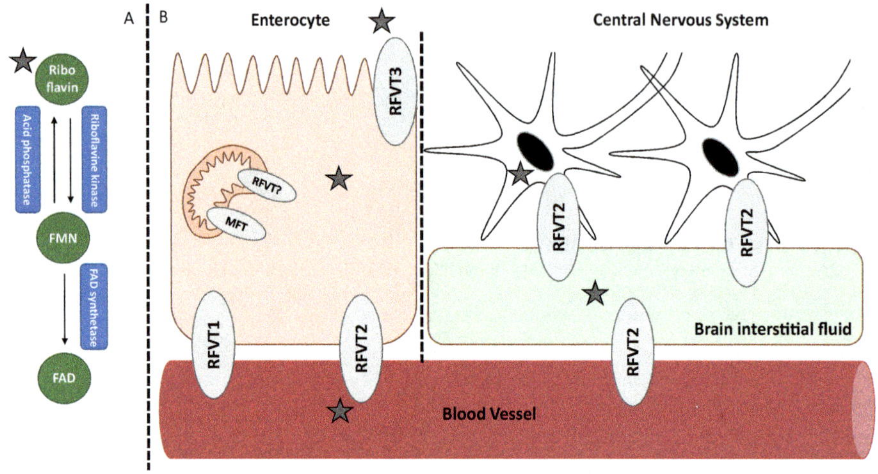

Fig. 1 Riboflavin metabolism and transport. (A) synthesis of FMN and FAD from riboflavin. (B) transport of riboflavin from intestinal lumen to the central nervous system.

chain, amino acids synthesis, metabolism of vitamins, glutathione metabolism, etc. The role of FMN/FAD as a cofactor is to undergo electron transfer, explaining why they are essentially used for oxido-reductions reactions. In total 90 proteins need FMN and/or FAD to function properly, mainly enzymes, defining what is named the "flavoproteome" (Lienhart et al., 2013). Riboflavin, as the primary and essential fuel for flavoproteome functioning, is therefore involved in a wide range of cellular functions.

Acquired riboflavin deficiency is rarely observed, and often in a malnutrition context where other nutritional deficiencies can also be observed (Plantone, Pardini, & Rinaldi, 2021). It is thought to be responsible for several skin, hair, and membranes disorders like glossitis, cheilosis, seborrheic dermatitis, and hair loss. No clear neurological disease is linked to acquired riboflavin deficiency, as it is the case for other vitamins like folate and cobalamin.

Interestingly this is not what is clinically observed in genetic diseases directly involving genes participating to riboflavin metabolism and/or transport (see Table 1). These diseases are (i) transporter deficiencies: *SLC52A2* and *SLC52A3* genes coding for riboflavin transporters and causing a lower motor neuron disease (also named motor neuronopathy) first described as the Brown-Vialetto-Van-Laere syndrome (BVVLS) (O'Callaghan, Bosch, & Houlden, 2019; Sathasivam, 2008), and *SLC25A32* gene, coding for a mitochondrial FAD transporter (MFT), causing the myopathic syndrome riboflavin responsive exercise intolerance (RREI) (Hellebrekers et al., 2017; Schiff et al., 2016); and (ii) FADS deficiency (*FLAD1* gene), causing also RREI and more severe phenotypes like early-onset and severe cardiac and respiratory insufficiency (Auranen et al., 2017; Olsen et al., 2016). These diseases are to be differentiated from other close riboflavin-related

Table 1 Riboflavin-related genetic diseases.

	Disease	Gene	Metabolic deficit	Clinical presentation	Biochemical test	Treatment
Disorders of riboflavin metabolism	Riboflavin transporter deficiency (part of Brown-Vialetto-Van Laere syndrome)	*SLCA52A2/ SLC52A3*	Riboflavin transporter deficiency	Motor neuronopathy or double neuronopathy with bulbar features, optic neuropathy, auditory neuropathy, cerebellar symptoms	Acylcarnitines	Riboflavin
	Lipid storage myopathy due to FAD synthetase deficiency (LSMFLAD)	*FLAD1*	Flavin adenine dinucleotide synthase 1 deficiency	Myopathy/exercise intolerance, cardiomyopathy and respiratory insufficiency	Acylcarnitines	Riboflavin
	Riboflavin-responsive exercise intolerance	*SLC25A32*	Mitochondrial flavine adenine dinucleotide transporter deficiency	Myopathy/exercise intolerance	Acylcarnitines	
Flavoenzymes deficits	Multiple acyl-coA deficiency (MADD)	*EFTA/ EFTB/ ETFDH*	Electron transport flavoproteins deficiency	Myopathy/exercise intolerance, rhabdomyolysis, encephalopathy, lactic acidosis	Acylcarnitines	Riboflavin
	Acyl-coA dehydrogenase 9 (ACAD9) deficiency	*ACAD9*	Acyl-coA dehydrogenase 9 deficiency	Myopathy/exercise intolerance	Acylcarnitines	Riboflavin

Continued

Table 1 Riboflavin-related genetic diseases—cont'd

	Disease	Gene	Metabolic deficit	Clinical presentation	Biochemical test	Treatment
	Combined Oxidative Phosphorylation Deficiency 6/ Charcot–Marie–Tooth Disease, X-Linked Recessive, 4	*AIFM1*	Apoptosis inducing factor mitochondria associated 1 deficiency	Infancy: encephalopathy, seizures, myopathy, sensorimotor neuropathy, hypertrophic cardiomyopathy, deafness, and respiratory insufficiency; adult: distal sensorimotor neuropathy (CMT-like), auditory neuropathy, cerebellar ataxia	None	Riboflavin

genetic neurological diseases with mutations in genes coding for flavo-proteins: the MADD (multiple acyl-CoA dehydrogenase deficiency) syndrome (*ETFA*, *ETFB*, and *ETFDH* genes), *ACAD9* deficiency, and *AIFM1* deficiency (Oette et al., 2009).

In all these diseases (except *AIFM1* deficiency) can be found in patients samples biochemical abnormalities that are the consequences of mitochondrial beta-oxidation defects. Indeed, several flavoenzymes participate to this pathway that shortens fatty acids to acetyl-CoA with subsequent production of NADH that will be used in other pathways for energy production. In routine investigations, blood acylcarnitine profile is used to detect mitochondrial beta oxidation defects, with specific patterns of increased acylcarnitines depending on the underlying deficit. For all riboflavin-related defects, the associated pattern is what was first observed in the MADD syndrome, i.e., increased acylcarnitines with different chain lengths (short, medium, and long) (Oette et al., 2009).

Among those diseases, *SLC52A2* and *SLC52A3* genes mutations (now referred as RTD for riboflavin transporters deficiencies) have a quite singular presentation, as they involve the lower motor neuron (LMN) and not the muscle (O'Callaghan et al., 2019). We will focus on these two diseases for the rest of this chapter and will explore possible links between amyotrophic lateral sclerosis (ALS), the more frequent cause of LMN degeneration (van Es et al., 2017), and riboflavin metabolism.

Riboflavin transporters deficiencies (RTD)

Riboflavin transporters (RFVT2 and RFVT3 from *SLC52A2* and *SLC52A3* genes, respectively) were recently shown to be involved in the well-described BVVL syndrome (Green et al., 2010; IJlst et al., 2012; Johnson, Gibbs, Van Maldergem, Houlden, & Singleton, 2010; Knoester et al., 2010; van Es et al., 2017). BVVL was identified for many years as a hereditary pediatric lower motor neuron disease of recessive inheritance, with associated hearing loss, and possible poor prognosis leading to death in a few years (Sathasivam, 2008). Exome sequencing revealed that some BVVL patients had mutations in *SLC52A3*, the first identified gene, or *SLC52A2* identified soon after. Phenotype of RTD, clinical evolution, diagnostic tests including biochemistry and genetics, and response to riboflavin supplementation were then clarified along the following years (Balasubramaniam, Christodoulou, & Rahman, 2019; Barile, Giancaspero, Leone, Galluccio, & Indiveri, 2016; Bosch et al., 2012; Jaeger & Bosch, 2016; O'Callaghan et al., 2019; Sathasivam, 2008).

Pathophysiology of RTD

RFVT2 and RFVT3 transporters riboflavin have differential tissue distribution: RFVT3 is mainly expressed in the intestine, therefore probably involved in intestinal absorption of riboflavin, whereas RFVT2 is widely expressed in many tissues, but especially in the brain and therefore thought to be involved in riboflavin transport from blood to the central nervous system (Jin et al., 2017; Patel, Vadlapatla, Pal, & Mitra, 2012; Teafatiller, Said, Sabui, Subramanian, & Bohl, 2017) (see Fig. 1). It is clear that indeed the riboflavin defect is responsible for clinical manifestations as the majority of patients do improve under riboflavin supplementation; however, it is not understood how these transporters defects induce a LMN degeneration. The exact odyssey of riboflavin from intestine to the LMN through the different membranes expressing the transporters is not documented (Tolomeo, Nisco, Leone, & Barile, 2020). The LMN could degenerate because of a lack of external riboflavin supply, and/or because its inability to capture riboflavin, and/or because its inability to correctly address riboflavin in determined subcellular localizations. Indeed subcellular localization of the transporters is not elucidated (Tolomeo et al., 2020). In addition, RTD could also impact glial cells (mainly astrocytes and microglia) that interact with and support the LMN, and could therefore participate to its degeneration. Indeed, in two patients, spinal cord histology found an inflammatory reaction restricted to microglial activation and macrophages infiltrates with no significant lymphocytic inflammation (Manole et al., 2017). Another important question is to understand what subpart of the flavoproteome is impacted by RTD, and what minimal altered subpart is sufficient for the triggering of LMN degeneration. Some studies have demonstrated some metabolic defect in different models (patients fibroblasts, flies, induced pluripotent stem cells or iPSC, and iPSC-derived LMN) like decreased mitochondrial electron chain transport (ETC) complexes I and II activities (with associated ultrastructural mitochondrial abnormalities), decreased peroxysomal activity, altered antioxidant systems (Colasuonno et al., 2020; Colasuonno, Bertini, Tartaglia, Compagnucci, & Moreno, 2020; Manole et al., 2017; Marioli, Magliocca, Petrini, Niceforo, & Borghi, 2020; Niceforo et al., 2021; Paul et al., 2017; Rizzo et al., 2017). However, it is not clear if these abnormalities represent an epiphenomenon or if they are an important physiopathological aspect of the disease. Probably that combined metabolic defects are needed for LMN degeneration triggering. Apart for metabolic defects, altered cell–cell contacts, decreased neurite length, and abnormal calcium homeostasis were found in RTD cellular models (Niceforo et al., 2021).

Clinics of riboflavin transporters deficiencies

Disease onset usually takes place during childhood with weakness from an LMN dysfunction, that can have several localizations, the most typical being facial and bulbar palsy (leading to dysarthria and swallowing disorder) (Niceforo et al., 2021; O'Callaghan et al., 2019). As any LMN disorder, weakness is associated with amyotrophy (tongue, limbs). Classically, there is no additional involvement of the upper motor neuron (UMN). Diaphragm weakness can also occur and is the main prognostic factor of the disease: patients may die a few years after disease onset from respiratory failure without artificial ventilation. Hearing loss caused by an auditory neuropathy is very often associated to the LMN disorder and often precedes motor symptoms (Menezes et al., 2016). Differences between the two involved transporters exist in clinical presentation (Carreau et al., 2021; Manole et al., 2017): in *SLC52A3* RTD facial and bulbar weakness are most often present; in *SLC52A2* RTD, axial weakness (dropped head) and limb weakness are more often present, with also possible other signs: visual loss from optic neuropathy and sensory ataxia from sensory neuronopathy. The diagnosis can be suggested by abnormalities of two blood tests, but their sensitivity is quite low, so normality does not exclude at all an RTD diagnosis: acylcarnitine profile that may have a MADD profile, and riboflavin level that may be low. Treatment is high dose of riboflavin (15 mg/kg), which usually allows improvement or at least stabilization of the disease, that concern motor but also sensorial symptoms. The first treated patient was successfully treated at 3 months of age, weaned from artificial ventilation due to diaphragmatic dysfunction, and was still asymptomatic at 8 years old, demonstrating the maintenance of the therapeutic effect of riboflavin supplementation over the years (O'Callaghan et al., 2019). There is no reported side effects of high-dose riboflavin treatment, probably because of urinary excretion of riboflavin in excess (Chastain & McCormick, 1987).

Adult-onset RTD and differential diagnoses

More recently, adult-onset RTD patients were reported, reviewed here (Carreau et al., 2021). Adult-onset RTD patients globally have the same disorder as children, but with more heterogeneous clinical presentations and a clearly better prognosis. Onset may be subacute, or more insidious with a chronic evolution that can last decades, since diaphragmatic dysfunction may not be present. Therefore, according to its clinical presentation, RTD may potentially mimic other more common neurological syndromes associated with an LMN disorder: acute motor neuropathy (a subtype of Guillain-Barré syndrome), multiple cranial neuropathies (as described in neurosarcoidosis for example), motor and sensory neuronopathy (possibly of paraneoplastic or genetic origin), ALS and distal hereditary motor neuropathy (Allison, Roncero, Forsyth, Coffman, & Le Pichon, 2017; Carreau et al., 2021).

Riboflavin and amyotrophic lateral sclerosis (ALS)

RTD is different but also has several clinical similarities with the more frequent neurological disease with LMN degeneration which is ALS (Carreau et al., 2021; Garg et al., 2017; van Es et al., 2017).

ALS clinics

ALS is characterized by onset of weakness caused by LMN degeneration with often associated UMN degeneration (van Es et al., 2017). Other clinical signs are also caused by MN degeneration: amyotrophy, fasciculations for LMN, and spasticity for UMN. Weakness often has a local onset with classical distinctions of bulbar onset (causing dysarthria, swallowing disorder, bucco-facial apraxia, tongue amyotrophy, rire et pleurer spasmodique), limb onset (which can be asymmetrical, concerning UL and/or LL), and respiratory onset that manifests with clinical signs of diaphragmatic dysfunction (orthopnea, exertion dyspnea, hyercapnea). As the disease progresses, weakness extends to other anatomical locations, increasing disabilities, that may at some point need some support like gastrostomy when food intake is not sufficient or noninvasive ventilation, which are two mainstays of current ALS therapeutic strategy. The third therapeutic option is a drug, Riluzole, prescribed for all ALS forms, that is the only drug that proved to be (moderately) efficient by slowing worsening of the disease, increasing life span of 3 months (whereas mean life span is 5 years). Riluzole is thought to act as limiting excito-toxicity (Chiò, Mazzini, & Mora, 2020).

Pathophysiology of ALS

ALS can be a sporadic (90% of cases) or a familial (10%) disease. Recently, the number of genes identified as causing ALS has greatly increased, with approximately 30 genes now reported (Chia, Chiò, & Traynor, 2018). This was taken as an opportunity to help understanding the pathophysiology of the disease, which is not currently fully understood, despite extensive research. This may be partly due to the fact that ALS is more a syndrome than a disease, with very heterogeneous clinical presentations related to hypothetical underlying heterogeneous pathophysiological cellular mechanisms (Swinnen & Robberecht, 2014). Many of them have been involved to explain LMN degeneration (including mitochondrial dysfunction, axonal transport, toxic protein aggregation, impaired protein degradation involving the proteasome or autophagy, prion-like spreading, excitotoxicity, decreased neurotrophic support from nonneuronal cells, oxidative stress, hypermetabolism, inflammation, RNA metabolism defects, and RNA toxicity), with some difficulties to identify primary from secondary mechanisms (Robberecht & Philips, 2013). Also a multistep accumulation of cellular defects is thought to precede the triggering of the pathological process leading to LMN degeneration (Al-Chalabi et al., 2014). A secondary riboflavin metabolism defect in some determined ALS patients could be one of these steps, for which specific therapeutic intervention could be conceivable (Timmerman & De Jonghe, 2014). No clear link has ever been suggested between ALS pathophysiology and riboflavin metabolism, but some observed mechanisms are shared between ALS and RTD, notably mitochondrial dysfunction. Riboflavin defect could for example occur in the context of aggregated proteins, or abnormal RNA processing, with consecutive trapping of proteins or abnormal RNA production from riboflavin-related genes. In addition, some ALS patients, for different reasons including a hypermetabolic state, lose some weight at the beginning of the disease, with an impact on the prognosis of the disease; as illustrated by RTD, it seems that acquired poor riboflavin supply in the context of malnutrition could have deleterious effects on disease evolution (D'amico et al., 2021). However, before ALS onset, nutritional risk factors, including vitamins deficiencies, were looked for, without evidence of association, on the basis of questionnaires but not biological assessments (D'amico et al., 2021).

Conclusion

RTD is a genetic disease characterized by LMN degeneration that may mimic other LMN disorders in adults, especially the frequent ALS. Pathophysiology of RTD is not understood, and future research aiming at improving this knowledge could also give broader insights on the LMN biology and its susceptibility to degeneration.

Summary points

- Riboflavin (also named vitamin B2) is a water-soluble vitamin mainly used as a substrate for the synthesis of its two derived co factors, FMN (flavin mononucleotide) and FAD (flavin adenine dinucleotide).
- FMN and FAD act as essential cofactors for numerous and various enzymes, named as flavoenzymes.
- Riboflavin transporter deficiency (RTD), caused by *SLC52A2* and *SLC52A3* genes mutations, is characterized by generalized weakness due to a lower motor neuron (LMN) degeneration, potentially leading to death by respiratory compromise.
- RTD may mimic other LMN disorders including the far more common amyotrophic lateral sclerosis (ALS).
- Pathophysiology of RTD is not understood, and future research aiming at improving this knowledge could also give broader insights on the LMN biology and its susceptibility to degeneration.

References

Al-Chalabi, A., Calvo, A., Chio, A., Colville, S., Ellis, C. M., Hardiman, O., et al. (2014). Analysis of amyotrophic lateral sclerosis as a multistep process: A population-based modelling study. *Lancet Neurology*, *13*(11), 1108–1113.

Allison, T., Roncero, I., Forsyth, R., Coffman, K., & Le Pichon, J.-B. (2017). Brown-Vialetto-Van Laere syndrome as a mimic of neuroimmune disorders: 3 cases from the clinic and review of the literature. *Journal of Child Neurology*, *32*(6), 528–532. [Internet]. Available from http://journals.sagepub.com/doi/10.1177/0883073816689517.

Auranen, M., Paetau, A., Piirilä, P., Pohju, A., Salmi, T., Lamminen, A., et al. (2017). Patient with multiple acyl-CoA dehydrogenation deficiency disease and FLAD1 mutations benefits from riboflavin therapy. *Neuromuscular Disorders*, *27*(6), 581–584.

Balasubramaniam, S., Christodoulou, J., & Rahman, S. (2019). Disorders of riboflavin metabolism. *Journal of Inherited Metabolic Disease*. [Internet]. [cited 2019 Mar 9]; Available from http://doi.wiley.com/10.1002/jimd.12058.

Barile, M., Giancaspero, T. A., Leone, P., Galluccio, M., & Indiveri, C. (2016). Riboflavin transport and metabolism in humans. *Journal of Inherited Metabolic Disease*, *39*(4), 545–557. [Internet]. Available from http://link.springer.com/10.1007/s10545-016-9950-0.

Bosch, A. M., Stroek, K., Abeling, N. G., Waterham, H. R., Ijlst, L., & Wanders, R. J. A. (2012). The Brown-Vialetto-Van Laere and Fazio Londe syndrome revisited: Natural history, genetics, treatment and future perspectives (Internet). *Orphanet Journal of Rare Diseases*, *7*, 1. Available from: Orphanet Journal of Rare Diseases.

Carreau, C., Benoit, C., Ahle, G., Cauquil, C., Roubertie, A., Lenglet, T., et al. (2021). Late-onset riboflavin transporter deficiency: A treatable mimic of various motor neuropathy aetiologies. *Journal of Neurology, Neurosurgery, and Psychiatry*, *92*(1), 27–35.

Chastain, J. L., & McCormick, D. B. (1987). Flavin catabolites: Identification and quantitation in human urine. *The American Journal of Clinical Nutrition*, *46*(5), 830–834. [Internet]. [cited 2021 Nov 8]. Available from: https://pubmed.ncbi.nlm.nih.gov/3673930/.

Chia, R., Chiò, A., & Traynor, B. J. (2018). Novel genes associated with amyotrophic lateral sclerosis: Diagnostic and clinical implications. *Lancet Neurology*, *17*(1), 94–102.

Chiò, A., Mazzini, L., & Mora, G. (2020). Disease-modifying therapies in amyotrophic lateral sclerosis. *Neuropharmacology*, *167*. [Internet]. May 1 [cited 2021 Nov 8]. Available from https://pubmed.ncbi.nlm.nih.gov/32062193/.

Colasuonno, F., Bertini, E., Tartaglia, M., Compagnucci, C., & Moreno, S. (2020). Mitochondrial abnormalities in induced pluripotent stem cells-derived motor neurons from patients with riboflavin transporter deficiency. *Antioxidants*, *9*(12), 1–10.

Colasuonno, F., Niceforo, A., Marioli, C., Fracassi, A., Stregapede, F., Massey, K., et al. (2020). Mitochondrial and Peroxisomal alterations contribute to energy dysmetabolism in riboflavin transporter deficiency. *Oxidative Medicine and Cellular Longevity*, *2020*.

D'amico, E., Grosso, G., Nieves, J. W., Zanghì, A., Factor-Litvak, P., & Mitsumoto, H. (2021). Metabolic abnormalities, dietary risk factors and nutritional management in amyotrophic lateral sclerosis. *Nutrients*, *13*. MDPI AG.

Garg, N., Park, S. B., Vucic, S., Yiannikas, C., Spies, J., Howells, J., et al. (2017). Differentiating lower motor neuron syndromes (Internet). *Journal of Neurology, Neurosurgery and Psychiatry*, *88*, 474–483. BMJ Publishing Group. [cited 2020 Mar 20]. Available from http://www.ncbi.nlm.nih.gov/pubmed/28003344.

Green, P., Wiseman, M., Crow, Y. J., Houlden, H., Riphagen, S., Lin, J.-P., et al. (2010). Brown-Vialetto-Van Laere syndrome, a ponto-bulbar palsy with deafness, is caused by mutations in C20orf54. *American Journal of Human Genetics*, *86*(3), 485–489. [Internet]. Available from https://doi.org/10.1016/j.ajhg.2010.02.006.

Hellebrekers, D. M. E. I., Sallevelt, S. C. E. H., Theunissen, T. E. J., Hendrickx, A. T. M., Gottschalk, R. W., Hoeijmakers, J. G. J., et al. (2017). Novel SLC25A32 mutation in a patient with a severe neuromuscular phenotype. *European Journal of Human Genetics*, *25*(7), 886–888.

IJlst, L., Wanders, R. J., Bosch, A. M., Stroek, K., Waterham, H. R., & Abeling, N. G. (2012). The Brown-Vialetto-Van Laere and Fazio Londe syndrome revisited: Natural history, genetics, treatment and future perspectives. *Orphanet Journal of Rare Diseases*, *7*(1), 83. [Internet]. [cited 2018 Dec 19]. Available from http://ojrd.biomedcentral.com/articles/10.1186/1750-1172-7-83.

Jaeger, B., & Bosch, A. M. (2016). Clinical presentation and outcome of riboflavin transporter deficiency: Mini review after five years of experience. *Journal of Inherited Metabolic Disease*, *39*(4), 559–564. [Internet]. [cited 2019 Mar 20]. Available from: http://doi.wiley.com/10.1007/s10545-016-9924-2.

Jin, C., Yao, Y., Yonezawa, A., Imai, S., Yoshimatsu, H., Otani, Y., et al. (2017). Riboflavin transporters RFVT/SLC52A mediate translocation of riboflavin, rather than FMN or FAD, across plasma membrane. *Biological and Pharmaceutical Bulletin*, *40*(11), 1990–1995. [Internet]. [cited 2018 Dec 19]. Available from: https://www.jstage.jst.go.jp/article/bpb/40/11/40_b17-00292/_article.

Johnson, J. O., Gibbs, J. R., Van Maldergem, L., Houlden, H., & Singleton, A. B. (2010). Exome sequencing in Brown-Vialetto-Van Laere syndrome. *American Journal of Human Genetics*, *87*, 567–569.

Bosch, A. M., Abeling, N. G. G. M., Ijlst, L., Knoester, H., van der Pol, W. L., Stroomer, A. E. M., et al. (2010). Brown-Vialetto-Van Laere and Fazio Londe syndrome is associated with a riboflavin transporter defect mimicking mild MADD: A new inborn error of metabolism with potential treatment. *Journal of Inherited Metabolic Disease*, *34*(1), 159–164.

Lienhart, W. D., Gudipati, V., & MacHeroux, P. (2013). The human flavoproteome (Internet). *Archives of Biochemistry and Biophysics*, *535*, 150–162. Elsevier Inc. Available from https://doi.org/10.1016/j.abb.2013.02.015.

Manole, A., Jaunmuktane, Z., Hargreaves, I., Ludtmann, M. H. R., Salpietro, V., Bello, O. D., et al. (2017). Clinical, pathological and functional characterization of riboflavin-responsive neuropathy. *Brain*, *140*(11), 2820–2837. [Internet]. [cited 2018 Dec 19]. Available from: https://academic.oup.com/brain/article/140/11/2820/4237466.

Marioli, C., Magliocca, V., Petrini, S., Niceforo, A., Borghi, R., Petrillo, S., et al. (2020). Antioxidant amelioration of riboflavin transporter deficiency in motoneurons derived from patient-specific induced pluripotent stem cells. *International Journal of Molecular Sciences*, *21*(19), 1–17.

Menezes, M. P., O'Brien, K., Hill, M., Webster, R., Antony, J., Ouvrier, R., et al. (2016). Auditory neuropathy in Brown–Vialetto–Van Laere syndrome due to riboflavin transporter RFVT2 deficiency. *Developmental Medicine and Child Neurology*, *58*(8), 848–854.

Niceforo, A., Marioli, C., Colasuonno, F., Petrini, S., Massey, K., Tartaglia, M., et al. (2021). Altered cytoskeletal arrangement in induced pluripotent stem cells and motor neurons from patients with riboflavin transporter deficiency. DMM. *Disease Models & Mechanisms*, *14*(2).

O'Callaghan, B., Bosch, A. M., & Houlden, H. (2019). An update on the genetics, clinical presentation, and pathomechanisms of human riboflavin transporter deficiency. *Journal of Inherited Metabolic Disease*, *42*(4), 598–607.

Oette, M., Stone, M. J., Scholl, H. P. N., Issa, P. C., Fleckenstein, M., et al. (2009). Multiple acyl-CoA dehydrogenase deficiency. In *Encyclopedia of molecular mechanisms of disease* (p. 1363).

Olsen, R. K. J., Koňaříková, E., Giancaspero, T. A., Mosegaard, S., Boczonadi, V., Mataković, L., et al. (2016). Riboflavin-responsive and non-responsive mutations in FAD synthase cause multiple acyl-CoA dehydrogenase and combined respiratory-chain deficiency. *American Journal of Human Genetics*, *98*(6), 1130–1145.

Patel, M., Vadlapatla, R. K., Pal, D., & Mitra, A. K. (2012). Molecular and functional characterization of riboflavin specific transport system in rat brain capillary endothelial cells. *Brain Research*, *1468*, 1–10. [Internet]. Available from https://doi.org/10.1016/j.brainres.2012.05.052.

Paul, A., Drecourt, A., Petit, F., Deguine, D. D., Vasnier, C., Oufadem, M., et al. (2017). FDXR mutations cause sensorial neuropathies and expand the Spectrum of mitochondrial Fe-S-synthesis diseases. *American Journal of Human Genetics*, *101*(4), 630–637.

Plantone, D., Pardini, M., & Rinaldi, G. (2021). Riboflavin in neurological diseases: A narrative review (Internet). *Clinical Drug Investigation*, *41*, 513–527. [cited 2021 Jul 14]. Available from https://doi.org/10.1007/s40261-021-01038-1.

Rizzo, F., Ramirez, A., Compagnucci, C., Salani, S., Melzi, V., Bordoni, A., et al. (2017). Genome-wide RNA-seq of iPSC-derived motor neurons indicates selective cytoskeletal perturbation in Brown-Vialetto disease that is partially rescued by riboflavin. *Scientific Reports*, 7(1), 46271. [Internet]. [cited 2018 Dec 19]. Available from: http://www.nature.com/articles/srep46271.

Robberecht, W., & Philips, T. (2013). The changing scene of amyotrophic lateral sclerosis. *Nature Reviews. Neuroscience*, *14*(4), 248–264.

Sathasivam, S. (2008). Brown-Vialetto-Van Laere syndrome. *Orphanet Journal of Rare Diseases*, *3*(1), 9. [Internet]. Available from http://www.ncbi.nlm.nih.gov/pubmed/18416855%0Ahttp://www.pubmedcentral.nih.gov/articlerender.fcgi?artid=PMC2346457.

Schiff, M., Veauville-Merllié, A., Su, C. H., Tzagoloff, A., Rak, M., de Baulny, H. O., et al. (2016). SLC25A32 mutations and riboflavin-responsive exercise intolerance. *The New England Journal of Medicine*, *374*(8), 795–797.

Swinnen, B., & Robberecht, W. (2014). The phenotypic variability of amyotrophic lateral sclerosis. *Nature Reviews. Neurology*, *10*(11), 661–670.

Teafatiller, T., Said, H. M., Sabui, S., Subramanian, V. S., & Bohl, J. A. (2017). Structure/functional aspects of the human riboflavin transporter-3 (SLC52A3): Role of the predicted glycosylation and substrate-interacting sites. *American Journal of Physiology. Cell Physiology*, *313*(2), C228–C238.

Timmerman, V., & De Jonghe, P. (2014). Promising riboflavin treatment for motor neuron disorder (Internet). *Brain*, *137*, 2–3. [cited 2018 Dec 19]. Available from: https://academic.oup.com/brain/article-lookup/doi/10.1093/brain/awt342.

Tolomeo, M., Nisco, A., Leone, P., & Barile, M. (2020). Development of novel experimental models to study flavoproteome alterations in human neuromuscular diseases: The effect of Rf therapy. *International Journal of Molecular Sciences*, *21*(15).

van Es, M. A., Hardiman, O., Chio, A., Al-Chalabi, A., Pasterkamp, R. J., Veldink, J. H., et al. (2017). Amyotrophic lateral sclerosis. *Lancet*, *390*(10107), 2084–2098.

PART III

Brain injury

CHAPTER 8

Vitamin C (L-ascorbic acid) in brain injury

Stefan W. Leichtle[a], Sulaiman Al Mohaish[b], Melissa Sandler[c], Vishal Yajnik[d], Dennis J. Rivet, II[e], Markos Kashiouris[f], Alpha A. (Berry) Fowler, III[f], and Gretchen M. Brophy[b]

[a]Division of Acute Care Surgical Services, Department of Surgery, Virginia Commonwealth University School of Medicine, Richmond, VA, United States
[b]Department of Pharmacotherapy & Outcomes Science, Virginia Commonwealth University School of Pharmacy, Richmond, VA, United States
[c]Department of Physical Medicine and Rehabilitation, Virginia Commonwealth University School of Medicine, Richmond, VA, United States
[d]Division of Anesthesia Critical Care, Department of Anesthesiology, Virginia Commonwealth University School of Medicine, Richmond, VA, United States
[e]Harold F. Young Neurosurgical Center, Department of Neurosurgery, Virginia Commonwealth University School of Medicine, Richmond, VA, United States
[f]Division of Pulmonary Disease and Critical Care Medicine, Virginia Commonwealth University School of Medicine, Richmond, VA, United States

Abbreviations

BBB blood–brain barrier
CNS central nervous system
CSF cerebrospinal fluid
CT computed tomography
DHA dehydroascorbic acid
G6PD glucose-6-phosphate dehydrogenase
IL interleukin
IV intravenous
MMP metalloproteinase
ROS reactive oxygen species
TBI traumatic brain injury
TNF tumor necrosis factor

Introduction

Vitamin C is a well-known nutrient, natural or added part of multiple foods, and frequently taken as a nutritional supplement. Historically, the systemic connective tissue disease scurvy was a feared consequence of nutritional vitamin C deficiency. Additionally, the important role of the nutrient in immune function and as potent antioxidant has become increasingly recognized. The potential therapeutic effects of high-dose, intravenous (IV) vitamin C administration in states of critical illness have become a promising area of research. Brain injury, discussed in this chapter with a focus on traumatic brain injury (TBI), is a major acute and chronic public health problem worldwide. Injury progression after the primary insult, the secondary brain injury, has deleterious ramifications.

Vitamins and Minerals in Neurological Disorders
https://doi.org/10.1016/B978-0-323-89835-5.00040-5

It is mediated by oxidative stress and characterized by inflammatory processes and edema. The potent antioxidant and antiinflammatory properties of vitamin C, in addition to its role in the synthesis of key neurotransmitters, make it a promising treatment strategy for the brain injury.

Vitamin C

Vitamin C (L-ascorbic acid) is a water-soluble vitamin with important functions as an antioxidant and as a co-factor in multiple biosynthetic reactions such as collagen and neurotransmitter production (Biesalski & McGregor, 2007; Ellis et al., 2000; Li & Schellhorn, 2007). As an essential vitamin, the human body cannot synthesize it and is dependent on adequate dietary intake; multiple foods, including citrus fruits and vegetables contain vitamin C. Severe deficiency has become rare in countries with adequate, year-round access to these natural sources of vitamin C. It remains, however, a relevant problem in low- and middle-income countries where large parts of the population suffer from malnutrition. Severe vitamin C deficiency leads to scurvy, manifesting as a connective tissue disease with symptoms of progressive fatigue, easy bruising, poor wound healing, and gum disease, ultimately leading to death. Historically, scurvy was a feared affliction for sailors until the implementation of strategies to provide vitamin C-rich foods during long journeys.

The recommended dietary allowance of vitamin C for healthy adults is 75 mg and 90 mg per day for women and men, respectively (National Institutes of Health Office of Dietary Supplements (ODS), 2020). Some guidelines recommend higher doses of up to 200 mg per day and higher daily intake may also be required for women during pregnancy and lactation, as well as for tobacco users of both genders. For children, lower doses are suggested. Importantly, these recommended doses apply only to healthy individuals and do not reflect the needs in both acute and chronic illness. Normal doses of vitamin C are readily absorbed from the intestinal tract when consumed orally, while amounts greater than 1000 mg per day are excreted renally (Berger, 2009; Levine et al., 1996). Dietary intake without additional supplementation is usually adequate to maintain homeostasis in healthy individuals.

Normal plasma vitamin C concentrations range from 21 μmol/L to 100 μmol/L (de Grooth et al., 2018; Levine et al., 1996; Levine, Rumsey, Daruwala, Park, & Wang, 1999). Due to the ready elimination of enteral doses above 1000 mg via the kidneys, peak plasma concentrations do not exceed 220 μmol/L even with high-dose oral administration of 3000 mg vitamin C every four hours per day. In contrast, IV administration of 1250 mg vitamin C results in peak plasma concentrations of up to 885 μmol/L and can reach 1760 μmol/L for IV administration of 3000 mg every four hours (Padayatty et al., 2004). In critically ill patients with multiple organ dysfunction, only IV doses of more than 10,000 mg per day or 200 mg per kg per day achieved plasma concentrations above 1000 μmol/L, which is the optimal concentration for free radical scavenging effects (de Grooth et al., 2018; Fowler et al., 2014; Jackson, Xu, Vita, & Keaney, 1998; Padayatty et al., 2004).

Therefore, while dietary intake of vitamin C is usually adequate to prevent symptoms of serious vitamin deficiency, substantially higher doses are required to harness potentially "therapeutic" effects of vitamin C. These include well-established potent antioxidant and free radical-scavenging properties, in addition to presumed stabilizing effects on cellular basal membranes and the microcirculation (Ashor, Lara, Mathers, & Siervo, 2014; Biesalski & McGregor, 2007). Based on decades of strong data from pre-clinical studies and promising though mixed more recent data from clinical trials (Fowler et al., 2014, 2019; Marik, Khangoora, Rivera, Hooper, & Catravas, 2017), vitamin C has re-emerged as a potential treatment strategy for patients with critical illness, including TBI and non-traumatic brain injury.

Traumatic brain injury

TBI is a leading public health problem worldwide: an estimated 69 million individuals are affected each year, with etiology, geographic distribution, severity, and quality of data collection varying widely (Dewan et al., 2018). In the United States, TBI alone results in 2.8 million Emergency Department visits and 56,000 deaths per year (Taylor, Bell, Breiding, & Xu, 2017). Already the leading cause of death and disability in children and adults up to age 44 years, the incidence of TBI continues to rise despite ongoing efforts in primary prevention (Taylor et al., 2017). TBI consists of a primary, direct brain injury caused by the initial traumatic insult, followed by secondary brain injury, which often results in even more damage than the initial trauma. Secondary brain injury occurs as a consequence of local tissue ischemia, brain edema, metabolic changes, and damage from reactive oxygen species (ROS) (Dearden, 1998; Ikeda & Long, 1990; Jones et al., 1994; Kontos & Povlishock, 1986; Lok et al., 2011; Povlishock & Kontos, 1992; Tyurin et al., 2002). The concept of secondary brain injury, its pathologic mechanisms, and the principles guiding its mitigating and treatment are shared between traumatic and non-traumatic brain injury. In the acute care setting, treatment strategies for TBI focus on preventing the "injury after the injury," i.e., secondary brain injury. However, these efforts are essentially limited to best general practices in the management of critically ill patients such as maintenance of adequate mean arterial pressures, provision of adequate cerebral perfusion pressures, prevention of hypoxia, and stabilization of patients' overall clinical condition. An effective treatment to actively provide neuroprotection or to promote neuro-regeneration after TBI does not exist (Andrews et al., 2015; Chang, Provencio, & Shah, 2021; Hutchinson et al., 2016).

Secondary brain injury follows the primary insult over the subsequent hours and days, mediated by local ischemia, cellular swelling, and reperfusion injury. Oxidative stress plays a crucial role in injury progression (Brau, García-Castiñeiras, & Rifkinson, 1984; Dearden, 1998; Ikeda & Long, 1990; Lok et al., 2011; Povlishock & Kontos, 1992), leading to an imbalance between neuroprotective factors and those causing additional harm. The former include ROS-scavengers and protective mediators such as nuclear factor erythroid 2 (NF-E2) (Zhao, Moore, Redell, & Dash, 2007). The latter are the result of upregulation of damaging agents such as metalloproteinase (MMP)-9, which disrupts the integrity of the blood–brain barrier (BBB) (Guo et al., 2008; Lin, Huang, Shen, Huang, & Liu, 2010; Sajja, Prasad, Tang, Kaisar, & Cucullo, 2017; Zhao et al., 2007), and ROS that cause direct cell damage through lipid peroxidation of the fatty acids in the cell and mitochondrial membranes (Smith, Andrus, Zhang, & Hall, 1994; Vagnozzi et al., 1999). Similar mechanisms involving the depletion or impairment of antioxidant dependent cellular defense mechanisms can be found in multiple states of critical illness (Carr et al., 2017; Polidori, Mecocci, & Frei, 2001) and numerous studies suggest an important role of exogenous administration of antioxidants such as vitamin C.

Therapeutic effects of vitamin C

Antiinflammatory and endothelial-stabilizing effects

The effect of different doses of vitamin C was studied extensively in animal models of severe physiologic derangement such as sepsis and hemorrhage. In rodent models of bacterial- and endotoxin-induced sepsis, high-dose (200 mg/kg per day) IV vitamin C administration resulted in potent attenuation of the systemic inflammatory response, decreasing the risk for multisystem organ dysfunction. Observed mechanisms included a stabilizing effect on the microvascular system, prevention of capillary leak, and mitigation of lung damage (Fisher et al., 2012, 2011). Similar effects were observed in a model of polytrauma and hemorrhage: high-dose (200 mg/kg per day) vitamin C lowered the extent of end-organ damage and severity of lung injury as seen on histopathologic exam and the systemic inflammatory response characterized by decreased expression of the proinflammatory markers interleukin (IL)-1β, IL-8, tumor necrosis factor (TNF)-α, plasminogen activator inhibitor-1, and tissue factor (Reynolds et al., 2018). Clinical studies corroborated the systemic antiinflammatory and lung-protective effects of supplemental vitamin C administration in patients with trauma and after major surgery (Berger et al., 2008; Nathens et al., 2002). The attenuating effect of vitamin C on capillary leak and fluid extravasation is also well-recognized in severe burn injuries and in both animal and clinical studies, where vitamin C significantly reduced resuscitation fluid requirements and the harmful effects associated therein (Barbosa et al., 2009; Dubick, Williams, Elgjo, & Kramer, 2005; Tanaka et al., 2000).

Therapeutic effects in brain injury

Conceptually, the beneficial effects of vitamin C administration demonstrated in states of systemic inflammation and critical illness can be expected in brain injury as well due to many pathologic mechanisms with similar characteristics. Ample animal and some clinical data on the effects of vitamin C and other compounds with antioxidant properties in brain injury exist. In animal models of ischemic stroke, higher doses of vitamin C alone or as component of an antioxidant "neuroprotective cocktail" mitigated ischemia reperfusion injury (Henry & Chandy, 1998; Huang et al., 2001; Mack et al., 2006). In models of TBI and intracranial hemorrhage, the beneficial effects of vitamin C included lower risk for vasospasm (Kawakami, Kodama, & Toda, 1991), improved cellular defense mechanisms, reduced tissue necrosis, and improved functional outcomes (Ishaq et al., 2013; Sajja et al., 2017; Yan et al., 2014). A major limitation of the existing animal data on vitamin C in brain injury from traumatic and nontraumatic causes is its extreme heterogeneity. Most animal models employ rodents, some nonhuman primates, and almost all have a low number of specimens. In many studies, vitamin C was administered enterally, in some in IV form; doses ranged from supplementary to one hundred times the RDA; and frequently, vitamin C was given as part of an antioxidant cocktail, making it impossible to determine the effect of vitamin C itself. Lastly, the outcomes examined in these studies included plasma concentrations of inflammatory markers; size of brain lesions on imaging; histopathologic changes in end-organs; few functional outcomes; and rarely, a comprehensive evaluation combining these assessments.

Extremely low concentrations of vitamin C and other antioxidants in plasma and cerebrospinal fluid (CSF) of patients with central nervous system (CNS) pathology provide an additional rationale for therapeutic and prophylactic vitamin C administration (Brau et al., 1984; Cherubini et al., 2000; Polidori et al., 2001; Sánchez-Moreno et al., 2004). In 28 patients with intracranial hemorrhage or TBI, plasma vitamin C concentrations within 24 h of the incident were almost half of those in healthy controls (29–30 μmol/L vs. 52–57 μmol/L, respectively) and inversely correlated with the severity of brain damage (Polidori et al., 2001). A 1984 study of CSF concentrations of vitamin C in 41 patients with CNS disorders demonstrated the lowest vitamin C concentrations of all patients to be in the five subjects suffering from TBI (77 μmol/L vs. 203 μmol/L in controls) (Brau et al., 1984).

Most interventional studies on vitamin C administration after brain injury are case reports and small observational studies, but there is some stronger evidence from the few prospective clinical trials. In 28 patients with ischemic stroke, enteral administration of 200 mg vitamin C per day together with 300 mg of aspirin over 90 days significantly lowered blood markers of lipid peroxidation as compared to the 31 patients receiving aspirin alone. Functional outcomes, however, were no different between the groups (Polidori et al., 2005). In 96 patients with ischemic stroke, enteral administration of antioxidant cocktails, one of which contained 500 mg of vitamin C, significantly lowered the concentration of C-reactive protein, a plasma marker of systemic inflammation (Ullegaddi, Powers, & Gariballa, 2006). There is only one randomized controlled trial of TBI patients that studied the IV administration of high-dose vitamin C. Twenty-three of 100 patients with severe TBI received vitamin C in doses of 10,000 mg on day 1,

10,000 mg on day 4, and 4000 mg daily for three more days. This group was found to have a significant decrease in brain edema surrounding the traumatic parenchymal lesions seen on computed tomography (CT) of the head, but there was no impact on clinical neurologic outcomes (Razmkon et al., 2011).

Vitamin C dosing and routes of administration in brain injury

Direct neuroprotective and antioxidant effects in the brain require adequate CSF concentrations of vitamin C. Dehydroascorbic acid (DHA), its oxidized form, is actively and preferentially transported across the BBB. Vitamin C concentrations are therefore higher in the CNS than in plasma (Harrison & May, 2009; Huang et al., 2001; May, 2012). Administration of DHA, however, is not FDA-approved in the US and therefore only used in experimental models. This makes high plasma concentrations of vitamin C the most important predictor of high CSF concentrations. Additionally, such high plasma concentrations may aid in attenuating the systemic state of inflammation caused by brain injury.

Animal and clinical data in TBI, nontraumatic brain injury, and critical illness suggest a dramatic difference in effect between "supplementary" doses of vitamin C, i.e., those aimed at restoring depleted levels and supratherapeutic, high-dose administration as IV medication. In acute injury and illness, the former may alleviate some of the adverse consequences of decreased antioxidant and defensive mechanisms in the CNS, but the latter is required to achieve the presumed microcirculatory and endothelial effects. The disparity between highly promising animal studies and limited clinic data may be explained herein, as animal studies mostly administered much larger doses than those provided via enteral administration in most clinical trials. Several recent, large clinical trials examining IV administration of vitamin C in septic patients (Fowler et al., 2019; Fujii et al., 2020; Sevransky et al., 2021) did not confirm the dramatic mortality reduction suggested in an earlier, smaller clinical study (Marik et al., 2017). However, while these patients received *high* doses of IV vitamin C (up to 1500 mg given four times per day), the dose remained substantially below the *high-dose* administration suggested in animal studies: therapeutic effects in the acute phase of brain injury will most likely require an IV treatment regimen of 200 mg/kg per day, i.e., daily doses in the range of 12,000–16,000 mg to ensure consistently supranormal, ROS-scavenging plasma concentrations. In addition to various dosing regimens and routes of administration, timing considerations introduce yet another challenge and element of heterogeneity. Secondary brain injury with capillary leak, intense inflammation, and brain edema occurs shortly after injury (Smith et al., 1994; Vagnozzi et al., 1999), and high plasma (and therefore CNS) vitamin C concentrations need to be achieved rapidly after the inciting event to achieve neuroprotective effects. Taken together, the complex pharmacokinetics of vitamin C may be the explanation for the discordance of results seen in animal and human studies (Lykkesfeldt & Tveden-Nyborg, 2019). Table 1 provides a selection of animal and clinical studies on the use of vitamin C in brain injury.

Table 1 A selection of studies on the use of vitamin C in brain injury.

Author/ year	Design	Sample	Interventions	Key findings
Henry and Chandy (1998)	Animal study: AA vs. placebo	Monkeys with ischemic stroke ($N=15$)	AA (500 mg/kg)	Mean infarct size in AA pretreated group significantly lower than placebo group
Huang et al. (2001)	Animal study: placebo vs. AA vs. DHA	Mice with ischemic stroke ($N=111$)	AA (250 mg/kg or 500 mg/kg), DHA (40 mg/kg, 250 mg/kg, or 500 mg/kg)	After reperfusion of cerebral ischemia, mean infarct volume reduced significantly from 53% and 59% in placebo and AA-treated animals, respectively, to 15% in 250 mg/kg DHA-treated animals
Mack et al. (2006)	Animal study: placebo vs. DHA	Mice with ischemic stroke ($N=24$)	DHA (50 mg/kg, 100 mg/kg, or 200 mg/kg)	DHA repletion of interstitial AA levels and reduction in cerebral lipid peroxidation associated with dose-dependent significant reductions in infarct volume
Yan et al. (2014)	Animal study: AA vs. placebo	Rats with spinal cord contusion ($N=30$)	AA (100 mg/kg or 200 mg/kg)	Reduced area of necrosis at injury site vs. control group on histological analysis in both normal- and high-dose AA groups
Polidori et al. (2001)	Observational study	Healthy adults ($N=40$), ICH hemorrhage ($N=13$), and TBI ($N=15$)	No intervention	Brain-injured patients had significantly lower plasma AA levels on day 1 compared with healthy control subjects
Razmkon et al. (2011)	RCT	Patients with severe TBI ($N=100$)	AA 500 mg/d IV, 10 g IV on 1st day, repeated on 4th day, followed by 4 g/d IV for remaining 3 days, vs. Vitamin E (400 IU/d IM) for 7 days	High-dose AA significantly stabilized or reduced diameter of perilesional edema in 68% of patients. GOS at discharge and follow-up significantly better for vitamin E group

Continued

Table 1 A selection of studies on the use of vitamin C in brain injury—cont'd

Author/ year	Design	Sample	Interventions	Key findings
Polidori et al. (2005)	RCT	Patients with ischemic stroke (N=59)	AA (200 mg/day) + aspirin (300 mg/day) or aspirin alone (300 mg/day)	Clinical functional outcome for both groups improved after three months of treatment, and vitamin C, significantly decreases ischemic stroke-related lipid peroxidation

AA, ascorbic acid; DHA, dehydroascorbic acid (oxidized form of ascorbic acid); ICH, intracranial hemorrhage; TBI, traumatic brain injury; RCT, randomized controlled trial; GOS, Glasgow outcome scale.

Risks and adverse effects

Vitamin C has an excellent safety profile. When taken orally, effective renal secretion essentially prevents any accumulation of toxic concentrations; minor gastrointestinal discomfort is the only side effect observed. Similarly, when administered IV and in higher doses, vitamin C is readily excreted in the urine, thus limiting the potential for systemic toxicity. Traditionally, direct renal damage and formation of oxalate stones in the setting of acute or chronic renal insufficiency were a frequently cited concern and an exclusion criteria in trials administering high-dose IV vitamin C (Buehner et al., 2016). However, none of the recent large randomized controlled trials administering up to 4500 mg IV vitamin C per day raised concern about serious adverse events (Fujii et al., 2020; Sevransky et al., 2021) related to vitamin C administration. In prior trials, some of which included very high doses or prolonged treatment courses, enteral and IV vitamin C was tolerated well without major side effects (Fowler et al., 2014; Hoffer et al., 2008; Muhlhofer et al., 2004; Nathens et al., 2002; Stephenson, Levin, Spector, & Lis, 2013). Few case reports of hemolysis in patients with glucose-6-phosphate dehydrogenase (G6PD) deficiency who received extremely high doses of IV vitamin C suggest caution or avoiding high-dose vitamin C administration in this patient group. Lastly, patients requiring glucose monitoring while receiving IV vitamin C need to undergo plasma glucose measurements as opposed to "finger stick" measurements with a glucometer. The latter may demonstrate falsely elevated glucose concentrations and "treatment" of this false positive finding of hyperglycemia could precipitate life-threatening hypoglycemia. Very high-dose (200 mg/kg per day and higher) administration should be limited to short treatment courses and while monitoring closely for the few known possible side effects.

Conclusions

Inflammatory processes, ROS, and endothelial dysfunction are well-established contributors to secondary brain injury from traumatic and nontraumatic causes. Numerous animal studies conducted over more than three decades achieved highly encouraging results with the administration of antioxidants and free radical scavengers in models of acute injury. These robust data contrast the limited and more ambiguous clinical results of vitamin C administration in brain injury. A major difference between preclinical and clinical studies of vitamin C were dose and route of administration as well as achieved plasma concentrations.

The well-documented depletion of vitamin C levels in states of systemic inflammation, stress, and brain injury demonstrates the importance of adequate vitamin C stores to strengthen the brain's baseline neuroprotective mechanisms and to enhance its resilience to oxidative stress. This is particularly important for those at risk for nutritional deficiencies like the elderly, who have lower baseline vitamin C levels, but also for those at higher risk for concussions and TBI such as athletes or military personnel. Harnessing therapeutic effects of vitamin C in the acute phase of brain injury and critical illness will require the administration of high doses of IV vitamin C. The future for research in this vital nutrient remains rich, as its properties and roles in cellular processes are vital and offer an opportunity to curb secondary injury in various disease processes.

Applications to other neurological conditions

This chapter focused on TBI to discuss the potential neuroprotective effects of vitamin C. Many, if not all, underlying considerations on pharmacokinetics, mechanisms of action, and comparison of preclinical and clinical data can be inferred to other situations of acute brain injury due to the "common pathway" of secondary injury. While prevention of primary traumatic- and nontraumatic brain injury is ideal, secondary injury is the mechanism most amenable to treatment and with the largest potential to improve long-term outcomes. A poorly studied but potentially interesting application of lower dose, vitamin C supplementation to ensure adequate plasma levels may be in patients at higher risk for brain injury or those with chronic CNS illnesses. The former may include athletes in contact sports or military personnel, the latter may include patients with seizure disorders or hydrocephalus. Such a benefit may be extrapolated from the findings of depleted CSF concentrations of vitamin C in patients with hydrocephalus, epilepsy, and other CNS disorders (Brau et al., 1984), but no clinical data on the effect of treatment for these conditions with vitamin C exists.

Other components of interest

A variety of nutritional supplements with antioxidant and antiinflammatory properties may be of benefit in brain injury. Indeed, several of the studies on vitamin C in brain injury administered an "antioxidant cocktail" to harness additional or synergistic effects of other antioxidants. Reliable evidence, however, is scarce. A promising antioxidant combination consists of vitamin C and E (alpha tocopherol). Vitamin E as individual compound has strong antioxidant properties and is an inhibitor of lipid peroxidation (Inci, Ozcan, & Kilinç, 1998), which is one of the primary pathologic mechanisms seen in secondary brain injury. When alpha tocopherol and ascorbic acid are combined, vitamin C has the ability to reduce the oxidized form of vitamin E, thus regenerating vitamin E and increasing its availability. Despite the use of this combination in brain injury (Barbosa et al., 2009; Razmkon et al., 2011), definitive conclusions about the role of vitamin E supplementation in brain injury cannot be made based on the existing evidence. Individuals taking vitamin K-dependent anticoagulants need to be aware of interactions with vitamin E supplements, leading to an increased bleeding risk. Additionally, a large meta-analysis on vitamin E supplementation raised concerns about increased all-cause mortality with high-dose vitamin E administration (Miller et al., 2005). Vitamins are not the only compounds with antioxidant properties studied in brain injury. *N*-acetyl cysteine (NAC) is a well-established medication for the treatment of acetaminophen-induced liver failure. Due to its potent free-radical scavenging properties and excellent safety profile, its use has expanded to the liver failure of other etiologies, radiographic contrast-induced acute kidney injury, among others. NAC also has neuroprotective properties. A clinical study on the use of NAC in military personnel with TBI from blast injury demonstrated faster resolution of TBI-related symptoms in those receiving NAC (M. E. Hoffer, Balaban, Slade, Tsao, & Hoffer, 2013). The existing data on NAC in brain injury, however, remain inadequate for strong recommendations.

Mini-dictionary of terms

- Primary brain injury: initial, direct damage to the brain parenchyma.
- Secondary brain injury: processes that cause additional or ongoing damage to the brain through inflammation, hypoperfusion, and other metabolic abnormalities.
- Reactive oxygen species: often also called "free radicals"; highly reactive oxygen compound with unpaired electrons, which can cause damage to cellular structures through various processes such as lipid peroxidation.
- Free radical scavenger: molecules that can eliminate free radicals by donating electrons and rendering them harmless.
- Capillary leak: increased capillary permeability seen in states of systemic or local inflammation such as trauma, sepsis, or critical illness results in intravascular volume loss and interstitial edema, both of which exacerbate hypoperfusion and inflammation.

Key facts

- As antioxidant and essential co-factor in multiple biosynthetic processes, vitamin C (L-ascorbic acid) has important functions in states of health and severe illness.
- Oxidative stress plays a crucial role in secondary brain injury, i.e., the brain damage sustained following the primary insult.
- The recommended dietary intake of vitamin C for healthy adults is 75 mg/day for women and 90 mg/day for men, but much larger doses are needed in severe illness.
- A large number of studies have demonstrated promising but mixed results of exogenous administration of high doses of antioxidants such as vitamin C.
- Vitamin C supplementation is generally safe, but it may cause kidney stones when given in very high doses over a long period of time.

Summary points

- Inflammatory processes, oxidative stress, and endothelial dysfunction are well-established pathologic mechanisms in brain injury.
- As a potent antioxidant and free radical scavenger, vitamin C demonstrated a strong potential to counteract these damaging processes in animal studies, but the results of several clinical studies are less definitive.
- The heterogeneity of existing research with regard to study setting, target population, vitamin C dose, route of administration, and examined outcomes make an assessment and definitive conclusions challenging.
- Preclinical studies with promising outcomes used vitamin C doses ten-fold higher than those administered in the clinical trials which had no significant beneficial effects, pointing toward dosing and pharmacokinetic considerations as a major reason for the discrepancy between animal and human data.
- High-dose IV vitamin C has a strong potential to be an effective treatment strategy in the acute phase of TBI by mitigating secondary brain injury due to its potent antioxidant, antiinflammatory, and cell membrane-stabilizing properties.
- Supplementation of vitamin C to replete baseline or acutely diminished vitamin C stores in individuals with brain injury or those at higher risk for it can be considered a low risk and potentially beneficial therapeutic strategy.

References

Andrews, P. J. D., Sinclair, H. L., Rodriguez, A., Harris, B. A., Battison, C. G., Rhodes, J. K. J., et al. (2015). Hypothermia for intracranial hypertension after traumatic brain injury. *The New England Journal of Medicine, 373*(25), 2403–2412. https://doi.org/10.1056/NEJMoa1507581.

Ashor, A. W., Lara, J., Mathers, J. C., & Siervo, M. (2014). Effect of vitamin C on endothelial function in health and disease: A systematic review and meta-analysis of randomised controlled trials. *Atherosclerosis, 235*(1), 9–20. https://doi.org/10.1016/j.atherosclerosis.2014.04.004.

Barbosa, E., Faintuch, J., Machado Moreira, E. A., Gonçalves da Silva, V. R., Lopes Pereima, M. J., Martins Fagundes, R. L., et al. (2009). Supplementation of vitamin E, vitamin C, and zinc attenuates oxidative stress in burned children: A randomized, double-blind, placebo-controlled pilot study. *Journal of Burn Care & Research: Official Publication of the American Burn Association, 30*(5), 859–866. https://doi.org/10.1097/BCR.0b013e3181b487a8.

Berger, M. M. (2009). Vitamin C requirements in parenteral nutrition. *Gastroenterology, 137*(5 Suppl), S70–S78. https://doi.org/10.1053/j.gastro.2009.08.012.

Berger, M. M., Soguel, L., Shenkin, A., Revelly, J.-P., Pinget, C., Baines, M., et al. (2008). Influence of early antioxidant supplements on clinical evolution and organ function in critically ill cardiac surgery, major trauma, and subarachnoid hemorrhage patients. *Critical Care (London, England), 12*(4), R101. https://doi.org/10.1186/cc6981.

Biesalski, H. K., & McGregor, G. P. (2007). Antioxidant therapy in critical care—Is the microcirculation the primary target? *Critical Care Medicine, 35*(9 Suppl), S577–S583. https://doi.org/10.1097/01.CCM.0000278598.95294.C5.

Brau, R. H., García-Castiñeiras, S., & Rifkinson, N. (1984). Cerebrospinal fluid ascorbic acid levels in neurological disorders. *Neurosurgery, 14*(2), 142–146.

Buehner, M., Pamplin, J., Studer, L., Hughes, R. L., King, B. T., Graybill, J. C., et al. (2016). Oxalate nephropathy after continuous infusion of high-dose vitamin C as an adjunct to burn resuscitation. *Journal of Burn Care & Research: Official Publication of the American Burn Association, 37*(4), e374–e379. https://doi.org/10.1097/BCR.0000000000000233.

Carr, A. C., Rosengrave, P. C., Bayer, S., Chambers, S., Mehrtens, J., & Shaw, G. M. (2017). Hypovitaminosis C and vitamin C deficiency in critically ill patients despite recommended enteral and parenteral intakes. *Critical Care (London, England), 21*(1), 300. https://doi.org/10.1186/s13054-017-1891-y.

Chang, C. W. J., Provencio, J. J., & Shah, S. (2021). Neurological critical care: The evolution of cerebrovascular critical care. *Critical Care Medicine.* https://doi.org/10.1097/CCM.0000000000004933.

Cherubini, A., Polidori, M. C., Bregnocchi, M., Pezzuto, S., Cecchetti, R., Ingegni, T., et al. (2000). Antioxidant profile and early outcome in stroke patients. *Stroke, 31*(10), 2295–2300. https://doi.org/10.1161/01.str.31.10.2295.

de Grooth, H.-J., Manubulu-Choo, W.-P., Zandvliet, A. S., Spoelstra-de Man, A. M. E., Girbes, A. R., Swart, E. L., et al. (2018). Vitamin C pharmacokinetics in critically ill patients: A randomized trial of four IV regimens. *Chest, 153*(6), 1368–1377. https://doi.org/10.1016/j.chest.2018.02.025.

Dearden, N. M. (1998). Mechanisms and prevention of secondary brain damage during intensive care. *Clinical Neuropathology, 17*(4), 221–228.

Dewan, M. C., Rattani, A., Gupta, S., Baticulon, R. E., Hung, Y.-C., Punchak, M., et al. (2018). Estimating the global incidence of traumatic brain injury. *Journal of Neurosurgery*, 1–18. https://doi.org/10.3171/2017.10.JNS17352.

Dubick, M. A., Williams, C., Elgjo, G. I., & Kramer, G. C. (2005). High-dose vitamin C infusion reduces fluid requirements in the resuscitation of burn-injured sheep. *Shock, 24*(2), 139–144.

Ellis, G. R., Anderson, R. A., Lang, D., Blackman, D. J., Morris, R. H., Morris-Thurgood, J., et al. (2000). Neutrophil superoxide anion—Generating capacity, endothelial function and oxidative stress in chronic heart failure: Effects of short- and long-term vitamin C therapy. *Journal of the American College of Cardiology, 36*(5), 1474–1482.

Fisher, B. J., Kraskauskas, D., Martin, E. J., Farkas, D., Wegelin, J. A., Brophy, D., et al. (2012). Mechanisms of attenuation of abdominal sepsis induced acute lung injury by ascorbic acid. *American Journal of Physiology. Lung Cellular and Molecular Physiology, 303*(1), L20–L32. https://doi.org/10.1152/ajplung.00300.2011.

Fisher, B. J., Seropian, I. M., Kraskauskas, D., Thakkar, J. N., Voelkel, N. F., Fowler, A. A., et al. (2011). Ascorbic acid attenuates lipopolysaccharide-induced acute lung injury. *Critical Care Medicine, 39*(6), 1454–1460. https://doi.org/10.1097/CCM.0b013e3182120cb8.

Fowler, A. A., Syed, A. A., Knowlson, S., Sculthorpe, R., Farthing, D., DeWilde, C., et al. (2014). Phase I safety trial of intravenous ascorbic acid in patients with severe sepsis. *Journal of Translational Medicine, 12*, 32. https://doi.org/10.1186/1479-5876-12-32.

Fowler, A. A., Truwit, J. D., Hite, R. D., Morris, P. E., DeWilde, C., Priday, A., et al. (2019). Effect of vitamin C infusion on organ failure and biomarkers of inflammation and vascular injury in patients with

Sepsis and severe acute respiratory failure: The CITRIS-ALI randomized clinical Trial. *JAMA*, *322*(13), 1261–1270. https://doi.org/10.1001/jama.2019.11825.

Fujii, T., Luethi, N., Young, P. J., Frei, D. R., Eastwood, G. M., French, C. J., et al. (2020). Effect of vitamin C, hydrocortisone, and thiamine vs hydrocortisone alone on time alive and free of vasopressor support among patients with septic shock: The VITAMINS randomized clinical Trial. *Journal of the American Medical Association*. https://doi.org/10.1001/jama.2019.22176.

Guo, M., Cox, B., Mahale, S., Davis, W., Carranza, A., Hayes, K., et al. (2008). Pre-ischemic exercise reduces matrix metalloproteinase-9 expression and ameliorates blood-brain barrier dysfunction in stroke. *Neuroscience*, *151*(2), 340–351. https://doi.org/10.1016/j.neuroscience.2007.10.006.

Harrison, F. E., & May, J. M. (2009). Vitamin C function in the brain: Vital role of the ascorbate transporter SVCT2. *Free Radical Biology & Medicine*, *46*(6), 719–730. https://doi.org/10.1016/j.freeradbiomed.2008.12.018.

Henry, P. T., & Chandy, M. J. (1998). Effect of ascorbic acid on infarct size in experimental focal cerebral ischaemia and reperfusion in a primate model. *Acta Neurochirurgica*, *140*(9), 977–980. https://doi.org/10.1007/s007010050201.

Hoffer, M. E., Balaban, C., Slade, M. D., Tsao, J. W., & Hoffer, B. (2013). Amelioration of acute sequelae of blast induced mild traumatic brain injury by N-acetyl cysteine: A double-blind, placebo controlled study. *PLoS One*, *8*(1), e54163. https://doi.org/10.1371/journal.pone.0054163.

Hoffer, L. J., Levine, M., Assouline, S., Melnychuk, D., Padayatty, S. J., Rosadiuk, K., et al. (2008). Phase I clinical trial of i.v. ascorbic acid in advanced malignancy. *Annals of Oncology: Official Journal of the European Society for Medical Oncology*, *19*(11), 1969–1974. https://doi.org/10.1093/annonc/mdn377.

Huang, J., Agus, D. B., Winfree, C. J., Kiss, S., Mack, W. J., McTaggart, R. A., et al. (2001). Dehydroascorbic acid, a blood-brain barrier transportable form of vitamin C, mediates potent cerebroprotection in experimental stroke. *Proceedings of the National Academy of Sciences of the United States of America*, *98*(20), 11720–11724. https://doi.org/10.1073/pnas.171325998.

Hutchinson, P. J., Kolias, A. G., Timofeev, I. S., Corteen, E. A., Czosnyka, M., Timothy, J., et al. (2016). Trial of decompressive craniectomy for traumatic intracranial hypertension. *The New England Journal of Medicine*, *375*(12), 1119–1130. https://doi.org/10.1056/NEJMoa1605215.

Ikeda, Y., & Long, D. M. (1990). The molecular basis of brain injury and brain edema: The role of oxygen free radicals. *Neurosurgery*, *27*(1), 1–11. https://doi.org/10.1227/00006123-199007000-00001.

Inci, S., Ozcan, O. E., & Kilinç, K. (1998). Time-level relationship for lipid peroxidation and the protective effect of alpha-tocopherol in experimental mild and severe brain injury. *Neurosurgery*, *43*(2), 330–335. discussion 335-336 https://doi.org/10.1097/00006123-199808000-00095.

Ishaq, G. M., Saidu, Y., Bilbis, L. S., Muhammad, S. A., Jinjir, N., & Shehu, B. B. (2013). Effects of α-tocopherol and ascorbic acid in the severity and management of traumatic brain injury in albino rats. *Journal of Neurosciences in Rural Practice*, *4*(3), 292–297. https://doi.org/10.4103/0976-3147.118784.

Jackson, T. S., Xu, A., Vita, J. A., & Keaney, J. F. (1998). Ascorbate prevents the interaction of superoxide and nitric oxide only at very high physiological concentrations. *Circulation Research*, *83*(9), 916–922.

Jones, P. A., Andrews, P. J., Midgley, S., Anderson, S. I., Piper, I. R., Tocher, J. L., et al. (1994). Measuring the burden of secondary insults in head-injured patients during intensive care. *Journal of Neurosurgical Anesthesiology*, *6*(1), 4–14.

Kawakami, M., Kodama, N., & Toda, N. (1991). Suppression of the cerebral vasospastic actions of oxyhemoglobin by ascorbic acid. *Neurosurgery*, *28*(1), 33–39. discussion 39-40 https://doi.org/10.1097/00006123-199101000-00006.

Kontos, H. A., & Povlishock, J. T. (1986). Oxygen radicals in brain injury. *Central Nervous System Trauma*, *3*(4), 257–263. https://doi.org/10.1089/cns.1986.3.257.

Levine, M., Conry-Cantilena, C., Wang, Y., Welch, R. W., Washko, P. W., Dhariwal, K. R., et al. (1996). Vitamin C pharmacokinetics in healthy volunteers: Evidence for a recommended dietary allowance. *Proceedings of the National Academy of Sciences of the United States of America*, *93*(8), 3704–3709.

Levine, M., Rumsey, S. C., Daruwala, R., Park, J. B., & Wang, Y. (1999). Criteria and recommendations for vitamin C intake. *JAMA*, *281*(15), 1415–1423.

Li, Y., & Schellhorn, H. E. (2007). New developments and novel therapeutic perspectives for vitamin C. *The Journal of Nutrition*, *137*(10), 2171–2184. https://doi.org/10.1093/jn/137.10.2171.

Lin, J.-L., Huang, Y.-H., Shen, Y.-C., Huang, H.-C., & Liu, P.-H. (2010). Ascorbic acid prevents blood-brain barrier disruption and sensory deficit caused by sustained compression of primary somatosensory cortex. *Journal of Cerebral Blood Flow and Metabolism: Official Journal of the International Society of Cerebral Blood Flow and Metabolism*, *30*(6), 1121–1136. https://doi.org/10.1038/jcbfm.2009.277.

Lok, J., Leung, W., Murphy, S., Butler, W., Noviski, N., & Lo, E. H. (2011). Intracranial hemorrhage: Mechanisms of secondary brain injury. *Acta Neurochirurgica. Supplement*, *111*, 63–69. https://doi.org/10.1007/978-3-7091-0693-8_11.

Lykkesfeldt, J., & Tveden-Nyborg, P. (2019). The pharmacokinetics of vitamin C. *Nutrients*, *11*(10). https://doi.org/10.3390/nu11102412.

Mack, W. J., Mocco, J., Ducruet, A. F., Laufer, I., King, R. G., Zhang, Y., et al. (2006). A cerebroprotective dose of intravenous citrate/sorbitol-stabilized dehydroascorbic acid is correlated with increased cerebral ascorbic acid and inhibited lipid peroxidation after murine reperfused stroke. *Neurosurgery*, *59*(2), 383–388. discussion 383-388 https://doi.org/10.1227/01.NEU.0000223496.96945.A7.

Marik, P. E., Khangoora, V., Rivera, R., Hooper, M. H., & Catravas, J. (2017). Hydrocortisone, vitamin C, and thiamine for the treatment of severe sepsis and septic shock: A retrospective before-after study. *Chest*, *151*(6), 1229–1238. https://doi.org/10.1016/j.chest.2016.11.036.

May, J. M. (2012). Vitamin C transport and its role in the central nervous system. *Sub-Cellular Biochemistry*, *56*, 85–103. https://doi.org/10.1007/978-94-007-2199-9_6.

Miller, E. R., Pastor-Barriuso, R., Dalal, D., Riemersma, R. A., Appel, L. J., & Guallar, E. (2005). Meta-analysis: High-dosage vitamin E supplementation may increase all-cause mortality. *Annals of Internal Medicine*, *142*(1), 37–46. https://doi.org/10.7326/0003-4819-142-1-200501040-00110.

Muhlhofer, A., Mrosek, S., Schlegel, B., Trommer, W., Rozario, F., Bohles, H., et al. (2004). High-dose intravenous vitamin C is not associated with an increase of pro-oxidative biomarkers. *European Journal of Clinical Nutrition*, *58*(8), 1151–1158. https://doi.org/10.1038/sj.ejcn.1601943.

Nathens, A. B., Neff, M. J., Jurkovich, G. J., Klotz, P., Farver, K., Ruzinski, J. T., et al. (2002). Randomized, prospective trial of antioxidant supplementation in critically ill surgical patients. *Annals of Surgery*, *236*(6), 814–822.

National Institutes of Health Office of Dietary Supplements (ODS). (2020, February 27). *Vitamin C: Fact sheet for health professionals*. https://ods.od.nih.gov/factsheets/VitaminC-HealthProfessional/#en3.

Padayatty, S. J., Sun, H., Wang, Y., Riordan, H. D., Hewitt, S. M., Katz, A., et al. (2004). Vitamin C pharmacokinetics: Implications for oral and intravenous use. *Annals of Internal Medicine*, *140*(7), 533–537.

Polidori, M. C., Mecocci, P., & Frei, B. (2001). Plasma vitamin C levels are decreased and correlated with brain damage in patients with intracranial hemorrhage or head trauma. *Stroke*, *32*(4), 898–902.

Polidori, M. C., Praticó, D., Ingegni, T., Mariani, E., Spazzafumo, L., Del Sindaco, P., et al. (2005). Effects of vitamin C and aspirin in ischemic stroke-related lipid peroxidation: Results of the AVASAS (aspirin versus ascorbic acid plus aspirin in stroke) study. *BioFactors*, *24*(1–4), 265–274. https://doi.org/10.1002/biof.5520240131.

Povlishock, J. T., & Kontos, H. A. (1992). The role of oxygen radicals in the pathobiology of traumatic brain injury. *Human Cell*, *5*(4), 345–353.

Razmkon, A., Sadidi, A., Sherafat-Kazemzadeh, E., Mehrafshan, A., Jamali, M., Malekpour, B., et al. (2011). Administration of vitamin C and vitamin E in severe head injury: A randomized double-blind controlled trial. *Clinical Neurosurgery*, *58*, 133–137 (Journal Article).

Reynolds, P. S., Fisher, B. J., McCarter, J., Sweeney, C., Martin, E. J., Middleton, P., et al. (2018). Interventional vitamin C: A strategy for attenuation of coagulopathy and inflammation in a swine multiple injuries model. *The Journal of Trauma and Acute Care Surgery*, *85*(1S Suppl. 2), S57–S67. https://doi.org/10.1097/TA.0000000000001844.

Sajja, R. K., Prasad, S., Tang, S., Kaisar, M. A., & Cucullo, L. (2017). Blood-brain barrier disruption in diabetic mice is linked to Nrf2 signaling deficits: Role of ABCB10? *Neuroscience Letters*, *653*, 152–158. https://doi.org/10.1016/j.neulet.2017.05.059.

Sánchez-Moreno, C., Dashe, J. F., Scott, T., Thaler, D., Folstein, M. F., & Martin, A. (2004). Decreased levels of plasma vitamin C and increased concentrations of inflammatory and oxidative stress markers after stroke. *Stroke*, *35*(1), 163–168. https://doi.org/10.1161/01.STR.0000105391.62306.2E.

Sevransky, J. E., Rothman, R. E., Hager, D. N., Bernard, G. R., Brown, S. M., Buchman, T. G., et al. (2021). Effect of vitamin C, thiamine, and hydrocortisone on ventilator- and vasopressor-free days in patients with sepsis: The VICTAS randomized clinical trial. *JAMA*, *325*(8), 742–750. https://doi.org/10.1001/jama.2020.24505.

Smith, S. L., Andrus, P. K., Zhang, J. R., & Hall, E. D. (1994). Direct measurement of hydroxyl radicals, lipid peroxidation, and blood-brain barrier disruption following unilateral cortical impact head injury in the rat. *Journal of Neurotrauma*, *11*(4), 393–404. https://doi.org/10.1089/neu.1994.11.393.

Stephenson, C. M., Levin, R. D., Spector, T., & Lis, C. G. (2013). Phase I clinical trial to evaluate the safety, tolerability, and pharmacokinetics of high-dose intravenous ascorbic acid in patients with advanced cancer. *Cancer Chemotherapy and Pharmacology*, *72*(1), 139–146. https://doi.org/10.1007/s00280-013-2179-9.

Tanaka, H., Matsuda, T., Miyagantani, Y., Yukioka, T., Matsuda, H., & Shimazaki, S. (2000). Reduction of resuscitation fluid volumes in severely burned patients using ascorbic acid administration: A randomized, prospective study. *Archives of Surgery*, *135*(3), 326–331.

Taylor, C. A., Bell, J. M., Breiding, M. J., & Xu, L. (2017). Traumatic brain injury-related emergency department visits, hospitalizations, and deaths—United States, 2007 and 2013. *Morbidity and Mortality Weekly Report. Surveillance Summaries*, *66*(9), 1–16. https://doi.org/10.15585/mmwr.ss6609a1.

Tyurin, V. A., Tyurina, Y. Y., Borisenko, G. G., Sokolova, T. V., Ritov, V. B., Quinn, P. J., et al. (2002). Oxidative stress following traumatic brain injury in rats: Quantitation of biomarkers and detection of free radical intermediates. *Journal of Neurochemistry*, *75*(5), 2178–2189. https://doi.org/10.1046/j.1471-4159.2000.0752178.x.

Ullegaddi, R., Powers, H. J., & Gariballa, S. E. (2006). Antioxidant supplementation with or without B-group vitamins after acute ischemic stroke: A randomized controlled trial. *JPEN Journal of Parenteral and Enteral Nutrition*, *30*(2), 108–114. https://doi.org/10.1177/0148607106030002108.

Vagnozzi, R., Marmarou, A., Tavazzi, B., Signoretti, S., Di Pierro, D., del Bolgia, F., et al. (1999). Changes of cerebral energy metabolism and lipid peroxidation in rats leading to mitochondrial dysfunction after diffuse brain injury. *Journal of Neurotrauma*, *16*(10), 903–913. https://doi.org/10.1089/neu.1999.16.903.

Yan, M., Yang, M., Shao, W., Mao, X.-G., Yuan, B., Chen, Y.-F., et al. (2014). High-dose ascorbic acid administration improves functional recovery in rats with spinal cord contusion injury. *Spinal Cord*, *52*(11), 803–808. https://doi.org/10.1038/sc.2014.135.

Zhao, J., Moore, A. N., Redell, J. B., & Dash, P. K. (2007). Enhancing expression of Nrf2-driven genes protects the blood brain barrier after brain injury. *The Journal of Neuroscience: The Official Journal of the Society for Neuroscience*, *27*(38), 10240–10248. https://doi.org/10.1523/JNEUROSCI.1683-07.2007.

CHAPTER 9

Antioxidants in brain injury with or without antibiotics

Santanu Kaity[a] **and Anoop Kumar**[b]
[a]Formulation Development Service, Mylan Laboratories Limited, Hyderabad, India
[b]Department of Pharmacology, Delhi Pharmaceutical Sciences & Research University (DPSRU), New Delhi, India

List of abbreviations

BBB blood-brain barrier
CBF cerebral blood flow
ER endoplasmic reticulum
GLT-1 glutamate transporter-1
GABA gamma-aminobutyric acid
ICP intracranial pressure
LP lipid peroxidation
mTOR Mammalian target of rapamycin
PN peroxy-nitrite
RNS reactive nitrogen species
ROS reactive oxygen species
TBI traumatic brain injury
WHO World Health Organization

Introduction

Traumatic brain injury (TBI) is the worst outcome of the fast-moving generation in the present scenario. TBI occurs after a severe or moderate impact of the brain against the skull which creates major consequences on the health and socioeconomic conditions of the victims because of physical and mental disabilities. According to WHO, a sharp rise of using a motor vehicle is one of the leading causes of the accident and this will increase TBI-related incidents globally (Finfer & Cohen, 2001). The young generation is mostly affected by TBI due to speeding vehicle accidents. The other affected population is geriatrics who experiences multiple falls because of age-related loss of muscle control, fear, slip, or any other sudden impact, which leads to loss of control and sudden fall. The other major causes of TBI are social violence, terrorist attack, firearms, and blast, and all such traumatic incidents, which affect a person practically or sometimes generate huge mental tension, lead to temporary or permanent damage of brain (Butcher, McHugh, Lu, et al., 2007).

Vitamins and Minerals in Neurological Disorders
https://doi.org/10.1016/B978-0-323-89835-5.00010-7

Epidemiology

As per the Centres for Disease Control and Prevention, USA, the yearly cases of hospital admissions and emergency department visits due to TBI are related to 403 individuals out of 100,000 affected persons and 85 individuals out of 100,000 incidents, respectively (Langlois, Rutland-Brown, & Thomas, 2006). According to the report of the European Union, the yearly total incidence of fatal TBI incidents count is approximately 235 individuals out of 100,000 incidents (Tagliaferri, Compagnone, Korsic, Servadei, & Kraus, 2006). Almost similar numbers of cases are found in Australia too (Hillier, Hiller, & Metzer, 1997). In India, annually, 1.5 to 2 million individuals die every year because of TBI as per a published report (Gururaj, 2013).

Mechanism of TBI propagation

The primary injury to the brain results in mechanical focal brain damage; however, serious injury could result in the damage of blood-brain barrier (BBB), hemorrhage, increased intracranial pressure (ICP), decreased cerebral blood flow (CBF), etc. These all-primary injury events in most of the cases further altered various normal brain signaling pathways such as alteration of glutamate and GABA levels, induction of calcium signaling pathways (phospholipase C as well as endoplasmic stress (ER) pathways), activation of oxidative stress signaling pathways (caspase-dependent as well as caspase-independent), inflammatory signaling pathways, alteration of mTOR and p53 signaling pathways, etc. (Rana, Singh, Sharma, & Kumar, 2019).

All these activated signaling pathways take time (from hours to days or months also) and result in neurochemical, cellular, and metabolic changes. All these changes together result in neuronal cell death and trigger multiple complications in the individual. The propagation of the damage varies differently among individuals based on the biological variation and extent of the impact.

Classification of TBI for targeted therapy

TBI can be classified in many ways; however, the targeted therapy-based classification has made it easy to initiate prompt action and therapy toward restoring the patient to normal condition and save the life. The categorization of TBI is summarized in Table 1.

Hospitalization and treatment

The primary treatment aims to prevent hypoxia and hypotension which are very common just after the impact in most of the patients. The primary treatment is also helpful to prevent secondary brain damage. Extensive care should be taken to establish the oxygen saturation level, blood pressure normalization, and to prevent the hypovolemic

Table 1 Categorization of TBI (Saatman et al., 2008).

SL. No.	Categorization of TBI	Diagnosis system	Description
1.	Different Clinical indices based on the severity of the injury	GCS neurologic injury severity scale for adults	This is a 15-point scale which evaluates respiratory expansion, respiratory rate, capillary refill, systolic blood pressure, and based on the observation, a comprehensive score is calculated. Patient with GCS score less than 8 diagnosed as severe
		Brussells Coma Grade	Correlation with symptoms: ($r=-0.216/P=0.0001$)
		Grady coma scale	A scale used to rate the level of consciousness, with five grades corresponding to confusion (Grade I), stupor (Grade II), deep stupor (Grade III), decerebrate rigidity (Grade IV), and coma (Grade V)
		Innsbruck Coma Scale	A score-based evaluation after evaluating certain factors like reaction to acoustic stimuli, eye movement and response, pupil size, pain reaction, body posture, position and movement of the eyeball, and automatisms of oral cavity
		FOUR score Scale	Coma scale based on 4 components (brainstem reflexes, eye response, respiration pattern, and motor response)
		Hunt and Hess Scale	Correlation with symptoms: $r=-0.197$ and $P=0.0001$
		World Federation of Neurosurgeons Score	Correlation with symptoms: $r=-0.185$ and $P=0.0001$
		Glasgow Coma Score Compared	Correlation with symptoms: ($r=0.236/P=0.0001$)

Continued

Table 1 Categorization of TBI (Saatman et al., 2008)—cont'd

SL. No.	Categorization of TBI	Diagnosis system	Description
2.	Pathoanatomic type, used most often to describe injuries for acute management. This kind of classification is based on the location and anatomical abnormalities that are targeted for the treatment	Scalp laceration and contusion Skull fracture Epidural hemorrhage Hemorrhage of Subdural region Hemorrhage of Subarachnoid region laceration and brain contusion Hemorrhage of intraparenchymal region Hemorrhage of intraventricular parts Axonal injury and Focal injury patterns	Severity of the injury is evaluated based on the report of various diagnostic tools like CT scan, MRI, and different imaging techniques. Based on the patient condition and severity, various treatment options can be implemented. However, sometimes, combination of some of the different pathoanatomic conditions makes the treatment options more complicated
3.	Physical mechanism (involves the force of the impact leading to the injury)	High speed impact Gunshot Blast Hitting on head Sudden fall	The severity, direction and the magnitude, and the extent can give a detailed idea about the damage

shock-related event. Depending on the severity of the impact, immediate hospitalization is recommended to establish the condition to the normal. Symptomatic treatments should be started based on the medical condition of the patient.

After hospitalization, a series of diagnosis-based treatments can be induced; for example, artificial ventilation, osmotherapy, sedation, craniectomy, and many more condition-based therapies are recommended.

Currently, various classes of drugs such as anti-depressants, anti-epileptics, anti-psychotics, etc. are being used to prevent the secondary impact of the primary injury. Among these available treatment options, antibiotics and antioxidant admiration are also widely recommended to restore the patient to normal condition. However, the exact neuroprotective mechanisms of antibiotics are still unclear. Thus, we have searched the literature and compiled the relevant information in the below sections.

Use of antibiotics in TBI

Antibiotics are widely used in the prevention and treatment of bacterial infections. Apart from their antibacterial actions, some of antibiotics have also shown neuroprotective actions in experimental TBI models. In the present situation of the COVID-19 pandemic, we are using antibiotics such as doxycycline, azithromycin, etc. as preventive and treatment strategies, not due to their antibacterial action but due to their antiviral actions also. The antibiotics belonging to β-lactam are generally recommended in TBI to accelerate glutamate clearance. There is a prominent role of glutamate in an auto-initiating cascade of neuronal injury and subsequent death of neuronal cells and resulting changes which results in an imbalance between the situation of excitatory and inhibitory responses, which leads to incidents like seizures and prolonged mnemonic, behavioral and cognitive changes weeks or months after the injury. The important reports related to the use of antibiotics in TBI and related events are compiled in Table 2. The individual antibiotics used against TBI are discussed below.

Table 2 Role of antibiotics in TBI.

SL. No.	Drug studied	Model studied	Major findings	References
Antibiotics in TBI and related events				
1	Ceftriaxone	Primary human astrocytes	Ceftriaxone promotes excitatory amino acid transporter-2 expression along with glutamate uptake in primary human astrocytes. It also increases EAAT2 mRNA levels inducing EAAT2 promoter activity through nuclear factor-κB (NF-κB) signaling pathway	Lee et al. (2008)
2	Ceftriaxone	Rodents and human cultured astrocytes	ceftriaxone in in-vitro model can induce Nrf2 and xCT which elicit neuroprotective action	Lewerenz et al. (2009)
3	Ceftriaxone	Mouse models of ischemic injury, motor neuron degeneration, and model of the fatal disease ALS	Glutamate transporter expression is increased to offer neuroprotection	Rothstein et al. (2005)
4.	Ceftriaxone	TBI rodent model	Stabilize GLT-1 expression in the lesional cortex to near normal levels and reduce posttraumatic astrogliosis, reduce seizure frequencies	Goodrich et al. (2013)

Continued

Table 2 Role of antibiotics in TBI—cont'd

SL. No.	Drug studied	Model studied	Major findings	References
5	Ceftriaxone	Cortical impact injured rats	Post-traumatic cerebral edema and inflammatory cytokine expression is reduced	Wei et al. (2012)
6	Amoxicillin	Alcohol-preferring rat model	In mesocorticolimbic regions, a sharp fall in ethanol intake and increase in GLT-1 expression along with AKT phosphorylation were observed	Goodwani, Rao, Bell, and Sari (2015)
7	Sulbactam	Ischemic rat model	Prevents CA1 pyramidal cells from ischemic cell death, helps to increase GLT-1 expression	Cui et al. (2015)
8	Doxycycline	Weight drop model in rats	significantly attenuated TBI altered behavioral, oxidative stress and inflammatory parameters	Rana, Singh, Deshmukh, and Kumar (2020)

Ceftriaxone

Ceftriaxone belongs to third-generation cephalosporin groups of antibiotics which are available in the market in injection form for the treatment of a wide range of bacterial infections like lower respiratory tract infections (LRTI), urinary tract infections (UTI), meningitis, etc. Various reports have indicated that ceftriaxone crosses the blood-brain barrier (BBB) and is useful in the treatment of bacterial meningitis (Dwivedi et al., 2009; Prasil, Buchta, Paterova, & Hanovcova, 2010). The neuroprotective action of ceftriaxone has also been reported in the literature through upregulation of glutamate transporter 1 (Gunduz, Oltulu, & Ulugol, 2011; Hameed et al., 2019; Mimura et al., 2011; Nicholson, Gilliland, & Winkelstein, 2014). Cui et al. (2014) have also reported the neuroprotective effect of ceftriaxone in the experimental TBI model through upregulation of GLT-1 and reduction in neuronal autophagy. After TBI, the treatment with ceftriaxone also reduces gliosis as well as post-traumatic seizures (Goodrich et al., 2013). In some cases, ceftriaxone enhances GLT-1 expression in hypoxic and ischemic brain injury and thereby positively impacts the cell viability of the hypoxic hippocampal region (Jagadapillai, Mellen, Sachleben, & Gozal, 2014). Intracisternal ceftriaxone also reduces the posttraumatic downregulation of GLT-1 after hemorrhage of the subarachnoid part, which helps to decrease extracellular glutamate accumulation and thereby enhance memory function, and prevents neuronal death of the hippocampal part. Overall, it can be concluded that ceftriaxone might be useful in TBI through upregulation of GLT-1, reduction in neuronal autophagy, and reduction in gliosis.

Amoxicillin

Amoxicillin is a wide spectrum β-lactam antibiotic effective against not only Gram-positive organisms but also effective against enterococci, haemophilus, and listeria. It is available in the market in the form of tablets as well as injections. It is tolerated well by most individuals. This antibiotic could also be useful in TBI. However, reports on amoxicillin in TBI are very less. Goodwani et al. (2015) have shown an increase in GLT-1 expression in the major part of nucleus accumbens, which in turn can enhance glutamate clearance and prevent neural damage in TBI. The exact role of amoxicillin in TBI is still unclear; thus further experimental studies are required to confirm its safety and efficacy as a neuroprotective agent in TBI.

Amoxicillin and clavulanic acid or sulbactam

As beta lactamase degrades the β-lactam antibiotics including amoxicillin, the activity of these antibiotics is increased when used in combination with beta-lactamase inhibitors like sulbactam or clavulanic acid. Thus, Goodwani et al. (2015) have also tested the combination of amoxicillin and clavulanate and reported the upregulation of GLT-1 expression which are down regulated by ethanol in the part of nucleus accumbens and in the major portion of the medial prefrontal cortex of alcohol-preferring rats. Recently, Alasmari, Alhaddad, Wong, Bell, and Sari (2020) have reported the attenuation of neuroinflammation in chronically exposed animals to ethanol after treatment with amoxicillin and sulbactam. The sulbactam treatment also prevents the reduction of GLT-1 expression in cerebral ischemia and restores CA1 pyramidal cells from neuronal damage and thereby delays ischemia mediated neuronal death (Cui et al., 2015). In the literature, reports are available regarding the neuroprotective effect of amoxicillin alone and in combination with clavulanic acid or sulbactam; however, reports on neuroprotective effects of these antibiotics specifically in TBI are very few. Overall, these antibiotics alone and in combination with beta lactamase inhibitors could be promising through the upregulation of GLT-1, which results in the reduction of glutamate as shown in Fig. 1.

Doxycycline

Doxycycline is a well-known bacteriostatic tetracycline antibiotic used in the treatment of a wide range of bacterial infections. Few reports are available in the literature, which has indicated the neuroprotective role of doxycycline in TBI. Wang, Xue, Jiao, Liu, and Wang (2012) have reported the protective effect of doxycycline against focal cerebral ischemia-reperfusion injury in rats. Rana et al. (2020) have shown the neuroprotective effect of doxycycline through anti-inflammatory properties. The possible neuroprotective mechanism of antibiotics against TBI through inhibition of inflammation is presented in Fig. 2.

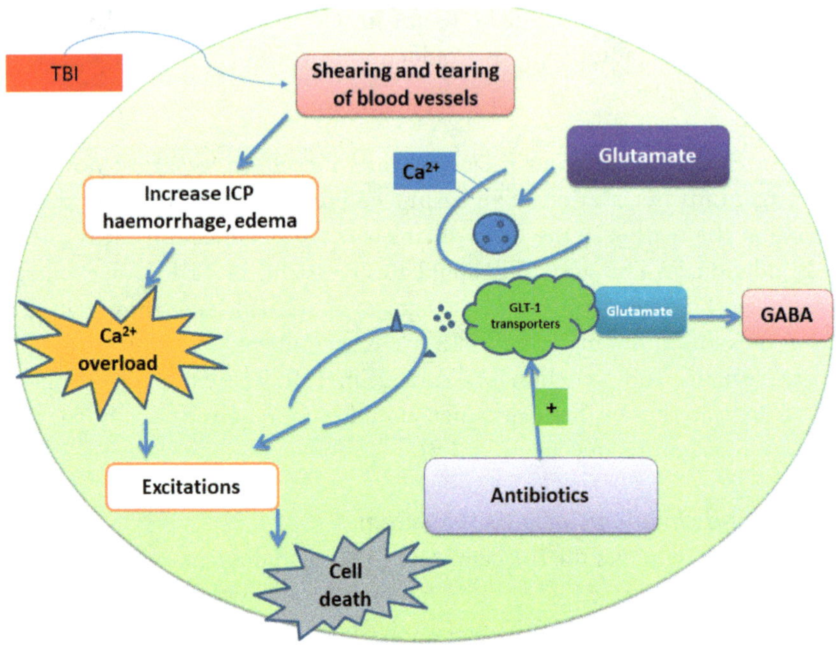

Fig. 1 Possible mechanism of antibiotics against TBI through upregulation of GLT-1 transporters.

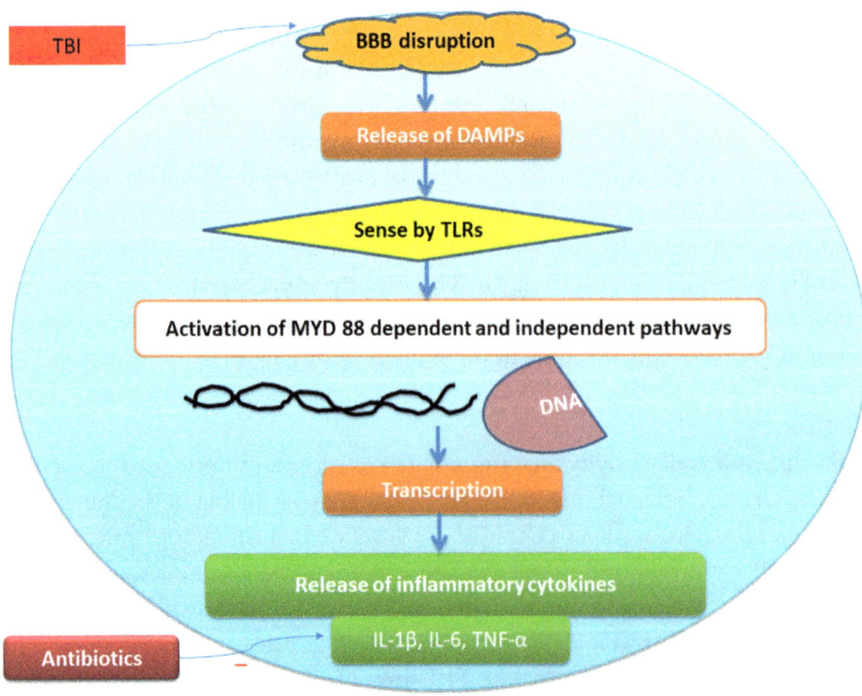

Fig. 2 Possible mechanism of antibiotics against TBI through inhibition of inflammatory signaling pathways.

Minocycline

Minocycline is another tetracycline antibiotic used against a wide range of bacterial infections. Elewa, Hilali, Hess, Machado, and Fagan (2006) have recommended the minocycline for short-term neuroprotection in acute neurologic injuries. The possible neuroprotective mechanism of minocycline is its anti-inflammatory, anti-apoptotic as well as anti-proteases activities. Further studies are required to investigate the exact role of minocycline in TBI.

Reactive oxygen species and antioxidant therapy

After mild or severe brain injury, the oxidative stress of the system plays an important role in the secondary sequences, and the observed variations enhance the complexity of clinical injury management. Formation of reactive nitrogen species (RNS) and reactive oxygen species (ROS) is a preliminary event in post injury, which occurs within few minutes of the incident. A prime factor in this event is lipid peroxidation, which is induced by peroxynitrite (PN). The formation of PN and lipid peroxidation damage protein functions and neuronal membrane lipids irreversibly, which hinders ion homeostasis, leading to mitochondrial respiratory failure and damage of microvascular parts. Antioxidant therapy acts by inhibition and/or scavenging free radicals associated with peroxynitrite, superoxide, or carbonyl compounds.

The strategies of antioxidant therapy can be classified into two major categories: (A) The molecules which act by inhibiting or scavenging the ROS and RNS prior to lipid peroxidation (LP) initiation, (B) the molecules which act by inhibiting the propagation of LP by inhibiting the lipid radicals and/or alkoxyl radicals. The first group of antioxidants possesses a narrow therapeutic window which sometimes fails to arrest the initial explosive production of post-traumatic free radical species which triggers the free radical chain reactions of oxidative damage. The second group of antioxidants acts by inhibiting the LP propagation chain reactions which act on neural cell membranes to block lipid-free radical interaction along with closely associated polyunsaturated fatty acids. Antioxidants are more promising and can be used with other therapies to enhance the treatment efficacy (Mona & Edward, 2012). The role of antioxidants in TBI and related events is summarized in Table 3, and a possible mechanism is presented in Fig. 3.

Antioxidants used in TBI

Dexamethasone

Dexamethasone can effectively reduce vasogenic pericontusional edema in TBI. In a recent study, Moll et al. (2020) reported that after treatment with dexamethasone for 10 days, there was a sharp reduction of edema volume, an increase in the fractional anisotropy value, and a reduction in the apparent diffusion coefficient value in TBI patients

Table 3 Role of antioxidants in TBI.

SL. No.	Drug studied	Model studied	Major findings	Reference
Antioxidants in TBI and related events				
1	Dexamethasone	Affected patients	Reduce cerebral edema, improve spinal cord function, synaptic transmission, and motor neuron excitability	Reulen and Schurmann (1972)
2	Methyl prednisolone	Cat spinal cord experimental model	Prevents free-radical chain reactions and lipid peroxidation in high intravenous administration	Hall and Braughler (1982)
3	Tirilazad	Phase III human clinical trial	Prevents lipid peroxidation Chain reaction by membrane stabilization and lipid free radical scavenging activity. However, not showed significant impact in severely injured TBI patients	Marshall et al. (1998)
4	U-83836E is a second-generation lazaroid	Mouse model	U-83836E decreases post traumatic lipid peroxidation and protein nitration. Respiratory function of mitochondria is preserved	Mustafa, Singh, Wang, Carrico, and Hall (2010)
5	Melatonin	Immature rats	Increases antioxidant levels in brain, improves cognitive function and reduces NF-kappaB activation	Ozdemir et al. (2005)
6	D-Penicillamine	Isolated rat brain mitochondria	Protect against PN-induced mitochondrial respiratory dysfunction, reduce 4-HNE levels	Singh, Sullivan, and Hall (2007)
7	Hydrazine compounds	Ex vivo spinal cord injury model	Protects compression and acrolein-mediated injuries	Hamann, Nehrt, Ouyang, Duerstock, and Shi (2008)
8	Tempol	Mouse traumatic brain injury model	Neuroprotection, peroxynitrite-derived free radicals scavenging activity	Deng-Bryant, Singh, Carrico, and Hall (2008)

Table 3 Role of antioxidants in TBI—cont'd

SL. No.	Drug studied	Model studied	Major findings	Reference
9	Resveratrol	Rat model of spinal cord injury	Promote neurologic and histopathologic improvements, reduce expression inflammatory cytokines and pro-apoptotic proteins, xanthine oxidase and nitric oxide free radicals reduce tissue lesion volume	Liu et al. (2011)
10	Cyclosporin A	Male Wistar rats	Reduce secondary mitochondrial dysfunction, increase survival, and ameliorate brain damage	Li, Kristian, He, and Siesjo (2000)
11	Tocopherol	Weight Drop Model in Rats	significantly attenuated TBI altered oxidative stress and inflammatory parameters	Rana et al. (2020)

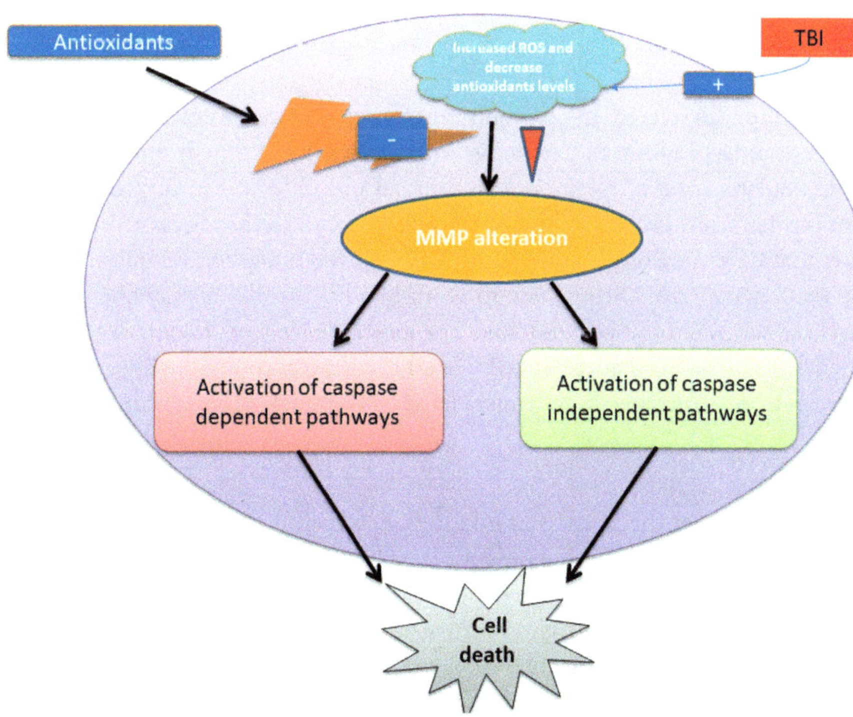

Fig. 3 Possible mechanism of antioxidants against TBI.

with pericontusional vasogenic edema. Glucocorticoid steroids like dexamethasone are efficient to reduce post-traumatic lipid peroxidation because of their lipophilic nature which allows them to invade through the cell membrane and effectively inhibit peroxide chain reaction propagation in the phospholipid bilayer and membrane lipids of the cell (Hall & Braughler, 1981, 1982).

Methyl prednisolone

Saul, Ducker, Salcman, and Carro (1981) carried out a clinical study on 100 TBI patients who were admitted within 6 h of injury and they were treated with placebo and methylprednisolone 250 mg bolus, followed by 125 mg every 6 h for 7–10 days of treatment. They observed improvement in Glasgow coma scale at 72-hour post treatment and improvement in Glasgow outcome scale scores at 6 months. Methyl prednisolone acts by inhibiting free-radical reactions and LP after a high dose intravenous administration. It also shows neuroprotective effects and efficiently inhibits posttraumatic ischemia of the spinal cord. It actively decreases calcium overload and supports aerobic energy metabolism in the intracellular region and reduces neurofilament loss mediated by calpain (Braughler & Hall, 1982). However, in a recent report, Hoshide, Cheung, Marshall, Kasper, and Chen (2016) showed that their corticosteroids do not have a significant benefit in the TBI population.

Lazaroid

Lazaroids were discovered by Upjohn researchers which were having a group of 21-aminosteroid compounds. Lazaroids are highly effective neuroprotective antioxidants. Particularly, one of the lazaroids, tirilazad mesylate was most promising among all from the Lazaroid group (Cahill & Hall, 2017). Lazaroids possess a very specialized structure that bears an alpha-tocopherol ring which is bonded to multiple amine groups. This special characteristic helps them to work in a bi-directional way like lipid peroxidation chain reaction inhibitor and lipid free radical scavenger. Lazaroids can inhibit an early initiation of events that are linked with the secondary injury pathways and provide neuroprotection at multiple biochemical levels (Hall, McCall, & Means, 1994).

Tirilazad

Tirilazad comes under the amino steroid category which is required in small dose to initiate the treatment. Unlike methyl prednisolone and dexamethasone, it showed very less steroid side effects like gastrointestinal bleeding, wound infections, and delayed healing (Bracken et al., 1990). This non-glucocorticoid lipid peroxidation inhibitor acts by inhibiting the peroxidation reaction chain propagation by membrane stabilization and scavenging peroxide free radicals. However, Roberts (2000) reported that the death risk in patients treated with tirilazad and with placebo was almost identical. Even, the risk of death and severe disability in patients treated with tirilazad and placebo was again almost similar. He concluded that there is no evidence that proves the beneficial effect of the routine use of aminosteroids in the management of TBI.

D-Penicillamine

D-Penicillamine covalently binds with peroxide-derived aldehydes and by that elicits carbonyl scavenging activity and have neuroprotective activity as researched in experimental trauma animal models. D-Penicillamine protects against mitochondrial respiratory dysfunction, decreases 4-HNE levels, and supports neurological recovery in the mouse concussive head injury model. (Hall, Kupina, & Althaus, 1999; Singh et al., 2007).

Melatonin

Melatonin is a hormone, made by the pineal gland, produced in the brain and binds to the MEL-specific receptors on neuronal membranes in multiple regions of the brain (Osier et al., 2018). Melatonin is highly lipophilic and can easily pass BBB with less side effects. Melatonin acts by showing free radicals scavenging activity and it has an indirect effect on the endogenous antioxidant enzyme expression (Reiter, Carneiro, & Oh, 1997). Wang et al. (2019) investigated the protective effects of ramelteon, a melatonin MT1/MT2 receptor agonist, and its probable mechanisms of action. Ramelteon (10 mg/kg) was administered daily at 10:00 a.m. which alleviated early brain damage on day 3 and long-term neurobehavioral outcomes on day 28 in C57BL/6 mice. Melatonin also helps to decrease lipid peroxidation, preserve the neuronal structure, and increase functional recovery after the injury. Melatonin helps to decrease NF-kappaB activation and increase brain antioxidant levels, and also helps to boost cognitive function in experimental animal models. Combined therapy with melatonin and dexamethasone, melatonin with minocycline also showed increased antioxidant level and histopathological or cognitive functional recovery in experimental animals (Beni, Kohen, Reiter, Tan, & Shohami, 2004; Kelso, Scheff, Scheff, & Pauly, 2011).

Tempol

Tempol is a stable membrane-permeable nitroxide, it has 1 h to several days treatment tolerance to achieve locomotor recovery and mitochondrial protection capacity (Hillard et al., 2004). Tempol helps to reduce post-traumatic lipid peroxidation and nitration-induced oxidative damage which generally originate from protein degradation. It helps to preserve mitochondrial bioenergetics, helps in the reduction of calpain-mediated cytoskeletal damage, mitochondrial impairment, and prevents neurodegeneration. It also decreased the chances of post-traumatic brain edema and improved neurological recovery in experimental animals (Deng-Bryant et al., 2008; Zhang et al., 1998).

Resveratrol

Resveratrol, which is having a polyphenolic structure, is widely available in grapes and red wine. This molecule is reported to help minimize oxidative stress, improve events of post-injury edema, intra and extra cellular ion mobility and promote neural recovery. Resveratrol also helps to decrease the expression of MDA, pro-apoptotic proteins, and inflammatory cytokines (Ates et al., 2006; Liu et al., 2011; Yang & Piao, 2003). In a recent study, Shi, Qiu, Xiao, Cheng, and Zhang (2018) reported that administration of resveratrol attenuated cognitive deficits of TBI rats. They reported that ROS generation and apoptosis in the brain of TBI rats were decreased markedly by the treatment. The resveratrol treatment also mediated the activation of the p38/Nrf2/HO1 signaling cascade.

Hydrazine

Hydrazine (—NH—NH2)-containing compounds like hydralazine, phenelzine, and iproniazid can interact with the carbonyl end of acrolein or 4-HNE, to prevent the reactive molecule from binding to amino acids part of proteins. This pathway mechanism helps widely in neuroprotection.

However, since hydralazine is a potent vasodilator, it is difficult to administer the molecule to a patient with TBI as severe hypotension is a major impact-oriented symptom with the affected individual. Other molecules with a hydrazine moiety, like iproniazid and phenelzine, do not hinder blood pressure as hydralazine, and can be a good alternative than the previously mentioned molecules (Galvani, Coatrieux, Elbaz, et al., 2008; Hamann et al., 2008).

Antibiotics in combination with antioxidants

Treatment with doxycycline alone and in combination with tocopherol can attenuate locomotor activity significantly. Such a treatment mode can improve memory recognition and can efficiently decrease neuroinflammation. Such combined therapy is efficient to preserve oxidative balance, and the neurotransmitter levels are well restored. The underlying mechanism of neuronal activity restoration might be triggered by free radical scavenging and anti-inflammatory activity of those molecules. The role of antibiotics in combination with other drugs is compiled in Table 4.

Concluding remarks

Traumatic brain injury (TBI) affects millions of people worldwide, specifically young adults. The current treatment options are symptomatic and not able to reduce secondary complications effectively. Thus, there is a need for new drugs against TBI. However, the

Table 4 Combination therapy in TBI and related events.

SL. No.	Drug studied	Model studied	Major findings	Reference
Combination therapy in TBI and related events				
1	Amoxicillin with clavulanate	Alcohol-preferring rat model	Reduce ethanol intake and increase GLT-1 expression, AKT phosphorylation in mesocorticolimbic regions	Goodwani et al. (2015)
2	Melatonin and dexamethasone	Experimental model of mouse SCI	Significant anti-inflammatory effects and tissue and motor recovery	Genovese et al. (2007)
3	Melatonin and minocycline	Rat CCI model	Failed to demonstrate promising histopathological or cognitive functional recovery, however, it was interpreted that treatment initiation, dosing, and duration played a major role in the negative outcome	Kelso et al. (2011)
4	Doxy with tocopherol	Weight drop model in rats	significantly attenuated TBI altered behavioral, oxidative stress and inflammatory parameters as well as altered neurotransmitter levels	Rana et al. (2020)

development of new chemical entities is a long and expensive process. The use of existing molecules against TBI could be a most promising approach due to the availability of data. Recent experimental studies have shown promising neuroprotective effects of different classes of antibiotics through various mechanisms in TBI models. The sufficient evidence is available regarding the involvement of oxidative stress signaling pathways in TBI. Thus, the combined administration of specific antibiotics (which have ability to cross BBB) with antioxidants could be a desirable combination in the treatment of TBI.

Applications to other neurological conditions

In this chapter, we have reviewed the preclinical and clinical evidence regarding the role of antibiotics alone and in combination with antioxidants against TBI. The neuroprotective role of selected antibiotics is due to upregulation of GLT-1, reduction in neuronal autophagy, gliosis, and inflammation, whereas antioxidants reduce the activation of undesired oxidative stress signaling pathways. The down regulation of GLT-1, increase in neuronal autophagy, gliosis, inflammation as well as increase in oxidative stress are observed in almost every neurodegenerative disease like Alzheimer's, Parkinson's, Huntington's, Amyotrophic lateral sclerosis, etc. Thus, this approach might be also helpful in neurodegenerative diseases. However, further studies are required to confirm the exact role of specific antibiotic alone and in combination with specific antioxidants.

Key facts

- The young generation is mostly affected by TBI due to speeding vehicle accidents followed by old age people who experience multiple falls because of age-related loss of muscle control, fear, slip, etc.
- Injury to brain could result in damage of blood-brain barrier (BBB), hemorrhage, increased intracranial pressure (ICP), decrease cerebral blood flow (CBF), etc.
- The treatment of primary injury is required as per the severity.
- It has been observed that most of the individuals experience many undesired symptoms after recovery from primary injury.
- Thus, primary injury could result in the activation of various undesired signaling pathways.
- Activation of oxidative stress signaling pathways is one of the undesired activated pathways after injury.
- Thus, antioxidants could be helpful in TBI.
- Antibiotics are used in the treatment of various types of bacterial infections.
- Emerging studies have shown beneficial effects of some antibiotics in TBI through different mechanisms.
- Thus, specific combination of antibiotics and antioxidants could help in the treatment of TBI associated neurological complications.

Summary points

- Antibiotics can reduce primary injury-induced secondary complications.
- Antibiotics could reduce the excess glutamate by increasing the expression of GLT-1 transporters in nearby astrocytes.
- Antibiotics could also inhibit TBI induced inflammatory signaling pathways.
- Antioxidants are also useful in TBI by inhibiting the activation of oxidative stress signaling pathways.
- Antibiotics in combination with antioxidants could be a useful combination against TBI.

Acknowledgments

Author, Anoop Kumar, is thankful to Vice-Chancellor, Professor (Dr.) R.K. Goyal for constant support and motivation for the completion of this work.

References

Alasmari, F., Alhaddad, H., Wong, W., Bell, R. L., & Sari, Y. (2020). Ampicillin/Sulbactam treatment modulates NMDA receptor NR2B subunit and attenuates neuroinflammation and alcohol intake in male high alcohol drinking rats. *Biomolecules*, *10*(7), 1030.

Ates, O., Cayli, S., Altinoz, E., Gurses, I., Yucel, N., Kocak, A., et al. (2006). Effects of resveratrol and methylprednisolone on biochemical, neurobehavioral and histopathological recovery after experimental spinal cord injury. *Acta Pharmacologica Sinica*, *27*, 1317–1325.

Beni, S. M., Kohen, R., Reiter, R. J., Tan, D. X., & Shohami, E. (2004). Melatonin-induced neuroprotection after closed head injury is associated with increased brain antioxidants and attenuated late-phase activation of NF-kappaB and AP-1. *The FASEB Journal*, *18*, 149–151.

Bracken, M. B., Shepard, M. J., Collins, W. F., Holford, T. R., Young, W., Baskin, D. S., et al. (1990). A randomized, controlled trial of methylprednisolone or naloxone in the treatment of acute spinal-cord injury. Results of the second national acute spinal cord injury study. *The New England Journal of Medicine*, *322*, 1405–1411.

Braughler, J. M., & Hall, E. D. (1982). Correlation of methylprednisolone levels in cat spinal cord with its effects on (Na+ + K+)-ATPase, lipid peroxidation, and alpha motor neuron function. *Journal of Neurosurgery*, *56*, 838–844.

Butcher, I., McHugh, G. S., Lu, J., et al. (2007). The prognostic value of cause of injury in traumatic brain injury: Results from the IMPACT study. *Journal of Neurotrauma*, *24*, 281–286.

Cahill, L., & Hall, E. D. (2017). Is it time to resurrect "lazaroids"? *Journal of Neuroscience Research*, *95*(1–2), 17–20.

Cui, C., Cui, Y., Gao, J., Sun, L., Wang, Y., Wang, K., et al. (2014). Neuroprotective effect of ceftriaxone in a rat model of traumatic brain injury. *Neurological Sciences*, *35*, 695–700.

Cui, X., Li, L., Hu, Y. Y., Ren, S., Zhang, M., & Li, W. B. (2015). Sulbactam plays neuronal protective effect against brain ischemia via upregulating GLT1 in rats. *Molecular Neurobiology*, *51*, 1322–1333.

Deng-Bryant, Y., Singh, I. N., Carrico, K. M., & Hall, E. D. (2008). Neuroprotective effects of tempol, a catalytic scavenger of peroxynitrite-derived free radicals, in a mouse traumatic brain injury model. *Journal of Cerebral Blood Flow and Metabolism*, *28*, 1114–1126.

Dwivedi, V. K., Chaudhary, M., Soni, A., Yadav, J., Tariq, A., Siddiqui, M. R., et al. (2009). Diffusion of sulbactomax and ceftriaxone into cerebrospinal fluid of meningitis induced rat model. *International Journal of Pharmacology*, *5*, 307–312.

Elewa, H. F., Hilali, H., Hess, D. C., Machado, L. S., & Fagan, S. C. (2006). Minocycline for short-term neuroprotection. Pharmacotherapy. *The Journal of Human Pharmacology and Drug Therapy*, *26*, 515–521.

Finfer, S. R., & Cohen, J. (2001). Severe traumatic brain injury. *Resuscitation*, *48*, 77–90.

Galvani, S., Coatrieux, C., Elbaz, M., et al. (2008). Carbonyl scavenger and antiatherogenic effects of hydrazine derivatives. *Free Radical Biology and Medicine*, *45*, 1457–1467.

Genovese, T., Mazzon, E., Crisafulli, C., Esposito, E., Di Paola, R., Muia, C., et al. (2007). Effects of combination of melatonin and dexamethasone on secondary injury in an experimental mice model of spinal cord trauma. *Journal of Pineal Research*, *43*, 140–153.

Goodrich, G. S., Kabakov, A. Y., Hameed, M. Q., Dhamne, S. C., Rosenberg, P. A., & Rotenberg, A. (2013). Ceftriaxone treatment after traumatic brain injury restores expression of the glutamate transporter, GLT-1, reduces regional gliosis, and reduces post-traumatic seizures in the rat. *Journal of Neurotrauma*, *30*, 1434–1441.

Goodwani, S., Rao, P. S., Bell, R. L., & Sari, Y. (2015). Amoxicillin and amoxicillin/clavulanate reduce ethanol intake and increase GLT-1 expression as well as AKT phosphorylation in meso-corticolimbic regions. *Brain Research*, *1622*, 397–408.

Gunduz, O., Oltulu, C., & Ulugol, A. (2011). Role of GLT-1 transporter activation in prevention of cannabinoid tolerance by the beta-lactam antibiotic, ceftriaxone, in mice. *Pharmacology Biochemistry and Behavior, 99*, 100–103.
Gururaj, J. (2013). Epidemiology of traumatic brain injuries: Indian scenario. *Neurological Research, 24*, 24–28.
Hall, E. D., & Braughler, J. M. (1981). Acute effects of intravenous glucocorticoid pretreatment on the in vitro peroxidation of cat spinal cord tissue. *Experimental Neurology, 73*, 321–324.
Hall, E. D., & Braughler, J. M. (1982). Glucocorticoid mechanisms in acute spinal cord injury: A review and therapeutic rationale. *Surgical Neurology, 18*, 320–327.
Hall, E. D., Kupina, N. C., & Althaus, J. S. (1999). Peroxynitrite scavengers for the acute treatment of traumatic brain injury. *Annals of the New York Academy of Sciences, 890*, 462–468.
Hall, E. D., McCall, J. M., & Means, E. D. (1994). Therapeutic potential of the lazaroids (21-aminosteroids) in acute central nervous system trauma, ischemia and subarachnoid hemorrhage. *Advances in Pharmacology, 28*, 221–268.
Hamann, K., Nehrt, G., Ouyang, H., Duerstock, B., & Shi, R. (2008). Hydralazine inhibits compression and acrolein-mediated injuries in ex vivo spinal cord. *Journal of Neurochemistry, 104*(3), 708–718.
Hameed, M. Q., Hsieh, T. H., Morales-Quezada, L., Lee, H. H., Damar, U., MacMullin, P. C., et al. (2019). Ceftriaxone treatment preserves cortical inhibitory interneuron function via transient salvage of GLT-1 in a rat traumatic brain injury model. *Cerebral Cortex, 29*, 4506–4518.
Hillard, V. H., Peng, H., Zhang, Y., Das, K., Murali, R., Etlinger, J. D., et al. (2004). Tempol, a nitroxide antioxidant, improves locomotor and histological outcomes after spinal cord contusion in rats. *Journal of Neurotrauma, 21*, 1405–1414.
Hillier, S., Hiller, J., & Metzer, J. (1997). Epidemiology of traumatic brain injury in South Australia. *Brain Injury, 11*, 649–659.
Hoshide, R., Cheung, V., Marshall, L., Kasper, E., & Chen, C. C. (2016). Do corticosteroids play a role in the management of traumatic brain injury? *Surgical Neurology International, 7*, 84.
Jagadapillai, R., Mellen, N. M., Sachleben, L. R., Jr., & Gozal, E. (2014). Ceftriaxone preserves glutamate transporters and prevents intermittent hypoxia-induced vulnerability to brain excitotoxic injury. *PLoS One, 9*, e100230.
Kelso, M. L., Scheff, N. N., Scheff, S. W., & Pauly, J. R. (2011). Melatonin and minocycline for combinatorial therapy to improve functional and histopathological deficits following traumatic brain injury. *Neuroscience Letters, 488*, 60–64.
Langlois, J. A., Rutland-Brown, W., & Thomas, K. E. (2006). *Traumatic brain injury in the United States: Emergency department visits, hospitalizations, and deaths*. Atlanta: Centers for Disease Control and Prevention, National Center for Injury Prevention and Control.
Lee, S. G., Su, Z. Z., Emdad, L., Gupta, P., Sarkar, D., Borjabad, A., et al. (2008). Mechanism of ceftriaxone induction of excitatory amino acid transporter-2 expression and glutamate uptake in primary human astrocytes. *Journal of Biological Chemistry, 283*, 13116–13123.
Lewerenz, J., Albrecht, P., Tien, M. L., Henke, N., Karumbayaram, S., Kornblum, H. I., et al. (2009). Induction of Nrf2 and xCT are involved in the action of the neuroprotective antibiotic ceftriaxone in vitro. *Journal of Neurochemistry, 111*, 332–343.
Li, P. A., Kristian, T., He, Q. P., & Siesjo, B. K. (2000). Cyclosporin a enhances survival, ameliorates brain damage, and prevents secondary mitochondrial dysfunction after a 30-minute period of transient cerebral ischemia. *Experimental Neurology, 165*, 153–163.
Liu, C., Shi, Z., Fan, L., Zhang, C., Wang, K., & Wang, B. (2011). Resveratrol improves neuron protection and functional recovery in rat model of spinal cord injury. *Brain Research, 1374*, 100–109.
Marshall, L. F., Maas, A. I., Marshall, S. B., Bricolo, A., Fearnside, M., Iannotti, F., et al. (1998). A multicenter trial on the efficacy of using tirilazad mesylate in cases of head injury. *Journal of Neurosurgery, 89*(4), 519–525.
Mimura, K., Tomimatsu, T., Minato, K., Jugder, O., Kinugasa-Taniguchi, Y., Kanagawa, T., et al. (2011). Ceftriaxone preconditioning confers neuroprotection in neonatal rats through glutamate transporter 1 upregulation. *Reproductive Sciences, 18*(12), 1193–1201.

Moll, A., Lara, M., Pomar, J., Orozco, M., Frontera, G., Llompart-Pou, J., et al. (2020). Effects of dexamethasone in traumatic brain injury patients with pericontusional vasogenic edema. *Medicine, 99*(43), e22879.

Mona, B., & Edward, D. H. (2012). Antioxidant therapies in traumatic brain and spinal cord injury. *Biochimica et Biophysica Acta, 1822*, 675–684.

Mustafa, A. G., Singh, I. N., Wang, J., Carrico, K. M., & Hall, E. D. (2010). Mitochondrial protection after traumatic brain injury by scavenging lipid peroxyl radicals. *Journal of Neurochemistry, 114*(1), 271–280.

Nicholson, K. J., Gilliland, T. M., & Winkelstein, B. A. (2014). Upregulation of GLT-1 by treatment with ceftriaxone alleviates radicular pain by reducing spinal astrocyte activation and neuronal hyperexcitability. *Journal of Neuroscience Research, 92*(1), 116–129.

Osier, N., McGreevy, E., Pham, L., Puccio, A., Ren, D., Conley, Y. P., et al. (2018). Melatonin as a therapy for traumatic brain injury: A review of published evidence. *International Journal of Molecular Sciences, 19*(5), 1539.

Ozdemir, D., Uysal, N., Gonenc, S., Acikgoz, O., Sonmez, A., Topcu, A., et al. (2005). Effect of melatonin on brain oxidative damage induced by traumatic brain injury in immature rats. *Physiological Research, 54*, 631–637.

Prasil, P., Buchta, V., Paterova, P., & Hanovcova, I. (2010). Penetration of ceftriaxone into the cerebrospinal fluid and its relationship to inflammatory markers during bacterial meningitis. *Klinickamikrobiologieainfekcnilekarstvi, 16*(2), 64–72.

Rana, A., Singh, S., Deshmukh, R., & Kumar, A. (2020). Pharmacological potential of tocopherol and doxycycline against traumatic brain injury-induced cognitive/motor impairment in rats. *Brain Injury, 34*(8), 1039–1050.

Rana, A., Singh, S., Sharma, R., & Kumar, A. (2019). Traumatic brain injury altered normal brain signaling pathways: Implications for novel therapeutics approaches. *Current Neuropharmacology, 17*(7), 614–629.

Reiter, R. J., Carneiro, R. C., & Oh, C. S. (1997). Melatonin in relation to cellular antioxidative defense mechanisms. *Hormone and Metabolic Research, 29*, 363–372.

Reulen, H. J., & Schurmann, K. (Eds.). (1972). *Steroids and brain edema*. Berlin: Springer-Verlag Inc.

Roberts, I. (2000). Aminosteroids for acute traumatic brain injury. *Cochrane Database of Systematic Reviews*, CD001527. https://doi.org/10.1002/14651858.CD001527 (Online).

Rothstein, J. D., Patel, S., Regan, M. R., Haenggeli, C., Huang, Y. H., Bergles, D. E., et al. (2005). Beta-lactam antibiotics offer neuroprotection by increasing glutamate transporter expression. *Nature, 433*, 73–77.

Saatman, K. E., Duhaime, A. C., Bullock, R., Maas, A. I., Valadka, A., Manley, G. T., et al. (2008). Classification of traumatic brain injury for targeted therapies. *Journal of Neurotrauma, 25*(7), 719–738.

Saul, T. G., Ducker, T. B., Salcman, M., & Carro, E. (1981). Steroids in severe head injury: A prospective randomized clinical trial. *Journal of Neurosurgery, 54*, 596–600.

Shi, Z., Qiu, W., Xiao, G., Cheng, J., & Zhang, N. (2018). Resveratrol attenuates cognitive deficits of traumatic brain injury by activating p38 signaling in the brain. *Medical Science Monitor: International Medical Journal of Experimental and Clinical Research, 24*, 1097–1103.

Singh, I. N., Sullivan, P. G., & Hall, E. D. (2007). Peroxynitrite-mediated oxidative damage to brain mitochondria: Protective effects of peroxynitrite scavengers. *Journal of Neuroscience Research, 85*, 2216–2223.

Tagliaferri, F., Compagnone, C., Korsic, M., Servadei, F., & Kraus, J. (2006). A systematic review of brain injury epidemiology in Europe. *Acta Neurochirurgica, 148*, 255–268.

Wang, J., Jiang, C., Zhang, K., Lan, X., Chen, X., Zang, W., et al. (2019). Melatonin receptor activation provides cerebral protection after traumatic brain injury by mitigating oxidative stress and inflammation via the Nrf2 signaling pathway. *Free Radical Biology & Medicine, 131*, 345–355.

Wang, Z., Xue, Y., Jiao, H., Liu, Y., & Wang, P. (2012). Doxycycline-mediated protective effect against focal cerebral ischemia-reperfusion injury through the modulation of tight junctions and PKCδ signaling in rats. *Journal of Molecular Neuroscience, 47*, 89–100.

Wei, J., Pan, X., Pei, Z., Wang, W., Qiu, W., Shi, Z., et al. (2012). The beta-lactam antibiotic, ceftriaxone, provides neuroprotective potential via anti-excitotoxicity and anti-inflammation response in a rat model of traumatic brain injury. *The Journal of Trauma and Acute Care Surgery, 73*, 654–660.

Yang, Y. B., & Piao, Y. J. (2003). Effects of resveratrol on secondary damages after acute spinal cord injury in rats. *Acta Pharmacologica Sinica, 24*, 703–710.
Zhang, R., Shohami, E., Beit-Yannai, E., Bass, R., Trembovler, V., & Samuni, A. (1998). Mechanism of brain protection by nitroxide radicals in experimental model of closed head injury. *Free Radical Biology & Medicine, 24*, 332–340.

CHAPTER 10

Vitamin D level and the outcome after traumatic brain injury

Ajay Choudhary[a], Ashok Kumar[a], Aakanksha Choudhary[b], and Kaviraj Kaushik[a]
[a]Department of Neurosurgery, Atal Bihari Vajpayi Institute of Medical Sciences, Dr. R.M.L. Hospital, New Delhi, India
[b]Institute of Home Economics, University of Delhi, New Delhi, India

Abbreviations

ACTH Adrenocorticotropic hormone
GCS Glasgow Coma Scale
GH growth hormone
PTSD post-traumatic stress disorder
TBI traumatic brain injury
TNFα tumor necrosis factor α
TSH thyroid stimulating hormone

Introduction

Dearth of vitamin D in traumatic brain injury patients is a common and lethal condition for which early detection and treatment are known to improve the survival and quality of life (Toman et al., 2017). Vitamin D deficiency in head injury is a common condition which can be a major cause of mortality and morbidity among the younger population worldwide. The incidence of vitamin D deficiency in TBI patients is snowballing and is associated with impaired cognitive function and psychiatric illness. Among TBI survivors, 30% to 50% of the patients develop one or more endocrinopathies, in which 35% to 40% possess some degree of hypopituitarism, which is associated with serious morbidity along with an estimated 20% experience a reduction in life expectancy. As an impact of TBI, alteration in vitamin D levels results in a cascade of neurobehavioral changes, physical disability, and endocrine function impairment (HOPE TBI, 2015). Due to the impairment of the pituitary gland after the injury, the afflicted patients may develop vitamin D deficiency. Since there are limited data and evidence available on the therapeutic role of vitamin D in predicting the outcomes in neurocritical patients after TBI, this review highlights the clinical co-relation of vitamin D with severity of traumatic brain injury and subsequent improvement in quality of life of the patient (Anjum, Jaffery, Fayyaz, Zarak, & Anjum, 2018).

Vitamins and Minerals in Neurological Disorders
https://doi.org/10.1016/B978-0-323-89835-5.00009-0

Vitamin D

Vitamin D is an important micronutrient required for the growth of the skeleton and regulation of calcium homeostasis. It is a secosteroid, primarily synthesized in the skin in the presence of sunlight and plays a dual role in skeletal escalation and neurodevelopment (Brown, Dusso, & Slatopolsky, 1999). Vitamin D acts as a key biomarker in predicting the outcome of recovery in TBI individuals. Clinical subjects with TBI have shown progressive vitamin D paucity and impaired neurocognitive functioning during indoor stay which could be due to the lack of sunlight exposure. Calcitriol, the active form of vitamin D, regulates the levels of calcium and phosphate in the body (Socci, Valvasone, Cecchi, & Bonizzoli, 2014). Vitamin D, its metabolites, and other neurosteroids bind to vitamin D receptors located in different regions of the brain, namely the hippocampus, thalamus, hypothalamus, cerebellum, and basal ganglia (Tabish & Syed, 2015). Thus, vitamin D has been considered a vital marker for predicting the extent of outcome in cognitive and neurophysiological dysfunction. It is noteworthy that its deficiency negatively affects the recovery and quality of life in TBI individuals (Rosenfeld, Maas, Bragge, Morganti-Kossmann, & Manley, 2012) (Table 1 and Fig. 1).

Traumatic brain injury and neuroendocrine disturbance

In current years, neuroendocrine assessments have emerged as a popular tool in predicting the extent of morbidity and mortality in individuals with traumatic brain injury. TBI results in endocrine disturbances and reduces the integrity of the blood-brain barrier (BBB) and thereby worsens the neuroinflammatory response. However, there is a small correlation between TBI severity and the level of endocrine impairment. A mild TBI may have significant endocrinopathies, while a moderate-to-severe TBI was shown to have minor to no endocrine abnormalities (Stocchetti & Zanier, 2016).

Table 1 Studies on clinical prevalence of vitamin D deficiency in patients after TBI.

Author/year	Injury severity	Units	Total sample size	Vitamin D insufficient	Vitamin D deficient
European Society of Endocrinology (2010)	Moderate to Severe	*n* (%)	90	0 (0%)	58 (65%)
Socci et al. (2014)	Moderate to Severe	*n* (%)	46	0 (0%)	34 (73.9%)
Jamall et al. (2016)	Moderate to Severe	*n* (%)	353	163 (74.5%)	119 (72.6%)
Toman et al. (2017)	Mild to Severe	*n* (%)	124	29 (23%)	(34%)

Value stated as *n* (%).

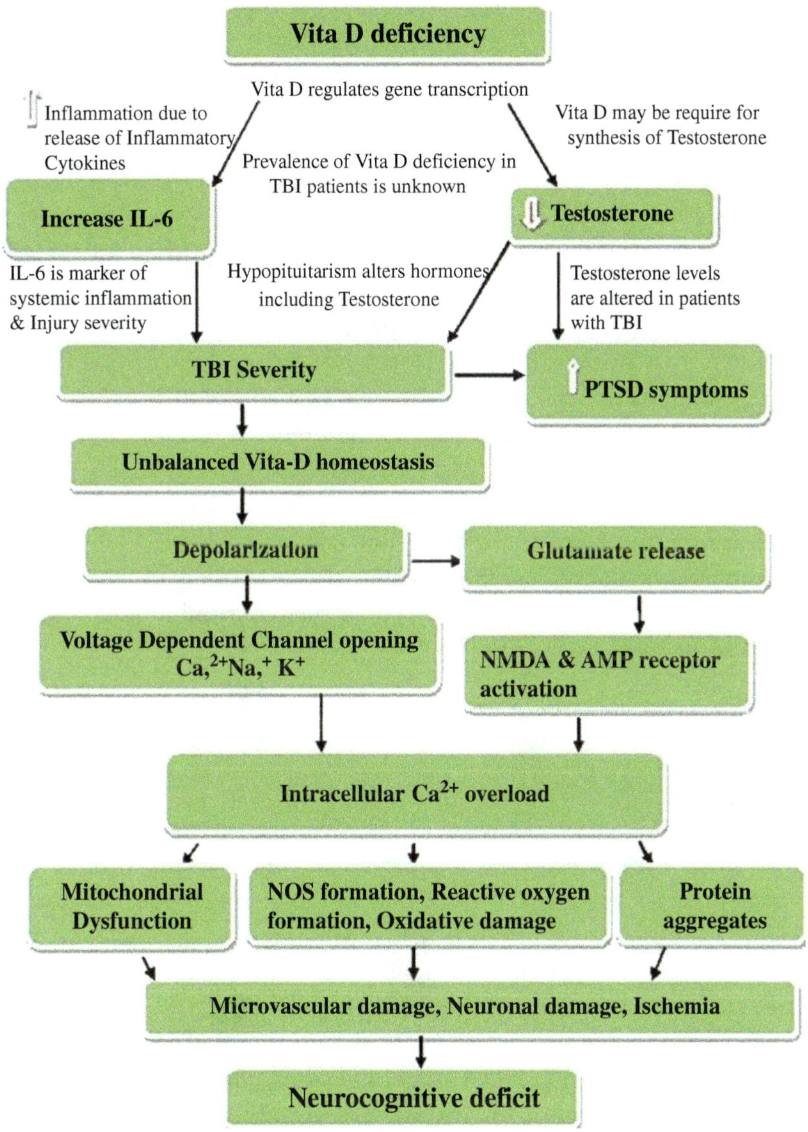

Fig. 1 Vitamin D role in post-traumatic neurocognitive illness. Role of vitamin D in regulation of inflammatory response and androgen synthesis associated with TBI severity with post-traumatic stress disorder and neurocognitive impairment.

In chronic cases of TBI, endocrine deficiency and pituitary malfunction are becoming more apparent (Ulutabanca, Hatipoglu, Tanriverdi, & Gökoğlu, 2014). Majority of the afflicted patients may develop endocrine deficiency as an immediate and delayed effect that may precipitate even months or years after the injury. Persistent vitamin D deficiency evokes baseline inflammation even prior to injury. After injury, this acute inflammatory phase response leads to an increased cell death and DNA damage, indicating a more severe secondary injury process after the initial injury even with progesterone therapy (Wentz, Eldred, Henry, & Berry-Caban, 2014). Neuroinflammation in TBI triggers neuroglial cells to initiate the phosphorylation of mitogen-activated protein kinase (*MAPK*) pathways, Jun-N-terminal (*JNK*) *MAPK*, extracellular signal-regulated kinase (*ERK*), *p38-MAPK*, and nuclear transcription factors (NTF) cascade to upregulate cellular activities like cellular inflammation and mitosis. In TBI patients, these pathways have been considered crucial for the synthesis of pro-inflammatory mediators and associated detrimental effects in brain. Thus, an apparent hyperactivation of microglial cells may exacerbate neuronal impairment via upregulation of glial pathways in injured individuals (Wehr, Pilz, Boehm, März, & Obermayer, 2010). Experimental evidence suggests that post-injury treatment with vitamin D along with progesterone decreases brain edema, attenuates free radical damage, and reduces neuronal loss in TBI animal models. In addition, vitamin D reduces the inflammatory cytokines, and nitric oxide (NO) attenuates neurological abnormalities after ischemia (Jennifer & Bess, 2008). Rate of endocrinopathies among TBI patients depicts a high prevalence. However, the cause of endocrine abnormalities after TBI is still not well understood. In acute phase of injury, these abnormalities remain serious but can be treated to avoid further complications of brain defacement (Eserian, 2013).

Vitamin D deficiency and androgen synthesis in TBI

Various studies have shown that vitamin D plays an important role in the regulation of mineral metabolism. The vitamin D receptor and enzymes regulating vitamin D expression have been reported in the synthesis of androgen in both the genders. The association between 25-hydroxy vitamin D, testosterone levels, sex-hormone binding globulin, and assessment of homeostatic insulin resistance model (HOMA-IR) in women surprisingly explains the favorable effect of vitamin D on fertility process (Smolders, Moen, Damoiseaux, Huitinga, & Holmoy, 2011). The association between vitamin D and testosterone has been clearly explained in both the sexes. Low levels of testosterone have been elucidated in conditions like cognitive disorders, Type-2 diabetes mellitus, hypertension, obesity, insulin resistance, and other metabolic problems. Therefore, the level of testosterone produced is directly proportional to the quality of life. Poor vitamin D level affects the testosterone synthesis, potentially lowers down the muscle mass, causes fatigue, which indirectly affects the performance of an individual along with elevated risk of post-traumatic stress disorder (PTSD) (Gallagher & Sai, 2010; Yang & Xu, 2007) (Table 2).

The optimal levels of vitamin D aids overcome the inflammatory response and testosterone levels which thereby cut down the risk for PTSD and associated neurocognitive disorder (Cekic, Saveed, & Stein, 2009).

Table 2 Clinical studies on association of neurocognitive disorder and neuroendocrine abnormalities with TBI.

Author/year	Sample size (*n*)	mean age group (years)	TBI (severity)	Neurocognitive dysfunction (%), Neuroendocrine abnormalities (%)	Follow up time (months)
Gil, Caspi, Zilberman, Koren, and Klein (2005)	120	31.4	Mild	Post-traumatic stress disorder (14.3%)	3 months
Tanriverdi et al. (2006)	49	35.9	Mild to Severe	Acute phase and at 12 months-TSH deficiency (5.8%, 5.8%), Gonadotropin deficiency (41.6%, 7.7%), ACTH deficiency (9.8%, 19.2%), GH deficiency (20.4%, 37.7%)	0–12 months
Schneider, Kreitschmann, Ghigo, Stalla, and Agha (2007)	246	36.2	Moderate to Severe	Hypopituitarism (21%), Total, partial and isolated Hypopituitarism (1%, 2% and 18%)	12 months
Klose, Watt, Brennum, and Rasmussen (2007)	104	41	Mild to Severe	ACTH deficiency (5%), TSH deficiency (2%), ADH deficiency (2%)	12 months
Xydakis, Robbins, and Grant (2008)	2525	30	Mild to Moderate	Loss of consciousness (4.9%), Altered mental status (10.3%), 43.9% met criteria for post-traumatic stress disorder	3–4 months
Behan, Phillips, Thompson, and Agha (2008)	26	11.47	Mild to Moderate	Pituitary impairment (61%), GH deficiency (42%)	30 months
Schneider et al. (2007)	90	37.4	Mild to Moderate	Pituitary hormone deficiency (29%), GH deficiency (24%), Gonadal hormone deficiency (10%)	0–12 months
Ghaffarzadegan, Ebrahimvandi, and Jalali (2016)	756	31.4	Mild	Post-Traumatic Stress Disorder (8% to 16%)	Data base series of 168 months

Tables show the clinical studies on neurocognitive disorder and neuroendocrine discrepancies with TBI.

Role of vitamin D in neuroprotection

Vitamin D plays a vital role in bone and tissue growth. The wide distribution of vitamin D receptors and enzymes (*1-alpha O Hase*) highlighted that vitamin D is associated with cognitive function and dementia. Most of the studies showed that vitamin D has a unique role in neurodevelopment and also possesses neuroprotective effects on dopaminergic cascade in the adult brain (Xydakis et al., 2008). Vitamin D regulates the expression levels of tyrosine hydroxylase and dopaminergic receptors. Since vitamin D with other nutrients reduces the inflammatory response (cytokines and nuclear factor kappa-light-chain-enhancer of activated B cells (NF_kB) in the body, its therapy becomes crucial in patients recovering from TBI (Matthews et al., 2012).

The inflammatory response is usually elevated in the post-concussion period. Vitamin D guards the brain damage and reduces the pro-inflammatory cytokine and NF_kB pathways. Vitamin D with other nutrients helps in lowering down the neuronal excitotoxicity of glutamate and aids in inhibition of reactive oxygen levels in the nerve cells (Aminmansour et al., 2012).

It is noticeable that vitamin D therapy exhibits shielding effect in TBI patients. Therefore, the vitamin can be labeled to be protective against neurodegenerative processes. Since vitamin D has favorable effect on dopamine and serotonin cascade in the brain, we can conclude that it plays a vital role in regulating the mood and energy levels in the body. Vitamin D works synergistically with other anti-inflammatory drugs to prevent secondary brain damage by restricting the inflammatory process. It acts as a neuroprotector that makes the neurons defiant to stress, hypothermia, ischemia, hypoglycaemia, hyperglycaemia, hypotension, and hypertension (Matthews et al., 2012). Vitamin D produces a neuroprotective action by modulating inflammatory cytokines, dipping oxidative stress, regulating concentration of neural calcium, and enhancing nerve conduction mechanism. Vitamin D in 25 (OH) D form easily surpasses the blood-brain barrier and is accessible in the central nervous system. Vitamin D targets neural genes and works as a gene transcription regulator which influences a variety of behavioral and biochemical processes. Recent clinical study revealed that vitamin D as a monotherapy reduces inflammation and prevents neuronal damage (Crista, 2018). In addition, it also exhibits synergistic effects when combined with progesterone (PROG) and improves the outcome in Patients with TBI (Jennifer & Bess, 2008). One more study highlighted that vitamin D hormone ameliorates key factors in pain severity, mobility, and routine activities in TBI patients (Brown et al., 1999).

Paucity of vitamin D in association with severity of TBI

Incidence of vitamin D deficiency is widespread among TBI patients. It is closely associated with impaired cognitive function and more severe depressive symptoms (Anjum et al., 2018). In acute injury, damage to the neurons initiates a cascade of neurodegenerative process, imbalance in cellular calcium homeostasis, tissue acidosis, and reactive oxygen species production (Aminmansour et al., 2012).

In acute injury, chronic vitamin D deficiency leads to a more intense pro-inflammatory type-1 reaction, which could exacerbate the processes of secondary damage to the initial traumatic impact.

Macrophage and microglial activity and TH1 response lead to the production of varied free radical species, which leads to oxidative stress and contribute to secondary injury (Bramlett & Dietrich, 2004). There is a direct co-relation between TBI severity and levels of vitamin D (Sharma et al., 2020). Individuals with vitamin D deficiency are found to have been more prone to severe head injury as low levels of vitamin D are insufficient to overcome an anti-inflammatory and neuroprotective response after the injury (Klose et al., 2007).

Vitamin D deficiency is commonly seen in the general population as well as in critically ill patients (Matthews et al., 2012). A study was conducted to determine the relationship between vitamin D levels in patients admitted to ICU and associated mortality. However, the author could not find any clear and possible conclusion pertaining to a relation between vitamin D levels in patients when they get admitted and when they get discharged from the hospital after the injury. The index of recovery in individuals with TBI is usually determined by the difference between the admission, Glasgow Comma Scale (GCS), and shifting of patients from ICU (Intensive care unit) to normal wards (Aminmansour et al., 2012). Though paucity of vitamin D is linked to a poor outcome in ICU patients, low vitamin D levels at the time of ICU admission showed a marked rise in the mortality rate, thus prolonging ICU stay (Atalan & Güçyetmez, 2017) (Table 3).

Applications to other areas of neurological conditions

Vitamin D deficiency is found to be associated with various diseases and is a risk factor for conditions with compelling morbidity. VDD is the most accepted cause of rickets in children and osteomalacia in adults. Vitamin D deficiency is widely associated with complications including muscle weakness, skeletal pain, and increased risk factor for fracture and skeletal abnormality, i.e., "squadron" and skull; "craniotabes." Vitamin D has also been found to be related with complex immune modulating system action, (Villamor, 2006) that has been linked with cardiovascular disease, which plays an important role in multiple types of cancer, and many studies have been related with other conditions including multiple sclerosis, preeclampsia, rheumatoid arthritis, diabetes mellitus, and dementia (Balvers, Brouwer-Brolsma, Endenburg, de Groot, et al., 2015).

The prevalence of vitamin D deficiency has been found in neuropsychiatric illnesses that include depressive disorders, Alzheimer's disease, Schizophrenia, Parkinson's disease, cognitive disorder, dementia, and autism. For these observations, reversal causation of brain autism has not been abandoned till date (Kraus et al., 1986).

Data from another prospective cohort studies revealed the reasons behind the cognitive deterioration in older adults more likely to be vitamin D deficient. Some prospective

Table 3 Clinical studies on the role of vitamin D in determining the outcome after CNS injury.

Author/year	Sample size (*n*)	Time to testing	Vitamin D status	GCS score	Vitamin D Treatment	Association with TBI severity	Outcome	Follow up (months)
Matthews et al. (2012)	3	8–24 h	17–29 ng/mL	3–9	Vitamin D3 (50,000 IU), progesterone (20 mg), omega-3 fatty acids (2 g (Loveza), and enteral glutamine (20 g) up to 6 weeks	Temporal subdural hematomas (SDH), left frontal subarachnoid hemorrhage (SAH), diffuse axonal injury (DAI), memory deficient	GCS-12–15, Extubate-9-12 days, improved memory	3
Aminmansour et al. (2012)	20	0–8 h	10–29 ng/mL	3–8	Progesterone (1 mg/kg), Vitamin D (5 μg/kg) up to 5 days	Brain Trauma and DAI	GCS-7–15, Improved GOS	3

Matthews et al. (2012), Aminmansour et al. (2012) showed vitamin D role in central nervous injury.

studies have also claimed the same type of evidence regarding the association of defined vitamin D insufficiency to enhance the risk of Alzheimer's disease and vascular dementia during 30 years of follow-up. Unsuccessful to note baseline cognitive data or to apply goals tools to judge cognitive capabilities has restricted the importance of the observations (Toffanello, Coin, Perissinotto, Zambon, et al., 2014).

Other component of interest

The eventual approach of TBI research is to avoid, analyze, and manage traumatic brain injuries in various situations. Many important parameters need to be investigated that include the cascade pathway by which primary injury triggers secondary injury, and the acuteness/specificity of various biomarkers determines the severity of injury. Apart from this, identifying and assessing novel remedies for pharmacological action in protection of TBI and synthesizing decent extrapolation methods to discover pharmacological feedback are necessary clinically. Other than this, the healthcare costs associated with TBI are sky touching in billions of dollars to discover in research and development centre; (Finkelstein, Corso, & Miller, 2006) thus, discovering novel pharmacological targets which may be impactful in the injury pathways evolution has to be the focus of primary research for better TBI management. Moreover, the accurate pathways for secondary injury development, i.e., the pathways of injury and its progression that speeds up in the human brain, are unknown. This non-regulative nature of the injury pathways limits the rationale behind TBI design pharmacotherapies; therefore, clinical studies have been unsuccessful in establishing the treatment's efficacy. All these unsuccessful stories are likely due to several reasons, involving the complexity, heterogeneity, and pathophysiology of TBI. There is a lack of understanding about ideal drugs, dosages, routes of administration, dosing regimens, and/or medication durations; and also there's less understanding of the pharmacokinetics (PK), dose-response association, and therapeutic windows (Margulies et al., 2016). Consequently, there is no "magic bullet" for reducing the progression of secondary brain injury recently.

During TBI, simple concussion and diffuse axonal injury occur due to neuronal mechanical disruption and disruption of cerebral micro-circulation that is the result of primary injury, which can further play a decisive role in the severity index. (Adams, Vinken, & Bruyn, 1975). A very little effort has been made to reduce its direct effects, as may be because of this, it has less interest among clinicians than the forensic experts. Over the last 25 years, "secondary brain insult" has received much attention. During the brain injury, patients who "talked and died," their death may be caused by anything other than devastating primary brain insult (Reilly, Adams, Graham, & Jennett, 1975). They suspected for this to increase intracranial pressure (ICP) as a result of hematoma or brain inflammation, and classified "unnecessary factors" which had caused or associated to the death. Evolution of hematomas undisciplined intracranial infection or epilepsy, systemic hypoxia, and hypercarbia has been potential causes for secondary damage (Gentleman & Jennett, 1981). Vitamin D may have promising role in the prevention or alleviation of secondary brain damage. The neurons that are mechanically injured are more vulnerable to hypoxic-ischemic insult and the intensity and time of the damage decide the amount and feasibility of this damage (Douglas, 1999).

All remedies which are being used for seizures treatment or any other head injury can only avoid secondary injury but not alleviate secondary injury because these medications are restricted in passing or infiltrating the Blood Brain Barrier (BBB).

Conclusion

Traumatic brain injury is a complex pathophysiological process rather than a true pathological injury, which could lead to an alteration in the mental state of an individual (Schneider et al., 2007). Its clinical complication, resulting in a secondary brain injury and symptom diversity, possesses unique challenges in TBI management (Klose et al., 2007). Furthermore, vitamin D deficiency is prevalent in a large segment of the population, especially the elderly and can significantly affect recovery after CNS injury. After TBI, many of the symptoms such as PTSD, psychomotor abilities, and cognitive decline are also associated with vitamin D deficiency (Karsy et al., 2019). Furthermore, these symptoms are associated with endocrine dysfunction and deficiencies of pituitary hormones, such as testosterone and growth hormone, which may be overcome by vitamin D therapy. Vitamin D insufficiency is linked to an altered post traumatic cognitive function and endocrine system abnormalities, owing to its role in regulating inflammatory process and hormone synthesis.

This review offers a new paradigm in the management of traumatic brain injury recovery outcome by highlighting the importance of vitamin D status in pre and post TBI and associated neurological outcome. Conventional therapy of TBI is limited to the use of anticonvulsants, antibiotics, anti-inflammatory agents, etc. However, the use of vitamin D along with these agents may improve clinical outcomes in TBI patients. Preliminary studies have indicated that vitamin D holds substantial potential to reduce neuro-inflammation and neurodegenerative process (Sharma et al., 2020; Tang et al., 2013). However, further studies are still required to establish vitamin D as a therapeutic candidate in improving clinical outcomes and quality of life in patients with TBI.

Mini-dictionary terms

Vitamin D

- A secosteroid
- Hormone
- A neuroprotecting agent

Post traumatic cognitive dysfunction

- Pre-Alzheimer's and dementia due to TBI
- Psychiatric impairment after TBI, affecting recovery outcome along with quality of life.

Post TBI complications

- Psychomotor disabilities
- Memory deficits
- Attention deficits
- Cognitive deficits with altered general fluid intelligence (Gf)
- Issues in understanding language
- Altered hearing and visual function

Key facts

Vitamin D deficiency

- Vitamin D deficiency is a comprehensive well-known dilemma with today's physical and mental wellness.
- Alone India has greater than 90% of allegedly healthy Indians and global prevalence rate is 80%–100% with 25(OH) D deficiency levels
- The VDD prevalence in the TBI is enhanced as correlated to the global predominance, and the local predominance noted in Indian region is $\geq$65%
- Vitamin D injection can later lead to painful granuloma, hemorrhage, and infections at the injection site, which further delays its absorption
- Oral route has been considered as the most obvious route of administration due to high patient compliance

Vitamin D in TBI

- Early detection of vitamin D deficiency after traumatic brain injury (TBI) has been established as an important factor in predicting survival rate and quality of life in an individual
- Post-injury treatment with vitamin D decreases brain edema, attenuates free radical damage, reduces neuronal loss in TBI animal models, and reduces the inflammatory cytokines TNF-α, IL-6, and nitric oxide and attenuates neurological abnormalities after ischemia
- Vitamin D3 supplementation in neurocritical care patients is suggested to have post TBI benefits.

Summary points

- Vitamin D deficiency directly influences the rate of recovery in acute TBI patients.
- Paucity of vitamin D in the injury potentiates the process of a secondary brain injury.

- Treatment with vitamin D reduces brain edema, free radical, and neuronal loss in TBI.
- Prolong administration of vitamin D improves neurological outcome in acute TBI patients.

References

Adams, H., Vinken, P. J., & Bruyn, G. (1975). The neuropathology of head injury. *Handbook of Clinical Neurology*, *23*(35).

Aminmansour, B., Nikbakht, H., Ghorbani, A., Rezvani, M., Rahmani, P., Torkashvand, M., et al. (2012). Comparison of the administration of progesterone versus progesterone and vitamin D in improvement of outcomes in patients with traumatic brain injury: A randomized clinical trial with placebo group. *Advanced Biomedical Research*, *1*, 1–5.

Anjum, I., Jaffery, S., Fayyaz, M., Zarak, S., & Anjum, S. (2018). The role of vitamin D in brain health: A mini literature review. *Cureus Journal of Medical Sciences*, *10*, 2960.

Atalan, K., & Güçyetmez, B. (2017). Serum vitamin D level at ICU admission and mortality. *Turkish Journal of Anaesthesiology & Reanimation*, *45*, 193–196.

Balvers, M. G., Brouwer-Brolsma, E. M., Endenburg, S., de Groot, L. C., et al. (2015). Recommended intakes of vitamin D to optimize health associated circulating 25- hydroxyvitamin D concentrations, and dosing regimens to treat deficiency: workshop report and overview of current literature. *Journal of Nutritional Science*, *4*, 23.

Behan, A., Phillips, J., Thompson, C., & Agha, A. (2008). Neuroendocrine disorders after traumatic brain injury. *Journal of Neurology, Neurosurgery, and Psychiatry*, *79*, 753–759.

Bramlett, H. M., & Dietrich, W. D. (2004). Pathophysiology of cerebral ischemia and brain trauma: Similarities and differences. *Journal of Cerebral Blood Flow & Metabolism*, *24*, 133–150.

Brown, J., Dusso, A., & Slatopolsky, E. (1999). Vitamin D. *The American Journal of Physiology*, *277*, 157–175.

Cekic, M., Saveed, I., & Stein, D. G. (2009). Combination treatment with progesterone and vitamin D may be more effective than monotherapy for nervous system injury and disease. *Frontiers in Endocrinology*, *30*, 158–172.

Crista, J. (2018). Endocrine effects of traumatic brain injury. *Journal of Natural Medicines*, *10*, 51–55.

Douglas, G. (1999). Improving outcome after traumatic brain injury—Progress and challenges. *British Medical Bulletin*, *55*(4), 910–926.

Eserian, J. (2013). Vitamin D as an effective treatment approach for drug abuse and addiction. *Medical Hypotheses*, 7, 35–39.

European Society of Endocrinology. (2010). *Vitamin D deficiency associated with chronic fatigue in brain injured patients*. Science Daily. www.sciencedaily.com/releases/2010/04/100427182609.htm.

Finkelstein, E., Corso, P., & Miller, T. (2006). *The incidence and economic burden of injuries in the United States*. New York (NY): Oxford University Press.

Gallagher, J. C., & Sai, A. J. (2010). Vitamin D insufficiency, deficiency, and bone health. *Journal of Clinical Endocrinology and Metabolism*, *95*(6), 2630–2633.

Gentleman, D., & Jennett, B. (1981). Hazards of inter-hospital transfer of comatose head injured patients. *Lancet*, 853–855.

Ghaffarzadegan, N., Ebrahimvandi, A., & Jalali, M. (2016). A dynamic model of post-traumatic stress disorder for military personnel and veterans. *PLoS One*, *10*, 0161405.

Gil, S., Caspi, Y., Zilberman, I., Koren, D., & Klein, E. (2005). Does memory of a traumatic event increase the risk for posttraumatic stress disorder in patients with traumatic brain injury? A prospective study. *The American Journal of Psychiatry*, *162*, 963–969.

HOPE TBI. (2015). *Vitamin D Deficiency post trauma*. Available from: https://hopetbi.com/vitamin-d-deficiency-post-trauma.

Jamall, A., Feeney, C., Zaw-Linn, J., Malik, A., Niemi, E. K., Jimenez, T., et al. (2016). Prevalence and correlates of vitamin D deficiency in adults after traumatic brain injury. *Clinical Endocrinology*, *4*, 636–644.

Jennifer, B., & Bess, H. (2008). Vitamin D and neurocognitive dysfunction. *Molecular Aspects of Medicine*, *29*, 415–422.

Karsy, M., Guan, J., Eli, I., Brock, A. A., Menacho, S. T., & Park, M. S. (2019). The effect of supplementation of vitamin D in neurocritical care patients: Randomized clinical trial of hypovitaminosis D (RECTIFY). *Journal of Neurosurgery*, 1–10.

Klose, M., Watt, T., Brennum, J., & Rasmussen, U. (2007). Posttraumatic hypopituitarism is associated with an unfavorable body composition and lipid profile, and decreased quality of life 12 months after injury. *The Journal of Clinical Endocrinology and Metabolism*, *92*(10), 3861–3868.

Kraus, J. F., Block, M. A., Hessol, L., Ley, P., Rokaw, W., Sullivan, C., et al. (1986). The incidence of acute brain injury and serious impairment in defined population. *American Journal of Public Health*, *119*, 186–201.

Margulies, S., Anderson, G., Atif, F., Badaut, J., Clark, R., Empey, P., et al. (2016). Combination therapies for traumatic brain injury: retrospective considerations. *Journal of Neurotrauma*, *33*, 101–112.

Matthews, L. R., Danner, O. K., Ahmed, Y. A., Dennis-Griggs, D. M., Frederick, A., Clark, C., et al. (2012). Combination therapy with vitamin D3, progesterone, omega-3 fatty acids and glutamine reverses coma and improves clinical outcomes in patients with severe traumatic brain injuries. *International Journal of Case Reports and Images*, *4*(3), 143–149.

Reilly, L., Adams, H., Graham, I., & Jennett, B. (1975). Patients with head injury who talk and die. *Lancet*, 375–377.

Rosenfeld, V., Maas, I., Bragge, P., Morganti-Kossmann, M., & Manley, T. (2012). Early management of severe traumatic brain injury. *Lancet*, *380*, 1088–1098.

Schneider, J., Kreitschmann, I., Ghigo, E., Stalla, K., & Agha, A. (2007). Hypothalamopituitary dysfunction following traumatic brain injury and aneurismal subarachnoid hemorrhage: A systematic review. *Journal of the American Medical Association*, *298*, 1429–1438.

Sharma, S., Kumar, A., Choudhary, A., Sharma, S., Khurana, L., Sharma, N., et al. (2020). Neuroprotective role of oral vitamin D supplementation on consciousness and inflammatory biomarkers in determining severity outcome in acute traumatic brain injury patients: A double-blind randomized clinical trial. *Clinical Drug Investigation*, *40*(4), 327–334.

Smolders, J., Moen, M., Damoiseaux, J., Huitinga, I., & Holmoy, T. (2011). Vitamin D in the healthy and inflamed central nervous system: Access and function. *Journal of Neurological Sciences*, *311*, 37–43.

Socci, F., Valvasone, S., Cecchi, A., & Bonizzoli, M. (2014). Vitamin D level could affect the recovery rate in traumatic brain injury: A retrospective study. *Critical Care*, *18*, 471.

Stocchetti, N., & Zanier, R. (2016). Chronic impact of traumatic brain injury on outcome and quality of life: A narrative review. *Critical Care*, *20*, 148.

Tabish, A., & Syed, N. (2015). Recent advances and future trends in traumatic brain injury. *Emergency Medicine Journal*, *5*, 2165–7548.

Tang, H., Hua, F., Wang, J., Sayeed, I., Wang, X., Chen, Z., et al. (2013). Progesterone and vitamin D: Improvement after traumatic brain injury in middle-aged rats. *Hormones and Behavior*, *64*, 527–538.

Tanriverdi, F., Senyurek, H., Unluhizarci, K., Selcuklu, A., Casanueva, F. F., & Kelestimur, F. (2006). High risk of hypopituitarism after traumatic brain injury: A prospective investigation of anterior pituitary function in the acute phase and 12 months after trauma. *Journal of Clinical Endocrinology and Metabolism*, *91*, 2105–2111.

Toffanello, E. D., Coin, A., Perissinotto, E., Zambon, S., et al. (2014). Vitamin D deficiency predicts cognitive decline in older men and women: The Pro.V.A. study. *Neurology*, *83*(24), 2292–2298.

Toman, E., Bishop, J., Davies, D., Su, Z., Criseno, S., Mason, A., et al. (2017). Vitamin D deficiency in traumatic brain injury and its relationship with severity of injury and quality of Life: A prospective, observational study. *Journal of Neurotrauma*, 1448–1456.

Ulutabanca, H., Hatipoglu, N., Tanriverdi, F., & Gökoğlu, A. (2014). Prospective investigation of anterior pituitary function in the acute phase and 12 months after pediatric traumatic brain injury. *Childs Nervous System*, *30*, 1021–1028.

Villamor, E. (2006). A potential role for vitamin D on HIV infection? *Nutrition Reviews*, *2006*(64), 226–233.

Wehr, E., Pilz, S., Boehm, B., März, W., & Obermayer, B. (2010). Association of vitamin D status with serum androgen levels in men. *Clinical Endocrinology*, *73*, 243–248.

Wentz, L. M., Eldred, J. D., Henry, M. D., & Berry-Caban, C. (2014). Clinical relevance of optimizing vitamin D status in soldiers to enhance physical and cognitive performance. *Journal of Special Operations Medicine*, *14*(1), 58–66.

Xydakis, M., Robbins, A., & Grant, G. A. (2008). Mild traumatic brain injury in U.S. soldiers returning from Iraq. *New England Journal of Medicine*, *358*, 2177–2179.

Yang, L., & Xu, F. (2007). Effect of dipeptide of glutamine and alanine on severe traumatic brain injury. *Chinese Journal of Traumatology*, *10*, 145–149.

CHAPTER 11

Zinc and traumatic brain injury

Sydney M.A. Juan*, Maria Daglas*, Phan H. Truong*, and Paul A. Adlard
Synaptic Neurobiology Laboratory, The Florey Institute of Neuroscience and Mental Health, The University of Melbourne, Melbourne, VIC, Australia

Abbreviations

AD	Alzheimer's disease
AMPA	α-amino-3-hydroxy-5-methyl-4-isoxazolepropionic acid
Ca-EDTA	calcium ethylenediaminetetraacetic acid
CCI	controlled cortical impact
CQ	clioquinol
CTE	chronic traumatic encephalopathy
DNA	deoxyribonucleic acid
FPI	fluid percussion injury
MT	metallothionein
NADPH	nicotinamide adenine dinucleotide phosphate oxidase
NMDA	*N*-methyl-D-aspartate
PD	Parkinson's disease
PP2A	protein phosphatase 2A
ROS	reactive oxygen species
TBI	traumatic brain injury
TPEN	N,N,N′,N′-tetrakis(2-pyridylmethyl) ethylenediamine
WDIA	weight drop impact acceleration
ZIP	zinc/iron transporter protein
ZnT3	zinc transporter 3

Introduction

Traumatic brain injury (TBI) typically occurs when there is an acute or chronic application of force to the head/skull of sufficient magnitude that it results in an injury to the enclosed brain. Such injuries can occur in an acute setting, such as when an individual knocks their head following a fall and experiences a closed-skull injury, through to cases of penetrating brain injury resulting from high-velocity events such as car accidents or military conflict. TBI can also occur in a chronic setting, such as in cases of self-injury behavior and in some contact sports.

Whatever the underlying cause, TBI remains a significant problem within the community and remains the leading cause of death/disability in developed nations, with a

* Equal contribution.

Vitamins and Minerals in Neurological Disorders
https://doi.org/10.1016/B978-0-323-89835-5.00039-9

lifetime incidence rate of ~40% (Gilchrist, Thomas, Xu, McGuire, & Coronado, 2011). In the USA, approximately 2.5 million people receive a diagnosis of TBI each year (Coronado et al., 2012), resulting in >50,000 deaths and often causing lifelong impairments in the survivors (e.g., cognitive deficits, impaired vision or hearing, disrupted motor function, and behavioral issues such as mood and anxiety disorders) (Faul, Xu, Wald, Coronado, & Dellinger, 2010). The associated social and economic impact of TBI is immense, with a lifetime cost per patient of $85,000–$3,000,000 in the USA, resulting in an estimated cost of > $70 billion/year in that country (Humphreys, Wood, Phillips, & Macey, 2013).

One of the significant issues is that TBI remains an unmet clinical need, with there currently being no approved therapeutics that prevent or reverse the symptoms of injury. As such, there is an urgent need for a greater understanding of the mechanistic drivers of the secondary injury cascades that occur following injury and which contribute to the symptomatic presentation of TBI. In this review, we will address the potential role of zinc in TBI.

Pathophysiology of TBI

TBI typically leads to irreversible brain damage such as axonal injury, neuronal death, and hemorrhage (Portbury & Adlard, 2015). Following this primary injury, secondary injury mechanisms occur within hours to weeks and result in cellular consequences, such as neuroinflammation, oxidative damage, excitotoxicity, protein aggregation, and metal dyshomeostasis (Daglas & Adlard, 2018; Portbury & Adlard, 2015). This cascade of events often persists for months to years post-TBI and can lead to patients experiencing cognitive, motor, and psychological impairments that often result in lifelong disability (Faul et al., 2010). Importantly, injury type, and severity will affect clinical outcomes post-TBI, and brain injury is a contributing factor in the development of neurological disorders including Alzheimer's disease (AD) (Djordjevic, Sabbir, & Albensi, 2016), Parkinson's disease (PD) (Delic, Beck, Pang, & Citron, 2020), and chronic traumatic encephalopathy (CTE) (Portbury & Adlard, 2015). A common pathway that may underlie these conditions, and may represent a therapeutic target, is altered zinc homeostasis.

Zinc in physiology

Dietary zinc

Zinc is an essential micronutrient for human development, naturally found in foods such as meats, fish and poultry, and also available as a dietary supplement. The complex form of zinc in food, rather than free ions, affects its bioavailability. In Australia, the National Institute of Health Recommended Dietary Intake is 14 mg of zinc/day for men and 8 mg for women. Under normal physiological conditions, zinc ions released from food

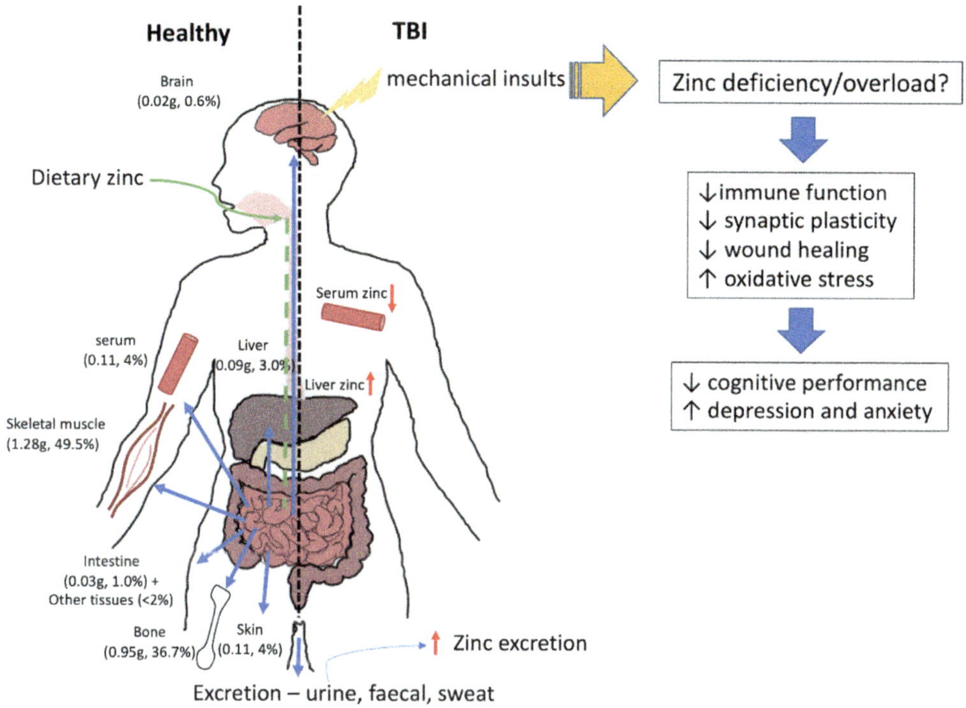

Fig. 1 Zinc absorption in health and TBI. The healthy human adult contains 2–3 g of zinc. Dietary zinc is absorbed in the duodenum and jejunum in the small intestine and distributed to peripheral tissues. Zinc is maintained in the skeletal muscle, bone, skin, serum, liver, intestine, brain, and can be lost through fecal, urine, and sweat excretion. Brain injury can decrease serum zinc, increase the release of zinc in the liver, and upregulate zinc excretion. The shuttling of systemic zinc results in zinc deficiency/overload, implicated in the secondary effects of brain injury such as cognitive/behavioral impairments.

are absorbed in the small intestine mainly in the duodenum and jejunum and are then distributed to peripheral tissues (Fig. 1). Zinc turnover in these tissues is slow, and unlike other micronutrients such as iron, the body has no storage of zinc and is dependent on regular dietary intake (Maares & Haase, 2020). Endogenous zinc is lost through a number of routes, including the feces, urine, and sweat. There are currently no zinc-dependent enzymes that can serve as a marker to specifically and sensitively measure zinc status; and while plasma zinc concentration is currently the most frequently used method for assessing zinc, this is only useful in severe cases of deficiency (Lowe, Fekete, & Decsi, 2009). Metabolic zinc is regulated through the equilibrium of zinc absorption during deficiency and secretion during repletion. Imbalances in nutritional intake and absorption could be an important factor in the pathological mechanisms of several neurological diseases including AD, PD, and TBI (Ayton, Finkelstein, Cherny, Bush, & Adlard, 2013; Portbury & Adlard, 2017).

Zinc in the brain

Zinc is the second most abundant metal in the body/brain and is essential for cognition, immunity, wound healing, DNA/protein synthesis, and synaptic transmission and is an antioxidant (Portbury & Adlard, 2017). It contributes to the transcription of >2000 proteins and is highly concentrated in the synaptic boutons, vesicles (containing 10%–15% of total zinc), and mossy fibers of the hippocampus (Portbury & Adlard, 2017). Zinc homeostasis is regulated by proteins that function to import, transport and export free zinc in the brain. These include metallothioneins (MT), zinc transporters (ZnT) and zinc/iron transporter proteins (ZIPs) (Portbury & Adlard, 2017). Zinc can exist in its free "chelatable" form as Zn^{2+} ions within synaptic vesicles, which can modulate both excitatory and inhibitory synapses, and is therefore essential for healthy neurotransmission and cognitive function (Choi, Hong, Choi, & Suh, 2020).

Implications of altered zinc homeostasis in TBI

Following TBI, neuronal hyperexcitability can cause excessive zinc to be released from synaptic vesicles, which can activate glutamate receptors such as *N*-methyl-D-aspartate (NMDA), kainate and α-amino-3-hydroxy-5-methyl-4-isoxazolepropionic acid (AMPA), and voltage-gated calcium channels (Levenson, 2020). Zinc can act on several other pathways involved in apoptosis/cell death (e.g., NF-κB, p38, and p53), neurogenesis (e.g., IGF-1, Akt, and Erk), inflammation (NF-κB), and promote the generation of reactive oxidative species (ROS) through a protein kinase C-dependent increase in nicotinamide adenine dinucleotide phosphate (NADPH) oxidase (Choi et al., 2020). Zinc can also modulate tau proteins that stabilize microtubules in axons, and tau pathology is associated with several neurodegenerative diseases including CTE that can develop from multiple head injuries (Portbury & Adlard, 2015). Zinc can promote tau hyperphosphorylation by activating kinases that phosphorylate tau (p70S6) and inhibiting phosphatases that dephosphorylate tau such as protein phosphatase 2A (PP2A) (Pei et al., 2006; Sun et al., 2012). Hence, altered zinc levels in the brain can have serious implications in TBI, potentially affecting crucial pathways which can contribute to neuropathology. However, it remains unknown to what extent zinc is beneficial or harmful in TBI, and whether this is influenced by the type, severity, or stage of injury.

Zinc changes following TBI

It has previously been shown that zinc deficiency occurs in TBI patients, as evidenced by significantly lower serum zinc levels (McClain et al., 1986). However, zinc accumulation has been documented in human cortical neurons and in rat hippocampal neurons following TBI (Sun et al., 2013). A preclinical TBI study using the weight drop impact acceleration (WDIA) model in rats found zinc accumulation in injured neurons and an associated deficiency of zinc in presynaptic boutons, indicating a presynaptic release of zinc at the injury site post-TBI (Suh et al., 2000). In another study utilizing the controlled cortical impact (CCI) model, no major changes in total zinc were observed in the cortex and hippocampus post-TBI over 28 days as assessed by laser-ablation inductively coupled plasma mass spectrometry (Portbury, Hare, Sgambelloni, Finkelstein, & Adlard, 2016). However, there was a slight increase in zinc at 7 days in the injured region and a trending accumulation of brain zinc over time, particularly in the hippocampus. Higher levels of zinc were more apparent in the hippocampus of aged mice compared to young adult mice post-TBI (Portbury et al., 2016, 2017). These changes in brain zinc in TBI are summarized in Table 1.

Mechanisms of zinc neurotoxicity and neuroprotection

The neurotoxic effect of zinc has been largely attributed to its excessive release from vesicles (Suh et al., 2000). Supporting this is a fluid percussion injury (FPI) study that found that degenerating hippocampal neurons are associated with increased levels of presynaptic zinc post-TBI. Inhibiting the excitotoxic functions of zinc decreased neuronal damage within the hippocampus (refer to Table 1) (Hellmich et al., 2007). Zinc may also mediate its toxic effects by inducing ubiquitin conjugation of neuronal proteins leading to neuronal injury (Sun et al., 2013). Another potential mechanism of zinc neurotoxicity is through its involvement in tau-regulatory pathways (Sun et al., 2012). Treatment with PBT2 (a copper/zinc chaperone), which increases tau phosphatase proteins and attenuates tau aggregation (Sedjahtera et al., 2018), reduced neuropathology and improved outcomes post-TBI supporting the notion that the enhancement of tau hyperphosphorylation by synaptic zinc can contribute to cell death in TBI (Portbury et al., 2019).

Other sources of zinc that contribute to neuronal damage/death have emerged in recent years, particularly through the use of zinc transporter-3 (ZnT3) knockout mice that lack vesicular zinc. ZnT3 loads zinc into synaptic vesicles at glutamatergic synapses. In a mouse model of subarachnoid hemorrhage, ZnT3 knockout mice exhibited improved behavioral outcomes and fewer apoptotic neurons compared to wildtype mice at 24h (Chen et al., 2018), confirming a detrimental role of zinc in the injured brain. However, other studies using these mice have revealed that zinc can still accumulate in hippocampal neurons and contribute to neuronal damage/death (Lee, Cole, Palmiter, & Koh, 2000). Potential sources of this excess zinc could be extracellular, mitochondrial, or via binding to MT, which are elevated post-TBI (Lee, Kim, Palmiter, & Koh, 2003; Penkowa, Giralt, Thomsen, Carrasco, & Hidalgo, 2001). The neurotoxicity of zinc has been intrinsically linked to MT alterations. For example, MT-3 null mice that experienced acute brain injury by kainate injection to induce seizures or by sodium nitroprusside, showed attenuated accumulation of cytoplasmic zinc, and reduced neuronal death in various brain regions including the hippocampus and thalamus. Double

Table 1 Clinical and preclinical evidence supporting zinc dyshomeostasis after TBI.

Model/patients	Zinc changes	Neuropathology/outcome	Time period (post-TBI)	Zn neurotoxic or protective?	Use of chelator/ chaperone	Study
TBI patients ($n = 26$)	↓ serum zinc ↑ urinary zinc	Correlated with injury severity	Up to 16 days	Possibly protective	No	McClain et al. (1986)
TBI patients ($n = 5$) and rats (WDIA model)	↑ zinc in human cortical neurons and rat hippocampal neurons	↑ ubiquitin-conjugated proteins and neurodegeneration	Up to 72 h (rats)	Neurotoxic	Yes (in rats), protective	Sun et al. (2013)
CCI, mice (young adult and aged)	↑ brain zinc only at day 7 and trending ↑ over 28 days (lesion site). Higher levels in aged mice (mainly hippocampus)	N/A	Up to 28 days	Unknown	No	Portbury et al. (2016, 2017)
FPI model, mice	↑ presynaptic zinc in degenerating neurons	Neuronal degeneration in hippocampus	4 and 24 h	Neurotoxic	No, however, lamotrigine (inhibits glutamate presynaptic release) and nicardipine (blocks voltage-gated calcium channels) protected neurons	Hellmich et al. (2007)
WDIA model, rats	↑ zinc in injured hippocampal neurons ↓ zinc in presynaptic boutons	Neuronal damage in hippocampus	6 and 24 h	Neurotoxic	Yes, protective	Suh et al. (2000)

Table outlining the clinical/experimental studies that have reported a change in zinc levels and its consequence post-TBI. Abbreviations: *N/A*, not applicable; *CCI*, controlled cortical impact; *FPI*, fluid percussion injury; *WDIA*, weight drop impact acceleration.

ZnT3/MT-3 knockout mice had a greater number of spared hippocampal neurons following kainate-induced injury (Lee et al., 2003).

On the other hand, the genetic depletion (ZnT3 knockout mice) and pharmacological targeting of vesicular zinc can also have negative effects on the injured brain as observed by an increase in necrotic and apoptotic neurons at 24 h following TBI (Doering et al., 2010). It has been suggested that the absence of zinc can mediate these effects by inducing ROS leading to oxidative stress, and impairing inflammation, which may be the result of MT changes in the brain after TBI (Penkowa et al., 2001). Hence, this suggests a protective role of zinc in TBI through its antioxidant and immune-modulatory functions, and possibly via promotion of neuronal survival and hippocampal neurogenesis (Choi et al., 2014).

These studies indicate that zinc can be neuroprotective acutely post-TBI but also neurotoxic (summarized in Fig. 2), and this may be dependent on timing or injury severity.

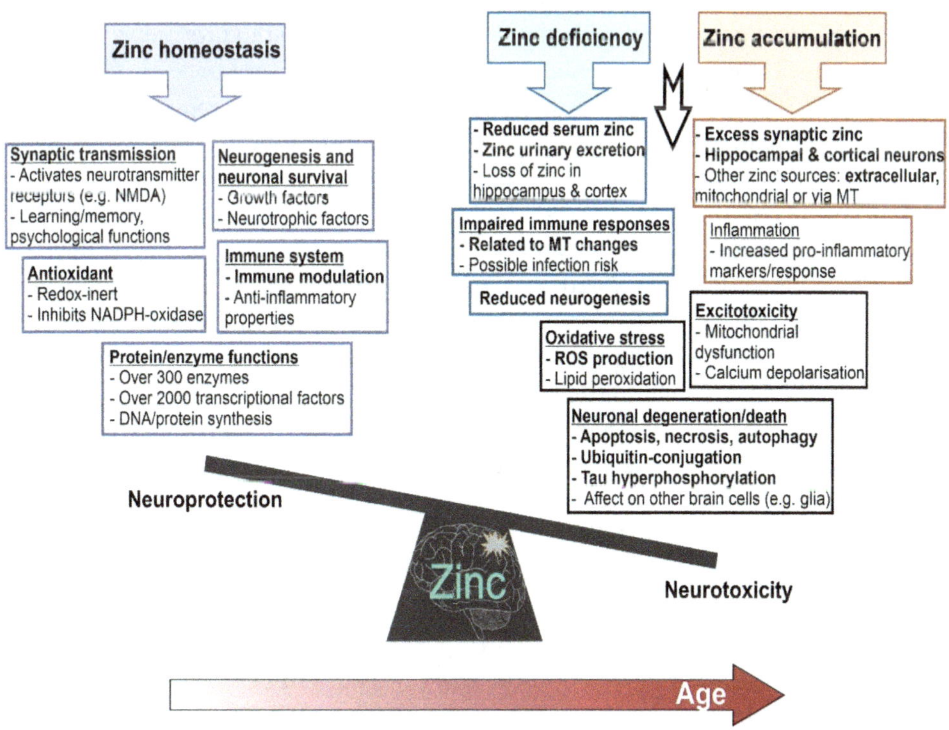

Fig. 2 Neuroprotective and neurotoxic effects of zinc in the brain and in TBI. Schematic depicts the mechanisms involved in zinc-mediated neuroprotection (promoted by zinc homeostasis) and the consequence of altered zinc balance (deficiency/accumulation) in the brain-promoting neurotoxicity. *Bold text* in boxes represents direct/indirect zinc effects reported in brain injury literature. *Red arrow* shows that zinc neurotoxicity increases with age. *MT*, metallothionein; *ROS*, reactive oxygen species; *NMDA*, *N*-methyl-D-aspartate; *NADPH*, nicotinamide adenine dinucleotide phosphate; *DNA*, deoxyribonucleic acid.

This strengthens the notion that both zinc accumulation and deficiency may be harmful to the brain and that a tightly balanced zinc concentration is needed for healthy brain functioning. It is perhaps not surprising that if this balance is tipped (Fig. 2), for example, by various factors such as aging, TBI, and neurodegenerative cascades, that this can exacerbate neurotoxicity and further impact learning/memory. Imbalanced zinc homeostasis, particularly zinc deficiency, has been strongly linked to the development of neurodegenerative diseases (e.g., AD and PD) and neurological conditions such as depression. The therapeutic potential of zinc supplementation or pharmacological targeting of zinc will be discussed in the following sections.

Dietary zinc in TBI

The mechanisms that maintain zinc homeostasis have proven to be pivotal in brain function. The regulation of zinc transport and its dynamics in protein binding and release in the intestinal epithelial cells for cellular distribution remain enigmatic. Possibly due to its regulatory process, where zinc regulates various signaling pathways in conjunction with other micronutrients such as iron, magnesium, calcium, chromium, and copper. Therefore, other dietary constituents can influence the efficiency of zinc absorption within the gastrointestinal environment (Maares & Haase, 2020). Zinc absorbed by enterocytes is either bound to MT or transported across the basolateral membrane via intracellular zinc transporters. When zinc consumption exceeds the normal dietary limit, intestinal MT limits its absorption by binding to excess. In contrast, when dietary zinc is low, MT synthesis is decreased, thus allowing for increased interactions of zinc and intracellular binding proteins. The major isoforms MT-1 and MT-2 are ubiquitously expressed, while MT-3 and MT-4 isoforms are typically found in neurons and stratified squamous epithelium, respectively (Santos, Martinho, Quintela, & Goncalves, 2012). Interestingly, MT-1/2 was found to provide neuroprotection following brain injury (Carrasco et al., 2003; Giralt, Penkowa, Lago, Molinero, & Hidalgo, 2002). Previous reports also demonstrate that the level of brain MT-1/2 expression is increased following TBI (Chung et al., 2004). However, the increase in MT-1/2 is not caused by increased urinary zinc excretion (Ott, McClain, Gillespie, & Young, 1994). Using mice, it was proposed that following brain injury, zinc is released from the liver, thereby causing an upregulation of MT-1/2 expression to restore hepatic zinc back to preinjury levels (Pankhurst et al., 2012). The function of MTs in zinc homeostasis in the brain during TBI and dietary deprivation remains to be fully elucidated.

Zinc deficiency in TBI

Zinc deficiency is a worldwide problem and has been implicated in many disorders of the brain. In childhood, it can cause the retardation of growth and intellectual disabilities

(Prasad, Schulert, Miale, Farid, & Sandstead, 1963). It also adversely affects the immune system, increases inflammation, and exacerbates oxidative stress (Portbury & Adlard, 2017). Zinc supplementation has shown benefits in brain disorders including depression (Ranjbar et al., 2014), ischemic stroke (Aquilani et al., 2009), autism (Vyas, Lee, Jung, & Montgomery, 2020), and AD (Constantinidis, 1992) (detailed in Table 2). In clinical studies, it has been identified that individuals are at risk of development of zinc deficiency postinjury. TBI patients had increased urinary zinc excretion with significantly reduced serum zinc concentrations directly proportional to severity of the trauma (McClain et al., 1986). Using rodent TBI models, it was observed that zinc deficiency can cause significant necrotic cell death both acutely and chronically post-TBI (Yeiser, Vanlandingham, & Levenson, 2002). Indeed, intestinal absorption of dietary zinc plays a crucial role in zinc homeostasis, thus making zinc supplementation a possible therapeutic adjunct for TBI.

Zinc supplementation in TBI

Dietary zinc supplementation has shown beneficial roles both in preclinical and clinical TBI studies. Preclinical TBI work in rats showed that preinjury zinc supplementation for 4 weeks could reduce anxiety, depression-like behaviors, and cognitive deficits following CCI (Cope et al., 2011). The same group also supplemented rats with either a zinc adequate or supplemented diet with some rats also receiving a zinc injection 1-h post-CCI. Animals treated with zinc supplement alone showed improvement in learning and memory deficits, while those given intraperitoneal and dietary zinc supplementation showed a significant reduction in cognitive impairment and depression, thus highlighting the potential importance of pre- and postinjury zinc administration following TBI (Cope et al., 2012).

In the clinic, TBI patients given zinc supplementation showed neurological recovery 2 weeks following treatment, which persisted over the 3-month study. Interestingly, the improved neurological outcomes occurred concomitantly with improved protein metabolism evidenced by increased retinol-binding protein and prealbumin concentrations, suggesting a beneficial effect of zinc supplementation in maintaining visceral protein and neurological functions (Young et al., 1996). Further supporting this, a recent study treated severely head-injured patients with various zinc diets (placebo, 120 mg or 528 mg) and found the 120 mg regime increased plasma zinc levels with promising effects on inflammatory factors and neurological recovery over 28 days postinjury and a nonsignificant improvement in survival rate (Khazdouz et al., 2018). These studies highlight a potential benefit of zinc supplementation for TBI patients (Table 2). However, special consideration should be taken to avoid off-target effects of administering high levels of dietary zinc such as zinc neurotoxicity and its interaction with other metal ions. Therefore, other therapeutic approaches should be explored and are discussed.

Table 2 Selective zinc supplementation studies in brain disorders.

Disorder	Study participants (animal/clinical)	Study description	Zinc dosage/route/duration	Outcome	References
TBI	Rats subjected to CCI (6 weeks old)	Rats supplemented with zinc diet 4 weeks preinjury, behavioral tests conducted 1 week postinjury	Zinc deficient (5 ppm), zinc adequate (30 ppm), or zinc supplemented (180 ppm) diet for 4 weeks preinjury	Increases in adrenal weight reduced depression-like behaviors	Cope, Morris, Scrimgeour, VanLandingham, and Levenson (2011)
TBI	Rats subjected to CCI	Rats subjected to CCI were either given or not given single i.p of zinc 1-h postinjury, then supplemented with zinc diet for 5 days	Group 1: Zinc adequate (30 ppm), zinc supplemented (180 ppm) for 5 days postinjury. Group 2: single i.p zinc (30 mg/kg) 1-h postinjury + zinc adequate diet (30 ppm), zinc supplemented (180 ppm) for 5 days postinjury	No improvement in TBI-induced anxiety for both groups Zinc supplement alone improve cognitive behavior Zinc supplement + i.p zinc group show improved depression-like behaviors	Cope, Morris, Scrimgeour, and Levenson (2012)
TBI	Sixty-eight patients (18–65 years, 13 women, 55 men) with blunt head injuries, 49 patients with diffuse injuries, and 19 focal injuries	Randomized, double-blind. Patients 24 h following admission were given standard and supplemented zinc diet and assessed after 1 month using GCS	Standard zinc (2.5 mg) or zinc supplemented (12 mg) daily for 15 days then oral zinc (168 mg zinc gluconate, 22 mg elemental zinc) for 3 months	Mortality rate reduced by ~50% Higher GCS score on Day 15 continued to Day 28 Higher urine zinc, higher prealbumin, and retinol-binding protein	Young et al. (1996)
TBI	One hundred patients (18–65 years) with severe head trauma	Randomized double-blind clinical trial. Zinc administered 24 h postinjury assessed over 28 days using SOFA, GOS score, and plasma	Zinc sulfate solution supplement (120 mg or 528 mg) or placebo (distilled water) daily for 15 days	↑plasma zinc, ↑urine zinc levels at Days 7 and 16, improved inflammatory factors,↓ SOFA score on Day 16, ↓mortality rate (17% vs. 22%) at Day 28 in 120 mg zinc supplemented group	Khazdouz et al. (2018)

AD	Five presenile AD (56–67 years old) Five senile AD (73–82 years old)	Psychometric tests conducted before and after treatment, no nootropic, or neuroleptic treatment given during the trial	Zinc hydrogen aspartate (50 mg/daily) + i.v zinc (30 mg) in three patients receive 3 months to 1 year	Eight patients reported improved memory, understanding, communication, and social interaction	Constantinidis (1992)
Severe depress-ion (DSM-IV)	Forty-four patients with DSM-IV (18–55 years, both sexes, Iran), zinc supplement ($n = 20$), placebo ($n = 17$)	Randomized, double-blind, measured using HDRS, plasma levels	Zinc sulfate (25 mg/day) given orally for 12 weeks + antidepressant medications	Zinc significantly reduced HDRS, reducing depressive symptoms No differences in plasma levels	Ranjbar et al. (2014)
Autism associated	Maternal Shank3-knockout mice Offspring juvenile (9–10 weeks), adult mice (16–17 weeks) mice	Maternal shank knockout mice treated with either zinc control or supplemented diet postnatal Day 21	Maternal Shank3 knockout treated with either zinc diet control (30 ppm), zinc supplemented (150 ppm)	Prevented ASD-associated deficits in social interaction and normalized anxiety behaviors in shank knockout offspring	Vyas et al. (2020)
Ischemic Stroke	Twenty-six patients subacute stroke (65–82, both sexes, Italy) Zinc group ($n = 13$) Placebo ($n = 13$)	Randomized, double-blind. Only patients with adequate energy protein intake selected	Zinc sulfate (10 mg/day) for 30 days	Improve neurological deficit, ↑energy protein and lipid intakes, ↑serum albumin levels	Aquilani et al. (2009)

Table describes selective studies of zinc supplementation in brain disorders highlighting the dosage and outcome postadministration. *DSM-IV*, diagnostic and statistical manual of mental disorders; *i.p*, intraperitoneal; *i.v*, intravenous; *CCI*, controlled cortical impact; *GCS*, Glasgow Coma Scale; SOFA, Sequential Organ Failure Assessment; *GOS*, Glasgow Outcome Score.

The potential for therapeutic approaches that target zinc

While the evidence for zinc deficiency post-TBI has historically led researchers to examine the benefits of zinc supplementation, the conflicting results have suggested the need for a more precise molecular targeting of zinc. As such, researchers started to study the effects of zinc chelation, which works by binding zinc and preventing it from participating in biological processes (Suh et al., 2000). However, chelation has also yielded contradictory results post-TBI, bringing about the consideration that another class of drugs known as zinc chaperones/ionophores, which effectively function to redistribute metal ions to other target proteins, may be beneficial in the outcomes post-TBI. These therapeutic approaches are discussed.

Zinc chelation

To our knowledge, zinc chelation clinical trials have yet to be undertaken in the context of TBI. However, an early preclinical study demonstrated that TBI induced the translocation of zinc from presynaptic boutons into injured neurons and that administration of the zinc chelator calcium ethylenediaminetetraacetic acid (Ca-EDTA) reduced the number of degenerating neurons in the hippocampus by 50%–75% (Suh et al., 2000), suggesting that removal of extracellular zinc post-TBI may be neuroprotective. Building on this work was a preclinical study that found that Ca-EDTA treatment upregulated mRNA and protein levels of neuroprotective genes which correlated with a reduction in apoptotic cell death (Hellmich et al., 2004). Indeed, the upregulation of neuroprotective genes suggests that zinc chelation may help alleviate the TBI-induced stress response. In another study, treatment with the zinc chelator N,N,N′,N′-tetrakis-(2-pyridylmethyl) ethylenediamine (TPEN) reduced excitotoxicity and improved markers of autophagy (Zhao, Liu, Ma, Zhang, & Liang, 2018). Importantly, the authors showed that TBI caused zinc translocation via mossy fiber sprouting in the hippocampus, which was inhibited by treatment with TPEN, suggesting that inhibiting zinc translocation post-TBI may be able to attenuate zinc-related neurotoxicity.

With mounting evidence demonstrating the cellular and molecular benefits of zinc chelation, the question remained whether chelation would improve behavioral outcomes post-TBI. To address this, a follow-up preclinical study reported a reduction in Fluoro-Jade-positive (injured) neurons following Ca-EDTA treatment in the hippocampus 24 h post-TBI (Hellmich et al., 2008). Despite rescue of injured neurons, a separate cohort of animals showed that chelation did not improve spatial memory performance, assessed using the water maze test at 15 days postinjury. In fact, zinc chelation worsened spatial memory function of control animals, which was not different to Ca-EDTA-treated animals. This might be expected because zinc chelation dysregulates physiological zinc homeostasis, which is required for normal cellular processes to take place. In fact, one study showed that Ca-EDTA administration in the hippocampus of healthy mice resulted in impairments in episodic memory (Daumas, Halley, & Lassalle, 2004).

Taken together, these data demonstrate that while zinc chelation may be beneficial in reducing a variety of TBI-mediated cellular and molecular outcomes following injury, so far it has not been demonstrated to improve TBI-induced cognitive deficits. One reason for this might be due to their binding affinity. Most zinc chelators have high-binding affinities and indeed several have additional affinities for copper, calcium, and magnesium ions (Grabrucker, Rowan, & Garner, 2011). Therefore, unintended side effects due to chelation of other metal ions have to be considered. Furthermore, the lack of synaptic zinc is known to impair cognition (Adlard, Parncutt, Finkelstein, & Bush, 2010), and as such, the continued blunt chelation of zinc, while initially of benefit postinjury, may ultimately impair/hamper memory processes due to a lack of zinc in the correct cellular locations. This highlights the need for a more sophisticated targeting of zinc within specific brain regions.

Zinc chaperones

A promising avenue in managing the outcomes post-TBI is metal chaperones, which bind metals but do not permanently chelate them out from "biological circulation." At a simplistic level then, these compounds can effectively redistribute metal ions, and in a disease context, this might translate to metals shifting from areas of "surplus" to areas of "deficiency," which can then result in zinc-mediated modulation of specific neuronal pathways (Adlard & Bush, 2018). In the case of zinc, several metal chaperone drugs have demonstrated neuroprotective effects in various models of aging and neurodegenerative disease (Adlard et al., 2011, 2014) and in clinical trials of AD (Faux et al., 2010). Based on the observed efficacy of these chaperone drugs, they have also been tested in preclinical models of TBI.

The zinc/copper chaperone clioquinol (CQ) has previously been tested in the context of AD; however, the literature following TBI is scarce. One preclinical TBI study confirmed that TBI alone induced hippocampal neurogenesis and that this was significantly reduced following treatment with CQ (Choi et al., 2014). The authors suggest that CQ may be either acting as a chelating agent to reduce zinc levels in the hippocampus or that alternatively, it may be acting as a chaperone by maintaining homeostatic zinc levels. So while not clear-cut, this study suggests that targeting zinc availability may play a role in modulating neurogenesis post-TBI. This also highlights the notion that these compounds are not "smart" and that their effect is impacted by the complexity of the cellular metal pools, their biological accessibility to those pools, and the differential regulation/requirements of various metals in health and disease at different stages of life.

Another preclinical TBI study found that mice treated with the zinc/copper chaperone PBT2 (Prana Biotechnology) had decreased lesioned area and astrocytosis, increased numbers of neurons and improved cognitive function at 26 days postinjury (Portbury et al., 2019). This suggests that PBT2 may improve the long-term cognitive deficits that are characteristic post-TBI. Importantly, this study is in line with some of our previous work demonstrating the beneficial effects of PBT2 treatment on cognitive function in models of normal aging and neurodegeneration (Adlard et al., 2008, 2014).

Conclusion

The past three decades of research have demonstrated that zinc is a key player in the maintenance of normal cellular function and that it is also critically involved in neurodegeneration. This has provided an opportunity for investigations into the mechanisms underlying neurodegenerative diseases such as AD as well as other conditions such as TBI. Importantly, it has also guided research into novel therapeutic approaches for these conditions. Indeed, several treatment options may be beneficial in modulating zinc levels, and this research is actively being pursued to better understand the mechanism of action of these drugs (Fig. 3). In the case of TBI, further studies are required to determine the dosage of these compounds and whether administration pre- versus postinjury can affect the cellular and behavioral outcomes post-TBI (Table 3). Further research is also required to more fully explore the role of zinc in different forms of TBI, such as mild concussion and repeat head trauma—issues which are becoming increasingly diagnosed and which may present with a different zinc profile to more severe forms of injury. These studies will be crucial in expanding our understanding of best practices when it comes to zinc-related pathology following brain injury and in other neurodegenerative diseases.

Applications to other neurological conditions

Aging changes brain metal levels, with studies reporting both increased zinc in healthy aging and zinc deficiency. Increased brain zinc levels were reported in aged cattle (Zatta et al., 2008), whereas they remained constant throughout life span in aged rodents (Takahashi et al., 2001). A clinical study showed that 44% of elderly patients were zinc-deficient, of which 20% were severe (Madej, Borowska, Bylinowska, Szybalska, & Pietruszka, 2013). Another study showed that zinc depletion progressed with age in the cortex/hippocampus (Vasudevaraju et al., 2010). In AD patients, increased cortical zinc levels are proportional to amyloid-beta load, correlating with poorer cognitive scores (Religa et al., 2006). Moreover, AD patients showed decreased serum zinc concentrations compared to healthy controls (Ventriglia et al., 2015). Indeed, zinc can bind/accumulate within amyloid-beta plaques (Bush, Pettingell Jr, Paradis, & Tanzi, 1994), possibly leading to cognitive impairment through depletion of synaptic zinc pools (Adlard et al., 2010). The lack of synaptic zinc in zinc transporter-3 knockout mice results in cognitive impairments, synaptic loss, and decreased hippocampal proteins involved in

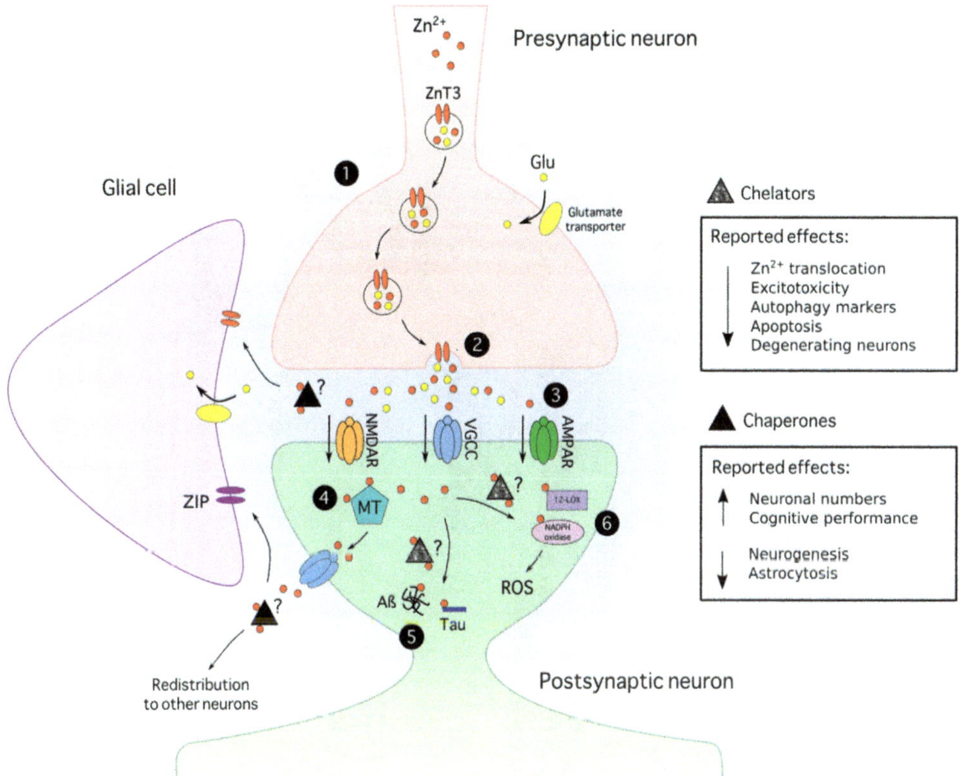

Fig. 3 Putative intracellular targets of zinc chelator/chaperones in the zinc-mediated excitotoxic cascade post-TBI. TBI-induced overactivation of excitatory signaling causes ZnT3-containing vesicles to transport Zn^{2+} along the axon (1). Zn^{2+}/glutamate are released into the synaptic cleft via exocytosis (2) where Zn^{2+} permeates through Zn^{2+}-permeable channels (3). Postsynaptic-Zn^{2+} is chaperoned by MT and released via oxidation/nitrosylation (4), leading to free-Zn^{2+} accumulation. Free-Zn^{2+} binds amyloid-beta/tau leading to conformational changes/aggregation (5). Zn^{2+} activates 12-LOX and NADPH oxidase generating ROS (6). Chaperones may act by redistributing Zn^{2+} between neurons/glia at the synapse or after postsynaptic permeation through VGCCs. Chelators may reduce the Zn^{2+} available to participate in cellular processes. *ZnT3*, zinc transporter 3; *VGCC*, voltage-gated calcium channels; *AMPAR*, a-amino-3-hydroxy-5-methyl-4-isoxazolepropionic acid receptor; *NMDAR*, *N*-methyl-D-aspartate receptor; *MT*, metallothionein; *ROS*, reactive oxygen species; *ZIP*, zinc/iron-regulated transporter protein.

neurogenesis and learning/memory, demonstrating that zinc is crucial for neurotransmission (Adlard et al., 2010). While the role of zinc remains conflicted, aging/neurodegeneration seems to shift homeostatic brain zinc levels, as is reported in brain injury. As previously discussed, therapeutic targeting of zinc has shown promising results (Adlard et al., 2011).

Table 3 A list of preclinical TBI studies investigating the effects of zinc chelators and chaperones.

Therapeutic approach	Drugs	TBI model	Dose	Timing/route of administration	Cellular/molecular outcomes	Behavioral outcomes	Reference
Zinc chelators	TPEN	CCI	20 mg/kg	Immediately post-TBI and twice daily for 48 h/i.p.	Reduced zinc translocation, excitotoxicity, autophagy markers	N/A	Zhao et al. (2018)
	Ca-EDTA	WDIA	100 mM	10 min pre-TBI/ i.c.v. injection	Reduced degenerating neurons	N/A	Suh et al. (2000)
		FPI	100 mM	30 min pre-TBI/i.c.v. injection	Reduced apoptotic cell death, upregulated neuroprotective genes	N/A	Hellmich et al. (2004)
		FPI	300 mM	30 min pre- and post-TBI/i.c.v. injection	Reduced degenerating neurons	No changes in cognitive function	Hellmich et al. (2008)
Zinc chaperones	PBT2	CCI	30 mg/kg	Initiated 5 h post-TBI, once daily for 26 days/oral gavage	Maintained neuronal integrity	Improved cognitive performance	Portbury et al. (2019)
	CQ	WDIA	30 mg/kg	Initiated 24 h post-TBI, twice daily for 1-week/i.p.	Reduced neurogenesis	N/A	Choi et al. (2014)

Table describes selective studies of zinc chaperone/chelator administration on outcomes postbrain injury. Abbreviations: *N/A*, not applicable; *i.p.*, intraperitoneal; *i.c.v.*, intracerebroventricular; *CCI*, controlled cortical impact; *FPI*, fluid percussion injury; *WDIA*, weight drop impact acceleration.

Other components of interest

Other metals of interest in TBI, aging, and neurodegenerative disorders include iron and copper. Iron is the most abundant transition metal in the body/brain and is essential for energy production and neuronal transmission, whereas copper, the third most abundant metal, is crucial for myelination and neurotransmitter synthesis. We have shown that iron progressively increases at the injury site over 28 days, whereas copper showed an acute reduction but thereafter increased until 28 days postinjury (Portbury et al., 2016). A follow-up study in aged mice similarly showed increased iron in the injury site, whereas copper was increased at 14 and 28 days only postinjury (Portbury et al., 2017). Like zinc, therapeutic targeting of copper is limited in TBI studies; however, promising results for iron chelation therapy on postinjury recovery have been reported in TBI models (Daglas & Adlard, 2018). Indeed, the zinc-copper chaperones PBT2 and clioquinol may also work by reducing copper accumulation (Choi et al., 2014; Portbury et al., 2019). This demonstrates the importance of regulating metal dyshomeostasis post-TBI and warrants further research.

Mini dictionary of terms

- Alzheimer's disease: a progressive neurological condition characterized by brain amyloid-beta/tau accumulation resulting in memory loss.
- Chaperones: a compound that binds one or more metals, facilitating cellular redistribution.
- Chelation: a compound that binds one or more metals, making them unavailable to the cell.
- Chronic traumatic encephalopathy: progressive neurodegenerative disorder characterized by brain protein tau accumulation linked to repetitive brain injury.
- Metal homeostasis/dyshomeostasis: process of maintaining balance of metal levels in the brain/body. Dyshomeostasis refers to the imbalance of metals, which may lead to disease.
- Neuronal hyperexcitability: process of abnormal neuronal firing as a result of overstimulation.
- Neurodegeneration: gradual process by which neurons are damaged leading to neuronal death.
- Neuroprotection: where a protein/compound acts to prevent pathology and protect brain cells.
- Neurotoxicity: where a protein/compound/process adversely affects a neuron leading to neuronal damage/death.
- TBI: brain insult typically caused by external forces disrupting normal brain function, often leading to chronic cognitive/functional deficits.

Key facts of TBI

- ~700,000 Australians live with a brain injury.
- TBI significantly increases the risk of morbidity/mortality.
- TBI leads to acute/chronic pathological events which may persist for months or years.
- Metal dyshomeostasis is a secondary process arising from TBI that may initiate/exacerbate the pathological sequence of events postinjury.

Key facts of zinc in the brain

- Zinc is an abundant and essential element for normal body/brain functions, absorbed through adequate dietary intake.
- Zinc plays a crucial role in synaptic transmission, cognition, neurogenesis, immune, and protein/enzyme functions.
- Zinc-regulatory proteins maintain zinc homeostasis.
- Zinc dyshomeostasis contributes to the pathogenesis of TBI, aging, and neurodegenerative/neurological conditions.
- Zinc modulation is a potential therapeutic approach to alleviate brain pathology and improve recovery for individuals with brain injury.

Summary points

- TBI causes a series of secondary processes, including metal dyshomeostasis, that exacerbate brain pathology and lead to short/long-term disabilities.
- Brain zinc imbalance, as occurs post-TBI, contributes to excitotoxicity, oxidative stress, inflammation, and neuronal damage/death.
- Pharmacological targeting of zinc is a potential therapeutic approach for TBI.
- Further investigation is warranted to better understand how zinc is protective/toxic in TBI.

References

Adlard, P. A., Bica, L., White, A. R., Nurjono, M., Filiz, G., et al. (2011). Metal ionophore treatment restores dendritic spine density and synaptic protein levels in a mouse model of Alzheimer's disease. *PLoS One*, *6*(3), e17669.

Adlard, P. A., & Bush, A. I. (2018). Metals and Alzheimer's disease: How far have we come in the clinic? *Journal of Alzheimer's Disease*, *62*(3), 1369–1379.

Adlard, P. A., Cherny, R. A., Finkelstein, D. I., Gautier, E., Robb, E., et al. (2008). Rapid restoration of cognition in Alzheimer's transgenic mice with 8-hydroxy quinoline analogs is associated with decreased interstitial Aβ. *Neuron*, *59*(1), 43–55.

Adlard, P. A., Parncutt, J. M., Finkelstein, D. I., & Bush, A. I. (2010). Cognitive loss in zinc transporter-3 knock-out mice: A phenocopy for the synaptic and memory deficits of Alzheimer's disease? *The Journal of Neuroscience*, *30*(5), 1631–1636. https://doi.org/10.1523/JNEUROSCI.5255-09.2010.

Adlard, P. A., Sedjahtera, A., Gunawan, L., Bray, L., Hare, D., et al. (2014). A novel approach to rapidly prevent age-related cognitive decline. *Aging Cell, 13*(2), 351–359.

Aquilani, R., Baiardi, P., Scocchi, M., Iadarola, P., Verri, M., et al. (2009). Normalization of zinc intake enhances neurological retrieval of patients suffering from ischemic strokes. *Nutritional Neuroscience, 12*(5), 219–225. https://doi.org/10.1179/147683009X423445.

Ayton, S., Finkelstein, D. I., Cherny, R. A., Bush, A. I., & Adlard, P. A. (2013). Zinc in Alzheimer's and Parkinson's diseases. In R. H. Kretsinger, V. N. Uversky, & E. A. Permyakov (Eds.), *Encyclopedia of metalloproteins* (pp. 2433–2441). New York, NY: Springer New York.

Bush, A. I., Pettingell, W. H., Jr., Paradis, M., & Tanzi, R. E. (1994). Modulation of A beta adhesiveness and secretase site cleavage by zinc. *Journal of Biological Chemistry, 269*(16), 12152–12158.

Carrasco, J., Penkowa, M., Giralt, M., Camats, J., Molinero, A., et al. (2003). Role of metallothionein-III following central nervous system damage. *Neurobiology of Disease, 13*(1), 22–36. https://doi.org/10.1016/s0969-9961(03)00015-9.

Chen, D., Nie, Z.-B., Chi, Z.-H., Wang, Z.-Y., Wei, X.-T., et al. (2018). Neuroprotective effect of ZnT3 knockout on subarachnoid hemorrhage. *Translational Neuroscience, 9*, 26–32. https://doi.org/10.1515/tnsci-2018-0006.

Choi, S., Hong, D. K., Choi, B. Y., & Suh, S. W. (2020). Zinc in the brain: Friend or foe? *International Journal of Molecular Sciences, 21*(23). https://doi.org/10.3390/ijms21238941.

Choi, B. Y., Kim, J. H., Kim, H. J., Lee, B. E., Kim, I. Y., et al. (2014). Zinc chelation reduces traumatic brain injury-induced neurogenesis in the subgranular zone of the hippocampal dentate gyrus. *Journal of Trace Elements in Medicine and Biology, 28*(4), 474–481. https://doi.org/10.1016/j.jtemb.2014.07.007.

Chung, R. S., Adlard, P. A., Dittmann, J., Vickers, J. C., Chuah, M. I., et al. (2004). Neuron-glia communication: Metallothionein expression is specifically up-regulated by astrocytes in response to neuronal injury. *Journal of Neurochemistry, 88*(2), 454–461.

Constantinidis, J. (1992). Treatment of Alzheimer's disease by zinc compounds. *Drug Development Research, 27*(1), 1–14.

Cope, E. C., Morris, D. R., Scrimgeour, A. G., & Levenson, C. W. (2012). Use of zinc as a treatment for traumatic brain injury in the rat: Effects on cognitive and behavioral outcomes. *Neurorehabilitation and Neural Repair*, 907.

Cope, E. C., Morris, D. R., Scrimgeour, A. G., VanLandingham, J. W., & Levenson, C. W. (2011). Zinc supplementation provides behavioral resiliency in a rat model of traumatic brain injury. *Physiology & Behavior, 104*(5), 942–947. https://doi.org/10.1016/j.physbeh.2011.06.007.

Coronado, V. G., McGuire, L. C., Sarmiento, K., Bell, J., Lionbarger, M. R., et al. (2012). Trends in traumatic brain injury in the U.S. and the public health response: 1995-2009. *Journal of Safety Research, 43*(4), 299–307. https://doi.org/10.1016/j.jsr.2012.08.011.

Daglas, M., & Adlard, P. A. (2018). The involvement of Iron in traumatic brain injury and neurodegenerative disease. *Frontiers in Neuroscience, 12*, 981. https://doi.org/10.3389/fnins.2018.00981.

Daumas, S., Halley, H., & Lassalle, J. M. (2004). Disruption of hippocampal CA3 network: Effects on episodic-like memory processing in C57BL/6J mice. *European Journal of Neuroscience, 20*(2), 597–600.

Delic, V., Beck, K. D., Pang, K. C. H., & Citron, B. A. (2020). Biological links between traumatic brain injury and Parkinson's disease. *Acta Neuropathologica Communications, 8*(1), 45. https://doi.org/10.1186/s40478-020-00924-7.

Djordjevic, J., Sabbir, M. G., & Albensi, B. C. (2016). Traumatic brain injury as a risk factor for Alzheimer's disease: Is inflammatory signaling a key player? *Current Alzheimer Research, 13*(7), 730–738. https://doi.org/10.2174/1567205013666160222110320.

Doering, P., Stoltenberg, M., Penkowa, M., Rungby, J., Larsen, A., et al. (2010). Chemical blocking of zinc ions in CNS increases neuronal damage following traumatic brain injury (TBI) in mice. *PLoS One, 5*(4), e10131. https://doi.org/10.1371/journal.pone.0010131.

Faul, M., Xu, L., Wald, M. M., Coronado, V., & Dellinger, A. M. (2010). Traumatic brain injury in the United States: National estimates of prevalence and incidence, 2002-2006. *Injury Prevention, 16*, A268. https://doi.org/10.1136/ip.2010.029215.951.

Faux, N. G., Ritchie, C. W., Gunn, A., Rembach, A., Tsatsanis, A., et al. (2010). PBT2 rapidly improves cognition in Alzheimer's disease: Additional phase II analyses. *Journal of Alzheimer's Disease, 20*(2), 509–516.

Gilchrist, J., Thomas, K. E., Xu, L. K., McGuire, L. C., & Coronado, V. (2011). Nonfatal traumatic brain injuries related to sports and recreation activities among persons aged <= 19 years-United States, 2001-2009 (Reprinted from MMWR, vol 60, pg 1337-1342, 2011). *JAMA-Journal of the American Medical Association, 306*(21), 2318–2320.

Giralt, M., Penkowa, M., Lago, N., Molinero, A., & Hidalgo, J. (2002). Metallothionein-1+2 protect the CNS after a focal brain injury. *Experimental Neurology, 173*(1), 114–128. https://doi.org/10.1006/exnr.2001.7772.

Grabrucker, A. M., Rowan, M., & Garner, C. C. (2011). Brain-delivery of zinc-ions as potential treatment for neurological diseases: Mini review. *Drug Delivery Letters, 1*(1), 13–23. https://doi.org/10.2174/2210303111101010013.

Hellmich, H. L., Eidson, K. A., Capra, B. A., Garcia, J. M., Boone, D. R., et al. (2007). Injured Fluoro-jade-positive hippocampal neurons contain high levels of zinc after traumatic brain injury. *Brain Research, 1127*(1), 119–126. https://doi.org/10.1016/j.brainres.2006.09.094.

Hellmich, H. L., Eidson, K., Cowart, J., Crookshanks, J., Boone, D. K., et al. (2008). Chelation of neurotoxic zinc levels does not improve neurobehavioral outcome after traumatic brain injury. *Neuroscience Letters, 440*(2), 155–159.

Hellmich, H. L., Frederickson, C. J., DeWitt, D. S., Saban, R., Parsley, M. O., et al. (2004). Protective effects of zinc chelation in traumatic brain injury correlate with upregulation of neuroprotective genes in rat brain. *Neuroscience Letters, 355*(3), 221–225.

Humphreys, I., Wood, R. L., Phillips, C. J., & Macey, S. (2013). The costs of traumatic brain injury: A literature review. *ClinicoEconomics and Outcomes Research, 5*, 281–287. https://doi.org/10.2147/CEOR.S44625 (default).

Khazdouz, M., Mazidi, M., Ehsaei, M. R., Ferns, G., Kengne, A. P., et al. (2018). Impact of zinc supplementation on the clinical outcomes of patients with severe head trauma: A double-blind randomized clinical trial. *Journal of Dietary Supplements, 15*(1), 1–10. https://doi.org/10.1080/19390211.2017.1304486.

Lee, J. Y., Cole, T. B., Palmiter, R. D., & Koh, J. Y. (2000). Accumulation of zinc in degenerating hippocampal neurons of ZnT3-null mice after seizures: Evidence against synaptic vesicle origin. *The Journal of Neuroscience, 20*(11), RC79.

Lee, J. Y., Kim, J. H., Palmiter, R. D., & Koh, J. Y. (2003). Zinc released from metallothionein-iii may contribute to hippocampal CA1 and thalamic neuronal death following acute brain injury. *Experimental Neurology, 184*(1), 337–347. https://doi.org/10.1016/s0014-4886(03)00382-0.

Levenson, C. W. (2020). Zinc and traumatic brain injury: From chelation to supplementation. *Medical Sciences (Basel), 8*(3). https://doi.org/10.3390/medsci8030036.

Lowe, N. M., Fekete, K., & Decsi, T. (2009). Methods of assessment of zinc status in humans: A systematic review. *The American Journal of Clinical Nutrition, 89*(6), 2040S–2051S. https://doi.org/10.3945/ajcn.2009.27230G.

Maares, M., & Haase, H. (2020). A guide to human zinc absorption: General overview and recent advances of in vitro intestinal models. *Nutrients, 12*(3), 762. https://doi.org/10.3390/nu12030762.

Madej, D., Borowska, K., Bylinowska, J., Szybalska, A., & Pietruszka, B. (2013). Dietary intakes of iron and zinc assessed in a selected group of the elderly: Are they adequate? *Roczniki Państwowego Zakładu Higieny, 64*(2).

McClain, C. J., Twyman, D. L., Ott, L. G., Rapp, R. P., Tibbs, P. A., et al. (1986). Serum and urine zinc response in head-injured patients. *Journal of Neurosurgery, 64*(2), 224–230. https://doi.org/10.3171/jns.1986.64.2.0224.

Ott, L., McClain, C. J., Gillespie, M., & Young, B. (1994). Cytokines and metabolic dysfunction after severe head injury. *Journal of Neurotrauma, 11*(5), 447–472. https://doi.org/10.1089/neu.1994.11.447.

Pankhurst, M. W., Gell, D. A., Butler, C. W., Kirkcaldie, M. T., West, A. K., et al. (2012). Metallothionein (MT) -I and MT-II expression are induced and cause zinc sequestration in the liver after brain injury. *PLoS One, 7*(2), e31185. https://doi.org/10.1371/journal.pone.0031185.

Pei, J.-J., An, W.-L., Zhou, X.-W., Nishimura, T., Norberg, J., et al. (2006). P70 S6 kinase mediates tau phosphorylation and synthesis. *FEBS Letters, 580*(1), 107–114.

Penkowa, M., Giralt, M., Thomsen, P. S., Carrasco, J., & Hidalgo, J. (2001). Zinc or copper deficiency-induced impaired inflammatory response to brain trauma may be caused by the concomitant metallothionein changes. *Journal of Neurotrauma, 18*(4), 447–463. https://doi.org/10.1089/089771501750171056.

Portbury, S. D., & Adlard, P. A. (2015). Traumatic brain injury, chronic traumatic encephalopathy, and Alzheimer's disease: Common pathologies potentiated by altered zinc homeostasis. *Journal of Alzheimer's Disease, 46*(2), 297–311.

Portbury, S. D., & Adlard, P. A. (2017). Zinc signal in brain diseases. *International Journal of Molecular Sciences, 18*(12), 2506.

Portbury, S. D., Hare, D. J., Sgambelloni, C. J., Bishop, D. P., Finkelstein, D. I., et al. (2017). Age modulates the injury-induced metallomic profile in the brain. *Metallomics, 9*(4), 402–410. https://doi.org/10.1039/c6mt00260a.

Portbury, S. D., Hare, D. J., Sgambelloni, C., Finkelstein, D. I., & Adlard, P. A. (2016). A time-course analysis of changes in cerebral metal levels following a controlled cortical impact. *Metallomics, 8*(2), 193–200. https://doi.org/10.1039/c5mt00234f.

Portbury, S. D., Sedjahtera, A., Perrones, K., Sgambelloni, C., Zhang, M., et al. (2019). Metal chaperones: A novel therapeutic strategy for brain injury? *Brain Injury, 33*(3), 305–312. https://doi.org/10.1080/02699052.2018.1552988.

Prasad, A. S., Schulert, A. R., Miale, A., Jr., Farid, Z., & Sandstead, H. H. (1963). Zinc and iron deficiencies in male subjects with dwarfism and hypogonadism but without ancylostomiasis, schistosomiasis or severe anemia. *The American Journal of Clinical Nutrition, 12*(6), 437–444. https://doi.org/10.1093/ajcn/12.6.437.

Ranjbar, E., Shams, J., Sabetkasaei, M., M-Shirazi, M., Rashidkhani, B., et al. (2014). Effects of zinc supplementation on efficacy of antidepressant therapy, inflammatory cytokines, and brain-derived neurotrophic factor in patients with major depression. *Nutritional Neuroscience, 17*(2), 65–71. https://doi.org/10.1179/1476830513Y.0000000066.

Religa, D., Strozyk, D., Cherny, R., Volitakis, I., Haroutunian, V., et al. (2006). Elevated cortical zinc in Alzheimer disease. *Neurology, 67*(1), 69–75.

Santos, C. R., Martinho, A., Quintela, T., & Goncalves, I. (2012). Neuroprotective and neuroregenerative properties of metallothioneins. *IUBMB Life, 64*(2), 126–135. https://doi.org/10.1002/iub.585.

Sedjahtera, A., Gunawan, L., Bray, L., Hung, L. W., Parsons, J., et al. (2018). Targeting metals rescues the phenotype in an animal model of tauopathy. *Metallomics, 10*(9), 1339–1347.

Suh, S. W., Chen, J. W., Motamedi, M., Bell, B., Listiak, K., et al. (2000). Evidence that synaptically-released zinc contributes to neuronal injury after traumatic brain injury. *Brain Research, 852*(2), 268–273. https://doi.org/10.1016/s0006-8993(99)02095-8.

Sun, X. Y., Wei, Y. P., Xiong, Y., Wang, X. C., Xie, A. J., et al. (2012). Synaptic released zinc promotes tau hyperphosphorylation by inhibition of protein phosphatase 2A (PP2A). *The Journal of Biological Chemistry, 287*(14), 11174–11182. https://doi.org/10.1074/jbc.M111.309070.

Sun, K. J., Zhu, L., Wang, H. D., Ji, X. J., Pan, H., et al. (2013). Zinc as mediator of ubiquitin conjugation following traumatic brain injury. *Brain Research, 1506*, 132–141. https://doi.org/10.1016/j.brainres.2013.02.011.

Takahashi, S., Takahashi, I., Sato, H., Kubota, Y., Yoshida, S., et al. (2001). Age-related changes in the concentrations of major and trace elements in the brain of rats and mice. *Biological Trace Element Research, 80*(2), 145–158.

Vasudevaraju, P., Bharathi, J. T., Shamasundar, N., Rao, K. S., Balaraj, B., et al. (2010). New evidence on iron, copper accumulation and zinc depletion and its correlation with DNA integrity in aging human brain regions. *Indian Journal of Psychiatry, 52*(2), 140.

Ventriglia, M., Brewer, G. J., Simonelli, I., Mariani, S., Siotto, M., et al. (2015). Zinc in Alzheimer's disease: A meta-analysis of serum, plasma, and cerebrospinal fluid studies. *Journal of Alzheimer's Disease, 46*(1), 75–87.

Vyas, Y., Lee, K., Jung, Y., & Montgomery, J. M. (2020). Influence of maternal zinc supplementation on the development of autism-associated behavioural and synaptic deficits in offspring Shank3-knockout mice. *Molecular Brain, 13*(1), 110. https://doi.org/10.1186/s13041-020-00650-0.

Yeiser, E. C., Vanlandingham, J. W., & Levenson, C. W. (2002). Moderate zinc deficiency increases cell death after brain injury in the rat. *Nutritional Neuroscience*, *5*(5), 345–352. https://doi.org/10.1080/1028415021000033811.

Young, B., Ott, L., Kasarskis, E., Rapp, R., Moles, K., et al. (1996). Zinc supplementation is associated with improved neurologic recovery rate and visceral protein levels of patients with severe closed head injury. *Journal of Neurotrauma*, 25.

Zatta, P., Drago, D., Zambenedetti, P., Bolognin, S., Nogara, E., et al. (2008). Accumulation of copper and other metal ions, and metallothionein I/II expression in the bovine brain as a function of aging. *Journal of Chemical Neuroanatomy*, *36*(1), 1–5.

Zhao, L., Liu, Q., Ma, S., Zhang, Y., & Liang, P. (2018). TPEN attenuates neural autophagy induced by synaptically-released zinc translocation and improves histological outcomes after traumatic brain injury in rats. *Annals of Clinical and Laboratory Science*, *48*(4), 446–452.

PART IV

Cerebral palsy

CHAPTER 12

Cerebral palsy: Evaluating vitamin D levels

Marcela Almeida Linden and Maria Angela Bellomo-Brandão*
Department of Pediatrics, School of Medical Sciences, University of Campinas, Campinas, São Paulo, Brazil

Abbreviations

$1,25(OH)_2D2$)	1,25-dihydroxyergocalciferol
$1,25(OH)_2D3$	1,25-dihydroxycolecalciferol
$1,25(OH)_2D3$	calcitriol
25[OH]D	25-hydroxyvitamin D
CYP27B1	1-α-hydroxylase
DBP	D-binding protein
O-25(OH)D3	calcidiol
PTH	parathyroid hormone

Introduction

First identified and classified as a nutrient in the early 20th century, the vitamin D is now considered as a hormone due to its role in the regulation of the calcium and phosphorus metabolism, together with parathyroid hormone (PTH) (Holick et al., 1980).

Vitamin D was initially described as a traditional vitamin, being essential for the functioning of the human body. This nutrient may be either obtained through food or produced by our organism, through a photosynthetic reaction in response to sun exposure. Vitamin D may occur in two forms: ergocalciferol or vitamin D2, synthesized from the plant steroid ergosterol via ultraviolet radiation; and cholecalciferol or vitamin D3, synthesized from cholesterol (Holick et al., 2011). Vitamin D deficiency may lead to rickets and osteomalacia, osteopenia, and osteoporosis, thus being essential for maintenance of the bone health in children and adults (Chung et al., 2009; Rovner et al., 2008). The biologically active forms of vitamin D3 are produced in the epidermis by a photolysis reaction of 7-dehydrocholesterol, whereby ultraviolet-B radiation induces the breakage of the precursor steroids B nucleus. Being directly correlated with vitamin D3, the production of the D2 form may occur in a similar manner (Hess, 1922; Holick et al., 2011).

* **In Absentia Contact Person:** Ms. Ana Monteiro, School of Medical Sciences, University of Campinas, Campinas, São Paulo, Brazil, Telephone: +55 19 3521 7082, E-mail: anapmon@unicamp.br.

Vitamins and Minerals in Neurological Disorders
https://doi.org/10.1016/B978-0-323-89835-5.00003-X

Table 1 Amount of vitamin D in natural sources.

• Exposure to ultraviolet B rays (5–10 min)	3000 IU
• Cod liver oil 13.6 g (1 col. soup)	400–1000 IU
• Wild salmon 100 g	451 IU
• 100 g canned sardines	300 IU
• Tuna in a 100 g can	230 IU
• Whole milk with added vitamin D 244 mL	124 IU
• Fresh shitake mushroom 100 g	100 IU
• Egg with yolk	20 IU
• *Mozzarella* 100 g	16 IU
• Whole milk without added vitamin D 244 mL	5 IU

Source: Adapted from Cesari, M., Incalzi, R., Zamboni, V., & Pahor, M. (2010). Vitamin D hormone: A multitude of actions potentially influencing the physical function decline in older persons. *Geriatrics & Gerontology International, 11*(2), 133–142 and Holick, M. (2007). Vitamin D deficiency. *New England Journal of Medicine, 357*(3), 266–281.

Diet is responsible for about 10%–20% of the vitamin D considered necessary for the adequate functioning of the human organism, whereas the other 80%–90% stems from endogenous synthesis through sun exposure (Holick et al., 2011). Cholecalciferol (vitamin D3) may be obtained from dietary sources of animal origins and ergocalciferol (vitamin D2) from plant origins (Table 1) (Cesari, Incalzi, Zamboni, & Pahor, 2010; Holick, 2007).

Vitamin D physiology

Cholecalciferol and ergocalciferol are transported into bloodstream bound to a D-binding protein (DBP), while vitamin D is absorbed by the intestinal mucosa. As protein-vitamin D complex, both cholecalciferol and ergocalciferol are transported to the liver, where cholecalciferol is hydroxylized at position 25 by the 25-hydroxylase enzyme in the hepatic microsomal system, resulting in 25-hydroxyvitamin D3 [25(OH)D3]. Through the same process, ergosterol becomes 25-hydroxyergocalciferol [25(OH)D2] (Windaus, Linsert, Lüttringhaus, & Weidlich, 1932).

The biologically activate form of vitamin D requires yet a second stage, whereby the product of the first reaction is bound to transcalciferrin (an alpha globulin synthesized by the liver with high affinity and specificity for this metabolite that functions as a carrier protein) and transported to the kidneys. The second hydrolyzation occurs at this site due to the catalytic action 1-α-hydroxylase (CYP27B1), resulting in 1,25-dihydroxycholecalciferol [1,25(OH)$_2$D3] and 1,25-dihydroxyergocalciferol [1,25 (OH)$_2$D2]. This enzyme of renal origin, also found in bones and in the placenta, can be activated either directly, by the parathyroid hormone (PTH) due to the decrease in serum phosphate concentrations, or indirectly, due to the decrease in plasma concentrations of calcium ions (Holick et al., 2011) (Fig. 1).

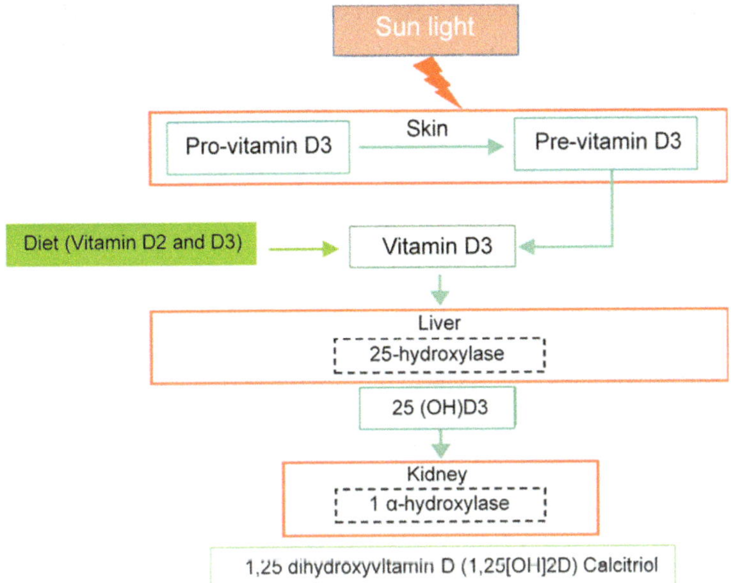

Fig. 1 Vitamin D metabolism.

Vitamin D activation is associated with nephrogenic cyclic AMP, estrogen, prolactin, and growth hormones, besides stimulating, together with PTH, osteoclastic bone resorption by increasing serum calcium concentrations. In the presence of hypocalcemia, 1,25 (OH)D3 plasma levels increase and CYP27B1 activity decreases. When compared to 1,25$(OH)_2$D3, or calcitriol (a very potent hormone), plasma concentrations of 25(OH)D3, or calcidiol, are about a thousand times higher. However, serum calcitriol concentrations function as an indicator of calcium homeostasis, so that its measurement is not reliable for indicating the real situation of plasma vitamin D levels, in which case the measurement of calcidiol is preferable (Chung et al., 2009; Holick et al., 1980; Holick et al., 2011).

Low levels of 25(OH)D may be associated with several internal and external factors, such as skin pigmentation, sunscreen use, clothing, latitude, season, air pollution, reduced food intake, malabsorption syndromes, obesity, abstinence from sun, pregnancy, age, and the kidney or liver failure (Holick, 2017; Misra, Pacaud, Petryk, Collett-Solberg, & Kappy, 2008).

Vitamin D deficiency is a major cause of rickets and osteomalacia, indicating the key role this vitamin plays in the bone health. Moreover, such hormone is crucial in the pathogenesis of osteoarthritic kidney disease, where its deficiency may accelerate bone loss and worsen osteoporosis in older adults (Bouillon et al., 2018).

In osteoblasts, the vitamin D receptor is responsible for activating osteoclast maturation (Kitazawa, Kajimoto, Kondo, & Kitazawa, 2003). By resorbing calcium and phosphate in the blood, osteoclasts have a bone-degrading function (Schlesinger, Blair, Teitelbaum, & Edwards, 1997). However, osteoclast activity is stimulated by the PTH only indirectly and inhibited by calcitonin—which acts against PTH main activities, associated with an increase in active vitamin D in the blood (Giustina et al., 2020; Zhong, Armbrecht, & Christakos, 2009).

Role of vitamin D

Besides promoting bone health and preventing rickets, vitamin D also actuates on the prevention of various immune-related diseases in children and adolescents, such as asthma, type 1 diabetes, infectious diseases, recurrent respiratory infections, and cardiovascular diseases (Christakos & DeLuca, 2011; Zhong et al., 2009). Aiming to prevent hypovitaminosis D, the Institute of Medicine (IOM) has established a recommended dietary allowance (RDA) of vitamin D equal to 400 IU/day until the first year of age, and of 600 IU/day for children aged between 1 and 18 years (American Academy of Pediatrics, 2012). Based on the results of a literature review, the IOM concluded that 25-hydroxyvitamin D [25(OH)D] concentration levels of at least 20 ng/mL are considered adequate for 97.5% of the population in all age groups, besides indicating the need for a consensus regarding the cut-off points for vitamin deficiency based on serum 25(OH)D, thus avoiding undertreatment and overtreatment issues related to hypovitaminosis D (Ross et al., 2011).

According to data from the IOM and the Endocrine Society, serum levels of 25(OH)D at 20–30 ng/mL (or 49.92 nmol/L) indicate an adequate amount of vitamin D. In this sense, 25(OH)D serum levels can be classified as follows in Table 2. Being 30 ng/mL is the threshold for bone health (Bischoff-Ferrari, Giovannucci, Willett, Dietrich, & Dawson-Hughes, 2006; Holick & Chen, 2008; Hossein-nezhad, Spira, & Holick, 2013; Rich-Edwards et al., 2011).

PTH levels are abnormal when 25(OH)D serum levels are lower than 20 ng/mL, in which case vitamin D deficiency may lead to skeletal consequences such as secondary hyperparathyroidism and bone loss, increased risk of cancer, impaired brain development in fetuses, oxidative stress in preeclampsia placentas, unregulated inflammation in placentas, and depression (Nowson et al., 2012).

For being highly prevalent and causing a great number of diseases, hypovitaminosis D is now recognized as a pandemic. About one billion people around the world are

Table 2 Classification of vitamin D and serum levels.

Vitamin D levels	Classification
Less than 20 ng/mL	Deficiency
Between 21 and 29 ng/mL	Insufficiency
Between 30 and 100 ng/mL	Sufficiency

Table 3 Average daily recommended amounts of vitamin D.

Ages	Recommended amount
Birth to 12 months	10 μg (400 IU)
1–70 years	15 μg (600 IU)
Pregnant and breastfeeding teens and women	15 μg (600 IU)
Adults 71 years and older	20 μg (800 IU)

estimated to present with vitamin D deficiency or insufficiency (Holick, 2017; Holick et al., 2011; Holick & Chen, 2008).

Average daily recommended amounts of vitamin D are listed below in micrograms (mcg) and International Units (IU, according NIH (National Institutes of Health) (Table 3) (Office of Dietary Supplements—Vitamin D, 2021).

Cerebral palsy and vitamin D

Cerebral palsy is a term used to classify movement and posture disorders attributed to nonprogressive disturbances that occurred in the developing fetal or infant brain, affecting the structure and functioning of the central nervous system (Bax et al., 2005). CP is diagnosed based on clinical criteria, including patients' medical history, physical and neurological evaluations, and motor manifestations—the main clinical characteristic of the disease (Koman, Smith, & Shilt, 2004).

In CP, damage caused to the CNS can occur by endogenous or exogenous factors in varying proportions. Genetic diseases are the endogenous factor, and among the exogenous factors, there is damage caused at different times such as the prenatal, perinatal, and postnatal periods. Evidence suggests that 70%–80% of CP cases are of prenatal origin. The causes of CP are presented in Table 4 (Causes and Risk Factors of Cerebral Palsy | CDC, 2021; Rotta, 2002).

Table 4 Possible etiologies of cerebral palsy.

Prenatal	Diseases such as Rubella, Cytomegalovirus, Toxoplasmosis, Human Immunodeficiency Virus, Von Willebrand disease; exposure to radiation, specific medications, alcohol and illicit drug abuse, severe abdominal trauma, as well as congenital malformation, multiple pregnancies and maternal conditions of mental retardation, hyperthyroidism, malnutrition, severe anemia.
Perinatal	Prematurity (which is associated with 25% of cases), low birth weight, acute or chronic asphyxia (may be related to placental insufficiency), hypoxic–ischemic lesion, intracranial hemorrhage, epilepsy, ischemic stroke, meconial aspiration, cesarean emergency.
Postnatal	Infections, cerebrovascular accidents, trauma, metabolic disorders (hyperglycemia, hypocalcemia, hypomagnesemia), postinfectious encephalitis, hyperbilirubinemia (due to maternal-fetal blood incompatibility), malnutrition, and vascular processes.

Table 5 Classification of gross motor function classification system.

GMFCS	Characteristic
Level I	Unrestricted walks
Level II	Patients able to walk without restrictions
Level III	Patients who require assistive mobility devices
Level IV	Patients with limited mobility who are transported
Level V	Patients who use technology to get around (for example, with wheelchairs)

As motor involvement varies greatly among patients affected with CP, health authorities have proposed the use of the Gross Motor Function Classification System (GMFCS) to assess and classify the severity of neuromotor impairment. This instrument groups patients according to severity levels (Table 5) (Palisano et al., 1997, 2000; Wood & Rosenbaum, 2000).

In addition to limited sunlight exposure, children affected by CP may present with oromotor dysfunction and gastrointestinal problems, hindering optimum nutrition and contributing to lower levels of vitamin D (Aggarwal, Chadha, & Pathak, 2015; Reading, 2012).

Cerebral palsy may co-occur with epilepsy, in which case antiepileptic drug treatment is necessary. The discharge of calcium in the urine and/or failure in calcium reabsorption triggered by the use of these drugs, as is the case of nonenzyme-induced antiepileptic drugs, may affect bone mineral density (BMD), thus increasing the risk of fractures within this population (Samaniego & Sheth, 2007). In turn, enzyme-induced antiepileptic drugs directly change vitamin D concentrations, increasing bone remodeling. Regardless of the above exposed, no protocol for vitamin D supplementation in patients using these drugs has been established up to date (Pack et al., 2005).

In another study, the authors observed that patients undergoing antiepileptic polytherapy had significantly lower values of vitamin D than those receiving monotherapy or in the control group. These data suggest that the use of these medications may inactivate vitamin D by different mechanisms, which are still poorly explored (Tosun et al., 2016).

Different GMFCS classifications (I, II, III, IV, and V) did not emerge as a clear independent risk factor for vitamin D deficiency in studies conducted in Australia (Ware, Whitelaw, Flett, & Parameswaran, 2013) and Iran (Toopchizadeh, Barzegar, Madinei, & Jahanjoo, 2017). In Brazil, vitamin D levels among children and adolescents with CP GMFCS classification V were evaluated and compared with a control group. The results of the study showed that even though Brazil is a sunny country for most of the year, the levels of insufficiency and/or deficiency of vitamin D were prevalent not only in children and adolescents with CP but also in healthy ones (Linden, Lomazi, Hessel, & Bellomo-Brandao, 2021).

Conclusion

Despite the numerous risks associated with the disease, systematic vitamin D supplementation in patients with CP is not recommended, but rather monitoring vitamin D levels, sun exposure, and regular intake of vitamin-rich foods. Vitamin D intoxication is extremely rare; effects related to overdose are vomiting, dehydration, pain, loss of appetite, hypercalcemia, and hyperphosphatemia (Galior, Grebe, & Singh, 2018; Holick, 2017), and frequently are related to errors in supplementation, as overdosing by patients or prescribers and manufacturing errors (Galior et al., 2018; Taylor & Davies, 2018). Therefore, supplementation is suggested with caution.

Applications to other neurological conditions

Other neurological disorders, as is the case with neuromuscular disorders and genetic conditions, may present with multifactorial malnutrition, including swallowing disorders, decreased food intake, gastroesophageal reflux, and intestinal constipation, contributing to the nutritional impairment in these individuals. These patients may also present mobility limitations, which may increase difficulties related to sun exposure and lead to bone deformities that will further affect bone health (Fong, Ong, Ong, Khoo, & Lee, 2020; Kaur & Chakrabarty, 2019).

Epilepsy syndromes requiring treatment with antiepileptic drugs such as phenytoin, phenobarbital, and carbamazepine that may interfere in vitamin D concentrations and osteoblast function, thus favoring a decrease in bone mineral density. In this scenario, monitoring vitamin D levels in high-risk populations is essential for maintaining satisfactory levels and preventing hypovitaminosis D, besides introducing screening and treatment measures (Shellhaas & Joshi, 2010).

Other components of interest

This chapter describes the importance of sun exposure, regular intake of foods rich in vitamin D, and the monitoring of vitamin D levels for maintaining satisfactory levels of this hormone in patients with cerebral palsy.

As motor impairment in patients affected with cerebral palsy varies greatly, children with less mobility may likewise be less exposed to sunlight, consequently presenting lower levels of vitamin D. In a study conducted in India by Seth et al., the authors found that children who walked and used anticonvulsants presented similar 25(OH)D levels than those who did not walk nor used anticonvulsants. However, the mean 25(OH)D level in the group of nonwalking children who received anticonvulsants was lower than other groups, and significantly lower among walking children and in the control group.

These data indicate that mobility limitations and the use of anticonvulsants affect 25(OH)D levels in children affected with cerebral palsy (Christakos & DeLuca, 2011; Seth et al., 2017).

Therefore, patients who have the association of factors, such as use of antiepileptic drugs, limited sun exposure, and malnutrition, represent the greatest risks for hypovitaminosis D and bone health.

Key facts of cerebral palsy (CP): Evaluating vitamin D levels

- Vitamin D may be either obtained through food or produced by our organism, through a photosynthetic reaction in response to sun exposure.
- Low levels of 25(OH)D may be associated with several internal and external factors, such as skin pigmentation, sunscreen use, clothing, latitude, season, air pollution, reduced food intake, malabsorption syndrome, and the kidney or liver failure.
- Vitamin D deficiency is a major cause of rickets and osteomalacia.
- In addition to limited sunlight exposure, children affected by CP may present with oromotor dysfunction and gastrointestinal problems, hindering optimum nutrition and contributing to lower levels of vitamin D.
- CP patients may experience co-occurring epilepsy syndromes, thus requiring treatment with antiepileptic drugs. However, the use of these medications affects bone mineral density (BMD), increasing the risk of bone fractures.

Dictionary of terms

- **Cerebral palsy (CP):** a chronic nonevolutionary encephalopathy that affects the structure and functioning of the central nervous system during its maturation phase. This condition implies the occurrence of motor dysfunctions, including disturbances in muscle tone, posture, and mobility. CP has several clinical presentations, classified according to the extent and intensity of motor dysfunction using the GMFCS.
- **Gross motor function classification system (GMFCS)** is a standardized instrument used to classify the motor functionality of patients according to levels I–V, where level V includes patients with highly compromised automobility, being transported by wheelchair.
- **IOM:** Institute of Medicine.
- **NIH (National Institutes of Health) RDA-(Recommended Dietary Allowance)**, where the RDA for vitamin D is equal to 400 IU/day for the first year of life, and 600 IU/day for children aged between 1 and 18 years.
- **European Society of Pediatric Gastroenterology, Hepatology and Nutrition (ESPGHAN) recommendations:** 25(OH)D $>$ 20 ng/mL: sufficiency, and $<$ 10 ng/mL as severe deficiency.

Summary points

- For being highly prevalent throughout the world, vitamin D deficiency is now recognized as a pandemic.
- Patients with cerebral palsy may present mobility limitations, which may increase difficulties related to sunlight exposure and lead to bone deformities that will further affect the bone health.
- Patients with cerebral palsy may present with multifactorial malnutrition, including swallowing disorders and decreased food intake, contributing to the reduced intake of foods rich in vitamin D.
- Despite the numerous risk factors associated with the condition, studies on systematic vitamin D supplementation in CP patients are still inconclusive.
- Treatment for CP patients includes the monitoring of vitamin D levels, sun exposure, and regular intake of foods high in vitamin D.

References

Aggarwal, S., Chadha, R., & Pathak, R. (2015). Nutritional status and growth in children with cerebral palsy: A review. *International Journal of Medical Science and Public Health*, *4*(6), 737.

American Academy of Pediatrics. (2012). Dietary reference intakes for calcium and vitamin D. *Pediatrics*, *130*(5), e1424.

Bax, M., Goldstein, M., Rosenbaum, P., Leviton, A., Paneth, N., Dan, B., et al. (2005). Proposed definition and classification of cerebral palsy, April 2005. *Developmental Medicine and Child Neurology*, *47*(8), 571–576.

Bischoff-Ferrari, H., Giovannucci, E., Willett, W., Dietrich, T., & Dawson-Hughes, B. (2006). Estimation of optimal serum concentrations of 25-hydroxyvitamin D for multiple health outcomes. *The American Journal of Clinical Nutrition*, *84*(1), 18–28.

Bouillon, R., Marcocci, C., Carmeliet, G., Bikle, D., White, J., Dawson-Hughes, B., et al. (2018). Skeletal and extraskeletal actions of vitamin D: Current evidence and outstanding questions. *Endocrine Reviews*, *40*(4), 1109–1151.

Centers for Disease Control and Prevention. (2021). *Causes and Risk Factors of Cerebral Palsy | CDC*. [online] Available at: https://www.cdc.gov/ncbddd/cp/causes.html>. (Accessed 1 July 2021).

Cesari, M., Incalzi, R., Zamboni, V., & Pahor, M. (2010). Vitamin D hormone: A multitude of actions potentially influencing the physical function decline in older persons. *Geriatrics & Gerontology International*, *11*(2), 133–142.

Christakos, S., & DeLuca, H. (2011). Minireview: Vitamin D: Is there a role in extraskeletal health? *Endocrinology*, *152*(8), 2930–2936. https://doi.org/10.1210/en.2011-0243.

Chung, M., Balk, E. M., Brendel, M., Ip, S., Lau, J., Lee, J., et al. (2009). Vitamin D and calcium: A systematic review of health outcomes. *Evidence Report/Technology Assessment*, (183), 1–420. [online] Available at: https://www.ncbi.nlm.nih.gov/pubmed/20629479. (Accessed 2 July 2021).

Fong, C. Y., Ong, F. N., Ong, L. C., Khoo, T. B., & Lee, M. L. (2020). Vitamin D deficiency and insufficiency in Malaysian children with spina bifida. *Spinal Cord*, *58*(9), 1030–1036.

Galior, K., Grebe, S., & Singh, R. (2018). Development of vitamin D toxicity from overcorrection of vitamin D deficiency: A review of case reports. *Nutrients*, *10*(8), 953.

Giustina, A., Bouillon, R., Binkley, N., Sempos, C., Adler, R., Bollerslev, J., et al. (2020). Controversies in vitamin D: A statement from the third international conference. *JBMR Plus*, *4*(12) (Giustina et al., 2020).

Hess, A. F. (1922). The prevention and cure of rickets by sunlight. *American Journal of Public Health*, *12*(2), 104–107.

Holick, M. (2007). Vitamin D Deficiency. *New England Journal of Medicine, 357*(3), 266–281.
Holick, M. (2017). The vitamin D deficiency pandemic: Approaches for diagnosis, treatment and prevention. *Reviews in Endocrine & Metabolic Disorders, 18*(2), 153–165.
Holick, M., & Chen, T. (2008). Vitamin D deficiency: A worldwide problem with health consequences. *The American Journal of Clinical Nutrition, 87*(4), 1080S–1086S.
Holick, M., MacLaughlin, J., Clark, M., Holick, S., Potts, J., Anderson, R., et al. (1980). Photosynthesis of previtamin D3 in human skin and the physiologic consequences. *Science, 210*(4466), 203–205.
Holick, M. S., Binkley, N. C., Bischoff-Ferrari, H. A., Gordon, C. M., Hanley, D. A., Heaney, R. P., et al. (2011). Evaluation, treatment, and prevention of vitamin D deficiency: An Endocrine Society clinical practice guideline. *The Journal of Clinical Endocrinology and Metabolism, 96*(7), 1911–1930. https://doi.org/10.1210/jc.2011-0385. In this issue.
Hossein-nezhad, A., Spira, A., & Holick, M. (2013). Influence of vitamin D status and vitamin D3 supplementation on genome wide expression of white blood cells: A randomized double-blind clinical trial. *PLoS One, 8*(3), e58725.
Kaur, P., & Chakrabarty, B. (2019). Vitamin D and neurological disorders: The conundrum continues. *The Indian Journal of Pediatrics, 86*(9), 771–772.
Kitazawa, S., Kajimoto, K., Kondo, T., & Kitazawa, R. (2003). Vitamin D3 supports osteoclastogenesis via functional vitamin D response element of human RANKL gene promoter. *Journal of Cellular Biochemistry, 89*(4), 771–777.
Koman, L., Smith, B., & Shilt, J. (2004). Cerebral palsy. *The Lancet, 363*(9421), 1619–1631.
Linden, M. F., Lomazi, E., Hessel, G., & Bellomo-Brandao, M. A. (2021). Vitamin D levels in children and adolescents with cerebral palsy: Cross-sectional study. *Brazilian Journal of Health Review, 4*(3), 11938–11949.
Misra, M., Pacaud, D., Petryk, A., Collett-Solberg, P., & Kappy, M. (2008). Vitamin D deficiency in children and its management: Review of current knowledge and recommendations. *Pediatrics, 122*(2), 398–417.
Nowson, C., McGrath, J., Ebeling, P., Haikerwal, A., Daly, R., Sanders, K., et al. (2012). Vitamin D and health in adults in Australia and New Zealand: A position statement. *Medical Journal of Australia, 196*(11), 686–687.
Office of Dietary Supplements—Vitamin D. (2021). *Office of dietary supplements—Vitamin D.* [online] Available at: https://ods.od.nih.gov/factsheets/VitaminD-Consumer>. (Accessed 1 July 2021).
Pack, A., Morrell, M., Marcus, R., Holloway, L., Flaster, E., Doñe, S., et al. (2005). Bone mass and turnover in women with epilepsy on antiepileptic drug monotherapy. *Annals of Neurology, 57*(2), 252–257.
Palisano, R., Hanna, S., Rosenbaum, P., Russell, D., Walter, S., Wood, E., et al. (1997). Development and reliability of a system to classify gross motor function in children with cerebral palsy. *Developmental Medicine and Child Neurology, 30*(4), 214–223.
Palisano, R., Hanna, S., Rosenbaum, P., Russell, D., Walter, S., Wood, E., et al. (2000). Validation of a model of gross motor function for children with cerebral palsy. *Physical Therapy, 80*(10), 974–985.
Reading, R. (2012). Feeding difficulties in children with cerebral palsy. *Child: Care, Health and Development, 38*(3), 455–456.
Rich-Edwards, J., Ganmaa, D., Kleinman, K., Sumberzul, N., Holick, M., Lkhagvasuren, T., et al. (2011). Randomized trial of fortified milk and supplements to raise 25-hydroxyvitamin D concentrations in schoolchildren in Mongolia. *The American Journal of Clinical Nutrition, 94*(2), 578–584.
Ross, A., Manson, J., Abrams, S., Aloia, J., Brannon, P., Clinton, S., et al. (2011). The 2011 dietary reference intakes for calcium and vitamin D: What dietetics practitioners need to know. *Journal of the American Dietetic Association, 111*(4), 524–527. This article is a summary of the institute of medicine report entitled dietary reference intakes for calcium and Vitamin D (Available at http://www.iom.edu/Reports/2010/Dietary-Reference-Intakes-for-Calcium-and-Vitamin-D.aspx) for dietetics practitioners; a similar summary for clinicians has also been published (Ross AC, Manson JE, Abrams SA, Aloia JF, Brannon PM, Clinton SK, Durazo-Arvizu RA, Gallagher JC, Gallo RL, Jones G, Kovacs CS, Mayne ST, Rosen CJ, Shapses SA. The 2011 report on Dietary Reference Intakes for calcium and vitamin D from the Institute of Medicine: What clinicians need to know. J Clin Endocrinol Metab. 2011;96:53–58).
Rotta, N. (2002). Cerebral palsy, new therapeutic possibilities. *Journal de Pediatria, 78*(7), 48–54.

Rovner, A., Stallings, V., Kawchak, D., Schall, J., Ohene-Frempong, K., & Zemel, B. (2008). High risk of vitamin D deficiency in children with sickle cell disease. *Journal of the American Dietetic Association, 108*(9), 1512–1516.

Samaniego, E., & Sheth, R. (2007). Bone consequences of epilepsy and antiepileptic medications. *Seminars in Pediatric Neurology, 14*(4), 196–200.

Schlesinger, P., Blair, H., Teitelbaum, S., & Edwards, J. (1997). Characterization of the osteoclast ruffled border Chloride channel and its role in bone resorption. *Journal of Biological Chemistry, 272*(30), 18636–18643.

Seth, A., Aneja, S., Singh, R., Majumdar, R., Sharma, N., & Gopinath, M. (2017). Effect of impaired ambulation and anti-epileptic drug intake on vitamin D status of children with cerebral palsy. *Paediatrics and International Child Health, 37*(3), 193–198.

Shellhaas, R., & Joshi, S. (2010). Vitamin D and bone health among children with epilepsy. *Pediatric Neurology, 42*(6), 385–393.

Taylor, P. N., & Davies, J. S. (2018). A review of the growing risk of vitamin D toxicity from inappropriate practice. *British Journal of Clinical Pharmacology, 84*(6), 1121–1127.

Toopchizadeh, V., Barzegar, M., Madinei, N., & Jahanjoo, F. (2017). A survey on growth and nutritional status of children with cerebral palsy in the Northwest Iran. *Iranian Journal of Child Neurology, 12*(2), 25–32.

Tosun, A., Erisen Karaca, S., Unuvar, T., Yurekli, Y., Yenisey, C., & Omurlu, I. (2016). Bone mineral density and vitamin D status in children with epilepsy, cerebral palsy, and cerebral palsy with epilepsy. *Child's Nervous System, 33*(1), 153–158.

Ware, T., Whitelaw, C., Flett, P., & Parameswaran, V. (2013). Vitamin D status in Tasmanian children with cerebral palsy. *Journal of Paediatrics and Child Health, 49*(4), E349–E350.

Windaus, A., Linsert, O., Lüttringhaus, A., & Weidlich, G. (1932). Über das krystallisierte Vitamin D2. *Justus Liebigs Annalen der Chemie, 492*(1), 226–241.

Wood, E., & Rosenbaum, P. (2000). The gross motor function classification system for cerebral palsy: A study of reliability and stability over time. *Developmental Medicine and Child Neurology, 42*(5), 292–296.

Zhong, Y., Armbrecht, H., & Christakos, S. (2009). Calcitonin, a regulator of the 25-hydroxyvitamin D3 1α-hydroxylase gene. *Journal of Biological Chemistry, 284*(17), 11059–11069.

PART V

Dietary neurotoxins

CHAPTER 13

Linking copper and neurotoxic activities to the CNS

Gesivaldo Santos[a], Marcos Avila-Rodriguez[b], Érika Pereira Rubio[c], Eduardo Galembeck[d], and Julita Maria Pereira Borges[c]
[a]Department of Biological Sciences, State University of Southwest of Bahia, Jequié, Bahia, Brazil
[b]Department of Clinical Science, University of Tolima, Tolima, Colombia
[c]Department of Science and Technology, State University of Southwest of Bahia, Jequié, Bahia, Brazil
[d]Institute of Biology, Department of Biochemistry, State University of Campinas-São Paulo, Campinas, São Paulo, Brazil

Abbreviations

AD	Alzheimer's disease
APP	amyloid precursor protein
ATOX1	transport metallochaperones protein
ATP	adenosine triphosphate
ATP7A	copper-transporting P-type ATPase
Aβ	β-amyloid
BBB	blood-brain barrier
CCS	cytosolic metallochaperones
CTR	copper transports
CUBD	copper-binding domain
Cu	copper
DJ-1	protein/nucleic acid deglycase
DMN	default mode network
DMT1	divalent metal transporter 1
DNA	deoxyribonucleic acid
EFSA	European Food and Safety Authority
ER	endoplasmic reticulum stress
ETC	electron transport chain
GPx	glutathione peroxidase enzymes
LRP1	density lipoprotein 1
LRRK2	leucine-rich repeat kinase 2
NADPH	nicotinamide and adenine dinucleotide phosphate
NFT	neurofibrillary tangles
NOX	NADPH oxidase
PARK2	Parkinson's protein mutations 2
PD	Parkinson's disease
PrPC	cellular prion protein
RAGE	advanced glycation end-products
ROS	reactive oxygen species
SOD1	superoxide dismutase 1
SUMO	small ubiquitin-like modifier
Zn	zinc

Vitamins and Minerals in Neurological Disorders
https://doi.org/10.1016/B978-0-323-89835-5.00035-1

Introduction

Copper (Cu) is a chemical element belonging to the group of transition metals that can present itself in two oxidation states: Cu (I) and Cu (II) (Palumaa, 2013; Rossi, Lombardo, Ciriolo, & Rotilio, 2004). It is an essential trace element from the diet that participates in important physiological processes, as a cofactor of several enzymes (Bisaglia & Bubacco, 2020; Pena, Lee, & Thiele, 1999). The highest concentrations of copper are present in tissues that have great metabolic activity, such as the brain and the liver (Lutsenko, Bhattacharjee, & Hubbard, 2010; Rossi et al., 2004).

Cu is involved with enzymes that take part in processes such as antioxidant defense of cells (superoxide dismutase), electron transport in mitochondria (cytochrome *c* oxidase), iron homeostasis (ceruloplasmin), melanin synthesis (tyrosinase), collagen crosslinking and elastin (lysyl oxidase), synthesis of neurotransmitters (dopamine β-hydroxylase), bioactivation of peptide hormones (peptidylglycine monooxygenase), induction of blood vessel formation and integrity (angiogenin) (Bisaglia & Bubacco, 2020; Pena et al., 1999; Rossi et al., 2004).

The production of adenosine triphosphate (ATP) in the cell respiration cycle that takes place in the mitochondria is catalyzed by the electron transport chain (ETC), which in turn acts in the presence of Cu ions, transporting electrons from cytochrome-c to oxygen (Ackerman, 2018; Bisaglia & Bubacco, 2020; Palumaa, 2013). The transport of Cu to the ETC is essential to produce energy (ATP) and cell survival, considering that the efficiency in the cell respiration process is dependent on the presence of this ion (Palumaa, 2013).

Due to its peculiar characteristic of transiting between different oxidation states (redox chemistry), Cu takes part in the primary antioxidant protection of cells, acting as a cofactor in the active site of the enzyme Cu/Zn-superoxide dismutase (SOD1). SOD1 mediates cell protection by catalyzing superoxide radicals in oxygen and hydrogen peroxide (Ackerman, 2018). Despite being essential in cell protection against reactive oxygen species (ROS), its redox chemistry causes the Cu homeostasis imbalance to be involved with the generation of highly reactive radicals, making it toxic when present in high concentrations to cellular functions (Mezzaroba, Alfieri, Colado Simao, & Vissoci Reiche, 2019). In other words, Cu's homeostatic equilibrium condition guarantees its protective activity as a cofactor in the enzyme Cu/Zn-superoxide dismutase (SOD1), in addition to not amplifying the production of ROS from the presence of high concentrations of copper (Culotta, Yang, & O'Halloran, 2006; Rossi et al., 2004).

Cu transport in the blood is carried out by proteins such as albumin and, mainly, ceruloplasmin (Cp), which has the capacity to transport up to six Cu ions (Bisaglia & Bubacco, 2020). The high reactivity of Cu in the free form interacts with these transport proteins also assume an important role in the control of Cu homeostasis (Rossi et al., 2004), being involved with mechanisms that protect cells from oxidative stress (Davies et al., 2013). Free Cu can produce and amplify the generation of reactive oxygen species (ROS) which, in turn, can cause damage to membranes (lipid peroxidation), proteins (direct oxidation), DNA and RNA (cleavage) (Pena et al., 1999), neurodegenerative disorders, aging, and even cancer (Davies et al., 2013; Lutsenko et al., 2010; Palumaa, 2013; Pena et al., 1999).

How do we get copper?

Copper and the diet

Almost all copper incorporated into our body comes from food and water in a concentration of which varies greatly depending on the type of food, as well as the soil and even the presence of copper in sources of water (Brewer, 2015; Hsu, Bondy, & Kitazawa, 2018). However, it is important to note that the richest sources of the mineral are of animal origin such as seafood and liver (Scheiber, Mercer, & Dringen, 2014). Another secondary natural sources of Cu in food are fish, peas, lentils, dried fruits, seeds, vegetables, and crustaceans. In addition, breast milk, despite being an abundant source of copper for the baby's diet, undergoes a varying concentration, according to the geographic region (Klein et al., 2017).

The lack of a physiological marker with high specificity for copper makes it difficult to adopt a dietary pattern to define a recommended daily amount of copper. Thus, these data usually are obtained indirectly by agencies, such as European Food and Safety Authority (EFSA), based on observation in healthy populations (EFSA, 2015). In addition, baseline levels of copper are strictly regulated in our bodies, so that excess copper levels tend to be excreted in the urine and feces. On the other hand, levels below the physiological pattern tend to increase Cu due to the upregulation of copper transport channels (Scheiber, Dringen, & Mercer, 2013).

According to the EFSA, the average daily intake of copper in developed countries is estimated between 1 and 5 mg/day varying, as seen previously, between different world regions. From this amount, the drinking water accounts for 0.1–1 mg/day. Yet EFSA estimates that a healthy daily intake of copper for adult is about 1.6 mg/day in man and 1.3 mg/day for woman. For children, depending on the age group, this variation is contained in the range of 1 mg/day for children under 3 years old and 1.1 mg/day for individuals under 18 years old (Harvey et al., 2003). Currently, studies have adopted the use of high throughput techniques to determine risk association of the disease's development related to copper metabolic control, such as Menke's and Wilson's disease. The analysis is based on the genomic and proteomic evaluation of healthy individuals, as well as those with copper pathologies (Conlan, Korula, & Tallentire, 1990).

Nutritional aspects and disease

Copper absorption and transport

Copper absorption in our body largely depends on the distribution of the Cu transporters (CTR1) family, which varies widely according to the tissue observed. However, in humans, the absorption of the mineral occurs between the stomach and the duodenum, from which gains the portal circulation by means of active transport mediated by copper-transporting P-type ATPase (ATP7A) (Petris, 2004). The absorption of cuprous (Cu^{+}) on the surface of intestinal microvilli is facilitated by the Cupper 1 (CTR1, (SLC31A1)) and cupric (Cu^{2+}) transporters by the divalent metal transporter 1 ((DMT1), DMT1-IRE, and DMT1-nonIRE) (Prohaska, 2008; Zheng et al., 2009). It is then quickly absorbed by hepatocytes, converting the liver into the main reservoir of copper, followed by the brain Fig. 1 (Bisaglia & Bubacco, 2020).

In neuronal cells, like that observed in enterocytes, the absorption of copper will be established according to the main transporters: copper transporter-1 or CTR1 and the DMT1, but first, it must be able to cross the blood-brain barrier, as seen in Fig. 2. CTR1 is commonly found in the apical portion of cell membranes and uptake Cu through endocytosis. While DMT1 is predominantly distributed in the intracellular region, and may eventually be found in the somma, depending on the Cu level in the diet, which also regulate CTR1 expression by negative feedback (Lutsenko, 2010). Inside the cell, copper transport is mediated by cytosolic metallochaperones (CCS), whose function is to transfer copper to superoxide dismutase, an enzyme present in mitochondrial intermembrane space. While another copper transporter, copper transport metallochaperones protein (ATOX1), will mediate the transport of copper to the Golgi complex from where it will be packaged into vesicles and secreted through active transport via ATP7A and ATP7B in opposite directions (Fig. 2) (Boal & Rosenzweig, 2009).

Homeostasis of copper in the brain

Copper and brain metabolism

Copper is an essential trace element in neurochemistry, involved in the synthesis of neurotransmitters, myelination, and activation of neuropeptides (Madsen & Gitlin, 2007). Like iron, copper has an enormous capacity to gain and donate electrons. After absorption and compartmentalization of copper in the body, the cycle of copper ions alternates between cupric Cu^{2+} (oxidized) and Cu^{+} (reduced) states (Kim, Nevitt, & Thiele, 2008).

It consists of prosthetic groups of various enzymes, such as ceruloplasmin (Cp), cytochrome-c oxidase, dopamine β-hydroxylase, tyrosinase, proteins involved in axonal targeting and neurite growth, in the modulation of signaling cascades induced by neurotrophic factors, with inhibitory and excitatory modification in neuronal transmission (Gromadzka, Tarnacka, Flaga, & Adamczyk, 2020; Magrì & La Mendola, 2018). Due to this, cellular regulation of metal trafficking, like copper, with such oxidizing power,

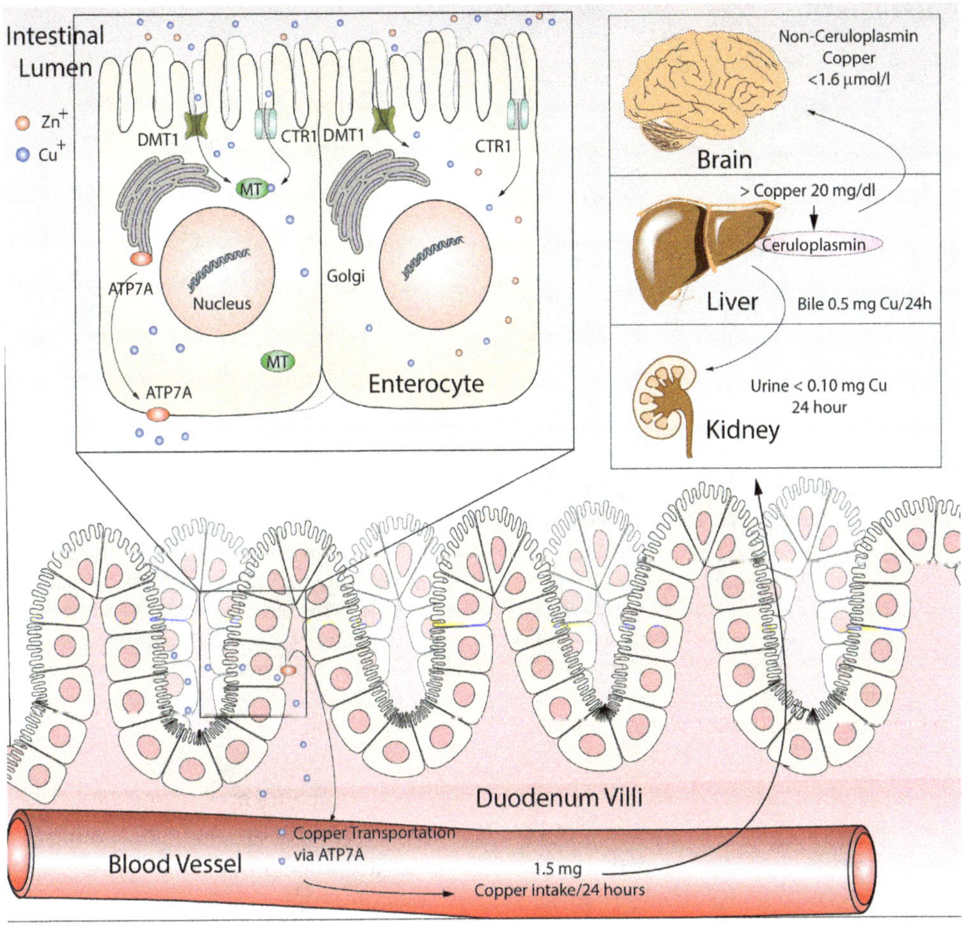

Fig. 1 In the figure above we have the schematic representation of the proposed process for the absorption of copper in the intestinal lumen. According to the model, copper would be captured in the apical region of the enterocytes by the transport proteins CTR1 and DMT1. Inside the cell, Cu follows two different pathways. The mitochondrial pathway by which COX17 is transported into the mitochondria and the trans-Golgi network pathway, whose transport is in charge of ATOX1, and is mainly destined for the kidneys, the liver, and the brain. Abbreviations: *CTR1*, copper transporter 1; *DMT1*, divalent metal transporter1; *ATOX1*, copper transport protein ATOX1.

must occur in a highly controlled manner, even though copper is constantly being ingested either through food or through water intake (Linder, 2020). In mammals, this balance is actively maintained by an enterohepatic recirculation process in which excess of copper is excreted mainly by the bile and feces (Shi, 2021). Acute copper dyshomeostasis occurs precisely when the net rate of Cu excreted, compared with Cu consumed, becomes unbalanced and, as seen previously, it is determined by environmental factors, such as high concentration of Cu present in water and foods (Squitti, Siotto, & Polimanti, 2014).

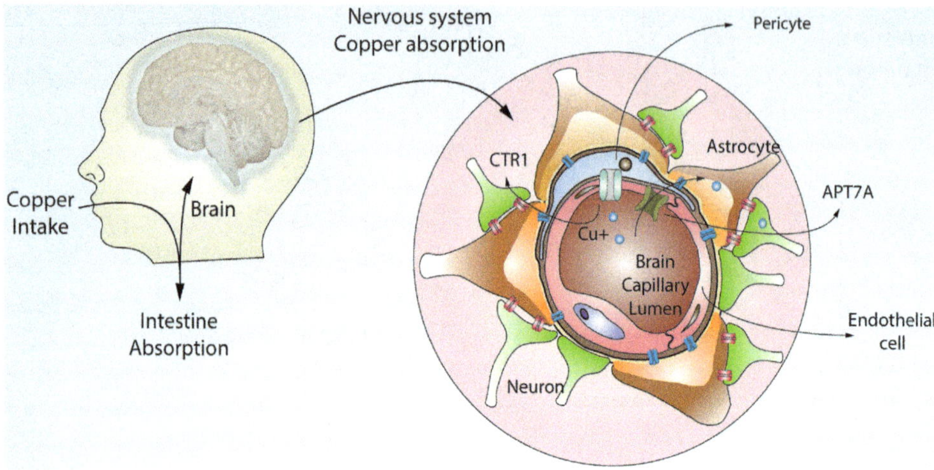

Fig. 2 Cu transport in the CNS is slightly different from the mechanism proposed for other tissues. As seen in the image above, the flow of Cu in the CNS would occur through blood capillaries of the BBB mediated in opposite directions by the CTR1 and ATP7B transporters, expressed in the luminal region of the endothelial cells. After being absorbed by CTR1, Cu would follow routes like those mentioned in Fig. 1, through which it would be expelled to ventricular regions by means of the ATP7A transporter. ATP7B would take part in the Cu excretion process, pumping it from the CNS back into the bloodstream. Abbreviations: *CTR1*, copper transporter1; *ATP7B*, copper-transporting ATPase2; *ATP7A*, copper-transporting ATPase1. *Adapted from Kaler, S. G. (2011). ATP7A-related copper transport diseases-emerging concepts and future trends.* Nature Reviews. Neurology, *7(1), 15–29.*

In the brain, additional Cu-binding proteins, such as amyloid precursor protein (APP) and the prion protein, bind to Cu and control copper homeostasis, being maintained in a small safety margin between the optimal physiological concentration and the onset of toxicity (Opazo, Greenough, & Bush, 2014). The metabolic disorder of copper is associated with different toxic effects, mainly represented by oxidative stress (Cheignon et al., 2018) and inflammation (Huat et al., 2019), being related to many neurodegenerative diseases, such as Alzheimer's disease (AD) and Parkinson's disease (PD) (Bandmann, Weiss, & Kaler, 2015; Brown et al., 1997; Rose, Hodak, & Bernholc, 2011; Xiao, Fan, Wang, & Zhou, 2013). In the chronic condition, by far the most frequent imbalance, the breakdown of copper homeostasis is determined by genetic factors such as mutations in metal transport proteins (Squitti et al., 2014). Two important examples of Cu imbalance are observed in Menke's and Wilson's diseases, in which changes in ATP7A and ATP7B transporters lead to abnormal copper levels imbalance (Kaler, Ferreira, & Yam, 2020; Vairo et al., 2019).

Copper and genetic disorder: Wilson's and Menke's disease

In Menke's disease, an X-linked recessive disorder, mutations in the gene responsible for coding P-type ATPase copper transporting protein (ATP7A), is characterized by neurodegeneration of the brain and the cerebellum. Manifesting in early onset of life causes a devastating effect on the nervous system like epilepsy, hypotonia, and neurodegeneration, even leading to death, if there is no adequate Cu supplementation. As seen in "Copper absorption and transport" section, after being captured by the enterocytes, part of the Cu follows to be excited in the bloodstream. The mutation in ATP7A hinders Cu's intestinal absorption, preventing it from reaching the systemic circulation, with the consequence of the low distribution of copper in all tissues (Fig. 3A) (Kaler et al., 2020; Vairo et al., 2019).

On the other hand, in Wilson's disease, an autosomal recessive and inherit condition, defects in the gene encoding the ATP7B protein, causes Cu levels to rise slowly, as this mutation hinders the bile excretion of copper (Chang & Hahn, 2017). ATP7B encodes the transport membrane protein, responsible for pumping Cu from the Golgi complex for its incorporation into ceruloplasmin (Cp) (Kaler, 2011; Squitti et al., 2014) (Fig. 3B). Thus, with the systemic spread of non-Cp-Cu, several tissues start to suffer damage, like the liver and the brain, in the latter, overload of the metal causes the patient to develop behavioral changes and a type of dyskinesia similar to Parkinson's, as we will see further up (Litwin et al., 2013).

Copper and cellular neurodegeneration

Bioinorganic complexes resulting from copper with protein also involve APP and cellular prion protein (PrP^C), with oxidizing and antioxidant action of neuronal structures, contributing to neurodegeneration reactions (Opazo, Barría, Ruiz, & Inestrosa, 2003; Salzano, Giachin, & Legname, 2019). Aβ levels in the brain are regulated by their rate of cerebral production of APP (Provias & Jeynes, 2014). The transport of Aβ across the blood-brain barrier (BBB) which involves specific proteins constitutes the endothelial barrier: density lipoprotein (LRP1), P-glycoprotein, and advanced glycation end-products (RAGE) (Deane et al., 2004; Provias & Jeynes, 2014). The accumulation of Cu in the cerebral endothelium seems to be an attempt to protect the brain from the actions of free Cu excess (Saido, 2013). However, the negative regulation of LRP1 in the endothelium can lead to the clearance of Aβ through the BBB, accumulating in the brain, leading to the increase of neuraldegenertion (Singh et al., 2013).

Copper and Alzheimer

Neurodegenerative pathogenesis of AD is caused by genetic and environmental factors, characterized by incorrect unfolding and aggregation of brain proteins, β-amyloid (Aβ) and tau, leading to the formation of neurofibrillary tangles (NFT) and plates. (Huat et al., 2019) It is a progressive disease associated with behavioral and psychological symptoms of dementia (Hsu et al., 2018).

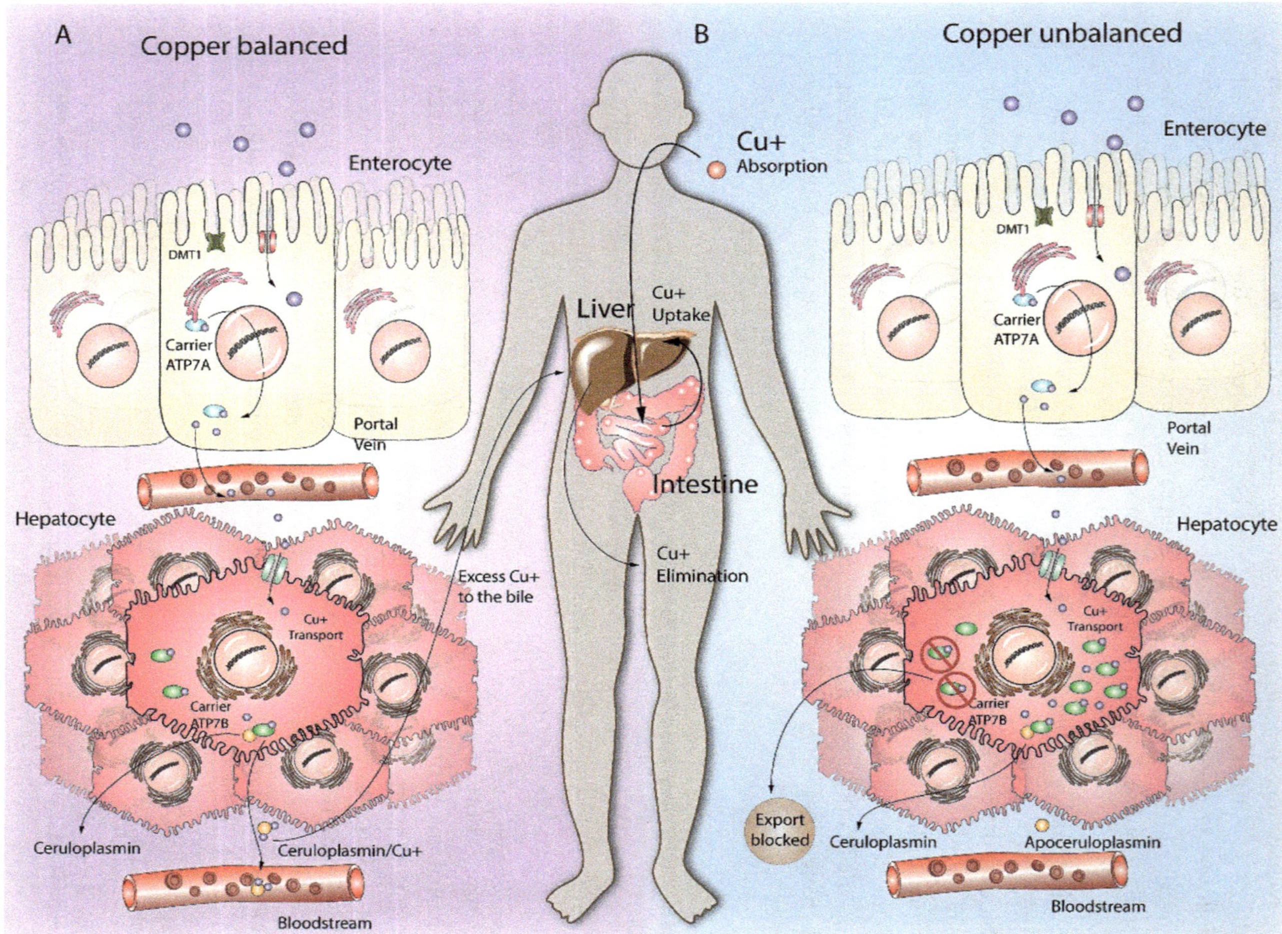

Fig. 3 (A) In Menkes disease, a mutation in ATP7A prevents the absorption of copper from the intestinal lumen. This change prevents copper from reaching the systemic circulation and, therefore, accumulates slowly in different tissues like the kidneys, but poorly in the liver and the brain. (B) In Wilson's disease, mutation in the gene encoding the ATP7B protein prevents the excretion of copper from the brain parenchyma back into the blood, causing an overload of the metal in the CNS. *Adapted from https://rare-liver.eu/healthcare-professionals/disease-specific-information/wilson-disease (Accessed 30 December 2022).*

Molecular interaction resulting from the affinity of copper to the extracellular domain of the transmembrane APP, due to excess copper or due to aberrant APP protein processing, in the neurons, has been shown to promote Aβ production and in vitro fibrillation (Huang et al., 1999). This interaction in the Aβ sequence, the redox-active Cu^{2+}, but not Cu^{+}, coordinated by histidine, promotes a change in conformation, with the release of reactive oxygen species involved in mitochondrial oxidative damage (Jiang et al., 2007). The N-terminal region of APP, E1, contains a copper binding domain (CUBD) through which it is cleaved by APP cleavage enzyme 1 at the beta site (BACE1) and γ-secretase (Kimura, Hata, & Suzuki, 2016; Yan, 2017).

Aβ peptide is subject to incorrect folding and aggregation, leading to eventual oligomerization and formation of AD (Weibull, Simonsen, Oksbjerg, Tiwari, & Hemmingsen, 2019). The dimerization of C-terminal APP fragments also seems to contribute to the packaging of Aβ peptides, in the formation of crossed β sheets of protofibril structures (Chen, Oh, Hinman, & Abraham, 2006).

Aβ plaques, which trigger a proinflammatory response from microglia and astrocytes, in Aβ also cause hyperphosphorylation of tau, leading to its dissociation from the microtubules and consequent destabilization within the neurons (Krasemann et al., 2017; Sala Frigerio et al., 2019). Hyperphosphorylated tau is also prone to aggregation, forming NFT, which correlates with neuronal loss and neurodegeneration (Chong, Ng, Koh, & Chye, 2018; Pîrşcoveanu et al., 2017; Sebastián-Serrano, de Diego-García, & Díaz-Hernández, 2018).

Rare mutations in APP (PSEN1 and PSEN2) alter the processing of this protein and are known to be strong genetic causes of AD (Lanoiselée et al., 2017). Other variations in genes related to lipid metabolism, endocytosis, and inflammatory responses, such as apoprotein E (APOE), TREM2, and CD33, are more common in the population, but confer moderate to low risk for AD (Huat et al., 2019; Kim, 2018).

No genetic mutations are observed involving Tau. However, abnormal posttransductional changes such as hyperosphorylation, small ubiquitin-like modifier (SUMO), and glycosylation may be synergistically involved in neurodegeneration in genesis of DA (Wang et al., 2020; Wang, Grundke-Iqbal, & Iqbal, 1996). Hyperphosphorylation of Tau appears to have an initial neuroprotective function (Wang & Liu, 2008). However, the accumulation of phosphorylated Tau leads to neurodegenerative disorder due to cellular dysfunctions such as deficits of mitophagy and autophagy, endoplasmic reticulum stress (ER), and synaptic transmission deficit (González, Farías, & Maccioni, 1998). The SUMO conjugation and binding to target proteins regulate epigenetic via pathways involved in the production, aggregation, and/or elimination of poorly folded proteins found in AD (Knock et al., 2018; Yau, Molina, & Courey, 2020). Other tau-modifying reactions involve glycation or nonenzymatic glycosylation. Glycation of tau does not seem to be the starting point for aggregation; however, it seems to promote the accumulation of tau aggregation (Alonso, Zaidi, Grundke-Iqbal, & Iqbal, 1994; Fontaine et al., 2015).

Copper and association with Parkinson's disease

The degenerative neurochemistry of PD has a multifactorial origin with achievements in the movement of the body and in the command of motor coordination. Among the causal factors, excess copper in the brain leads to the production of highly damaging free hydroxyl radicals by participating in Fenton or Haber-Weiss reactions, resulting in oxidative stress and cytotoxicity involved in the large loss of dopamine-producing neurons in the brain area called pars compacta of the substantia nigra. (Pal, Kumar, & Prasad, 2014; Pohanka, 2019; Strausak, Mercer, Dieter, Stremmel, & Multhaup, 2001)

The binding of copper to alpha-synuclein results in protein fibrillation and increased oxidative stress (Meloni & Vašák, 2011). Soluble α-synuclein-Cu (II) complexes have dopamine oxidase activity involved in the production of o-quinone and hydroxydopamine in the presence of biological reducing agents via the redox cycle Cu (II)/Cu (I). (Calvo, Mulpuri, Dao, Qazi, & Meloni, 2020) To protect neurons from the damage induced by 6-hydroxydopamine, glial cells are able to release the extracellular isoform of superoxide dismutase (Takano, Tanaka, Kawabe, Moriyama, & Nakamura, 2013). However, the persistence of oxidative stress causes DNA cleavage, protein oxidation, and lipid peroxidation, which results in poorly folded protein aggregates (mainly α-synuclein) and oxidized protein fragments, known as Lewy bodies. The binding of copper with α-synuclein is the main component of Lewy bodies. Oxidative DNA damage, when not repaired, appears to play an important role in Parkinson's protein mutations 2 (PARK2), leucine-rich repeat kinase 2 (LRRK2), phosphatase and tensin homolog (putative kinase 1 induced by PTEN) (PINK1), and DP protein 7 (autosomal recessive, early onset) (DJ-1) (Pan-Montojo et al., 2010; Rappold & Tieu, 2010; Wang et al., 2011). Genetic products involved in DP also contribute to increase oxidative stress, mainly due to mitochondrial dysfunction.

Parkinson's neuroprotection evolves an active detoxifying system (glutathione peroxidase enzymes (GPx; NADPH oxidase (NOX)), based on a constant antioxidant defense, modulated mainly in glial cells and system for removing unwanted damaged proteins (ubiquitin-proteasome and mitophagy) (Dias, Junn, & Mouradian, 2013; Senkevich & Gan-Or, 2020). In the accumulation of copper and during neurodegeneration reactions, the antioxidant defense is decreased. Antioxidant status, natural vitamins, antioxidant enzymes and peptides, antioxidant signaling pathways are the main factors to neutralize ROS preventing oxidative damage.

Copper supplementation

Copper and zinc supplementation

Several studies have demonstrated the existence of an inverse correlation of plasma inorganic copper levels with cognitive deficiencies, especially those observed in AD (Brewer,

2014). This evidence is so striking that some authors like Sparks and Schreurs (2003) have even postulated that the ingestion of free inorganic copper in water could be understood as the main environmental risk factor associated with the development of the Alzheimer's disease. However, what makes this fact even more impressive is that such level of copper is at least ten times lower than those found in the diet consumed by humans in different parts of the world (Singh et al., 2013).

As previously seen, the brain makes use of copper's powerful versatility in mediating redox reactions. However, this copper becomes associated with proteins, whose main function is to supply the high neuronal metabolic demand in different parts of the brain, mainly those related to default mode network (DMN). In these regions, the free copper levels tend to increase with age, reaching quite high levels in patients with AD. On the other hand, zinc levels are reduced in these same individuals with cognitive impairment, mainly because circulating Zn tends to be strongly captured by the amyloid protein present in senile plaques (Adlard, Parncutt, Finkelstein, & Bush, 2010; Sandusky-Beltran, Manchester, & McNay, 2017).

Zinc is undoubtedly one of the most important micronutrients, playing a considerable role in the oxidation state of copper and, consequently, in the conformation of its molecular structure when associated with other compounds such as SOD1 and A-Beta peptide (Sandusky-Beltran et al., 2017). It has also been shown that Zn improves short- and long-term memory formation, one of the most severe processes impacted by Alzheimer's progression according to Takeda et al. (2014); this fact must be related to the reduction in serum copper levels promoted by the presence of Zn. Taken together, it would be natural to infer that zinc food supplementation could be established as an alternative treatment to reduce the progression of Alzheimer's condition, since low levels of Zn are identified as a risk factor to AD development, as it has been demonstrated by some experiments (Brewer, 2014).

Conclusion

As we have seen along the chapter, copper is an essential trace element obtained from the diet, which participates in significant physiological processes, such as in the primary antioxidant protection of cells, acting as a cofactor in the active site of the enzyme Cu/Zn-superoxide dismutase (SOD1) (Bisaglia & Bubacco, 2020; Lutsenko et al., 2010; Pena et al., 1999; Rossi et al., 2004).

The highest concentrations of copper are present in tissues that have great metabolic activity, like the brain and the liver (Lutsenko et al., 2010; Rossi et al., 2004). In the brain, copper is primarily associated with synthesis of neurotransmitter and myelin formation (Madsen & Gitlin, 2007), and due to its high oxidizing power and narrow variation in oxidative state, copper dyshomeostasis can lead to severe neurological disorders, almost always associated with inflammatory processes, capable to lead to development of neurodegeneration, epilepsy, and motor disorders.

Studies show that the incorporation of copper in our body varies greatly with environmental factors such as food, water sources, and even geographic location. Thus, a lack of specific biological markers for copper makes it difficult to determine a standard concentration for its daily consumption. However, given its metabolic importance, copper intake levels are closely monitored by government agencies, which adopt limited values for Cu intake, based on the average intake of their populations (EFSA, 2015; Harvey et al., 2003).

Mini-dictionary of terms

Alpha-synuclein: Alpha-synuclein is an abundant neuronal protein that is highly enriched in presynaptic nerve terminals.

Alzheimer's disease: Alzheimer's disease is a progressive neurodegenerative disease that starts with mild memory loss and possibly leads to a loss of the ability to carry on a conversation and respond to the environment.

BACE1: Beta-secretase 1 is the amyloid precursor protein cleavage enzyme.

B-BB: Blood-brain barrier consists of a set of cell membranes, including endothelium and astrocytes, which together control the influx and efflux of substances into the brain environment.

Beta-amyloid: Beta amyloids are peptides with 36–43 amino acids resulting from cleavage by β- and γ-secretases in a transmembrane protein, the amyloid precursor protein (APP), being the main constituent of the amyloid plaques observed in the brain of patients with Alzheimer's disease.

Cytosolic metallochaperones (CCS): CCS is an essential cytosolic factor that maintains copper homeostasis in living cells.

Cu/Zn: Copper/zinc are two essential minerals for important biochemical functions in maintaining homeostasis throughout life.

DMN: Default mode network has been identified as an automatic brain module, since its activity seems to correlate negatively with tasks that require high selectivity, such as the development of attentional activities.

ETC: Electron transport chain consists of a complex of enzymes, located in the inner membrane of the mitochondrial, whose function is to gradually lower the energy level of electrons taken from organic species, such as sugar derivatives, culminating in the production of ATP.

GSH: Glutation is an antioxidant tripeptide, acts as a free radical scavenger and a detoxifying and agent important as a cofactor for the enzyme glutathione peroxidase.

Lewy bodies: Lewy bodies are intracellular inclusions rich in alpha-synuclein (alpha-sin) aggregates characteristic of neurodegenerative diseases.

NADPH: NADPH is nicotinamide adenine dinucleotide phosphate the reduced form of NADP+, is membrane-bound enzyme that catalyzes the transfer of one electron from NADH or NADPH to oxygen to generate superoxide.

Neurofibrillary tangles (NFT): NFT are paired helical filaments formed from hyperphosphorylation of TAU.

PD: Parkinson's disease is part of the neurodegenerative diseases, whose progression leads to neuromuscular impairment, characterized by tremors and bradykinesia.

PrPC: Cellular prion protein is one of the isoforms of the PrP protein expressed in several tissues; however, it has gained more notoriety in the nervous system because it is associated with mad cow disease.

PSEN1 and PSEN2: Presenilin-1 (PSEN-1) and presenilin-2 (PSEN-2) are amyloid precursor protein (APP) genes.

ROS: Reactive oxygen species is molecules derived from molecular oxygen, which are formed by reduction-oxidation (redox) reactions or by electronic excitation.

SOD1: Superoxide dismutase 1 is at the forefront of the body's antioxidant defenses. Its main function is to convert superoxide radicals into oxygen and H_2O_2.

TREM2: Triggering receptor expressed on myeloid cells 2 is a receptor for lipids expressed in microglia.

Key facts

Copper in a free state can induce the development of neurodegenerative diseases and aging:

- In part by the production and proliferation of ROS.
- For damage to cell membranes.

High copper concentrations tend to have a high impact on the development of Alzheimer's disease:

- Due to aberrant APP protein processing.
- Due to molecular interaction and high affinity of copper and transmembrane APP.

Summary points

- The highest concentrations of copper are present in the liver and in the brain.
- In the brain, copper homeostasis is mainly regulated by APP and the prion protein.
- Imbalance of copper can be resumed in oxidative stress and inflammation.
- Free copper levels tend to increase with age.
- Zinc food supplementation may be established as an alternative treatment to AD.

References

EFSA. (2015). Scientific opinion on dietary reference values for copper. *EFSA Journal*, 4253.

Ackerman, C. M. (2018). CopperSignaling in the brain na beyond. In: Chang, CJ, editors *Journal of Biological Chemistry*, 4628–4635.

Adlard, P. A., Parncutt, J. M., Finkelstein, D. I., & Bush, A. I. (2010). Cognitive loss in zinc transporter-3 knock-out mice: A phenocopy for the synaptic and memory deficits of Alzheimer's disease? *The Journal of Neuroscience*, *30*(5), 1631–1636.

Alonso, A. C., Zaidi, T., Grundke-Iqbal, I., & Iqbal, K. (1994). Role of abnormally phosphorylated tau in the breakdown of microtubules in Alzheimer disease. *Proceedings of the National Academy of Sciences of the United States of America*, *91*(12), 5562–5566.

Bandmann, O., Weiss, K. H., & Kaler, S. G. (2015). Wilson's disease and other neurological copper disorders. *Lancet Neurology*, *14*(1), 103–113.

Bisaglia, M., & Bubacco, L. (2020). Copper ions and Parkinson's disease: Why is homeostasis so relevant? *Biomolecules*, *10*(2).

Boal, A. K., & Rosenzweig, A. C. (2009). Structural biology of copper trafficking. *Chemical Reviews*, *109*(10), 4760–4779.

Brewer, G. J. (2014). Alzheimer's disease causation by copper toxicity and treatment with zinc. *Frontiers in Aging Neuroscience*, *6*, 92.

Brewer, G. J. (2015). Copper-2 ingestion, plus increased meat eating leading to increased Copper absorption, are major factors behind the current epidemic of Alzheimer's disease. *Nutrients*, 7(12), 10053–10064.
Brown, D. R., Qin, K., Herms, J. W., Madlung, A., Manson, J., Strome, R., et al. (1997). The cellular prion protein binds copper in vivo. *Nature*, *390*(6661), 684–687.
Calvo, J. S., Mulpuri, N. V., Dao, A., Qazi, N. K., & Meloni, G. (2020). Membrane insertion exacerbates the α-Synuclein-Cu(II) dopamine oxidase activity: Metallothionein-3 targets and silences all α-synuclein-Cu(II) complexes. *Free Radical Biology & Medicine*, *158*, 149–161.
Chang, I. J., & Hahn, S. H. (2017). The genetics of Wilson disease. *Handbook of Clinical Neurology*, *142*, 19–34.
Cheignon, C., Tomas, M., Bonnefont-Rousselot, D., Faller, P., Hureau, C., & Collin, F. (2018). Oxidative stress and the amyloid beta peptide in Alzheimer's disease. *Redox Biology*, *14*, 450–464.
Chen, C. D., Oh, S. Y., Hinman, J. D., & Abraham, C. R. (2006). Visualization of APP dimerization and APP-Notch2 heterodimerization in living cells using bimolecular fluorescence complementation. *Journal of Neurochemistry*, *97*(1), 30–43.
Chong, F. P., Ng, K. Y., Koh, R. Y., & Chye, S. M. (2018). Tau proteins and Tauopathies in Alzheimer's disease. *Cellular and Molecular Neurobiology*, *38*(5), 965–980.
Conlan, D., Korula, R., & Tallentire, D. (1990). Serum copper levels in elderly patients with femoral-neck fractures. *Age and Ageing*, *19*(3), 212–214.
Culotta, V. C., Yang, M., & O'Halloran, T. V. (2006). Activation of superoxide dismutases: Putting the metal to the pedal. *Biochimica et Biophysica Acta*, *1763*(7), 747–758.
Davies, K. M., Hare, D. J., Cottam, V., Chen, N., Hilgers, L., Halliday, G., et al. (2013). Localization of copper and copper transporters in the human brain. *Metallomics*, *5*(1), 43–51.
Deane, R., Wu, Z., Sagare, A., Davis, J., Du Yan, S., Hamm, K., et al. (2004). LRP/amyloid beta-peptide interaction mediates differential brain efflux of Abeta isoforms. *Neuron*, *43*(3), 333–344.
Dias, V., Junn, E., & Mouradian, M. M. (2013). The role of oxidative stress in Parkinson's disease. *Journal of Parkinson's Disease*, *3*(4), 461–491.
Fontaine, S. N., Sabbagh, J. J., Baker, J., Martinez-Licha, C. R., Darling, A., & Dickey, C. A. (2015). Cellular factors modulating the mechanism of tau protein aggregation. *Cellular and Molecular Life Sciences*, *72*(10), 1863–1879.
González, C., Farías, G., & Maccioni, R. B. (1998). Modification of tau to an Alzheimer's type protein interferes with its interaction with microtubules. *Cellular and Molecular Biology (Noisy-le-Grand, France)*, *44*(7), 1117–1127.
Gromadzka, G., Tarnacka, B., Flaga, A., & Adamczyk, A. (2020). Copper Dyshomeostasis in neurodegenerative diseases-therapeutic implications. *International Journal of Molecular Sciences*, *21*(23).
Harvey, L. J., Majsak-Newman, G., Dainty, J. R., Lewis, D. J., Langford, N. J., Crews, H. M., et al. (2003). Adaptive responses in men fed low- and high-copper diets. *The British Journal of Nutrition*, *90*(1), 161–168.
Hsu, H. W., Bondy, S. C., & Kitazawa, M. (2018). Environmental and dietary exposure to Copper and its cellular mechanisms Linking to Alzheimer's disease. *Toxicological Sciences*, *163*(2), 338–345.
Huang, X., Cuajungco, M. P., Atwood, C. S., Hartshorn, M. A., Tyndall, J. D., Hanson, G. R., et al. (1999). Cu(II) potentiation of alzheimer abeta neurotoxicity. Correlation with cell-free hydrogen peroxide production and metal reduction. *The Journal of Biological Chemistry*, *274*(52), 37111–37116.
Huat, T. J., Camats-Perna, J., Newcombe, E. A., Valmas, N., Kitazawa, M., & Medeiros, R. (2019). Metal toxicity links to Alzheimer's disease and Neuroinflammation. *Journal of Molecular Biology*, *431*(9), 1843–1868.
Jiang, D., Men, L., Wang, J., Zhang, Y., Chickenyen, S., Wang, Y., et al. (2007). Redox reactions of copper complexes formed with different beta-amyloid peptides and their neuropathological [correction of neuropathalogical] relevance. *Biochemistry*, *46*(32), 9270–9282.
Kaler, S. G. (2011). ATP7A-related copper transport diseases-emerging concepts and future trends. *Nature Reviews. Neurology*, 7(1), 15–29.
Kaler, S. G., Ferreira, C. R., & Yam, L. S. (2020). Estimated birth prevalence of Menkes disease and ATP7A-related disorders based on the genome aggregation database (gnomAD). *Molecular Genetics and Metabolism Reports*, *24*, 100602.
Kim, J. H. (2018). Genetics of Alzheimer's disease. *Dementia and Neurocognitive Disorders*, 17(4), 131–136.

Kim, B. E., Nevitt, T., & Thiele, D. J. (2008). Mechanisms for copper acquisition, distribution and regulation. *Nature Chemical Biology*, *4*(3), 176–185.

Kimura, A., Hata, S., & Suzuki, T. (2016). Alternative selection of β-site APP-cleaving enzyme 1 (BACE1) cleavage sites in amyloid β-protein precursor (APP) harboring protective and pathogenic mutations within the Aβ sequence. *The Journal of Biological Chemistry*, *291*(46), 24041–24053.

Klein, L. D., Breakey, A. A., Scelza, B., Valeggia, C., Jasienska, G., & Hinde, K. (2017). Concentrations of trace elements in human milk: Comparisons among women in Argentina, Namibia, Poland, and the United States. *PLoS One*, *12*(8), e0183367.

Knock, E., Matsuzaki, S., Takamura, H., Satoh, K., Rooke, G., Han, K., et al. (2018). SUMO1 impact on Alzheimer disease pathology in an amyloid-depositing mouse model. *Neurobiology of Disease*, *110*, 154–165.

Krasemann, S., Madore, C., Cialic, R., Baufeld, C., Calcagno, N., El Fatimy, R., et al. (2017). The TREM2-APOE pathway drives the transcriptional phenotype of dysfunctional microglia in neurodegenerative diseases. *Immunity*, *47*(3). 566–581.e9.

Lanoiselée, H. M., Nicolas, G., Wallon, D., Rovelet-Lecrux, A., Lacour, M., Rousseau, S., et al. (2017). APP, PSEN1, and PSEN2 mutations in early-onset Alzheimer disease: A genetic screening study of familial and sporadic cases. *PLoS Medicine*, *14*(3), e1002270.

Linder, M. C. (2020). Copper homeostasis in mammals, with emphasis on secretion and excretion. A review. *International Journal of Molecular Sciences*, *21*(14).

Litwin, T., Gromadzka, G., Szpak, G. M., Jabłonka-Salach, K., Bulska, E., & Członkowska, A. (2013). Brain metal accumulation in Wilson's disease. *Journal of the Neurological Sciences*, *329*(1–2), 55–58.

Lutsenko, S. (2010). Human copper homeostasis: A network of interconnected pathways. *Current Opinion in Chemical Biology*, *14*(2), 211–217.

Lutsenko, S., Bhattacharjee, A., & Hubbard, A. L. (2010). Copper handling machinery of the brain. *Metallomics*, *2*(9), 596–608.

Madsen, E., & Gitlin, J. D. (2007). Copper and iron disorders of the brain. *Annual Review of Neuroscience*, *30*, 317–337.

Magrì, A., & La Mendola, D. (2018). Copper binding features of tropomyosin-receptor-kinase-a fragment: Clue for neurotrophic factors and metals link. *International Journal of Molecular Sciences*, *19*(8).

Meloni, G., & Vašák, M. (2011). Redox activity of α-synuclein-Cu is silenced by Zn_7-metallothionein-3. *Free Radical Biology & Medicine*, *50*(11), 1471–1479.

Mezzaroba, L., Alfieri, D. F., Colado Simao, A. N., & Vissoci Reiche, E. M. (2019). The role of zinc, copper, manganese and iron in neurodegenerative diseases. *Neurotoxicology*, *74*, 230–241.

Opazo, C., Barría, M. I., Ruiz, F. H., & Inestrosa, N. C. (2003). Copper reduction by copper binding proteins and its relation to neurodegenerative diseases. *Biometals*, *16*(1), 91–98.

Opazo, C. M., Greenough, M. A., & Bush, A. I. (2014). Copper: From neurotransmission to neuroproteostasis. *Frontiers in Aging Neuroscience*, *6*, 143.

Pal, A., Kumar, A., & Prasad, R. (2014). Predictive association of copper metabolism proteins with Alzheimer's disease and Parkinson's disease: A preliminary perspective. *Biometals*, *27*(1), 25–31.

Palumaa, P. (2013). Copper chaperones. The concept of conformational control in the metabolism of copper. *FEBS Letters*, *587*(13), 1902–1910.

Pan-Montojo, F., Anichtchik, O., Dening, Y., Knels, L., Pursche, S., Jung, R., et al. (2010). Progression of Parkinson's disease pathology is reproduced by intragastric administration of rotenone in mice. *PLoS One*, *5*(1), e8762.

Pena, M. M., Lee, J., & Thiele, D. J. (1999). A delicate balance: Homeostatic control of copper uptake and distribution. *The Journal of Nutrition*, *129*(7), 1251–1260.

Petris, M. J. (2004). The SLC31 (Ctr) copper transporter family. *Pflügers Archiv*, *447*(5), 752–755.

Pîrşcoveanu, D. F. V., Pirici, I., Tudorică, V., Bălşeanu, T. A., Albu, V. C., Bondari, S., et al. (2017). Tau protein in neurodegenerative diseases - A review. *Romanian Journal of Morphology and Embryology*, *58*(4), 1141–1150.

Pohanka, M. (2019). Copper and copper nanoparticles toxicity and their impact on basic functions in the body. *Bratislavské Lekárske Listy*, *120*(6), 397–409.

Prohaska, J. R. (2008). Role of copper transporters in copper homeostasis. *The American Journal of Clinical Nutrition*, *88*(3), 826s–829s.

Provias, J., & Jeynes, B. (2014). The role of the blood-brain barrier in the pathogenesis of senile plaques in Alzheimer's disease. *International Journal of Alzheimer's Disease*, *2014*, 191863.

Rappold, P. M., & Tieu, K. (2010). Astrocytes and therapeutics for Parkinson's disease. *Neurotherapeutics*, 7(4), 413–423.

Rose, F., Hodak, M., & Bernholc, J. (2011). Mechanism of copper(II)-induced misfolding of Parkinson's disease protein. *Scientific Reports*, *1*, 11.

Rossi, L., Lombardo, M. F., Ciriolo, M. R., & Rotilio, G. (2004). Mitochondrial dysfunction in neurodegenerative diseases associated with copper imbalance. *Neurochemical Research*, *29*(3), 493–504.

Saido, T. C. (2013). Metabolism of amyloid β peptide and pathogenesis of Alzheimer's disease. *Proceedings of the Japan Academy. Series B, Physical and Biological Sciences*, *89*(7), 321–339.

Sala Frigerio, C., Wolfs, L., Fattorelli, N., Thrupp, N., Voytyuk, I., Schmidt, I., et al. (2019). The major risk factors for Alzheimer's disease: Age, sex, and genes modulate the microglia response to Aβ plaques. *Cell Reports*, *27*(4). 1293–1306.e6.

Salzano, G., Giachin, G., & Legname, G. (2019). Structural consequences of Copper binding to the prion protein. *Cell*, *8*(8).

Sandusky-Beltran, L. A., Manchester, B. L., & McNay, E. C. (2017). Supplementation with zinc in rats enhances memory and reverses an age-dependent increase in plasma copper. *Behavioural Brain Research*, *333*, 179–183.

Scheiber, I., Dringen, R., & Mercer, J. F. (2013). Copper: Effects of deficiency and overload. *Metal Ions in Life Sciences*, *13*, 359–387.

Scheiber, I. F., Mercer, J. F., & Dringen, R. (2014). Metabolism and functions of copper in brain. *Progress in Neurobiology*, *116*, 33–57.

Sebastián-Serrano, Á., de Diego-García, L., & Díaz-Hernández, M. (2018). The Neurotoxic role of extracellular tau protein. *International Journal of Molecular Sciences*, *19*(4).

Senkevich, K., & Gan-Or, Z. (2020). Autophagy lysosomal pathway dysfunction in Parkinson's disease; evidence from human genetics. *Parkinsonism & Related Disorders*, *73*, 60–71.

Shi, B. O. (2021). In I. Lu, X. Hu, M. B. Betancor, M. Zhao, D. R. Tocher, I. Zhou, et al. (Eds.), *Dietary copper improves growth and regulates energy generation by mediating lipolysis and autophagy in hepatopancreas of Pacific white shrimp (Litopenaeus vannamei)* (p. 736505).

Singh, I., Sagare, A. P., Coma, M., Perlmutter, D., Gelein, R., Bell, R. D., et al. (2013). Low levels of copper disrupt brain amyloid-beta homeostasis by altering its production and clearance. *Proceedings of the National Academy of Sciences of the United States of America*, *110*(36), 14771–14776.

Sparks, D. L., & Schreurs, B. G. (2003). Trace amounts of copper in water induce beta-amyloid plaques and learning deficits in a rabbit model of Alzheimer's disease. *Proceedings of the National Academy of Sciences of the United States of America*, *100*(19), 11065–11069.

Squitti, R., Siotto, M., & Polimanti, R. (2014). Low-copper diet as a preventive strategy for Alzheimer's disease. *Neurobiology of Aging*, *35*(Suppl 2), S40–S50.

Strausak, D., Mercer, J. F., Dieter, H. H., Stremmel, W., & Multhaup, G. (2001). Copper in disorders with neurological symptoms: Alzheimer's, Menkes, and Wilson diseases. *Brain Research Bulletin*, *55*(2), 175–185.

Takano, K., Tanaka, N., Kawabe, K., Moriyama, M., & Nakamura, Y. (2013). Extracellular superoxide dismutase induced by dopamine in cultured astrocytes. *Neurochemical Research*, *38*(1), 32–41.

Takeda, A., Tamano, H., Ogawa, T., Takada, S., Nakamura, M., Fujii, H., et al. (2014). Intracellular Zn(2+) signaling in the dentate gyrus is required for object recognition memory. *Hippocampus*, *24*(11), 1404–1412.

Vairo, F. P. E., Chwal, B. C., Perini, S., Ferreira, M. A. P., de Freitas Lopes, A. C., & Saute, J. A. M. (2019). A systematic review and evidence-based guideline for diagnosis and treatment of Menkes disease. *Molecular Genetics and Metabolism*, *126*(1), 6–13.

Wang, H. L., Chou, A. H., Wu, A. S., Chen, S. Y., Weng, Y. H., Kao, Y. C., et al. (2011). PARK6 PINK1 mutants are defective in maintaining mitochondrial membrane potential and inhibiting ROS formation of substantia nigra dopaminergic neurons. *Biochimica et Biophysica Acta*, *1812*(6), 674–684.

Wang, J. Z., Grundke-Iqbal, I., & Iqbal, K. (1996). Glycosylation of microtubule-associated protein tau: An abnormal posttranslational modification in Alzheimer's disease. *Nature Medicine*, *2*(8), 871–875.

Wang, J. Z., & Liu, F. (2008). Microtubule-associated protein tau in development, degeneration and protection of neurons. *Progress in Neurobiology, 85*(2), 148–175.
Wang, L., Yin, Y. L., Liu, X. Z., Shen, P., Zheng, Y. G., Lan, X. R., et al. (2020). Current understanding of metal ions in the pathogenesis of Alzheimer's disease. *Translational Neurodegeneration, 9*, 10.
Weibull, M. G. M., Simonsen, S., Oksbjerg, C. R., Tiwari, M. K., & Hemmingsen, L. (2019). Effects of Cu(II) on the aggregation of amyloid-β. *Journal of Biological Inorganic Chemistry, 24*(8), 1197–1215.
Xiao, G., Fan, Q., Wang, X., & Zhou, B. (2013). Huntington disease arises from a combinatory toxicity of polyglutamine and copper binding. *Proceedings of the National Academy of Sciences of the United States of America, 110*(37), 14995–15000.
Yan, R. (2017). Physiological functions of the β-site amyloid precursor protein cleaving enzyme 1 and 2. *Frontiers in Molecular Neuroscience, 10*, 97.
Yau, T. Y., Molina, O., & Courey, A. J. (2020). SUMOylation in development and neurodegeneration. *Development, 147*, 6.
Zheng, W., Xin, N., Chi, Z. H., Zhao, B. L., Zhang, J., Li, J. Y., et al. (2009). Divalent metal transporter 1 is involved in amyloid precursor protein processing and Abeta generation. *The FASEB Journal, 23*(12), 4207–4217.

PART VI

Epilepsy

CHAPTER 14

Vitamin E and neuroprotection in epilepsy

Rita Marinelli[a,*], Desirée Bartolini[a,b,*], Michele Betti[c], Anna Migni[a], Linda Zatini[a], Giada Marcantonini[a], Mario Rende[b], and Francesco Galli[a]
[a]Department of Pharmaceutical Sciences, University of Perugia, Perugia, Italy
[b]Department of Medicine and Surgery, Unit of Human, Clinical and Forensic Anatomy, University of Perugia, Perugia, Italy
[c]Department of Biomolecular Science, Urbino University, Urbino, Italy

Introduction

The term "epilepsy" originated from the Greek verb "epilambanein" (επιλαμϖανειν), i.e., "to be attacked or to be seized," is the definition for a chronic noncommunicable disease of the brain characterized by frequent seizures, transient symptoms, and/or signs due to abnormal and simultaneous neuronal activity of a population of neurons in the brain (Thijs et al., 2019). This ailment seriously affects the quality of life of around 50 million people worldwide thus representing a global health priority. Epilepsy is a major health concern of the pediatric age with incidence and prevalence ranging from 0.5 to 1.0/1000 and 4 to 40/1000, respectively, moreover, between 20% and 30% of children with epilepsy are refractory to conventional antiepileptic therapy (Camfield & Camfield, 2015). For these reasons, there is urgent need to develop more efficient and cost-effective treatments, which can substantially reduce epilepsy-derived morbidity, disability, and mortality.

Epilepsy classification is based on the type of seizures, including focal, generalized, and unknown onset seizures, and type of epilepsy, which includes generalized and focalized epilepsy that involves both hemispheres or only one hemisphere, respectively, and the combined form (Scheffer et al., 2017). Temporal lobe epilepsy (TLE) is the most common focal epilepsy; it is often preceded by a latent period without seizures during which maladaptive network reorganization occurs, followed by the appearance of spontaneous recurrent seizures that characterize the chronic stage of the disease.

The causes of this abnormal brain activity can be different and include structural, genetic, metabolic, infectious, and immune mechanisms. However, idiopathic forms are very frequent at clinical evaluation.

Animal models have been developed to simulate this form of human epilepsy, with pharmacological models in rodents that represent the most reliable tool to recapitulate the

* Equally contribution.

Vitamins and Minerals in Neurological Disorders
https://doi.org/10.1016/B978-0-323-89835-5.00023-5

human form of TLE at the experimental level (*vide infra*). Mechanistic studies performed on these models indicate that glutamate-mediated excitotoxicity, neuroinflammation, and oxidative stress characterize the neurobiology of this and other types of epilepsy (Ambrogini et al., 2019). These processes identify the so-called pathogenic "triad" of epilepsy, but they are also present in many other diseases of the brain at clinical and experimental evaluation. Further mechanistic insights demonstrate that maladaptive plasticity, which is the result of gliosis as well as of glial alterations and neuroinflammation, contributes to the development of neurodegeneration and to the chronicity of epilepsy. Unfortunately, most of the antiepileptic drugs (AEDs) commonly utilized to treat epilepsy are reported to further reduce the mitochondrial function, thus exacerbating brain oxidative stress (Mehvari et al., 2016).

In this context, neuroprotective agents have gained increased interest as either first-line treatment to protect the brain tissue at the disease onset or as an add-on therapy to increase the efficacy of antiepileptic drugs. These include a series of functional nutrients or nutraceuticals [that have recently been reviewed elsewhere (Geronzi, Lotti, & Grosso, 2018; Kim & Cho, 2019)]. In this respect, results obtained in experimental models of epilepsy and preliminary evidence coming from clinical trials, support the notion that vitamin E—under the form of alpha-tocopherol (α-TOH)—represents a promising neuroprotectant in epilepsy and a convulsive behavior attenuation agent (Ambrogini et al., 2019). α-TOH is the main form of this vitamin in human tissues and its role in neuroprotection is supported by the physiological function of this vitamin as lipid homeostasis agent of the cellular membrane (Galli et al., 2017). This role is particularly important in the brain tissue since it is rich in polyunsaturated fatty acids, especially the ω-6 arachidonic acid (AA) and the ω-3 docosahexaenoic acid (DHA) (Ambrogini et al., 2019; Pourmasoumi et al., 2018). This lipid environment requires high levels of homeostatic control of enzymatic peroxidations and membrane phospholipase activity. In this context, it is particularly important to maintain efficient antioxidant protection against unwanted and harmful lipoperoxyl radicals that can be directly scavenged by the H atom donation function of vitamin E (Torquato et al., 2019a). According to such an important role, vitamin E deficiency in humans generates major neurological symptoms. A severe form of ataxia is associated with a genetic defect of the hepatic uptake mechanism of vitamin E, and defects in learning, memory, and emotive functions have been reported to occur in secondary (diet- or malabsorption-related) forms of deficiency (Ulatowski & Manor, 2015). Aside of lipid peroxidation modulation, emerging physiological functions of this vitamin are associated with direct and indirect (vitamin E metabolites-mediated) effects of modulation on genes and proteins important in the metabolism of xenobiotics and lipids (Bartolini et al., 2021), including arachidonic acid-derived eicosanoids involved in the activation of inflammatory pathways (Jiang et al., 2008; Pein et al., 2018), and lipophilic peptides such as the β-amyloid peptide (Aβ) (Marinelli et al., 2020).

These aspects of vitamin E metabolism and function will be described in this chapter as far as their roles in neuroprotection during pre-epileptic lesions and status epilepticus (SE) are concerned. Furthermore, the available evidence on the utilization of this vitamin in the clinical management of epilepsy will be critically evaluated.

The pathogenic triad

Glutamate-mediated excitotoxicity

A neurodegeneration mechanism of SE is the excitotoxicity induced by the persistent discharges of glutamate pathway (Mattson, 2019). Glutamate is the main excitatory neurotransmitter in the brain. Normally, the astrocytes convert the glutamate released in the synaptic space in glutamine by the activity of glutamine synthase. Glutamine is a nonexcitotoxic amino acid used for the synthesis of glutamate and GABA that is an inhibitory neurotransmitter. For this reason, this glutamine synthase activity is considered a pharmacological target of epilepsy.

Neuroinflammation

Neuroinflammation is a key player in the pathogenesis of epilepsy because it is responsible for excessive neuroglial activation and expression of inflammatory mediators, such as IL-1β and TNF-α (Ambrogini et al., 2018). Under normal conditions, neuroinflammation is a natural intrinsic response of the brain to various insults (i.e., tissue damage, seizures, etc.), but in epilepsy this response is protracted and excessive thus leading to chronic damage. In fact, this inflammatory response if not properly regulated, can contribute to seizures, neuronal cell loss, maladaptive synaptic plasticity, blood-brain barrier dysfunction, and development of comorbidity of epilepsy (Vezzani, Balosso, & Ravizza, 2019).

Oxidative stress

The term oxidative stress describes a condition of altered homeostasis that involves the cellular redox eventually leading to insufficient control of physiological oxidations and accumulation of oxidative damage on structural and signaling protein, membrane lipids, and nucleic acids (Galli et al., 2005; Jones, 2008). These changes ultimately produce the accumulation of molecular lesions and loss of physiological functions, including the capability to control the cellular redox and its regulatory function on signal transduction and gene transcription pathways important to adapt to endogenous and exogenous stimuli, to control the cellular metabolism, as well as cell death and autophagy pathways, which may result in cellular loss and tissue degeneration (Sies & Jones, 2020). Mechanistically, oxidative stress develops when the flux or reactivity of cellular and extracellular reactive oxygen species (ROS) increase over the capability of homeostatic processes to prevent or repair their adverse effects on the molecular components and functions of cells and tissues.

ROS can be generated through different mechanisms including the activation of cellular oxidoreductases, such as the membrane complex of NADPH-oxidase of inflammatory cells, and the stimulation and/or damage of organelles with oxygen-activating properties such as mitochondria, peroxisomes, and the endoplasmic reticulum. Conditions of cellular stress can also derive from an excess or impaired utilization of reducing equivalents (Jones, 2008), which respond to the definition of reductive stress. Such condition can derive from the hyperstimulation or poor control of stress response and antioxidant pathways, including those associated with cellular production or uptake of reduced pyridine nucleotides and thiols (Bartolini et al., 2020).

The brain is highly susceptible to oxidative stress (Ambrogini et al., 2019; Geronzi et al., 2018). In this organ, an intense oxygen metabolism is associated with one of the highest levels of polyunsaturated fatty acids (PUFA) esterified to membrane phospholipids. These are highly oxidizable biomolecules and a preferential substrate for the generation of lipid metabolites involved in cellular signaling (Cobley, Fiorello, & Bailey, 2018). The most abundant PUFA of membrane phospholipids is arachidonic acid, a 20-carbon atoms ω-6 fatty acid with 4 double bonds. This is the substrate of enzymatic reactions responsible for the generation of a wide series of inflammatory and vasoactive eicosanoids. Nonenzymatic oxidations of this and other membrane PUFA are involved in the formation of specific classes of eicosanoids and oxylipins, but most importantly, these are responsible of phospholipid damage-induced membrane phospholipase activation, further increasing the availability of free AA to form inflammatory eicosanoids by the activity of cellular lipoxygenases, cyclooxygenases, and cytochrome P-450 isoenzymes (Ambrogini et al., 2019). Also, the brain is one of the tissues richest in ω-3 fatty acids, particularly DHA, which play important structural roles in neuronal membranes and neuroprotection function being involved in neurotransmitter signaling and inflammatory pathway modulation (Pourmasoumi et al., 2018). An insufficient intake of ω-3 with the diet and an abnormal peroxidation of these and other PUFA in the brain tissue are believed to increase the risk of neuroinflammation and epileptogenesis.

Therefore, to control the flux and reactivity of cellular ROS and other prooxidants, such as transition metals, is essential to preserve the integrity and function of neuronal membranes. Main cellular sources of ROS are 5-lipoxygenase, xanthine oxidase, NADPH oxidase, cytochrome P450 isoenzymes, and mitochondrial cytochromes (Lin et al., 2020). In particular, NADPH oxidase isoenzymes have been reported to significantly contribute to seizure-induced ROS generation (Shekh-Ahmad et al., 2019). The overproduction of these reactive species can lead to extensive damage to cellular components and activation of death programs that may lead to neuronal loss and irreversible brain damage. For example, the reactive nitrogen species (RNS) peroxynitrite, that forms by the reaction of NO_2 with superoxide anion, is a pro-oxidant metabolite capable to alter the structure and function of neuronal cells, which is produced in excess during inflammatory cell activation and tissue degeneration processes of

SE (Lin et al., 2020). Again, the ROS-induced alteration of Na/K-ATPase causes excitotoxicity by impairing the efficient reversal of neuronal depolarization-induced electrochemical gradients (Kim & Cho, 2019). Enzymatic and nonenzymatic pathways of lipid peroxidation are activated during cellular oxidative stress and affect membrane phospholipids generating lipoperoxides and other lipid metabolites with important role as cellular messengers and signaling molecules. Some of these metabolites have been identified to sustain the ferroptotic program of neuronal cell death (Wenzel et al., 2017), a process that may sustain brain tissue degeneration in SE. According with this, ferroptosis inhibition in kainic acid-induced temporal lobe epilepsy was demonstrated to ameliorate cognitive impairment in rats (Ye et al., 2019).

Vitamin E: General characteristics

Usually, the term vitamin E is intended to describe a group of eight lipophilic plant molecules. These are grouped in the subfamilies of tocopherols (TOHs) and tocotrienols (T3), that are available to different extent in the diet and in the lipid compartments of human tissues. α-Tocopherol is by far the most biologically active and abundant form of vitamin E in human tissues, followed by γ-tocopherol, whereas T3 are present in traces and with biological roles distinct from TOHs (Azzi, 2021; Galli et al., 2017). Specific symptoms of deficiency have only been described for α-TOH, thus confirming for this form the definition of essential nutrient. These symptoms include neurological abnormalities that have been reported with different degrees of severity in both the primary (or genetic) and secondary forms of deficiency described so far in humans (Ulatowski & Manor, 2015), and recapitulated in different experimental models (Takahashi et al., 2019). Main sources of this vitamin are edible oils, nuts, and seeds, such as corn oil, wheat germ, hemp, peanuts, sesame, almonds, and walnuts (Torquato et al., 2020a). Other tocopherols and tocotrienols are usually present in traces in human tissues, and their biological activities are different from, and sometimes alternative to, those of the vitamin α-TOH, suggesting a definition of "nonvitamin" forms for these plant metabolites (Azzi, 2019). As introduced earlier, vitamin E, as α-TOH, is one of the most important lipophilic antioxidants found in nature and available to all human tissues to prevent lipid peroxidation of cellular membranes and circulating lipoproteins (Torquato et al., 2019b). Main sources of this vitamin are edible oils, nuts, and seeds, such as corn oil, wheat germ, hemp, peanuts, sesame, almonds, and walnuts (Torquato et al., 2020a).

However, together with the specific role of fat-soluble antioxidant, vitamin E is recognized to have other biological functions (Azzi, 2021; Galli et al., 2017). It modulates the activity of enzymes, signal transduction pathways, and transcription factors (such as NFkB), and lipoxygenase-dependent and cyclooxygenase-dependent pathways important for the immune function and inflammatory response of tissues (Torquato et al., 2020b), including the brain (Ambrogini et al., 2019). Moreover, similar to the metabolites produced by the fat-soluble vitamins A and D, the long-chain metabolites of vitamin E, namely α-13′OH and α-13′COOH, show immunomodulation function, antiinflammatory properties and hormone-like activity (Galli et al., 2017; Schubert et al., 2018). More in detail, α-13′COOH has been demonstrated to behave as a potent allosteric inhibitor of 5-LO activity, a proinflammatory enzyme involved in arachidonic acid activation cascade (Pein et al., 2018). Garcinoic acid, a plant analog of the delta form of 13-COOH, has been demonstrated to be one of the most potent ligands with agonist function of the nuclear receptor PXR. This receptor control groups of genes involved in the metabolism of xenobiotics and inflammatory eicosanoid, and more in general in metabolism and detoxification of lipophilic metabolites, especially steroids (Bartolini et al., 2021; Marinelli et al., 2020). Important enough is the evidence that garcinoic acid and its PXR agonist function are available to the mouse brain to detoxify the lipophilic β-amyloid peptide (Aβ), thus preventing its aggregation and deposition that are considered main pathogenic processes in AD. Besides to PXR activation, this Aβ scavenging activity of garcinoic acid in the mouse brain was observed to involve both PXR and PPAR-gamma nuclear receptors, and their transcriptional induction of astrocyte ApoE gene (Marinelli et al., 2020). These pieces of evidence suggest a physiological role of vitamin E and its long-chain metabolites in the detoxification of lipophilic compounds of endogenous and exogenous origin that may promote neurodegenerative and neuroinflammatory processes of the brain. These considerations also support the utilization of this vitamin as an adjuvant treatment to antiepileptic drug therapy, with the aim of preventing neuroinflammation and brain toxicity complications.

VE and its effects on the pathogenic triad of epilepsy

Animal studies

Experimental models of epilepsy have been used to explore the therapeutic efficacy of vitamin E against the pathogenic triad of epilepsy. Most striking evidence on the neuroprotection and anticonvulsant function of vitamin E has been obtained in animal models of pharmacological epilepsy, including KA epilepsy (Ambrogini et al., 2018; Kiasalari et al., 2012) (Fig. 1), pentylenetetrazole-induced epileptic seizures in acute hyperglycemic rats (Kaya et al., 2002), and ferrous chloride seizures (Levy, Burnham, & Hwang, 1990).

However, in a systematic comparison of experimental models of epilepsy, Levy and colleagues found that the anticonvulsant effects of d-α-TOH observed in ferrous chloride and penicillin-induced seizures, as well as in hyperbaric oxygen seizures, were not replicated in standard animal models used for anticonvulsant drug screening, such as the maximal electroshock and threshold pentylenetetrazol tests, metrazol threshold model, the kindling model (well-established seizures), and the KA model (Levy et al., 1992).

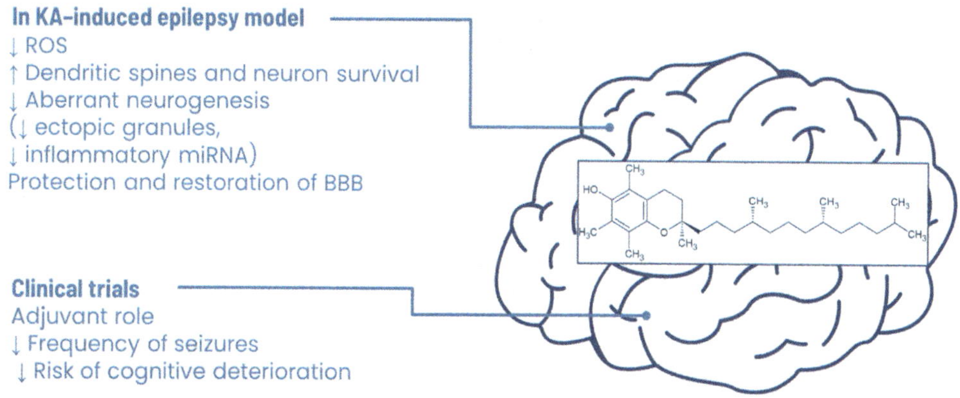

Fig. 1 Schematic representation of the neuroprotection function of vitamin E in the rat model of kainic acid epilepsy and as an add-on therapy in clinical trials. The insert presents the molecular structure of alpha-tocopherol. *(Modified from Ambrogini, P., et al. (2019). Excitotoxicity, neuroinflammation and oxidant stress as molecular bases of epileptogenesis and epilepsy-derived neurodegeneration: The role of vitamin E.* Biochimica et Biophysica Acta - Molecular Basis of Disease, 1865*(6), 1098–1112.)*

Again, in rats submitted to the pilocarpine model of epilepsy, Pansani et al. (2018) reported that long-term monotherapy with vitamin E reduces oxidative stress in the hippocampus, but not the frequency of seizures.

These apparently discording findings suggest that the studies on the efficacy of vitamin E to hindering the epileptogenic processes can be influenced by the type of experimental model and by the presence of favorable conditions for this vitamin to promote a therapeutic effect, especially through its antioxidant function. Apparently, these conditions depend on the study design and severity of the epileptogenic insult, and on the dosage of the vitamin and its compatibility with the levels of neuroinflammation and oxidative stress induced in the model.

In fact, considering the efficacy of α-TOH in delaying the onset of electrographic seizures of the intracerebral ferrous chloride model, compared with the absence of effect in other models of epilepsy (Levy et al., 1990), one may evoke the role of the transition metal iron as a potent inducer of the Fenton reaction (Galli et al., 2005) and of the lipid peroxidation process (Kahn-Kirby et al., 2019; Zhang et al., 2021), which provides the ideal target for this vitamin to exert its antioxidant and neuroprotection function (Ambrogini et al., 2019). In keeping with this mechanistic interpretation, the comparison with catalase treatment suggests an antioxidant mechanism for the BBB protection effect of α-TOH in pentylenetetrazole-induced epilepsy of acute hyperglycemic rats (Kaya et al., 2002).

Again, the available literature on KA epilepsy (recently reviewed in Ambrogini et al., 2018) confirms that oxidative stress, and especially lipid peroxidation, is present in this model and is effectively reduced during a successful therapy with vitamin E. Unfortunately, these molecular aspects dealing with the antiperoxidative function of vitamin E have not been verified in other studies that failed to verify its efficacy as anticonvulsant in this model (Levy et al., 1992).

Also, our studies carried out in the rat model of KA seizures suggested the importance of a timely intervention with α-TOH in preventing the development of chronic epilepsy affecting oxidative stress and the other components of the triad early in their pathogenic action (Ambrogini et al., 2018) (Fig. 1). However, using a different therapeutic approach, we found that even a late administration of vitamin E (4 days after the KA-induced status epilepticus) can be beneficial in this model preventing postseizure inflammation-induced brain damage (Ambrogini et al., 2014). Therefore, such pathogenic components of the triad appear to exert their neurodegenerative effects throughout the entire course of the disease, providing an opportunity to intervene with vitamin E both in the earliest and in the more advanced steps of SE. Important enough was the observation that the efficacy of a late administration of vitamin E was associated with the restoration of the seizure-induced inhibition of glutamine synthetase (GS), an enzyme that reduces the synaptic levels of glutamate thus preserving the glutamate/GABA ratio and avoiding repeated depolarization and repolarization cycle in glutamate terminals that may cause postsynaptic neuron damage by increased calcium afflux, *N*-methyl-D-aspartate (NMDA) levels, and NADPH-oxidase activity; the latter is a key player of cellular ROS production during seizures (Ambrogini et al., 2019). Therefore, the results obtained in KA epilepsy suggest that even if administered after seizures, α-TOH can intervene in (1) preventing key neurodegeneration and neuroinflammation processes of SE, (2) preventing GS inhibition to recover glutamate-glutamine-GABA cycle function of the hippocampus that is defective in patients with temporal lobe epilepsy, and (3) promoting synaptogenesis and synapses protection (further discussed elsewhere in Ambrogini et al., 2018; Ambrogini et al., 2019).

Glutamate can further cause excitotoxicity by preventing the entry into the neuronal cells of cystine. The reduction of this semi-essential amino acid to cysteine is important in the cell for the incorporation in cellular proteins as well as for the de novo synthesis of cellular glutathione. The inhibition of cystine uptake causes oxidative stress in the brain by depleting intracellular glutathione (Cobley et al., 2018). Therefore, the possibility that vitamin E might help to prevent the consequences of this oxidative stress of the brain is another point in favor of the use of this vitamin as a chemoprotection and/or prevention agent in epilepsy. In fact, this vitamin is recognized to represent an efficient inhibitor of ferroptosis, a cell death program observed in mitochondrial disease-dependent epilepsy and some forms of drug-induced epilepsy (Kahn-Kirby et al., 2019; Wenzel et al., 2017; Zhang et al., 2021). This cell death program depends on the formation of specific death signaling molecules consisting of arachidonic acid-derived formed during the alteration of the cellular iron metabolism (the prooxidant trigger of ferroptosis), or by the defect of cellular thiols and membrane glutathione peroxidase activity (the anti-oxygenic and ferroptosis prevention system).

Beside GS activity and glutamate pathways restoration, other mechanisms for this vitamin to prevent epilepsy-related neurodegeneration described in murine models include: (1) the inhibition of protein kinase C (PKCδ) activity (the activity and signaling function of this enzyme reduce the expression of GS) (Ambrogini et al., 2019); (2) a reduced availability of arachidonic acid to enter the catalytic site of 12-LOX enzyme, a key mediator of glutamate-induced neurodegeneration (Khanna et al., 2003); (3) the restoration of claudin levels, a molecule that forms tight junctions of the blood brain barrier, the levels of which are markedly decreased by the effects of neuroinflammation (Ambrogini et al., 2018); and (4) the control the aberrant expression in the hippocampus of epileptic patients of proinflammatory microRNAs, such as miR-124, miR-126, and miR-146a, with the latter that is proposed to regulate the inflammatory response mediated by the astroglia (Ambrogini et al., 2019).

As far as the neuroinflammation process is concerned, protective effects have consistently been demonstrated for this vitamin in neurodegenerative diseases characterized by by a pro-inflammatory phenotype, such as AD and Parkinson's disease (Browne et al., 2019; Fariss & Zhang, 2003), but not specifically in epilepsy.

A mechanistic linkage between lipid peroxidation and neuroinflammation has been suggested in experimental models of neurodegeneration, thus representing an important aspect in the pathobiology of epilepsy and a direct target for the antioxidant function of vitamin E. Furthermore, α-tocopherylquinone (α-TQ), the peroxidation metabolite of α-TOH (Torquato et al., 2019b), has been reported to suppress NF-κB signaling and inflammatory mediators, such as IL-6 and IL-1β, leading to reduced microglial activation and oxidative stress in both in vitro and in vivo models of AD (Wang et al., 2016). Moreover, experimental condition of VE deficiency led to impaired cognitive function and increased inflammation of the CNS (Takahashi et al., 2019).

Again, VE has been suggested to be effective in the protection and restoration of the BBB in several models of neurodegeneration and neuroinflammation, which is an important endpoint in the treatment of epilepsy. α-TOH has been demonstrated to protect the BBB function reducing the brain edema after ischemic stroke (Haghnejad Azar et al., 2017). Moreover, α-tocopheryl succinate, a synthetic analog of α-TOH, has been demonstrated to restore the BBB in experimental cerebral malaria (Kume et al., 2018). The BBB is also important to make α-TOH available to the brain tissue. Its trafficking across such barrier can occur through different transporters, including SRB1 (scavenger receptor class B type 1), TTP (tocopherol transport protein), and ABCA/ABCG (ATP-binding cassette A/G) (Lee & Ulatowski, 2019). Therefore it is conceivable to assume that the protection of BBB deriving from α-TOH treatment is related to its capability to access the brain to decreasing neuroinflammation and oxidative stress in the whole organ (Haghnejad Azar et al., 2017).

A limited and number of in vivo studies have been performed to evaluate the effects of VE on the neuroinflammation of epilepsy: the results of these studies, however, look promising. In the KA-epilepsy model of the rat, that causes glial activation in the hippocampus (Ambrogini et al., 2018, 2019; Betti et al., 2011) and increased ferroptotic neuronal death (Ye et al., 2019), VE treatment reduced all the characteristic cues of neuroinflammation stimulated by the treatment with this proconvulsant, including microglial activation, cytokines expression, and reactive astrocytosis consisting in the atypical increase of the number of astrocytes after the disruption of close by neurons. In this experimental model, three different protocols of vitamin E treatment were evaluated that differed by the time and dose of treatment. Important enough was the evidence that all these protocols showed similar results. In the first case, α-TOH was administered with the animal diet (as 10 g RRR-α-T/Kg pellet) before the induction of SE; the treatment resulted in the reduced expression of the glial fibrillary acidic protein (GFAP) and major histocompatibility complex II (MHC II), two astrocytic and microglial antigens, as well as of the proinflammatory cytokines IL-1β and TNF-α (Betti et al., 2011). These effects were associated with reduced levels of lipid peroxidation and neuronal cell death. In a second protocol, the treatment was performed for 4 days by intraperitoneal injection of pharmacological doses of α-TOH (250 mg/kg body weight) as a postictal intervention that started soon after SE, considering this as the timepoint of the transition phase in which the neuroinflammatory response is still active (Ravizza et al., 2008). This acute intervention significantly decreased astrocytosis and microglia activation in the hippocampus, in association with reduced lipid oxidation, neuronal degeneration, and spine loss of this area (Ambrogini et al., 2014). In a third protocol, a long-lasting treatment was compared with acute administration of VE. Treatments were performed by intraperitoneal injection soon after SE and the chronic treatment lasted for 15 days with a dose of 250 mg RRR-α-TOH/kg body weight for the first 4 days followed by 2 mg/kg for the remaining days. This schedule of treatment affects the latent period of the response to KA treatment at the end of which spontaneous recurrent seizures occur, a peculiarity of the rat model of KA-induced SE (Olney, Rhee, & Ho, 1974; Turski et al., 1989). The latent period is characterized by neuroinflammation representing a risk factor for the development of chronic epilepsy. After this long-lasting treatment protocol of treatment with VE, the number of spikelets in pyramidal neurons was restored to control levels and the death of neurons in the hippocampus was almost completely inhibited; moreover, the production of neuroinflammatory cytokines such as TNF-α, IL-1β, GFAP, IBA-1 was significantly reduced compared to untreated animals (Ambrogini et al., 2018). All these findings confirm the hypothesis that α-TOH treatment is effective in reducing brain inflammation both during and after KA-induced SE. Also, these in vivo studies suggested that nutritional interventions with α-TOH are as effective as i.p. treatments in treating epilepsy symptoms of neuroinflammation, including brain oxidative stress and neuroglial overactivation. Moreover, these findings suggest that an application of α-TOH in the secondary prevention of postseizure inflammatory events in the brain is worth investigating.

Clinical trials

Disturbance of the antioxidant defense systems and increased production of reactive species have been reported in patients affected by different forms of epilepsy (Costello & Delanty, 2004; Geronzi et al., 2018), and a significant association between epilepsy and increased lipid peroxidation has recently been confirmed in a meta-analysis study also highlighting the additional role that some antiepileptic drugs may play in this context (Martinc, Grabnar, & Vovk, 2014). Since uncontrolled membrane lipid peroxidation and ferroptotic signaling of lipid peroxides are considered important pathogenic mechanisms of neuronal damage in different forms of epilepsy (Cai & Yang, 2021; Chen et al., 2020; Petrillo et al., 2021), it is reasonable to propose that fat-soluble antioxidants, such as vitamin E, could provide useful neuroprotective effects (Fig. 1). This may consist of less severe structural damages of the brain tissue during epileptic seizures, reduced epileptogenesis, and lower risk of developing milder cognitive deterioration.

According with the postulate and with the preclinical evidence obtained in animal models of epilepsy presented earlier in this chapter, chronic vitamin E therapy has proven to be a useful adjuvant to anticonvulsant drugs (Fig. 1). The frequency of seizures was reduced in three different randomized placebo-controlled clinical trials performed both in pediatric and adult patients (reviewed in Kim & Cho, 2019). Improvements have also been reported in observational studies in patients with complex partial seizures, which are often resistant to drug therapy (Levy et al., 1992). However, another and more recent randomized, double-blind, cross-over clinical trial failed to report a significant effect of vitamin E as an add-on therapy to reduce seizure frequency in uncontrolled epilepsy (Raju et al., 1994).

Therapeutic applications of vitamin E may also include a role as adjuvant to ω-3 therapy in human epilepsy (Omrani et al., 2019 and references therein), which is supported by the antioxidant function of this vitamin described earlier in this chapter and by the evidence accumulated in literature concerning nutritional studies (EFSA Panel on Dietetic Products, Nutrition, and Allergies (NDA), 2015, 2016). These essential polyunsaturated fatty acids are abundantly represented in neuronal cells—especially DHA that represents from 10% to 20% of total brain fatty acids, whereas other forms of ω-3 collectively represent less than 2%; they represent structural and functional components of the neuronal membrane, also serving as precursor of antiinflammatory eicosanoids, signaling molecules important in neurotransmitter function (Pourmasoumi et al., 2018). These neuronal PUFA are highly susceptible to lipid peroxidation (Torquato et al., 2021) and vitamin E has been demonstrated to be important to protect these fatty acids (Galli et al., 2017).

Conclusion

In conclusion, considering the goal of identifying nutritional supplements or nutraceuticals that may support the efficacy and safety of drug therapy of epilepsy, and given the established safety of vitamin E at dosages that have been demonstrated to be efficient in the pediatric and adult form of the disease to decrease the frequency of seizures (400 UI/day for 3–6 months) (reviewed elsewhere in Kim & Cho, 2019), vitamin E—under the form of α-TOH—can be recommended as an add-on treatment for the neuroprotection and secondary prevention of neuronal damage in epileptic seizures and to reduce epileptogenesis and the risk of cognitive deterioration.

However, clinical data on the use of this antioxidant vitamin in human epilepsy remain insufficient to conclusively asses its efficacy. More carefully planned randomized clinical trials for the evaluation of vitamin E therapy in patients with epilepsy are warranted.

Mini dictionary

Epileptogenesis: the process occurring in the brain after an injury or insult resulting in spontaneous recurrent seizure or epilepsy.

Maladaptive plasticity: plasticity in the nervous system that leads to a disruption of its functionality resulting in a disease state.

Oxidative stress: a condition of altered homeostasis that involves the cellular redox eventually leading to insufficient control of physiological oxidations and accumulation of oxidative damages on structural and signaling protein, membrane lipids, and nucleic acids.

Kainic acid-induced status epilepticus: one of the most used experimental models of epilepsy in rat.

Key facts of vitamin E

- Vitamin E is a fat-soluble essential micronutrient.
- α-TOH is the main form of this vitamin in human tissues and the form with proven role of essential nutrient.
- Vitamin E is a fat-soluble antioxidant active in protecting lipid molecules of cells and body fluids.
- Vitamin E is capable to cross the blood-brain barrier to exert its antioxidant and signaling functions in the brain tissue.
- The long-chain metabolites of this vitamin are bioactive molecules active in xenobiotics metabolism, detoxification and antiinflammatory processes associated with neurodegeneration and neuroinflammation.

- Vitamin E holds potential as an adjuvant treatment to antiepileptic drug therapy, with the aim of preventing neuroinflammation and brain toxicity complications.

Summary points

- Epilepsy is a neurological disease mainly characterized by seizures and brain maladaptive morphofunctional changes.
- The pathogenic triad of glutamate excitoxicity, neuroinflammation, and oxidative stress characterizes the neurobiology of epilepsy.
- Vitamin E is a neuroprotectant contributing to lipid homeostasis processes of cellular membranes and extracellular fluids.
- In animal model of epileptogenesis, vitamin E prevents recurrent seizure and neurodegenerative lesions.
- In clinical trials, the adjuvant activity of vitamin E decreases the frequency of seizure and the risk of cognitive deterioration.

References

Ambrogini, P., et al. (2014). Post-seizure α-tocopherol treatment decreases neuroinflammation and neuronal degeneration induced by status epilepticus in rat hippocampus. *Molecular Neurobiology*, *50*(1), 246–256.

Ambrogini, P., et al. (2018). Neurobiological correlates of alpha-tocopherol Antiepileptogenic effects and MicroRNA expression modulation in a rat model of Kainate-induced seizures. *Molecular Neurobiology*, *55*(10), 7822–7838.

Ambrogini, P., et al. (2019). Excitotoxicity, neuroinflammation and oxidant stress as molecular bases of epileptogenesis and epilepsy-derived neurodegeneration: The role of vitamin E. *Biochimica et Biophysica Acta - Molecular Basis of Disease*, *1865*(6), 1098–1112.

Azzi, A. (2019). Tocopherols, tocotrienols and tocomonoenols: Many similar molecules but only one vitamin E. *Redox Biology*, *26*, 101259.

Azzi, A. (2021). Reflections on a century of vitamin E research: Looking at the past with an eye on the future. *Free Radical Biology & Medicine*, *175*, 155–160.

Bartolini, D., et al. (2020). Glutathione S-transferase P influences the Nrf2-dependent response of cellular thiols to seleno-compounds. *Cell Biology and Toxicology*, *36*(4), 379–386.

Bartolini, D., et al. (2021). Alpha-tocopherol metabolites (the vitamin E metabolome) and their Interindividual variability during supplementation. *Antioxidants (Basel, Switzerland)*, *10*(2), 173.

Betti, M., et al. (2011). Dietary supplementation with α-tocopherol reduces neuroinflammation and neuronal degeneration in the rat brain after kainic acid-induced status epilepticus. *Free Radical Research*, *45*(10), 1136–1142.

Browne, D., et al. (2019). Vitamin E and Alzheimer's disease: What do we know so far? *Clinical Interventions in Aging*, *14*, 1303–1317.

Cai, Y., & Yang, Z. (2021). Ferroptosis and its role in epilepsy. *Frontiers in Cellular Neuroscience*, *15*, 696889.

Camfield, P., & Camfield, C. (2015). Incidence, prevalence and aetiology of seizures and epilepsy in children. *Epileptic Disorders*, *17*(2), 117–123.

Chen, S., et al. (2020). Iron metabolism and Ferroptosis in epilepsy. *Frontiers in Neuroscience*, *14*, 601193.

Cobley, J. N., Fiorello, M. L., & Bailey, D. M. (2018). 13 reasons why the brain is susceptible to oxidative stress. *Redox Biology*, *15*, 490–503.

Costello, D. J., & Delanty, N. (2004). Oxidative injury in epilepsy: Potential for antioxidant therapy? *Expert Review of Neurotherapeutics*, *4*(3), 541–553.

EFSA Panel on Dietetic Products, Nutrition, and Allergies (NDA). (2015). Scientific opinion on dietary reference values for vitamin E as α -tocopherol. *EFSA Journal*, *13*(7), 72.

EFSA Panel on Dietetic Products, Nutrition, and Allergies (NDA). (2016). Vitamin E and protection of DNA, proteins and lipids from oxidative damage: evaluation of a health claim pursuant to Article 14 of Regulation (EC) No 1924/2006. *EFSA Journal*, *14*(10), 4588.

Fariss, M. W., & Zhang, J. G. (2003). Vitamin E therapy in Parkinson's disease. *Toxicology*, *189*(1–2), 129–146.

Galli, F., et al. (2005). Oxidative stress and reactive oxygen species. *Cardiovascular Disorders in Hemodialysis*, *149*, 240–260.

Galli, F., et al. (2017). Vitamin E: Emerging aspects and new directions. *Free Radical Biology & Medicine*, *102*, 16–36.

Geronzi, U., Lotti, F., & Grosso, S. (2018). Oxidative stress in epilepsy. *Expert Review of Neurotherapeutics*, *18*(5), 427–434.

Haghnejad Azar, A., et al. (2017). Alpha-tocopherol reduces brain edema and protects blood-brain barrier integrity following focal cerebral ischemia in rats. *Medical Principles and Practice*, *26*(1), 17–22.

Jiang, Q., et al. (2008). Long-chain carboxychromanols, metabolites of vitamin E, are potent inhibitors of cyclooxygenases. *Proceedings of the National Academy of Sciences of the United States of America*, *105*(51), 20464–20469.

Jones, D. P. (2008). Radical-free biology of oxidative stress. *American Journal of Physiology. Cell Physiology*, *295*(4), C849–C868.

Kahn-Kirby, A. H., et al. (2019). Targeting ferroptosis: A novel therapeutic strategy for the treatment of mitochondrial disease-related epilepsy. *PLoS One*, *14*(3), e0214250.

Kaya, M., et al. (2002). Catalase and alpha-tocopherol attenuate blood-brain barrier breakdown in pentylenetetrazole-induced epileptic seizures in acute hyperglycaemic rats. *Pharmacological Research*, *45*(2), 129–133.

Khanna, S., et al. (2003). Molecular basis of vitamin E action: Tocotrienol modulates 12-lipoxygenase, a key mediator of glutamate-induced neurodegeneration. *The Journal of Biological Chemistry*, *278*(44), 43508–43515.

Kiasalari, Z., et al. (2012). Antiepileptic effect of vitamin E on Kainic acid-induced temporal lobe epilepsy in rats. *Thrita Studied Journal of Medical Sciences*, *1*(1), 27–29.

Kim, J. E., & Cho, K. O. (2019). Functional nutrients for epilepsy. *Nutrients*, *11*, 6.

Kume, A., et al. (2018). α-Tocopheryl succinate-suppressed development of cerebral malaria in mice. *Parasitology Research*, *117*(10), 3177–3182.

Lee, P., & Ulatowski, L. M. (2019). Vitamin E: Mechanism of transport and regulation in the CNS. *IUBMB Life*, *71*(4), 424–429.

Levy, S. L., Burnham, W. M., & Hwang, P. A. (1990). An evaluation of the anticonvulsant effects of vitamin E. *Epilepsy Research*, *6*(1), 12–17.

Levy, S. L., et al. (1992). The anticonvulsant effects of vitamin E: A further evaluation. *The Canadian Journal of Neurological Sciences*, *19*(2), 201–203.

Lin, T.-K., et al. (2020). Seizure-induced oxidative stress in status epilepticus: Is antioxidant beneficial? *Antioxidants (Basel, Switzerland)*, *9*(11), 1029.

Marinelli, R., et al. (2020). Garcinoic acid prevents β-amyloid (Aβ) deposition in the mouse brain. *The Journal of Biological Chemistry*, *295*(33), 11866–11876.

Martinc, B., Grabnar, I., & Vovk, T. (2014). Antioxidants as a preventive treatment for epileptic process: A review of the current status. *Current Neuropharmacology*, *12*(6), 527–550.

Mattson, M. P. (2019). Excitotoxicity. In G. Fink (Ed.), *Stress: Physiology, biochemistry, and pathology* (pp. 125–134). Academic Press.

Mehvari, J., et al. (2016). Effects of vitamin E on seizure frequency, electroencephalogram findings, and oxidative stress status of refractory epileptic patients. *Advanced Biomedical Research*, *5*, 36.

Olney, J. W., Rhee, V., & Ho, O. L. (1974). Kainic acid: A powerful neurotoxic analogue of glutamate. *Brain Research*, *77*(3), 507–512.

Omrani, S., et al. (2019). The effect of omega-3 fatty acids on clinical and paraclinical features of intractable epileptic patients: A triple blind randomized clinical trial. *Clinical and Translational Medicine*, *8*(1), 3.

Pansani, A. P., et al. (2018). Long-term monotherapy treatment with vitamin E reduces oxidative stress, but not seizure frequency in rats submitted to the pilocarpine model of epilepsy. *Epilepsy & Behavior*, *88*, 301–307.

Pein, H., et al. (2018). Endogenous metabolites of vitamin E limit inflammation by targeting 5-lipoxygenase. *Nature Communications*, *9*(1), 3834.

Petrillo, S., et al. (2021). Imbalance of systemic redox biomarkers in children with epilepsy: Role of Ferroptosis. *Antioxidants (Basel)*, *10*(8).

Pourmasoumi, M., et al. (2018). Association of Omega-3 fatty acid and epileptic seizure in epileptic patients: A systematic review. *International Journal of Preventive Medicine*, *9*, 36.

Raju, G. B., et al. (1994). Randomized, double-blind, placebo-controlled, clinical trial of D-alpha-tocopherol (vitamin E) as add-on therapy in uncontrolled epilepsy. *Epilepsia*, *35*(2), 368–372.

Ravizza, T., et al. (2008). Interleukin converting enzyme inhibition impairs kindling epileptogenesis in rats by blocking astrocytic IL-1β production. *Neurobiology of Disease*, *31*(3), 327–333.

Scheffer, I. E., et al. (2017). ILAE classification of the epilepsies: Position paper of the ILAE Commission for Classification and Terminology. *Epilepsia*, *58*(4), 512–521.

Schubert, M., et al. (2018). Long-chain metabolites of vitamin E: Metabolic activation as a general concept for lipid-soluble vitamins? *Antioxidants (Basel, Switzerland)*, 7(1), 10.

Shekh-Ahmad, T., et al. (2019). Reactive oxygen species in status epilepticus. *Epilepsy & Behavior*, *101*(Pt B), 106410.

Sies, H., & Jones, D. P. (2020). Reactive oxygen species (ROS) as pleiotropic physiological signalling agents. *Nature Reviews Molecular Cell Biology*, *21*(7), 363–383. https://doi.org/10.1038/s41580-020-0230-3. Epub 2020 Mar 30.

Takahashi, K., et al. (2019). Vitamin C and vitamin E double-deficiency increased neuroinflammation and impaired conditioned fear memory. *Archives of Biochemistry and Biophysics*, *663*, 120–128.

Thijs, R. D., et al. (2019). Epilepsy in adults. *Lancet*, *393*(10172), 689–701.

Torquato, P., et al. (2019a). Nutritional and lipidomics biomarkers of docosahexaenoic acid-based multivitamin therapy in pediatric NASH. *Scientific Reports*, *9*(1), 2045.

Torquato, P., et al. (2019b). Increased plasma levels of the lipoperoxyl radical-derived vitamin E metabolite alpha-tocopheryl quinone are an early indicator of lipotoxicity in fatty liver subjects. *Free Radical Biology & Medicine*, *131*, 115–125.

Torquato, P., et al. (2020a). Vitamin E: Nutritional aspects. In V. B. Patel (Ed.), *Elsevier. Molecular nutrition: Vitamins* (pp. 447–485).

Torquato, P., et al. (2020b). Vitamin E: Metabolism and molecular aspects. In V. B. Patel (Ed.), *Elsevier. Molecular nutrition: Vitamins* (pp. 487–518).

Torquato, P., et al. (2021). Pre-analytical monitoring and protection of oxidizable lipids in human plasma (vitamin E and ω-3 and ω-6 fatty acids): An update for redox-lipidomics methods. *Free Radical Biology & Medicine*, *176*, 142–148.

Turski, L., et al. (1989). Review: Cholinergic mechanisms and epileptogenesis. The seizures induced by pilocarpine: A novel experimental model of intractable epilepsy. *Synapse*, *3*(2), 154–171.

Ulatowski, L. M., & Manor, D. (2015). Vitamin E and neurodegeneration. *Neurobiology of Disease*, *84*, 78–83.

Vezzani, A., Balosso, S., & Ravizza, T. (2019). Neuroinflammatory pathways as treatment targets and biomarkers in epilepsy. *Nature Reviews Neurology*, *15*(8), 459–472.

Wang, S.-w., et al. (2016). Alpha-tocopherol quinine ameliorates spatial memory deficits by reducing beta-amyloid oligomers, neuroinflammation and oxidative stress in transgenic mice with Alzheimer's disease. *Behavioural Brain Research*, *296*, 109–117.

Wenzel, S. E., et al. (2017). PEBP1 wardens Ferroptosis by enabling lipoxygenase generation of lipid death signals. *Cell*, *171*(3), 628–641.e26.

Ye, Q., et al. (2019). Inhibition of ferroptosis processes ameliorates cognitive impairment in kainic acid-induced temporal lobe epilepsy in rats. *American Journal of Translational Research*, *11*(2), 875–884.

Zhang, X., et al. (2021). Vitamin E exerts neuroprotective effects in Pentylenetetrazole kindling epilepsy via suppression of Ferroptosis. *Neurochemical Research*, *47*(3), 739–747.

CHAPTER 15

Selenium use in epilepsy

Haitham S. Mohammed
Biophysics Department, Faculty of Science, Cairo University, Giza, Egypt

List of abbreviations

DIO	iodothyronine deiodinase
GPx	glutathione peroxidase
GST	glutathione S transferase
PTZ	pentylenetetrazole
RDA	recommended dietary allowance
Sec	selenocysteine
SeMet	selenomethionine
SeNPs	selenium nanoparticles
TXNRD	thioredoxin reductase

Introduction

Since its discovery in 1817, selenium was recognized as a toxic substance. Only during the fifties of the last century and exactly in 1957, scientists discovered its significant role in the health of humans and animals (Brown & Arthur, 2001; Schwarz & Foltz, 1957). Selenium is a trace mineral that exists in nature in organic forms, selenomethionine (SE-Met) and selenocysteine (Sec), and inorganic forms, selenate (SeO_4), selenite (SeO_3), and selenide (Se^{2-}) (Puzanowska-Tarasiewicz, Kuzmicka, & Tarasiewicz, 2009). The amount of selenium in the soil and consequently in the plants and the human body varies with the geographic location (Skalny et al., 2019).

Selenium concentration in the body was also found to vary with age. Blood selenium levels were negatively correlated with age in healthy adults. It was found that selenium is concentrated in some of the body organs more than others. Brain, reproductive, and endocrine organs are the main organs that act as a priority pathway of selenium during inadequate selenium intake. Furthermore, Ejima, Watanabe, Koyama, Matsuno, and Satoh (1996) reported that selenium concentrations varied in different adult brain regions.

A recommended dietary allowance (RDA) of selenium for humans has been established by The National Research Council to be 55 μg/day for adult men and 70 μg/day for adult women (National Research Council, 1989). However, smaller limits were put for younger ages (Table 1). Selenium is obtained from food such as cereals, fish, meat, and dairy products (Table 2) (Klein, 2004). Selenomethionine (SeMet) is the chemical form of dietary selenium which is metabolized into Sec followed by the biosynthesis of

Vitamins and Minerals in Neurological Disorders
https://doi.org/10.1016/B978-0-323-89835-5.00017-X

Table 1 Recommended dietary allowances (RDAs) for selenium.

Age in years	Selenium intake limit in µg/day
0.5–3	20
4–8	30
9–13	40
14–>50	55

selenoproteins (Labunskyy, Hatfield, & Gladyshev, 2014). Selenium, as an essential component of many enzymes in the human body, has several roles such as antioxidant (Rayman, 2000), antiinflammatory (Malhotra, Welling, Mantri, & Desai, 2016), anticancer (Stolzoff & Webster, 2016), and antibacterial (Cihalova et al., 2015).

Human contains 25 genes for selenoproteins; however, not all functions of selenoproteins are known. Selenoproteins of known function have been named according to the function they are involved in (Brigelius-Flohe & Flohe, 2017). Glutathione peroxidases (GPx1–4 and 6), thioredoxin reductase (TXNRD1–3), iodothyronine deiodinase (DIO1–3), methionine sulfoxide reductase (MSRB1), and selenophosphate synthetase (SEPHS2) are the known-functional selenoproteins. Others are named as selenoproteins followed by a letter (e.g., Selenoprotein P or Selenop) (Gladyshev et al., 2016).

Selenium as an antioxidant

The main biological role played by selenium is attributed to its incorporation into proteins to make selenoproteins that act as antioxidant enzymes. Synthesis of selenoproteins is accomplished via the incorporation of the amino acid selenocysteine (Sec, the so-called

Table 2 Selenium in food.

Food	Amount of selenium in mcg per serving
Braizl nuts	544
Fish	92
Ham	42
Shrimp	40
Macaroni	37
Beef	33
Turkey	31
Chicken	22
Brown rice	19
Hard-boiled egg	15
Whole-wheat bread	13
Low fat milk	8
Potato	1

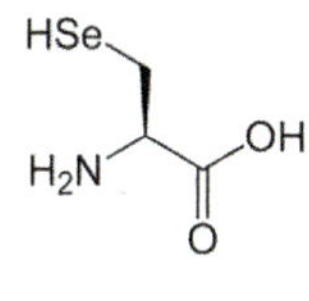

Fig. 1 Selenocysteine.

21st amino acid, Fig. 1) into proteins. This reaction is mainly encoded by the UGA codon (Birringer, Pilawa, & Flohé, 2002). The main group of selenoproteins that is involved in the regulation of the cellular redox state is the glutathione peroxidase (GPx) group which is characterized by the presence of Sec in the N-terminal of the functional domain of the protein (Behn et al., 1995). The antioxidant property of selenoproteins protects the biological tissues from the oxidative damage of free radicals (Lu & Holmgren, 2009). GPx isozymes help in maintaining low levels of hydrogen peroxide, which is responsible for the lipid peroxidation within the cells.

Glutathione peroxidase

Selenium has been associated with a selenoenzyme GPx in the neuronal tissues. GPx is the first identified selenoenzyme that consists of four subunits. Selenium is thought to be present at the active site of GPx in its selenolate form as selenocysteine (Wajner, Latini, Wyse, & Dutra-Filho, 2004). Two isoforms of GPx have been detected in the human brain, GPx1 and GPx4 (Lubos, Loscalzo, & Handy, 2011). GPx1 is a selenoenzyme expressed in the astroglial cells and is responsible for antioxidant activity against the hydrogen peroxides and the alkyl hydroperoxides (Lubos et al., 2011). Inhibition in the activity of this enzyme could elevate the brain's vulnerability to oxidative stress. GPx1 knockout mice suffer from increased susceptibility to oxidative stress (Fu, Cheng, Porres, Ross, & Lei, 1999). GPx4 is expressed in the cerebral cortex, hippocampus, and cerebellum, and it is involved in the elimination of phospholipid hydroperoxides (Squires, Stoytchev, Forry, & Berry, 2007). The estimates of the amounts of selenium in a rat's brain have indicated that GPx may account for only a fifth of the total selenium found in the brain (Larsen & Berry, 1995). Most of the selenium is bound to proteins and not to amino acids or nucleic acids (Wrobel, Power, & Toborek, 2016).

$$H_2O_2 + 2GSH \xrightarrow{GPx} GSSG + 2H_2O \text{ (antioxidant activity of GPx)}$$

Thioredoxin reductase

Three types of thioredoxin reductase have been identified, TR1, TR2, and TR3. The main function of these selenium-containing compounds is controlling the intracellular redox potential via their antioxidant activity. TR1 is located in the cellular cytoplasm and nucleus, while TR2 is located mainly in the mitochondria, and TR3 is located in the testes (Ducros & Favier, 2004; Fairweather-Tait, Collings, & Hurst, 2010).

Selenium deficiency

Insufficient supply of selenium has been associated with growth retardation (Moreno-Reyes, Egrise, Neve, Pasteels, & Schoutens, 2001), thyroid function abnormalities (Winther, Rayman, Bonnema, & Hegedus, 2020), cardiomyopathy (Chen, 2012), and epilepsy (Seven, Basaran, Cengiz, Unal, & Yuksel, 2012). Selenium deficiency leads to a decrease in glutathione peroxidase (GPx) activity and an increase in the hepatic glutathione S transferase (GST) activity (Burk, Hill, Awad, Morrow, & Lyons, 1995). It causes also a decrease in the plasma concentration of homocysteine which is considered a risk factor for cardiovascular disease (Uthus, Yokoi, & Davis, 2002).

Selenium deficiency could also cause an increase in the toxicity of some drugs and attenuation in the efficiency of others (Davis & Uthus, 2002). It was shown that selenium deficiency exposes the neurons to glutamate-induced excitotoxicity due to the absence of the inhibitory effect of selenium on the glutamate-induced NF-kB and AP-1 activation (Santamaría et al., 2005).

Some diets are used as a therapeutic tool for some diseases such as a high-fat ketogenic diet that is used in several treatment modalities including intractable epilepsy (Rogovik & Goldman, 2010). However, this treatment protocol may lead to serum selenium deficiency with its related health complications (Arslan, Kose, & Guzel, 2017).

Selenium toxicity

Although selenium in small concentration is crucial for human and animal health, the large concentration of selenium is toxic and can lead to many health disorders such as liver damage, heart muscle atrophy, stiffness and irritation of limbs, anemia, dyspnea, hair loss, emaciation, and depression. An excessive intake of selenium leads to the inhibition of cellular proliferation, DNA replication, protein synthesis, and oxidative stress (Dodig & Cepelak, 2004; Kim & Mahan, 2001; Tinggi, 2005; Zachary & McGavin, 2014). It was reported that a blood selenium level of more than 100 μg/dL leads to selenium toxicity or selenosis (Goldhaber, 2003).

High levels of selenium induce nitric oxide synthase and activation of caspase-3. Furthermore, high doses of selenium induce oxidative stress which resulted in the degeneration of neuronal cells (Vinceti et al., 2014). A characteristic marker for selenium toxicity in humans is the odor of garlic during exhaling due to the presence of dimethyl selenide as a volatile compound (Kieliszek & Błażejak, 2013). The acute poisoning by selenium causes hypotension, tachycardia, and tremor (Aldosary, Sutter, Schwartz, & Morgan, 2012).

The toxic symptoms may occur in individuals taking more than 5 mg selenium per day or persons who are exposed to air with a selenium concentration of more than 0.2 mg/m^3 (Bampidis et al., 2019). Convulsions and motor weakness have been reported due to chronic overexposure to selenium from food and water (Li, Xiao, & Zheng, 2012).

Selenium and neurological disorders

The brain is the most metabolically active organ in the body and contains a high concentration of unsaturated fatty acids. This makes the brain highly vulnerable to lipid peroxidation. Oxidative stress is considered a central mechanism in the pathogenesis of several neurological disorders including epilepsy. Therefore, the role played by antioxidant compounds is pivotal in combating such diseases. It was shown that selenium is essential for the proper development of neuronal tissue and that selenium deficiency could lead to some neurological disorders (Fig. 2). It was reported that a decline in the cognitive function of the human with age is related to the decrease in selenium level in the brain (Smorgon et al., 2004; Soerensen et al., 2008). Therefore, selenium and organo-selenium compounds have been shown to have therapeutic potential against neurophathological diseases. It was reported that treatment with sodium selenite could prevent oxidative damage, calcium overload, and cell death in the dorsal root ganglion of rats (Uguz & Naziroglu, 2012).

Selenium and epilepsy

Oxidative stress and the generation of reactive oxygen species are strongly implicated in epilepsy. Free radicals and lipid peroxidation reactions seem to be involved in epileptic

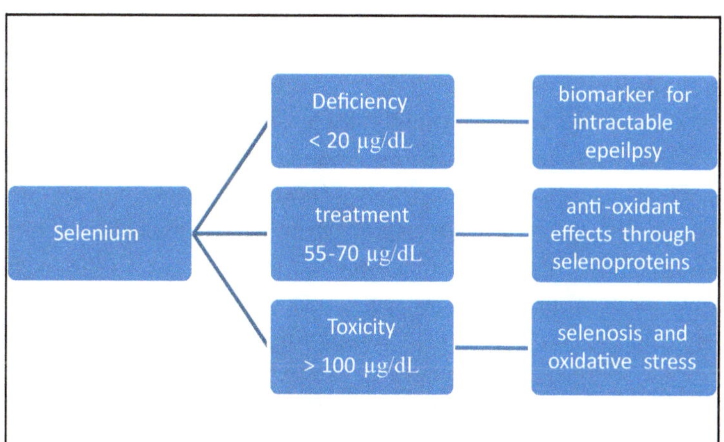

Fig. 2 Schematic diagram showing selenium effects and uses and dependence on its dose.

seizure development. The initiation and propagation of seizure activity and its accompanying induced neuronal damage are linked to the increase in the free radicals (Halliwell, 2001). Frequent seizures as the main manifestation of epilepsy shift the redox status of the neurons and lead to oxidative stress that eventually resulted in neuronal loss and reactive gliosis (Lovell, Xie, Gabbita, & Markesbery, 2000). The oxidative stress, in turn, increases the severity and frequency of epileptic seizures which creates a self-perpetuating vicious cycle.

Selenium as an antioxidant element could play an important role in protecting cells from oxidative damage and help in stopping this vicious cycle. It was suggested that the depletion of selenium from epileptic patients may constitute a key factor for the initiation of seizures (Savaskan et al., 2003). Furthermore, recurrent seizures have been attributed to the activation of glutamate receptors due to selenium depletion, which in turn leads to the failure of the antiepileptic drugs which act mainly through the GABAergic receptors (Savaskan et al., 2003).

An early study was behind the relationship between epilepsy and selenium deficiency (Ramaekers, Calomme, Vanden Berghe, & Makropoulos, 1994). In this study, two children have intractable epilepsy and low plasma selenium level. Selenium supplementation was capable to control the seizure activity, and discontinuation of selenium caused a return to the disease with symptomatic seizure activity.

Selenium has been involved in the protection of the neuronal tissues through selenoproteins (Wang et al., 2010) and showed a synergistic effect when combined with the anticonvulsant medication (Mohammed, Aboul Ezz, Zedan, & Ali, 2020). Furthermore, neurons frequently have a shift in their redox status during seizure activity due to the large influx of Ca^{2+}. These neurons are subjected to apoptosis and cell death if their redox equilibrium is not restored. It has been shown that selenium through its function in the glutathione peroxidase (GP) or thioredoxin reductase (TR) can neutralize the redox shift that is generated due to Ca^{2+} influx and prevent neuronal cell death (Xiong, Markesbery, Shao, & Lovell, 2007).

The notion of selenium relationship to seizures was supported with the animal studies that showed an increase in susceptibility to epileptic seizures in animals kept on a selenium-deficient diet and exposed to seizure evoked chemical such as kainate (Schweizer, Schomburg, & Savaskan, 2004). Moreover, GPx4 knockout mice displayed a tendency to develop epileptic seizures (Schweizer, Brauer, Kohrle, Nitsch, & Savaskan, 2004). Animal studies on epileptic models revealed that seizures could have several biochemical consequences such as activation of membrane phospholipases and proteases which lead to degradation of membrane phospholipids and lyses of the cytoskeleton proteins (Walz et al., 2000) that eventually contribute to neuronal injury and death (Patel, 2004). On the other hand, selenium-treated animal models of epilepsy showed promising results represented in inhibition of oxidative stress and regulation of calcium-dependent processes (Schweizer, Michaelis, Kohrle, & Schomburg, 2004). Additionally, selenium treatment has been able to reduce the number of seizures induced by pentylenetetrazole (PTZ) injection in animals (Schweizer et al., 2005).

Human studies

Selenium as biomarker

Several reports of human studies have focused on selenium as a biomarker for epilepsy and have correlated between selenium deficiency and different types of epilepsy in humans (Table 3). Reports showed that epilepsy patients have a significantly lower level of serum selenium compared to their healthy counterparts. A correlation between seizure activity and serum selenium has been elaborated. The importance of the measurement of serum selenium in patients with intractable epilepsy has been emphasized (Ashrafi et al., 2007).

Selenium as treatment

Selenium has been used as a therapeutic or adjunct therapy for epilepsy in humans (Table 3). It was found that selenium is capable of attenuating oxidative toxicity in the blood of epileptic patients (Yürekli & Nazıroğlu, 2013). Several reports demonstrated the importance of selenium supplements with some epilepsy treating modalities. Valproic acid and carbamazepine antiepileptic drugs produce highly reactive oxygen species that consume selenium resources in epilepsy patients, which necessitates the use of selenium supplements to avoid selenium deficiency. A high-fat ketogenic diet-based therapy for intractable epilepsy requires selenium supplements as an adjunct therapy to avoid health complications of selenium deficiency (Arslan et al., 2017).

Animal studies

Animal studies are carried out mainly to investigate the relationship between selenium deficiency and epileptic seizures and to use selenium as a treating agent in an animal model of epilepsy (Table 3). Transgenic mouse model studies have emphasized the role played by selenium in preventing seizures and other neurological disorders (Schweizer, Schomburg, et al., 2004). Anitseizure and neuroprotective effects of selenium have been reported in animal models of epilepsy (Dabhade et al., 2013; Tawfik et al., 2018).

The main observation on studies that used selenium in treating epilepsy in animal models is that most studies have been carried out on the pentylenetetrazole model of epilepsy and this could be a direct and easy way to provoke seizures in rats and mice. However, other models with different face validity (e.g., pilocarpine and kainic acid models) with human symptoms could also shed more light on the use of selenium in the treatment of epilepsy. Furthermore, studies have been carried out mainly on male animals and there

Table 3 Human and animal epilepsy studies use selenium.

Study	Subjects	Main results	Reference
Selenium as a biomarker in human studies			
Intractable epilepsy	90 children (30 patients treated with AED; 30 patients treated with ketogenic diet + AED; 30 control)	Selenium levels were significantly lower in epileptic patients and markedly lower in the ketogenic group	El-Rashidy et al. (2020)
Intractable epilepsy	70 patients and 60 healthy children	Intractable patients have a significantly lower level of serum selenium	Seven et al. (2012)
Febrile seizures	30 children with simple febrile seizures and 30 children without seizures	Simple febrile seizures children have a significantly lower level of serum selenium	Mahyar, Ayazi, Fallahi, and Javadi (2010)
Intractable epilepsy	80 patients and 80 healthy children	Significant lower selenium in the patient group compared with the control group No association between selenium and intractable seizures	Ashrafi, Shabanian, et al. (2007)
Patients with epilepsy	53 epileptic and 57 healthy children	Significant lower serum selenium in the patient group compared to healthy control ($P < 0.001$)	Ashrafi, Shams, et al. (2007)

Selenium as treatment in human patients			
Children with intractable epilepsy	2 children	Two weeks oral supplements of 3–5 μg/kg body weight resulted in a reduction of seizures and improvement of EEG	Ramaekers et al. (1994)
Patients with epilepsy and refractory epilepsy	15 epileptic patients and 19 controls 8 refractory epilepsy patients and 15 controls	0.1 mg sodium selenite twice daily for 45 days. Selenium and topiramate caused protective effects on epilepsy and refractory epilepsy-induced oxidative injury	Yürekli and Nazıroğlu (2013)
Selenium as treatment in animal models			
Pentylenetetrazole model	60 adult male Swiss mice (25–32 g)	5 mg/kg, oral dose of SeNPs for 10 days. SeNPs delayed the onset and decreased the duration of seizures and provided neuroprotection	Yuan et al. (2020)
Pentylenetetrazole model	50 adult male Wistar rats	Sodium selenite (0.3 mg/kg i.p.). Selenium ameliorates the adverse effects of the chemoconvulsant PTZ	Mohammed et al. (2020)

Continued

Table 3 Human and animal epilepsy studies use selenium—cont'd

Study	Subjects	Main results	Reference
Pentylenetetrazole model	70 male Swiss albino mice	0.2 mg/kg for 21 days every other day. Neuroprotective effects of selenium and sildenafil against PTZ-kindled model	Tawfik, Moustafa, and El-Azab (2018)
Pentylenetetrazole and electroshock models	36 male adult Wistar rats	Sodium selenite 100, 200, 300, and 400 μg/kg/day for 1 week. Protective effects of sodium selenite against seizures induced by PTZ and electroshock	Dabhade, Rizvi, and Kumar (2013)
Pentylenetetrazole model	Male Swiss albino mice	Sodium selenite 1, 3, and 10 mg/kg i.p. Selenium induced reduction in seizures and exerts synergistic anticonvulsant effect with celecoxib	Rehni and Singh (2013)

is limited information available about female animals in response to treatment. Interestingly, some early reports have linked sexual hormones and epilepsy. The proconvulsant effects of estrogen and the inhibitory effects of progesterone have been reported (Bäckström, Zetterlund, Blom, & Romano, 1984; Hall, 1977).

Nanotechnology has been implemented in selenium treatment of epilepsy through the conversion of selenium native compounds into nanoparticles. Selenium nanoparticles (SeNPs) have been shown to be more biocompatible, bioavailable, less toxic, and more biologically active than their native counterparts (Yuan et al., 2020).

Conclusions

Selenium is a trace element found in food and water and is essential for human health. It executes its action via incorporation in selenoproteins which primarily have strong antioxidant properties. Studies on selenium have revealed the important role it played in epilepsy. Selenium may be considered a double-edged sword for epilepsy. Between its deficiency and toxicity lies its appropriate usage for preventing and controlling seizure disorders. More studies should be carried out for the complete understanding of selenium effects on neurological disorders.

Applications to other neurological conditions

Since oxidative stress is involved in the pathogenesis of neurodegenerative diseases, selenium as a diet nutrient with antioxidant activity plays an important role in the treatment of Parkinson's and Alzheimer's diseases. Some studies showed that selenium could aid in preventing Parkinson's disease (PD) if used before the emergence of disease or when patients suffer from early stages of the disease (Ellwanger et al., 2015). Other studies showed that selenium is capable of partially reversing the toxic effects caused by monoamine-depleting agents used in animal models of PD (Khan, 2010). It was found that a daily dose of selenium could improve the locomotor dysfunction accompanying PD (Zafar et al., 2003). On the other hand, selenium deficiency can modify the dopaminergic neurons function (Romero-Ramos, Venero, Cano, & Machado, 2000) and selenium overdose can worsen the conditions of the PD patients (Vinceti et al., 2016). Fathi et al. (2013) found a relationship between serum level of selenium and PD using the nuclear magnetic resonance (NMR) metabonomics study. Recently, Zhang et al. (2021) have carried out a meta-analysis study to correlate selenium level and PD. They found that selenium levels do not differ in blood but they were, despite the small number of studies included, significantly higher in the cerebrospinal fluid (CSF) in PD patients than normal individuals.

Other components of interest

Besides selenium, as a nutritional active component involved in the diagnosis and treatment of epilepsy, other essential trace elements that the human body gets from food intake, and related to seizures and epilepsy are copper, zinc, and magnesium. These elements play an important role in the pathogenesis of epilepsy due to their involvement and effects on synaptogenesis, ion channels, neurotransmitters, and membrane lipid peroxidation. Copper is involved in the electron transport chain and the intracellular oxidation reaction. Zinc is the cofactor for several enzymes in the human body, and magnesium is involved in the synaptic transmission between neurons (Salwen, 2007). Zinc plays a central role in regulating the synthesis of gamma-aminobutyric acid (GABA) which is the main inhibitory neurotransmitter in the brain and closely related to seizure disorders (Mathie, Sutton, Clarke, & Veale, 2006). Kheradmand et al. (2014) have correlated zinc and copper levels with intractable epilepsy. They concluded that zinc serum analysis should be considered in intractable epilepsy patients, and zinc supplements could be useful in epilepsy treatment. Furthermore, long-term therapy with antiepileptic drugs could have an effect on the levels of zinc and copper in the epileptic patients (Talat, Ahmed, & Mohammed, 2015).

Mini-dictionary of terms

Selenosis: A health state that resulted from chronic selenium toxicity
Codon: A sequence of DNA or RNA that corresponds to specific amino acid
Selenoproteins: Proteins that contain selenocysteine residues
Excitotoxicity: Overactivation of excitatory receptors (glutamate) leads to damage and death of the cells
Ketogenic diet: A very low carbohydrate, moderate protein, and high-fat diet
Dyspnea: Shortness of breath
Emaciation: Extreme weight loss
Tachycardia: Heart rate elevation >100 beats/min
Redox shift: Change in the oxidation/reduction state of the cells
Intractable epilepsy: Uncontrolled seizure disorder that doesn't respond to drug treatment
Adjunct therapy: A secondary treatment that is used to assist the primary treatment

Key facts of selenium

- Selenium is an essential trace element and it is very important to human health.
- Selenium is an integral part of many antioxidant enzymes that protect the human body from oxidative stress.

- Selenium level in human serum is considered a biomarker for an epilepsy diagnosis.
- Selenium supplements in appropriate doses are considered as a treatment or adjuvant agent in epilepsy treatment.
- Selenium toxicity originates from higher doses of selenium intake.

Summary

- Selenium is an integral part of human health and disease states.
- Selenium through selenoproteins plays a vital role in the redox state of cells.
- Selenium deficiency, as well as toxicity, has serious health complications in humans.
- Ketogenic diets as a treatment strategy in intractable epilepsy lead to selenium deficiency.
- Animal models of epilepsy are needed for a complete understanding of the roles played by selenium in this disease.

References

Aldosary, B. M., Sutter, M. E., Schwartz, M., & Morgan, B. W. (2012). Case series of selenium toxicity from a nutritional supplement. *Clinical Toxicology*, *50*, 57–64.

Arslan, N., Kose, E., & Guzel, O. (2017). The effect of ketogenic diet on serum selenium levels in patients with intractable epilepsy. *Biological Trace Element Research*, *178*(1), 1–6.

Ashrafi, M. R., Shabanian, R., Abbaskhanian, A., Nasirian, A., Ghofrani, M., & Mohammadi, M. (2007). Selenium and intractable epilepsy: Is there any correlation? *Pediatric Neurology*, *36*(1), 25–29. https://doi.org/10.1016/j.pediatrneurol.2006.09.001.

Ashrafi, M. R., Shams, S., Nouri, M., Mohseni, M., Shabanian, R., Yekaninejad, M. S., et al. (2007). A probable causative factor for an old problem: Selenium and glutathione peroxidase appear to play important roles in epilepsy pathogenesis. *Epilepsia*, *48*(9), 1750–1755.

Bäckström, T., Zetterlund, B., Blom, S., & Romano, M. (1984). Effects of intravenous progesterone infusions on the epileptic discharge frequency in women with partial epilepsy. *Acta Neurologica Scandinavica*, *69*(4), 240–248.

Bampidis, V., Azimonti, G., Bastos, M. D. L., Christensen, H., Dusemund, B., Kouba, M., et al. (2019). Assessment of the application for renewal of authorisation of selenomethionine produced by Saccharomyces cerevisiae NCYC R397 for all animal species. *EFSA Journal*, *17*, e05539.

Behn, D., Weiss-Nowak, C., Kalcklösch, M., Westphal, C., Gessner, H., & Kyriakopoulos, A. (1995). Studies on the distribution and characteristics of new mammalian selenium-containing proteins. *Analyst*, *120*(3), 823–825.

Birringer, M., Pilawa, S., & Flohé, L. (2002). Trends in selenium biochemistry. *Natural Product Reports*, *19*(6), 693–718.

Brigelius-Flohe, R., & Flohe, L. (2017). Selenium and redox signaling. *Archives of Biochemistry and Biophysics*, *617*, 48–59.

Brown, K. M., & Arthur, J. R. (2001). Selenium, selenoproteins and human health: A review. *Public Health Nutrition*, *4*, 593–599.

Burk, R. F., Hill, K. E., Awad, J. A., Morrow, J. D., & Lyons, P. R. (1995). Liver and kidney necrosis in selenium-deficient rats depleted of glutathione. *Laboratory Investigation*, *72*(6), 723–730.

Chen, J. (2012). An original discovery: Selenium deficiency and keshan disease (an endemic heart disease). *Asia Pacific Journal of Clinical Nutrition*, *21*(3), 320–326.

Cihalova, K., Chudobova, D., Michalek, P., Moulick, A., Guran, R., Kopel, P., et al. (2015). Staphylococcus aureus and MRSA growth and biofilm formation after treatment with antibiotics and SeNPs. *International Journal of Molecular Sciences*, *16*, 24656–24672.

Dabhade, D. C., Rizvi, W., & Kumar, A. (2013). Beneficial role of selenium supplementation against experimental seizures in rats. *International Journal of Pharmaceutical and Phytopharmacological Research*, *2*(4), 272–276.

Davis, C. D., & Uthus, E. O. (2002). Dietary selenite and azadeoxycytidine treatments affect dimethylhydrazine-induced aberrant crypt formation in rat colon and DNA methylation in HT-29 cells. *The Journal of Nutrition*, *132*(2), 292–297.

Dodig, S., & Cepelak, I. (2004). The facts and controversies about selenium. *Acta Pharmaceutica*, *54*, 261–276.

Ducros, V., & Favier, A. (2004). Selenium metabolism. *EMC - Endocrinologie*, *1*, 19–28.

Ejima, A., Watanabe, C., Koyama, H., Matsuno, K., & Satoh, H. (1996). Determination of selenium in the human brain by graphite furnace atomic absorption spectrometry. *Biological Trace Element Research*, *54*, 9–21.

Ellwanger, J. H., Molz, P., Dallemole, D. R., Pereira dos Santos, A., Müller, T. E., Cappelletti, L., et al. (2015). Selenium reduces bradykinesia and DNA damage in a rat model of Parkinson's disease. *Nutrition (Burbank, Los Angeles County, Calif.)*, *31*(2), 359–365.

El-Rashidy, O. F., Youssef, M. M., Elgendy, Y. G., Mohsen, M. A., Morsy, S. M., Dawh, S. A., et al. (2020). Selenium and antioxidant levels in children with intractable epilepsy receiving ketogenic diet. *Acta Neurologica Belgica*, *120*(2), 375–380.

Fairweather-Tait, S. J., Collings, R., & Hurst, R. (2010). Selenium bioavailability: Current knowledge and future research requirements. *The American Journal of Clinical Nutrition*, *91*, 1484S–1491S.

Fathi, F., Kyani, A., Darvizeh, F., Mehrpour, M., Tafazzoli, M., & Shahidi, G. (2013). Relationship between serum level of selenium and metabolites using ^{1}HNMR-based Metabonomics in Parkinson's disease. *Applied Magnetic Resonance*, *44*, 721–734.

Fu, Y., Cheng, W. H., Porres, J. M., Ross, D. A., & Lei, X. G. (1999). Knockout of cellular glutathione peroxidase gene renders mice susceptible to diquat-induced oxidative stress. *Free Radical Biology & Medicine*, *27*(5–6), 605–611.

Gladyshev, V. N., Arner, E. S., Berry, M. J., Brigelius-Flohe, R., Bruford, E. A., Burk, R. F., et al. (2016). Selenoprotein gene nomenclature. *The Journal of Biological Chemistry*, *291*, 24036–24040.

Goldhaber, S. B. (2003). Trace element risk assessment: Essentiality vs. toxicity. *Regulatory Toxicology and Pharmacology*, *38*(2), 232–242.

Hall, S. M. (1977). Treatment of menstrual epilepsy with a progesterone-only oral contraceptive. *Epilepsia*, *18*(2), 235–236.

Halliwell, B. (2001). Role of free radicals in the neurodegenerative diseases: Therapeutic implications for antioxidant treatment. *Drugs & Aging*, *18*(9), 685–716.

Khan, H. A. (2010). Selenium partially reverses the depletion of striatal dopamine and its metabolites in MPTP-treated C57BL mice. *Neurochemistry International*, *57*(5), 489–491.

Kheradmand, Z., Yarali, B., Zare, A., Pourpak, Z., Shams, S., & Ashrafi, M. R. (2014). Comparison of serum zinc and copper levels in children and adolescents with intractable and controlled epilepsy. *Iranian Journal of Child Neurology*, *8*(3), 49–54.

Kieliszek, M., & Błazejak, S. (2013). Selenium: Significance, and outlook for supplementation. *Nutrition*, *29*, 713–718.

Kim, Y. Y., & Mahan, D. C. (2001). Comparative effects of high dietary levels of organic and inorganic selenium on selenium toxicity of growing-finishing pigs. *Journal of Animal Science*, *79*, 942–948.

Klein, E. A. (2004). Selenium: Epidemiology and basic science. *Urology Journal*, *171*, S50–S53.

Labunskyy, V. M., Hatfield, D. L., & Gladyshev, V. N. (2014). Selenoproteins: Molecular pathways and physiological roles. *Physiological Reviews*, *94*, 739–777.

Larsen, P. R., & Berry, M. J. (1995). Nutritional and hormonal regulation of thyroid hormone deiodinases. *Annual Review of Nutrition*, *15*, 323–352.

Li, S., Xiao, T., & Zheng, B. (2012). Medical geology of arsenic, selenium and thallium in China. *The Science of the Total Environment*, *421–422*, 31–40.

Lovell, M. A., Xie, C., Gabbita, S. P., & Markesbery, W. R. (2000). Decreased thioredoxin and increased thioredoxin reductase levels in Alzheimer's disease brain. *Free Radical Biology & Medicine*, *28*(3), 418–427.

Lu, J., & Holmgren, A. (2009). Selenoproteins. *Journal of Biological Chemistry*, *284*(2), 723–727.

Lubos, E., Loscalzo, J., & Handy, D. E. (2011). Glutathione peroxidase-1 in health and disease: From molecular mechanisms to therapeutic opportunities. *Antioxidants & Redox Signaling*, *15*(7), 1957–1997.
Mahyar, A., Ayazi, P., Fallahi, M., & Javadi, A. (2010). Correlation between serum selenium level and febrile seizures. *Pediatric Neurology*, *43*(5), 331–334. https://doi.org/10.1016/j.pediatrneurol.2010.05.024.
Malhotra, S., Welling, M. N., Mantri, S. B., & Desai, K. (2016). In vitro and in vivo antioxidant, cytotoxic, and anti-chronic inflammatory arthritic effect of selenium nanoparticles. *Journal of Biomedical Materials Research. Part B, Applied Biomaterials*, *104*, 993–1003.
Mathie, A., Sutton, G. L., Clarke, C. E., & Veale, E. L. (2006). Zinc and copper: Pharmacological probes and endogenous modulators of neuronal excitability. *Pharmacology & Therapeutics*, *111*(3), 567–583.
Mohammed, H. S., Aboul Ezz, H. S., Zedan, A., & Ali, M. A. (2020). Electrophysiological and neurochemical assessment of selenium alone or combined with carbamazepine in an animal model of epilepsy. *Biological Trace Element Research*, *195*, 579–590.
Moreno-Reyes, R., Egrise, D., Neve, J., Pasteels, J. L., & Schoutens, A. (2001). Selenium deficiency-induced growth retardation is associated with an impaired bone metabolism and osteopenia. *Journal of Bone and Mineral Research*, *16*(8), 1556–1563.
National Research Council. (1989). *Recommended dietary allowances* (10th ed.). Washington, DC: Matinal Academy Press.
Patel, M. (2004). Mitochondrial dysfunction and oxidative stress: Cause and consequence of epileptic seizures. *Free Radical Biology & Medicine*, *37*, 1951–1962.
Puzanowska-Tarasiewicz, H., Kuzmicka, L., & Tarasiewicz, M. (2009). Biological function of some elements and their compounds. II. Selenium, selenate, selenium organic compounds. *Polski Merkuriusz Lekarski*, *27*(159), 249–252.
Ramaekers, V. T., Calomme, M., Vanden Berghe, D., & Makropoulos, W. (1994). Selenium deficiency triggering intractable seizures. *Neuropediatrics*, *25*(4), 217–223.
Rayman, M. P. (2000). The importance of selenium to human health. *Lancet*, *356*, 233–241.
Rehni, A. K., & Singh, T. G. (2013). Selenium induced anticonvulsant effect: A potential role of prostaglandin E (1) receptor activation linked mechanism. *Journal of Trace Elements in Medicine and Biology*, *27*, 31–39.
Rogovik, A. L., & Goldman, R. D. (2010). Ketogenic diet for treatment of epilepsy. *Canadian Family Physician*, *56*(6), 540–542.
Romero-Ramos, M., Venero, J. L., Cano, J., & Machado, A. (2000). Low selenium diet induces tyrosine hydroxylase enzyme in nigrostriatal system of the rat. *Brain Research. Molecular Brain Research*, *84*(1–2), 7–16.
Salwen, M. J. (2007). Vitamis and trace elements. In R. A. Pherson, & M. R. Pincus (Eds.), *Henry's clinical diagnosis and management by laboratory methods* (pp. 379–389). Saunders.
Santamaría, A., Vázquez-Román, B., La Cruz, V. P., González-Cortés, C., Trejo-Solís, M. C., Galván-Arzate, S., et al. (2005). Selenium reduces the proapoptotic signaling associated to NF-kappa B pathway and stimulates glutathione peroxidase activity during excitotoxic damage produced by quinolinate in rat corpus striatum. *Synapse*, *58*(4), 258–266.
Savaskan, N. E., Brauer, A. U., Kuhbacher, M., Eyupoglu, I. Y., Kyriakopoulos, A., Ninnemann, O., et al. (2003). Selenium deficiency increases susceptibility to glutamate-induced excitotoxicity. *FASEB Journal*, *17*(1), 112–114.
Schwarz, K., & Foltz, C. M. (1957). Selenium as an integral part of factor 3 against dietary necrotic liver degeneration. *Journal of the American Chemical Society*, *79*(12), 3292–3293.
Schweizer, U., Brauer, A. U., Kohrle, J., Nitsch, R., & Savaskan, N. E. (2004). Selenium and brain function: A poorly recognized liaison. *Brain Research. Brain Research Reviews*, *45*(3), 164–178.
Schweizer, U., Michaelis, M., Kohrle, J., & Schomburg, L. (2004). Efficient selenium transfer from mother to offspring in selenoprotein-P deficient mice enables dose-dependent rescue of phenotypes associated with selenium deficiency. *The Biochemical Journal*, *378*(1), 21–26.
Schweizer, U., Schomburg, L., & Savaskan, N. E. (2004). The neurobiology of selenium: Lessons from transgenic mice. *The Journal of Nutrition*, *134*(4), 707–710.
Schweizer, U., Streckfuss, F., Pelt, P., Carlson, B. A., Hatfield, D. L., Kohrle, J., et al. (2005). Hepatically derived selenoprotein P is a key factor for kidney but not for brain selenium supply. *The Biochemical Journal*, *386*(2), 221–226.

Seven, M., Basaran, S. Y., Cengiz, M., Unal, S., & Yuksel, A. (2012). Deficiency of selenium and zinc as a causative factor for idiopathic intractable epilepsy. *Epilepsy Research*, *104*(1–2), 35–39.

Skalny, A. V., Burtseva, T. I., Salnikova, E. V., Ajsuvakova, O. P., Skalnaya, M. G., Kirichuk, A. A., et al. (2019). Geographic variation of environmental, food, and human hair selenium content in an industrial region of Russia. *Environmental Research*, *171*, 293–301.

Smorgon, C., Mari, E., Atti, A. R., Dalla Nora, E., Zamboni, P. F., Calzoni, F., et al. (2004). Trace elements and cognitive impairment: An elderly cohort study. *Archives of Gerontology and Geriatrics. Supplement*, (9), 393–402.

Soerensen, J., Jakupoglu, C., Beck, H., Forster, H., Schmidt, J., Schmahl, W., et al. (2008). The role of thioredoxin reductases in brain development. *PLoS One*, *3*(3), e1813.

Squires, J. E., Stoytchev, I., Forry, E. P., & Berry, M. J. (2007). SBP2 binding affinity is a major determinant in differential selenoprotein mRNA translation and sensitivity to nonsense-mediated decay. *Molecular and Cellular Biology*, *27*(22), 7848–7855.

Stolzoff, M., & Webster, T. J. (2016). Reducing bone cancer cell functions using selenium nanocomposites. *Journal of Biomedial Materials Research Part B Applied Biomaterials*, *104*, 476–482.

Talat, M. A., Ahmed, A., & Mohammed, L. (2015). Serum levels of zinc and copper in epileptic children during long-term therapy with anticonvulsants. *Neurosciences (Riyadh, Saudi Arabia)*, *20*(4), 341–345.

Tawfik, K. M., Moustafa, Y. M., & El-Azab, M. F. (2018). Neuroprotective mechanisms of sildenafil and selenium in PTZ-kindling model: Implications in epilepsy. *European Journal of Pharmacology*, *833*, 131–144.

Tinggi, U. (2005). Selenium toxicity and its adverse health effects. In R. Preedy, & R. R. Watson (Eds.), *Reviews in food and nutrition toxicity* (pp. 29–55). Boca Raton, FL: Taylor & Francis.

Uguz, A. C., & Naziroglu, M. (2012). Effects of selenium on calcium signaling and apoptosis in rat dorsal root ganglion neurons induced by oxidative stress. *Neurochemical Research*, *37*(8), 1631–1638.

Uthus, E. O., Yokoi, K., & Davis, C. D. (2002). Selenium deficiency in Fisher-344 rats decreases plasma and tissue homocysteine concentrations and alters plasma homocysteine and cysteine redox status. *The Journal of Nutrition*, *132*(6), 1122–1128.

Vinceti, M., Ballotari, P., Steinmaus, C., Malagoli, C., Luberto, F., Malavolti, M., et al. (2016). Long-term mortality patterns in a residential cohort exposed to inorganic selenium in drinking water. *Environmental Research*, *150*, 348–356.

Vinceti, M., Mandrioli, J., Borella, P., Michalke, B., Tsatsakis, A., & Finkelstein, Y. (2014). Selenium neurotoxicity in humans: Bridging laboratory and epidemiologic studies. *Toxicology Letters*, *230*(2), 295–303.

Wajner, M., Latini, A., Wyse, A. T. S., & Dutra-Filho, C. S. (2004). The role of oxidative damage in the neuropathology of organic acidurias: Insights from animal studies. *Journal of Inherited Metabolic Disease*, *27*(4), 427–448.

Walz, R., Morreira, J. C. F., Benfato, M. S., Quevedo, J., Schorer, N., & Vianna, M. M. R. (2000). Lipid peroxidation in hippocampus early and late after status epilepticus induced by pilocarpine of kainic acid in Wistar rats. *Neuroscience Letters*, *291*, 179–182.

Wang, G. S., Geng, D. Q., Wang, Y. W., Chen, X. D., Yang, T. H., & Chang, C. H. (2010). Protective effect of Na2SeO3 against cerebral ischemia-reperfusion injury to the hippocampal neurons in rats. *Nan Fang Yi Ke Da Xue Xue Bao*, *30*(10), 2336–2339.

Winther, K. H., Rayman, M. P., Bonnema, S. J., & Hegedus, L. (2020). Selenium in thyroid disorders—Essential knowledge for clinicians. *Nature Reviews. Endocrinology*, *16*, 165–176.

Wrobel, J. K., Power, R., & Toborek, M. (2016). Biological activity of selenium: Revisited. *IUBMB Life*, *68*(2), 97–105.

Xiong, S., Markesbery, W. R., Shao, C., & Lovell, M. A. (2007). Seleno-L-methionine protects against beta-amyloid and iron/hydrogen peroxide-mediated neuron death. *Antioxidants & Redox Signaling*, *9*(4), 457–467.

Yuan, X., Fu, Z., Ji, P., Guo, L., Al-Ghamdy, A. O., Alkandiri, A., et al. (2020). Selenium nanoparticles pre-treatment reverse behavioral, oxidative damage, neuronal loss and neurochemical alterations in pentylenetetrazole-induced epileptic seizures in mice. *International Journal of Nanomedicine*, *15*, 6339–6353.

Yürekli, V. A., & Nazıroğlu, M. (2013). Selenium and topiramate attenuates blood oxidative toxicity in patients with epilepsy: A clinical pilot study. *Biological Trace Element Research*, *152*(2), 180–186.

Zachary, J. F., & McGavin, M. D. (2014). *Pathologic basis of veterinary disease* (5th ed.). St. Louis, MO: Mosby-Elsevier (Chapter 20).

Zafar, K. S., Siddiqui, A., Sayeed, I., Ahmad, M., Salim, S., & Islam, F. (2003). Dose-dependent protective effect of selenium in rat model of Parkinson's disease: Neurobehavioral and neurochemical evidences. *Journal of Neurochemistry, 84*(3), 438–446.

Zhang, Y. J., Sun, H. L., Wang, T., Liu, X. X., Liu, C., Li, W. J., et al. (2021). Selenium level does not differ in blood but increased in cerebrospinal fluid in Parkinson's disease: A meta-analysis. *The International Journal of Neuroscience, 131*(1), 95–101.

CHAPTER 16

Correlation of reduced vitamin C and minerals with epilepsy

Md. Abdul Aziz and Mohammad Safiqul Islam*
Department of Pharmacy, Noakhali Science and Technology University, Noakhali, Bangladesh

Abbreviations

CNS Central nervous system
PNS Peripheral nervous system
Cu Copper
GABA Gamma-aminobutyric acid
Mn Manganese
NMDA *N*-methyl-D-aspartate
Zn Zinc

Introduction

Epilepsy is a common chronic neurological disorder characterized by spontaneous and periodic unprovoked seizures (International League Against Epilepsy Consortium on Complex Epilepsies, 2018; Pfisterer et al., 2020). It is commonly defined as an unprovoked seizure with a strong recurrence risk (>60%), and/or two unprovoked repeated seizures in less than 24h, and/or diagnosis of any epileptic symptom (Thijs, Surges, O'Brien, & Sander, 2019). Epilepsy is caused by cerebral cortex or hippocampus dysfunction in the human brain (Kathiravan, Kavitha, & Shanthi, 2021). Over 70 million people worldwide are affected by this disorder, which has a prevalence rate of nearly 7% (Kathiravan, Kavitha, & Shanthi, 2021; Saxena & Li, 2017). Epilepsy affects people of all ages, impairing their health, behavior, social and economic balance (Guerreiro, 2016). Approximately 80% of people with this disorder come from low- and middle-income countries, and in most cases, people do not get proper treatment due to familial or social stigma. Moreover, almost 75% of patients remain untreated, which results in a major epilepsy treatment gap in these countries (Saxena & Li, 2017).

Epilepsy is not only a simple disorder but also a symptom-complex condition with diverse risk factors. It may result from various factors starting with inherited mutations

* In Absentia Contact Person: Md. Shahid Sarwar, Department of Pharmacy, Noakhali Science and Technology University, Noakhali-3814, Bangladesh, Mobile: +8801818280905, Email: pharma_sarwar@yahoo.compharma_sarwar@yahoo.com.

Vitamins and Minerals in Neurological Disorders
https://doi.org/10.1016/B978-0-323-89835-5.00001-6

to structural abnormalities of the brain. It is a detrimental condition that causes havoc in the lives of victims and their families by imposing physical, mental, and psychosocial burdens (Negraes et al., 2021; Pfisterer et al., 2020). It generates cognitive, neurological, and psychological barriers to a patient's life (Devinsky et al., 2018).

Vitamin C is a nonenzymatic water-soluble vitamin with antioxidant properties that plays a crucial role in neuroprotective activities (Du, Cullen, & Buettner, 2012). According to reports, vitamin C is related to a fundamental function in maintaining the appropriate balance of the central nervous system (CNS). Any alteration in the optimal concentration of vitamin C leads to structural and functional damage of the neurotransmission system (Das et al., 2019; Harrison & May, 2009). Previous observations showed that a depleted level of vitamin C in the human body results in epilepsy (Shaikh et al., 2018).

Minerals and trace elements have an imperative role in developing the acquired immune system and are directly correlated with antioxidant defense mechanisms in the human body. Deficiency in minerals leads to the development of neuropsychiatric and neurological disorders (Shohag, Ullah, Qusar, Rahman, & Hasnat, 2012). To maintain a healthy and balanced central nervous system, it is a prerequisite to keep the concentrations of minerals in a balanced state (Wojciak, Mojs, Stanislawska-Kubiak, & Samborski, 2013). Investigations have revealed that a reduced level of minerals and trace elements leads to the recurrence of seizures or epilepsy (Hamed, Abdellah, & El-Melegy, 2004; Wojciak et al., 2013). In the present chapter, neuronal distortions caused by epilepsy and the relation of reduced vitamin C and minerals in epilepsy will be discussed.

Neuronal distortion in epilepsy

Epilepsy occurs due to epileptogenesis, a process by which a normal brain converts into an epileptic one which causes repeated seizures spontaneously (Thijs et al., 2019). This theory is conceptualized based on the distortion of balance between excitatory and inhibitory function of the brain (Stafstrom et al., 2011). Due to epileptogenesis, the neuronal circuit becomes highly synchronous that disrupts the normal activity of the brain and sometimes affects other neuronal networks. Primarily, abnormal activity in cortical neurons results in seizures, after which axons and glial cells in the white matter get involved (Thijs et al., 2019). Loss of excitatory and inhibitory neurons, sprouting of axons, change of glial cell functions, and reorganization of synapses are some commonly diagnosed conditions of the epileptic brain (Kobow et al., 2012).

Berkovic (2015) described that the imbalance between excitatory and inhibitory function results from attenuation of brain activity at different levels, including genes, subcellular signaling pathways, and neuronal networks. The study demonstrated that excitatory and inhibitory alteration factors may be due to genetic causes or acquired after birth. The genetic causes can affect anywhere in the brain, including circuit, receptor, or ion channel level. On the other hand, acquired causes can change circuit activity (Berkovic, 2015). Moreover, a number of physiological factors make a normally developing brain prone to epileptic seizures. For instance, the excitatory synaptic activity develops in a developing brain before the inhibitory activity, which increases the possibility of seizure. It is also evident that the gamma-aminobutyric acid (GABA), a critical neurotransmitter, shows the excitatory function in early life rather than inhibitory function (Pitkänen, Lukasiuk, Dudek, & Staley, 2015; Stafstrom & Carmant, 2015). A recent study by Kobylarek et al. (2019) summarized the possible mechanisms that occurred during the epileptogenesis as shown in Fig. 1.

Classification of epileptic seizures

Based on the etiology, epilepsy is categorized into three major types, such as idiopathic epilepsy, acquired epilepsy, and developmental or genetic epilepsy. Each class of epilepsy comes with its own features. Idiopathic epilepsy, for example, is described as the absence of any neurological symptoms and begins in childhood. Juvenile myoclonic epilepsy and childhood absence epilepsy are two common idiopathic epilepsy. Again, structural damages or neuronal distortions in brain are considered as acquired type. This epilepsy occurs due to infection or trauma or tumor in the cerebrum, immunological dysfunctions in the cerebrum, sclerosis in hippocampus, and infantile or perinatal disorders. Examples of acquired epilepsy are viral meningitis epilepsy, cerebral infarction epilepsy, and epilepsy due to open head surgery. Developmental or genetic epilepsy, on the other hand, is caused by almost 977 genes. These genes may be associated with enzymatic activities, ion channel regulation, cellular transportation, or cellular adhesion (Anwar et al., 2020; Valton et al., 2020).

Again, based on the symptoms, the International League Against Epilepsy (ILAE) classified epilepsy into three groups: focal onset epilepsy, generalized epilepsy, and unknown onset epilepsy. The latest lucid classification of epilepsy was revised by ILAE in 2017, in which more clear and more accurate types were included into these three types based on their onset, diagnosis, seizure type, and electroencephalography (Fisher, 2017). Table 1 represents the latest classification of epileptic seizures provided by the ILAE, depending on the etiology and symptoms.

Reduced vitamin C in epilepsy

Vitamin C, known as ascorbic acid, is a necessary hydrophilic vitamin associated with numerous beneficial effects on the human body, most notably on the brain. This vital vitamin commonly acts as an antioxidant and deposits in the CNS. However, human body cannot synthesize vitamin C, making them dependent on dietary intake (Berger, 2009; Du et al., 2012; Traber & Stevens, 2011). The recommended dose of vitamin

Table 1 Classification of epileptic seizures.

Criteria	Classes	Subclasses
Etiology		
	Idiopathic epilepsy Acquired epilepsy Developmental or genetic epilepsy	
Symptoms		
	Focal onset epilepsy	Impaired awareness Motor onset Nonmotor onset Focal to bilateral tonic–clonic
	Generalized epilepsy	Motor Nonmotor
	Unknown onset	Motor Nonmotor Unclassified

This table represents the latest classification of epileptic seizures provided by the ILAE in 2017, depending on the etiology and symptoms.

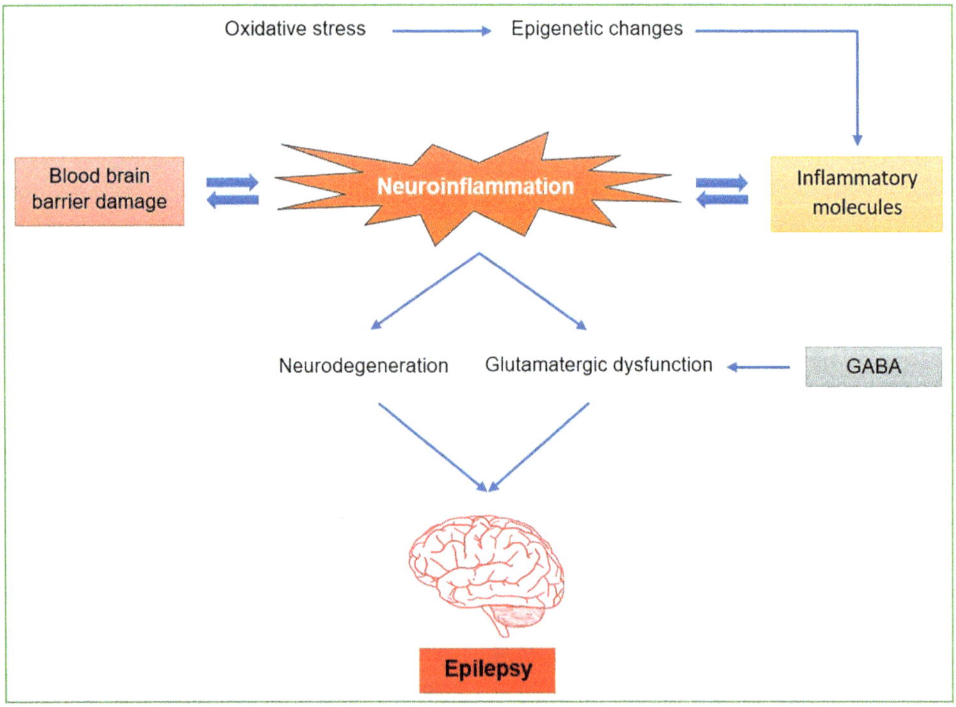

Fig. 1 Possible mechanisms in epileptogenesis. This figure shows the different pathways by which epilepsy occurs in human body, as described by Kobylarek et al. (2019).

C for women is 75 mg/day and for man is 90 mg/day (Hart, Cota, Makhdom, & Harvey, 2015; Waly, Al-Attabi, & Guizani, 2015).

Vitamin C has numerous functions in human brain development and shows a prominent activity in neuromodulation. It is also linked with reactive oxygen species scavenging and substantially depletes harmful oxidants (Covarrubias-Pinto, Acuña, Beltrán, Torres-Díaz, & Castro, 2015). By reducing lipid peroxidation and associated oxidative stress, vitamin C acts as a neuroprotector in brain. Moreover, reacting with alpha-tocopherol, this vitamin decreases massive brain injury during seizures (Sawicka-Glazer & Czuczwar, 2014). Vitamin C also exerts anticonvulsant activities through the expression of various glutamate transporter-related genes (Domith et al., 2018) and activation of neurotransmitter GABA (Ullah, Badshah, Naseer, Lee, & Kim, 2015), which may play a major therapeutic function in reducing the effect of epilepsy.

The contribution of vitamin C in the brain is extensively studied. It is found that vitamin C is transported to cerebrospinal fluid from blood through the sodium-dependent transporter and subsequently from the extracellular fluid to the neurons (Sawicka-Glazer & Czuczwar, 2014). Previous reports explicated that the deficiency of this vitamin enhances the susceptibility of different neurological disorders such as Alzheimer's disease, anxiety, depressive disorders, Huntington's disease, Parkinson's disease, schizophrenic condition, and sclerosis (Kocot, Luchowska-Kocot, Kiełczykowska, Musik, & Kurzepa, 2017). An investigation by Martinc, Grabnar, and Vovk (2014) demonstrated that ascorbic acid pretreatment before pilocarpine-induced seizure could reduce seizure-induced hippocampus damage by almost 60% (Martinc et al., 2014). Additionally, vitamin C can reduce penicillin-induced (Ayyildiz, Coskun, Yildirim, & Agar, 2007) or $FeCl_3$-induced epileptic seizures (Yamamoto, Kabuto, Matsumoto, Ogawa, & Yokoi, 2002). A recent analysis reported that the optimum level of ascorbic acid is reduced in epileptic patients than in the normal patients (Shaikh et al., 2018).

Preclinical studies reported that treatment with vitamin C by increasing concentrations of progesterone hormone shows anticonvulsant activity against pentylenetetrazol-induced seizures (González-Ramírez et al., 2010). Another study depicted enhanced anticonvulsant effect in strychnine-induced seizures in rats when vitamin C was co-administered with a low dose of sodium valproate (Zalkhani, Moazedi, & Najafzadehvarzi, 2018). A similar formula was also applied in male mice by Alhassan et al. (2017), which showed synergistic anticonvulsive activity by prolonging the latency period and decreasing seizure duration in pentylenetetrazol-induced seizures. Vitamin C pretreatment increases the concentration of hippocampal enzymes like superoxide dismutase and catalase, reduces lipid peroxidation, enhances the latency period of seizures, represses seizure episodes, and reduces hippocampal lesions (Santos et al., 2009; Shin et al., 2011).

Reduced minerals in epilepsy

Minerals are vital elements for healthy brain growth and development, like vitamins. The sensitivity of neurons for excitability is also dependent on the availability of minerals. Their movement between cell membranes or extra- and intra-cellular fluid helps to properly function the brain (Chen, Zhao, Sun, Wang, & Ni, 2019; Heffernan, Horner, De Vito, & Conway, 2019; Shohag et al., 2012). Any alteration in the concentration of minerals in the brain was found to increase the likelihood and progression of seizures in animals with genetically determined epilepsy (Hamed et al., 2004). Microminerals are significantly associated with immune system development and defense mechanism. Reduction of microminerals concentration results in neurological and neuropsychiatric disorders (Shohag et al., 2012). For sustaining the homeostasis of the CNS and neuronal integrity, a balanced state of these elements is mandatory (Wojciak et al., 2013). Previous research studies have indicated that altered minerals level has an imperative function in the pathophysiology and incidence of epileptic seizures (Seven, Basaran, Cengiz, Unal, & Yuksel, 2013; Wojciak et al., 2013).

Voltage-gated calcium channels are required for a fast influx of calcium ions, which depolarizes the membrane potential. Again, membrane potential has a very significant part in the regulation of neurotransmitters release. Any change in their expression level leads to the alteration of brain pathophysiology, resulting in epileptic seizures (Cain & Snutch, 2012). Another study demonstrated that the concentration of serum calcium levels leads to the development of epileptic convulsions (Prasad et al., 2009).

Zinc (Zn) is one of the most common minerals investigated considerably for its engagement in neuronal activities in epileptic seizures. A depleted amount of Zn results in central nervous system defects and peripheral neuropathy. Zn deficiency leads to hippocampal dysfunction, changes in emotional behavior, and learning disorder (Chen, Zhou, & Zhu, 2019; Oki et al., 2012). Altered metabolism of Zn and lower Zn levels in serum may play a critical role in inducing seizures by means of *N*-methyl-D-aspartate (NMDA) receptors activation (Prasad et al., 2009). In addition, a reduced Zn level is associated with febrile seizure (Das et al., 2019; Wojciak et al., 2013). Multiple analyses have demonstrated that Zn modulates GABA receptors, leading to the inhibition of seizure (Amiri, Farzin, Moassesi, & Sajadi, 2010; André et al., 2010). These outcomes support the contribution of Zn deficiency to the pathogenesis of epilepsy.

Another common crucial element is copper (Cu) that mainly behaves as a cofactor during various enzymatic reactions (Nahar et al., 2010). Cu acts on several ion channels like voltage-gated ion channels and NMDA receptors that are correlated with the excitability of neurons and synaptic transmission (Saghazadeh et al., 2015). Particularly, the release of Cu during synaptic activities in normal conditions modulates neuronal excitability but decreases serum Cu results in epileptiform discharges (Hamed & Abdellah, 2004; Talat, Ahmed, & Mohammed, 2015). Several studies described that depleted level of Cu in serum is significantly correlated with the susceptibility and progression of epileptic seizures (Nazıroğlu & Yürekli, 2013; Saad, Hammad, Hassan, & Badry, 2014). An investigation by Das et al. (2019) also reported that lower serum Cu level leads to epileptogenesis.

A fundamental micromineral for the appropriate development and activity of brain and the neuronal system is manganese (Mn) (Horning, Caito, Tipps, Bowman, & Aschner, 2015; Miah et al., 2020). It is associated with many metabolic pathways of brain. It has a major impact on glutamine synthesis and metabolism because it is a metalloenzyme, and it also influences the Na^{+}-K^{+} ATPase (Das et al., 2019). Alterations in serum Mn concentration attenuate the dopamine, GABA, and glutamate regulations (Gonzalez-Reyes, Gutierrez-Alvarez, & Moreno, 2007). The imbalance of Mn level in human can seriously disturb the pathophysiological status of the CNS. Several researchers showed that Mn deficiency leads to the development of a convulsive disorder, that is, epilepsy (Balachandran et al., 2020; Gonzalez-Reyes et al., 2007).

Conclusion

Epilepsy is a frequent and disabling neurobiological disorder that occurs due to unusual exaggeration or hyper-synchronization of neuronal function in human brain. It affects people from all age groups. However, imbalances in vitamins like vitamin C and minerals like Ca, Cu, Mn, and Zn lead to alterations of neuronal activities that ultimately result in epilepsy. A proper diagnosis of seizures is crucial in epileptic disorder to reduce any incorrect diagnosis besides keeping the balance of vitamins and minerals.

Applications to other neurological conditions

We have explained the role of low vitamin C and minerals in epileptic disorders in this chapter. However, as vitamin C is directly correlated with brain function, we suggest the role of this vitamin in other neurological conditions such as Alzheimer's disease, Parkinson's disease, Huntington's disease, multiple sclerosis, and amyotrophic lateral sclerosis. This is because vitamin C plays a crucial role in α-synuclein modification, antioxidant mechanism, neurotransmission, cognitive functions, collagen synthesis, amyloid accumulation, tau protein pathology, dopaminergic and glutaminergic systems, all of which may be associated with structural and functional damage and lead to the development and progression of neurodegenerative disorders (Kocot et al., 2017).

Again, the optimum concentration of minerals including Ca, Zn, Cu, Mn, Mg is necessary for proper maintenance of the neurotransmission system in the CNS and PNS (peripheral nervous system). These minerals regulate excitatory-inhibitory signaling systems such as glutamatergic system and control membrane stability in both nervous and muscular system. For instance, Ca is associated with membrane stability, Mg activates different intracellular enzymes, Zn regulates GABA transmission system, Cu acts as cofactors, and Mn regulates several neurotransmission systems. Any alterations in their concentrations may lead to other neurodevelopmental disorders, including Alzheimer's disease, Parkinson's disease in addition to epilepsy (Agrawal, Habib, & Emanuele, 2014; Kirkland, Sarlo, & Holton, 2018). Thus, it is imperative to maintain a balanced level of vitamin C and minerals to combat other neurological disorders as well.

Other components of interest

In the present chapter, we observed that reduced levels in vitamin C and minerals may be linked to the development of epileptic seizures. However, evidence shows that other vitamins, like vitamin B complex, vitamin D, and vitamin E, also have a prominent role in neurological disorders, specifically epilepsy. Vitamin B complex plays a prominent function in both CNS and PNS. Common B vitamins are B1, B6, and B12 which play a critical role in antioxidant mechanisms, cellular homeostasis, neurotransmission system, and neuroprotection. Deficiencies in these vitamins lead to several neurological disorders (Calderón-Ospina & Nava-Mesa, 2020).

It was previously observed that vitamin D has a role in free radical scavenging, antioxidant activity, muscle performance and acts as a neurosteroid or neurotrophic factor. Lower serum levels of vitamin D elevate the risk of epilepsy, Parkinson's disease, Alzheimer's disease, sclerosis, and other neurodegenerative diseases (Di Somma et al., 2017). Vitamin E, on the other hand, has shown to be effective in free radical scavenging as well as the synthesis of prostaglandins and nucleic acids. It possesses anti-inflammatory properties and is associated with reduced cholesterol levels essential for maintaining brain health. Reduced status of vitamin E in the body is implicated in developing neurological disorders (Lloret, Esteve, Monllor, Cervera-Ferri, & Lloret, 2019). Based on this evidence, it is transparent that neurological disorders may be developed due to different vitamins, which need to be investigated more wisely to get the best insight into disease etiology and select the best possible strategy for the management as well as treatment.

Mini-dictionary of terms

- **Epilepsy:** A common chronic neurological disorder that is associated with spontaneous and periodic unprovoked seizures.
- **Epileptogenesis:** A process by which a normal brain converts into an epileptic one which causes repeated seizures spontaneously.
- **Vitamin C:** A necessary water-soluble vitamin that is associated with numerous beneficial effects on the human body, most notably on the brain.
- **GABA:** A neurotransmitter that is released during synaptic transmission and affects neurological activities.
- **NMDA:** Receptors that are correlated with the excitability of neurons and synaptic transmission.

Key facts of reduced vitamin C and minerals in epilepsy

- Epileptic seizures seriously affect more than 70 million people worldwide.
- Epilepsy is a symptom-complex condition leading to damage in the neurological, cognitive, and psychological systems.
- Imbalance between the excitatory and inhibitory neurotransmission system of the brain leads to epilepsy development.
- Vitamins, especially vitamin C play a crucial role in neuroprotective activities.
- Minerals also have prominent activity in maintaining brain homeostasis.

Summary points

- Epilepsy is a prevalent chronic neurological disorder associated with spontaneous and periodic unprovoked seizures.
- Epilepsy is a symptom-complex condition with diverse risk factors which generate cognitive, neurological, and psychosocial barriers in a patient's life.
- A depleted level of vitamin C in the human body results in epilepsy.
- Deficiency in minerals leads to the development of neurological disorders, especially epilepsy.
- A proper balance of vitamin C and minerals is a prerequisite to tackle epilepsy.

References

Agrawal, L., Habib, Z., & Emanuele, N. V. (2014). Neurologic disorders of mineral metabolism and parathyroid disease. *Handbook of Clinical Neurology*, *120*, 737–748.

Alhassan, A. W., Mohammed, D. K., Saleh, M. A., Mshelia, P. P., Yarube, I. U., & Lawan, I. (2017). Combined effect of sodium valproate and ascorbic acid on pentylenetetrazoleinduced seizures in mice. *Bayero Journal of Pure and Applied Sciences*, *10*(1), 47–50.

Amiri, M., Farzin, L., Moassesi, M. E., & Sajadi, F. (2010). Serum trace element levels in febrile convulsion. *Biological Trace Element Research*, *135*(1–3), 38–44.

André, V. M., Cepeda, C., Vinters, H. V., Huynh, M., Mathern, G. W., & Levine, M. S. (2010). Interneurons, GABAA currents, and subunit composition of the GABAA receptor in type I and type II cortical dysplasia. *Epilepsia*, *51 Suppl 3*(Suppl 3), 166–170.

Anwar, H., Khan, Q. U., Nadeem, N., Pervaiz, I., Ali, M., & Cheema, F. F. (2020). Epileptic seizures. *Discoveries (Craiova)*, *8*(2), e110.

Ayyildiz, M., Coskun, S., Yildirim, M., & Agar, E. (2007). The effects of ascorbic acid on penicillin-induced epileptiform activity in rats. *Epilepsia*, *48*(7), 1388–1395.

Balachandran, R. C., Mukhopadhyay, S., McBride, D., Veevers, J., Harrison, F. E., Aschner, M., et al. (2020). Brain manganese and the balance between essential roles and neurotoxicity. *The Journal of Biological Chemistry*, *295*(19), 6312–6329.

Berger, M. M. (2009). Vitamin C requirements in parenteral nutrition. *Gastroenterology*, *137*(5 Suppl), S70–S78.

Berkovic, S. F. (2015). Genetics of epilepsy in humans. *Cold Spring Harbor Perspectives in Medicine*. https://doi.org/10.1101/cshperspect.a022400.

Cain, S. M., & Snutch, T. P. (2012). Voltage-gated calcium channels in epilepsy. In J. L. Noebels, et al. (Eds.), *Jasper's basic mechanisms of the epilepsies* (4th ed., pp. 66–84). US: National Center for Biotechnology Information.

Calderón-Ospina, C. A., & Nava-Mesa, M. O. (2020). B Vitamins in the nervous system: Current knowledge of the biochemical modes of action and synergies of thiamine, pyridoxine, and cobalamin. *CNS Neuroscience & Therapeutics*, *26*(1), 5–13.

Chen, N. N., Zhao, D. J., Sun, Y. X., Wang, D. D., & Ni, H. (2019). Long-term effects of zinc deficiency and zinc supplementation on developmental seizure-induced brain damage and the underlying GPR39/ZnT-3 and MBP expression in the hippocampus. *Frontiers in Neuroscience*, *13*, 920.

Chen, W., Zhou, X., & Zhu, W. (2019). Trace elements homeostatic imbalance in psoriasis: A meta-analysis. *Biological Trace Element Research*, *191*(2), 313–322.

Covarrubias-Pinto, A., Acuña, A. I., Beltrán, F. A., Torres-Díaz, L., & Castro, M. A. (2015). Old things new view: Ascorbic acid protects the brain in neurodegenerative disorders. *International Journal of Molecular Sciences*, *16*(12), 28194–28217.

Das, A., Sarwar, M. S., Hossain, M. S., Karmakar, P., Islam, M. S., Hussain, M. E., et al. (2019). Elevated serum lipid peroxidation and reduced vitamin C and trace element concentrations are correlated with epilepsy. *Clinical EEG and Neuroscience*, *50*(1), 63–72.

Devinsky, O., Vezzani, A., O'Brien, T. J., Jette, N., Scheffer, I. E., de Curtis, M., et al. (2018). Epilepsy. *Nature Reviews. Disease Primers*, *4*, 18024.

Di Somma, C., Scarano, E., Barrea, L., Zhukouskaya, V. V., Savastano, S., Mele, C., et al. (2017). Vitamin D and neurological diseases: An endocrine view. *International Journal of Molecular Sciences*, *18*(11), 2482.

Domith, I., Socodato, R., Portugal, C. C., Munis, A. F., Duarte-Silva, A. T., & Paes-de-Carvalho, R. (2018). Vitamin C modulates glutamate transport and NMDA receptor function in the retina. *Journal of Neurochemistry*, *144*(4), 408–420.

Du, J., Cullen, J. J., & Buettner, G. R. (2012). Ascorbic acid: Chemistry, biology and the treatment of cancer. *Biochimica et Biophysica Acta*, *1826*(2), 443–457.

Fisher, R. S. (2017). The new classification of seizures by the international league against epilepsy 2017. *Current Neurology and Neuroscience Reports*, *17*, 48.

González-Ramírez, M., Razo-Juárez, L. I., Sauer-Ramírez, J. L., González-Trujano, M. E., Salgado-Ceballos, H., & Orozco-Suarez, S. (2010). Anticonvulsive effect of vitamin C on pentylenetetrazol-induced seizures in immature rats. *Pharmacology, Biochemistry, and Behavior*, *97*(2), 267–272.

Gonzalez-Reyes, R. E., Gutierrez-Alvarez, A. M., & Moreno, C. B. (2007). Manganese and epilepsy: A systematic review of the literature. *Brain Research Reviews*, *53*(2), 332–336.

Guerreiro, C. A. (2016). Epilepsy: Is there hope? *The Indian Journal of Medical Research*, *144*(5), 657–660.

Hamed, S. A., & Abdellah, M. M. (2004). Trace elements and electrolytes homeostasis and their relation to antioxidant enzyme activity in brain hyperexcitability of epileptic patients. *Journal of Pharmacological Sciences*, *96*(4), 349–359.

Hamed, S. A., Abdellah, M. M., & El-Melegy, N. (2004). Blood levels of trace elements, electrolytes, and oxidative stress/antioxidant systems in epileptic patients. *Journal of Pharmacological Sciences*, *96*(4), 465–473.

Harrison, F. E., & May, J. M. (2009). Vitamin C function in the brain: Vital role of the ascorbate transporter SVCT2. *Free Radical Biology & Medicine*, *46*(6), 719–730.

Hart, A., Cota, A., Makhdom, A., & Harvey, E. J. (2015). The role of vitamin C in orthopedic trauma and bone health. *American Journal of Orthopedics*, *44*(7), 306–311.

Heffernan, S. M., Horner, K., De Vito, G., & Conway, G. E. (2019). The role of mineral and trace element supplementation in exercise and athletic performance: A systematic review. *Nutrients*, *11*(3), 696.

Horning, K. J., Caito, S. W., Tipps, K. G., Bowman, A. B., & Aschner, M. (2015). Manganese is essential for neuronal health. *Annual Review of Nutrition*, *35*, 71–108.

International League Against Epilepsy Consortium on Complex Epilepsies. (2018). Genome-wide mega-analysis identifies 16 loci and highlights diverse biological mechanisms in the common epilepsies. *Nature Communications*, *9*(1), 5269.

Kathiravan, M., Kavitha, S., & Shanthi, R. (2021). To determine the effect of long-term antiepileptic drug on the serum folate and vitamin B12 among epileptic patients. *Scientific Reports*, *11*(1), 4393.

Kirkland, A. E., Sarlo, G. L., & Holton, K. F. (2018). The role of magnesium in neurological disorders. *Nutrients*, *10*(6), 730.

Kobow, K., Auvin, S., Jensen, F., Löscher, W., Mody, I., Potschka, H., et al. (2012). Finding a better drug for epilepsy: Antiepileptogenesis targets. *Epilepsia*, *53*(11), 1868–1876.

Kobylarek, D., Iwanowski, P., Lewandowska, Z., Limphaibool, N., Szafranek, S., Labrzycka, A., et al. (2019). Advances in the potential biomarkers of epilepsy. *Frontiers in Neurology*, *10*, 685.

Kocot, J., Luchowska-Kocot, D., Kiełczykowska, M., Musik, I., & Kurzepa, J. (2017). Does vitamin C influence neurodegenerative diseases and psychiatric disorders? *Nutrients*, *9*(7), 659.

Lloret, A., Esteve, D., Monllor, P., Cervera-Ferri, A., & Lloret, A. (2019). The effectiveness of vitamin E treatment in Alzheimer's disease. *International Journal of Molecular Sciences*, *20*(4), 879.

Martinc, B., Grabnar, I., & Vovk, T. (2014). Antioxidants as a preventive treatment for epileptic process: A review of the current status. *Current Neuropharmacology*, *12*(6), 527–550.

Miah, M. R., Ijomone, O. M., Okoh, C., Ijomone, O. K., Akingbade, G. T., Ke, T., et al. (2020). The effects of manganese overexposure on brain health. *Neurochemistry International*, *135*, 104688.

Nahar, Z., Azad, M. A., Rahman, M. A., Rahman, M. A., Bari, W., Islam, S. N., et al. (2010). Comparative analysis of serum manganese, zinc, calcium, copper and magnesium level in panic disorder patients. *Biological Trace Element Research*, *133*(3), 284–290.

Nazıroğlu, M., & Yürekli, V. A. (2013). Effects of antiepileptic drugs on antioxidant and oxidant molecular pathways: Focus on trace elements. *Cellular and Molecular Neurobiology*, *33*(5), 589–599.

Negraes, P. D., Trujillo, C. A., Yu, N. K., Wu, W., Yao, H., Liang, N., et al. (2021). Altered network and rescue of human neurons derived from individuals with early-onset genetic epilepsy. *Molecular Psychiatry*. https://doi.org/10.1038/s41380-021-01104-2.

Oki, G., Wada, T., Iba, K., Aiki, H., Sasaki, K., Imai, S. I., et al. (2012). Metallothionein deficiency in the injured peripheral nerves of complex regional pain syndrome as revealed by proteomics. *Pain*, *153*(3), 532–539.

Pfisterer, U., Petukhov, V., Demharter, S., Meichsner, J., Thompson, J. J., Batiuk, M. Y., et al. (2020). Identification of epilepsy-associated neuronal subtypes and gene expression underlying epileptogenesis. *Nature Communications*, *11*(1), 5038.

Pitkänen, A., Lukasiuk, K., Dudek, F. E., & Staley, K. J. (2015). Epileptogenesis. *Cold Spring Harbor Perspectives in Medicine*, *5*(10), a022822.

Prasad, R., Singh, A., Das, B. K., Upadhayay, R. S., Singh, T. B., & Mishra, O. P. (2009). Cerebrospinal fluid and serum Zn, Cu, Mg and Ca levels in children with idiopathic seizures. *Journal of Clinical and Diagnostic Research*, *3*(6), 1841–1846.

Saad, K., Hammad, E., Hassan, A. F., & Badry, R. (2014). Trace element, oxidant, and antioxidant enzyme values in blood of children with refractory epilepsy. *The International Journal of Neuroscience*, *124*(3), 181–186.

Saghazadeh, A., Mahmoudi, M., Meysamie, A., Gharedaghi, M., Zamponi, G. W., & Rezaei, N. (2015). Possible role of trace elements in epilepsy and febrile seizures: A meta-analysis. *Nutrition Reviews*, *73*(11), 760–779.

Santos, I. M., Tomé, A., Saldanha, G. B., Ferreira, P. M., Militão, G. C., & Freitas, R. M. (2009). Oxidative stress in the hippocampus during experimental seizures can be ameliorated with the antioxidant ascorbic acid. *Oxidative Medicine and Cellular Longevity*, *2*(4), 214–221.

Sawicka-Glazer, E., & Czuczwar, S. J. (2014). Vitamin C: A new auxiliary treatment of epilepsy? *Pharmacological Reports*, *66*(4), 529–533.

Saxena, S., & Li, S. (2017). Defeating epilepsy: A global public health commitment. *Epilepsia Open*, *2*(2), 153–155.

Seven, M., Basaran, S. Y., Cengiz, M., Unal, S., & Yuksel, A. (2013). Deficiency of selenium and zinc as a causative factor for idiopathic intractable epilepsy. *Epilepsy Research*, *104*(1–2), 35–39.

Shaikh, A. S., Guo, X., Li, Y., Cao, L., Liu, X., Li, P., et al. (2018). The impact of antiepileptic drugs on vitamin levels in epileptic patients. *Current Pharmaceutical Biotechnology*, *19*(8), 674–681.

Shin, E. J., Jeong, J. H., Chung, Y. H., Kim, W. K., Ko, K. H., Bach, J. H., et al. (2011). Role of oxidative stress in epileptic seizures. *Neurochemistry International*, *59*(2), 122–137.

Shohag, H., Ullah, A., Qusar, S., Rahman, M., & Hasnat, A. (2012). Alterations of serum zinc, copper, manganese, iron, calcium, and magnesium concentrations and the complexity of interelement relations in patients with obsessive-compulsive disorder. *Biological Trace Element Research*, *148*(3), 275–280.

Stafstrom, C. E., Arnason, B. G., Baram, T. Z., Catania, A., Cortez, M. A., Glauser, T. A., et al. (2011). Treatment of infantile spasms: Emerging insights from clinical and basic science perspectives. *Journal of Child Neurology*, *26*(11), 1411–1421.

Stafstrom, C. E., & Carmant, L. (2015). Seizures and epilepsy: An overview for neuroscientists. *Cold Spring Harbor Perspectives in Medicine*, *5*(6), a022426.

Talat, M. A., Ahmed, A., & Mohammed, L. (2015). Serum levels of zinc and copper in epileptic children during long-term therapy with anticonvulsants. *Neurosciences*, *20*(4), 341–345.

Thijs, R. D., Surges, R., O'Brien, T. J., & Sander, J. W. (2019). Epilepsy in adults. *Lancet*, *393*(10172), 689–701.

Traber, M. G., & Stevens, J. F. (2011). Vitamins C and E: Beneficial effects from a mechanistic perspective. *Free Radical Biology & Medicine*, *51*(5), 1000–1013.

Ullah, I., Badshah, H., Naseer, M. I., Lee, H. Y., & Kim, M. O. (2015). Thymoquinone and vitamin C attenuates pentylenetetrazole-induced seizures via activation of GABAB1 receptor in adult rats cortex and hippocampus. *Neuromolecular Medicine*, *17*(1), 35–46.

Valton, L., Benaiteau, M., Denuelle, M., Rulquin, F., Hachon Le Camus, C., Hein, C., et al. (2020). Etiological assessment of status epilepticus. *Revue Neurologique*, *176*(6), 408–426.

Waly, M. I., Al-Attabi, Z., & Guizani, N. (2015). Low nourishment of vitamin C induces glutathione depletion and oxidative stress in healthy young adults. *Preventive Nutrition and Food Science*, *20*(3), 198–203.

Wojciak, R. W., Mojs, E., Stanislawska-Kubiak, M., & Samborski, W. (2013). The serum zinc, copper, iron, and chromium concentrations in epileptic children. *Epilepsy Research*, *104*(1–2), 40–44.

Yamamoto, N., Kabuto, H., Matsumoto, S., Ogawa, N., & Yokoi, I. (2002). Alpha-Tocopheryl-L-ascorbate-2-O-phosphate diester, a hydroxyl radical scavenger, prevents the occurrence of epileptic foci in a rat model of post-traumatic epilepsy. *Pathophysiology*, *8*(3), 205–214.

Zalkhani, R., Moazedi, A. A., & Najafzadehvarzi, H. (2018). Investigation the effect of sodium valproate in the presence of vitamin C and E on seizure induced by strychnine in adult rat. *Journal of Sabzevar University of Medical Sciences*, *26*(2), 195–201.

CHAPTER 17

Thiamine, transporters, and epilepsy

Sunita Bijarnia-Mahay[a], Arushi Gahlot Saini[b], and Harjit Singh Mahay[c]
[a]Department of Clinical & Metabolic Genetics, Institute of Medical Genetics & Genomics, Sir Ganga Ram Hospital, New Delhi, India
[b]Department of Pediatrics, Postgraduate Institute of Medical Education and Research (PGIMER), Chandigarh, India
[c]Department of Critical Care, Fortis Hospital, New Delhi, India

Abbreviations

ALP	alkaline/acid phosphatase
ATTP	adenosine thiamine triphosphate
BCKAD	branched-chain keto acid dehydrogenase
BTBGD	biotin-thiamine-responsive basal ganglia disease
HeMT	human extraneuronal monoamine transporter
ITD	infantile thiamine deficiency
MRI	magnetic resonance imaging
NADPH	nicotinamide adenine dinucleotide phosphate
OCT	organic cationic transporters
SLC19A3	solute carrier family 19 member 3
TDD	thiamine deficiency disorders
THTR	thiamine membrane transporter
TKT	transketolase
TMP	thiamine monophosphate
TPK	thiamine phosphokinase
TPP	thiamine pyrophosphate
TTP	thiamine triphosphate
WKS	Wernicke-Korsakoff syndrome

Introduction

At the turn of the last century and the late 19th century, the story of the mysterious disease, *beriberi,* was unfolding, with reports from Southeast Asia, a primarily rice-eating region. Around the same time, Wernicke-Korsakoff syndrome (WKS) was also being discovered as a separate entity associated with alcoholism (Eijkman, 1990; Thomson, 2000). A Dutch physician Christian Eijkman attributed the occurrence of beriberi to a poor diet consisting mainly of polished rice. Hopkins hypothesized the presence of some essential micronutrients in the diet, which were later given the name "*vitamine*" by Punk in 1922 (Eijkman, 1990; Funk, 1975; Hopkins, 1912). The "e" in the term was later deleted when the theory of all vitamins being "amines" was proven incorrect. Thiamine deficiency was implicated in the causation of WKS. The pivotal discoveries of Eijkman and Hopkins not only led to their Nobel prize in 1929 but also to isolation of

Vitamins and Minerals in Neurological Disorders
https://doi.org/10.1016/B978-0-323-89835-5.00020-X

thiamine by Jansen and Donath, the first water-soluble B vitamin (vitamin B1) in 1926 (Jansen & Donath, 1926; Thomson, 2000).

The initial cases of thiamine deficiency were reported in either malnourished individuals or chronic alcoholics with isolated thiamine deficiency causing WKS (Beltramo, Mazzeo, & Porta, 2021; Bose, 1949). The neuropathy noted in Wernicke's encephalopathy had similarity with that in diabetes (Bose, 1949). After a quiet period of about 3–4 decades, the vitamin came back into focus owing to modernization of dietary practices with abundance of processed foods, reducing the nutritional quality of foods, thus leading to this vitamin's deficiency even in the apparently "well-nourished" population. Thiamine deficiency also finds a mention in celiac disease, where it occurs due to impaired absorption similar to what happens in chronic alcoholism.

Thiamine is involved in over 20 biochemical reactions in the body, carrying out multiple functions, mainly, though not limited to, brain and energy metabolism. Though required in very small quantities, it is rapidly excreted from the body owing to its water-soluble nature and hence needs to be replenished on a daily basis. There is very little storage in the body, except in the liver where it can maintain replenishment for up to 2 weeks (Bettendorff, 2021). As an inherent feature of a vitamin, de novo synthesis of thiamine only occurs in bacteria, fungi, and plants (Bettendorff, 2021). For humans, the various dietary sources of thiamine include meat, dairy, fish, wheat bran, rice, other grains, oils, beans, etc. Meat, especially pork, whole grains, yeast, and legumes are the richest source. After ingestion, it is taken up by the body using various transport mechanisms illustrated below to reach the active site for use.

Deficiency of thiamine, either dietary/nutritional or due to an underlying genetic defect, is responsible for a variety of clinical features ranging from common symptoms such as lethargy, listlessness, forgetfulness, and muscle cramps to severe infantile epilepsy or heart failure. The extreme ages of life, i.e., infancy and old age are especially vulnerable to dietary thiamine deficiency states. Nutritional thiamine deficiency disorder beriberi has been traditionally classified as wet beriberi, pertaining to fluid accumulation in lungs and extremities secondary to cardiac failure, or dry beriberi encompassing neurological symptoms of peripheral neuropathy (WKS). Since many genetically determined disorders have now also been identified, mostly related to transport mechanism for thiamine, leading to symptoms more or less similar to that of nutritional deficiency or worse, the disorders are now referred to as thiamine deficiency disorders (TDDs). TDDs present with a wide range of clinical symptoms with multiorgan system involvement including neurologic, respiratory, cardiac, gastrointestinal, musculoskeletal, and energy metabolism. Early recognition of thiamine deficiency is paramount as it is evident that supplementation of thiamine can be life-saving, reverting symptoms as early as in hours, even in the severely affected individuals.

In this chapter, we aim to describe thiamine absorption and transportation and its crucial role in metabolic pathways leading to energy production, with a special focus on neurological functions, and then describe the disorders related to thiamine deficiency, either nutritionally or due to genetic determinants.

Chemical structure and biological forms of thiamine

The thiamine molecule is composed of two rings—a sulfur-containing thiazole ring connected to a pyrimidine ring through a methylene group (Fig. 1). It exists in the body in one of four physiological states, depending upon its phosphorylation status. The most abundant form is the thiamine pyrophosphate (TPP), also known as thiamine diphosphate, making up >80% of total body thiamine. Thiamine triphosphate (TTP) makes up 10%–15% of total thiamine (Lu & Frank, 2008). Thiamine monophosphate (TMP) is the least abundant, but crucial state since it facilitates active transport using transporters to intracellular compartments. TPP is the precursor for two triphosphorylated derivatives, thiamine triphosphate and adenosine thiamine triphosphate, the function of which is clearly known (Bettendorff, 2021). TTP plays a role in membrane conductance properties through activation of chloride channels and functions as a phosphate group donor in phosphorylation reactions in energy metabolism. The unphosphorylated form is a cationic form of thiamine (T^{+}) and is the only form that can passively diffuse across membranes and be spontaneously absorbed. Free thiamine also has antioxidant function.

There is no actual storage of thiamine in the body, but liver containing about 10%–20% of body's thiamine as TPP, can act as a reservoir for other organs, especially the brain. This may allow body to remain healthy for about 2 weeks only (Bettendorff, 2021). The normal levels of thiamine and thiamine pyrophosphate in blood are in Table 1 (Gangolf et al., 2010). The recommended daily intake or dietary reference intake of thiamine varies from 0.2–0.3 mg in infancy to 1–1.2 mg in adults (Whitfield, Smith, Rohner, Wieringa, & Green, 2021). The requirements vary according to body's physiological demands, and it is more in pregnant and lactating women, individuals performing hard physical labor and in those taking relative high carbohydrates, and is shown in Table 2 (Smith et al., 2021). The European Food Safety Authority (EFSA) revised the guidelines for dietary reference intake of thiamine in 2016, which was measured in terms of energy intake, with levels of 0.4 mg/1000 kcal (0.1 mg/MJ) taken as reference intake per day. The EFSA expert panel proposed the same requirement and reference intake for the adult population, expressed in mg/MJ or mg/1000 kcal, that for infants 7–11 months, children 1 to <18 years and during pregnancy and lactation, assuming that the ratio

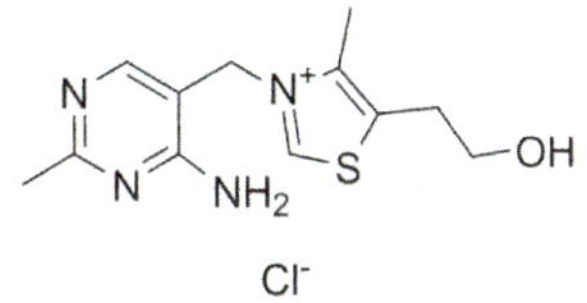

Fig. 1 Chemical structure of thiamine.

Table 1 Normal levels of various forms of thiamine in blood.[a]

Thiamine derivative	Whole blood levels (nmol/l)	Plasma levels (nmol/l)
Free thiamine (T+)	4±3	11±3
Thiamine monophosphate (TMP)	10±4	5±2
Thiamine diphosphate/pyrophosphate (TPP)	138±33	Not detected
Thiamine triphosphat (TTP)	13±4	Not detected

[a]Based on Gangolf et al., levels detected in whole blood and plasma in 7 and 3 individuals, respectively. Majority of thiamine is in erythrocytes as depicted in samples.

Table 2 Dietary reference intake of thiamine in mg/day (EFSA NDA Panel, 2016).

Age group	Female	Male
0–6 months	0.2	
7–12 months	0.3	
1–3 years	0.5	
4–6 years	0.6	
7–9 years	0.9	
≥10 years including adult	1.1	1.2
Pregnancy	1.4	
Lactation	1.5	

between the requirement of thiamine and that of energy is the same in all population groups (EFSA NDA Panel, 2016). No known toxicity has been reported with a higher intake of thiamine, it being excreted rapidly by the kidneys. Additionally, oral thiamine absorption via enterocytes is blunted after a concentration of thiamine >2.5 mg in the intestine (Whitfield et al., 2021).

Absorption and transport of thiamine

As there is no mechanism of de novo synthesis of thiamine in human body, the two main sources of thiamine are the diet and gut bacterial flora. The dietary thiamine is absorbed from the proximal small intestine, mainly the jejunum. The thiamine made available by the intestinal bacteria is absorbed in the colon too and is used mainly for the colonic health. After the food is digested in the stomach and intestine, thiamine is released in free-state (T+) by hydrolysis by enzymes. The phosphorylated thiamine is further reduced to release free thiamine by phosphorylases. The free thiamine is then absorbed from the brush border of jejunum and parts of colon. The absorption process takes place by two methods, depending upon the availability of thiamine in the gut. In the state of abundance, i.e., when the concentration is higher than 1–2 μmol, the absorption takes place via spontaneous passive diffusion across the luminal border of the small intestine, called diffusive process (Rindi & Laforenza, 2000). This passive absorption takes in the cationic thiamine (T+). For active absorption, saturable thiamine/H+ antiport system is required. This is via specific thiamine transporters, THTR1 and THTR2, via active Na+-independent, pH-dependent, amiloride-sensitive, carrier-mediated exchange process across the human intestinal brush border membrane (Dudeja, Tyagi, Kavilaveettil, Gill, & Said, 2001). The two transporters are part of *SLC19A* family of transporters. While absorption of thiamine on the luminal side occurs by both THTR1 and THTR2 across the brush border membrane (BBM), it is majorly done by THTR1 on the basolateral border (BLB) from where thiamine exits the enterocyte to enter the blood stream (Fig. 2). In addition, in the colonic absorption, a small contribution is also made by isoform 3 of SLC44A4, a nonneuronal choline carrier protein, which mediates the absorption of microbiota-generated TPP, especially in infants (Nabokina et al., 2017).

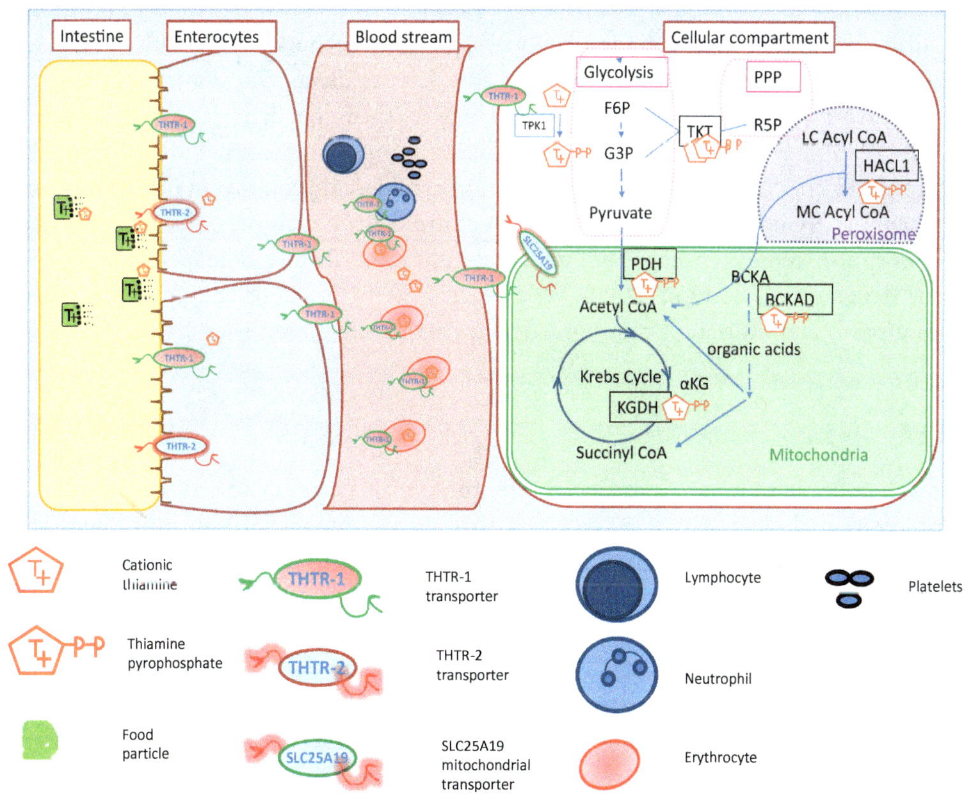

Fig. 2 Intestinal absorption, transport via blood stream and intracellular metabolic function of thiamine. *BCKA*, branched chain keto acids; *BCKADH*, branched chain keto acid dehydrogenase; *F6P*, fructose-6-phosphate; *G3P*; glyceraldehyde-3-phosphate; *HACL1*, hydroxy acyl coenzyme A lyase 1; *KGDH*, alpha ketoglutarate dehydrogenase; *LC*, acyl CoA long chain acyl coenzyme A; *MC*, acyl CoA medium chain acyl coenzyme A; *PDH*, pyruvate dehydrogenase; *PPP*, pentose phosphate pathway; *R5P*, ribulose-5-phosphate; *SLC25A19*, mitochondrial thiamine transporter; *T+*, cationic thiamine; *THTR1/2*, thiamine receptors; *TKT*, cytosolic transketolase; *TPK1*, thiamine phosphokinase enzyme; *TPP*, thiamine pyrophosphate; *αKG*, alpha ketoglutarate.

In the blood stream, >90% of the thiamine is transported via blood cells where it is bound to transketolase in the form of TPP (Talwar, Davidson, Cooney, & St, 2000). While mostly present in the erythrocytes (>75% of blood thiamine content) and leukocytes (15%), a small amount is also circulating free in plasma (<10%) and platelets (Casirola, Patrini, Ferrari, & Rindi, 1990). Studies have shown that both types of uptakes, active (energy-dependent, via transporter) as well as passive diffusion occur in all the tissues at the same time, as free cationic T+ or neutral TMP. The transportation of either TPP or TTP requires their conversion to TMP or T+ (Manzetti, Zhang, & van der Spoel, 2014).

Factors affecting intestinal absorption (Rindi & Laforenza, 2000)

Thiamine uptake is enhanced by thiamine deficiency and reduced by thyroid hormone and diabetes. Ingestion of thiamine antagonists, such as betel nuts and tea, and thiaminase-containing fish, ferns, and African silkworm larvae affects the absorption process (Whitfield et al., 2021). Other factors leading to thiamine deficiency include heat, tannin-containing or food rich in caffeine, theobromine, and theophylline (such as coffee, tea, and chocolate), and drugs such as furosemide, amprolium, pyrithiamine, and oxythiamine (Rindi & Laforenza, 2000). In chronic alcohol abusers, decreased intake, malabsorption, reduced storage, and impaired utilization all contribute to functional thiamine deficiency (Thomson, 2000). Other factors and conditions hampering thiamine consumption and utilization causing deficiency will be discussed later.

Thiamine transporters

Cross membrane transport is carried out by three major carriers described later. Other transporters include SLC25A19, required to transport intracellular thiamine into the mitochondria, the SLC21 organic anion transporter, and the SLC44A4 isoform 3 non-neuronal choline transporter in the colon (Nabokina et al., 2017).

1. The SLC19 transporter protein superfamily: Three proteins, namely SLC19A1 (reduced folate carrier or RFC), SLC19A2 (thiamine transporter-1 or THTR-1), and SCL19A3 (thiamine transporter-2) exist, each encoded by genes of the same name. All the three proteins have structural similarity with 12 transmembrane domains consisting of approximately 500 amino acids and at least 40% of homology between them, but with different thiamine-binding affinities. The main features of these proteins are summarized in Table 3.

   Proteins, THTR1 and 2 are more specific for thiamine, accounting for total carrier-mediated thiamine uptake in the intestine, as determined by using gene-specific siRNAs to selectively knock down the individual transporter (Said, Balamurugan, Subramanian, & Marchant, 2004). Both are ubiquitous in their

Table 3 Comparative description of three SLC19 Transporters.

S.No.	Characteristics	SLC19A1	SLC19A2	SLC19A3
1	Name	Reduced folate carrier or RFC	Thiamine transporter 1 or THTR1	Thiamine transporter 2 or THTR2
2	Transport function	TMP TPP TTP Folates	T+ No folates	T+ No folates
3	Type of carrier	Anion carrier	Cation carrier	Cation carrier
4	Localisation	Intestinal epithelial cells, placenta, and kidney.	Skeletal muscles, systemic tissues including intestine	Intestines
	Intestinal location	Upper half of the villi, apical, brush border	Apical/brush border, basolateral	Apical/ brush border
5	Thiamine binding affinity	Lesser than folates	Influx constant K_t 2.5 uM	Influx constant K_t 25 mM, most affinity for thiamine

distribution, although THTR1 is most abundant in the intestine, skeletal muscle, nervous system, and eye, whereas THTR2 is located mostly in adipose tissue, breast tissue, liver, lymphocytes, spleen, gallbladder, placenta, pancreas, and brain (Ganapathy, Smith, & Prasad, 2004). Within the intestinal mucosa, while the expression of THTR1 occurs at both the BBM and BLM domains of the polarized enterocyte, expression of the THTR2 protein is restricted to the BBM domain of these cells (Said, 2011). The process of upregulation of expression in response to thiamine deficiency, mediated via RNA transcriptional modification, occurs only for THTR2 transporters and not THTR1 (Reidling, Lambrecht, Kassir, & Said, 2010).

2. Human extraneuronal monoamine transporter (heMT). This large family of transporters, known as SLC22A, has eight families, two of which are known to transport thiamine. These two families of nonneuronal transporters staking responsibility are the organic cationic transporters (OCT1,2,3 and 6) and the OCT variants, OCTN1 and 2. Much like the SLC19A superfamily, transporters in both families consist of approximately 550 amino acids, have high sequence homology, and have 12 transmembrane domains including at least one extracellular domain and one intracellular domain for binding and release of thiamine (Engelhart et al., 2020). With an established function as drug transporters, they have a polyspecific nature of binding to proteins, functioning at variable efficiencies, depending upon the co-substrate being transported simultaneously. The unique antiporter transporters simultaneously bind multiple substrates

to either potentiate or inhibit transport of thiamine. For example, the OCT1 is known to function as a major drug transporter in the liver. Recent studies have found that OCT1 is a high-capacity thiamine transporter and as such plays an important regulatory role in carbohydrate and lipid metabolism in hepatocytes. Animal experiments have implicated that thiamine deficiency in OCT1-deficient mice is the reason behind reduced hepatic steatosis (Chen, Yee, & Giacomini, 2015). The results need to be substantiated, but there is growing knowledge of these transporters having an increasingly significant role in regulation, intracellular signaling, organ crosstalk, and interorganismal communication (Chen et al., 2015; Engelhart et al., 2020).

3. Alkaline/acid phosphatase (ALP) transport system. They are the third group of proteins associated with cross membrane transport of thiamine. These are essentially membrane-bound metalloproteins performing a crucial function to transport thiamine in both directions to inside the cell, as well as export to outside the cell. ALPs facilitate the transport by dephosphorylating all the phosphorylated forms of thiamine to T+. Being involved in cross membrane transport of many other substances, such as lipids and phosphates, they are present in many tissues, remaining particularly active in exchange surfaces. With respect to thiamine, they are present in the intestine and plasma membrane, while also being expressed in liver, kidney, bone, heart, and placenta (Manzetti et al., 2014).

Functions of thiamine in metabolic processes

Thiamine plays an indispensable role in various metabolic processes of the body, both as a coenzyme as well as in a noncoenzymatic way.

Co-enzyme function of thiamine

Thiamine is involved in at least 5 bioenergetic reactions within the cell which are illustrated in Fig. 2. The enzymes requiring thiamine as a cofactor are mainly linked in the process of energy production while being situated in cytosol (transketolase or TKT), mitochondria (PDH, A-KG, and BCKAD) and the peroxisome (2-hydroxyacyl-CoA lyase) (Fig. 2). While the role of T+ and TMP is mainly in transport across cellular membranes, it is the TPP which is the most active and abundant form, performing coenzymatic functions. The role of triphosphorylated derivatives of thiamine, thiamine triphosphate (TTP), and adenosine thiamine triphosphate (ATTP) is not clearly defined, although postulated to be in facilitating neurotransmission either directly or via enabling generation of acetyl-CoA (Bettendorff, 2021).

1. As thiamine enters the cell, either at T+ or TMP, it is swiftly converted into TPP by thiamine pyrophosphokinase (TPK), which then acts as a phosphate donor in the bioenergetic pathways with ultimate aim of ATP production. The first such reaction occurs in the cytosol where it is utilized by enzyme, cytosolic transketolase, to

produce ribose-5-phosphate. This is a key step in the pentose-phosphate pathway as it helps in generating nicotinamide adenine dinucleotide phosphate (NADPH), which is required as a reducing agent in various reactions including fatty acid synthesis, antioxidant enzyme glutathione peroxidase-reductase system among others. The ribose-5-phosphate is also required in the biosynthesis of DNA and RNA.

2. TPP is also required as a cofactor by pyruvate dehydrogenase complex in converting pyruvate to acetyl-CoA within the mitochondria to kick-start the citric acid cycle. This step lies at the boundaries of aerobic (citric acid cycle→ respiratory chain→ ATP production) and anaerobic energy production via lactate. TPP deficiency would lead to a switch in anaerobic energy production and elevation of lactate.
3. Within the citric acid cycle, the role of TPP again comes in as a cofactor to enzyme alpha-ketoglutarate dehydrogenase, in the conversion of alpha-ketoglutarate to succinyl CoA. This step also leads to the production of NADH which enters the respiratory chain to directly produce ATP. The crucial role of TPP at this step is highlighted by evidence of neurodegeneration occurring due to decreased αKGHD activity in TPP deficient states, thus leading to reduced energy production and accumulation of glutamate in the cell with neurodegeneration as a consequence (Gibson, Blass, Beal, & Bunik, 2005).
4. The peroxisomal enzyme, 2-hydroxyacyl-CoA lyase, functions to remove the methyl group in branched chain fatty acid, phytanic acid, a process called alpha-oxidation. This crucial step in the oxidation of branched chain fatty acids requires TPP as a cofactor. Deficiency of TPP leads to accumulation of triglycerides and causes symptoms similar to Refsum's disease which itself is caused by deficiency of enzyme HACL1 or phytanoyl-CoA lyase (Wierzbicki, 2007).
5. The second crucial step in branched chain amino acid catabolism carried out by the enzyme complex, BCKAD, requires thiamine as a cofactor. Deficiency of the enzyme complex leads to accumulation of branched chain amino acids, especially leucine which is neurotoxic, causing maple syrup urine disease. Studies have shown that deficiency of thiamine leads to reduced activity of BCKAD enzyme in the medial thalamus, causing toxic accumulation of leucine and leading to progressive damage (Navarro, Zwingmann, & Butterworth, 2008).

Role of thiamine other than as a cofactor

There is growing knowledge of a noncoenzymatic role of thiamine in living beings (Manzetti et al., 2014). Thiamine triphosphate is important for facilitating neurotransmission, possibly via stimulating acetylcholine synthesis. Thiamine also has an important role as an antioxidant and other immune-mediated mechanisms. In diabetes, for example, supplementation of thiamine helps in reducing the formation of reactive oxygen species (ROS) by normalizing the metabolic pathways that may be overactive in high glucose environment (Beltramo et al., 2021).

Table 4 Factors causing thiamine deficiency (Dhir et al., 2019).

Acquired deficiency

- Excessive consumption of processed food (high calorie malnutrition)
- Increased carbohydrate intake (increases thiamine demand in the body)
- High rates of white rice consumption (industrial rice milling and processing)
- Ingesting diets contaminated with thiaminase enzyme
- Thiamine inactivation in food by heat
- Heavy consumption of tannin-containing (such as coffee, chocolate, tea)
- Aging
- Poor socio-economic status
- Eating disorders
- Medical or surgical conditions affecting the gastrointestinal tract
- Parental nutrition
- Bariatric surgery
- Diabetes
- Alcohol abuse
- Lactation, pregnancy, and increased physical activity

Inherited deficiency

Thiamine transporter defects

- *SLC19A2* gene associated thiamine-responsive megaloblastic anemia or thiamine metabolism dysfunction syndrome 1
- *SLC19A3* gene associated biotin–thiamine responsive basal ganglia disease or thiamine metabolism dysfunction syndrome 2
- *SLC25A19* gene associated Amish lethal microcephaly (thiamine metabolism dysfunction syndrome 3) and bilateral striatal degeneration and progressive polyneuropathy type (thiamine metabolism dysfunction syndrome 4)

Enzymatic defects

- Thiamine pyrophosphokinase 1 (TPK1) deficiency or Thiamine metabolism dysfunction syndrome 5 or episodic encephalopathy type

Thiamine deficiency states

As thiamine is water-soluble, it is not stored in the body and rapidly gets excreted (half-life of only 9–18 days). Thus, thiamine-rich tissues in body like skeletal muscles, heart, liver, kidney, and brain bear the brunt of its deficiency. The common factors leading to thiamine deficiency are listed in Table 4. Chronic alcoholism is perhaps the most frequent cause of thiamine deficiency in the adults leading to Wernicke's encephalopathy. Particularly vulnerable are the extremes of age—infancy and old age—pregnant and lactating women and the critically ill patients with prolonged hospitalization.

Thiamine deficiency in critically ill patients

Patients with a critical illness are at high risk for thiamine deficiency owing to an imbalance between demand and supply. There is increased oxidative stress and heightened catabolism on the one hand and inadequate supplementation owing to nutritional deprivation on the other. This may be further compromised by patients with preexisting medical conditions such as diabetes, chronic alcoholism, and dialysis-dependent renal conditions. Sepsis forms a major subset of patients admitted to intensive care units. A recent trial on HAT therapy (hydrocortisone, ascorbic acid, and thiamine) in critically ill patients with septic shock and other smaller trials later showed a marked improvement in organ function and reduction in mortality, thus highlighting the crucial role of thiamine (Marik, Khangoora, Rivera, Hooper, & Catravas, 2017; Raj et al., 2021). However, more large-scale randomized controlled trials are being recommended.

Pediatric disorders due to acquired nutritional deficiency of thiamine

Nutritional deficiency of thiamine may cause neurological and cardiovascular disease in the vulnerable population. The underlying pathogenesis involves thiamine deficiency-induced mitochondrial dysfunction, ensuing lactic acidosis, secondary neurotransmitter abnormalities, and energy crisis in several vital organs such as brain, heart, and muscles (Hiffler, Rakotoambinina, Lafferty, & Martinez Garcia, 2016). The common pediatric disorders associated with acquired thiamine deficiency are discussed later:

- ***Infantile thiamine deficiency (ITD)*:** It predominantly affects infants who are exclusively breast-fed by thiamine-deficient mothers or with thiamine-deficient formulas. It manifests after 2–3 weeks of thiamine deprivation starting with milder symptoms such as refusal to eat, vomiting, and irritability (Dhir, Tarasenko, Napoli, & Giulivi, 2019). It may progress to alteration in sensorium, diminished muscle stretch reflexes, ataxia, quadriparesis, and even death. Severely deficient newborns develop infantile beriberi, associated with neurological and cardiac manifestations. Three phenotypes have been classically described: the cardiac form or "wet beriberi," the aphonic form, and the neurological form or "dry beriberi." Shoshin beriberi is the more severe form associated with cardiac failure and lactic acidosis. Typical neuroimaging findings in acute-onset infantile beriberi encephalopathy include bilateral symmetrical signal changes in putamen, followed by caudate nuclei, thalami, globi pallidi, and less commonly in the cerebral cortex (Wani, Qureshi, Jehangir, Ahmad, & Ahmad, 2016). There is often associated restricted diffusion and increased lactate peak on spectroscopy. Early recognition of these changes may help in a timely diagnosis and thiamine supplementation. In developing countries, infantile beriberi is associated with consumption of food sources with deficient thiamine, such as milled white cereals, polished rice and wheat flour, and lesser intake of thiamine-rich foods (such as meat, fish, and vegetables). In developed countries, thiamine deficiency often occurs in outbreaks due to the use of thiamine-deficient soy formula. Long-term effects of thiamine deficiency include developmental problems, epilepsy, intellectual impairment, and variable motor dysfunction (Mimouni-Bloch et al., 2014). Management includes thiamine supplementation at 10–50 mg/day IV or IM or 10–50 mg/day orally followed by 5–10 mg/day for 1 month.

- ***Wernicke's encephalopathy in children*:** It is commonly seen in predisposed children with underlying malnutrition, malignancies, or malabsorptive states such as post-gastric surgeries, prolonged total parenteral nutrition with low contents of thiamine, or pancreatitis (Ortigoza-Escobar et al., 2017; Perko et al., 2012). Similar to adults, children present with altered mental status, ataxia, ophthalmoplegia, and seizures. Involvement of efferent sympathetic outflow may result in hypotension and tachycardia, and involvement of hypothalamus may result in hypothermia. The diagnosis may be missed due to nonspecific presentation and the outcome may be fatal in up to one-fifth of the cases. Magnetic resonance imaging (MRI) often gives an early clue due to the peculiar involvement of brain areas (Kobayashi, Kaneko, & Matsunaga, 2011; Kohnke & Meek, 2021). These include bilaterally symmetrical signal changes in the medial thalami, mammillary bodies, tectal plate, and periaqueductal area of the midbrain (Perko et al., 2012). Management includes thiamine supplementation in doses at 100–500 mg/day. Clinical symptoms resolve in hours to days and brain changes resolve in days to weeks (Ho, York, Rustad, Felde, & Stern, 2021).

Pediatric disorders due to genetic deficiency of thiamine

The inherited disorders of thiamine metabolism result from pathogenic variations in the genes encoding for enzymes and transporters in the thiamine metabolic pathways. The onset of these disorders is often early in life. The clinical presentation mimics the nutritional thiamine deficiency and mitochondria disorders, thus highlighting the key role of thiamine in energy metabolism in the body. Currently, there are five genetic phenotypes referred to as thiamine metabolism dysfunction syndromes 1–5, caused by mutations in *SLC19A2, SLC19A3, SLC25A19,* and *TPK1* genes (Table 4) (Marce-Grau, Marti-Sanchez, Baide-Mairena, Ortigoza-Escobar, & Perez-Duenas, 2019).

Thiamine transporter defects

- ***Thiamine-responsive megaloblastic anemia or thiamine metabolism dysfunction syndrome 1 or Rogers syndrome*:** This early-onset, rare, autosomal recessive disorder is associated with pathogenic mutations in the thiamine transporter gene *SLC19A2,* encoding protein THTR1 transporter. Clinical features include megaloblastic anemia, macrocytosis, nonautoimmune diabetes mellitus, and sensorineural deafness. Other features include optic atrophy, stroke, epilepsy, pigmentary retinopathy, short stature, cytopenia, and congenital heart defects (Ganesh, Ezhilarasi, Vasanthi, Gowrishankar, & Rajajee, 2009; Oishi & Diaz, 1993). The condition partly responds to high doses of thiamine (50–100 mg/day).
- ***Biotin-thiamine-responsive basal ganglia disease (BTBGD) or thiamine transporter-2 deficiency or thiamine metabolism dysfunction syndrome 2*:** This rare neurometabolic, autosomal-recessive disorder is associated with biallelic pathogenic

mutations in the *SLC19A3* gene. First described by Ozand et al. in 1998 as a childhood encephalopathy responsive to biotin, (Ozand et al., 1998) the disorder is now termed 'biotin-thiamine-responsive basal ganglia disease' to denote the spectrum of inherited neurological illnesses associated with thiamine transport in the body and bilateral basal ganglia lesions (Alfadhel et al., 2013). The encoded protein is thiamine membrane transporter 2 (THTR2), which is either incorrectly localized in the cells or has a reduced affinity for thiamine, or fails to up regulate the gene expression during periods of metabolic stress, thus restricting the transport of thiamine into the cells (Schanzer et al., 2014). Thiamine supplementation in high doses tries to surpass the transporter block and biotin supplementation helps in the optimal expression of the gene encoding the transporter.

The three main clinical presentations are (Saini & Sharma, 2021):

(1) *Early childhood encephalopathy*: This is the classical form that presents in early or late childhood between 3 and 10 years of age with episodic encephalopathy triggered by an intercurrent illness or stress, recurrent seizures, extrapyramidal involvement, and bilateral lesions of the striatum, cerebral cortex and brainstem (Alfadhel et al., 2013)

(2) *Early infantile form*: This is more severe and presents as a rapidly progressing encephalopathy, seizures, lactic acidosis, poor response to thiamine and early death, Leigh-like presentation (Kevelam et al., 2013; Perez-Duenas et al., 2013) Some infants present with typical infantile spasms, severe psychomotor retardation and progressive brain atrophy (Yamada et al., 2010). The two types of presentations Leigh-like and infantile spasms are not mutually exclusive as shown in a recent study (Mishra et al., 2021). Patients may present with infantile spasms on follow-up after an acute crisis. The epilepsy may evolve into Lennox–Gastaut syndrome with multiple seizure types (Sremba, Chang, Elbalalesy, Cambray-Forker, & Abdenur, 2014).

(3) *Late-onset Wernicke-like encephalopathy*: Often presents in the second decade with acute-onset neurological symptoms such as epilepsy, nystagmus, ophthalmoplegia, and ataxia. This presentation has been associated with compound heterozygous variations in the *SLC19A3* and shows reversal of symptoms with high-dose thiamine supplementation (Kono et al., 2009). The condition responds variably to timely and high doses of biotin (5–10 mg/kg/day) and thiamine (300–800 mg/day). (Alfadhel et al., 2013; Saini & Sharma, 2021)

- ***Amish lethal microcephaly or thiamine metabolism dysfunction syndrome-3*:** This severe, rare, autosomal-recessive disorder is associated with pathogenic variations in the thiamine transporter encoding gene *SLC25A19*. The encoded protein is a mitochondrial TPP carrier. Clinical features include severe microcephaly at birth, brain malformations, profound developmental delay, and episodic encephalopathy with lactic acidosis and α-ketoglutaric aciduria. (Kelley, Robinson, Puffenberger,

Strauss, & Morton, 2002) Treatment includes high-fat, low carbohydrate diet which utilizes the alternative mitochondrial β-oxidation pathway for ATP generation via Krebs cycle. (Siu, Ratko, Prasad, Prasad, & Rupar, 2010) A less severe variant of the disease is known as thiamine metabolism dysfunction syndrome 4, presenting with flaccid weakness, progressive axonal polyneuropathy, episodic encephalopathy and bilateral striatal necrosis in patients with normal psychomotor development. (Spiegel et al., 2009) The condition responds to high-dose thiamine, but residual weakness may persist.

Thiamine enzymatic pathway defects

- ***Thiamine pyrophosphokinase 1 (TPK1) enzyme deficiency*:** The condition is also known as thiamine metabolism dysfunction syndrome 5. The enzyme deficiency results into failure of phosphorylation of thiamine to TPP, the active co-factor. Children present with episodic, acute-onset encephalopathy, lactatemia, progressive motor dysfunction, gait abnormalities, dystonia, and spasticity. Brain MRI abnormalities described in TPK1-deficient patients are involvement of cerebellum, dentate nuclei, basal ganglia and spinal cord. (Invernizzi et al., 2017) Treatment includes supplementation with thiamine, niacin, biotin, α-lipoic acid, and ketogenic diet. (Fraser et al., 2014)

Applications to other neurological conditions

In this chapter, we have reviewed the role of thiamine in various bio-energetic metabolic processes in the body. There is growing knowledge of a nonco-enzymatic role of thiamine in living beings. There is evidence of thiamine-binding proteins having specific function in the nervous system, especially in facilitation of neurotransmission (Bettendorff, 2021). Thiamine supplementation has been known to improve symptoms in patients with Alzheimer's disease and Parkinson's disease, thus highlighting the role of the nutrient in brain function. Thiamine is also known to affect function of our immune system, such as in diabetes (Bose, 1949), in anti-oxidation as described with glutathione function and in septic shock, in both adults and children (Raj et al., 2021). Much less is known about adenylated derivatives of thiamine, at least in human beings (Bettendorff, 2021). An inverse association has been shown between thiamine levels and symptoms of depression in adults. Positive mood changes have been linked with thiamine supplementation in healthy elderly and younger women, and in postpartum women with postpartum depression. Thiamine-based treatments are gaining momentum for several neurodevelopmental and pediatric neurological disorders linked to mitochondrial dysfunction such as autism spectrum disorder, behavioral and attention disorders, learning disorders, pediatric migraine, and psychiatric disorders.

Other components of interest

In this chapter, we described the role of thiamine in the functioning of the nervous system as a cofactor and a resurgence of deficiency states in recent times, all treatable with thiamine supplementation. Such vitamin-responsive conditions are important to recognize early as they are eminently treatable. Pediatric movement disorders responsive to vitamin therapy include infantile tremor syndrome, biotinidase deficiency, pyruvate dehydrogenase deficiency, aromatic amino acid decarboxylase deficiency, ataxia with vitamin E deficiency, abetalipoproteinemia, cerebral folate deficiency, and cobalamin metabolism defects. Metabolic epilepsy associated with other vitamins include pyridoxine dependent epilepsy, pyridoxal-5-phosphate dependent epilepsy, folinic acid responsive epilepsies, biotinidase deficiency, and infantile B12 deficiency (Reddy & Saini, 2020).

Mini-dictionary of terms

- *Thiamine pyrophosphate*: One of four physiological states of thiamine in the body that is most active and abundant, responsible for >80% of total body thiamine.
- *Thiamine transporters*: Family of solute carrier proteins transporting thiamine across membranes.
- *Nutritional thiamine deficiency disorder*: Includes the traditionally described disorders of wet beriberi (pertaining to fluid accumulation in lungs and extremities secondary to cardiac failure), or dry beriberi (pertaining to the neurological symptoms of peripheral neuropathy as in Wernicke-Korsakoff syndrome)
- *Thiamine deficiency disorders*: Umbrella term used for both acquired and genetically-determined disorders which lead to symptoms more or less similar to that of nutritional deficiency.
- *Biotin-thiamine-responsive basal ganglia disease*: A type of inherited thiamine metabolic defect characterized by progressive encephalopathy, epilepsy, tone and gait abnormalities, bilateral basal ganglia involvement and response to high doses of biotin and thiamine.

Key facts of thiamine metabolism

- Thiamine or vitamin B1 is a water-soluble vitamin of the vitamin B complex family
- Thiamine is not endogenously synthesized in the human body, hence it is an essential vitamin for humans
- Thiamine-rich tissues in the human body are the skeletal muscles, heart, liver, kidney and brain, which bear the brunt of its deficiency
- The recommended dietary allowance of thiamine is age-dependent, ranging from 0.2 to 0.6 mg per day in children to 1–1.2 mg/day in adults. A dietary reference value

calculated in mg/1000 kcal is now utilized, with normal value being 0.4 mg/1000 kcal, which is uniform across all ages.
- The common pediatric disorders associated with acquired thiamine deficiency are infantile thiamine deficiency and Wernicke's encephalopathy, previously termed wet and dry Beriberi.
- The inherited disorders of thiamine metabolism include five genetic phenotypes caused by mutations in *SLC19A2*, *SLC19A3*, *SLC25A19*, and *TPK1* genes
- Early recognition of thiamine deficiency is paramount as supplementation of thiamine can be life-saving, reverting symptoms as early as in hours, even in the severely affected individuals.

Summary points

- The water-soluble vitamin thiamine or vitamin B1 has historically been associated with beriberi and Wernicke-Korsakoff syndrome in a deficiency state, largely due to malnutrition.
- Thiamine is back into focus owing to the modernisation of dietary practices with abundance of processed foods, reducing the nutritional quality of food and leading to thiamine deficiency even in the apparently 'well nourished' population.
- Thiamine is involved in over 20 biochemical reactions in the body, mainly involving the energy metabolism in brain, heart, muscle.
- Though required in very small quantities, it is rapidly excreted from the body owing to its water soluble nature, and hence needs to be replenished on a daily basis.
- In the chapter, we aim to describe thiamine absorption and transportation, and its crucial role in metabolic pathways leading to energy production, with special focus on neurological functions and related disorders.
- Early recognition of thiamine deficiency is paramount as supplementation of thiamine can be life-saving, reverting symptoms as early as in hours, even in the severely affected individuals.

References

Alfadhel, M., Almuntashri, M., Jadah, R. H., Bashiri, F. A., Al Rifai, M. T., Al Shalaan, H., et al. (2013). Biotin-responsive basal ganglia disease should be renamed biotin-thiamine-responsive basal ganglia disease: A retrospective review of the clinical, radiological and molecular findings of 18 new cases. *Orphanet Journal of Rare Diseases*, *8*, 83. https://doi.org/10.1186/1750-1172-8-83.

Beltramo, E., Mazzeo, A., & Porta, M. (2021). Thiamine and diabetes: Back to the future? *Acta Diabetologica*, *58*(11), 1433–1439. https://doi.org/10.1007/s00592-021-01752-4.

Bettendorff, L. (2021). Update on thiamine triphosphorylated derivatives and metabolizing enzymatic complexes. *Biomolecules*, *11*(11). https://doi.org/10.3390/biom11111645.

Bose, J. P. (1949). The role of vitamin B1 in the treatment of diabetes. *Indian Medical Gazette*, *83*(12), 554–556.

Casirola, D., Patrini, C., Ferrari, G., & Rindi, G. (1990). Thiamin transport by human erythrocytes and ghosts. *The Journal of Membrane Biology*, *118*(1), 11–18. https://doi.org/10.1007/BF01872201.

Chen, L., Yee, S. W., & Giacomini, K. M. (2015). OCT1 in hepatic steatosis and thiamine disposition. *Cell Cycle*, *14*(3), 283–284. https://doi.org/10.1080/15384101.2015.1006532.

Dhir, S., Tarasenko, M., Napoli, E., & Giulivi, C. (2019). Neurological, psychiatric, and biochemical aspects of thiamine deficiency in children and adults. *Frontiers in Psychiatry*, *10*, 207. https://doi.org/10.3389/fpsyt.2019.00207.

Dudeja, P. K., Tyagi, S., Kavilaveettil, R. J., Gill, R., & Said, H. M. (2001). Mechanism of thiamine uptake by human jejunal brush-border membrane vesicles. *American Journal of Physiology. Cell Physiology*, *281*(3), C786–C792. https://doi.org/10.1152/ajpcell.2001.281.3.C786.

EFSA NDA Panel (EFSA Panel on Dietetic Products, Nutrition and Allergies), et al. (2016). Scientific opinion on dietary reference values for thiamin. *EFSA Journal*, *14*(12), 55. https://doi.org/10.2903/j.efsa.2016.4653.

Eijkman, C. (1990). *Polyneuritis in chickens, or the origins of vitamin research* (DG van der Heij, translator). Basel: Hoffman-La Roche.

Engelhart, D. C., Granados, J. C., Shi, D., Saier, M. H., Jr., Baker, M. E., Jr., Abagyan, R., et al. (2020). Systems biology analysis reveals eight SLC22 transporter subgroups, including OATs, OCTs, and OCTNs. *International Journal of Molecular Sciences*, *21*(5). https://doi.org/10.3390/ijms21051791.

Fraser, J. L., Vanderver, A., Yang, S., Chang, T., Cramp, L., Vezina, G., et al. (2014). Thiamine pyrophosphokinase deficiency causes a Leigh disease like phenotype in a sibling pair: Identification through whole exome sequencing and management strategies. *Molecular Genetics and Metabolism Reports*, *1*, 66–70. https://doi.org/10.1016/j.ymgmr.2013.12.007.

Funk, C. (1975). The journal of State Medicine. Volume XX: 341-368, 1912. The etiology of the deficiency diseases, Beri-beri, polyneuritis in birds, epidemic dropsy, scurvy, experimental scurvy in animals, infantile scurvy, ship beri-beri, pellagra. *Nutrition Reviews*, *33*(6), 176–177. https://doi.org/10.1111/j.1753-4887.1975.tb05095.x.

Ganapathy, V., Smith, S. B., & Prasad, P. D. (2004). SLC19: The folate/thiamine transporter family. *Pflügers Archiv*, *447*(5), 641–646. https://doi.org/10.1007/s00424-003-1068-1.

Ganesh, R., Ezhilarasi, S., Vasanthi, T., Gowrishankar, K., & Rajajee, S. (2009). Thiamine responsive megaloblastic anemia syndrome. *Indian Journal of Pediatrics*, *76*(3), 313–314. https://doi.org/10.1007/s12098-009-0058-5.

Gangolf, M., Czerniecki, J., Radermecker, M., Detry, O., Nisolle, M., Jouan, C., et al. (2010). Thiamine status in humans and content of phosphorylated thiamine derivatives in biopsies and cultured cells. *PLoS One*, *5*(10), e13616. https://doi.org/10.1371/journal.pone.0013616.

Gibson, G. E., Blass, J. P., Beal, M. F., & Bunik, V. (2005). The alpha-ketoglutarate-dehydrogenase complex: A mediator between mitochondria and oxidative stress in neurodegeneration. *Molecular Neurobiology*, *31*(1–3), 43–63. https://doi.org/10.1385/MN:31:1-3:043.

Hiffler, L., Rakotoambinina, B., Lafferty, N., & Martinez Garcia, D. (2016). Thiamine deficiency in tropical pediatrics: New insights into a neglected but vital metabolic challenge. *Frontiers in Nutrition*, *3*, 16. https://doi.org/10.3389/fnut.2016.00016.

Ho, P. A., York, A., Rustad, J. K., Felde, A., & Stern, T. A. (2021). Non-alcohol-related Wernicke's encephalopathy: Diagnosis and treatment. *Primary Care Companion to CNS Disorders*, *23*(5). https://doi.org/10.4088/PCC.21f02968.

Hopkins, F. G. (1912). Feeding experiments illustrating the importance of accessory factors in normal dietaries. *The Journal of Physiology*, *44*(5–6), 425–460. https://doi.org/10.1113/jphysiol.1912.sp001524.

Invernizzi, F., Panteghini, C., Chiapparini, L., Moroni, I., Nardocci, N., Garavaglia, B., et al. (2017). Thiamine-responsive disease due to mutation of tpk1: Importance of avoiding misdiagnosis. *Neurology*, *89*(8), 870–871. https://doi.org/10.1212/WNL.0000000000004270.

Jansen, B. C. P., & Donath, W. F. (1926). On the isolation of the anti-beri-beri vitamin. *Proc Konink Akad van Wetensch Amsterdam*, *107*, 1390–1400.

Kelley, R. I., Robinson, D., Puffenberger, E. G., Strauss, K. A., & Morton, D. H. (2002). Amish lethal microcephaly: A new metabolic disorder with severe congenital microcephaly and 2-ketoglutaric aciduria. *American Journal of Medical Genetics*, *112*(4), 318–326. https://doi.org/10.1002/ajmg.10529.

Kevelam, S. H., Bugiani, M., Salomons, G. S., Feigenbaum, A., Blaser, S., Prasad, C., et al. (2013). Exome sequencing reveals mutated SLC19A3 in patients with an early-infantile, lethal encephalopathy. *Brain, 136*(Pt 5), 1534–1543. https://doi.org/10.1093/brain/awt054.

Kobayashi, M., Kaneko, A., & Matsunaga, T. (2011). MRI findings in nonalcoholic Wernicke's encephalopathy. *Internal Medicine, 50*(19), 2245–2246. https://doi.org/10.2169/internalmedicine.50.5726.

Kohnke, S., & Meek, C. L. (2021). Don't seek, don't find: The diagnostic challenge of Wernicke's encephalopathy. *Annals of Clinical Biochemistry, 58*(1), 38–46. https://doi.org/10.1177/0004563220939604.

Kono, S., Miyajima, H., Yoshida, K., Togawa, A., Shirakawa, K., & Suzuki, H. (2009). Mutations in a thiamine-transporter gene and Wernicke's-like encephalopathy. *The New England Journal of Medicine, 360*(17), 1792–1794. https://doi.org/10.1056/NEJMc0809100.

Lu, J., & Frank, E. L. (2008). Rapid HPLC measurement of thiamine and its phosphate esters in whole blood. *Clinical Chemistry, 54*(5), 901–906. https://doi.org/10.1373/clinchem.2007.099077.

Manzetti, S., Zhang, J., & van der Spoel, D. (2014). Thiamin function, metabolism, uptake, and transport. *Biochemistry, 53*(5), 821–835. https://doi.org/10.1021/bi401618y.

Marce-Grau, A., Marti-Sanchez, L., Baide-Mairena, H., Ortigoza-Escobar, J. D., & Perez-Duenas, B. (2019). Genetic defects of thiamine transport and metabolism: A review of clinical phenotypes, genetics, and functional studies. *Journal of Inherited Metabolic Disease, 42*(4), 581–597. https://doi.org/10.1002/jimd.12125.

Marik, P. E., Khangoora, V., Rivera, R., Hooper, M. H., & Catravas, J. (2017). Hydrocortisone, vitamin C, and thiamine for the treatment of severe sepsis and septic shock: A retrospective before-after study. *Chest, 151*(6), 1229–1238. https://doi.org/10.1016/j.chest.2016.11.036.

Mimouni-Bloch, A., Goldberg-Stern, H., Strausberg, R., Brezner, A., Heyman, E., Inbar, D., et al. (2014). Thiamine deficiency in infancy: Long-term follow-up. *Pediatric Neurology, 51*(3), 311–316. https://doi.org/10.1016/j.pediatrneurol.2014.05.010.

Mishra, R., Bijarnia-Mahay, S., Kumar, P., Buxi, T. B. S., Kulshrestha, S., Kuldeep, J., et al. (2021). Early infantile thiamine transporter-2 deficiency with epileptic spasms—A phenotypic spectrum with a novel mutation. *Journal of Pediatric Epilepsy, 10*(4), 168–174. https://doi.org/10.1055/s-0041-1731018.

Nabokina, S. M., Inoue, K., Subramanian, V. S., Valle, J. E., Yuasa, H., & Said, H. M. (2017). Molecular identification and functional characterization of the human colonic thiamine pyrophosphate transporter. *The Journal of Biological Chemistry, 292*(40), 16526. https://doi.org/10.1074/jbc.A113.528257.

Navarro, D., Zwingmann, C., & Butterworth, R. F. (2008). Impaired oxidation of branched-chain amino acids in the medial thalamus of thiamine-deficient rats. *Metabolic Brain Disease, 23*(4), 445–455. https://doi.org/10.1007/s11011-008-9105-6.

Oishi, K., & Diaz, G. A. (1993). Thiamine-responsive megaloblastic anemia syndrome. In M. P. Adam, H. H. Ardinger, R. A. Pagon, S. E. Wallace, L. J. H. Bean, G. Mirzaa, & A. Amemiya (Eds.), *GeneReviews((R))*. Seattle (WA).

Ortigoza-Escobar, J. D., Alfadhel, M., Molero-Luis, M., Darin, N., Spiegel, R., de Coo, I. F., et al. (2017). Thiamine deficiency in childhood with attention to genetic causes: Survival and outcome predictors. *Annals of Neurology, 82*(3), 317–330. https://doi.org/10.1002/ana.24998.

Ozand, P. T., Gascon, G. G., Al Essa, M., Joshi, S., Al Jishi, E., Bakheet, S., et al. (1998). Biotin-responsive basal ganglia disease: A novel entity. *Brain, 121*(Pt 7), 1267–1279. https://doi.org/10.1093/brain/121.7.1267.

Perez-Duenas, B., Serrano, M., Rebollo, M., Muchart, J., Gargallo, E., Dupuits, C., et al. (2013). Reversible lactic acidosis in a newborn with thiamine transporter-2 deficiency. *Pediatrics, 131*(5), e1670–e1675. https://doi.org/10.1542/peds.2012-2988.

Perko, R., Harreld, J. H., Helton, K. J., Sabin, N. D., Haidar, C. E., & Wright, K. D. (2012). What goes around comes around? Wernicke encephalopathy and the nationwide shortage of intravenous multivitamins revisited. *Journal of Clinical Oncology, 30*(31), e318–e320. https://doi.org/10.1200/JCO.2012.42.7237.

Raj, K. M., Baranwal, A. K., Attri, S. V., Jayashree, M., Kumar, M. P., Patial, A., et al. (2021). Thiamine status in children with septic shock from a developing country: A prospective case-control study. *Journal of Tropical Pediatrics, 67*(1). https://doi.org/10.1093/tropej/fmaa107.

Reddy, C., & Saini, A. G. (2020). Metabolic epilepsy. *Indian Journal of Pediatrics*. https://doi.org/10.1007/s12098-020-03510-w.

Reidling, J. C., Lambrecht, N., Kassir, M., & Said, H. M. (2010). Impaired intestinal vitamin B1 (thiamin) uptake in thiamin transporter-2-deficient mice. *Gastroenterology, 138*(5), 1802–1809. https://doi.org/10.1053/j.gastro.2009.10.042.

Rindi, G., & Laforenza, U. (2000). Thiamine intestinal transport and related issues: Recent aspects. *Proceedings of the Society for Experimental Biology and Medicine, 224*(4), 246–255. https://doi.org/10.1046/j.1525-1373.2000.22428.x.

Said, H. M. (2011). Intestinal absorption of water-soluble vitamins in health and disease. *The Biochemical Journal, 437*(3), 357–372. https://doi.org/10.1042/BJ20110326.

Said, H. M., Balamurugan, K., Subramanian, V. S., & Marchant, J. S. (2004). Expression and functional contribution of hTHTR-2 in thiamin absorption in human intestine. *American Journal of Physiology. Gastrointestinal and Liver Physiology, 286*(3), G491–G498. https://doi.org/10.1152/ajpgi.00361.2003.

Saini, A. G., & Sharma, S. (2021). Biotin-thiamine-responsive basal ganglia disease in children: A treatable neurometabolic disorder. *Annals of Indian Academy of Neurology, 24*(2), 173–177. https://doi.org/10.4103/aian.AIAN_952_20.

Schanzer, A., Doring, B., Ondrouschek, M., Goos, S., Garvalov, B. K., Geyer, J., et al. (2014). Stress-induced upregulation of SLC19A3 is impaired in biotin-thiamine-responsive basal ganglia disease. *Brain Pathology, 24*(3), 270–279. https://doi.org/10.1111/bpa.12117.

Siu, V. M., Ratko, S., Prasad, A. N., Prasad, C., & Rupar, C. A. (2010). Amish microcephaly: Long-term survival and biochemical characterization. *American Journal of Medical Genetics. Part A, 152A*(7), 1747–1751. https://doi.org/10.1002/ajmg.a.33373.

Smith, T. J., Johnson, C. R., Koshy, R., Hess, S. Y., Qureshi, U. A., Mynak, M. L., et al. (2021). Thiamine deficiency disorders: A clinical perspective. *Annals of the New York Academy of Sciences, 1498*(1), 9–28. https://doi.org/10.1111/nyas.14536.

Spiegel, R., Shaag, A., Edvardson, S., Mandel, H., Stepensky, P., Shalev, S. A., et al. (2009). SLC25A19 mutation as a cause of neuropathy and bilateral striatal necrosis. *Annals of Neurology, 66*(3), 419–424. https://doi.org/10.1002/ana.21752.

Sremba, L. J., Chang, R. C., Elbalalesy, N. M., Cambray-Forker, E. J., & Abdenur, J. E. (2014). Whole exome sequencing reveals compound heterozygous mutations in SLC19A3 causing biotin-thiamine responsive basal ganglia disease. *Molecular Genetics and Metabolism Reports, 1*, 368–372. https://doi.org/10.1016/j.ymgmr.2014.07.008.

Talwar, D., Davidson, H., Cooney, J., & St, J. R. D. (2000). Vitamin B(1) status assessed by direct measurement of thiamin pyrophosphate in erythrocytes or whole blood by HPLC: Comparison with erythrocyte transketolase activation assay. *Clinical Chemistry, 46*(5), 704–710.

Thomson, A. D. (2000). Mechanisms of vitamin deficiency in chronic alcohol misusers and the development of the Wernicke-Korsakoff syndrome. *Alcohol and Alcoholism. Supplement, 35*(1), 2–7. https://doi.org/10.1093/alcalc/35.supplement_1.2.

Wani, N. A., Qureshi, U. A., Jehangir, M., Ahmad, K., & Ahmad, W. (2016). Infantile encephalitic beriberi: Magnetic resonance imaging findings. *Pediatric Radiology, 46*(1), 96–103. https://doi.org/10.1007/s00247-015-3437-2.

Whitfield, K. C., Smith, T. J., Rohner, F., Wieringa, F. T., & Green, T. J. (2021). Thiamine fortification strategies in low- and middle-income settings: A review. *Annals of the New York Academy of Sciences, 1498*(1), 29–45. https://doi.org/10.1111/nyas.14565.

Wierzbicki, A. S. (2007). Peroxisomal disorders affecting phytanic acid alpha-oxidation: A review. *Biochemical Society Transactions, 35*(Pt 5), 881–886. https://doi.org/10.1042/BST0350881.

Yamada, K., Miura, K., Hara, K., Suzuki, M., Nakanishi, K., Kumagai, T., et al. (2010). A wide spectrum of clinical and brain MRI findings in patients with SLC19A3 mutations. *BMC Medical Genetics, 11*, 171. https://doi.org/10.1186/1471-2350-11-171.

PART VII

Headaches and migraines

CHAPTER 18

Riboflavin usage in pediatric migraine

Gaku Yamanaka
Department of Pediatrics and Adolescent Medicine, Tokyo Medical University, Tokyo, Japan

Abbreviations

CGRP Calcitonin gene-related peptide
CIMs Complementary and integrative medicines
PUFAs Polyunsaturated fatty acids
RCT Randomized controlled trial

Introduction

Migraine headaches are common in children, adolescents, and adults worldwide, occurring in 3%–5% of young children and up to 18% of adolescents (Lewis, Diamond, Scott, & Jones, 2004). Because of their substantial negative impact, these headaches can affect the quality of life in a manner similar to that of childhood cancer, heart disease, and rheumatic disease (Powers, Patton, Hommel, & Hershey, 2003). Long-term migraine outcomes in children can be improved with early diagnosis and multidisciplinary interventions (Charles et al., 2009; Kabbouche et al., 2005) consisting of both pharmacologic and non-pharmacologic approaches (such as biobehavioral and biofeedback therapy) (Esparham et al., 2018; Kabbouche et al., 2005; Soee, Skov, Skovgaard, & Thomsen, 2013). Prophylactic treatment is necessary to avoid excessive drug intake and improve the quality of life of children with migraines and their families (Lewis et al., 2004; Silberstein, 2000), which is an important part of treatment. However, to date, there has been no consensus regarding migraine prophylaxis in children and adolescents, even in recently published practice guidelines from the United States (Oskoui et al., 2019).

In 2017, the largest and most methodologically rigorous trial on the efficacy of pharmacological interventions for the prevention of migraine in children and adolescents (the Childhood and Adolescent Migraine Prevention Trial; CHAMP) denied the usefulness of both amitriptyline and topiramate, which are widely used as preventive treatments for migraine in children (Powers et al., 2017). On the basis of the results of the CHAMP trial, many clinicians now use nutritional supplements, which cause minimal side effects, as their first-line treatment prior to the use of pharmaceuticals (Patterson-Gentile & Szperka, 2018). The following complementary and integrative medicines (CIMs) are now attracting attention for preventing migraine: riboflavin, coenzyme Q10,

https://doi.org/10.1016/B978-0-323-89835-5.00026-0

magnesium, melatonin, polyunsaturated fatty acids (PUFAs), feverfew, butterbur, vitamin D, and ginkgolide B, and various current reviews on the effectiveness of these agents have been published (Orr, 2018; Wells, Beuthin, & Granetzke, 2019; Yamanaka et al., 2021). In most cases, these agents have been associated with fewer side effects than pharmaceuticals; however, while butterbur has been reported to be effective in preventing migraine in the pediatric population (Oelkers-Ax et al., 2008; Pothmann & Danesch, 2005), postmarketing data have shown that it is associated with hepatotoxicity (Prieto, 2014). In this context, riboflavin can be considered to be a relatively safe CIM that can be used to prevent pediatric migraines, although classifying drugs as CIMs does not necessarily mean that they are safe.

As a migraine prophylactic agent in children, riboflavin is gradually gaining research interest and is being frequently used clinically. Therefore, in this chapter, I have attempted to summarize the evidence for the efficacy and safety of riboflavin in the prevention of pediatric migraines.

Underlying research on riboflavin

Riboflavin plays a crucial role in the production of adenosine triphosphate since it is a precursor for two coenzymes that are required for oxidation-reduction reactions during mitochondrial energy production (Taylor, 2011). Previous studies have described the role of mitochondrial dysfunction in migraine pathophysiology (Reyngoudt, Paemeleire, Descamps, De Deene, & Achten, 2011; Younis, Hougaard, Vestergaard, Larsson, & Ashina, 2017). Magnetic resonance spectroscopy studies have revealed reduced interictal rates of mitochondrial oxidative phosphorylation in the brains of patients with migraine (Barbiroli et al., 1992; Montagna, Cortelli, & Barbiroli, 1994; Younis et al., 2017). Another recent study reported the neuroprotective potential of riboflavin to ameliorate oxidative stress, neuroinflammation, and glutamine excitotoxicity through various mechanisms (Marashly & Bohlega, 2017) (Fig. 1).

Clinical studies of riboflavin

Seven previous studies (four randomized controlled trials [RCTs] and three retrospective studies) have evaluated the role of riboflavin in the prevention of migraine, as summarized in Tables 1 (positive studies) and 2 (negative studies). Among these studies, five—two RCTs and three retrospective studies—reported that riboflavin was effective (Athaillah et al., 2012; Das & Qubty, 2020; Talebian et al., 2018; Yamanaka et al., 2020), while the other two studies concluded that there was a lack of evidence to support the use of riboflavin for pediatric migraine (Bruijn et al., 2010; MacLennan et al., 2008).

These studies evaluated a wide range of riboflavin doses (10–400 mg). Some studies have shown that higher doses are more effective (Talebian et al., 2018), while others have

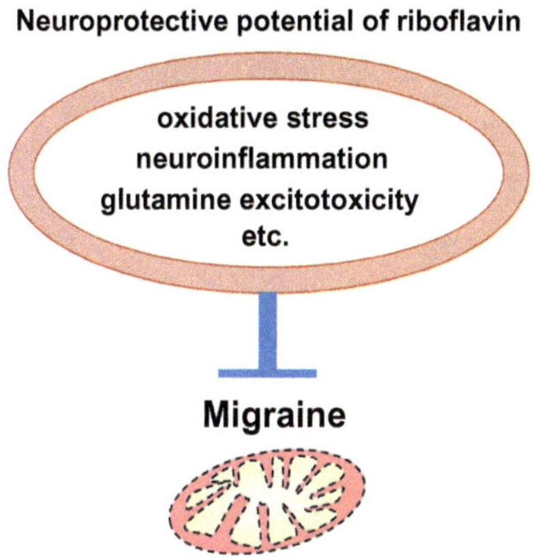

Fig. 1 Neuroprotective potential of riboflavin. The role of mitochondrial dysfunction in the pathophysiology of migraine has been reported, and riboflavin may contribute to the improvement of migraine pathology by neuroprotective effects against oxidative stress, neuroinflammation, and glutamine excitotoxicity through various mechanisms.

Table 1 Summary of positive studies on complementary and integrative medicines for pediatric migraine.

Reference no: Study design	N/patients	Intervention	Comparison	Outcomes	Adverse effects
Retrospective chart review (Condo, Posar, Arbizzani, & Parmeggiani, 2009)	41 children and adolescents with a variety of headache disorders	Riboflavin (200 mg or 400 mg daily) for three, four, or 6 months	N/A	Significant reduction in headache frequency after treatment for three or 4 months ($P < .01$), which was not sustained at 6 months ($P > .05$)	Vomiting, increased appetite without weight gain, temporary yellow-orange coloration of urine
RCT (Athaillah et al., 2012)	98 adolescents with migraine	Riboflavin (400 mg daily) for 3 months ($n = 50$)	Placebo for 3 months ($n = 48$)	Headache frequency decreased from the 1st month to the 2nd month to the 3rd month (3.7 per month) and headache duration decreased ($P = .012$ and	Polyuria, diarrhea

Continued

Table 1 Summary of positive studies on complementary and integrative medicines for pediatric migraine—cont'd

Reference no: Study design	N/patients	Intervention	Comparison	Outcomes	Adverse effects
				P=.001, respectively) vs placebo. Disability, as measured by the PedMIDAS, also decreased (P=.001)	
RCT (Talebian et al., 2018)	90 children and adolescents with migraine	Riboflavin (200 mg or 400 mg daily) for 3 months (n=30, and 30, respectively)	Placebo for 3 months (n=30)	The riboflavin 400 mg group showed a greater reduction in the headache frequency and duration than the placebo (P=.00 and P=.00, respectively)	None
Retrospective observational study (Yamanaka et al., 2020)	68 children and adolescents with migraine	Riboflavin (10 or 40 mg daily) for 3 months (n=13 and 55, respectively)	N/A	Significant overall reduction in the median frequency of headache episodes from the baseline to 3 months (P=.00) was detected	None
Retrospective observational study (Das & Qubty, 2020)	42 children and adolescents with a variety of headache disorders	Riboflavin (100 and 200 mg for children weighing 20–40 kg and greater than 40 kg, respectively)	N/A	There was a significant decrease in the frequency of headache days after 2–4 months compared to baseline. Mean headache intensity decreased (P<.001), and headache duration also significantly decreased (P<.001)	None

RCT, randomized controlled trial; MIDAS, Migraine Disability Assessment; N/A, not available. N/D: not described.

Table 2 Summary of negative studies on complementary and integrative medicines for pediatric migraine.

Reference no: Study design	N/patients	Intervention	Comparison	Outcomes	Adverse effects
RCT (MacLennan et al., 2008)	48 children and adolescents with migraine	Riboflavin (200 mg daily) for 12 weeks ($n=27$)	Placebo for 12 weeks ($n=21$)	No difference was found between the groups in terms of the proportion of participants with a 50% or greater reduction of migraine frequency ($P=.125$)	Change in urine color
Crossover RCT (Bruijn et al., 2010)	42 children with migraine	Riboflavin (50 mg daily) for 4 months ($n=20$)	Placebo for 4 months ($n=22$)	No difference between groups in terms of change in migraine frequency ($P=.44$); the riboflavin group showed a greater reduction in the frequency of tension-type headaches as compared to the placebo ($P=.04$)	None

RCT, randomized controlled trial; MIDAS, Migraine Disability Assessment; N/A, not available. N/D: not described.

shown that small doses of riboflavin (25 mg or 100 mg) have been shown to have pharmacological action even in adults (Maizels, Blumenfeld, & Burchette, 2004; Nambiar, Aiyappa, & Srinivasa, 2011). The exact riboflavin requirement is not clear even in adults with migraine; a dose of 400 mg was chosen in a previous adult study since similar high doses were previously used to treat mitochondrial disorders and due to riboflavin's lack of toxicity (Schoenen, Jacquy, & Lenaerts, 1998). A prior pharmacokinetic study conducted among adults did not find significant differences in blood concentrations, maximal serum concentrations, or areas under the curve with 20-, 40-, and 60-mg riboflavin supplementation (Zempleni, Galloway, & McCormick, 1996).

The estimated average requirements and recommended dietary allowances for riboflavin for men and women between the ages of 19 and 70 years are 0.9–1.1 mg/day and 1.1–1.3 mg/day (Institute of Medicine Committee on Use of Dietary Reference Intakes in Nutrition, 2003), respectively, and those for children aged 1–9 years and adolescents aged 10–18 years range from 0.5 to 0.6 mg/day and from 0.9 to 1.3 mg/day, respectively (Pinto & Zempleni, 2016). Our recent univariate analysis also showed no difference between 10 and 40 mg, so it is possible that smaller doses may have exerted pharmacological effects in children (Yamanaka et al., 2020).

The two studies that did not show any effectiveness of riboflavin as a preventive therapy for pediatric migraines, as shown in Table 2 may have been influenced by the small number of cases included, the amount of riboflavin used, and the background of the patients (Bruijn et al., 2010; MacLennan et al., 2008). The following factors have been reported to likely inhibit the effects of riboflavin: the presence of comorbid headaches, male sex, and age under 12 years (Condo et al., 2009; Yamanaka et al., 2020). However, there is no solid evidence, and further studies with larger numbers of patients and longer observational periods are required to assess the effectiveness of riboflavin-based prophylaxis for pediatric migraines.

Although there is a widespread perception that riboflavin has no side effects, riboflavin administration has been reported to cause some adverse effects, as shown in the Tables. In particular, studies utilizing high doses reported orange discoloration of the urine, diarrhea, vomiting, and an increase in appetite without weight gain (Athaillah et al., 2012; Condo et al., 2009; MacLennan et al., 2008), although studies using low doses of the riboflavin supplement did not report adverse events (Bruijn et al., 2010; Yamanaka et al., 2020). No serious side effects have been reported to date. Factors that may interfere with the effects of riboflavin include concomitant tension-type headache and other conditions, and age 12 years and younger have been reported (Condo et al., 2009; Yamanaka et al., 2020).

Applications to other neurological conditions

Although the American Academy of Neurology guidelines do not mention CIMs, including riboflavin (Oskoui et al., 2019), there are several reports, including four RCTs, with high evidence levels for the efficacy of riboflavin for migraine in adults (Boehnke et al., 2004; Nambiar et al., 2011; Rahimdel, Zeinali, Yazdian-Anari, Hajizadeh, & Arefnia, 2015; Schoenen et al., 1998; Schoenen, Lenaerts, & Bastings, 1994). Riboflavin may also exert neuroprotective effects in several neurological diseases, including Parkinson's disease (Ferraz et al., 2004; Marashly & Bohlega, 2017; Murakami et al., 2010) and multiple sclerosis (Evans, Piccio, & Cross, 2018; Parks, Jackson-Tarlton, Vacchi, Merdad, & Johnston, 2020), through its role in several pathways hypothesized to be impaired in neurological diseases, such as antioxidation, myelin formation, mitochondrial function, and iron metabolism (Saedisomeolia & Ashoori, 2018).

Other components of interest

Nutritional supplements such as coenzyme Q10, polyunsaturated fatty acids (PUFAs), feverfew, butterbur, vitamin D, and ginkgolide B have also been used to prevent migraine. In most cases, these drugs have shown fewer adverse effects than pharmaceutical drugs. However, while butterbur has been reported to be effective in migraine prophylaxis in children (Oelkers-Ax et al., 2008), postmarketing data have shown that it is associated with hepatotoxicity (Prieto, 2014). Therefore, the classification of a drug as a CIM does not necessarily mean that it is safe, and individual efficacy and mechanisms of action may vary widely.

Coenzyme Q10

Since mitochondrial energy is depleted in migraine (Reyngoudt et al., 2011; Younis et al., 2017), supplementation with coenzyme Q10 may prevent migraine by restoring mitochondrial energy stores. Current data show that in migraine patients, coenzyme Q10 supplementation leads to reduced levels of calcitonin gene-related peptide (CGRP) (Dahri, Tarighat-Esfanjani, Asghari-Jafarabadi, & Hashemilar, 2019). Although no serious adverse events have occurred with coenzyme Q10 treatment, there is not enough evidence to draw definitive conclusions about its efficacy (Slater et al., 2011).

PUFAs

PUFAs, in addition to their established antivascular pressure effects (Mori et al., 2000), are necessary for the optimal functioning of the nervous system (Crawford et al., 1997) and are known to have antiinflammatory effects (Gutiérrez, Svahn, & Johansson, 2019; Yan et al., 2013). Several studies have examined the benefits of PUFAs (Fayyazi, Khajeh, Ghazavi, & Sangestani, 2016; Harel et al., 2002), but there is insufficient evidence to draw definitive conclusions.

Vitamin D

A single-blind, randomized clinical trial in children with migraine suggested that the combination of topiramate and vitamin D3 was more effective in preventing migraine than topiramate alone (Fallah, Sarraf Yazd, & Sohrevardi, 2020). Furthermore, in a current meta-analysis of the serum levels of 25-hydroxyvitamin D3 in migraine patients, preliminary evidence confirmed low levels in both adults and children (Liampas, Siokas, Brotis, & Dardiotis, 2020). The potential efficacy of vitamin D suggests that CIMs with an attractive safety profile can be used in combination regimens with vitamin D.

Feverfew

Feverfew, also known as *Tanacetum parthenium*, is a medicinal herb that has been used for centuries to treat a variety of ailments. It has been reported to affect the processes involved in the pathogenesis of migraine (Benemei, Fusi, Trevisan, & Geppetti, 2014) and has shown antimigraine effects, such as inhibition of NO synthesis, cytokine production, CGRP induction, and subsequent serotonin release from platelets (Sangermani & Boncimino, 2017). A recent RCT evaluating the use of combination therapy (feverfew, coenzyme Q10, riboflavin, and magnesium) showed positive results (Moscano et al., 2019). Treatment was discontinued in 4.4% of patients due to gastrointestinal symptoms (nausea and diarrhea), but the study did not report whether these adverse effects were due to the use of feverfew.

Ginkgolide B

Ginkgolide B is a natural component that modulates the action of glutamate in the central nervous system (Nogami, Hirashima, Endo, & Takaku, 1997). In an open-label study on migraine prophylaxis in children, ginkgolide B (in combination with various other CIMs) appeared to produce favorable results without adverse events (Esposito, Ruberto, Pascotto, & Carotenuto, 2012).

Summary

The use of various combinations of feverfew, vitamin D, and ginkgolide B for pediatric migraine prevention appears to be effective without serious adverse effects. However, due to the lack of standalone studies, the individual effects of each integrative agent cannot be determined.

Mini-dictionary of terms

- **Complementary and integrative medicines (CIMs)**. Formerly known as complementary and alternative medicine (CAM), CIMs are based on evidence-based methods to support patient health and healing. CIMs typically include nutritional supplements and exercise therapy and are becoming increasingly recognized for their migraine prevention benefits.
- **Calcitonin gene-related peptide (CGRP)**. CGRP, a member of the calcitonin family, is hypothesized to play an essential role in the pathophysiology of migraine. Blockade of CGRP ligands or their receptors has been demonstrated to treat and prevent migraine.

- **Coenzyme Q10**. This enzyme is a dietary supplement required for cellular energy production due to its role as an electron carrier in the electron transport chain of mitochondria.
- **Polyunsaturated fatty acids (PUFAs)**. PUFAs, including eicosapentaenoic acid and docosahexaenoic acid, are essential dietary components that humans cannot synthesize.
- **Ginkgolide B**. Ginkolide B is a herbal constituent extracted from *Ginkgo biloba* tree leaves and is a natural modulator.

Summary points

- The level of evidence for using riboflavin to treat pediatric migraine was relatively high.
- Riboflavin may be useful for migraine prophylaxis in children and adolescents.
- It may be useful for children older than 13 years of age, especially those without concomitant conditions such as tension-type headaches.
- Since riboflavin showed no serious adverse effects, it may be used as a first-line option for preventing migraine in children.
- Coenzyme Q10, PUFAs, feverfew, butterbur, vitamin D, and ginkgolide B might also be effective in preventing pediatric migraine, but the level of evidence is limited compared to riboflavin.
- There is no solid evidence to prove the efficacy of riboflavin in the treatment of migraine in children, and further studies with a larger number of patients and longer observation periods are needed.

References

Athaillah, A., Dimyati, Y., Johannes, H. S., Bistok, S., Hakimi, H., et al. (2012). Riboflavin as migraine prophylaxis in adolescents. *Paediatrica Indonesiana*, *52*(3), 132–137.

Barbiroli, B., Montagna, P., Cortelli, P., Funicello, R., Iotti, S., Monari, L., et al. (1992). Abnormal brain and muscle energy metabolism shown by 31P magnetic resonance spectroscopy in patients affected by migraine with aura. *Neurology*, *42*(6), 1209–1214. https://doi.org/10.1212/wnl.42.6.1209.

Benemei, S., Fusi, C., Trevisan, G., & Geppetti, P. (2014). The TRPA1 channel in migraine mechanism and treatment. *British Journal of Pharmacology*, *171*(10), 2552–2567. https://doi.org/10.1111/bph.12512.

Boehnke, C., Reuter, U., Flach, U., Schuh-Hofer, S., Einhaupl, K. M., & Arnold, G. (2004). High-dose riboflavin treatment is efficacious in migraine prophylaxis: An open study in a tertiary care centre. *European Journal of Neurology*, *11*(7), 475–477. https://doi.org/10.1111/j.1468-1331.2004.00813.x.

Bruijn, J., Duivenvoorden, H., Passchier, J., Locher, H., Dijkstra, N., & Arts, W. F. (2010). Medium-dose riboflavin as a prophylactic agent in children with migraine: A preliminary placebo-controlled, randomised, double-blind, cross-over trial. *Cephalalgia*, *30*(12), 1426–1434. https://doi.org/10.1177/0333102410365106.

Charles, J. A., Peterlin, B. L., Rapoport, A. M., Linder, S. L., Kabbouche, M. A., & Sheftell, F. D. (2009). Favorable outcome of early treatment of new onset child and adolescent migraine-implications for disease modification. *Journal of Headache and Pain*, *10*(4), 227–233. https://doi.org/10.1007/s10194-009-0133-3.

Condo, M., Posar, A., Arbizzani, A., & Parmeggiani, A. (2009). Riboflavin prophylaxis in pediatric and adolescent migraine. *Journal of Headache and Pain*, *10*(5), 361–365. https://doi.org/10.1007/s10194-009-0142-2.

Crawford, M. A., Costeloe, K., Ghebremeskel, K., Phylactos, A., Skirvin, L., & Stacey, F. (1997). Are deficits of arachidonic and docosahexaenoic acids responsible for the neural and vascular complications of preterm babies? *American Journal of Clinical Nutrition*, *66*(4 Suppl), 1032s–1041s. https://doi.org/10.1093/ajcn/66.4.1032S.

Dahri, M., Tarighat-Esfanjani, A., Asghari-Jafarabadi, M., & Hashemilar, M. (2019). Oral coenzyme Q10 supplementation in patients with migraine: Effects on clinical features and inflammatory markers. *Nutritional Neuroscience*, *22*(9), 607–615. https://doi.org/10.1080/1028415X.2017.1421039.

Das, R., & Qubty, W. (2020). Retrospective observational study on riboflavin prophylaxis in child and adolescent migraine. *Pediatric Neurology*, *114*, 5–8. https://doi.org/10.1016/j.pediatrneurol.2020.09.009.

Esparham, A., Herbert, A., Pierzchalski, E., Tran, C., Dilts, J., Boorigie, M., et al. (2018). Pediatric headache clinic model: Implementation of integrative therapies in practice. *Children (Basel)*, *5*(6). https://doi.org/10.3390/children5060074.

Esposito, M., Ruberto, M., Pascotto, A., & Carotenuto, M. (2012). Nutraceutical preparations in childhood migraine prophylaxis: Effects on headache outcomes including disability and behaviour. *Neurological Sciences*, *33*(6), 1365–1368. https://doi.org/10.1007/s10072-012-1019-8.

Evans, E., Piccio, L., & Cross, A. H. (2018). Use of vitamins and dietary supplements by patients with multiple sclerosis: A review. *JAMA Neurology*, *75*(8), 1013–1021. https://doi.org/10.1001/jamaneurol.2018.0611.

Fallah, R., Sarraf Yazd, S., & Sohrevardi, S. M. (2020). Efficacy of Topiramate alone and Topiramate plus vitamin D3 in the prophylaxis of pediatric migraine: A randomized clinical trial. *Iranian Journal of Child Neurology*, *14*(4), 77–86.

Fayyazi, A., Khajeh, A., Ghazavi, A., & Sangestani, M. (2016). Omega 3 in childhood migraines: A double blind randomized clinical trial. *Iranian Journal of Child Neurology*, *10*(1), 9–13.

Ferraz, H. B., Quagliato, E. A., Rieder, C. R., Silva, D. J., Teive, H. A., Barbosa, E. R., et al. (2004). Comments on the paper "High doses of riboflavin and the elimination of dietary red meat promote the recovery of some motor functions in Parkinson's disease patients. C.G. Coimbra and V.B.C. Junqueira. Brazilian Journal of Medical and Biological Research, 36: 1409–1417, 2003". *Brazilian Journal of Medical and Biological Research*, *37*(9), 1297–1299. discussion 1299–1302 https://doi.org/10.1590/s0100-879x2004000900002.

Gutiérrez, S., Svahn, S. L., & Johansson, M. E. (2019). Effects of omega-3 fatty acids on immune cells. *International Journal of Molecular Sciences*, *20*(20). https://doi.org/10.3390/ijms20205028.

Harel, Z., Gascon, G., Riggs, S., Vaz, R., Brown, W., & Exil, G. (2002). Supplementation with omega-3 polyunsaturated fatty acids in the management of recurrent migraines in adolescents. *Journal of Adolescent Health*, *31*(2), 154–161. https://doi.org/10.1016/s1054-139x(02)00349-x.

Institute of Medicine Committee on Use of Dietary Reference Intakes in Nutrition, L. (2003). Dietary reference intakes: Guiding principles for nutrition labeling and fortification. In *Dietary reference intakes: Guiding principles for nutrition labeling and fortification*. Washington (DC): National Academies Press (US). Copyright 2003 by the National Academy of Sciences. All rights reserved.

Kabbouche, M. A., Powers, S. W., Vockell, A. L., LeCates, S. L., Ellinor, P. L., Segers, A., et al. (2005). Outcome of a multidisciplinary approach to pediatric migraine at 1, 2, and 5 years. *Headache*, *45*(10), 1298–1303. https://doi.org/10.1111/j.1526-4610.2005.00261.x.

Lewis, D., Ashwal, S., Hershey, A., Hirtz, D., Yonker, M., & Silberstein, S. (2004). Practice parameter: Pharmacological treatment of migraine headache in children and adolescents: Report of the American Academy of Neurology Quality Standards Subcommittee and the Practice Committee of the Child Neurology Society. *Neurology*, *63*(12), 2215–2224. https://doi.org/10.1212/01.wnl.0000147332.41993.90.

Lewis, D. W., Diamond, S., Scott, D., & Jones, V. (2004). Prophylactic treatment of pediatric migraine. *Headache*, *44*(3), 230–237. https://doi.org/10.1111/j.1526-4610.2004.04052.x.

Liampas, I., Siokas, V., Brotis, A., & Dardiotis, E. (2020). Vitamin D serum levels in patients with migraine: A meta-analysis. *Revue Neurologique*, *176*(7–8), 560–570. https://doi.org/10.1016/j.neurol.2019.12.008.

MacLennan, S. C., Wade, F. M., Forrest, K. M., Ratanayake, P. D., Fagan, E., & Antony, J. (2008). High-dose riboflavin for migraine prophylaxis in children: A double-blind, randomized, placebo-controlled trial. *Journal of Child Neurology*, *23*(11), 1300–1304. https://doi.org/10.1177/0883073808318053.

Maizels, M., Blumenfeld, A., & Burchette, R. (2004). A combination of riboflavin, magnesium, and feverfew for migraine prophylaxis: A randomized trial. *Headache*, *44*(9), 885–890. https://doi.org/10.1111/j.1526-4610.2004.04170.x.

Marashly, E. T., & Bohlega, S. A. (2017). Riboflavin has neuroprotective potential: Focus on Parkinson's disease and migraine. *Frontiers in Neurology*, *8*, 333. https://doi.org/10.3389/fneur.2017.00333.

Montagna, P., Cortelli, P., & Barbiroli, B. (1994). Magnetic resonance spectroscopy studies in migraine. *Cephalalgia*, *14*(3), 184–193. https://doi.org/10.1046/j.1468-2982.1994.014003184.x.

Mori, T. A., Watts, G. F., Burke, V., Hilme, E., Puddey, I. B., & Beilin, L. J. (2000). Differential effects of eicosapentaenoic acid and docosahexaenoic acid on vascular reactivity of the forearm microcirculation in hyperlipidemic, overweight men. *Circulation*, *102*(11), 1264–1269. https://doi.org/10.1161/01.cir.102.11.1264.

Moscano, F., Guiducci, M., Maltoni, L., Striano, P., Ledda, M. G., Zoroddu, F., et al. (2019). An observational study of fixed-dose *Tanacetum parthenium* nutraceutical preparation for prophylaxis of pediatric headache. *Italian Journal of Pediatrics*, *45*(1), 36. https://doi.org/10.1186/s13052-019-0624-z.

Murakami, K., Miyake, Y., Sasaki, S., Tanaka, K., Fukushima, W., Kiyohara, C., et al. (2010). Dietary intake of folate, vitamin B6, vitamin B12 and riboflavin and risk of Parkinson's disease: A case-control study in Japan. *British Journal of Nutrition*, *104*(5), 757–764. https://doi.org/10.1017/s0007114510001005.

Nambiar, N. J., Aiyappa, C., & Srinivasa, R. (2011). Oral riboflavin versus oral propranolol in migraine prophylaxis: An open label randomized controlled trial. *Neurology Asia*, *16*(3), 223–229.

Nogami, K., Hirashima, Y., Endo, S., & Takaku, A. (1997). Involvement of platelet-activating factor (PAF) in glutamate neurotoxicity in rat neuronal cultures. *Brain Research*, *754*(1–2), 72–78. https://doi.org/10.1016/s0006-8993(97)00047-4.

Oelkers-Ax, R., Leins, A., Parzer, P., Hillecke, T., Bolay, H. V., Fischer, J., et al. (2008). Butterbur root extract and music therapy in the prevention of childhood migraine: An explorative study. *European Journal of Pain (London, England)*, *12*(3), 301–313. https://doi.org/10.1016/j.ejpain.2007.06.003.

Orr, S. L. (2018). The evidence for the role of nutraceuticals in the management of pediatric migraine: A review. *Current Pain and Headache Reports*, *22*(5), 37. https://doi.org/10.1007/s11916-018-0692-6.

Oskoui, M., Pringsheim, T., Billinghurst, L., Potrebic, S., Gersz, E. M., Gloss, D., et al. (2019). Practice guideline update summary: Pharmacologic treatment for pediatric migraine prevention: Report of the guideline development, dissemination, and implementation subcommittee of the American academy of neurology and the American headache society. *Headache*, *59*(8), 1144–1157. https://doi.org/10.1111/head.13625.

Parks, N. E., Jackson-Tarlton, C. S., Vacchi, L., Merdad, R., & Johnston, B. C. (2020). Dietary interventions for multiple sclerosis-related outcomes. *Cochrane Database of Systematic Reviews*, *5*(5), Cd004192. https://doi.org/10.1002/14651858.CD004192.pub4.

Patterson-Gentile, C., & Szperka, C. L. (2018). The changing landscape of pediatric migraine therapy: A review. *JAMA Neurology*, *75*(7), 881–887. https://doi.org/10.1001/jamaneurol.2018.0046.

Pinto, J. T., & Zempleni, J. (2016). Riboflavin. *Advances in Nutrition*, 7(5), 973–975. https://doi.org/10.3945/an.116.012716.

Pothmann, R., & Danesch, U. (2005). Migraine prevention in children and adolescents: Results of an open study with a special butterbur root extract. *Headache*, *45*(3), 196–203. https://doi.org/10.1111/j.1526-4610.2005.05044.x.

Powers, S. W., Coffey, C. S., Chamberlin, L. A., Ecklund, D. J., Klingner, E. A., Yankey, J. W., et al. (2017). Trial of amitriptyline, topiramate, and placebo for pediatric migraine. *New England Journal of Medicine*, *376*(2), 115–124. https://doi.org/10.1056/NEJMoa1610384.

Powers, S. W., Patton, S. R., Hommel, K. A., & Hershey, A. D. (2003). Quality of life in childhood migraines: Clinical impact and comparison to other chronic illnesses. *Pediatrics*, *112*(1 Pt 1), e1–e5. https://doi.org/10.1542/peds.112.1.e1.

Prieto, J. M. (2014). Update on the efficacy and safety of Petadolex®, a butterbur extract for migraine prophylaxis. *Botanics: Targets and Therapy*. https://doi.org/10.2147/btat.S54023.

Rahimdel, A., Zeinali, A., Yazdian-Anari, P., Hajizadeh, R., & Arefnia, E. (2015). Effectiveness of vitamin B2 versus sodium valproate in migraine prophylaxis: a randomized clinical trial. *Electronic Physician*, 7(6), 1344–1348. https://doi.org/10.14661/1344.

Reyngoudt, H., Paemeleire, K., Descamps, B., De Deene, Y., & Achten, E. (2011). 31P-MRS demonstrates a reduction in high-energy phosphates in the occipital lobe of migraine without aura patients. *Cephalalgia*, *31*(12), 1243–1253. https://doi.org/10.1177/0333102410394675.

Saedisomeolia, A., & Ashoori, M. (2018). Riboflavin in human health: A review of current evidences. *Advances in Food and Nutrition Research*, *83*, 57–81. https://doi.org/10.1016/bs.afnr.2017.11.002.

Sangermani, R., & Boncimino, A. (2017). The use of nutraceutics in children's and adolescent's headache. *Neurological Sciences*, *38*(Suppl. 1), 121–124. https://doi.org/10.1007/s10072-017-2896-7.

Schoenen, J., Jacquy, J., & Lenaerts, M. (1998). Effectiveness of high-dose riboflavin in migraine prophylaxis. A randomized controlled trial. *Neurology*, *50*(2), 466–470. https://doi.org/10.1212/wnl.50.2.466.

Schoenen, J., Lenaerts, M., & Bastings, E. (1994). High-dose riboflavin as a prophylactic treatment of migraine: Results of an open pilot study. *Cephalalgia*, *14*(5), 328–329. https://doi.org/10.1046/j.1468-2982.1994.1405328.x.

Silberstein, S. D. (2000). Practice parameter: Evidence-based guidelines for migraine headache (an evidence-based review): Report of the quality standards subcommittee of the American academy of neurology. *Neurology*, *55*(6), 754–762. https://doi.org/10.1212/wnl.55.6.754.

Slater, S. K., Nelson, T. D., Kabbouche, M. A., LeCates, S. L., Horn, P., Segers, A., et al. (2011). A randomized, double-blinded, placebo-controlled, crossover, add-on study of CoEnzyme Q10 in the prevention of pediatric and adolescent migraine. *Cephalalgia*, *31*(8), 897–905. https://doi.org/10.1177/0333102411406755.

Soee, A. B., Skov, L., Skovgaard, L. T., & Thomsen, L. L. (2013). Headache in children: Effectiveness of multidisciplinary treatment in a tertiary paediatric headache clinic. *Cephalalgia*, *33*(15), 1218–1228. https://doi.org/10.1177/0333102413490349.

Talebian, A., Soltani, B., Banafshe, H. R., Moosavi, G. A., Talebian, M., & Soltani, S. (2018). Prophylactic effect of riboflavin on pediatric migraine: A randomized, double-blind, placebo-controlled trial. *Electronic Physician*, *10*(2), 6279–6285. https://doi.org/10.19082/6279.

Taylor, F. R. (2011). Nutraceuticals and headache: The biological basis. *Headache*, *51*(3), 484–501. https://doi.org/10.1111/j.1526-4610.2011.01847.x.

Wells, R. E., Beuthin, J., & Granetzke, L. (2019). Complementary and integrative medicine for episodic migraine: An update of evidence from the last 3 years. *Current Pain and Headache Reports*, *23*(2), 10. https://doi.org/10.1007/s11916-019-0750-8.

Yamanaka, G., Kanou, K., Takamatsu, T., Takeshita, M., Morichi, S., Suzuki, S., et al. (2021). Complementary and integrative medicines as prophylactic agents for pediatric migraine: A narrative literature review. *Journal of Clinical Medicine*, *10*(1). https://doi.org/10.3390/jcm10010138.

Yamanaka, G., Suzuki, S., Takeshita, M., Go, S., Morishita, N., Takamatsu, T., et al. (2020). Effectiveness of low-dose riboflavin as a prophylactic agent in pediatric migraine. *Brain Dev*. https://doi.org/10.1016/j.braindev.2020.04.002.

Yan, Y., Jiang, W., Spinetti, T., Tardivel, A., Castillo, R., Bourquin, C., et al. (2013). Omega-3 fatty acids prevent inflammation and metabolic disorder through inhibition of NLRP3 inflammasome activation. *Immunity*, *38*(6), 1154–1163. https://doi.org/10.1016/j.immuni.2013.05.015.

Younis, S., Hougaard, A., Vestergaard, M. B., Larsson, H. B. W., & Ashina, M. (2017). Migraine and magnetic resonance spectroscopy: A systematic review. *Current Opinion in Neurology*, *30*(3), 246–262. https://doi.org/10.1097/wco.0000000000000436.

Zempleni, J., Galloway, J. R., & McCormick, D. B. (1996). Pharmacokinetics of orally and intravenously administered riboflavin in healthy humans. *American Journal of Clinical Nutrition*, *63*(1), 54–66. https://doi.org/10.1093/ajcn/63.1.54.

CHAPTER 19

Vitamin B12, folate, and migraine

Ioannis Liampas[a,*], Vasileios Siokas[a], Christos Bakirtzis[b], and Efthimios Dardiotis[a]
[a]Department of Neurology, Faculty of Medicine, School of Health Sciences, University of Thessaly, Larissa, Greece
[b]Multiple Sclerosis Center, B' Department of Neurology, AHEPA University Hospital, Aristotle University of Thessaloniki, Thessaloniki, Greece

Abbreviations

CM	chronic migraine
CSD	cortical spreading depression
GABA	gamma-aminobutyric acid
Hcy	homocysteine
MA	migraine with aura
MMA	methylmalonic acid
MO	migraine without aura
NMDA(R)	*N*-methyl-D-aspartic (receptor)
NO	nitric oxide
NSAIDs	nonsteroidal antiinflammatory drugs

Introduction

Migraine is the second commonest primary headache disorder following tension-type headache, with more than 1 billion individuals suffering from migrainous headaches worldwide. Its prevalence registers a peak during the 4th decade of life, with progressively withering numbers towards the two ends of the lifespan. An overall female predominance is observed, with women being affected almost twice as commonly as men. Migraine constitutes the second cause of disability worldwide following low back pain, while it appears particularly burdensome among women of reproductive age (GBD 2016 Headache Collaborators, 2018).

There are three major subtypes of migraine: migraine with aura (MA), migraine without aura (MO), and chronic migraine (CM). MA is differentiated from MO by the presence of transitory focal neurologic manifestations that precede or accompany headaches. Visual symptoms are the most common (90%), but sensory, speech, motor, brainstem (e.g., dysarthria, vertigo, tinnitus, diplopia, and so on), and retinal (monocular symptoms) aura may be encountered as well. In some individuals, aura is associated

* **In Absentia Contact Person:** Katerina Satira, Faculty of Medicine, School oh Health Sciences, University of Thessaly, Telephone [1]: +30 2413502792 Telephone [2]: +30 6987974380, Email: ksat@uth.gr.

Vitamins and Minerals in Neurological Disorders
https://doi.org/10.1016/B978-0-323-89835-5.00042-9

with indistinct headache or not succeeded-accompanied by headache at all. CM is defined irrespective of the presence of aura, based on the emergence of migraine for 15 or more days per month. The diagnosis of these discrete entities is performed according to the International Headache Society criteria and has important therapeutic and prognostic implications (Headache Classification Committee of the International Headache Society, 2018).

Aura affects about one-third of migraine patients (Russell, Rasmussen, Thorvaldsen, & Olesen, 1995). Of note, clinical evidence suggests that MA is associated with elevated cardiovascular risk, especially in females under the age of 45 (Øie, Kurth, Gulati, & Dodick, 2020). Endothelial dysfunction and vasospasm, as well as hypercoagulability and microembolism, are among the potential connecting mechanisms between migraine and increased cardiovascular risk (Gollion et al., 2020; Gryglas & Smigiel, 2017; Kumar, Kijima, West, & Tobis, 2019; Sen et al., 2018; Tietjen & Collins, 2018). Regardless of the underlying link, MA is considered to confer an important risk towards cardiovascular events, which appears to be independent of the traditional vascular risk factors (Øie et al., 2020).

The pathogenesis of migraine

Cortical spreading depression (CSD), a spreading depolarization of neurons and glia accompanied by altered cerebral blood flow (a short-lasting cerebral spreading hyperemia followed by a long-lasting spreading oligemia) and metabolic abnormalities, is considered the neurophysiological substrate of aura (Dreier & Reiffurth, 2015). Glutamate-mediated activation of the *N*-methyl-D-aspartic receptors (NMDARs) may contribute to the development and propagation of CSD (Close, Eftekhari, Wang, Charles, & Russo, 2019). CSD can be induced by hypoxic-ischemic conditions and lead to neuronal death depending on the local energy equilibrium (Dreier & Reiffurth, 2015). Intriguingly, CSD has been reproduced by microembolization in animal models, without causing permanent tissue damage (Nozari et al., 2010).

On the other hand, the trigeminovascular system is of pivotal importance in the functional anatomy of migraine. It has been theorized that CSD could activate the trigeminovascular system and cause migrainous headaches (Zhang et al., 2010). The activation of meningeal nociceptors by CSD could initiate the process and activate the peripheral trigeminovascular constituent. Dural neurogenic inflammation and vasodilation, involving the release of vasoactive substances, including calcitonin gene-related peptide (mainly), nitric oxide (NO), and substance P, are considered to play an important role in the activation and sensitization of the meningeal nociceptors. In turn, peripheral trigeminovascular neurons activate the spinal trigeminal nucleus, which projects to multiple thalamic and hypothalamic nuclei, (among other functionally important central nervous system components) that followingly project to the clinically relevant cortical areas. However, given its complex functional neurocorrelates, the exact pathophysiological cascade of migraine is expectantly not completely understood to date (Ashina et al., 2019).

Serum homocysteine, folate, and vitamin B12 in migraine

Vitamin B12 and folate (pyridoxine as well) are two important coenzymes in the metabolism of homocysteine (Hcy) (Froese, Fowler, & Baumgartner, 2019). Specifically, the process of remethylation of Hcy to methionine (the main metabolic pathway of Hcy) requires both vitamin B12 and folate, while hyperhomocysteinemia is often a result of either folate and/or B12 deficiency. Hcy and methylmalonic acid (MMA) are regarded as superior indicators of the functional vitamin B12 status, compared to B12 itself, with normal serum B12 values often accompanying relative functional B12 deficiencies reflected in higher Hcy or MMA values (Hannibal et al., 2016).

Genetic evidence suggests that disturbances in the Hcy-methionine circle are associated with migraine (Stuart, Cox, Lea, & Griffiths, 2012). Complementarily, pooled evidence from observational studies indicates that adults with migraine have higher Hcy values in comparison with healthy controls (Liampas, Siokas, Aloizou, Tsouris, et al., 2020). Specifically, a mean difference of about 1.64 μmol/L was determined between individuals with migraine and healthy adults, whereas a direct comparison of patients with MA and MO revealed that this difference was mainly driven by the MA participants (the MA group presented about 1.20 μmol/L higher Hcy values than the MO group). Regardless, subgroup comparisons demonstrated that both MA and MO are differentially associated with elevated serum Hcy. Of interest, most published studies reported that mean Hcy values were within the "normal" range (<15 μmol/L) for the group of migraine individuals (similar to the healthy control group). However, the conventionally "normal" Hcy concentrations have been positively correlated with the overall cardiovascular risk (Held et al., 2008). Moreover, in accordance with the migraine-stroke demographics, increased serum Hcy is considered to be an important cardiovascular risk factor mainly among middle-aged adults (Madonna et al., 2002; Zhang et al., 2020). It is, therefore, possible that migraine patients, especially MA individuals, present clinically important elevations in serum Hcy compared to healthy adults. Intriguingly, a cerebrospinal fluid-based investigation has, additionally, estimated increased Hcy levels in the cerebrospinal fluid of subjects with MA (Isobe & Terayama, 2010).

In accordance with serum Hcy, serum folate has been determined lower among migraine patients (MA and MO) compared to healthy controls, whereas no differences have been revealed for serum B12 (Liampas, Siokas, Mentis, Aloizou, et al., 2020). As mentioned before, the latter finding could be attributed to the inferior value of vitamin B12 as a marker of its own functional status. Two case-control studies measuring serum and urine MMA (superior B12 index, similar to Hcy) demonstrated that MMA is higher in patients with migraine, additionally advocating for a functional deficiency of vitamin

B12 in migraine individuals (İpçioğlu, Özcan, Gültepe, Tekeli, & Şenol, 2008; Togha, Razeghi Jahromi, Ghorbani, Martami, & Seifishahpar, 2019).

Considering the whole body of evidence, folate and B12 deficiencies may be implicated in migraine, and their effects could be exerted through an increase in serum (or even CSF) Hcy. Despite the limited pediatric data, there is some published evidence of lower folate and B12 levels in children with migraine compared to healthy children (Aydin, Bucak, & Geyik, 2020; Bottini et al., 2006). Nonetheless, additional studies are required in pediatric populations in order to reach safer conclusions.

Folate and vitamin B12 supplementation for migraine prophylaxis

Menon et al. (2015) examined the effect of dietary folate on MA and revealed that increased folate consumption (also reflected in serum concentrations) has protective properties against migraine (reduces the frequency of migraine). Complementarily, Sadeghi, Maghsoudi, Khorvash, Ghiasvand, and Askari (2016) determined a lower dietary folate intake among patients with MA compared to healthy individuals. The results of diet-based studies were additionally supported by the results of clinical trials exploring the efficacy of Hcy-lowing vitamins in the prevention of migraine (Liampas, Siokas, Aloizou, Tsouris, et al., 2020).

Specifically, published evidence from clinical trials involving adults with MA suggests that the combination of folate (5 mg) and pyridoxine (80 mg) for 3 months, as well as the combination of folate (2 mg), vitamin B12 (400 μg), and pyridoxine (25 mg) for 6 months, may ameliorate the overall migraine burden (frequency, severity, duration, and induced disability). Moreover, the safety profile of these vitamins was very attractive, with occasional gastrointestinal events being recorded as the most frequent adverse event (namely gastroesophageal reflux or epigastralgia). However, folate supplementation alone (5 mg for 3 months) has not proven beneficial against migraine. Furthermore, contrary to the 2-mg folate regimens, the combination of 1-mg folate with vitamin B12 (400 μg) and pyridoxine (25 mg) for 6 months has not provided encouraging results. It seems that the interplay between genetic (related to the metabolic pathway of Hcy) and environmental factors is important in the determination of the migraine subgroup that ultimately benefits from the supplementation of folate and B12 and may partly explain these contradictory results (Lea, Colson, Quinlan, Macmillan, & Griffiths, 2009; Menon et al., 2012). Regardless, clinical trials have demonstrated that the combination of Hcy-lowering vitamins might exert a positive effect on MA, in adults.

On the other hand, the management of acute migraine attacks with folate or B12 supplementation has been long abandoned (Kopjas, 1969). In addition, the prophylactic use of folate and vitamin B12 in adult MO, as well as nonadult populations (MO and MA), is studied by only 3 uncontrolled, open-label trials. One of them assessed folate (5 mg) supplementation in children ($n = 16$) with MO (Di Rosa et al., 2007), while the second evaluated the intranasal administration of hydroxocobalamin (1 mg) to 19 adults with migraine (both MO and MA) (van der Kuy, Merkus, Lohman, ter Berg, & Hooymans, 2002). Finally, Jahromi and Togha (2016) assessed the use of Vitamin B12 (1 mg), along with pyridoxine (100 mg) and vitamin B1 (100 mg) in menstrually related migraine ($n = 30$). Despite the encouraging results, on the basis of the substantial placebo effect in migraine research (especially regarding children), it would not be prudent to extract any conclusions from these uncontrolled trials (Evers, Marziniak, Frese, & Gralow, 2009; van der Kuy & Lohman, 2002).

Homocysteine-induced vascular risk—Clinical implications

Migraine, most notably MA, is firmly linked to an elevated cardiovascular risk. Endothelial dysfunction and vasospasm, as well as hypercoagulability and microembolism, are among the potential mechanisms that explain this relationship (Gollion et al., 2020; Gryglas & Smigiel, 2017; Kumar et al., 2019; Sen et al., 2018; Tietjen & Collins, 2018). Hcy is considered to have a substantial contribution to the overall cardiovascular risk, especially among younger individuals (Madonna et al., 2002; Zhang et al., 2020). This age-group coincides with the age-group that is primarily affected by the MA-induced cardiovascular risk (Øie et al., 2020). The migraine-stroke demographics are, therefore, in agreement with a potential implication of Hcy in the increased vascular risk of individuals with migraine.

The elevated vascular risk conferred by Hcy is associated with the induction of endothelial injury, platelet aggregation, and atherosclerosis (Harker, Ross, Slichter, & Scott, 1976; Wall, Harlan, Harker, & Striker, 1980). Folate and vitamin B12 (pyridoxine as well) can lower Hcy levels and protect from its detrimental vascular effects (Kaye et al., 2020; Spence, 2016). Of note, elevated serum Hcy should be routinely treated with B12 supplementation, despite the presence of normal folate and B12 serum levels (Spence, 2016). Given the overall protective effect of Hcy-lowering vitamins in the prevention of stroke, the value of the routine supplementation of Hcy-lowering vitamins to individuals with MA should be evaluated in parallel with the modification of the long-term vascular risk (Martí-Carvajal, Solà, Lathyris, & Dayer, 2017). Additionally, the role of serum Hcy screening and individualized supplementation of Hcy-lowering vitamins in patients with MA should be investigated as it may be related to superior outcomes. As mentioned earlier, genetic-environmental interactions appear to play an important part in the determination of the migraine subgroup that benefits from the supplementation of Hcy-lowering vitamins and should be taken into account in the translation of the results.

Folate-B12 deficiency and homocysteine in the pathogenesis of migraine

The potential underlying mechanisms connecting Hcy, as well as folate and B12 deficiency, with migraine are summarized in Fig. 1. CSD, the neurophysiological equivalent

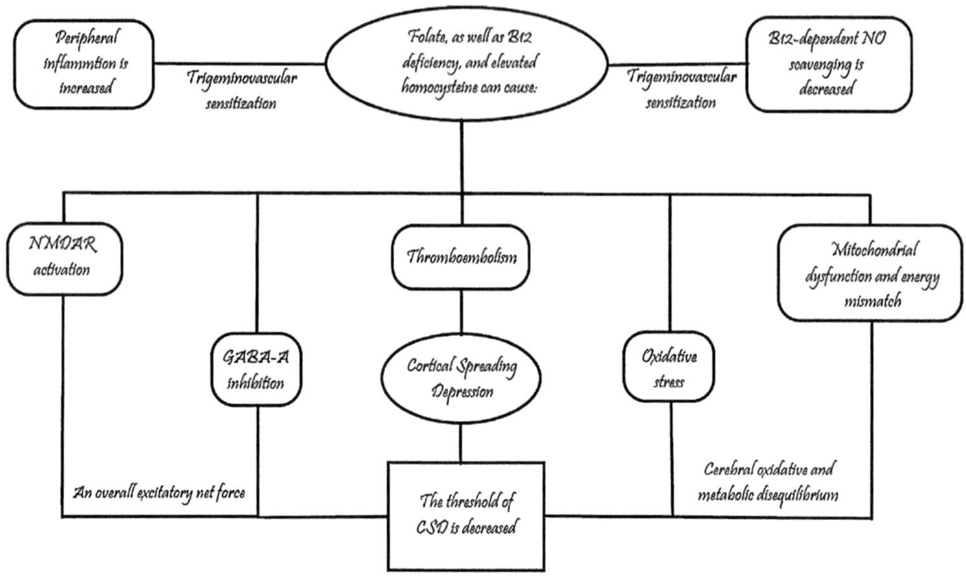

Fig. 1 Schematic representation depicting the potential role of folate and vitamin B12 deficiency in the pathogenesis of migraine. *CSD*: cortical spreading depression, *GABA*: gamma-aminobutyric acid, *NMDAR*: *N*-methyl-D-aspartic receptor, *NO*: nitric oxide.

of aura, can be reproduced by microembolization (Nozari et al., 2010). Therefore, hyperhomocysteinemia could be implicated in the thromboembolic induction of CSD which theoretically ignites the pathophysiological cascade of a migraine episode. In this context, Hcy could be part of a multifactorial panel of prothrombotic and microembolic conditions which may account for the overall thrombophilic predisposition of individuals with MA, compared to the general population (Gollion et al., 2020; Kumar et al., 2019; Liampas, Mylonas, et al., 2021; Sen et al., 2018).

Glutamate-activated NMDARs are important in the initiation and propagation of CSD (Close et al., 2019). Homocysteic acid and cysteine sulfinic acid are two neurotoxic oxidation by-products of the Hcy-methionine metabolic pathway. Both molecules are increased in hyperhomocysteinemia and constitute potent glutamic acid analogs, with NMDAR-activating properties (Bhatia & Singh, 2015). Therefore, elevated serum Hcy could contribute to the NMDA-mediated generation and spreading of CSD. On the other hand, both folate and vitamin B12 have proved beneficial in the amelioration of the NMDA-mediated excitotoxicity, which might in part explain their beneficial effects on MA (Akaike, Tamura, Sato, & Yokota, 1993; Lin, Desbois, Jiang, & Hou, 2004). Of interest, the disturbed homeostasis of these excitatory amino acids could in part account for the substantial psychiatric burden of migraine, as well (Santhosh-Kumar, Hassell, Deutsch, & Kolhouse, 1994).

Hcy, additionally, inhibits the gamma-aminobutyric acid (GABA)-A receptor (Lominadze, Tyagi, Sen, Ovechkin, & Tyagi, 2012). The activation of GABA-A receptors counteracts the excitatory neurotransmission; therefore, the inhibitory effect of Hcy on GABA-A ultimately produces an excitatory-excitotoxic net force (analogous to the activation of NMDARs). It has been argued that GABA-ergic neurotransmission modulates cortical excitability, especially in the visual cortex (which is most often implicated in MA) and may, therefore, be involved in the inhibition of CSD (Holland, Akerman, & Goadsby, 2010; Palmer, Chronicle, Rolan, & Mulleners, 2000). The activation of the GABA-A receptors by some efficient preventive medications in migraine, additionally, implies their potential involvement in its pathogenesis. Overall, the NMDA-GABA mediated cortical excitatory effects of Hcy could explain its risk-conferring nature based on the facilitation of CSD (Kuzniecky et al., 2002; Palmer et al., 2000).

Accumulating evidence, also, indicates that migraine may be a sequela of cerebral energy deprivation and oxidative stress (Gross, Lisicki, Fischer, Sándor, & Schoenen, 2019). It is hypothesized that cerebral metabolic and oxidative homeostasis is involved in migraine by affecting the threshold for the induction of CSD. In this context, the oxidative properties of Hcy could be involved in the induction of CSD (Guest et al., 2015; Moretti & Caruso, 2019). At the same time, the Hcy-dependent, as well as -independent, antioxidant properties of folate and vitamin B12 may account for their prophylactic effects (Joshi, Adhikari, Patro, Chattopadhyay, & Mukherjee, 2001; van de Lagemaat, de Groot, & van den Heuvel, 2019). Folate and vitamin B12 (pyridoxine as well) could also prevent migraine episodes by improving mitochondrial dysfunction and ameliorating energy metabolism (Depeint, Bruce, Shangari, Mehta, & O'Brien, 2006).

Regarding the peripheral trigeminovascular constituent of migraine, neurogenic inflammation and vasodilation are considered of central importance (Ashina et al., 2019). Among others, NO is an important vasoactive peptide that assumes a key role in these processes. Circulating B12 in the form of hydroxocobalamin presents substantial scavenging properties against NO (Rajanayagam, Li, & Rand, 1993). Additionally, Hcy itself has inflammatory properties and could potentially increase the individual susceptibility towards migraine (Moretti & Caruso, 2019). On the other hand, vitamin B12 has substantial antinociceptive properties against orofacial pain of inflammatory origin, further supporting the potential importance of the peripheral actions of Hcy-lowering vitamins in migraine (Erfanparast et al., 2014).

Finally, it has also been argued that the elevation of serum Hcy may be indirectly associated with migraine, on the basis of nonsteroidal antiinflammatory drug (NSAID) abuse (Gabriel, Jaakkimainen, & Bombardier, 1991). Gastric mucus damage and intrinsic factor deficiency caused by NSAIDs could account for a B12-dependent Hcy elevation. Nonetheless, the beneficial effects of Hcy-lowering vitamins supplementation on migraine come in contradiction with the aforementioned narrative of an indirect association.

Applications to other primary headache disorders

Folate and B12 are two crucial vitamins with multidimensional effects on the nervous system. Cognitive impairment, cerebrovascular disease, psychiatric, and sensory disorders are among the various neurologic complications of folate and B12 deficiency (Moretti & Caruso, 2019). In this chapter, we reviewed the evidence for the connection between folate, vitamin B12, and migraine, while the potential underlying mechanisms of this connection were described. Published evidence was indicative of lower folate and functional B12 (i.e., higher Hcy and MMA) levels among patients with migraine compared to healthy controls (a difference mainly driven by MA), while the supplementation of Hcy-lowering vitamins to individuals with MA could have prophylactic properties. The stronger relationship of hyperhomocysteinemia with MA might partly account for MA's prominent vascular aspects, compared to MO.

On the other hand, despite the differential association of MO with higher serum Hcy compared to healthy individuals (a weaker association in comparison with MA), the supplementation of Hcy-lowering vitamins for MO has been substantially understudied (Liampas, Siokas, Aloizou, Tsouris, et al., 2020; Liampas, Siokas, Mentis, Aloizou, et al., 2020). However, the complex hypothesized pathogenetic background of migraine, involving peripheral inflammation and vasodilation, as well as cerebral metabolic and energy disequilibrium, warrants further investigation of folate-B12 supplementation to patients with MO. Of note, although retrieved evidence for non-adult populations was exceptionally limited (MA and MO), there are a few studies that support the potential extrapolation of these findings in children too (Aydin et al., 2020; Bottini et al., 2006).

Aside from migraine, preliminary evidence suggests the involvement of vitamin B12 deficiency in a pediatric tension-type headache which consists of the commonest primary headache disorder (Calik et al., 2018). In parallel, there is evidence for the involvement of Hcy, folate, and B12 in psychiatric disorders (Esnafoglu & Ozturan, 2020). Although the neurophysiological background of tension-type headache remains vague, an important psychogenic component is implicated, while a bidirectional relationship is theorized between migraine and psychological disturbances, as well (Liampas, Siokas, et al., 2021; Liampas et al., 2022). This two-way association of headache and psychogenic disorders may be in part related to folate and B12 deficiencies, and the investigation of Hcy-lowering vitamins supplementation to patients with headache and psychological comorbidities might provide superior results. In any case, the value of individualized approaches based on the combination of clinical criteria (presence of aura, psychogenic comorbidities, and so on) with laboratory indices (Hcy and MMA levels, folate and B12 values or genetic biomarkers)in the prevention of primary headache disorders with Hcy-lowering vitamins remains to be determined by future studies.

Other components of interest

Aside from folate and B12, a great number of additional nutrients, nutraceuticals, and dietary approaches have proven useful in the prevention of migraine. For starters, patients with migraine frequently report the emergence of an episode following the exposure to specific dietary triggers such as caffeine, alcohol (most importantly red wine), food seasoning, dairy products, chocolate, and skipping regular meals. Avoiding such triggers is often an efficient strategy to reduce the number of migraine attacks. Moreover, dietary patterns appear to affect the burden of the disease. Fixed scheduling of meals distributed throughout the day (without important fasting intervals), late night-time eating, and "healthy" eating habits present a positive impact, while shifting eating hours, western dietary patterns, as well as consumption of fried food, have the opposite effect. Finally, dietary interventions can ameliorate migraine parameters with low-lipid, ketogenic, and elimination diets constituting efficacious protective approaches (Hindiyeh et al., 2020). Overall, considering the interlink between migraine, obesity, inflammation, energy metabolism, and oxidative stress, dietary approaches (lifestyle interventions as well) are probably warranted in the management of migraine.

At the same time, nutraceuticals are becoming an increasingly popular option, partly owing to their favorable safety profiles. Improvement of the mitochondrial function and energy metabolism, modulation of inflammatory responses, and attenuation of oxidative stress are among the potential protective mechanisms. Riboflavin, coenzyme Q10, magnesium, vitamin D, melatonin, herbal preparations, and several other less-studied complementary approaches are among the complementary approaches that tend to be integrated into the management of migrainous headaches (D'Onofrio, Raimo, Spitaleri, Casucci, & Bussone, 2017; Liampas, Siokas, Brotis, & Dardiotis, 2020; Liampas, Siokas, Brotis, Vikelis, & Dardiotis, 2020). In view of their relatively rare and mild adverse actions, alternative treatments could be also examined as part of combination regimens with other traditional antimigraine medications.

Mini-dictionary of terms

- **Homocysteine.** A sulfur-containing amino acid that is converted into methionine or cysteine through folate-B12 or pyridoxine-dependent pathways, respectively.
- **CGRP and Substance P.** Vasoactive neuropeptides with important nociceptive properties, implicated in the transmission of pain.
- **Nitric oxide.** A highly reactive, endothelium-derived free radical with transient vasodilatory autocrine and paracrine properties.
- **Folate.** Also known as vitamin B9, an essential water-soluble nutrient mainly obtained from fruits and vegetables, in the form of folate, or from fortified food, in the synthetic form of folic acid.

- **Vitamin B12.** Also known as cobalamin, an essential water-soluble nutrient mainly obtained from animal-sourced food, dietary supplements, or parenteral formulations (hydroxocobalamin and cyanocobalamin).
- **Glutamate.** The anion of the amino acid glutamic acid which acts as an excitatory neurotransmitter.
- **NMDA receptors.** A class of glutamate receptors that act as cation channels. Their activation leads to an excitatory effect.
- **GABA-A receptors.** Gamma-aminobutyric acid receptors that act as anion channels. Their activation leads to an inhibitory effect.

Key facts of migraine

- Migraine has two major types: migraine without aura and migraine with aura, which is usually preceded by the transient focal neurological symptoms, occurring minutes to hours before the headache.
- Migraine attacks usually last for 4–72h when untreated (or unsuccessfully treated).
- Migraine usually manifests with unilateral headaches of pulsating quality, moderate to severe intensity and are worsened by routine physical activity. Nausea and/or vomiting, photophobia, and phonophobia often concur.
- Prodromal symptoms are subtle changes that may herald an upcoming migraine episode, occurring hours to days before the headache and including (among others) fatigue, depression, irritability, cravings for particular foods, repetitive yawning, neck stiffness and/or pain, light and sound sensitivity.
- Following the recession of a migraine attack, a postdrome phase (migraine hangover) may occur with fatigue, depressed or euphoric mood, or cognitive dysfunction.

Summary points

- Folate and functional B12 deficiencies (elevated serum homocysteine and methylmalonic acid) appear to be associated with migraine, especially with aura.
- The supplementation of combination regimens of homocysteine-lowering vitamins has demonstrated a prophylactic effect against migraine with aura.
- The prothrombotic properties of homocysteine may contribute to the microembolic induction of cortical spreading depression and the elevated cardiovascular risk of patients suffering from migraine with aura.
- Important homocysteine–NMDA and GABA-A receptor interactions may facilitate the emergence and propagation of cortical depression.
- Homocysteine-induced mitochondrial dysfunction and oxidative stress could lower the threshold of cortical depression.

- Folate and vitamin B12 supplementation may counteract the above-listed effects of homocysteine and, additionally, confer several homocysteine-independent benefits.
- Peripheral actions including the NO scavenging properties of cobalamin, as well as the antiinflammatory effect of homocysteine-lowering vitamins, may in part account for the beneficial effect of folate-B12 supplementation.

References

Akaike, A., Tamura, Y., Sato, Y., & Yokota, T. (1993). Protective effects of a vitamin B12 analog, methylcobalamin, against glutamate cytotoxicity in cultured cortical neurons. *European Journal of Pharmacology, 241*(1), 1–6.

Ashina, M., Hansen, J. M., Do, T. P., Melo-Carrillo, A., Burstein, R., & Moskowitz, M. A. (2019). Migraine and the trigeminovascular system-40 years and counting. *Lancet Neurology, 18*(8), 795–804.

Aydin, H., Bucak, I. H., & Geyik, M. (2020). Vitamin B12 and folic acid levels in pediatric migraine patients. *Acta Neurologica Belgica, 121*(6), 1741–1744.

Bhatia, P., & Singh, N. (2015). Homocysteine excess: Delineating the possible mechanism of neurotoxicity and depression. *Fundamental & Clinical Pharmacology, 29*(6), 522–528.

Bottini, F., Celle, M. E., Calevo, M. G., Amato, S., Minniti, G., Montaldi, L., et al. (2006). Metabolic and genetic risk factors for migraine in children. *Cephalalgia, 26*(6), 731–737.

Calik, M., Aktas, M. S., Cecen, E., Piskin, I. E., Ayaydın, H., Ornek, Z., et al. (2018). The association between serum vitamin B_{12} deficiency and tension-type headache in Turkish children. *Neurological Sciences, 39*(6), 1009–1014.

Close, L. N., Eftekhari, S., Wang, M., Charles, A. C., & Russo, A. F. (2019). Cortical spreading depression as a site of origin for migraine: Role of CGRP. *Cephalalgia, 39*(3), 428–434.

Depeint, F., Bruce, W. R., Shangari, N., Mehta, R., & O'Brien, P. J. (2006). Mitochondrial function and toxicity: Role of B vitamins on the one-carbon transfer pathways. *Chemico-Biological Interactions, 163*(1–2), 113–132.

Di Rosa, G., Attinà, S., Spanò, M., Ingegneri, G., Sgrò, D. L., Pustorino, G., et al. (2007). Efficacy of folic acid in children with migraine, hyperhomocysteinemia and MTHFR polymorphisms. *Headache, 47*(9), 1342–1344.

D'Onofrio, F., Raimo, S., Spitaleri, D., Casucci, G., & Bussone, G. (2017). Usefulness of nutraceuticals in migraine prophylaxis. *Neurological Sciences, 38*(Suppl 1), 117–120.

Dreier, J. P., & Reiffurth, C. (2015). The stroke-migraine depolarization continuum. *Neuron, 86*(4), 902–922.

Erfanparast, A., Escort, M., Tamaddonfard, E., Maroufi, S., Kazemi-Shojaei, S., Dabbaghi, M., et al. (2014). Systemic and local peripheral injections of vitamin B12 suppressed orofacial nociception induced by formalin in rats. *Drug Research, 64*(2), 85–90.

Esnafoglu, E., & Ozturan, D. D. (2020). The relationship of severity of depression with homocysteine, folate, vitamin B12, and vitamin D levels in children and adolescents. *Child and Adolescent Mental Health, 25*(4), 249–255.

Evers, S., Marziniak, M., Frese, A., & Gralow, I. (2009). Placebo efficacy in childhood and adolescence migraine: An analysis of double-blind and placebo-controlled studies. *Cephalalgia, 29*(4), 436–444.

Froese, D. S., Fowler, B., & Baumgartner, M. R. (2019). Vitamin B_{12}, folate, and the methionine remethylation cycle-biochemistry, pathways, and regulation. *Journal of Inherited Metabolic Disease, 42*(4), 673–685.

Gabriel, S. E., Jaakkimainen, L., & Bombardier, C. (1991). Risk for serious gastrointestinal complications related to use of nonsteroidal anti-inflammatory drugs. A meta-analysis. *Annals of Internal Medicine, 115*(10), 787–796.

GBD 2016 Headache Collaborators. (2018). Global, regional, and national burden of migraine and tension-type headache, 1990-2016: A systematic analysis for the global burden of disease study 2016. *Lancet Neurology, 17*(11), 954–976.

Gollion, C., Gazagnes, J., Fabry, V., Barbieux-Guillot, M., Lerebours, F., & Larrue, V. (2020). Atrial fibrillation and migraine with aura in young adults with ischemic stroke. *Cephalalgia*, *41*(3), 375–382.

Gross, E. C., Lisicki, M., Fischer, D., Sándor, P. S., & Schoenen, J. (2019). The metabolic face of migraine—From pathophysiology to treatment. *Nature Reviews. Neurology*, *15*(11), 627–643.

Gryglas, A., & Smigiel, R. (2017). Migraine and stroke: What's the link? What to do? *Current Neurology and Neuroscience Reports*, *17*(3), 22.

Guest, J., Bilgin, A., Hokin, B., Mori, T. A., Croft, K. D., & Grant, R. (2015). Novel relationships between B12, folate and markers of inflammation, oxidative stress and NAD(H) levels, systemically and in the CNS of a healthy human cohort. *Nutritional Neuroscience*, *18*(8), 355–364.

Hannibal, L., Lysne, V., Bjørke-Monsen, A. L., Behringer, S., Grünert, S. C., Spiekerkoetter, U., et al. (2016). Biomarkers and algorithms for the diagnosis of vitamin B12 deficiency. *Frontiers in Molecular Biosciences*, *3*, 27.

Harker, L. A., Ross, R., Slichter, S. J., & Scott, C. R. (1976). Homocystine-induced arteriosclerosis. The role of endothelial cell injury and platelet response in its genesis. *The Journal of Clinical Investigation*, *58*(3), 731–741.

Headache Classification Committee of the International Headache Society (IHS). (2018). The international classification of headache disorders, 3rd edition. *Cephalalgia*, *38*(1), 1–211.

Held, C., Sumner, G., Sheridan, P., McQueen, M., Smith, S., Dagenais, G., et al. (2008). Correlations between plasma homocysteine and folate concentrations and carotid atherosclerosis in high-risk individuals: Baseline data from the homocysteine and atherosclerosis reduction trial (HART). *Vascular Medicine*, *13*(4), 245–253.

Hindiyeh, N. A., Zhang, N., Farrar, M., Banerjee, P., Lombard, L., & Aurora, S. K. (2020). The role of diet and nutrition in migraine triggers and treatment: A systematic literature review. *Headache*, *60*(7), 1300–1316.

Holland, P. R., Akerman, S., & Goadsby, P. J. (2010). Cortical spreading depression-associated cerebral blood flow changes induced by mechanical stimulation are modulated by AMPA and GABA receptors. *Cephalalgia*, *30*(5), 519–527.

İpçioğlu, O. M., Özcan, Ö., Gültepe, M., Tekeli, H., & Şenol, M. G. (2008). Functional vitamin B12 deficiency represented by elevated urine methylmalonic acid levels in patients with migraine. *Turkish Journal of Medical Sciences*, *38*, 409–414.

Isobe, C., & Terayama, Y. (2010). A remarkable increase in total homocysteine concentrations in the CSF of migraine patients with aura. *Headache*, *50*(10), 1561–1569.

Jahromi, S. R., & Togha, M. (2016). Vitamin B compound could be effective in menstrual related migraine. *Cephalalgia*, *36*(1S), 105–106.

Joshi, R., Adhikari, S., Patro, B. S., Chattopadhyay, S., & Mukherjee, T. (2001). Free radical scavenging behavior of folic acid: Evidence for possible antioxidant activity. *Free Radical Biology & Medicine*, *30*(12), 1390–1399.

Kaye, A. D., Jeha, G. M., Pham, A. D., Fuller, M. C., Lerner, Z. I., Sibley, G. T., et al. (2020). Folic acid supplementation in patients with elevated homocysteine levels. *Advances in Therapy*, *37*(10), 4149–4164.

Kopjas, T. L. (1969). The use of folic acid in vascular headache of the migraine type. *Headache*, *8*(4), 167–170.

Kumar, P., Kijima, Y., West, B. H., & Tobis, J. M. (2019). The connection between patent foramen Ovale and migraine. *Neuroimaging Clinics of North America*, *29*(2), 261–270.

Kuzniecky, R., Ho, S., Pan, J., Martin, R., Gilliam, F., Faught, E., et al. (2002). Modulation of cerebral GABA by topiramate, lamotrigine, and gabapentin in healthy adults. *Neurology*, *58*(3), 368–372.

Lea, R., Colson, N., Quinlan, S., Macmillan, J., & Griffiths, L. (2009). The effects of vitamin supplementation and MTHFR (C677T) genotype on homocysteine-lowering and migraine disability. *Pharmacogenetics and Genomics*, *19*(6), 422–428.

Liampas, I., Bourlios, S., Siokas, V., Aloizou, A.-M., Dervenis, P., Nasios, G., et al. (2022). Vitamin D and tension-type headache: Causal association or epiphenomenon? *International Journal of Neuroscience*, 1–11. https://doi.org/10.1080/00207454.2022.2110495.

Liampas, I., Mylonas, K. S., Brotis, A., Dervenis, P., Siokas, V., Mentis, A. A., et al. (2021). Serum lipid abnormalities in migraine: A meta-analysis of observational studies. *Headache*. https://doi.org/10.1111/head.14039 (Epub ahead of print).

Liampas, I. N., Siokas, V., Aloizou, A. M., Tsouris, Z., Dastamani, M., Aslanidou, P., et al. (2020). Pyridoxine, folate and cobalamin for migraine: A systematic review. *Acta Neurologica Scandinavica*, *142*(2), 108–120.

Liampas, I., Siokas, V., Brotis, A., & Dardiotis, E. (2020). Vitamin D serum levels in patients with migraine: A meta-analysis. *Revue Neurologique*, *176*(7–8), 560–570. 2020.

Liampas, I., Siokas, V., Brotis, A., Mentis, A.-A., Aloizou, A.-M., Dastamani, M., et al. (2021). Endogenous melatonin levels and therapeutic use of exogenous melatonin in tension type headache: A systematic review. *Revue Neurologique (Paris)*, *177*(8), 871–880. https://doi.org/10.1016/j.neurol.2021.01.014.

Liampas, I., Siokas, V., Brotis, A., Vikelis, M., & Dardiotis, E. (2020). Endogenous melatonin levels and therapeutic use of exogenous melatonin in migraine: Systematic review and meta-analysis. *Headache*, *60*(7), 1273–1299.

Liampas, I., Siokas, V., Mentis, A. A., Aloizou, A. M., Dastamani, M., Tsouris, Z., et al. (2020). Serum homocysteine, pyridoxine, folate, and vitamin B12 levels in migraine: Systematic review and meta-analysis. *Headache*, *60*, 1508–1534.

Lin, Y., Desbois, A., Jiang, S., & Hou, S. T. (2004). Group B vitamins protect murine cerebellar granule cells from glutamate/NMDA toxicity. *Neuroreport*, *15*(14), 2241–2244.

Lominadze, D., Tyagi, N., Sen, U., Ovechkin, A., & Tyagi, S. C. (2012). Homocysteine alters cerebral microvascular integrity and causes remodelling by antagonizing GABA-A receptor. *Molecular and Cellular Biochemistry*, *371*(1–2), 89–96.

Madonna, P., de Stefano, V., Coppola, A., Cirillo, F., Cerbone, A. M., Orefice, G., et al. (2002). Hyperhomocysteinemia and other inherited prothrombotic conditions in young adults with a history of ischemic stroke. *Stroke*, *33*(1), 51–56.

Martí-Carvajal, A. J., Solà, I., Lathyris, D., & Dayer, M. (2017). Homocysteine-lowering interventions for preventing cardiovascular events. *Cochrane Database of Systematic Reviews*, *8*(8), CD006612.

Menon, S., Lea, R. A., Ingle, S., Sutherland, M., Wee, S., Haupt, L. M., et al. (2015). Effects of dietary folate intake on migraine disability and frequency. *Headache*, *55*(2), 301–309.

Menon, S., Lea, R. A., Roy, B., Hanna, M., Wee, S., Haupt, L. M., et al. (2012). Genotypes of the MTHFR C677T and MTRR A66G genes act independently to reduce migraine disability in response to vitamin supplementation. *Pharmacogenetics and Genomics*, *22*(10), 741–749.

Moretti, R., & Caruso, P. (2019). The controversial role of homocysteine in neurology: From labs to clinical practice. *International Journal of Molecular Sciences*, *20*(1), 231.

Nozari, A., Dilekoz, E., Sukhotinsky, I., Stein, T., Eikermann-Haerter, K., Liu, C., et al. (2010). Microemboli may link spreading depression, migraine aura, and patent foramen ovale. *Annals of Neurology*, *67*(2), 221–229.

Øie, L. R., Kurth, T., Gulati, S., & Dodick, D. W. (2020). Migraine and risk of stroke. *Journal of Neurology, Neurosurgery, and Psychiatry*, *91*(6), 593–604.

Palmer, J. E., Chronicle, E. P., Rolan, P., & Mulleners, W. M. (2000). Cortical hyperexcitability is cortical under-inhibition: Evidence from a novel functional test of migraine patients. *Cephalalgia*, *20*(6), 525–532.

Rajanayagam, M. A., Li, C. G., & Rand, M. J. (1993). Differential effects of hydroxocobalamin on NO-mediated relaxations in rat aorta and anococcygeus muscle. *British Journal of Pharmacology*, *108*(1), 3–5.

Russell, M. B., Rasmussen, B. K., Thorvaldsen, P., & Olesen, J. (1995). Prevalence and sex-ratio of the subtypes of migraine. *International Journal of Epidemiology*, *24*(3), 612–618.

Sadeghi, O., Maghsoudi, Z., Khorvash, F., Ghiasvand, R., & Askari, G. (2016). Assessment of pyridoxine and folate intake in migraine patients. *Advanced Biomedical Research*, *5*, 47.

Santhosh-Kumar, C. R., Hassell, K. L., Deutsch, J. C., & Kolhouse, J. F. (1994). Are neuropsychiatric manifestations of folate, cobalamin and pyridoxine deficiency mediated through imbalances in excitatory sulfur amino acids? *Medical Hypotheses*, *43*(4), 239–244.

Sen, S., Androulakis, X. M., Duda, V., Alonso, A., Chen, L. Y., Soliman, E. Z., et al. (2018). Migraine with visual aura is a risk factor for incident atrial fibrillation: A cohort study. *Neurology*, *91*(24), e2202–e2210.

Spence, J. D. (2016). Homocysteine lowering for stroke prevention: Unravelling the complexity of the evidence. *International Journal of Stroke*, *11*(7), 744–747.

Stuart, S., Cox, H. C., Lea, R. A., & Griffiths, L. R. (2012). The role of the MTHFR gene in migraine (2012). *Headache, 52*(3), 515–520.

Tietjen, G. E., & Collins, S. A. (2018). Hypercoagulability and migraine. *Headache, 58*(1), 173–183.

Togha, M., Razeghi Jahromi, S., Ghorbani, Z., Martami, F., & Seifishahpar, M. (2019). Serum vitamin B12 and methylmalonic acid status in migraineurs: A case-control study. *Headache, 59*, 1492–1503.

van de Lagemaat, E. E., de Groot, L. C. P. G. M., & van den Heuvel, E. G. H. M. (2019). Vitamin B_{12} in relation to oxidative stress: A systematic review. *Nutrients, 11*(2), 482.

van der Kuy, P. H., & Lohman, J. J. (2002). A quantification of the placebo response in migraine prophylaxis. *Cephalalgia, 22*(4), 265–270.

van der Kuy, P. H., Merkus, F. W., Lohman, J. J., ter Berg, J. W., & Hooymans, P. M. (2002). Hydroxocobalamin, a nitric oxide scavenger, in the prophylaxis of migraine: An open, pilot study. *Cephalalgia, 22*(7), 513–519.

Wall, R. T., Harlan, J. M., Harker, L. A., & Striker, G. E. (1980). Homocysteine-induced endothelial cell injury in vitro: A model for the study of vascular injury. *Thrombosis Research, 18*(1–2), 113–121.

Zhang, X., Levy, D., Noseda, R., Kainz, V., Jakubowski, M., & Burstein, R. (2010). Activation of meningeal nociceptors by cortical spreading depression: Implications for migraine with aura. *The Journal of Neuroscience, 30*(26), 8807–8814.

Zhang, N., Zhang, L., Wang, Q., Zhao, J., Liu, J., & Wang, G. (2020). Cerebrovascular risk factors associated with ischemic stroke in a young non-diabetic and non-hypertensive population: A retrospective case-control study. *BMC Neurology, 20*(1), 424.

CHAPTER 20

Vitamin D and migraine

Parisa Gazerani[a,b]
[a]Department of Life Sciences and Health, Faculty of Health Sciences, Oslo Metropolitan University, Oslo, Norway
[b]Department of Health Science and Technology, Faculty of Medicine, Aalborg University, Aalborg, Denmark

Abbreviations

1,25(OH)2D 1,25-dihydroxyvitamin D
25(OH)D 25-hydroxyvitamin D
ALZ Alzheimer's disease
BMI body mass index
CGRP calcitonin gene-related peptide
Ca calcium
CLIA chemiluminescence immunoassays
CM chronic migraine
CH cluster headache
EIA enzyme-linked immunosorbent assay
EM episodic migraine
GABA gamma aminobutyric acid
HIT-6 headache impact test-6
HSE high solar-exposure
HPLC-UV high-performance liquid chromatography-ultraviolet
HD hypovitaminosis D
iNOS inducible nitric oxide synthase
ICHD-3 International Classification of Headache Disorders version 3
IU International unit
LC–MS/MS liquid chromatography coupled with tandem mass spectrometry
LSE low solar-exposure
MMPs matrix metalloproteinases
MOH medication overuse headache
MIDAS migraine disability assessment test
MIGSEV migraine severity scale
MS multiple sclerosis
NO nitric oxide
PD Parkinson's disease
RRMS relapsing-remitting multiple sclerosis
TTH tension-type headaches
VDR vitamin D receptor

Vitamins and Minerals in Neurological Disorders
https://doi.org/10.1016/B978-0-323-89835-5.00016-8

Introduction

Primary headaches

Headache is a common complaint and a frequent reason for seeking medical care. According to the third edition of the International Classification of Headache Disorders (ICHD-3), primary headaches are defined as those without a definitive underlying cause, while secondary headaches result from a specific etiology (Headache Classification Committee of the International Headache Society (IHS), 2018). Migraine (14.4%) and tension-type headaches (TTH, 26.1%) are the two most prevalent types of primary headaches (Collaborators, 2018). Migraine is manifested as a severe unilateral, pulsing headache that lasts 4–72h and often with accompanying symptoms such as nausea, sensitivity to light and sound, and cutaneous allodynia (Collaborators, 2018). TTH, however, is sensed mostly as a pressure-like headache bilaterally around the head, often without other symptoms. Cluster headache (CH) is a short-lasting severe primary headache with typical dysautonomia features such as eye tearing and ptosis (Wei, Yuan Ong, & Goadsby, 2018). Treatment of primary headache disorders is challenging and currently consists of abortive (i.e., to stop headaches) and prophylactic therapy (i.e., to reduce headache frequency and severity, or disease progression). Besides pharmacotherapy, several other strategies have been introduced, including neurostimulation, cognitive therapy, acupuncture, supplementary treatments with, e.g., minerals and vitamins (Nowaczewska, Wicinski, Osinski, & Kazmierczak, 2020). Modification of lifestyle, diet, smoking, sleep, and exercise habits have been promoted before or alongside the treatment strategies (Gazerani, 2020; Martin & Vij, 2016; Nattagh-Eshtivani et al., 2018; Orr, 2018).

Migraine

Recurrent migraines affect 6% of men and 18% of women (Nattagh-Eshtivani et al., 2018). Migraine has a significant impact on patient's quality of life (Pradeep, Nemichandra, Harsha, & Radhika, 2020), is ranked number 2 among the causes of disability, and is the first among young females (Steiner et al., 2020). Intractable forms of migraine are often resistant to treatment, and other approaches, including the nutritional approach, have been proposed to overcome the challenges (Puledda, Messina, & Goadsby, 2017). In addition, a transition from acute to chronic type and complexity of migraine that often accompanies by the medication overuse headache (MOH) (Goadsby, 2016) have emphasized the value of nondrug strategies, e.g., considering lifestyle modification (Bigal & Lipton, 2006; Buse, Greisman, Baigi, & Lipton, 2019).

The diagnosis of migraine remains mostly clinical, and it is based on the latest version of the headache classification, where subclassification of migraine (e.g., chronic migraine (CM), episodic migraine (EM), migraine with aura, migraine without aura) is also present. Aura occurs in one-third of patients with migraines, and it is manifested with transient neurological symptoms, e.g., visual aura. Migraine with aura may respond differently to acute and preventive therapies (Hansen & Charles, 2019).

The pathophysiology of migraine is not completely understood; however, neurogenic inflammation, altered pain processing, cortical hyperexcitability, and central sensitization have been proposed (Charles, 2018; Pozo-Rosich, Coppola, Pascual, & Schwedt, 2020; Ramachandran, 2018; Schwedt, 2014). Progress has been made in the identification of neurotransmitters, hormones, and other biochemical elements in the pathophysiology of migraine, and potential biomarkers for diagnosis of migraine and response to treatments have been suggested (Gazerani, 2019). For example, the novel CGRP-based therapeutics (Do, Guo, & Ashina, 2019) that are targeting the CGRP pathway in migraine (Edvinsson, 2018) were based on the identification of the key molecule, CGRP, that permitted for the identification of targets and development of mechanism-based treatments.

Besides genetic factors and several environmental elements have been known to associate with migraine (Mulder et al., 2003), including nutrients, vitamins, and minerals (Nattagh-Eshtivani et al., 2018). Among those, vitamin D has been investigated following the recognition of its important roles in physiological activities that might be impaired or altered in migraine (Patel et al., 2019). Therefore, it was assumed that lack or insufficiency of vitamin D can influence migraine negatively and supplementation of vitamin D would affect it positively (Nowaczewska et al., 2020).

Vitamin D

Vitamin D (Deluca, 2014) is a fat-soluble molecule (Holick, 2007) that is known as a prohormone produced following skin exposure to sunlight (Nair & Maseeh, 2012) or taken via diet (Polzonetti, Pucciarelli, Vincenzetti, & Polidori, 2020). Naturally, vitamin D can be found as ergocalciferol (D2, mainly present in plants or their products (Japelt & Jakobsen, 2013)) or cholecalciferol (D3, in animal sources (Schmid & Walther, 2013)) (Wacker & Holick, 2013). Vitamin D plays a critical role in calcium (Ca) homeostasis and bone health (Antonucci, Locci, Clemente, Chicconi, & Antonucci, 2018). Vitamin D is hydroxylated to 25-hydroxyvitamin D (25(OH)D) in the liver and to the active form 1,25-dihydroxyvitamin D (1,25(OH)2D) in the kidneys (Jorde & Grimnes, 2018). 25(OH)D is used as a marker of vitamin D status in an individual (Holick, 2009). Seasonal alterations for vitamin D, increased in summer and decreased in winter, have been reported (Klingberg, Olerod, Konar, Petzold, & Hammarsten, 2015). This variation might depend on the latitude, with higher vitamin D in latitudes close to the equator (Ladizesky et al., 1995). Sun exposure or ultraviolet B (UVB) light helps in the absorption of vitamin D. This might be an issue for Nordic countries and other areas with limited UVB exposure (Virtanen et al., 2017). However, vitamin D deficiency has also been reported in sunny regions (Fields, Trivedi, Horton, & Mechanick, 2011).

The binding of vitamin D to vitamin D receptor (VDR), a member of the nuclear transcription factors (Moon et al., 2013), initiates its effect that is classified as genomic effects via VDR, and nongenomic effects by activating some intracellular pathways, e.g., cyclic AMP, protein kinase-A, phospholipase-C, and MAP kinase (Fernandes de Abreu, Eyles, & Feron, 2009). Basic animal research has demonstrated a diverse localization of VDR, for example, in central and peripheral nervous systems (Eyles, Smith, Kinobe, Hewison, & McGrath, 2005), in neurons, glial cells, and macrophages in areas such as hippocampus, hypothalamus, amygdala, cortex, thalamus, and cerebellum in the brain (Brewer et al., 2001).

Vitamin D deficiency

Vitamin D deficiency is a common condition worldwide (Amrein et al., 2020; Arabi, El Rassi, & El-Hajj Fuleihan, 2010; Hilger et al., 2014; Yousef et al., 2019). Several factors influence an individual's vitamin D status, which includes variation in sun exposure (latitude, season, time of day, atmospheric components, clothing, sunscreen use, and skin pigmentation), age, diet, body composition and obesity, and presence of chronic disorders (Jorde & Grimnes, 2018; Tsiaras & Weinstock, 2011). Low serum 25(OH)D levels are associated with several diseases (e.g., cancer, infections, autoimmune diseases, metabolic syndrome, diabetes), risk factors for diseases (e.g., muscle weakness and falls, hypertension, and other cardiovascular diseases), and an increased rate of mortality (Bouillon, 2018; Holick et al., 2011; Melamed, Michos, Post, & Astor, 2008). Vitamin D has also been associated with several neurological disorders (Llewellyn et al., 2010; Moretti, Morelli, & Caruso, 2018; van den Bos et al., 2013) such as multiple sclerosis (MS), Alzheimer's disease (ALZ), Parkinson's disease (PD), and stroke (Brondum-Jacobsen, Nordestgaard, Schnohr, & Benn, 2013; DeLuca, Kimball, Kolasinski, Ramagopalan, & Ebers, 2013).

A relationship between vitamin D and pain, for example, chronic musculoskeletal pain (Straube, Andrew Moore, Derry, & McQuay, 2009), has also been investigated (Habib, Nagi, Thillaiappan, Sukumaran, & Akhtar, 2020; Shipton & Shipton, 2015; Turner et al., 2008; Wu, Malihi, Stewart, Lawes, & Scragg, 2018). Pain comorbidities, in association with vitamin D, for example, depression (Anglin, Samaan, Walter, & McDonald, 2013), have also been studied. The current status and rationale for the beneficial effects of vitamin D supplementation, either alone, or in combination with analgesics, for better control of chronic pain have been reported (Habib et al., 2020). The most promising results have been presented in patients suffering from fibromyalgia (Wepner et al., 2014).

Few recent studies have demonstrated that migraine (and other primary headaches, such as TTH and CH) may be related to low vitamin D serum levels (Liampas, Siokas, Brotis, & Dardiotis, 2020; Nowaczewska et al., 2020). However, a clear relationship between vitamin D and headaches, because of the low quality of the reviewed studies, has not been established (Liampas et al., 2020; Nowaczewska et al., 2020). Most of the studies in the literature include the adult population with migraine; however, few reports are also available for pediatric migraine (Cayir, Turan, & Tan, 2014; Donmez, Orun, & Sonmez, 2018; Kilic & Kilic, 2019; O'Brien, Meteyer, Reisen, Ip, & Brown, 2010). One study has also looked into the migraines in the elderly (Virtanen et al., 2017) that demonstrates a potential link between migraine and vitamin D.

Vitamin D and migraine

Three main domains have been investigated with a focus on vitamin D and migraine association. One avenue has been related to the identification of vitamin D receptor polymorphism that was found to be more prevalent in migraine patients than in controls and the polymorphisms found were related to the severity of headaches (Motaghi et al., 2013). Several studies have looked into the identification of lower serum concentration of 25(OH)D and whether it was related to clinical features of migraine (Hussein, Fathy, & Abd Elkareem, 2019), such as the severity of headaches, or higher frequency of attacks, and migraine complications, for example, rate of hospitalization (Patel et al., 2019). In line with this, several studies have investigated vitamin D supplementation to identify if this approach could improve clinical features of migraine, for example, by reducing headache severity or frequency (Buettner et al., 2015; Cayir et al., 2014; Gazerani et al., 2019; Thys-Jacobs, 1994a). Among these types of studies, some also determined vitamin D levels (i.e., for a potential insufficiency or deficiency) prior and following the supplementation period (i.e., searching for a normalization of vitamin D levels).

The significant interest in vitamin D and migraine association was generated following a couple of reports (Prakash & Shah, 2009; Thys-Jacobs, 1994a, 1994b). Other studies on seasonal (Cugini, Romit, Di Palma, & Giacovazzo, 1990; Soriani et al., 2006) and latitude-related migraine (Prakash et al., 2010) also highlighted a possible relationship. A number of literature reviews on the topic were then followed (Lippi, Cervellin, & Mattiuzzi, 2014; Nattagh-Eshtivani et al., 2018; Orr, 2018; Wells, Beuthin, & Granetzke, 2019), but the controversial outcome prevented a clear conclusion and the overall assessment did not show a relationship (Lippi et al., 2014). The most updated status, however, is presented recently (Liampas et al., 2020; Nowaczewska et al., 2020).

The comprehensive effort in the meta-analysis (Liampas et al., 2020) compared serum level of vitamin D in migraine patients with healthy controls, and collectively, it became evident that 25(OH)D concentration was lower in patients with migraine. In the primary analysis of the 8 included studies, serum levels of 25(OH)D were determined significantly lower in migraine patients compared with healthy controls. Secondary analysis revealed no difference between patients with migraine and patients with other types of primary headaches, and no difference between TTH and controls (Liampas et al., 2020).

The comprehensive review (Nowaczewska et al., 2020) collected 22 studies that investigated serum vitamin D levels in association with primary headaches. Analysis showed that 18 studies found a link between serum vitamin D levels and headaches, with the strongest connection for migraine headaches (Nowaczewska et al., 2020). The effect of vitamin D supplementation was investigated in 8 studies, and collectively, it was summarized that in some patients with migraine, the supplementation seems to reduce the frequency of headaches, especially in those with vitamin D deficiency (Nowaczewska et al., 2020). However, the authors indicated that enough evidence yet to be accumulated before recommending vitamin D supplementation. In addition, due to heterogeneity among the interventional studies, the optimal dose of vitamin D remained to be determined (Nowaczewska et al., 2020).

Overall, the updated current evidence encourages larger studies and identification of parameters that can influence the outcome, such as age, gender, comorbid diseases with migraine, and co-treatments with other agents (Hussein et al., 2019; Kilic & Kilic, 2019; Virtanen et al., 2017). Supplementation of vitamin D in patients with migraine has been either on its own (Gazerani et al., 2019; Kilic & Kilic, 2019; Mottaghi, Askari, Khorvash, & Maracy, 2015) or in combination with other drugs, such as simvastatin (Buettner et al., 2015) or amitriptyline (Cayir et al., 2014). These trials have provided encouraging results, at least for several migraine-related parameters. Among the parameters, such as headache frequency and severity, some studies have focused on other outcomes, for example, improvement in MIDAS score (Ghorbani et al., 2020). In addition, mechanisms underlying the effects have been reported, for example, lower CGRP level following the vitamin D supplementation (Ghorbani et al., 2020).

A question has remained open as to whether vitamin D supplementation must be offered for those migraine patients with low levels of vitamin D or could this be a general recommendation. In 2019, a trial of vitamin D3 supplementation in adult patients with migraine (Gazerani et al., 2019) reported beneficial effects of vitamin D supplementation on reducing the number of headache days. 25(OH)D levels increased significantly for the treatment group during the first 12 weeks, but no significant change in 1,25(OH)2D was identified (Gazerani et al., 2019). In another study, vitamin D supplementation reduced the headache frequency in migraine patients (Mottaghi et al., 2015). It seemed that the effects of vitamin D supplementation were not related to vitamin D deficiency, and this group (Mottaghi et al., 2013) reported a nonsignificant correlation between serum vitamin D and migraine severity. The beneficial effect of combination therapy by vitamin D and simvastatin in patients with EM was reported as lowering the number of migraine days (Buettner et al., 2015), and severity of migraine was reduced in patients with high serum levels of 25(OH)D (Buettner & Burstein, 2015). Therefore, the current picture shows that vitamin D can serve as a primary drug or as an adjunct therapy

(Al-Quliti & Assaedi, 2016), related (Buettner & Burstein, 2015), or not associated (Mottaghi et al., 2013) with the level of vitamin D in migraine.

Another open question is whether and how hypovitaminosis D (HD) in patients with migraine can influence the responsiveness of patients to other pharmacological treatments. One study (Iannacchero et al., 2015) has presented that migraine patients appear to have a high rate of hypovitaminosis D and vitamin D may interfere with the absorption of the medication and reduce the efficacy of pharmacological agents for the management of migraine. Hence, restoring vitamin D levels would enhance the efficacy of antimigraine drugs (Iannacchero et al., 2015).

Generally, vitamin D is considered safe for supplementation. However, a recent basic study (Hausler et al., 2019) has shown that high but not medium-dose vitamin D could cause mild hypercalcemia, which can make T cells more prone to a proinflammatory state and this might trigger active phase of a disease such as multiple sclerosis (MS) (Hausler et al., 2019). Interestingly, a recent study (Hamamci et al., 2020) has looked into reasons for the coexistence of MS and migraine to better understand the etiology of this co-existence. The vitamin D and VDR values of the MS patients with migraine were found lower than the patients without migraine. The results of this study (Hamamci et al., 2020) indicated that the status of vitamin D might influence comorbidities, for example, migraine and MS with lower vitamin D levels. The value of these findings and implications in the clinic for treatment strategies remains for further investigation.

Vitamin D and pediatric migraine

Both migraine and vitamin D deficiency have also been reported in children. One study (Kumar, Muntner, Kaskel, Hailpern, & Melamed, 2009) has indicated that 70% of children and adolescents in the United States have suboptimal (9% deficiency, 61% insufficiency) vitamin D status. A case study on 5–16 years patients with migraine or TTH and healthy controls was conducted in 2018 (Donmez et al., 2018). Each group was then divided into high solar exposure (HSE) and low solar exposure (LSE). Several biofactors were measured, including the levels of 25(OH)D. Vitamin D deficiency was defined as 25(OH)D < 20 ng/mL, insufficiency as 25(OH)D of 20–29.9 ng/mL, and sufficiency as 25(OH)D of ≥30 ng/mL (Holick et al., 2011). Results from this study (Donmez et al., 2018) revealed that the levels of 25(OH)D were significantly lower in migraine and TTH compared to the controls (Donmez et al., 2018). Based on the findings, the authors concluded that there might be a relationship between vitamin D deficiency and children's primary headaches, with a particular significance of LSE. Another study (Güngör, Güngör, & Acıpayam, 2020) has demonstrated that in children with migraine, low levels of vitamin D were associated with the severity of headaches. Interestingly, in this study, no relationship was evident between the levels of calcium, phosphorus, alkaline phosphatase, and the severity of headaches, when it was evaluated by the Migraine Disability Assessment Score (MIDAS).

In pediatric migraine, nutraceuticals are commonly considered. A review (Orr, 2018) has summarized all forms of available data (collected from 21 studies) on the efficacy and safety of nutraceuticals (e.g., vitamin D, riboflavin, coenzyme Q10, magnesium, butterbur, and polyunsaturated fatty acids) for the management of migraine in children and adolescents. Collectively, due to the low quality and sample size of the studies, no definite conclusion could be made (Orr, 2018).

Vitamin D and other types of primary headaches

TTH is the most common primary headache disorder. Some studies have reported that TTH is associated with vitamin D deficiency. However, collective data show that sufficient evidence is lacking to draw a sharp conclusion for an association between TTH and vitamin D (Nowaczewska et al., 2020).

CH shows an association with diurnal and seasonal variations (Barloese, 2015; Lee et al., 2014; Rozen & Fishman, 2012) that have been linked to sun exposure and vitamin D metabolism. The study by Sohn et al. is the first to examine serum vitamin D levels in patients with CH compared with those of age- and sex-matched controls (Sohn, Chu, Park, Ahn, & Cho, 2018). This study (Sohn et al., 2018) enrolled 28 patients with CH, 36 migraineurs, and 36 normal controls. Serum 25(OH)D levels showed that 92.8% of CH had vitamin D deficiency. There was no sex-related difference in the serum levels of 25(OH)D between cluster attack and remission time, first and next attack, daily or seasonal periods, or 3-month recurrence. A trend, however, was observed with regard to the lower serum 25(OH)D in 14 patients with the seasonal periodicity of winter to spring compared with summer to autumn periodicity (Sohn et al., 2018). The findings suggest that vitamin D deficiency is common in CH, but the significance of this finding remains to be established.

Proposed mechanisms underlying vitamin D and headache relationship

Several hypotheses have been proposed to explain how vitamin D deficiency and headaches are associated and that restoring it would be beneficial (Nowaczewska et al., 2020; Straube et al., 2009). For example, vitamin D deficiency, via elevation of iNOS, may promote inflammation (Garcion, Nataf, Berod, Darcy, & Brachet, 1997; Straube et al., 2009). The role of NO in the pathogenesis of migraine has been known (Neeb & Reuter, 2007) and NOS inhibitors have been investigated for the treatment of migraine (Pradhan, Bertels, & Akerman, 2018). Vitamin D inhibits iNOS expression and reduces NO production and hence can block the development of headaches such as migraine. It has also been proposed that due to the role of vitamin D in Ca regulation (Brewer et al., 2001), its deficiency can cause neuronal hyperexcitability, which is also known for the pathogenesis of migraine (Welch, D'Andrea, Tepley, Barkley, & Ramadan, 1990). Another mechanism is the contribution of vitamin D in the synthesis of serotonin, and hence, its deficiency can dysregulate this key molecule, which is known in the pathogenesis of, for example, depression, autism, and migraine (Nowaczewska et al., 2020; Straube et al., 2009). A basic study (Krisanova et al., 2019) has presented recently that vitamin D deficiency can alter glutamate-GABA transporter expressions and, consequently, the balance between the states of excitation and inhibition, which might have an impact on the pathogenesis of, for example, migraine. Fig. 1 depicts several proposed mechanisms and potential reasons that can link vitamin D deficiency and some aspects of migraine pathogenesis. It is not known whether and how features of migraine pathogenesis might influence vitamin D levels and efficiency.

Vitamin D measurement

Usually, the laboratory values of vitamin D are used for the estimation of diseases' burden (Reid, 2016). A key point, therefore, seems to be a proper measure of vitamin D (LeFevre & LeFevre, 2018). Variations in the measurement and precision of analytical methods can lead to incorrect levels and associated interpretation. A study has looked into the measurement of 25(OH)D by using different methods in 220 patients with migraine of both sexes, above the age of 10 years (Gallelli et al., 2019). 120 healthy volunteers served as controls. Liquid chromatography coupled with tandem mass spectrometry (LC–MS/MS)

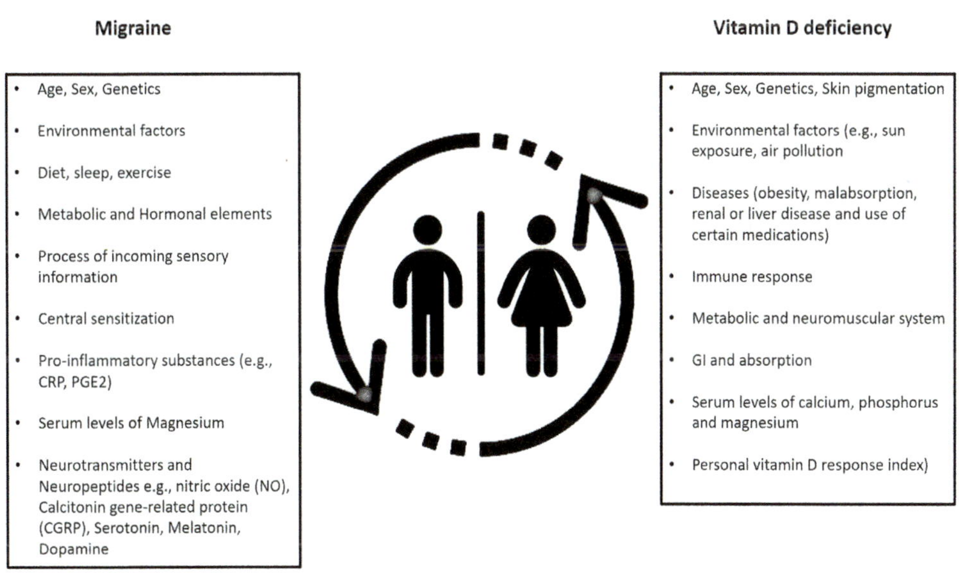

Fig. 1 Proposed mechanisms and potential reasons to link vitamin D deficiency and migraine pathogenesis (CRP: C reactive protein, PGE2: prostaglandin E2, GI: gastrointestinal).

evaluation demonstrated that participants in the migraine and control groups had lower serum vitamin D levels (15.4 ng/mL), compared with the normal range (30–100 ng/mL) and no sex difference was found. When samples were measured with high-performance liquid chromatography-ultraviolet (HPLC-UV), enzyme-linked immunosorbent assay (EIA), and chemiluminescence immunoassays (CLIA), no significant differences were found compared with the mean values from the reference method (LC–MS/MS).

Applications to other neurological conditions

Vitamin D deficiency has been linked to both neurological diseases and neuropsychological disorders (Di Somma et al., 2017; Moretti et al., 2018). Mood disorders and several psychiatric conditions, such as depression (Menon, Kar, Suthar, & Nebhinani, 2020), have also been investigated for a link between vitamin D deficiency and the potential advantages of the supplementation. Current evidence does not definitively demonstrate that vitamin D deficiency is a cause of or risk for developing these conditions nor that if vitamin D is an effective therapy for all patients affected by these disorders. It is a general view that vitamin D supplementation in neurological diseases needs to be optimized in terms of dosing and timing to generate a therapeutic potential. Recent studies suggest that vitamin D supplementation should be individualized depending on baseline vitamin D status and patients' responsiveness. Confounding factors and VDR genotypes should also be appropriately addressed (Carlberg, 2016). Large, randomized, well-controlled clinical trials are still awaited to provide high-quality evidence on the safety and tolerability of vitamin D as an adjunct therapy in a wide range of neurological and neuropsychiatric disorders (Di Somma et al., 2017).

Other components of interest

Vitamin D functions in a coherent way with other components; therefore, evaluation of other vitamins and minerals is important. The determinants of vitamin D levels must be recognized and considered, including the amount of ultraviolet exposure, dietary intake (e.g., enriched or fortified food), location, season, vitamin D supplementation, BMI, physical activity, socioeconomic status, and lifestyle (Kimlin et al., 2014). Age, ethnic background, type of disease, other therapeutics, sleep pattern, and disease comorbidities are also important to include. It is also highly important to obtain baseline levels of 25(OH) D, for comparisons with the schedule of the treatment. Dose and interval of supplementation must be justified. In addition to attempts for using proper supplementation of vitamin D, the identification of potential mechanisms is interesting (Ghorbani et al., 2020). Most studies are focused on headache frequency and severity; however, it is also important that other migraine features can be taken into consideration, for example, migraine disability (Ghorbani et al., 2020), or aura, phonophobia, photophobia, autonomic manifestations, allodynia, and resistance to medications (Hussein et al., 2019). Even though vitamin D is considered safe, any side effects or toxicity must also be monitored and reported. Coexisting migraine with other diseases should not be neglected, for example, migraine and epilepsy, or migraine and MS, depression, cardiovascular diseases, or neurodegenerative disorders such as ALZ and PD, in which also vitamin D seems to play a role.

Summary points

- Population-based studies around the globe report low levels of vitamin D.
- Lifestyle changes and adequate supplementation have been recommended to achieve optimal vitamin D levels.
- The current evidence highlights an association between a low level of vitamin D and clinical features of migraine and associated symptoms.
- There is still not sufficient and high-quality evidence to recommend vitamin D supplementation for migraine patients; however, promising results are being accumulated.
- Future basic and clinical investigations need to consider an optimized design, rational dose, and supplementation schedule for interventional studies, a wide and diverse range of influencing factors, and potential underlying mechanisms.

References

Al-Quliti, K. W., & Assaedi, E. S. (2016). New advances in prevention of migraine. Review of current practice and recent advances. *Neurosciences (Riyadh)*, *21*(3), 207–214. https://doi.org/10.17712/nsj.2016.3.20150506.

Amrein, K., Scherkl, M., Hoffmann, M., Neuwersch-Sommeregger, S., Kostenberger, M., Tmava Berisha, A., et al. (2020). Vitamin D deficiency 2.0: An update on the current status worldwide. *European Journal of Clinical Nutrition*, *74*(11), 1498–1513. https://doi.org/10.1038/s41430-020-0558-y.

Anglin, R. E., Samaan, Z., Walter, S. D., & McDonald, S. D. (2013). Vitamin D deficiency and depression in adults: Systematic review and meta-analysis. *The British Journal of Psychiatry*, *202*, 100–107. https://doi.org/10.1192/bjp.bp.111.106666.

Antonucci, R., Locci, C., Clemente, M. G., Chicconi, E., & Antonucci, L. (2018). Vitamin D deficiency in childhood: Old lessons and current challenges. *Journal of Pediatric Endocrinology & Metabolism*, *31*(3), 247–260. https://doi.org/10.1515/jpem-2017-0391.

Arabi, A., El Rassi, R., & El-Hajj Fuleihan, G. (2010). Hypovitaminosis D in developing countries-prevalence, risk factors and outcomes. *Nature Reviews. Endocrinology*, *6*(10), 550–561. https://doi.org/10.1038/nrendo.2010.146.

Barloese, M. C. (2015). Neurobiology and sleep disorders in cluster headache. *The Journal of Headache and Pain*, *16*, 562. https://doi.org/10.1186/s10194-015-0562-0.

Bigal, M. E., & Lipton, R. B. (2006). Modifiable risk factors for migraine progression. *Headache*, *46*(9), 1334–1343. https://doi.org/10.1111/j.1526-4610.2006.00577.x.

Bouillon, R. (2018). Extra-skeletal effects of vitamin D. *Frontiers of Hormone Research*, *50*, 72–88. https://doi.org/10.1159/000486072.

Brewer, L. D., Thibault, V., Chen, K. C., Langub, M. C., Landfield, P. W., & Porter, N. M. (2001). Vitamin D hormone confers neuroprotection in parallel with downregulation of L-type calcium channel expression in hippocampal neurons. *The Journal of Neuroscience*, *21*(1), 98–108. https://www.ncbi.nlm.nih.gov/pubmed/11150325.

Brondum-Jacobsen, P., Nordestgaard, B. G., Schnohr, P., & Benn, M. (2013). 25-hydroxyvitamin D and symptomatic ischemic stroke: An original study and meta-analysis. *Annals of Neurology*, *73*(1), 38–47. https://doi.org/10.1002/ana.23738.

Buettner, C., & Burstein, R. (2015). Association of statin use and risk for severe headache or migraine by serum vitamin D status: A cross-sectional population-based study. *Cephalalgia*, *35*(9), 757–766. https://doi.org/10.1177/0333102414559733.

Buettner, C., Nir, R. R., Bertisch, S. M., Bernstein, C., Schain, A., Mittleman, M. A., et al. (2015). Simvastatin and vitamin D for migraine prevention: A randomized, controlled trial. *Annals of Neurology*, *78*(6), 970–981. https://doi.org/10.1002/ana.24534.

Buse, D. C., Greisman, J. D., Baigi, K., & Lipton, R. B. (2019). Migraine progression: A systematic review. *Headache*, *59*(3), 306–338. https://doi.org/10.1111/head.13459.

Carlberg, C. (2016). Molecular approaches for optimizing vitamin D supplementation. *Vitamins and Hormones*, *100*, 255–271. https://doi.org/10.1016/bs.vh.2015.10.001.

Cayir, A., Turan, M. I., & Tan, H. (2014). Effect of vitamin D therapy in addition to amitriptyline on migraine attacks in pediatric patients. *Brazilian Journal of Medical and Biological Research*, *47*(4), 349–354. https://doi.org/10.1590/1414-431x20143606.

Charles, A. (2018). The pathophysiology of migraine: Implications for clinical management. *Lancet Neurology*, *17*(2), 174–182. https://doi.org/10.1016/S1474-4422(17)30435-0.

Collaborators, G. B. D. H. (2018). Global, regional, and national burden of migraine and tension-type headache, 1990-2016: A systematic analysis for the global burden of disease study 2016. *Lancet Neurology*, *17*(11), 954–976. https://doi.org/10.1016/S1474-4422(18)30322-3.

Cugini, P., Romit, A., Di Palma, L., & Giacovazzo, M. (1990). Common migraine as a weekly and seasonal headache. *Chronobiology International*, *7*(5–6), 467–469. https://doi.org/10.3109/07420529009059158.

Deluca, H. F. (2014). History of the discovery of vitamin D and its active metabolites. *Bonekey Reports*, *3*, 479. https://doi.org/10.1038/bonekey.2013.213.

DeLuca, G. C., Kimball, S. M., Kolasinski, J., Ramagopalan, S. V., & Ebers, G. C. (2013). Review: The role of vitamin D in nervous system health and disease. *Neuropathology and Applied Neurobiology*, *39*(5), 458–484. https://doi.org/10.1111/nan.12020.

Di Somma, C., Scarano, E., Barrea, L., Zhukouskaya, V. V., Savastano, S., Mele, C., et al. (2017). Vitamin D and neurological diseases: An Endocrine view. *International Journal of Molecular Sciences*, *18*(11). https://doi.org/10.3390/ijms18112482.

Do, T. P., Guo, S., & Ashina, M. (2019). Therapeutic novelties in migraine: New drugs, new hope? *The Journal of Headache and Pain*, *20*(1), 37. https://doi.org/10.1186/s10194-019-0974-3.

Donmez, A., Orun, E., & Sonmez, F. M. (2018). Vitamin D status in children with headache: A case-control study. *Clinical nutrition ESPEN*, *23*, 222–227. https://doi.org/10.1016/j.clnesp.2017.09.010.

Edvinsson, L. (2018). The CGRP pathway in migraine as a viable target for therapies. *Headache*, *58*(Suppl 1), 33–47. https://doi.org/10.1111/head.13305.

Eyles, D. W., Smith, S., Kinobe, R., Hewison, M., & McGrath, J. J. (2005). Distribution of the vitamin D receptor and 1 alpha-hydroxylase in human brain. *Journal of Chemical Neuroanatomy*, *29*(1), 21–30. https://doi.org/10.1016/j.jchemneu.2004.08.006.

Fernandes de Abreu, D. A., Eyles, D., & Feron, F. (2009). Vitamin D, a neuro-immunomodulator: implications for neurodegenerative and autoimmune diseases. *Psychoneuroendocrinology*, *34*(Suppl 1), S265–S277. https://doi.org/10.1016/j.psyneuen.2009.05.023.

Fields, J., Trivedi, N. J., Horton, E., & Mechanick, J. I. (2011). Vitamin D in the Persian Gulf: Integrative physiology and socioeconomic factors. *Current Osteoporosis Reports*, *9*(4), 243–250. https://doi.org/10.1007/s11914-011-0071-2.

Gallelli, L., Michniewicz, A., Cione, E., Squillace, A., Colosimo, M., Pelaia, C., et al. (2019). 25-Hydroxy vitamin D detection using different analytic methods in patients with migraine. *Journal of Clinical Medicine*, *8*(6). https://doi.org/10.3390/jcm8060895.

Garcion, E., Nataf, S., Berod, A., Darcy, F., & Brachet, P. (1997). 1,25-Dihydroxyvitamin D3 inhibits the expression of inducible nitric oxide synthase in rat central nervous system during experimental allergic encephalomyelitis. *Brain Research. Molecular Brain Research*, *45*(2), 255–267. https://doi.org/10.1016/s0169-328x(96)00260-4.

Gazerani, P. (2019). Current evidence on potential uses of MicroRNA biomarkers for migraine: From diagnosis to treatment. *Molecular Diagnosis & Therapy*, *23*(6), 681–694. https://doi.org/10.1007/s40291-019-00428-8.

Gazerani, P. (2020). Migraine and Diet. *Nutrients*, *12*(6). https://doi.org/10.3390/nu12061658.

Gazerani, P., Fuglsang, R., Pedersen, J. G., Sorensen, J., Kjeldsen, J. L., Yassin, H., et al. (2019). A randomized, double-blinded, placebo-controlled, parallel trial of vitamin D3 supplementation in adult patients with migraine. *Current Medical Research and Opinion*, *35*(4), 715–723. https://doi.org/10.1080/03007995.2018.1519503.

Ghorbani, Z., Rafiee, P., Fotouhi, A., Haghighi, S., Rasekh Magham, R., Ahmadi, Z. S., et al. (2020). The effects of vitamin D supplementation on interictal serum levels of calcitonin gene-related peptide (CGRP) in episodic migraine patients: Post hoc analysis of a randomized double-blind placebo-controlled trial. *The Journal of Headache and Pain*, *21*(1), 22. https://doi.org/10.1186/s10194-020-01090-w.

Goadsby, P. J. (2016). Bench to bedside advances in the 21st century for primary headache disorders: Migraine treatments for migraine patients. *Brain*, *139*(Pt 10), 2571–2577. https://doi.org/10.1093/brain/aww236.

Güngör, O., Güngör, S., & Acıpayam, C. (2020). The relationship between serum vitamin D levels and pain severity in children with migraine. *Trends in Pediatrics*, *1*(1), 22–26. https://doi.org/10.5222/TP.2020.54264.

Habib, A. M., Nagi, K., Thillaiappan, N. B., Sukumaran, V., & Akhtar, S. (2020). Vitamin D and its potential interplay with pain signaling pathways. *Frontiers in Immunology*, *11*, 820. https://doi.org/10.3389/fimmu.2020.00820.

Hamamci, M., Gocmen, A. Y., Say, B., Alpua, M., Badem, N. D., Ergun, U., et al. (2020). Why do multiple sclerosis and migraine coexist? *Multiple Sclerosis and Related Disorders*, *40*. https://doi.org/10.1016/j.msard.2020.101946, 101946.

Hansen, J. M., & Charles, A. (2019). Differences in treatment response between migraine with aura and migraine without aura: Lessons from clinical practice and RCTs. *The Journal of Headache and Pain*, *20*(1), 96. https://doi.org/10.1186/s10194-019-1046-4.

Hausler, D., Torke, S., Peelen, E., Bertsch, T., Djukic, M., Nau, R., et al. (2019). High dose vitamin D exacerbates central nervous system autoimmunity by raising T-cell excitatory calcium. *Brain*, *142*(9), 2737–2755. https://doi.org/10.1093/brain/awz190.

Headache classification Committee of the International Headache Society (IHS). (2018). The international classification of headache disorders, 3rd edition. *Cephalalgia*, *38*(1), 1–211. https://doi.org/10.1177/0333102417738202.

Hilger, J., Friedel, A., Herr, R., Rausch, T., Roos, F., Wahl, D. A., et al. (2014). A systematic review of vitamin D status in populations worldwide. *The British Journal of Nutrition*, *111*(1), 23–45. https://doi.org/10.1017/S0007114513001840.

Holick, M. F. (2007). Vitamin D deficiency. *The New England Journal of Medicine*, *357*(3), 266–281. https://doi.org/10.1056/NEJMra070553.

Holick, M. F. (2009). Vitamin D status: Measurement, interpretation, and clinical application. *Annals of Epidemiology*, *19*(2), 73–78. https://doi.org/10.1016/j.annepidem.2007.12.001.

Holick, M. F., Binkley, N. C., Bischoff-Ferrari, H. A., Gordon, C. M., Hanley, D. A., Heaney, R. P., et al. (2011). Evaluation, treatment, and prevention of vitamin D deficiency: An Endocrine Society clinical practice guideline. *The Journal of Clinical Endocrinology and Metabolism*, *96*(7), 1911–1930. https://doi.org/10.1210/jc.2011-0385.

Hussein, M., Fathy, W., & Abd Elkareem, R. M. (2019). The potential role of serum vitamin D level in migraine headache: A case-control study. *Journal of Pain Research*, *12*, 2529–2536. https://doi.org/10.2147/JPR.S216314.

Iannacchero, R., Costa, A., Squillace, A., Gallelli, L., Cannistra, U., & De Sarro, G. (2015). P060. Vitamin D deficiency in episodic migraine, chronic migraine and medication-overuse headache patients. *The Journal of Headache and Pain*, *16*(Suppl 1), A184. https://doi.org/10.1186/1129-2377-16-S1-A184.

Japelt, R. B., & Jakobsen, J. (2013). Vitamin D in plants: A review of occurrence, analysis, and biosynthesis. *Frontiers in Plant Science*, *4*, 136. https://doi.org/10.3389/fpls.2013.00136.

Jorde, R., & Grimnes, G. (2018). Serum cholecalciferol may be a better marker of vitamin D status than 25-hydroxyvitamin D. *Medical Hypotheses*, *111*, 61–65. https://doi.org/10.1016/j.mehy.2017.12.017.

Kilic, B., & Kilic, M. (2019). Evaluation of vitamin D levels and response to therapy of childhood migraine. *Medicina (Kaunas, Lithuania)*, *55*(7). https://doi.org/10.3390/medicina55070321.

Kimlin, M. G., Lucas, R. M., Harrison, S. L., van der Mei, I., Armstrong, B. K., Whiteman, D. C., et al. (2014). The contributions of solar ultraviolet radiation exposure and other determinants to serum 25-hydroxyvitamin D concentrations in Australian adults: The AusD study. *American Journal of Epidemiology*, *179*(7), 864–874. https://doi.org/10.1093/aje/kwt446.

Klingberg, E., Olerod, G., Konar, J., Petzold, M., & Hammarsten, O. (2015). Seasonal variations in serum 25-hydroxy vitamin D levels in a Swedish cohort. *Endocrine*, *49*(3), 800–808. https://doi.org/10.1007/s12020-015-0548-3.

Krisanova, N., Pozdnyakova, N., Pastukhov, A., Dudarenko, M., Maksymchuk, O., Parkhomets, P., et al. (2019). Vitamin D3 deficiency in puberty rats causes presynaptic malfunctioning through alterations in exocytotic release and uptake of glutamate/GABA and expression of EAAC-1/GAT-3 transporters. *Food and Chemical Toxicology*, *123*, 142–150. https://doi.org/10.1016/j.fct.2018.10.054.

Kumar, J., Muntner, P., Kaskel, F. J., Hailpern, S. M., & Melamed, M. L. (2009). Prevalence and associations of 25-hydroxyvitamin D deficiency in US children: NHANES 2001-2004. *Pediatrics*, *124*(3), e362–e370. https://doi.org/10.1542/peds.2009-0051.

Ladizesky, M., Lu, Z., Oliveri, B., San Roman, N., Diaz, S., Holick, M. F., et al. (1995). Solar ultraviolet B radiation and photoproduction of vitamin D3 in central and southern areas of Argentina. *Journal of Bone and Mineral Research*, *10*(4), 545–549. https://doi.org/10.1002/jbmr.5650100406.

Lee, Y. J., Chen, Y. T., Ou, S. M., Li, S. Y., Yang, A. C., Tang, C. H., et al. (2014). Temperature variation and the incidence of cluster headache periods: A nationwide population study. *Cephalalgia*, *34*(9), 656–663. https://doi.org/10.1177/0333102413520083.

LeFevre, M. L., & LeFevre, N. M. (2018). Vitamin D screening and supplementation in community-dwelling adults: Common questions and answers. *American Family Physician*, *97*(4), 254–260. https://www.ncbi.nlm.nih.gov/pubmed/29671532.

Liampas, I., Siokas, V., Brotis, A., & Dardiotis, E. (2020). Vitamin D serum levels in patients with migraine: A meta-analysis. *Revue Neurologique (Paris)*, *176*(7–8), 560–570. https://doi.org/10.1016/j.neurol.2019.12.008.

Lippi, G., Cervellin, G., & Mattiuzzi, C. (2014). No evidence for an association of vitamin D deficiency and migraine: A systematic review of the literature. *BioMed Research International*, *2014*. https://doi.org/10.1155/2014/827635, 827635.

Llewellyn, D. J., Lang, I. A., Langa, K. M., Muniz-Terrera, G., Phillips, C. L., Cherubini, A., et al. (2010). Vitamin D and risk of cognitive decline in elderly persons. *Archives of Internal Medicine*, *170*(13), 1135–1141. https://doi.org/10.1001/archinternmed.2010.173.

Martin, V. T., & Vij, B. (2016). Diet and headache: Part 2. *Headache*, *56*(9), 1553–1562. https://doi.org/10.1111/head.12952.

Melamed, M. L., Michos, E. D., Post, W., & Astor, B. (2008). 25-hydroxyvitamin D levels and the risk of mortality in the general population. *Archives of Internal Medicine*, *168*(15), 1629–1637. https://doi.org/10.1001/archinte.168.15.1629.

Menon, V., Kar, S. K., Suthar, N., & Nebhinani, N. (2020). Vitamin D and depression: A critical appraisal of the evidence and future directions. *Indian Journal of Psychological Medicine*, *42*(1), 11–21. https://doi.org/10.4103/IJPSYM.IJPSYM_160_19.

Moon, M., Song, H., Hong, H. J., Nam, D. W., Cha, M. Y., Oh, M. S., et al. (2013). Vitamin D-binding protein interacts with Abeta and suppresses Abeta-mediated pathology. *Cell Death and Differentiation*, *20*(4), 630–638. https://doi.org/10.1038/cdd.2012.161.

Moretti, R., Morelli, M. E., & Caruso, P. (2018). Vitamin D in neurological diseases: A rationale for a pathogenic impact. *International Journal of Molecular Sciences*, *19*(8). https://doi.org/10.3390/ijms19082245.

Motaghi, M., Haghjooy Javanmard, S., Haghdoost, F., Tajadini, M., Saadatnia, M., Rafiee, L., et al. (2013). Relationship between vitamin D receptor gene polymorphisms and migraine without aura in an Iranian population. *BioMed Research International*, *2013*. https://doi.org/10.1155/2013/351942, 351942.

Mottaghi, T., Askari, G., Khorvash, F., & Maracy, M. R. (2015). Effect of vitamin D supplementation on symptoms and C-reactive protein in migraine patients. *Journal of Research in Medical Sciences: The Official Journal of Isfahan University of Medical Sciences*, *20*(5), 477–482. https://doi.org/10.4103/1735-1995.163971.

Mottaghi, T., Khorvash, F., Askari, G., Maracy, M. R., Ghiasvand, R., Maghsoudi, Z., et al. (2013). The relationship between serum levels of vitamin D and migraine. *Journal of Research in Medical Sciences: The Official Journal of Isfahan University of Medical Sciences*, *18*(Suppl. 1), S66–S70. https://www.ncbi.nlm.nih.gov/pubmed/23961291.

Mulder, E. J., Van Baal, C., Gaist, D., Kallela, M., Kaprio, J., Svensson, D. A., et al. (2003). Genetic and environmental influences on migraine: A twin study across six countries. *Twin Research*, *6*(5), 422–431. https://doi.org/10.1375/136905203770326420.

Nair, R., & Maseeh, A. (2012). Vitamin D: The "sunshine" vitamin. *Journal of Pharmacology and Pharmacotherapeutics*, *3*(2), 118–126. https://doi.org/10.4103/0976-500X.95506.

Nattagh-Eshtivani, E., Sani, M. A., Dahri, M., Ghalichi, F., Ghavami, A., Arjang, P., et al. (2018). The role of nutrients in the pathogenesis and treatment of migraine headaches: Review. *Biomedicine & Pharmacotherapy*, *102*, 317–325. https://doi.org/10.1016/j.biopha.2018.03.059.

Neeb, L., & Reuter, U. (2007). Nitric oxide in migraine. *CNS & Neurological Disorders Drug Targets*, *6*(4), 258–264. https://doi.org/10.2174/187152707781387233.

Nowaczewska, M., Wicinski, M., Osinski, S., & Kazmierczak, H. (2020). The role of vitamin D in primary headache-from potential mechanism to treatment. *Nutrients*, *12*(1). https://doi.org/10.3390/nu12010243.

O'Brien, V. A., Meteyer, C. U., Reisen, W. K., Ip, H. S., & Brown, C. R. (2010). Prevalence and pathology of West Nile virus in naturally infected house sparrows, western Nebraska, 2008. *The American Journal of Tropical Medicine and Hygiene*, *82*(5), 937–944. https://doi.org/10.4269/ajtmh.2010.09-0515.

Orr, S. L. (2018). The evidence for the role of nutraceuticals in the Management of Pediatric Migraine: A review. *Current Pain and Headache Reports*, *22*(5), 37. https://doi.org/10.1007/s11916-018-0692-6.

Patel, U., Kodumuri, N., Malik, P., Kapoor, A., Malhi, P., Patel, K., et al. (2019). Hypocalcemia and vitamin D deficiency amongst migraine patients: A Nationwide retrospective study. *Medicina (Kaunas, Lithuania)*, *55*(8). https://doi.org/10.3390/medicina55080407.

Polzonetti, V., Pucciarelli, S., Vincenzetti, S., & Polidori, P. (2020). Dietary intake of vitamin D from dairy products reduces the risk of osteoporosis. *Nutrients*, *12*(6). https://doi.org/10.3390/nu12061743.

Pozo-Rosich, P., Coppola, G., Pascual, J., & Schwedt, T. J. (2020). How does the brain change in chronic migraine? Developing disease biomarkers. *Cephalalgia*. https://doi.org/10.1177/0333102420974359, 333102420974359.

Pradeep, R., Nemichandra, S. C., Harsha, S., & Radhika, K. (2020). Migraine disability, quality of life, and its predictors. *Annals of Neurosciences*, *27*(1), 18–23. https://doi.org/10.1177/0972753120929563.

Pradhan, A. A., Bertels, Z., & Akerman, S. (2018). Targeted nitric oxide synthase inhibitors for migraine. *Neurotherapeutics*, *15*(2), 391–401. https://doi.org/10.1007/s13311-018-0614-7.

Prakash, S., Mehta, N. C., Dabhi, A. S., Lakhani, O., Khilari, M., & Shah, N. D. (2010). The prevalence of headache may be related with the latitude: A possible role of vitamin D insufficiency? *The Journal of Headache and Pain*, *11*(4), 301–307. https://doi.org/10.1007/s10194-010-0223-2.

Prakash, S., & Shah, N. D. (2009). Chronic tension-type headache with vitamin D deficiency: Casual or causal association? *Headache*, *49*(8), 1214–1222. https://doi.org/10.1111/j.1526-4610.2009.01483.x.

Puledda, F., Messina, R., & Goadsby, P. J. (2017). An update on migraine: Current understanding and future directions. *Journal of Neurology*, *264*(9), 2031–2039. https://doi.org/10.1007/s00415-017-8434-y.

Ramachandran, R. (2018). Neurogenic inflammation and its role in migraine. *Seminars in Immunopathology*, *40*(3), 301–314. https://doi.org/10.1007/s00281-018-0676-y.

Reid, I. R. (2016). What diseases are causally linked to vitamin D deficiency? *Archives of Disease in Childhood*, *101*(2), 185–189. https://doi.org/10.1136/archdischild-2014-307961.

Rozen, T. D., & Fishman, R. S. (2012). Cluster headache in the United States of America: Demographics, clinical characteristics, triggers, suicidality, and personal burden. *Headache*, *52*(1), 99–113. https://doi.org/10.1111/j.1526-4610.2011.02028.x.

Schmid, A., & Walther, B. (2013). Natural vitamin D content in animal products. *Advances in Nutrition, 4*(4), 453–462. https://doi.org/10.3945/an.113.003780.

Schwedt, T. J. (2014). Chronic migraine. *BMJ, 348*. https://doi.org/10.1136/bmj.g1416, g1416.

Shipton, E. A., & Shipton, E. E. (2015). Vitamin D and pain: Vitamin D and its role in the Aetiology and maintenance of chronic pain states and associated comorbidities. *Pain Research and Treatment, 2015*. https://doi.org/10.1155/2015/904967, 904967.

Sohn, J. H., Chu, M. K., Park, K. Y., Ahn, H. Y., & Cho, S. J. (2018). Vitamin D deficiency in patients with cluster headache: A preliminary study. *The Journal of Headache and Pain, 19*(1), 54. https://doi.org/10.1186/s10194-018-0886-7.

Soriani, S., Fiumana, E., Manfredini, R., Boari, B., Battistella, P. A., Canetta, E., et al. (2006). Circadian and seasonal variation of migraine attacks in children. *Headache, 46*(10), 1571–1574. https://doi.org/10.1111/j.1526-4610.2006.00613.x.

Steiner, T. J., Stovner, L. J., Jensen, R., Uluduz, D., Katsarava, Z., The, L., et al. (2020). Migraine remains second among the world's causes of disability, and first among young women: Findings from GBD2019. *The Journal of Headache and Pain, 21*(1), 137. https://doi.org/10.1186/s10194-020-01208-0.

Straube, S., Andrew Moore, R., Derry, S., & McQuay, H. J. (2009). Vitamin D and chronic pain. *Pain, 141*(1–2), 10–13. https://doi.org/10.1016/j.pain.2008.11.010.

Thys-Jacobs, S. (1994a). Alleviation of migraines with therapeutic vitamin D and calcium. *Headache, 34*(10), 590–592. https://doi.org/10.1111/j.1526-4610.1994.hed3410590.x.

Thys-Jacobs, S. (1994b). Vitamin D and calcium in menstrual migraine. *Headache, 34*(9), 544–546. https://doi.org/10.1111/j.1526-4610.1994.hed3409544.x.

Tsiaras, W. G., & Weinstock, M. A. (2011). Factors influencing vitamin D status. *Acta Dermato-Venereologica, 91*(2), 115–124. https://doi.org/10.2340/00015555-0980.

Turner, M. K., Hooten, W. M., Schmidt, J. E., Kerkvliet, J. L., Townsend, C. O., & Bruce, B. K. (2008). Prevalence and clinical correlates of vitamin D inadequacy among patients with chronic pain. *Pain Medicine, 9*(8), 979–984. https://doi.org/10.1111/j.1526-4637.2008.00415.x.

van den Bos, F., Speelman, A. D., van Nimwegen, M., van der Schouw, Y. T., Backx, F. J., Bloem, B. R., et al. (2013). Bone mineral density and vitamin D status in Parkinson's disease patients. *Journal of Neurology, 260*(3), 754–760. https://doi.org/10.1007/s00415-012-6697-x.

Virtanen, J. K., Giniatullin, R., Mantyselka, P., Voutilainen, S., Nurmi, T., Mursu, J., et al. (2017). Low serum 25-hydroxyvitamin D is associated with higher risk of frequent headache in middle-aged and older men. *Scientific Reports*, 7, 39697. https://doi.org/10.1038/srep39697.

Wacker, M., & Holick, M. F. (2013). Vitamin D - effects on skeletal and extraskeletal health and the need for supplementation. *Nutrients, 5*(1), 111–148. https://doi.org/10.3390/nu5010111.

Wei, D. Y., Yuan Ong, J. J., & Goadsby, P. J. (2018). Cluster headache: Epidemiology, pathophysiology, clinical features, and diagnosis. *Annals of Indian Academy of Neurology, 21*(Suppl 1), S3–S8. https://doi.org/10.4103/aian.AIAN_349_17.

Welch, K. M., D'Andrea, G., Tepley, N., Barkley, G., & Ramadan, N. M. (1990). The concept of migraine as a state of central neuronal hyperexcitability. *Neurologic Clinics, 8*(4), 817–828. https://www.ncbi.nlm.nih.gov/pubmed/1979655.

Wells, R. E., Beuthin, J., & Granetzke, L. (2019). Complementary and integrative medicine for episodic migraine: An update of evidence from the last 3 years. *Current Pain and Headache Reports, 23*(2), 10. https://doi.org/10.1007/s11916-019-0750-8.

Wepner, F., Scheuer, R., Schuetz-Wieser, B., Machacek, P., Pieler-Bruha, E., Cross, H. S., et al. (2014). Effects of vitamin D on patients with fibromyalgia syndrome: A randomized placebo-controlled trial. *Pain, 155*(2), 261–268. https://doi.org/10.1016/j.pain.2013.10.002.

Wu, Z., Malihi, Z., Stewart, A. W., Lawes, C. M., & Scragg, R. (2018). The association between vitamin D concentration and pain: A systematic review and meta-analysis. *Public Health Nutrition, 21*(11), 2022–2037. https://doi.org/10.1017/S1368980018000551.

Yousef, S., Elliott, J., Manuel, D., Colman, I., Papadimitropoulos, M., Hossain, A., et al. (2019). Study protocol: Worldwide comparison of vitamin D status of immigrants from different ethnic origins and native-born populations-a systematic review and meta-analysis. *Systematic Reviews, 8*(1), 211. https://doi.org/10.1186/s13643-019-1123-4.

PART VIII

Multiple sclerosis

CHAPTER 21

Iron, zinc, and multiple sclerosis patients

Alaa Matar and Jamilah Borjac*
Department of Biological Sciences, Faculty of Science, Beirut Arab University, Debbieh, Lebanon

Abbreviations

1,25 $(OH)_2$ vitD	1,25-Dihydroxy-vitamin D
12-LOX	12-Lipoxygenase
25-OH-vitD	25-Hydroxy-vitamin D
AD	Alzheimer's disease
ATP	Adenosine triphosphate
AMPA	Alpha-amino-3-hydroxy-5-methyl-4-isoxazole propionic acid
BBB	Blood-brain barrier
BCSFB	Blood-cerebrospinal fluid barrier
Ca^{2+}	Calcium ion
CD163	Cluster of differentiation 163
CD	Cluster of differentiation
CP	Ceruloplasmin
CNS	Central nervous system
CSF	Cerebrospinal fluid
CYP450	Cytochrome P450
DMT1	Divalent metal transporter 1
DNA	Deoxyribonucleic acid
EAE	Experimental autoimmune encephalomyelitis
ERK	1/2 Signal-regulated kinase 1/2
Fe	Iron
FPN1	Ferroportin 1
HEPH	Hephaestin
HFE	Homeostatic iron regulator
iNOS	Inducible nitric oxide synthase
LIP	Labile iron pool
MBP	Myelin basic protein
MMP	Matrix metalloproteinase
MMP-9	Matrix metallopeptidase-9
MRI	Magnetic resonance imaging
MS	Multiple sclerosis
MT	Metallothionein
NADPH	Nicotinamide adenine dinucleotide phosphate
NAWM	Normal appearing white matter
NRAMP1	Natural resistance-associated macrophage protein 1

* **In Absentia Contact Person:** Wassim Kanbar, Telephone: +961 7 985858 ext 3317, Fax [1]: +961 7 985070, Email: wassim.kanbar@bau.edu.lb.

Vitamins and Minerals in Neurological Disorders
https://doi.org/10.1016/B978-0-323-89835-5.00011-9

NTBI	Nontransferrin bound iron
OPC	Oligodendrocyte progenitor cell
PARP-1	Poly(ADP-ribose) polymerase 1
PD	Parkinson's disease
RBC	Red blood cell
ROS	Reactive oxygen species
Tf	Transferrin
Tf-Fe	Transferrin-bound iron
TfR	Transferrin receptor
TGF-β	Tumor growth factor-beta
TIM-1	T-cell immunoglobulin and mucin domain-1
TMPRSS6	Transmembrane serine protease 6
TNF-α	Tumor necrosis factor-alpha
Zn	Zinc
Zn^{2+}	Zinc ion
ZnT3	Zinc transporter 3

Introduction

Multiple sclerosis (MS) is a chronic neurodegenerative disease of the central nervous system (CNS). Metals such as iron (Fe) and zinc (Zn) are believed to play an important role in MS pathogenesis and progression. Iron is an important element used by different cells of the body including those of the CNS, for various basic functions. In the brain, iron is a cofactor for diverse proteins and enzymes involved in the synthesis of neurotransmitters, mitochondrial respiration, myelin production, and nucleic acid repair. It is also needed for the proliferation of inflammatory cells, crucial players in MS. Moreover, its involvement in oxygen transport and glucose metabolism is very well established. Given its association with diverse functions in the human nervous system, alteration in its metabolism is accompanied with serious problems and impairments such as neurological disorders including MS (Jouini et al., 2021; Knyszynska et al., 2020; Tai et al., 2020).

On the other hand, zinc is an important trace element for all living beings as it is a cofactor for more than 300 enzymes and proteins including the matrix metalloproteinases (MMPs) and the myelin basic protein (MBP). It is also abundantly present in the brain especially in presynaptic vesicles of neurons (Portbury & Adlard, 2017). Zinc levels perturbation has negative consequences on the nervous and immune system particularly (Sensi, Granzotto, Siotto, & Squitti, 2018) as high levels of zinc may induce neurotoxicity (Li, Du, & Ni, 2019) and low levels of zinc are able to promote neurons' apoptosis (Seth et al., 2015).

Hence, zinc and iron play crucial roles in the brain's homeostasis and are crucial for human health (Bredholt & Frederiksen, 2016; Schweser et al., 2020).

Brain iron metabolism

The brain, a highly metabolically active organ, consumes approximately 20% of the energy produced by the entire human body. Continuous iron supply is needed to secure this high-energy need. Iron is one of the major metals needed by the brain where it serves as a cofactor for distinct physiological processes inside the CNS such as myelination and neurotransmitters biosynthesis, and its levels are tightly regulated. The passage of iron from the blood to the brain is controlled mainly by the blood-brain barrier and somehow by the blood-cerebrospinal fluid barrier (BCSFB). Transferrin-bound iron (Tf-Fe) is unable to traverse directly the BBB and the process of iron entry to the brain is not well established. Two potential models for iron transport have been suggested involving Tf-Fe and nontransferrin bound iron (NTBI).

The Tf-Fe pathway is thought to be the main way for iron transport through the luminal membrane of the capillary endothelium. It has been suggested that ferritin (Tf) binds to ferritin receptor (TfR) at the luminal side of the brain capillaries and then the complex crosses by endocytosis. The acidic environment allows the release of ferric iron from Tf and its reduction to ferrous iron by endosomal reductase. The subsequent steps are not entirely well clear. It has been hypothesized that ferrous iron is transported from the endosome to the cytosol by the divalent metal transporter 1 (DMT1) and then reaches the intracellular labile iron pool (LIP). Subsequently, it may be used by the endothelial cells, stored in endothelial cell ferritin, or transported to the mitochondria by TfR2 and mitoferrins. Export proteins called ferroportin 1 (FPN1), mainly expressed by neurons, astrocytes, and the capillary endothelium of BBB, may also allow its release into the extracellular fluid, where it is reoxidized to ferric iron through ceruloplasmin (CP) and the ferroxidases hephaestin (HEPH).

The second potential mechanism is based on the fact that the endosome containing Tf-TfR1 complex liberates ferrous iron between astrocyte end-foot processes and endothelial cells where it can be oxidized to ferric iron by HEPH or CP present on the end-foot processes. Then, iron binds to apo-Tf circulating within the brain. Iron may also join the brain through the epithelial cells of the choroid plexus. Next, the iron is released just the same way as from the BBB endothelial cells, via ferroxidases, DMT1 and FPN1. When reaching the cerebrospinal fluid (CSF), iron binds to Tf and supports CNS cells that express TfR1.

Iron is needed by various cell types inside the brain where it is used in different pathways. Since neurons present high amounts of TfR1, Tf is meant to be their major source of iron. NTBI can also be uptaken by neurons where it represents the main source of iron for astrocytes and oligodendrocytes. Additionally, ferrous iron present in the brain interstitium can bind to citrate or adenosine triphosphate (ATP) in order to be transported to astrocytes as well as to oligodendrocytes, the glial cells responsible in myelin sheath synthesis in the CNS, the same way as NTBI. Oligodendrocytes get NTBI by the T-cells immunoglobulin and mucin domain-1 (TIM-1) protein, which is a ferritin receptor specifically expressed in oligodendrocytes (Fig. 1).

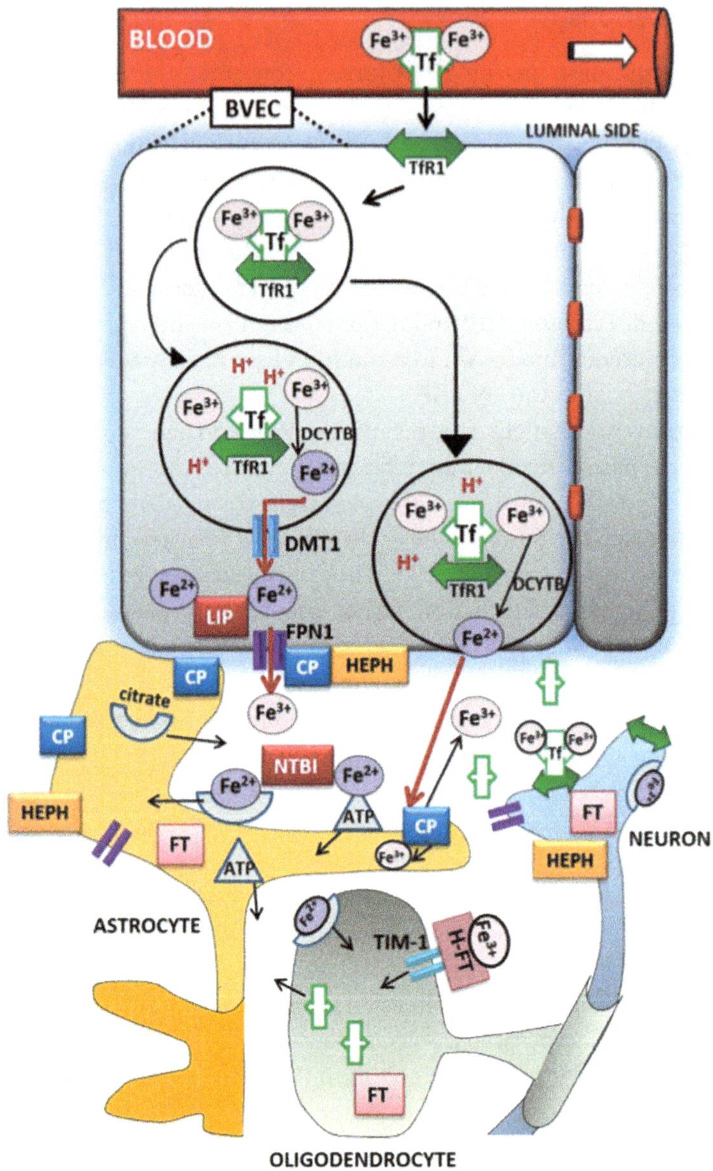

Fig. 1 Iron transport inside the brain. Scheme proposing the transferrin and nontransferrin bound iron transport pathways in the brain. Abbreviations: *BVEC*: blood vascular endothelial cell, *Tf*: transferrin, *TfR1*: transferrin receptor 1, Fe^{3+}: ferric iron, Fe^{2+}: ferrous iron, *DCYTB*: duodenal cytochrome *b*, *DMT1*: divalent metal transporter 1, *FPN1*: ferroportin1, *LIP*: labile iron pool, *CP*: ceruloplasmin, *HEPH*: hephaestin, *NTBI*: nontransferrin bound iron, *FT*: ferritin, *H-FT*: H-ferritin, *TIM-1*: T-cell immunoglobulin and mucin domain-1. *(Source: Grubić Kezele, T. & Ćurko-Cofek, B. (2020). Age-related changes and sex related differences in brain iron metabolism.* Nutrients, *12, 2601. https://doi.org/10.3390/nu12092601.)*

Iron deposition in the brain of MS patients

During MS, a perturbation in iron homeostasis occurs within the gray and white matters of the brain and the remyelinated plaques (Hametner et al., 2013). Using magnetic resonance imaging (MRI) techniques, elevated iron levels were detected in the proximity of lesions and in various gray matter structures such as globus pallidus, caudate, putamen, and the thalamus of MS patients (Al-Radaideh et al., 2013) and these levels were correlated with the evolvement of the disease and disability progression. However, reduced iron levels were noticed in the remyelinated plaques as well as in the white matter in association with inflammatory lesions (Mehta et al., 2013).

These differentially dysregulated levels of iron inside distinct regions of the brain suggest iron shuttling during MS disease course or its redistribution between the different regions of the brain (Al-Radaideh et al., 2013).

Histochemical localization of iron deposits in both MS patients and in the experimental autoimmune encephalomyelitis (EAE), an animal MS model, tracked it around vessels within macrophages/microglia as a potential consequence of microhemorrhaging (Bagnato et al., 2011; Williams et al., 2011). Staining of pathological iron deposit was also detected in punctate structures of neurons that may be the result of mitochondrial injury. In the white matter, iron accumulation was found in macrophages/microglia, astrocytes, and in hemosiderin of red blood cells (RBCs) of the erupted BBB as well as in the injured axons (Hametner et al., 2013; Williams et al., 2011).

In the normal appearing white matter (NAWM), iron accumulation was basically observed in oligodendrocytes and myelin sheath (Hametner et al., 2013). In this regard, iron chelation therapy may present a beneficial approach once the precise contribution of iron in MS is elucidated. This strategy is generally used to treat iron overload in several disorders such as thalassemia, where deferiprone drug is commonly used. Deferiprone is able to cross the BBB (Roy, Preston, Hider, & Ma, 2010) where it can impede the functions of infiltrating immune cells in the CNS (Sweeney et al., 2011) through the binding of free iron released from the degenerated neuronal cells or located in the perivascular vessels (Hametner et al., 2013; Williams, Buchheit, Berman, & LeVine, 2012), limiting thereby the oxidative stress progress. It could also reduce the pro-inflammatory cytokines secretion through iron chelation, limiting hence its uptake by macrophages/microglia (Mehta et al., 2013). It can manage iron redistribution as well within cells of the deep gray matter in MS by importing free iron outside the mitochondria. Table 1 summarizes the main therapeutic mechanisms shown to be putatively involved in iron chelation in MS.

To sum up, iron dysregulation may be considered as a double-edged sword in MS etiology; from one hand, high iron levels contribute to free radical formation that enhances oxidative stress, resulting in myelin loss and axonal injury leading subsequently to neurodegeneration. On the other hand, low iron levels could worsen the immune system response and reduce energy production due to mitochondrial membrane potential loss, resulting in neurodegeneration.

Table 1 Suggested therapeutic mechanisms of iron chelation in MS.

Pathological target	Potential mechanism
Activated or iron-enriched microglia, macrophage or astrocyte	Slowing down ROS and pro-inflammatory cytokine production
Demyelination and axonal damage that contribute to the liberation of iron-containing products	Slowing down tissue damage enhancement
Excessive iron deposit in the deep gray matter	Decreasing iron uptake by the brain, preventing iron induced inhibition of DNA repair and protecting against oxidative stress
Perivascular iron deposit	Preventing iron catalyzed damage to the BBB and the myelin sheath
B- and T-cells	Blocking cell proliferation in the periphery and potentially in the CNS
Cytokines	Modulating cytokine liberation in the periphery and potentially in the CNS

Different proposed therapeutic mechanisms contribute in iron chelation in MS. Abbreviations: ROS: reactive oxygen species; DNA: deoxyribonucleic acid; BBB: blood-brain barrier; CNS: central nervous system.

These findings that originate from an inadequate iron homeostasis (low or high) may be due to differences in nutritional behavior, biochemical background, age, and sex-related factors between individuals, making iron treatment an individualized approach in MS. The onset of MS was shown to be early in females but worse in males. Gender-related changes in the brain iron metabolism were observed.

A higher ferritin serum level with a lower serum iron level and iron accumulation was found in females in comparison to males with low sex hormones in both genders (Table 2).

Table 2 Sex-related differences in brain iron metabolism.

	Female	Male
MS disease onset	Early	Worse
Serum ferritin level	High	Low
Serum iron level	Low	High
Iron accumulation in CNS	Low	High
Sexual hormone	Low estrogen	Low testosterone

Differences between female and male MS patients in terms of state of MS disease onset, serum ferritin and iron level, iron accumulation in CNS variations as well as sexual hormone changes.

Iron and oligodendrocytes

Iron is the crucial key in the protection of oligodendrocytes against oxidizing agents. Particularly, iron is essential for these cells due to the high energy cost needed for the production and the maintenance of the myelin sheath and for the maturation of oligodendrocyte progenitor cells (OPCs) (Rosko, Smith, Yamazaki, & Huang, 2019). Low iron levels were shown to threaten the viability of oligodendrocytes.

During remyelination, OPCs are driven to the injured axonic lesions where they differentiate into mature oligodendrocytes in order to participate in their remyelination. Taking into consideration the requirement of iron-containing enzymes in all these stages, it is safe to conclude that iron levels in oligodendrocytes have a great impact on neuronal repair and remyelination processes.

Iron-deficient oligodendrocytes may result from iron trafficking disturbance (Grishchuk et al., 2015), overall iron deficiency (Chen et al., 2019) or iron export from astrocytes (Schulz, Kroner, & David, 2012), resulting in a decreased OPCs proliferation and thus a lack of remyelination.

Oligodendrocytes are sensitive to oxidation, and antioxidants such as glutathione are required especially during MS relapses that are characterized by elevated levels of inflammatory cells and free radicals (Kuhn, Gritti, Crooks, & Dombrowski, 2019). Iron is required for adenosine triphosphate (ATP) generation, which, in turn, is crucial for nicotinamide adenine dinucleotide phosphate (NADPH) synthesis that is considered as a major intracellular reducing agent involved in diverse metabolic pathways (Todorich, Pasquini, Garcia, Paez, & Connor, 2009). Furthermore, NADPH is required for cholesterol synthesis, a main lipid component of the oligodendrocytes membrane. It is also a cofactor for cytochrome P450 enzymes (CYP450) involved in the synthesis of active form of vitamin D from 7-dehydrocholesterol. Active vitamin D_3 binds to its receptor that is highly expressed in oligodendrocytes, resulting in improved OPCs differentiation. Thus, adequate amount of iron is needed for cytochrome P450 enzymes synthesis implicated in oligodendrocytes maturation and hence MS.

Possible sources of iron deposit

Distinct causes of abnormal iron deposition inside the brain of MS patients have been suggested such as BBB dysfunction, reduced iron clearance as a consequence of inflammation and axonal injury, or alteration of iron transport proteins due to inflammation (Stankiewicz et al., 2007).

There are a number of sources for the accumulated iron in the brain. In pathological conditions, iron deposit may originate from the release of protein-containing iron from damaged structures such as axons, oligodendrocytes, and myelin. Moreover, perivascular inflammation, a basic process occurring in MS, induces vessel walls injury, resulting in

RBCs eruption inside the CNS, which releases hemoglobin. Hemoglobin breakdown liberates heme whose degradation leads to perivascular iron accumulation (Williams et al., 2011). Generally, heme is required for distinct proteins in neurons and oligodendrocytes. However, during MS inflammation, these cells degenerate and release heme-containing proteins leading to pathological iron accumulation.

Iron is released from heme by heme oxygenase 1 and stored intracellularly as ferritin, limiting by this its pro-oxidant effect and therefore reducing its effect on MS. However, if excessive iron is released, improper storage of iron occurs resulting in iron-induced oxidative stress that may be a major cause of MS exacerbation, contributing potentially in axonal injury and demyelination (Haider et al., 2014).

In addition, excess iron deposition in macrophages/microglia is believed to stimulate pro-inflammatory cytokines release including tumor necrosis factor alpha (TNF-α) and interleukin 1 beta. These cytokines, along with the ROS released by microglia and the released iron that may react with the hydrogen peroxide produced, inhibit glutamate uptake by astrocytes, resulting in elevated extracellular levels of glutamate and an enhanced neuronal damage (Matute, 2011).

So altogether, heme and iron are able to catalyze reactions that lead to the production of reactive oxygen species, harmful products to the CNS.

Moreover, mitochondrial dysfunction may enhance MS pathology during the inflammatory process and neurodegeneration progression (Su, Bourdette, & Forte, 2013). In vitro, pro-inflammatory cytokines such as TNF-α and transforming growth factor-beta1 (TGF-β1) were shown to promote iron uptake by microglia and astrocytes (Rathore, Redensek, & David, 2012). The accumulated iron leads to mitochondrial damage due to oxidative stress (Wang et al., 2019). Herein, a feedback loop may occur given that mitochondrial damage may stimulate mitochondrial iron uptake. So, iron can act independently inside the mitochondria of neuronal cells and cause oxidative damage that could trigger demyelination, axonal damage, and neuronal attack.

Genetics of iron metabolism

The implication of iron in MS will be fully understood once taking into consideration the influence of genetic variants related to iron metabolism. The regulation of iron uptake and transport processes is strongly complicated due to the involvement of different biochemical players. Two major genes implicated in iron transport and absorption are of high interest: the natural resistance-associated macrophage protein 1 (*NRAMP1*) and the homeostatic iron regulator (*HFE*) genes. Genetic variations in both genes were associated with iron deficiency in MS patients (Table 3). In addition, hepcidin, an iron regulatory protein, plays a role in the regulation of dietary iron absorption, distribution, and storage inside the body post binding to the ferroportin receptor. Higher levels of hepcidin were detected during MS progression, leading to a decline in iron absorption and

Table 3 Correlation between MS susceptibility and genetic variants in two iron-related genes: *NRAMP1* and *HFE*.

Gene	Population	Study size	Conclusion on the statistical significance of the genes association with MS
NRAMP1	South African	104 MS; 522 HC	Yes
	Sardinian	66 MS; 60 HC	Yes
	Turkish	100 MS; 100 HC	No
HFE	Tasmanian	166 MS families; 489 MS individuals; 104 HC families	No
	Slovenian	314 MS; 400 HC	No
	Portuguese	373 MS	No

This table summarizes a number of studies performed on different populations in an attempt to investigate the association between genetic variations in two iron-related genes (*NRAMP1* and *HFE*) and MS disease susceptibility. Abbreviations: *NRAMP1*: natural resistance-associated macrophage protein 1; *HFE*: homeostatic iron regulator; MS: multiple sclerosis; HC: healthy control.

consequently, to its accumulation inside macrophages (Bsteh et al., 2019). Differential expression of hepcidin and its receptor during infection and inflammation minimizes iron accessibility needed to get rid of infectious agents. Moreover, it leads to iron deficiency inside cells during inflammatory episodes of MS. Additionally, variations in the transmembrane serine protease 6 (*TMPRSS6*) gene may upregulate hepcidin expression resulting in iron deficiency (Cau, Melis, Congiu, & Galanello, 2010).

In white matter lesions and periplaque white matter, the expression of ferritin gene that encodes for the iron storage protein was found to be upregulated while the expression of frataxin gene that encodes for the protein responsible in iron export from mitochondria was downregulated (Hametner et al., 2013). However, the transferrin receptor, needed for iron transfer to cells, was found to be upregulated in NAWM of MS patients. Adding to this, proteins engaged in iron import/export or those acting as ferrireductases or ferrioxidases are also shown to be upregulated (Hametner et al., 2013), and proteins like hephaestin that facilitates intracellular iron transport were found in oligodendrocytes and astrocytes of NAWM and in the white matter, implying an active export state of iron during MS (Hametner et al., 2013).

In conclusion, genetic variations of genes implicated in iron metabolism may directly affect iron levels inside the body and thus implicated in MS outcome/progression.

The molecular mechanisms of zinc-induced neurotoxicity and the possible role of zinc in MS

Vesicular zinc release is thought to be a key mediator in neuronal death during stressful circumstances like hypoglycemia, where it contributes to neuronal toxicity through a series of mechanisms. Glucose reperfusion induces nitric oxide synthesis that promotes vesicular zinc liberation from presynaptic neurons. Reaching the postsynaptic neurons, the high intracellular zinc levels lead to ROS overproduction initiating neuronal death and imposing oxidative stress on DNA. This stress may lead to additional release of zinc bound to metallothionein (MT) activating the poly(ADP-ribose) polymerase-1 (PARP-1) leading to loss of mitochondrial permeability, followed by the release of apoptotic factors and ending up by neuronal toxicity. Chelation of zinc after hypoglycemia and glucose reperfusion decrease ROS, PARP-1 hyperactivation, and neurotoxicity. Knocking out the zinc transporter 3 gene (*ZnT3*) was shown to result in vesicular zinc depletion, reduction of ROS levels, and consequently impeding neuronal death. Both *ZnT3* deletion and zinc chelation have been shown to ameliorate EAE clinical score and to induce a lower rate of demyelination, inflammation, and infiltration of encephalitogenic immune cells such as the CD4+ and CD8+ T-cells inside the spinal cord. Moreover, motor perturbations were shown to be remarkably less pronounced, accompanied by a reduced BBB disruption and matrix metallopeptidase-9 (MMP-9) activation, minimizing subsequently immune cells infiltration and preventing myelin damage (Choi et al., 2016; Choi, Lee, Denney, & Lynch, 2011). These studies show that zinc plays an important role in MS pathogenesis.

Several studies demonstrated differential serum or CSF zinc level and tried to correlate it to MS. However, the results were conflicting (Table 4).

Table 4 Serum zinc levels in MS patients in comparison to healthy controls.

Country	Study size	Conclusion on the statistical significance of Zn serum level in MS
Lebanon	28 MS; 28 HC	Nonsignificant
Germany	151 MS; 50 HC	Significantly lowered
Poland	101 MS; 68 HC	Significantly lowered
Iran	58 MS; 39 HC	Nonsignificant
Iran	60 MS; 60 HC	Significantly lowered
Iraq	32 MS; 32 HC	Significantly lowered
Italy	60 MS; 124 HC	Significantly lowered
Greece	15 MS; 28 HC	Nonsignificant
USA	68 MS; 62 HC	Nonsignificant
Sweden	50 MS; 50 HC	Significantly lowered

This table represents a comparison in serum zinc levels between MS patients and controls of different populations where some studies demonstrated an association between reduced zinc serum levels and MS. Abbreviations: Zn: zinc; MS: multiple sclerosis; HC: healthy control.

Role of zinc in oligodendrocyte death

The release of intracellular zinc may be a crucial key in oligodendrocyte death, resulting in demyelination, the initial pathological feature of MS. Oligodendrocytes are known as the glial cells responsible in myelin sheath synthesis in the CNS. Oligodendrocytes are prone to oxidative stress as previously mentioned (Bradl & Lassmann, 2010). The AMPA receptors (AMPAR) located on the surface of these cells increase their susceptibility to glutamate excitoxicity, particularly in the absence of GluR2, which is an AMPAR subunit, resulting in Ca^{2+} accumulation and intracellular zinc release. Evidence shows that intracellular zinc release is responsible for oligodendrocytes death in MS. Mato et al. showed that intracellular zinc liberation triggers glutamate-induced excitotoxic damage to oligodendrocytes. Zn^{2+} was reported to cause mitochondrial damage and ROS generation inside oligodendrocytes with subsequent activation of biochemical cascades that may lead to apoptosis such as PARP-1-mediated cell death pathway (Mato, Sanchez-Gomez, Bernal-Chico, & Matute, 2013). Moreover, it has been proposed that peroxynitrite that is generated through the reaction between nitric oxide and superoxide is produced as a consequence of intracellular zinc release and may be a key mediator in oligodendrocytes death in MS (Li, Vana, Ribeiro, & Zhang, 2011). The effect of peroxynitrite in MS patients is mediated through inducible nitric oxide activation (iNOS) inside lesions. MS patients were shown to present high levels of iNOS and nitrotyrosine that are directly involved in peroxynitrite synthesis. Peroxynitrite plays a critical role in zinc release from its stock inside mitochondria and from MT that binds zinc and acts as zinc buffer. Subsequently, zinc induces extracellular activation of 12-lipoxygenase (12-LOX), signal-regulated kinase 1/2 (ERK 1/2) beside ROS generation, causing oligodendrocyte death. Adding to this, peroxynitrite-induced neurotoxicity may be initiated through protein kinase C (PKC), which mediates neurotoxic zinc release. Peroxynitrite toxicity may be alleviated through zinc chelation treatment such as TPEN (N,N,N_0,N_0-tetrakis(2-pyridylmethyl)ethylenediamine), highlighting the role of zinc in peroxynitrite-induced toxicity (Zhang et al., 2006) (Fig. 2).

Applications to other neurological conditions

The deep gray matter structures are targets of iron deposition in MS subjects and, thereby, the most susceptible areas of neurodegeneration (Stankoff & Louapre, 2018). However, iron deposition is also a hallmark in Parkinson's disease (PD) and Alzheimer's disease (AD) at neurodegenerated areas that are also implicated in MS. Furthermore, mitochondrial dysfunction, a feature observed in MS, is also observed in the neurological disorders PD and AD (Picca et al., 2020). This might be due to the decreased blood pressure in the cerebrum of MS patients leading to low oxygen level and thus disrupting the electron transport chain and ATP production. Low oxygen level poses stress and negatively

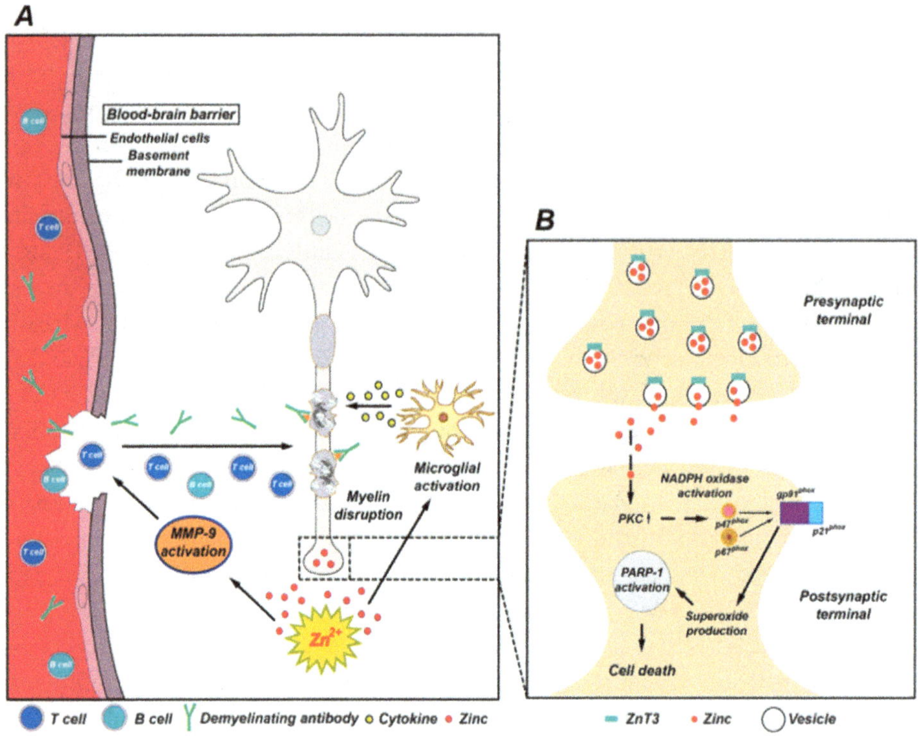

Fig. 2 Possible association of zinc with MS pathogenesis. (A) Schematic drawing indicating several ways in which zinc may influence MS pathogenesis. Released zinc may induce matrix metallopeptidase-9 (MMP-9)-dependent blood-brain barrier (BBB) disruption and enhancing the migration of autoreactive immune cells and demyelinating antibodies by degrading the extracellular matrix. In addition, the released zinc may activate microglia and then pro-inflammatory cytokines released by activated microglia, causing damage to the myelin sheath. (B) Mechanism of zinc-induced neuronal damage in MS. Cytoplasmic influx of synaptically released zinc stimulates protein kinase C (PKC), NADPH oxidases, and superoxide-generating enzymes. Superoxide production activates poly(ADP-ribose) polymerase-1 (PARP-1) in the nucleus leading to cell death. *(Source: Choi, B. Y., Jung, J. W., & Suh, S. W. (2017). The emerging role of zinc in the pathogenesis of multiple sclerosis.* International Journal of Molecular Sciences, 18*(10), 2070. https://doi.org/10.3390/ijms18102070.)*

impacts the cerebral metabolism. Hypoxia may lead to mitochondrial damage and excessive release of ROS induces a pro-oxidative state inside neurons leading subsequently to their degeneration (Pelizzoni et al., 2011). A high iron level may be accompanied with increased ROS release through the Fenton reaction that may induce an oxidative damage to the nucleic acids, proteins, lipids, and specifically to myelin. Iron may also impede the activity of enzymes engaged in base excision repair for DNA damage of neurons in vitro (Li, Swiercz, & Englander, 2009).

On the other hand, Zn^{2+} is shown to exert an effect on beta-amyloid metabolism in AD leading consequently to oxidative stress. It has also been hypothesized that oxidative stress and zinc dyshomeostasis participate in the AD brain damage where few studies have shed the light on the putative implication of Zn^{2+} dysregulation in the structural and functional damage of AD synapses. Of interest are also studies indicating that Zn^{+} can worsen learning and memory performance processes. They were shown to be impaired in AD.

In conclusion, both iron and zinc could play an important role in neurodegeneration through their direct involvement in neuronal tissue damage.

Other components of interest

Differences in latitude are directly related to the prevalence of the disease, giving that MS occurrence is more frequently observed in regions located far away from the equator (Ghareghani, Reiter, Zibara, & Farhadi, 2018). Several factors are correlated with latitude, but one of the best associates is the intensity and duration of sunlight radiation (Gallagher et al., 2019). Various ecological studies have indicated an inverse association between sunlight exposure and MS prevalence. It has been suggested that ultraviolet radiation may exert immunosuppressive effects at the time where it is essential for the cutaneous initiation of vitamin D_3 production that terminates in the kidneys (Hart & Norval, 2018).

Vitamin D is a fat-soluble hormone that is a member of the steroid hormone family. It possesses pleiotropic function: it is vital for the maintenance of calcium homeostasis and bone metabolism; it has a detectable influence on the regulation of the immune response as well as on neuroprotective processes as it promotes the differentiation of oligodendrocytes. Adequate vitamin D level can be achieved either through dietary supply (including food and pharmaceutical supplies) or through its synthesis in the body. The endogenous metabolism provides 80–90% of the vitamin target, and thus it stands as the major source, while only 10–20% of vitamin D is supplied via diet.

In addition to its modulatory effects on immune cells, active vitamin D exerts a neuroprotective function on brain cells as it supports remyelination by enhancing neural stem proliferation and survival, as well as by promoting oligodendrocyte differentiation. Several studies showed a correlation between low serum 25-OH-vitamin D levels and MS occurrence (Duan et al., 2014; Jacobs, Noyce, Giovannoni, & Dobson, 2020; Mokry et al., 2015; Munger, Levin, Hollis, Howard, & Ascherio, 2006). The variability in sunlight exposure and vitamin D intake during seasons make sunlight and/or vitamin D investigation challenging, but it is probable that interactions between genetic factors modulating the effects of vitamin D and exposure to sunlight may be the reason under the geographic differences in MS risk (Akturk et al., 2019).

Key facts

- Low iron levels may be accompanied with depression.
- Fertile women requirement of iron is higher than men due to its loss during menstruation.
- High zinc supplementation can cause copper deficiency that may lead to myelopathy, a status with neurological symptoms comparable to those in MS.
- Food enriched in antioxidants such as fruits and vegetables is recommended for MS patients.
- Unsaturated fat can help in decreasing inflammation in MS. However, saturated fat intake may be directly linked to the risk of MS relapses and disease disability.
- Cow dairy products are to be avoided in MS since they increase disease activity due to butyrophilin, a cow protein that may trigger antibodies production against it. These antibodies may recognize oligodendrocyte glycoprotein on nerve cells causing their damage.

Mini-dictionary of terms

Oxidative stress: A phenomenon due to an imbalance between free radicals and antioxidants inside the body, resulting in cells and tissues damage.

Heme: An iron-containing molecule that forms the nonprotein part in hemoglobin. It is known to transport oxygen in the bloodstream and other cytochrome proteins needed in electron transport chain and in xenobiotics detoxification.

Normal appearing white matter: It refers to the normal, yet infected tissue around the white matter in magnetic resonance imaging.

Reactive oxygen species: Unstable free radical molecules containing oxygen that can easily react with other compounds inside a cell, leading to nucleic acids and proteins damage.

Perivascular inflammation: A cluster of inflammatory cells surrounding the blood vessels, which occurs in different conditions like MS.

AMPA receptors: Glutamate receptors coupled with cation channels that are involved in cell excitability at different postsynaptic membranes.

Glutamate excitotoxicity: A pathological process of neuronal cell death due to excessive release of glutamate.

Summary points

- Iron dyshomeostasis may be considered as a double-edged sword in MS etiology: high levels of iron may stimulate free radical production and low levels of iron may perturbate the immune system function. However, both dysregulation status end up by neurodegeneration.

- Iron is essential for oligodendrocytes as it allows their maturation in order to achieve their task in myelin synthesis around axons of the CNS.
- Iron can act independently inside the mitochondrial neuronal cells and induces neuronal damage in MS.
- In stressful conditions, high intracellular zinc release results in an excessive ROS production and a reduced mitochondrial membrane permeability, leading consequently to neuronal toxicity.
- High intracellular zinc levels play a role in oligodendrocytes death in MS.

References

Akturk, T., Turan, Y., Tanik, N., Karadag, M. E., Sacmaci, H., & Inan, L. E. (2019). Vitamin D, vitamin D binding protein, vitamin D receptor levels and cardiac dysautonomia in patients with multiple sclerosis: A cross-sectional study. *Arquivos de Neuro-Psiquiatria*, 77(12), 848–854. https://doi.org/10.1590/0004-282X20190182.

Al-Radaideh, A. M., Wharton, S. J., Lim, S. Y., Tench, C. R., Morgan, P. S., Bowtell, R. W., et al. (2013). Increased iron accumulation occurs in the earliest stages of demyelinating disease: An ultra-high field susceptibility mapping study in clinically isolated syndrome. *Multiple Sclerosis*, *19*(7), 896–903. https://doi.org/10.1177/1352458512465135.

Bagnato, F., Hametner, S., Yao, B., van Gelderen, P., Merkle, H., Cantor, F. K., et al. (2011). Tracking iron in multiple sclerosis: A combined imaging and histopathological study at 7 tesla. *Brain*, *134*(Pt 12), 3602–3615. https://doi.org/10.1093/brain/awr278.

Bradl, M., & Lassmann, H. (2010). Oligodendrocytes: Biology and pathology. *Acta Neuropathologica*, *119*(1), 37–53. https://doi.org/10.1007/s00401-009-0601-5.

Bredholt, M., & Frederiksen, J. L. (2016). Zinc in multiple sclerosis: A systematic review and meta-analysis. *ASN Neuro*, *8*(3). https://doi.org/10.1177/1759091416651511.

Bsteh, G., Haschka, D., Tymoszuk, P., Berek, K., Petzer, V., Hegen, H., et al. (2019). Serum hepcidin levels in multiple sclerosis. *Multiple Sclerosis Journal – Experimental, Translational and Clinical*, *5*(4). https://doi.org/10.1177/2055217319885984. 2055217319885984.

Cau, M., Melis, M. A., Congiu, R., & Galanello, R. (2010). Iron-deficiency anemia secondary to mutations in genes controlling hepcidin. *Expert Review of Hematology*, *3*(2), 205–216. https://doi.org/10.1586/ehm.10.2.

Chen, Z., Jiang, R., Chen, M., Zheng, J., Chen, M., Braidy, N., et al. (2019). Multi-copper ferroxidase deficiency leads to iron accumulation and oxidative damage in astrocytes and oligodendrocytes. *Scientific Reports*, *9*(1), 9437. https://doi.org/10.1038/s41598-019-46019-9.

Choi, B. Y., Kim, I. Y., Kim, J. H., Kho, A. R., Lee, S. H., Lee, B. E., et al. (2016). Zinc transporter 3 (ZnT3) gene deletion reduces spinal cord white matter damage and motor deficits in a murine MOG-induced multiple sclerosis model. *Neurobiology of Disease*, *94*, 205–212. https://doi.org/10.1016/j.nbd.2016.06.018.

Choi, I. Y., Lee, S. P., Denney, D. R., & Lynch, S. G. (2011). Lower levels of glutathione in the brains of secondary progressive multiple sclerosis patients measured by 1H magnetic resonance chemical shift imaging at 3 T. *Multiple Sclerosis*, *17*(3), 289–296. https://doi.org/10.1177/1352458510384010.

Duan, S., Lv, Z., Fan, X., Wang, L., Han, F., Wang, H., et al. (2014). Vitamin D status and the risk of multiple sclerosis: A systematic review and meta-analysis. *Neuroscience Letters*, *570*, 108–113. https://doi.org/10.1016/j.neulet.2014.04.021.

Gallagher, L. G., Ilango, S., Wundes, A., Stobbe, G. A., Turk, K. W., Franklin, G. M., et al. (2019). Lifetime exposure to ultraviolet radiation and the risk of multiple sclerosis in the US radiologic technologists cohort study. *Multiple Sclerosis*, *25*(8), 1162–1169. https://doi.org/10.1177/1352458518783343.

Ghareghani, M., Reiter, R. J., Zibara, K., & Farhadi, N. (2018). Latitude, vitamin D. melatonin. And gut microbiota act in concert to initiate multiple sclerosis: A new mechanistic pathway. *Frontiers in Immunology*, *9*, 2484. https://doi.org/10.3389/fimmu.2018.02484.

Grishchuk, Y., Pena, K. A., Coblentz, J., King, V. E., Humphrey, D. M., Wang, S. L., et al. (2015). Impaired myelination and reduced brain ferric iron in the mouse model of mucolipidosis IV. *Disease Models & Mechanisms*, *8*(12), 1591–1601. https://doi.org/10.1242/dmm.021154.

Haider, L., Simeonidou, C., Steinberger, G., Hametner, S., Grigoriadis, N., Deretzi, G., et al. (2014). Multiple sclerosis deep grey matter: The relation between demyelination, neurodegeneration, inflammation and iron. *Journal of Neurology, Neurosurgery, and Psychiatry*, *85*(12), 1386–1395. https://doi.org/10.1136/jnnp-2014-307712.

Hametner, S., Wimmer, I., Haider, L., Pfeifenbring, S., Bruck, W., & Lassmann, H. (2013). Iron and neurodegeneration in the multiple sclerosis brain. *Annals of Neurology*, *74*(6), 848–861. https://doi.org/10.1002/ana.23974.

Hart, P. H., & Norval, M. (2018). Ultraviolet radiation-induced immunosuppression and its relevance for skin carcinogenesis. *Photochemical & Photobiological Sciences*, *17*(12), 1872–1884. https://doi.org/10.1039/c7pp00312a.

Jacobs, B. M., Noyce, A. J., Giovannoni, G., & Dobson, R. (2020). BMI and low vitamin D are causal factors for multiple sclerosis: A Mendelian randomization study. *Neurology Neuroimmunology & Neuroinflammation*, *7*(2). https://doi.org/10.1212/NXI.0000000000000662.

Jouini, N., Saied, Z., Ben Sassi, S., Nebli, F., Messaoud, T., Hentati, F., et al. (2021). Impacts of Iron metabolism dysregulation on Alzheimer's disease. *Journal of Alzheimer's Disease*. https://doi.org/10.3233/JAD-201250.

Knyszynska, A., Radecka, A., Zabielska, P., Luczak, J., Karakiewicz, B., & Lubkowska, A. (2020). The role of Iron metabolism in fatigue, depression, and quality of life in multiple sclerosis patients. *International Journal of Environmental Research and Public Health*, *17*(18). https://doi.org/10.3390/ijerph17186818.

Kuhn, S., Gritti, L., Crooks, D., & Dombrowski, Y. (2019). Oligodendrocytes in development, myelin generation and beyond. *Cell*, *8*(11). https://doi.org/10.3390/cells8111424.

Li, X., Du, X., & Ni, J. (2019). Zn (2+) aggravates tau aggregation and neurotoxicity. *International Journal of Molecular Sciences*, *20*(3). https://doi.org/10.3390/ijms20030487.

Li, H., Swiercz, R., & Englander, E. W. (2009). Elevated metals compromise repair of oxidative DNA damage via the base excision repair pathway: Implications of pathologic iron overload in the brain on integrity of neuronal DNA. *Journal of Neurochemistry*, *110*(6), 1774–1783. https://doi.org/10.1111/j.1471-4159.2009.06271.x.

Li, S., Vana, A. C., Ribeiro, R., & Zhang, Y. (2011). Distinct role of nitric oxide and peroxynitrite in mediating oligodendrocyte toxicity in culture and in experimental autoimmune encephalomyelitis. *Neuroscience*, *184*, 107–119. https://doi.org/10.1016/j.neuroscience.2011.04.007.

Mato, S., Sanchez-Gomez, M. V., Bernal-Chico, A., & Matute, C. (2013). Cytosolic zinc accumulation contributes to excitotoxic oligodendroglial death. *Glia*, *61*(5), 750–764. https://doi.org/10.1002/glia.22470.

Matute, C. (2011). Glutamate and ATP signalling in white matter pathology. *Journal of Anatomy*, *219*(1), 53–64. https://doi.org/10.1111/j.1469-7580.2010.01339.x.

Mehta, V., Pei, W., Yang, G., Li, S., Swamy, E., Boster, A., et al. (2013). Iron is a sensitive biomarker for inflammation in multiple sclerosis lesions. *PLoS One*, *8*(3), e57573. https://doi.org/10.1371/journal.pone.0057573.

Mokry, L. E., Ross, S., Ahmad, O. S., Forgetta, V., Smith, G. D., Goltzman, D., et al. (2015). Vitamin D and risk of multiple sclerosis: A Mendelian randomization study. *PLoS Medicine*, *12*(8), e1001866. https://doi.org/10.1371/journal.pmed.1001866.

Munger, K. L., Levin, L. I., Hollis, B. W., Howard, N. S., & Ascherio, A. (2006). Serum 25-hydroxyvitamin D levels and risk of multiple sclerosis. *JAMA*, *296*(23), 2832–2838. https://doi.org/10.1001/jama.296.23.2832.

Pelizzoni, I., Macco, R., Morini, M. F., Zacchetti, D., Grohovaz, F., & Codazzi, F. (2011). Iron handling in hippocampal neurons: Activity-dependent iron entry and mitochondria-mediated neurotoxicity. *Aging Cell*, *10*(1), 172–183. https://doi.org/10.1111/j.1474-9726.2010.00652.x.

Picca, A., Calvani, R., Coelho-Junior, H. J., Landi, F., Bernabei, R., & Marzetti, E. (2020). Mitochondrial dysfunction, oxidative stress, and Neuroinflammation: Intertwined roads to neurodegeneration. *Antioxidants (Basel)*, *9*(8). https://doi.org/10.3390/antiox9080647.

Portbury, S. D., & Adlard, P. A. (2017). Zinc signal in brain diseases. *International Journal of Molecular Sciences*, *18*(12). https://doi.org/10.3390/ijms18122506.

Rathore, K. I., Redensek, A., & David, S. (2012). Iron homeostasis in astrocytes and microglia is differentially regulated by TNF-alpha and TGF-beta1. *Glia*, *60*(5), 738–750. https://doi.org/10.1002/glia.22303.

Rosko, L., Smith, V. N., Yamazaki, R., & Huang, J. K. (2019). Oligodendrocyte bioenergetics in health and disease. *The Neuroscientist*, *25*(4), 334–343. https://doi.org/10.1177/1073858418793077.

Roy, S., Preston, J. E., Hider, R. C., & Ma, Y. M. (2010). Glucosylated deferiprone and its brain uptake: Implications for developing glucosylated hydroxypyridinone analogues intended to cross the blood-brain barrier. *Journal of Medicinal Chemistry*, *53*(15), 5886–5889. https://doi.org/10.1021/jm100380k.

Schulz, K., Kroner, A., & David, S. (2012). Iron efflux from astrocytes plays a role in remyelination. *The Journal of Neuroscience*, *32*(14), 4841–4847. https://doi.org/10.1523/JNEUROSCI.5328-11.2012.

Schweser, F., Hagemeier, J., Dwyer, M. G., Bergsland, N., Hametner, S., Weinstock-Guttman, B., et al. (2020). Decreasing brain iron in multiple sclerosis: The difference between concentration and content in iron MRI. *Human Brain Mapping*. https://doi.org/10.1002/hbm.25306.

Sensi, S. L., Granzotto, A., Siotto, M., & Squitti, R. (2018). Copper and zinc dysregulation in Alzheimer's disease. *Trends in Pharmacological Sciences*, *39*(12), 1049–1063. https://doi.org/10.1016/j.tips.2018.10.001.

Seth, R., Corniola, R. S., Gower-Winter, S. D., Morgan, T. J., Jr., Bishop, B., & Levenson, C. W. (2015). Zinc deficiency induces apoptosis via mitochondrial p53- and caspase-dependent pathways in human neuronal precursor cells. *Journal of Trace Elements in Medicine and Biology*, *30*, 59–65. https://doi.org/10.1016/j.jtemb.2014.10.010.

Stankiewicz, J., Panter, S. S., Neema, M., Arora, A., Batt, C. E., & Bakshi, R. (2007). Iron in chronic brain disorders: Imaging and neurotherapeutic implications. *Neurotherapeutics*, *4*(3), 371–386. https://doi.org/10.1016/j.nurt.2007.05.006.

Stankoff, B., & Louapre, C. (2018). Can we use regional grey matter atrophy sequence to stage neurodegeneration in multiple sclerosis? *Brain*, *141*(6), 1580–1583. https://doi.org/10.1093/brain/awy114.

Su, K., Bourdette, D., & Forte, M. (2013). Mitochondrial dysfunction and neurodegeneration in multiple sclerosis. *Frontiers in Physiology*, *4*, 169. https://doi.org/10.3389/fphys.2013.00169.

Sweeney, M. E., Slusser, J. G., Lynch, S. G., Benedict, S. H., Garcia, S. L., Rues, L., et al. (2011). Deferiprone modulates in vitro responses by peripheral blood T cells from control and relapsing-remitting multiple sclerosis subjects. *International Immunopharmacology*, *11*(11), 1796–1801. https://doi.org/10.1016/j.intimp.2011.07.007.

Tai, S., Zheng, Q., Zhai, S., Cai, T., Xu, L., Yang, L., et al. (2020). Alpha-lipoic acid mediates clearance of Iron accumulation by regulating Iron metabolism in a Parkinson's disease model induced by 6-OHDA. *Frontiers in Neuroscience*, *14*, 612. https://doi.org/10.3389/fnins.2020.00612.

Todorich, B., Pasquini, J. M., Garcia, C. I., Paez, P. M., & Connor, J. R. (2009). Oligodendrocytes and myelination: The role of iron. *Glia*, *57*(5), 467–478. https://doi.org/10.1002/glia.20784.

Wang, Z. B., Liu, J. Y., Xu, X. J., Mao, X. Y., Zhang, W., Zhou, H. H., et al. (2019). Neurodegeneration with brain iron accumulation: Insights into the mitochondria dysregulation. *Biomedicine & Pharmacotherapy*, *118*, 109068. https://doi.org/10.1016/j.biopha.2019.109068.

Williams, R., Buchheit, C. L., Berman, N. E., & LeVine, S. M. (2012). Pathogenic implications of iron accumulation in multiple sclerosis. *Journal of Neurochemistry*, *120*(1), 7–25. https://doi.org/10.1111/j.1471-4159.2011.07536.x.

Williams, R., Rohr, A. M., Wang, W. T., Choi, I. Y., Lee, P., Berman, N. E., et al. (2011). Iron deposition is independent of cellular inflammation in a cerebral model of multiple sclerosis. *BMC Neuroscience*, *12*, 59. https://doi.org/10.1186/1471-2202-12-59.

Zhang, Y., Wang, H., Li, J., Dong, L., Xu, P., Chen, W., et al. (2006). Intracellular zinc release and ERK phosphorylation are required upstream of 12-lipoxygenase activation in peroxynitrite toxicity to mature rat oligodendrocytes. *The Journal of Biological Chemistry*, *281*(14), 9460–9470. https://doi.org/10.1074/jbc.M510650200.

CHAPTER 22

Considerations on vitamin D supplementation in multiple sclerosis

Joost Smolders[a,b] and Jan Damoiseaux[c]
[a]Departments of Neurology and Immunology, MS Center ErasMS, Erasmus Medical Center, Rotterdam, The Netherlands
[b]Neuroimmunology Research Group, Netherlands Institute for Neuroscience, Amsterdam, The Netherlands
[c]Central Diagnostic Laboratory, Maastricht University Medical Center, Maastricht, The Netherlands

Abbreviations

$1,25(OH)_2D$	1,25-dihydroxyvitamin D
25(OH)D	25-hyodrxyvitamin D
BMI	body mass index
CIS	clinically isolated syndrome
CNS	central nervous system
CYP24A1	cytochrome P2A1
CYP27B1	cytochrome P27B1
EAE	experimental autoimmune encephalomyelitis
HLA	human leukocyte antigen
MRI	magnetic resonance imaging
MS	multiple sclerosis
RRMS	relapsing remitting multiple sclerosis
VDR	vitamin D receptor

Introduction

Multiple sclerosis (MS) is an inflammatory disease of the central nervous system (CNS) (Thompson, Baranzini, Geurts, Hemmer, & Ciccarelli, 2018), and with an estimated age-standardized prevalence of 127/100,000 in western Europe, the leading cause of nontraumatic neurological disability among young adults (Wallin et al., 2019). MS starts in most patients with attacks of temporal neurological disability (i.e., relapsing remitting (RR)MS), followed by a gradual accumulation of disability later in its course (i.e., progressive MS). A combination of genetic variations and environmental risk-factors is believed to cause this disease. Single-nucleotide polymorphisms within risk-alleles point toward a critical role of lymphocytes and myeloid cells as mediators of the genetic risk of MS (Patsopoulos et al., 2019). A limited exposure to vitamin D has been identified as an environmental risk factors for MS. A limited dietary intake of vitamin D supplements (Munger et al., 2004), low circulating 25-hydroxyvitamin D (25(OH)D) levels (Munger, Levin, Hollis, Howard, & Ascherio, 2006), low maternal 25(OH)D levels

Vitamins and Minerals in Neurological Disorders
https://doi.org/10.1016/B978-0-323-89835-5.00012-0

during intrauterine development and directly after birth (Munger et al., 2016; Nielsen et al., 2017), and an increased load of genetic variants associated with lower 25(OH)D levels (Rhead et al., 2016) have been associated with a higher risk of developing MS. The antiinflammatory properties of vitamin D have been postulated as the mediator of these associations (Smolders, Damoiseaux, Menheere, & Hupperts, 2008).

Besides putative effects on MS onset, vitamin D supplementation has received much scientific attention as possible treatment of relapsing remitting and progressive forms of MS. As such, several clinical trials explored the efficacy of vitamin D supplementation in MS. In this chapter, we will focus on clinical trials on vitamin D_3 supplementation in MS as has been published thus far. We will discuss the effects on MS-related endpoints and safety. Furthermore, we will discuss considerations in the interpretation of these data and several considerations on the potential benefits of vitamin D_3 supplements for people with MS. A discussion on vitamin D_3 supplements and possible prevention of MS falls beyond the scope of this chapter.

Vitamin D metabolism and cellular effects

Vitamin D is a cytochrome P 450 (CYP450) superfamily enzymes-mediated metabolite of cholesterol, and a precursor of a steroid hormone (Holick, 2007). Within this metabolism-cascade, exposure of the skin to sunlight forms a critical step in the formation of cholecalciferol (vitamin D_3). In plants, this photosynthesis step results in a slightly different vitamin D molecule, ergocalciferol (vitamin D_2). Only a small fraction of vitamin D in the human body is vitamin D_2, with the majority being vitamin D_3 acquired by sun exposure of the skin or, to a lesser extent, dietary intake. The most abundant and stable circulating metabolite of vitamin D is 25(OH)D. This metabolite is accepted to reflect an individual overall vitamin D status best and can be converted to the bioactive metabolite of vitamin D, 1,25-dihydroxyvitamin D (1,25$(OH)_2$D). The latter molecule serves pleiotropic roles in calcium homeostasis and immune regulation and its metabolism is therefore tightly controlled. Although the kidneys are the major source of circulating 1,25$(OH)_2$D, many cells and tissues express the CYP27B1 enzyme essential for the generation of this molecule. These include myeloid cells, lymphocytes, but also the CNS (Smolders & Damoiseaux, 2011; Smolders, Moen, Damoiseaux, Huitinga, & Holmøy, 2011). Interestingly, CYP27B1 expression and/or 1,25$(OH)_2$D synthesis are upregulated in lymphocytes and myeloid cells upon activation, but also in association with MS white matter lesions (Smolders et al., 2013), and in spinal cords of mice with the experimental autoimmune encephalitis (EAE) model of neuroinflammation (Nashold, Spach, Spanier, & Hayes, 2009). Active 1,25$(OH)_2$D binds to the nuclear vitamin D receptor (VDR), which regulates the transcription of a multitude of vitamin D responsive genes. Lymphocytes and myeloid cells express VDR (Smolders & Damoiseaux, 2011), but also neurons, microglia, astrocytes, and oligodendrocytes (Smolders et al., 2011, 2013). Dependent on cell type and stimulus provided, 1,25$(OH)_2D$ can have pro- or anti-inflammatory properties in vitro. In B- or T-cells stimulated polyclonally, antigen specific, or with mitogens, antiinflammatory properties have been attributed to vitamin D. Accordingly, treatment with vitamin D or 1,25 $(OH)_2D$ resulted in a prevention of disease onset or attenuation of severity in the T- and B-cell-driven EAE model (Cantorna, Hayes, & DeLuca, 1996). These inhibiting properties of vitamin D and its metabolites on B- and T-cell-driven neuroinflammation have been argued to underlie the association of low vitamin D status with an active disease course of MS.

Vitamin D status and supplement use associate with MS activity

Low circulating levels of vitamin D have been associated with several clinical endpoints in MS. In general, low circulating 25(OH)D levels have been reported in MS cohorts (Smolders, Menheere, Kessels, Damoiseaux, & Hupperts, 2008; van der Mei et al., 2007), most notably in patients with a progressive disease course (Smolders, Menheere, et al., 2008).

In cross-sectional studies, lower 25(OH)D levels have been measured in patients during relapses compared to remission (Soilu-Hänninen et al., 2005, 2008), or in patients without compared to with relapses the previous 2 years (Smolders, Menheere, et al., 2008). In several prospective cohort studies in adult and pediatric RRMS, lower 25(OH)D levels were associated with a higher risk of subsequent relapses (Ascherio et al., 2014; Mowry et al., 2010; Runia, Hop, de Rijke, Buljevac, & Hintzen, 2012; Simpson et al., 2010). These were mostly studies in early MS after a first clinical presentation (i.e., *clinically isolated syndrome*). In another large but more heterogeneous cohort of CIS patients, low 25(OH)D levels did predict a second relapse of RRMS only in the unadjusted but not the analysis adjusted for potential confounders (Kuhle et al., 2015). In the largest cohort of RRMS patients with a longer disease course, no association of relapses with vitamin D status was found (Fitzgerald et al., 2015). In other studies, associations were only found in subjects with a younger age (Muris et al., 2016), or carriers of the HLA-DRB1*1501 risk-allele (Mowry et al., 2012). Therefore, although associations of low 25(OH)D levels with a higher relapse risk in early MS, the generalizability of this association for the entire MS population is uncertain. There have been several small-scale randomized placebo-controlled clinical trials on vitamin D supplementation performed in MS (Table 1), of which several addressed the endpoints annualized relapse rate (ARR) or relapse-free status as primary or secondary endpoints. In the two largest trials, CHOLINE and SOLAR, a nonsignificant lower point-estimate of ARR was reported as primary (21% reduction, $P=.379$) and secondary endpoints (30% reduction; $P=.17$), respectively (Camu et al., 2019; Hupperts et al., 2019). In addition, SOLAR reported no difference in relapse-free proportion between treatment arms, while CHOLINE

Table 1 Overview of controlled clinical trials in MS.

Study (first author)	Cohort	Number included	Number completed	Duration (months)	Intervention	Comparator	Primary endpoint	Primary endpoint met
Stein et al. (2011)	RRMS	23	20	6	High-dose vitamin D_2 aiming on 130–175 nmol/L 25(OH)D	1000 IU/day vitamin D_2	MRI (Gd^+ T1 and new T2 lesions)	No
Kampman, Steffensen, Mellgren, and Jørgensen (2012)	RRMS	71	68	24	20,000 IU/week vitamin D_3	Placebo	Relapse	No
Shaygannejad, Janghorbani, Ashtari, and Dehghan (2012)	RRMS	50	50	12	0.5 μg/day 1,25 $(OH)_2D$	Placebo	EDSS increase	Yes
Soilu-Hänninen et al. (2012)	RRMS	66	62	12	20,000 IU/week vitamin D_3	Placebo	MRI (T2 lesion volume increase)	No
Derakhshandi et al. (2013)	Optic neuritis	30	24	12	50,000 IU/week vitamin D_3	Placebo	Conversion to RMS	No
Golan et al. (2013)	RRMS	45	30	12	4370 IU/day vitamin D_3	800 IU/day vitamin D_3	Flu-like symptoms	No
Achiron et al. (2015)	RRMS	158	143	6	1 μg/day 1(OH)D	Placebo	Fatigue	Yes
Salari et al. (2015)	Optic neuritis	74	52	5	50,000 IU/week vitamin D_3	Placebo	Retinal nerve fiber layer thickness	No
Sotirchos et al. (2016)	RRMS	40	35	6	10,400 IU/day vitamin D_3	800 IU/day vitamin D_3	CD4 T-cell characteristics	Yes

O'Connell et al. (2017)	CIS	32	29	5	10,000 and 5000 IU/day vitamin D_3	Placebo	CD4 T-cell characteristics	No
Camu et al. (2019)	RRMS	129	90	21	100,000 IU 2x/ month vitamin D_3	Placebo	Relapse rate	No
Hupperts et al. (2019)	RRMS	229	183	12	14,000 IU/day vitamin D_3	Placebo	Absence of relapses, EDSS-progression, and new MRI lesions	No
Dörr et al. (2020)	CIS/RRMS	53	41	18	20,400 IU/2 days vitamin D_3	400 IU/2 days vitamin D_3	MRI (T2 lesion number increase)	No

reported in the completers of the full 96-week study protocol a reduced ARR in the vitamin D-arm (60% reduction, $P=.012$). In other (exploratory) studies on vitamin D supplements, no significant effects on these endpoints have been reported (Dörr et al., 2020; Golan et al., 2013; Kampman et al., 2012; Soilu-Hänninen et al., 2012; Sotirchos et al., 2016; Stein et al., 2011). This is except for a small study on optic neuritis, which can be considered a CIS, where 5/11 in the placebo and 0/13 patients in the vitamin D group experienced a relapse of MS during 12-months follow-up ($P=.007$) (Derakhshandi et al., 2013).

A high level of chronic MS-related disability, as reflected by a high score on the commonly used clinometric expanded disability status scale (EDSS), correlated in several cross-sectional studies with low 25(OH)D levels (Smolders, Menheere, et al., 2008; van der Mei et al., 2007). In a cohort of CIS patients, disability progression was less pronounced in participants with high 25(OH)D levels at the start of follow-up (Ascherio et al., 2014). Accordingly, low 25(OH)D-levels have been reported in diagnostic blood samples of an extreme subgroup progressing rapidly into a progressive disease stage compared to matched controls (Muris et al., 2016). However, in cohorts with more advanced MS, progression of disability was not predicted by 25(OH)D levels (Mowry et al., 2012; Muris, Smolders, et al., 2016; University of California, San Francisco MS-EPIC Team, et al., 2016; van der Mei et al., 2007), nor time of reaching a progressive disease stage during follow-up (Muris, Rolf, et al., 2016). The association of 25(OH)D levels with MS-related disability is therefore even more uncertain than the association with relapses. In clinical supplementation studies, no effects of supplements on EDSS progression were noticed (Camu et al., 2019; Hupperts et al., 2019; Kampman et al., 2012; Soilu-Hänninen et al., 2012). Notably, Stein et al. reported in a study group of $N=23$ MS patients, after adjusting for entry EDSS, a higher exit-EDSS score in the high-dose vs low-dose vitamin D group ($P=.05$), where four relapses were recorded versus two in the low-dose group ($P=.04$) (Stein et al., 2011).

Several biomarkers are used in clinical practice to monitor the disease process of MS, most notably magnetic resonance imaging (MRI). New hyperintense T2 or gadolinium-enhancing T1 lesions are believed to reflect inflammatory disease activity of MS (Barkhof et al., 2012). In several cohorts of CIS and RRMS, low 25(OH)D levels predicted a higher risk of such lesions on MRI (Ascherio et al., 2014; Fitzgerald et al., 2015; Loken-Amsrud et al., 2012; Mowry et al., 2012; University of California, San Francisco MS-EPIC Team, et al., 2016). In the SOLAR trial, the composite primary-endpoint of absence of new/enhancing MRI lesions, relapses, and disability progression was not met, yet as secondary endpoint, a lower number of new hyperintense T2 or gadolinium-enhancing T1 MRI-lesions were recorded in the vitamin D vs placebo group (32% reduction, $P=.0045$) (Hupperts et al., 2019). Accordingly, the increase in T2 lesion-volume was also attenuated in the vitamin D vs. placebo group (3.57% vs 6.07%; $P=.035$). In a Finnish study, a reduced number of gadolinium-enhancing lesions was found in the vitamin D vs placebo group, with a trend toward a reduced T2 lesion-volume increase (Soilu-Hänninen et al., 2012). For the CHOLINE study, new hyperintense T2 and gadolinium enhancing T1 lesions did not differ between vitamin D and placebo group in intention to treat and completers subgroup (Camu et al., 2019). However, occurrence and volume of T1 lesions were reduced in the vitamin D arm. Other (exploratory) studies neither reported any effects on the new MRI lesions (Dörr et al., 2020; O'Connell et al., 2017; Stein et al., 2011).

The progression of brain atrophy was predicted by low 25(OH)D levels in a cohort of CIS patients (Ascherio et al., 2014), but not in RRMS (Fitzgerald et al., 2015). No effect of vitamin D supplementation on brain volume changes were found in the CHOLINE and SOLAR trials (Camu et al., 2019; Hupperts et al., 2019). Additionally, circulating levels of neurofilament light chain (NfL) associate with inflammation in MS (Gafson et al., 2020). No associations between natural variation in NfL and 25(OH)D levels have been reported in MS (Røsjø et al., 2020). In three small clinical trials, circulating NfL levels were assessed and were not affected by vitamin D supplements (Hänninen, Jääskeläinen, Herukka, & Soilu-Hänninen, 2020; Holmøy et al., 2018; Smolders et al., 2020).

Although most clinical trials supplemented various doses of vitamin D_3, with the study of Stein et al. being the exemption in supplementing vitamin D_2 (Stein et al., 2011), two studies investigated the effect of 1(OH)D or 1,25$(OH)_2$D treatment in MS. Achiron et al. recorded in a 6-month study among 158 MS patients with fatigue 8 relapses in the subgroup supplemented with 1 μg/day 1(OH)D, vs 25 in the placebo group (P=.006) (Achiron et al., 2015). Shaygannejad et al. found in a study among 50 RRMS patients no difference in relapse rate reduction between groups treated with up to 0.5 μg/day 1,25$(OH)_2$D vs placebo (Shaygannejad et al., 2012).

In short, several endpoints of MS activity associate with low circulating levels of 25(OH)D, although there is substantial heterogeneity between studies. Notably, none of the clinical trials on vitamin D supplements thus far reached its primary endpoint but showed in secondary or exploratory analyses attenuation of some inflammatory parameters of MS. Therefore, despite several trials being performed, there remains uncertainty whether vitamin D supplements suppress MS activity as reflected by relapses, MRI parameters, and neurofilament levels. An effect of vitamin D supplements on disability progression is even more uncertain.

Heterogeneity in causality and confounding

These mixed results and negative endpoints could be primarily the result of underpowered clinical trials and association studies. However, additional explanations for the considerable heterogeneity in results of observational studies and clinical trials could be reverse causality and confounders.

First, a more active disease process of MS could lower circulating 25(OH)D levels. In both dogs as well as healthy human volunteers, experimental endotoxemia with lipopolysaccharide reversibly lowers within hours circulating 25(OH)D levels (Holowaychuk et al., 2012; Smolders, van den Ouweland, Geven, Pickkers, & Kox, 2020). Accordingly, multiple inflammatory and infectious diseases associate with lower 25(OH)D levels. Notably, elective knee surgery has also been associated with an inflammatory response as well as overnight lowering of circulating 25(OH)D levels (Reid et al., 2011). Additionally, remission of active Crohn's disease has been shown to precede recovery of lowered 25(OH)D levels (Haifer et al., 2019). Possibly, the reduction of 25(OH)D levels in association with inflammation in general is mediated by an upregulation of CYP27B1 in activated immune cells (Smolders & Damoiseaux, 2011) as has also been observed in association with active MS white matter lesions (Smolders et al., 2013). These observations do not prove that a chronic inflammatory response in MS lowers 25(OH)D levels, but reveals this as a plausible alternative explanation for the observed associations in MS.

Second, sunlight exposure has been propagated to have immune-suppressive properties by itself independent of vitamin D (Lucas, Byrne, Correale, Ilschner, & Hart, 2015). Since sunlight is the main inducer of circulating 25(OH)D levels, independent effects of sunlight on MS activity could be a significant confounder. Indeed, sunlight exposure predicts a second clinical attack in early MS independent of serum 25(OH)D levels (Simpson et al., 2018). In an exploratory study on artificial exposure of the skin to narrowband UVB light in CIS, first signals were observed of a protective effect on relapses (Hart et al., 2018), and transient shifts in circulating subsets of lymphocytes and myeloid cells (Trend et al., 2019). Nevertheless, since this intervention was also associated with elevated circulating 25(OH)D levels, an entirely independent effect of vitamin D cannot be concluded (Hart et al., 2018). A recent nationwide study among MS patients in Germany found a clear association between sunlight exposure and disease course of MS, but was not able to fully separate vitamin D-dependent and -independent effects (Ostkamp et al., 2021).

Third, a more active disease course of MS is associated with a higher amount of MS-related disability during relapses. The higher levels of MS-related disability as reflected by EDSS score are associated with a reduced sunlight exposure (van der Mei et al., 2007). Therefore, a more active disease course of MS may induce sun-avoiding behavior in patients and hereby a reduction in circulating 25(OH)D levels.

Fourth, a higher bodyweight or obesity at age 18 has been associated with an increased risk of developing MS (Munger, Chitnis, & Ascherio, 2009). Accordingly, genes associated with a higher childhood bodyweight are enriched in people with MS (Harroud et al., 2021). A higher bodyweight at age 20 also associated with smokers with a higher risk of reaching a progressive MS course (Manouchehrinia et al., 2018) and a higher actual bodyweight correlated with a higher rate of brain atrophy on MRI (Mowry et al., 2018). A higher bodyweight is associated with lower circulating 25(OH)D levels (Wortsman,

Matsuoka, Chen, Lu, & Holick, 2000), and also with an attenuated serological response to high-dose vitamin D_3 supplementation in multiple sclerosis (Bhargava et al., 2016; Smolders, Mimpen, et al., 2020). The explanation of these observations can be found in vitamin D being a lipophilic molecule that may be stored in the fat tissue. Although the interaction between bodyweight and vitamin D has been argued to be limited regarding the risk on developing MS based on genetic studies (Harroud et al., 2021), this interaction could be substantial regarding the disease course of MS.

The positive effects of vitamin D supplements on several MS endpoint suggest that there is an effect of vitamin D on disease activity in MS. Accordingly, an enrichment of genes associated with lower 25(OH)D levels has been associated with a higher relapse activity in children with MS (Graves et al., 2020). However, the association of serum 25(OH)D and MS endpoint could be a complex interaction of several contributing mechanisms. This complex interaction could also explain that effects of vitamin D supplements in controlled trials on specific endpoints could be less pronounced than suggested by association studies (Smolders, Torkildsen, Camu, & Holmøy, 2019).

Heterogeneity in participant selection

Another explanation for the heterogeneity in study results could be the selection of different subgroups of participants.

On the whole, association studies in patients very early in their disease course of MS showed more pronounced associations between 25(OH)D levels and MS activity, compared to patients with more advanced MS. The most illustrative examples are the data sets analyzed by Ascherio et al. (2014) which included participants of the BENEFIT study on interferon beta treatment in CIS, and by Fitzgerald et al. (2015) which included participants of the BEYOND study on interferon beta (IFNβ) treatment in RRMS. Both large studies report associations between 25(OH)D levels and MRI endpoints, while associations with relapse risk and EDSS progression were only evident in the CIS data set. Likewise, associations of 25(OH)D levels with relapse risk were also most clear in subjects with a younger age (Muris, Smolders, et al., 2016). Therefore, inclusion of patients of different age and disease duration could influence results.

Besides from CIS, both observational studies and clinical trials in RRMS included patients with MS treated with disease-modifying therapies. These therapies may mask potential effects of vitamin D_3 supplements on the disease course of MS. Participants were mostly treated with IFNβ in the largest observational studies (Ascherio et al., 2014; Fitzgerald et al., 2015) and clinical trials (Camu et al., 2019; Hupperts et al., 2019; Soilu-Hänninen et al., 2012) on MS and vitamin D. Loken-Amsrud et al. (2012) reported in an earlier study that the association of low 25(OH)D levels with new T2 and gadolinium-enhancing lesions only was observed in untreated patients and not after starting IFNβ therapy. Alternatively, Stewart et al. (2012) found the association between vitamin D and relapse risk most clear in MS patients treated with IFNβ compared to no disease-modifying therapy. In cohorts treated with other disease-modifying therapy, associations of low 25(OH)D levels with increased risk of disease activity have also been reported, including glatiramer acetate, fingolimod, and natalizumab (Laursen, Søndergaard, Sørensen, Sellebjerg, & Oturai, 2016; Rotstein et al., 2015; Scott, Hackett, Dworek, & Schramke, 2013). Although results of an vitamin D add-on study to glatiramer acetate are being expected (Bhargava et al., 2014), no other controlled trials assessed combination therapy of vitamin D supplements with disease-modifying therapies other than IFNβ.

Genetic variation of included patients could also contribute. Mowry et al. observed in a cohort of CIS and RRMS patients an association between low 25(OH)D levels and new T2- and gadolinium-enhancing lesions, while an association with relapse risk was restricted to carriers of the HLA DRB1*1501 risk-allele (Mowry et al., 2012). HLA-DRB1*1501 has the largest effect size of all risk-alleles identified for MS (Patsopoulos et al., 2019). Interestingly, the HLA-DRB1 region only contains a vitamin D response element in its promoter region when the HLA-DRB1*15 risk-allele for MS is present (Ramagopalan et al., 2009). Accordingly, expression of HLA-DRB1 by lymphoblastoid cells in vitro was in the latter study only promoted by $1,25(OH)_2D$ in cells expressing HLA-DRB1*15. Although some association studies controlled for presence of this major risk-allele, this was not the case for clinical trials performed thus far. Several other risk-alleles contain a vitamin D response element in their promotors, of which transcription of IL-2 receptor alpha chain (CD25) and T-Cell Activation RhoGTPase Activating Protein (TAGAP) is also regulated by $1,25(OH)_2D$ (Berge et al., 2016). Interestingly, in the latter study, higher 25(OH)D levels correlate also with a higher $CD4^+$ T-cell mRNA expression of the IL-2 receptor alpha chain directly ex vivo in MS. Additionally, genome-wide association studies identified the genes encoding the CYP27B1 and CYP24A1, enzymes which regulate metabolism of $1,25(OH)_2D$, as risk-alleles for MS (Patsopoulos et al., 2019). Other sets of genes associated with both lower 25(OH)D levels and a higher risk of MS (Rhead et al., 2016) could also interact with associations between 25(OH)D levels and clinical endpoints in MS. Accordingly, 25(OH)D levels reached by high-dose vitamin D supplements in Dutch participants of the SOLAR trial were affected by single-nucleotide polymorphisms near the D-binding protein (DBP or group component) and CYP27B1 genes (Mimpen et al., 2021). Studies on vitamin D supplementation in MS thus far did not take interaction of clinical or radiological endpoints with these multiple putative genetic interactors into account.

Heterogeneity in dosing

Lastly, supplementation studies on vitamin D used a wide variety of dosages ranging from 20,000 IU weekly (Kampman et al., 2012; Soilu-Hänninen et al., 2012) to 100,000 IU twice monthly in the CHOLINE trial (Camu et al., 2019), and 14,000 IU/day in the

SOLAR trial (Hupperts et al., 2019). This resulted in average circulating 25(OH)D levels in the intervention arms of 110–123 nmol/L, 157 nmol/L, and 215 nmol/L, respectively. Clearly, these trials targeted high physiological to supra physiological concentrations of vitamin D. It remains unsolved whether there is any gain in supplementation toward supraphysiological levels rather than preventing deficient or insufficient values (<50 nmol/L; vide infra). Clearly, point estimate of effect sizes on MS endpoints did not differ dramatically between individual trials performed thus far. An important study by Pierrot-Deseilligny, Rivaud-Péchoux, Clerson, de Paz, and Souberbielle (2012) observed a protective effect of vitamin D on relapse risk up to circulating 25(OH)D levels of 110 nmol/L above which a plateau effect was reached. Additionally, observational studies did not investigate the association of 25(OH)D levels with MS endpoints above the physiological range of about 150 nmol/L (Ascherio et al., 2014; Fitzgerald et al., 2015; Simpson et al., 2010). The rationale of targeting these high levels is therefore mostly theoretically based on presumed working mechanisms of vitamin D, and not supported by observational or trial data. Although no safety issues were encountered in the clinical trials performed, caution for hypercalcemia was raised in EAE studies supplementing these high levels of vitamin D (Häusler et al., 2019).

Alternatively, if therapeutic approaches should be aimed on preventing the lowest rather than reaching the highest circulating concentration of vitamin D, the target level which should be crossed is uncertain. For endpoints in calcium metabolism (see below), levels below 50 nmol/L have been argued to be insufficient, while levels exceeding 80–100 nmol/L have been argued to be optimal (Holick, 2007). Whether similar levels hold for MS-related endpoints is uncertain. Regarding risk of developing MS, levels below 50 nmol/L were associated with a significantly higher risk of MS compared to levels above 100 nmol/L (Munger et al., 2006). When extrapolating these data to MS activity of MS, they can be interpreted as a call to avoid levels below 50 nmol/L, or rather a call to reach levels above 100 nmol/L. The data of Pierrot-Deseilligny et al. (2012) provide some support for the latter interpretation.

Ultimately, metaanalyses of randomized controlled trials taking into account baseline 25(OH)D levels and 25(OH)D levels reached should be able to provide a more definitive answer to this dosing question. However, it is to be said that current data provide insufficient support for targeting 25(OH)D levels exceeding 100–150 nmol/L.

Benefits of vitamin D supplements beyond MS

Although this chapter has a focus mostly on vitamin D and the disease process of MS, there are several other vitamin D-related endpoints relevant for people with MS.

People with MS are at an increased risk of falls, both early after first symptoms of MS (Moen et al., 2011) as well as in rehabilitation setting (Nilsagård, Lundholm, Denison, & Gunnarsson, 2009). Additionally, an increased risk of fractures in people with MS has been noted in several national registries (Bazelier, Bentzen, et al., 2012; Bazelier, van Staa, et al., 2012). One of the contributing factors to this risk could be the loss of bone mineral density frequently encountered in people with MS (Huang, Qi, Du, Chen, & Yan, 2015; Simonsen et al., 2016). In community dwelling elderly, low circulating 25(OH) levels are associated with loss of bone mineral density and an increased risk of falls and fractures (Bischoff-Ferrari, Dietrich, Orav, & Dawson-Hughes, 2004). Besides loss of bone mineral density, a loss of muscle strength is also associated with low 25(OH) D levels (Halfon, Phan, & Theta, 2015). Especially in MS, where muscular strength and control can be impaired, this can be of particular relevance. Additionally, supplementation of vitamin D in these cohorts is associated with a reduced risk of falls and fractures and promotion of bone mineral density (Bischoff-Ferrari et al., 2004, 2005; Tang, Eslick, Nowson, Smith, & Bensoussan, 2007). Since low circulating 25(OH)D levels are common in MS (Smolders, Menheere, et al., 2008; van der Mei et al., 2007), preventing these low concentration with vitamin D supplements could be worthwhile to improve these endpoints. Based on decompensation of calcium metabolism based on elevation of parathyroid hormone levels and population studies, levels below 50 nmol/L are generally deemed insufficient for maintaining a healthy calcium metabolism (Holick, 2007).

Conclusion

Low circulating 25(OH)D levels are associated with several adverse MS endpoints, and supplementation studies on vitamin D showed several signals of improvement on these endpoints. Inconsistent results of scientific research are most likely the result of underpowered studies, reverse causation, confounding, variation in included patient subgroups, variation in genetic constitution of included patients, and variation in dosing of vitamin D. Although there are several items uncertain, there is probably more to gain than to lose for people with MS in supplementing low circulating 25(OH)D levels. This accounts both for disease activity of MS as well as the general bone health. Although there may be additional benefits in targeting 25(OH)D levels exceeding 100 nmol/L, preventing levels below 50 nmol/L is the lowest reasonable target of supplementation. Future clinical studies and metaanalyses should reveal which doses of vitamin D supplements could be of benefit for which (subgroups of) MS patients.

Applications on other neurological conditions

Although most extensively studied in MS, low circulating 25(OH)D levels have been associated in several other neurological diseases. These include other neuroinflammatory disease, but also neurodegenerative diseases, such as amyotrophic lateral sclerosis (ALS), Alzheimer's disease, and in conditions such as delirium (Smolders, Hiller, & Camu, 2021). Studies in these diseases encounter similar challenges as observations in the MS field. Although the theoretic framework to postulate an effect of vitamin D supplements in these diseases is quite solid, association studies on 25(OH)D-levels are difficult to interpret and well-powered trials on vitamin D supplements are mostly lacking. Nevertheless, regardless of any effects on neurological endpoints, these diseases all comprise vulnerable patients which may benefit from optimization of 25(OH)D levels for the sake of optimizing calcium metabolism.

Other components of interest

Physicians treating people with MS will probably recognize that diet and MS are often a hot topic during patient consultations. A high prudent diet factor score has been associated with a lower risk of subsequent relapses in early MS (Simpson-Yap et al., 2020). Accordingly, a more healthy diet has been associated with lesser disability in MS (Fitzgerald et al., 2018). The exact nutrients mediating these associations are not well defined. Experimental studies pinpointed a high salt intake as possibly detrimental in MS, yet this factor was not associated with MS course in observational studies (Fitzgerald et al., 2017). A study on omega-3 fatty acid supplementation did not make its primary endpoint in MS (Torkildsen et al., 2012). Additionally, many supplements or dietary regimens are promoted in layman's literature and within patient communities with any solid scientific basis and clinical trials lacking. Therefore, it remains difficult to provide evidence-based recommendations to patients with MS. On the whole, promotion of a healthy bodyweight, active lifestyle, and an accordingly healthy diet appear reasonable advises for people with MS to promote long-term well-being and mobility.

Mini dictionary of terms

25(OH)D. Most abundant and stable vitamin D metabolite in the circulation. This molecule reflects an individual's vitamin D status best. Synonym: calcediol or calcifediol.

1,25(OH)$_2$*D.* Biologically active metabolite of vitamin D. Synonym: calcitriol.

Clinically isolate syndrome. First attack of demyelinating disease in an individual presenting as a syndrome of focal neurological deficits. In patients with MS, this is usually the first clinical presentation of disease.

Relapsing remitting MS. Most frequent clinical phenotype of MS, in which subacute attacks of temporal neurological disability occur.

Progressive MS. Phenotype in MS in which disability gradually accumulates during follow-up. Usually, a relapsing remitting disease course evolves in a secondary progressive phenotype. A small proportion of patients have a primary progressive disease course from diagnosis onward.

Vitamin D$_2$*.* Vitamin D variant derived from a vegetal diet. Synonym: ergocalciferol.

Vitamin D_3. Vitamin D variant synthesized in the skin upon sunlight (ultraviolet B) exposure and additionally derived from an animal diet, in particular, fatty fish. Synonym: cholecalciferol.

Summary points

- Vitamin D is the precursor of a steroid hormone with immune-modulating properties.
- Low circulating levels of 25-hydroxyvitamin D have been associated with a higher risk of disease activity in patients suffering from multiple sclerosis (MS).
- Several randomized controlled clinical trials in MS, although mostly negative on primary endpoints, provided some indications for positive effects of vitamin D supplements on these endpoints.
- Results of individual association studies and trials in MS are heterogeneous, which may be due to underpowered studies, causality of associations, confounders, variation in included patient subgroups, variation in genetic constitution of included patients, and variation in dosing of vitamin D.
- Targeting with vitamin D supplementation circulating 25-hydroxyvitamin D levels of above 50 nmol/L is likely of benefit, with possibly additional benefit of targeting levels above 100 nmol/L.

References

Achiron, A., Givon, U., Magalashvili, D., Dolev, M., Liraz Zaltzman, S., Kalron, A., et al. (2015). Effect of Alfacalcidol on multiple sclerosis-related fatigue: A randomized, double-blind placebo-controlled study. *Multiple Sclerosis*, *21*, 767–775.

Ascherio, A., Munger, K. L., White, R., Köchert, K., Simon, K. C., Polman, C. H., et al. (2014). Vitamin D as an early predictor of multiple sclerosis activity and progression. *JAMA Neurology*, *71*, 306.

Barkhof, F., Simon, J. H., Fazekas, F., Rovaris, M., Kappos, L., De Stefano, N., et al. (2012). MRI monitoring of immunomodulation in relapse-onset multiple sclerosis trials. *Nature Reviews. Neurology*, *8*, 13–21.

Bazelier, M. T., Bentzen, J., Vestergaard, P., Stenager, E., Leufkens, H. G., van Staa, T.-P., et al. (2012). The risk of fracture in incident multiple sclerosis patients: The Danish National Health Registers. *Multiple Sclerosis Journal*, *18*, 1609–1616.

Bazelier, M. T., van Staa, T.-P., Uitdehaag, B. M. J., Cooper, C., Leufkens, H. G. M., Vestergaard, P., et al. (2012). Risk of fractures in patients with multiple sclerosis: A population-based cohort study. *Neurology*, *78*, 1967–1973.

Berge, T., Leikfoss, I. S., Brorson, I. S., Bos, S. D., Page, C. M., Gustavsen, M. W., et al. (2016). The multiple sclerosis susceptibility genes TAGAP and IL2RA are regulated by Vitamin D in CD4 + T cells. *Genes and Immunity*, *17*, 118–127.

Bhargava, P., Cassard, S., Steele, S. U., Azevedo, C., Pelletier, D., Sugar, E. A., et al. (2014). The vitamin D to Ameliorate Multiple Sclerosis (VIDAMS) trial: Study design for a multicenter, randomized, double-blind controlled trial of vitamin D in multiple sclerosis. *Contemporary Clinical Trials*, *39*, 288–293.

Bhargava, P., Steele, S. U., Waubant, E., Revirajan, N. R., Marcus, J., Dembele, M., et al. (2016). Multiple sclerosis patients have a diminished serologic response to Vitamin D supplementation compared to healthy controls. *Multiple Sclerosis*, *22*, 753–760.

Bischoff-Ferrari, H. A., Dawson-Hughes, B., Willett, W. C., Staehelin, H. B., Bazemore, M. G., Zee, R. Y., et al. (2004). Effect of vitamin D on falls: A meta-analysis. *Journal of the American Medical Association, 291*, 1999–2006.

Bischoff-Ferrari, H. A., Dietrich, T., Orav, E. J., & Dawson-Hughes, B. (2004). Positive association between 25-hydroxy vitamin D levels and bone mineral density: A population-based study of younger and older adults. *The American Journal of Medicine, 116*, 634–639.

Bischoff-Ferrari, H. A., Willett, W. C., Wong, J. B., Giovannucci, E., Dietrich, T., & Dawson-Hughes, B. (2005). Fracture prevention with vitamin D supplementation: A meta-analysis of randomized controlled trials. *Journal of the American Medical Association, 293*, 2257–2264.

Camu, W., Lehert, P., Pierrot Deseilligny, C., Hautecoeur, P., Besserve, A., Jean Deleglise, A., et al. (2019). Cholecalciferol in relapsing–remitting MS. A randomized clinical trial (CHOLINE). *Neurology Neuroimmunology & Neuroinflammation, 6*, e597.

Cantorna, M. T., Hayes, C. E., & DeLuca, H. F. (1996). 1,25-Dihydroxyvitamin D3 reversibly blocks the progression of relapsing encephalomyelitis, a model of multiple sclerosis. *Proceedings of the National Academy of Sciences of the United States of America, 93*, 7861–7864.

Derakhshandi, H., Etemadifar, M., Feizi, A., Abtahi, S.-H., Minagar, A., Abtahi, M.-A., et al. (2013). Preventive effect of vitamin D3 supplementation on conversion of optic neuritis to clinically definite multiple sclerosis: A double blind, randomized, placebo-controlled pilot clinical trial. *Acta Neurologica Belgica, 113*, 257–263.

Dörr, J., Bäcker-Koduah, P., Wernecke, K.-D., Becker, E., Hoffmann, F., Faiss, J., et al. (2020). High-dose vitamin D supplementation in multiple sclerosis – results from the randomized EVIDIMS (efficacy of vitamin D supplementation in multiple sclerosis) trial. *Multiple Sclerosis Journal – Experimental, Translational and Clinical, 6*, 205521732090347.

Fitzgerald, K. C., Munger, K. L., Hartung, H. P., Freedman, M., Montalbán, X., Edan, G., et al. (2017). Sodium intake and multiple sclerosis activity and progression in BENEFIT. *Annals of Neurology, 82*, 20–29.

Fitzgerald, K. C., Munger, K. L., Köchert, K., Arnason, B. G. W., Comi, G., Cook, S., et al. (2015). Association of vitamin D levels with multiple sclerosis activity and progression in patients receiving interferon beta-1b. *JAMA Neurology, 72*, 1458.

Fitzgerald, K. C., Tyry, T., Salter, A., Cofield, S. S., Cutter, G., Fox, R., et al. (2018). Diet quality is associated with disability and symptom severity in multiple sclerosis. *Neurology, 90*, E1–E11.

Gafson, A. R., Barthélemy, N. R., Bomont, P., Carare, R. O., Durham, H. D., Julien, J. P., et al. (2020). Neurofilaments: Neurobiological foundations for biomarker applications. *Brain, 143*, 1975–1998.

Golan, D., Halhal, B., Glass-Marmor, L., Staun-Ram, E., Rozenberg, O., Lavi, I., et al. (2013). Vitamin D supplementation for patients with multiple sclerosis treated with interferon-beta: A randomized controlled trial assessing the effect on flu-like symptoms and immunomodulatory properties. *BMC Neurology, 13*, 60.

Graves, J. S., Barcellos, L. F., Krupp, L., Belman, A., Shao, X., Quach, H., et al. (2020). Vitamin D genes influence MS relapses in children. *Multiple Sclerosis Journal, 26*, 894–901.

Haifer, C., Lawrance, I. C., Center, J. R., Clarke, M. W., Hart, P. H., Eisman, J. A., et al. (2019). Vitamin D metabolites are lower with active Crohn's disease and spontaneously recover with development of remission. *Therapeutic Advances in Gastroenterology, 12*.

Halfon, M., Phan, O., & Theta, D. (2015). Vitamin D: A review on its effects on muscle strength, the risk of fall, and frailty. *BioMed Research International, 2015*.

Hänninen, K., Jääskeläinen, O., Herukka, S. K., & Soilu-Hänninen, M. (2020). Vitamin D supplementation and serum neurofilament light chain in interferon-beta-1b-treated MS patients. *Brain and Behavior: A Cognitive Neuroscience Perspective*.

Harroud, A., Manousaki, D., Butler-Laporte, G., Mitchell, R. E., Davey Smith, G., Richards, J. B., et al. (2021). The relative contributions of obesity, vitamin D, leptin, and adiponectin to multiple sclerosis risk: A Mendelian randomization mediation analysis. *Multiple Sclerosis*, 1352458521995484.

Harroud, A., Mitchell, R. E., Richardson, T. G., Morris, J. A., Forgetta, V., Davey Smith, G., et al. (2021). Childhood obesity and multiple sclerosis: A Mendelian randomization study. *Multiple Sclerosis Journal*.

Hart, P. H., Jones, A. P., Trend, S., Cha, L., Fabis-Pedrini, M. J., Cooper, M. N., et al. (2018). A randomized, controlled clinical trial of narrowband UVB phototherapy for clinically isolated syndrome: The PhoCIS study. *Multiple Sclerosis Journal – Experimental, Translational and Clinical, 4*, 205521731877311.

Häusler, D., Torke, S., Peelen, E., Bertsch, T., Djukic, M., Nau, R., et al. (2019). High dose vitamin D exacerbates central nervous system autoimmunity by raising T-cell excitatory calcium. *Brain*.

Holick, M. F. (2007). Vitamin D deficiency. *The New England Journal of Medicine, 357*, 266–281.

Holmøy, T., Røsjø, E., Zetterberg, H., Blennow, K., Lindstrøm, J. C., Steffensen, L. H., et al. (2018). Vitamin D supplementation and neurofilament light chain in multiple sclerosis. *Acta Neurologica Scandinavica*.

Holowaychuk, M. K., Birkenheuer, A. J., Li, J., Marr, H., Boll, A., & Nordone, S. K. (2012). Hypocalcemia and hypovitaminosis D in dogs with induced endotoxemia. *Journal of Veterinary Internal Medicine, 26*, 244–251.

Huang, Z., Qi, Y., Du, S., Chen, G., & Yan, W. (2015). BMI levels with MS bone mineral density levels in adults with multiple sclerosis: A meta-analysis. *The International Journal of Neuroscience, 125*, 904–912.

Hupperts, R., Smolders, J., Vieth, R., Holmøy, T., Marhardt, K., Schluep, M., et al. (2019). Randomized trial of daily high-dose vitamin D3 in patients with RRMS receiving subcutaneous interferon beta-1a. *Neurology, 93*, 1–19.

Kampman, M. T., Steffensen, L. H., Mellgren, S. I., & Jørgensen, L. (2012). Effect of vitamin D 3 supplementation on relapses, disease progression, and measures of function in persons with multiple sclerosis: Exploratory outcomes from a double-blind randomized controlled trial. *Multiple Sclerosis Journal, 18*, 1144–1151.

Kuhle, J., Disanto, G., Dobson, R., Adiutori, R., Bianchi, L., Topping, J., et al. (2015). Conversion from clinically isolated syndrome to multiple sclerosis: A large multicentre study. *Multiple Sclerosis Journal, 21*, 1013–1024.

Laursen, J. H., Søndergaard, H. B., Sørensen, P. S., Sellebjerg, F., & Oturai, A. B. (2016). Vitamin D supplementation reduces relapse rate in relapsing–remitting multiple sclerosis patients treated with natalizumab. *Multiple Sclerosis and Related Disorders, 10*, 169–173.

Loken-Amsrud, K. I., Holmoy, T., Bakke, S. J., Beiske, A. G., Bjerve, K. S., Bjornara, B. T., et al. (2012). Vitamin D and disease activity in multiple sclerosis before and during interferon- treatment. *Neurology, 79*, 267–273.

Lucas, R. M., Byrne, S. N., Correale, J., Ilschner, S., & Hart, P. H. (2015). Ultraviolet radiation, vitamin D and multiple sclerosis. *Neurodegenerative Disease Management, 5*, 413–424.

Manouchehrinia, A., Hedström, A. K., Alfredsson, L., Olsson, T., Hillert, J., & Ramanujam, R. (2018). Association of pre-disease body mass index with multiple sclerosis prognosis. *Frontiers in Neurology, 9*, 232.

Mimpen, M., Rolf, L., Poelmans, G., van den Ouweland, J., Hupperts, R., Damoiseaux, J., et al. (2021). Vitamin D related genetic polymorphisms affect serological response to high-dose vitamin D supplementation in multiple sclerosis. *PLoS One, 16*(12), e0261097.

Moen, S. M., Celius, E. G., Sandvik, L., Nordsletten, L., Eriksen, E. F., & Holmoy, T. (2011). Low bone mass in newly diagnosed multiple sclerosis and clinically isolated syndrome. *Neurology*, 77, 151–157.

Mowry, E. M., Azevedo, C. J., McCulloch, C. E., Okuda, D. T., Lincoln, R. R., Waubant, E., et al. (2018). Body mass index, but not vitamin D status, is associated with brain volume change in MS. *Neurology*. https://doi.org/10.1212/WNL.0000000000006644.

Mowry, E. M., Krupp, L. B., Milazzo, M., Chabas, D., Strober, J. B., Belman, A. L., et al. (2010). Vitamin D status is associated with relapse rate in pediatric-onset MS. *Annals of Neurology, 67*.

Mowry, E. M., Waubant, E., McCulloch, C. E., Okuda, D. T., Evangelista, A. A., Lincoln, R. R., et al. (2012). Vitamin D status predicts new brain magnetic resonance imaging activity in multiple sclerosis. *Annals of Neurology, 72*, 234–240.

Munger, K. L., Åivo, J., Hongell, K., Soilu-Hänninen, M., Surcel, H.-M., & Ascherio, A. (2016). Vitamin D status during pregnancy and risk of multiple sclerosis in offspring of women in the finnish maternity cohort. *JAMA Neurology, 73*, 515.

Munger, K. L., Chitnis, T., & Ascherio, A. (2009). Body size and risk of MS in two cohorts of US women. *Neurology, 73*, 1543–1550.

Munger, K. L., Levin, L. I., Hollis, B. W., Howard, N. S., & Ascherio, A. (2006). Serum 25-hydroxyvitamin D levels and risk of multiple sclerosis. *JAMA, 296*, 2832.

Munger, K. L., Zhang, S. M., O'Reilly, E., Hernán, M. A., Olek, M. J., Willett, W. C., et al. (2004). Vitamin D intake and incidence of multiple sclerosis. *Neurology, 62*, 60–65.

Muris, A.-H., Rolf, L., Broen, K., Hupperts, R., Damoiseaux, J., & Smolders, J. (2016). A low vitamin D status at diagnosis is associated with an early conversion to secondary progressive multiple sclerosis. *The Journal of Steroid Biochemistry and Molecular Biology, 164*, 254–257.

Muris, A.-H., Smolders, J., Rolf, L., Klinkenberg, L. J., van der Linden, N., Meex, S., et al. (2016). Vitamin D status does not affect disability progression of patients with multiple sclerosis over three year follow-up. *PLoS One, 11*, e0156122.

Nashold, F. E., Spach, K. M., Spanier, J. A., & Hayes, C. E. (2009). Estrogen controls vitamin D 3 -mediated resistance to experimental autoimmune encephalomyelitis by controlling vitamin D 3 metabolism and receptor expression. *Journal of Immunology, 183*, 3672–3681.

Nielsen, N. M., Munger, K. L., Koch-Henriksen, N., Hougaard, D. M., Magyari, M., Jørgensen, K. T., et al. (2017). Neonatal vitamin D status and risk of multiple sclerosis. *Neurology, 88*, 44–51.

Nilsagård, Y., Lundholm, C., Denison, E., & Gunnarsson, L. G. (2009). Predicting accidental falls in people with multiple sclerosis - A longitudinal study. *Clinical Rehabilitation, 23*, 259–269.

O'Connell, K., Sulaimani, J., Basdeo, S. A., Kinsella, K., Jordan, S., Kenny, O., et al. (2017). Effects of vitamin D3 in clinically isolated syndrome and healthy control participants: A double-blind randomized controlled trial. *Multiple Sclerosis Journal – Experimental, Translational and Clinical, 3*, 2055217317727296.

Ostkamp, P., Salmen, A., Pignolet, B., Görlich, D., Andlauer, T. F. M., Schulte-Mecklenbeck, A., et al. (2021). Sunlight exposure exerts immunomodulatory effects to reduce multiple sclerosis severity. *Proceedings of the National Academy of Sciences of the United States of America, 118*.

Patsopoulos, N. A., Baranzini, S. E., Santaniello, A., Shoostari, P., Cotsapas, C., Wong, G., et al. (2019). Multiple sclerosis genomic map implicates peripheral immune cells and microglia in susceptibility. *Science (80-)*, 365.

Pierrot-Deseilligny, C., Rivaud-Péchoux, S., Clerson, P., de Paz, R., & Souberbielle, J.-C. (2012). Relationship between 25-OH-D serum level and relapse rate in multiple sclerosis patients before and after vitamin D supplementation. *Therapeutic Advances in Neurological Disorders, 5*, 187–198.

Ramagopalan, S. V., Maugeri, N. J., Handunnetthi, L., Lincoln, M. R., Orton, S. M., Dyment, D. A., et al. (2009). Expression of the multiple sclerosis-associated MHC class II allele HLA-DRB1*1501 is regulated by vitamin D. *PLoS Genetics, 5*.

Reid, D., Toole, B. J., Knox, S., Talwar, D., Harten, J., O'Reilly, D. S. J., et al. (2011). The relation between acute changes in the systemic inflammatory response and plasma 25-hydroxyvitamin D concentrations after elective knee arthroplasty. *The American Journal of Clinical Nutrition, 93*, 1006–1011.

Rhead, B., Bäärnhielm, M., Gianfrancesco, M., Mok, A., Shao, X., Quach, H., et al. (2016). Mendelian randomization shows a causal effect of low vitamin D on multiple sclerosis risk. *Neurology Genetics, 2*, e97.

Røsjø, E., Lindstrøm, J. C., Holmøy, T., Myhr, K. M., Varhaug, K. N., & Torkildsen, Ø. (2020). Natural variation of vitamin D and neurofilament light chain in relapsing–remitting multiple sclerosis. *Frontiers in Neurology, 11*.

Rotstein, D. L., Healy, B. C., Malik, M. T., Carruthers, R. L., Musallam, A. J., Kivisakk, P., et al. (2015). Effect of vitamin D on MS activity by disease-modifying therapy class. *Neurology Neuroimmunology & Neuroinflammation, 2*.

Runia, T. F., Hop, W. C. J., de Rijke, Y. B., Buljevac, D., & Hintzen, R. Q. (2012). Lower serum vitamin D levels are associated with a higher relapse risk in multiple sclerosis. *Neurology, 79*, 261–266.

Salari, M., Janghorbani, M., Etemadifar, M., Dehghani, A., Razmjoo, H., & Naderian, G. (2015). Effects of vitamin D on retinal nerve fiber layer in vitamin D deficient patients with optic neuritis: Preliminary findings of a randomized, placebo-controlled trial. *Journal of Research in Medical Sciences: The Official Journal of Isfahan University of Medical Sciences, 20*, 372–378.

Scott, T. F., Hackett, C. T., Dworek, D. C., & Schramke, C. J. (2013). Low vitamin D level is associated with higher relapse rate in natalizumab treated MS patients. *Journal of the Neurological Sciences, 330*, 27–31.

Shaygannejad, V., Janghorbani, M., Ashtari, F., & Dehghan, H. (2012). Effects of adjunct low-dose vitamin D on relapsing–remitting multiple sclerosis progression: Preliminary findings of a randomized placebo-controlled trial. *multiple sclerosis International, 2012*, 1–7.

Simonsen, C. S., Celius, E. G., Brunborg, C., Tallaksen, C., Eriksen, E. F., Holmøy, T., et al. (2016). Bone mineral density in patients with multiple sclerosis, hereditary ataxia or hereditary spastic paraplegia after at least 10 years of disease - A case control study. *BMC Neurology, 16*, 252.

Simpson, S., Taylor, B., Blizzard, L., Ponsonby, A.-L., Pittas, F., Tremlett, H., et al. (2010). Higher 25-hydroxyvitamin D is associated with lower relapse risk in MS. *Annals of Neurology, 68*.

Simpson, S., Van Der Mei, I., Lucas, R. M., Ponsonby, A. L., Broadley, S., Blizzard, L., et al. (2018). Sun exposure across the life course significantly modulates early multiple sclerosis clinical course. *Frontiers in Neurology, 9*.

Simpson-Yap, S., Oddy, W. H., Taylor, B., Lucas, R. M., Black, L. J., Ponsonby, A. L., et al. (2020). High Prudent diet factor score predicts lower relapse hazard in early multiple sclerosis. *Multiple Sclerosis Journal*.

Smolders, J., & Damoiseaux, J. (2011). Vitamin D as a T-cell modulator in multiple sclerosis. *Vitamins & Hormones, 86*, 401–428.

Smolders, J., Damoiseaux, J., Menheere, P., & Hupperts, R. (2008). Vitamin D as an immune modulator in multiple sclerosis, a review. *Journal of Neuroimmunology, 194*, 7–17.

Smolders, J., Hiller, A., & Camu, W. (2021). Editorial: Vitamin D in neurological diseases: From pathophysiology to therapy. *Frontiers in Neurology, 12*, 614900.

Smolders, J., Menheere, P., Kessels, A., Damoiseaux, J., & Hupperts, R. (2008). Association of vitamin D metabolite levels with relapse rate and disability in multiple sclerosis. *Multiple Sclerosis, 14*, 1220–1224.

Smolders, J., Mimpen, M., Oechtering, J., Damoiseaux, J., Ouweland, J., Hupperts, R., et al. (2020). Vitamin D 3 supplementation and neurofilament light chain in multiple sclerosis. *Acta Neurologica Scandinavica, 141*, 77–80.

Smolders, J., Moen, S. M., Damoiseaux, J., Huitinga, I., & Holmøy, T. (2011). Vitamin D in the healthy and inflamed central nervous system: Access and function. *Journal of the Neurological Sciences, 311*, 37–43.

Smolders, J., Schuurman, K. G., Van Strien, M. E., Melief, J., Hendrickx, D., Hol, E. M., et al. (2013). Expression of vitamin D receptor and metabolizing enzymes in multiple sclerosis-affected brain tissue. *Journal of Neuropathology and Experimental Neurology, 72*, 91–105.

Smolders, J., Torkildsen, Ø., Camu, W., & Holmøy, T. (2019). An update on vitamin D and disease activity in multiple sclerosis. *CNS Drugs, 33*, 1187–1199.

Smolders, J., van den Ouweland, J., Geven, C., Pickkers, P., & Kox, M. (2020). Letter to the editor: Vitamin D deficiency in COVID-19: Mixing up cause and consequence. *Metabolism, 154434*.

Soilu-Hänninen, M., Airas, L., Mononen, I., Heikkilä, A., Viljanen, M., & Hänninen, A. (2005). 25-Hydroxyvitamin D levels in serum at the onset of multiple sclerosis. *Multiple Sclerosis, 11*, 266–271.

Soilu-Hänninen, M., Aivo, J., Lindström, B.-M., Elovaara, I., Sumelahti, M.-L., Färkkilä, M., et al. (2012). A randomized, double blind, placebo controlled trial with vitamin D3 as an add on treatment to interferon β-1b in patients with multiple sclerosis. *Journal of Neurology, Neurosurgery, and Psychiatry, 83*, 565–571.

Soilu-Hänninen, M., Laaksonen, M., Laitinen, I., Erälinna, J. P., Lilius, E. M., & Mononen, I. (2008). A longitudinal study of serum 25-hydroxyvitamin D and intact parathyroid hormone levels indicate the importance of vitamin D and calcium homeostasis regulation in multiple sclerosis. *Journal of Neurology, Neurosurgery, and Psychiatry, 79*, 152–157.

Sotirchos, E. S., Bhargava, P., Eckstein, C., Van Haren, K., Baynes, M., Ntranos, A., et al. (2016). Safety and immunologic effects of high- vs low-dose cholecalciferol in multiple sclerosis. *Neurology, 86*, 382–390.

Stein, M. S., Liu, Y., Gray, O. M., Baker, J. E., Kolbe, S. C., Ditchfield, M. R., et al. (2011). A randomized trial of high-dose vitamin D2 in relapsing–remitting multiple sclerosis. *Neurology, 77*, 1611–1618.

Stewart, N., Simpson, S., Van Der Mei, I., Ponsonby, A. L., Blizzard, L., Dwyer, T., et al. (2012). Interferon-β and serum 25-hydroxyvitamin D interact to modulate relapse risk in MS. *Neurology, 79*, 254–260.

Tang, B. M., Eslick, G. D., Nowson, C., Smith, C., & Bensoussan, A. (2007). Use of calcium or calcium in combination with vitamin D supplementation to prevent fractures and bone loss in people aged 50 years and older: A meta-analysis. *Lancet, 370*, 657–666.

Thompson, A. J., Baranzini, S. E., Geurts, J., Hemmer, B., & Ciccarelli, O. (2018). Multiple sclerosis. *Lancet, 391*, 1622–1636.
Torkildsen, Ø., Wergeland, S., Bakke, S., Beiske, A. G., Bjerve, K. S., Hovdal, H., et al. (2012). ω-3 Fatty acid treatment in multiple sclerosis (OFAMS study): A randomized, double-blind, placebo-controlled trial. *Archives of Neurology, 69*, 1044–1051.
Trend, S., Jones, A. P., Cha, L., Cooper, M. N., Geldenhuys, S., Fabis-Pedrini, M. J., et al. (2019). Short-term changes in frequencies of circulating leukocytes associated with narrowband UVB phototherapy in people with clinically isolated syndrome. *Scientific Reports, 9*.
University of California, San Francisco MS-EPIC Team, BAC, C., Gourraud, P.-A., Oksenberg, J. R., Bevan, C., Crabtree-Hartman, E., et al. (2016). Long-term evolution of multiple sclerosis disability in the treatment era. *Annals of Neurology, 80*, 499–510.
van der Mei, I. A. F., Ponsonby, A.-L., Dwyer, T., Blizzard, L., Taylor, B. V., Kilpatrick, T., et al. (2007). Vitamin D levels in people with multiple sclerosis and community controls in Tasmania, Australia. *Journal of Neurology, 254*, 581–590.
Wallin, M. T., Culpepper, W. J., Nichols, E., Bhutta, Z. A., Gebrehiwot, T. T., Hay, S. I., et al. (2019). Global, regional, and national burden of multiple sclerosis 1990–2016: A systematic analysis for the Global Burden of Disease Study 2016. *Lancet Neurology, 18*, 269–285.
Wortsman, J., Matsuoka, L. Y., Chen, T. C., Lu, Z., & Holick, M. F. (2000). Decreased bioavailability of vitamin D in obesity. *The American Journal of Clinical Nutrition, 72*, 690–693.

PART IX

Neuroinflammation

CHAPTER 23

Neuroinflammation: Role of magnesium

Renu Agarwal[a] and Igor Iezhitsa[a,b]
[a]Department of Pharmacology and Therapeutics, School of Medicine, International Medical University, Kuala Lumpur, Malaysia
[b]Department of Pharmacology and Bioinformatics, Volgograd State Medical University, Volgograd, Russia

Neuroinflammation

Neuroinflammation has now been recognized as the key pathophysiological mechanism underlying several neurological conditions affecting CNS such as multiple sclerosis, Alzheimer's disease, stroke, migraine, and epilepsy, and those affecting neural tissue elsewhere such as retina and peripheral nerves. Neuroinflammation refers to inflammatory responses of nervous tissue to various types of insults such as ischemia, invasion by pathogens, or other stimuli. The consequences of neuroinflammatory response depend on the nature of initial insult as well as the severity and duration of inflammatory reaction. A controlled and brief inflammatory response promotes healing and is beneficial to the tissue. Such responses also provide preconditioning to innate immune system toward a phenotype with greater neuroprotective abilities. However, chronic uncontrolled inflammatory responses associated with excessive production of cytokines and reactive oxygen and nitrogen species cause accumulation of neutrophils and macrophages at the site of injury and consequent tissue damage. Several brief and repeated episodes of inflammation may also result in monocyte and macrophage accumulation and tissue damage. A low level of chronic inflammation may result in impaired neuronal plasticity and cognitive functions whereas more intense chronic inflammation is known to underlie several neurodegenerative diseases (DiSabato, Quan, & Godbout, 2016).

Magnesium and neurological disorders

Magnesium (Mg) is the second most abundant intracellular cation after potassium and overall is the fourth most abundant cation (Iezhitsa & Spasov, 2008). This bivalent cation is involved in wide-ranging cellular functions due to its regulatory role in several biochemical and metabolic reactions (Durlach, 1988). It acts as a cofactor for more than 300 enzymes that regulate diverse biochemical reactions such as protein synthesis, blood glucose control, muscle, and nerve function (Durlach, 1988; Iezhitsa & Spasov, 2008). It is complexed with ATP and participates in a variety of redox reactions. Its regulatory functions are not limited to enzymatic activities but also extend to regulation of ion channels and intracellular signaling (Wolf & Trapani, 2008). In fact, Mg has been shown to couple cell surface receptors on T lymphocytes with intracellular effectors (Li et al., 2011).

Vitamins and Minerals in Neurological Disorders
https://doi.org/10.1016/B978-0-323-89835-5.00027-2

Mg deficiency is prevalent mainly due to intestinal and urinary losses in certain diseases such as alcoholism (Elisaf, Merkourpoulos, Tsianos, & Siamopoulos, 1995) and diabetes (Gommers, Hoenderop, Bindels, & de Baaij, 2016). Inadequate Mg intake may be a contributory factor. Drug therapy such as use of diuretics, aminoglycosides, digoxin, cisplatin, cyclosporin, etc., may also result in Mg deficiency (Iezhitsa, 2005; Liamis, Hoorn, Florentin, & Milionis, 2021). The crucial role of Mg in maintaining immune functions is well described and Mg deficiency is known to induce inflammatory responses. Epidemiological studies have shown association of low Mg status with low-grade inflammation and increased level of C-reactive protein (CRP), an acute phase protein. Supplementation with Mg rich preparations exerts an antiinflammatory effect. Considering that the inflammation underlies the pathophysiology of vast majority of neurological conditions, it is not surprising that Mg deficiency has been shown to be associated with several neurological conditions. For example, in patients with migraine, significantly lower Mg levels were detected in serum, saliva, and the brain (Ramadan et al., 1989; Sarchielli et al., 1992). Patients with Parkinson's disease, Alzheimer's disease, stroke, epilepsy, and psychiatric symptoms have also been found to have low Mg level (Kirkland, Sarlo, & Holton, 2018). Physiologically, cerebrospinal fluid (CSF) concentration of Mg is higher than its concentration in plasma ultrafiltrate as Mg is actively transported across blood-brain barrier. Deficiency of Mg causes depletion of Mg concentration in CSF; however, fall in CSF Mg level lags and is less pronounced compared to reduction in its plasma concentration. Reduction in CSF Mg level correlates with similar changes in extracellular brain Mg level and these changes are quickly restored by Mg supplementation (Morris, 1992). In fact, in AD brains, the Mg levels were found to be significantly low compared to healthy controls (Cilliler, Ozturk, & Ozbakir, 2007). One of the preclinical studies showed that Mg-L-threonate effectively prevents synaptic loss and alleviates cognitive decline in aging rodents (Liu, Weinger, Lu, Xue, & Sadeghpour, 2015).

Mg deficiency and neuroinflammation

Although, in human, Mg deficiency alone may not induce inflammation, it is known to predispose to exaggerated inflammatory response to injury and may potentiate the preexisting inflammatory reactions (King, 2009). Hence, Mg-deficient state sufficient to induce inflammation in human volunteers for experimental purposes is unlikely to be achieved. Moreover, there are ethical implications due to adverse effects of Mg deficiency including cardiac adverse events. In population-based studies, effects of low Mg levels resulting from dietary Mg deficiency are generally not sufficiently pronounced. In patients who develop significant Mg depletion due to disease conditions, it is difficult to associate pathological changes with Mg depletion alone due to complexities of pathogenetic processes underlying various diseases. Hence, it is pertinent to use animal models to understand the whole range of implications of Mg-deficient state.

In animals, the clinical symptoms of inflammation were first observed in Mg-deficient rats as early as in 1930s (Kruse, Orent, & McCollum, 1932), and subsequently, Mg-deficient animals were used for experimental purposes. Under experimental conditions, the Mg contents of diet and pathophysiological condition of animals determine the severity and rapidity of the development of Mg-deficient state (Mazur et al., 2007). The features of inflammation resulting from Mg deficiency often manifest as hyperemia of ears and paws, oedema, and enlarged spleen are particularly pronounced in rats compared to other animal species. These animals also show marked leucocytosis, increased level of proinflammatory cytokines, neuropeptides such as substance P, inflammatory mediators, and acute phase proteins. Notably, specificity of Mg deficiency as a cause of inflammation in these experimental models is supported by absence of any pathogen or lipopolysaccharide in the blood and organs and relief of symptoms by restoring normal Mg levels after supplementation (Mazur, Libako, Nowacki, & Maier, 2013). The use of such animal models has played a key role in enhancing our understanding of the molecular mechanisms associated with Mg deficiency.

Mechanisms of neuroinflammation in Mg deficiency

Mg was defined as natural physiological calcium (Ca) antagonist. The ability of Mg to cause nonspecific Ca channel blockade explains its several cellular mechanisms of action that translate into antiinflammatory effect. The key pathophysiological mechanisms underlying neuroinflammation resulting from Mg deficiency include:

i. Priming and activation of phagocytic cells
ii. Activation of *N*-methyl-D-aspartate (NMDA) receptors
iii. Activation of proinflammatory signaling pathways and increased expression of proinflammatory cytokines and neuropeptides
iv. Increased oxidative and nitrosative stress

(i) *Mg and phagocyte priming and activation*

The role of phagocytes in defending the host organism against injurious agents is pivotal. The hallmark feature of the phagocytic activity is the "respiratory burst" which refers to a multistep process enabling phagocytes to rapidly produce and release reactive oxygen species (ROS). The respiratory burst by phagocytes, largely leucocytes, is preceded by their "priming" due to exposure to priming agents such as cytokines (e.g., tumor necrosis factor alpha, TNF-α), lipid mediators (e.g., platelet activating factor, PAF), interferon gamma (IFN-γ), or bacterial products (e.g., lipopolysaccharide, LPS). Priming transforms quiescent phagocytes into the primed phagocytes that are ready to respond to a stimulus (Coffer & Koenderman, 1997; Kitchen, Rossi, Condliffe, Haslett, & Chilvers, 1996; Sample & Czuprynski, 1991). Priming of phagocytes is a reversible phenomenon, which means that primed cells may be de-primed and de-primed cells may be re-primed by the same or different priming stimulus (Kitchen et al., 1996). The role of low Mg status as a priming factor was first observed in vitro. It was noted that exposure of resting human neutrophils to a chemotactic tripeptide *N*-formyl-methionyl-leucyl-phenylalanine causes influx of Ca^{2+} via Na^{+}/Ca^{2+} exchange mechanism. A transient increase in free intracellular Ca^{2+} concentration primes granulocytes and drives respiratory burst. Further, it was shown that low extracellular Mg^{2+} levels promote oxidative activity and its high extracellular concentration markedly inhibits the same in human polymorphonuclear cells (Bussière, Zimowska, Gueux, Rayssiguier, & Mazur, 2002). Further support for the role of Mg in leucocyte activation comes from in vivo studies, which showed that leucocytes from Mg deficient rats produce larger quantities of ROS even without stimulation compared to control rats (Malpuech-Brugère et al., 2000). Additionally, it was noted that neutrophils from Mg-deficient rats overexpressed a phenotypic marker of primed cell, vimentin, which is a cytoskeletal protein related to cell adhesion, migration, and integrity (Bussière, Zimowska, et al., 2002). Although precise mechanism of extracellular Mg^{2+} on leukocyte activation remains unclear, evidence suggests that Mg^{2+} competitively inhibits intracellular Ca^{2+} rise and consequently respiratory burst (Simchowitz, Foy, & Cragoe Jr, 1990). Moreover, production of proinflammatory cytokines and neuromediators also promotes priming and activation of leucocytes, which is inhibited by Mg (Bussière, Mazur, et al., 2002).

(ii) *Mg and activation of NMDA receptors*

NMDA receptors are excitatory glutamatergic receptors that play an important role in regulating physiological functions of neurons. Their excessive activation, however, is involved in the pathophysiological progression of various neurodegenerative diseases. Glutamate, the major excitatory neurotransmitter in CNS, mediates its actions via ionotropic and metabotropic receptors. NMDA receptors are the ionotropic type of glutamatergic receptors. There is sufficient evidence to suggest that glutamate-mediated excitotoxicity, particularly involving NMDA receptors, is associated with increased expression of proinflammatory markers such as nuclear factor kappa B (NF-κB), interleukin 1 beta (IL-1β), TNF-α, ROS, and nitrite resulting in neuroinflammation in neurodegenerative conditions such as Alzheimer's disease. Blockade of NMDA receptors prevents neuroinflammatory reactions (Ju Hwang, Choi, Park, & Hong, 2019; Sil, Ghosh, & Ghosh, 2016). NMDA receptors are also expressed on the cell bodies and peripheral terminals of primary afferent nerves in the gut and are known to play an important role in normal visceral pain transmission (McRoberts et al., 2001). Furthermore, activated neutrophils add to enhance inflammatory reaction by releasing glutamate and D-serine, NMDA receptor co-agonist, which in turn perpetuate neuronal injury (Del Arroyo et al., 2019). The exposure of rat retina to NMDA causes increased expression of proinflammatory mediators and results in neuronal death (Arfuzir, Agarwal, Iezhitsa, Agarwal, & Ismail, 2020). In fact, under various neurodegenerative conditions, NMDA receptors are upregulated and excessively activated, initiating Ca^{2+}

influx and intracellular Ca^{2+} overloading, which in turn initiates various Ca-dependent intracellular signaling pathways leading to inflammatory responses.

In order to understand the role of Mg in regulating activity of NMDA receptors, it is important to look into their structure and binding affinity. NMDA receptors incorporate an ion channel possessing several receptor and modulatory sites, which recognize individual regulatory substances. It possesses a site for binding with Mg^{2+} in a voltage-dependent manner. Depolarization of neurons by glutamatergic activation overcomes the Mg^{2+} block and allows inflow of ions through the cation channel that is highly permeable to Ca^{2+} (Fig. 1).

Mg deficiency increases neuronal excitability by decreasing threshold of excitation by glutamate on NMDA receptors, increasing the Ca^{2+} current in pre- and postsynaptic membranes and alteration of the Na^{+}/K^{+} ATPases (Papadopol & Nechifor, 2011). Blockade of NMDA receptors by Mg counteracts the activation of NMDA receptors and subsequent inflammatory responses. In our previous studies, pretreatment with Mg acetyltaurate protected rat retinas and optic nerve against ischemic and excitotoxic injury (Arfuzir et al., 2016; Lambuk et al., 2017). These effects of Mg acetyltaurate were associated with improved ERG recordings and visual functions (Arfuzir et al., 2020; Lambuk, Iezhitsa, Agarwal, Agarwal, & Ismail, 2021). In fact, exposure to both endothelin-1 and NMDA inflicts a dose-dependent injury to retina and optic nerve (Arfuzir, Agarwal, Iezhitsa, Agarwal, & Ismail, 2019; Lambuk et al., 2019) and Mg acetyltaurate protects against both the endothelin-1 and NMDA-induced retinal and optic nerve injury, which at least partially could be attributed to NMDA receptor blockade by Mg (Arfuzir et al., 2018, 2016; Iezhitsa & Agarwal, 2018; Jafri et al., 2018; Lambuk et al., 2017). The protective effect of Mg acetyltaurate against NMDA-induced retinal damage was

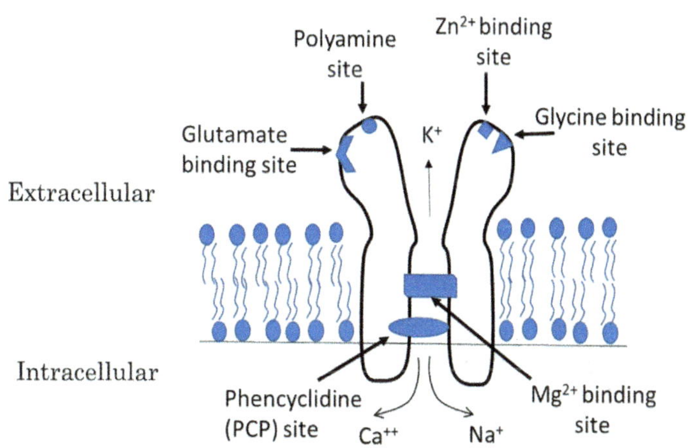

Fig. 1 Diagrammatic representation of NMDA receptor showing Mg^{2+} binding site within the ion channel.

also found to be associated with restoration of changes in minerals and trace elements homeostasis (Jafri et al., 2017). These studies clearly indicate the protective effect of Mg against NMDA-induced excitotoxic injury to neurons.

In addition to suppressing NMDA-mediated excessive postsynaptic glutamatergic transmission, Mg also reduces presynaptic release of glutamate and enhances release of inhibitory neurotransmitter gamma-aminobutyric acid (GABA) by modulating presynaptic metabotropic glutamate receptors. The direct presynaptic effect of Mg decreases catecholamine release (Papadopol & Nechifor, 2011). It is also noteworthy that NMDA receptors are also distributed on astrocytes, oligodendrocytes, and polydendrocytes. Since these cells are capable of producing proinflammatory cytokines upon stimulation, their blockade by Mg may further contribute to suppression of neuroinflammation (Guo et al., 2007).

(iii) Activation of proinflammatory signaling pathways and increased expression of proinflammatory cytokines and neuropeptides

An earlier study (Weglicki & Phillips, 1992) demonstrated that Mg-deficient diet in rats led to an increase in serum levels of inflammatory cytokines, such as interleukin-1 (IL-1), interleukin-6 (IL-6), and TNF-α. Mg-deficiency was also found to promote substance P-associated neurogenic inflammation leading to enhanced free radical production and oxidation of proteins and lipids (Weglicki et al., 1996). The precise cellular mechanisms underlying the protective effect of supraphysiological Mg concentrations against injury induced by neuroinflammation remain unclear; however, several pathways that could be modulated by Mg have been investigated.

In neuroinflammation, activated microglia play a central role by migrating to the site of injury and producing inflammatory mediators. Additionally, astrocytes, endothelial cells, and perivascular macrophages also contribute to inflammatory responses (Dunn, Swiergiel, Zhang, & Quan, 2006; Norden, Trojanowski, Villanueva, Navarro, & Godbout, 2016). Glial cell activation and neuronal injury subsequent to cellular insult are initiated by the release of pathogen-associated molecular patterns (PAMPs) or damage-associated molecular patterns (DAMPs) from the injured cells. DAMPs include several proteins such as high-mobility group box 1 (HMGB1), mitochondrial components, uric acid, and others (Shen, Kreisel, & Goldstein, 2013), and interact with two types of pattern recognition receptors (PRRs) that include toll-like receptors (TLRs) and nod-like receptors (NLRs) on neurons and glial cells. TLRs are membrane-spanning receptors whereas NLRs are cytoplasmic receptors. Activation of TLRs by DAMPs initiates intracellular signaling through mitogen-activated protein kinases (MAPKs) causing activation of proinflammatory transcription factors including NF-κB, AP-1, and c-Jun/c-Fos, increased expression of inflammatory cytokines such as IL-1β, IL-6, TNF-α, and activation of inflammasomes (Kawasaki & Kawai, 2014). Inflammasomes are intracellular, multimeric protein complexes that are formed after detection of DAMPs by TLRs. Inflammasome activation stimulates expression of proinflammatory cytokines, which in turn stimulate production of nitric oxide (NO) and ROS and initiate the caspase cascade resulting in neuronal apoptosis (Chakraborty, Kaushik, Gupta, & Basu, 2010). TLR stimulation on glial cell causes their activation and promotes cytokine production via MAPK and NF-κB (Kawasaki & Kawai, 2014). The cytokines released by reactive glial cells initiate signaling in neurons via NF-κB and c-Jun and promote neuronal apoptosis. Therefore, the inflammatory cascade can be summarized to include inducers/sensors (TLRs), transducers (MAPKs, NF-κB), and amplifiers (proinflammatory cytokines: IL-1β, IL-6, TNF-α).

Suppression of the activation of proinflammatory transcription factors is associated with reduced expression of inflammatory cytokines and reduced neuroinflammatory response. Studies have shown that Mg can interrupt the chain of events leading to inflammatory responses. In one of the studies, Mg isoglycyrrhizinate exerted antiinflammatory effects against isoprenaline-induced myocardial fibrosis in mice by inhibiting the TLR4/NF-κB (p65) signaling pathway (Ma et al., 2018). Mg sulfate was shown to inhibit lipopolysaccharide-induced activation of NLR3 and subsequent upregulation of IL-1β in the human monocytic leukemia cell line by suppressing Ca^{2+} currents (Chang et al., 2018). In our studies, pretreatment with Mg acetyltaurate inhibited endothelin-1-induced retinal ganglion cell apoptosis by suppressing activation of NF-κB in rat retinas. This effect of Mg acetyltaurate was associated with reduced retinal expression of IL-1β, IL-6, and TNF-α (Arfuzir et al., 2020). Further, we observed that Mg acetyltaurate also reduces NMDA-induced retinal ganglion cell apoptosis by downregulating transcriptional activity of NF-κB, p53, and AP-1-mediated c-Jun/c-Fos (Lambuk et al., 2021). Noteworthy that studies have also shown that pretreatment with Mg acetyltaurate prevents NMDA-induced changes in the expression of Ca-regulated proteins including CAMKII, cabin-1, and calpain-1 in rat retinas indicating that Mg prevents Ca^{2+} influx induced by exposure to NMDA (Jafri, Agarwal, Iezhitsa, Agarwal, & Ismail, 2020). In amyloid beta precursor protein/presenilin 1 transgenic mice, Mg-suppressed neuroinflammation when administered as Mg-L-threonate. It suppressed the expression of IL-1β in the mouse brain D1A glial cells by stimulating the activity of extracellular signal-regulated protein kinases 1 and 2 (ERK1/2) and peroxisome proliferator-activated receptor gamma (PPARγ) signaling pathways (Wang et al., 2017). In experimental animals, attenuation of memory impairment by restoration of Mg^{2+} level was also found to be associated with activation of protein kinase C (PKC) (Libien, Sacktor, & Kass, 2005).

Cytokine production by glial cells is also suppressed by Mg. NF-κB activation and cytokine expression induced by LPS in microglia was found to be suppressed after treatment with Mg sulfate (Gao et al., 2013). In microglia, nuclear enzyme poly(ADP-ribose) polymerase-1 (PARP-1) is a key signaling hub driving proinflammatory responses. The activity of PARP-1 can be stimulated via ERK1/2 activation in a Ca^{2+} influx-dependent manner. Furthermore, Ca^{2+}-permeable transient receptor potential melastatin 2 (TRPM2) channels are activated downstream of PARP-1. Interestingly, NMDA receptor stimulation in primary cultured microglia was found to induce their proliferation and activation and caused release of proinflammatory mediators by recruiting PARP-1 and TRPM2 channels (Raghunathan, Vosoughi, Kauppinen, & Jackson, 2020). Mg^{2+} was shown to inhibit a variety of TRPM channels hence preventing Ca^{2+} influx and inhibiting the inflammatory responses (Bouron, Kiselyov, & Oberwinkler, 2015).

Mg deficiency is also associated with the release of neuropeptides such as substance P, which stimulate tachykinin receptors and mediate pain. Animal studies have shown that in the early phase after the initiation of Mg deficient diet, the blocking effect of Mg on NMDA channels is lost resulting in their opening and release of neuropeptides such as calcitonin gene-related peptide and substance P from the sensory C fibers. The increased circulating levels of these neuropeptides initiate "neurogenic inflammation," which is characterized by increased activity of inflammatory cells, increased expression of proinflammatory cytokines, and increased production of ROS and reactive nitrogen species (RNS) (Kramer et al., 2009; Tejero-Taldo, Kramer, Mak, Komarov, & Weglicki, 2006) (Figs. 2 and 3).

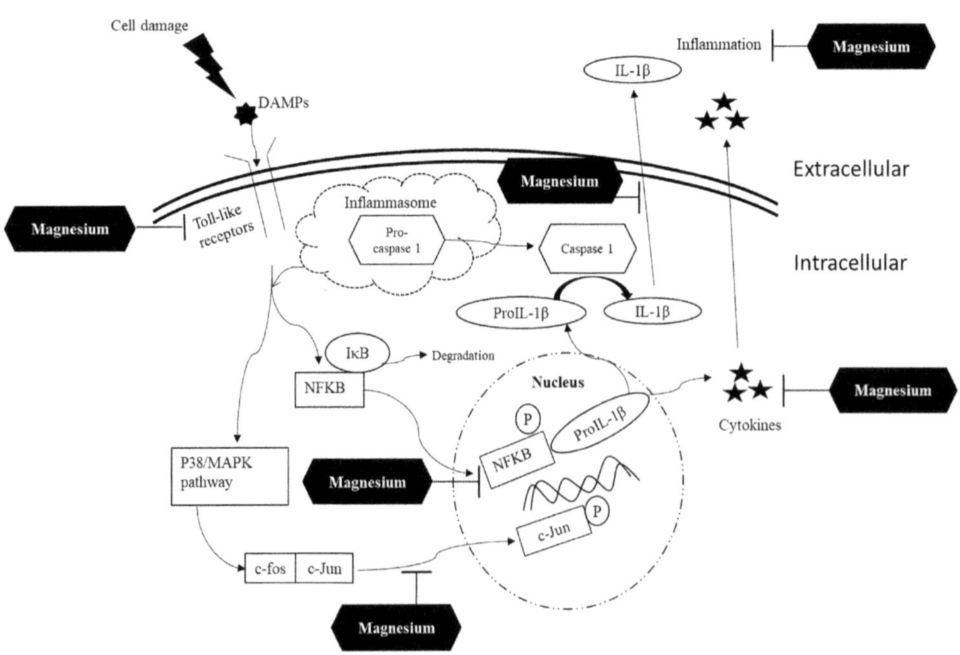

Fig. 2 A diagrammatic representation of toll-like receptor-mediated activation of transcription factors and expression of inflammatory cytokines. Potential sites of inhibition by Mg are shown.

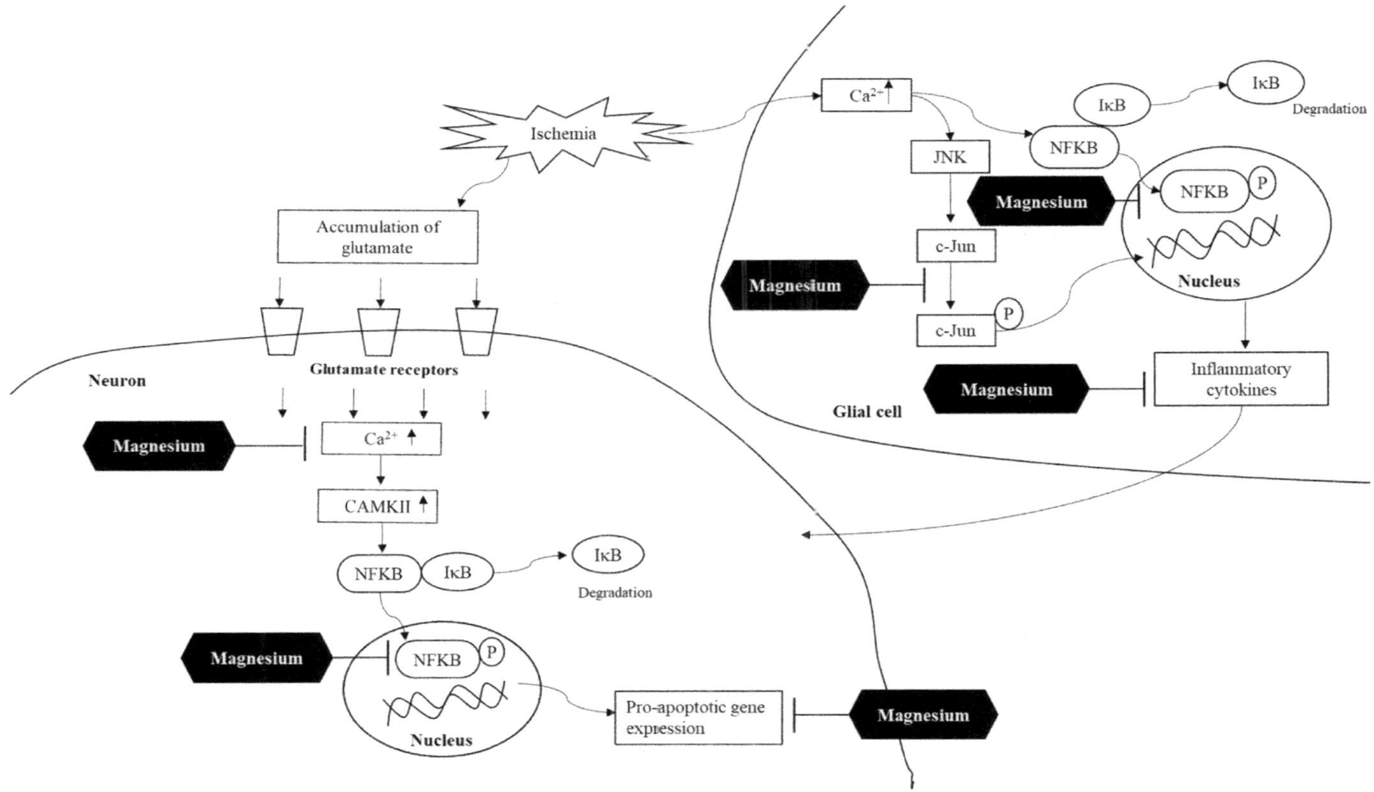

Fig. 3 A diagrammatic representation of intracellular pathways causing neuroinflammation as a result of neuron-glia interaction in the presence of ischemia and/or excitotoxicity. Potential sites of inhibition by Mg are shown. NF-κB is used as a representative transcription factor.

(iv) Oxidative and nitrosative stress

Oxidative stress is an integral part of the neuroinflammatory conditions. A short-lasting inflammatory response is beneficial and promotes healing; however, severe and long-lasting inflammation promotes neurodegenerative conditions such as Alzheimer's disease and Parkinson's disease. Oxidative stress results from an imbalance between the amount of free radicals produced and tissue's antioxidant capacity. Chronic loss of the redox balance causes intracellular accumulation of ROS which include superoxide anion ($O\cdot^{-2}$), hydrogen peroxide (H_2O_2), and hydroxyl radical ($\cdot OH$). RNS include nitric oxide (NO) and peroxynitrite (ONOO −). Accumulation of ROS and RNS damages proteins, lipids, and DNA by causing their oxidation and nitration. It also dysregulates signal transduction and leads to activation of microglia and increased expression of DAMPs and proinflammatory cytokines and chemokines. In turn, cytokines such as IL-6, IL-1β, TNF-α, and interferons (IFNs) induce the generation of ROS in a variety of cells. The free radicals derived from mitochondria diffuse into the cytoplasm and activate cytosolic targets including redox-sensitive transcription factors such as hypoxia-inducible factor 1 alpha (HIF-1α) and NF-κB, which in turn increase the production of proinflammatory cytokines (Solleiro-Villavicencio & Rivas-Arancibia, 2018). Suppression of oxidative stress, therefore, is one of the potential therapeutic strategies in neuroinflammatory conditions.

Mg deficiency is known to be associated with oxidative stress. Serum Mg level is negatively correlated with markers of oxidative stress such as plasma superoxide anions and malondialdehyde (MDA) (Cernak et al., 2000). MDA is a product of lipid peroxidation and its increased levels are observed in the presence of oxidative stress. A similar correlation has been observed between RBC Mg levels and the ratio of GSH/GSSG concentration (circulating reduced/oxidized glutathione) (Barbagallo, Dominguez, Tagliamonte, Resnick, & Paolisso, 1999). GSH serves to remove free radicals and is a part of antioxidant defense system of the body. Enhanced oxidative stress has also been observed in rats fed with Mg-deficient diet (Hans, Chaudhary, & Bansal, 2002). Mg-deficient diet led to 1.4-fold rise of plasma protein carbonyl concentration in rats, the marker of protein oxidative damage, compared to control group (Kharitonova, Zheltova, Iezhitsa, Ozerov, & Spasov, 2013). Mg deficiency promotes release of catecholamines perhaps due to increased Ca^{2+}/Mg^{2+} ratio, which in turn increase the oxidative stress. For the same reason phagocytes and neutrophils are activated and produce increasing quantities of free radicals. Mg deficiency also causes uncoupling of oxidative phosphorylation in mitochondria resulting in production of excessive quantity of free radicals. In fact, mitochondrial dysfunction is the point of crosstalk between neuroinflammation and oxidative stress. Continued exposure to ROS leads to mitochondrial DNA (mtDNA) mutation, accumulation of which in turn enhances

ROS production, establishing a positive feedback loop. The sources of ROS production in microglia include intracellular peroxidase, membrane surface NADPH oxidase (NOX), and mitochondrial oxidation. Increased ROS production in glial cells results in increased expression of proinflammatory genes and perpetuates neuroinflammation (He, Zhu, Wang, & Zhang, 2020). Furthermore, Mg deficiency is associated with endothelial dysfunction. It has been observed that expression of inducible nitric oxide synthase (iNOS) is upregulated in the presence of Mg deficiency, which causes production of excessive quantities of NO. Reaction of NO with superoxide produces peroxynitrite, which is a highly reactive species and damages biomolecules (Zheltova, Kharitonova, Iezhitsa, & Spasov, 2016). In our studies, exposure to NMDA and endothelin-1 was found to cause increased peroxynitrite levels in the rat retinas and this was suppressed by pretreatment with Mg acetyltaurate (Arfuzir et al., 2018; Jafri et al., 2018). Mg acetyltaurate also reduced the retinal oxidative stress in rat retinas (Arfuzir et al., 2016; Iezhitsa & Agarwal, 2018; Lambuk et al., 2017). Neuroinflammation and oxidative stress in Alzheimer's disease are known to be associated with a lower Mg status. Similarly, plasma Mg level was positively correlated with the activity of catalase and glutathione peroxidase and with concentrations of IL-6 and TNF-α in patients with preeclampsia (de Sousa Rocha, Della Rosa, Ruano, Zugaib, & Colli, 2015). These observations clearly indicate an association between Mg status, oxidative stress, and inflammation (Fig. 4).

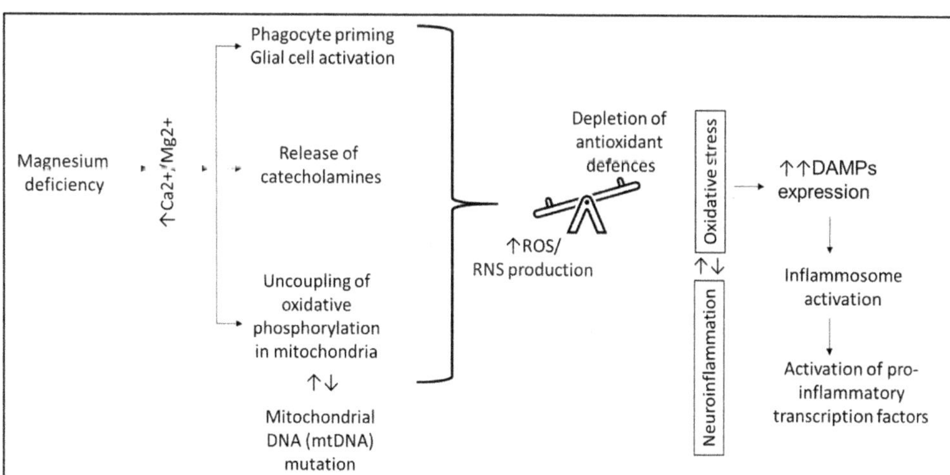

Fig. 4 A diagrammatic representation of role of Mg deficiency in the development of oxidative stress and crosstalk between oxidative stress and neuroinflammation.

Conclusions

- Deficiency of Mg is known to be associated with neuroinflammation, which is the major pathophysiological process underlying several neurodegenerative conditions.
- Since Mg is a natural Ca antagonist, most of the effects of its deficiency on cellular processes result from increased intracellular Ca levels.
- The major effects of these ionic imbalance include increased priming and activation of phagocytic cells, activation of NMDA receptors, activation of proinflammatory transcription factors, cytokines, chemokines and neuropeptides, and oxidative and nitrosative stress.
- Hence, correction of Mg deficiency in the presence of neuroinflammation may at least partially be of therapeutic importance.

Applications to other neurological conditions

In this chapter, we have reviewed the role of magnesium in neuroinflammation. Various pathways through which Mg modifies inflammatory responses are discussed. Since alteration of these pathways is associated with not only the neurodegenerative conditions but also several other neurological conditions such as metabolic disturbances and malignancy, wider implications of Mg deficiency are possible and need further exploration.

Key facts

- Magnesium is the second most abundant intracellular cation after potassium, and overall, it is the fourth most abundant cation.
- Magnesium serves as a cofactor for over 600 enzymes and may act as an activator for additional 200 enzymes playing important physiological role particularly in the brain, the heart, and the skeletal muscles.
- Recommended dietary allowance for magnesium is 4.5 mg/kg, which is a total daily allowance of 350–400 mg for adult men and 280–300 mg for adult women.
- Low magnesium intake is linked to chronic (neuro)inflammation, which is one of the drivers of neurodegeneration.
- Supraphysiological levels of magnesium exhibit protective effect against excitotoxic neuronal cell death (excitotoxicity) and has been implicated in neuroprotection of multiple neurological disorders.

Mini-dictionary of terms

Neuroinflammation is an activation of innate immune system in neuronal tissue in response to the inflammatory challenges and is characterized by cellular and molecular changes within the brain and nervous tissues.

Neurodegeneration refers to progressive irreversible neuronal damage and death.
Cytokines are cell-signaling molecules released by cells and serve as molecular messengers for cellular communication in regulating immune responses.
Tumor necrosis factor α (TNFα) is a cytokine that plays an important role in diverse cellular events such as cell survival, proliferation, differentiation, and death.
Interleukins are cytokines that act as signaling molecules to regulate inflammation and immune system modulation via promoting development and differentiation of natural killer cells, T- and B-lymphocytes, and hematopoietic stem cells.
Nuclear factor kappa B (NF-κB) is a family of inducible transcription factors, which regulates a large array of genes responsible for regulation of immune responses.
Oxidative stress is an imbalance between free radical production and activity of innate antioxidant system.
Nitrosative stress is overproduction of nitric oxide resulting in the formation of peroxynitrite and other reactive nitrogen species.
Excitotoxicity is neuronal death-induced by excessive simulation of neuronal excitatory receptors.

Other components of interest

Degree of antiinflammatory and neuroprotective effects of Mg may considerably vary depending on the nature of anions, ligands, or complexing agents used in Mg preparations. Observed differences are likely to be a result of either different bioavailability of Mg salts or synergistic pharmacological effects between Mg and specific anions or complexing agents (Kharitonova et al., 2015; Spasov et al., 2010, 2011). As an example, a combined salt of Mg and taurine (magnesium acetyltaurate) seems to produce greater neuroprotective effect in preventing NMDA and endothelin-1 (ET-1)-induced retinal and optic nerve damage compared to taurine alone (Arfuzir et al., 2018; Jafri et al., 2018). The relatively newly developed Mg L-threonate, unlike other Mg compounds, significantly increases Mg^{2+} levels in the CSF (Slutsky et al., 2010). Biochemical relationships between pyridoxine and Mg are interesting as well. Although the precise mechanism of the interaction between vitamin pyridoxine and Mg is not completely known, it seems an established fact (Durlach, 1969) that vitamin pyridoxine speeds up the transport of Mg through the cellular membranes and is essential for storing Mg.

It is also noteworthy that several micronutrients including minerals and vitamins play a crucial role in regulating the immune homeostasis. In this chapter, the role of only Mg in neuroinflammation is discussed; however, other nutrients also have a role to play in neuroinflammation. Iron deficiency is a common condition particularly in pregnant and growing children. Low serum concentration of iron and ferritin is associated with inflammatory responses (Williams et al., 2019). In fact, insufficient iron intake causes increased production of cytokines and other mediators of inflammation, hence promoting inflammatory responses. Similarly, zinc is also known to act as an inflammatory regulator. Its dietary deficiency results in increased proinflammatory signal transduction and cytokine production resulting in an inflammatory response (Bordeleau, Fernández de Cossío, Chakravarty, & Tremblay, 2021). Several vitamins including calcium, chromium, and vitamins A, E, D, folate, B_6, and B_{12} also play a key role in maintaining body's immune responses (Shenkin, 2006). Hence, it is important to consider dietary sufficiency of these micronutrients as well to account for inflammation when assessing overall nutritional status.

Summary points

- Deficiency of Mg is known to be associated with neuroinflammation, which is the major pathophysiological process underlying several neurodegenerative conditions.
- Since Mg is a natural Ca antagonist, most of the effects of its deficiency on cellular processes result from increased intracellular Ca levels.
- The major effects of these ionic imbalance include increased priming and activation of phagocytic cells, activation of NMDA receptors, activation of proinflammatory transcription factors, cytokines, chemokines and neuropeptides, and oxidative and nitrosative stress.
- Hence, correction of Mg deficiency in the presence of neuroinflammation may at least partially be of therapeutic importance.

References

Arfuzir, N. N. N., Agarwal, R., Iezhitsa, I., Agarwal, P., & Ismail, N. M. (2019). Dose-dependent effects of endothelin-1 on retinal and optic nerve morphology in Sprague Dawley rats. *Neurochemical Journal, 13*, 73–80. https://doi.org/10.1134/S1819712419010045.

Arfuzir, N. N. N., Agarwal, R., Iezhitsa, I., Agarwal, P., & Ismail, N. M. (2020). Magnesium acetyltaurate protects against endothelin-1 induced RGC loss by reducing neuroinflammation in *Sprague dawley* rats. *Experimental Eye Research, 194*, 107996. https://doi.org/10.1016/j.exer.2020.107996.

Arfuzir, N. N. N., Agarwal, R., Iezhitsa, I., Agarwal, P., Sidek, S., Spasov, A., et al. (2018). Effect of magnesium acetyltaurate and taurine on endothelin1-induced retinal nitrosative stress in rats. *Current Eye Research, 43*(8), 1032–1040. https://doi.org/10.1080/02713683.2018.1467933.

Arfuzir, N. N. N., Lambuk, L., Jafri, A. J. A., Agarwal, R., Iezhitsa, I., Sidek, S., et al. (2016). Protective effect of magnesium acetyltaurate against endothelin-induced retinal and optic nerve injury. *Neuroscience, 325*, 153–164. https://doi.org/10.1016/j.neuroscience.2016.03.041.

Barbagallo, M., Dominguez, L. J., Tagliamonte, M. R., Resnick, L. M., & Paolisso, G. (1999). Effects of glutathione on red blood cell intracellular magnesium: Relation to glucose metabolism. *Hypertension, 34*(1), 76–82. https://doi.org/10.1161/01.hyp.34.1.76.

Bordeleau, M., Fernández de Cossío, L., Chakravarty, M. M., & Tremblay, M.È. (2021). From maternal diet to neurodevelopmental disorders: A story of neuroinflammation. *Frontiers in Cellular Neuroscience, 14*, 612705. https://doi.org/10.3389/fncel.2020.612705. PMID: 33536875.

Bouron, A., Kiselyov, K., & Oberwinkler, J. (2015). Permeation, regulation and control of expression of TRP channels by trace metal ions. *Pflügers Archiv/European Journal of Physiology, 467*(6), 1143–1164. https://doi.org/10.1007/s00424-014-1590-3.

Bussière, F. I., Mazur, A., Fauquert, J. L., Labbe, A., Rayssiguier, Y., & Tridon, A. (2002). High magnesium concentration in vitro decreases human leukocyte activation. *Magnesium Research, 15*(1–2), 43–48.

Bussière, F. I., Zimowska, W., Gueux, E., Rayssiguier, Y., & Mazur, A. (2002). Stress protein expression cDNA array study supports activation of neutrophils during acute magnesium deficiency in rats. *Magnesium Research, 15*(1–2), 37–42.

Cernak, I., Savic, V., Kotur, J., Prokic, V., Kuljic, B., Grbovic, D., et al. (2000). Alterations in magnesium and oxidative status during chronic emotional stress. *Magnesium Research, 13*(1), 29–36.

Chakraborty, S., Kaushik, D. K., Gupta, M., & Basu, A. (2010). Inflammasome signaling at the heart of central nervous system pathology. *Journal of Neuroscience Research, 88*(8), 1615–1631.

Chang, Y. Y., Kao, M. C., Lin, J. A., Chen, T. Y., Cheng, C. F., Wong, C. S., et al. (2018). Effects of $MgSO_4$ on inhibiting Nod-like receptor protein 3 inflammasome involve decreasing intracellular calcium. *The Journal of Surgical Research, 221*, 257–265.

Cilliler, A. E., Ozturk, S., & Ozbakir, S. (2007). Serum magnesium level and clinical deterioration in Alzheimer's disease. *Gerontology, 53*, 419–422.

Coffer, P. J., & Koenderman, L. (1997). Granulocyte signal transduction and priming: Cause without effect? *Immunology Letters, 57*(1–3), 27–31.

de Sousa Rocha, V., Della Rosa, F. B., Ruano, R., Zugaib, M., & Colli, C. (2015). Association between magnesium status, oxidative stress and inflammation in preeclampsia: A case-control study. *Clinical Nutrition (Edinburgh), 34*(6), 1166–1171. https://doi.org/10.1016/j.clnu.2014.12.001.

Del Arroyo, A. G., Hadjihambi, A., Sanchez, J., Turovsky, E., Kasymov, V., Cain, D., et al. (2019). NMDA receptor modulation of glutamate release in activated neutrophils. *eBioMedicine, 47*, 457–469.

DiSabato, D. J., Quan, N., & Godbout, J. P. (2016). Neuroinflammation: The devil is in the details. *Journal of Neurochemistry, 139*(Suppl. 2), 136–153. https://doi.org/10.1111/jnc.13607.

Dunn, A. J., Swiergiel, A. H., Zhang, H., & Quan, N. (2006). Reduced ingestion of sweetened milk induced by interleukin-1 and lipopolysaccharide is associated with induction of cyclooxygenase-2 in brain endothelia. *Neuroimmunomodulation, 13*(2), 96–104.

Durlach, J. (1969). Données actuelles sur les mécanismes de synergie entre vitamine B6 et magnésium. *Journal de Médecine de Besançon, 5*, 349–359 (in French).

Durlach, J. (1988). *Magnesium in clinical practice*. London, Paris: John Libbey & Co.

Elisaf, M., Merkourpoulos, M., Tsianos, E. V., & Siamopoulos, K. S. (1995). Pathogenetic mechanisms of hypomagnesemia in alcoholic patients. *Journal of Trace Elements in Medicine and Biology, 9*(4), 210–214. https://doi.org/10.1016/S0946-672X(11)80026-X.

Gao, F., Ding, B., Zhou, L., Gao, X., Guo, H., & Xu, H. (2013). Magnesium sulfate provides neuroprotection in lipopolysaccharide-activated primary microglia by inhibiting NF-κB pathway. *The Journal of Surgical Research, 184*(2), 944–950. https://doi.org/10.1016/j.jss.2013.03.034.

Gommers, L. M., Hoenderop, J. G., Bindels, R. J., & de Baaij, J. H. (2016). Hypomagnesemia in type 2 diabetes: A vicious circle? *Diabetes, 65*(1), 3–13. https://doi.org/10.2337/db15-1028.

Guo, W., Wang, H., Watanabe, M., Shimizu, K., Zou, S., LaGraize, S. C., et al. (2007). Glial-cytokine-neuronal interactions underlying the mechanisms of persistent pain. *The Journal of Neuroscience, 27*(22), 6006–6018. https://doi.org/10.1523/JNEUROSCI.0176-07.2007.

Hans, C. P., Chaudhary, D. P., & Bansal, D. D. (2002). Magnesium deficiency increases oxidative stress in rats. *Indian Journal of Experimental Biology, 40*(11), 1275–1279.

He, J., Zhu, G., Wang, G., & Zhang, F. (2020). Oxidative stress and neuroinflammation potentiate each other to promote progression of dopamine neurodegeneration. *Oxidative Medicine and Cellular Longevity, 2020*, 6137521. https://doi.org/10.1155/2020/6137521.

Iezhitsa, I., & Agarwal, R. (2018). Magnesium acethyltaurate as a potential agent for retinal and optic nerve protection in glaucoma. *Neural Regeneration Research, 13*(5), 807–808. https://doi.org/10.4103/1673-5374.232470.

Iezhitsa, I. (2005). Potassium and magnesium depletions in congestive heart failure—Pathophysiology, consequences and replenishment. *Clinical calcium, 15*(11), 123–133.

Iezhitsa, I. N., & Spasov, A. A. (2008). Potassium magnesium homeostasis: Physiology, pathophysiology, clinical consequences of deficiency and pharmacological correction. *Uspekhi Fiziologicheskikh Nauk, 39*(1), 23–41. PMID: 18314767.

Jafri, A. J. A., Agarwal, R., Iezhitsa, I., Agarwal, P., & Ismail, N. M. (2020). Protective effect of magnesium acetyltaurate against NMDA-induced retinal ganglion cell loss in rats involves calcium-regulated proteins. *Molecular Vision, 26*, 392–408. http://www.molvis.org/molvis/v26/392.

Jafri, A. J. A., Agarwal, R., Iezhitsa, I., Agarwal, P., Spasov, A., Ozerov, A., et al. (2018). Protective effect of magnesium acetyltaurate and taurine against NMDA-induced retinal damage involves reduced nitrosative stress. *Molecular Vision, 24*, 495–508. http://www.molvis.org/molvis/v24/495.

Jafri, A. J. A., Arfuzir, N. N. N., Lambuk, L., Iezhitsa, I., Agarwal, R., Agarwal, P., et al. (2017). Protective effect of magnesium acetyltaurate against NMDA-induced retinal damage involves restoration of minerals and trace elements homeostasis. *Journal of Trace Elements in Medicine and Biology, 39*, 147–154. https://doi.org/10.1016/j.jtemb.2016.09.005.

Ju Hwang, C., Choi, D. Y., Park, M. H., & Hong, J. T. (2019). NF-κB as a key mediator of brain inflammation in Alzheimer's disease. *CNS & Neurological Disorders Drug Targets, 18*(1), 3–10. https://doi.org/10.2174/1871527316666170807130011.

Kawasaki, T., & Kawai, T. (2014). Toll-like receptor signaling pathways. *Frontiers in Immunology, 5*, 461. https://doi.org/10.3389/fimmu.2014.00461.

Kharitonova, M., Iezhitsa, I., Zheltova, A., Ozerov, A., Spasov, A., & Skalny, A. (2015). Comparative angioprotective effects of magnesium compounds. *Journal of Trace Elements in Medicine and Biology, 29*, 227–234. https://doi.org/10.1016/j.jtemb.2014.06.026.

Kharitonova, M. V., Zheltova, A. A., Iezhitsa, I. N., Ozerov, A. A., & Spasov, A. A. (2013). Effect of magnesium salts on oxidative protein modification in rats fed low magnesium diet. *Acta Pharmacologica Sinica, 34*(Suppl), 151. https://doi.org/10.1038/aps.2013.118.

King, D. E. (2009). Inflammation and elevation of C-reactive protein: Does magnesium play a key role? *Magnesium Research, 22*(2), 57–59.

Kirkland, A. E., Sarlo, G. L., & Holton, K. F. (2018). The role of magnesium in neurological disorders. *Nutrients, 10*(6), 730. https://doi.org/10.3390/nu10060730.

Kitchen, E., Rossi, A. G., Condliffe, A. M., Haslett, C., & Chilvers, E. R. (1996). Demonstration of reversible priming of human neutrophils using platelet-activating factor. *Blood, 88*(11), 4330–4337.

Kramer, J. H., Spurney, C., Iantorno, M., Tziros, C., Mak, I. T., Tejero-Taldo, M. I., et al. (2009). Neurogenic inflammation and cardiac dysfunction due to hypomagnesemia. *The American Journal of the Medical Sciences, 338*(1), 22–27. https://doi.org/10.1097/MAJ.0b013e3181aaee4d.

Kruse, H. D., Orent, E. R., & McCollum, E. V. (1932). Studies on magnesium deficiency in animals: I. symptomatology resulting from magnesium deprivation. *Journal of Biological Chemistry, 96*(2), 519–539. https://doi.org/10.1016/S0021-9258(18)76290-X.

Lambuk, L., Iezhitsa, I., Agarwal, R., Agarwal, P., & Ismail, N. M. (2021). Magnesium acetyltaurate prevents retinal damage and visual impairment in rats through suppression of NMDA-induced upregulation of NF-κB, p53 and AP-1 (c-Jun/c-Fos). *Neural Regeneration Research, 16*(11), 2330–2344. https://doi.org/10.4103/1673-5374.310691.

Lambuk, L., Jafri, A. J., Arfuzir, N. N., Iezhitsa, I., Agarwal, R., Rozali, K. N., et al. (2017). Neuroprotective effect of magnesium acetyltaurate against NMDA-induced excitotoxicity in rat retina. *Neurotoxicity Research, 31*(1), 31–45. https://doi.org/10.1007/s12640-016-9658-9.

Lambuk, L., Jafri, A. J. A., Iezhitsa, I., Agarwal, R., Bakar, N. S., Agarwal, P., et al. (2019). Dose-dependent effects of NMDA on retinal and optic nerve morphology in rats. *International Journal of Ophthalmology, 12*(5), 746–753. https://doi.org/10.18240/ijo.2019.05.08.

Li, F. Y., Chaigne-Delalande, B., Kanellopoulou, C., Davis, J. C., Matthews, H. F., Douek, D. C., et al. (2011). Second messenger role for Mg^{2+} revealed by human T-cell immunodeficiency. *Nature, 475*(7357), 471–476. https://doi.org/10.1038/nature10246.

Liamis, G., Hoorn, E. J., Florentin, M., & Milionis, H. (2021). An overview of diagnosis and management of drug-induced hypomagnesemia. *Pharmacology Research & Perspectives, 9*(4), e00829. https://doi.org/10.1002/prp2.829.

Libien, J., Sacktor, T. C., & Kass, I. S. (2005). Magnesium blocks the loss of protein kinase C, leads to a transient translocation of PKCα and PKCε, and improves recovery after anoxia in rat hippocampal slices. *Molecular Brain Research, 136*(1–2), 104–111. https://doi.org/10.1016/j.molbrainres.2005.01.005.

Liu, G., Weinger, J. G., Lu, Z. L., Xue, F., & Sadeghpour, S. (2015). Efficacy and safety of MMFS-01, a synapse density enhancer, for treating cognitive impairment in older adults: A randomized, double-blind, placebo-controlled trial. *Journal of Alzheimer's Disease, 49*, 971–990.

Ma, D., Zhang, J., Zhang, Y., Zhang, X., Han, X., Song, T., et al. (2018). Inhibition of myocardial hypertrophy by magnesium isoglycyrrhizinate through the TLR4/NF-κB signaling pathway in mice. *International Immunopharmacology, 55*, 237–244.

Malpuech-Brugère, C., Nowacki, W., Daveau, M., Gueux, E., Linard, C., Rock, E., et al. (2000). Inflammatory response following acute magnesium deficiency in the rat. *Biochimica et Biophysica Acta, 1501*(2–3), 91–98.

Mazur, A., Libako, P., Nowacki, W., & Maier, J. A. (2013). Magnesium and inflammation. In R. H. Kretsinger, V. N. Uversky, & E. A. Permyakov (Eds.), *Encyclopedia of metalloproteins* (pp. 1232–1237). New York: Springer. https://doi.org/10.1007/978-1-4614-1533-6_281.

Mazur, A., Maier, J. A., Rock, E., Gueux, E., Nowacki, W., & Rayssiguier, Y. (2007). Magnesium and the inflammatory response: Potential physiopathological implications. *Archives of Biochemistry and Biophysics, 458*(1), 48–56.

McRoberts, J. A., Coutinho, S. V., Marvizón, J. C., Grady, E. F., Tognetto, M., Sengupta, J. N., et al. (2001). Role of peripheral N-methyl-D-aspartate (NMDA) receptors in visceral nociception in rats. *Gastroenterology, 120*(7), 1737–1748.

Morris, M. E. (1992). Brain and CSF magnesium concentrations during magnesium deficit in animals and humans: Neurological symptoms. *Magnesium Research, 5*(4), 303–313.

Norden, D. M., Trojanowski, P. J., Villanueva, E., Navarro, E., & Godbout, J. P. (2016). Sequential activation of microglia and astrocyte cytokine expression precedes increased Iba-1 or GFAP immunoreactivity following systemic immune challenge. *Glia, 64*(2), 300–316.

Papadopol, V., & Nechifor, M. (2011). Magnesium in neuroses and neuroticism. In R. Vink, & M. Nechifor (Eds.), *Magnesium in the central nervous system [internet]*. Adelaide: University of Adelaide Press. https://www.ncbi.nlm.nih.gov/books/NBK507254/.

Raghunathan, P., Vosoughi, A., Kauppinen, T. M., & Jackson, M. F. (2020). Microglial NMDA receptors drive pro-inflammatory responses *via* PARP-1/TRMP2 signaling. *Glia, 68*(7), 1421–1434. https://doi.org/10.1002/glia.23790.

Ramadan, N. M., Halvorson, H., Vande-Linde, A., Levine, S. R., Helpern, J. A., & Welch, K. M. (1989). Low brain magnesium in migraine. *Headache, 29*(9), 590–593. https://doi.org/10.1111/j.1526-4610.1989.hed2909590.x.

Sample, A. K., & Czuprynski, C. J. (1991). Priming and stimulation of bovine neutrophils by recombinant human interleukin-1 alpha and tumor necrosis factor alpha. *Journal of Leukocyte Biology, 49*(2), 107–115.

Sarchielli, P., Coata, G., Firenze, C., Morucci, P., Abbritti, G., & Gallai, V. (1992). Serum and salivary magnesium levels in migraine and tension-type headache. Results in a group of adult patients. *Cephalalgia, 12*(1), 21–27. https://doi.org/10.1046/j.1468-2982.1992.1201021.x.

Shen, H., Kreisel, D., & Goldstein, D. R. (2013). Processes of sterile inflammation. *Journal of Immunology, 191*(6), 2857–2863.

Shenkin, A. (2006). Micronutrients in health and disease. *Postgraduate Medical Journal, 82*(971), 559–567. https://doi.org/10.1136/pgmj.2006.047670.

Sil, S., Ghosh, T., & Ghosh, R. (2016). NMDA receptor is involved in neuroinflammation in intracerebroventricular colchicine-injected rats. *Journal of Immunotoxicology, 13*(4), 474–489.

Simchowitz, L., Foy, M. A., & Cragoe, E. J., Jr. (1990). A role for Na^{+}/Ca^{2+} exchange in the generation of superoxide radicals by human neutrophils. *The Journal of Biological Chemistry, 265*(23), 13449–13456.

Slutsky, I., Abumaria, N., Wu, L.-J., Haung, C., Zhang, L., Li, B., et al. (2010). Enhancement of learning and memory by elevating brain magnesium. *Neuron, 65*, 165–177. https://doi.org/10.1016/j.neuron.2009.12.026.

Solleiro-Villavicencio, H., & Rivas-Arancibia, S. (2018). Effect of chronic oxidative stress on neuroinflammatory response mediated by $CD4^{+}$ T cells in neurodegenerative diseases. *Frontiers in Cellular Neuroscience, 12*, 114. https://doi.org/10.3389/fncel.2018.00114.

Spasov, A. A., Ozerov, A. A., Iezhitsa, I. N., Kharitonova, M. V., Kravchenko, M. S., & Zheltova, A. A. (2011). Correction of furosemide-induced magnesium deficiency with different stereoisomers of organic magnesium salts: A comparative study. *Bulletin of Experimental Biology and Medicine, 151*(3), 333–335. https://doi.org/10.1007/s10517-011-1323-y.

Spasov, A. A., Petrov, V. I., Iezhitsa, I. N., Kravchenko, M. S., Kharitonova, M. V., & Ozerov, A. A. (2010). Comparative study of magnesium salts bioavailability in rats fed a magnesium-deficient diet. *Vestnik Rossiĭskoĭ akademii meditsinskikh nauk*, (2), 29–37.

Tejero-Taldo, M. I., Kramer, J. H., Mak, I., Komarov, A. M., & Weglicki, W. B. (2006). The nerve-heart connection in the pro-oxidant response to mg-deficiency. *Heart Failure Reviews*, *11*(1), 35–44. https://doi.org/10.1007/s10741-006-9191-7.

Wang, P., Yu, X., Guan, P. P., Guo, J. W., Wang, Y., Zhang, Y., et al. (2017). Magnesium ion influx reduces neuroinflammation in Aβ precursor protein/Presenilin 1 transgenic mice by suppressing the expression of interleukin-1β. *Cellular & Molecular Immunology*, *14*(5), 451–464. https://doi.org/10.1038/cmi.2015.93.

Weglicki, W. B., Mak, I. T., Kramer, J. H., Dickens, B. F., Cassidy, M. M., Stafford, R. E., et al. (1996). Role of free radicals and substance P in magnesium deficiency. *Cardiovascular Research*, *31*(5), 677–682. https://doi.org/10.1016/S0008-6363(95)00196-4.

Weglicki, W. B., & Phillips, T. M. (1992). Pathobiology of magnesium deficiency: A cytokine/neurogenic inflammation hypothesis. *American Journal of Physiology*, *263*(3 Pt 2), R734–R737. https://doi.org/10.1152/ajpregu.1992.263.3.R734.

Williams, A. M., Ladva, C. N., Leon, J. S., Lopman, B. A., Tangpricha, V., Whitehead, R. D., et al. (2019). Changes in micronutrient and inflammation serum biomarker concentrations after a norovirus human challenge. *The American Journal of Clinical Nutrition*, *110*(6), 1456–1464. https://doi.org/10.1093/ajcn/nqz201. PMID: 31504095; PMCID: PMC6885472.

Wolf, F. I., & Trapani, V. (2008). Cell (patho)physiology of magnesium. *Clinical Science (London)*, *114*(1), 27–35. https://doi.org/10.1042/CS20070129.

Zheltova, A. A., Kharitonova, M. V., Iezhitsa, I. N., & Spasov, A. A. (2016). Magnesium deficiency and oxidative stress: An update. *BioMedicine (Taiwan)*, *6*(4), 8–14. https://doi.org/10.7603/s40681-016-0020-6. https://link.springer.com/article/10.7603/s40681-016-0020-6.

CHAPTER 24

Vitamin B12 as a neuroprotectant in neuroinflammation

Larissa Marcely Gomes Cassiano[a,b], Marina da Silva Oliveira[b], and Roney Santos Coimbra[b]
[a]Graduate Program of Biochemistry and Immunology, Federal University of Minas Gerais, Belo Horizonte, Brazil
[b]Neurogenomics/Immunopatology, René Rachou Institute, Oswaldo Cruz Foundation (FIOCRUZ), Belo Horizonte, Brazil

Abbreviations

2-OGDD	2-oxoglutarate-dependent dioxygenase
3-MST	3-mercaptopyruvate sulfurtransferase
AdoCbl	adenosylcobalamin
ALS	amyotrophic lateral sclerosis
AMPA	α-amino-3-hydroxy-5-methyl-4-isoxazolepropionic acid
AP-1	activator protein-1
Apo $E^{-/-}$	apolipoprotein E deficient
APP	amyloid precursor protein
ATP	adenosine triphosphate
Aβ	β-amyloid peptide
BACE	beta-site APP cleaving enzyme
Cbl	cobalamin
CBS	cystathionine beta-synthase
CCL3	chemokine (C-C motif) ligand 3
CGL	cystathionine gamma-lyase
CNCbl	cyanocobalamin
CNS	central nervous system
DNMT	DNA methyltransferase
$FADH_2$	flavin adenine dinucleotide
GSCbl	glutathionylcobalamin
GSH	glutathione
GSSG	glutathione disulfide
GTP	guanosine triphosphate
H_2O_2	hydrogen peroxide
H_2OCbl	aquacobalamin
H_2S	hydrogen sulfide
HAT	histone acetyltransferase
HCY	homocysteine
HDAC	histone deacetylase
HDM	histone demethylase
HHCY	hyperhomocysteinemia
HIF	hypoxia inducible factor
HMT	histone methyltransferase

Vitamins and Minerals in Neurological Disorders
https://doi.org/10.1016/B978-0-323-89835-5.00038-7

HoCbl	hydroxycobalamin
HS^-	hydrosulfide anion
IL	interleukin
JMJD	jumonji C domain-containing
MCM	methylmalonyl-CoA mutase
MeCbl	methylcobalamin
MET	methionine
MS	methionine synthase
MTHFR	5,10-methylenetetrahydrofolate reductase
NADH	nicotinamide adenine dinucleotide
NF-κB	nuclear factor kappa-light-chain-enhancer of activated B cells
NMDA	*N*-methyl D-aspartate
NMDAR	*N*-methyl D-aspartate receptor
NO	nitric oxide
O_2^-	superoxide radical
PHD	prolyl-hydroxylases
PKC	protein kinase C
PP2A	protein phosphatase 2A
PSEN1	presenilin 1
RNS	reactive nitrogen species
ROS	reactive oxygen species
SAM	*S*-adenosylmethionine
SARS-CoV-2	severe acute respiratory syndrome coronavirus 2
TC	transcobalamin
TET	ten-eleven translocation methylcytosine dioxygenase
TNF-α	tumor necrosis factor alpha

Introduction

The term vitamin B12 refers to a set of water-soluble organic compounds with a cobalamin (Cbl) in their structures. These vitamers are adenosylcobalamin (AdoCbl), cyanocobalamin (CNCbl), hydroxycobalamin (HoCbl), and methylcobalamin (MeCbl). Humans acquire B12 through diet, where it is usually found in animal products such as meat, eggs, milk, and dairy products. In plasma, vitamers form a complex with transcobalamin (TC) protein types I, II, or III. Although it is not the most abundant plasma TC, type II (TCII) is essential for transferring Cbl into cells through interaction with specific receptors. After endocytosis of the complex, TCII is degraded in lysosomes and Cbl is released to the cytoplasm, where it can be transformed into its active forms AdoCbl and MeCbl. These vitamers act as cofactors of mitochondrial methylmalonyl-CoA mutase (MCM) and cytosolic methionine synthase (MS), respectively. MCM catalyzes the conversion of L-methylmalonyl-CoA into succinyl-CoA, an intermediate in the Krebs cycle, which is pivotal for energy metabolism, and MS, in the sulfur amino acids pathway, transfers methyl between 5-methyl-tetrahydrofolate and homocysteine (HCY) forming methionine (MET), which is required for synthesis of proteins, DNA, and neurotransmitters. MET is then converted into *S*-adenosylmethionine

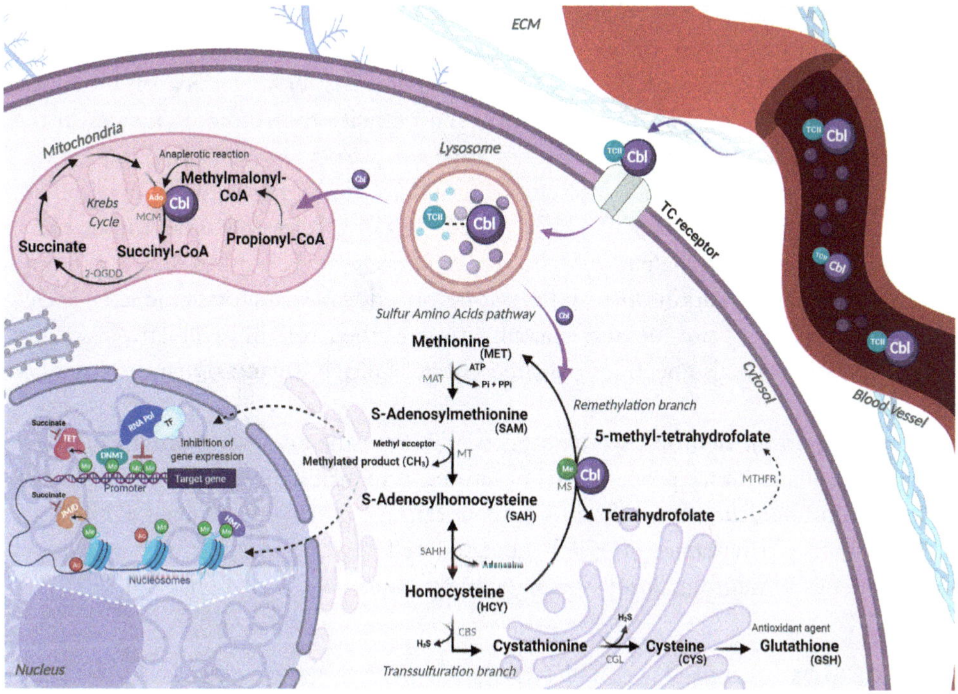

Fig. 1 Cobalamin and cell metabolism. Endocytosis of cobalamin (Cbl) bound to its blood carrier transcobalamin (TCII) occurs through the transcobalamin receptor (TC receptor). In the lysosome, TCII is degraded and free Cbl is released into the cytosol for metabolization. In the mitochondria, Cbl undergoes adenosylation to form adenosylcobalamin (AdoCbl). Methylcobalamin (MeCbl) originates from the degradation of methionine synthase (MS) in the sulfur amino acids pathway. The methyl (Me) products generated by the conversion of *S*-adenosylmethionine (SAM) to *S*-adenosylhomocysteine (SAH) fuel DNA and histone methylation reactions. 2-OGDD, 2-oxoglutarate-dependent dioxygenase; Ac, acetyl; ATP, adenosine triphosphate; CBS, cystathionine beta-synthase; CGL, cystathionine gamma-lyase; DNMT, DNA methyltransferase; ECM, extracellular matrix; H_2S, hydrogen sulfide; HMT, histone methyltransferases; JMJD = Jumonji C domain-containing (histone demethylase); MAT, methionine adenosyltransferase; MCM, methylmalonyl-CoA mutase; MT, methyltransferase; MTHFR, 5,10-methylenetetrahydrofolate reductase; Pi, phosphate (inorganic); PPi, pyrophosphate; RNA pol, RNA polymerase; SAHH, adenosylhomocysteinase; TET, ten-eleven translocation methylcytosine dioxygenases (DNA demethylase); and TF, transcription factor. *(Created with http://Biorender.com.)*

(SAM), the universal methyl donor (Fig. 1). Furthermore, Cbl is also involved in processes such as myelination and neurotransmitter synthesis, combat to oxidative stress and immune response regulation (Peterson, Rodionov, Osterman, & Peterson, 2020). Therefore, maintaining adequate levels of this vitamin is essential for proper functioning of the nervous system and its deficiency can lead to neuroinflammation-associated disorders (Hunt, Harrington, & Robinson, 2014).

Inflammation in the central nervous system (CNS) is named neuroinflammation and comprises complex cellular and molecular processes triggered in response to danger signals. The main agent of this response is the microglia, whose activation involves changes in gene expression, cellular morphology, and behavior. Proinflammatory and neurotrophic factors released by activated microglia trigger a variety of biological processes, ranging from oxidative stress and excitotoxicity to profound changes in the epigenetic cell landscape. These processes involve also astrocytes, neurons, oligodendrocytes, vascular pericytes, and infiltrated peripheral immune cells (Ransohoff & Engelhardt, 2012; Ransohoff, Kivisäkk, & Kidd, 2003). Uncontrolled neuroinflammatory response is pivotal in the pathophysiology of neurological disorders such as multiple sclerosis, Alzheimer's, Parkinson's, and Huntington's diseases, amyotrophic lateral sclerosis (ALS), meningitis and meningoencephalitis, and their role in psychiatric disorders such as depression has also been investigated (Aktas, Ullrich, Infante-Duarte, Nitsch, & Zipp, 2007).

In this chapter, the neuroprotective aspects of B12 will be discussed from the perspective of neuroinflammatory processes. Its functions in regulating energy metabolism and activating immune cells, responding to oxidative stress, glutamate-induced excitotoxicity, hyperhomocysteinemia (HHCY)-mediated cell death, and epigenetic regulation of pivotal genes in neuroinflammation signaling pathway will be emphasized.

B12 regulates energy metabolism and immune response

The Krebs cycle is a mitochondrial pathway of energy metabolism that converts the end product of glycolysis into several intermediates in a closed loop of biochemical reactions generating GTP/ATP and reduced forms of NADH and $FADH_2$ which provide electrons to the oxidative phosphorylation. Multiple substrates can feed the Krebs cycle, in addition to acetyl-CoA and oxaloacetate, both derived from pyruvate, the last product of glycolysis. Concomitantly, as the cycle runs, its intermediates are exported to the cytosol to become building blocks for anabolic reactions. The cycle is replenished by anaplerotic reactions, such as the conversion, within the mitochondria, of methylmalonyl-CoA into succinyl-CoA by MCM, which requires AdoCbl as cofactor (Nelson & Cox, 2021).

In response to proinflammatory stimuli, macrophages and their closely related immunocompetent cells of the CNS, the microglia, change from resting to activated state. To supply their increased energy demand, activated macrophages and microglia switch from oxidative phosphorylation to aerobic glycolysis. This metabolic shift is mediated by the inhibition of Krebs cycle and oxidative phosphorylation pathways and upregulation of glycolysis-associated genes (Ghosh, Castillo, Frias, & Swanson, 2018; Kramer, Ravi, Chacko, Johnson, & Darley-Usmar, 2014). Speculatively, Cbl, by stimulating the anaplerotic production of succinyl-CoA to replenish the Krebs cycle, may contribute to reactivate aerobic respiration, favoring the inactive phenotype of immune cells. However, it is worth noting that excess succinyl-CoA can also slow down the Krebs cycle by inhibiting citrate synthase and α-ketoglutarate dehydrogenase (Martínez-Reyes & Chandel, 2020). Other circulating immune cells, such as lymphocytes and neutrophils, can be recruited from blood to the CNS during neuroinflammation. However, activated lymphocytes increase both glycolysis and mitochondrial respiration, and neutrophils, which have few mitochondria, use glycolysis to sustain the energy demand for their activities such as chemotaxis (Kramer et al., 2014).

In the Krebs cycle, succinyl-CoA converts to succinate, which also has nonmetabolic signaling functions, such as regulation of the innate immune response (Mills et al., 2016; Tannahill et al., 2013). For instance, the increase in succinate concentration stimulates the production of reactive oxygen species (ROS) and inhibits prolyl-hydroxylases (PHD) preventing HIF-α hydroxylation and allowing it to migrate to the nucleus where it induces the expression of glycolytic genes and *IL-1β*, among other proinflammatory cytokines and chemokines (Selak et al., 2005; Tannahill et al., 2013). Succinate also plays an important role in epigenetic modulation of gene transcription. Indeed, succinate is the product of 2-oxoglutarate-dependent dioxygenase (2-OGDD) enzymes, a phylogenetically conserved family that includes histone (JMJD) and DNA (TET) demethylases. When succinate accumulates, it inhibits 2-OGDD and, consequently, histone and DNA demethylation (Letouzé et al., 2013). Thus, it is conceivable that supplemental B12, by enhancing the production of succinyl-CoA, the precursor of succinate, could also promote downregulation of neuroinflammatory genes.

B12 is antioxidant and modulates the oxidative stress response

The brain is responsible for about 20% of the body basal O_2 consumption. Furthermore, it has a reduced antioxidant capacity and a high concentration of polyunsaturated fatty acids prone to lipid peroxidation, which makes it an organ especially vulnerable to oxidative stress (Rink & Khanna, 2011). This process is characterized by the unbalance between the antioxidant response and the production of reactive species that cause oxidative damage, especially those derived from oxygen (ROS) and nitrogen (RNS). One of the most important cell sources of reactive species is the mitochondria, arising from the incomplete reduction of O_2 to H_2O. In this case, O_2 is reduced with one electron producing the superoxide radical (O_2^-), which can be converted into peroxynitrite by reacting with nitric oxide (NO), or to hydrogen peroxide (H_2O_2) by the action of superoxide dismutase. Excessive mitochondrial ROS production can lead to functional and structural changes in the organelle and DNA damage, impairing energy production and other relevant cellular processes, such as fatty acid oxidation and amino acid metabolism. This cascade of events culminates in triggering proapoptotic signaling and cell death (Murphy, 2009). Oxidative stress is implicated in the pathophysiology of several neurological diseases, such as Alzheimer's (Butterfield & Halliwell, 2019) and Parkinson's (Guo, Zhao,

Li, Li, & Liu, 2018), and is also a hallmark of the host response to CNS infections (Koedel & Pfister, 1999; Percário et al., 2012).

Oxidative stress is an inhibitor of MS activity. The Cbl present in this enzyme is an intermediate carrier of methyl groups. The cobalt that makes up Cbl can present different oxidation states in Cbl(I), Cbl(II), or Cbl(III). Cbl(I) in MS is an excellent nucleophile and acts as a methyl acceptor, forming MeCbl, which is the methyl donor to the conversion of HCY into MET by MS (Fig. 2). Alternatively, under oxidative stress conditions, Cbl(I) can act as an antioxidant by undergoing oxidation to Cbl(II). This partial oxidation temporarily inhibits MS by interrupting the conversion of HCY into MET and, consequently, favoring the production of glutathione (GSH) by the transsulfuration branch of the sulfur amino acids pathway (Banerjee & Matthews, 1990) (Fig. 1). GSH is an important cellular antioxidant due to its role as a cofactor for the enzyme glutathione peroxidase, which catalyzes the conversion of H_2O_2 into O_2 and H_2O. Moreover, GSH disulfide (GSSG) generated by oxidation of two GSH is a RNS and ROS scavenger (Aoyama, 2021; Pritsos & Pardini, 1983). Aquacobalamin (H_2OCbl) is abundant in cells and can readily combine with reduced GSH to form glutathionylcobalamin (GSCbl), an intermediary in the biosynthesis of MeCbl and AdoCbl (Pezacka, Green, & Jacobsen, 1990; Xia, Cregan, Berben, & Brasch, 2004).

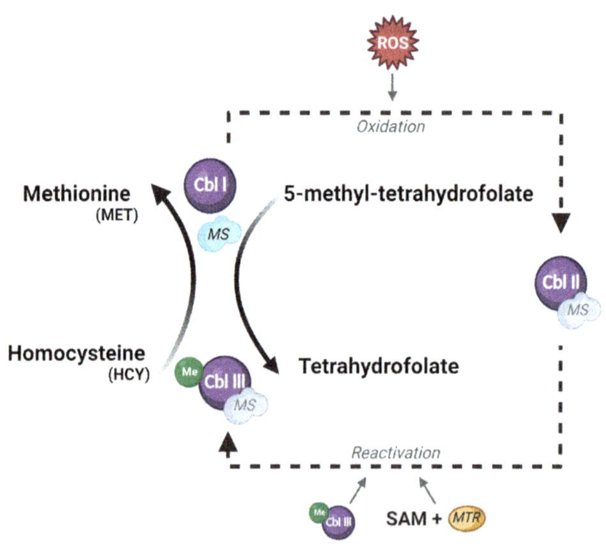

Fig. 2 Effect of ROS in Cobalamin oxidation. During the primary turnover cycle of methionine synthase (MS), Cob(I)alamin (Cbl I) is occasionally oxidized by reactive oxygen species (ROS) and electrophilic xenobiotic metabolites to Cob(II)alamin (Cbl II). This reaction inhibits MS. Reactivation of the enzyme and subsequent formation of methylcobalamin (MeCbl) depends on the supply of the methyl group by SAM and methionine synthase reductase (MTR) or the replacement of oxidized Cbl(II). Cbl(III), Coba(III)amin; Me, methyl; MS, methionine synthase; ROS, reactive oxygen species; and SAM, *S*-adenosylmethionine. *(Created with http://Biorender.com.)*

Thus, Cbl and MS are sensors and modulators of the CNS cell response to oxidative stress, aiding to preserve a healthy redox intracellular state and, consequently, acting antiinflammatory.

B12 boosts the synthesis of H_2S within the CNS

Hydrogen sulfide (H_2S) is a gasotransmitter mainly produced from the desulfhydration of cysteine carried out by the enzymes CBS, CGL, and 3-MST (Xie, Liu, & Bian, 2016). Vitamin B6-dependent CBS and CGL participate in the transsulfuration branch of the sulfur amino acids pathway, converting HCY into cystathionine and this into cysteine and alpha-ketobutyrate, respectively (Fig. 1).

H_2S acts antiinflammatory and prevents oxidative damage through several enzymatic and nonenzymatic mechanisms, such as by enhancing cystine/cysteine transporters to increase intracellular GSH production and maintaining its reduced status. Furthermore, the hydrosulfide anion (HS^-), resulting from the dissociation of H_2S, is a potent reducing agent against ROS (Xie et al., 2016).

It has been reported that the deficit of H_2S in patients with Alzheimer's disease correlates with low levels of SAM (McCarty, O'Keefe, & DiNicolantonio, 2019). This is probably due to the allosteric activation of CBS, the chief enzyme which synthesizes H_2S in the brain parenchyma, by SAM, whose production by MS requires MeCbl (Fig. 1).

Therefore, B12 plays a role in maintaining adequate levels of H_2S, which has a protective role in neuroinflammation. There is also an interplay between B12 and H_2S in shaping the epigenetic cell landscape that may account for their neuroprotective effects in neuroinflammation.

B12 prevents NMDAR-mediated glutamate excitotoxicity

Proinflammatory cytokines, such as IL-1β and TNF-α, can interfere with glutamatergic transmission by increasing glutamate release, inhibiting its reuptake and modulating the expression of NMDA receptors, which are involved in excitotoxicity processes (Viviani, Boraso, Marchetti, & Marinovich, 2014). NMDAR is the main receptor of the glutamatergic system due to its fundamental role in mediating functions such as cognition, memory, neural plasticity, and neurotoxicity (Kugaya & Sanacora, 2005). Activated NMDAR opens nonselective ion channels for cations, allowing the influx of Na^+ and Ca^{2+} and the efflux of K^+. Hyperactivation of NMDAR and, subsequently, excessive influx of Ca^{2+} can lead to excitotoxicity. This mechanism, which induces the production of free radicals, mitochondrial dysfunction, and cell death, is implicated in various neuropsychiatric diseases (Paoletti, Bellone, & Zhou, 2013).

CNCbl inhibited NMDAR-mediated glutamate excitotoxicity through a mechanism involving the inhibition of PKC, which, in its turn, suppresses the voltage-dependent Ca^{2+} channel activity and decreases the Ca^{2+} influx into synaptosomes isolated from the prefrontal cortex of rats (Hung, Wang, Huang, & Wang, 2009). Furthermore, glutamate excitotoxicity was prevented when rat cortical neurons were cultured in a medium containing MeCbl (Akaike, Tamura, Sato, & Yokota, 1993).

Taken together, these evidences that B12 can prevent NMDAR-mediated excitotoxicity in neural cells indicate that this vitamin has a great potential as an adjuvant therapy for some neuropsychiatric diseases whose physiopathology involves imbalances in the homeostasis of excitatory amino acids.

B12 rescues neurons from homocysteine-mediated cell death

HCY is synthesized from MET as an intermediary in the sulfur amino acids pathway and this enzymatic reaction catalyzed by MS depends on MeCbl as a cofactor (Fig. 1). Indeed, reduced MS or 5,10-methylenetetrahydrofolate reductase (MTHFR) activity, due to MeCbl or folate and tetrahydrofolate deficiency, respectively, can lead to HHCY (Kim, Kim, Roh, & Kwon, 2018). HCY regulates the levels of NMDAR and AMPA (Liu, Wen, Qiao, & An, 2013) and, when in excess, causes oxidative stress (Guest et al., 2015) and impaired mitochondrial function (Zhang et al., 2020) in CNS cells, contributing to synaptic dysfunction and memory deficits (Wyse et al., 2020; Zhang, Ma, Jing, Wang, & Zhang, 2020). HHCY has been associated with increased risk for several neuropsychiatric conditions, such as stroke, epilepsy, multiple sclerosis, Parkinson's and Alzheimer's diseases, dementia, cognitive impairment, and depression (Ansari, Mahta, Mallack, & Luo, 2014).

In vitro and in vivo experimental models have been used to demonstrate the mechanisms of HCY neurotoxicity and to test whether they can be mitigated by supplementing B12. MeCbl provided neuroprotection and neurorescue in a dose-dependent manner against ROS-independent apoptosis induced by high levels of HCY in NSC-34D cells, a hybrid cell line of motor neurons. In this experimental model, MeCbl prevented the HCY activation of caspases 3 and 7 (Hemendinger, Armstrong, & Brooks, 2011). Additionally, Tripathi et al. (2016) have demonstrated the central role of endoplasmic reticulum stress, lysosomal dysfunction, and impairment of autophagy induced by excess HCY in human and mouse primary astrocyte cultures as well as in a diet-induced mouse model of HHCY. Interestingly, B12 and folate cotreatment restored these functions. B12 and folate supplementation also reduced mitochondrial damage, autophagy, and neuronal apoptosis in apolipoprotein *E*-deficient (Apo $E^{-/-}$) mice with HHCY-enhanced brain damage (Zhang et al., 2017).

Therefore, the neuroprotectant effect of B12 in neuroinflammation involves preventing some noxious effects of excess HCY in CNS cells, such as endoplasmic reticulum stress, lysosomal dysfunction, impaired autophagy, and ROS-independent apoptosis.

B12 modulates the expression of inflammatory genes through cell epigenetic landscape changings

DNA methylation is an epigenetic mechanism that, in mammals, occurs mainly in CpG dinucleotides. When occurring in regulatory regions, it can affect the accessibility of transcription factors and RNA polymerases to DNA, thus modulating gene expression. The transfer of a methyl group to a cytosine residue is catalyzed by a family of DNMT and requires SAM as a methyl donor. In general, DNA hypomethylation is associated with active gene expression, while its hypermethylation is related to transcriptional downregulation (Fuks, 2005) (Fig. 1).

A second level of epigenetic regulation comprises a set of posttranslational modifications in amino acids that form the amino terminal tails of histone proteins. Chromatin is composed of nucleosome repeats made up of a protein octamer containing histones H2A, H2B, H3, and H4, which is physically associated with a sequence of 147 nucleotides. The histone N-terminal tails contain amino acids that are targets of a variety of covalent modifications, such as methylation, acetylation, phosphorylation, and ubiquitination. Such posttranslational modifications of specific residues in histones alter the chromatin state at the vicinity and regulate the transcription of associated genes: the euchromatin allows transcription, while the heterochromatin silences genes. This so-called histone code is established and maintained by dozens of chromatin-modifying enzymes, including the histone methyltransferases (HMT), demethylases (HDM), acetyltransferases (HAT), and deacetylases (HDAC), which target specific histone residues and genome loci, in a complex regulatory hierarchy for the control of global gene expression (Zhang et al., 2021).

DNA and histone methylation reactions depend on the bioavailability of SAM. Accordingly, adjuvant therapy with CNCbl increased genome-wide DNA methylation and reduced expression of proinflammatory genes including *IL-1β* and *CCL3* in infant rats with pneumococcal meningitis (de Queiroz et al., 2020). In addition to this mechanism involving MS and the sulfur amino acids pathway, Cbl also regulates gene expression through MCM, which produces succinyl-CoA in mitochondria from L-methylmalonyl-CoA (Fig. 1). In the Krebs cycle, succinyl-CoA is converted into succinate by 2-OGDD. When accumulated, succinate acts as an antagonist of histone and DNA demethylation reactions catalyzed by enzymes of the 2-OGDD family (Letouzé et al., 2013).

B12 also indirectly changes the epigenetic landscape by regulating oxidative stress and inflammation. Oxidative stress inactivates the histone deacetylating enzyme HDAC2, which can contribute to the formation of euchromatin regions where DNA is more accessible to the transcription machinery (Rahman, Marwick, & Kirkham, 2004). However, oxidative stress also activates signaling pathways that increase the expression of the DNA methylating enzyme DNMT1 (Molognoni, de Melo, da Silva, & Jasiulionis, 2013), which can lead to inhibition of gene transcription. The net balance of the epigenetic changes resulting from oxidative stress alleviation by B12 remains to be determined. Inflammation also has a direct influence on DNA methylation and vice versa. Gene promoters encoding cytokines are rapidly demethylated (~6 h) after activation of immune cells (Li et al., 2012), and cytokines such as TNF-α, IL-1β, and other inflammatory molecules, such as caspases, recruited during inflammatory processes can induce histone demethylation through the activation of JMJD3 (Bayarsaihan, 2011; Greer & Shi, 2012).

B12 modulates different mechanisms of chromatin remodeling that may affect gene expression. However, the net balance of the orchestrated contributions of all these mechanisms in determining the up- or downregulation of neuroinflammatory genes is still unknown. Further research is still needed to elucidate this intriguing and complex aspect of B12 mode of action.

B12 regulates pivotal neuroinflammation signaling pathway-associated genes

Among various genes and proteins regulated by B12, APP, Tau, NF-κB, TNF-α, and AP-1 play a central role in the regulation of neuroinflammation signaling (Fig. 3).

APP is a particularly abundant protein membrane at neuron synapses and a precursor of the β-amyloid peptide (Aβ), whose fibrillar form is a primary component of amyloid plaques. The accumulation of APP leads to increased activity of γ- and β-secretases, proteolytic enzymes that cleave APP generating the Aβ peptide. Also, extracellular deposits of Aβ peptide and intracellular neurofibrillary tangles, consisting mainly of hyperphosphorylated Tau protein, are observed in the brain tissue of patients affected by Alzheimer's disease or other neurodegenerative diseases. Together, these mechanisms intensify local oxidative stress and trigger a neuroinflammatory cascade (Sastre, Walter, & Gentleman, 2008).

Cbl, through SAM production, regulates the methylation state of CpG islands of *PSEN1* promoter, a gene encoding one of the four main proteins of the γ-secretase complex. There is also evidence that *BACE*, that codes a β-secretase, is regulated by promoter methylation (Fuso, Seminara, Cavallaro, D'Anselmi, & Scarpa, 2005). HHCY induced by dietary B12 suppression increased transcriptional and protein levels of *PSEN1* and *BACE*, in addition to increasing APP phosphorylation and Aβ production in the rat brain (Fuso et al., 2008; Zhang et al., 2009). B12 deficiency also causes inactivation of PP2A and subsequently hyperphosphorylation and aggregation of Tau protein (Rafiee, Asadollahi, Riazi, Ahmadian, & Saboury, 2017). In line with these findings, B12

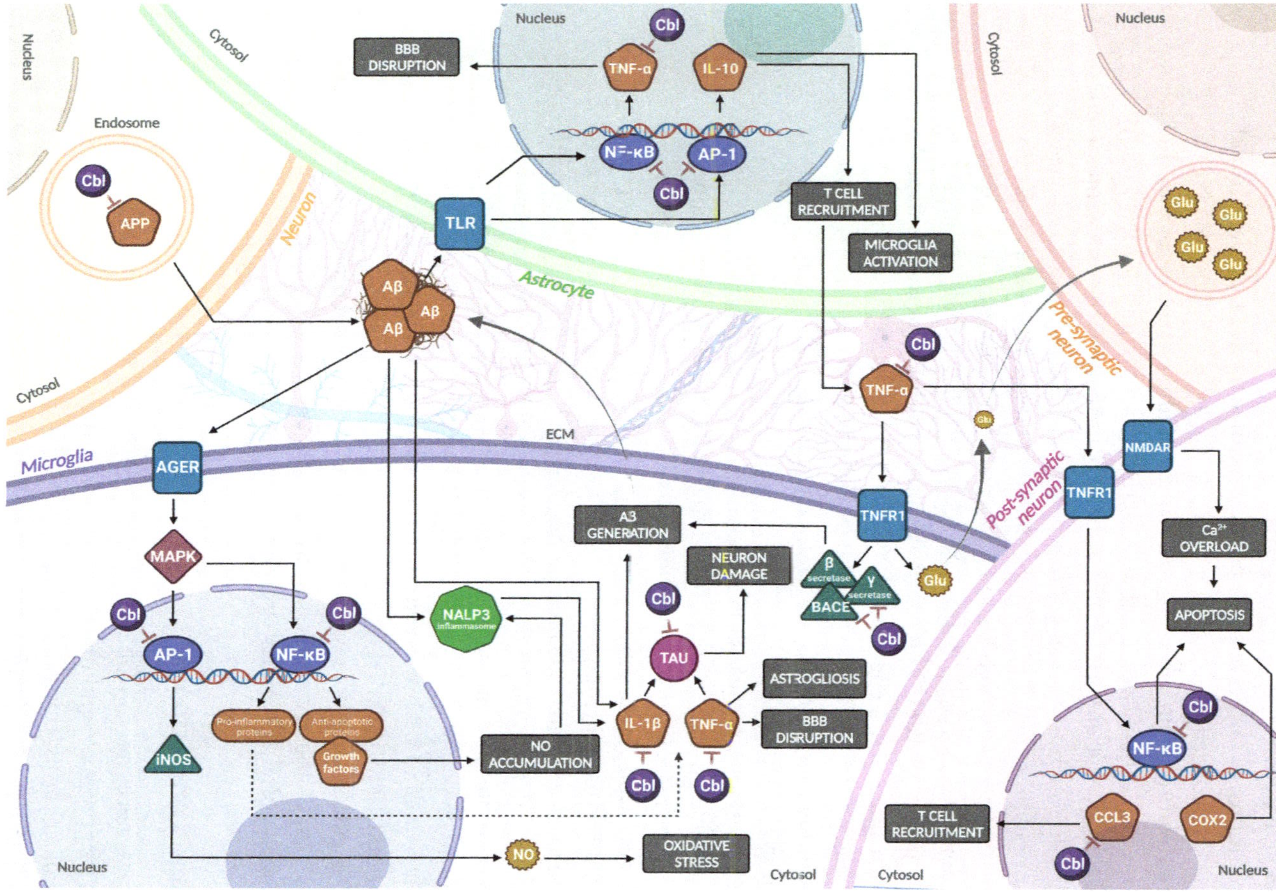

Fig. 3 B12 regulates pivotal components of the neuroinflammation signaling pathway. At the early stages of neuroinflammation, accumulated APP is processed to form Aβ peptides that aggregate in the extracellular matrix (ECM). The deposit of Aβ activates membrane receptors (advanced glycosylation end-product-specific receptor—AGER and toll-like receptor—TLR) of microglia and astrocytes, starting the inflammatory response cascade. In microglia, AGER signaling triggers the activation of AP-1 and NF-κB via MAPK. This mechanism culminates in NO accumulation, increased oxidative stress, blood-brain barrier (BBB) disruption, astrogliosis, neuronal damage, and generation of Aβ in a positive feedback loop. In astrocytes, the activation of AP-1 and NF-κB via TLR induces the production of proinflammatory mediators, contributing to BBB disruption, microglia activation, and T cell recruitment. TNF-α plays an important role in T-cell recruitment induced by neurons and in the production of glutamate (Glu). Excess glutamate hyperactivates NMDAR triggering neuronal apoptosis. Highlighted are the main molecules negatively regulated by Cbl. MAPK, mitogen-activated protein kinase; iNOS, inducible nitric oxide synthase; TNFR, tumor necrosis factor receptor 1; and COX2, cyclooxygenase 2. *(Created with http://Biorender.com.)*

inhibited Tau fibrillization in vitro (Rafiee et al., 2017), and combined B12 and folate supplementation prevented Tau hyperphosphorylation and HHCY-induced memory deficits in rats (Wei et al., 2011).

NF-κB is an evolutionarily conserved transcription factor activated by several stimuli and that regulates a complex gene network not yet fully elucidated. Among the NF-κB activators, there are neurotoxic proteins (e.g., Aβ) and, mainly, cytokines such as IL-1 and TNF-α that are also upregulated by the transcription factor in a feedback loop (Kaltschmidt, Widera, & Kaltschmidt, 2005). Cbl also modulates the inflammatory response and its resolution by regulating NF-κB activity. Among the mechanisms involved, the direct or indirect removal of ROS stands out. Cbl increases the production and preservation of GSH which is a RNS and ROS scavenger. Cbl also inhibits Aβ production (Alam, Siddiqi, Chaturvedi, Zaman, & Khan, 2017; Andrade, Loureiro, & Pereira, 2021) and downregulates TNF-α (Peracchi, Bamonti Catena, Pomati, De Franceschi, & Scalabrino, 2001; Veber et al., 2008). B12 deficiency has actually been associated with increased CNS TNF-α expression (Scalabrino et al., 2004). Accordingly, Cbl supplementation downregulated NF-κB and, consequently, TNF-α in the peripheral nervous system (Veber et al., 2008).

The JUN protein, associated with FOS, forms the AP-1 heterodimer. As well as NF-κB, AP-1 is an essential transcription factor for the maintenance of the immune system, participating in the regulation of tissue homeostasis, proliferation, differentiation, apoptosis, and cell migration (Hess, Angel, & Schorpp-Kistner, 2004). AP-1 regulates transcription of genes encoding cytokines and chemokines in addition to other proteins involved in T-cell recruitment and ROS production (Wei et al., 2019). The functionality of AP-1 is strongly dependent on the phosphorylation of its JUN and FOS subunits, the combinatorial patterns of these components, and the interaction with other transcription factors (Ameyar, Wisniewska, & Weitzman, 2003). It has been demonstrated that *Jun* mRNA levels are increased in B12-deficient mice (Liu et al., 2011), but the underlying mechanism remains to be elucidated.

Therefore, B12 regulates gene expression and the activity of central proteins in neuroinflammatory signaling, either by acting directly on these molecules or their activators or indirectly, through methyl-dependent epigenetic mechanisms.

Final remarks and conclusion

Neuroinflammation is necessary to maintain a healthy environment in the nervous system. However, an inappropriate and unbalanced immune response can cause or aggravate various neurological or psychiatric conditions. Cbl, besides its central role in MET production, which is required for synthesis of proteins, DNA, and neurotransmitters, also acts pleiotropically preventing or attenuating several mediators of neuroinflammation, such as the imbalance in redox homeostasis, accumulation of HCY, neuronal excitotoxicity mediated by NMDAR, also ensuring adequate levels of H_2S. In addition, it acts regulating the expression and activity of pivotal transcription factors and other components of the neuroinflammation signaling cascade. This plethora of regulatory mechanisms is mostly mediated by succinyl-CoA and SAM, whose production depends directly or indirectly on reactions catalyzed by the enzymes MCM and MS that require Cbl as cofactor. Furthermore, B12 plays a key role in the control of methyl-dependent mechanisms that modify the epigenetic cell landscape regulating gene transcription. Taken together, these properties make Cbl a strong candidate medication for new adjuvant neuroprotectant therapies to neuropsychiatric diseases involving neuroinflammation.

Applications to other neurological conditions

B12 has a safe pharmacological profile and concomitantly controls several components of neuroinflammation. For these reasons, B12 supplementation has been evaluated and proposed as a neuroprotectant adjuvant therapy for various neuropsychiatric disorders with neuroinflammation involved in their pathophysiology, such as ALS, Parkinson's disease, Alzheimer's disease, stroke, epilepsy, acute bacterial meningitis, depression, among others. However, there is still room for investigating the therapeutic potential of B12 in several other diseases related to neuroinflammation, which include viral meningitis and meningoencephalitis, cerebral malaria, and infections by emerging pathogens such as dengue, the Zika virus, and, why not, SARS-CoV-2.

Other components of interest

Much of the antiinflammatory effect of Cbl is mediated by metabolites of the Krebs cycle and the sulfur amino acid pathway. Other steps in these pathways are catalyzed by enzymes that require thiamine (B1), pyridoxine (B6), and folate (B9). Therefore, it is possible that the combined supplementation of these B-complex vitamins may potentiate the neuroprotectant effect of Cbl in neuroinflammation.

Mini-dictionary of terms

- **Anaplerotic reaction**—Chemical reactions that replenish metabolic pathways, producing intermediates. One example is the conversion of methylmalonyl-CoA into succinyl-CoA by MCM to feed the Krebs cycle.
- **Autophagy**—Natural catabolic process by which cytoplasmic components are degraded and recycled in the cell lysosome. It is a protective mechanism for cleaning damaged cells. However, autophagic dysfunction is associated with neuroinflammation, neurodegeneration, and many other disease-related processes.
- **Gasotransmitter**—Gaseous signaling molecules that can chemically modify their targets and regulate various intra- and extracellular functions, such as inflammation, cellular respiration, and apoptosis. They are produced endogenously, when required,

or acquired exogenously. These gaseous neurotransmitters are beneficial in physiologic concentrations but may be toxic when elevated. Well-known examples are nitric oxide, carbon monoxide, and hydrogen sulfide.

- **Epigenetic landscape—**Term coined by Conrad H. Waddington that refers to a conceptual model in which environmental factors may interact with genes determining cellular fate and modifying cell function and phenotype. According to Waddington's visual concept, epigenetic landscape consists of a series of ridges and valleys (environmental factors) that a cell can traverse on its way to a final fate (function). Later on, it has been proved that the cell epigenetic landscape is shaped with chemical modifications of specific residues in DNA, RNA, and histones, as well as by noncoding RNAs.
- **The histone code—**Proposed hypothesis in which multiple chemical modifications in histone proteins combined in different ways, primarily those of histone tails, may regulate DNA transcription, thus participating in the epigenetic code. Histones are proteins that associate with DNA to form nucleosomes, helping in the assembly of nuclear DNA to form the chromosome.

Applications to other areas

B12 supplementation has the potential as an adjuvant therapy for other clinical conditions that do not have the CNS as a primary site and whose pathophysiology involves oxidative stress, excessive cytokine, and chemokine production and imbalance in HCY and GSH production, such as influenza infections, sepsis, and others.

Summary points

- Cbl may favor the resting state of microglia and macrophages by feeding the Krebs cycle with succinyl-CoA and, consequently, contributing to reactivate the aerobic respiration to the detriment of glycolysis.
- Cbl and MS are sensors and modulators of the cell response to oxidative stress.
- B12 plays a role in maintaining adequate levels of H_2S, which has a protective role in neuroinflammation.
- There is an interplay between B12 and H_2S in shaping the epigenetic cell landscape.
- B12 can prevent NMDAR-mediated excitotoxicity in neural cells.
- B12 prevents the noxious effects of excess HCY in CNS cells that include endoplasmic reticulum stress, lysosomal dysfunction, impaired autophagy, and ROS-independent apoptosis.
- B12 modulates different mechanisms of chromatin remodeling that may contribute to downregulating neuroinflammatory genes.

References

Akaike, A., Tamura, Y., Sato, Y., & Yokota, T. (1993). Protective effects of a vitamin B12 analog, methylcobalamin, against glutamate cytotoxicity in cultured cortical neurons. *European Journal of Pharmacology*, *241*(1), 1–6. https://doi.org/10.1016/0014-2999(93)90925-8.

Aktas, O., Ullrich, O., Infante-Duarte, C., Nitsch, R., & Zipp, F. (2007). Neuronal damage in brain inflammation. *Archives of Neurology*, *64*(2), 185–189. https://doi.org/10.1001/archneur.64.2.185.

Alam, P., Siddiqi, M. K., Chaturvedi, S. K., Zaman, M., & Khan, R. H. (2017). Vitamin B12 offers neuronal cell protection by inhibiting Aβ-42 amyloid fibrillation. *International Journal of Biological Macromolecules*, *99*, 477–482. https://doi.org/10.1016/j.ijbiomac.2017.03.001.

Ameyar, M., Wisniewska, M., & Weitzman, J. B. (2003). A role for AP-1 in apoptosis: The case for and against. *Biochimie*, *85*(8), 747–752. https://doi.org/10.1016/j.biochi.2003.09.006.

Andrade, S., Loureiro, J. A., & Pereira, M. C. (2021). Vitamin B12 inhibits Aβ fibrillation and disaggregates preformed fibrils in the presence of synthetic neuronal membranes. *ACS Chemical Neuroscience*. https://doi.org/10.1021/acschemneuro.1c00210.

Ansari, R., Mahta, A., Mallack, E., & Luo, J. J. (2014). Hyperhomocysteinemia and neurologic disorders: A review. *Journal of Clinical Neurology*, *10*(4), 281–288. https://doi.org/10.3988/jcn.2014.10.4.281.

Aoyama, K. (2021). Glutathione in the brain. *International Journal of Molecular Sciences*, *22*(9). https://doi.org/10.3390/ijms22095010.

Banerjee, R. V., & Matthews, R. G. (1990). Cobalamin-dependent methionine synthase. *The FASEB Journal*, *4*(5), 1450–1459. https://doi.org/10.1096/fasebj.4.5.2407589.

Bayarsaihan, D. (2011). Epigenetic mechanisms in inflammation. *Journal of Dental Research*, *90*(1), 9–17. https://doi.org/10.1177/0022034510378683.

Butterfield, D. A., & Halliwell, B. (2019). Oxidative stress, dysfunctional glucose metabolism and Alzheimer disease. *Nature Reviews. Neuroscience*, *20*(3), 148–160. https://doi.org/10.1038/s41583-019-0132-6.

de Queiroz, K. B., Cavalcante-Silva, V., Lopes, F. L., Rocha, G. A., D'Almeida, V., & Coimbra, R. S. (2020). Vitamin B12 is neuroprotective in experimental pneumococcal meningitis through modulation of hippocampal DNA methylation. *Journal of Neuroinflammation*, *17*(1), 96. https://doi.org/10.1186/s12974-020-01763-y.

Fuks, F. (2005). DNA methylation and histone modifications: Teaming up to silence genes. *Current Opinion in Genetics & Development*, *15*(5), 490–495. https://doi.org/10.1016/j.gde.2005.08.002.

Fuso, A., Nicolia, V., Cavallaro, R. A., Ricceri, L., D'Anselmi, F., Coluccia, P., et al. (2008). B-vitamin deprivation induces hyperhomocysteinemia and brain S-adenosylhomocysteine, depletes brain S-adenosylmethionine, and enhances PS1 and BACE expression and amyloid-beta deposition in mice. *Molecular and Cellular Neurosciences*, *37*(4), 731–746. https://doi.org/10.1016/j.mcn.2007.12.018.

Fuso, A., Seminara, L., Cavallaro, R. A., D'Anselmi, F., & Scarpa, S. (2005). S-adenosylmethionine/homocysteine cycle alterations modify DNA methylation status with consequent deregulation of PS1 and BACE and beta-amyloid production. *Molecular and Cellular Neurosciences*, *28*(1), 195–204. https://doi.org/10.1016/j.mcn.2004.09.007.

Ghosh, S., Castillo, E., Frias, E. S., & Swanson, R. A. (2018). Bioenergetic regulation of microglia. *Glia*, *66*(6), 1200–1212. https://doi.org/10.1002/glia.23271.

Greer, E. L., & Shi, Y. (2012). Histone methylation: A dynamic mark in health, disease and inheritance. *Nature Reviews. Genetics*, *13*(5), 343–357. https://doi.org/10.1038/nrg3173.

Guest, J., Bilgin, A., Hokin, B., Mori, T. A., Croft, K. D., & Grant, R. (2015). Novel relationships between B12, folate and markers of inflammation, oxidative stress and NAD(H) levels, systemically and in the CNS of a healthy human cohort. *Nutritional Neuroscience*, *18*(8), 355–364. https://doi.org/10.1179/1476830515Y.0000000041.

Guo, J. D., Zhao, X., Li, Y., Li, G. R., & Liu, X. L. (2018). Damage to dopaminergic neurons by oxidative stress in Parkinson's disease (Review). *International Journal of Molecular Medicine*, *41*(4), 1817–1825. https://doi.org/10.3892/ijmm.2018.3406.

Hemendinger, R. A., Armstrong, E. J., & Brooks, B. R. (2011). Methyl Vitamin B12 but not methylfolate rescues a motor neuron-like cell line from homocysteine-mediated cell death. *Toxicology and Applied Pharmacology*, *251*(3), 217–225. https://doi.org/10.1016/j.taap.2011.01.003.

Hess, J., Angel, P., & Schorpp-Kistner, M. (2004). AP-1 subunits: Quarrel and harmony among siblings. *Journal of Cell Science, 117*(Pt 25), 5965–5973. https://doi.org/10.1242/jcs.01589.

Hung, K. L., Wang, C. C., Huang, C. Y., & Wang, S. J. (2009). Cyanocobalamin, vitamin B12, depresses glutamate release through inhibition of voltage-dependent Ca2+ influx in rat cerebrocortical nerve terminals (synaptosomes). *European Journal of Pharmacology, 602*(2–3), 230–237. https://doi.org/10.1016/j.ejphar.2008.11.059.

Hunt, A., Harrington, D., & Robinson, S. (2014). Vitamin B12 deficiency. *BMJ, 349*, g5226. https://doi.org/10.1136/bmj.g5226.

Kaltschmidt, B., Widera, D., & Kaltschmidt, C. (2005). Signaling via NF-kappaB in the nervous system. *Biochimica et Biophysica Acta, 1745*(3), 287–299. https://doi.org/10.1016/j.bbamcr.2005.05.009.

Kim, J., Kim, H., Roh, H., & Kwon, Y. (2018). Causes of hyperhomocysteinemia and its pathological significance. *Archives of Pharmacal Research, 41*(4), 372–383. https://doi.org/10.1007/s12272-018-1016-4.

Koedel, U., & Pfister, H. W. (1999). Oxidative stress in bacterial meningitis. *Brain Pathology, 9*(1), 57–67. https://doi.org/10.1111/j.1750-3639.1999.tb00211.x.

Kramer, P. A., Ravi, S., Chacko, B., Johnson, M. S., & Darley-Usmar, V. M. (2014). A review of the mitochondrial and glycolytic metabolism in human platelets and leukocytes: Implications for their use as bioenergetic biomarkers. *Redox Biology, 2*, 206–210. https://doi.org/10.1016/j.redox.2013.12.026.

Kugaya, A., & Sanacora, G. (2005). Beyond monoamines: Glutamatergic function in mood disorders. *CNS Spectrums, 10*(10), 808–819. https://doi.org/10.1017/s1092852900010403.

Letouzé, E., Martinelli, C., Loriot, C., Burnichon, N., Abermil, N., Ottolenghi, C., et al. (2013). SDH mutations establish a hypermethylator phenotype in paraganglioma. *Cancer Cell, 23*(6), 739–752. https://doi.org/10.1016/j.ccr.2013.04.018.

Li, Y., Chen, G., Ma, L., Ohms, S. J., Sun, C., Shannon, M. F., et al. (2012). Plasticity of DNA methylation in mouse T cell activation and differentiation. *BMC Molecular Biology, 13*, 16. https://doi.org/10.1186/1471-2199-13-16.

Liu, Z., Ciappio, E. D., Crott, J. W., Brooks, R. S., Nesvet, J., Smith, D. E., et al. (2011). Combined inadequacies of multiple B vitamins amplify colonic Wnt signaling and promote intestinal tumorigenesis in BAT-LacZxApc1638N mice. *The FASEB Journal, 25*(9), 3136–3145. https://doi.org/10.1096/fj.11-184143.

Liu, H., Wen, L. M., Qiao, H., & An, S. C. (2013). Modulation of hippocampal glutamate and NMDA/AMPA receptor by homocysteine in chronic unpredictable mild stress-induced rat depression. *Sheng Li Xue Bao, 65*(1), 61–71.

Martínez-Reyes, I., & Chandel, N. S. (2020). Mitochondrial TCA cycle metabolites control physiology and disease. *Nature Communications, 11*(1), 102. https://doi.org/10.1038/s41467-019-13668-3.

McCarty, M. F., O'Keefe, J. H., & DiNicolantonio, J. J. (2019). A diet rich in taurine, cysteine, folate, B12 and betaine may lessen risk for Alzheimer's disease by boosting brain synthesis of hydrogen sulfide. *Medical Hypotheses, 132*, 109356. https://doi.org/10.1016/j.mehy.2019.109356.

Mills, E. L., Kelly, B., Logan, A., Costa, A. S. H., Varma, M., Bryant, C. E., et al. (2016). Succinate dehydrogenase supports metabolic repurposing of mitochondria to drive inflammatory macrophages. *Cell, 167*(2), 457–470.e413. https://doi.org/10.1016/j.cell.2016.08.064.

Molognoni, F., de Melo, F. H., da Silva, C. T., & Jasiulionis, M. G. (2013). Ras and Rac1, frequently mutated in melanomas, are activated by superoxide anion, modulate Dnmt1 level and are causally related to melanocyte malignant transformation. *PLoS One, 8*(12), e81937. https://doi.org/10.1371/journal.pone.0081937.

Murphy, M. P. (2009). How mitochondria produce reactive oxygen species. *The Biochemical Journal, 417*(1), 1–13. https://doi.org/10.1042/BJ20081386.

Nelson, D. L., & Cox, M. M. (2021). *Lehninger principles of biochemistry* (8th ed.). W. H. Freeman.

Paoletti, P., Bellone, C., & Zhou, Q. (2013). NMDA receptor subunit diversity: Impact on receptor properties, synaptic plasticity and disease. *Nature Reviews. Neuroscience, 14*(6), 383–400. https://doi.org/10.1038/nrn3504.

Peracchi, M., Bamonti Catena, F., Pomati, M., De Franceschi, M., & Scalabrino, G. (2001). Human cobalamin deficiency: Alterations in serum tumour necrosis factor-alpha and epidermal growth factor. *European Journal of Haematology, 67*(2), 123–127. https://doi.org/10.1034/j.1600-0609.2001.t01-1-00507.x.

Percário, S., Moreira, D. R., Gomes, B. A., Ferreira, M. E., Gonçalves, A. C., Laurindo, P. S., et al. (2012). Oxidative stress in malaria. *International Journal of Molecular Sciences*, *13*(12), 16346–16372. https://doi.org/10.3390/ijms131216346.

Peterson, C. T., Rodionov, D. A., Osterman, A. L., & Peterson, S. N. (2020). B vitamins and their role in immune regulation and cancer. *Nutrients*, *12*(11). https://doi.org/10.3390/nu12113380.

Pezacka, E., Green, R., & Jacobsen, D. W. (1990). Glutathionylcobalamin as an intermediate in the formation of cobalamin coenzymes. *Biochemical and Biophysical Research Communications*, *169*(2), 443–450. https://doi.org/10.1016/0006-291x(90)90351-m.

Pritsos, C. A., & Pardini, R. S. (1983). A role for glutathione disulfide as a scavenger of oxygen radicals produced by 2,3-dichloro-1,4-naphthoquinone. *Research Communications in Chemical Pathology and Pharmacology*, *42*(2), 271–280.

Rafiee, S., Asadollahi, K., Riazi, G., Ahmadian, S., & Saboury, A. A. (2017). Vitamin B12 inhibits tau fibrillization via binding to cysteine residues of tau. *ACS Chemical Neuroscience*, *8*(12), 2676–2682. https://doi.org/10.1021/acschemneuro.7b00230.

Rahman, I., Marwick, J., & Kirkham, P. (2004). Redox modulation of chromatin remodeling: Impact on histone acetylation and deacetylation, NF-kappaB and pro-inflammatory gene expression. *Biochemical Pharmacology*, *68*(6), 1255–1267. https://doi.org/10.1016/j.bcp.2004.05.042.

Ransohoff, R. M., & Engelhardt, B. (2012). The anatomical and cellular basis of immune surveillance in the central nervous system. *Nature Reviews. Immunology*, *12*(9), 623–635. https://doi.org/10.1038/nri3265.

Ransohoff, R. M., Kivisäkk, P., & Kidd, G. (2003). Three or more routes for leukocyte migration into the central nervous system. *Nature Reviews. Immunology*, *3*(7), 569–581. https://doi.org/10.1038/nri1130.

Rink, C., & Khanna, S. (2011). Significance of brain tissue oxygenation and the arachidonic acid cascade in stroke. *Antioxidants & Redox Signaling*, *14*(10), 1889–1903. https://doi.org/10.1089/ars.2010.3474.

Sastre, M., Walter, J., & Gentleman, S. M. (2008). Interactions between APP secretases and inflammatory mediators. *Journal of Neuroinflammation*, *5*, 25. https://doi.org/10.1186/1742-2094-5-25.

Scalabrino, G., Carpo, M., Bamonti, F., Pizzinelli, S., D'Avino, C., Bresolin, N., et al. (2004). High tumor necrosis factor-alpha [corrected] levels in cerebrospinal fluid of cobalamin-deficient patients. *Annals of Neurology*, *56*(6), 886–890. https://doi.org/10.1002/ana.20325.

Selak, M. A., Armour, S. M., MacKenzie, E. D., Boulahbel, H., Watson, D. G., Mansfield, K. D., et al. (2005). Succinate links TCA cycle dysfunction to oncogenesis by inhibiting HIF-alpha prolyl hydroxylase. *Cancer Cell*, *7*(1), 77–85. https://doi.org/10.1016/j.ccr.2004.11.022.

Tannahill, G. M., Curtis, A. M., Adamik, J., Palsson-McDermott, E. M., McGettrick, A. F., Goel, G., et al. (2013). Succinate is an inflammatory signal that induces IL-1β through HIF-1α. *Nature*, *496*(7444), 238–242. https://doi.org/10.1038/nature11986.

Tripathi, M., Zhang, C. W., Singh, B. K., Sinha, R. A., Moe, K. T., DeSilva, D. A., et al. (2016). Hyperhomocysteinemia causes ER stress and impaired autophagy that is reversed by Vitamin B supplementation. *Cell Death & Disease*, *7*(12), e2513. https://doi.org/10.1038/cddis.2016.374.

Veber, D., Mutti, E., Tacchini, L., Gammella, E., Tredici, G., & Scalabrino, G. (2008). Indirect downregulation of nuclear NF-kappaB levels by cobalamin in the spinal cord and liver of the rat. *Journal of Neuroscience Research*, *86*(6), 1380–1387. https://doi.org/10.1002/jnr.21599.

Viviani, B., Boraso, M., Marchetti, N., & Marinovich, M. (2014). Perspectives on neuroinflammation and excitotoxicity: A neurotoxic conspiracy? *Neurotoxicology*, *43*, 10–20. https://doi.org/10.1016/j.neuro.2014.03.004.

Wei, C. J., Li, Y. L., Zhu, Z. L., Jia, D. M., Fan, M. L., Li, T., et al. (2019). Inhibition of activator protein 1 attenuates neuroinflammation and brain injury after experimental intracerebral hemorrhage. *CNS Neuroscience & Therapeutics*, *25*(10), 1182–1188. https://doi.org/10.1111/cns.13206.

Wei, W., Liu, Y. H., Zhang, C. E., Wang, Q., Wei, Z., Mousseau, D. D., et al. (2011). Folate/vitamin-B12 prevents chronic hyperhomocysteinemia-induced tau hyperphosphorylation and memory deficits in aged rats. *Journal of Alzheimer's Disease*, *27*(3), 639–650. https://doi.org/10.3233/JAD-2011-110770.

Wyse, A. T. S., Sanches, E. F., Dos Santos, T. M., Siebert, C., Kolling, J., & Netto, C. A. (2020). Chronic mild hyperhomocysteinemia induces anxiety-like symptoms, aversive memory deficits and hippocampus atrophy in adult rats: New insights into physiopathological mechanisms. *Brain Research*, *1728*, 146592. https://doi.org/10.1016/j.brainres.2019.146592.

Xia, L., Cregan, A. G., Berben, L. A., & Brasch, N. E. (2004). Studies on the formation of glutathionylcobalamin: Any free intracellular aquacobalamin is likely to be rapidly and irreversibly converted to glutathionylcobalamin. *Inorganic Chemistry*, *43*(21), 6848–6857. https://doi.org/10.1021/ic040022c.

Xie, Z. Z., Liu, Y., & Bian, J. S. (2016). Hydrogen sulfide and cellular redox homeostasis. *Oxidative Medicine and Cellular Longevity*, *2016*, 6043038. https://doi.org/10.1155/2016/6043038.

Zhang, T., Huang, D., Hou, J., Li, J., Zhang, Y., Tian, M., et al. (2020). High-concentration homocysteine inhibits mitochondrial respiration function and production of reactive oxygen species in neuron cells. *Journal of Stroke and Cerebrovascular Diseases*, *29*(10), 105109. https://doi.org/10.1016/j.jstrokecerebrovasdis.2020.105109.

Zhang, J. W., Ma, Y. M., Jing, L., Wang, Y. L., & Zhang, J. Z. (2020). Synaptic remodeling and reduced expression of the transcription factors, HES1 and HES5, in the cortex neurons of cognitively impaired hyperhomocysteinemic mice. *Pathology, Research and Practice*, *216*(6), 152953. https://doi.org/10.1016/j.prp.2020.152953.

Zhang, Y., Sun, Z., Jia, J., Du, T., Zhang, N., Tang, Y., et al. (2021). Overview of histone modification. *Advances in Experimental Medicine and Biology*, *1283*, 1–16. https://doi.org/10.1007/978-981-15-8104-5_1.

Zhang, C. E., Wei, W., Liu, Y. H., Peng, J. H., Tian, Q., Liu, G. P., et al. (2009). Hyperhomocysteinemia increases beta-amyloid by enhancing expression of gamma-secretase and phosphorylation of amyloid precursor protein in rat brain. *The American Journal of Pathology*, *174*(4), 1481–1491. https://doi.org/10.2353/ajpath.2009.081036.

Zhang, J. W., Yan, R., Tang, Y. S., Guo, Y. Z., Chang, Y., Jing, L., et al. (2017). Hyperhomocysteinemia-induced autophagy and apoptosis with downregulation of hairy enhancer of split 1/5 in cortical neurons in mice. *International Journal of Immunopathology and Pharmacology*, *30*(4), 371–382. https://doi.org/10.1177/0394632017740061.

CHAPTER 25

Vitamin D3 (VD3) effects on neuroinflammation and neurodegenerative disorders: Focus on Alzheimer's and Parkinson's diseases

Iana Bantim Felicio Calou[a], Milena de Barros Viana[b], Daniel Araki Ribeiro[b], Débora Estadella[b], and Glauce Socorro de Barros Viana[c]
[a]Coordenação de Nutrição, Rua Cícero Duarte, Federal University of Piauí (UFPi), Picos, Piauí, Brazil
[b]Departamento de Biociência, Rua Silva Jardim, Federal University of São Palo (UNIFESP), Santos, São Paulo, Brazil
[c]Departamento de Fisiologia e Farmacologia, Rua Coronel Nunes de Melo, Federal University of Ceará (UFC), Fortaleza, Brazil

Abbreviations

AD	Alzheimer's disease
Aβ	amyloid-β
COX-2	cyclooxygenase-2
DA	dopamine
IL	interleukin
iNOS	inducible nitric oxide synthase
LB	Lewy body
MHC	major histocompatibility complex
MPTP	1-methyl-4-phenyl-1,2,3,6-tetrahydropyridine
NFTs	neurofibrillary tangles
NRF2	nuclear factor erythroid 2-related factor 2
OS	oxidative stress
PD	Parkinson's disease
RA	retinoic acid
RANK	receptor activator of nuclear factor kappa-B
RARs	retinoic acid receptors
RNS	reactive nitrogen species
ROS	reactive oxygen species
RXR	retinoid X receptor
TH	tyrosine hydroxylase
TNF-α	tumor necrosis factor-α
UPS	ubiquitin-proteasomal system
VD3	vitamin D3
VD3R	VD3 receptor
VDREs	vitamin D response elements
α-syn	α-synuclein
1α,25-$(OH)_2$ D_3	1α,25-dihydroxyvitamin D3

Vitamins and Minerals in Neurological Disorders
https://doi.org/10.1016/B978-0-323-89835-5.00024-7

Introduction

This review will discuss the literature data on the possible neuroprotective properties of VD3, focusing on the two most prevalent neurological disorders, like Alzheimer's and Parkinson's diseases. Besides, the association of VD3 deficiency with other neurological conditions and the value of other vitamins/nutrients as beneficial for neurodegenerative disorders will be discussed as well.

Vitamin D3 and physiological features

Vitamin D3 (VD3), or cholecalciferol, is a prohormone that, after being synthesized by the skin when exposed to sunlight, is metabolically processed into calcitriol, responsible for its main action. Cholecalciferol, the physiologically inactive form of VD3, binds to VD3 binding protein and is transported to the liver, where it is converted into the major circulating metabolite of VD3, 25-hydroxyvitamin D, or calcifediol, through a hydroxylation mediated by 25-hydroxylase (CYP2R1). A further hydroxylation mediated by 1-alpha-hydroxylase (CYP27B1), occurring mainly in the kidney, is required for obtaining the fully active metabolite 1,25(OH)2D3, also known as calcitriol (Bikle, 2014; Cao et al., 2017) (see Fig. 1).

The connection of calcitriol to the membrane and nuclear vitamin D3 receptors (VD3R) triggers molecular cascades, important in serum calcium and phosphorus regulation, autoimmunity, hypertension control, obesity, infectious diseases, and some types of cancers. Although it was initially considered a nutrient, it has been recognized that the molecules derived from the VD3 metabolism are best considered as a complex endocrine system (Gallieni et al., 2009). In the last decades, VD3 has become a hot topic in endocrinology research, because of its nonclassical actions involving immunomodulation, lung and muscle function, cardiovascular health, infectious disease prevention, neurodegenerative diseases, and even antiaging effects (Christakos et al., 2013; Xu et al., 2020). Besides acting to promote calcium absorption in the gut and maintain serum calcium and phosphate concentrations, VD3 also affects bone growth and reorganization through its action on osteoblasts and osteoclasts cells (Gil, Plaza-Diaz, & Mesa, 2018).

VD3 works through two types of receptors: (i) the nuclear VD3R to induce genomic action and (ii) the putative membrane receptor MARRS (membrane-associated, rapid response steroid-binding) to induce nongenomic actions. Similar to the receptors of other steroid hormones, VD3R belongs to the nuclear receptor superfamily that can bind their ligands at very low concentrations in a specific ligand-binding domain where it can interact with gene response elements, thereby modulating gene transcription. VD3R binds to $1\alpha,25\text{-}(OH)_2\,D_3$ and then reaches the nucleus to form a heterodimer with retinoid X receptor (RXR) (Carlberg & Molnár, 2015).

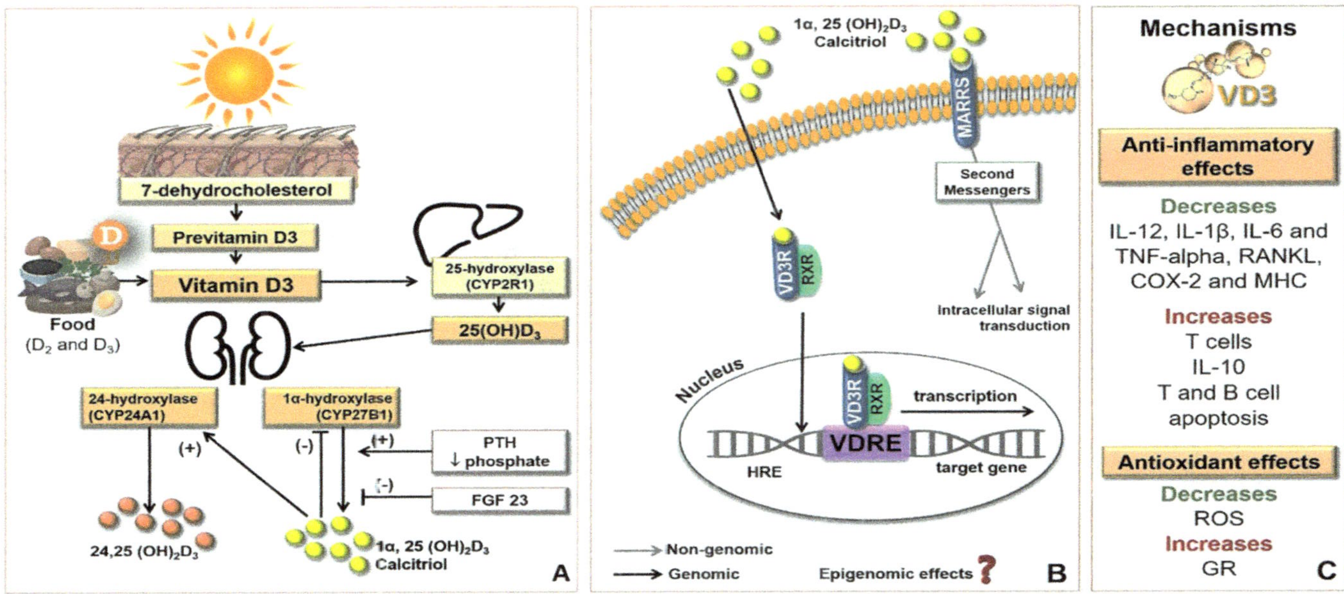

Fig. 1 (A) Vitamin D3 (VD3) synthesis from the sun and dietary sources. (B) Vitamin D3 genomic and nongenomic pathways. (C) Vitamin D3 antiinflammatory and antioxidant mechanisms. The VD3 system comprises hormone precursors, active metabolites, carriers, enzymes, and receptors involved in genomic and nongenomic effects. *FGF 23*, fibroblast growth factor 23; *GR*, glutathione reductase; *HRE*, hormone response element in target genes; *IL*, interleukin; *MARRS*, membrane-associated, rapid response steroid-binding; *MHC*, major histocompatibility complex; *PTH*, parathyroid hormone; *RANKL*, receptor activator of nuclear factor kappa-B ligand; *RXR*, retinoid X receptor; *TNF-alpha*, tumor necrosis factor-alpha; *VD3R*, nuclear vitamin D3 receptor; *VDRE*, vitamin D response element.

VD3R is a ligand-responsive transcription factor that forms active, heterodimeric complexes with the 9-*cis* retinoic acid receptor (RXR) on vitamin D response elements (VDREs). VD3 and retinoic acid (RA) receptors bind as heterodimers with the retinoid X receptor (RXR) to hormone response elements (HREs) in target genes. Therefore, the transcriptional response to RA and VD3 depends on a complex combinatory pattern of interaction among different receptors with DNA and coactivators (Jimenez-Lara & Aranda, 2000) (see Fig. 1).

VD3 exerts autocrine and paracrine effects in different cells of the immune system, presenting antiinflammatory effects by suppressing the innate as well as the adaptive immune response (Bouillon & Bikle, 2019; Colotta, Jansson, & Bonelli, 2017). VD3 deficiency is strongly associated with an increased risk of infections, diseases with a high inflammatory chronic pathophysiologic process, and autoimmune diseases (Yin & Agrawal, 2014). Inflammatory and immune cells not only convert inactive VD3 metabolites into calcitriol, the active form of VD3, but also express the nuclear receptor of VD3 (VD3R) that modulates differentiation, activation and proliferation of these cells. In vitro, calcitriol upregulates different antiinflammatory pathways and downregulates molecules that activate immune and inflammatory cells (Colotta et al., 2017), as shown in Fig. 1. The antiinflammatory cytokine IL-10 production and T- and B-cell apoptosis are increased by VD3 (van Halteren, Tysma, van Etten, Mathieu, & Roep, 2004). Also, an antioxidant effect has been observed, mediated by increased glutathione reductase (GR) and a decrease in reactive oxygen species (ROS) (Martens, Gysemans, Verstuyf, & Mathieu, 2020).

The antiinflammatory effects mediated by VD3 have also been demonstrated in the CNS: it increases calcium uptake by neurons, increases phagocytic activity, and reduces nitric oxide synthase (iNOS) expression in the microglia (Garcion, Nataf, Berod, Darcy, & Brachet, 1997). Lastly, VD3 stimulates oligodendrocyte maturation and astrocyte activation (Nystad et al., 2014). We have recently (Leal et al., 2020) demonstrated that VD3 presents antiinflammatory and antioxidative effects, manifested at peripheral and central sites as shown in this study for the first time. Therefore, after discussing VD3 actions and their physiological features, the second chapter deals with the effects of VD3 on neuroinflammation and oxidative stress (OS).

Vitamin D: Neuroinflammation and OS

As illustrated in Fig. 2, accumulating evidence suggests that neurodegeneration occurs, in part, because the environment is affected during disease in a cascade of processes collectively termed neuroinflammation. These observations indicate that therapies targeting glial cells might provide benefits for those afflicted by neurodegenerative disorders (Ransohoff, 2016). Aging is the primary risk factor for most neurodegenerative diseases, including Parkinson's disease (PD) and Alzheimer's disease (AD). The central biological

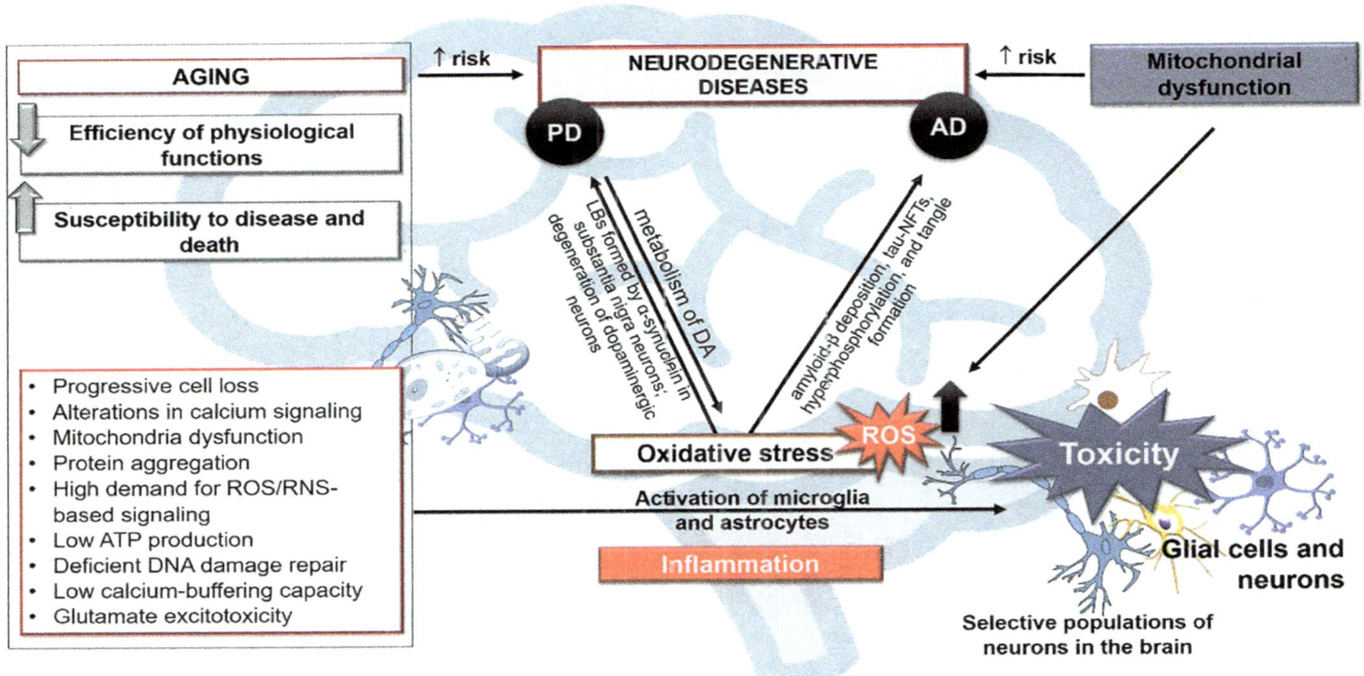

Fig. 2 Parkinson's disease (PD) is a progressive neurodegenerative disorder associated with a selective loss of dopamine (DA) neurons in the substantia nigra *pars compacta*. Alzheimer's disease (AD) is characterized by the abnormal deposition of amyloid-β (Aβ) peptide, and intracellular accumulation of neurofibrillary tangles of hyperphosphorylated tau protein and dementia. One of the disease hallmarks is oxidative stress (OS), which interconnects with the pathological AD and PD processes. *ATP*, adenosine triphosphate; *DNA*, deoxyribonucleic acid; *LBs*, Lewy bodies; *NFTs*, neurofibrillary tangles; *RNS*, reactive nitrogen species; *ROS*, reactive oxygen species.

mechanisms of aging and their potential as targets for the development of novel therapies for neurodegenerative diseases include NAD^+ precursors, mitophagy inducers, and inhibitors of cellular senescence (Ransohoff, 2016).

Aging is characterized by a progressive decline in the efficiency of physiological function and by increased susceptibility to disease and death. Currently, one of the most plausible and acceptable explanations for the mechanistic basis of aging is the "free radical theory of aging." This theory postulates that aging and its related diseases are the consequences of free radical-induced damage to cellular macromolecules and the inability to counterbalance these changes by endogenous antioxidant defenses (Gemma, Vila, Bachstetter, & Pickford, 2007).

Aging is a driving factor in various age-related diseases, including neurodegenerative diseases (Li et al., 2021). It is a complex process and many factors contribute to aging, such as the time-dependent accumulation of macromolecular damage, including DNA damage. Links between damage to the genetic code and other pillars of aging support the notion that DNA damage could be the root of aging (Yousefzadeh et al., 2021). Furthermore, the mechanisms of aging include a loss of proteostasis, mitochondrial dysfunction, altered nutrient sensing, telomere shortening, genomic instability, cellular senescence, stem cell exhaustion, epigenetic alterations, and altered intercellular communication (Keshavarz, Xie, Schaaf, Bano, & Ehninger, 2022).

Neurodegenerative diseases are morphologically featured by progressive cell loss in specific vulnerable neuronal cells, often associated with cytoskeletal protein aggregates, forming inclusions in neurons and/or glial cells. Deposition of abnormal aggregated proteins and disruption of metal ions homeostasis are highly associated with OS. These events occur as a result of neuroinflammatory cascades, involving alterations in the crosstalk between glial cells and neurons as a consequence of the activation of microglia and astrocytes (Guzman-Martinez et al., 2019).

While many brain neurons can cope with a rise in OS, there are select populations of neurons in the brain that are vulnerable. Because of their selective vulnerability, these neurons are usually the first to exhibit functional decline and cell death during normal aging, or in age-associated neurodegenerative diseases, such as AD and PD (Uttara, Singh, Zamboni, & Mahajan, 2009). Among the major underlying factors are high intrinsic OS, high demand for ROS/reactive nitrogen species (RNS)-based signaling, low ATP production, mitochondrial dysfunction, and high inflammatory response in vulnerable neurons. Other intrinsic and extrinsic factors contributing to the selective vulnerability of neurons to OS are deficient DNA damage repair, low calcium-buffering capacity, and glutamate excitotoxicity (Wang & Michaelis, 2010).

Brain cells normally respond adaptively to OS or bioenergetic challenges, resulting from ongoing activity in neuronal circuits. As the nervous system ages, neuron metabolism may change, inducing glucose hypometabolism, impaired transport of critical substrates underlying metabolism, alterations in calcium signaling, and mitochondrial dysfunction. Moreover, in neuronal aging, impaired and aggregated proteins in the cytoplasm and mitochondria accumulate, resulting in OS, reduced antioxidant defenses, and/or an increase in ROS. These changes lead to greater vulnerability of neurons in various regions of the brain and increased susceptibility to several diseases (Castelli et al., 2019).

Increased ROS production and OS have been implicated in the pathogenesis of numerous neurodegenerative conditions including, among others, AD and PD. The endogenous antioxidant response pathway protects cells from OS, by increasing the expression of cytoprotective enzymes and is regulated by the transcription factor nuclear factor erythroid 2-related factor 2 (Nrf2). In addition to regulating the expression of antioxidant genes, Nrf2 has also been shown to exert antiinflammatory effects besides modulating mitochondrial function, emerging as an antioxidant therapy and as a promising therapeutic target to be considered for the management of these diseases (Brandes and Gray, 2020; Liu et al., 2018).

AD is the most common cause of disability in individuals aged >65 years worldwide and is characterized by the abnormal deposition of amyloid-β (Aβ) peptides and intracellular accumulation of neurofibrillary tangles of hyperphosphorylated tau protein and dementia. One of the disease hallmarks is OS, which interconnects with other processes, such as amyloid-β deposition, tau hyperphosphorylation, and tangle formation occurring particularly in the brain, a vulnerable organ, highly affected by ROS (Huang, Zhang, & Chen, 2016) (see Fig. 3). Oxidative stress (OS) connects the different hypotheses and mechanisms of AD. It is a process causing neuronal damage and occurring in various pathways. Besides, OS plays a critical role in AD and can be considered a crucial factor in the pathogenesis of AD (Bai, Guo, Ye, Xie, & Xie, 2022). Blood biomarkers indicating elevated amyloid-β (Aβ) pathology in AD are needed for facilitating the initial screening process of participants in disease-modifying trials. Plasma p-tau231 and p-tau217 better capture the earliest cerebral Aβ changes, before Aβ plaque pathology is present, and are promising blood biomarkers to enrich a preclinical population for Alzheimer's disease clinical trials (Milà-Alomà et al., 2022).

Following this line of reasoning, OS is increasingly recognized as a central event contributing to the degeneration of dopaminergic neurons in the pathogenesis of Parkinson's disease (PD). Although reactive oxygen species (ROS) production is implicated as a causative factor in PD, the cellular and molecular mechanisms linking OS with dopaminergic neuron death are complex and not well characterized (Guo, Zhao, Li, Li, & Liu, 2018).

PD is a progressive neurodegenerative disorder associated with a selective loss of the dopaminergic (DA) neurons in the substantia nigra *pars compacta* (*SNpc*). Because there is currently no therapy that delays the neurodegenerative process, modification of the disease course by neuroprotective therapy is an important unmet clinical need. Thus, understanding cellular mechanisms that render the nigral neurons particularly vulnerable has been a subject of intensive research. Increasing evidence suggests that OS plays a major role. The metabolism of DA itself contributes to OS, resulting in the modification of intracellular macromolecules whose functions are important for cell survival.

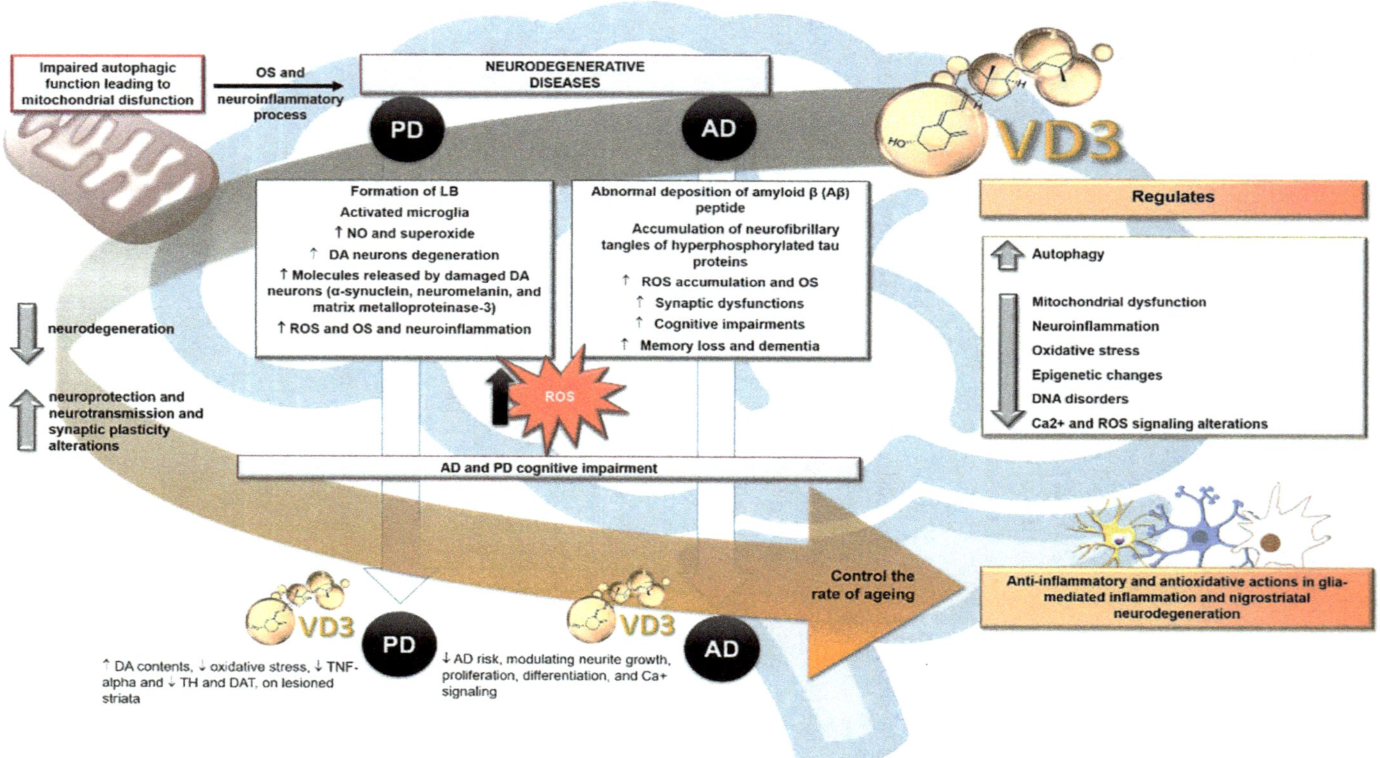

Fig. 3 The effects of vitamin D3 (VD3) on neuroinflammation and oxidative stress in the central nervous system: it increases calcium uptake by neurons, increases phagocytic activity, reduces nitric oxide synthase (iNOS) expression in the microglia, stimulates oligodendrocyte maturation and astrocyte activation, and presents antiinflammatory and antioxidative effects, observed at peripheral and central sites. VD3 deficiency is associated with an increased risk of age-related chronic diseases, including Alzheimer's (AD) and Parkinson's diseases (PD). VD3 neuroprotective effects result from its antiinflammatory and antioxidant actions. *DA*, dopamine; *DAT*, dopamine transporter; *LB*, Lewy bodies; *NO*, nitric oxide; *OS*, oxygen stress; *ROS*, reactive oxygen species; *TH*, tyrosine hydroxylase; *TNF-alpha*, tumor necrosis factor-alpha.

Mitochondrial dysfunction and the consequent increase in ROS also trigger a sequence of events that lead to cell demise. Besides, activated microglia produce nitric oxide and superoxide during neuroinflammatory responses, and this is aggravated by the molecules released by damaged DA neurons, such as α-synuclein, neuromelanin, and matrix metalloproteinase-3 (Dias, Junn, & Mouradian, 2013) (see Fig. 3).

The hallmark of PD pathology is the formation of Lewy bodies and the deposition of α-synuclein (α-syn). The mechanisms responsible for the progressive feature of DA neurodegeneration are not fully illustrated. Recently, OS and neuroinflammation have been pointed out as key factors, due to external influences or changes in the internal environment, such as the accumulation of ROS, exposure to a toxic environment, and the transformation of systemic inflammation into neuroinflammation (He, Zhu, Wang, & Zhang, 2020).

Microglia-mediated neuroinflammation has been described as a common hallmark of PD and is believed to further exacerbate the progressive degeneration of dopaminergic neurons. It has been shown that VD3, administered orally in a 1-methyl-4-phenyl-1,2,3,6-tetrahydropyridine (MPTP)-induced preclinical animal model of PD, protects against glia-mediated inflammation and nigrostriatal neurodegeneration. VD3 significantly attenuated the MPTP-induced loss of tyrosine hydrolase (TH)-positive neuronal cells, microglial cell activation (Iba1-immunoreactive), inducible nitric oxide synthase (iNOS), and TLR-4 expression, typical hallmarks of the proinflammatory activation of microglia. Additionally, VD3 was able to decrease proinflammatory cytokine mRNA expression in distinct brain areas of the MPTP mouse. Collectively, these results demonstrate that VD3 exhibits substantial neuroprotective effects in this PD animal model, by attenuating proinflammatory and upregulating antiinflammatory processes (Calvello et al., 2017).

We also showed that pre- or posttreatment with VD3 prevented behavioral changes and improved the decreased DA contents of the 6-OHDA group of hemiparkinsonian rats. Besides, VD3 reduced OS, increased TH and DAT, and reduced TNF-alpha immunostainings in the lesioned striata. While significant decreases in VD3R immunoreactivity were observed, after the 6-OHDA lesion, these changes were blocked after VD3 pre- or posttreatments. Furthermore, VD3 offers neuroprotection, decreasing behavioral changes, DA depletion, and OS and reverses partially or completely TH, DAT, TNF-alpha and VD3R alterations of immunoreactivities, in this 6-OHDA model of PD (Lima et al., 2018). Importantly, we also recently demonstrated VD3 neuroprotective effects on brain mitochondria from hemiparkinsonian rats (Araújo de Lima, Oliveira Cunha, Felício Calou, Tavares Neves, & Socorro de Barros Viana, 2022).

Processes such as autophagy, mitochondrial dysfunction, inflammation, OS, epigenetic changes, DNA disorders, and alterations in Ca^{2+} and reactive oxygen species (ROS) signaling are all known to be regulated by vitamin D (Berridge, 2017). Furthermore, the importance of decreasing the neuroinflammatory and OS processes associated with neurodegeneration points out the value of VD3 for supplementation, to be used for prevention or even as an adjuvant therapeutic strategy in neurodegenerative diseases such as AD and PD.

VD3 and neurodegenerative diseases

As discussed earlier, neurodegenerative diseases are a major global burden affecting millions of people worldwide. The prevalence of disorders such as AD and PD is increasing throughout the world, in part by the increased population lifespan. Age, as well as mitochondrial dysfunction, one of the main features of the aging process, is the main risk factor for a neurodegenerative process (Cenini, Lloret, & Cascella, 2019).

Currently, there is no cure for these diseases. However, new and more effective therapeutic strategies are in development. Models focused on cell-based systems or animal models have proven to be useful tools for helping researchers to shed light on the mechanisms underlying neurodegenerative diseases, and these advances have now begun to provide promising therapeutic avenues (Gitler, Dhillon, & Shorter, 2017).

One of the main goals of the research is to enhance our understanding of the mechanisms underlying proteotoxicity. As shown in Fig. 3, specific proteins are associated with these neurodegenerative diseases, as determined by both immunohistochemical studies on *postmortem* tissue and genetic screening, where protein misfolding and aggregation are key hallmarks. Many of these proteins are shown to misfold and aggregate into soluble nonnative oligomers and large insoluble protein deposits (fibrils and plaques) and are considered to be an underlying mechanism mediating the pathogenesis of late-onset neurodegenerative diseases (McAlary, Plotkin, & Cashman, 2019; Ruz, Alcantud, Montero, Duran, & Bandes-Ciga, 2020).

Furthermore, protein homeostasis or proteostasis is a balance of cellular protein levels mediated by a network of biochemical pathways that regulate different steps of protein quality control, from synthesis to degradation. Protein misfolding and/or alterations during protein synthesis result in inactive or toxic proteins. Thus, the maintenance of a balanced proteome, preventing the formation of impaired proteins, is accomplished by two major catabolic routes: the ubiquitin-proteasomal system and the autophagy-lysosomal system. The proteostasis network is particularly important for neurons, as its failure is implicated in the development of neurodegenerative diseases, such as AD and PD, among others (Höhn, Tramutola, & Cascella, 2020).

Moreover, aggregation of brain proteins into filamentous lesions is emerging as a common mechanistic theme in sporadic and hereditary neurodegenerative disorders, including AD and PD. For example, in the clinical setting, numerous telencephalic β-amyloid peptide-rich senile plaques and τ-rich neurofibrillary tangles (NFTs) are diagnostics of AD, while Lewy bodies (LBs) formed by α-synuclein (α-syn) in *substantia nigra* neurons are signatures of PD (Trojanowski, 2000).

The identification of the exact molecular mechanisms involved in these diseases would facilitate the discovery of earlier pathophysiological markers along with substantial therapies, which may consist of mitochondria-targeted antioxidant therapy, mitochondrial dynamic modulators, epigenetic modulators, and neural stem cell therapy. Therefore, all these therapies may hold particular assurance, as influential neuroprotective therapies in the treatment of neurodegenerative diseases (Patel & Mandal, 2019).

AD is the most common cause of disability in 65 years of age or older individuals and is clinically characterized by progressive deficits of cognitive functions. Pathological hallmarks of AD include histological changes induced by the extracellular deposition of amyloid-β peptides, forming senile plaques and intracellular neurofibrillary tangle of hyperphosphorylated tau proteins in the brain. However, these phenomena are mainly initiated and enhanced by OS, a process referring to an imbalance between antioxidants and oxidants in favor of oxidants (Cioffi, Hassan, Adam, & Broersen, 2019; Huang et al., 2016). Studies show that low serum concentrations of VD3 can substantially increase the risk of AD. In addition to modulating neurite growth, proliferation, differentiation, and calcium signaling, VD3 has also been implicated in neuroprotection and may alter neurotransmission and synaptic plasticity, important processes involved in neurodegeneration (Lauer, Janitschke, Hartmann, Grimm, & Grimm, 2019).

PD prevalence is rapidly increasing in an aging global population and this increase exponentially rises social and economic costs, emphasizing the immediate need for effective disease-modifying treatments (Trist, Hare, & Double, 2019). PD is characterized by a progressive neurodegenerative movement disorder, associated with a selective loss of dopaminergic neurons in the *SNpc* and degeneration of projecting nerve fibers to the striatum. Evidence implicates the protein α-synuclein (α-Syn), in the pathogenesis of PD, and as such, there is intense interest in targeting α-Syn for potential disease modification (Grosso Jadutkar et al., 2022, Henderson et al., 2022). Increasing evidence suggests that OS plays a major role and the metabolism of DA itself contributes to this. Mitochondrial dysfunction and the consequent increase in ROS also trigger a sequence of events that leads to cell demise. In addition, activated microglia produce nitric oxide and superoxide during neuroinflammatory responses, and this is aggravated by the molecules released by damaged DA neurons, such as α-synuclein, neuromelanin, and matrix metalloproteinase (Hwang, 2013).

Furthermore, these two prevalent neurodegenerative diseases are characterized by the progressive appearance of abnormal proteinaceous assemblies in the nervous system. Studies in experimental systems indicate that the assemblies originated from the prion-like seeded aggregation of specific misfolded proteins form the intracellular and/or extracellular lesions typical of each disorder. The characteristics of cells and regions in the affected network govern their vulnerability and thereby influence the neuropathological and clinical attributes of the disease (Jucker & Walker, 2018).

Aging can occur at different rates, but what controls these variable rates is unknown and a hypothesis that VD3 may act to control the rate of aging was developed. Processes such as autophagy, mitochondrial dysfunction, inflammation, OS, epigenetic changes, DNA disorders, and alterations in Ca^{2+} and ROS signaling are known to be regulated by VD3. The activity of these processes will be enhanced in individuals that are deficient in VD3. Not only will this increase the rate of aging, but it will also increase the probability of developing age-related diseases, such as AD and PD. For an individual with normal VD3 levels, these aging-related processes will occur at lower rates, resulting in a reduced rate of aging and enhanced protection against these age-related diseases (Berridge, 2017).

The clinical and preclinical exploration of the therapeutic properties of VD3 has significantly increased in the past decade, owing to the growing associative evidence suggesting vitamin D is neuroprotective. However, whether depletion of VD3 contributes to the onset of neurological disorders or is a symptom of neurological disease has yet to be defined. The current data supporting neuroprotective benefits from VD3 are based on preclinical and observational studies (Iacopetta et al., 2020). All these findings emphasize the therapeutic potential of VD3 as a neuroprotective agent with a great number of possibilities for the treatment of neurodegenerative diseases, such as AD and PD, due to VD3 antioxidant and antiinflammatory actions, among others.

The autophagic process mitophagy is the only known mechanism for mitochondrial turnover and it has been speculated that its dysfunction may result in cellular stress. Investigations have provided indications of how mitophagy is associated with cellular OS and its impact on neurodegeneration. This impaired autophagic function, leading to mitochondria dysfunction, may be considered a possible mechanism in the pathogenesis of several neurodegenerative disorders, including AD and PD. Therefore, mitochondrial protective agents represent a unique direction for the development of drug candidates that can modify the pathogenesis of neurodegeneration (Kamat, Kalani, Kyles, Tyagi, & Tyagi, 2014).

In AD, mitochondria progressively lose their proper functions that are ultimately responsible for their accumulation and removal via the autophagic process, which further worsens the progression of the disease. Preclinical and clinical studies have suggested that mitochondrial dysfunction along with mitophagy significantly contribute to the accumulation of amyloid-beta (Aβ) fibrils and hyperphosphorylated tau protein tangles which lead to synaptic dysfunctions and cognitive impairments, such as memory loss through reactive oxygen species (ROS)-mediated pathway (Rai, Singh, Singh, Singh, & Singh, 2020). Despite their specific primary etiologies, neurodegenerative diseases present mitochondrial dysfunction with consequent OS accumulation. As these two events occur early during the neurodegenerative process, they could be considered important therapeutic targets (Pezzini, Mattoli, & Ciofani, 2017).

VD3R mediates many genomic and nongenomic effects of VD3. Recently, the mitochondrial effects of VD3 have been characterized in many cell types. In the long run, the absence of the receptor caused impairment of mitochondrial integrity and, finally, cell death. VD3R plays a central role in protecting cells from excessive respiration and the production of ROS that leads to cell damage (Ricca et al., 2018).

Neurodegenerative diseases such as AD and PD have several common pathological signals and are greatly affected by OS and neuroinflammatory processes that are also related to mitochondrial dysfunction. Antioxidant and antiinflammatory effects of VD3 have been demonstrated in vivo and in vitro systems (Leal et al., 2020). Additionally, VD3 presents important neuroprotective effects on hemiparkinsonian rats (Costa et al., 2017), which makes VD3 an important therapeutic strategy and target to be considered for the treatments of neurodegenerative diseases, such as AD and PD.

In conclusion, VD3 deficiency is associated with an increased risk of age-related chronic diseases, including AD and PD cognitive impairments. VD3 improves protein homeostasis and slows aging, highlighting the importance of maintaining appropriate VD3 levels and explaining why such a wide variety of human age-related diseases are associated with VD3 deficiency (Kamat et al., 2014). Taken together, VD3 neuroprotective effects could result from its antiinflammatory and antioxidant actions and its actions on VD3R (Lima et al., 2018). All these findings should stimulate translational research toward the VD3 potential for the prevention or treatment of neurodegenerative diseases, such as PD and AD (see Fig. 3). Additionally, these data point out the importance of this micronutrient on the diet, especially in the elderly. It certainly may contribute beneficially to a healthy brain, improving cognitive functions and delaying or avoiding the development of neurodegenerative diseases like AD and PD.

Applications of VD3 to other neurological/pathological conditions

The role of VD3 is commonly focused on the maintenance of skeletal health in older adults. However, with the discovery of its receptor (VD3R) in the nervous, cardiovascular, and endocrine systems, the role of VD3 and its impact on these systems have become an important area of research. Epidemiologic evidence indicates an association between low levels of VD3 and diseases related to aging, such as cognitive decline, depression, osteoporosis, cardiovascular disease, hypertension, type 2 diabetes, and cancer (Meghan & Penckofer, 2014).

In the elderly, an association has been observed between VD3 deficiency and neurologic disorders, indicating that VD3 may have an important role in physiological and pathological processes of the CNS, where VD3 is involved in the regulation of calcium-mediated neuronal excitotoxicity, in the reduction of OS, and the induction of synaptic structural proteins, neurotrophic factors, and deficient neurotransmitters (Moretti, Morelli, & Caruso, 2018).

Increasing evidence highlights the impact of VD3 deficiency in various neurological diseases, especially multiple sclerosis and neurodegenerative diseases, such as amyotrophic lateral sclerosis and multiple sclerosis, besides Parkinson's disease and Alzheimer's disease (Amrein et al., 2020; Koduah, Paul, & Dorr, 2017). There is increasing evidence indicating that lower vitamin D levels are associated with an increased risk of MS, thus in a recent systematic review, a great number of studies, dated between the years 2010 and 2017, were predominantly on multiple sclerosis (Caballero-Villaraso et al., 2019; Sintzel, Rametta, & Reder, 2018).

The role of vitamin D in mammalian aging is also well documented. Since vitamin D is essential in bone formation and mineralization, its deficiency results in impaired bone mineralization, such as rickets in children, osteomalacia in adults, and osteoporosis in the aged population (Lanske & Razzaque, 2007). The recent "Consensus statement from 2nd International Conference on Controversies in Vitamin D" (Giustina et al., 2020) suggested that several other conditions, besides those related to chronic degenerative neurological disorders, are impacted by VD3 deficiency, as shown below.

Thus, VD3 is essential to prevent rickets and osteomalacia, and its supplementation with adequate calcium intake can decrease the incidence of fractures in elderly, VD3-deficient subjects. Published RCTs indicate that vitamin D supplementation did not significantly reduce cancer risk, but did significantly improve cancer survival. Besides, both obesity and diabetes are known to be associated with VD3 deficiency, however the mechanisms involved have not yet been elucidated. Besides, VD3 has not been shown to improve outcome measures of diabetes and obesity (Giustina et al., 2020).

Furthermore, VD3 deficiency might be involved in the pathogenesis of celiac disease, in line with its potential role in the immune response, as it occurs in other autoimmune diseases. As already indicated, increasing evidence suggests that several environmental factors including inadequate VD3 levels are associated with the progression of MS. Individuals with adequate levels of vitamin D appear to have reduced prevalence, activity, and progression of MS (Giustina et al., 2020).

Finally, VD3 may play a role in aging and age-related cognitive decline, and its deficiency is associated with muscle dysfunction and falls in the elderly. VD3 supplementation can reduce or increase the risk of falls in individuals who are moderately deficient in VD3. The concentration of 30 nmol/L is considered as the threshold for hypovitaminosis D, and the range of VD3 intake of between 4000 and 10,000 IU/day will probably remain as a useful safe buffer zone that can be used in the short term.

Other components of interest for neurological conditions

An association between food habits and the individual lifestyle with neurodegeneration has been manifested, thereby proposing the role of nutraceuticals as a prophylactic treatment for neurological interventions. Furthermore, emerging evidence is showing nutrition as a crucial factor in the high prevalence and incidence of neurodegenerative disorders. Preventive interventions on neuroinflammation seem to be able to interfere with neurodegeneration and supplementation of essential nutrients, minimizing inflammation, enhancing antioxidative defense and lowering the risk and incidence of age-related diseases, such as cardiovascular diseases and neurodegenerative diseases. Thus, dietary supplementation-based strategies have been demonstrated to be effective in subjects with mild cognitive impairment, while weaker results have been obtained in patients with advanced neurodegenerative disease (Businaro et al., 2021).

Inflammatory processes and vascular dysfunctions appear to play important roles in the pathogenesis of age-associated pathologies. Several neurodegenerative disorders, such as Alzheimer's disease (AD), Parkinson's disease (PD), amyotrophic lateral sclerosis (ALS), tardive dyskinesia, Huntington's disease (HD), and multiple sclerosis are known to be associated with OS manifested by lipid peroxidation, protein oxidation and other markers (Stephenson, Nutma, van der Valk, & Amor, 2018).

A large body of evidence shows that both vitamins E and C are important for the central nervous system and that a decrease in their concentrations causes structural and functional damage to the cells (Butterfield, Castegna, Drake, Scapagnini, & Calabrese, 2002; Martin, Youdim, Szprengiel, Shukitt-Hale, & Joseph, 2002). This fact, together with a growing body of evidence indicating that neurodegenerative processes are associated with OS, leads to the convincing idea that several neurological disorders may be prevented and/or cured by the antioxidant properties of vitamin E (Ricciarelli, Argellati, Pronzato, & Domenicotti, 2007).

The demands for vitamin B12 are particularly high in nervous tissue. Hypovitaminosis is accompanied by pathological lesions, both in brain tissue where several types of neurological manifestations are described as subacute, combined degeneration of spinal cord, sensorimotor polyneuropathy, optic nerve neuropathy, cognitive disorders (Pavlov, Damulin, Shulpekova, & Andreev, 2019). Vitamin B12 deficiency causes other neurological impairments, including cognitive and psychiatric disturbances, gait instability, neuropathy, and autonomic dysfunction, as well as in neurodegenerative disorders (Luthra, Marcus, Hills, et al., 2020).

Despite all these studies on the importance of other vitamins in neurological conditions, the literature data seem to be more straightforward for VD3. Thus, accumulating evidence shows that VD3 is important for accurate brain functioning, and many important neurological diseases are related to low levels of VD3. Unfortunately, up to now, the results for its supplementation in neurological disorders are inconclusive (Moretti et al., 2018) and there is the need for more and better translational studies and well-designed randomized clinical trials, in order to establish the efficacy and benefits of VD3 for the treatment of neurodegenerative diseases. VD3 is essential to maintain important functions of the body, such as calcium homeostasis, maintenance of skeleton integrity, and neurodevelopment, so that its deficiency has been linked to many problems, such as dementia, depression, diabetes mellitus, autism, and schizophrenia. Finally, measures for correcting the VD3 deficiency state can help prevent many negative health consequences (Anjum, Jaffery, Fayyaz, Samop, & Anjum, 2018).

Concluding remarks

VD3 is essential for the maintenance of healthy bones, through the regulation of calcium homeostasis, and it is also implicated in the regulation of neuronal integrity and many brain functions (Morello et al., 2020). VD3 deficiency is common among the elderly and has an important role in the mechanism of neuroinflammation, which is the basis of several aging conditions, including cognitive decline and neurodegeneration (Farghali, Ruga, Morsanuto, & Uberti, 2020).

Some of the effects of VD3 deficiency include the development of dementia, caused by the increase of cerebral soluble and insoluble amyloid-β (Aβ) peptides, and a decrease of its antiinflammatory/antioxidant properties (Anjum et al., 2018). Furthermore, VD3R and 1α-hydroxylase (the enzyme that converts VD3 to its active form) are highly expressed in the *substantia nigra*, what led to the hypothesis that inadequate levels of VD3 may cause dysfunction or cell death within the *SNPc*, strengthening the relationship between VD3 deficiency and PD (Fullard & Duda, 2020).

Mini-dictionary of terms

Vitamin D3 receptor (VD3R): Belongs to the nuclear hormone receptor superfamily and acts as a ligand-inducible transcription factor. MARRS (membrane-associated, rapid response steroid) is a second VD3R.

OS: Oxidative stress is an imbalance between free radicals and antioxidants.

α-Synuclein: Presynaptic neuronal protein genetically and neuropathologically linked to Parkinson's disease (PD).

Amyloid-beta: The main component of amyloid plaques found in the brains of Alzheimer's disease patients.

Proteostasis: Dynamic regulation of a balanced, functional proteome. Its failure is implicated with the development of neurodegenerative diseases.

Key facts of vitamin D3 (VD3) effects on neuroinflammation and neurodegenerative disorders: Focus on Alzheimer's and Parkinson's diseases

1. Vitamin D3 (VD3) and physiological features
2. Vitamin D: neuroinflammation and oxidative stress (OS)
3. Vitamin D3 (VD3) and neurodegenerative diseases
4. Applications of vitamin D3 (VD3) to other neurological/pathological conditions
5. Other components of interest

Summary points

VD3 presents antiinflammatory and antioxidative effects, manifested at peripheral and central sites.

Epidemiological and clinical studies found reduced VD3 levels associated with an increased risk of several neurodegenerative disorders, such as AD and PD.

VD3 could be used in the prevention of cognitive impairment in dementia.

VD3 is neuroprotective, as shown in animal models of AD and DP.

VD3 could be also used in the prevention and treatment of neurodegenerative diseases, such as AD and PD.

References

Amrein, K., Scherkl, M., Hoffmann, M., Neuwersch-Sommeregger, S., Köstenberger, M., Berisha, A. T., et al. (2020). Vitamin D deficiency 2.0: An update on the current status worldwide. *European Journal of Clinical Nutrition, 74*, 1498–1513.

Anjum, I., Jaffery, S. S., Fayyaz, M., Samop, Z., & Anjum, S. (2018). The role of vitamin D in brain health: A mini literature review. *Cureus, 10*(7), e2960.

Araújo de Lima, L., Oliveira Cunha, P. L., Felício Calou, I. B., Tavares Neves, K. R., & Socorro de Barros Viana, G. (2022). Effects of vitamin D (VD3) supplementation on brain mitochondrial function of male rats, in the 6-OHDA-induced model of Parkinson's disease. *Neurochemistry International, 154*, 105280.

Bai, R., Guo, J., Ye, X. Y., Xie, Y., & Xie, T. (2022). Oxidative stress: The core pathogenesis and mechanism of Alzheimer's disease. *Ageing Research Reviews*, 101619.

Berridge, M. J. (2017). Vitamin D deficiency accelerates ageing and age-related diseases: A novel hypothesis. *Journal of Physiology, 597*(22), 6825–6836.

Bikle, D. D. (2014). Vitamin D metabolism, mechanism of action, and clinical applications. *Chemistry & Biology, 21*, 319–329.

Bouillon, R., & Bikle, D. (2019). Vitamin D metabolism revised: Fall of dogmas. *Journal of Bone and Mineral Research, 34*, 1985–1992.

Brandes, M. S., & Gray, N. E. (2020). NRF2 as a therapeutic target in neurodegenerative diseases. *ASN Neuro, 12*. https://doi.org/10.1177/1759091419899782. Jan–Dec. 175909. PMID: 31964153; PMCID: PMC6977098.

Businaro, R., Vauzour, D., Sarris, J., Münch, G., Gyengesi, E., Brogelli, L., et al. (2021). Therapeutic opportunities for food supplements in neurodegenerative disease and depression. *Frontiers in Nutrition, 8*, 669846. https://doi.org/10.3389/fnut.2021.669846.

Butterfield, D. A., Castegna, A., Drake, J., Scapagnini, G., & Calabrese, V. (2002). Vitamin E and neurodegenerative disorders associated with oxidative stress. *Nutritional Neuroscience, 5*(4), 229–239.

Caballero-Villaraso, J., Jiménez-Jiménez, M. J., Escribano, B. M., Agüera, E., Santamaria, A., & Túnez, I. (2019). Implications of vitamin D in multiple sclerosis and other neurodegenerative processes: Bibliometric analysis and systematic review. *CNS & Neurological Disorders Drug Targets, 18*(6), 478–490.

Calvello, R., Cianciulli, A., Nicolardi, G., DeNuccio, F., Giannoti, L., Salvatore, R., et al. (2017). Vitamin D treatment attenuates neuroinflammation and dopaminergic neurodegeneration in an animal model of Parkinson's disease, shifting M1 to M2 microglia responses. *Journal of Neuroimmune Pharmacology, 12*(2), 327–339.

Cao, H., Xu, Y., Necochea-Campion, R., Baylink, D. J., Payne, K. J., Tanq, X., et al. (2017). Application of vitamin D and vitamin D analogs in acute myelogenous leukemia. *Experimental Hematology, 50*, 1–12. https://doi.org/10.1016/j.exphem.2017.01.007.

Carlberg, C., & Molnár, F. (2015). Vitamin D receptor signaling and its therapeutic implications: Genome-wide and structural view. *Canadian Journal of Physiology and Pharmacology*, *93*(5), 311–318.

Castelli, V., Benedetti, E., Antonosante, A., Catanesi, M., Pitari, G., Ippoliti, R., et al. (2019). Neuronal cells rearrangement during aging and neurodegenerative disease: Metabolism, oxidative stress and organelles dynamic. *Frontiers in Molecular Neuroscience*. https://doi.org/10.3389/fnmol.2019.00132.

Cenini, G., Lloret, A., & Cascella, R. (2019). Oxidative stress in neurodegenerative diseases: From a mitochondrial point of view. *Oxidative Medicine and Cellular Longevity*, *2019*, 2105607.

Christakos, S., Hewison, M., Gardner, D. G., Wagner, C. L., Sergeev, I. N., Rutten, E., et al. (2013). Vitamin D: Beyond bone. *Annals of the New York Academy of Sciences*, *1287*(1), 45–58.

Cioffi, F., Hassan, R., Adam, I., & Broersen, K. (2019). Molecular mechanisms and genetics of oxidative stress in Alzheimer's disease. *Journal of Alzheimer's Disease*, *72*(4), 981–1017.

Colotta, F., Jansson, B., & Bonelli, F. (2017). Modulation of inflammatory and immune responses by vitamin D. *Journal of Autoimmunity*, *85*, 78–97.

Costa, R. O., Gadelha-Filho, C. V. J., Costa, A. E. M., Feitosa, M. L., Araújo, D. P., Lucena, J. D., et al. (2017). Exercise protects against dopaminergic neuron loss and brain oxidative stress in parkinsonian rats. *Oxidative Medicine and Cellular Longevity*, *2017*, 2138169. https://doi.org/10.1155/2017/2138169.

Dias, V., Junn, E., & Mouradian, M. M. (2013). The role of oxidative stress in Parkinson's disease. *Journal of Parkinson's Disease*, *3*(4), 461–491.

Farghali, M., Ruga, S., Morsanuto, V., & Uberti, F. (2020). Can brain health be supported by vitamin D-based supplements? A critical review. *Brain Sciences*, *10*, 0660. https://doi.org/10.3390/brainsci10090660.

Fullard, M. E., & Duda, J. E. (2020). A review of the relationship between vitamin D and Parkinson disease symptoms. *Frontiers in Neurology*. https://doi.org/10.3389/fneur.2020.00454.

Gallieni, M., Cozzolino, M., Fallabrino, G., Pasho, S., Olivi, L., & Brancaccio, D. (2009). Vitamin D: Physiology and pathophysiology. *International Journal of Artificial Organs*, *32*(2), 87–94. https://doi.org/10.1177/039139880903200205, 19363780

Garcion, E., Nataf, S., Berod, A., Darcy, F., & Brachet, P. (1997). 1,25-Dihydroxyvitamin D3 inhibits the expression of inducible nitric oxide synthase in rat central nervous system during experimental allergic encephalomyelitis. *Molecular Brain Research*, *45*, 255–267.

Gemma, C., Vila, J., Bachstetter, A., & Pickford, P. C. (2007). Oxidative stress and the aging brain: From theory to prevention. In D. R. Riddle (Ed.), *Brain aging: Models, methods, and mechanisms* (p. 2007). Boca Raton, FL: CRC Press/Taylor & Francis. Chapter 15. Available from: https://www.ncbi.nlm.nih.gov/books/NBK3869/.

Gil, A., Plaza-Diaz, J., & Mesa, M. D. (2018). Vitamin D: Classic and novel actions. *Annals of Nutrition & Metabolism*, *72*(2), 87–95.

Gitler, A. D., Dhillon, P., & Shorter, J. (2017). Neurodegenerative disease: Models, mechanisms, and a new hope. *Disease Models & Mechanisms*, *10*(5), 499–502.

Giustina, A., Adler, R. A., Binkley, N., Bollerslev, J., Bouillon, R., Dawson-Hughes, B., et al. (2020). Consensus statement from 2nd international conference on controversies in vitamin D. *Reviews in Endocrine & Metabolic Disorders*, *21*, 89–116.

Grosso Jasutkar, H., Oh, S. E., & Mouradian, M. M. (2022). Therapeutics in the pipeline targeting α-synuclein for Parkinson's disease. *Pharmacological Reviews*, 207–237. https://doi.org/10.1124/pharmrev.120.000133.

Guo, J.-D., Zhao, X., Li, Y., Li, G.-R., & Liu, X.-L. (2018). Damage to dopaminergic neurons by oxidative stress in Parkinson's disease. *International Journal of Molecular Medicine*. https://doi.org/10.3892/ijmm.2018.3406:1817-1825.

Guzman-Martinez, L., Maccioni, R. B., Andrade, V., Navarrete, L. P., Pastor, M. G., & Ramos-Escobar, N. (2019). Neuroinflammation as a common feature of neurodegenerative disorders. *Frontiers in Pharmacology*, *10*, 1008. https://doi.org/10.3389/fphar.2019.01008. eCollection 2019.

He, J., Zhu, G., Wang, G., & Zhang, F. (2020). Oxidative stress and neuroinflammation potentiate each other to promote progression of dopamine neurodegeneration. *Oxidative Medicine and Cellular Longevity*, *2020*, 6137521. https://doi.org/10.1155/2020/6137521. eCollection 2020.

Henderson, M. X., Henrich, M., Geibl, F. F., Oertel, W. H., Brundin, P., & Surmeier, D. J. (2022). The roles of connectivity and neuronal phenotype in determining the pattern of α-synuclein pathology in Parkinson's disease. *Neurobiology of Disease*, 105687. https://doi.org/10.1016/j.nbd.2022.105687.
Höhn, A., Tramutola, A., & Cascella, R. (2020). Proteostasis failure in neurodegenerative diseases: Focus on oxidative stress. *Oxidative Medicine and Cellular Longevity*, 5497046. https://doi.org/10.1155/2020/5497046.
Huang, W.-J., Zhang, X., & Chen, W.-W. (2016). Role of oxidative stress in Alzheimer's disease. *Biomedical Reports*, *4*(5), 519–522.
Hwang, O. (2013). Role of oxidative stress in Parkinson's disease. *Experimental Neurobiology*, *22*(1), 11–17.
Iacopetta, K., Collins-Praino, L. E., Buisman-Pijlman, F. T. A., Liu, J., Hutchinson, A. D., & Hutchinson, M. R. (2020). Are the protective benefits of vitamin D in neurodegenerative disease dependent on route of administration? A systematic review. *Nutritional Neuroscience*, *23*(4), 251–280.
Jimenez-Lara, A. M., & Aranda, A. (2000). Interaction of vitamin D and retinoid receptors on regulation of gene expression. *Hormone Research*, *54*(5–6), 301–305. https://doi.org/10.1159/000053276.
Jucker, M., & Walker, L. C. (2018). Propagation and spread of pathogenic protein assemblies in neurodegenerative diseases. *Nature Neuroscience*, *21*, 1341–1349.
Kamat, P. K., Kalani, A., Kyles, P., Tyagi, S. C., & Tyagi, N. (2014). Autophagy of mitochondria: A promising therapeutic target for neurodegenerative disease. *Cell Biochemistry and Biophysics*, *70*, 707–719. https://doi.org/10.1007/s12013-014-0006-5. PMID: 24807843; PMCID: PMC4184921.
Keshavarz, M., Xie, K., Schaaf, K., Bano, D., & Ehninger, D. (2022). Targeting the "hallmarks of aging" to slow aging and treat age-related disease: Fact or fiction? *Molecular Psychiatry*. https://doi.org/10.1038/s41380-022-01680-x.
Koduah, P., Paul, F., & Dorr, J.-M. (2017). Vitamin D in the prevention, prediction and treatment of neurodegenerative and neuroinflammatory diseases. *EPMA Journal*, *8*(4), 313–325.
Lanske, B., & Razzaque, M. S. (2007). Vitamin D and aging: Old concepts and new insights. *Journal of Nutritional Biochemistry*, *18*(12), 771–777.
Lauer, A. A., Janitschke, D., Hartmann, T., Grimm, H. S., & Grimm, M. O. W. (2019). In J. Fedotova (Ed.), *The effects of vitamin D deficiency on neurodegenerative diseases, vitamin D deficiency* IntechOpen. https://doi.org/10.5772/intechopen.89160.
Leal, L. K. A. M., Lima, L. A., Aquino, P. E. A., Sousa, J. A. C., Gadelha-Filho, C. V. J., Calou, I. B. F., et al. (2020). Vitamin D (VD3) antioxidative and anti-inflammatory activities: Peripheral and central effects. *European Journal of Pharmacology*, *879*, 173099. https://doi.org/10.1016/j.ejphar.2020.173099.
Li, Z., Zhang, Z., Ren, Y., Fang, J., Yue, H., Ma, S., et al. (2021). Aging and age-related diseases: From mechanisms to therapeutic strategies. *Biogerontology*, 165–187. https://doi.org/10.1007/s10522-021-09910-5.
Lima, L. A. R., Lopes, M. J. P., Costa, R. O., Lima, F. A. V., Neves, K. R. T., Calou, I. B. F., et al. (2018). Vitamin D protects dopaminergic neurons against neuroinflammation and oxidative stress in hemiparkinsonian rats. *Journal of Neuroinflammation*, *15*, 249.
Liu, W., Zhang, L., Xu, H.-J., Li, Y., Hu, C.-M., Yang, J.-Y., et al. (2018). The anti-inflammatory effects of vitamin D in tumorigenesis. *International Journal of Molecular Sciences*, *19*(9), 2736.
Luthra, N. S., Marcus, A. H., Hills, N. K., et al. (2020). Vitamin B12 measurements across neurodegenerative disorders. *Journal of Clinical Movement Disorders*, 7, 3. https://doi.org/10.1186/s40734-020-00085-8.
Martens, P. J., Gysemans, C., Verstuyf, A., & Mathieu, C. (2020). Vitamin D's effect on immune function. *Nutrients*, *12*, 1248.
Martin, A., Youdim, K., Szprengiel, A., Shukitt-Hale, B., & Joseph, J. (2002). Roles of vitamins E and C on neurodegenerative diseases and cognitive performance. *Nutrition Reviews*, *60*(10 Pt 1), 308–326.
McAlary, L., Plotkin, S. S., & Cashman, N. R. (2019). Emerging developments in targeting proteotoxicity in neurodegenerative diseases. *CNS Drugs*, *33*, 883–904.
Meghan, M., & Penckofer, S. (2014). The role of vitamin D in the aging adult. *Journal of Aging and Gerontology*, *2*(2), 60–71. https://doi.org/10.12974/2309-6128.2014.02.02.1. PMID: 25893188; PMCID: PMC4399494.

Milà-Alomà, M., Ashton, N. J., Shekari, M., Salvadó, G., Ortiz-Romero, P., Montoliu-Gaya, L., et al. (2022). Plasma p-tau231 and p-tau217 as state markers of amyloid-β pathology in preclinical Alzheimer's disease. *Nature Medicine*, 1–5.

Morello, M., Pieri, M., Zenobi, R., Talamo, A., Stephan, D., & Landel, V. (2020). The influence of vitamin D on neurodegeneration and neurological disorders: A rationale for its physio-pathological actions. *Current Pharmaceutical Design*, *26*(21), 2475–2491.

Moretti, R., Morelli, M. E., & Caruso, P. (2018). Vitamin D in neurological diseases: A rationale for a pathogenic impact. *International Journal of Molecular Sciences*, *19*, 2245. https://doi.org/10.3390/ijms19082245.

Nystad, A. E., Wergeland, S., Aksnes, L., Myhr, K. M., Bø, L., & Torkildsen, Ø. (2014). Effect of high-dose 1.25 dihydroxyvitamin D3 on remyelination in the cuprizone model. *APMIS*, *122*, 1178–1186.

Pavlov, C. S., Damulin, I. V., Shulpekova, Y. O., & Andreev, E. A. (2019). Neurological disorders in vitamin B12 deficiency. *Terapevticheskiĭ Arkhiv*, *91*(4), 122–129.

Pezzini, I., Mattoli, V., & Ciofani, G. (2017). Mitochondria and neurodegenerative diseases: The promising role of nanotechnology in targeted drug delivery. *Expert Opinion on Drug Delivery*, *14*(4), 513–523.

Rai, A. N., Singh, C., Singh, A., Singh, M. P., & Singh, B. K. (2020). Mitochondrial dysfunction: A potential therapeutic target to treat Alzheimer's disease. *Molecular Neurobiology*, *57*(7), 3075–3088.

Ransohoff, R. M. (2016). How neuroinflammation contributes to neurodegeneration. *Science*, *353*(6301), 777–783. https://doi.org/10.1126/science.aag2590. PMID: 27540165.

Ricca, C., Aillon, A., Bergandi, L., Alotto, D., Castagnoli, C., & Silvagno, F. (2018). Vitamin D receptor is necessary for mitochondrial function and cell health. *International Journal of Molecular Sciences*, *19*(6), 1672.

Ricciarelli, R., Argellati, F., Pronzato, M. A., & Domenicotti, C. (2007). Vitamin E and neurodegenerative diseases. *Molecular Aspects of Medicine*, *85*(5–6), 591–606.

Ruz, C., Alcantud, J. L., Montero, F. V., Duran, R., & Bandes-Ciga, S. (2020). Proteotoxicity and neurodegenerative diseases. *International Journal of Molecular Sciences*, *21*(16), 5646. https://doi.org/10.3390/ijms21165646.

Sintzel, M. B., Rametta, M., & Reder, A. T. (2018). Vitamin D and multiple sclerosis: A comprehensive review. *Neurology and Therapy*, 7(1), 59–85.

Stephenson, J., Nutma, E., van der Valk, P., & Amor, S. (2018). Inflammation in CNS neurodegenerative diseases. *Immunology*, *154*(2), 204–219.

Trist, B. G., Hare, D. J., & Double, K. L. (2019). Oxidative stress in the aging substantia nigra and the etiology of Parkinson's disease. *Aging Cell*, *18*(6), e13031.

Trojanowski, J. (2000). Biochemical markers of neurodegenerative diseases: τ and synucleins. *Archives of Neurology*, *57*(8), 1236.

Uttara, B., Singh, A. V., Zamboni, P., & Mahajan, R. T. (2009). Oxidative stress and neurodegenerative diseases: A review of upstream and downstream antioxidant therapeutic options. *Current Neuropharmacology*, 7(1), 65–74.

van Halteren, A. G., Tysma, O. M., van Etten, E., Mathieu, C., & Roep, B. O. (2004). 1alpha,25-dihydroxyvitamin D3 or analogue treated dendritic cells modulate human autoreactive T cells via the selective induction of apoptosis. *Journal of Autoimmunity*, *23*(3), 233–239. https://doi.org/10.1016/j.jaut.2004.06.004. PMID: 15501394.

Wang, X., & Michaelis, E. K. (2010). Selective neuronal vulnerability to oxidative stress in the brain. *Frontiers in Aging Neuroscience*, *2*, 12. https://doi.org/10.3389/fnagi.2010.00012. PMID: 20552050; PMCID: PMC2874397.

Xu, Y., Baylink, D. J., Chen, C.-S., Reeves, M. E., Xiao, J., Lacy, C., et al. (2020). The importance of vitamin d metabolism as a potential prophylactic, immunoregulatory and neuroprotective treatment for COVID-19. *Journal of Translational Medicine*, *18*(1), 322. https://doi.org/10.1186/s12967-020-02488-5.

Yin, K., & Agrawal, D. K. (2014). Vitamin D and inflammatory diseases. *Journal of Inflammation Research*, 7, 69–87.

Yousefzadeh, M. J., Flores, R. R., Zhu, Y., Schmiechen, Z. C., Brooks, R. W., Trussoni, C. E., et al. (2021). An aged immune system drives senescence and ageing of solid organs. *Nature*, 100–105. https://doi.org/10.1038/s41586-021-03547-7.

Further reading

Bouillon, R., Lieben, L., Mathieu, C., Verstuyf, A., & Carmeliet, G. (2013). Vitamin D action: Lessons from VDR and Cyp27b1 null mice. *Pediatric Endocrinology Reviews*, (Suppl. 2), 354–366.

Carlberg, C., & Campbell, M. J. (2013). Vitamin D receptor signaling mechanisms: Integrated actions of a well-defined transcription factor. *Steroids*, *78*(2), 127–136.

Feige, J., Moser, T., Bieler, L., Schwenker, K., Hauer, L., & Sellner, J. (2020). Vitamin D supplementation in multiple sclerosis: A critical analysis of potentials. *Nutrients*, *12*, 783. https://doi.org/10.3390/nu12030783.

Hii, C. S., & Ferrante, A. (2016). The non-genomic actions of vitamin D. *Nutrients*, *8*, 135.

Hou, Y., Dan, X., Babbar, M., Wei, Y., Hasselbalch, S. G., & Croteau, D. L. (2019). Ageing as a risk factor for neurodegenerative disease. *Nature Reviews. Neurology*, *15*, 565–581.

Kamat, P. K., Kalani, A., Rai, S., Swarnkar, S., Tota, S., Nath, C., et al. (2016). Mechanism of oxidative stress and synapse dysfunction in the pathogenesis of Alzheimer's disease: Understanding the therapeutics strategies. *Molecular Neurobiology*, *53*, 648–661.

Kim, G. H., Kim, J. E., Rhie, S. J., & Yon, S. (2015). The role of oxidative stress in neurodegenerative disease. *Experimental Neurobiology*, *24*, 325–340.

Moreira, P. I., Zhu, X., Wang, X., Lee, H.-G., Nunomura, A., Petersen, R. B., et al. (2010). Mitochondria: A therapeutic target in neurodegeneration. *Biochimica et Biophysica Acta*, *1802*(1), 212–220.

Patel, F., & Mandal, P. (2019). Neurodegenerative diseases and their therapeutic approaches. In G. E. A. Abreu, & M. E. H. Aguilar (Eds.), *Neurons – Dendrites and axons* IntechOpen. https://doi.org/10.5772/intechopen.82129 2.

CHAPTER 26

Vitamin C and neuroinflammation

Amanda L. Marino[a], **David C. Consoli**[a], **and Fiona E. Harrison**[b]
[a]Division of Diabetes, Endocrinology, and Metabolism, Vanderbilt University Medical Center, Nashville, TN, United States
[b]Division of Diabetes, Endocrinology, and Metabolism, Department of Medicine, Vanderbilt University Medical Center, Nashville, TN, United States

Abbreviations

5hmC	5-hydroxymethylcytosine
5mC	5-methylcytosine
AD	Alzheimer's disease
ASC	ascorbate/vitamin C
ASC	ascorbic acid
C1q	complement component 1q
Cav-1	c-Src–caveolin-1
CNS	central nervous system
DHA	dehydroascorbic acid
IFNγ	interferon gamma
IKK	IκB/a kinase complex
IL-1β	interleukin 1 beta
IL-6	interleukin-6
NF-κB	nuclear factor-kB
PD	Parkinson's disease
ROS	reactive oxygen species
SLC23	solute carrier family 23
SNP	single-nucleotide polymorphism
SVCT1	sodium-dependent vitamin C transporter 1
SVCT2	sodium-dependent vitamin C transporter 2
TET	ten-eleven translocation methylcytosine dioxygenases
TNFα	tumor necrosis factor-alpha

Introduction

Neuroinflammation is characterized by inflammation of brain or spinal cord tissue within the central nervous system (CNS). Depending on the cause, neuroinflammation can be transient or persist for an extended period of time. Acute triggers of neuroinflammation include traumatic brain and spinal cord injury, bacterial and viral infections, and environmental toxins. Chronic neuroinflammation is also a hallmark feature of many neurological disorders and neurodegenerative diseases as well as normal aging. Neuroinflammation occurs through activation of microglia and other inflammatory immune cells through

Vitamins and Minerals in Neurological Disorders
https://doi.org/10.1016/B978-0-323-89835-5.00028-4

coordinated release of cytokines and reactive oxygen species (ROS). While ROS act as important signaling molecules, they can also compromise protein and membrane integrity through oxidation of amino acid residues. Vitamin C (ascorbic acid, ASC) is a potent antioxidant and is particularly abundant in the brain. It can rapidly scavenge reactive oxygen species (ROS) through spontaneous electron donation, including to transition metals such as iron and manganese which are intimately involved in active sites of many enzymes. Therefore, ASC serves as an important enzymatic cofactor for neuromodulation including synthesis of dopamine (dopamine beta-hydroxylase and tyrosine hydroxylase), and the 10,11 translocation (TET) enzymes which are important in DNA methylation. Due to these distinguishing features, ASC deficiency may aggravate or drive neuroinflammation whereas maintaining appropriately replete ASC in brain may serve as a potential therapeutic for neuroinflammation and help to protect against some of the consequences of neuroinflammatory conditions including cognitive and other behavioral impairment. A fundamental role of ASC in neuroinflammation would render it applicable to multiple neurological disorders where neuroinflammation is present.

Neuroinflammation overview

Neuroinflammation refers to the coordinated immune response carried out by resident immune cells specifically in the brain and spinal cord and differs from peripheral inflammation due to the clear separation provided by the blood-brain barrier (BBB). The BBB is composed of the neurovascular unit, which includes endothelial cells that form tight junctions that tightly regulate infiltration of cells and other molecules from the peripheral immune system. To defend against infections or injury and maintain neuron homeostasis, the brain depends on its own type of immune system comprised of multiple glial cell types: microglia, astrocytes, and oligodendrocytes. Microglia provide the primary immune support in the brain. Following an insult or injury, the brain undergoes a broad range of changes to activate and recruit microglia including release of cytokines, reactive oxygen species, and chemokines (DiSabato, Quan, & Godbout, 2016). The main immune response to an injury or infection in the brain is by microglial cells. As the resident macrophages of the brain, they are able to migrate, proliferate, and respond to changes in their microenvironment (Shabab, Khanabdali, Moghadamtousi, Kadir, & Mohan, 2017). Increased cytokines and chemokines during neuroinflammation recruit peripheral monocytes to the brain to aid in recovery following illness or injury (DiSabato et al., 2016). Under normal conditions, microglia provide support to neurons in their local environment as they extend their processes to constantly survey neurons and maintain homeostasis. In particular "activation" of microglia results in a shift from the normal, ramified state to an activated phagocytic state which includes further release of proinflammatory mediators (Lyman, Lloyd, Ji, Vizcaychipi, & Ma, 2014). Astrocytes provide the primary metabolic, structural, and neuromodulatory support for neurons especially within the synapse, but astrocytes also assume a reactive morphology and are attracted to the insult site in response to cytokines and chemokines. These reactive astrocytes display changes in glutamate signaling, ion homeostasis, and energy metabolism that could alter connectivity (Palpagama, Waldvogel, Faull, & Kwakowsky, 2019). Though the main response to neuroinflammatory conditions is by microglia and astrocytes, oligodendrocytes may also have an instrumental role. Oligodendrocytes primarily maintain myelination, axonal integrity and aid in neuronal survival (Bankston, Mandler, & Feng, 2013). Under neuroinflammatory conditions, oligodendrocytes have been shown to secrete many mediators known to modulate microglia activation (Balabanov et al., 2007). For example, under stress, oligodendrocytes secrete chemoattractants, which suggests a role for attracting microglia to the site of injury or infection (Balabanov et al., 2007). Together, microglia, astrocytes, and oligodendrocytes coordinate the neuroinflammatory immune response to injury in the brain. Neuroinflammation is a necessary defensive and protective response in the CNS, promoting recovery through phagocytosis of cellular debris, axonal regrowth, tissue repair, and cellular signaling

Microglia and neuroinflammatory response

In a normal environment, microglia have highly ramified processes and constantly survey their surroundings to be able to quickly detect and respond to any debris or damaged neurons (Sousa, Biber, & Michelucci, 2017). Microglia with this morphology are often referred to as surveying microglia or "M0." The first indication of neuroinflammation is usually evident through changes in microglia morphology. Though microglial morphology is dynamic, classically microglial activation has been divided into two categories "M1" and "M2" microglia (Fig. 1). M1 microglia shift from a normal surveying, ramified state "M0" into an amoeboid and mobile state (Li & Barres, 2018). This activation is a salient feature of all inflammation. In this activated state, M1 microglia are activated by interferon gamma (IFNγ) and tumor necrosis factor-alpha (TNFa) and secrete proinflammatory cytokines such as IL (interleukin)-1β, IL-6, and complement component 1q (C1q) among others (DiSabato et al., 2016; Li & Barres, 2018). M1 microglia engulf and phagocytose damaged cells and axons, in an effort to maintain homeostasis at the insult site (DiSabato et al., 2016). M2 microglia are in contrast, antiinflammatory in nature, activated by of IL-4 and IL-13 cytokines following insult or injury (Aguilera, Colín-González, Rangel-López, Chavarría, & Santamaría, 2018). They themselves secrete arginase and antiinflammatory cytokines such as transforming growth factor-beta (TGF-β), leading to the downregulation of the M1 phenotype (Aguilera et al., 2018). M2 microglia are considered antiinflammatory because they promote tissue repair and neuroprotection by clearing disease pathology. Together, a successfully mounted neuroinflammatory response coordinates M1 and M2 microglial activation to maintain or return to normal brain function.

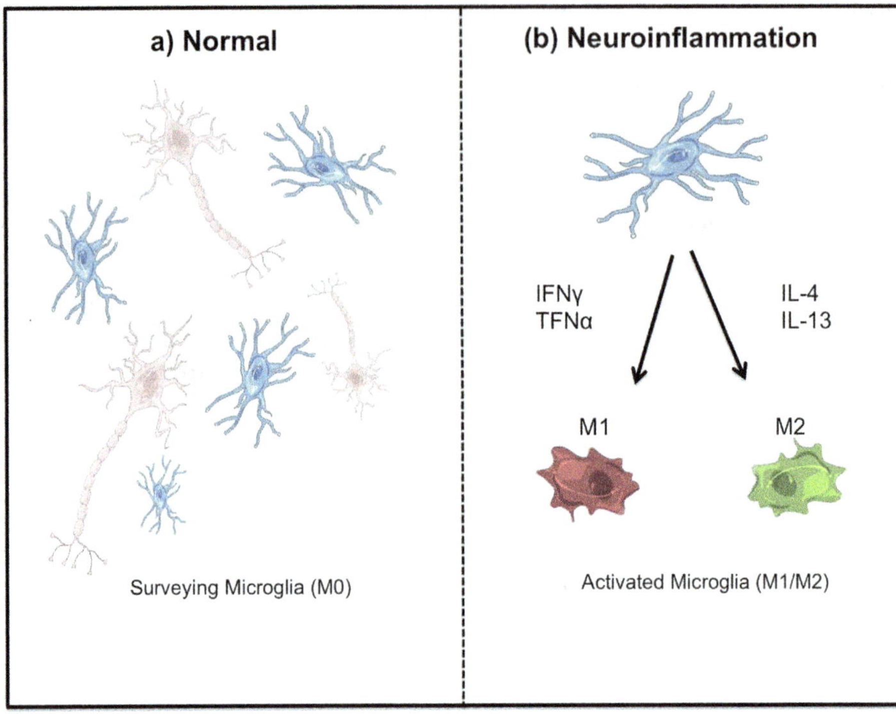

Fig. 1 Microglia changes under homeostatic and neuroinflammatory conditions. (A) Under normal conditions, microglia maintain a surveying phenotype with highly ramified processes and provide support to neurons. (B) Under neuroinflammatory conditions, microglia shift into an activated state with an amoeboid morphology. M1 are activated by IFNγ and TNFα, and M2 are activated by IL-4 and IL-13. This is a simplified summary; for more in-depth information including genetic and molecular markers for each type, see Lan, Han, Li, Yang, and Wang (2017).

Acute versus chronic neuroinflammation

An acute neuroinflammatory response is critical to mitigate damage in the brain. Immediately following a neuroinflammatory trigger, microglia and astrocytes activate in response to cellular mediators and mobilize to the site of injury. In the acute stage, activation of microglia and astrocytes is transient and upregulation of cytokine release is temporary (Norden & Godbout, 2013). Activated microglia play an important role in synaptic pruning and phagocytosing debris following an insult (DiSabato et al., 2016). Microglia and astrocyte mobilization during the acute phase of injury minimizes brain damage and neuronal loss in an effort to stabilize the microenvironment of the brain. However, when neuroinflammation persists over long periods of time, this leads to neuronal damage and death that can cause impairments in plasticity and cognition (DiSabato et al., 2016). The exact mechanism of perpetuating neuroinflammation into a chronic state is not currently known, but chronic neuroinflammation is associated with negative behavioral effects and many neurological diseases including Alzheimer's disease (AD), Parkinson's disease (PD), anxiety, and depression (Lyman et al., 2014; Zheng et al., 2021). The severity of the resulting neuroinflammation response is typically directly related to the context of the injury or infection, such that greater damage leads to a larger inflammatory response (DiSabato et al., 2016). Under chronic neuroinflammatory conditions, M1 microglia become the dominant cells, with M2 microglia being downregulated (Kigerl et al., 2009). Prolonged microglial activation can also directly induce reactivity in astrocytes (Li et al., 2020; Liddelow et al., 2017). In this activated state, astrocytes lose their normal functions of promoting neuronal outgrowth and synaptogenesis and instead can induce neuronal death (Liddelow et al., 2017). Chronic neuroinflammation is also associated with significant upregulation in cytokine and chemokine production, BBB breakdown and permeability, and infiltration of peripheral immune cells (DiSabato et al., 2016; Hawkins & Davis, 2005; Michael et al., 2016; Monahan, Warren, & Carvey, 2008). These chronic and prolonged neuroinflammatory conditions are perpetuated by reactive microglia, and under these chronic conditions, the neuroinflammation may never resolve.

Vitamin C overview

ASC has many roles in the brain as an antioxidant, an enzymatic cofactor, and a neuromodulator (Ballaz & Rebec, 2019; Buettner, 1993; Harrison & May, 2009) (Fig. 2). As an antioxidant, ASC can rapidly donate electrons to free radicals to stabilize them, as well as scavenge ROS, thereby contributing to a state of redox homeostasis in the body and brain (Birben, Sahiner, Sackesen, Erzurum, & Kalayci, 2012). When ASC quenches ROS, the immediate oxidation product of ASC is dehydroascorbic acid (DHA). ASC is also an important cofactor for many enzymes in the body including collagen. ASC is necessary component in the synthesis of hydroxyproline and hydroxylysine during collagen formation, which stabilizes the collagen triple helix keeping it structurally stable and forms the crosslinking that gives collagen its strength (Murad et al., 1981). ASC also preserves antioxidant potential in other molecules such as through efficient reduction of the antioxidant vitamin E free radicals back to vitamin E (Birben et al., 2012; DePhillipo et al., 2018). ASC has properties as a neuromodulator through its intrinsic links with glutamate release and reuptake (Mi et al., 2018; Wilson, Peters, Sitar, Daoust, & Gelb, 2000). Presence of extracellular glutamate stimulates ASC release from astrocytes in culture, and following seizure, trauma, and cerebral ischemia in order to counterbalance mitochondrial free radical production from glutamatergic activation (Wilson et al., 2000).

ASC must be consumed through the diet by humans and nonhuman primates because we lack a functional copy of the gulonolactone oxidase gene making us unable to synthesize ASC, unlike most other mammals. Serum ASC levels that fall below ~23 μM are considered to reflect hypovitaminosis, and when plasma levels fall below 1 μM, humans

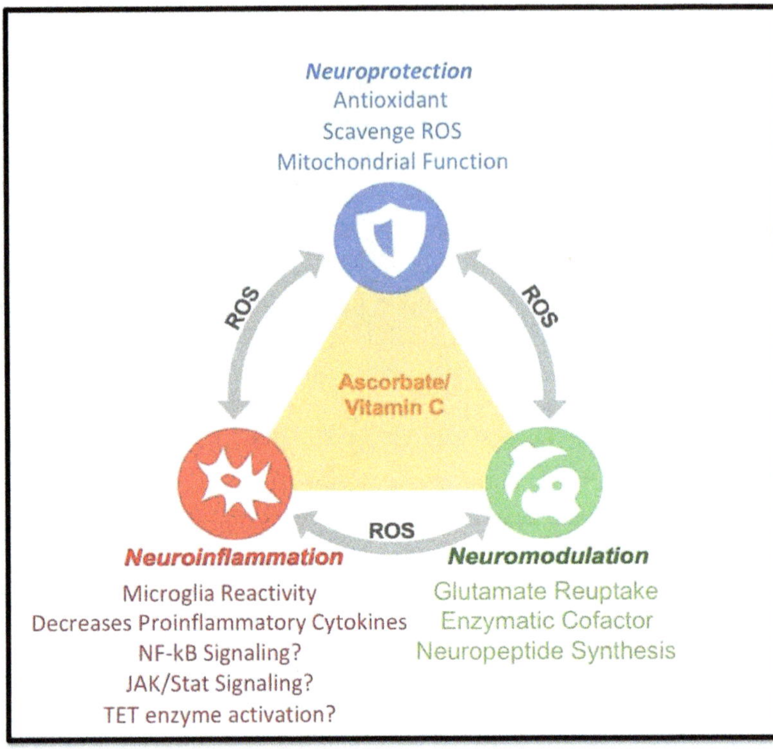

Fig. 2 Simplified roles for ASC in the brain and body. There are many well-established roles for ASC in neuroprotection and neuromodulation, mostly due to antioxidant potential and roles as an enzymatic cofactor. Several direct roles for ASC in neuroinflammation have been identified more recently.

can develop scurvy which can be fatal (Levine et al., 1996; Rowe & Carr, 2020). The highest concentrations of ASC in the body are in metabolically active tissues such as the brain. Within the brain tissue, concentrations can be as high as 10 mM within neurons and highest brain concentrations are observed in neuronally dense areas such as the hippocampus and cortex (Travica et al., 2020). Comparatively, organs such as liver tend to have levels around fourfold lower (Harrison, Green, Dawes, & May, 2010).

Since ASC is obtained from food sources, uptake into tissues is dependent on an efficient transport system throughout the body. First, sodium-dependent vitamin C transporter (SVCT) 1 located in intestinal cells transports dietary ASC into the blood stream. SVCT1 expressed in kidneys recycles ASC and prevents excretion and is thus primarily responsible for maintaining whole body ASC homeostasis. ASC in the blood is then taken up into metabolically active tissues via SVCT2 (Savini, Rossi, Pierro, Avigliano, & Catani, 2008). Recently a novel exporter, GLUT12, has been identified in the active transport of ASC from blood into the cerebrospinal fluid (CSF) (Miyata et al., 2022). SVCT2 transports ASC from blood into the choroid plexus epithelial cells, then GLUT12 exports vitamin c from the choroid plexus and into the CSF where it is dispersed to the brain parenchyma (Miyata et al., 2022). SVCT2 is also present on neurons to allow further concentration into these cells. In astrocytes which do not typically express SVCT2, the oxidized form of ASC, DHA can be transported via GLUT1 and GLUT3 and, once within the cell, can be reduced back to ASC (Harrison & May, 2009).

Vitamin C and neuroinflammation—Where the research stands

Growing evidence supports a protective role of ASC in the attenuation of neuroinflammation. As an antioxidant, ASC is particularly effective at mediating ROS, which is one of the contributing factors to neuroinflammation, particularly with regard to proinflammatory cytokines. In one model of oxidative stress induced following pilocarpine-induced seizures, ASC administration was effective in reducing ROS and suppressed convulsions (Xavier et al., 2007). Pretreatment of ASC has also been shown to reduce levels of proinflammatory cytokines that are produced following acute and excessive exercise paradigms in humans, and this may be due to the close relationship between changes in redox potential and proinflammatory cytokines (Nieman, Peters, Henson, Nevines, & Thompson, 2000; Yfanti et al., 2012). What remains unknown are the exact mechanisms through which ASC moderates neuroinflammation including ROS, cytokine expression, and microglial activation. Current research has also focused more strongly on protective roles of added ASC, rather than potential pathways impacted by ASC deficiency.

In an immunohistochemical study of SVCT2 immunoreactivity throughout the brain, SVCT2 was identified as predominantly located in neuronal cell bodies but also notably in microglia (Mun et al., 2006). SVCT2 seems to be necessary for microglia to activate properly under inflammatory conditions (Portugal et al., 2017). SVCT2 is not expressed in astrocytes in homeostatic states but can be induced under pathophysiological conditions such as ischemia in vivo and in primary cell culture environments (Berger et al., 2003; Salazar et al., 2018). SVCT2 also becomes internalized following an inflammatory challenge. Following treatment with lipopolysaccharide (LPS), SVCT2 in rodent microglia was significantly decreased in the plasma membrane within 30 min following internalization and degradation through the c-Src–caveolin-1 (Cav-1) signaling complex (Portugal et al., 2017). This internalization is associated with a reactive microglial phenotype (Portugal et al., 2017). SVCT2 on microglia is strongly suggestive of a required role of ASC in the neuroinflammatory response but this is an understudied area and has not been well described in rodent models or human samples.

Potential mechanisms for ASC attenuation of neuroinflammation

Though research is ongoing to identify the exact mechanisms through which ASC mediates neuroinflammation, several candidate pathways have been identified. ASC has an inhibitory relationship with transcription factor nuclear factor-kB (NF-kB), which

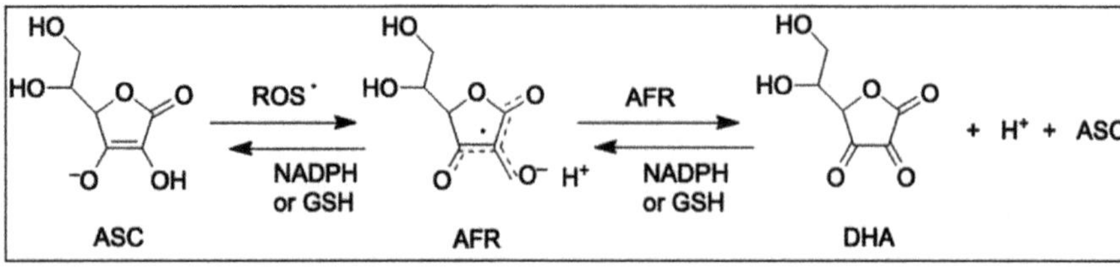

Fig. 3 Ascorbic acid (ASC) oxidizes readily through electron donation to radical oxygen species (ROS) or enzymatic active sites to form the ascorbate free radical (AFR). Two AFR molecules readily react with one another to produce reduced ASC and oxidized DHA.

regulates genes involving the innate and adaptive immune response. Under normal conditions, the IKK (IκB kinase) complex keeps NF-kB in the cytoplasm rather than translocating to the nucleus. However, in the presence of proinflammatory cytokines, the IKKβ protein is phosphorylated which signals for proteasome degradation of the IKK complex and allows NF-kB translocation to the nucleus. This translocation of NF-kB into the nucleus induces gene transcription changes, including many proinflammatory genes (Engelmann, Weih, & Haenold, 2014). DHA has been shown to regulate IKKa/B by directly binding to the kinase complex and inhibiting IκBα phosphorylation, preventing NF-kB translocation and subsequent upregulation of proinflammatory and cell death genes (Cárcamo et al., 2004). Under inflammatory conditions with sufficient ASC, ROS accumulates in the cell ASC is more likely to be present as DHA, its oxidized form, following electron donation (Fig. 3). With DHA accumulating in the cell, NF-kB can remain blocked from translocating and mediating proinflammatory gene changes and potential cell death (Ferrada et al., 2021) (Fig. 4).

A second potential pathway for ASC mediation of neuroinflammation is through an interaction between SVCT2 and the Janus kinase (JAK) signal transducer and activators of transcription (STAT) signaling pathway. The JAK/STAT signaling pathway is activated by many cytokines and interferons and is essential for innate and adaptive immunity. The abnormal activation of the JAK/STAT pathway is present in a variety of neuroinflammatory diseases including Parkinson's disease and multiple sclerosis (Yan, Gibson, Buckley, Qin, & Benveniste, 2018). SVCT2 directly interacts with JAK2. Coupled with ASC transport, SVCT2 induces the autophosphorylation of JAK2, which subsequently recruits and phosphorylates STAT2. Activated STAT2 dimerizes and translocates to the nucleus where it alters gene transcription, though these exact gene changes have yet to be fully elucidated (Fig. 5). Therefore, SVCT2 might act as a receptor-like transporter with direct effects on epigenetic regulation (Han et al., 2021). Though these findings need to be replicated in vivo, it is possible the regulation of JAK/STAT by SVCT2 attenuates the neuroinflammatory response.

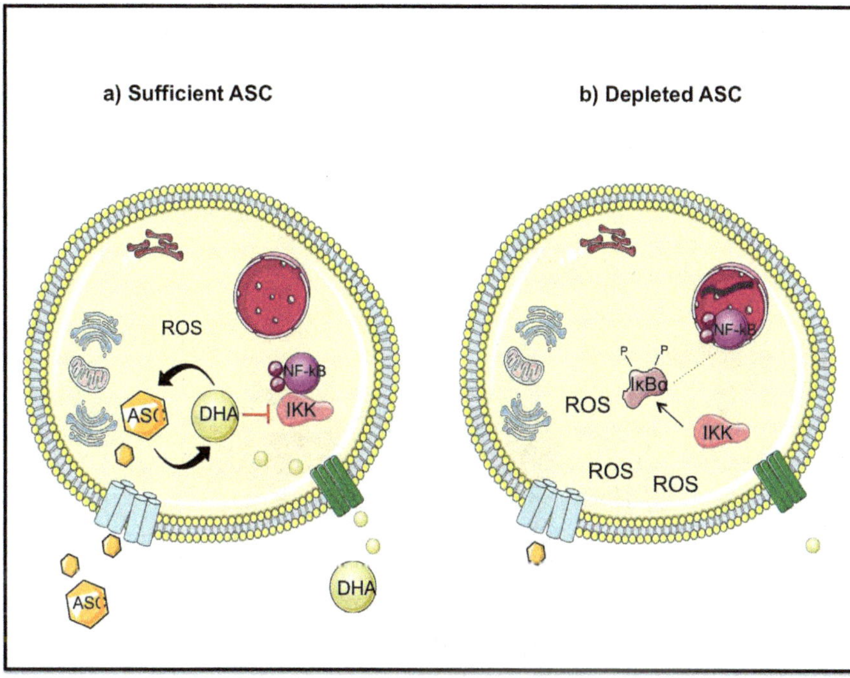

Fig. 4 Proposed mechanism of ASC-mediated NF-kB signaling. (A) Under ASC-sufficient conditions, DHA inhibits IKK from phosphorylation and degradation, thus preventing NF-kB from translocating to the nucleus. (B) Under depleted ASC conditions, IKK will be phosphorylated and degraded allowing NF-kB to translocate to the nucleus and initiating proinflammatory gene transcription changes (Cárcamo et al., 2004).

ASC is also a cofactor for ten-eleven translocation (TET) enzymes 1–3 which catalyze the conversion of 5-methylcytosine (5mC) to 5-hydroxymethylcytosine (5hmC) leading to demethylation of DNA (Ma, Hung, & Svaren, 2016; Varela-Rey et al., 2014). In primary cultures, 5hmC generation increases proportionally to treatment with ASC (Huff et al., 2020). The activation of both TET enzymes and 5hmC has roles in remyelination by Schwann cells following peripheral injury (Huff et al., 2020). 5hmC and TET enzymes have important roles in suppressing proinflammatory cytokines like IL-1B and in the proper functioning of peripheral immune cells (Lio & Rao, 2019). Interestingly, SVCT2 deficiency has been shown to impair remyelination following a nerve crush injury in SVCT2$^{+/-}$ mice, leading to reduced myelin thickness compared to wild-type mice and worse functional recovery (Röhr, Halfter, Schulz, Young, & Gess, 2017). This role for ASC in epigenetic modulation through TET enzymes and 5hmC is becoming clearer in the peripheral nervous system, and further research is needed to clarify a role in the CNS immune response.

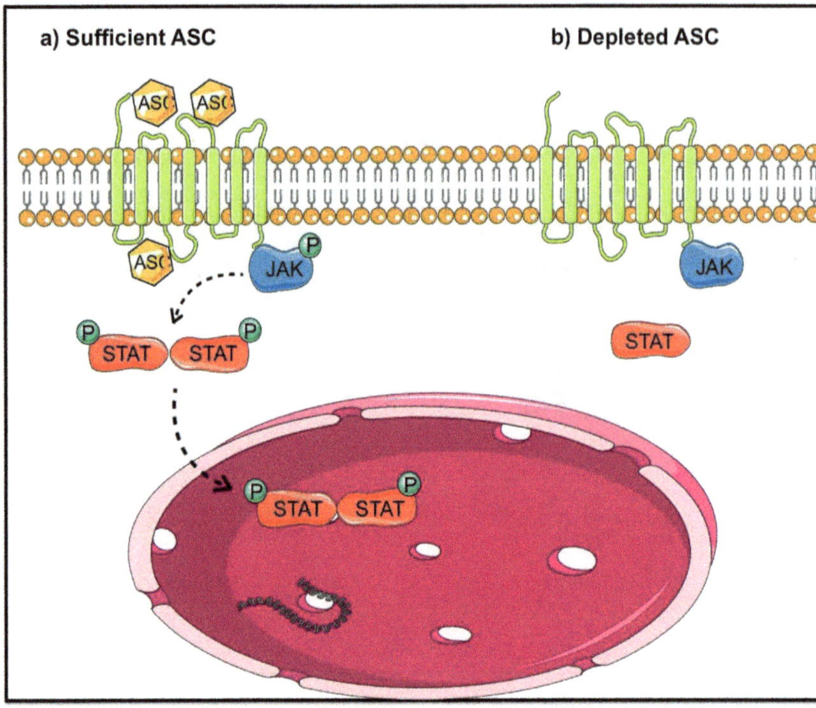

Fig. 5 Proposed mechanism of ASC-mediated JAK/STAT signaling. (A) Under ASC-sufficient conditions, SVCT2 acts as both a receptor and transporter. Coupled with SVCT2-mediated ASC influx into the cell, JAK is phosphorylated triggering recruitment, dimerization and phosphorylation of STAT. STAT then translocates into the nucleus where neuroprotective gene transcription is upregulated. (B) Under low ASC conditions, ASC is more likely to be retained by neurons and is therefore unavailable within the extracellular space to support this signaling process (Han et al., 2021).

Application to other neurological conditions

This discussion may be broadly applicable to other neurological conditions that present with neuroinflammation and ASC deficiency, such as depression, schizophrenia, Alzheimer's disease (AD), and diabetes (Lopes da Silva et al., 2014; Pullar, Carr, Bozonet, & Vissers, 2018; Subotičanec, Folnegović-Šmalc, Korbar, Meštrović, & Buzina, 1990; Will & Byers, 2009) (Fig. 6). For example, in depression, some small clinical studies found that ASC deficiency is linked to depressive-like symptoms (Kinsman & Hood, 1971; Pullar et al., 2018). In schizophrenic patients, one study found significantly lower levels of ASC compared to other hospital patients who received the same diet (Subotičanec et al., 1990). In one study in mouse models of ASC brain depletion and AD, low ASC supplementation accelerated AD pathology and increased oxidative stress in the brain in the early stages of disease development compared to mice with high ASC supplementation (Dixit et al., 2015). Patients with AD also tend to have lower plasma ASC concentrations than healthy individuals (Charlton, Rabinowitz, Geffen, & Dhansay,

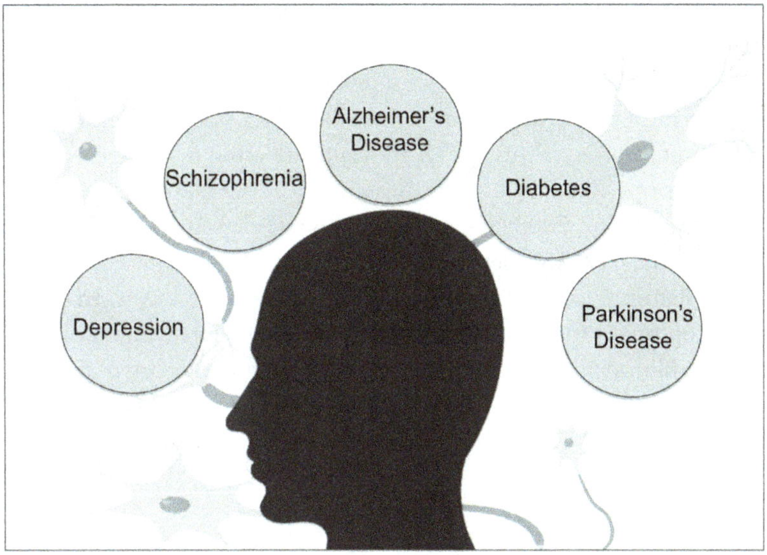

Fig. 6 Multiple neurological disorders have been associated with ASC deficiency and neuroinflammation.

2004). This suggests that ASC deficiency may accelerate some neuropathological features of AD. In AD and depression, chronic microglial activation is present (Brites & Fernandes, 2015; Mandrekar-Colucci & Landreth, 2010). In schizophrenia, several cases correcting deficiency or ASC supplementation were associated with an improvement in positive symptoms (Arvindakshan, 2003). These associations between ASC deficiency and neurological conditions with high neuroinflammation reinforce a greater oxidative demand in these conditions and suggest that efforts to maintain adequate ASC concentrations in the brain could ease disease progression and pathological burden.

Other areas of interest

Of other interest are combination treatments of antioxidants, such as ASC and vitamin E together. Vitamin E (alpha-tocopherol) is a fat-soluble antioxidant that can scavenge hydroperoxyl radical species in lipids throughout the body (Lee & Han, 2018). Humans take in vitamin E through our diet through foods such as vegetable oils and nuts. Vitamin E has been shown to modulate immune response in animal models. In a study looking at young and aged mice exposed to bacterial pneumonia, mice who received a diet without vitamin E showed uncontrolled neutrophil-driven pulmonary inflammation compared to mice supplemented with vitamin E (Bou Ghanem et al., 2015). ASC has the ability to recycle vitamin E through reduction of tocopheroxyl radicals back to α-tocopherol. Combination treatment of ASC and vitamin E in mice unable to synthesize ASC was protective against ROS damage following a traumatic optic neuropathy blast injury

(Bernardo-Colón et al., 2018). The decreased visual evoked potentials and axonal damage that were exacerbated in the mice on a low-ASC diet were attenuated by vitamin E (Bernardo-Colón et al., 2018). These findings have provided the basis for human clinical trials supplementing ASC and vitamin E to benefit a variety of health conditions; however, significant improvements in patient health are often not observed in large portions of treated groups (Retana-Ugalde, Casanueva, Altamirano-Lozano, González-Torres, & Mendoza-Núñez, 2008; Sesso, 2008). Nevertheless, in many cases, participants of these trials may not have had depleted levels of ASC or vitamin E. There is usually some benefit to vitamin treatment if correcting from a deficient status. In contrast, most vitamins have an optimally functional level, but when provided in excess of that amount, they provide no additional benefit and may even be harmful (Morris, 2011).

Mini-dictionary terms

Antioxidant—A compound that can neutralize radical species through donation of electrons.

Cytokine—Small proteins that are secreted and are important for singling in various immune responses.

Neuroinflammation—An inflammatory state of nervous tissue due to insult or injury.

Microglia—A type of glial cell in the central nervous system and acts as a part of the immune response in the brain.

SVCT1/2—Sodium-dependent vitamin C transporters that transport vitamin C throughout the body and brain.

Vitamin C—Ascorbic acid. A vitamin found in various fruits and vegetables and functions as a potent antioxidant in the brain and body.

Vitamin E—Alpha-tocopherol. A lipid-soluble vitamin that functions as an antioxidant in the body.

Key facts

Key facts of neuroinflammation:
- Neuroinflammation is inflammation of brain tissue caused by insult or injury
- Chronic neuroinflammation can contribute to cognitive decline
- Chronic neuroinflammation is a key feature in many neurological disorders

Key facts of Vitamin C:
- Vitamin C is also known as ascorbic acid
- Vitamin C is a powerful antioxidant in the brain
- It scavenges free radicals, is an enzymatic cofactor, and a neuromodulator

Summary points

- This chapter focuses on vitamin C (ASC) and the role it may play in neuroinflammatory conditions
- ASC is ingested through the diet in humans from various fruits, vegetables, and also supplements
- ASC concentrations are highest in brain tissue and may attenuate damage caused by neuroinflammation
- Key features of neuroinflammation include reactive glial cells, an increase in cytokine signaling, ROS production, and recruitment of peripheral monocytes and lymphocytes
- New data suggest a specific role for ASC in modulating different signaling pathways of neuroinflammation including through NF-kB or Jak/STAT
- Neuroinflammation in a variety of neurological conditions may benefit from proper ASC repletion

References

Aguilera, G., Colín-González, A. L., Rangel-López, E., Chavarría, A., & Santamaría, A. (2018). Redox signaling, neuroinflammation, and neurodegeneration. *Antioxidants & Redox Signaling*, *28*(18), 1626–1651. https://doi.org/10.1089/ars.2017.7099.

Arvindakshan, M. (2003). Supplementation with a combination of ω-3 fatty acids and antioxidants (vitamins E and C) improves the outcome of schizophrenia. *Schizophrenia Research*, *62*(3), 195–204. https://doi.org/10.1016/S0920-9964(02)00284-0.

Balabanov, R., Strand, K., Goswami, R., McMahon, E., Begolka, W., Miller, S. D., et al. (2007). Interferon-gamma-oligodendrocyte interactions in the regulation of experimental autoimmune encephalomyelitis. *Journal of Neuroscience*, *27*(8), 2013–2024. https://doi.org/10.1523/JNEUROSCI.4689-06.2007.

Ballaz, S. J., & Rebec, G. V. (2019). Neurobiology of vitamin C: Expanding the focus from antioxidant to endogenous neuromodulator. *Pharmacological Research*, *146*, 104321. https://doi.org/10.1016/j.phrs.2019.104321.

Bankston, A. N., Mandler, M. D., & Feng, Y. (2013). Oligodendroglia and neurotrophic factors in neurodegeneration. *Neuroscience Bulletin*, *29*(2), 216–228. https://doi.org/10.1007/s12264-013-1321-3.

Berger, U. V., Lu, X.-C. M., Liu, W., Tang, Z., Slusher, B. S., & Hediger, M. A. (2003). Effect of middle cerebral artery occlusion on mRNA expression for the sodium-coupled vitamin C transporter SVCT2 in rat brain: Ascorbate transporter expression after ischemia. *Journal of Neurochemistry*, *86*(4), 896–906. https://doi.org/10.1046/j.1471-4159.2003.01891.x.

Bernardo-Colón, A., Vest, V., Clark, A., Cooper, M. L., Calkins, D. J., Harrison, F. E., et al. (2018). Antioxidants prevent inflammation and preserve the optic projection and visual function in experimental neurotrauma. *Cell Death & Disease*, *9*(11), 1097. https://doi.org/10.1038/s41419-018-1061-4.

Birben, E., Sahiner, U. M., Sackesen, C., Erzurum, S., & Kalayci, O. (2012). Oxidative stress and antioxidant defense. *World Allergy Organization Journal*, *5*(1), 9–19. https://doi.org/10.1097/WOX.0b013e3182439613.

Bou Ghanem, E. N., Clark, S., Du, X., Wu, D., Camilli, A., Leong, J. M., et al. (2015). The α-tocopherol form of vitamin E reverses age-associated susceptibility to *Streptococcus pneumoniae* lung infection by modulating pulmonary neutrophil recruitment. *Journal of Immunology*, *194*(3), 1090–1099. https://doi.org/10.4049/jimmunol.1402401.

Brites, D., & Fernandes, A. (2015). Neuroinflammation and depression: Microglia activation, extracellular microvesicles and microRNA dysregulation. *Frontiers in Cellular Neuroscience, 9*. https://doi.org/10.3389/fncel.2015.00476.

Buettner, G. R. (1993). The pecking order of free radicals and antioxidants: Lipid peroxidation, α-tocopherol, and ascorbate. *Archives of Biochemistry and Biophysics, 300*(2), 535–543. https://doi.org/10.1006/abbi.1993.1074.

Cárcamo, J. M., Pedraza, A., Bórquez-Ojeda, O., Zhang, B., Sanchez, R., & Golde, D. W. (2004). Vitamin C is a kinase inhibitor: Dehydroascorbic acid inhibits IκBα kinase β. *Molecular and Cellular Biology, 24*(15), 6645–6652. https://doi.org/10.1128/MCB.24.15.6645-6652.2004.

Charlton, K. E., Rabinowitz, T. L., Geffen, L. N., & Dhansay, M. A. (2004). Lowered plasma vitamin C, but not vitamin E, concentrations in dementia patients. *Journal of Nutrition, Health & Aging, 8*(2), 99–107.

DePhillipo, N. N., Aman, Z. S., Kennedy, M. I., Begley, J. P., Moatshe, G., & LaPrade, R. F. (2018). Efficacy of vitamin C supplementation on collagen synthesis and oxidative stress after musculoskeletal injuries: A systematic review. *Orthopaedic Journal of Sports Medicine, 6*(10). https://doi.org/10.1177/2325967118804544. 232596711880454.

DiSabato, D. J., Quan, N., & Godbout, J. P. (2016). Neuroinflammation: The devil is in the details. *Journal of Neurochemistry, 139*(S2), 136–153. https://doi.org/10.1111/jnc.13607.

Dixit, S., Bernardo, A., Walker, J. M., Kennard, J. A., Kim, G. Y., Kessler, E. S., et al. (2015). Vitamin C deficiency in the brain impairs cognition, increases amyloid accumulation and deposition, and oxidative stress in APP/PSEN1 and normally aging mice. *ACS Chemical Neuroscience, 6*(4), 570–581. https://doi.org/10.1021/cn500308h.

Engelmann, C., Weih, F., & Haenold, R. (2014). Role of nuclear factor kappa B in central nervous system regeneration. *Neural Regeneration Research, 9*(7), 707. https://doi.org/10.4103/1673-5374.131572.

Ferrada, L., Magdalena, R., Barahona, M. J., Ramírez, E., Sanzana, C., Gutiérrez, J., et al. (2021). Two distinct faces of vitamin C: AA vs. DHA. *Antioxidants, 10*(2), 215. https://doi.org/10.3390/antiox10020215.

Han, Z., Zhang, Z., Guan, Y., Chen, B., Yu, M., Zhang, L., et al. (2021). New insights into Vitamin C function: Vitamin C induces JAK2 activation through its receptor-like transporter SVCT2. *International Journal of Biological Macromolecules, 173*, 379–398. https://doi.org/10.1016/j.ijbiomac.2021.01.120.

Harrison, F. E., Green, R. J., Dawes, S. M., & May, J. M. (2010). Vitamin C distribution and retention in the mouse brain. *Brain Research, 1348*, 181–186. https://doi.org/10.1016/j.brainres.2010.05.090.

Harrison, F. E., & May, J. M. (2009). Vitamin C function in the brain: Vital role of the ascorbate transporter SVCT2. *Free Radical Biology and Medicine, 46*(6), 719–730. https://doi.org/10.1016/j.freeradbiomed.2008.12.018.

Hawkins, B. T., & Davis, T. P. (2005). The blood-brain barrier/neurovascular unit in health and disease. *Pharmacological Reviews, 57*(2), 173–185. https://doi.org/10.1124/pr.57.2.4.

Huff, T. C., Sant, D. W., Camarena, V., Van Booven, D., Andrade, N. S., Mustafi, S., et al. (2020). Vitamin C regulates Schwann cell myelination by promoting DNA demethylation of pro-myelinating genes. *Journal of Neurochemistry*. https://doi.org/10.1111/jnc.15015.

Kigerl, K. A., Gensel, J. C., Ankeny, D. P., Alexander, J. K., Donnelly, D. J., & Popovich, P. G. (2009). Identification of two distinct macrophage subsets with divergent effects causing either neurotoxicity or regeneration in the injured mouse spinal cord. *Journal of Neuroscience, 29*(43), 13435–13444. https://doi.org/10.1523/JNEUROSCI.3257-09.2009.

Kinsman, R. A., & Hood, J. (1971). Some behavioral effects of ascorbic acid deficiency. *American Journal of Clinical Nutrition, 24*(4), 455–464. https://doi.org/10.1093/ajcn/24.4.455.

Lan, X., Han, X., Li, Q., Yang, Q.-W., & Wang, J. (2017). Modulators of microglial activation and polarization after intracerebral haemorrhage. *Nature Reviews Neurology, 13*(7), 420–433. https://doi.org/10.1038/nrneurol.2017.69.

Lee, G., & Han, S. (2018). The role of vitamin E in immunity. *Nutrients, 10*(11), 1614. https://doi.org/10.3390/nu10111614.

Levine, M., Conry-Cantilena, C., Wang, Y., Welch, R. W., Washko, P. W., Dhariwal, K. R., et al. (1996). Vitamin C pharmacokinetics in healthy volunteers: Evidence for a recommended dietary allowance.

Proceedings of the National Academy of Sciences of the United States of America, 93(8), 3704–3709. https://doi.org/10.1073/pnas.93.8.3704.

Li, Q., & Barres, B. A. (2018). Microglia and macrophages in brain homeostasis and disease. *Nature Reviews Immunology, 18*(4), 225–242. https://doi.org/10.1038/nri.2017.125.

Li, T., Liu, T., Chen, X., Li, L., Feng, M., Zhang, Y., et al. (2020). Microglia induce the transformation of A1/A2 reactive astrocytes via the CXCR7/PI3K/Akt pathway in chronic post-surgical pain. *Journal of Neuroinflammation, 17*(1), 211. https://doi.org/10.1186/s12974-020-01891-5.

Liddelow, S. A., Guttenplan, K. A., Clarke, L. E., Bennett, F. C., Bohlen, C. J., Schirmer, L., et al. (2017). Neurotoxic reactive astrocytes are induced by activated microglia. *Nature, 541*(7638), 481–487. https://doi.org/10.1038/nature21029.

Lio, C.-W. J., & Rao, A. (2019). TET enzymes and 5hmC in adaptive and innate immune systems. *Frontiers in Immunology, 10*, 210. https://doi.org/10.3389/fimmu.2019.00210.

Lopes da Silva, S., Vellas, B., Elemans, S., Luchsinger, J., Kamphuis, P., Yaffe, K., et al. (2014). Plasma nutrient status of patients with Alzheimer's disease: Systematic review and meta-analysis. *Alzheimer's & Dementia, 10*(4), 485–502. https://doi.org/10.1016/j.jalz.2013.05.1771.

Lyman, M., Lloyd, D. G., Ji, X., Vizcaychipi, M. P., & Ma, D. (2014). Neuroinflammation: The role and consequences. *Neuroscience Research, 79*, 1–12. https://doi.org/10.1016/j.neures.2013.10.004.

Ma, K. H., Hung, H. A., & Svaren, J. (2016). Epigenomic regulation of Schwann cell reprogramming in peripheral nerve injury. *Journal of Neuroscience, 36*(35), 9135–9147. https://doi.org/10.1523/JNEUROSCI.1370-16.2016.

Mandrekar-Colucci, S., & Landreth, G. E. (2010). Microglia and inflammation in Alzheimer's disease. *CNS & Neurological Disorders Drug Targets, 9*(2), 156–167. https://doi.org/10.2174/187152710791012071.

Mi, D. J., Dixit, S., Warner, T. A., Kennard, J. A., Scharf, D. A., Kessler, E. S., et al. (2018). Altered glutamate clearance in ascorbate deficient mice increases seizure susceptibility and contributes to cognitive impairment in APP/PSEN1 mice. *Neurobiology of Aging, 71*, 241–254. https://doi.org/10.1016/j.neurobiolaging.2018.08.002.

Michael, B. D., Griffiths, M. J., Granerod, J., Brown, D., Keir, G., Wnęk, C., et al. (2016). The interleukin-1 balance during encephalitis is associated with clinical severity, blood-brain barrier permeability, neuroimaging changes, and disease outcome. *Journal of Infectious Diseases, 213*(10), 1651–1660. https://doi.org/10.1093/infdis/jiv771.

Miyata, H., et al. (2022). Identification of an exporter that regulates vitamin C supply from blood to the brain. *iScience, 117*(31).

Monahan, A. J., Warren, M., & Carvey, P. M. (2008). Neuroinflammation and peripheral immune infiltration in Parkinson's disease: An autoimmune hypothesis. *Cell Transplantation, 17*(4), 363–372. https://doi.org/10.3727/096368908784423328.

Morris, M. C. (2011). A potential design flaw of randomized trials of vitamin supplements. *JAMA, 305*(13), 1348. https://doi.org/10.1001/jama.2011.383.

Mun, G. H., Kim, M. J., Lee, J. H., Kim, H. J., Chung, Y. H., Chung, Y. B., et al. (2006). Immunohistochemical study of the distribution of sodium-dependent vitamin C transporters in adult rat brain. *Journal of Neuroscience Research, 83*(5), 919–928. https://doi.org/10.1002/jnr.20751.

Murad, S., Grove, D., Lindberg, K. A., Reynolds, G., Sivarajah, A., & Pinnell, S. R. (1981). Regulation of collagen synthesis by ascorbic acid. *Proceedings of the National Academy of Sciences of the United States of America, 78*(5), 2879–2882. https://doi.org/10.1073/pnas.78.5.2879.

Nieman, D. C., Peters, E. M., Henson, D. A., Nevines, E. I., & Thompson, M. M. (2000). Influence of vitamin C supplementation on cytokine changes following an ultramarathon. *Journal of Interferon & Cytokine Research, 20*(11), 1029–1035. https://doi.org/10.1089/10799900050198480.

Norden, D. M., & Godbout, J. P. (2013). Review: Microglia of the aged brain: Primed to be activated and resistant to regulation: Increased microglial reactivity with age. *Neuropathology and Applied Neurobiology, 39*(1), 19–34. https://doi.org/10.1111/j.1365-2990.2012.01306.x.

Palpagama, T. H., Waldvogel, H. J., Faull, R. L. M., & Kwakowsky, A. (2019). The role of microglia and astrocytes in Huntington's disease. *Frontiers in Molecular Neuroscience, 12*, 258. https://doi.org/10.3389/fnmol.2019.00258.

Portugal, C. C., Socodato, R., Canedo, T., Silva, C. M., Martins, T., Coreixas, V. S. M., et al. (2017). Caveolin-1-mediated internalization of the vitamin C transporter SVCT2 in microglia triggers an inflammatory phenotype. *Science Signaling*, *10*(472). https://doi.org/10.1126/scisignal.aal2005.

Pullar, J., Carr, A., Bozonet, S., & Vissers, M. (2018). High vitamin C status is associated with elevated mood in male tertiary students. *Antioxidants*, 7(7), 91. https://doi.org/10.3390/antiox7070091.

Retana-Ugalde, R., Casanueva, E., Altamirano-Lozano, M., González-Torres, C., & Mendoza-Núñez, V. M. (2008). High dosage of ascorbic acid and alpha-tocopherol is not useful for diminishing oxidative stress and DNA damage in healthy elderly adults. *Annals of Nutrition and Metabolism*, *52*(2), 167–173. https://doi.org/10.1159/000129652.

Röhr, D., Halfter, H., Schulz, J. B., Young, P., & Gess, B. (2017). Sodium-dependent Vitamin C transporter 2 deficiency impairs myelination and remyelination after injury: Roles of collagen and demethylation: RöHR et al. *Glia*, *65*(7), 1186–1200. https://doi.org/10.1002/glia.23152.

Rowe, S., & Carr, A. C. (2020). Global vitamin C status and prevalence of deficiency: A cause for concern? *Nutrients*, *12*(7), 2008. https://doi.org/10.3390/nu12072008.

Salazar, K., Martínez, F., Pérez-Martín, M., Cifuentes, M., Trigueros, L., Ferrada, L., et al. (2018). SVCT2 expression and function in reactive astrocytes is a common event in different brain pathologies. *Molecular Neurobiology*, *55*(7), 5439–5452. https://doi.org/10.1007/s12035-017-0762-5.

Savini, I., Rossi, A., Pierro, C., Avigliano, L., & Catani, M. V. (2008). SVCT1 and SVCT2: Key proteins for vitamin C uptake. *Amino Acids*, *34*(3), 347–355. https://doi.org/10.1007/s00726-007-0555-7.

Sesso, H. D. (2008). Vitamins E and C in the prevention of cardiovascular disease in men: The Physicians' Health Study II randomized controlled trial. *JAMA*, *300*(18), 2123. https://doi.org/10.1001/jama.2008.600.

Shabab, T., Khanabdali, R., Moghadamtousi, S. Z., Kadir, H. A., & Mohan, G. (2017). Neuroinflammation pathways: A general review. *International Journal of Neuroscience*, *127*(7), 624–633. https://doi.org/10.1080/00207454.2016.1212854.

Sousa, C., Biber, K., & Michelucci, A. (2017). Cellular and molecular characterization of microglia: A unique immune cell population. *Frontiers in Immunology*, *8*. https://doi.org/10.3389/fimmu.2017.00198.

Subotičanec, K., Folnegović-Šmalc, V., Korbar, M., Meštrović, B., & Buzina, R. (1990). Vitamin C status in chronic schizophrenia. *Biological Psychiatry*, *28*(11), 959–966. https://doi.org/10.1016/0006-3223(90)90061-6.

Travica, N., Ried, K., Hudson, I., Sali, A., Scholey, A., & Pipingas, A. (2020). The contribution of plasma and brain vitamin C on age and gender-related cognitive differences: A mini-review of the literature. *Frontiers in Integrative Neuroscience*, *14*, 47. https://doi.org/10.3389/fnint.2020.00047.

Varela-Rey, M., Iruarrizaga-Lejarreta, M., Lozano, J. J., Aransay, A. M., Fernandez, A. F., Lavin, J. L., et al. (2014). S-Adenosylmethionine levels regulate the Schwann cell DNA methylome. *Neuron*, *81*(5), 1024–1039. https://doi.org/10.1016/j.neuron.2014.01.037.

Will, J. C., & Byers, T. (2009). Does diabetes mellitus increase the requirement for vitamin C? *Nutrition Reviews*, *54*(7), 193–202. https://doi.org/10.1111/j.1753-4887.1996.tb03932.x.

Wilson, J. X., Peters, C. E., Sitar, S. M., Daoust, P., & Gelb, A. W. (2000). Glutamate stimulates ascorbate transport by astrocytes. *Brain Research*, *858*(1), 61–66. https://doi.org/10.1016/s0006-8993(99)02433-6.

Xavier, S. M., Barbosa, C. O., Barros, D. O., Silva, R. F., Oliveira, A. A., & Freitas, R. M. (2007). Vitamin C antioxidant effects in hippocampus of adult Wistar rats after seizures and status epilepticus induced by pilocarpine. *Neuroscience Letters*, *420*(1), 76–79. https://doi.org/10.1016/j.neulet.2007.04.056.

Yan, Z., Gibson, S. A., Buckley, J. A., Qin, H., & Benveniste, E. N. (2018). Role of the JAK/STAT signaling pathway in regulation of innate immunity in neuroinflammatory diseases. *Clinical Immunology*, *189*, 4–13. https://doi.org/10.1016/j.clim.2016.09.014.

Yfanti, C., Fischer, C. P., Nielsen, S., Åkerström, T., Nielsen, A. R., Veskoukis, A. S., et al. (2012). Role of vitamin C and E supplementation on IL-6 in response to training. *Journal of Applied Physiology*, *112*(6), 990–1000. https://doi.org/10.1152/japplphysiol.01027.2010.

Zheng, Z.-H., Tu, J.-L., Li, X.-H., Hua, Q., Liu, W.-Z., Liu, Y., et al. (2021). Neuroinflammation induces anxiety- and depressive-like behavior by modulating neuronal plasticity in the basolateral amygdala. *Brain, Behavior, and Immunity*, *91*, 505–518. https://doi.org/10.1016/j.bbi.2020.11.007.

CHAPTER 27

Vitamin A and all-trans retinoic acid (ATRA) on neuroinflammatory changes in the brain

Saritha S. Nair[a], S.H. Priyanka[b], and M. Indira[b]
[a]R&D Division, Sipwise Beverages Pvt. Ltd., Chennai, India
[b]Department of Biochemistry, University of Kerala, Thiruvananthapuram, India

Abbreviations

9-CRA	9-*cis*-retinoic acid
AMPA	α-amino-3-hydroxy-5-methyl-4-isoxazolepropionic acid
APL	acute promyelocytic leukemia
APO E	apolipoprotein E
ATRA	all-trans retinoic acid
BBB	blood-brain barrier
BMVEC	brain microvascular endothelial cells
CIR	cerebral ischemic injury
CNS	central nervous system
LPS	lipopolysaccharide
MCA	middle cerebral artery
MS	multiple sclerosis
MVAD	marginal vitamin A deficiency
NETS	neutrophil extracellular traps
NMDARS	*N*-methyl D-aspartate receptors
NO	nitric oxide
NOS-2	nitric oxide synthase 2
PBMC	peripheral blood mononuclear cell
RA	retinoic acid
RAR	retinoic acid receptor
ROS	reactive oxygen species
RXR	retinoid receptor X
SVZ	subventricular zone
TJP	tight junction proteins

Introduction

Nutraceuticals and nutraceutical supplementation as a modality for treatment support have come into the fore presently in the disease management for many diseases. This strategy combined with therapy has been found to contribute positively to enhance

Vitamins and Minerals in Neurological Disorders
https://doi.org/10.1016/B978-0-323-89835-5.00019-3

the quality of life and health of the patients and hence has been supported by nutritionists and physicians worldwide. Nutrition was found to play a crucial role in the pathogenesis and evolution of neurodegenerative diseases as well. Certain vitamin deficiencies, especially vitamin A, have been documented to have a negative impact on brain health and cause neurodegenerative changes. Explorations of the biochemical role of vitamin A pointed out that most of its actions are manifested through its active metabolite all-trans retinoic acid (ATRA). So, in this review, even though we have begun with the role of vitamin A in neurodegenerative and neuroinflammation, our focus is on ATRA. We have attempted to paint their current picture in brain health. Our recent data on the impact of ATRA in the brain of alcoholic rat models have been added to reinforce the beneficial role of ATRA.

Vitamin A

The term "vitamin A" relates to a group of unsaturated organic compounds, which includes retinoids, including retinol, retinal, and retinyl esters (Johnson & Russell, 2010; Ross, 2010). It has long been recognized as a vitamin essential for visual process. But later, scientists could attribute more functions to this fat-soluble vitamin. It is necessary for normal differentiation of epithelial tissues, reproduction, and for the optimal maintenance of both innate and adaptive immune system. Its deficiency has been also linked with a variety of disorders such as ocular disorders, developmental or growth disorders, and compromised immunity. This is substantiated by the findings of a decrease in serum retinol concentration during infection and disease (Huang, Liu, Qi, Brand, & Zheng, 2018).

Vitamin A metabolism

Retinoid metabolism is complex and involves many different retinoid forms, including retinyl esters, retinol, retinal, retinoic acid (RA), and oxidized and conjugated metabolites of both retinol and retinoic acid.

The different retinoids are often modified on the end group of the side chain, which can be oxidized to form retinaldehyde and then further oxidized to form ATRA (Fig. 1). Dietary vitamin A from meat in the form of retinyl esters is hydrolyzed to retinol, or from plants as beta-carotene is cleaved to retinaldehyde and then reduced to retinol. Vitamin A and its precursors are absorbed in the intestine by intestinal epithelium cells. The precursors are esterified by the lecithin retinol acyltransferase. They are further loaded to chylomicrons and transported to hepatocytes, wherein after hydrolysis to retinol are mostly stored in hepatic stellate cells. Under conditions when RA is essential for the organism, the formed retinol binds retinol-binding protein in the liver and is carried through the bloodstream (Azaïs-Braesco & Pascal, 2000; Oliveira, Teixeira, & Sato, 2018). RA can be generated in multiple forms as all-trans, 9-*cis*, and 13-*cis* RA

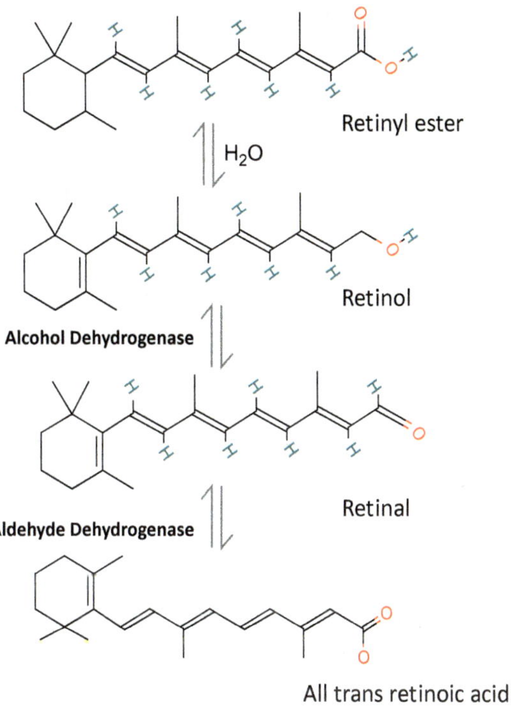

Fig. 1 Vitamin A metabolism.

(Campo-Paysaa, Marlétaz, Laudet, & Schubert, 2008). ATRA is the most abundant and functionally most active (Arnold, Amory, Walsh, & Isoherranen, 2012; Oliveira et al., 2018). RA interacts with nuclear receptors, such as the retinoic acid receptor (RAR) and retinoid receptor X (RXR), to regulate the transcription of a variety of target genes (Schuster, Kenyon, & Stephensen, 2008) by binding the retinoic acid-responsive elements in DNA (Balmer & Blomhoff, 2002; Oliveira et al., 2018).

Brain, vitamin A, and its metabolites

Vitamin A and vitamin A deficiency

Vitamin A is associated with the development of both eye and brain and exerts its action on the development of central nervous system (CNS) through its active metabolite RA. Its role in embryonic CNS includes the induction of neurogenesis and control of neuron patterning (Shearer, Stoney, Morgan, & McCaffery, 2012). Retinoids were demonstrated to modulate the maturation and maintenance of the adult brain and associated behaviors, primarily due to the abundance of retinoid receptors in brain and its activity at transcriptional levels itself (Lane & Bailey, 2005). But its role in adult brain is not fully understood. Dietary vitamin A supplementation improves learning and memory in vitamin A-deficient rodents and can ameliorate cognitive declines associated with normal aging. Cocco et al. (2002) reported that vitamin A deficiency (VAD) at early stages in the life cycle of rodents contributes to irreversible effects in the brain, but adult animals are more tolerant to retinoid malnutrition. VAD induces inflammation and aggravates existing inflammatory states. But supplementation with vitamin A has been found to be of benefit in certain cases (Reifen, 2002). Marginal vitamin A deficiency (MVAD) is associated with Alzheimer's disease. Zeng et al. (2017) stated that MVAD is approximately twofold more prevalent than VAD in the elderly. Mostly, prenatal MVAD promotes neuritic plaque formation and aggravates memory deficits in Alzheimer's disease models. Supplementing a therapeutic dose of vitamin A ameliorated the MVAD-induced memory deficits. This is also supported by studies indicating the therapeutic efficiencies of vitamin A and retinol derivatives in certain psychiatric and neurological disorders like schizophrenia and Alzheimer's disease (Tafti & Ghyselinck, 2007).

Role of the active metabolite ATRA in developmental and adult brain

ATRA directs embryonic development in the CNS, and retinoic acid signaling declines beyond the developmental stages. Hence, earlier studies were concentrated on retinoid signaling in the nervous system during developmental stages of embryo or early postnatal brain. But later studies have pointed out that it is more regionally focused on the adult brain and has acquired newer functions (Rhinn & Dollé, 2012). The metabolites of retinoids are found in specific regions of the adult, indicating their synthesis on discrete regions of the brain. The distribution of retinoid receptors in the adult nervous system is different from those in developmental stages. So, retinoid signaling may have a physiological role in different brain regions. Several neuronal specific genes having recognition sequences for the retinoid receptors have also been identified. They may be directly regulated by retinoids. Studies causing the disruption of retinoid signaling pathways in rodent models revealed their involvement in regulating synaptic plasticity and associated learning and memory behaviors. The various processes of neuroplasticity through hippocampus, the olfactory bulb, and the hypothalamus involvement have been demonstrated to be mediated through retinoid involvement. In hypothalamus, retinoids have been suggested to be involved in the process of seasonal switch in hibernating animals (Bremner, Shearer, & McCaffery, 2012; Helfer et al., 2012). They are also involved in the pathophysiology of Alzheimer's disease, schizophrenia, and depression. Retinoids were demonstrated as a neuroprotective agent, efficacious in facilitating regeneration (Lane & Bailey, 2005) and having antiinflammatory potential (van Neerven et al., 2010).

Alzheimer's disease is characterized by dementia that affects reasoning, behavior, and memory. The two distinct features of Alzheimer's disease are amyloid plaques and the other neurofibrillary tangles, which impair neurons and causes dementia. RA facilitates neuroprotective action in Alzheimer's disease models by reducing the accumulation of amyloid beta fibrils, inhibiting their formation, and destabilizing existing fibrils (Fragoso, Stoney, & McCaffery, 2014; Sahin, Karauzum, Perry, Smith, & Aliciguzel, 2005). Impaired α-amino-3-hydroxy-5-methyl-4-isoxazolepropionic acid (AMPA)-mediated synaptic transmission and defects in long-term potentiation (LTP) and long-term depression (LTD) have been observed in knockdown of RAR expression and in VAD, affecting social and spatial memory (Fragoso et al., 2014; Nomoto et al., 2012). Age-related memory deficits have also been linked to decreased retinoid signaling and were found to improve upon its upregulation (Dumetz et al., 2020). Both in vivo and in vitro studies on the administration of RA to experimental models of Alzheimer's disease showed beneficial effects. It inhibited hippocampal microglial activation, improved stem cell proliferation (Oliveira et al., 2018; Takamura, Watamura, Nikkuni, & Ohshima, 2017), and increased the synthesis of apolipoprotein E (Apo E) in human macrophages (Clemens, Regen, Le Bret, Heuser, & Hellmann-Regen, 2018). Apo E shields microglia, from the neurotoxic effects of amyloid β, and maintains neuronal homeostasis (Qin et al., 2006). Oral coadministration of RAR-α/β agonist, Am80 and RXR agonist, HX630 diminished the level of Aβ peptide in the brain by inducing the differentiation of IL-4-responsive M2-like microglia (Kawahara et al., 2014).

Investigations of Behairi et al. (2016) suggested that dietary or pharmacological ATRA consumption could potentially delay in the development of Alzheimer's disease in the elderly. The neuroinflammation model characterizing cerebral amyloid genesis showed that aging aggravated lipopolysaccharide (LPS)-induced neuroinflammation, thereby altering brain tissue integrity with resultant neuronal degeneration in both the cerebral cortex and the hippocampus as seen in Alzheimer's disease. They demonstrated that ATRA prevented the LPS-dependent nitric oxide (NO) production and upregulated cerebral nitric oxide synthase 2 (NOS2). The study demonstrated that ATRA pretreatment preserved memory function in LPS-treated rats by improving the integrity of cerebral tissue. ATRA downregulated iNOS expression and thereby hampered spontaneous NO production, causing a reduction in IL-17A production and increased IL-10 in peripheral blood mononuclear cell (PBMC) from patients of Alzheimer's disease.

Beneficial effects of ATRA in brain injury and neuroinflammatory conditions

Hummel et al. (2020) demonstrated the therapeutic value of posttraumatic ATRA treatment in experimental models. ATRA gave long-term protection to hippocampal granule cell layer integrity and reduced brain lesion size, reactive astrogliosis, and axonal injury. Li et al. (2018) showed that ATRA pretreatment could alleviate cerebral ischemic injury (CIR)-induced neurological deficits, inhibit MMP-9 protein expression and activity, increase BBB permeability, infarct volume, and degradation of tight junction proteins (TJPs). ATRA may be inhibiting the JNK/P38 MAPK pathway to ameliorate BBB disruption and improving CIR outcomes.

Investigations conducted by Mizee et al. (2014) suggested that RA may be a regulator of BBB function in multiple sclerosis (MS). MS lesions are characterized by activated astrocytes and their proinflammatory secretions. The expression of retinaldehyde dehydrogenase 2, a key enzyme for RA synthesis, is generally high in reactive astrocytes. In vitro modeling studies revealed a protective role for astrocyte-derived RA on BBB function during inflammation. They could demonstrate that RA reduced oxidative stress through an Nrf2-mediated pathway.

Treatment with ATRA alone or in combination with calcitriol in murine autoimmune encephalomyelitis, an experimental model of MS, increased the expression of the Foxp3 and TGF-β genes in splenocytes while reducing RORγt gene expression (Oliveira et al., 2018). Supplementation of vitamin A for 6 months to patients with MS revealed that there was an upregulation of TGF-β and Foxp3 gene expression in their PBMC (Saboor-Yaraghi et al., 2015) and a reduction in IFN-γ and T-bet gene expression (Mohammadzadeh Honarvar et al., 2016). RA treatment also suppressed Tγδ cell pathogenic activity by decreasing IL-17 production (Raverdeau, Breen, Misiak, & Mills, 2016). Coadministration of ATRA and atorvastatin, a lipid-lowering drug with pleotropic effects, resulted in the regression of the clinical and neuropathological features of encephalomyelitis (Abtahi Froushani, Delirezh, Hobbenaghi, & Mosayebi, 2014).

Effects of RA on neurogenesis are controversial. Some results suggest that RA regulates postnatal neurogenesis in the murine subventricular zone olfactory (SVZ) bulb pathway (Wang, Zhang, & Parent, 2005). Also, in SVZ neurosphere stem cell and parasagittal slice cultures from postnatal mouse brain, RA exposure enhanced neurogenesis by improving the proliferation and neuronal differentiation of forebrain SVZ neuroblasts. Prolonged oral RA administration also increased adult rat SVZ cell proliferation. Contrary to these reports, Crandall et al. (2004) documented that chronic exposure to 13-*cis*-RA reduced cell proliferation in the hippocampus and SVZ.

9-*Cis*-retinoic acid (9cRA), the easily interconvertible metabolite of ATRA, reduced ischemic brain injury in rodents by upregulation of bone morphogenetic protein7 (Shen et al., 2009). The combative effect of 9cRA in an animal model for stroke was examined in the study. 9cRA-pretreated rats were subjected to transient focal ischemia induced by ligation of the middle cerebral artery (MCA). Rats pretreated with 9cRA had betterment in locomotor activity and lesser neurological deficits 2 days after MCA ligation. It also reduced cerebral infarction, thus pointing the effectiveness of 9cRA pretreatment in strokes.

The therapeutic efficacy of ATRA in ischemic stroke by attenuating neural inflammation was reported by Cai et al. (2019). ATRA treatment prevented neutrophil accumulation, promoted N2 polarization, and thereby blocked neutrophil extracellular traps formation in ischemic lesion. STAT1 signaling played a decisive role in the mechanisms of ATRA-mediated regulation of neutrophil. Prophylactic administration of ATRA reduced brain injury in stroke as neuronal viability and BBB integrity were improved and glia cell activation was controlled. RA also has neuroprotective activity and is capable of reinstating barrier tightness in human-induced pluripotent stem cell-derived brain endothelial cells by RARα, RARγ, and RXRα activation (Stebbins et al., 2018).

Liao et al. (2015) reported the contradictory effects of ATRA on lupus-like disease in the MRL/lpr mouse model. It suggests that while treatment with ATRA improved glomerular pathology, it also caused severe inflammation in other peripheral organs. The increased tissue inflammation is attributable to ATRA-mediated augmentation of B-cell responses. This was supported by Theus, Sparks, Liao, Ren, and Luo (2017) that ATRA induced a paradoxical effect on immune activation during the development of autoimmune lupus.

In a recent study, Alatshan et al. (2020) demonstrated that ATRA can enhance the production of IL-1β secretion of LPS-activated human monocyte-derived macrophages via the NLRP3 inflammasome-mediated pathway by enhancing both activation and priming signals.

All these studies imply the beneficial effects of ATRA in mitigating neuroinflammation, which is a major factor in the development of neurodegenerative diseases. The pathological changes associated with various neurological problems are similar; hence, a therapeutic effect found in one case can be extrapolated to other neurodegenerative diseases.

ATRA and alcoholic neurotoxicity

Our group has been concentrating over the last few years on ameliorating potential of various micronutrients in alcohol-induced toxicity. Although most of the ethanol ingested gets metabolized by liver, it causes intoxication in the brain (Calabrese et al., 2002). Chronic alcohol intake results in neuroinflammation (Vallés, Blanco, Pascual, & Guerri, 2004) and neurodegeneration in both humans and animal models (Banerjee, 2014). Hence, we focused on neurotoxicity and studied the impact of supplementation of ATRA on both coadministration with alcohol and its action on the regression of alcohol-induced toxicity.

Our quest for an economical and noninvasive therapy for alcoholic injury showed that ATRA could reduce alcohol-induced brain injury in both chronic and regression alcohol models (Nair, Prathibha, Syam Das, Kavitha, & Indira, 2015; Priyanka, Syam Das, Thushara, Rauf, & Indira, 2018; Priyanka, Thushara, Rauf, & Madambath, 2019). Our studies have shown that ATRA has antioxidant and antiinflammatory potential as evidenced by the reduction in the markers of free radicals and mRNA expression of lipoxygenase and cyclooxygenase. Neuroinflammatory changes were recorded in alcohol-administered rats since we observed the elevated expression of downstream cytokines IL-2 and IL-10. Administration of ATRA had a protective effect in all these cases. Supplementation of ATRA reduced the expression and translocation of NFκB to nucleus, indicating the role NFκB-mediated mechanism. Alcohol exposure also reduced the expression of SIRT1, which may be due to enhanced oxidative stress and lower NAD+ on alcohol administration. SIRT1 is activated by increased levels of NAD+ and directly binds and deactivates NFκB RelA/p65 through deacetylation and proteasome degradation (Lei et al., 2012). But SIRT expression was enhanced on ATRA supplementation with a concomitant decrease in the activity of alcohol dehydrogenase.

There are contradictory reports on the action of ATRA on glutathione, an endogenous antioxidant. ATRA at low concentration was shown to inhibit Nrf2, which is the master regulator of glutathione, and other target antioxidant genes (Nefedova et al., 2007; Yin, Zhou, Wu, Chen, & Bao, 2015). The action of ATRA on Nrf2 is concentration dependent. High concentration of ATRA (15–30 mg/kg) was shown to activate Nrf2 in mice (Wang, Hayes, Henderson, & Wolf, 2007). However, lower concentration was shown to inhibit Nrf2 (Tan, Kosuge, Yang, & Ito, 2008).

Effects of ATRA on neurodegenerative alterations in chronic alcohol model

Ethanol can bring about neurotoxicity by the compensatory upregulation of *N*-methyl D-aspartate receptors (NMDARs) and related alterations that are involved in several physiological and pathophysiological processes, including synaptic plasticity, ischemia, and neurodegeneration (Ridge, Ho, Innes, & Dodd, 2008). KLF11-MAO-B cell death cascade and its contribution to chronic alcohol-induced brain damage were reported earlier by Udemgba et al. (2014). Hence, we probed NMDAR expression and activation of KLF-11-MAO-B cascade on coadministration of alcohol and ATRA in rats (Nair et al., 2015). We assessed the mRNA expressions of the obligatory subunit NR1 and regulatory subunit NR2B and found that both NR1 and NR2 subunits of the NMDA receptor were upregulated on chronic ethanol intoxication and the elevated glutamate levels in the brain further substantiated the findings. These coincided with a marked suppression in the expression of GABBR1 receptor and overexpression of 5-HTR1A, which were antagonized on ATRA supplementation. Chronic ethanol exposure significantly upregulated KLF11, a MAO-B transcriptional activator, with consequent upregulation of both MAO-A and MAO-B. ATRA supplementation downregulated the expression of KLF11 and thereby MAO-A and MAO-B expressions in the brain, eventually decreasing MAO activity. This caused elevation of serotonin and dopamine on exposure to ATRA and reduction of glutamate.

In short, we could demonstrate that low-dose ATRA supplementation can mitigate ethanol-induced alterations in the brain by substantially reducing oxidative stress with subsequent modulation of NMDAR and KL-11 expression and causing enhancement in the catabolism of glutamate and decrease in the catabolism of serotonin and dopamine.

Effects of ATRA on blood-brain barrier on chronic alcoholism

BBB is vital in maintaining homeostasis within the brain to safeguard optimal neuronal function (Mizee et al., 2014). BBB dysfunction may lead to apoptosis of endothelium. Loss of BBB integrity is a critical event in the development and progression of neurological disorders such as Alzheimer's disease, human immunodeficiency virus-1 (HIV-1) encephalitis, MS, stroke, and traumatic brain injury (Sweeney, Sagare, & Zlokovic, 2018). Neurodegenerative changes and neuronal death were also observed on neuropathologic examination of brain tissue from chronic alcoholics. Increased activity and expression of the ethanol-metabolizing enzymes, CYP2E1 and alcohol dehydrogenase, in brain microvascular endothelial cells (BMVEC) on chronic ethanol insult (Haorah et al., 2005) was shown to be the underlying cause of BBB impairment. Ethanol, being polar, and ATRA, being a small molecule, both can cross the BBB and exert their action. Hence, we channeled our studies to investigate the alterations in mRNA expression in TJPs in BMVEC, both in chronic alcoholic and in regression models. In order to investigate the impact of ATRA in chronic alcoholic conditions, male rats were grouped as follows: I—Control, II—Ethanol (4g/kg b.w.), III ATRA (100μg/kg b.w.), IV—Ethanol (4g/kg b.w.), and +ATRA (100μg/kg b.w.). Rats were sacrificed on 91st day, and various parameters were assayed.

ATRA could downregulate ethanol-induced upregulation of MMP-9, which mediates both inflammatory and vascular changes in brain. This might have caused the diminution of total occludin, ZO-1, and claudin contents, probably due to an MMP-9-induced activation of MMP-3/9. This was altered by ATRA since there was upregulation of TJPs, which are crucial in maintaining BBB integrity (Fig. 2).

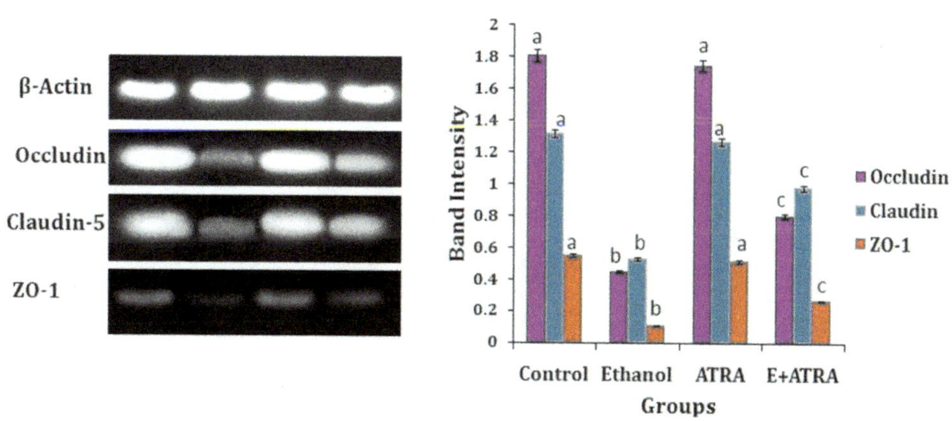

Fig. 2 mRNA expression of TJPs in BMVEC. The relative amount of mRNA was estimated by semiquantitative RT-PCR. The PCR products were quantified by densitometry and standardized to their respective β-actin controls. Results are expressed as average of quadruplicates ± SEM. *Different letters* indicate the values statistically different at $P < .05$.

The ethanol-induced excitotoxicity has been linked to the process of neurodegeneration, and ATRA-mediated alterations to NMDAR might have reduced neurodegeneration (Fig. 3).

The ethanol-mediated neuroinflammatory processes were found to be lowered by ATRA as evident in the downregulation of proinflammatory NFκB and its target genes TNFα, MCP-1, and IL-6. Also, these studies hint at the potential of ATRA to prevent neuronal apoptosis mediated by TNFα activation and stimulation of IL-6 (Fig. 4).

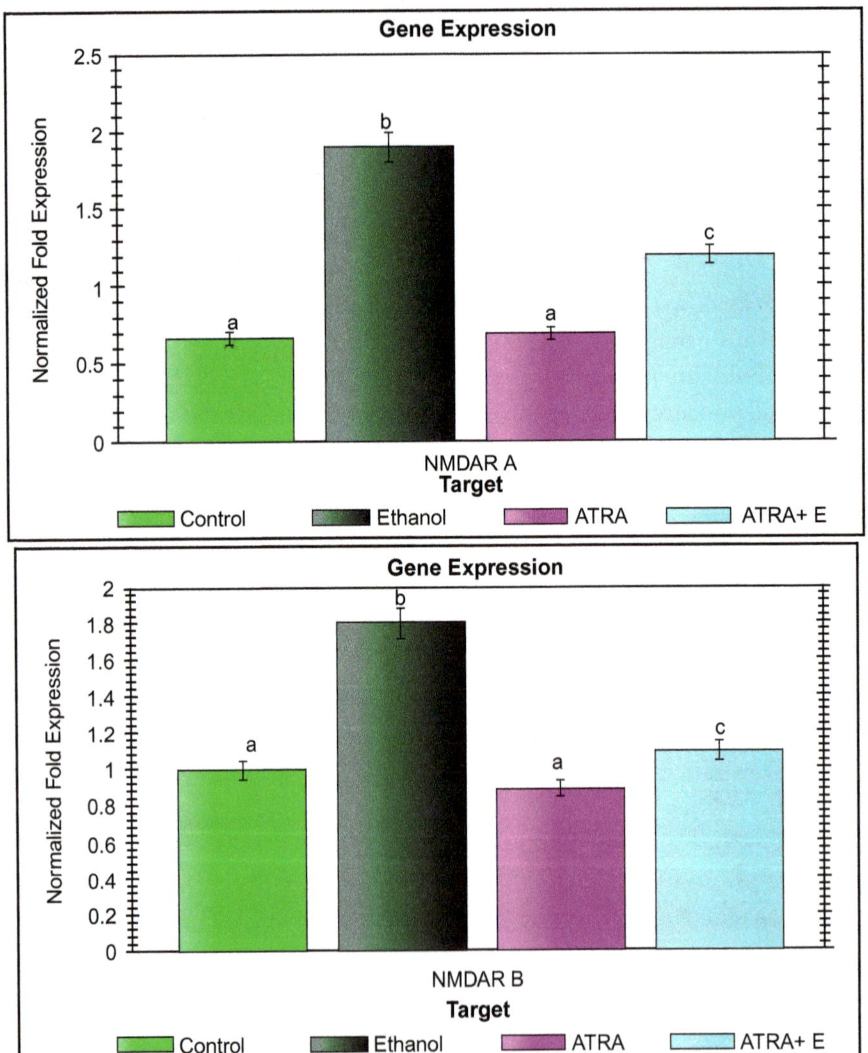

Fig. 3 qPCR of NMDAR expression in the brain tissue. mRNA were quantified by qPCR and the products quantified against their respective β-actin controls. Results are expressed as average of triplicates ± SEM. *Different letters* indicate the values statistically different at $P < .05$.

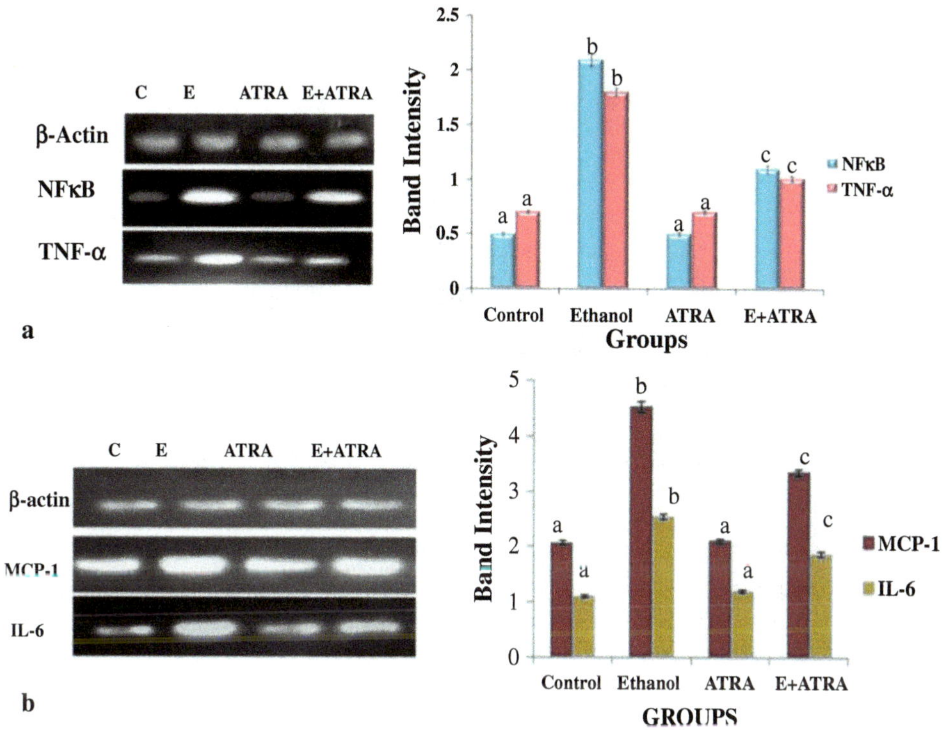

Fig. 4 mRNA expression NFκB, TNF-α, MCP-1, and IL-6 in brain. The relative amount of mRNA was estimated by semiquantitative RT-PCR. The PCR products were quantified by densitometry and standardized to their respective β-actin controls. Results are expressed as average of quadruplicates ± SEM. *Different letters* indicate the values statistically different at $P < .0$.

This study suggested ATRA to be an effective modality of nutritional intervention against ethanol-induced BBB dysfunction and neuroinflammation, in chronic alcoholism models.

Effects of ATRA on blood-brain barrier on alcohol regression

Endotoxemia has been well documented in alcoholic patients (Kirpich et al., 2017). Alcohol-induced intestinal barrier dysfunction may lead to intestinal hyperpermeability and cause endotoxemia. These endotoxins may penetrate to major organs, including brain and liver, to cause alcoholic injury. Hence, in a regression study for alcohol model, we furthered our investigations on the impact of supplementation of ATRA on BBB. For this regression model, rats were divided into two groups and were maintained for 90 days as control (C) and ethanol groups (E) (4 g/kg b.w.). Ethanol administration was stopped after 90 days, and animals in the control group were divided into two groups and then maintained for 30 days as the control (C) and ATRA (100 μg/kg b.w.) and those in the ethanol group as ethanol abstention (A) and ATRA + abstention (ATRA + A) (100 μg/kg b.w.). At the end of experimental period, animals were sacrificed and various biochemical parameters were assayed.

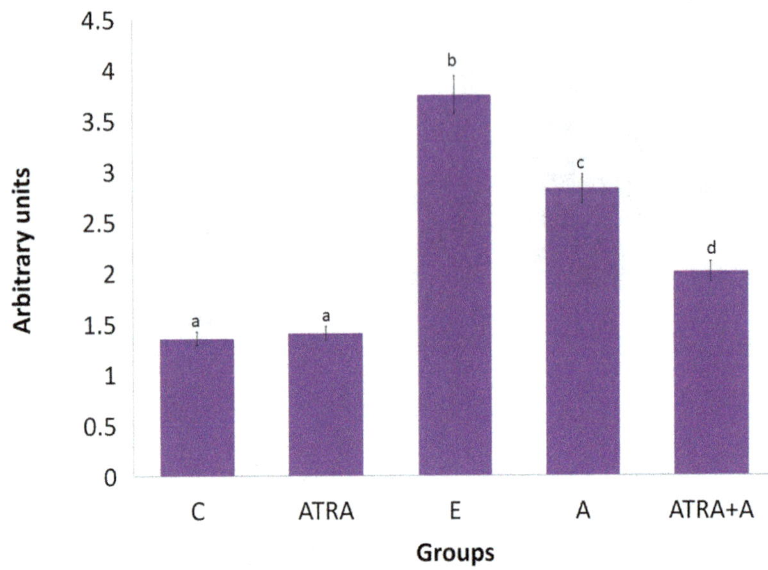

Fig. 5 qPCR of 16S rRNA in intestine. 16S rRNA were quantified by qPCR and the products standardized against their respective β-actin controls. Results are expressed as average of six rats per group ± SEM. *Different superscripts* indicate the values statistically significant at $P < .05$.

Priyanka et al. (2019) observed enhanced endotoxins level in the intestinal fluid in ethanol group. Supplementation of ATRA reduced endotoxemia better than abstention. ATRA decreased the hyperpermeability of intestine better than abstention as evidenced by the expression of 16S rRNA (Fig. 5).

Endotoxemia and intestinal barrier defect lead to BBB dysfunction through the disruption of TJPs. TJPs are found both in intestine and in BBB. The decreased mRNA expressions of claudin, occludin, and ZO-1 and increased endotoxins in ethanol-fed rats indicated intestinal barrier and BBB defect. But ATRA administration reduced bacterial overgrowth, endotoxemia, and intestinal barrier defect than alcohol abstention (Fig. 6).

The alterations in the TJPs cause barrier defects both in intestine and in BBB, releasing endotoxins into the circulation, transgressing into the brain, through structurally compromised BBB. Our results too demonstrate this fact, since we observed defective BBB and enhanced endotoxemia in the brain. But ATRA could partially alter this situation. ATRA has been reported to have an antibacterial effect (Wheelwright et al., 2014). Hence, ATRA may be directly inhibiting small intestinal bacterial overgrowth, thus reducing intestinal alterations caused by alcohol. Since gut-brain-microbiota has been implicated in various other neurodegenerative disorders (Osadchiy, Martin, & Mayer, 2019), this finding is particularly important in the elucidation of mechanistic aspects of the process.

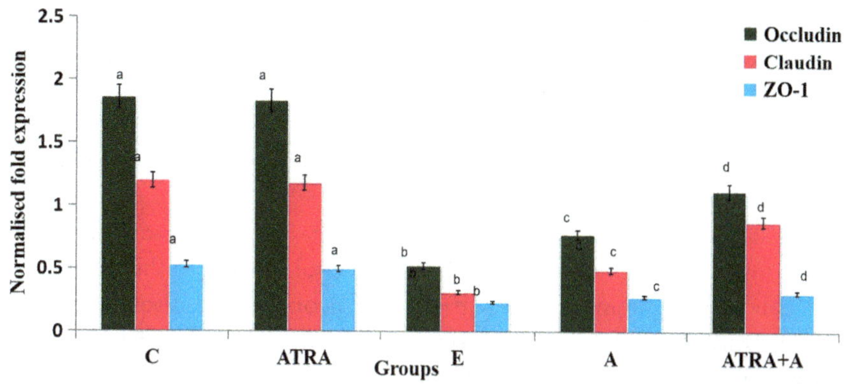

Fig. 6 qPCR of TJPs in BMVEC. Occludin, claudin, and ZO-1 mRNAs were quantified by qPCR and the products standardized against their respective β-actin controls. Fold change (*y*-axis) represents the expression of the target gene mRNAs relative to that of the control group. Values are expressed as mean ± SEM of six rats per group. Different superscripts indicate the values statistically significant at $P < .05$.

Effects of ATRA on neuroinflammation and on NLP3 inflammasomes in regression models

We have documented the impact of ATRA on inflammasomes and neurodegeneration in alcohol regression models (Priyanka et al., 2019). One of the main players triggering acute or chronic inflammation is the activation of inflammasomes. The oxidative stress induced by ethanol can potentially lead to inflammation via NFκB activation. We probed the impact of ATRA supplementation on the NLRP3 inflammasome activation. Our studies revealed the increased gut permeability and plasma endotoxin level in rats chronically fed alcohol. But ATRA administration reduced ROS, endotoxemia, and decreased intestinal permeability than alcohol abstention.

We also observed elevated expression of caspase 1, which may be responsible for the elevated levels of IL-1β, since it is involved in the processing of pro-IL-1β to IL-1β. These markers of NLRP3 activation were decreased on ATRA supplementation. Overall, these cascades of cellular events due to ATRA supplementation would have reduced pyroptosis. NLRP3 activation was higher in abstention in comparison with ATRA-supplemented group, thus pointing out the therapeutic potential of ATRA.

Conclusion

A myriad of studies on retinoids and ATRA has emphasized its role in embryonic developmental stages and differentiation. But most importantly, the evolved and localized functions of retinoids in brain have been revealed, and they suggest a very distinct neuroprotective effect for ATRA supplementation. Administration of ATRA has been shown to delay the development of various diseases like Alzheimer's disease and MS. It also helps in recovery from cerebral ischemic injury, reduces brain tissue damage, and prevents BBB disruption. ATRA was also reported to be generating endogenously in response to neuroinflammatory insult. All these studies have shown the therapeutic potential of ATRA on neuroinflammatory events, eventually contributing to the prevention of neurodegenerative processes.

Our quest for a noninvasive and economical therapy for alleviating alcoholic brain injury has shown beneficial effects of ATRA in alcohol models, in the instances of both chronic alcohol of consumption and regression of alcohol-induced toxicity. Our studies have outlined the effects of ATRA on BBB restoration at molecular levels. Besides, it also demonstrated how ATRA ameliorates alcohol-induced neuroinflammatory and neurodegenerative changes via various pathways. The alterations in the brain due to alcohol-induced toxicity have similarities to other neurodegenerative diseases. Hence, our findings may be extrapolated to other neurological problems. But these observations should be validated in relevant models. The dose response must be studied in detail for extrapolation since vitamin A and retinoids are redox-active molecules, which exert an antioxidant action at low concentration and pro-oxidant effects at high vitamin concentration (Lee et al., 2009; Siddikuzzaman & Grace, 2013). Our study involved the supplementation of ATRA at low doses. Hence, it seems ATRA can be developed into nutritional modality to support the recovery or prevention of neuroinflammatory/neurodegenerative conditions. Moreover, ATRA is an approved drug for the treatment of acute promyelocytic leukemia (APL), and so it can be easily repurposed for the mitigation of neuroinflammation.

Applications to other neurological conditions

The alterations in the brain due to alcohol-induced toxicity have similarities to other neurodegenerative diseases, and they share almost the same pathways for the progression and development of diseased conditions. Alcohol regression models have similarities to brain recovery after injury or trauma. KLF-11-mediated pathways have been elucidated in neuroprotection (Harris et al., 2015; Yin et al., 2013), and we have also characterized the same in chronic alcoholism and regression models. We have studied the alterations to TJPs in BBB, and the same has been reported in many studies related to brain injury and neuroinflammation. Hence, our findings might find applications in other neurogenerative, neuroinflammatory, and recovery from brain injury milieu. But these findings need to be substantiated in relevant models. This is in tune with the observations that supplementation of ATRA delays the development of various diseases like Alzheimer's disease and MS. Hence, ATRA can be developed into a nutritional modality to prevent the progression of neuroinflammatory alterations.

Other components of interest

Several other nutrients also alleviate neuroinflammation by interacting through gut-microbiome-brain axis by mitigating the endotoxemia. Riccio and Rossano (2018) have documented that certain antiinflammatory dietary nutrients—omega-3 (n-3) polyunsaturated fatty acids, vitamins (D, A, B12, C, E, PP), oligoelements (selenium, zinc, magnesium) thiolic compounds (lipoic acid, *N*-acetylcysteine), polyphenols (quercetin, catechins, ...), nonflavonoids (resveratrol, curcumin, hydroxytyrosol)—reduce the low-grade inflammation found in MS.

McGrattan et al. (2019) reported that dietary components of the Mediterranean diet (omega-3 fatty acids, antioxidants, and polyphenols) can inhibit neuroinflammation associated with Alzheimer's disease. Similarly, pathogenesis of Parkinson's disease can also be alleviated by many natural compounds, phytochemicals, vitamins, and minerals (Ciulla, Marinelli, Cacciatore, & Stefano, 2019).

Recently, numerous antiinflammatory compounds have been isolated from marine algae with therapeutic potential against neuroinflammation (Barbalace et al., 2019).

Ascorbic acid has been found to ameliorate the pathogenesis of Alzheimer's disease. Its mechanism of action includes its antioxidant, antiinflammatory properties and the chelation of iron, copper, and zinc on the suppression of amyloid plaques (Monacelli, Acquarone, Giannotti, Borghi, & Nencioni, 2017). Khan, Muhammad, Ikram, and Kim (2019) observed that curcumin may be a therapeutic candidate for LPS-induced ROS-mediated neurotoxicity and memory deficits in an adult rat model. The same group (Khan, Ikram, Park, Park, & Kim, 2020) has also observed the potential of anthocyanin in alleviating the induction of Alzheimer's disease. Hesperetin, a citrus flavonoid, attenuates LPS-induced neuroinflammation (Muhammad, Ikram, Ullah, Rehman, & Kim, 2019).

We have also observed a reduction of neurotoxicity by selenium (Sreekala & Indira, 2009) and *Sida cordifolia* Linn (Swathy et al., 2010) in quinolinic acid-induced toxicity models in rats. Our studies also pointed out that atorvastatin improved Y-maze learning behavior in nicotine-treated male rats (Das, Nair, Kavitha, Febi, & Indira, 2015).

All these studies suggest that a combination of different dietary supplements may give a synergistic action to mitigate neuroinflammation.

Key facts of ATRA

- ATRA is an anticancer ("antineoplastic" or "cytotoxic") chemotherapy drug, used in the treatment of APL.
- It has a hepatoprotective effect.
- It is important in early embryonic neuronal development.
- It has region-specific functions in adult brain too.

Mini-dictionary of terms

- **Abstention:** Restraining from the consumption of alcohol.
- **Endotoxin:** LPS found in the cell wall of gram-negative bacteria induces various biological reactions.
- **Supplementation of ATRA:** In addition to what is included in the diet.
- **Alcohol regression models:** Initially, induction of alcoholic toxicity by administering alcohol for a definite period. Then, alcohol administration is stopped and monitored the reduction of toxicity after supplementing the micronutrient/ drug.
- **Intestinal permeability:** The gut epithelial wall loses its integrity and allows materials from the lumen to translocate into the bloodstream and other organs.

Summary points

- Vitamin A is essential for visual process and necessary for normal differentiation of epithelial tissues, reproduction, and for the optimal maintenance of both innate and adaptive immune system.
- Vitamin A deficiencies cause a myriad of cognitive disorders, and its active metabolite ATRA is crucial in developmental stages as well as maintenance and regeneration of adult neurons, in a region-specific manner.
- ATRA holds therapeutic potential for the brain recovering from injuries and neuroinflammatory conditions.
- Our investigations on alcoholism have characterized molecular pathways of the beneficial effects of ATRA, in relation to BBB disruption and other neurodegenerative alterations due to ethanol exposure.
- Alcoholic neurotoxicity is mechanistically similar to many other neurodegenerative and neuroinflammatory disorders, and hence, findings relevant to alcoholic neuropathy may find its applications in any such disorders, as well.
- Hence, these insights may serve for a better understanding and application of ATRA in neurological disorders having a mechanistic similarity to alcoholic neuropathy.

References

Abtahi Froushani, S. M., Delirezh, N., Hobbenaghi, R., & Mosayebi, G. (2014). Synergistic effects of atorvastatin and all-trans retinoic acid in ameliorating animal model of multiple sclerosis. *Immunological Investigations*, *43*(1), 54–68.

Alatshan, A., Kovács, G. E., Aladdin, A., Czimmerer, Z., Tar, K., & Benkő, S. (2020). All-trans retinoic acid enhances both the signaling for priming and the glycolysis for activation of NLRP3 inflammasome in human macrophage. *Cells*, *9*(7), 1591.

Arnold, S. L., Amory, J. K., Walsh, T. J., & Isoherranen, N. (2012). A sensitive and specific method for measurement of multiple retinoids in human serum with UHPLC-MS/MS. *Journal of Lipid Research*, *53*(3), 587–598.

Azaïs-Braesco, V., & Pascal, G. (2000). Vitamin A in pregnancy: Requirements and safety limits. *American Journal of Clinical Nutrition*, *71*(5 Suppl), 1325S–1333S.

Balmer, J. E., & Blomhoff, R. (2002). Gene expression regulation by retinoic acid. *Journal of Lipid Research*, *43*(11), 1773–1808.

Banerjee, N. (2014). Neurotransmitters in alcoholism: A review of neurobiological and genetic studies. *Indian Journal of Human Genetics*, *20*(1), 20–31.

Barbalace, M. C., Malaguti, M., Giusti, L., Lucacchini, A., Hrelia, S., & Angeloni, C. (2019). Anti-inflammatory activities of marine algae in neurodegenerative diseases. *International Journal of Molecular Sciences*, *20*(12), 3061.

Behairi, N., Belkhelfa, M., Rafa, H., Labsi, M., Deghbar, N., Bouzid, N., et al. (2016). All-trans retinoic acid (ATRA) prevents lipopolysaccharide-induced neuroinflammation, amyloidogenesis and memory impairment in aged rats. *Journal of Neuroimmunology*, *300*, 21–29.

Bremner, J. D., Shearer, K. D., & McCaffery, P. J. (2012). Retinoic acid and affective disorders: The evidence for an association. *Journal of Clinical Psychiatry*, *73*(1), 37–50.

Cai, W., Wang, J., Hu, M., Chen, X., Lu, Z., Bellanti, J. A., et al. (2019). All trans-retinoic acid protects against acute ischemic stroke by modulating neutrophil functions through STAT1 signaling. *Journal of Neuroinflammation*, *16*(1), 175.

Calabrese, V., Scapagnini, G., Latteri, S., Colombrita, C., Ravagna, A., & Catalano, C. (2002). Long-term ethanol administration enhances age-dependent modulation of redox state in different brain regions in the rat: Protection by acetyl carnitine. *International Journal of Tissue Reactions*, *24*(3), 97–104.

Campo-Paysaa, F., Marlétaz, F., Laudet, V., & Schubert, M. (2008). Retinoic acid signaling in development: Tissue-specific functions and evolutionary origins. *Genesis (New York, N.Y.: 2000)*, *46*(11), 640–656.

Ciulla, M., Marinelli, L., Cacciatore, I., & Stefano, A. D. (2019). Role of dietary supplements in the management of Parkinson's disease. *Biomolecules*, *9*(7), 271.

Clemens, V., Regen, F., Le Bret, N., Heuser, I., & Hellmann-Regen, J. (2018). Retinoic acid enhances apolipoprotein E synthesis in human macrophages. *Journal of Alzheimer's Disease*, *61*(4), 1295–1300.

Cocco, S., Diaz, G., Stancampiano, R., Diana, A., Carta, M., Curreli, R., et al. (2002). Vitamin A deficiency produces spatial learning and memory impairment in rats. *Neuroscience*, *115*(2), 475–482.

Crandall, J., Sakai, Y., Zhang, J., Koul, O., Mineur, Y., Crusio, W. E., et al. (2004). 13-cis-retinoic acid suppresses hippocampal cell division and hippocampal-dependent learning in mice. *Proceedings of the National Academy of Sciences of the United States of America*, *101*(14), 5111–5116.

Das, S. S., Nair, S. S., Kavitha, S., Febi, J., & Indira, M. (2015). Atorvastatin improves Y-maze learning behaviour in nicotine treated male albino rats. *Pharmacology, Biochemistry, and Behavior*, *138*, 117–122.

Dumetz, F., Buré, C., Alfos, S., Bonneu, M., Richard, E., Touyarot, K., et al. (2020). Normalization of hippocampal retinoic acid level corrects age-related memory deficits in rats. *Neurobiology of Aging*, *85*, 1–10.

Fragoso, Y. D., Stoney, P. N., & McCaffery, P. J. (2014). The evidence for a beneficial role of vitamin A in multiple sclerosis. *CNS Drugs*, *28*(4), 291–299.

Haorah, J., Heilman, D., Knipe, B., Chrastil, J., Leibhart, J., Ghorpade, A., et al. (2005). Ethanol-induced activation of myosin light chain kinase leads to dysfunction of tight junctions and blood-brain barrier compromise. *Alcoholism, Clinical and Experimental Research*, *29*(6), 999–1009.

Harris, S., Johnson, S., Duncan, J. W., Udemgba, C., Meyer, J. H., Albert, P. R., et al. (2015). Evidence revealing deregulation of the KLF11-MAO A pathway in association with chronic stress and depressive disorders. *Neuropsychopharmacology: Official Publication of the American College of Neuropsychopharmacology*, *40*(6), 1373–1382.

Helfer, G., Ross, A. W., Russell, L., Thomson, L. M., Shearer, K. D., Goodman, T. H., et al. (2012). Photoperiod regulates vitamin A and Wnt/β-catenin signaling in F344 rats. *Endocrinology*, *153*(2), 815–824.

Huang, Z., Liu, Y., Qi, G., Brand, D., & Zheng, S. G. (2018). Role of vitamin A in the immune system. *Journal of Clinical Medicine*, *7*(9), 258.

Hummel, R., Ulbrich, S., Appel, D., Li, S., Hirnet, T., Zander, S., et al. (2020). Administration of all-trans retinoic acid after experimental traumatic brain injury is brain protective. *British Journal of Pharmacology*, *177*(22), 5208–5223.

Johnson, E. J., & Russell, R. M. (2010). Beta-carotene. In P. M. Coates, J. M. Betz, M. R. Blackman, et al. (Eds.), *Encyclopedia of dietary supplements* (2nd ed., pp. 115–120). London and New York: Informa Healthcare.

Kawahara, K., Suenobu, M., Ohtsuka, H., Kuniyasu, A., Sugimoto, Y., Nakagomi, M., et al. (2014). Cooperative therapeutic action of retinoic acid receptor and retinoid x receptor agonists in a mouse model of Alzheimer's disease. *Journal of Alzheimer's Disease: JAD*, *42*(2), 587–605.

Khan, M. S., Ikram, M., Park, J. S., Park, T. J., & Kim, M. O. (2020). Gut microbiota, its role in induction of Alzheimer's disease pathology, and possible therapeutic interventions: Special focus on anthocyanins. *Cell*, *9*(4), 853.

Khan, M. S., Muhammad, T., Ikram, M., & Kim, M. O. (2019). Dietary supplementation of the antioxidant curcumin halts systemic LPS-induced neuroinflammation-associated neurodegeneration and memory/synaptic impairment via the JNK/NF-κB/Akt signaling pathway in adult rats. *Oxidative Medicine and Cellular Longevity*, *2019*, 7860650.

Kirpich, I. A., McClain, C. J., Vatsalya, V., Schwandt, M., Phillips, M., Falkner, K. C., et al. (2017). Liver injury and endotoxemia in male and female alcohol-dependent individuals admitted to an alcohol treatment program. *Alcoholism, Clinical and Experimental Research*, *41*(4), 747–757.

Lane, M. A., & Bailey, S. J. (2005). Role of retinoid signalling in the adult brain. *Progress in Neurobiology*, *75*(4), 275–293.

Lee, H. P., Casadesus, G., Zhu, X., Lee, H. G., Perry, G., Smith, M. A., et al. (2009). All-trans retinoic acid as a novel therapeutic strategy for Alzheimer's disease. *Expert Review of Neurotherapeutics*, *9*(11), 1615–1621.

Lei, M., Wang, J. G., Xiao, D. M., Fan, M., Wang, D. P., Xiong, J. Y., et al. (2012). Resveratrol inhibits interleukin 1β-mediated inducible nitric oxide synthase expression in articular chondrocytes by activating SIRT1 and thereby suppressing nuclear factor-κB activity. *European Journal of Pharmacology*, *674*(2–3), 73–79.

Li, M., Tian, X., An, R., Yang, M., Zhang, Q., Xiang, F., et al. (2018). All-trans retinoic acid ameliorates the early experimental cerebral ischemia-reperfusion injury in rats by inhibiting the loss of the blood-brain barrier via the JNK/P38MAPK signaling pathway. *Neurochemical Research*, *43*(6), 1283–1296.

Liao, X., Ren, J., Wei, C. H., Ross, A. C., Cecere, T. E., Jortner, B. S., et al. (2015). Paradoxical effects of all-trans-retinoic acid on lupus-like disease in the MRL/lpr mouse model. *PLoS One*, *10*(3), e0118176.

McGrattan, A. M., McGuinness, B., McKinley, M. C., Kee, F., Passmore, P., Woodside, J. V., et al. (2019). Diet and inflammation in cognitive ageing and Alzheimer's disease. *Current Nutrition Reports*, *8*(2), 53–65.

Mizee, M. R., Nijland, P. G., van der Pol, S. M., Drexhage, J. A., van Het Hof, B., Mebius, R., et al. (2014). Astrocyte-derived retinoic acid: A novel regulator of blood-brain barrier function in multiple sclerosis. *Acta Neuropathologica*, *128*(5), 691–703.

Mohammadzadeh Honarvar, N., Harirchian, M. H., Abdolahi, M., Abedi, E., Bitarafan, S., Koohdani, F., et al. (2016). Retinyl palmitate supplementation modulates T-bet and interferon gamma gene expression in multiple sclerosis patients. *Journal of Molecular Neuroscience: MN*, *59*(3), 360–365.

Monacelli, F., Acquarone, E., Giannotti, C., Borghi, R., & Nencioni, A. (2017). Vitamin C, aging and Alzheimer's disease. *Nutrients*, *9*(7), 670.

Muhammad, T., Ikram, M., Ullah, R., Rehman, S. U., & Kim, M. O. (2019). Hesperetin, a citrus flavonoid, attenuates LPS-induced neuroinflammation, apoptosis and memory impairments by modulating TLR4/NF-κB signaling. *Nutrients*, *11*(3), 648.

Nair, S. S., Prathibha, P., Syam Das, S., Kavitha, S., & Indira, M. (2015). All trans retinoic acid (ATRA) mediated modulation of N-methyl D-aspartate receptor (NMDAR) and Kruppel like factor 11 (KLF11) expressions in the mitigation of ethanol induced alterations in the brain. *Neurochemistry International*, *83–84*, 41–47.

Nefedova, Y., Fishman, M., Sherman, S., Wang, X., Beg, A. A., & Gabrilovich, D. I. (2007). Mechanism of all-trans retinoic acid effect on tumor-associated myeloid-derived suppressor cells. *Cancer Research*, *67*(22), 11021–11028.

Nomoto, M., Takeda, Y., Uchida, S., Mitsuda, K., Enomoto, H., Saito, K., et al. (2012). Dysfunction of the RAR/RXR signaling pathway in the forebrain impairs hippocampal memory and synaptic plasticity. *Molecular Brain*, *5*, 8.

Oliveira, L. M., Teixeira, F., & Sato, M. N. (2018). Impact of retinoic acid on immune cells and inflammatory diseases. *Mediators of Inflammation, 2018*, 3067126.

Osadchiy, V., Martin, C. R., & Mayer, E. A. (2019). The gut-brain Axis and the microbiome: Mechanisms and clinical implications. *Clinical Gastroenterology and Hepatology: The Official Clinical Practice Journal of the American Gastroenterological Association, 17*(2), 322–332.

Priyanka, S. H., Syam Das, S., Thushara, A. J., Rauf, A. A., & Indira, M. (2018). All trans retinoic acid attenuates markers of neuroinflammation in rat brain by modulation of SIRT1 and NFκB. *Neurochemical Research, 43*(9), 1791–1801.

Priyanka, S. H., Thushara, A. J., Rauf, A., & Madambath, I. (2019). Alcohol induced NLRP3 inflammasome activation in the brain of rats is attenuated by ATRA supplementation. *Brain, Behavior, & Immunity – Health, 2*, 100024.

Qin, S., Colin, C., Hinners, I., Gervais, A., Cheret, C., & Mallat, M. (2006). System Xc- and apolipoprotein E expressed by microglia have opposite effects on the neurotoxicity of amyloid-beta peptide 1–40. *Journal of Neuroscience: The Official Journal of the Society for Neuroscience, 26*(12), 3345–3356.

Raverdeau, M., Breen, C. J., Misiak, A., & Mills, K. H. (2016). Retinoic acid suppresses IL-17 production and pathogenic activity of γδ T cells in CNS autoimmunity. *Immunology and Cell Biology, 94*(8), 763–773.

Reifen, R. (2002). Vitamin A as an anti-inflammatory agent. *Proceedings of the Nutrition Society, 61*(3), 397–400.

Rhinn, M., & Dollé, P. (2012). Retinoic acid signalling during development. *Development (Cambridge, England), 139*(5), 843–858.

Riccio, P., & Rossano, R. (2018). Diet, gut microbiota, and vitamins D + A in multiple sclerosis. *Neurotherapeutics: The Journal of the American Society for Experimental NeuroTherapeutics, 15*(1), 75–91.

Ridge, J. P., Ho, A. M., Innes, D. J., & Dodd, P. R. (2008). The expression of NMDA receptor subunit mRNA in human chronic alcoholics. *Annals of the New York Academy of Sciences, 1139*, 10–19.

Ross, C. A. (2010). Vitamin A. In P. M. Coates, J. M. Betz, M. R. Blackman, et al. (Eds.), *Encyclopedia of dietary supplements* (2nd ed., pp. 778–791). Informa Healthcare.

Saboor-Yaraghi, A. A., Harirchian, M. H., Mohammadzadeh Honarvar, N., Bitarafan, S., Abdolahi, M., Siassi, F., et al. (2015). The effect of vitamin A supplementation on FoxP3 and TGF-β gene expression in Avonex-treated multiple sclerosis patients. *Journal of Molecular Neuroscience, 56*(3), 608–612.

Sahin, M., Karauzum, S. B., Perry, G., Smith, M. A., & Aliciguzel, Y. (2005). Retinoic acid isomers protect hippocampal neurons from amyloid-beta induced neurodegeneration. *Neurotoxicity Research*, 7(3), 243–250.

Schuster, G. U., Kenyon, N. J., & Stephensen, C. B. (2008). Vitamin A deficiency decreases and high dietary vitamin A increases disease severity in the mouse model of asthma. *Journal of immunology (Baltimore, Md.: 1950), 180*(3), 1834–1842.

Shearer, K. D., Stoney, P. N., Morgan, P. J., & McCaffery, P. J. (2012). A vitamin for the brain. *Trends in Neurosciences, 35*(12), 733–741.

Shen, H., Luo, Y., Kuo, C. C., Deng, X., Chang, C. F., Harvey, B. K., et al. (2009). 9-Cis-retinoic acid reduces ischemic brain injury in rodents via bone morphogenetic protein. *Journal of Neuroscience Research, 87*(2), 545–555.

Siddikuzzaman, & Grace, V. (2013). Antioxidant potential of all-trans retinoic acid (ATRA) and enhanced activity of liposome encapsulated ATRA against inflammation and tumor-directed angiogenesis. *Immunopharmacology and Immunotoxicology, 35*(1), 164–173.

Sreekala, S., & Indira, M. (2009). Impact of co administration of selenium and quinolinic acid in the rat's brain. *Brain Research, 1281*, 101–107.

Stebbins, M. J., Lippmann, E. S., Faubion, M. G., Daneman, R., Palecek, S. P., & Shusta, E. V. (2018). Activation of RARα, RARγ, or RXRα increases barrier tightness in human induced pluripotent stem cell-derived brain endothelial cells. *Biotechnology Journal, 13*(2). https://doi.org/10.1002/biot.201700093.

Swathy, S. S., Panicker, S., Nithya, R. S., Anuja,, M. M., Rejitha, S., & Indira, M. (2010). Antiperoxidative and antiinflammatory effect of *Sida cordifolia* Linn. on quinolinic acid induced neurotoxicity. *Neurochemical Research, 35*(9), 1361–1367.

Sweeney, M. D., Sagare, A. P., & Zlokovic, B. V. (2018). Blood-brain barrier breakdown in Alzheimer disease and other neurodegenerative disorders. *Nature Reviews Neurology, 14*(3), 133–150.

Tafti, M., & Ghyselinck, N. B. (2007). Functional implication of the vitamin A signaling pathway in the brain. *Archives of Neurology, 64*(12), 1706–1711.

Takamura, R., Watamura, N., Nikkuni, M., & Ohshima, T. (2017). All-trans retinoic acid improved impaired proliferation of neural stem cells and suppressed microglial activation in the hippocampus in an Alzheimer's mouse model. *Journal of Neuroscience Research, 95*(3), 897–906.

Tan, K. P., Kosuge, K., Yang, M., & Ito, S. (2008). NRF2 as a determinant of cellular resistance in retinoic acid cytotoxicity. *Free Radical Biology & Medicine, 45*(12), 1663–1673.

Theus, M. H., Sparks, J. B., Liao, X., Ren, J., & Luo, X. M. (2017). All- trans-retinoic acid augments the histopathological outcome of neuroinflammation and neurodegeneration in lupus-prone MRL/lpr mice. *Journal of Histochemistry and Cytochemistry: Official Journal of the Histochemistry Society, 65*(2), 69–81.

Udemgba, C., Johnson, S., Stockmeier, C. A., Luo, J., Albert, P. R., Wang, J., et al. (2014). The expression of KLF11 (TIEG2), a monoamine oxidase B transcriptional activator in the prefrontal cortex of human alcohol dependence. *Alcoholism, Clinical and Experimental Research, 38*(1), 144–151.

Vallés, S. L., Blanco, A. M., Pascual, M., & Guerri, C. (2004). Chronic ethanol treatment enhances inflammatory mediators and cell death in the brain and in astrocytes. *Brain Pathology (Zurich, Switzerland), 14*(4), 365–371.

van Neerven, S., Regen, T., Wolf, D., Nemes, A., Johann, S., Beyer, C., et al. (2010). Inflammatory chemokine release of astrocytes in vitro is reduced by all-trans retinoic acid. *Journal of Neurochemistry, 114*(5), 1511–1526.

Wang, T. W., Zhang, H., & Parent, J. M. (2005). Retinoic acid regulates postnatal neurogenesis in the murine subventricular zone-olfactory bulb pathway. *Development (Cambridge, England), 132*(12), 2721–2732.

Wang, X. J., Hayes, J. D., Henderson, C. J., & Wolf, C. R. (2007). Identification of retinoic acid as an inhibitor of transcription factor Nrf2 through activation of retinoic acid receptor alpha. *Proceedings of the National Academy of Sciences of the United States of America, 104*(49), 19589–19594.

Wheelwright, M., Kim, E. W., Inkeles, M. S., De Leon, A., Pellegrini, M., Krutzik, S. R., et al. (2014). All-trans retinoic acid-triggered antimicrobial activity against *Mycobacterium tuberculosis* is dependent on NPC2. *Journal of Immunology, 192*(5), 2280–2290.

Yin, K. J., Fan, Y., Hamblin, M., Zhang, J., Zhu, T., Li, S., et al. (2013). KLF11 mediates PPARγ cerebrovascular protection in ischemic stroke. *Brain: A Journal of Neurology, 136*(Pt 4), 1274–1287.

Yin, X. P., Zhou, J., Wu, D., Chen, Z. Y., & Bao, B. (2015). Effects of that ATRA inhibits Nrf2-ARE pathway on glial cells activation after intracerebral hemorrhage. *International Journal of Clinical and Experimental Pathology, 8*(9), 10436–10443.

Zeng, J., Chen, L., Wang, Z., Chen, Q., Fan, Z., Jiang, H., et al. (2017). Marginal vitamin A deficiency facilitates Alzheimer's pathogenesis. *Acta Neuropathologica, 133*(6), 967–982.

CHAPTER 28

Fluoride excess and neuroinflammation

Namrata Dudha[a], Lomas Kumar Tomar[b], Charu Tyagi[a], Yugandhar P. Reddy[c], and Varun Kumar Sharma[a]

[a]Department of Biotechnology and Microbiology, School of Sciences, Noida International University-NIU, Gautam Budh Nagar, Uttar Pradesh, India
[b]Department of Biotechnology, VSPG(PG) College, CCS University, Meerut, Uttar Pradesh, India
[c]Department of Zoology, Government Degree College, Kurnool, Andhra Pradesh, India

Abbreviations

4-HNE 4-hydroxynonenal
% percentage
AA arachidonic acid
CNS central nervous system
COX cyclooxygenase
DHA docosahexaenoic acid
ERKs extracellular signal-regulated kinases
HSP70 heat shock protein 70
IL interleukin
IQ intelligence quotient
iNOS inducible nitric oxide synthase
JNKs Jun amino-terminal kinases
MAPK mitogen-activated protein kinase
Mg milligram
PPM parts per million
PKC protein kinase C
PGE2 prostaglandin E2
ROS reactive oxygen species
RNS reactive nitrogen species
SOD superoxide dismutase
TNF-α tumor necrosis factor
VEGF vascular endothelial growth factor
WHO World Health Organization

Introduction

Fluoride is a natural mineral, which is present in various foods and drinking water. It plays a pivotal role in developmental process; however, excess intake of fluoride is associated with various pathological conditions, including fluorosis, infertility among women, premature aging, cancer as well as certain neurological conditions. Evidence indicates that

Vitamins and Minerals in Neurological Disorders
https://doi.org/10.1016/B978-0-323-89835-5.00005-3

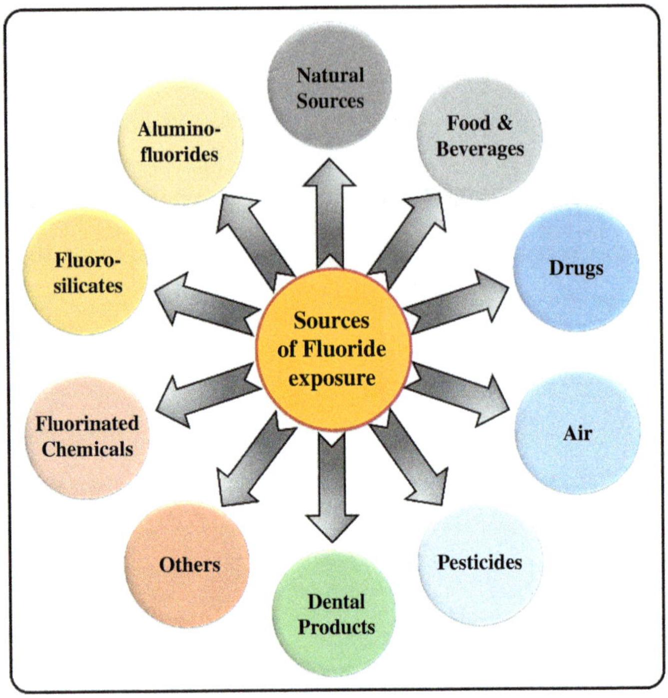

Fig. 1 Major sources of fluoride exposure. Fluoride naturally exists in a wide variety of natural components (air and water) or manmade products (drugs, dental products, and manufacturing processes) that are the major sources of fluoride exposure.

excessive fluoride exposure may be due to various metabolic diseases and neurological damage leading to neurodegeneration (Dec et al., 2017; Goschorska et al., 2018). Fig. 1 gives a snapshot of the various sources, which may lead to fluoride exposure. Utilization of drinking water with excess fluoride (1–12 ppm) is linked to dental and skeletal fluorosis as well as damaging the central nervous system of suffering patients (WHO, 1999, 2001). Interestingly, limited amount of fluoride (about 0.7 mg/L) in drinking water prevents tooth decay. Therefore, many countries such as Australia, Brazil, Canada, Chile, Ireland, New Zealand, United Kingdom, and the United State of America added fluoride to many community water supplies (Grandjean, 2019; USEPA, 2010). Further, observation based on the fluoride-exposed rat model revealed that fluoride effects nerve cells in animal model (Mahaboob & Saumya, 2013). Previous observations suggest that sodium fluoride exposure induces hyperactivity and cognitive deficits along with hippocampus damage, and also promotes neurotoxicity and neuroinflammation via modulating various signaling pathways (Mullenix, Denbesten, Schunior, & Kernan, 1995; Shanmugam, Abdulla, Yakulasamy, Selvaraj, & Mathan, 2018).

Sources of fluoride exposure

Natural sources

Fluoride is naturally present in the form of fluorspar, fluorapatite, and cryolite, which is mainly distributed in the lithosphere (rocky outer layer of earth), and 13th most frequent element in the earth's crust. It is also naturally found in water, viz., sea water (1.2–1.4 mg/L), ground water (about 67 mg/L), and surface water (less than <0.1 mg/L). Hence, fluoride is found in a wide range of food and drinking items, including fish and tea (Näsman, Ekstrand, Granath, Ekbom, & Fored, 2013; WHO, 2001).

Fluoride in food and beverages

Almost, all foodstuffs contain at least trace amounts of fluoride, as it is ubiquitous in nature (Lakdawala & Punekar, 1973). Fluoride enters human food-and-beverage chain in increasing amounts through the consumption of tea, wheat, spinach, cabbage, carrots, and other food items. The fluoride in these food items presumably results from the use of soil or fertilizer-borne fluoridated water for food and beverage processing (Sengupta & Pal, 1937). For example, fluoride concentration ranging from 3.2 to 400 mg/kg has been observed in leaves of the tea plant; however, in general, tea contains fluoride about 8.6 mg/L, which mainly depends on infusion time, and the concentration of fluoride could vary with different brands prepared using distilled water, and the observed average fluoride concentration is about 1.5 mg/L (Malinowska, Inkielewicz, Czarnowski, & Szefer, 2008). Additionally, tea plants possess high fluoride uptake capacity, and 97% of it gets accumulated in leaves. The fluoride content of tea leaves is about 1000 times the soluble fluoride content of soil and 2–7 times the total fluoride content in soil (Gupta & Sandesh, 2012; Ruan, Ma, Shi, & Han, 2004).

Drugs

A set of drugs contain a trace of fluoride that helps in the treatment of different pathological diseases, in particular bone-associated damages, such as bone fracture, osteoporosis, and arthritis cases. For instance, sodium-fluoride (treatment for osteoporosis), niflumic acid (treatment for rheumatoid arthritis), Proflo (fluoride-containing mouth rinse), and other drugs contain fluoride traces (Ghosh, Mukherjee, Ghosh, & Saha, 2013; Pak, Sakhaee, Piziak, & Peterson, 1994). Prolonged consumption of these drugs increases the risk of fluoride-associated chronic disorders.

Air

The traces of fluoride are also identified in nature air in the form of hydrogen fluoride, particulate fluorides, or fluorine gas, which is released to the atmosphere by natural sources such as volcanoes and by a number of anthropogenic sources. Additionally, excessive fluoride in air can also be released during industrial manufacture process, such as inorganic fluoride compounds used in the production of aluminum, glassfiber, steel, and phosphate fertilizers. Additional production unit-associated construction sector plants also release fluoride in environment during the production of tiles, ceramics, and bricks (Ghosh et al., 2013; Jagtap, Yenkie, Labhsetwar, & Rayalu, 2012).

Pesticides

Pesticides such as herbicides, fungicides, and insecticides have played an important role in minimizing crop loss and promoting crop yields. Moreover, pesticides serve to protect human health from parasitic (vector borne) diseases such as malaria, dengue, and other disease. Fluoro-DDT was one of the faster-acting congeners of DDT used to control malaria during the 20th century (Ogawa, Tokunaga, Kobayashi, Hirai, & Shibata, 2020). Additionally, cryolite and sulfuryl fluoride are the two pesticides that are regulated for their contribution to the residue of inorganic fluoride in foods. Fluoride ions are also the degradation products of sulfuryl fluoride (EPA-U.S. Environmental Protection Agency, 1992). Thus, the recent evaluation of the dietary risk of sulfuryl fluoride use on food takes into account the additional exposure to fluoride from cryolite. Sulfuryl fluoride is also regulated as a compound with its own toxicologic characteristics.

Dental products and supplements

Fluoridated dental products include dentifrices (toothpastes, powders, liquids, and other preparations for cleaning teeth) for home use and various gels and other topical applications for use in dental health. Highly significant associations were found between estimated fluoride ingestion from toothpaste and fluorosis. In general, the presence of fluoride could vary in different products such as fluoride content from the range of 1000–1500 mg/kg present in toothpaste, in the form of sodium fluoride and sodium monofluorophosphate. Contaminated raw materials like calcium carbonate, talc and chalk-like products that have been used for manufacturing toothpaste, may have high fluoride concentration that can lead to presence of fluoride content as high as 1000 ppm in the product as well (Fawell et al., 2006; Ghosh et al., 2013).

Fluorinated organic chemicals

Organic fluorine is present in many pharmaceuticals, consumer products, and pesticides. Interestingly, organic fluorine is not as easily displaced from the alkyl carbon and is much more lipophilic than the hydrogen substitutes. Naturally in biological condition, organofluoride compounds only rarely release fluorine (fluoride ions, F^{-}) and seldom causes fluoride poisoning. Fluoride ions is the major cause of fluoride poising; organo-fluoride compounds (e.g., methoxyfluoride) contain strong bond between carbon and fluorine ion and does not release free fluoride ions. As a result, organo-fluoride compounds have been used in the production of drugs and other products such as Teflon and chlorofluorocarbon (Ogawa et al., 2020; Okazoe, 2009).

Aluminofluorides, fluoroaluminate, and beryllofluorides

Aluminofluorides and fluoroaluminate are complexes of aluminum and fluoride, which are present in the form of AlF3 or AlF-4. Beryllofluorides contain beryllium and fluoride and denominate as BeF-3. The origin of these metallocomplexes of fluoride is possible when these metals exist in the same environment as fluoride (Strunecka & Patocka, 2002). Cryolite (Na3AlF6) is one of the most common forms of the inorganic salt of aluminum and fluoride. In the late 1800s and during the 1900s Industrial development across the world led to increasing air pollution by factory effluents. This resulted in adverse effects on the environment and a higher prevalence of acid rain that led to increased concentration of metal ions such as aluminum not only in the soil but the overall environment. Aluminofluorides is one of the key sources of fluoride and aluminum, which enters the ecosystem and is ingested during the consumption of water, tea, and other daily use products like medications, deodorants, cosmetics, and glassware. Similarly, fluorosilicates such as fluorosilicic acid (H2SiF6) and sodium fluorosilicate (Na2SiF6) are also a major source of fluoride in drinking water (Strunecka & Patocka, 2002; Wong, Fung, & Carr, 2003).

Other sources

Fluoride source in drinking water is also commercially available in the form of bottled mineral water, which is a most common source of excessive fluoride ingestion. It has been observed that the majority of bottled mineral water producers do not display the fluoride content on their labels even though it contains high concentrations of fluoride. Hence, it is recommended that sanitary regulatory system for the quality control of the level of fluoride in the bottled mineral waters marketed is needed (Villena, Borges, & Cury, 1996).

Fluoride exposure-associated major health burden

Fluorosis: Dental and skeletal fluorosis

Prolonged excessive fluoride intake has been linked to cause fluorosis and endemic in different region of the world. Dental fluorosis in children and skeletal fluorosis in adults are most frequent. Chronic fluoride toxicity denotes a major hazard to human health. Excessive intake increases the risk of fluorosis and other disease, during the ingestion of fluoride-contaminated food and water; above the 1.5 ppm of fluoride increases the severity of diseases and one of the major public health problems. Fluorosis can be classified on the basis of specific target site of fluorosis in suffering patients, such as dental fluorosis, skeletal classes, and non-skeletal fluorosis (Fig. 2; DenBesten & Li, 2011; Grandjean, 2019).

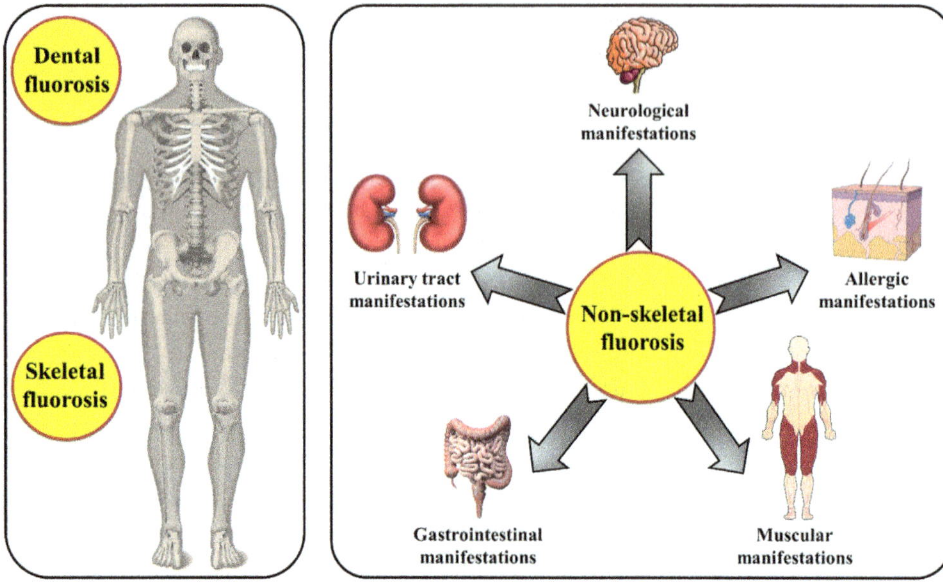

Fig. 2 Different physiological manifestations of fluorosis. Excessive fluoride intake induces different pathological conditions, include fluorosis-associated dental, skeletal, and non-skeletal complications.

Dental fluorosis is mainly observed in the early childhood years and damages the tooth-forming cells, leading to a defect in the enamel. Severe form of dental fluorosis manifests as heavily stained, pitted, and friable enamel that can result in the loss of dental function (DenBesten & Li, 2011). Skeletal fluorosis is a result of excessive intake of sodium fluoride, which leads to the development of musculoskeletal disease with a high morbidity. Severity of skeletal fluorosis-associated pathologies includes severe pain in the backbone, joints, hip region, stiffness of the backbone, immobile/stiff joints, increased density of bones, besides calcification of ligaments, constriction of vertebral canal and intervertebral foramen pressure on nerves, and paralysis (Srivastava & Flora, 2020).

Additionally, induced intake of fluoride is also a major cause of non-skeletal-associated disease, which affects different physiological systems in the human body such as urinary system, reproductive system, gastric system, and central nervous system, and other manifestation associated with muscles and skin, including skin allergies (Kumar, Sood, Arora, Singh, et al., 2010; Valdez-Jiménez, Soria Fregozo, Miranda, Gutiérrez, & Pérez, 2011; Whitford, 2011).

Impact of fluoride in neurotoxicity and neurobehavioral

Fluoride contamination is associated with neurodevelopmental disorders, including learning disabilities, autism spectrum disorder, and attention deficit hyperactivity disorder, intellectual disabilities, and developmental delays. Observation based on studies in animals and human population revealed that excessive fluoride has adverse impacts on developing brain and induces neurotoxicity. A hypothesis proposed by Isaacson and his research group proposes that fluoride can modulate the structure and function of the brain, resulting in the suppression of metabolic activities (Isaacson, 2008). Other observations based on epidemiological and animal-based observations indicate that fluoride damages intelligence capabilities, reduces the learning potential, and impairs memories among children and animals (Mullenix et al., 1995). High fluoride content in drinking water badly impacts on nervous system without affecting primary symptom linked to fluorosis.

Evidence indicates that fluoride can get transferred from mother to infant through placenta; also, small amount of fluoride transfer takes place from mother to baby via breastfeeding (Green, Lanphear, Hornung, Flora, et al., 2019; Till et al., 2020). Accumulation of fluoride in brain alters the expression profile of specific proteins and neurotransmitters in the central nervous system, which badly impacts on learning and memory processes. Meta-analysis of different epidemiologic studies indicates that fluoride exposure from drinking water has been associated with children's intelligence. Additionally, mean intelligence quotient (IQ) level of students exposed to high-fluoride-level drinking water was significantly lower than that of the students exposed to a lower-fluoride-level drinking water. It concludes that high fluoride exposure is directly associated with lowering the IQ level in children (Xiang et al., 2003). Children from the reason of endemic fluorosis area have more time higher risk of developing lower IQ level compared to those who live in a low-fluoridated area region (Tang, Du, Ma, Jiang, & Zhou, 2008).

Mechanism involved in fluoride toxicity

Fluoride toxicity is directly and indirectly involved in different biological activities. Fluoride is able to freely migrate through the biological membranes via passive diffusion. It is rapidly absorbed into the blood and bonds with plasma by which it can reach in any tissues of the body. Finally, it accumulates in teeth, bone, and other soft tissues. Excessive fluoride induces oxidative stress, which damages different tissues (Thangapandiyan & Miltonprabu, 2013). Thangapandiyan and his colleagues reported that fluoride interacts and binds Ca^+ and reduces calcium in host, called as hypocalcemia, and promotes osteoid formation. Further, it leads to impair various cellular processes such as oxidative phosphorylation, glycolysis, coagulation, and neurotransmission. The mechanism of fluoride toxicity in oxidative stress is demonstrated in Fig. 3.

Fluoride influences various enzyme regulation and endocrine gland functions, which control the oxidant or antioxidant system. Negatively charged fluoride promotes genotoxicity by direct interaction with DNA and induces free radical production, resulting in DNA damages. In addition to this, it also suppresses DNA polymerase and negatively controls DNA replication or repair process (Thangapandiyan & Miltonprabu, 2013; Wang et al., 2004).

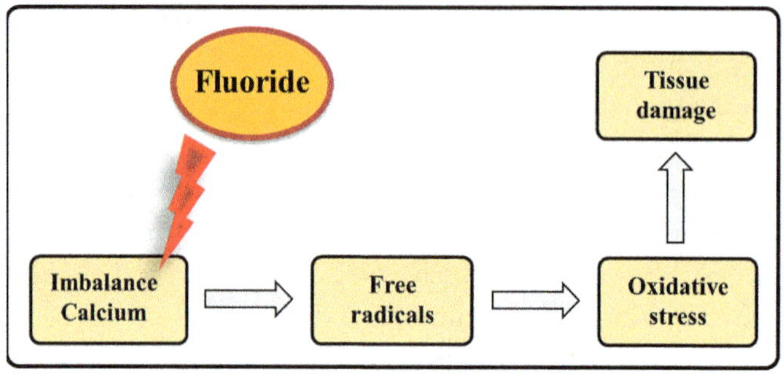

Fig. 3 Flowchart explaining fluoride-mediated toxicity in tissues. Fluoride induces free radical production via modulating the regulation of calcium ions (Ca^{2+}) ion and promotes tissue damage.

Neurochemical effects

Prolonged excessive fluoride exposure exerts long-term synergistic function on central nervous system (CNS) and promotes neuronal destruction and synaptic injury, by various mechanisms such as free radical production and lipid peroxidation (Dec et al., 2017). The exact mechanism of fluoride-mediated neurotoxic effect and complete brain damage process has not been fully elucidated, even though recent research evidence elaborated that fluoride plays an important role in neuropathological conditions in human and animals. Various neurotoxic metals, including mercury, lead, aluminum, and iron, damage neural elements in the central nervous system by an excitotoxic mechanism. Pro-inflammatory cytokines and excitotoxic amino acids are regulated by these metals and activate microglial cells (a resident macrophages of the brain). Similarly, the common mechanism is also involved in fluoride-mediated neurotoxicity by promoting the generation of free radicals, suppressing the regulation of antioxidant enzymes, mitochondrial energy enzymes, and glutamate transporters (Dec et al., 2017; Farina, Avila, Rocha, & Aschner, 2013; Tan, Norhaizan, Liew, & Rahman, 2008).

Free radical generation

Fluoride induces free radicals production and plays an active role as neurotoxic agent. Excessive production of fluoride-mediated free radicals production process is known as excitotoxicity. Excitotoxicity is also promoted by lipid peroxidation and the release of free radicals and malondialdehyde, which promote impaired dendrites and synaptic connections and lead to neuronal destruction. Additionally, free radicals also regulate other processes that trigger excitotoxicity; free radicals impair glutamate removal process in microglia, and excessive deposition of glutamate regulates microglia activation. Another mechanism is that fluoride might increase brain free radical generation in the brain and regulate protein kinase C activation (Fig. 4). Activation of protein kinase C regulates glutamate mediated excitotoxicity (Shuhua, Ziyou, Ling, Fei, & Sun, 2012; Yan et al., 2013).

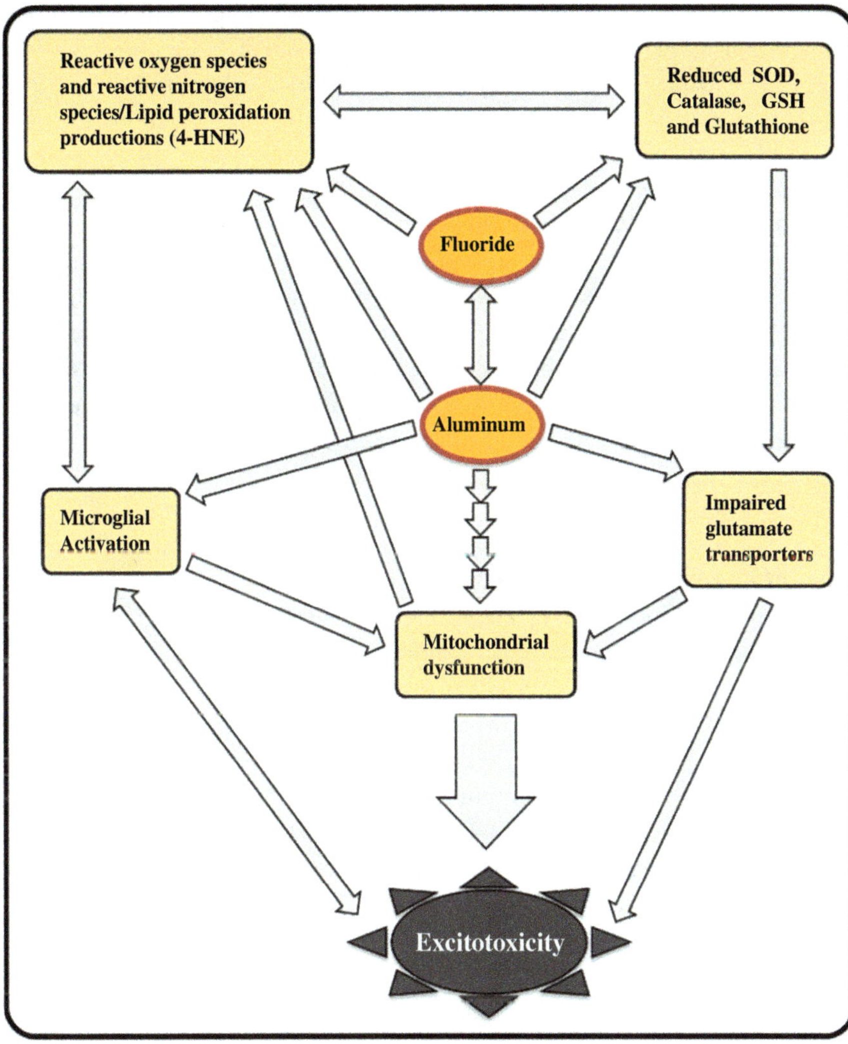

Fig. 4 Fluoride and aluminum promotes neurodegeneration via inducing excitotoxicity. The flowchart highlights fluoride-aluminum promoting excitotoxicity via inducing oxidative stress, impaired glutamate regulation, mitochondrial dysfunction, and microglial activation. *(Adopted and modified from Blaylock, R. L., & Ridgeland, M. S. (2004). Excitotoxicity: A possible central mechanism in fluoride neurotoxicity.* Fluoride, 37, *301–304.)*

Inhibition of antioxidant enzymes

Fluoride promotes the depletion of antioxidant enzymes and is responsible for fluoride toxicity. Vani and Reddy have reported that fluoride suppresses antioxidant molecule and enzymes such as superoxide dismutase (SOD), glutathione reductase, glutathione peroxidase, catalase, and glutathione. Fluoride not only induces free radical-mediated injuries, but also promotes excitotoxicity by triggering the excitotoxic process by producing reactive oxygen species, nitrogen species, and lipid peroxidation, and by suppressing antioxidant enzyme activities. Hence, fluoride-mediated suppression of antioxidant enzyme enhances neurotoxicity. Fluoride and other excitotoxins impose the harmful impact in brain by suppressing hormonal regulation (Goschorska et al., 2018; Vani & Reddy, 2000).

Inhibition of mitochondrial energy enzymes

Mitochondrial enzymes regulation is important in cell survival and its function. Fluoride has been identified to suppress the cellular energy-producing enzymes, including mitochondrial electron transport enzymes. These enzyme play an important role in glycolytic and Kreb's cycle. Therefore, fluoride can control mitochondrial function. Observation indicates that 20 mg/kg of fluoride concentration suppresses the antioxidant enzymes and energy-generating enzymes in female mice. Even low concentration of fluoride can inhibit both enzymes and play an active role in excitotoxicity (Mahaboob & Saumya, 2013; Vani & Reddy, 2000).

Inhibition of glutamate transporters

As discussed earlier, fluoride-mediated free radical production impairs glutamate removal process in microglia and excessive deposition of glutamate regulates microglial activation, which triggers excitotoxicity. Glutamate transport protein plays an important role in central nervous system. Impairment of these proteins diminishing the function of nervous system leads to neurodevelopmental disorders and neurodegenerative diseases. Various neurotoxins such as mercury, aluminum, iron, cytokines, eicosanoids (prostaglandin E2-PGE2), and 4-hydroxynonenal (4-HNE) can alter the neurological function. García-López and his colleagues observed that fluoride exposure reduces glutamine uptake and mobilization in glia cells by a disruption of glutamatergic transmission (García-López, Hernández-Castillo, Hernández-Kelly, Olivares-Bañuelos, & Ortega, 2020). Fluoride promotes oxidative stress, glial activation, and inflammation in the central nervous system and, leads to neurodegeneration, impaired learning ability, memory dysfunction, and behavioral imbalance (Dec et al., 2017).

Neuroinflammation

Neuroinflammation is a host protective mechanism that helps to neutralize the cellular insult and promotes the restoration of normal structure and function of brain tissues from uninjured area, destroys affected cells, and repairs the extracellular matrix and is one of the hallmarks of major central nervous system-related diseases (Correale & Villa, 2004). Brain tissues contain different types of cells such as glial cells, the microglia, astrocytes, and oligodendrocytes, which cover about 70% of cell population. These cells actively participate in neuroinflammatory response in brain. Neuroinflammation is characterized by the activation of astrocytes and microglial cells (Takeuchi et al., 2005). During the brain tissue injury or infection, activation and transformation of microglial cells take place, which includes cell migration, cell proliferation, cytokine/chemokine production, leads to the stimulation of phospholipases A2 and cyclooxygenases. These process lead to the breakdown of membrane glycerophospholipids with the release of arachidonic acid (AA) and docosahexaenoic acid (DHA). Neural membrane glycerophospholipid is a precursor of prostaglandins and leukotrienes, which are regulated by multiple forms of PLA2, cyclooxygenases (COX-1 and COX-2), and lipoxygenases (LOX) (Bachiller, Jiménez-Ferrer, Paulus, Yang, et al., 2018; Harry & Kraft, 2008).

During neuroinflammation, cyclooxygenase enzymes oxidize arachidonic acid to prostaglandin H2, which is major precursor of several prostaglandins, thromboxanes, and prostacyclins (Fig. 5). In an alternative pathway, lipoxygenases enzymes oxidize arachidonic acid to leukotrienes A4, which is a major precursor of leukotrienes subtypes B4,

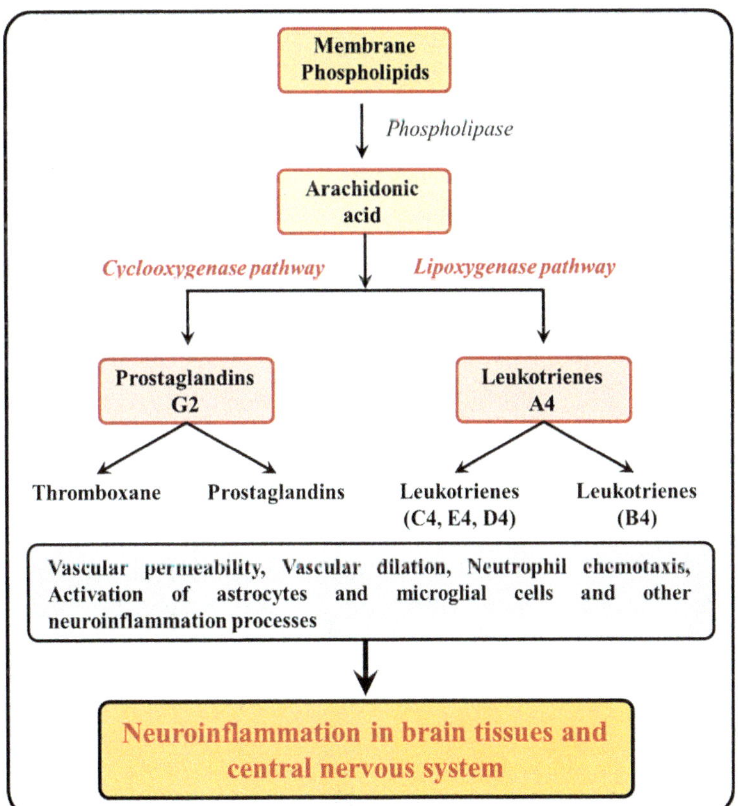

Fig. 5 Mechanism of neuroinflammation in brain tissues and central nervous system. **The flowchart explains the mechanism of neuroinflammation, in which** glycerophospholipid is the source of prostaglandins, thromboxane, and leukotrienes, which regulated by cyclooxygenases and lipoxygenases. Induced production of prostaglandins, thromboxane, and leukotrienes facilitates vascular permeability, neutrophil activation, and chemotaxis, which lead to neuroinflammation in brain tissues and central nervous system.

C4, E4, and D4. Activation of microglial cells via these signaling pathways rapidly modulates the microglial function, which promotes the migration of these cells toward site of injury and promotes the elimination of dead cell via phagocytosis. Additionally, during this process, the modulation of complement proteins, adhesion molecules, and inflammatory cytokines also occurs to promote neuroinflammation (Farooqui, Ong, & Horrocks, 2006). Inflammation promotes leukocytic infiltration processes by which polymorphonuclear cells (neutrophils) in acute inflammation and different mononuclear cells (macrophages, lymphocytes, plasma cells) migrate to damage site in the brain. This process compromises the function of brain tissues and highlights the role of astrocytes and microglia in the initiation of central nervous system inflammation (Bachiller et al., 2018; Farooqui et al., 2006; Hanna & Hafez, 2018; Harry & Kraft, 2008).

Effect of sodium fluoride on free radical enzymes and oxidative stress

As discussed in the previous section, fluoride mediates free radical production and suppression of oxidative enzyme function, which promotes oxidative stress in brain tissues and central nervous system. Fluoride-mediated oxidative stress induces the neurotoxicity by destroying the homeostasis in central nervous system. Generation of free radicals molecules promotes oxidation process and generates oxidative stress condition (oxidative injury) by producing reactive oxygen species (ROS). ROS damages the cell and tissues by promoting oxidation process, results in the imbalance between formation and neutralization of pro-oxidants. Oxidative stress regulates various cellular and molecular damages in different tissues, including brain tissues, by inducing the lipid peroxidation process, thus leading to neurodegeneration. In contrast, antioxidant molecule helps to maintain the balance between the free oxygen production and consumption, which helps to reduce oxidative stress in healthy central nervous system (Nita & Grzybowski, 2016; Yan et al., 2013).

Generation of ROS is an essential cellular process that helps to maintain regular functions of cells such as mitochondrial respiratory chain, phagocytosis, arachidonic acid metabolism, ovulation, and fertilization. In different pathological conditions, it has been observed induced ROS production and hyperactivation of these cellular processes. Furthermore, hydrogen peroxide, alkoxy radical peroxyl, hydroxyl radicals, and hypochlorous acid are the key super reactive oxygen species. Some of the other non-oxygen species, including reactive nitrogen species (RNS) such as peroxynitrite and nitric oxide (NO), also show similar free radical toxicity as ROS in both animal-based observations and in human living in areas of endemic fluorosis region. Observation indicates that excessive fluoride exposure also promotes induced production of ROS and RNS in microglial cells (Nita & Grzybowski, 2016; Shuhua et al., 2012; Yamaguti, Ganzerla, Nogueira, Nesadal, & Nicolau, 2008).

Observations point that fluoride is a highly reactive ionized element which regulates oxygen metabolism and encourages the oxygen free radical production. Excessive fluoride exposure endorses the production of ROS and free radicals by which fluoride suppresses the various physiological functions in brain and develops neurodegenerative conditions (Yamaguti et al., 2008; Yan et al., 2013). ROS act as secondary messenger, which modulate various intracellular signaling pathways such as JNK, NF-κB, MAPK, and other pathways (Fig. 6).

Mitogen-activated protein kinase (MAPK) pathway is also known as MAPK/ERK pathway or Ras–Raf–MEK–ERK pathway. MAPK is a complex protein, which is a group of three proteins, namely, extracellular signal-regulated kinases (ERKs), Jun amino-terminal kinases (JNKs), and stress-activated protein kinases (p38/SAPKs). In neuronal, growth factor activates ERK1/2 and promotes cellular processes such as cell

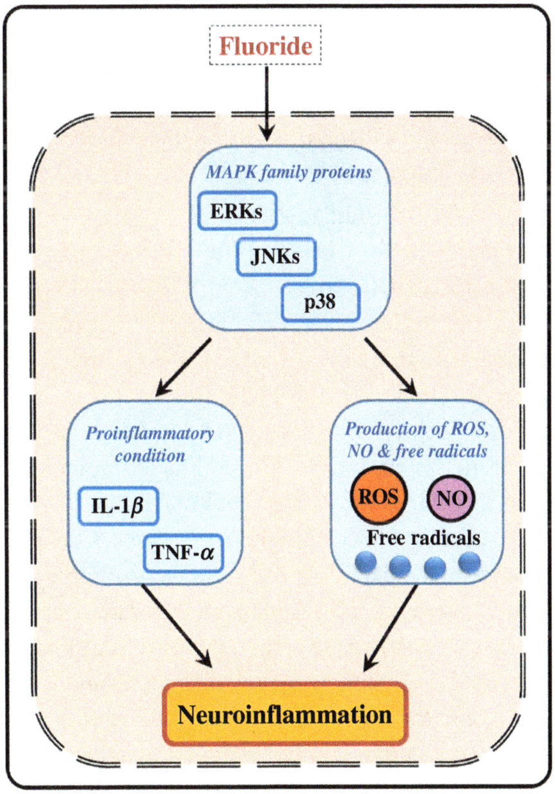

Fig. 6 Fluoride regulates neuroinflammation via modulating MAPK signaling pathways. Diagram demonstrates the fluoride-mediated aberrant regulation of ERKs, JNK, and p38, which indicate induced MAPK signaling pathways. Overall induction in MAPK pathways promotes ROS and NOS production, and also regulates IL-1β and TNF-α, which together leads to regulate inflammatory conditions in brain tissues and central nervous system.

proliferation, differentiation, and development. In spite of this, JNK and p38 signaling cascades are activated during environmental stress and inflammatory cytokines. Activation of these MAPK family proteins promote cell apoptosis, which leads to neuronal cell death. Additionally, MAPKs also induce the production of free radicals and ROS. Observations indicate that fluoride induces phosphorylation of the MAPK family proteins and promotes neuroinflammation in microglial cell, thus resulting in neuronal cell damages (Cargnello & Roux, 2011; Chen et al., 2017; Yan et al., 2013).

Microglia serve as a critical role in neural development and its maintenance. Iba-1 is a marker that is involved in microglia activation. Earlier observations indicate that fluoride induces intracellular Iba-1 expression level and promotes the activation of microglial cells. In the continuation of the observation, fluoride-mediated activated microglial cells produce and release pro-inflammatory cytokines (TNF-α, IL-1β, and IL-6), which promotes neuroinflammation by disrupting the brain–blood barrier (Chen et al., 2017; Yan et al., 2013). Exposure of fluoride has also been identified to regulate NF-κB signaling pathway in microglial cells and release a set of different pro-inflammatory mediators, including iNOS, IL-1β, TNF-α, and IL-6 (Chen et al., 2017). Additionally, fluoride-mediated oxidative stress can also promote NF-κB signaling by activating IκB kinases and promoting NF-κB p52 translocation into the nucleus of microglial cells, which promote neuroinflammation in the brain.

In spite of this, in general, neuroinflammation is a protective mechanism at certain limit that isolates the damaged brain tissue from uninjured area, destroys affected cells, and repairs the extracellular matrix (Correale & Villa, 2004). Microglia, astrocytes, neurons, oligodendrocytes, and other brain cells critically participate in inflammatory responses; however, the hallmark of neuroinflammation is the activation of microglial cells. As discussed earlier in detail, microglial cells initiate a rapid response that involves cell migration, proliferation, release of cytokines/chemokines, and trophic and/or toxic effects. A variety of immune system modulators include complement proteins, adhesion molecules, inflammatory cytokines such as interleukin (IL)-1α, IL-1β, IL-3, IL-6, TNF-α (Drew, Storer, Xu, & Chavis, 2005; Minghetti, Ajmone-Cat, De Berardinis, & De, 2005; Noda, Kettenmann, & Wada, 2006; Sun, Newman, Perry, & Weller, 2004).

Furthermore, neuronal damage and fluoride concentration are interlinked with an increase in fluoride exposure causing its increased deposition in brain tissues. The neurotoxic effect of fluoride could be due to metabolic perturbations at the subcellular levels, most commonly related to excitotoxicity by free radical generation and lipid peroxidation. As discussed in the previous section, 4-hydeoxynonenal (4-HNE), a lipid peroxidation by-product, can specifically impair synaptic functions and inhibit glutamate removal by the transporter proteins (Mattson, 2009). Additionally, it has also been shown that nitric oxide synthase activity is enhanced upon exposure to NaF, which acts by damaging mitochondrial energy production, inhibiting glutamate reuptake, and stimulating lipid peroxidation (García-López et al., 2020; Goschorska et al., 2018; Shuhua et al., 2012).

Recent study based on animal-based observation identified that fluoride induces neuroinflammation by measuring the levels of specific biomolecules expression such as iNOS, COX-2, and TNF-1 along with PKC, VEGF, and HSP70, which seemed to play a role in the induction of inflammatory responses (Allan & Rothwell, 2003; Reddy, Tiwari, Tomar, Desai, & Sharma, 2021). Another observations showed that rats with traumatic brain injury showed elevated iNOS protein, suggesting a role of this enzyme in cerebrovascular disturbances due to chemicals with neurotoxic potential. Induced expression of COX-2 acts as a rate-limiting enzyme in the synthesis of prostaglandins, which have been implicated in various inflammatory pathologies (Nørregaard, Kwon, & Frøkiær, 2015). Further, evidence indicates that fluoride-mediated elevated expression of PKC in brain tissue plays as a potential role in neurodegeneration. More importantly, an increased VEGF expression occurs in a fluoride dose-dependent manner and protects different cells in the brain such as hippocampal, cortical, cerebellar granule, and dopaminergic neurons, under hypoxic or oxidative conditions (Storkebaum & Carmeliet, 2004). Furthermore, upregulated expression of HSP70 in mice model is under the exposure of higher doses of fluoride. HSP70, an endogenous factor and an inflammatory mediator, is known for protecting cell and tissue injury under various pathological conditions (Lin et al., 2001). Therefore, the overexpression of HSP70 like VEGF could be a mechanism to protect the system from various stress-induced injuries. Hsp70 and other heat shock proteins may thus play an essential role in mediating iNOS inhibition and in the protection of cells against stress-induced cell death and apoptosis in brain cells (Bernardini, Zannoni, Turba, Fantinati, et al., 2005; Manucha, 2017).

Fluoride induced neuronal alterations in the sciatic nerve, the brain, and the spinal cord were scattered, and there were low rough endoplasmic reticulum, swollen mitochondria, compressed Golgi cisternae, crenulated nucleus, as was observed with fluoride exposure in previous studies (Reddy et al., 2021). In addition, cell organelle degeneration, alteration in the cell morphology, and other changes indicate that fluoride caused extensive damage to neurons in the brain, spinal cord, and sciatic nerve of the mice model, which may further lead to paralysis and brain dysfunction. Morphological alterations in the brain including effects on neurons and cerebrovasculature in mice chronically administered with NaF (Reddy, Tiwari, Tomar, Desai, & Sharma, 2021; Varner, Jensen, Horvath, & Isaacson, 1998). However, this study reported alterations in brain; the present study sought to include additional tissues of the brain. Moreover, observation also indicates the permissible limit of fluoride (<1.0 ppm) in regulating the metabolic activities of the body, and overdoses of it can cause several neurological, neuroinflammatory, and neurodegenerative complications.

Mini-dictionary of terms

- **Free Radicals.** A highly reactive and transitory uncharged molecule having a valency electron that is unpaired

- **Oxidative stress**. It is a physiological accumulation of free radicals in the body owing to unequal neutralization by antioxidants. It may lead to other disease conditions like inflammation, cancer, neurological disorders, etc.
- **Chlorofluorocarbons.** These are complexes of chlorine, fluorine, and carbon and at times hydrogen. These complex molecules have very low toxicity, reactivity, and flammability because they are preferentially used as refrigerants, cleaning solvents, and aerosol propellants in medical applications.
- **Excitotoxicity.** It is degeneration of dendrites leading to cell death caused by the activation of glutamate receptor by a complex mechanism.
- **Neurodegeneration.** Degeneration and death of neuronal cell in brain and spinal cord that results in its damage and dysfunction owing to multifactorial process.

Key facts of neuroinflammation

- Excessive fluoride exposure is associated with major health burden related to neuroinflammation.
- Neuroinflammation is defined as an inflammatory response within the brain or spinal cord.
- Microglia are central players and focal points for any discussion of neuroinflammation.
- Neuroinflammatory responses are mediated by several key pro-inflammatory cytokines (IL-1β, IL-6, and TNF-α), chemokines (ccl2, ccl5, cxcl1), secondary messengers, (NO and prostaglandins) and reactive oxygen species (ROS).
- Neurodegenerative disorders are distinguished by the accumulation of misfolded proteins, such as α-synuclein (α-syn) protein in dementia.

Summary points

- Excessive intake of fluoride is linked to dental, skeletal fluorosis, and damaging factor to central nervous system.
- Excessive exposure induces hyperactivity and cognitive deficits along with hippocampus damage and also promotes neurotoxicity via modulating different signaling pathways
- Excessive exposure of fluoride directly affects the intelligent quotient (IQ) of the young age group.
- Interaction of fluoride with calcium ion leads to impaired different cellular processes such as oxidative phosphorylation, glycolysis, and neurotransmission.
- The permissible limit of fluoride intake (<1 ppm) is required in the regulation of metabolic activity, while excess to it can cause several neurological neuroinflammatory and neurodegenerative complications.

References

Allan, S. M., & Rothwell, N. J. (2003). Inflammation in central nervous system injury. *Philosophical Transactions of the Royal Society of London. Series B, Biological Sciences*, *358*, 1669–1677.

Bachiller, S., Jiménez-Ferrer, I., Paulus, A., Yang, Y., et al. (2018). Microglia in neurological diseases: A road map to brain-disease dependent-inflammatory response. *Frontiers in Cellular Neuroscience*, *12*, 480–488.

Bernardini, C., Zannoni, A., Turba, M. E., Fantinati, P., et al. (2005). Heat shock protein 70, heat shock protein 32, and vascular endothelial growth factor production and their effects on lipopolysaccharide-induced apoptosis in porcine aortic endothelial cells. *Cell Stress & Chaperones*, *10*, 340–348.

Cargnello, M., & Roux, P. P. (2011). Activation and function of the MAPKs and their substrates, the MAPK-activated protein kinases. *Microbiology and Molecular Biology Reviews*, *75*, 50–83.

Chen, R., Zhao, L. D., Liu, H., Li, H. H., Ren, C., Zhang, P., et al. (2017). Fluoride induces neuroinflammation and alters Wnt signaling pathway in BV2 microglial cells. *Inflammation*, *40*, 1123–1130.

Correale, J., & Villa, A. (2004). The neuroprotective role of inflammation in nervous system injuries. *Journal of Neurology*, *251*, 1304–1316.

Dec, K., Łukomska, A., Maciejewska, D., Jakubczyk, K., Baranowska-Bosiacka, I., et al. (2017). The influence of fluorine on the disturbances of homeostasis in the central nervous system. *Biological Trace Element Research*, *177*, 224–234.

DenBesten, P., & Li, W. (2011). Chronic fluoride toxicity: Dental fluorosis. *Monographs in Oral Science*, *22*, 81–96.

Drew, P. D., Storer, P. D., Xu, J. H., & Chavis, J. A. (2005). Hormone regulation of microglial cell activation: Relevance to multiple sclerosis. *Brain Research Reviews*, *48*, 322–327.

EPA-U.S. Environmental Protection Agency. (1992). National primary drinking water regulations: Fluoride, final rule. *Fed. Regist.*, *50*, 47142–47155.

Farina, M., Avila, D. S., Rocha, J. B. T., & Aschner, M. (2013). Metals, oxidative stress and neurodegeneration: A focus on iron, manganese and mercury. *Neurochemistry International*, *62*, 575–594.

Farooqui, A. A., Ong, W. Y., & Horrocks, L. A. (2006). Inhibitors of brain phospholipase A2 activity: Their neuropharmacologic effects and therapeutic importance for the treatment of neurologic disorders. *Pharmacological Reviews*, *58*, 591–620.

Fawell, J., Bailey, K., Chilton, J., Dahi, E., Fewtrell, L., & Magara, Y. (2006). *Fluoride in drinking-water* (pp. 1–47). London: World Health Organization.

García-López, A. L., Hernández-Castillo, J., Hernández-Kelly, L. C., Olivares-Bañuelos, T. N., & Ortega, A. (2020). Fluoride exposure affects glutamine uptake in Müller glia cells. *Neurotoxicity Research*, *38*, 765–774.

Ghosh, A., Mukherjee, K., Ghosh, S. K., & Saha, B. (2013). Sources and toxicity of fluoride in the environment. *Research on Chemical Intermediates*, *39*, 2881–2915.

Goschorska, M., Baranowska-Bosiacka, I., Gutowska, I., Metryka, E., Skórka-Majewicz, M., & Chlubek, D. (2018). Potential role of fluoride in the etiopathogenesis of Alzheimer's disease. *International Journal of Molecular Sciences*, *19*, 1–23.

Grandjean, P. (2019). Developmental fluoride neurotoxicity: An updated review. *Environmental Health*, *18*, 110.

Green, R., Lanphear, B., Hornung, R., Flora, D., et al. (2019). Association between maternal fluoride exposure during pregnancy and IQ scores in offspring in Canada. *JAMA Pediatrics*, *173*, 940–948.

Gupta, P., & Sandesh, N. (2012). Estimation of fluoride concentration in tea infusions, prepared from different forms of tea, commercially available in Mathura city. *Journal of International Society of Preventive and Community Dentistry*, *2*, 64–68.

Hanna, V. S., & Hafez, E. A. A. (2018). Synopsis of arachidonic acid metabolism: A review. *Journal of Advanced Research*, *13*, 23–32.

Harry, G. J., & Kraft, A. D. (2008). Neuroinflammation and microglia: Considerations and approaches for neurotoxicity assessment. *Expert Opinion on Drug Metabolism & Toxicology*, *10*, 1265–1277.

Isaacson, R. L. (2008). Fluoride and the brain: Dangers and diversities. *Fluoride*, *41*, 233–258.

Jagtap, S., Yenkie, M. K., Labhsetwar, N., & Rayalu, S. (2012). Fluoride in drinking water and defluoridation of water. *Chemical Reviews*, *112*, 2454–2466.

Kumar, N., Sood, S., Arora, B., Singh, M., et al. (2010). Effect of duration of fluoride exposure on the reproductive system in male rabbits. *Journal of Human Reproductive Sciences*, *3*, 148–152.

Lakdawala, D. R, & Punekar, B. D. (1973). Fluoride content of water and commonly consumed foods in Bombay and a study of dietary intake. *Indian Journal of Medical Research*, *16*, 1679–1687.

Lin, B. L., Wang, J. S., Liu, H. C., Chen, R. W., Meyer, Y., Barakat, A., & Delseny, M. (2001). Genomic analysis of the Hsp70 superfamily in Arabidopsis thaliana. *Cell Stress Chaperones*, *6*(3), 201–208. https://doi.org/10.1379/1466-1268(2001)006<0201:gaoths>2.0.co;2.

Mahaboob, B. P., & Saumya, S. M. (2013). Suppression of mitochondrial oxidative phosphorylation and TCA enzymes in discrete brain regions of mice exposed to high fluoride: Amelioration by *Panax ginseng* (Ginseng) and *Lagerstroemia speciosa* (Banaba) extracts. *Cellular and Molecular Neurobiology*, *33*, 453–464.

Malinowska, E., Inkielewicz, I., Czarnowski, W., & Szefer, P. (2008). Assessment of fluoride concentration and daily intake by human from tea and herbal infusions. *Food and Chemical Toxicology*, *46*, 1055–1061.

Manucha, W. (2017). Mitochondrial dysfunction associated with nitric oxide pathways in glutamate neurotoxicity. *Clínica e Investigación en Arteriosclerosis*, *29*, 92–97.

Mattson, M. P. (2009). Roles of the lipid peroxidation product 4-hydroxynonenal in obesity, the metabolic syndrome, and associated vascular and neurodegenerative disorders. *Experimental Gerontology*, *44*, 625–633.

Minghetti, L., Ajmone-Cat, M. A., De Berardinis, M. A., & De Simon, R. (2005). Microglial activation in chronic neurodegenerative diseases: Roles of apoptotic neurons and chronic stimulation. *Brain Research Reviews*, *48*, 251–256.

Mullenix, P. J., Denbesten, P. K., Schunior, A., & Kernan, W. J. (1995). Neurotoxicity of sodium fluoride in rats. *Neurotoxicology and Teratology*, *17*, 169–177.

Näsman, P., Ekstrand, J., Granath, F., Ekbom, A., & Fored, C. M. (2013). Estimated drinking water fluoride exposure and risk of hip fracture: A cohort study. *Journal of Dental Research*, *92*, 1029–1034.

Nita, M., & Grzybowski, A. (2016). The role of the reactive oxygen species and oxidative stress in the pathomechanism of the age-related ocular diseases and other pathologies of the anterior and posterior eye segments in adults. *Oxidative Medicine and Cellular Longevity*, *2016*, 3164734. https://doi.org/10.1155/2016/3164734.

Noda, M., Kettenmann, H., & Wada, K. (2006). Anti-inflammatory effects of kinins via microglia in the central nervous system. *Biological Chemistry*, *387*, 167–171.

Nørregaard, R., Kwon, T. H., & Frøkiær, J. (2015). Physiology and pathophysiology of cyclooxygenase-2 and prostaglandin E2 in the kidney. *Kidney Research and Clinical Practice*, *34*, 194–200.

Ogawa, Y., Tokunaga, E., Kobayashi, O., Hirai, K., & Shibata, N. (2020). Current contributions of organofluorine compounds to the agrochemical industry. *iScience.*, *23*, 101467.

Okazoe, T. (2009). Overview on the history of organofluorine chemistry from the viewpoint of material industry. *Proceedings of the Japan Academy. Series B, Physical and Biological Sciences*, *85*, 276–289.

Pak, C. Y., Sakhaee, K., Piziak, V., & Peterson, R. D. (1994). Slow-release sodium fluoride in the management of postmenopausal osteoporosis. A randomized controlled trial. *Annals of Internal Medicine*, *120*, 625–632.

Reddy, Y. P., Tiwari, S., Tomar, L. K., Desai, N., & Sharma, V. K. (2021). Fluoride-induced expression of neuroinflammatory markers and neurophysiological regulation in the brain of Wistar rat model. *Biological Trace Element Research*, *199*, 2621–2626.

Ruan, J., Ma, L., Shi, Y., & Han, W. (2004). The impact of pH and calcium on the uptake of fluoride by tea plants (*Camellia sinensis* L.). *Annals of Botany*, *93*, 97–105.

Sengupta, S. R., & Pal, B. (1937). Iodine and fluoride contents of food stuffs. *The Indian Journal of Nutrition and Dietetics*, *8*, 66–71.

Shanmugam, T., Abdulla, S., Yakulasamy, V., Selvaraj, M., & Mathan, R. (2018). A mechanism underlying the neurotoxicity induced by sodium fluoride and its reversal by epigallocatechin gallate in the rat hippocampus: Involvement of NrF2/Keap-1 signaling pathway. *The Journal of Basic and Applied Zoology*, *79*, 1–19.

Shuhua, X., Ziyou, L., Ling, Y., Fei, W., & Sun, G. (2012). A role of fluoride on free radical generation and oxidative stress in BV-2 microglia cells. *Mediators of Inflammation*, *2012*, 102954.

Srivastava, S., & Flora, S. J. S. (2020). Fluoride in drinking water and skeletal fluorosis: A review of the global impact. *Current Environmental Health Reports*, *7*, 140–146.

Storkebaum, E., & Carmeliet, P. (2004). VEGF: A critical player in neurodegeneration. *The Journal of Clinical Investigation*, *113*, 14–28.

Strunecka, A., & Patocka, J. (2002). Aluminofluoride complexes in the etiology of Alzheimer's disease. *Structure and Bonding*, *104*, 239–280.

Sun, D., Newman, T. A., Perry, V. H., & Weller, R. O. (2004). Cytokine induced enhancement of autoimmune inflammation in the brain and spinal cord: Implications for multiple sclerosis. *Neuropathology and Applied Neurobiology*, *30*, 374–384.

Takeuchi, H., Mizuno, T., Zhang, G. Q., Wang, J. Y., Kawanokuchi, J., Kuno, R., & Suzumura, A. (2005). Neuritic beading induced by activated microglia is an early feature of neuronal dysfunction toward neuronal death by inhibition of mitochondrial respiration and axonal transport. *The Journal of Biological Chemistry*, *280*, 10444–10454.

Tan, B. L., Norhaizan, M. E., Liew, W. P., & Rahman, H. S. (2008). Antioxidant and oxidative stress: A mutual interplay in age-related diseases. *Frontiers in Pharmacology*, *16*, 1162.

Tang, Q. Q., Du, J., Ma, H. H., Jiang, S. J., & Zhou, X. J. (2008). Fluoride and children's intelligence: A meta-analysis. *Biological Trace Element Research*, *126*, 115–120.

Thangapandiyan, S., & Miltonprabu, S. (2013). Molecular mechanism of fluoride induced oxidative stress and its possible reversal by chelation therapy. *Research and Reviews: A Journal of Toxicology*, *3*, 1–26.

Till, C., Green, R., Flora, D., Hornung, R., Martinez-Mier, E. A., et al. (2020). Fluoride exposure from infant formula and child IQ in a Canadian birth cohort. *Environment International*, *134*, 105315.

USEPA U.S. Environmental Protection Agency. (2010). *Fluoride: Exposure and relative source contribution analysis*. Washington, DC: Health and Ecological Criteria Division, Office of Water, U.S. EPA.

Valdez-Jiménez, L., Soria Fregozo, C., Miranda, M. L. B., Gutiérrez, O., & Pérez, M. I. V. (2011). Effects of the fluoride on the central nervous system. *Neurología*, *26*, 297–300.

Vani, M. L., & Reddy, K. P. (2000). Effects of fluoride accumulation on some enzymes of brain and gastrocnemius muscle of mice. *Fluoride*, *33*, 17–26.

Varner, J. A., Jensen, K. F., Horvath, W., & Isaacson, R. L. (1998). Chronic administration of aluminum-fluoride or sodium-fluoride to rats in drinking water: Alterations in neuronal and cerebrovascular integrity. *Brain Research*, *784*(1–2), 284–298. https://doi.org/10.1016/s0006-8993(97)01336-x.

Villena, R. S., Borges, D. G., & Cury, J. A. (1996). Evaluation of fluoride content of bottled drinking waters in Brazil. *Rev-Saude-Publica*, *30*, 512–518.

Wang, A. G., Xia, T., Chu, Q. L., Zhang, M., Liu, F., et al. (2004). Effects of fluoride on lipid peroxidation, DNA damage and apoptosis in human embryo hepatocytes. *Biomedical and Environmental Sciences*, *17*, 217–222.

Whitford, G. M. (2011). Acute toxicity of ingested fluoride. *Monographs in Oral Science*, *22*, 66–80.

Wong, M. H., Fung, K. F., & Carr, H. P. (2003). Aluminum and fluoride content in tea, with emphasis on brick tea, and their health implications. *Toxicology Letters*, *137*, 111–120.

World Health Organization. (1999). *Guidelines for drinking-water quality. Vol. 2* (2nd ed.). Geneva: World Health Organization.

World Health Organization. (2001). Prepared for World Water Day 2001. In *Reviewed by staff and experts from Oral Health Programme (ORH), and Water, Sanitation and Health Programme (WSH)*. Geneva: World Health Organization (WHO).

Xiang, Q., Liang, Y., Chen, L., Wang, C., Chen, B., Chen, X., et al. (2003). Effect of fluoride in drinking water on children's intelligence. *Fluoride*, *36*, 84–94.

Yamaguti, P. M., Ganzerla, E., Nogueira, F. N., Nesadal, D., & Nicolau, J. (2008). Salivary gland response to fluoride induced oxidative stress and lipid peroxidation. *Fluoride*, *41*, 233–258.

Yan, L., Liu, S., Wang, C., Wang, F., Song, Y., Yan, N., et al. (2013). JNK and NADPH oxidase involved in fluoride-induced oxidative stress in BV-2 microglia cells. *Mediators of Inflammation*, *2013*, 895975.

PART X

Parkinson's disease

CHAPTER 29

Vitamin D and Parkinson's disease

Valeria C. Gonçalves[a], Fulvio A. Scorza[a], Antônio-Carlos G. de Almeida[b], and Carla A. Scorza[a]
[a]Departamento de Neurologia e Neurocirurgia, Escola Paulista de Medicina, Universidade Federal de São Paulo, São Paulo, Brazil
[b]Departamento de Engenharia de Biossistemas, Laboratório de Neurociências Experimental e Computacional, Universidade Federal de São João del-Rei, São João del-Rei, Minas Gerais, Brazil

Abbreviations

1,25-(OH)2D	1α,25-dihydroxyvitamin D
25-(OH)D	25-hydroxycholecalciferol
CYP27B1	extrarenal enzyme 1α-hydroxylase
MS	multiple sclerosis
PD	Parkinson's disease
ROS	reactive oxygen species
SNCA	cells with α-synuclein
SNpc	*Substantia nigra pars compacta*
UVB	ultraviolet B
VDR	vitamin D receptor
Vitamin D2	ergocalciferol
Vitamin D3	cholecalciferol

Vitamin D metabolism

The discovery of vitamin D is linked to the history of rickets, a troubling condition that damages tissue and bone structure in childhood and an intractable medical problem with an elusive cause until the twentieth century (Chang & Lee, 2019). After World War I, scientists were working to solve the puzzle of rickets when they came to the conclusion that the consumption of cod liver oil and improved sunlight exposure could be the cure (Chick, Dalyell, Hume, et al., 1922). In this scenario, McCollum, Simmonds, Becker, et al. (1922), after destroying the vitamin A in cod liver oil, suggested that the presence of a fat-soluble factor found in this preparation was responsible for healing rickets. Further research was needed to identify the structure of this substance with powerful calcemic actions when Askew, Bourdillon, Bruce, et al. (1930), isolated the vitamin from a mixture of its precursor, ergosterol, and subsequently, the term "vitamin D" was coined and it came to the light.

First classified as a fat-soluble vitamin crucial for calcium homeostasis, vitamin D is currently considered as a steroid hormone with multifaceted actions. Produced in the presence of the solar ultraviolet B (UVB) radiation, ergocalciferol (vitamin D2) and

Vitamins and Minerals in Neurological Disorders
https://doi.org/10.1016/B978-0-323-89835-5.00031-4

cholecalciferol (vitamin D3) are the prime forms of vitamin D and both function as pro-hormones; therefore, they are biologically inactive compounds. Vitamin D2 is derived from plants, such as fungi exposed to UVB, while vitamin D3 is found in animal-sourced foods (such as fishes with a high-fat content (salmon and tuna), egg yolk and fish oil) or can be photochemically produced in the human skin. Since few foods naturally contain vitamin D, humans depend on sunlight exposure to meet their vitamin D needs. Sun's UVB photons (295–310 nm) absorbed by the skin act on 7-dehydrocholesterol, resulting in the pre-vitamin D, which requires a thermoconversion to give rise to the vitamin D3 (cholecalciferol) (Chang & Lee, 2019). In turn, diet-derived vitamin D is absorbed in the small intestine and incorporated into the chylomicrons to be taken to the liver. Then, two hydroxylation steps are required to convert vitamins D2 and D3 into the biologically active hormone. The first step, performed by the P450 enzyme 25-hydroxylase in the liver, produces the 25-hydroxycholecalciferol (25-(OH)D). The second one finally gives rise to the active form 1α,25-dihydroxyvitamin D (1,25-(OH)2D) or calcitriol by the action of the enzyme 1α-hydroxylase in the kidney, while the renal 1,25-dihydroxyvitamin D 24-hydroxylase is responsible for 1,25-(OH)2D degradation, hence controlling its levels. The active hormone 1,25-(OH)2D is released into the blood stream linked to vitamin D-binding protein and then reaches various target organs (Fig. 1). Vitamin D metabolites have the ability to cross the blood–brain barrier, reaching the brain (Hribar, Cobbold, & Church, 2020). The hormone calcitriol binds to a nuclear receptor, the vitamin D receptor (VDR), and controls the expression of hundreds of genes linked to cell proliferation, differentiation, adipogenesis, angiogenesis, inflammation, adaptive and innate immunity control, among others (Szymczak-Pajor, Drzewoski, & Śliwińska, 2020). Further, it is now firmly established that many cells and tissues (including the brain) present in the extrarenal enzyme 1α-hydroxylase (CYP27B1) and, as a result, are competent to produce active 1,25-(OH)2D3, a hormone capable of acting in an autocrine or paracrine manner (Szymczak-Pajor et al., 2020). Vitamin D receptors are found throughout the body, meaning that 1,25-(OH)2D exerts key biological roles beyond calcium metabolism, and scientists are vigorously investigating the extraskeletal effects of vitamin D.

Prospective studies indicate an association between low concentrations of vitamin D and a wide range of diseases, including cardiovascular, inflammatory, infectious and autoimmune diseases, cancer, cognitive deficits, neurological and neurodegenerative diseases, as well as mortality in general (Hribar et al., 2020). Although scientific results have not effectively proven that vitamin D supplementation can minimize or reverse the effects of these diseases, interventional and observational studies have suggested that major drop-off in levels of vitamin D should be an alert to health problems (Autier, Boniol, Pizot, et al., 2014). Harmfully, vitamin D deficiency has reached global epidemic proportions. The level of the circulating vitamin D metabolite 25(OH)D is a prime indicator of vitamin D status, as it safely correlates the amount of vitamin D obtained from

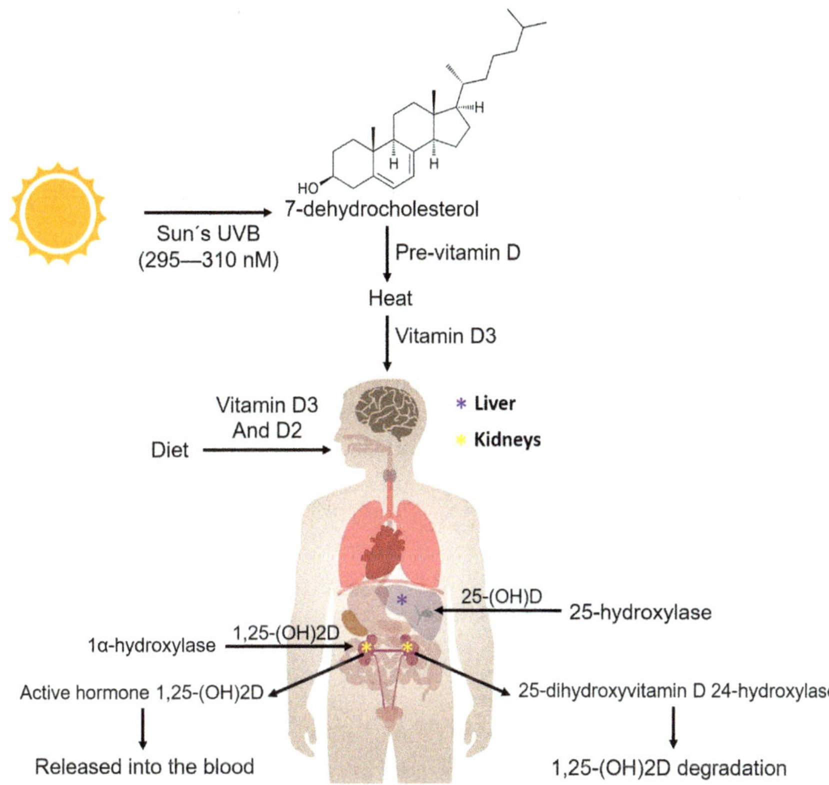

Fig. 1 Vitamin D metabolism. The pro-hormone vitamin D is obtained directly from the diet or produced in the skin from 7-dehydrocholesterol, by means of sunlight exposure. The vitamin Dis found in two biologically inactive forms (vitamin D3 (cholecalciferol) and vitamin D2 (ergocalciferol)) and is activated by a sequential two-step hydroxylation process, the first in the liver (25-hydroxylation produces 25-(OH)D) and the second in the kidneys (25-(OH)D is converted to 1.25-(OH)2D by the enzyme 1α-hydroxylase). The 1,25-(OH)2D degradation also takes place in the kidneys by the action of the enzyme 1,25-dihydroxyvitamin D 24-hydroxylase, regulating the hormone circulating levels. Released in the blood circulation, the active hormone 1,25-(OH)2D reaches the target organs.

dietary intake and epidermal production from sunlight exposure, as well as the conversion from fatty deposits in the liver. Currently, there is no consensus on the ideal 25(OH) D3 concentration. However, according to mainstream medical guidelines, blood levels lower than 40–60 nmol/L are deficient and associated with unfavorable skeletal outcomes (Hribar et al., 2020).

Hypovitaminosis D is a global health problem and affects mainly pregnant women, people with higher melanin content, obese subjects, the elderly, and those individuals who are not directly exposed to the sun. It is estimated that, worldwide, 60% of the population have a vitamin D deficiency or insufficiency, in all age groups (Holick, 2017).

Elderly people are at a singularly elevated risk of vitamin D deficiency as a result of lack of exposure to sunlight, reduced skin synthesis, and low dietary intake of vitamin D (Chang & Lee, 2019). Indeed, it is estimated that about 90% of the elderly people have vitamin D hypovitaminosis (Lauer, Janitschke, Hartmann, et al., 2020). Certainly, aging is the greatest known risk factor for the majority of neurodegenerative diseases, such as Parkinson's disease (PD). As well, inadequate levels of vitamin D are copiously assumed as a risk factor for neurodegenerative diseases (Lauer et al., 2020). PD, in comparison with other brain disorders surveyed by the Neurological Disorders Collaborator Group, was designated as "the fastest growing in disability and deaths" (GBD, 2018). Over the last two decades, the number of individuals with PD has more than doubled (Dorsey, Sherer, Okun, et al., 2018; GBD, 2018). The significant increase in average life expectancy throughout the world and the dramatic age-related disease burden have triggered considerable interest in studying the vitamin D and its impact on body systems in view of the essential pleiotropic functions of this secosteroid hormone. This chapter focuses on vitamin D and PD.

Introduction to Parkinson's disease

Parkinson's disease (PD) is a complex and common age-related incapacitating neurodegenerative disorder affecting more than 6 million people in the world, reaching 1% of the population over 65% and 3% of the population over 85 (GBD, 2018). Annually, the incidence of PD varies between 8 and 18 new cases per hundred thousand people, being the second most common neurodegenerative disease in the elderly (Balestrino & Schapira, 2020). The typical neuropathological hallmarks of PD include α-synuclein-containing Lewy bodies and prominent death of dopamine-producing neurons in the *substantia nigra pars compacta* (SNpc), leading to poor facilitation of voluntary movements. PD is clinically diagnosed from classical parkinsonian motor manifestations, including bradykinesia, tremor, and rigidity (Postuma, Berg, Stern, et al., 2015). When motor features emerge, individuals with PD have already lost more than 60% of dopaminergic neurons, indicating that pathological alterations may start long ahead the occurrence of motor symptoms. Indeed, PD is associated with clinically significant non-motor manifestations that frequently appear during a protracted prodromal period as much as a decade before motor dysfunction (Balestrino & Schapira, 2020). The broad spectrum of non-motor features includes hyposmia, pain, REM sleep behavior disorder, constipation, and autonomic dysfunction, among others (Kalia & Lang, 2015). Not only the nigrostriatal dopaminergic system, but many other brain neurotransmitter systems and regions are also involved and account for the presence of motor and non-motor features of the disease. PD treatment is uniquely focused on the dopaminergic system to control and relief motor signs; however, therapeutic approaches to slow the neurodegenerative process are non-existent. Abnormal α-synuclein deposits are not exclusive to the brain; nevertheless, they are also found in the esophagus, stomach, small intestine, colon, skin, and even in submandibular salivary gland in living PD patients (Tsukita, Sakamaki-Tsukita, Tanaka, et al., 2019). Indeed, there is increasing evidence that disturbances and Lewy pathology may even start in the enteric nervous system (Tan, Chao, West, et al., 2020). It is becoming clear that the called "gut-brain axis" with bidirectional interactions between the central nervous system and the gastrointestinal tract is critical in the etiopathogenesis of PD. In addition to neural connections, this axis also includes immune and endocrine pathways (Houser & Tansey, 2017). Overwhelming evidence implicates immune system abnormalities and aberrant inflammation as key pieces of PD development and progression (Tan et al., 2020). Current evidence suggested the contribution of the adaptive immune system in PD pathogenesis, with the detection of T-cell in the basal ganglia of PD patients and T-cell reactivity to α-synuclein peptides (Arlehamn, Dhanwani, Pham, et al., 2020). A number of key molecular events are at center stage in PD pathogenesis, including mitochondrial dysfunction, oxidative stress, protein aggregation, impaired clearance of proteins, neuroinflammation, defective regulation of cell cycle and death, among several others (Balestrino & Schapira, 2020; Tan et al., 2020). Vitamin D is a crucial direct or indirect regulator of these pathological processes underlying PD and other age-related neurodegenerative diseases (Lauer et al., 2020). Cutting-edge data linking etiological risk factors for PD point to multiple kinds of cellular dysregulation (Tran, Anastacio, & Bardy, 2020). The largest majority of PD cases (~90%) are idiopathic, and disease appears to be an intricate combination of aging, genetic and environmental factors compromising a myriad of elementary cellular processes (Ball, Teo, Chandra, et al., 2019). Scientists are just starting to understand the many factors and cellular events underlying PD and, as the same way are starting to design novel strategies to retard or restrain the unavoidable disease progression. The challenge is to translate the laboratory findings to effective and safe clinical treatment.

Parkinson's disease and vitamin D

Compelling evidence from epidemiological studies supports the association between hypovitaminosis D and a miscellany of psychiatric disorders and neurological/neurodegenerative diseases (Lauer et al., 2020). Vitamin D insufficiency appears to be common in Parkinson's disease (PD) patients, and despite studies suggesting the association between low vitamin D levels and PD incidence and progression, evidence is controversial and not fully understood; therefore, further research is needed (Homann, Ivanic, Homann, et al., 2020). The study by Sato and collaborators was a pioneer in pointing out a relationship between vitamin D and PD (Sato, Kikuyama, & Oizumi, 1997). Currently, a growing body of research indicates the higher prevalence of hypovitaminosis D in individuals with PD when compared to age-matched controls (Rimmelzwaan, van Schoor, Lips, et al., 2016). Ding and colleagues identified the association between plasma levels of 25-hydroxy-vitamin D3 and the prevalence of PD, and suggested that vitamin D deficiency reaches thousands of PD patients in North America and has gone largely underrated (Ding, Dhima, Lockhart, et al., 2013). A small number of longitudinal studies have addressed the association between low vitamin D status and PD incidence. A major investigation on this topic was carried out with a Finnish database, selecting data from 3173 people aged between 50 and 79 years and following them for 3 decades. It has been found that 50 people developed PD and data showed that people with higher levels of vitamin D had a 65% lower chance of developing PD (Knekt, Kilkkinen, & Rissanen, 2010). On the other hand, the investigation conducted by Shrestha, Lutsey, Alonso, et al. (2016) reached a different conclusion. Their prospective study followed 12,700 individuals between 45 and 64 years of age without PD for 17 years and detected the incidence of 67 cases of the disease, but no association between serum vitamin D concentrations and PD risk was found (Shrestha et al., 2016). The literature also shows that higher serum levels of vitamin D are associated with better neuropsychiatric function and cognitive performance in PD patients without dementia (Zhou, Zhou, Zhang, et al., 2019).

The vitamin D receptor (VDR) gene affects susceptibility for PD, and studies have suggested a link between the functional VDR Fokl polymorphism and cognitive impairments in patients with PD (Lv, Tan, Peng, et al., 2020). Interestingly, brain tissue expression of VDR in the midbrain is earliest detected on gestational day 12, coinciding with the development of the SNpc dopamine cells; therefore, prenatal vitamin D hypovitaminosis influences nigrostriatal pathways with lifelong outcomes (Cui, Pelekanos, & Liu, 2013). A key finding is that VDR and the enzyme that converts vitamin D into its active form are both strongly expressed in the SNpc, reinforcing the hypothesis that inadequate levels of vitamin D can be implicated in dopaminergic cell death in this brain region (Fullard & Duda, 2020). In rats, vitamin D3 appears to act positively in the synthesis of dopamine in the basal ganglia and also protects the dopaminergic neurons from the actions of the neurotoxin 6-hydroxydopamine that selectively destroys these dopamine cells (Pertile, Cui, & Eyles, 2016; Wang, Wu, Cherng, et al., 2001). Vitamin D also appears to support the growth and survival of developing or mature neurons and protects against cytotoxicity (Fullard & Duda, 2020). Animal studies have suggested a range of benefits of vitamin D treatment against the key molecular and cellular pathophysiological processes behind PD such as neuroinflammation, oxidative stress, glial activation, endothelial vascular impairment, mitochondrial dysfunction, and cell death among others (Bayo-Olugbami, Nafiu, Amin, et al., 2020; Kim, Shin, Lee, et al., 2020). Attenuation of motor and cognitive impairments has been described in PD rodents, treated with vitamin D (Bayo-Olugbami et al., 2020).

Mitochondrial dysfunction is a central piece in the etiopathogenesis of PD and is suggested as the primary defect implicated in dopaminergic cell loss. Findings from genetic studies enrolling patients with monogenic and sporadic forms of PD have reinforced the implication of mitochondrial deficiency in the disease (Billingsley, Barbosa, Bandrés-Ciga, et al., 2019). The impairment of the electron transport chain complex I activity in PD results in elevated production of reactive oxygen species (ROS) and subsequent oxidative stress causing a massive depletion of cellular energy levels (Billingsley et al., 2019). PD patients show elevated oxidative stress in peripheral blood mononuclear cells expressing dopamine receptors and tyrosine hydroxylase (Leite, Lima, Marino, et al., 2016). It has been shown that the vitamin D receptor not only protects cells from excessive mitochondrial respiration-elicited ROS production, but also is essential for mitochondrial activity and integrity and for cell health (Szymczak-Pajor et al., 2020). In this line, a study showed that the active metabolite of vitamin D, 1,25-dihydroxyvitamin D3, protects dopaminergic cells from ROS and glutamate-induced excitotoxicity (Ibi, Sawada, Nakanishi, et al., 2001). Even though dopamine turned out to be the cornerstone neurotransmitter in PD, there are downstream consequences for other transmitter systems, for instance, glutamate and GABA. Hypovitaminosis D also affects the network balance between excitatory and inhibitory neurotransmission associated with increased ROS production and elevated concentration of intracellular presynaptic calcium and a proinflammatory shift (Kasatkina, Tarasenko, Krupko, et al., 2019).

A number of cellular processes are regulated by the intracellular concentration of free calcium. Yamada and colleagues showed that nigral dopaminergic neurons expressing calbindin-D28K, a vitamin D-dependent calcium-binding protein, are relatively spared when compared to calbindin-D28K-negative neurons in PD (Yamada, McGeer, Baimbridge, et al., 1990). Elevated intracellular free calcium concentration and oxidative stress mutually induce the accumulation of α-synuclein; however, it was shown that calcipotriol, a vitamin D3 analog, was able to inhibit this accumulation in human neuroblastoma cells by inducing calbindin-D28K (Lauer et al., 2020).

Although the neuronal changes identified in PD are not limited to nigrostriatal dopaminergic neurons, these cells are lost to a greater extent in the disease, raising the possibility that the increased dopamine metabolism may also contribute to neurodegeneration (Haddad & Nakamura, 2016). The dopamine neurons in the SNpc of PD patients show a higher proportion of cells with α-synuclein (SNCA) gains than in controls (Mokretar, Pease, & Taanman, 2018). Mutations in the SNCA gene, which encodes the protein α-synuclein, are implicated in PD pathogenesis and are tightly associated with the early-onset familial form of the disease. A recent study showed that the SNCA gene is altered in the brain of rodents with vitamin D inadequate status (Grimm, Lauer, Grosgen, et al., 2019).

Neuroinflammation orchestrated by immune cells, including glial cells, cytokines, and chemokines, has been also considered as a crucial component of PD etiopathogenesis and progression (Troncoso-Escudero, Parra, Nassif, et al., 2018). Studies in animal models of PD have shown that inflammatory stimuli facilitate neurodegeneration of the nigral dopaminergic cells, and in this regard, available data has suggested that systemic inflammation can also be a potential risk factor for PD (Koprich, Reske-Nielsen, Mithal, et al., 2008; Villarán, Espinosa-Oliva, Sarmiento, et al., 2010). Numerous studies have detected an inverse association between vitamin D levels and inflammatory markers in PD patients, but it is still controversial whether inflammation is lowered by vitamin D or whether lower concentrations of vitamin D are a consequence of the underlying inflammation (Autier et al., 2014; Lee, Eisman, & Center, 2009). Vitamin D is a negative acute-phase reactant, that is, its concentration can be reduced as a consequence of the inflammatory process (Chen, Luo, Zhao, et al., 2015).

T cells and dendritic cells are key players in the inflammatory processes, and both express vitamin D receptors. In the presence of inflammation, dendritic cells pick up indications of infection or tissue damage and activate T cells, pointing out where the identified antigens are assisting T cells in the body's immune defense (Tan et al., 2020). Inadequate vitamin D levels lead to the wrong functioning of the immune system (Hribar et al., 2020). PD patients have more T cells expressed in the brain stem, and recent studies indicate that these cells may be naturally autoreactive leading to an autoimmune response to α-synuclein proteins (Garretti, Agalliu, LindestamArlehamn, et al., 2019). Errors during the folding process of α-synuclein cause the agglutination of this protein, generating inclusions known as Lewy bodies, which compromise the functioning of mitochondria and cause damage to the plasma membranes of neurons. Literature adds mounting evidence that PD exhibits features of an autoimmune disease (Garretti et al., 2019). The immune system is influenced by vitamin D, which increases the expression of the antimicrobials that intervene in the bacterial cell membranes, making a link between the immune memory and the adaptive immune response (Hribar et al., 2020). Impaired vitamin D status triggers immune-mediated processes in animal models of autoimmune diseases that are alleviated by vitamin D supplementation (Szodoray, Nakken, Gaal, et al., 2008). To better understand vitamin D signaling in immune-mediated processes, researchers point out the need for long-term studies and the urgency of examining models of vitamin D metabolism.

Gut dysfunction in Parkinson's disease and vitamin D

The earliest stages of PD may involve the gastrointestinal tract. Observed in up to 80% of PD patients, constipation is a precocious and frequent gastrointestinal disturbance that commonly precedes motor manifestations for more than a decade and worsens as the disease progresses (Adams-Carr, Bestwick, Shribman, et al., 2015). The majority of PD subjects exhibit gastrointestinal dysfunctions, such as gut motility disorders, increased intestinal permeability, and altered microbiota composition; however, the reasons of these disturbances and their role in PD pathogenesis are far from clear (Yang, Zhao, Ali Shah, et al., 2019). Current state of knowledge of PD pathophysiology suggests the abundant α-synuclein inclusions in the pathogenesis of disease. Although the cause of sporadic PD is still unidentified, it has been recently postulated that α-synuclein aggregates in the intestinal enteric nerves spread out to the brain, suggesting that the gastrointestinal tract might be the starting point for the whole ethiopathogenic process of PD (Chalazonitis & Rao, 2018; Hawkes, Del Tredici, & Braak, 2007). Gut microbiota-immune interactions are recognized as crucial controllers of the gut-brain communication in the context of PD (Houser & Tansey, 2017). Gut inflammation seems to be a remarkably important driver of PD pathogenesis. Intestinal hyperpermeability and dysbiosis precociously found in PD associated with gut proinflammatory immune activity may contribute to α-synuclein pathology (Yang et al., 2019). Elevated vascular endothelial growth factor receptor 1 and C-reactive protein as well as enhanced levels of inflammatory cytokines released by activating immune cells, including tumor necrosis factor α, interleukin (IL)-1β, IL-6, and interferon-γ, have been found in PD patients and are negatively associated with PD duration (Hasegawa, Goto, Tsuji, et al., 2015; Houser & Tansey, 2017). There is growing interest in establishing how the gut-brain axis influences and contributes to the pathogenesis of PD, as well as understanding the molecules that can substantially affect the patient's gut milieu and metabolism, but these research topics are still in their infancy. Vitamin D deficiency is a significant public health problem in the world and has been linked with gut dysbiosis and inflammation (Singh, Rawat, Alwakeel, et al., 2020). The benefits of vitamin D on the gut microbiome have been widely assessed in people affected by chronic conditions (Singh et al., 2020).

The literature shows that vitamin D regulates intestinal homeostasis and microbiota by maintaining butyrate-producing bacteria, inhibiting and regulating intestinal permeability. Butyrate, in addition to acting in the supply of energy to the colon and regulating T cells, can also play a role in other organs, such as the brain. When there is an imbalance in the intestinal bacterial flora, there are implications for neurological pathologies (Silva, Bernardi, & Frozza, 2020). Studies undertaken in inflammatory bowel disease have demonstrated the beneficial effects of vitamin D on inflammation and intestinal permeability, suggesting oral vitamin D supplementation as a potential adjunctive therapy (Raftery, Martineau, & Greiller, 2015). On the other hand, deficient levels of vitamin D led to gut inflammation and intestinal epithelial barrier dysfunction (Assa, Vong, & Pinnell, 2014). Several multiple cross-sectional studies have reported the alterations of gut microbiota composition in PD patients in comparison with age-matched controls, but the therapeutic potential of the intestinal microbiota remains to be firmly established (Baldini, Hertel, Sandt, et al., 2020; Barichella, Severgnini, Cilia, et al., 2018). It has been recently revealed that the human VDR gene, which encodes vitamin D receptor, significantly influences host's gut microbiome at the genetic level (Wang, Thingholm, Skiecevičienė, et al., 2016). The interplay between vitamin D levels and the gut microbiome continues to elude us; however, available data suggest that gut dysbiosis might even affect the ability to respond to vitamin D supplementation (Singh et al., 2020). Furthermore, there is compelling evidence that vitamin D levels are decreased during acute and chronic inflammatory processes, which in part clarify the association between low vitamin D and a wide range of inflammatory conditions, including PD (Autier et al., 2014). It is a very exciting time to study the effects of vitamin D in PD-associated intestinal microbiota alterations, still an unexplored research topic. Broadening our understanding of the roles of vitamin D in Parkinson's disease may conduce to beneficial therapeutic approaches for improving patient's condition and quality of life.

Applications to other neurological conditions

In this chapter, we discuss the role of vitamin D in PD, mainly the improvement of immune system function associated with adequate levels of vitamin D as well as the beneficial actions of vitamin D in the molecular pathophysiological processes associated with neuroinflammation and mitochondrial disorders present in PD. The benefits of vitamin D are not limited to PD, but have also been linked to other neuropathologies. Research over the years provides evidence that elevated circulating concentrations of vitamin D are associated with lower risk for developing multiple sclerosis (MS), a chronic autoimmune inflammatory and demyelinating neurological disorder. The risk for getting MS is governed by a combination of genetic, epigenetic, and environmental factors. One of the latter is hypovitaminosis D (Feige, Moser, Bieler, et al., 2020). MS has a distinct latitudinal pattern, and disease is more prevalent in medium- or high-latitude countries, suggesting that individuals living in these places may not get enough solar ultraviolet radiation, impacting vitamin D production; however, other factors may also play a role. Studies have shown that vitamin D is a direct and indirect T-cell regulator in MS (Andrei, Maud, Andrew, & Robert, 2020). Evidence points that low concentrations of vitamin D are implicated in more T-cell activation. Experimental autoimmune encephalomyelitis is an animal model widely used in MS research that partially mimics clinical, histopathological, and immunological aspects of human disease and has contributed to a better understanding of the pathophysiological mechanisms of MS and potential protective effects of therapies. In addition to the potential role of vitamin D as an adjunct therapy to conventional treatment of multiple sclerosis with immunotherapy, recent preclinical data have also suggested vitamin D as a way of boosting steroid efficacy for treatment of acute relapse (Andrei et al., 2020). The literature is clear about the role of vitamin D supplementation in the prevention and treatment of MS. However, it should be taken into account that several drugs are commonly prescribed to patients with chronic diseases; therefore, drug interactions associated with high levels of vitamin D and other substances may have undesirable effects that need to be better known. High levels of vitamin D toxicity can camouflage other symptoms of MS, such as fatigue and muscle weakness, so it is up to doctors and researchers to monitor the patient and investigate the best vitamin D dosage and supplementation profile for each individual (Feige et al., 2020).

Other components of interest

Vitamin D, reviewed in the context of Parkinson's disease in this chapter, has numerous benefits at the mitochondrial level. Another substance that also acts against oxidative stress and cellular inflammation is melatonin, a major hormone secreted by the pineal gland, in response to darkness. Besides regulating the circadian rhythm and mitochondrial bioenergetics, melatonin also has potent antioxidant properties (Li, Zhao, Ma, et al., 2020). The administration of melatonin at adequate levels inhibits pathways associated with cell death, autophagy, the formation of Lewy bodies, oxidative stress, and dopaminergic loss, and stimulates the reduction of some non-motor symptoms in PD (Tamtaji, Reiter, & Alipoor, 2020). Most PD patients have sleep disturbances, and studies have suggested that dopaminergic pathways are modulated by melatonin. Concentrations of melatonin fall during the day and dopamine levels are increased. While melatonin levels are up at night and dopamine content decreases. However, a recent study showed that PD patients have higher circulating early-morning concentrations of melatonin than subjects without PD, which are linked to non-motor symptoms in PD. Dopamine replacement therapy may increase circulating melatonin levels (Li et al., 2020). Melatonin has been suggested as a potential adjunctive treatment to dopamine-related disorders; however, more research is needed to reveal the relationship between melatonin and PD.

Mini-dictionary of terms

- **Cholecalciferol**: Also known as vitamin D3, it is a form of vitamin D, synthesized by the skin directly exposed to solar ultraviolet light and acts as a precursor to calcitriol.
- **Ergocalciferol**: Also known as vitamin D2. It is an inactivated vitamin D analog. It works by helping the body to make better use of the calcium supplied by foods and supplements.
- **Calcitriol**: The active form of vitamin D, also known as 1,25-dihidroxicholecalciferol. it acts on calcium homeostasis throughout the body, but most importantly on the gastrointestinal tract.
- **25-Hydroxycholecalciferol**: Calcifediol is a type of vitamin D produced in the liver.
- **Hypovitaminosis**: Condition caused by the deficiency of a vitamin.
- **Dopamine**: Is a monoaminergic neurotransmitter involved in mood, motivation, memory, attention, pleasure, reward, and also in motor control.
- **α-Synuclein**: A protein, encoded by the SCNA gene, that regulates the release of neurotransmitters. It is abundantly expressed in the brain, while small amounts are found in other tissues.
- **Lewy body**: When errors occur in α-synuclein folding process, the protein aggregates into clumps known as Lewy bodies, which can lead to cell death.

- **Reactive oxygen species**: Highly reactive molecules that play important physiological roles as cell signaling molecules. They are free radicals and derivatives of molecular oxygen that, when produced in excess, can cause oxidative damage and cell death.

Key facts of Parkinson's disease

- Parkinson's disease was first described as a nervous disorder in 1817, by James Parkinson, who called it "Shaking Palsy."
- It is the second most common neurodegenerative disease affecting 1% of the population in general and 3% of the population over 85 years. The cardinal symptoms of Parkinson's disease are rest tremor, bradykinesia, and muscle stiffness.
- The diagnosis usually occurs when the motor features of the disease appear, and when there is already a dopaminergic neuronal loss of approximately 60% and 80% less dopamine release.
- The available medications treat motor symptoms; however, they are not able to stop or reverse the condition.
- Supplementation of vitamin D in adequate doses, as adjunctive therapy to polypharmacy in Parkinson's disease, has a beneficial effect and seems to have no side effects.
- Non-motor symptoms of Parkinson's disease, such as impaired sense of smell, sleep disorders and constipation, appear to be predictive of the onset of the disease or help to identify it in its early stages.

Key facts of hypovitaminosis D

- Vitamin D is linked to rickets, an intractable problem until the twentieth century. The apparent solution at the time was the use of cod oil and sun exposure, which years later would lead to the discovery of vitamin D.
- Vitamin D was first classified as a fat-soluble vitamin necessary for calcium homeostasis; however, it is currently considered a steroid hormone with multifaceted actions.
- Low concentrations of vitamin D are associated with a range of diseases in all systems of the body, as well as mortality in general. Inadequate levels of vitamin D can be considered a risk factor for these diseases.
- Hypovitaminosis D is a global health problem and affects mainly pregnant women, people with higher melanin content, obese subjects, the elderly, and those individuals who are not directly exposed to the sun.
- It is estimated that approximately 60% of the world population has vitamin D deficiency or insufficiency. In the elderly groups, this percentage increases to 90%.

Summary points

- Vitamin D can cross the blood–brain barrier, and its insufficiency has been associated with several diseases, such as neurodegenerative ones.
- Parkinson's disease triggers several molecular events, mainly linked to the accumulation of proteins, inflammation, and oxidative stress.
- Studies suggest benefits of vitamin D treatment against the molecular and cellular pathophysiological processes behind Parkinson's disease, such as neuroinflammation, oxidative stress, glial activation, endothelial vascular impairment, mitochondrial dysfunction, and cell death, among others.
- Vitamin D is a crucial direct or indirect regulator of pathological processes. There is a higher prevalence of hypovitaminosis D in individuals with PD when compared to age-matched controls.
- Vitamin D appears to support the growth and survival of developing or mature neurons and protects against cytotoxicity.
- The enzyme that converts vitamin D into its active form is very expressed in substantia nigra pars compacta, reinforcing the hypothesis of hypovitaminosis D in dopaminergic death in this region, aggravating or favoring Parkinson's disease.
- It has recently been discovered that α-synuclein aggregates in intestinal enteric nerves spread to the brain, suggesting that the gastrointestinal tract may be the starting point for the entire etiopathogenic process of Parkinson's disease.
- The literature shows that vitamin D regulates intestinal homeostasis and microbiota by maintaining butyrate-producing bacteria, inhibiting and regulating intestinal permeability.

References

Adams-Carr, K. L., Bestwick, J. P., Shribman, S., et al. (2015). Constipation preceding Parkinson's disease: A systematic review and meta-analysis. *Journal of Neurology, Neurosurgery, and Psychiatry*, *87*(7), 710–716.

Andrei, M., Maud, B., Andrew, C., & Robert, H. (2020). A brief review of the effects of vitamin D on multiple sclerosis. *Frontiers in Immunology*, *11*, 781.

Arlehamn, C. S. L., Dhanwani, R., Pham, J., et al. (2020). α-Synuclein-specific T cell reactivity is associated with preclinical and early Parkinson's disease. *Nature Communications*, *11*(1).

Askew, F. A., Bourdillon, R. B., Bruce, H. M., et al. (1930). The distillation of vitamin D. *Proceedings of the Royal Society B: Biological Sciences*, *107*(748), 76–90.

Assa, A., Vong, L., & Pinnell, L. J. (2014). Vitamin D deficiency promotes epithelial barrier dysfunction and intestinal inflammation. *The Journal of Infectious Diseases*, *210*(8), 1296–1305.

Autier, P., Boniol, M., Pizot, C., et al. (2014). Vitamin D status and ill health: A systematic review. *The Lancet Diabetes and Endocrinology*, *2*(1), 76–89.

Baldini, F., Hertel, J., Sandt, E., et al. (2020). Parkinson's disease-associated alterations of the gut microbiome predict disease-relevant changes in metabolic functions. *BMC Biology*, *18*(1).

Balestrino, R., & Schapira, A. H. V. (2020). Parkinson disease. *European Journal of Neurology*, *27*(1), 27–42.

Ball, N., Teo, W.-P., Chandra, S., et al. (2019). Parkinson's disease and the environment. *Frontiers in Neurology*, *10*, 218.

Barichella, M., Severgnini, M., Cilia, R., et al. (2018). Unraveling gut microbiota in Parkinson's disease and atypical parkinsonism. *Movement Disorders*, *34*, 396–405.
Bayo-Olugbami, A., Nafiu, A. B., Amin, A., et al. (2020). Vitamin D attenuated 6-OHDA-induced behavioural deficits, dopamine dysmetabolism, oxidative stress, and neuro-inflammation in mice. *Nutritional Neuroscience*, 1–12.
Billingsley, K. J., Barbosa, I. A., Bandrés-Ciga, S., et al. (2019). Mitochondria function associated genes contribute to Parkinson's disease risk and later age at onset. *NPJ Parkinson's Disease*, *5*(1).
Chalazonitis, A., & Rao, M. (2018). Enteric nervous system manifestations of neurodegenerative disease. *Brain Research*, *1693*, 207–213.
Chang, S.-W., & Lee, H.-C. (2019). Vitamin D and health—The missing vitamin in humans. *Pediatrics and Neonatology*, *60*(3), 237–244.
Chen, Z., Luo, Z., Zhao, X., et al. (2015). Association of vitamin D status of septic patients in intensive care units with altered procalcitonin levels and mortality. *The Journal of Clinical Endocrinology and Metabolism*, *100*(2), 516–523.
Chick, H., Dalyell, E. J., Hume, M., et al. (1922). The aetiology of rickets in infants: Prophylactic and curative observations at the Vienna University Kinderklinik. *Lancet*, *200*, 7–11.
Cui, X., Pelekanos, M., & Liu, P.-Y. (2013). The vitamin D receptor in dopamine neurons; its presence in human substantia nigra and its ontogenesis in rat midbrain. *Neuroscience*, *236*, 77–87.
Ding, H., Dhima, K., Lockhart, K. C., et al. (2013). Unrecognized vitamin D3 deficiency is common in Parkinson disease: Harvard Biomarker Study. *Neurology*, *81*(17), 1531–1537.
Dorsey, E. R., Sherer, T., Okun, M. S., et al. (2018). The emerging evidence of the Parkinson pandemic. *Parkinson's Disease*, *8*, 3–8.
Feige, J., Moser, T., Bieler, L., et al. (2020). Vitamin D supplementation in multiple sclerosis: A critical analysis of potentials and threats. *Nutrients*, *12*(3), 783.
Fullard, M. E., & Duda, J. E. (2020). A review of the relationship between vitamin D and Parkinson disease symptoms. *Frontiers in Neurology*, *11*, 454.
Garretti, F., Agalliu, D., LindestamArlehamn, C. S., et al. (2019). Autoimmunity in Parkinson's disease: The role of α-synuclein-specific T cells. *Frontiers in Immunology*, *10*, 303.
GBD—Parkinson's Disease Collaborators. (2018). Global, regional, and national burden of Parkinson's disease, 1990–2016: A systematic analysis for the global burden of disease study 2016. *Lancet Neurology*, *17*(11), 939–953.
Grimm, M. O. W., Lauer, A. A., Grosgen, S., et al. (2019). Profiling of Alzheimer's disease related genes in mild to moderate vitamin D hypovitaminosis. *The Journal of Nutritional Biochemistry*, *67*, 123–137.
Haddad, D., & Nakamura, K. (2016). Understanding the susceptibility of dopamine neurons to mitochondrial stressors in Parkinson's disease. *FEBS Letters*, *589*, 3702–3713.
Hasegawa, S., Goto, S., Tsuji, H., et al. (2015). Intestinal Dysbiosis and lowered serum lipopolysaccharide-binding protein in Parkinson's disease. *PLoS One*, *10*(11), e0142164.
Hawkes, C. H., Del Tredici, K., & Braak, H. (2007). Parkinson's disease: A dual-hit hypothesis. *Neuropathology and Applied Neurobiology*, *33*(6), 599–614.
Holick, M. F. (2017). The vitamin D deficiency pandemic: Approaches for diagnosis, treatment and prevention. *Reviews in Endocrine & Metabolic Disorders*, *18*(2), 153–165.
Homann, C. N., Ivanic, G., Homann, B., et al. (2020). Vitamin D and hyperkinetic movement disorders: A systematic review. *Tremor and Other Hyperkinetic Movements*, *10*, 32.
Houser, M. C., & Tansey, M. G. (2017). The gut-brain axis: Is intestinal inflammation a silent driver of Parkinson's disease pathogenesis? *NPJ Parkinson's Disease*, *3*(1).
Hribar, C. A., Cobbold, P. H., & Church, F. C. (2020). Potential role of vitamin D in the elderly to resist COVID-19 and to slow progression of Parkinson's disease. *Brain Sciences*, *10*(5), 284.
Ibi, M., Sawada, H., Nakanishi, M., et al. (2001). Protective effects of 1α,25-(OH)2D3 against the neurotoxicity of glutamate and reactive oxygen species in mesencephalic culture. *Neuropharmacology*, *40*(6), 761–771.
Kalia, L. V., & Lang, A. E. (2015). Parkinson's disease. *Lancet*, *386*(9996), 896–912.

Kasatkina, L. A., Tarasenko, A. S., Krupko, O. A., et al. (2019). Vitamin D deficiency induces the excitation/inhibition brain imbalance and the proinflammatory shift. *The International Journal of Biochemistry & Cell Biology, 105665.*

Kim, H., Shin, J.-Y., Lee, Y.-S., et al. (2020). Brain endothelial P-glycoprotein level is reduced in Parkinson's disease via a vitamin D receptor-dependent pathway. *International Journal of Molecular Sciences*, *21*(22), 8538.

Knekt, P., Kilkkinen, A., & Rissanen, H. (2010). Serum vitamin D and the risk of Parkinson disease. *Archives of Neurology*, *67*, 808–811.

Koprich, J. B., Reske-Nielsen, C., Mithal, P., et al. (2008). Neuroinflammation mediated by IL-1β increases susceptibility of dopamine neurons to degeneration in an animal model of Parkinson's disease. *Journal of Neuroinflammation*, *5*(1), 8.

Lauer, A. A., Janitschke, D., Hartmann, T., et al. (2020). *The effects of vitamin D deficiency on neurodegenerative diseases. Vitamin D deficiency*. IntechOpen.

Lee, P., Eisman, J. A., & Center, J. R. (2009). Vitamin D deficiency in critically ill patients. *The New England Journal of Medicine*, *360*, 1912–1914.

Leite, F., Lima, M., Marino, F., et al. (2016). Dopaminergic receptors and tyrosine hydroxylase expression in peripheral blood mononuclear cells: A distinct pattern in central obesity. *PLoS One*, *11*(1).

Li, L., Zhao, Z., Ma, J., et al. (2020). Elevated plasma melatonin levels are correlated with the non-motor symptoms in Parkinson's disease: A cross-sectional study. *Frontiers in Neuroscience*, *14*.

Lv, L., Tan, X., Peng, X., et al. (2020). The relationships of vitamin D, vitamin D receptor gene polymorphisms, and vitamin D supplementation with Parkinson's disease. *Translational Neurodegeneration*, *9*(1).

McCollum, E. V., Simmonds, N., Becker, J. E., et al. (1922). Studies on the experimental demonstration of the existence of a vitamin which promotes calcium deposition. *The Journal of Biological Chemistry*, *53*, 293–312.

Mokretar, K., Pease, D., & Taanman, J.-W. (2018). Somatic copy number gains of α-synuclein (SNCA) in Parkinson's disease and multiple system atrophy brains. *Brain*, *141*(8), 2419–2431.

Pertile, R. A. N., Cui, X., & Eyles, D. W. (2016). Vitamin D signaling and the differentiation of developing dopamine systems. *Neuroscience*, *333*, 193–203.

Postuma, R. B., Berg, D., Stern, M., et al. (2015). MDS clinical diagnostic criteria for Parkinson's disease. *Movement Disorders*, *30*(12), 1591–1601.

Raftery, T., Martineau, A. R., & Greiller, C. L. (2015). Effects of vitamin D supplementation on intestinal permeability, cathelicidin and disease markers in Crohn's disease: Results from a randomised double-blind placebo-controlled study. *United European Gastroenterology Journal*, *3*(3), 294–302.

Rimmelzwaan, L. M., van Schoor, N. M., Lips, P., et al. (2016). Systematic review of the relationship between vitamin D and Parkinson's disease. *Journal of Parkinson's Disease*, *6*(1), 29–37.

Sato, Y., Kikuyama, M., & Oizumi, K. (1997). High prevalence of vitamin D deficiency and reduced bone mass in Parkinson's disease. *Neurology*, *49*, 1273–1278.

Shrestha, S., Lutsey, P. L., Alonso, A., et al. (2016). Serum 25-hydroxyvitamin D concentrations in mid-adulthood and Parkinson's disease risk. *Movement Disorders*, *31*(7).

Silva, Y. P., Bernardi, A., & Frozza, R. L. (2020). The role of short-chain fatty acids from gut microbiotain gut-brain communication. *Frontiers in Endocrinology*, *11*, 25.

Singh, P., Rawat, A., Alwakeel, M., et al. (2020). The potential role of vitamin D supplementation as a gut microbiota modifier in healthy individuals. *Scientific Reports*, *10*, 21641.

Szodoray, P., Nakken, B., Gaal, J., et al. (2008). The complex role of vitamin D in autoimmune diseases. *Scandinavian Journal of Immunology*, *68*, 261–269.

Szymczak-Pajor, I., Drzewoski, J., & Śliwińska, A. (2020). The molecular mechanisms by which vitamin D prevents insulin resistance and associated disorders. *International Journal of Molecular Sciences*, *21*(18), 6644.

Tamtaji, O. R., Reiter, R. J., Alipoor, R., et al. (2020). Melatonin and Parkinson disease: Current status and future perspectives for molecular mechanisms. *Cellular and Molecular Neurobiology*, *40(1)*, 15–23.

Tan, E.-K., Chao, Y.-X., West, A., et al. (2020). Parkinson disease and the immune system—Associations, mechanisms and therapeutics. *Nature Reviews. Neurology*, *16*, 303–318.
Tran, J., Anastacio, H., & Bardy, C. (2020). Genetic predispositions of Parkinson's disease revealed in patient-derived brain cells. *Parkinson's Disease*, *6*, 8.
Troncoso-Escudero, P., Parra, A., Nassif, M., et al. (2018). Outside in: Unraveling the role of neuroinflammation in the progression of Parkinson's disease. *Frontiers in Neurology*, *9*.
Tsukita, K., Sakamaki-Tsukita, H., Tanaka, K., et al. (2019). Value of in vivo α-synuclein deposits in Parkinson's disease: A systematic review and meta-analysis. *Movement Disorders*, *34*(10), 1452–1463.
Villarán, R. F., Espinosa-Oliva, A. M., Sarmiento, M., et al. (2010). Ulcerative colitis exacerbates lipopolysaccharide-induced damage to the nigral dopaminergic system: Potential risk factor in Parkinson's disease. *Journal of Neurochemistry*, *114*(6), 1687–1700.
Wang, J., Thingholm, L. B., Skiecevičienė, J., et al. (2016). Genome-wide association analysis identifies variation in vitamin D receptor and other host factors influencing the gut microbiota. *Nature Genetics*, *48*(11), 1396–1406.
Wang, J. Y., Wu, J. N., Cherng, T. L., et al. (2001). Vitamin D (3) attenuates 6-hydroxydopamine-induced neurotoxicity in rats. *Brain Research*, *904*(1), 67–75.
Yamada, T., McGeer, P. L., Baimbridge, K. G., et al. (1990). Relative sparing in Parkinson's disease of substantia nigra dopamine neurons containing calbindin-D28K. *Brain Research*, *526*(2), 303–307.
Yang, D., Zhao, D., Ali Shah, S. Z., et al. (2019). The role of the gut microbiota in the pathogenesis of Parkinson's disease. *Frontiers in Neurology*, *10*.
Zhou, Z., Zhou, R., Zhang, Z., et al. (2019). The association between vitamin D status, vitamin D supplementation, sunlight exposure, and Parkinson's disease: A systematic review and meta-analysis. *Medical Science Monitor*, *25*, 666–674.

CHAPTER 30

Tocotrienols and Parkinson's disease: In vitro and in vivo modeling

Tatsuya Matsura[a], Kazuhiro Nakaso[b], and Yosuke Horikoshi[b]
[a]Department of Nutritional Sciences, Faculty of Human Ecology, Yasuda Women's University, Hiroshima, Japan
[b]Division of Biochemistry, Department of Pathophysiological and Therapeutic Sciences, School of Medicine, Faculty of Medicine, Tottori University, Yonago, Japan

Abbreviations

AD	Alzheimer's disease
ER	estrogen receptor
L·	lipid radical
LOO·	lipid peroxyl radical
LOOH	lipid hydroperoxide
MAPK	mitogen-activated protein kinase
MPP$^+$	1-methyl-4-phenylpyridinium
MPTP	1-methyl-1,2,3,6-tetrahydropyridine
OH·	hydroxyl radical
PD	Parkinson's disease
PH	pleckstrin homology
PHLPP1	pleckstrin homology domain leucine-rich repeat protein phosphatase 1
PI3K	phosphatidylinositol 3-kinase
PPAR	peroxisome proliferator-activated receptor
PUFA	polyunsaturated fatty acid
ROS	reactive oxygen species
siRNA	small interfering ribonucleic acid
T3	tocotrienol
Toc	tocopherol
TTP	tocopherol transfer protein

Introduction

Parkinson's disease (PD) is the second most common progressive neurodegenerative disorder after Alzheimer's disease. It is characterized by the loss of dopaminergic neurons in the substantia nigra (Nussbaum & Ellis, 2003).

In Europe, the incidence and prevalence rates of PD are estimated to be approximately 11–19 and 108–275 per 100,000 persons/year, respectively (Balestrino & Schapira, 2020). In the Western Pacific Region, including Japan, the WHO age-standardized incidence rates of PD are 6.7–26.9 per 100,000 persons/year (Lim et al., 2019). In Japan, seven record-based studies and one door-to-door survey carried out

Vitamins and Minerals in Neurological Disorders
https://doi.org/10.1016/B978-0-323-89835-5.00002-8

between 1980 and 2007 across different metropolitan and rural areas showed consistently lower prevalence rates (35.2–75.8 per 100,000 persons) compared to the studies conducted during a similar period in Europe and North America (Lim et al., 2019). Age is the greatest risk factor for the development of PD. The prevalence and incidence of PD increase nearly exponentially with age reaching a peak after 80 years (Kalia & Lang, 2015). Male sex is considered to be a moderate risk factor (global age-standardized prevalence 1.4 times higher in males). However, female predominance has been reported in Japan and Korea, possibly because of the greater disparities between sexes in life expectancy and smoking rates in these countries (Lim et al., 2019). The clinical features of PD include motor symptoms such as resting tremor, muscular rigidity, bradykinesia, and postural anomalies. Nonmotor symptoms, including impaired olfaction, constipation, depression, and rapid eye movement sleep behavior disorder, are also frequently observed in PD before the onset of motor symptoms (Kalia & Lang, 2015; Klockgether, 2004). These symptoms are closely related to a decreased quality of life. The main pathological hallmarks of PD are the loss of dopaminergic neurons in the substantia nigra, and accumulation of Lewy bodies (Kalia & Lang, 2015). Neuroinflammation associated with activated astrocytes and microglia is also recognized as a pathological feature of PD (Tansey & Goldberg, 2010). The pathogenesis of PD remains unknown, but has been linked to environmental factors, including the exposure to pesticides such as paraquat (Brooks, Chadwick, Gelbard, Cory-Slechta, & Federoff, 1999) and rotenone (Testa, Sherer, & Greenamyre, 2005), which induce oxidative stress, mitochondrial dysfunction, and genetic factors related to α-synuclein, parkin, and DJ-1 (Kalia & Lang, 2015).

Since oxidative stress has been implicated in the pathogenesis of PD, antioxidants are potential candidates for neuroprotective therapy against PD.

Vitamin E is one of the most potent fat-soluble antioxidants in the human body. Evidence from cell and animal models has revealed the potential protective effects of vitamin E from diseases associated with oxidative stress (Niki, 2014). In this chapter, we detailed the neuroprotective effects of vitamin E, especially tocotrienols (T3s), and demonstrated their nonantioxidant activities in in vitro and in vivo PD models.

Chemical structure and bioavailability of vitamin E

Vitamin E was first discovered in 1922 as an antisterility factor based on its role in rat reproduction (Evans & Bishop, 1922), and was isolated from wheat germ oil in 1936 (Evans, Emerson, & Emerson, 1936). Vitamin E was named "tocopherol" based on Greek words tocos meaning "childbirth," phero "to bear," and ending with ol to denote the alcohol group in its structure.

Vitamin E is classified into tocopherol (Toc) and tocotrienol (T3) based on the differences in the sidechain structure bound to the chromanol ring. Toc has a saturated

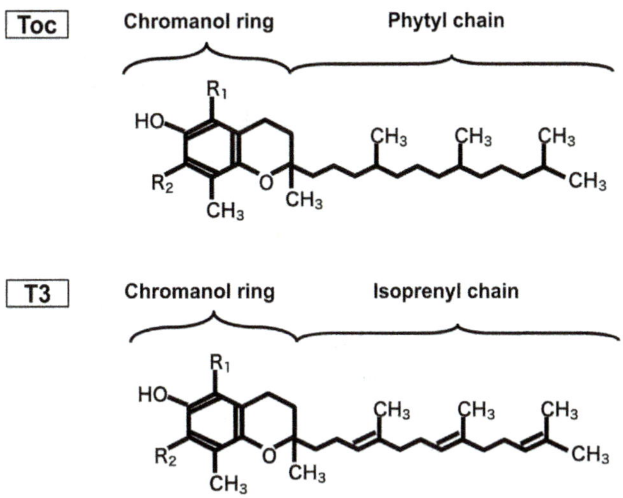

Homologs	α	β	γ	δ
R_1	CH_3	CH_3	H	H
R_2	CH_3	H	CH_3	H

Fig. 1 Chemical structures of tocopherols (Tocs) and tocotrienols (T3s), two classes of vitamin E. Toc has a saturated phytyl side chain, and T3 has an unsaturated isoprenyl side chain. Toc and T3 have four derivatives (i.e., α, β, γ, and δ) according to the position and number of methyl groups on their chromanol rings.

phytyl-like side chain, while T3 has an unsaturated isoprenoid side chain. Both Toc and T3 have four derivatives (namely, α, β, γ, and δ) according to the position and number of methyl groups on their chromanol rings (Fig. 1). Tocs and T3s are abundantly found in vegetable oils and nuts, and in palm oil, wheat germ, and rice bran, respectively (Miyazawa, Burdeos, Itaya, Nakagawa, & Miyazawa, 2019). As humans cannot synthesize vitamin E, it must be obtained from diet or via supplements. α-Toc and γ-Toc are more abundant in food than other forms of Tocs or T3s. However, despite the variations in food sources, it has long been observed that α-Toc is the predominant form of vitamin E in the body. Since α-tocopherol transfer protein (α-TTP) in the liver has a high binding affinity for α-Toc, α-Toc preferentially transfers to very-low-density lipoprotein via α-TTP, and is taken up by the peripheral tissues (Jiang, 2014). Vitamin E activity is generally considered to be mostly related to the antioxidant properties of its chemical structure. The free hydroxyl group on the chromanol ring is responsible for its antioxidant properties (Fig. 1).

Antioxidant activity of vitamin E

The membranes that surround cells and subcellular organelles are highly susceptible to oxidative attack due to the presence of polyunsaturated fatty acid (PUFA) moieties in their membrane phospholipids. Lipid peroxidation can be initiated by the interaction of reactive oxygen species (ROS) with an esterified PUFA to generate lipid radical (L·). Specifically, ROS such as hydroxyl radical (OH·) abstract a hydrogen atom from the methylene group of PUFA to form L·. Following abstraction, the carbon-centered radical is stabilized by the molecular rearrangement to a conjugated diene, which rapidly reacts with O_2 to form lipid peroxyl radical (LOO·). LOO· has sufficient oxidizing potential to attack an adjacent PUFA to form the lipid hydroperoxide (LOOH) and a new L·. This propagation stage of lipid peroxidation newly forms L·, which can react with O_2 to form new LOO·; thus, the chain reaction of lipid peroxidation continues until LOO· meets chain-breaking antioxidants such as vitamin E or until the chain is terminated by the collision of two radical species to form nonradical products (Fig. 2). The antioxidant activity of vitamin E is conferred by scavenging for active free radicals, especially LOO·, produced during the propagation stage of lipid peroxidation, to yield a non-radical product (LOOH in this case) and vitamin E radical. In organic solvents, the rate of scavenging LOO· by vitamin E is approximately 10,000 times faster than that of the abstraction of a hydrogen atom from PUFA by LOO·; vitamin E can thus effectively terminate the chain reaction (Niki, 2014). Tocs and corresponding T3s have the same free radical scavenging capacity (Niki, 2014). The relative reactivity of α, β, γ, and δ forms toward oxygen radicals decreases in the order of $\alpha > \beta = \gamma > \delta$ (Niki, 2014).

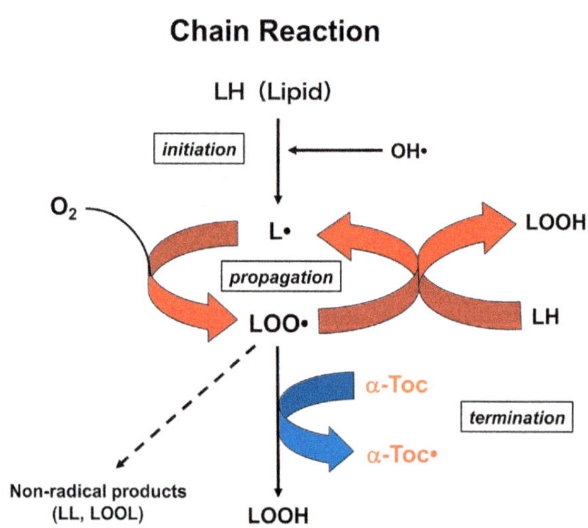

Fig. 2 Initiation and propagation of lipid peroxidation chain reactions and termination by α-tocopherol (α-Toc). Lipid peroxidation can be initiated by the interaction of reactive oxygen species such as hydroxyl radical (OH·) with an esterified polyunsaturated fatty acid to generate lipid radical (L·), which rapidly reacts with O_2 to form lipid peroxyl radical (LOO·). α-Toc scavenges LOO· before it attacks an adjacent lipid (LH) to yield lipid hydroperoxide (LOOH) and a new L·, which propagates chain oxidation. α-Toc·, α-tocopheroxyl radical.

Nonantioxidant activities of vitamin E

Recently, the nonantioxidant activities of vitamin E have received substantial attention. These activities include the modulation of signal transduction, and gene expression, and alteration of membrane lipid domains. In most cases, the nonantioxidant function of vitamin E was assessed based on the experimental results that other antioxidants did not have significant effects under the same conditions, or that vitamin E exhibited a beneficial effect without the reduction of lipid peroxide content. It has been reported that α-Toc and γ-Toc do not function as ligands of peroxisome proliferator-activated receptor-γ (PPAR-γ) but induce the transcription and activation of PPAR-γ via an increase in intracellular concentrations of a PPAR-γ endogenous ligand 15-deoxy-Δ12,14-prostaglandin J2 in 3T3-L1 cells, leading to the upregulation of adiponectin expression, independent of their antioxidant activities. In addition, γ-Toc feeding induced adiponectin in mice at both the mRNA and protein levels (Landrier et al., 2009). Another in vitro study has shown that α-Toc and γ-Toc directly bind to protein kinase C (PKC)α, where the former inhibits nonoxidative cofactor-dependent PKCα activity and the latter enhances it (McCary et al., 2012). α-Toc has been reported to directly modulate the catalytic activity of several enzymes, including that of phosphatidylinositol 3-kinase (PI3K), which is involved in cell signal transduction (Azzi, 2018; Zingg, 2015). It was also demonstrated that α-Toc promotes the cell polarization and migration in human keratinocyte HaCaT cells after wounding via the PI3K/atypical PKC signaling cascade, independently of its antioxidant activity, resulting in accelerated wound repair (Horikoshi et al., 2018). It has been reported that α-Toc and, to a greater extent, γ-Toc, can mediate Akt inactivation in prostate cancer cells due to their unique ability to facilitate the membrane co-localization of Akt and pleckstrin homology (pH) domain leucine-rich repeat protein phosphatase 1 (PHLPP1), a Ser473-specific Akt protein phosphatase. This is achieved through interactions of the polar head group in the chromanol ring with the PH domain of Akt and PHLPP1 (Yan et al., 2015).

Regarding the alteration of membrane lipid domains, it has been suggested that the effect of α-Toc on membrane proteins and enzymes, signaling cascades, and gene expression may be related to its contribution to membrane structure and dynamics, independent of its antioxidant activity (Atkinson, Harroun, Wassall, Stillwell, & Katsaras, 2010). In HER2-positive human breast cancer cells, HER2 receptors in lipid rafts dimerize and result in the autophosphorylation and activation of mitogenic signal transduction. However, in γ-T3-treated cells, γ-T3 accumulates in the lipid rafts and interferes with HER2 dimerization, resulting in the suppression of HER2 activation and subsequent HER2 oncogenic mitogenic signaling (Alawin, Ahmed, Ibrahim, Briski, & Sylvester, 2016).

Clinical studies on effects of α-Toc treatment for PD

There are conflicting findings on the protective effects of vitamin E (especially α-Toc) against PD. Some studies showed that the intake of vitamin E reduced the risk of PD (Etminan, Gill, & Samii, 2005; Ghani, Stevens, Weiss, & Rosenbaum, 2002; Miyake et al., 2011; Yang, Wolk, Hakansson, Pedersen, & Wirdefeldt, 2017). It has been shown that a higher dietary intake of vitamin E is inversely associated with PD occurrence, independent of age and sex (Schirinzi et al., 2019). A randomized, double-blind, placebo-controlled clinical trial conducted in 60 patients with PD showed a significant improvement in Unified Parkinson's Disease Rating Stage (UPDRS) scores reported for patients supplemented with 400 IU vitamin E in combination with 1000 mg omega-3 fatty acids (Taghizadeh et al., 2017). In contrast, it has been reported that vitamin E intake does not substantially affect the risk of PD in 1036 PD cases identified from two large, prospective, cohort studies, the Nurses' Health Study and the Health Professionals Follow-up Study (Hughes et al., 2016). Furthermore, there was no correlation between the severity or duration of PD and the concentration of vitamin E (King, Playfer, & Roberts, 1992). A multicenter, double-blind, randomized clinical trial (DATATOP) in 202 untreated patients with PD showed no beneficial effects of α-Toc (2000 IU/day) on disease progression (Parkinson Study Group, 1993). In addition, antioxidant therapy for PD has not been successful in a clinical setting (Yacoubian & Standaert, 2009). These findings suggest that the nonantioxidant activities of vitamin E other than the antioxidant activities provided by α-Toc, may be beneficial for PD prevention. In this context, T3s (or in some cases, β-, γ-, and δ-Tocs) may represent promising therapeutic agents for PD prevention.

Neuroprotective effects of T3s in PD models in vitro and in vivo via nonantioxidant activities

For the nonantioxidant activities of vitamin E, we found a novel cytoprotective function of T3s involving the binding of T3s to the membrane receptor using in vitro and in vivo PD models (Nakaso et al., 2014, 2016). We examined whether T3s exhibit neuroprotective effects in PD models and determined whether the effect is mediated by their antioxidant activity or other mechanisms. We used a mitochondrial dysfunction-related PD model to examine the neuroprotective effects of T3 (Nakaso et al., 2014). SH-SY5Y cells were incubated with MPP^+ (2 mM) and/or T3s (10 nM, 100 nM, or 1 μM) for 48 h, and cell viability was measured. MPP^+ reduced cell viability to approximately 40%. Simultaneous treatment with T3s significantly increased cell viability in a dose-dependent manner. Next, we examined whether T3s affected MPP^+-induced oxidative stress. T3s did not significantly reduce the production of lipid hydroperoxides and protein carbonyl by MPP^+. Furthermore, we examined the effect of T3s on oxidative stress-independent PD-related neurotoxicity (i.e., proteasome inhibitor or endoplasmic reticulum stress-induced toxicities). T3s, especially γ- and δ-T3s, significantly protected cells against these neurotoxicities. These findings suggest that T3s protect cells against neurotoxicity, independent of their antioxidant activities. Thus, we hypothesized that the cytoprotective effect of T3s is dependent on intracellular signal transduction and examined the activation of the MAPK and PI3K/Akt pathways. Akt was phosphorylated 2 h after treatment with T3s (1 μM), especially in response to γ-T3 and δ-T3. As an index of Akt activation, we examined the translocation of Akt from the cytosolic area to the plasma membrane area. γ-T3 and δ-T3 induced Akt translocation. These results suggest that γ-T3 and δ-T3 can activate the PI3K/Akt signaling pathway. Based on these results, we examined whether kinase inhibitors can inhibit the cytoprotective effects of γ-T3 and δ-T3. PI3K inhibitors dramatically blocked T3-dependent cytoprotection. Our study thus revealed that PI3K/Akt signaling is essential for T3-mediated cytoprotection. As several receptors have been reported as upstream mediators of PI3K/Akt signaling in neuronal cells, we investigated the association of several receptors with T3-mediated cytoprotection using chemical inhibitors. Among the inhibitors, only the estrogen receptor (ER) inhibitor tamoxifen inhibited the cytoprotective effects of T3s. This result indicates that ER signaling is associated with the cytoprotective effects of T3. In silico docking simulations and in vitro binding analyses indicated the high-affinity binding of γ-T3 and δ-T3 to ERβ but not to ERα (Camitato et al., 2009). Therefore, we measured the binding activity between T3s and ERα/ERβ using a radiometric competitive inhibition method. The direct binding of γ-T3 and δ-T3 to the ERβ was observed. The binding of both T3s to ERα was weak. Next, we examined the translocation of ERβ with T3 treatment. In SH-SY5Y cells, ERβ is localized mainly in extranuclear space, such as in the cytosol and plasma membrane. Treatment with γ-T3 and δ-T3 induced ERβ translocation to the perinuclear space. To confirm whether the cytoprotective effect of T3 is due to ERβ, we depleted the ERβ using siRNAs. ERβ silencing abrogated T3-mediated cytoprotection and PI3K/Akt signaling activation. As ER-dependent genomic signaling-related molecules, cathepsin D and Egr-1 were not induced by γ- or δ-T3 treatment, the cytoprotective effect of T3 seems to be due to its nongenomic action via the direct interaction of T3 with membranous ERβ. To characterize the immediate signaling events triggered by γ-T3 treatment, we performed an antibody microarray. Four of the nine upregulated proteins were classified as endocytosis-related or actin/microtubule-related proteins. These findings suggest potential endocytotic actions, such as caveola formation during γ-T3/ERβ signaling. Knockdown of caveolin-1 and/or caveolin-2 prevented cytoprotection by γ-T3 and δ-T3. However, it did not inhibit Akt phosphorylation mediated by γ-T3 and δ-T3. This finding suggests that PI3K/Akt activation may be an upstream mediator of caveola formation. Taken together, γ-T3 and δ-T3 bind to ERβ directly, and this signal induces the activation of the PI3K/Akt signaling pathway, which triggers caveola formation including ERβ (Fig. 3).

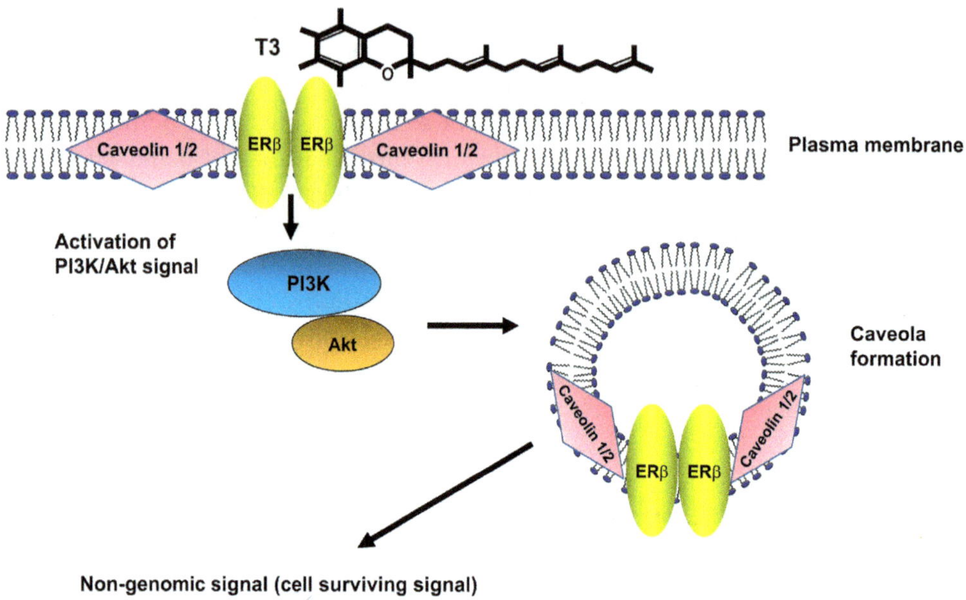

Fig. 3 Proposed underlying mechanisms for T3-mediated neuroprotection in a Parkinson's disease model of SH-SY5Y cells. T3s bind to membranous ERβ directly, and this signal induces the activation of the PI3K/Akt pathway, which triggers caveola formation including ERβ, leading to cell survival (nongenomic signaling pathway).

To examine the ERβ-mediated neuroprotective effects of δ-T3 in vivo, we used the MPTP mouse PD model (Nakaso et al., 2016). The neurotoxicant MPTP is metabolized by monoamine oxidase in astrocytes to generate the active metabolite MPP^+, which was also used in the in vitro model above. MPP^+ is taken up by dopaminergic neurons, resulting in their degeneration. First, we confirmed the expression of ERβ and, to a lesser extent, ERα in the substantia nigra, using immunoblotting. Immunohistochemistry demonstrated that ERβ expression in neuronal cells in the substantia nigra was observed in the cell nuclei as well as extracellularly. In addition, we confirmed that δ-T3 was detected in the mouse brain after daily oral administration of δ-T3. MPTP (20 mg/kg) was injected intraperitoneally four times on day 0, and the oral administration of α-T3, δ-T3, α-Toc, or δ-Toc (100 μg/kg) was carried out every day from day −2 to day 6. Tamoxifen (10 μg/kg) was injected intraperitoneally every 2 days from day −2 to day 6. Wheel running activity and rotarod performance were measured on days 0, 3, and 7. Brain sampling was performed on day 7. We evaluated the voluntary performance in the wheel-running activity. MPTP administration decreased the performance. δ-T3-treated MPTP mice recovered voluntary performance, and tamoxifen canceled the δ-T3-induced recovery. Furthermore, we evaluated motor coordination and balance using the rotarod test. There was no significant difference in the time on the rotarod in mice receiving different treatments. Immunohistochemistry showed a loss of dopaminergic neurons in the substantia nigra of the MPTP-administered mouse brain. δ-T3 protected against MPTP-induced neurotoxicity, while tamoxifen canceled the positive effects of δ-T3. These results suggest that δ-T3 exhibits neuroprotective effects via the ERβ in a mouse PD model.

In conclusion, the results from our studies suggest that T3s (especially γ- and δ-T3) exhibit the neuroprotective effects via the ERβ/PI3K/Akt signaling pathway in cellular and murine PD models. However, whether similar mechanisms also occur in humans remains to be elucidated.

Applications to other neurological conditions

In this chapter, we have reviewed the effects of α-Toc on PD in humans. Since studies on the effects of T3 on PD are lacking in the clinical setting as well as in cellular and animal models, we presented the neuroprotective effect of T3 via its direct interaction with membranous ERβ in cellular and murine PD models. Oxidative stress is known to be involved in the development of Alzheimer's disease (AD), the most prevalent age-related neurodegenerative disease. However, clinical trials with antioxidant therapies, including vitamin E (α-Toc), have not produced consistent results (Tönnies & Trushina, 2017). Diarylpropionitrile, a selective ERβ agonist, protected rat primary hippocampal neurons against glutamate-induced cell death via the enhancement of Bcl-2 expression in the neurons (Zhao, Wu, & Brinton, 2004). In addition, an in vivo study demonstrated that the activation of ERβ could regulate hippocampal plasticity and improve hippocampus-dependent cognition (Liu et al., 2008). As ERβ is highly expressed in the brain and little or no expression is observed in the breast or uterus (Zhao et al., 2004), ERβ-selective molecules could provide a strategy for achieving the beneficial effects of estrogen on the brain without affecting the reproductive organs. As shown in our studies on PD, since T3s (especially γ- or δ-T3) likely have a much higher binding affinity to ERβ than to ERα to drive ER-dependent nongenomic signaling, they may also exert neuroprotective effects in AD. Further studies are required in this area.

Other components of interest

In this chapter, we describe the neuroprotective effect of T3 via its interaction with ERβ in cellular and murine PD models. Lecomte et al. (2017) investigated the selective ER modulator activity of major natural dietary compounds that have previously been shown to bind to ERs. Among them, resveratrol, a polyphenol with antioxidant and anti-inflammatory activities found in grapes and red wine, has a powerful effect on neuritogenesis in the PC12 cells stably transfected with ERα, while it showed a slight or no effect on the proliferation of breast cancer cells. However, the neuroprotective effects of resveratrol via ERβ were not investigated. An in vivo study demonstrated that resveratrol exerts neuroprotective effects via both ERα and ERβ in rat cerebral ischemia–reperfusion injury (Saleh, Connell, & Saleh, 2013). It has been indicated that the use of ERβ-selective agonists for brain disorders is safer than using ERα agonists or estradiol since it cannot stimulate the proliferation of breast or endometrial tissue (Cue, Diaz, Briegel, Patel, & Raval, 2015). Although resveratrol has been reported to provide neuroprotective effects via both ERα and ERβ, given the high expression of ERβ in the brain, it may also interact with ERβ directly leading to neuroprotection through nongenomic pathways similar to T3, in PD models in vitro and in vivo. Future studies should further investigate this possibility.

Mini-dictionary of terms

Oxidative stress: A disturbance in the balance of the production of reactive oxygen species (ROS) and antioxidant defense systems leads to ROS-mediated biomolecular damage.

MPTP and MPP^+: 1-Methyl-1,2,3,6-tetrahydropyridine (MPTP), a neurotoxin, is converted to its toxic metabolite 1-methyl-4-phenylpyridinium (MPP^+) by monoamine oxidase in astrocytes. MPP^+ is taken up by dopaminergic neurons, and it inhibits mitochondrial electron transport chain complex I, leading to dopaminergic neuron degeneration and Parkinson's disease-like behavioral deficits.

Tocotrienol: Another class of vitamin E, besides tocopherol. In addition to antioxidant activity, the following physiological effects of tocotrienols are known: anti-hypercholesterolemic, anticancer, anti-inflammatory, anti-angiogenic, and neuroprotective effects.

Estrogen receptor: Estrogen receptor (ER)α and ERβ belong to the steroid/thyroid hormone superfamily of nuclear receptors. Upon stimulation of ERα and ERβ by estrogen, the receptors translocate to the nucleus and act as transcription factors that modify the activity of target genes (genomic pathway).

Nongenomic pathway: In addition to the genomic pathway, estrogen also utilizes plasma membrane ERs and the downstream signaling pathway referred to as the nongenomic pathway.

Key facts of Parkinson's disease

Parkinson's disease (PD) is a neurodegenerative disorder that results in the death of dopaminergic neurons in the substantia nigra.

PD is recognized as the most common neurodegenerative disorder after Alzheimer's disease.

Clinical manifestations of PD include resting tremor, muscular rigidity, bradykinesia, and postural anomalies.

Environmental factors such as oxidative stress, mitochondrial dysfunction, abnormal protein degradation, and endoplasmic reticulum stress are implicated in the pathogenesis of PD.

Genetic factors are also implicated in PD pathogenesis.

Key facts of vitamin E

Vitamin E is a fat-soluble vitamin that includes two classes: tocopherol (Toc) and tocotrienol (T3).

Tocs and T3s have four derivatives (α, β, γ, and δ).

Tocs and T3s are abundantly found in vegetable oils and nuts, and in palm oil, wheat germ, and rice bran, respectively.

Vitamin E is one of the most potent antioxidants in the body.

The nonantioxidant activities of vitamin E have attracted substantial attention in recent years.

Summary points

This chapter focuses on the neuroprotective effect of T3, independent of its antioxidant activity, in PD models in vitro and in vivo.

T3s prevented MPP^+-induced neurotoxicity in SH-SY5Y cells.

T3s (γ-T3 and δ-T3) activated PI3K/Akt signaling in SH-SY5Y cells.

The PI3K inhibitors dramatically reduced the cytoprotective effects of γ-T3 and δ-T3.

γ-T3 and δ-T3 are directly bound to ERβ in vitro.

Knockdown of ERβ abrogated T3-mediated cytoprotection and PI3K/Akt signal activation.

Knockdown of caveolin abrogated T3-mediated cytoprotection, but not PI3K/Akt signal activation.

Oral administration of δ-T3 inhibited MPTP-induced neurotoxicity in mice and improved the performance of PD mice in wheel-running activity.

ER inhibitor tamoxifen disrupted the neuroprotective effects of T3.

These results suggest that T3 elicits the cytoprotective effects via the ERβ/PI3K/Akt signaling pathway in cellular and murine PD models.

References

Alawin, O. A., Ahmed, R. A., Ibrahim, B. A., Briski, K. P., & Sylvester, P. W. (2016). Antiproliferative effects of γ-tocotrienol are associated with lipid raft disruption in HER2-positive human breast cancer cells. *Journal of Nutritional Biochemistry*, *27*, 266–277.

Atkinson, J., Harroun, T., Wassall, S. R., Stillwell, W., & Katsaras, J. (2010). The location and behavior of a-tocopherol in membranes. *Molecular Nutrition and Food Research*, *54*, 641–651.
Azzi, A. (2018). Many tocopherols, one vitamin E. *Molecular Aspects of Medicine*, *61*, 92–103.
Balestrino, R., & Schapira, A. H. V. (2020). Parkinson disease. *European Journal of Neurology*, *27*, 27–42.
Brooks, A. I., Chadwick, C. A., Gelbard, H. A., Cory-Slechta, D. A., & Federoff, H. J. (1999). Paraquat elicited neurobehavioral syndrome caused by dopaminergic neuron loss. *Brain Research*, *823*, 1–10.
Camitato, R., Nesaretnam, K., Leoni, G., Ambra, R., Canali, R., Bolli, A., Marino, M., & Virgili, F. (2009). A novel mechanism of natural vitamin E tocotrienol activity: Involvement of ERβ signal transduction. *American Journal of Physiology—Endocrinology and Metabolism*, *297*, E427–E437.
Cue, L., Diaz, F., Briegel, K. J., Patel, H. H., & Raval, A. P. (2015). Periodic estrogen receptor-beta activation: A novel approach to prevent ischemic brain damage. *Neurochemical Research*, *40*, 2009–2017.
Etminan, M., Gill, S. S., & Samii, A. (2005). Intake of vitamin E, vitamin C, and carotenoids and the risk of Parkinson's disease: A meta-analysis. *The Lancet Neurology*, *4*, 362–365.
Evans, H. M., & Bishop, K. S. (1922). On the existence of a hitherto unrecognized dietary factor essential for reproduction. *Science*, *56*, 650–651.
Evans, H. M., Emerson, O. H., & Emerson, G. A. (1936). The isolation from wheat germ oil of an alcohol, α-tocopherol, having the properties of vitamin E. *Journal of Biological Chemistry*, *113*, 319–332.
Ghani, H., Stevens, D., Weiss, J., & Rosenbaum, R. (2002). Vitamins and the risk of Parkinson's disease. *Neurology*, *59*, E8–E9.
Horikoshi, Y., Kamizaki, K., Hanaki, T., Morimoto, M., Kitagawa, Y., Nakaso, K., Kusumoto, C., & Matsura, T. (2018). A-tocopherol promotes HaCaT keratinocyte wound repair through the regulation of polarity proteins leading to the polarized cell migration. *BioFactors*, *44*, 180–191.
Hughes, K., Gao, X., Kim, I. Y., Rimm, E. B., Wang, M., Weisskopf, M. G., Schwarzschild, M. A., & Ascherio, A. (2016). Intake of antioxidant vitamins and risk of Parkinson's disease. *Movement Disorders*, *31*, 1909–1914.
Jiang, Q. (2014). Natural forms of vitamin E: Metabolism, antioxidant, and anti-inflammatory activities and their role in disease prevention and therapy. *Free Radical Biology and Medicine*, *72*, 76–90.
Kalia, L. V., & Lang, A. E. (2015). Parkinson's disease. *The Lancet*, *386*, 896–912.
King, D., Playfer, J. R., & Roberts, N. B. (1992). Concentrations of vitamins a, C and in elderly patients with Parkinson's disease. *Postgraduate Medical Journal*, *68*, 634–637.
Klockgether, T. (2004). Parkinson's disease: Clinical aspects. *Cell and Tissue Research*, *318*, 115–120.
Landrier, J. F., Gouranton, E., Yazidi, C. E., Malezet, C., Balaguer, P., Borel, P., & Amiot, J. M. J. (2009). Adiponectin expression is induced by vitamin E via a peroxisome proliferator-activated receptor γ-dependent mechanism. *Endocrinology*, *150*, 5318–5325.
Lecomte, S., Lelong, M., Bourgine, G., Efstathiou, T., Saligaut, C., & Pakdel, F. (2017). Assessment of the potential activity of major dietary compounds as selective estrogen receptor modulators in two distinct cell models for proliferation and differentiation. *Toxicology and Applied Pharmacology*, *325*, 61–70.
Lim, S. Y., Tan, A. H., Ahmad-Annuar, A., Klein, C., Tan, L. C. S., Rosales, R. L., Bhidayasiri, R., Wu, Y. R., Shang, H. F., Evans, A., Pal, P. K., Hattori, N., Tan, C. T., Jeon, B., Tan, E. K., & Lang, A. E. (2019). Parkinson's disease in the Western Pacific region. *The Lancet Neurology*, *184*, 865–879.
Liu, F., Day, M., Muñiz, L. C., Bitran, D., Arias, R., Revilla-Sanchez, R., Grauer, S., Zhang, G., Kelley, C., Pulito, V., Sung, A., Mervis, R. F., Navarra, R., Hirst, W. D., Reinhart, P. H., Marquis, K. L., Moss, S. J., Pangalos, M. N., & Brandon, N. J. (2008). Activation of estrogen receptor-b regulates hippocampal synaptic plasticity and improves memory. *Nature Neuroscience*, *11*, 334–343.
McCary, C. A., Yoon, Y., Panagabko, C., Cho, W., Atkinson, J., & Cook-Mills, J. M. (2012). Vitamin E isoforms directly bind PKCα and differentially regulate activation of PKCα. *Biochemical Journal*, *441*, 189–198.
Miyake, Y., Fukushima, W., Tanaka, K., Sasaki, S., Kiyohara, C., Tsuboi, Y., Yamada, T., Oeda, T., Miki, T., Kawamura, N., Sakae, N., Fukuyama, H., Hirota, Y., Nagai, M., & Fukuoka Kinki Parkinson's Disease Study Group. (2011). Dietary intake of antioxidant vitamins and risk of Parkinson's disease: A case-control study in Japan. *European Journal of Neurology*, *18*, 106–113.
Miyazawa, T., Burdeos, G. C., Itaya, M., Nakagawa, K., & Miyazawa, T. (2019). Vitamin E: Regulatory redox interactions. *International Union of Biochemistry and Molecular Biology*, *71*, 430–441.

Nakaso, K., Horikoshi, Y., Takahashi, T., Hanaki, T., Nakasone, M., Kitagawa, Y., Koike, T., & Matsura, T. (2016). Estrogen receptor-mediated effect of δ-tocotrienol prevents neurotoxicity and motor deficit in the MPTP mouse model of Parkinson's disease. *Neuroscience Letters*, *610*, 117–122.

Nakaso, K., Tajima, N., Horikoshi, Y., Nakasone, M., Hanaki, T., Kamizaki, K., & Matsura, T. (2014). The estrogen receptor β-PI3K/Akt pathway mediates the cytoprotective effects of tocotrienol in a cellular Parkinson's disease model. *Biochimica et Biophysica Acta*, *1842*, 1303–1312.

Niki, E. (2014). Role of vitamin E as a lipid-soluble peroxyl radical scavenger: In vitro and in vivo evidence. *Free Radical Biology and Medicine*, *66*, 3–12.

Nussbaum, R. L., & Ellis, C. E. (2003). Alzheimer's disease and Parkinson's disease. *The New England Journal of Medicine*, *348*, 1356–1364.

Parkinson Study Group. (1993). Effect of tocopherol and deprenyl on the progression of disability in early Parkinson's disease. *The New England Journal of Medicine*, *328*, 176–183.

Saleh, M. C., Connell, B. J., & Saleh, T. M. (2013). Resveratrol induced neuroprotection is mediated via both estrogen receptor subtypes, ERα and ERβ. *Neuroscience Letters*, *548*, 217–221.

Schirinzi, T., Martella, G., Imbriani, P., Lazzaro, G. D., Franco, D., Colona, V. L., Alwardat, M., Salimei, P. S., Mercuri, N. B., Pierantozzi, M., & Pisani, A. (2019). Dietary vitamin E as a protective factor for Parkinson's disease: Clinical and experimental evidence. *Frontiers in Neurology*, *10*, 148.

Taghizadeh, M., Tamtaji, O. R., Dadgostar, E., Kakhaki, R. D., Bahmani, F., Abolhassani, J., Arabi, M. H., Kouchaki, E., Memarzadeh, R., & Asemi, Z. (2017). The effects of omega-3 fatty acids and vitamin E co-supplementation on clinical and metabolic status in patients with Parkinson's disease: A randomized, double-blind, placebo-controlled trial. *Neurochemistry International*, *108*, 183–189.

Tansey, M. G., & Goldberg, M. S. (2010). Neuroinflammation in Parkinson's disease: Its role in neuronal death and implications for therapeutic intervention. *Neurobiology of Disease*, *37*, 510–518.

Testa, C. M., Sherer, T. B., & Greenamyre, J. T. (2005). Rotenone induces oxidative stress and dopaminergic neuron damage in organotypic substantia nigra cultures. *Molecular Brain Research*, *134*, 109–118.

Tönnies, E., & Trushina, E. (2017). Oxidative stress, synaptic dysfunction, and Alzheimer's disease. *Journal of Alzheimer's Disease*, *57*, 1105–1121.

Yacoubian, T. A., & Standaert, D. G. (2009). Targets for neuroprotection in Parkinson's disease. *Biochimica et Biophysica Acta*, *1792*, 676–687.

Yan, R., Chuang, H. C., Kapuriya, N., Chou, C. C., Lai, P. T., Chang, H. W., Yang, C. N., Kulp, S. K., & Chen, C. S. (2015). Exploitation of the ability of γ-tocopherol to facilitate membrane co-localization of Akt and PHLPP1 to develop PHLPP1-targeted Akt inhibitors. *Journal of Medicinal Chemistry*, *58*, 2290–2298.

Yang, F., Wolk, A., Hakansson, N., Pedersen, N. L., & Wirdefeldt, K. (2017). Dietary antioxidants and risk of Parkinson's disease in two population-based cohorts. *Movement Disorders*, *32*, 1631–1636.

Zhao, L., Wu, T. W., & Brinton, D. (2004). Estrogen receptor subtypes alpha and beta contribute to neuroprotection and increased Bcl-2 expression in primary hippocampal neurons. *Brain Research*, *1010*, 22–34.

Zingg, J. M. (2015). Modulation of signal transduction and gene expression by vitamin E via PI3Kgamma/PKB and hTAP1/SEC14L2-mediated lipid exchange. *Journal of Nutritional Science and Vitaminology*, *61*, S76–S77.

CHAPTER 31

B vitamins: Pyridoxal phosphate and parkinsonism

Olakunle J. Onaolapo[a] and Adejoke Y. Onaolapo[b]
[a]Department of Pharmacology, Behavioural Neuroscience Unit, Neuropharmacology Subdivision, Ladoke Akintola University of Technology, Ogbomoso, Oyo, Nigeria
[b]Department of Anatomy, Behavioural Neuroscience Unit, Neurobiology Subdivision, Ladoke Akintola University of Technology, Ogbomoso, Oyo, Nigeria

Abbreviations

DNA deoxyribonucleic acid
GABA gamma amino butyric acid
LCIG levodopa-carbidopa intestinal gel
RNA ribonucleic acid
PD Parkinson's disease
PDXK pyridoxal kinase gene
PLP pyridoxal phosphate

Introduction

Vitamins are defined as a group of closely related organic compounds, which in humans are required in small quantities for normal body metabolism and growth. Since they cannot be synthesized endogenously, vitamins must be sourced from dietary intake; however, in conditions of inadequate dietary intake, deficiencies develop especially in the elderly and infants. Traditionally, vitamins are categorized into fat-soluble (A, D, E, and K) or water-soluble (B and C), which determines their site of action and function. While vitamins that are water-soluble are found in high concentrations in the extracellular fluid or cell cytosol, and act mainly as co-factors for a number of enzymatic reactions (Chawla & Kvarnberg, 2014), fat-soluble vitamins are able to bind to cellular nuclear receptors, influence gene expression (Sánchez-Hernández et al., 2016), and also protect cell membranes from free radical injury.

The B family of vitamins or B complex vitamins are a group of eight water-soluble vitamins, which include thiamine (B1), riboflavin (B2), niacin (B3), pantothenic acid (B5), pyridoxine (B6), biotin (B7), folate (B9), and cobalamin (B12) (Sauberlich, 1984; Zhao et al., 2019). These group of vitamins act as co-enzymes in a number of enzymatic reactions and are also involved in cellular functioning (Kennedy, 2016). Singly or collectively, they impact energy metabolism, DNA/RNA repair, genomic and non-

Vitamins and Minerals in Neurological Disorders
https://doi.org/10.1016/B978-0-323-89835-5.00029-6

genomic methylation, and brain function (Kennedy, 2016; Mikkelsen, Stojanovska, Tangalakis, Bosevski, & Apostolopoulos, 2016). Most of the vitamins in the B group (Table 1) with the exception of niacin (which can be synthesized from tryptophan) are available to the body only through dietary intake (Zhao et al., 2019). The B complex vitamins are also called "neurotropic" vitamins largely because a number

Table 1 B vitamins, sources, and deficiencies.

Vitamins	Sources	Deficiencies
Thiamin B1	Whole grains and fortified bread, cereal, pasta, brown rice, pork, trout, mussels, acorn squash, legumes, such as black beans and soybeans, seeds, and nuts	Deficiencies causes weight loss, loss of appetite, cognitive deficit, confusion, cardiac disease, tingling and numbness in the extremities, loss of muscle mass, poor reflexes, and Wernicke-Korsakoff syndrome
• Riboflavin B2	Organ meats, salmon, eggs broccoli cereals,yogurt, milk, oatmeal, almonds, and mushrooms	Skin disorders, sores at the corners of the mouth, swelling of the mouth and throat, swollen, cracked lips, alopecia, red, itchy eyes, anemia, and cataracts
• Niacin B3	Meat, poultry, fish, nuts, legumes, grains peanut, liver and brown rice	Pellagra, headache, fatigue, depression, brown discoloration on skin exposed to sunlight, patches of skin with a rough appearance, bright red tongue, vomiting, diarrhea, and constipation
• Pantothenic acid B5	Beef liver, mushrooms, avocados, vegetables, nut, shiitake mushrooms, sunflower seeds, chicken, tuna, and fortified breakfast cereals	Irritability numbness and burning of the extremities, headache, loss of appetite, restlessness, and poor sleep
• Pyridoxine B6	Trout, blue berries, molasses, poultry meat, Organ meats, chickpeas, tuna, salmon, potatoes, and fortified cereals	Anemia, scaling on the lips, cracks at corners of the mouth, swollen tongue, weakened immune system, confusion, and depression
• Biotin B7	Egg yolk, organ meat, milk, barley, royal jelly, yeast, salmon, pork, beef, and sunflower seeds	Thinning of the hair, a scaly rash around eyes, nose, and mouth, brittle nails, depression, and fatigue

Table 1 B vitamins, sources, and deficiencies—cont'd

Vitamins	Sources	Deficiencies
• Folate or "folic acid" B9	Berries, salmon spinach, eggs, dark green leafy vegetables, beef liver, avocado, pawpaw, orange juice, beans, and nuts	Weakness, headache, palpitations, irritability, sores on the tongue or in the mouth, and changes in the skin, hair, and nails
• Cyanocobalamin B12	Trout, liver, beef, eggs, salmon, milk, yogurt, shellfish, and clams	Megaloblastic anemia, fatigue, weight loss, constipation, loss of appetite, numbness in the extremities, memory impairment, and depression

Source: Authors.

of them such as vitamins B1, B6, and B12 are essential for the normal functioning of the peripheral and central nervous systems (Kennedy, 2016; Mikkelsen et al., 2016; Wendołowicz, Stefańska, & Ostrowska, 2018). Deficiencies of B vitamins such as folate, B6, and B12, which are involved in one-carbon metabolism pathway, have been associated with the development of cognitive decline that has been linked to oxidative stress resulting from increased homocysteine levels (An et al., 2019).

Vitamin B6 occurs in a variety of related forms, which include pyridoxal, pyridoxine, pyridoxamine, and their phosphorylated forms such as pyridoxal phosphate (PLP) (Dakshinamurti, Dakshinamurti, & Czubryt, 2017). Rich food sources of vitamin B6 include cereals, beans, vegetables, liver, meat, and eggs. A few decades ago, there were reports suggesting that pyridoxine hydrochloride, an older formulation of vitamin B6 (pyridoxine hydrochloride), was associated with an improvement in Parkinson's disease (PD) symptoms (Olofinnade et al., 2020; Shen, 2015); however, its use was limited by increasing incidence of side effects observed when higher doses were administered. In the last four decades, research has continued to examine the effects of vitamin B6 alone or as an adjunct in PD management (Klawans, Ringel, & Shenker, 1971; Mars, 1974; Olofinnade et al., 2020). While a deficiency of PLP may play a role in the etiology of parkinsonism, drugs that are used in managing parkinsonism such as the levodopa/carbidopa combination had been reported to be associated with lowered plasma PLP in patients. The relationship between parkinsonism and PLP may be a complex one, and the understanding of it may bring more benefits to patient management. In this chapter, we review extant literature for the effects of PLP in PD, with emphasis on its impact on pathogenesis, disease progression, and treatment. In the course of this chapter, we will attempt to shed more light on the probably evolving roles of PLP in parkinsonism while trying to decipher the links involving PLP deficiency, parkinsonism, and antiparkinsonian drugs. Finally, we will try to determine whether PLP is likely to have a recognized role in the clinical management of parkinsonism.

Brief history of vitamin B6

In the early 1930s, the discovery by Rudolf Peters that young rats placed on a semisynthetic diet to which only thiamine and riboflavin were added developed severe cutaneous lesions known as "rat acrodynia" prompted extensive research into the possible causes of these lesions (Rosenberg, 2012). A few years later, Paul György reported that the administration of vitamin B6 to these rats was associated with amelioration of the skin lesions; this fact was corroborated by other studies that had demonstrated that deficiency of vitamin B6 resulted in the development of microcytic anemia, convulsions, and skin lesions in animals (Rosenberg, 2012). In 1938, the isolation and crystallization of vitamin B6 was carried out by Samuel Lepkovsky, while in 1939, Leslie Harris, Karl Folkers, and Richard Kuhn independently demonstrated that vitamin B6 was a pyridine derivative, 3-hydroxy-4,5-dihydroxy-methyl-2-methyl-pyridine. The name "pyridoxine" was proposed by Paul György for this pyridine derivative. Shortly after, other derivatives such as pyridoxamine and pyridoxal were described (Hayashi, Tanase, & Yagi, 2010; Rosenberg, 2012), with reports showing that in animals, these three derivatives had approximately equal activities, owing their vitamin activity to the ability of the organism to convert them into the enzymatically active form, which is pyridoxal-5-phosphate (Ramos et al., 2019).

Although practically all living organisms require vitamin B6 for survival, only microorganisms and plants have the capacity to synthesize vitamin B6 de novo (Fitzpatrick et al., 2007). In humans and other mammals, B6 vitamers are sourced from diet following which they are converted to PLP through the vitamin B6 salvage pathways (Ramos et al., 2019).

Pyridoxal phosphate

Pyridoxal phosphate, the enzymatically active form of vitamin B6, is an important co-factor for several enzymes and proteins involved in vast aspects of cellular processes and metabolism (Olofinnade et al., 2020). Reports show that it is involved in the activities of approximately 140 enzyme systems (Percudani & Peracchi, 2003) while also contributing significantly to the synthesis of some neurotransmitters (Clayton, 2006). Currently available reports show that in humans, over 56 PLP-dependent enzymes exist (Percudani & Peracchi, 2009), suggesting that in humans, it is a crucial co-factor for the maintenance of health and prevention of disease.

In humans, the importance of PLP in the brain has been examined severally. Reports show that it improves cognition and verbal fluency in health and in disease conditions like mild cognitive impairment, chronic alcohol abuse, and Alzheimer's disease (Jannusch et al., 2017). It possibly influences brain function through its ability to participate in the synthesis and proper functioning of brain neurotransmitters, including dopamine, norepinephrine, serotonin, gamma aminobutyric acid, and acetylcholine through the activities of PLP-dependent enzymatic reactions (Ramos et al., 2019). It is involved in transamination, decarboxylation, racemization, deamination, and desulfhydration reactions, which are crucial to amino acid synthesis, while also acting as a co-factor in the metabolism of proteins, lipids, and carbohydrates (Ueland, Ulvik, Rios-Avila, Midttun, & Gregory, 2015). In addition to causing anemia and growth retardation, PLP deficiency has also been linked with the development of seizures, hyperirritability, neuropathies, and hyperexcitability. The importance of PLP in the brain is further reinforced by reports that have demonstrated the impact of nutritional deficiencies of vitamin B6 on the growth and development of the brain and its neuronal processes. Also, inborn errors of vitamin B6 metabolism have also been associated with the development of mental retardation. More recently, because of its link to neurotransmitter synthesis and function, there has been increased research into its possible effect on the pathogenesis and management of central nervous system disorders, including neurologic, neurodegenerative, and neuropsychiatric disorders.

Pyridoxal phosphate and the brain in health and disease

The effects of vitamin B6 on the brain in health (Table 2) have been examined in preclinical and clinical studies (Erickson et al., 2008; Nassiri-Asl et al., 2012). Results from one study demonstrated that in rats, pretreatment with vitamin B6 was associated with a dose-dependent improvement in memory retrieval in the passive avoidance task (Nassiri-Asl et al., 2012). In humans, there have also been reports that an increased intake of vitamin B6 among community-dwelling, healthy aging population was associated with an increase in the volume of the gray matter (Erickson et al., 2008). However, the memory-enhancing effects of vitamin B6 have been attributed to its ability to inhibit the release of glutamate from the cortical synaptosomes of the rat through the suppression of the presynaptic voltage-dependent entry of Ca^{2+} and protein kinase C activity (Nassiri-Asl et al., 2012); a systematic review of studies in humans did not show any significant short-term benefits of vitamin B6 supplementation in health (Malouf & Grimley Evans, 2003). There have also been reports associating severe forms of convulsions and epileptic encephalopathy with a reduced PLP level in the cell, which is believed to occur as a result of inborn errors of salvage pathway enzymes/other metabolic pathways, or the interactions of drugs with PLP. Neurological disorders, such as autism, Alzheimer's disease, schizophrenia, epilepsy, and Parkinson's disease, have also been linked to inadequate levels of PLP in cells (di Salvo, Safo, & Contestabile, 2012).

Vitamin B6 is also an essential co-factor in the re-methylation of homocysteine, and homocysteine is an important risk factor for the development of cerebrovascular disease,

Table 2 Vitamin B6 and the brain in health and disease.

Study type	Species	Effect	Reference
Preclinical	Rats	Pretreatment with vitamin B6 was associated with a dose-dependent improvement in memory retrieval in the passive avoidance task	Nassiri-Asl et al. (2012)
A voxel-based morphometry study	Humans	an increased intake of vitamins B6 among community-dwelling, healthy aging population was associated with an increase in the volume of the gray matter	Erickson et al. (2008)
Systematic review	Humans	No significant short-term benefits of vitamin B6 supplementation in healthy individuals. There is, however, a decrease in pyridoxal phosphate levels with increasing age	Malouf and Grimley Evans (2003)
Clinical trial	Human	Polyneuropathy, seizures, and elevated levels of homocysteine arising in individuals with advanced-stage PD following treatment with LCIG infusion due to vitamin B6 deficiencies	Skodda and Müller (2013)
Meta-analysis	Human	high dietary intake of vitamin B6 was associated with a reduced risk of developing PD compared with other B vitamins	Shen (2015)
Clinical study	Human	Pyridoxine administration reversed the therapeutic benefits of L-DOPA but failed to do so when L-DOPA was co-administered with carbidopa	Klawans et al. (1971) and Mars (1974)
Clinical study	Human	Intravenous administration of pyridoxal phosphate stabilized polyneuropathy due to levodopa carbidopa intestinal gel	Urban et al. (2010)
Clinical study	Human	Intravenous administration of pyridoxal phosphate reversed motor but not sensory deficits due to levodopa carbidopa intestinal gel	Klostermann et al. (2012)
Preclinical	Mice	Pyridoxal phosphate supplementation was associated with a reversal of PD-related motor symptoms, improvements in spatial working memory, and a decrease in anxiety-related behaviors	Olofinnade et al. (2020)
Preclinical	Mice	Dietary pyridoxal phosphate supplementation was also associated with increased γ-amino butyric acid levels, improved antioxidant status, reduction in oxidative stress, and reduction in mortality	Olofinnade et al. (2020)

Source: Authors.
Legend: L-Dopa: Levodopa, PD: Parkinson's disease, LCIG: levodopa-carbidopa intestinal gel.

neurodegenerative disorders, and dementias (Malouf & Grimley Evans, 2003). There have also been reports associating B6 deficiency with an increase in blood homocysteine levels, which is reversed following vitamin B6 supplementation. The results of epidemiological studies have also demonstrated a decline in vitamin B6 status with increasing age (Malouf & Grimley Evans, 2003). There are also indications that vitamin B6 requirements in the elderly are higher as a result of decreased absorption, increased catabolism, and impaired phosphorylation (Kjeldby, Fosnes, Ligaarden, & Farup, 2013; Morris, Picciano, Jacques, & Selhub, 2008; Ribaya-Mercado, Russell, Sahyoun, Morrow, & Gershoff, 1991). Due to the decreased dietary intake and increased requirement, there is an increased risk of vitamin B6 deficiency with increasing age (Kjeldby et al., 2013). While we understand the risk that exists with vitamin deficiencies, how vitamin B6 supplementation specifically impacts aging and aging-related neurodegenerative disorders like Parkinson's disease is still being examined.

Pyridoxal phosphate and parkinsonism

The increasing aging population has resulted in neurodegenerative disorders, becoming a leading cause of morbidity and mortality worldwide (GBD 2015 Neurological Disorders Collaborator Group, 2017). Neurodegenerative disorders including Alzheimer's disease and PD are conditions in which genetics and the environment are believed to play important roles in etiology and disease susceptibility. Significant advances in our understanding of the possible etiology, risk factors, and treatment options in Parkinson's disease and parkinsonism have occurred since the first description of the shaking palsy by James Parkinson (Del Rey et al., 2018).

Parkinson's disease is reported to affect approximately1–2 persons per 1000 of the population. Also, its prevalence is reported to increase with age, affecting about 1% of persons aged 60 years and older. Although it is more common in the elderly, there are increasing reports of PD affecting persons of younger age groups. The core neuropathological hallmarks of PD include the presence of α-synuclein-containing Lewy bodies and loss of dopaminergic neurons in the substantia nigra, which have been associated with the appearance of hypokinesia, rigidity, and tremor. Although the conservation of striatal dopamine is the mainstay of pharmacological management, the worsening of symptoms over time, and the search for more permanent, less toxic treatment options, has continued to fuel neuroscience research.

In the last decade or more, the beneficial effects of nutritional status and diet, and the use of vitamin supplements like PLP in reducing the incidence and burden of PD have been examined (Olofinnade et al., 2020; Onaolapo et al., 2019). The suggestions that vitamin B6 could be beneficial in PD have arisen from reports from genetic studies that show that mutations or variations in PLP-dependent genes can increase the risk of developing PD; and clinical studies that showed that PLP deficiency which results from prolonged use of conventional PD therapy worsens outcome and prognosis in persons with advanced PD. Also, high dietary intake of vitamin B6 has been linked with a decreased risk of PD (Shen, 2015).

Vitamin B6 and Parkinson's disease pathogenesis

Parkinson's disease is generally regarded as a multifactorial disease, with both environmental and genetic factors playing important roles. While factors like age are considered the single most important risk factor in PD, factors such as nutritional status and nutritional deficiencies are gradually becoming recognized as important in the pathogenesis of PD (Onaolapo, Odeniyi, & Onaolapo, 2021).

The importance of diets rich in vitamins and/or vitamin supplements in PD pathogenesis has been demonstrated by studies that reported that an increased consumption of fish, fruits, and vegetables decreased the risk of developing PD (Muralidhara & Yenisetti, 2017; Seidl, Santiago, Bilyk, & Potashkin, 2014). Also, deficiencies of B vitamins, including vitamin B6, have also been associated with the precipitation and worsening of neurodegenerative disorders (Skodda & Müller, 2013; Tsagalioti, Trifonos, Morari, Vadikolias, & Giaginis, 2018), including PD (Muralidhara & Yenisetti, 2017). Vitamin B6 is an important co-factor in homocysteine metabolism, and its deficiencies have also been linked with the development of homocysteinemia, which is now considered a risk factor in dementia (Bonetti, Brombo, & Zuliani, 2016). Genetic mutations of sequences that control the biosynthesis and metabolism of the B vitamins have been linked to the development of PD (Elstner et al., 2009; M'Angale & Staveley, 2017), with mutations in genes involved in PLP/dopamine metabolism (PDXK) also being associated with an increased risk of developing PD (Elstner et al., 2009). In preclinical studies, mutations in these genes have been linked with the development of parkinsonian-like symptoms in Drosophilae (M'Angale & Staveley, 2017). Overall, these findings suggest that vitamin B6 and vitamers such as PLP could be beneficial in reducing the risk of developing PD or retarding disease progression.

Vitamin B6 and parkinsonism therapy

Vitamin B6 and its vitamers have been beneficial in the management of diseases through their use as supplements, to target select enzymes in the vitamin B6 biosynthetic pathways, and also by targeting PLP-dependent enzymes. The role of vitamin B6 and PLP as possible therapeutic agents in neurologic disorders (Adelufosi, Abayomi, & Ojo, 2015) and neurodegenerative disorders like PD has been examined severally (Olofinnade et al., 2020; Onaolapo et al., 2021; Skodda & Müller, 2013). There have been reports suggesting that PLP could be beneficial in mitigating the severity of neuroleptic-induced tardive dyskinesia (a chronic debilitating movement disorder) among individuals with schizophrenia (Adelufosi et al., 2015). Suggestions of the possible therapeutic benefits of PLP are supported by clinical reports of polyneuropathy, seizures, and elevated levels of homocysteine arising in individuals with advanced-stage PD following treatment with levodopa-carbidopa intestinal gel (LCIG) infusion, features that were attributed to vitamin B6 deficiencies (Skodda & Müller, 2013), and in most cases resolved following vitamin B6 supplementation. The antioxidant effects of pyridoxine in neurons have also been reported, and it has been shown to increase glutathione level through the conversion of homocysteine to cysteine, a glutathione precursor (Segura-Aguilar, 2015). Also, the results of a meta-analysis had shown that high dietary intake of vitamin B6 was associated with a reduced risk of developing PD compared with other B vitamins (Shen, 2015). Suggestions that exogenous pyridoxine had the ability to antagonize the neuropharmacologic activity of levodopa (without carbidopa) in PD (Klawans et al., 1971; Mars, 1974) initially limited its use, but it also prompted research into the possible mechanism responsible for this effect (Mars, 1974). Pyridoxine had been shown to accelerate the systemic metabolism of levodopa in 15 persons who had received long-term levodopa therapy; however, this was retarded in individuals on levodopa/carbidopa combination therapy (Mars, 1974; Onaolapo et al., 2021).

In recent times, the search for more effective anti-parkinsonism therapies and increasing understanding of the crucial roles of vitamins in the prevention or management of neurodegenerative disorders is increasing research into the therapeutic benefits of PLP in PD (Table 2). Olofinnade et al. (2020) examined the effects of PLP supplementation on neurobehavioral changes, antioxidant status, and mortality in a mouse model of PD, with the report that PLP supplementation was associated with a reversal of PD-related motor symptoms, improvements in spatial working memory, and a decrease in anxiety-related behaviors. Dietary PLP supplementation was also associated with increased γ-amino butyric acid levels, improved antioxidant status, reduction in oxidative stress, and reduction in mortality (Olofinnade et al., 2020). Although research is ongoing to delineate the mechanisms behind the beneficial effects of PLP, there have been suggestions that PLP could exert its effect through its antioxidant potential, which has been shown to improve neuronal metabolism, reduce neuronal excitotoxicity, and protect against neuronal loss (Olofinnade et al., 2020). There have also been suggestions that PLP plays an important role in mitigating inflammation through its ability to act as a co-factor in inflammatory pathways such as those responsible for vitamin B6 catabolism, sphingosine 1-phosphate metabolism, the kynurenine pathway, serine and glycine metabolism, and the trans-sulfuration pathways (Ueland, McCann, Midttun, & Ulvik, 2017). In humans, the result of individual case reports demonstrated that intravenous administration of PLP stabilized polyneuropathy due to LCIG according to one report (Urban et al., 2010); however, in another report (Klostermann, Jugel, Müller, & Marzinzik, 2012), it reversed motor but not sensory deficits, despite continuation of LCIG therapy (Müller et al., 2013). There is a dearth of clinical and preclinical studies evaluating the possible benefits of PLP in

PD; however, inferences from available studies show that while dietary PLP supplementation (or its intravenous administration) may not be a replacement for levodopa-carbidopa therapies, it may be useful in reversing some of the unwanted effects of levodopa-carbidopa therapy, thereby improving the quality of life of individuals with PD.

Side effects of pyridoxine/pyridoxal phosphate therapy

Although it is generally believed that toxicities occurring as a result of supplementation with water-soluble vitamins are rare, there have been reports of a few adverse reactions to vitamin B6 or pyridoxal phosphate supplementation. Neurotoxicity and photosensitivity have been observed following the administration of pyridoxine/pyridoxal phosphate at above 500 mg/day (Ronis, Pedersen, & Watt, 2018). There have been suggestions that because the morphologic changes that occur due to genetic mutations of pyridoxal kinase (PDXK) which causes axonal neuropathy are very similar to those seen with excess pyridoxine/pyridoxal phosphate ingestion, they could possibly share a common mechanism where in this case excessive pyridoxine intake inhibits the activity of PDXK, resulting in neuronal degeneration and loss, due to excitotoxicity. There have been suggestions that a decrease in GABA neurotransmission could also be responsible for these changes, because there are reports that compounds that inhibit PDXK are associated with the development of convulsions and decreased GABA levels in the nervous system (Hadtstein & Vrolijk, 2021). There have also been reports of polyneuropathy in some individuals taking pyridoxal phosphate at doses as low as 0.5 mg/day. These variations in sensitivity to pyridoxine/PLP have been attributed to interindividual difference in vitamin kinetics (Vrolijk, Hageman, van de Koppel, van Hunsel, & Bast, 2020). In the pediatric age group, there have been reports of asymptomatic increases in liver enzymes following the use of PLP at 10–30 mg/kg/day (Cortès-Saladelafont, Molero-Luis, Hsjd Working Group, Artuch, and García-Cazorla, 2016). In another case report, there was evidence of recurrent hemarthrosis that coincided with an increase in the dosing of PLP in an 8-year-old boy who had previously been on long-term PLP therapy for epilepsy (Borst & Tchapyjnikov, 2018).

Conclusion

The possible relationships that link pyridoxine/PLP and parkinsonism may be a complex one, which science is yet to uncover in its entirety. However, we now have proof that while pyridoxine/PLP may be involved in the pathogenesis of parkinsonism, giving them as nutritional supplements is not likely to be a magic bullet. Also, it is becoming increasingly obvious that pyridoxine/PLP supplementation may not be a straightforward endeavor due to possible wide variations in sensitivity and expression of toxicity.

Therefore, while it is possible that they may be useful in parkinsonism prevention or management, a lot of research still has to be conducted in this area.

Applications to other areas of neurological conditions

In this chapter, we have examined the role(s) of vitamin B6 and its active component pyridoxal phosphate in the pathogenesis and management of Parkinson's disease (PD). Clinical studies have reported that low levels of vitamin B6 were associated with an increased risk of developing PD. In particular, increasing understanding of the crucial role played by pyridoxal phosphate-dependent enzymes in brain development and aging has led to suggestions that vitamin B6 or pyridoxal phosphate could be possible therapeutic agent in neurologic disorders and neurodegenerative disorders like PD. The results of a study carried out in our laboratory, revealed that dietary pyridoxal phosphate supplementation could ameliorate changes in open-field behaviors, cognitive deficit, anxiety-related behaviors, catalepsy, oxidative stress, apoptosis, and neurochemical changes in a rodent model of PD (Olofinnade et al., 2020). However, there is a need for more studies, particularly clinical studies, to examine its role a sole or adjunctive agent in PD management.

Other components of interest

In this chapter, we discussed the potential benefit of pyridoxal phosphate a vitamer of vitamin in Parkinson's disease pathogenesis and therapy. Of interest are studies that have also demonstrated the possible benefits of other B vitamins including nicotinamide (Xu et al., 2012), folate, and cobalamin (Schaffner et al., 2019) in animal models of Parkinson's disease. Clinical studies have also demonstrated that low levels of folic acid and cobalamin are associated with the development of depression and cognitive deficits, respectively, in levodopa-treated Parkinson's disease or in patients with advanced-stage Parkinson's disease (Triantafyllou et al., 2008). However, low levels of vitamin B12 were associated with gait instability and cognitive deficits in early Parkinson's disease (Christine et al., 2018; Dietiker, Kim, Zhang, & Christine, 2019). Also, supplementation with these vitamins was associated with reversal of Parkinson's disease symptoms. More recently, results from a study by Schaffner et al. (2019) have provided evidence that treatment with 5′-deoxyadenosylcobalamin (a physiological form of vitamin B12) has the ability to inhibit the activity Leucine-Rich Repeat Kinase 2 (LRRK2). Treatment with 5′-deoxyadenosylcobalamin inhibits LRRK2 activity in cultured cells and brain tissue and also protects against the development of neurotoxicity in culture mouse brain cells. In vivo, it sustains dopamine availability in mouse models of Parkinson's disease. This suggests that vitamin B12 could be harnessed in the development of novel therapies especially in LRRK2-based Parkinson's disease.

Mini-dictionary of terms

1. **Vitamin:** any of a group of organic compounds which are essential for normal growth and nutrition and are required in small quantities in the diet because they cannot be synthesized by the body.
2. **Parkinsonism:** any condition that causes a combination of the movement abnormalities seen in **Parkinson's disease**
3. **Neurodegeneration:** degeneration of the nervous system, especially of neurons in the brain.
4. **Polyneuropathy:** a general degeneration of peripheral nerves that spreads toward the center of the body
5. **Supplementation:** the addition of an extra element or amount to something

Key facts of B vitamins: Pyridoxal phosphate and parkinsonism

Key facts of Parkinson's disease

1. Parkinson's disease is named after James Parkinson who first described the shaking palsy
2. Parkinson's disease affects approximately1–2 persons per 1000 of the population.
3. Prevalence increases with age, affecting approximately 1% of persons ≥60 years
4. Neuropathological hallmarks of the disease include the presence of α Lewy bodies and loss of dopaminergic neurons in the substantia nigra
5. Till date, the conservation of striatal dopamine is the mainstay of pharmacological management

Key facts of vitamins

1. Vitamins are a group of organic compounds required in small quantities by humans for normal body metabolism and growth
2. Vitamins are classified as either fat-soluble or water-soluble
3. In conditions of inadequate dietary intake, deficiencies develop
4. B vitamins act as co-enzymes in a number of enzymatic reactions
5. A number of B vitamins are involved in one-carbon metabolism pathways

Key facts of vitamin B6

1. Vitamin B6 occurs as pyridoxal, pyridoxine, pyridoxamine, and their phosphorylated forms
2. Food sources include cereals, beans, vegetables, liver, meat, and eggs.
3. Deficiency of vitamin B6 causes microcytic anemia
4. Pyridoxal phosphate (PLP) is the enzymatically active form of vitamin B6
5. PLP is involved in the activities of over 140 enzyme systems

Summary points

1. Parkinsonism is a neurodegenerative condition, for which science is yet to get a cure
2. Although pharmacotherapy retards disease progression, halting disease progression has not been achieved.
3. Research into the etiology of parkinsonism has revealed the possible roles of vitamins
4. The relationship between parkinsonism, vitamin B6, and pyridoxal phosphate (PLP) appears to be complex
5. This chapter tries to shed more light on the probably evolving roles of PLP in parkinsonism while trying to decipher the links involving PLP deficiency, parkinsonism, and antiparkinsonian drugs

References

Adelufosi, A. O., Abayomi, O., & Ojo, T. M. (2015). Pyridoxal 5 phosphate for neuroleptic-induced tardive dyskinesia. *The Cochrane Database of Systematic Reviews*, *4*. https://doi.org/10.1002/14651858.CD010501.pub2, CD010501.

An, Y., Feng, L., Zhang, X., Wang, Y., Wang, Y., Tao, L., Qin, Z., & Xiao, R. (2019). Dietary intakes and biomarker patterns of folate, vitamin B_6, and vitamin B_{12} can be associated with cognitive impairment by hypermethylation of redox-related genes NUDT15 and TXNRD1. *Clinical Epigenetics*, *11*(1), 139. https://doi.org/10.1186/s13148-019-0741-y.

Bonetti, F., Brombo, G., & Zuliani, G. (2016). The relationship between hyperhomocysteinemia and neurodegeneration. *Neurodegenerative Disease Management*, *6*(2), 133–145. https://doi.org/10.2217/nmt-2015-0008.

Borst, A. J., & Tchapyjnikov, D. (2018). B_6 and bleeding: A case report of a novel vitamin toxicity. *Pediatrics*, *141*(Suppl 5), S430–S433. https://doi.org/10.1542/peds.2017-2039.

Chawla, J., & Kvarnberg, D. (2014). Hydrosoluble vitamins. In *Vol. 120. Handbook of clinical neurology* (pp. 891–914). https://doi.org/10.1016/B978-0-7020-4087-0.00059-0.

Christine, C. W., Auinger, P., Joslin, A., Yelpaala, Y., Green, R., & Parkinson Study Group-DATATOP Investigators. (2018). Vitamin B12 and homocysteine levels predict different outcomes in early Parkinson's disease. *Movement Disorders: Official Journal of the Movement Disorder Society*, *33*(5), 762–770. https://doi.org/10.1002/mds.27301.

Clayton, P. T. (2006). B6-responsive disorders: A model of vitamin dependency. *Journal of Inherited Metabolic Disease*, *29*(2–3), 317–326. https://doi.org/10.1007/s10545-005-0243-2.

Cortès-Saladelafont, E., Molero-Luis, M., Hsjd Working Group, Artuch, R., & García-Cazorla, À. (2016). Pyridoxal phosphate supplementation in neuropediatric disorders. *Seminars in Pediatric Neurology*, *23*(4), 351–358. https://doi.org/10.1016/j.spen.2016.11.009.

Dakshinamurti, K., Dakshinamurti, S., & Czubryt, M. P. (2017). Vitamin B6: Effects of deficiency, and metabolic and therapeutic functions. In V. Preedy, & V. Patel (Eds.), *Handbook of famine, starvation, and nutrient deprivation*. Cham: Springer. https://doi.org/10.1007/978-3-319-40007-5_81-1.

Del Rey, N. L., Quiroga-Varela, A., Garbayo, E., Carballo-Carbajal, I., Fernández-Santiago, R., Monje, M., Trigo-Damas, I., Blanco-Prieto, M. J., & Blesa, J. (2018). Advances in Parkinson's disease: 200 years later. *Frontiers in Neuroanatomy*, *12*, 113. https://doi.org/10.3389/fnana.2018.00113.

di Salvo, M. L., Safo, M. K., & Contestabile, R. (2012). Biomedical aspects of pyridoxal 5′-phosphate availability. *Frontiers in Bioscience (Elite Edition)*, *4*, 897–913.

Dietiker, C., Kim, S., Zhang, Y., & Christine, C. W. (2019). Characterization of vitamin B12 supplementation and correlation with clinical outcomes in a large longitudinal study of early Parkinson's disease. *Journal of Movement Disorders*, *12*(2), 91–96. https://doi.org/10.14802/jmd.18049.

Elstner, M., Morris, C. M., Heim, K., Lichtner, P., Bender, A., Mehta, D., Schulte, C., Sharma, M., Hudson, G., Goldwurm, S., Giovanetti, A., Zeviani, M., Burn, D. J., McKeith, I. G., Perry, R. H., Jaros, E., Krüger, R., Wichmann, H. E., Schreiber, S., ... Turnbull, D.M. (2009). Single-cell expression profiling of dopaminergic neurons combined with association analysis identifies pyridoxal kinase as Parkinson's disease gene. *Annals of Neurology, 66*(6), 792–798. https://doi.org/10.1002/ana.21780.

Erickson, K. I., Suever, B. L., Prakash, R. S., Colcombe, S. J., McAuley, E., & Kramer, A. F. (2008). Greater intake of vitamins B6 and B12 spares gray matter in healthy elderly: A voxel-based morphometry study. *Brain Research, 1199*, 20–26.

Fitzpatrick, T. B., Amrhein, N., Kappes, B., Macheroux, P., Tews, I., & Raschle, T. (2007). Two independent routes of de novo vitamin B6 biosynthesis: Not that different after all. *The Biochemical Journal, 407*(1), 1–13. https://doi.org/10.1042/BJ20070765.

GBD 2015 Neurological Disorders Collaborator Group. (2017). Global, regional, and national burden of neurological disorders during 1990-2015: A systematic analysis for the global burden of disease study 2015. *The Lancet. Neurology, 16*(11), 877–897. https://doi.org/10.1016/S1474-4422(17)30299-5.

Hadtstein, F., & Vrolijk, M. (2021). Vitamin B-6-induced neuropathy: Exploring the mechanisms of pyridoxine toxicity. *Advances in Nutrition, 2021*. https://doi.org/10.1093/advances/nmab033, nmab033.

Hayashi, H., Tanase, S., & Yagi, T. (2010). Esmond E. Snell–the pathfinder of B vitamins and cofactors. *Journal of Biochemistry, 147*(4), 451–457. https://doi.org/10.1093/jb/mvq016.

Jannusch, K., Jockwitz, C., Bidmon, H. J., Moebus, S., Amunts, K., & Caspers, S. (2017). A complex interplay of vitamin B1 and B6 metabolism with cognition, brain structure, and functional connectivity in older adults. *Frontiers in Neuroscience, 11*, 596. https://doi.org/10.3389/fnins.2017.00596.

Kennedy, D. O. (2016). B vitamins and the brain: Mechanisms, dose and efficacy—A review. *Nutrients, 8*(2), 68. https://doi.org/10.3390/nu8020068.

Kjeldby, I. K., Fosnes, G. S., Ligaarden, S. C., & Farup, P. G. (2013). Vitamin B6 deficiency and diseases in elderly people–a study in nursing homes. *BMC Geriatrics, 13*, 13. https://doi.org/10.1186/1471-2318-13-13.

Klawans, H. L., Ringel, S. P., & Shenker, D. M. (1971). Failure of vitamin B6 to reverse the L-dopa effect in patients on a dopa decarboxylase inhibitor. *Journal of Neurology, Neurosurgery, and Psychiatry, 34*(6), 682–686. https://doi.org/10.1136/jnnp.34.6.682.

Klostermann, F., Jugel, C., Müller, T., & Marzinzik, F. (2012). Malnutritional neuropathy under intestinal levodopa infusion. *Journal of Neural Transmission (Vienna, Austria: 1996), 119*(3), 369–372. https://doi.org/10.1007/s00702-011-0689-3.

Malouf, R., & Grimley Evans, J. (2003). The effect of vitamin B6 on cognition. *The Cochrane Database of Systematic Reviews, 4*. https://doi.org/10.1002/14651858.CD004393, CD004393.

M'Angale, P. G., & Staveley, B. E. (2017). A loss of Pdxk model of Parkinson disease in Drosophila can be suppressed by Buffy. *BMC Research Notes, 10*(1), 205. https://doi.org/10.1186/s13104-017-2526-8.

Mars, H. (1974). Levodopa, carbidopa, and pyridoxine in Parkinson disease. Metabolic interactions. *Archives of Neurology, 30*(6), 444–447. https://doi.org/10.1001/archneur.1974.00490360020005.

Mikkelsen, K., Stojanovska, L., Tangalakis, K., Bosevski, M., & Apostolopoulos, V. (2016). Cognitive decline: A vitamin B perspective. *Maturitas, 93*, 108–113. https://doi.org/10.1016/j.maturitas.2016.08.001.

Morris, M. S., Picciano, M. F., Jacques, P. F., & Selhub, J. (2008). Plasma pyridoxal 5′-phosphate in the US population: The National Health and nutrition examination survey, 2003-2004. *The American Journal of Clinical Nutrition, 87*(5), 1446–1454. https://doi.org/10.1093/ajcn/87.5.1446.

Müller, T., van Laar, T., Cornblath, D. R., Odin, P., Klostermann, F., Grandas, F. J., Ebersbach, G., Urban, P. P., Valldeoriola, F., & Antonini, A. (2013). Peripheral neuropathy in Parkinson's disease: Levodopa exposure and implications for duodenal delivery. *Parkinsonism & Related Disorders, 19*(5), 501. https://doi.org/10.1016/j.parkreldis.2013.02.006.

Muralidhara, Y. S. V., & Yenisetti, S. C. (2017). Current understanding on the beneficial role of nutrition in Parkinson's disease –An overview. *Journal of Aging Science, 5*, 1. https://doi.org/10.4172/2329-8847.1000177.

Nassiri-Asl, M., Sarookhan, M.-R., Abbasi, E., Zangivand, A.-A., Shakiba, E., Sedighi, A., & Rahbari, M. (2012). The effects of pre-treatment with vitamin B6 on memory retrieval in rats. *Food Chemistry, 132*, 1046–1048.

Olofinnade, A. T., Onaolapo, T. M., Oladimeji, S., Fatoki, A. M., Balogun, C. I., Onaolapo, A. Y., & Onaolapo, O. J. (2020). An evaluation of the effects of pyridoxal phosphate in chlorpromazine induced parkinsonism using mice. *Central Nervous System Agents in Medicinal Chemistry*, *20*(1), 13–25. https://doi.org/10.2174/1871524920666200120142508.

Onaolapo, O. J., Odeniyi, A. O., Jonathan, S. O., Samuel, M. O., Amadiegwu, D., Olawale, A., Tiamiyu, A. O., Ojo, F. O., Yahaya, H. A., Ayeni, O. J., & Onaolapo, A. Y. (2019). An investigation of the anti-parkinsonism potential of co-enzyme Q10 and co-enzyme Q10/levodopa-carbidopa combination in mice. *Current Aging Science*. https://doi.org/10.2174/1874609812666191023153724. Advance online publication.

Onaolapo, O. J., Odeniyi, A. O., & Onaolapo, A. Y. (2021). Parkinson's disease: Is there a role for dietary and herbal supplements? *CNS & Neurological Disorders Drug Targets*. https://doi.org/10.2174/1871527320666210218082954. Advance online publication. https://doi.org/10.2174/1871527320666210218082954.

Percudani, R., & Peracchi, A. (2003). A genomic overview of pyridoxal-phosphate-dependent enzymes. *EMBO Reports*, *4*(9), 850–854. https://doi.org/10.1038/sj.embor.embor914.

Percudani, R., & Peracchi, A. (2009). The B6 database: A tool for the description and classification of vitamin B6-dependent enzymatic activities and of the corresponding protein families. *BMC Bioinformatics*, *10*, 273. https://doi.org/10.1186/1471-2105-10-273.

Ramos, R. J., Albersen, M., Vringer, E., Bosma, M., Zwakenberg, S., Zwartkruis, F., Jans, J., & Verhoeven-Duif, N. M. (2019). Discovery of pyridoxal reductase activity as part of human vitamin B6 metabolism. *Biochimica et Biophysica Acta. General Subjects*, *1863*(6), 1088–1097. https://doi.org/10.1016/j.bbagen.2019.03.019.

Ribaya-Mercado, J. D., Russell, R. M., Sahyoun, N., Morrow, F. D., & Gershoff, S. N. (1991). Vitamin B-6 requirements of elderly men and women. *The Journal of Nutrition*, *121*(7), 1062–1074. https://doi.org/10.1093/jn/121.7.1062.

Ronis, M., Pedersen, K. B., & Watt, J. (2018). Adverse effects of nutraceuticals and dietary supplements. *Annual Review of Pharmacology and Toxicology*, *58*, 583–601. https://doi.org/10.1146/annurev-pharmtox-010617-05284.

Rosenberg, I. H. (2012). A history of the isolation and identification of vitamin B(6). *Annals of Nutrition & Metabolism*, *61*(3), 236–238. https://doi.org/10.1159/000343113.

Sánchez-Hernández, D., Anderson, G. H., Poon, A. N., Pannia, E., Cho, C. E., Huot, P., & Kubant, R. (2016). Maternal fat-soluble vitamins, brain development, and regulation of feeding behavior: An overview of research. *Nutrition Research (New York, N.Y.)*, *36*(10), 1045–1054. https://doi.org/10.1016/j.nutres.2016.09.009.

Sauberlich, H. E. (1984). Implications of nutritional status on human biochemistry, physiology, and health. *Clinical Biochemistry*, *17*(2), 132–142. https://doi.org/10.1016/s0009-9120(84)90344-8.

Schaffner, A., Li, X., Gomez-Llorente, Y., Leandrou, E., Memou, A., Clemente, N., Yao, C., Afsari, F., Zhi, L., Pan, N., Morohashi, K., Hua, X., Zhou, M. M., Wang, C., Zhang, H., Chen, S. G., Elliott, C. J., Rideout, H., Ubarretxena-Belandia, I., & Yue, Z. (2019). Vitamin B_{12} modulates Parkinson's disease LRRK2 kinase activity through allosteric regulation and confers neuroprotection. *Cell Research*, *29*(4), 313–329. https://doi.org/10.1038/s41422-019-0153-8.

Segura-Aguilar, J. (2015). A new mechanism for protection of dopaminergic neurons mediated by astrocytes. *Neural Regeneration Research*, *10*(8), 1225–1227. https://doi.org/10.4103/1673-5374.162750.

Seidl, S. E., Santiago, J. A., Bilyk, H., & Potashkin, J. A. (2014). The emerging role of nutrition in Parkinson's disease. *Frontiers in Aging Neuroscience*, *6*, 36. https://doi.org/10.3389/fnagi.2014.0003.

Shen, L. (2015). Associations between B vitamins and Parkinson's disease. *Nutrients*, 7(9), 7197–7208. https://doi.org/10.3390/nu7095333.

Skodda, S., & Müller, T. (2013). Refractory epileptic seizures due to vitamin B6 deficiency in a patient with Parkinson's disease under duodopa® therapy. *Journal of Neural Transmission (Vienna, Austria: 1996)*, *120*(2), 315–318. https://doi.org/10.1007/s00702-012-0856-1.

Triantafyllou, N. I., Nikolaou, C., Boufidou, F., Angelopoulos, E., Rentzos, M., Kararizou, E., Evangelopoulos, M. E., & Vassilopoulos, D. (2008). Folate and vitamin B12 levels in levodopa-treated Parkinson's disease patients: Their relationship to clinical manifestations, mood and cognition. *Parkinsonism & Related Disorders*, *14*(4), 321–325. https://doi.org/10.1016/j.parkreldis.2007.10.002.

Tsagalioti, E., Trifonos, C., Morari, A., Vadikolias, K., & Giaginis, C. (2018). Clinical value of nutritional status in neurodegenerative diseases: What is its impact and how it affects disease progression and management? *Nutritional Neuroscience*, *21*(3), 162–175. https://doi.org/10.1080/1028415X.2016.12615296.
Ueland, P. M., McCann, A., Midttun, Ø., & Ulvik, A. (2017). Inflammation, vitamin B6 and related pathways. *Molecular Aspects of Medicine*, *53*, 10–27. https://doi.org/10.1016/j.mam.2016.08.00.
Ueland, P. M., Ulvik, A., Rios-Avila, L., Midttun, Ø., & Gregory, J. F. (2015). Direct and functional biomarkers of vitamin B6 status. *Annual Review of Nutrition*, *35*, 33–70. https://doi.org/10.1146/annurev-nutr-071714-034330.
Urban, P. P., Wellach, I., Faiss, S., Layer, P., Rosenkranz, T., Knop, K., & Weis, J. (2010). Subacute axonal neuropathy in Parkinson's disease with cobalamin and vitamin B6 deficiency under duodopa therapy. *Movement Disorders: Official Journal of the Movement Disorder Society*, *25*(11), 1748–1752. https://doi.org/10.1002/mds.23342.
Vrolijk, M. F., Hageman, G. J., van de Koppel, S., van Hunsel, F., & Bast, A. (2020). Inter-individual differences in pharmacokinetics of vitamin B6: A possible explanation of different sensitivity to its neuropathic effects. *PharmaNutrition*, *12*, 100188.
Wendołowicz, A., Stefańska, E., & Ostrowska, L. (2018). Influence of selected dietary components on the functioning of the human nervous system. *Roczniki Panstwowego Zakladu Higieny*, *69*(1), 15–21.
Xu, J., Xu, S. Q., Liang, J., Lu, Y., Luo, J. H., & Jin, J. H. (2012). Zhejiang da xue xue bao. *Yi xue ban = Journal of Zhejiang University. Medical Sciences*, *41*(2), 146–152.
Zhao, X., Zhang, M., Li, C., Jiang, X., Su, Y., & Zhang, Y. (2019). Benefits of vitamins in the treatment of Parkinson's disease. *Oxidative Medicine and Cellular Longevity*, *2019*, 9426867. https://doi.org/10.1155/2019/9426867.

CHAPTER 32

Antioxidant vitamins in Parkinson's disease: Vitamins A, C, E

Ariel F. Ying[a] and Woon-Puay Koh[b]
[a]Health Services and Systems Research Programme, Duke-National University of Singapore Medical School, Singapore, Singapore
[b]Healthy Longevity Translational Research Programme, Yong Loo Lin School of Medicine, National University of Singapore, Singapore, Singapore

Abbreviations

6-OHDA	6-hydroxydopamine
DATATOP	Deprenyl And Tocopherol Antioxidative Therapy Of Parkinsonism
DJ-1	daisuke-junko-1, PARK7
LRRK2	leucine-rich repeat kinase 2, PARK8
MAPK	mitogen-activated protein kinase
MPTP	1-methyl-4-phenyl-1,2,3,6-tetrahydropyridine
PD	Parkinson's disease
PINK1	Phosphatase and tensin homolog-induced kinase 1

Introduction

Parkinson's disease (PD) is a progressive neurodegenerative disease characterized by substantial death of the dopaminergic neurons in the substantia nigra, resulting in a deficiency of dopamine within the basal ganglia (Kalia & Lang, 2015). Although there are several hypotheses as to its pathogenesis, studies have generally agreed that one of the key mechanisms underlying the development of PD is oxidative stress (Jankovic & Tan, 2020).

With the rise of PD-mediated disability worldwide (G.B.D. 2016 Neurology Collaborators, 2019), the search for modifiable risk factors that can reduce PD risk is also increasingly pertinent (Belvisi et al., 2020). Since PD is a chronic progressive disease with an extensive prodromal period that can last decades (Hawkes, Del Tredici, & Braak, 2010), one area that has long been of interest is the midlife diet before the onset of symptoms at older age. In particular, due to the "oxidative stress" hypothesis, one major focus of research has been on dietary nutrients with antioxidant properties, such as vitamins A, C, and E (Ascherio & Schwarzschild, 2016).

In this chapter, we have studied evidence from the literature, including in vitro and in vivo experimental studies (Table 1), as well as human observational studies (Table 2), to

Vitamins and Minerals in Neurological Disorders
https://doi.org/10.1016/B978-0-323-89835-5.00006-5

Table 1 Evidence for the protective effects of antioxidant vitamins on experimental models of Parkinson's disease.

Study	Model	Treatment (T) versus (vs.) Control (C)	Outcomes	Main Findings
Tanaka, Sotomatsu, & Hirai, 1992	Rat brain tissues	T: vitamin E-deficient rat vs. C: normal rat	Lipid peroxidation in substantia nigra of brain slice	Vitamin E deficiency greatly increased lipid peroxidation in substantia nigra
Offen, Ziv, Sternin, Melamed, & Hochman, 1996	Rat pheochromocytoma cell	T: 0.3 mM dopamine plus varying concentrations of vitamins C and E vs. C: 0.3 mM dopamine only	Cell proliferation	Vitamin C showed a mild protective effect at 0.1 mM but vitamin E did not show any protective effect at all concentrations; both were cytotoxic at high concentrations
Roghani & Behzadi, 2001	Rat	T: intramuscular injection of 24 IU/kg D-α-tocopheryl acid succinate, 3 times/week vs. C: none	Neuron death in substantia nigra after 6-OHDA lesion	Vitamin E protected against 6-OHDA-mediated cell death in substantia nigra
Pasbakhsh et al., 2008	Rat	T: intramuscular injection of 24 IU/kg D-α-tocopheryl acid succinate, every 2 days for 8 weeks vs. C: none	Neuron death after 6-OHDA lesion	Vitamin E injections protected against 6-OHDA-mediated cell death in locus coeruleus
di Matteo et al., 2009	Rat	T: high β-carotene content tomato in diet vs. C: no tomato	Striatal dopamine levels Plasma β-carotene levels	β-Carotene did not protect against 6-OHDA-mediated dopaminergic neuron death
Niso-Santano et al., 2010	Human neuroblastoma cell	T: preincubation with vitamin E vs. C: none	Paraquat-induced phosphorylation of MAPK	Preincubation with vitamin E prevented paraquat-induced phosphorylation of MAPK
Khan et al., 2012	Drosophila with neurons that express α-synuclein	T: L-ascorbic acid in diet at varying concentrations vs. C: normal diet	Climbing ability	L-Ascorbic acid in the diet delayed the loss of climbing ability in flies
Yin et al., 2012	Rat dopaminergic neurons Rat	T: 9-cis-retinoic acid injection vs. C: vehicle	Cell: cell death after 6-OHDA lesion Rat: death of nigrostriatal dopaminergic neurons	Early treatment with 9-cis-retinoic acid after lesion protected against 6-OHDA-mediated cell death and death of nigrostriatal dopaminergic neurons in rats

Table 1 Evidence for the protective effects of antioxidant vitamins on experimental models of Parkinson's disease—cont'd

Study	Model	Treatment (T) versus (vs.) Control (C)	Outcomes	Main Findings
Nakaso et al., 2014	Human neuroblastoma cell	T: tocotrienols vs. C: none	Cell death	Tocotrienols protected against oxidative stress toxin-mediated cell death
Nakaso et al., 2016	Mouse	T: 100 μg/kg of δ-tocotrienol administered orally vs. C: vehicle	Motor function tests dopaminergic neuron death in substantia nigra	δ-Tocotrienol protected against MPTP-mediated motor function deficits and death of dopaminergic neurons in substantia nigra
Wulansari et al., 2017	Rat neural stem cells	T: supplementation of vitamin C or E during cell passage vs. C: none	Loss of dopaminergic neurons	Vitamin C, but not vitamin E, protected against loss of dopaminergic neurons during cell passage
Gaba et al., 2019	Rat	T: intranasal vitamin E emulsion vs. C: none	Motor function tests Oxidative stress levels	Vitamin E protected against 6-OHDA-mediated motor function deficits and oxidative stress
Schirinzi et al., 2019	PINK1-knockout mouse	T: Vitamins A, C, or E, or β-carotene, preincubated in parasagittal brain slices (ex vivo) or intraperitoneal injections (in vivo) vs. C: none	Electrophysiology studies	Only vitamin E restored synaptic plasticity in PINK1-knockout mice

summarize the current understanding of the biological plausibility underlying possible protective effects of antioxidant vitamins in the pathogenesis of PD, conclude whether dietary antioxidant vitamins may affect PD risk in the population, and provide potential future research directions on the subject.

Table 2 Observational studies of the association between antioxidant vitamins and the risk of developing Parkinson's disease.

Study	Population/Follow-up time	Exposures	Main findings
Cross-sectional studies			
de Rijk et al., 1997	5342 Dutch men and women	Vitamins C and E, and β-carotene	Null: Vitamin C and β-carotene Lower risk: Vitamin E
Silva et al., 2015	34 Brazilian men and women	Vitamins A, C, and E	Null: Vitamins A and C Lower risk: Vitamin E
Case-control studies			
Hellenbrand et al., 1996	684 German men and women	Vitamins C and E, and β-carotene	Null: Vitamins C and E, and β-carotene
Logroscino et al., 1996	397 U.S. residents in a community	Vitamins A, C, and E	Null: Vitamins A, C, and E
Morens et al., 1996	420 Japanese-American men	Vitamin E	Null: Vitamin E
Scheider et al., 1997	107 U.S. residents	Vitamins C and E, α-carotene, β-carotene, and xanthins	Null: Vitamins C and E, α-carotene, β-carotene, and xanthins
Anderson et al., 1999	259 U.S. residents	Vitamins A, C, and E	Null: Vitamins A, C, and E
Johnson, Gorell, Rybicki, Sanders, & Peterson, 1999	558 U.S. residents from a population-based sample	Vitamins A, C, and E, and β-carotene	Null: Vitamins A, C, and E, and β-carotene
Paganini-Hill, 2001	2715 U.S. residents from a population-based cohort	Vitamins A, C, and E, and β-carotene	Null: Vitamins A, C, and E, and β-carotene
Powers et al., 2003	638 U.S. residents	Vitamins C and E, and β-carotene	Null: Vitamins C and E, and β-carotene
Ådén et al., 2011	115 Swedish men and women	Vitamins A, C, and E	Null: Vitamins A, C, and E
Miyake et al., 2011	617 Japanese men and women	Vitamins C and E, α-carotene, β-carotene, and cryptoxanthin	Null: Vitamin C, α-carotene, and cryptoxanthin Lower risk: Vitamin E and β-carotene
Schirinzi et al., 2019	200 Italian men and women	Vitamin E	Lower risk: Vitamin E
Cohort studies			
Zhang et al., 2002	124,221 U.S. men and women/12 years for men, 14 years for women	Vitamins C and E, α-carotene, β-carotene, and β-cryptoxanthin	Null: Vitamin C, α-carotene, β-carotene, and β-cryptoxanthin Lower risk: Vitamin E

Table 2 Observational studies of the association between antioxidant vitamins and the risk of developing Parkinson's disease—cont'd

Study	Population/Follow-up time	Exposures	Main findings
Abbott et al., 2003	8006 Japanese-American men/30 years	Vitamins A and C, α-tocopherol, and β-carotene	Null: Vitamins A and C, α-tocopherol, and β-carotene
Hughes et al., 2016	129,422 U.S. men and women/24 years	Vitamins C and E, α-carotene, β-carotene, and β-cryptoxanthin	Null: Vitamins C and E, α-carotene, β-carotene, and β-cryptoxanthin
Yang, Wolk, Håkansson, Pedersen, & Wirdefeldt, 2017	84,774 Swedish men and women/14.9 years	Vitamins C and E, and β-carotene	Null: Vitamins C and E in men Lower risk: β-carotene, and Vitamins C and E in women
Ying et al., 2020	60,249 Chinese men and women/19.4 years	Vitamins A, C, and E, α carotene, β-carotene, and β-cryptoxanthin	Null: Vitamins A, C, and E, α-carotene, β-carotene, and β-cryptoxanthin
Hantikainen et al., 2021	43,865 Swedish men and women/17.6 years	Vitamins C and E, and β-carotene	Null: β-carotene Lower risk: Vitamins C and E
Meta-analyses			
Etminan, Gill, & Samii, 2005	1 cross-sectional, 6 case-control, and 1 cohort study	Vitamins C and E, and β-carotene	Null: Vitamin C and β-carotene Lower risk: Vitamin E
Takeda et al., 2014	1 cross-sectional, 11 case-control, and 1 cohort study	Vitamin A, α-carotene, β-carotene, and β-cryptoxanthin	Null: Vitamin A, α-carotene, β-carotene, and β-cryptoxanthin

The "oxidative stress" hypothesis

The idea that the pathogenesis of PD stemmed from oxidative stress is not a new one. As early as the 1980s, neuropathological studies have suggested the existence of a possible link between the generation of free radicals and the development of PD (Perry, Godin, & Hansen, 1982), possibly due to an increase in lipid peroxidation in the dopaminergic cells (Saggu et al., 1989). There are multiple molecular mechanisms that point to the plausibility of this hypothesis as well. The metabolism pathways for dopamine involve its conversion into several toxic compounds, namely, (1) dopamine-quinones, which have been implicated in the formation of Lewy bodies (Conway, Rochet, Bieganski, & Lansbury Jr, 2001); (2) 6-OHDA, which is a neurotoxin (Jenner, Dexter, Sian, Schapira, & Marsden, 1992); and (3) hydrogen peroxide, which causes lipid peroxidation (Jenner et al., 1992). Moreover, several of the genetic mutations associated with PD, such as LRRK2 and DJ-1, have been found to induce oxidative stress in neurons (Nguyen et al., 2011; Parsanejad et al., 2014).

Similarly, as the "oxidative stress" hypothesis in PD gains traction, studies that examined whether dietary vitamin antioxidants could delay or even prevent PD have also garnered increasing interest. While early studies on diet (Golbe, Farrell, & Davis, 1988) only examined food groups, with further development of food frequency questionnaire and creation of food composition tables (Butrum, 1992), it has become possible for observational studies to examine the effects of individual antioxidant vitamins in the diet on PD risk.

Vitamin A

Vitamin A is a family of fat-soluble compounds that are converted to retinol in the liver (Blomhoff, Green, Berg, & Norum, 1990). Preformed vitamin A is obtained from animal sources such as dairy, eggs, oily fish, and liver products in the form of retinyl esters (Benzie & Choi, 2014), while provitamin A precursors are obtained from yellow and green vegetables such as carrots, spinach, and sweet potatoes in the form of carotenoids, chiefly α-carotene, β-carotene, and β-cryptoxanthin (Block, 1994). The antioxidant properties of vitamin A are related to its ability to prevent lipid peroxidation (Kartha & Krishnamurthy, 1977) and scavenge for free radicals, especially in a lipid-based environment (Tesoriere et al., 1993). More pertinent to PD, a study found that the addition of vitamin A could prevent the formation of α-synuclein fibrils in a dose-dependent manner, and also destabilize preformed α-synuclein fibrils (Ono & Yamada, 2006).

Evidence from experimental studies

Evidence on the effect of vitamin A on PD risk in experimental models is scant and conflicting. Yin et al. (2012) found that early treatment with 9-*cis*-retinoic acid was able to significantly reduce 6-OHDA-mediated neuron death both in rat dopaminergic neurons in vitro and in rat models in vivo. On the other hand, Schirinzi et al. (2019) reported that neither preincubation of brain slices with vitamin A or β-carotene, nor intraperitoneal injections of vitamin A or β-carotene in their PINK1 mouse knockout model, significantly improved synaptic plasticity measured via electrophysical recordings. Another study using a high β-carotene diet (di Matteo et al., 2009) also observed that β-carotene did not protect against 6-OHDA-mediated dopaminergic neuron death in rats.

Evidence from observational studies

A total of five case-control studies (Ådén et al., 2011; Anderson et al., 1999; Johnson et al., 1999; Logroscino et al., 1996; Paganini-Hill, 2001), one cross-sectional study (Silva et al., 2015), and two cohort studies (Abbott et al., 2003; Ying et al., 2020) have assessed the effects of vitamin A intake on PD risk, and all have reported null results. A meta-analysis (Takeda et al., 2014) also found no association between vitamin A intake and PD risk.

Another six case-control studies (Hellenbrand et al., 1996; Johnson et al., 1999; Miyake et al., 2011; Paganini-Hill, 2001; Powers et al., 2003; Scheider et al., 1997), one cross-sectional study (de Rijk et al., 1997), and six cohort studies (Abbott et al., 2003; Hantikainen et al., 2021; Hughes et al., 2016; Yang et al., 2017; Ying et al., 2020; Zhang et al., 2002) have assessed the effects of dietary provitamin A precursors on PD risk. Although a case-control study in Japan (Miyake et al., 2011) and the Swedish Mammography Cohort and the Cohort of Swedish Men (Yang et al., 2017) have reported that β-carotene intake could reduce PD risk, most other studies on β-carotene, including an early meta-analysis conducted by Etminan et al. (2005), found no association between β-carotene intake and PD risk (Abbott et al., 2003; de Rijk et al., 1997; Hantikainen et al., 2021; Hellenbrand et al., 1996; Hughes et al., 2016; Johnson et al., 1999; Paganini-Hill, 2001; Powers et al., 2003; Scheider et al., 1997; Ying et al., 2020; Zhang et al., 2002). All studies on α-carotene (Hughes et al., 2016; Miyake et al., 2011; Scheider et al., 1997; Ying et al., 2020; Zhang et al., 2002) and β-cryptoxanthin (Hughes et al., 2016; Miyake et al., 2011; Scheider et al., 1997; Ying et al., 2020; Zhang et al., 2002) also did not find any association between intake levels and PD risk. Finally, the meta-analysis conducted by Takeda et al. (2014) concurred that there was no significant association between intake of these provitamin A carotenoids and PD risk.

Vitamin C

Also known as ascorbic acid, vitamin C is a water-soluble antioxidant found in many fruits (Benzie & Choi, 2014), with neuroprotective and neuromodulating effects (Ballaz & Rebec, 2019). More pertinently, ascorbate was noted to be able to directly interact with the α-synuclein-Cu^{2+} complex (Wang, Liu, Zhang, Peng, & Zhou, 2010), and hence, it is biologically plausible that vitamin C may play a role in mitigating the development of PD.

Evidence from experimental studies

Evidence on the effect of vitamin C on PD risk in experimental models has been conflicting. In in vitro studies, Offen et al. (1996) reported that lower vitamin C concentrations promoted cell proliferation in rat pheochromocytoma cells, while Wulansari et al. (2017) reported that vitamin C protected against dopaminergic neuron death during the passage of rat neural stem cells. In in vivo studies, Khan et al. (2012) noted that L-ascorbic acid in the diet could delay the development of motor dysfunction in *Drosophila* engineered to express α-synuclein, but Schirinzi et al. (2019) reported that in their PINK1-knockout mice, neither preincubation of brain slices nor intraperitoneal injections of vitamin C improved synaptic plasticity.

Evidence from observational studies

A total of nine case-control studies (Ådén et al., 2011; Anderson et al., 1999; Hellenbrand et al., 1996; Johnson et al., 1999; Logroscino et al., 1996; Miyake et al., 2011; Paganini-Hill, 2001; Powers et al., 2003; Scheider et al., 1997), two cross-sectional studies (de Rijk et al., 1997; Silva et al., 2015), and six cohort studies (Abbott et al., 2003; Hantikainen et al., 2021; Hughes et al., 2016; Yang et al., 2017; Ying et al., 2020; Zhang et al., 2002) have assessed the association between vitamin C intake and PD risk. Interestingly, only two large cohort studies from Sweden have reported inverse associations between vitamin C intake and PD risk, namely, the Swedish Mammography Cohort and the Swedish National March Cohort (Hantikainen et al., 2021; Yang et al., 2017). All other large cohort studies, including the Cohort of Swedish Men, the Singapore Chinese Health Study, the Honolulu-Asia Aging Study, the Nurse's Health Study, and the Health Professionals Follow-Up Study (Abbott et al., 2003; Hughes et al., 2016; Yang et al., 2017; Ying et al., 2020; Zhang et al., 2002), as well as an earlier meta-analysis conducted by Etminan et al. (2005), have reported null associations.

Vitamin E

Widely acknowledged as one of the most potent antioxidant nutrients in the diet, Vitamin E comprises a family of tocopherol and tocotrienol compounds found in plant oils, nuts, and seeds (Benzie & Choi, 2014; Morrissey & Sheehy, 1999). It is a potent antioxidant that acts not only as a scavenger of free radicals (Iannone, Rota, Bergamini, Tomasi, & Canfield, 1998; Tesoriere et al., 1993) but may also reduce lipid peroxidation and modulate signal transduction pathways to affect response to oxidative stress (Morrissey & Sheehy, 1999).

Evidence from experimental studies

Experimental studies on the effect of vitamin E have been conducted as early as three decades ago, when Tanaka et al. (1992) reported that vitamin *E*-deficient rats had significantly higher lipid peroxidation in the substantia nigra than control rats. In in vitro studies, although some investigators did not observe neuroprotective properties associated with vitamin E in dopamine-producing cells of rodent origin (Offen et al., 1996; Wulansari et al., 2017), other studies on human neuroblastoma cells reported that vitamin E attenuated cytotoxin-mediated molecular changes and protected against cell death (Nakaso et al., 2014; Niso-Santano et al., 2010). In in vivo studies, investigators using rodent models have generally concurred that vitamin E is able to protect against neurotoxin-mediated neuron death (Gaba et al., 2019; Nakaso et al., 2016; Pasbakhsh et al., 2008; Roghani & Behzadi, 2001) as well as restore synaptic plasticity in PINK1-knockout mice (Schirinzi et al., 2019).

Evidence from observational studies

A total of 11 case-control studies (Ådén et al., 2011; Anderson et al., 1999; Hellenbrand et al., 1996; Johnson et al., 1999; Logroscino et al., 1996; Miyake et al., 2011; Morens et al., 1996; Paganini-Hill, 2001; Powers et al., 2003; Scheider et al., 1997; Schirinzi et al., 2019), two cross-sectional studies (de Rijk et al., 1997; Silva et al., 2015), and six cohort studies (Abbott et al., 2003; Hantikainen et al., 2021; Hughes et al., 2016; Yang et al., 2017; Ying et al., 2020; Zhang et al., 2002) have assessed the effects of vitamin E intake on PD risk. Like the experimental studies, results from epidemiologic studies have also been conflicting. Although two case-control studies (Miyake et al., 2011, Schirinzi et al., 2019) and an early meta-analysis (Etminan et al., 2005) reported inverse associations between vitamin E and PD risk, most of the other case-control studies concurred that there was no significant association (Ådén et al., 2011; Anderson et al., 1999; Hellenbrand et al., 1996; Johnson et al., 1999; Logroscino et al., 1996; Morens et al., 1996; Paganini-Hill, 2001; Powers et al., 2003; Scheider et al., 1997). Interestingly, the same two Swedish cohort studies that had reported inverse associations between vitamin C and PD risk also noted that vitamin E intake reduced PD risk in their cohorts (Hantikainen et al., 2021; Yang et al., 2017), suggesting that there could be residual confounding from another dietary component often consumed with antioxidant vitamins. This was the case in the Singapore Chinese Health Study, which initially reported an inverse association between vitamin E and PD risk (Tan et al., 2008), but later noted that this association was attenuated and no longer significant after further adjustment for other dietary factors found to be protective in the cohort (Ying et al., 2020). Three other large USA-based cohorts, namely, the Honolulu-Asia Aging Study, the Nurse's Health Study, and the Health Professionals Follow-Up Study, have all reported null associations between dietary vitamin E and PD risk (Abbott et al., 2003; Hughes et al., 2016).

Interventional studies

Few interventional studies have examined the effect of antioxidant vitamins in PD patients. Of these, the most well-known is the Deprenyl And Tocopherol Antioxidative

Therapy Of Parkinsonism (DATATOP) trial, which was a randomized, double-blind, placebo-controlled trial involving 800 untreated early-stage PD patients followed up for 24 months (Parkinson Study Group, 1993). Of the two compounds studied in this trial, selegiline was found to be effective at delaying the onset of motor symptoms, but vitamin E had no effect. Given the null association with vitamin E from this large, well-designed trial, interest in clinical PD trials involving dietary antioxidants has somewhat waned, although there has been some recent curiosity in testing the effect of vitamin E in combination with other antioxidant supplements such as omega-3 fatty acids (Tamtaji et al., 2019).

Conclusions

Despite strong biological plausibility and some promising results in experimental models, there is little evidence from human studies that antioxidant vitamins in the diet have a significant effect on PD risk. Given that animals are often directly injected with antioxidants or fed highly concentrated extracts in experimental models, one possibility for the discrepancy with human studies is that antioxidant vitamins are not consumed in sufficient quantities in most diets to have measurable effects in reducing PD risk that can be detected in observational studies based in the population. Further research in this area could focus on examining combinations of antioxidant supplements at higher doses in well-conducted randomized controlled trials.

Applications to other neurological conditions

In this chapter, we have described the "oxidative stress" hypothesis as applicable to PD, and summarized the current state of evidence for the potential of dietary antioxidant vitamins in the prevention or treatment of PD. Of note, many other neurodegenerative diseases, such as Alzheimer's disease, amyotrophic lateral sclerosis, and tardive dyskinesia have also been linked to oxidative stress (Kamat et al., 2008). As such, the effect of antioxidant vitamins, particularly vitamin E, has also been investigated in other neurodegenerative diseases.

Preclinical studies in these neurodegenerative diseases have yielded promising results. In cellular models of Alzheimer's disease, vitamin E has been found to prevent amyloid β-peptide-induced neurotoxicity (Butterfield, Koppal, Subramaniam, & Yatin, 1999). In a rodent model of amyotrophic lateral sclerosis, vitamin E has been found to slow the progression of motor symptoms (Gurney et al., 1996). In addition, vitamin E has also been examined in clinical studies, although the results vary from disease to disease. A meta-analysis conducted by Boomershine, Shelton, and Boomershine (1999) assessed 18 clinical trials that evaluated the efficacy of vitamin E as treatment for tardive dyskinesia, and found that vitamin E was beneficial particularly in early stage of tardive dyskinesia. A double-blinded, randomized, placebo-controlled trial of selegiline and α-tocopherol was conducted in patients with moderate Alzheimer's disease, and α-tocopherol was found to slow the progression of Alzheimer's disease (Sano et al., 1997). Nevertheless, many observational studies on Alzheimer's disease risk have reported null results (Kamat et al., 2008). As such, the evidence for the effect of antioxidant vitamins in other neurodegenerative diseases remains inconclusive.

Other components of interest

In this chapter, we have described the use of antioxidant vitamins, namely, vitamins A, C, and E, in the prevention or treatment of PD. However, these vitamins are not the only dietary antioxidants that have been examined in PD. Notably, catechins in green tea, such as epigallocatechin-3-gallate, have also been studied in relation to PD. In experimental models, catechins have been found to inhibit α-synuclein aggregation, reduce nigrostriatal dopamine loss after lesion, and improve motor function (Chiu, Venkatakrishnan, & Wang, 2020). However, results have been more equivocal in observational studies. While a study in Japan observed an inverse association between Japanese and Chinese teas and PD risk (Tanaka et al., 2011), two other studies conducted in Singapore and China, respectively, reported no statistically significant association between green tea consumption and PD risk (Chen et al., 2015; Tan et al., 2008).

Data on other antioxidants are less comprehensive. Studies on coenzyme Q10, an antioxidant found in the body, had yielded encouraging results in experimental studies (Chiu et al., 2020), and coenzyme Q10 was indeed found to be reduced in the serum of PD patients compared to controls (Jin et al., 2014). Unfortunately, a large phase III randomized controlled trial of high-dose coenzyme Q10 for PD treatment (Parkinson Study Group QE3 Investigators et al., 2014), as well as a meta-analysis of five randomized controlled trials (Negida et al., 2016), has failed to produce evidence of any clinical benefit. There have also been many promising experimental studies on other naturally occurring antioxidants such as curcumin from turmeric and resveratrol from fruits (Chiu et al., 2020; Wang et al., 2017), but their efficacy has not been established in clinical studies.

Key facts of antioxidant vitamins

- There are three main antioxidant vitamins in the diet: vitamins A, C, and E.
- Vitamin A is a family of fat-soluble compounds obtained from the diet either as preformed vitamin A from animal sources like liver or dairy products, or as provitamin A carotenoids from yellow-green vegetables.
- Vitamin C is a water-soluble compound obtained from fruits and vegetables.
- Vitamin E is a family of fat-soluble tocopherols and tocotrienols obtained from plant oils and nuts.

- All three vitamins exhibit free radical scavenging properties and reduce lipid peroxidation under experimental conditions, and may thus protect cells against oxidative stress damage.
- Maintaining adequate intake of these vitamins in the diet has been recommended for the prevention of chronic diseases and promotion of health.

Mini-dictionary of terms

- **Ascorbic acid**. Also known as vitamin C, this is a water-soluble compound with many neuroprotective and neuromodulating effects.
- **Daisuke-junko-1 (DJ-1)**. Also known as PARK7, DJ-1 is a peptidase that protects neurons against oxidative stress. Autosomal recessive mutations in DJ-1 cause early-onset PD.
- **Leucine-rich repeat kinase 2 (LRRK2)**. Also known as PARK8, LRRK2 is an enzyme involved in autophagy and mitophagy. Autosomal dominant mutations in LRRK2 cause familial PD.
- **PTEN-induced kinase 1 (PINK1)**. PINK1 is an enzyme that protects against oxidative stress-induced mitochondrial dysfunction. Autosomal recessive mutations in PINK1 cause early-onset PD.
- **Retinol**. The active form of vitamin A, this is obtained from the diet from two different sources: preformed vitamin A and provitamin A carotenoids.
- **Tocopherol**. Part of a family of fat-soluble compounds known collectively as vitamin E, this is one of the most potent dietary antioxidant nutrients.

Summary points

- One potential molecular mechanism underpinning the development of PD is oxidative stress.
- Studies have focused on finding modifiable risk factors that can prevent the development of PD.
- One area of interest has been the diet, particularly the presence of dietary nutrients with antioxidant properties, such as the vitamins A, C, and E.
- Although positive findings have not been consistently reported, some experimental studies on antioxidant vitamins have reported promising results in both cellular and animal models of PD.
- Observational studies in both Western and Asian populations have largely concurred that higher intake of antioxidant vitamins in the diet is not associated with a reduction in PD risk.

- The largest randomized controlled trial on antioxidant vitamins in PD, DATATOP, has reported that daily high-dose supplementation of vitamin E could not delay the onset of functional symptoms.

References

Abbott, R. D., Ross, G. W., White, L. R., Sanderson, W. T., Burchfiel, C. M., Kashon, M., et al. (2003). Environmental, life-style, and physical precursors of clinical Parkinson's disease: Recent findings from the Honolulu-Asia aging study. *Journal of Neurology*, *250*(Suppl 3), III30–III39. https://doi.org/10.1007/s00415-003-1306-7.

Ådén, E., Carlsson, M., Poortvliet, E., Stenlund, H., Linder, J., Edström, M., et al. (2011). Dietary intake and olfactory function in patients with newly diagnosed Parkinson's disease: A case-control study. *Nutritional Neuroscience*, *14*(1), 25–31. https://doi.org/10.1179/174313211X12966635733312.

Anderson, C., Checkoway, H., Franklin, G. M., Beresford, S., Smith-Weller, T., & Swanson, P. D. (1999). Dietary factors in Parkinson's disease: The role of food groups and specific foods. *Movement Disorders: Official Journal of the Movement Disorder Society*, *14*(1), 21–27. https://doi.org/10.1002/1531-8257(199901)14:1<21::aid-mds1006>3.0.co;2-y.

Ascherio, A., & Schwarzschild, M. A. (2016). The epidemiology of Parkinson's disease: Risk factors and prevention. *The Lancet. Neurology*, *15*(12), 1257–1272. https://doi.org/10.1016/S1474-4422(16)30230-7.

Ballaz, S. J., & Rebec, G. V. (2019). Neurobiology of vitamin C: Expanding the focus from antioxidant to endogenous neuromodulator. *Pharmacological Research*, *146*, 104321. https://doi.org/10.1016/j.phrs.2019.104321.

Belvisi, D., Pellicciari, R., Fabbrini, G., Tinazzi, M., Berardelli, A., & Defazio, G. (2020). Modifiable risk and protective factors in disease development, progression and clinical subtypes of Parkinson's disease: What do prospective studies suggest? *Neurobiology of Disease*, *134*, 104671. https://doi.org/10.1016/j.nbd.2019.104671.

Benzie, I. F., & Choi, S. W. (2014). Antioxidants in food: Content, measurement, significance, action, cautions, caveats, and research needs. *Advances in Food and Nutrition Research*, *71*, 1–53. https://doi.org/10.1016/B978-0-12-800270-4.00001-8.

Block, G. (1994). Nutrient sources of provitamin a carotenoids in the American diet. *American Journal of Epidemiology*, *139*(3), 290–293. https://doi.org/10.1093/oxfordjournals.aje.a116996.

Blomhoff, R., Green, M. H., Berg, T., & Norum, K. R. (1990). Transport and storage of vitamin A. *Science (New York, N.Y.)*, *250*(4979), 399–404. https://doi.org/10.1126/science.2218545.

Boomershine, K. H., Shelton, P. S., & Boomershine, J. E. (1999). Vitamin E in the treatment of tardive dyskinesia. *The Annals of Pharmacotherapy*, *33*(11), 1195–1202. https://doi.org/10.1345/aph.17025.

Butrum, R. R. (1992). Overview of international food composition data bases. *World Review of Nutrition and Dietetics*, *68*, 1–14. https://doi.org/10.1159/000420936.

Butterfield, D. A., Koppal, T., Subramaniam, R., & Yatin, S. (1999). Vitamin E as an antioxidant/free radical scavenger against amyloid beta-peptide-induced oxidative stress in neocortical synaptosomal membranes and hippocampal neurons in culture: Insights into Alzheimer's disease. *Reviews in the Neurosciences*, *10*(2), 141–149. https://doi.org/10.1515/revneuro.1999.10.2.141.

Chen, H., Ding, D., Wang, J., Zhao, Q., Meng, H., Li, H., et al. (2015). Parkinson's disease research in a prospective cohort in China. *Parkinsonism & Related Disorders*, *21*(10), 1200–1204. https://doi.org/10.1016/j.parkreldis.2015.08.020.

Chiu, H. F., Venkatakrishnan, K., & Wang, C. K. (2020). The role of nutraceuticals as a complementary therapy against various neurodegenerative diseases: A mini-review. *Journal of Traditional and Complementary Medicine*, *10*(5), 434–439. https://doi.org/10.1016/j.jtcme.2020.03.008.

Conway, K. A., Rochet, J. C., Bieganski, R. M., & Lansbury, P. T., Jr. (2001). Kinetic stabilization of the alpha-synuclein protofibril by a dopamine-alpha-synuclein adduct. *Science (New York, N.Y.)*, *294*(5545), 1346–1349. https://doi.org/10.1126/science.1063522.

de Rijk, M. C., Breteler, M. M., den Breeijen, J. H., Launer, L. J., Grobbee, D. E., van der Meché, F. G., et al. (1997). Dietary antioxidants and Parkinson disease. The Rotterdam Study. *Archives of Neurology, 54*(6), 762–765. https://doi.org/10.1001/archneur.1997.00550180070015.

di Matteo, V., Pierucci, M., Di Giovanni, G., Dragani, L. K., Murzilli, S., Poggi, A., et al. (2009). Intake of tomato-enriched diet protects from 6-hydroxydopamine-induced degeneration of rat nigral dopaminergic neurons. *Journal of Neural Transmission. Supplementum, 73*, 333–341. https://doi.org/10.1007/978-3-211-92660-4_28.

Etminan, M., Gill, S. S., & Samii, A. (2005). Intake of vitamin E, vitamin C, and carotenoids and the risk of Parkinson's disease: A meta-analysis. *The Lancet: Neurology, 4*(6), 362–365. https://doi.org/10.1016/S1474-4422(05)70097-1.

G.B.D. 2016 Neurology Collaborators. (2019). Global, regional, and national burden of neurological disorders, 1990-2016: A systematic analysis for the global burden of disease study 2016. *The Lancet: Neurology, 18*(5), 459–480. https://doi.org/10.1016/S1474-4422(18)30499-X.

Gaba, B., Khan, T., Haider, M. F., Alam, T., Baboota, S., Parvez, S., et al. (2019). Vitamin E loaded Naringenin Nanoemulsion via intranasal delivery for the Management of Oxidative Stress in a 6-OHDA Parkinson's disease model. *BioMed Research International, 2019*, 2382563. https://doi.org/10.1155/2019/2382563.

Golbe, L. I., Farrell, T. M., & Davis, P. H. (1988). Case-control study of early life dietary factors in Parkinson's disease. *Archives of Neurology, 45*(12), 1350–1353. https://doi.org/10.1001/archneur.1988.00520360068014.

Gurney, M. E., Cutting, F. B., Zhai, P., Doble, A., Taylor, C. P., Andrus, P. K., et al. (1996). Benefit of vitamin E, riluzole, and gabapentin in a transgenic model of familial amyotrophic lateral sclerosis. *Annals of Neurology, 39*(2), 147–157. https://doi.org/10.1002/ana.410390203.

Hantikainen, E., Trolle Lagerros, Y., Ye, W., Serafini, M., Adami, H. O., Bellocco, R., et al. (2021). Dietary antioxidants and the risk of Parkinson disease: The Swedish National March Cohort. *Neurology, 96*(6), e895–e903. https://doi.org/10.1212/WNL.0000000000011373.

Hawkes, C. H., Del Tredici, K., & Braak, H. (2010). A timeline for Parkinson's disease. *Parkinsonism & Related Disorders, 16*(2), 79–84. https://doi.org/10.1016/j.parkreldis.2009.08.007.

Hellenbrand, W., Boeing, H., Robra, B. P., Seidler, A., Vieregge, P., Nischan, P., et al. (1996). Diet and Parkinson's disease. II. A possible role for the past intake of specific nutrients. Results from a self-administered food-frequency questionnaire in a case-control study. *Neurology, 47*(3), 644–650. https://doi.org/10.1212/wnl.47.3.644.

Hughes, K. C., Gao, X., Kim, I. Y., Rimm, E. B., Wang, M., Weisskopf, M. G., et al. (2016). Intake of antioxidant vitamins and risk of Parkinson's disease. *Movement Disorders: Official Journal of the Movement Disorder Society, 31*(12), 1909–1914. https://doi.org/10.1002/mds.26819.

Iannone, A., Rota, C., Bergamini, S., Tomasi, A., & Canfield, L. M. (1998). Antioxidant activity of carotenoids: An electron-spin resonance study on beta-carotene and lutein interaction with free radicals generated in a chemical system. *Journal of Biochemical and Molecular Toxicology, 12*(5), 299–304. https://doi.org/10.1002/(sici)1099-0461(1998)12:5<299::aid-jbt6>3.0.co;2-g.

Jankovic, J., & Tan, E. K. (2020). Parkinson's disease: Etiopathogenesis and treatment. *Journal of Neurology, Neurosurgery, and Psychiatry, 91*(8), 795–808. https://doi.org/10.1136/jnnp-2019-322338.

Jenner, P., Dexter, D. T., Sian, J., Schapira, A. H., & Marsden, C. D. (1992). Oxidative stress as a cause of nigral cell death in Parkinson's disease and incidental Lewy body disease. The Royal Kings and Queens Parkinson's disease research group. *Annals of Neurology, 32*(Suppl), S82–S87. https://doi.org/10.1002/ana.410320714.

Jin, H., Kanthasamy, A., Ghosh, A., Anantharam, V., Kalyanaraman, B., & Kanthasamy, A. G. (2014). Mitochondria-targeted antioxidants for treatment of Parkinson's disease: Preclinical and clinical outcomes. *Biochimica et Biophysica Acta, 1842*(8), 1282–1294. https://doi.org/10.1016/j.bbadis.2013.09.007.

Johnson, C. C., Gorell, J. M., Rybicki, B. A., Sanders, K., & Peterson, E. L. (1999). Adult nutrient intake as a risk factor for Parkinson's disease. *International Journal of Epidemiology, 28*(6), 1102–1109. https://doi.org/10.1093/ije/28.6.1102.

Kalia, L. V., & Lang, A. E. (2015). Parkinson's disease. *Lancet, 386*(9996), 896–912. https://doi.org/10.1016/S0140-6736(14)61393-3.

Kamat, C. D., Gadal, S., Mhatre, M., Williamson, K. S., Pye, Q. N., & Hensley, K. (2008). Antioxidants in central nervous system diseases: Preclinical promise and translational challenges. *Journal of Alzheimer's disease, 15*(3), 473–493. https://doi.org/10.3233/jad-2008-15314.

Kartha, V. N., & Krishnamurthy, S. (1977). Antioxidant function of vitamin A. *International Journal for Vitamin and Nutrition Research, 47*(4), 394–401.

Khan, S., Jyoti, S., Naz, F., Shakya, B., Rahul, Afzal, M., et al. (2012). Effect of L-ascorbic acid on the climbing ability and protein levels in the brain of Drosophila model of Parkinson's disease. *The International Journal of Neuroscience, 122*(12), 704–709. https://doi.org/10.3109/00207454.2012.709893.

Logroscino, G., Marder, K., Cote, L., Tang, M. X., Shea, S., & Mayeux, R. (1996). Dietary lipids and antioxidants in Parkinson's disease: A population-based, case-control study. *Annals of Neurology, 39*(1), 89–94. https://doi.org/10.1002/ana.410390113.

Miyake, Y., Fukushima, W., Tanaka, K., Sasaki, S., Kiyohara, C., Tsuboi, Y., et al. (2011). Dietary intake of antioxidant vitamins and risk of Parkinson's disease: A case-control study in Japan. *European Journal of Neurology, 18*(1), 106–113. https://doi.org/10.1111/j.1468-1331.2010.03088.x.

Morens, D. M., Grandinetti, A., Waslien, C. I., Park, C. B., Ross, G. W., & White, L. R. (1996). Case-control study of idiopathic Parkinson's disease and dietary vitamin E intake. *Neurology, 46*(5), 1270–1274. https://doi.org/10.1212/wnl.46.5.1270.

Morrissey, P. A., & Sheehy, P. J. (1999). Optimal nutrition: Vitamin E. *The Proceedings of the Nutrition Society, 58*(2), 459–468. https://doi.org/10.1017/s0029665199000609.

Nakaso, K., Horikoshi, Y., Takahashi, T., Hanaki, T., Nakasone, M., Kitagawa, Y., et al. (2016). Estrogen receptor-mediated effect of δ-tocotrienol prevents neurotoxicity and motor deficit in the MPTP mouse model of Parkinson's disease. *Neuroscience Letters, 610*, 117–122. https://doi.org/10.1016/j.neulet.2015.10.062.

Nakaso, K., Tajima, N., Horikoshi, Y., Nakasone, M., Hanaki, T., Kamizaki, K., et al. (2014). The estrogen receptor β-PI3K/Akt pathway mediates the cytoprotective effects of tocotrienol in a cellular Parkinson's disease model. *Biochimica et Biophysica Acta, 1842*(9), 1303–1312. https://doi.org/10.1016/j.bbadis.2014.04.008.

Negida, A., Menshawy, A., El Ashal, G., Elfouly, Y., Hani, Y., Hegazy, Y., et al. (2016). Coenzyme Q10 for patients with Parkinson's disease: A systematic review and meta-analysis. *CNS & Neurological Disorders Drug Targets, 15*(1), 45–53. https://doi.org/10.2174/1871527314666150821103306.

Nguyen, H. N., Byers, B., Cord, B., Shcheglovitov, A., Byrne, J., Gujar, P., et al. (2011). LRRK2 mutant iPSC-derived DA neurons demonstrate increased susceptibility to oxidative stress. *Cell Stem Cell, 8*(3), 267–280. https://doi.org/10.1016/j.stem.2011.01.013.

Niso-Santano, M., González-Polo, R. A., Bravo-San Pedro, J. M., Gómez-Sánchez, R., Lastres-Becker, I., Ortiz-Ortiz, M. A., et al. (2010). Activation of apoptosis signal-regulating kinase 1 is a key factor in paraquat-induced cell death: Modulation by the Nrf2/Trx axis. *Free Radical Biology & Medicine, 48*(10), 1370–1381. https://doi.org/10.1016/j.freeradbiomed.2010.02.024.

Offen, D., Ziv, I., Sternin, H., Melamed, E., & Hochman, A. (1996). Prevention of dopamine-induced cell death by thiol antioxidants: Possible implications for treatment of Parkinson's disease. *Experimental Neurology, 141*(1), 32–39. https://doi.org/10.1006/exnr.1996.0136.

Ono, K., & Yamada, M. (2006). Vitamin a potently destabilizes preformed alpha-synuclein fibrils in vitro: Implications for Lewy body diseases. *Neurobiology of Disease, 25*(2), 446–454. https://doi.org/10.1016/j.nbd.2006.10.010.

Paganini-Hill, A. (2001). Risk factors for Parkinson's disease: The leisure world cohort study. *Neuroepidemiology, 20*(2), 118–124. https://doi.org/10.1159/000054770.

Parkinson Study Group. (1993). Effects of tocopherol and deprenyl on the progression of disability in early Parkinson's disease. *The New England Journal of Medicine, 328*(3), 176–183. https://doi.org/10.1056/NEJM199301213280305.

Parkinson Study Group QE3 Investigators, Beal, M. F., Oakes, D., Shoulson, I., Henchcliffe, C., Galpern, W. R., et al. (2014). A randomized clinical trial of high-dosage coenzyme Q10 in early Parkinson disease: No evidence of benefit. *JAMA Neurology, 71*(5), 543–552. https://doi.org/10.1001/jamaneurol.2014.131.

Parsanejad, M., Bourquard, N., Qu, D., Zhang, Y., Huang, E., Rousseaux, M. W., et al. (2014). DJ-1 interacts with and regulates paraoxonase-2, an enzyme critical for neuronal survival in response to oxidative stress. *PLoS One*, *9*(9), e106601. https://doi.org/10.1371/journal.pone.0106601.

Pasbakhsh, P., Omidi, N., Mehrannia, K., Sobhani, A. G., Ragerdi Kashani, I., Abbasi, M., et al. (2008). The protective effect of vitamin E on locus coeruleus in early model of Parkinson's disease in rat: Immunoreactivity evidence. *Iranian Biomedical Journal*, *12*(4), 217–222.

Perry, T. L., Godin, D. V., & Hansen, S. (1982). Parkinson's disease: A disorder due to nigral glutathione deficiency? *Neuroscience Letters*, *33*(3), 305–310. https://doi.org/10.1016/0304-3940(82)90390-1.

Powers, K. M., Smith-Weller, T., Franklin, G. M., Longstreth, W. T., Jr., Swanson, P. D., & Checkoway, H. (2003). Parkinson's disease risks associated with dietary iron, manganese, and other nutrient intakes. *Neurology*, *60*(11), 1761–1766. https://doi.org/10.1212/01.wnl.0000068021.13945.7f.

Roghani, M., & Behzadi, G. (2001). Neuroprotective effect of vitamin E on the early model of Parkinson's disease in rat: Behavioral and histochemical evidence. *Brain Research*, *892*(1), 211–217. https://doi.org/10.1016/s0006-8993(00)03296-0.

Saggu, H., Cooksey, J., Dexter, D., Wells, F. R., Lees, A., Jenner, P., et al. (1989). A selective increase in particulate superoxide dismutase activity in parkinsonian substantia nigra. *Journal of Neurochemistry*, *53*(3), 692–697. https://doi.org/10.1111/j.1471-4159.1989.tb11759.x.

Sano, M., Ernesto, C., Thomas, R. G., Klauber, M. R., Schafer, K., Grundman, M., et al. (1997). A controlled trial of selegiline, alpha-tocopherol, or both as treatment for Alzheimer's disease. The Alzheimer's disease cooperative study. *The New England Journal of Medicine*, *336*(17), 1216–1222. https://doi.org/10.1056/NEJM199704243361704.

Scheider, W. L., Hershey, L. A., Vena, J. E., Holmlund, T., Marshall, J. R., & Freudenheim. (1997). Dietary antioxidants and other dietary factors in the etiology of Parkinson's disease. *Movement Disorders: Official Journal of the Movement Disorder Society*, *12*(2), 190–196. https://doi.org/10.1002/mds.870120209.

Schirinzi, T., Martella, G., Imbriani, P., Di Lazzaro, G., Franco, D., Colona, V. L., et al. (2019). Dietary vitamin E as a protective factor for Parkinson's disease: Clinical and experimental evidence. *Frontiers in Neurology*, *10*, 148. https://doi.org/10.3389/fneur.2019.00148.

Silva, M. Z. C., Fritzen, N. C., de Oliveira, M., da Silva, M. P., Petterle, R. R., Teive, H. A., et al. (2015). Protein intake, nitrogen balance and nutritional status in patients with Parkinson's disease; time for a change? *Nutricion Hospitalaria*, *31*(6), 2764–2770. https://doi.org/10.3305/nh.2015.31.6.8938.

Takeda, A., Nyssen, O. P., Syed, A., Jansen, E., Bueno-de-Mesquita, B., & Gallo, V. (2014). Vitamin A and carotenoids and the risk of Parkinson's disease: A systematic review and meta-analysis. *Neuroepidemiology*, *42*(1), 25–38. https://doi.org/10.1159/000355849.

Tamtaji, O. R., Taghizadeh, M., Aghadavod, E., Mafi, A., Dadgostar, E., Daneshvar Kakhaki, R., et al. (2019). The effects of omega-3 fatty acids and vitamin E co-supplementation on gene expression related to inflammation, insulin and lipid in patients with Parkinson's disease: A randomized, double-blind, placebo-controlled trial. *Clinical Neurology and Neurosurgery*, *176*, 116–121. https://doi.org/10.1016/j.clineuro.2018.12.006.

Tan, L. C., Koh, W. P., Yuan, J. M., Wang, R., Au, W. L., Tan, J. H., et al. (2008). Differential effects of black versus green tea on risk of Parkinson's disease in the Singapore Chinese health study. *American Journal of Epidemiology*, *167*(5), 553–560. https://doi.org/10.1093/aje/kwm338.

Tanaka, K., Miyake, Y., Fukushima, W., Sasaki, S., Kiyohara, C., Tsuboi, Y., et al. (2011). Intake of Japanese and Chinese teas reduces risk of Parkinson's disease. *Parkinsonism & Related Disorders*, *17*(6), 446–450. https://doi.org/10.1016/j.parkreldis.2011.02.016.

Tanaka, M., Sotomatsu, A., & Hirai, S. (1992). Aging of the brain and vitamin E. *Journal of Nutritional Science and Vitaminology*, 240–243. https://doi.org/10.3177/jnsv.38.special_240.

Tesoriere, L., Ciaccio, M., Bongiorno, A., Riccio, A., Pintaudi, A. M., & Livrea, M. A. (1993). Antioxidant activity of all-trans-retinol in homogeneous solution and in phosphatidylcholine liposomes. *Archives of Biochemistry and Biophysics*, *307*(1), 217–223. https://doi.org/10.1006/abbi.1993.1581.

Wang, C., Liu, L., Zhang, L., Peng, Y., & Zhou, F. (2010). Redox reactions of the α-synuclein-Cu(2+) complex and their effects on neuronal cell viability. *Biochemistry*, *49*(37), 8134–8142. https://doi.org/10.1021/bi1010909.

Wang, X. S., Zhang, Z. R., Zhang, M. M., Sun, M. X., Wang, W. W., & Xie, C. L. (2017). Neuroprotective properties of curcumin in toxin-base animal models of Parkinson's disease: A systematic experiment literatures review. *BMC Complementary and Alternative Medicine*, *17*(1), 412. https://doi.org/10.1186/s12906-017-1922-x.

Wulansari, N., Kim, E. H., Sulistio, Y. A., Rhee, Y. H., Song, J. J., & Lee, S. H. (2017). Vitamin C-induced epigenetic modifications in donor NSCs establish midbrain marker expressions critical for cell-based therapy in Parkinson's disease. *Stem Cell Reports*, *9*(4), 1192–1206. https://doi.org/10.1016/j.stemcr.2017.08.017.

Yang, F., Wolk, A., Håkansson, N., Pedersen, N. L., & Wirdefeldt, K. (2017). Dietary antioxidants and risk of Parkinson's disease in two population-based cohorts. *Movement Disorders: Official Journal of the Movement Disorder Society*, *32*(11), 1631–1636. https://doi.org/10.1002/mds.27120.

Yin, L. H., Shen, H., Diaz-Ruiz, O., Bäckman, C. M., Bae, E., Yu, S. J., et al. (2012). Early post-treatment with 9-cis retinoic acid reduces neurodegeneration of dopaminergic neurons in a rat model of Parkinson's disease. *BMC Neuroscience*, *13*, 120. https://doi.org/10.1186/1471-2202-13-120.

Ying, A. F., Khan, S., Wu, Y., Jin, A., Wong, A., Tan, E. K., et al. (2020). Dietary antioxidants and risk of Parkinson's disease in the Singapore Chinese health study. *Movement Disorders: Official Journal of the Movement Disorder Society*, *35*(10), 1765–1773. https://doi.org/10.1002/mds.28173.

Zhang, S. M., Hernán, M. A., Chen, H., Spiegelman, D., Willett, W. C., & Ascherio, A. (2002). Intakes of vitamins E and C, carotenoids, vitamin supplements, and PD risk. *Neurology*, *59*(8), 1161–1169. https://doi.org/10.1212/01.wnl.0000028688.75881.12.

CHAPTER 33

Role of calcium in Parkinson's disease

Payam Saadat[a] and Kiarash Saleki[b,c]
[a]Mobility Impairment Research Center, Health Research Institute, Babol University of Medical Sciences, Babol, Iran
[b]Student Research Committee, Babol University of Medical Sciences, Babol, Iran
[c]USERN Office, Babol University of Medical Sciences, Babol, Iran

Abbreviations

AD	Alzheimer's disease
AFSSA	The French Food Safety Agency
AI	adequate intake
ALS	amyotrophic lateral sclerosis
BMD	bone mineral density
EFSA	European Food Safety Authority
EOPD	early-onset PD
FAO	Food and Agriculture Organization
FSH	follicle-stimulating hormone
GWAS	Genome-Wide Association Study
HCN	Health Council of the Netherlands
IOM	USA Institute of Medicine
LH	luteinizing hormone
MS	multiple sclerosis
NCM	Nordic Council of Ministers
PD	Parkinson's disease
PPN	pedunculopontine nucleus
PTH	parathyroid hormone
TH	tyrosine hydroxylase
WHO	World Health Organization

Introduction

Calcium is a mineral that plays key roles in numerous physiological processes. The role of calcium intake has been verified in bone disorders such as osteoporosis. Recently, the effects of both dietary and supplemental calcium have been expanded to other health issues (Ross et al., 2011; WHO, 2004). Calcium is mostly found in dairy products, but can also be found in significant amounts in vegetables, supplements, and traditional herbs (Cormick & Belizán, 2019; Service, U. D. o. A. A. R, 2019). Moreover, sources and levels of calcium intake are different across the globe due to varied patterns of diet and socioeconomic status. For instance, dairy products form about 14% of overall dietary energy intake in more advanced regions, while they form solely about 4% of total energy

Vitamins and Minerals in Neurological Disorders
https://doi.org/10.1016/B978-0-323-89835-5.00021-1

intake in less developed regions (Silanikove, Leitner, & Merin, 2015). Of note, these differences effect the calcium levels in addition to factors like hormonal profile, pregnancy, genetic differences, and dietary components that affect calcium absorption (e.g., phytic acid and oxalic acid) (Amalraj & Pius, 2017; Kovacs, 2000).

Parkinson's disease (PD) follows Alzheimer's disease (AD) among the most prevalent neurodegenerative illnesses globally. Although nonmotor presentations may be found, the classical clinical presentations include motor malfunction that presents with bradykinesia, tremor at rest, and rigidity, with postural inconsistency developing at a more advanced stage. PD arises due to the loss of dopaminergic neurons of the substantia nigra pars compacta, with a definitive hallmark being cellular accumulation of α-synuclein, Lewy bodies, or Lewy neurites. To date, etiology of PD has not been determined clearly. However, several genetic, environmental, and lifestyle risk elements have been identified. Influence of lifestyle factors including diet, smoking, intake of caffeine, and pesticide exposure has been speculated to affect the prospect of developing PD (Kouli, Torsney, & Kuan, 2018).

Recent clinical and preclinical research has explored the link between dietary calcium intake and several neurodegenerative conditions, including PD (Samavarchi Tehrani et al., 2020), AD (Kern et al., 2016), and amyotrophic lateral sclerosis (ALS) (Yasui, Ota, & Yoshida, 1997). Furthermore, many studies have shown that the imbalance of calcium levels contributes to the development of PD (Samavarchi Tehrani et al., 2020; Surmeier, Guzman, Sanchez-Padilla, & Schumacker, 2011; Surmeier & Schumacker, 2013). Even so, due to inconsistent results among studies, it remains unclear whether dietary calcium or calcium levels can predict PD. This chapter goes on to explain the dietary and supplemental sources of calcium, factors that influence calcium levels, and the processes through which calcium or related minerals (e.g., magnesium) could affect PD. Also, we discuss factors that lead to inconsistencies in identifying this linkage in order to direct future studies in the field.

Parkinson's disease: Influence of diet in neurodegenerative mechanisms

First diagnosed by Dr. James Parkinson, PD is now recognized by symptoms like resting tremor, muscle rigidity, speech abnormality, postural instability, bradykinesia, abnormal gait, skin issues, urinary incontinence, constipation, and sleep disruption. The presentations of PD involve accelerated aging, oxidation imbalance, and mainly degeneration of dopaminergic nerve cells in the substantia nigra (Hermanns, Deal, & Haas, 2012). Two, direct and indirect, striatal pathways from substantia nigra have been found. The direct pathway mediates its action via the D1 receptor, which is not affected by dopamine deprivation. However, the indirect pathway exerts its function via the D2 receptor and is heavily affected by dopamine deprivation through degeneration of the dendritic spine. Spiny synapses have been presented as major clues in the pathology of neurological disorders, because they regulate many key factors such as postsynaptic scaffold and adhesion proteins. Additionally, Lewy body aggregates are made up of abnormal α-synuclein; accumulate in the substantia nigra, medulla oblongata, pons, and the neocortex; and help in neurodegeneration (Gerfen, 2006). A notable aspect of neuropathology in PD is its locally selective effects. Specific deterioration of substantia nigra pars compacta dopaminergic nerve cells causes motor dysfunction due to their vulnerability. Exclusive vulnerability in these neurons is due to (i) dependence on dopamine as a neuronal transmitter; (ii) a markedly branched and extensive axonal arbor; and (iii) a discrete pacemaker phenotype which relies on Ca^{2+} penetration by Cav1 channels and causes oxidative stress on basal mitochondria (Surmeier et al., 2017). A study has shown that the pedunculopontine nucleus (PPN) is entailed in the pathophysiology of PD by causing gait freeze and falls (Thevathasan et al., 2011). Cortical thinning has been found as a pathophysiological mechanism for PD and a possible explanation for PD-induced cognitive impairment (Colloby, Watson, Blamire, O'Brien, & Taylor, 2020). In addition to catecholamine mis-metabolism, a significant noradrenergic mis-metabolism could be found that could be explained by the accumulation of aldehyde intermediate products. These mis-metabolisms could contribute to behavioral alterations associated with PD (Salvatore, Disbrow, & Emborg, 2014). PD could cause deterioration in pragmatic communication (verbal and nonverbal social communications). For example, cases may present diminished turn taking, vocal tone, facial expression, and eye contact as signs of cognitive impairment (Hall, Ouyang, Lonnquist, & Newcombe, 2011). Altogether, these arguments drive the line from basic pathophysiology to more clinical aspects.

Currently, evidence highlights that interaction between genome, environment (e.g., insecticides and pesticides), and lifestyle (e.g., diet and exercise) affects the initiation and progression of PD (Gopalakrishna & Alexander, 2015). For example, doing certain types of physical activity (e.g., Taichi) may increase neuroplasticity and improve PD symptoms. In addition, it is advised that the diet of PD patients is checked thoroughly to avoid processed food, gluten, heavy metals, and agents that cause sensitivity and intolerance. PD pathophysiology substantially relies on enhanced neuroinflammation. Therefore, research has suggested the consumption of calcium, turmeric, flavonoids, and omega-three fatty acids for their antiinflammatory antioxidative potential. Other changes to lifestyle such as sleep pattern and stress reduction have been shown to improve the symptoms of PD (Lister, 2020).

Nutritional sources of calcium

Calcium-rich sources and recommended dietary intake

Foods

Calcium consumption is mainly related to the use of dairy food including milk, yogurt, and cheese, because these contain high calcium concentrations. Among dairy products, hard cheese has 1 g of calcium per 100 g, compared to 100–180 mg provided by milk and yogurt. Cereals provide approximately 30 mg per 100 g, but, given that they are fortified, the concentration of 180 mg per 100 g can be reached. Also, seeds and nuts possess high concentrations of calcium, particularly almonds, sesame, and chia that have 250–600 mg per 100 g. Vegetables high in calcium are kale, broccoli, and watercress, which comprise 100–150 mg per 100 g (Cormick & Belizán, 2019; Service, U. D. o. A. A. R, 2019). Importantly, the influence of these products on total calcium intake is related to the nutritional algorithms in each population. While dairy foods make up approximately 14% of combined dietary energy input in more advanced countries, they represent only around 4% of total energy intake in developing regions. This is due to that certain Asian regions obtain a more significant proportion of total calcium intake from nonanimal foods such as vegetables, legumes, and grains than from dairy products, though they also have a much lower calcium intake (Silanikove et al., 2015). In the United states, 72% of calcium resource is achieved by the consumption of dairy foods, compared to 58% in the Netherlands, and 7% in China, in which it is mostly supplied by vegetables (30.2%) and legumes (16.7%) (Huang et al., 2018; Willemse et al., 2020). Fortified products like cereals or juices may also be considered as valuable sources of calcium.

Herbals and supplements

Supplements are also rich calcium sources in specific populations. Levels of calcium in supplements, which are accessible without prescription, can reach 1 g of calcium per tablet, meeting what is considered the nutritional need for most adults. But, the use of supplements differs among regions. In the United States and Canada, around 40% of the adult population claimed to have consumed calcium supplements in the month prior to the study, and this figure reached 70% in the aged group of women (Bailey et al., 2010; Silanikove et al., 2015). However, in Argentina and the Netherlands, women, even when pregnant, rarely took supplemental calcium (Cormick et al., 2014). A recent study showed that herbals, including chamomile, mint, sage, St John's wort, and nettle, could be utilized as infusions of calcium. In this study, varying levels of calcium were detected according to the origin and raw material of herbs (Suliburska & Kaczmarek, 2012).

Different national and global guidelines, including UK (SACN), USA and Canada (IOM), European (EFSA), and World Health Organization (WHO)/Food and Agriculture Organization (FAO), report different optimal intakes of calcium by age group. Among these, WHO/FAO recommends the highest intake of calcium, especially in age groups 7–10 and 50 years old (1000–1300 mg) (EFSA Panel on Dietetic products, Nutrition, and Allergies (NDA), 2010; Heaney, 2006). Table 1 summarizes the reference calcium intake values in various national and global guidelines.

Table 1 Overview of dietary reference for calcium in guidelines.

Guideline	FAO/WHO		European (EFSA)		USA and Canada (IOM)		New Zealand and Australia		HCN	NCM	AFSSA
Age	Estimated average requirement (mg/day)	Recommended nutrient intake (mg/day)	Average requirement (mg/day)	Population reference intake (mg/day)	Estimated average requirement (mg/day)	Recommended dietary allowance (mg/day)	Estimated average requirement (mg/day)	Recommended dietary intake (mg/day)	Population reference intake (mg/day)	Population reference intake (mg/day)	Population reference intake (mg/day)
0–6 month	240–300	300–400				200 (AI)		210 (AI)			
6–12 month	240–300	300–400		280(AI)		260 (AI)		270 (AI)	450	540	
1–3 year		500	390	450	500	700	360	500	500	600	500
4–6 year	440	600	680	800	800	1000	520[a]	700[a]	700[a]	600[c]	700
7–10 year		1300	680	800	800	1000				700[d]	900[f]
Male											
11–14 year	1040	1300	960	1150	1100	1300	1050[b]	1300[b]	1200	900[e]	1200[g]
15–18 year	1040	1300	960	1150	1100	1300	1050	1300	1200	900	1200
19–20 year	840	1000	860	1000	800	1000	840	1000	1000	900	900
21–24	840	1000	860	1000	800	1000	840	1000	1000	800	900
25–50 year	840	1000	750	950	800	1000	840	1000	1000	800	900
50 year	840	1000/1300	750	950	800	1000	840	1000	1100	800	900
Females											
11–14 year	1040	1300	960	1150	1100	1300	1050	1300	1200	900[e]	1200[g]
15–18 year	1040	1300	960	1150	1100	1300	1050	1300	1200	900	1200
19–20 year	840	1000	860	1000	800	1000	340	1000	1000	900	900
21–24 year	840	1000	860	1000	800	1000	340	1000	1000	800	900
25–50 year	840	1000	750	950	800	1000	340	1000	1000	800	900
50 year	840	1000	750	950	1000	1200	1100	1300	1100	800	900
Pregnancy											
14 to 18 year			Same as nonpregnant	Same as nonpregnant	1100	1300	1050	1300	Same as nonpregnant	900	1000 (third semester)
19 year and older	940	1200	Same as nonpregnant	Same as nonpregnant	800	1000	840	1000	Same as nonpregnant	900	1000 (third semester)
Lactation	1040	1000	Same as nonlactating	Same as nonlactating	1100/800	1300/1000	Same as nonpregnant	Same as nonpregnant	Same as nonpregnant	900	1000
Reference	(Heaney, 2006)		(Cormick & Belizán, 2019; World Health, 2005)		(EFSA Panel on Dietetic products, Nutrition, and Allergies (NDA), 2010)		(Capra, 2006)		(EFSA Panel on Dietetic products, Nutrition, and Allergies (NDA), 2015)		

Abbreviations: Afssa, The French Food Safety Agency; AI, Adequate intake; EFSA, European Food Safety Authority; HCN, Health Council of the Netherlands; IOM, USA Institute of Medicine; NCM, Nordic Council of Ministers.

[a]For 4–8 years.
[b]For 12–14 years.
[c]For 4–5 years.
[d]For 6–9 years.
[e]For 10–14 years.
[f]For 7–9 years.
[g]For 10–14 years.

Table adapted from Cormick, G., & Belizán, J. M. (2019). Calcium intake and health. *Nutrients, 11*(7), 1606.

Determinants of calcium levels

Sources that affect absorption of dietary calcium

The efficacy of absorption is associated with biological demands for calcium and is dose-reliant. Nutritional suppressors of calcium absorption consist of agents that produce complexes in the intestine. Phytates, regarded as major zinc antagonists, are prominent components in grains or cereal seeds, vegetables, legumes, fruits, roots, tubers, nuts, and protein and oil-producing plants (Konietzny & Greiner, 2003). Sodium can alter bioavailability of calcium because elevated concentrations enhance urinary calcium loss. While this co-occurs with an elevation in calcium absorption by the intestine, the overall outcome can be a decrease in the relative amount of dietary calcium that is biologically usable for the body, (e.g., diminished biological availability). Conversely, lactose increases calcium absorption. Currently, promoters of calcium absorption are being explored, with the aim of developing dietary products for the prophylaxis of osteoporosis (Fairweather-Tait & Southon, 2003). These efforts should be utilized considering their influence on severe calcium-related neurological conditions such as PD.

Genetics and hormonal disease

Absorption of calcium is influenced by an interlink between calcium intake, genetics, and diet. Polymorphisms and genetic syndromes affect calcium levels. Genetic syndromes should be considered in cases with hypercalcemia or hypocalcemia (Moreira, Brás, Lopes, & Januário, 2018), in particular in the youth and those who have first-degree relatives with hypercalcemia; or have had a history of tumor of endocrine pancreas, thyroid, pituitary, adrenal gland, and jaw bone. Also, polymorphisms such as *Fok1* in the VDR gene have been markedly associated with calcium absorption. In *Fok1* polymorphism, subjects with the FF genotype show the highest calcium absorption efficacy (Lewis, Laing, & Weaver, 2018).

Hypercalcemia could be explored according to the serum concentration of parathyroid hormone (PTH). Genetic types are a result categorized by existence or lack of a linkage to a parathyroid gland illness. During the raise in PTH concentration or when it is not diminished despite the hypercalcemia, findings that suggest potential familial history of hypercalcemia due to a genetic cause include syndromic manifestations in the patient or family members, or parathyroid malignancy, multiple or recurring parathyroid cancers, family background of primary hyperparathyroidism, and the diagnosis of primary hyperthyroidism in cases less than 50 years old. A precise assessment of clinical and paraclinical results can define whether hypercalcemia can be attributed to genetic disease. The advances in genetic evaluations have aided research on the phenotypes and control of genetic hypercalcemia or hypocalcemia (Cormier, 2019; Moreira et al., 2018). Also, therapeutic herbals can affect hormonal profile along with calcium levels. *Eurycoma longifolia* is a medicinal plant known for its protective effects on menopause. Its effect was evaluated on rats at incremental doses compared with testosterone as positive control, and showed increased calcium levels in the treatment group. In addition, following administration of this herbal, concentrations of luteinizing hormone (LH), follicle-stimulating hormone (FSH), and other sex hormones were found to be balanced (Chinnappan, George, Ashok, & Choudhary, 2020).

Other than described factors, a large body of evidence explains physiological conditions (e.g., pregnancy and aging) and various diseases alter calcium demands, and therefore, their role should be considered as a substantial determinant of calcium levels (Kumar & Kaur, 2017). Together, dietary intake of calcium should be linked to serum levels, only in light of such determinants, that affect its absorption and metabolism.

Calcium and Parkinson's disease

Role of food intake in Parkinson's disease

A link has been found between calcium intake and the risk of being diagnosed with PD later in life. In contrast, part of the literature shows no association between dietary calcium intake and PD; for example, no correlation has been found between calcium and dairy intake with the diagnosis of PD (Park et al., 2005). In a cohort, after adjusting for calcium intake and other factors, the femoral bone mineral density (BMD) of female PD patients was lower compared to controls (Lam, Li, Mok, Hui, & Woo, 2010). Collected data regarding calcium and dairy intake from PD patients and a control group showed no correlation between the risk of PD and calcium or dairy intake, even considering the gender factor (Miyake et al., 2011).

Both healing and harmful associations between calcium intake and PD have been investigated. The evaluation of several PD patients through a questionnaire and searching for hobbies like regular exercise, smoking, drinking, and consuming calcium supplements or dairy products showed that taking preventive measures like exercising and consuming dairy products is quintessential for PD patients' health. The examination of the daily calcium intake in PD patients with osteoporosis or osteopenia showed that higher levels of calcium intake and sunlight exposure were correlated with higher BMD, which was correlated with lower PD severity (Bezza et al., 2008). On the other hand, the harmful effects of calcium consumption on PD status have been demonstrated in a prospective cohort that experimented the association between dairy consumption and PD risk. The role of source of calcium and gender has been recently highlighted. There was a positive association between dairy calcium consumption and PD risk in men but not in women, though nondairy calcium sources were not associated with increased PD risk (Chen, Zhang, Hernán, Willett, & Ascherio, 2002). Factors affecting the role of calcium in PD and pathology of PD are shown in Fig. 1.

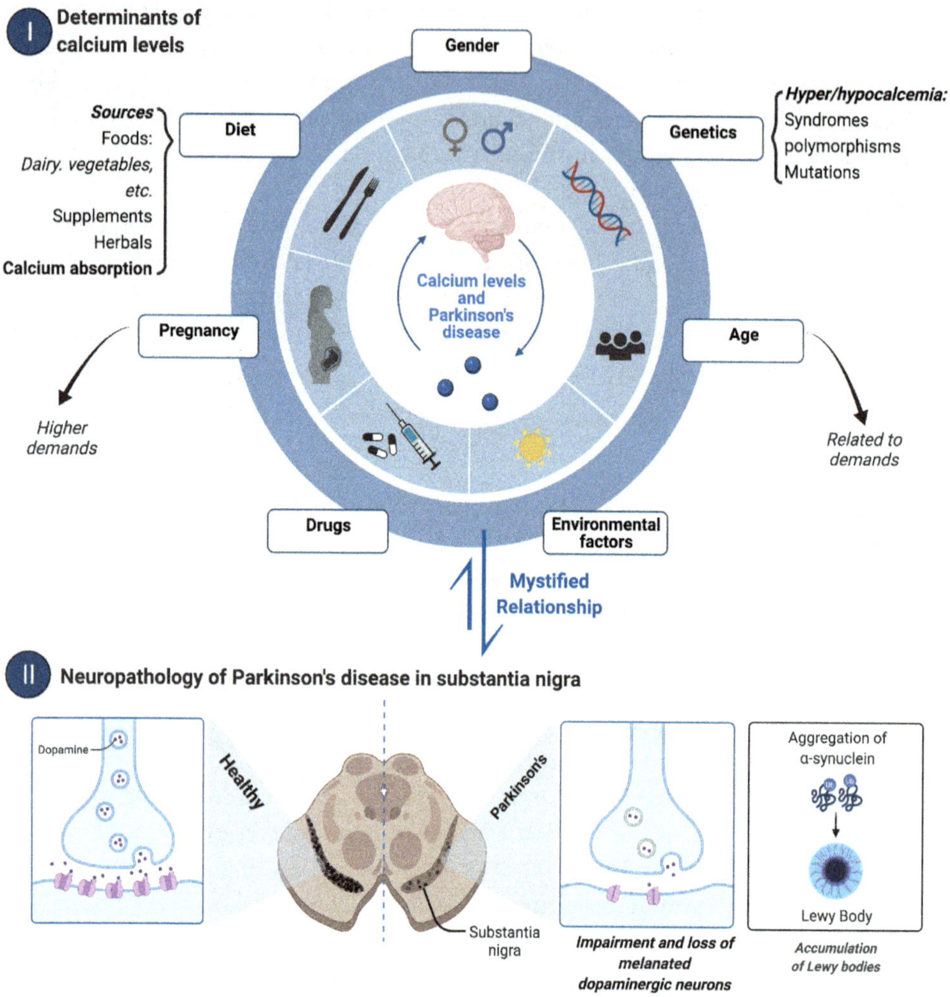

Fig. 1 Determinants of the possible role of nutritional calcium in Parkinson's disease. Many factors modify calcium absorption and demands. The role of dietary calcium in PD should be interpreted in light of these determinants. Also, exploring the linkage between calcium levels and PD pathogenesis has shown very heterogeneous results in different contexts (e.g., populations, study designs, methodology, sample types). Created with BioRender.com.

The link between calcium concentrations and Parkinson's disease

In addition to dietary intake, calcium levels have been linked to PD. This is important as dairy intake does not directly translate to calcium levels due to multiple factors affecting its absorption and body demands. A Genome-Wide Association Study (GWAS) found no relation between PD and genetically amplified levels of calcium in the European population. This study used the Mendelian randomization approach for eight gene variants on the basis of three presumptions, including that the variants influenced calcium concentrations, exerted their effect on PD only through altering calcium concentrations, and were not influenced by confounders. The authors avoided the use of dependent gene variants to reduce linking disequilibrium influences. The study utilized serum calcium and PD GWAS data from European population, decreasing stratification bias. Therefore, these results could not be established for other regions of different descents (Wang et al., 2020). In an Iranian population, however, we demonstrated that calcium levels, but not vitamin D levels, were higher in patients with PD compared to healthy subjects, significantly (Samavarchi Tehrani et al., 2020). These results warrant international studies with a wider spectrum of participants to confirm the potential effect of genetics.

In a novel research, the imbalances of the electrolytes such as calcium and magnesium and their relation with biochemical features of PD cases at various age classes (15–75 years) and disease phases were explored. The collected data of PD cases were assessed in comparison with healthy referents of similar age groups. The biological samples included scalp hair, serum, and plasma and were retrieved from PD patients, classified at three intervals, early (15–30 years), moderate (30–60 years), and critical (61–75 years). Results showed that calcium was associated with PD in different stages as well as compared to healthy referents. While disruptions in electrolyte levels are not generally known to result in neurodegeneration, this study established their role as a potential risk element (Rajput, Afridi, Kazi, Talpur, & Baig, 2020). Researches exploring the association between dietary calcium consumption and calcium levels are provided in Table 2.

Conclusions

The evidence that has studied the significance of calcium levels in the development of neurological disease accomplished many interesting results. However, in order to clearly define the role of calcium in PD, further studies should direct focus on the effect of confounding factors and apply a more robust methodology internationally. Defining the role of calcium in neurodegenerative conditions can help guidelines and decision makers to reduce the burden of disease across the globe by addressing shortcomings in the current therapeutic arsenal against PD.

Applications to other neurological conditions

In this chapter, we have reviewed the role of calcium in Parkinson's disease. Interestingly, the role of dietary calcium consumption is also implied in the pathogenesis of other neurodegenerative illnesses like ALS and multiple sclerosis (MS).

MS is a progressive neurodegenerative illness with autoimmune etiology. Diet is believed to affect disease course in MS. In a clinical study, the intake of calcium was compared between the normal population and MS patients. It was shown that higher calcium

Table 2 Studies exploring association between calcium and Parkinson's disease.

Type of calcium measurement	Other minerals/ factors measured	Location	Sample size	Study design	Summary	References
Serum calcium level	Phosphorus, alkaline phosphatase, vitamin D, PTH	Iran	239	Case-control	Mean level of calcium was found higher in PD compared to healthy individuals	(Samavarchi Tehrani et al., 2020)
Serum calcium level		Europe	12,760	Mendelian randomization trial, GWAS	The higher levels of serum calcium were not significantly linked to PD	(Wang et al., 2020)
Calcium concentration in subregions of the brain	Al, Mg	Japan		Case-control	The study showed that calcium concentration in brain was not associated with PD	(Yasui, Kihira, & Ota, 1992)
Calcium concentration in scalp hair, serum, and plasma	Mg, Na, K	Pakistan	243	Case-control	The higher concentrations of calcium were found in scalp hair sample of PD patients	(Rajput et al., 2020)
Calcium content of hair	Cu, Fe, Mg, Si, Zn	Italy	98	Case-control	Calcium contents of hair were not associated with PD	(Forte et al., 2005)
Calcium content of hair	Fe, Zn	Brazil	59	Case-control	Lower concentrations of calcium were found in hair sample of PD patients compared to healthy individuals	(Dos Santos et al., 2018)
Serum calcium level		China	152	Case-control	A correlation exists between serum calcium concentration and cognitive impairment in PD cases.	(Liu et al., 2016)

Calcium intake in diary food (milk, yogurt, cheese, or ice cream)	Cholesterol, dietary glycemic index, vitamin E, β-carotene, vitamin B(6), caffeine, iron, and alcohol	Japan	617	Case-control	There were no significant associations between calcium consumption and PD	(Miyake et al., 2011)
Calcium intake by milk consumption		Republic of Korea	7504	Cohort	Findings suggested that milk consumption is related to an increased risk of PD	(Park et al., 2005)

Abbreviations: GWAS, Genome-Wide Association Study; PTH, parathyroid hormone.

intake along with higher vitamin D intake has protective effects against MS progression (Ramsaransing, Mellema, & De Keyser, 2009).

ALS is the most prevalent progressive neurodegenerative illness in the adult population that impacts the motor neurons in the lower and upper parts of CNS. Disease illness seems to be associated with oxidative imbalance. In a clinical study, the association between total, animal, and plant calcium intake and the risk of ALS was evaluated. Interestingly, plant calcium showed protective effects against ALS symptoms, but total and animal calcium exacerbated the course of ALS (Jin et al., 2014). In contrast, in another clinical study, the dietary routines of the ALS patients were recorded for 5 years; and outcomes indicated no relationship between calcium intake and the risk of developing ALS (Longnecker et al., 2000). In an animal study, rats were fed on a low-calcium/magnesium and high-aluminum diet and their CNS and skin specimen were collected. The skin of rats showed pathological changes including atrophy, amorphous material aggregation, and collagen separation. Also, the spinal cord and cortex specimen showed atrophy of the CNS (Kihira, Yoshida, Kondo, Yase, & Ono, 2004). In a similar study, rats were fed on a low-calcium, high-aluminum diet and their trabecular bone and CNS tissue magnesium levels were evaluated. The results showed lower levels of magnesium in the CNS and trabecula, meaning a higher degeneration in the CNS and lower calcification in the bones due to abnormal magnesium levels (Yasui & Ota, 1998). To conclude, studies generally found a positive impact of calcium intake on ALS and MS pathophysiology; though, exceptions were also present. The ongoing coronavirus pandemic has shown us that a single agent can cause unexpected, vast, and potentially long-lasting neurological damage. Role of nutrients in neurological disorders is therefore better explored considering that they are multifaceted players in the nervous system (Saleki, Banazadeh, Saghazadeh, & Rezaei, 2020).

Other components of interest: Role of magnesium in Parkinson's disease

In this chapter, we described the use of calcium in PD and how its diagnosis and stages could be linked to calcium levels. Additionally, research has shown that magnesium could offer protective effects against PD (Miyake et al., 2011). Magnesium is an important player in cellular physiology. It adjusts potassium and calcium transport and regulates the cell cycle by modulating protein and nucleotide production, along with structural coherence in the mitochondria. Magnesium is proposed to protect against toxin-induced PD and α-synuclein aggregation (Agim & Cannon, 2015). It is also proposed to be a natural blocker of calcium in cells (McCarty & Lerner, 2020).

Magnesium has been explored along with calcium in multiple studies. In an animal study, rats were exposed to magnesium deficiency and both magnesium and calcium deficiency groups. Surprisingly, the solely magnesium-deficient rats showed more severe pathological symptoms resembling PD-like substantia nigra degeneration, formation of Lewy bodies, and fibrillary tangles (Oyanagi, 2005). In another study, rats were deprived of calcium/magnesium for 6 weeks resulting in catalepsy that could be alleviated by dopamine enhancers such as amantadine, bromocriptine, and L-DOPA. Also, tyrosine hydroxylase (TH) was lowest after 6 weeks of deprivation. These results highlight the crucial role of magnesium and/or calcium in dopamine metabolism (Taniguchi et al., 2013). Surprisingly, high levels of magnesium could also lead to damaged dopaminergic transmission through the induction of mitochondrial damage (Rasheed, Tripathi, Mishra, & Singh, 2017). The concentration of calcium, potassium, magnesium, and sodium have been determined in the scalp of PD patients vs healthy controls; other than calcium, higher levels of magnesium and also lower levels of sodium and potassium have been detected in the scalp samples of PD patients based on the stage of disease (Rajput et al., 2020). In conclusion, although the literature on the function of magnesium in PD etiology remains unclear, it could be implied that too low as well as too high magnesium intake would detrimentally affect dopamine homeostasis in the substantia nigra possibly leading to increased propensity of being diagnosed with PD.

Mini-dictionary of terms

- **Parkinson's disease**. A progressive neurodegenerative disease of the CNS characterized by shaking, postural instability, slowness, or other abnormalities in movement.
- **α-synuclein.** A protein encoded by the SNCA gene that regulates the trafficking of vesicles and ultimately the secretion of neurotransmitters in the CNS. It is an essential element in classic pathology of PD.
- **Parathyroid hormone.** A hormone that parathyroid glands secrete to adjust calcium homeostasis by means of altering the physiology of the intestines, bones, and the kidneys.
- **Mendelian randomization.** Using genetic variants as a model to estimate the impact of a risk factors on disease propensity.
- **Eurycoma longifolia.** A slender shrub used as an aphrodisiac in the traditional medicine of Indonesia and Indochina areas.
- **Genome-Wide Association Studies**. The study of the entire genome in individuals with different traits of the same quality.
- **Linkage disequilibrium.** When the association of two alleles is not random in a way that alters the expected haplotype production.

Key facts

- Calcium is found in dietary products, vegetables, supplements, and various herbs.
- In addition to calcium signaling, dietary calcium is implicated in PD.

- The effectiveness of dietary calcium is affected by hormonal profile, nutritional suppressors, genetic factors.
- The double-edged role of dietary calcium in PD as evidenced by recent research indicates the need for further research.

Summary points

- Reference values for calcium are mostly based on bone disease, and differ among national and global guidelines
- It is recommended that reference values are determined considering possible effect of calcium on neurological conditions.
- Calcium intake through diet and supplements can be translated into levels in body fluids in light of genetics, environmental, and absorption-altering compounds.
- Defining the clear function of dietary calcium in PD in different populations needs further investigation.
- Studies found positive association between either calcium intake or levels (in body fluids or scalp samples) and PD, while other studies could not establish such relations.
- PD has been found in case reports of patients with rare genetics syndromes with hypocalcemia.
- Calcium levels also play a role in other neurodegenerative conditions, such as ALS and AD.

References

Agim, Z. S., & Cannon, J. R. (2015). Dietary factors in the etiology of Parkinson's disease. *BioMed Research International, 2015*. https://doi.org/10.1155/2015/672838, 672838.

Amalraj, A., & Pius, A. (2017). Relative contribution of oxalic acid, phytate and tannic acid on the bioavailability of calcium from various calcium salts-an in vitro study. *International Food Research Journal, 24*(3), 1278.

Bailey, R. L., Dodd, K. W., Goldman, J. A., Gahche, J. J., Dwyer, J. T., Moshfegh, A. J., et al. (2010). Estimation of total usual calcium and vitamin D intakes in the United States. *The Journal of Nutrition, 140*(4), 817–822.

Bezza, A., Ouzzif, Z., Naji, H., Achemlal, L., Mounach, A., Nouijai, M., et al. (2008). Prevalence and risk factors of osteoporosis in patients with Parkinson's disease. *Rheumatology International, 28*(12), 1205–1209. https://doi.org/10.1007/s00296-008-0632-6.

Capra, S. (2006). New nutrient reference values for Australia and New Zealand: Implementation issues for nutrition professionals. *Nutrition & Dietetics: The Journal of the Dietitians Association of Australia, 63*(2), 64–66.

Chen, H., Zhang, S. M., Hernán, M. A., Willett, W. C., & Ascherio, A. (2002). Diet and Parkinson's disease: A potential role of dairy products in men. *Annals of Neurology: Official Journal of the American Neurological Association and the Child Neurology Society, 52*(6), 793–801.

Colloby, S. J., Watson, R., Blamire, A. M., O'Brien, J. T., & Taylor, J.-P. (2020). Cortical thinning in dementia with Lewy bodies and Parkinson disease dementia. *Australian & New Zealand Journal of Psychiatry, 54*(6), 633–643.

Cormick, G., & Belizán, J. M. (2019). Calcium intake and health. *Nutrients, 11*(7), 1606.

Cormick, G., Zhang, N. N., Andrade, S. P., Quiroga, M. J., Di Marco, I., Porta, A., et al. (2014). Gaps between calcium recommendations to prevent pre-eclampsia and current intakes in one hospital in Argentina. *BMC Research Notes*, 7(1), 1–6.

Cormier, C. (2019). Genetic hypercalcemia. *Joint, Bone, Spine*, *86*(4), 459–466.

Dos Santos, A. B., Kohlmeier, K. A., Rocha, M. E., Barreto, G. E., Barreto, J. A., de Souza, A. C. A., et al. (2018). Hair in Parkinson's disease patients exhibits differences in calcium, Iron and zinc concentrations measured by flame atomic absorption spectrometry—FAAS. *Journal of Trace Elements in Medicine and Biology*, *47*, 134–139.

EFSA Panel on Dietetic products, Nutrition, and Allergies (NDA). (2010). Scientific opinion on principles for deriving and applying dietary reference values. *EFSA Journal*, *8*(3), 1458.

EFSA Panel on Dietetic products, Nutrition, and Allergies (NDA). (2015). Scientific opinion on dietary reference values for calcium. *EFSA Journal*, *13*(5), 4101.

Fairweather-Tait, S. J., & Southon, S. (2003). Bioavailability of nutrients. In B. Caballero (Ed.), *Encyclopedia of food sciences and nutrition* (2nd ed., pp. 478–484). Oxford: Academic Press.

Forte, G., Alimonti, A., Violante, N., Di Gregorio, M., Senofonte, O., Petrucci, F., et al. (2005). Calcium, copper, iron, magnesium, silicon and zinc content of hair in Parkinson's disease. *Journal of Trace Elements in Medicine and Biology*, *19*(2–3), 195–201.

Gerfen, C. R. (2006). Indirect-pathway neurons lose their spines in Parkinson disease. *Nature Neuroscience*, *9*(2), 157–158.

Gopalakrishna, A., & Alexander, S. A. (2015). Understanding Parkinson disease: A complex and multifaceted illness. *Journal of Neuroscience Nursing*, *47*(6), 320–326.

Hall, D., Ouyang, B., Lonnquist, E., & Newcombe, J. (2011). Pragmatic communication is impaired in Parkinson disease. *The International Journal of Neuroscience*, *121*(5), 254–256. https://doi.org/10.3109/00207454.2010.550389.

Heaney, R. P. (2006). Absorbability and utility of calcium in mineral waters. *The American Journal of Clinical Nutrition*, *84*(2), 371–374.

Hermanns, M., Deal, B., & Haas, B. (2012). Biopsychosocial and spiritual aspects of Parkinson disease: An integrative review. *Journal of Neuroscience Nursing*, *44*(4), 194–205.

Huang, F., Wang, Z., Zhang, J., Du, W., Su, C., Jiang, H., et al. (2018). Dietary calcium intake and food sources among Chinese adults in CNTCS. *PLoS One*, *13*(10), e0205045.

Jin, Y., Oh, K., Oh, S. I., Baek, H., Kim, S. H., & Park, Y. (2014). Dietary intake of fruits and beta-carotene is negatively associated with amyotrophic lateral sclerosis risk in Koreans: A case-control study. *Nutritional Neuroscience*, *17*(3), 104–108. https://doi.org/10.1179/1476830513y.0000000071.

Kern, J., Kern, S., Blennow, K., Zetterberg, H., Waern, M., Guo, X., et al. (2016). Calcium supplementation and risk of dementia in women with cerebrovascular disease. *Neurology*, *87*(16), 1674–1680. https://doi.org/10.1212/wnl.0000000000003111.

Kihira, T., Yoshida, S., Kondo, T., Yase, Y., & Ono, S. (2004). ALS-like skin changes in mice on a chronic low-Ca/Mg high-Al diet. *Journal of the Neurological Sciences*, *219*(1–2), 7–14. https://doi.org/10.1016/j.jns.2003.11.010.

Konietzny, U., & Greiner, R. (2003). PHYTIC ACID | nutritional impact. In B. Caballero (Ed.), *Encyclopedia of food sciences and nutrition* (2nd ed., pp. 4555–4563). Oxford: Academic Press.

Kouli, A., Torsney, K. M., & Kuan, W.-L. (2018). *Parkinson's disease: Etiology, neuropathology, and pathogenesis* (pp. 3–26). Exon Publications.

Kovacs, C. S. (2000). Calcium and phosphate metabolism and related disorders during pregnancy and lactation. In K. R. Feingold, B. Anawalt, A. Boyce, G. Chrousos, W. W. de Herder, K. Dhatariya, … D. P. Wilson (Eds.), *Endotext*. South Dartmouth (MA): MDText.com, Inc. Copyright © 2000–2021, MDText.com, Inc.

Kumar, A., & Kaur, S. (2017). Calcium: A Nutrient in Pregnancy. *Journal of Obstetrics and Gynaecology of India*, *67*(5), 313–318. https://doi.org/10.1007/s13224-017-1007-2.

Lam, K., Li, M., Mok, V., Hui, A., & Woo, J. (2010). A case control study on bone mineral density in Chinese patients with Parkinson's disease. *Parkinsonism & Related Disorders*, *16*(7), 471–474.

Lewis, R., Laing, E., & Weaver, C. (2018). Chapter 41—Adolescence and acquisition of peak bone mass. In D. Feldman (Ed.), *Vitamin D* (4th ed., pp. 731–753). Academic Press.

Lister, T. (2020). Nutrition and lifestyle interventions for managing Parkinson's disease: A narrative review. *Journal of Movement Disorders, 13*(2), 97–104. https://doi.org/10.14802/jmd.20006.

Liu, J., Zhou, X. P., Zhang, L., Zhang, Q., Liu, C. F., & Luo, W. F. (2016). Correlation analysis of serum calcium level and cognition in the patients with Parkinson's disease. *Zhonghua Yi Xue Za Zhi, 96*(41), 3284–3288. https://doi.org/10.3760/cma.j.issn.0376-2491.2016.41.002.

Longnecker, M. P., Kamel, F., Umbach, D. M., Munsat, T. L., Shefner, J. M., Lansdell, L. W., et al. (2000). Dietary intake of calcium, magnesium and antioxidants in relation to risk of amyotrophic lateral sclerosis. *Neuroepidemiology, 19*(4), 210–216. https://doi.org/10.1159/000026258.

McCarty, M. F., & Lerner, A. (2020). Nutraceuticals targeting generation and oxidant activity of Peroxynitrite may aid prevention and control of Parkinson's disease. *International Journal of Molecular Sciences, 21*(10). https://doi.org/10.3390/ijms21103624.

Miyake, Y., Tanaka, K., Fukushima, W., Sasaki, S., Kiyohara, C., Tsuboi, Y., et al. (2011). Lack of association of dairy food, calcium, and vitamin D intake with the risk of Parkinson's disease: A case-control study in Japan. *Parkinsonism & Related Disorders, 17*(2), 112–116.

Moreira, F., Brás, A., Lopes, J. R., & Januário, C. (2018). Parkinson's disease with hypocalcaemia: Adult presentation of 22q11.2 deletion syndrome. *BMJ Case Reports, 2018*. https://doi.org/10.1136/bcr-2017-223751. bcr-2017-223751.

Oyanagi, K. (2005). The nature of the parkinsonism-dementia complex and amyotrophic lateral sclerosis of Guam and magnesium deficiency. *Parkinsonism & Related Disorders, 11*(Suppl 1), S17–S23. https://doi.org/10.1016/j.parkreldis.2005.02.010.

Park, M., Ross, G. W., Petrovitch, H., White, L. R., Masaki, K. H., Nelson, J. S., et al. (2005). Consumption of milk and calcium in midlife and the future risk of Parkinson disease. *Neurology, 64*(6), 1047–1051. https://doi.org/10.1212/01.Wnl.0000154532.98495.Bf.

Rajput, K., Afridi, H. I., Kazi, T. G., Talpur, F. N., & Baig, J. A. (2020). Sodium, potassium, calcium, and magnesium in the scalp hair and blood samples related to the clinical stages of the Parkinson's disease. *Biological Trace Element Research*. https://doi.org/10.1007/s12011-020-02399-y.

Ramsaransing, G. S., Mellema, S. A., & De Keyser, J. (2009). Dietary patterns in clinical subtypes of multiple sclerosis: An exploratory study. *Nutrition Journal, 8*, 36. https://doi.org/10.1186/1475-2891-8-36.

Rasheed, M. S. U., Tripathi, S., Mishra, S., & Singh, M. P. (2017). Coherent and contradictory facts, feats and fictions associated with metal accumulation in Parkinson's disease: Epicenter or outcome, Yet a Demigod Question. *Molecular Neurobiology, 54*(6), 4738–4755. https://doi.org/10.1007/s12035-016-0016-y.

Ross, A. C., Manson, J. E., Abrams, S. A., Aloia, J. F., Brannon, P. M., Clinton, S. K., et al. (2011). The 2011 report on dietary reference intakes for calcium and vitamin D from the Institute of Medicine: What clinicians need to know. *Journal of Clinical Endocrinology and Metabolism, 96*(1), 53–58. https://doi.org/10.1210/jc.2010-2704.

Chinnappan, S. M., George, A., Ashok, G., & Choudhary, Y. K. (2020). Effect of herbal extract *Eurycoma longifolia* (Physta(®)) on female reproductive hormones and bone biochemical markers: An ovariectomised rat model study. *BMC Complementary Medicine and Therapies, 20*(1), 31. https://doi.org/10.1186/s12906-020-2814-z.

Saleki, K., Banazadeh, M., Saghazadeh, A., & Rezaei, N. (2020). The involvement of the central nervous system in patients with COVID-19. *Reviews in the Neurosciences, 31*(4), 453–456. https://doi.org/10.1515/revneuro-2020-0026.

Salvatore, M. F., Disbrow, E. A., & Emborg, M. E. (2014). Peripheral and cognitive signs: Delineating the significance of impaired catecholamine metabolism in Parkinson's disease progression. *Journal of Neurochemistry, 131*(2), 129–133. https://doi.org/10.1111/jnc.12796.

Samavarchi Tehrani, S., Sarfi, M., Yousefi, T., Ahmadi Ahangar, A., Gholinia, H., Mohseni Ahangar, R., et al. (2020). Comparison of the calcium-related factors in Parkinson's disease patients with healthy individuals. *Caspian Journal of Internal Medicine, 11*(1), 28–33. https://doi.org/10.22088/cjim.11.1.28.

Service, U. D. o. A. A. R. (2019). *USDA nutrient database for standard reference, release 14: Nutrient data laboratory home page*. US Department of Agriculture, Agricultural Research Service.

Silanikove, N., Leitner, G., & Merin, U. (2015). The interrelationships between lactose intolerance and the modern dairy industry: Global perspectives in evolutional and historical backgrounds. *Nutrients, 7*(9), 7312–7331. https://doi.org/10.3390/nu7095340.

Suliburska, J., & Kaczmarek, K. (2012). Herbal infusions as a source of calcium, magnesium, iron, zinc and copper in human nutrition. *International Journal of Food Sciences and Nutrition*, *63*(2), 194–198. https://doi.org/10.3109/09637486.2011.617359.

Surmeier, D. J., Guzman, J. N., Sanchez-Padilla, J., & Schumacker, P. T. (2011). The role of calcium and mitochondrial oxidant stress in the loss of substantia nigra pars compacta dopaminergic neurons in Parkinson's disease. *Neuroscience*, *198*, 221–231.

Surmeier, D. J., & Schumacker, P. T. (2013). Calcium, bioenergetics, and neuronal vulnerability in Parkinson's disease. *Journal of Biological Chemistry*, *288*(15), 10736–10741.

Surmeier, D. J., Schumacker, P. T., Guzman, J. D., Ilijic, E., Yang, B., & Zampese, E. (2017). Calcium and Parkinson's disease. *Biochemical and Biophysical Research Communications*, *483*(4), 1013–1019. https://doi.org/10.1016/j.bbrc.2016.08.168.

Taniguchi, R., Nakagawasai, O., Tan-no, K., Yamadera, F., Nemoto, W., Sato, S., et al. (2013). Combined low calcium and lack magnesium is a risk factor for motor deficit in mice. *Bioscience, Biotechnology, and Biochemistry*, 77(2), 266–270. https://doi.org/10.1271/bbb.120671.

Thevathasan, W., Coyne, T. J., Hyam, J. A., Kerr, G., Jenkinson, N., Aziz, T. Z., et al. (2011). Pedunculopontine nucleus stimulation improves gait freezing in Parkinson disease. *Neurosurgery*, *69*(6), 1248–1254.

Wang, Y., Gao, L., Lang, W., Li, H., Cui, P., Zhang, N., et al. (2020). Serum calcium levels and Parkinson's disease: A Mendelian randomization study. *Frontiers in Genetics*, *11*, 824. https://doi.org/10.3389/fgene.2020.00824.

WHO. (2004). *Vitamin and mineral requirements in human nutrition*. World Health Organization.

Willemse, J. P. M. M., Meertens, L. J. E., Scheepers, H. C. J., Achten, N. M. J., Eussen, S. J., van Dongen, M. C., et al. (2020). Calcium intake from diet and supplement use during early pregnancy: The expect study I. *European Journal of Nutrition*, *59*(1), 167–174. https://doi.org/10.1007/s00394-019-01896-8.

World Health Organization. (2005). *Vitamin and mineral requirements in human nutrition* (2nd ed.). Geneva: World Health Organization.

Yasui, M., Kihira, T., & Ota, K. (1992). Calcium, magnesium and aluminum concentrations in Parkinson's disease. *Neurotoxicology*, *13*(3), 593–600.

Yasui, M., & Ota, K. (1998). Aluminum decreases the magnesium concentration of spinal cord and trabecular bone in rats fed a low calcium, high aluminum diet. *Journal of the Neurological Sciences*, *157*(1), 37–41. https://doi.org/10.1016/s0022-510x(98)00075-6.

Yasui, M., Ota, K., & Yoshida, M. (1997). Effects of low calcium and magnesium dietary intake on the central nervous system tissues of rats and calcium-magnesium related disorders in the amyotrophic lateral sclerosis focus in the Kii peninsula of Japan. *Magnesium Research*, *10*(1), 39–50.

CHAPTER 34

Micronutrient levels in Parkinson's disease

Hassan Imran Afridi
National Centre of Excellence in Analytical Chemistry, University of Sindh, Jamshoro, Pakistan

Abbreviations

(%)	percentage
°C	degree Celsius
μg	microgram
AAS	atomic absorption spectrometry
AD	Alzheimer's disease
ANOVA	analysis of variance
BMI	body mass index
Ca	calcium
CPD	chronic Parkinson's disease
FAAS	flame atomic absorption spectrometry
Hb	hemoglobin
HCl	hydrochloric acid
HNO_3	nitric acid
K	potassium
L	liter
MD	microwave-assisted acid digestion method
Mg	magnesium
MPD	mild Parkinson's disease
Na	sodium
PD	Parkinson's disease
ppb	part per billion
ppm	part per million
rpm	rotation per minute
RSD	relative standard deviation
Triton X-114	octylphenoxypolyethoxyethanol
TX-100	polyoxyethylene octyl phenyl ether
TX-114	Triton X-114
WHO	World Health Organization

Neurodegeneration

Neurodegeneration is complex and multifaceted, leading to many chronic diseases like Parkinson's disorder. Electrolyte imbalance has a significant role in the pathophysiology

Vitamins and Minerals in Neurological Disorders
https://doi.org/10.1016/B978-0-323-89835-5.00034-X

of neurological disorders, which might serve as a bioindicator of the neurological problems (Drago, Bolognin, & Zatta, 2008). Neurological disorders are considered as active inflammation process, and inside brain parenchyma, protein deposits accumulate, which leads to neuron loss and damage (Yokel, Crossgrove, & Bukaveckas, 2003). Parkinson's disease (PD), the second most common neurodegenerative disorder after Alzheimer's disease (AD), remains difficult to diagnose clinically due to its overlapping symptoms with other diseases, such as essential tremor, multiple system atrophy, and progressive supranuclear palsy, particularly during the early course of the disease (Emre et al., 2007; Jellinger, 2003). AD and PD are defined by the progressive loss of specific neuronal cell populations and are associated with protein aggregates (Barnham, Masters, & Bush, 2004). In Pakistan alone, there are estimated to be 450,000 persons afflicted from the disease (Thakur et al., 2013). AD is a devastating late-life dementia that produces progressive loss of memory and mental faculties in elderly people. More than 5 million people in the United States have AD, leading to a total healthcare cost of more than $50 billion USD annually. There is significant interest in biomarkers that can identify the earliest evidence of AD to allow for early initiation of treatment, quantitatively evaluate new interventions, and track a patient's response to treatment (Huitrón-Ramírez, Ricárdez-Salinas, & Camacho-Ferre, 2009).

Neurological disorders are an important cause of mortality and constitute 12% of total deaths globally. Neurological disorders constitute 16.8% of the total deaths in lower-middle-income countries compared with 13.2% of the total deaths in high-income countries. Neurological disorders contribute to 92 million DALYs in 2005 projected to increase to 103 million in 2030 (approximately a 12% increase). However, AD and other dementias are projected to show a 66% increase from 2005 to 2030. Dementia is one of the causes of disability and premature death in elderly adults and is a high-priority public health concern worldwide (Litvan et al., 2007).

Types of neurodegeneration

Parkinson's disease

PD is the second most common neurodegenerative disorder in humans after AD. PD patients present motor symptoms such as resting tremor, bradykinesia, and postural rigidity. However, the appearance of other disturbances such as constipation, sleep disorders, olfactory deficit, apathy, pain, sexual difficulties, and in some case cognitive decline is currently observed to anticipate motor deficits in many patients (De Lau & Breteler, 2006; Surmeier, Guzman, Sanchez-Padilla, & Schumacker, 2011). It is also associated with a diversity of nonmotor symptoms, together with late-onset motor symptoms (such as postural instability and falls, freezing of gait, speech, and swallowing difficulties) (Arvanitakis & Bennett, 2019). Age and dietary habits, as well as environmental and occupational factors, can favor the onset of this pathology, while less than 1% of PD cases seem to have a genetic origin. Neurochemical observations have suggested a role of metals in PD (Witholt, Gwiazda, & Smith, 2000). The view of etiological factors in PD has changed from one of a purely sporadic basis to the view that both genetic and environmental factors contribute to the development of the disease, while increasingly genetic predisposition is probably a major contributor leading to molecular mechanisms of neurodegeneration (Verstraeten, Theuns, & Van Broeckhoven, 2015). Nonmotor complications in many cases are not directly related to the involvement of dopaminergic pathways and may therefore develop even in patients where motor symptoms are well controlled. Sleep disorders are frequent in PD. PD is a common neurodegenerative illness. A combination of genetic and environmental factors is likely to be important in producing abnormal protein aggregation within select groups of neurons, leading to cell dysfunction and then death (Arvanitakis & Bennett, 2019).

Dementia

Many patients with PD develop dementia, typically 10 years or more after the onset of motor symptoms. Dementia in PD may be related to a number of pathologies. However, it seems to be the development of cortical Lewy bodies and/or Alzheimer pathology, which are most relevant (Arvanitakis & Bennett, 2019). Dementia is a disorder of the brain in which a person experiences decreasing cognitive abilities. Dementia is caused by brain disease or injury. It can be caused by medical conditions that start in other parts of the body. One of the most common is AD, which affects about 5 million people in the United States (Schneider, Arvanitakis, Leurgans, & Bennett, 2009). Dementia mainly affects older people: only 2% of cases start before the age of 65 years. After this, the prevalence doubles with every 5-year increment in age. Dementia is one of the major causes of disability in later life (Arvanitakis & Bennett, 2019).

Alzheimer's disease

AD is the most prevalent form of dementia in older people. This condition is characterized by cognitive and memory deterioration, progressive impairment in the activities of daily living, and other disabling neuropsychiatric and nutritional problems (Abbott, 2011; Bianchetti, Ranieri, Margiotta, & Trabucchi, 2006). It is a neurodegenerative disorder characterized by the presence of neurofibrillary tangles, dystrophic neurites, and senile plaques composed predominantly of amyloid beta peptide (Aβ) derived from the amyloid precursor protein (Mattson, 2004). AD is associated with impaired glutamate clearance and depressed Na^+/K^+ ATPase levels in AD brain that might lead to a cellular ion imbalance. AD, the most common form of dementia, affects more than 5 million Americans, placing a tremendous strain on our healthcare system (Mattson, 2004; Ringman, 2017), Studies of a community-based Japanese cohort investigated the relationship of self-reported dietary intake of potassium, calcium, and magnesium with the occurrence of dementia over a 17-year period. They found that increased dietary intake of potassium, calcium, and magnesium reduced the risk of all-cause dementia, especially vascular dementia. The risk of AD tended to decrease with higher reported dietary mineral intake, but there was no significant linear progression. Through the use of an animal AD model, Ozawa et al. (2012) found that higher potassium levels resulted in lower levels of oxidative stress and inflammation as well as enhanced cognition. Additionally, potassium was related to a change in the aggregation pattern of Aβ peptide and a reduction in tau phosphorylation (Cisternas et al., 1852). There is significant interest in biomarkers that can identify the earliest evidence of AD to allow for early initiation of treatment, quantitatively evaluate new interventions, and track a patient's response to treatment (Mellon et al., 2009; Mielke et al., 2006).

Role of micronutrients in pathogenesis of neurological disorders

The function of electrolytes in neuroscience has been designated that they have a role in neurological disorders/diseases. Electrolyte disturbances occur commonly and are associated with a variety of characteristic neurologic manifestations involving both the central and peripheral nervous systems. Electrolyte disturbances are essentially always secondary processes. Effective management requires the identification and treatment of the underlying primary disorder. Since neurological symptoms of electrolyte disorders are generally functional rather than structural, the neurologic manifestations of electrolyte disturbances are typically reversible (Riggs, 2002). Electrolyte disruptions are significantly always secondary processes. Effectual administration are needed for identification and cure of the underlying most important disorder (Aminoff & Josephson, 2014). Majority of electrolyte disturbances in neurologic patients, except dysnatremia, display an equivalent regularity with sick person from further disorder and primarily affected by hospital comorbidities (Kim & Joo, 2009). Abnormality of electrolyte may influence several organs, tissues, and the brain. In addition, neurological disorders or diseases may possibly emerge with seizures, and it quickly progressive neurologic symptoms that why need emergency management (Rose, 2001). As imbalance of electrolytes results in neurological problems in function instead of structural defects, the neurological symptoms due to imbalance of electrolytes are mostly reversible. The neurologic expressions might be a change due to the disturbance in electrolytes in blood serum (Augusto, Morais, Santana, & Almeida, 2019).

Calcium

It is known that significant changes in serum calcium levels, especially hypocalcemic states, can lead to neurological disorders, among them, parkinsonism (De Rosa et al.,

2011). Parkinsonism refers to a group of diseases characterized by tremors, gait disorders, stiffness, slowness, muscle pain, paresthesia, generalized fatigue, depression, nervousness, and others (Augusto et al., 2019). The function of calcium regulation in PD has been of attention for a short time and is another case of an issue that cuts across a number of the customary pathways that show way to neurodegeneration. Calcium is vital to normal cell physiology and the function of excitable cells (Surmeier & Schumacker, 2013). It is known that calcium is necessary for neuromuscular stimulation and also in neurotransmitters (Luján-Martínez, Sánchez-Cifuentes, Peña-Ros, Albarracín-Marín-Blázquez, & Candel-Arenas, 2018). Therefore, altered calcium levels affect the transmission of nerve impulse, which could lead to partial or total damage of dopaminergic receptors (Luján-Martínez et al., 2018). Moreover, the calcium concentration in blood serum decreases in PD patients. The suggested reasons for a decrease in levels of serum calcium of PD patients are inactivity, reducing sun exposure, and calcium and vitamin D consumption (Luján-Martínez et al., 2018). Calcium is chemical substance that is present in the human body and is originated within the cells of human body and extracellular fluids like blood plasma. Neurons use Ca^{2+} signals to control membrane excitability; trigger the release of neurotransmitters; mediate activity-dependent changes in gene expression; and modulate neuronal growth, differentiation, and transition to apoptosis (Bezprozvanny & Mattson, 2008).

It is well documented that alterations in the levels of serum Ca^{2+} are responsible for the initiation of convulsions (Roitg et al., 2004). Neonatal hypocalcaemia and hypomagnesaemia have also been reported to be the cause of convulsions. Serum Ca^{2+} levels were reported to be unaltered in generalized and unspecified seizures (Musolino et al., 1992). Rutter et al. reported normal serum levels in children with febrile convulsion (Rutter & Smales, 1976).

Magnesium

Magnesium is the fourth most abundant mineral and the second most abundant intracellular divalent cation and has been recognized as a cofactor for >300 metabolic reactions in the body (Takaya, Higashino, & Kobayashi, 2004). Magnesium also plays a critical role in maintaining normal nerve and muscle function, cardiac excitability (normal heart rhythm), neuromuscular conduction, muscular contraction, vasomotor tone, normal blood pressure, bone integrity, and glucose and insulin metabolism (Abou-Raya, Helmii, & Abou-Raya, 2009; Barbagallo et al., 2003; Bohl & Volpe, 2002; Chubanov, Gudermann, & Schlingmann, 2005; Guerrero-Romero & Rodrı´guez-Morán, 2000; He et al., 2006; Murakami, Okubo, & Sasaki, 2005; Newhouse & Finstad, 2000). The magnesium is a crucial electrolyte for normal physiological activity of human health, especially the functions of the heart, skeletal muscles, and nervous system. In this regard, magnesium deficiency has been associated with a number of chronic diseases, including migraine headaches, ADs, cerebrovascular accident (stroke), hypertension, cardiovascular disease (De Baaij, Hoenderop, & Bindels, 2015; Sanghvi, Kellerman, & Nanovic, 2007; Schlingmann et al., 2002; Walder et al., 2002). AD is the most widespread reason for dementia (Sanghvi et al., 2007). AD is the sixth leading cause of death in the United States, with >79,000 deaths per year. Barbagallo et al. examined magnesium balance in patients with mild to moderate AD (Barbagallo et al., 2003). Generally, Mg has antiinflammatory characteristics and performs as antagonist to Ca (Sterns, 2015). It was investigated that Parkinson's patients have low Mg in different parts of nervous system such as in white matter, brain stem cortex, and basal ganglia. In Parkinson's, the depleted Mg as well as Ca termed as hypomagnesemia and hypocalcemia (Schlingmann et al., 2002), which is illustrated by tremendously low levels of Mg in blood serum as 0.1–0.3 mM, along with depleted levels of Ca in serum, which creates rigorous complications in muscular and neurological problems involving mental and seizure disorders. (De Baaij et al., 2015; Sanghvi et al., 2007; Schlingmann et al., 2002; Walder et al., 2002).

Serum levels of Mg^{2+} were found to be lowered in epileptic patients (Rosenbauer, Koschmieder, Steidl, & Tenen, 2005a). Papierkowski et al., reported lowered serum Mg^{2+} concentrations in children with febrile convulsions. A variety of pathological conditions are associated with low serum Mg^{2+} levels in humans. Hypocalcemia and hypokalemia frequently accompany magnesium deficiency in humans (Schlingmann et al., 2002). Mg^{2+} is an essential element with a role in neuronal excitability. It inhibits the facilitating effect of calcium (Ca^{2+}) on synaptic transmission; Mg^{2+} alters Ca^{2+} mobilization and may stabilize excitable membranes and also exerts a voltage-dependent blockage of the NMDA receptor channel (Lewis & Sibai, 1997). It has been shown previously that low Mg^{2+} level-induced epileptiform activity in rat entorhinal cortex slices changes with time from a pattern of serial seizure-like events (SLEs) to a state of continuously recurring epileptiform activity. Low magnesium levels are observed in patients given anticonvulsant drugs designed for the clinical problem of pharmacoresistant epilepsy. In addition, efficacy of magnesium sulfate in the prevention and control of eclamptic convulsions has been validated in randomized controlled trials performed worldwide (Lewis & Sibai, 1997). Steidl et al., stated that various AEDs result in deficiency of red blood cells (RBCs) Mg^{2+} and serum Zn^{2+}, and therapy of magnesium lactate produces significantly higher RBC Mg^{2+} and serum Zn^{2+} with improved clinical EEG and biochemical findings (Rosenbauer, Koschmieder, Steidl, & Tenen, 2005b). Iihan et al. reported reduced serum level of Mg^{2+} in epileptic patients whether treated or not and also regardless of the type of AEDs utilized (Adjei, 2011). Shah et al., reported a significant decrease in serum Mg^{2+} levels in all seizure groups in comparison with controls, and they attributed this decrease to be responsible for hyperexcitability encountered for uncontrolled seizures on AEDs (Shah et al., 2011). However, many studies reported unchanged levels of Mg^{2+} in patients receiving AEDs. In contrast, Hamed et al., reported normal levels of serum Mg^{2+} among all groups of epileptics (treated or untreated) and irrespective of the degree of control on AEDs (Shah et al., 2011).

Sodium

The brain accomplishes in an extremely convoluted environment, which requires accurate balance of electrolytes. Tight control in excess of their concentrations across cellular partitions is crucial, and when these relationships are distressed, neurologic problems may build up. Perturbations of sodium are the electrolyte imbalances that mostly lead to neurologic diseases. Changes in extracellular liquid sodium levels produce water alters that causes brain to damage, contraction or enlargement (Arvanitakis & Bennett, 2019). Hyponatremia is the most common electrolyte imbalance in neurologic patients. It can be affected not only by the patient's own medical situation, but by neurologic disease itself though the disruption of arginine vasopressin (AVP) regulation. Hyponatremia is defined as serum sodium level less than 135 mEq/L 3105 ppm, which is known to occur in up to 20% of hospitalized patients and is associated with 1.5-fold higher in-hospital mortality rate than in patients without hyponatremia. Hyponatremia is more common in patients with neurologic disorders than in the general in-hospital population; the brain has a limited capacity to adapt to changes in serum osmolarity; the propensity of hyponatremia and hypernatremia to produce neurologic symptoms depends on the rapidity with which the sodium disturbance develops (Kim & Joo, 2009). The diagnosis of PD stays fundamentally a clinical one, and it is necessary to recognize the early stages together with symptoms and signs suggesting other causes of Parkinsonism. The exact diagnosis of seizures insignificant to electrolyte abnormalities initiates with a complete serum chemistry estimation, including determinations of electrolytes, particularly sodium. This diagnostic examination of patients should remain part of the first stage of workup in younger age group patients with initial seizures (Davie, 2008; U. Parkinson's, 2012).

In general, no abnormality was noticed in serum Na^{+} in epileptic patients during and after seizure activity (Natelson, Miletich, Seals, Visintine, & Albrecht, 1979). The observed hyponatremia leading to grand mal seizures in some cases of epileptic children treated with CBZ was associated with water intoxication, which is a rare side effect of CBZ (Natelson et al., 1979). Biochonski reported hypernatremia and hypokalemia in epileptic children. White et al., reported an elevation of K^{+} levels during periods of intense seizure activity (White, Chow, Yen-Chow, & Woodbury, 1992).

Potassium

Potassium (K) is a very significant body mineral, important to both cellular and electrical functions. It is one of the main blood minerals called "electrolytes" (the others are Na and Cl), which means it carries a tiny electrical charge (potential). K is the primary positive ion (cation) found within the cells, where 98% of the 120 g of K contained in the body is found (Baks et al., 2006). Imbalances in concentration of potassium hardly create problems in the central nervous system but it is possible that it is associated with weaknesses of muscles and tremors as the main clinical problems. Disorders of electrolytes mostly cause seizures, and seizures are mostly found in Parkinson's patients suffering from sodium disorders (Shi et al., 2010; Victor, Ropper, & Adams, 2001).

Nishimoto et al. (2005) have investigated that though the body readily conserves Na, there is no effective method for K conservation (Nishimoto et al., 2005). The elevations or depletions of this important mineral can cause problems and, in the extreme, even death. Maintaining consistent levels of K in the blood and cells is vital to body function (Lakatos et al., 2010). Bavaresco, Giachino, and Pezzutto (2003) have concluded that the deficiency of K is much more common, especially with aging or chronic disease. Some common problems that have been associated with low K levels include hypertension, congestive heart failure, cardiac arrhythmia, fatigue, depression, and other mood changes (Bavaresco et al., 2003). It was reported that decreased K level observed in both serum and liver tissue may be attributed to the pathophysiology of hepatic fibrosis and the underlying disease process (Zillich, Garg, Basu, Bakris, & Carter, 2006).

In contrast to sodium, the neurologic manifestations of potassium disturbance, whether hypokalemia or hyperkalemia, rarely involve the CNS. About 98% of total body potassium is located intracellularly, and 60% of intracellular potassium is within muscle. This distribution may, in part, account for the predominance of muscle symptoms associated with either hypokalemia or hyperkalemia. Hypokalemia is the most frequent electrolyte disorder encountered in clinical practice and is produced by a variety of mechanisms, including inadequate potassium intake or excessive renal or gastrointestinal potassium loss.

Evaluate the correlation of electrolytes with biochemical parameters in biological samples of Parkinson's disease patients at different stages

The macro- and microminerals are needed for progression and physiology of the organism. The current hospital-based research was performed to find out the different amount (concentrations) of Ca, Mg, Na, and K in the specimen (scalp hair, blood serum, and blood plasma) of male referents and PD patients. This study consists of two groups, one is chronic Parkinson's disease (CPD) patients (66–75 years), and other is mild Parkinson's disease (MPD) patients, age ranged 50–65 years. Of the selected 95 MPD patients, 50% patients initially have depression, 30% have palpitation, while 20% suffer from insomnias as well as other two physiological disorders. However, the CPD patients suffered from somatic and psychiatric comorbidities, such as pain in body (52.1%), palpitations (70.0%), poor appetite (80.2%), fatigue (67.7%), insomnia (56%), and

Table 1 Clinical and biochemical characteristics of referents and MPD and CPD patients.

Parameters	Normal range	55–65		66–75	
		Referents	MPD	Referents	CPD
BMI (kg/m^2)	–	24.9 ± 0.42	23.3 ± 0.65	23.5 ± 0.50	21.8 ± 0.35
SBP (mmHg)	–	127.9 ± 4.60	145.3 ± 5.82	129.8 ± 7.50	163.5 ± 5.95
DBP (mmHg)	–	83.6 ± 6.15	95.5 ± 7.35	85.2 ± 6.15	109 ± 8.05
Heart rate (min)	72–80	76.9 ± 7.55	87.9 ± 6.60	78.5 ± 5.05	92.7 ± 5.38
Hb (g/dl)	13.5–17.5	14.0 ± 0.12	13.0 ± 0.15	12.3 ± 1.34	10.6 ± 0.45
RBC (× 10^{12}/L)	4.28–5.81	4.92 ± 0.16	4.15 ± 0.33	4.10 ± 0.32	3.05 ± 0.39
Vitamin D (ng/mL)	25–30	24.8 ± 0.63	8.65 ± 1.62	22.5 ± 0.40	6.38 ± 1.62

Key: Hb, hemoglobin; SBP, systolic blood pressure; DBP, diastolic blood pressure; RBC, red blood cells; BMI, body mass index.

psychomotor retardation (64.6%). All the analyzed samples were grouped into two types of PD patients and referents.

The biochemical parameters of PD patients are indicated in Table 1 Demographic data such as sex, age, body mass index (BMI), systolic/diastolic blood pressure, vitamin D (ng/mL), hemoglobin (Hb), RBCs, and heart rate/min were measured in OPD and concerned pathological laboratories of hospitals for all patients.

The BMI of each patient and healthy participant was calculated by dividing the height value by its square (kg/m^2). It was significantly lower values of BMI, vitamin D, hemoglobin, and RBCs, in MPD and CPD patients. However, most of the PD patients have irregular blood pressure (systolic blood pressure and diastolic blood pressure) in patients as compared to healthy control.

Concentration of electrolytes in scalp hair samples of Parkinson's disease patients

The concentration of Ca in scalp hair samples of male healthy donors and MPD patients' age ranged 50–65 years, were found to be at 95% confidence intervals with median values, 347 [CI: 298, 386] and 437 [Cl: 315, 519] μg/g, respectively, which describes 20% Ca higher in scalp hair of MPD than the referent subjects. However, in higher age group (66–75 years), the Ca concentrations in scalp hair samples of male referent and CPD patients were found to 222 [CI: 162, 330] μg/g and 320 [CI: 238, 404] μg/g, which shows about 27% Ca concentration was greater in scalp hair of CPD than in the age-matched referent subjects. The median Mg contents in scalp hair samples of both male nondiseased subjects and MPD patients were detected to be 40.2 [CI: 22.6, 66.0] and 48.9 [CI: 25.6, 71.5] μg/g. However, in the higher age group, Mg contents in scalp hair samples of male non-diseased subjects and CPD patients were 38.9 [CI: 29.7, 52.4] and 50.4 [CI: 35.6, 68.2] μg/g; the resulted data indicated that MPD patients have >10% Mg than age-matched nondiseased subjects. Although the concentration of Mg was 20% greater in scalp hair of CPD patients than in that of age-matched referents. The median Na contents in scalp hair samples of male nondiseased subjects and MPD patients were found to be 309 [CI: 263, 345] and 268 [CI: 214, 304] μg/g, respectively; in the higher age group of male referent and CPD patients, the median Na contents were 408 [CI: 352, 456] and 251 [CI: 223, 288] μg/g. The resulted data indicated that MPD patients have <15% Na in scalp hair than referents, whereas the concentration of Na in scalp hair of CPD patients was <38.4% lower than age-matched controls. The median K contents in scalp hair samples of male referent and MPD patients were observed to be 68.1 [CI: 42.4, 92.3] and 31.5[CI:20.4,46.7] μg/g, respectively, which indicates that 52% K was higher in scalp hair of MPD than referents. However, in the higher age group, the K contents in scalp hair samples of male referent and CPD patients were 85.77 [CI:64.3136] and 32.7 [CI:25.3, 41.6] μg/g, which shows 62% K higher in scalp hair of CPD than that of referents as shown in Table 2 and Fig. 1.

Concentration of electrolytes in blood serum samples of Parkinson's disease patients

The concentration of Ca in blood serum samples of male referent and MPD patients aged 50–65 years was found to be 88 [CI: 76.3, 103] and 72.5 [CI: 63.2, 88.3] mg/L, respectively. However, in higher age group, the level of Ca in blood serum samples of referent and CPD patients were found as 85.1 [CI: 69.8, 103] and 77.9 [CI: 65.1, 89.3] mg/L, respectively. It was found that the levels of Ca were found to be lower in MPD and CPD patients (9.2%–16.9%, respectively) than in age-matched nondiseased male subjects. It was observed that the levels of Mg were found to be 14.2% and 11.3% lower in blood serum samples of MPD and CPD patients, respectively, than in those of age-

Table 2 Electrolyte concentrations (μg/g) in the scalp hair of the male referents and PD patients.

Age groups	Elements	Referents	Mild PD	*P*-value	Referents	Chronic PD	*P*-value
		(52–65) years			(66–75) years		
Scalp hair	Ca	350±34.5 (317–387)	435±36.7 (399–472)	0.010	242±44.1 (218–265)	324±31.7 (302–342)	0.001
	Mg	43.4±9.52 (33.5–50.5)	47.9±7.93 (40.2–53.5)	0.062	38.7±6.02 (30–45)	50.3±6.54 (46.5–53.4)	0.006
	Na	310±21.9 (285–330)	267±16.4 (250–280)	0.008	410.2±23.2 (385–432)	252±15.2 (237–266)	0.001
	K	64.87±12.7 (50–76)	31.2±4.50 (26–37)	0.001	87.5±13.2 (73–99)	32.8±3.15 (28–37)	0.001

Key: PD, Parkinson's disease; Ca, calcium; Mg, magnesium; Na, sodium; K, potassium.

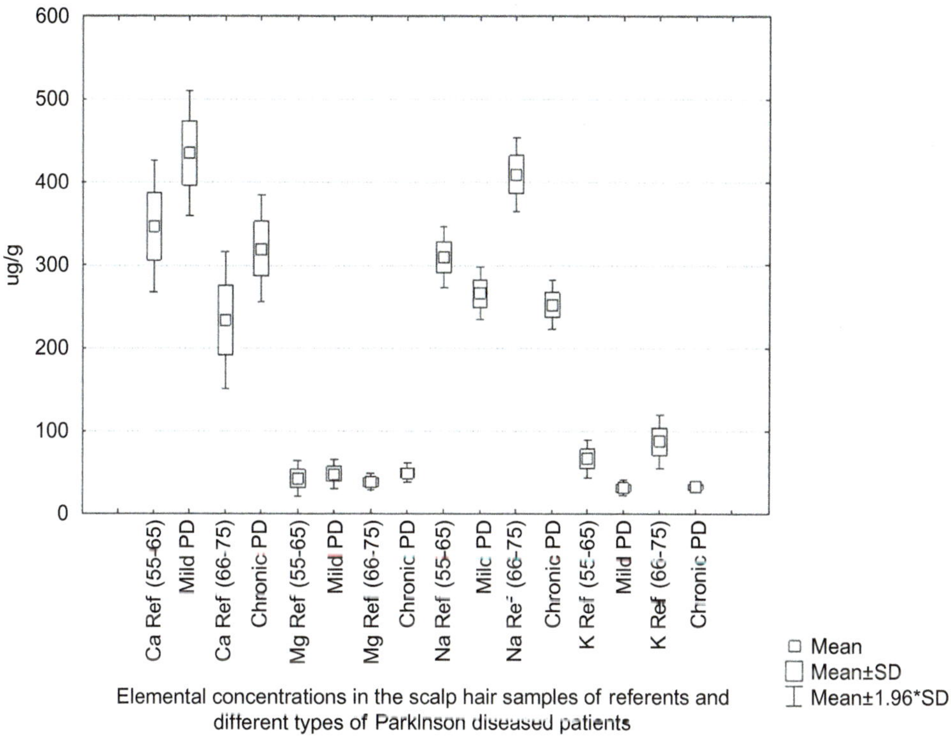

Fig. 1 Elemental concentrations in the scalp hair samples of referents versus PD patients.

matched referent subjects of both age groups. The Na level in blood serum samples of control and MPD patients was obtained to be 3201 [CI: 3048, 3365] and 2655 [CI: 2560,2778] mg/L; correspondingly, the concentration of Na in blood serum samples of nondiseased subjects is found to be greater [CI: 3039, 3257] mg/L, as compared to CPD patients [CI: 2429, 2670] mg/L. It was observed that the concentration of K was found to be 24% and 52% lower in blood serum samples of mild and CPD patients, respectively, than in blood serum samples of age-matched referent subjects of both age groups, which is shown in Table 3 and Fig. 2.

Concentration of electrolytes in blood plasma samples of Parkinson's disease patients

The Ca concentration in blood plasma samples of referent and MPD patients was observed to be [CI: 79.2116] and [CI: 48.9, 92.8] mg/L, while the concentration of Ca in blood plasma samples of male referent and CPD patients was found to be [CI: 72.7102] and [CI: 60.2, 85.0] mg/L, correspondingly. It was noted that the levels of Ca were found to be 22.1%–17.6% lower in blood plasma of MPD and CPD patients, respectively, than those of age-matched referent subjects. It was observed that 16% and

Table 3 Electrolyte concentrations (mg/L) in the blood serum of the male referents and PD patients.

Age groups	Elements	Referents	Mild PD	*P*-value	Referents	Chronic PD	*P*-value
		(52–65) years			(66–75) years		
Blood serum	Ca	88.9±5.10 (82–94)	73.4±5.02 (68–77)	0.001	84.3±7.09 (80.6–87.9)	77.1±5.62 (73.9–80.4)	0.001
	Mg	18.2±1.42 (16.7–19.6)	15.5±0.34 (15.16–15.84)	0.001	16.6±1.25 (15.9–17.2)	15.0±1.10 (14.5–15.5)	0.001
	Na	3194±57.5 (3135–3252)	2655±47.6 (2608–2700)	0.001	3160±60.2 (3100 – 3220)	2570±50.6 (2530–2610)	0.001
	K	174±17.5 (155–190)	132.5±13.5 (145–120)	0.001	158±12.5 (170–145)	78.9±9.86 (73.9–84.0)	0.001

Key: PD, Parkinson's disease; Ca, calcium; Mg, magnesium; Na, sodium; K, potassium.

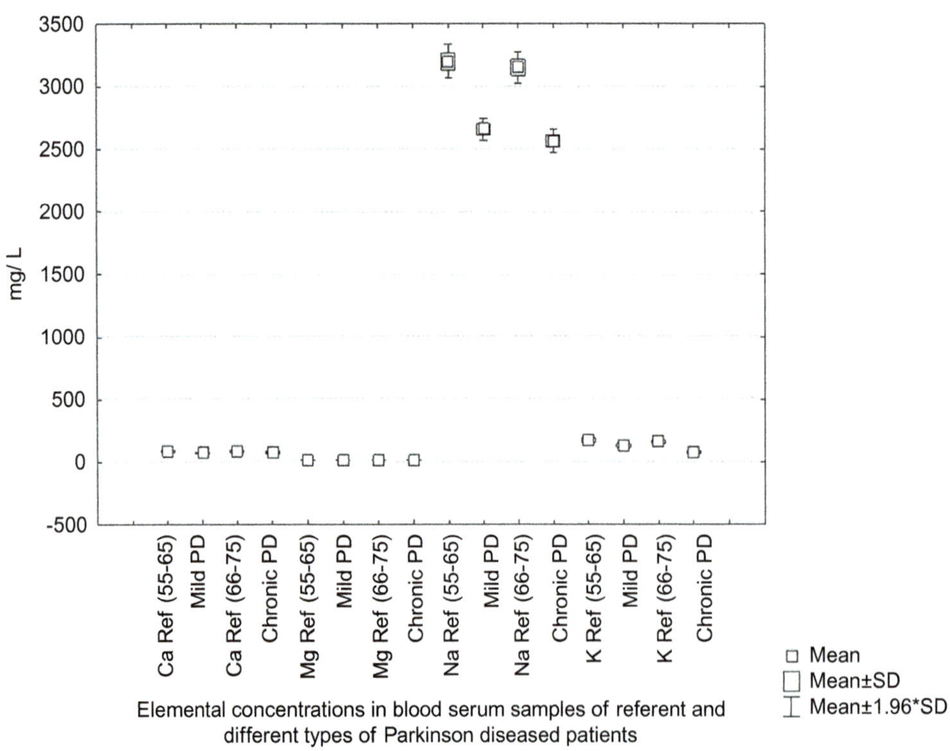

Fig. 2 Elemental concentrations in the blood serum samples of referents versus PD patients.

18% Na levels were obtained to be lower in MPD and CPD patients, respectively, than in the same age-grouped referent male subjects. However, 4.6% and 2.5% Na were found to be lower in blood plasma samples of MPD and CPD patients. It was also observed that 41.2% and 38% K concentrations were obtained to be lower in blood plasma samples of mild and CPD patients, respectively, than in those of age-matched nondiseased male subjects. The unpaired Student's *t*-test between male mild/chronic PD patients and nondiseased healthy subjects at different degrees of freedom was calculated at different probabilities. Our calculated *t*-value exceeds that of *t*-critical value at 95% confidence intervals, which showed that the difference among mean concentration of electrolytes in healthy subjects and mild/chronic PD patients showed significance ($P < 0.019$), as shown in Table 4 and Fig. 3.

Correlation of electrolytes versus different biochemical parameters in referents and PD patients

The interelemental correlation (r) among biochemical parameters included systolic blood pressure; diastolic blood pressure; hemoglobin; RBCs versus electrolytes in different types of Parkinson's disease patients (MPD/CPD); and normal persons in scalp hair, blood serum, and blood plasma is shown in Table 5.

Discussion

This research study gives information on electrolytes in biological (scalp hair, serum and plasma) samples, gained from two classes of PD (mild and chronic) sufferers and those of nondiseased subjects of age groups (55–65 and 66–75) years, and we showed the function of electrolytes in neurological disorder patients. We also correlated the biochemical

Table 4 Electrolyte concentrations (mg/L) in the blood plasma of the male referents and PD patients.

Age groups	Elements	Referents	Mild PD	*P*-value	Referents	Chronic PD	*P*-value
		(52–65) years			(66–75) years		
Blood plasma	Ca	97.5 ± 7.10 (90–105)	77.5 ± 9.02 (66.5–87.2)	0.001	90.3 ± 9.20 (84.3–96.9)	71.9 ± 5.90 (65.9–75.8)	0.001
	Mg	23.5 ± 3.5 (20.9–30.2)	15.3 ± 0.51 (14.9–15.7)	0.001	20.2 ± 4.2 (16.9–23.8)	10.7 ± 0.33 (10.5–10.95)	0.001
	Na	3015 ± 120 (2895–3137)	2890 ± 95.3 (2794–2987)	0.072	3165 ± 67.8 (3095–3230)	3085 ± 83.6 (3000–3168)	0.060
	K	197 ± 13.7 (182–210)	115 ± 6.35 (109–120)	0.001	176 ± 16.7 (160–192)	107 ± 7.29 (103.7–112)	0.001

Key: PD, Parkinson's disease; Ca, calcium; Mg, magnesium; Na, sodium; K, potassium.

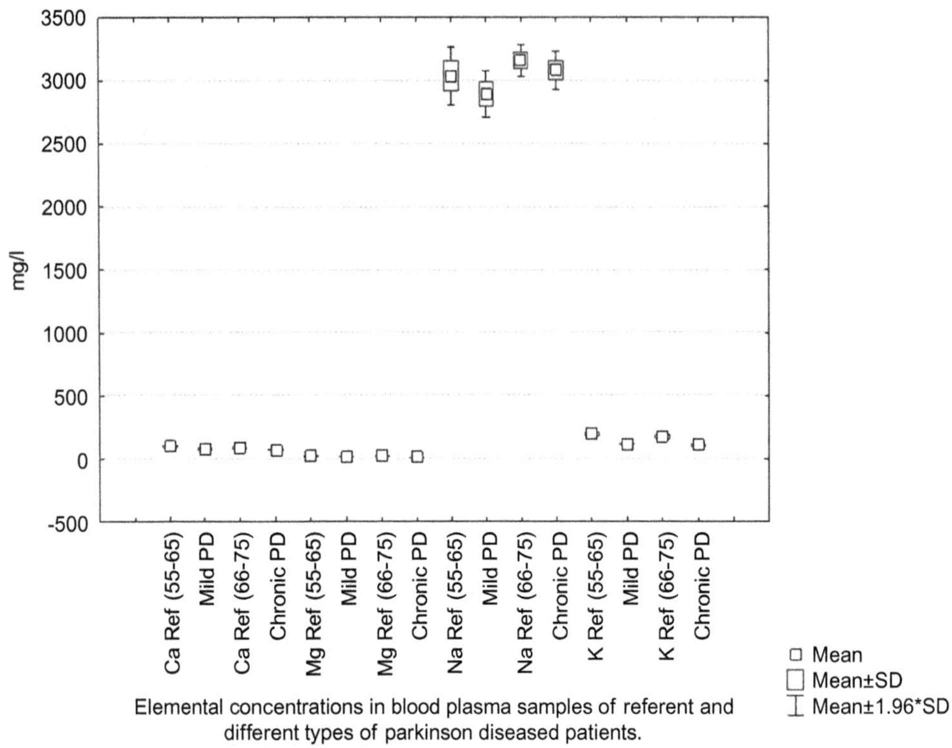

Fig. 3 Elemental concentrations in the blood plasma samples of referents versus PD patients.

parameters of both types of PD patients with electrolytes. The electrolytes (Ca and Mg) concentrations were obtained to be significantly higher in the scalp hair samples of mild/chronic PD patients than in controls. The mean values of Ca content in scalp hair samples of male PD patients were higher than in referents, while the difference was significant ($P<0.01$). The mean amount of Mg contents in scalp hair samples of both types of MPD and CPD patients was greater than that of control subjects, but in MPD group, the difference was not statistically significant ($P>0.05$). However, in CPD group, the statistically significant difference was observed ($P<0.05$). When comparing MPD and CPD patients, the Ca mean concentrations were a little higher in MPD patients than in CPD patients. However, the Mg content was slightly lower in MPD patients than in the CPD group ($P<0.05$). A reverse link was monitored with Na and K, which displayed a significantly lower concentrations in both types of PD groups as compared to control group levels ($P<0.05$ for both).

PD, the most common neurodegenerative disorder after AD, stays hard to diagnose clinically due to having common characteristics symptoms with other neurological diseases, for example, vital tremor, multiple system atrophy, and progressive supra nuclear palsy, mainly for the period of early course of the disease (Jellinger, 2015; Jellinger &

Table 5 Linear regression and Pearson's coefficient for electrolytes versus different biochemical parameters in referents and PD patients.

Age groups	Elements	Biochemical parameter	Referents	Mild PD	Referents	Chronic PD
			(52–65) years		(66–75) years	
Scalp hair	Ca	SBP	$y=7.765\times-644.5$ $r=0.84$	$y=5.965\times-434.2$ $r=0.86$	$y=4.619\times-374.0$ $r=0.82$	$y=4.197\times-366.4$ $r=0.87$
		DBP	$y=4.55\times-31.27$ $r=0.81$	$y=4.03\times+48.71$ $r=0.88$	$y=5.391\times-235.0$ $r=0.88$	$y=3.826\times-93.61$ $r=0.87$
		Hb	$y=-285.5\times+4340$ $r=0.86$	$y=205.1\times-2234$ $r=0.83$	$y=21.71\times-28.10$ $r=0.83$	$y=57.91\times-291.3$ $r=0.85$
		RBC	$y=201.6\times-643.7$ $r=0.79$	$y=78.84\times+107.8$ $r=0.78$	$y=78.84\times+107.8$ $r=0.77$	$y=89.56\times-133.3$ $r=0.82$
		Vitamin D	$y=109.0\times-2225$ $r=0.72$	$y=21.89\times+241.4$ $r=0.84$	$y=140.9\times-2945$ $r=0.93$	$y=13.45\times+235.0$ $r=0.75$
	Mg	SBP	$y=-1.964\times+293.1$ $r=0.77$	$y=-1.407\times+253.4$ $r=0.85$	$y=-0.517\times+106.4$ $r=0.75$	$y=-0.743\times+171.4$ $r=0.85$
		DBP	$y=1.170\times-54.69$ $r=0.76$	$y=0.942\times-42.05$ $r=0.76$	$y=0.620\times-15.43$ $r=0.83$	$y=-0.673\times+122.5$ $r=0.85$
		Hb	$y=76.23\times-1024$ $r=0.84$	$y=-47.30\times+663.7$ $r=0.81$	$y=-2.658\times+70.52$ $r=0.81$	$y=9.819\times-53.72$ $r=0.80$
		RBC	$y=42.90\times-168.2$ $r=0.61$	$y=20.25\times-35.69$ $r=0.84$	$y=-10.71\times+82.38$ $r=0.81$	$y=-12.63\times+87.44$ $r=0.88$
		Vitamin D	$y=-15.34\times+422.8$ $r=0.81$	$y=5.317\times+1.344$ $r=0.87$	$y=14.69\times-292.8$ $r=0.79$	$y=3.235\times+29.98$ $r=0.86$
	Na	SBP	$y=2.752\times-42$ $r=0.65$	$y=1.271\times+80.59$ $r=0.44$	$y=1.912\times+158.2$ $r=0.63$	$y=-1.148\times+439.8$ $r=0.52$
		DBP	$y=1.722\times+166.2$ $r=0.67$	$y=-0.902\times+352.4$ $r=0.47$	$y=2.119\times+225.4$ $r=0.64$	$y=-0.951\times+354.7$ $r=0.47$
		Hb	$y=101.5\times-1110$ $r=0.47$	$y=-53.57\times+963.0$ $r=0.51$	$y=-8.945\times+517.3$ $r=0.62$	$y=-16.19\times+422.9$ $r=0.51$
		RBC	$y=-79.00\times+697.1$ $r=0.68$	$y=22.55\times+172.3$ $r=0.53$	$y=-37.91\times+564.9$ $r=0.64$	$y=16.71\times+202.3$ $r=0.46$
		Vitamin D	$y=21.54\times-224.7$ $r=0.66$	$y=5.246\times+219.5$ $r=0.48$	$y=-51.73\times+1576$ $r=0.63$	$y=4.794\times+222.6$ $r=0.25$

Continued

Table 5 Linear regression and Pearson's coefficient for electrolytes versus different biochemical parameters in referents and PD patients—cont'd

Age groups	Elements	Biochemical parameter	Referents	Mild PD	Referents	Chronic PD
			(52–65) years		(66–75) years	
	K	SBP	$y=-1.166\times+215.4$ $r=0.43$	$y=0.331\times-16.76$ $r=0.39$	$y=0.977\times-40.84$ $r=0.44$	$y=0.182\times+3.027$ $r=0.37$
		DBP	$y=0.725\times+6.310$ $r=0.45$	$y=-0.237\times+54.28$ $r=0.43$	$y=1.132\times-10.75$ $r=0.47$	$y=-0.156\times+49.71$ $r=0.35$
		Hb	$y=-48.89\times+750.4$ $r=0.51$	$y=12.08\times-125.8$ $r=0.41$	$y=-4.434\times+141.0$ $r=0.42$	$y=2.782\times+3.453$ $r=0.41$
		RBC	$y=-36.3\times+244.7$ $r=0.49$	$y=-4.981\times+52.12$ $r=0.41$	$y=18.15\times+13.33$ $r=0.43$	$y=-2.825\times+41.21$ $r=0.35$
		Vitamin D	$y=-8.571\times+278.9$ $r=0.43$	$y=1.241\times+20.49$ $r=0.40$	$y=9.015\times-156.9$ $r=0.45$	$y=-24.51\times+640.4$ $r=0.42$
Blood serum	Ca	SBP	$y=0.732\times-5.200$ $r=0.54$	$y=-0.498\times+145.8$ $r=0.57$	$y=-0.568\times+159.8$ $r=0.56$	$y=-0.48\times+155.9$ $r=0.61$
		DBP	$y=0.493\times+47.35$ $r=0.59$	$y=0.493\times+47.35$ $r=0.59$	$y=-0.623\times+139.2$ $r=0.56$	$y=0.415\times+32.61$ $r=0.58$
		Hb	$y=26.04\times-276.0$ $r=0.53$	$y=-17.97\times+307.1$ $r=0.57$	$y=2.708\times+52.43$ $r=0.55$	$y=6.642\times+7.359$ $r=0.59$
		RBC	$y=22.23\times-20.85$ $r=0.59$	$y=7.511\times+42.11$ $r=0.58$	$y=-11.02\times+130.1$ $r=0.55$	$y=-7.657\times+100.2$ $r=0.59$
		Vitamin D	$y=-6.763\times+255.9$ $r=0.67$	$y=-1.866\times+89.76$ $r=0.57$	$y=15.93\times-274.3$ $r=0.57$	$y=0.738\times+1.643$ $r=0.73$
	Mg	SBP	$y=0.790\times+3.796$ $r=0.79$	$y=0.027\times+11.54$ $r=0.47$	$y=-0.090\times+28.64$ $r=0.61$	$y=-0.054\times+23.78$ $r=0.40$
		DBP	$y=-0.153\times+30.83$ $r=0.75$	$y=-0.022\times+17.66$ $r=0.57$	$y=-0.119\times+27.07$ $r=0.74$	$y=0.065\times+7.750$ $r=0.53$
		Hb	$y=-8.864\times+142.1$ $r=0.74$	$y=-0.931\times+27.66$ $r=0.45$	$y=0.467\times+11.07$ $r=0.66$	$y=1.048\times+3.754$ $r=0.54$
		RBCs	$y=-6.183\times+48.47$ $r=0.68$	$y=0.419\times+13.79$ $r=0.49$	$y=-2.443\times+26.71$ $r=0.85$	$y=-1.224\times+18.46$ $r=0.54$
		Vitamin D	$y=-1.932\times+66.00$ $r=0.78$	$y=-0.116\times+16.56$ $r=0.53$	$y=2.866\times-47.93$ $r=0.72$	$y=0.279\times+13.10$ $r=0.47$
	Na	SBP	$y=41.83\times+2442$ $r=0.88$	$y=-4.198\times+3269$ $r=0.52$	$y=-7.283\times+4107$ $r=0.86$	$y=1.731\times+2281$ $r=0.24$
		DBP	$y=-6.118\times+3708$ $r=0.63$	$y=3.000\times+2369$ $r=0.56$	$y=6.612\times+2575$ $r=0.73$	$y=-3.446\times+2936$ $r=0.53$

		Hb	$y=392.1\times-2284$ $r=0.69$	$y=-153.0\times-4648$ $r=0.53$	$y=-32.07\times+3536$ $r=0.64$	$y=-57.58\times+3171$ $r=0.56$
		RBCs	$y=317.3\times+1642$ $r=0.73$	$y=60.32\times+2407$ $r=0.51$	$y=-135.6\times+3705$ $r=0.83$	$y=59.53\times+2387$ $r=0.51$
		Vitamin D	$y=-82.56\times+5246$ $r=0.70$	$y=-18.65\times+2822$ $r=0.61$	$y=-63.73\times+1846$ $r=0.78$	$y=-4.859\times+281.9$ $r=0.51$
	K	SBP	$y=-9.684\times+349.5$ $r=0.78$	$y=-0.802\times+247.8$ $r=0.31$	$y=-1.177\times+316.0$ $r=0.77$	$y=-0.670\times+186.0$ $r=0.48$
		DBP	$y=2.114\times-1.310$ $r=0.84$	$y=0.791\times+55.00$ $r=0.45$	$y=1.337\times+45.03$ $r=0.81$	$y=-0.483\times+128.7$ $r=0.38$
		Hb	$y=105.4\times-1301$ $r=0.71$	$y=-42.55\times+684.6$ $r=0.45$	$y=-5.563\times+228.2$ $r=0.76$	$y=9.013\times-18.57$ $r=0.46$
		RBC	$y=-85.59\times+594.3$ $r=0.755$	$y=-21.12\times+218.5$ $r=0.54$	$y=19.34\times+82.02$ $r=0.66$	$y=11.22\times+43.13$ $r=0.49$
		Vitamin D	$y=23.63\times-411.6$ $r=0.77$	$y=3.332\times+101.4$ $r=0.39$	$y=-28.76\times+809.9$ $r=0.71$	$y=-2.845\times+93.99$ $r=0.47$
Blood plasma	Ca	SBP	$y=-0.971\times+221.3$ $r=0.59$	$y=0.890\times-54.01$ $r=0.57$	$y=-0.657\times+174.7$ $r=0.61$	$y=0.481\times-5.964$ $r=0.37$
		DBP	$y=0.649\times+43.40$ $r=0.63$	$y=0.622\times+16.09$ $r=0.61$	$y=0.672\times+29.78$ $r=0.57$	$y=-0.403\times+116.3$ $r=0.57$
		Hb	$y=33.90\times-376.9$ $r=0.56$	$y=32.33\times-344.9$ $r=0.58$	$y=3.233\times+49.29$ $r=0.621$	$y=-6.928\times+145.8$ $r=0.62$
		RBCs	$y=26.21\times-31.43$ $r=0.564$	$y=-13.59\times+132.1$ $r=0.590$	$y=-11.94\times+137.1$ $r=0.57$	$y=-7.043\times+93.66$ $r=0.55$
		Vitamin D	$y=-7.430\times+281.4$ $r=0.59$	$y=-3.673\times+108.2$ $r=0.392$	$y=26.61\times-511.7$ $r=0.59$	$y=-1.251\times+80.39$ $r=0.37$
	Mg	SBP	$y=-1.020\times+227.6$ $r=0.61$	$y=0.035\times+10.03$ $r=0.44$	$y=-0.426\times+75.99$ $r=0.75$	$y=0.023\times+6.854$ $r=0.44$
		DBP	$y=0.248\times+2.494$ $r=0.49$	$y=0.023\times+13.00$ $r=0.44$	$y=-0.436\times+57.81$ $r=0.70$	$y=-0.023\times+13.27$ $r=0.48$
		Hb	$y=-20.21\times+305.8$ $r=0.68$	$y=1.468\times-3.872$ $r=0.51$	$y=-1.963\times+43.55$ $r=0.72$	$y=-0.431\times+15.28$ $r=0.57$

Continued

Table 5 Linear regression and Pearson's coefficient for electrolytes versus different biochemical parameters in referents and PD patients—cont'd

Age groups	Elements	Biochemical parameter	Referents (52–65) years	Mild PD (52–65) years	Referents (66–75) years	Chronic PD (66–75) years
		RBCs	$y = -17.20\times + 107.5$ $r = 0.75$	$y = 0.562\times + 12.90$ $r = 0.47$	$y = 9.205\times - 17.79$ $r = 0.83$	$y = 0.417\times + 9.497$ $r = 0.48$
		Vitamin D	$y = -4.701\times + 139.6$ $r = 0.76$	$y = 0.157\times + 13.84$ $r = 0.52$	$y = 12.41\times - 260.0$ $r = 0.80$	$y = -0.115\times + 11.44$ $r = 0.50$
	Na	SBP	$y = -20.70\times + 5679$ $r = 0.75$	$y = -8.973\times + 4201$ $r = 0.52$	$y = -6.313\times + 3991$ $r = 0.73$	$y = -5.771\times + 4024$ $r = 0.51$
		DBP	$y = -14.23\times + 4218$ $r = 0.85$	$y = 5.785\times + 2339$ $r = 0.54$	$y = 7.001\times + 2552$ $r = 0.74$	$y = 5.552\times + 2481$ $r = 0.54$
		Hb	$y = 708.5\times - 6874$ $r = 0.72$	$y = -346.1\times + 7398$ $r = 0.57$	$y = 33.32\times + 2759$ $r = 0.80$	$y = -89.06\times + 4020$ $r = 0.56$
		RBCs	$y = -571.7\times + 5844$ $r = 0.76$	$y = 135.2\times + 2333$ $r = 0.54$	$y = -128.7\times + 3688$ $r = 0.59$	$y = 98.74\times + 2787$ $r = 0.53$
		Vitamin D	$y = 85.27\times + 1086$ $r = 0.72$	$y = 35.71\times + 2578$ $r = 0.56$	$y = 185.1\times - 1013$ $r = 0.79$	$y = 26.08\times + 2921$ $r = 0.53$
	K	SBP	$y = -2.138\times + 469.4$ $r = 0.78$	$y = -0.474\times + 184.6$ $r = 0.42$	$y = -1.606\times + 385.9$ $r = 0.73$	$y = 0.473\times + 29.33$ $r = 0.43$
		DBP	$y = 1.412\times + 79.39$ $r = 0.85$	$y = 0.345\times + 82.28$ $r = 0.46$	$y = -1.751\times + 326.9$ $r = 0.73$	$y = 0.451\times + 57.97$ $r = 0.45$
		Hb	$y = 86.60\times - 1014$ $r = 0.84$	$y = -17.61\times + 344.6$ $r = 0.44$	$y = -8.434\times + 276.3$ $r = 0.65$	$y = -6.184\times + 172$ $r = 0.40$
		RBC	$y = -53.59\times + 459.7$ $r = 0.71$	$y = 6.962\times + 86.54$ $r = 0.42$	$y = -34.19\times + 314.7$ $r = 0.81$	$y = -8.931\times + 133.3$ $r = 0.49$
		Vitamin D	$y = -14.81\times + 563.8$ $r = 0.73$	$y = -1.559\times + 129.2$ $r = 0.37$	$y = -44.31\times + 1173$ $r = 0.75$	$y = -1.862\times + 118.1$ $r = 0.39$

Key: SBP, systolic blood pressure; DBP, diastolic blood pressure; Hb, hemoglobin RBC, red blood cells.

Lantos, 2010). PD is a multiorgan disease functionalized by progressive deterioration of the dopaminergic neuronal system, many other organs and neuronal systems responsible for the foundation of motor symptoms (like akinesis, rigidity, tremor, and postural instability) and a variety of nonmotor and neuropsychological problems that influence the patient's life style (Gallegos, Pacheco, Peters, Opazo, & Aguayo, 2015). The resulted data indicated that higher contents of Ca and Mg in cytosol caused neurotoxicity and also promoted free radical generation. Usually, this observable fact is limited by a voltage-dependent magnesium (Mg) blockade (Olanow & Arendash, 1994). Further, an unbalanced metabolism of Ca and Mg in the central nervous system and an raised oxidative damage due to the shortage of Mg have been noticed in PD (Berg, Youdim, & Riederer, 2004).

Calcium is chemical substance that is present in the human body and is present within the body's cells and extracellular liquids such as blood plasma. The electrolyte (serum calcium) is of greatest importance to cellular functions that positively charged ions such as calcium. These ions are crucial for nerve messages conduction, muscle fiber contraction, and care of cell membrane permeability. We assessed the most common yet the major electrolyte (serum calcium), which helps in neuronal development and function, maintains the resting membrane potential, and causes the generation of action potential across the excitable tissue along the course of nerve, nerve transmission, intracellular signaling, and hormonal secretion. In all types of neurodegenerative disorders patients, there are alterations in Ca^{2+} homeostasis, which contribute to cell death (Brini, 2003; Nicholls, Vesce, Kirk, & Chalmers, 2003). The calcium homeostasis is affected near the beginning of the pathological process and consequence in major neuronal demise (Toescu & Verkhratsky, 2003). In neurodegenerative disorders, it is the disturbance in endoplasmic reticulum. Calcium homoeostasis leads to mitochondrial dysfunction and cell death. Increased intracellular calcium can predispose to Aβ aggregation and fibril formation in AD. Alterations in Ca^{2+} maintaining system-mediated signal transduction processes at the cellular and subcellular levels play a vital role in normal aging and age-related neuronal dysfunction in neurodegenerative diseases (Bezprozvanny, 2009; Farooqui & Farooqui, 2009; Mattson, Furukawa, Bruce, Mark, & Blanc, 1997). Neuronal populations with high levels of calcium-binding proteins (CaBPs) are relatively spared in neurodegenerative diseases, whereas those with reduced levels of CaBP are severely affected (Overgaard-Steensen & Ring, 2013). Human body holds about 24 g Mg, of which 99% is accumulated in bone, other soft tissues, and muscle. Mg is fundamentally important for the function of every organ in the human body. Furthermore, insufficiency of Mg is related to a broad series of diseases, and Mg supplementation is considered as potential treatment in many of diseases (Lunøe & Overgaard-Steensen, 2015). Deficiency of blood serum magnesium concentration caused different neurological diseases such as depression, migraine, and Parkinson's. Neuronal magnesium level plays a wide role in the regulation of *N*-methyl-D-aspartate (NMDA) receptor excitability. All NMDA receptors contain four subunits, and each one has different biochemical characteristics. Hyperexcitable NMDA receptors are due to deficiency of magnesium. Overall, magnesium plays an important role in the maintenance and release of neuropeptide, which may have serious consequences in neuronal disease.

Hyponatremia and hypernatremia are frequent in the critically ill patient and associated with significant morbidity and mortality. A rapid change in serum sodium concentration can result in brain damage and death resulting from brain edema or osmotic demyelination (Overgaard-Steensen & Ring, 2013). However, it is well known that a rapid decrease in serum sodium causes brain edema, and delayed treatment can result in brain damage and death (Lunøe & Overgaard-Steensen, 2015). On the other hand, sodium and potassium regulation have an effect on serum sodium, which are found by the difference between the input and output of Na^+ and K^+. Similarly, water balance is reviewed the serum sodium (input of water minus output, e.g., urine), for that reason SNa always should be corrected for hyperglycemia (SNa decreases approximately 0.4 mmol/L per mmol/L raise in plasma glucose concentration above 5 mmol/L) (Lunøe & Overgaard-Steensen, 2015). Sodium is lost together with water in urine, gastrointestinal fluids, bleeding, wounds, drainage of extracellular fluids (e.g., pleural effusions, ascites, CSF, dialysis), and sweat. Hypokalemia is generally visible as a mild form associated with hypochloremic alkalosis and largely combined with pharmaceutical treatments such as diuretics, intravenous fluid, and iatrogenic hyperventilation (Lunøe & Overgaard-Steensen, 2015; Overgaard-Steensen & Ring, 2013).

The electrolyte levels in blood serum were found to be lower in both types of PD patient groups than in healthy groups; the difference was significant ($P<0.01$). When comparing MPD patients with CPD patients group, all the electrolyte level was found to be slightly higher in MPD patients than in CPD patients ($P<0.01$). It was observed that electrolytes concentration in blood plasma was found lower in both groups of patients ($P<0.01$). The correlation (r) between electrolytes (Ca, Mg, Na, K) concentration in biological samples (scalp hair, blood serum, and blood plasma) and biochemical parameters (SBP, DBP, Hb, RBCs, vitamin D) of referents and PD patients subjects of age range (55–75 years) was observed from the resulted data. The correlation (r) of calcium in biological samples concentration versus biochemical parameter of referents shows higher ($r = 0.72$–0.93), while in the PD patients, the correlation was found to be lower ($r = 0.37$–0.87), respectively. In case of magnesium, the correlation (r) values in biological samples concentration versus biochemical parameter of referents show ($r=0.49$–0.85), while in the PD patients, the correlation was found to be ($r=0.40$–0.88) of both age groups. In Na, the correlation (r) values in biological samples concentration versus biochemical parameter of referents show ($r=0.47$–0.88), while in the PD patients, the correlation was found to be in the range of ($r=0.24$–0.57) of both age groups. It was observed from the obtained result that in K the correlation (r) values in biological samples concentration versus biochemical parameter of referents show

($r=0.42$–0.85), while in the PD patients, the correlation was found to be in the range of ($r=0.31$–0.54) of both age groups. This indicated that the imbalances in concentration of electrolytes are highly associated with biochemical parameters and the resulted data showed a strong correlation with calcium and magnesium in both categories.

With data extracted from the current work and knowledge attained, we recommend in the future referring this work for other elements that need to be monitored and regulated for general public. Future research is needed to determine these electrolyte concentration impact on clinical outcomes and to establish whether routinely providing supplements will result in improved neurological treatment outcomes and/or life quality. The investigation of concentrations of the electrolytes Ca, Mg, Na, and K and their interaction with biochemical parameters may help in studying the development of neurological diseases. The increase or decrease in the concentrations of these electrolytes may be associated with typical symptoms of PD, AD, and dementia. Therefore, measurements of electrolytes can be used as a tool to assess the susceptibility to earlier stage of disease.

Summary

It can be summarized that imbalance of electrolytes plays a crucial role in many neurodegenerative disease pathogenesis such as Alzheimer's, Parkinson's, and dementia disorders. The work focuses on the assessment of electrolytes in scalp hair, blood serum, and blood plasma samples. The correlation study was also carried out between electrolytes concentration and different biochemical parameters.

Summary points

- Neurodegeneration is complex and multifaceted, leading to many chronic diseases like Parkinson's disorder, dementia, and AD.
- Electrolytes imbalance has a significant role in the pathophysiology of neurological disorders, which might be serve as a bioindicator of the neurological problems.
- The disturbances of electrolytes (calcium, magnesium, sodium, and potassium) and its correlation with biochemical parameters in patients with PD at different stages is an important aspect.

References

Abbott, A. (2011). A problem for our age. *Nature*, *475*, S2.
Abou-Raya, S., Helmii, M., & Abou-Raya, A. (2009). Bone and mineral metabolism in older adults with Parkinson's disease. *Age and Ageing*, *38*, 675–680.
Adjei, P. (2011). *Autonomic function in epilepsy*. UCL (University College London).
Aminoff, M. J., & Josephson, S. A. (2014). *Aminoff's neurology and general medicine*. Elsevier.
Arvanitakis, Z., & Bennett, D. A. (2019). What is dementia? *JAMA*, *322*, 1728.

Augusto, C. M. G., Morais, N., Santana, R. P., & Almeida, M. (2019). *Parkinsonism as an atypical manifestation of primary hyperparathyroidism: A case report*. AACE Clinical Case Reports.
Baks, T., van Geuns, R.-J., Biagini, E., Wielopolski, P., Mollet, N. R., Cademartiri, F., et al. (2006). Effects of primary angioplasty for acute myocardial infarction on early and late infarct size and left ventricular wall characteristics. *Journal of the American College of Cardiology*, *47*, 40–44.
Barbagallo, M., Dominguez, L. J., Galioto, A., Ferlisi, A., Cani, C., Malfa, L., et al. (2003). Role of magnesium in insulin action, diabetes and cardio-metabolic syndrome X. *Molecular Aspects of Medicine*, *24*, 39–52.
Barnham, K. J., Masters, C. L., & Bush, A. I. (2004). Neurodegenerative diseases and oxidative stress. *Nature Reviews Drug Discovery*, *3*, 205.
Bavaresco, L., Giachino, E., & Pezzutto, S. (2003). Grapevine rootstock effects on lime-induced chlorosis, nutrient uptake, and source-sink relationships. *Journal of Plant Nutrition*, *26*, 1451–1465.
Berg, D., Youdim, M. B., & Riederer, P. (2004). Redox imbalance. *Cell and Tissue Research*, *318*, 201–213.
Bezprozvanny, I. (2009). Calcium signaling and neurodegenerative diseases. *Trends in Molecular Medicine*, *15*, 89–100.
Bezprozvanny, I., & Mattson, M. P. (2008). Neuronal calcium mishandling and the pathogenesis of Alzheimer's disease. *Trends in Neurosciences*, *31*, 454–463.
Bianchetti, A., Ranieri, P., Margiotta, A., & Trabucchi, M. (2006). Pharmacological treatment of Alzheimer's disease. *Aging Clinical and Experimental Research*, *18*, 158–162.
Bohl, C. H., & Volpe, S. L. (2002). Magnesium and exercise. *Critical Reviews in Food Science and Nutrition*, *42*, 533–563.
Brini, M. (2003). Ca^{2+} signalling in mitochondria: Mechanism and role in physiology and pathology. *Cell Calcium*, *34*, 399–405.
Chubanov, V., Gudermann, T., & Schlingmann, K. P. (2005). Essential role for TRPM6 in epithelial magnesium transport and body magnesium homeostasis. *Pflügers Archiv*, *451*, 228–234.
Cisternas, P., Lindsay, C. B., Salazar, P., Silva-Alvarez, C., Retamales, R. M., Serrano, F. G., et al. (1852). The increased potassium intake improves cognitive performance and attenuates histopathological markers in a model of Alzheimer's disease. *Biochimica et Biophysica Acta (BBA)—Molecular Basis of Disease*, *2015*, 2630–2644.
Davie, C. A. (2008). A review of Parkinson's disease. *British Medical Bulletin*, *86*, 109–127.
De Baaij, J. H., Hoenderop, J. G., & Bindels, R. J. (2015). Magnesium in man: Implications for health and disease. *Physiological Reviews*, *95*, 1–46.
De Lau, L. M., & Breteler, M. M. (2006). Epidemiology of Parkinson's disease. *The Lancet Neurology*, *5*, 525–535.
De Rosa, A., Rinaldi, C., Tucci, T., Pappatà, S., Rossi, F., Morra, V., et al. (2011). *Co-existence of primary hyperparathyroidism and Parkinson's disease in three patients: An incidental finding?* (179 views).
Drago, D., Bolognin, S., & Zatta, P. (2008). Role of metal ions in the Aβ oligomerization in Alzheimer's disease and in other neurological disorders. *Current Alzheimer Research*, *5*, 500–507.
Emre, M., Aarsland, D., Brown, R., Burn, D. J., Duyckaerts, C., Mizuno, Y., et al. (2007). Clinical diagnostic criteria for dementia associated with Parkinson's disease. *Movement Disorders: Official Journal of the Movement Disorder Society*, *22*, 1689–1707.
Farooqui, T., & Farooqui, A. A. (2009). Aging: An important factor for the pathogenesis of neurodegenerative diseases. *Mechanisms of Ageing and Development*, *130*, 203–215.
Gallegos, S., Pacheco, C., Peters, C., Opazo, C. M., & Aguayo, L. G. (2015). Features of alpha-synuclein that could explain the progression and irreversibility of Parkinson's disease. *Frontiers in Neuroscience*, *9*, 59.
Guerrero-Romero, F., & Rodrı´guez-Morán, M. (2000). Hypomagnesemia is linked to low serum HDL-cholesterol irrespective of serum glucose values. *Journal of Diabetes and its Complications*, *14*, 272–276.
He, K., Liu, K., Daviglus, M. L., Morris, S. J., Loria, C. M., Van Horn, L., et al. (2006). Magnesium intake and incidence of metabolic syndrome among young adults. *Circulation*, *113*, 1675–1682.
Huitrón-Ramírez, M. V., Ricárdez-Salinas, M., & Camacho-Ferre, F. (2009). Influence of grafted watermelon plant density on yield and quality in soil infested with melon necrotic spot virus. *HortScience*, *44*, 1838–1841.
Jellinger, K. A. (2003). α-Synuclein pathology in Parkinson's and Alzheimer's disease brain: Incidence and topographic distribution—A pilot study. *Acta Neuropathologica*, *106*, 191–202.
Jellinger, K. A. (2015). *How close are we to revealing the etiology of Parkinson's disease?*. Taylor & Francis.

Jellinger, K. A., & Lantos, P. L. (2010). Papp-Lantos inclusions and the pathogenesis of multiple system atrophy: An update. *Acta Neuropathologica, 119*, 657–667.
Kim, D. K., & Joo, K. W. (2009). Hyponatremia in patients with neurologic disorders. *Electrolytes & Blood Pressure, 7*, 51–57.
Lakatos, L., Kiss, L. S., David, G., Pandur, T., Erdelyi, Z., Mester, G., et al. (2010). Incidence, disease phenotype at diagnosis, and early disease course in inflammatory bowel diseases in Western Hungary, 2002–2006. *Inflammatory Bowel Diseases, 17*, 2558–2565.
Lewis, R., & Sibai, B. (1997). Recent advances in the management of preeclampsia. *Journal of Maternal-Fetal Medicine, 6*, 6–15.
Litvan, I., Halliday, G., Hallett, M., Goetz, C. G., Rocca, W., Duyckaerts, C., et al. (2007). The etiopathogenesis of Parkinson disease and suggestions for future research. Part I. *Journal of Neuropathology & Experimental Neurology, 66*, 251–257.
Luján-Martínez, D., Sánchez-Cifuentes, Á., Peña-Ros, E., Albarracín-Marín-Blázquez, A., & Candel-Arenas, M. F. (2018). Parkinsonism as presenting symptom of primary hyperparathyroidism: Improvement after surgery. *Cirugia y Cirujanos, 86*, 105–107.
Lunøe, M., & Overgaard-Steensen, C. (2015). Prevention of hospital-acquired hyponatraemia: individualised fluid therapy. *Acta Anaesthesiologica Scandinavica, 59*, 975–985.
Mattson, M. P. (2004). Pathways towards and away from Alzheimer's disease. *Nature, 430*, 631.
Mattson, M. P., Furukawa, K., Bruce, A. J., Mark, R. J., & Blanc, E. (1997). Calcium homeostasis and free radical metabolism as convergence points in the pathophysiology of dementia. In *Molecular mechanisms of dementia* (pp. 103–143). Springer.
Mellon, E., Pilkinton, D., Clark, C., Elliott, M., Witschey, W., Borthakur, A., et al. (2009). Sodium MR imaging detection of mild Alzheimer disease: Preliminary study. *American Journal of Neuroradiology, 30*, 978–984.
Mielke, M. M., Zandi, P. P., Blennow, K., Gustafson, D., Sjögren, M., Rosengren, L., et al. (2006). Low serum potassium in mid life associated with decreased cerebrospinal fluid Aβ42 in late life. *Alzheimer Disease & Associated Disorders, 20*, 30–36
Murakami, K., Okubo, H., & Sasaki, S. (2005). Effect of dietary factors on incidence of type 2 diabetes: A systematic review of cohort studies. *Journal of Nutritional Science and Vitaminology, 51*, 292–310.
Musolino, R., De, P. D., Marino, D., Labate, C., Serra, S., Silvestri, R., et al. (1992). Epilepsy associated with hypocalcemia: Description of a family. *Functional Neurology, 7*, 215–222.
Natelson, S., Miletich, D., Seals, C. F., Visintine, D., & Albrecht, R. (1979). Clinical biochemistry of epilepsy. I. Nature of the disease and a review of the chemical findings in epilepsy. *Clinical Chemistry, 25*, 889–897.
Newhouse, I. J., & Finstad, E. W. (2000). The effects of magnesium supplementation on exercise performance. *Clinical Journal of Sport Medicine, 10*, 195–200.
Nicholls, D. G., Vesce, S., Kirk, L., & Chalmers, S. (2003). Interactions between mitochondrial bioenergetics and cytoplasmic calcium in cultured cerebellar granule cells. *Cell Calcium, 34*, 407–424.
Nishimoto, N., Kanakura, Y., Aozasa, K., Johkoh, T., Nakamura, M., Nakano, S., et al. (2005). Humanized anti-interleukin-6 receptor antibody treatment of multicentric Castleman disease. *Blood, 106*, 2627–2632.
Olanow, C., & Arendash, G. (1994). Metals and free radicals in neurodegeneration. *Current Opinion in Neurology, 7*, 548–558. Find this article online.
Overgaard-Steensen, C., & Ring, T. (2013). Clinical review: Practical approach to hyponatraemia and hypernatraemia in critically ill patients. *Critical Care, 17*, 206.
Ozawa, M., Ninomiya, T., Ohara, T., Hirakawa, Y., Doi, Y., Hata, J., et al. (2012). Self-reported dietary intake of potassium, calcium, and magnesium and risk of dementia in the Japanese: The Hisayama study. *Journal of the American Geriatrics Society, 60*, 1515–1520.
Parkinson's UK. (2012). A quick introduction to Parkinson's: Parkinson's UK. London: Parkinson's UK.
Riggs, J. E. (2002). Neurologic manifestations of electrolyte disturbances. *Neurologic Clinics, 20*, 227–239.
Ringman, J. M. (2017). Update on Alzheimer's and the dementias: Introduction. *Neurologic Clinics, 35*, 171–174.
Roitg, I., Jamard, B., Constantin, A., Cantagrel, A., Mazières, B., & Laroche, M. (2004). Severe refractory hypocalcemia in patients with enteropathy: Consider hypomagnesemia. *Joint, Bone, Spine: Revue du Rhumatisme, 71*, 252.

Rose, B. (2001). *"Post TW," clinical physiology of acid-base and electrolyte disorders* (5th ed., pp. 701–707). New York: McGraw-Hill.

Rosenbauer, F., Koschmieder, S., Steidl, U., & Tenen, D. G. (2005a). Effect of transcription-factor concentrations on leukemic stem cells. *Blood, 106*, 1519–1524.

Rosenbauer, F., Koschmieder, S., Steidl, U., & Tenen, D. G. (2005b). Effect of transcription-factor concentrations on leukemic stem cells. *Blood, 106*, 1519–1524.

Rutter, N., & Smales, O. (1976). Calcium, magnesium, and glucose levels in blood and CSF of children with febrile convulsions. *Archives of Disease in Childhood, 51*, 141–143.

Sanghvi, S. R., Kellerman, P. S., & Nanovic, L. (2007). Beer potomania: An unusual cause of hyponatremia at high risk of complications from rapid correction. *American Journal of Kidney Diseases, 50*, 673–680.

Schlingmann, K. P., Weber, S., Peters, M., Nejsum, L. N., Vitzthum, H., Klingel, K., et al. (2002). Hypomagnesemia with secondary hypocalcemia is caused by mutations in TRPM6, a new member of the TRPM gene family. *Nature Genetics, 31*, 166.

Schneider, J. A., Arvanitakis, Z., Leurgans, S. E., & Bennett, D. A. (2009). The neuropathology of probable Alzheimer disease and mild cognitive impairment. *Annals of Neurology: Official Journal of the American Neurological Association and the Child Neurology Society, 66*, 200–208.

Shah, N. C., Liu, J.-P., Iqbal, J., Hussain, M., Jiang, X.-C., Li, Z., et al. (2011). Mg deficiency results in modulation of serum lipids, glutathione, and NO synthase isozyme activation in cardiovascular tissues: Relevance to de novo synthesis of ceramide, serum Mg^{2+} and atherogenesis. *International Journal of Clinical and Experimental Medicine, 4*, 103.

Shi, M., Zabetian, C. P., Hancock, A. M., Ginghina, C., Hong, Z., Yearout, D., et al. (2010). Significance and confounders of peripheral DJ-1 and alpha-synuclein in Parkinson's disease. *Neuroscience Letters, 480*, 78–82.

Sterns, R. H. (2015). Disorders of plasma sodium—Causes, consequences, and correction. *New England Journal of Medicine, 372*, 55–65.

Surmeier, D. J., Guzman, J. N., Sanchez-Padilla, J., & Schumacker, P. T. (2011). The role of calcium and mitochondrial oxidant stress in the loss of substantia nigra pars compacta dopaminergic neurons in Parkinson's disease. *Neuroscience, 198*, 221–231.

Surmeier, D. J., & Schumacker, P. T. (2013). Calcium, bioenergetics, and neuronal vulnerability in Parkinson's disease. *Journal of Biological Chemistry, 288*, 10736–10741.

Takaya, J., Higashino, H., & Kobayashi, Y. (2004). Intracellular magnesium and insulin resistance. *Magnesium Research, 17*, 126–136.

Thakur, K. T., Motta, M., Asemota, A. O., Kirsch, H. L., Benavides, D. R., Schneider, E. B., et al. (2013). Predictors of outcome in acute encephalitis. *Neurology, 81*, 793–800.

Toescu, E. C., & Verkhratsky, A. (2003). Neuronal ageing from an intraneuronal perspective: Roles of endoplasmic reticulum and mitochondria. *Cell Calcium, 34*, 311–323.

Verstraeten, A., Theuns, J., & Van Broeckhoven, C. (2015). Progress in unraveling the genetic etiology of Parkinson disease in a genomic era. *Trends in Genetics, 31*, 140–149.

Victor, M., Ropper, A. H., & Adams, R. D. (2001). Adams & Victor's principles of neurology (7th ed.). McGraw-Hill Professional Hardcover Paperback.

Walder, R. Y., Landau, D., Meyer, P., Shalev, H., Tsolia, M., Borochowitz, Z., et al. (2002). Mutation of TRPM6 causes familial hypomagnesemia with secondary hypocalcemia. *Nature Genetics, 31*, 171.

White, H. S., Chow, S. Y., Yen-Chow, Y., & Woodbury, D. M. (1992). Effect of elevated potassium on the ion content of mouse astrocytes and neurons. *Canadian Journal of Physiology and Pharmacology, 70*, S263–S268.

Witholt, R., Gwiazda, R., & Smith, D. (2000). The neurobehavioral effects of subchronic manganese exposure in the presence and absence of pre-parkinsonism. *Neurotoxicology and Teratology, 22*, 851–861.

Yokel, R. A., Crossgrove, J. S., & Bukaveckas, B. L. (2003). Manganese distribution across the blood-brain barrier. II. Manganese efflux from the brain does not appear to be carrier mediated. *Neurotoxicology, 24*, 15–22.

Zillich, A. J., Garg, J., Basu, S., Bakris, G. L., & Carter, B. L. (2006). Thiazide diuretics, potassium, and the development of diabetes: A quantitative review. *Hypertension, 48*, 219–224.

CHAPTER 35

Genes and dietary metals in Parkinson's disease

Shiek Fareeth Ahmed, Jency Roshni, S. Mahima, and V. Janakiraman
Faculty of Allied Health Sciences, Chettinad Hospital and Research Institute [CHRI], Chettinad Academy of Research and Education [CARE], Kelambakkam, Tamilnadu, India

Abbreviations

AD Alzheimer's disease
CNS central nervous system
OxS oxidative stress
PD Parkinson's disease
RDA recommended dietary allowance
ROS reactive oxygen species
SeP selenophosphate
SN substantia nigra
SYN synuclein

Introduction

Dr. James Parkinson was the first person to discover Parkinson's disease (PD) in 1817. PD affects people between the age group of 55 and 65. After Alzheimer's disease (AD), PD is the second most common neurodegenerative disease (DeMaagd & Philip, 2015). PD arises due to the death of neurons of substantia nigra (SN) region in the brain. The death of neurons results due to the accumulation of Lewy's bodies in the dopamine-producing neurons. So far, the pathogenesis of PD remains unclear. Globally, more than 10 million were affected with PD. The neurodegeneration occurs when the neurons of SN, a structure in the midbrain, are affected (Saeed et al., 2017). The SN region regulates the balance and movement of the body. Impairment in this region gives rise to motor and nonmotor symptoms in PD. Other symptoms like resting tremor, postural instability, bradykinesia, and rigidity are clinically common. The neuron impairment occurs either in familial (early onset) or in sporadic (late onset) form (Ferguson, Rajput, & Rajput, 2016). Both forms of PD were known to arise due to genetic mutations. Genetic mutations of PD were observed in *SNCA*, *LRRK2*, *PARK7*, *PINK1*, *PRKN*, and *ATP13A2* genes. *SNCA* and *LRRK2* gene mutations result in autosomal dominant form of PD, while mutated *PARK7*, *PINK1*, *PRKN*, and *ATP13A2* genes contribute to autosomal recessive PD (Nuytemans, Theuns, Cruts, & Van Broeckhoven, 2010). Furthermore, several

Vitamins and Minerals in Neurological Disorders
https://doi.org/10.1016/B978-0-323-89835-5.00018-1

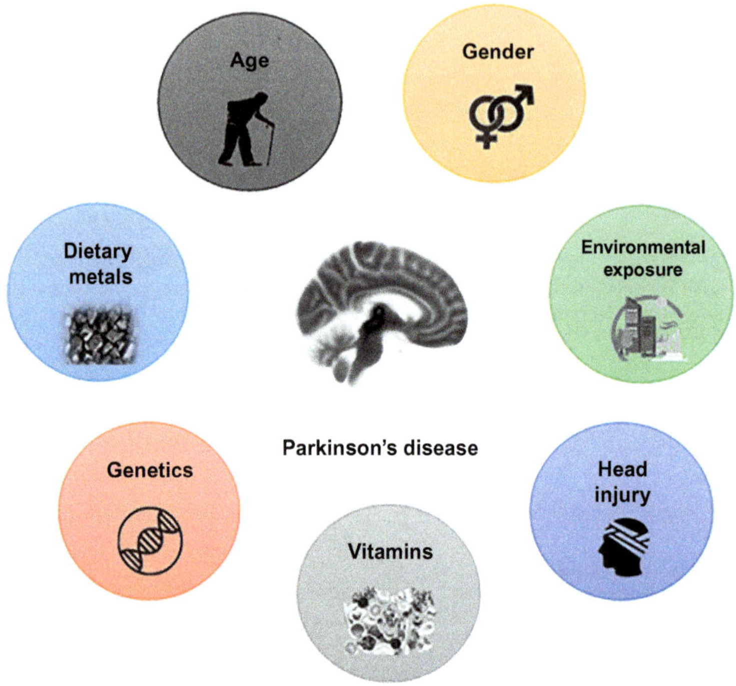

Fig. 1 Etiological factors of Parkinson's disease. Parkinson's disease is caused by several factors like aging, gender, environmental agents (such as pesticides, chemicals, vitamins), dietary metals, genes, and trauma.

other factors like gender, age, vitamins, environmental agents, and dietary metals also progress PD (Tanner et al., 1999) (Fig. 1). However, dietary habits have been shown to modify the risk of PD. Metals at both deficient and excess conditions were observed to influence PD pathogenesis. Several studies provide an association between metals and their interaction with PD risk genes such as *SNCA*, *LRRK2*, and *DJ-1*. This chapter focuses on major dietary metals (like arsenic, calcium, cobalt, chlorine, chromium, copper, iron, iodine, magnesium, manganese, molybdenum, selenium, sodium, potassium, vanadium, and zinc) and their role in genetically regulating PD.

Dietary metals and minerals in PD

Metals play a key role in the functioning of central nervous system (CNS). Several metals are found to be involved in maintaining homeostasis and also, in contrary, promoting dyshomeostasis. Trace elements assist in interaction of DNA and proteins binding, catalytic activity of enzymatic process, and regulation of specific genes (Cousins, 1994). Metal plays a major role in various biological processes, which include modification of proteins, neurotransmitter synthesis, electron and oxygen transportation, immunological response, carbohydrate and protein metabolism, and redox reaction. Especially,

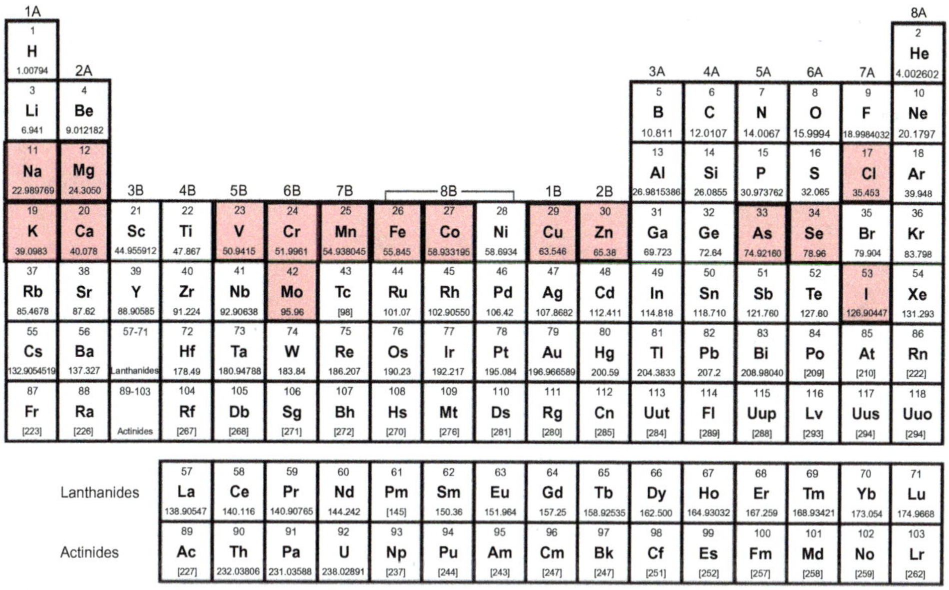

Fig. 2 Dietary metals and minerals associated with Parkinson's disease. Dietary metals play a key role in several functions of the nervous system. Arsenic, calcium, cobalt, chlorine, chromium, copper, iron, iodine, magnesium, manganese, molybdenum, selenium, sodium, potassium, vanadium, and zinc either promote or reduce the risk of PD.

accumulation of metals in brain expresses their importance in neurological function; moreover, their deficiency leads to various neurodegenerative diseases, such as pediatric stroke, pseudotumor cerebri, restless leg syndrome, cranial nerve palsy, and breath-holding spells (Chen, Parmalee, & Aschner, 2014; Cracan & Banerjee, 2013). In brain, metallotoxicity may affect the neurotransmission and leads to various neurological diseases like Huntington's disease, AD, PD, magnesium multiple sclerosis, Wilson's disease, autism spectrum disorder, amyotrophic lateral sclerosis, and Guillain-Barre disease (Authier et al., 2001; Chen, Chakraborty, Peres, Bowman, & Aschner, 2015; Okuda, Iwamoto, Tachibana, & Sugita, 1997; Shaw & Tomljenovic, 2013). Dietary metals have a significant pathological association with PD. Deficient conditions or overdose contributes to severe impact on PD, the dietary metals and minerals along with their dietary source, which are associated with PD are shown in Figs. 2 and 3.

Iron

Iron (Fe) is the first abundant essential transition metal in the human body. The recommended dietary allowance (RDA) intake of Fe for adults is 12–18 mg/day. Dietary iron is divided into three categories: (1) heme iron, which is found in red meat and is readily absorbed by the human body; (2) nonheme iron, which is found in plants like spinach and grains/cereal and is poorly absorbed by the body; and (3) supplementation (Mochizuki, Choong, & Baba, 2020). Inflammation, in combination with Fe

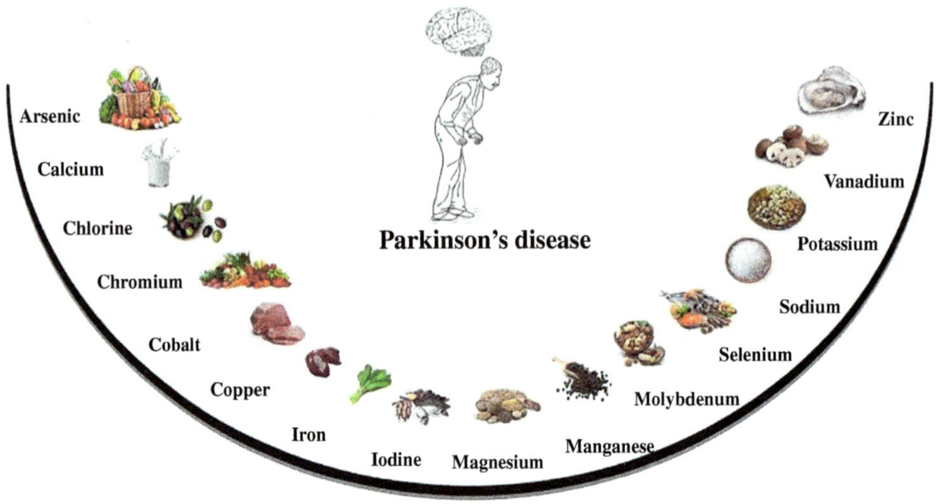

Fig. 3 Dietary metals and minerals along with their sources. Dietary arsenic, calcium, cobalt, chlorine, chromium, copper, iron, iodine, magnesium, manganese, molybdenum, selenium, sodium, potassium, vanadium, and zinc can be obtained from fruits, vegetables, meat, cereals, nuts, and dairy products.

dyshomeostasis and OxS, is a key factor in the progression of many neurodegenerative diseases like PD. Reactive oxygen species (ROS) observed in PD catalyze the oxidation of Fe to its ferric state. Fe^{3+} and Fe^{+2} bind directly to alpha SYN, a hallmark of PD, worsening SYN aggregation (Hare et al., 2017).

Zinc

Zinc (Zn) is one of the most abundant elements in the earth's crust. Also, it is known to be the second most common trace element in the human body and perform a wide range of physiological activities. The average intake of Zn is 2.5–10.0 mg/day. Animal-based diet, oysters, lean red meat, beef liver, turkey meat, skimmed milk powder, egg yolk, and cheddar cheese are the major sources of zinc (Mocchegiani et al., 2013). Studies show that Zn prevents oxidative stress (OxS) by influencing the activity of antioxidant enzymes and signaling pathways. In deficient conditions, nigrostriatal dopaminergic system undergoes degeneration by intranigral fusion. Besides, excess intracellular Zn^{2+} also causes OxS. Evidence suggests that synaptic Zn^{2+} signaling abnormalities have a role in PD (Sikora, Kieffer, Paoletti, & Ouagazzal, 2020).

Copper

Copper (Cu) is a necessary transition metal in the human body. Cereals, fresh fruits, and vegetables are the main sources of Cu. The average amount of Cu stored in the adult human body ranges from 50 to 120 mg (Erdman, Macdonald, & Zeisel, 2012). The brain has a high metabolic rate and high concentration of Cu after the liver and is prone to OxS.

However, Cu concentration varies in different areas of the brain due to physiological and structural heterogeneity. Cu content and distribution in the brain alter during development and neurodegenerative disorders. A 50% drop in Cu concentration was observed in the SN region of PD patients. Cu dyshomeostasis results in the formation of free radicals and degeneration of SN (Hare et al., 2017).

Manganese

Manganese (Mn) is an important minor element present naturally in foods, which acts as a cofactor for various enzymatic activities such as arginase, pyruvate carboxylase, and manganese superoxide dismutase. About 10–20 mg of Mn is present in human body, which is stored in kidney, brain, pancreas, and liver (Nielsen, 2012). Mn is involved in protein metabolism, bone mineralization, metabolic regulation, and cellular protection from ROS. Dietary Mn is obtained from whole grains, clams, oysters, mussels, almonds, soybeans and other legumes, rice, green vegetables, coffee, tea, and black pepper. Manganese toxicity primarily affects the CNS, and it establishes parkinsonism symptoms like tremors, muscle spasms, hearing loss, insomnia, depression, and headaches. Mn is found to block mitochondrial complexes I and II of the electron transport chain, and contributes to mitochondrial dysfunction. It also lowers mitochondrial membrane potential by raising intracellular ROS levels (Ullah et al., 2021).

Magnesium

In human body, magnesium (Mg) is the most abundant divalent cation. About 25 g of Mg is present in human body and contained mostly in soft tissues and bones. The major sources of Mg are whole grains, legumes, spinach, seeds, and nuts (Erdman et al., 2012). Mg is required for a wide range of neural activities, including energy production, neuromuscular transmission, and release of neurotransmitter. Mg acts as a cofactor for many critical enzymes. Disturbances in Mg homeostasis have been linked to a variety of aging-related neurodegenerative illnesses, including PD and AD. The aggregation of alpha SYN is directly inhibited by physiological Mg levels. Also, Mg is involved in autophagy that is observed in PD (Anirudhan et al., 2021; Yamanaka, Shindo, & Oka, 2019).

Calcium

Calcium (Ca^{2+}) is the most abundantly present in human body, which involves in the formation of teeth and bones. Ca^{2+} is enriched in food items like organ meats, whole grain, chocolate, shell fish, nuts, and seeds. Average dietary intake of Ca in adults is 1000 μg/day (Del Valle, Yaktine, Taylor, & Ross, 2011). The disruption of calcium homeostasis is a pathogenic characteristic of PD caused by alpha SYN aggregation.

Mitochondrial influx/efflux dysfunction also affects Ca^{2+} homeostasis (Zaichick, McGrath, & Caraveo, 2017). Ca^{2+} dyshomeostasis in PD is caused by SYN, Parkin, and DJ-1 mutations (Goldberg et al., 2012; Lashuel et al., 2002).

Iodine

Iodine (I) is a well-known essential metal found in foods. The dietary sources of iodine are seaweed, fish, and eggs. The RDA intake of iodine is 150–250 μg/day and varies according to age group (Trumbo, Yates, Schlicker, & Poos, 2001). It is also added in salt as a dietary supplement in many countries. Iodine is necessary for the regulation of thyroxine and triiodothyronine hormones of the thyroid. Iodine-producing thyroid hormones have an impact on cell migration, myelination, differentiation, synaptogenesis, synaptic plasticity, dendrite structure, and transcriptional regulation (Koibuchi, 2008; Rivas & Naranjo, 2007). These hormones also play an important role in enzymatic activity and protein synthesis. It also helps in the development of the fetus, especially in the development of CNS and skeletal system. Hypothyroidism causes some parkinsonism features like hypomimia and bradykinesia. On the other hand, hyperthyroidism increases the risk of dyskinesia and physical movement disability (García-Moreno & Chacón, 2002). Iodine imbalance may influence the pathogenesis of PD in various conditions related to hormonal changes.

Chlorine

Chlorine (Cl^-) is the major anionic element of our body and has 70% of total negative ions (Yunos, Bellomo, Story, & Kellum, 2010). The reference value for adult dietary intake of chloride is 2300 mg/day. Cl^- is mostly present in the foods, which include rye, lettuce, olives, seaweeds, table salt, and tomatoes. Chloride plays a crucial role in various activities of the body, including acid-base balance, water movement across fluid compartments, muscular activity, and osmotic pressure. Inhibition of synapses is crucial for maintaining neural representation. Migration of chloride into postsynaptic neurons leads to a rapid inhibition of the CNS. Dysregulation of Cl^- is concerned in a few psychiatric and neurological disorders (Doyon, Vinay, Prescott, & De Koninck, 2016). Increased mortality of dopaminergic neurons is due to hypochlorite-dependent conversion of cytotoxic redox cycling product, which in turn causes the OxS and might be a causing factor of PD (Mehta et al., 2016).

Chromium

Chromium (Cr) is a trace element present naturally in various foods and tends to have two trivalent (+3) and hexavalent (+6) forms. Trivalent (+3) Cr is the common dietary supplement and required in a smaller amount. However, the other hexavalent (+6) Cr is a secondary toxic product of stainless steel (Anderson & Cefalu, 2010). Cr is found in vegetables, fruits, nuts, grain products, brewer's yeast, species, wine, meat, and beer. The daily intake of Cr is about 1–13 μg per serving (Thor, Harnack, King, Jasthi, & Pettit, 2011). Cr plays a crucial role in maintaining lipid, protein, and glucose metabolism. In a case study, total parenteral nutrition patients showed some adverse effects like hyperglycemia, confusion, apparent weight loss, glycosuria, and glucose intolerance (Brown, Forloines-Lynn, Cross, & Heizer, 1986). Those neurological effects are decreased with adequate amount of chromium. Studies report that Cr may help to reduce the risk of PD.

Sodium

Sodium (Na) is the chief transporter of active solutes in the extracellular niche (Adrogué & Madias, 2000). The reference value for adult dietary intake of Na ranges between 920 and 2300 mg/day. The cell membrane integrity is maintained by numerous ion channels and a sodium-potassium pump, which is responsible for electrochemical and osmotic transportation. Brain homeostasis is maintained by stabilized gradient concentration by neurotransmission, membrane potential, glucose transport, pH regulation, calcium regulation, and cell volume control (Madelin, Klinc, Walvick, & Regatte, 2014). One of the frequently encountered electrolyte disorders is hyponatremia, which is denoted by a low concentration of Na in serum (Gankam Kengne & Decaux, 2017). The primary defense mechanism of the brain due to hypotonicity is the flow of water to the cerebrospinal fluid from parenchyma, and soon after into systemic circulation (Melton & Nattie, 1983). Neuronal dopamine-associated transmission via ion channels, including *N*-methyl-D-aspartate receptors, Na channels, G-protein-coupled channels, and potassium channels, might be involved in dyskinesia of PD (Dragicevic, Schiemann, & Liss, 2015).

Molybdenum

Molybdenum (Mo)-containing diets include legumes, grain products, and nuts. Its concentration in plant-based foods varies depending on the type of soil in which they are grown, but it is negligible in water. The established RDA for adults was set at 45 g/day (Novotny & Peterson, 2018). Mo is transported by plasma proteins like α-macroglobulin. Protein-bound molybdenum constitutes 83%–97% of total amount in erythrocytes. After digestion, Mo having low-order toxicity is eliminated via kidney within few weeks. In humans, Mo gains catalytic activity when complexed with special cofactors (pterin, dithiolene, and pyran (Moco)) (Mendel & Kruse, 2012). Studies have identified only four Mo-based enzymes, namely, xanthine oxidase, mitochondrial amidoxime-reducing component (mARC), aldehyde oxidase, and sulfite oxidase. Xanthine oxidase forms xanthine from hypoxanthine and further to uric acid. mARC with both type B and NAD(H) of cytochrome *b*5 reduces N-hydroxylated substrate groups. Aldehyde oxidase plays an important role in drug metabolism. Sulfite oxidase catalyzes oxidation of sulfur-containing amino acids. Mo-deficient diet from Mo-deficient soils can lead to neurodegenerative disorder like PD (Mendel & Kruse, 2012). Recent studies in sheep model exhibit that Mo deficiency parallels with purine (adenine and guanine) ingestion and consequently neurodegeneration in PD. Due to the lack of molybdoenzyme activity, these purines do not catabolize at the gastrointestinal tract or liver. They are loaded into the central nervous system and get accumulated. This accumulation primarily leads to astrocyte along with its receptors dysfunction that is localized abundantly in nigrostriatum (Bourke, 2018).

Selenium

Selenium (Se), a ubiquitous trace element, serves as a cofactor for antioxidant enzymes (Vinceti, Filippini, & Wise, 2018). Milk, nuts, meat, and seafood provide the main source of the metal. The RDA for selenium is 55 mg/day, and the Institute of Medicine has advised 400 mg/day as tolerable upper intake level. Dietary Se is first converted to selenide (Se2) and selenophosphate by the action of selenophosphate synthase-2 (SPS-2). Selenophosphate is incorporated into SeP for its activation. The SeP has diverse activities such as selenium transport, apoptosis regulation, immunomodulation, and control of redox state (Cardoso, Roberts, Bush, & Hare, 2015). There are 25 SePs identified so far, and they belong to thioredoxin reductase (TrxRs), iodothyronine deiodinase (DIOs), and glutathione peroxidase (GPxs) (Solovyev, 2015). The antioxidant activity of Se is exercised through active form of SePs that have selenocysteine (Sec) residue in their active site. During inadequate Se intake in PD, premature termination of SeP occurs. This inactivation of SeP may result in vulnerability of dopaminergic neurons to OxS, decreased PV-interneuron expression, and degeneration (Cardoso et al., 2015). Evidence shows that Oxs-induced neurotoxicity is attenuated after Se intake by evaluating the dopamine depletion in the substantia nigra (Ellwanger, Franke, Bordin, Prá, & Henriques, 2016). On contrary, it has also been found that Se at higher doses promotes OxS in cholinergic neurons of *Caenorhabditis elegans* and cognitive impairment in zebrafish (Estevez et al., 2012).

Vanadium

Vanadium (V) can be obtained through spinach, black pepper, mushrooms, parsley, seafood, fennel seeds, and grains. Almost, 9 μg can be consumed by adults per day. The absorption of the micronutrient is 1%–10% due to its instability in the stomach pH. It diffuses across cell membrane as vanadyl cations or remains in extracellular fluids in the form of metavanadate (Gruzewska, Michno, Pawelczyk, & Bielarczyk, 2014). OxS and apoptosis are identified as significant mediators of neurodegeneration in PD. Vanadium compounds have been found to induce oxidative stress and apoptosis in dopaminergic neurons, cytochrome *c* release, and proteolytic cleavage of PKCδ (AfesehNgwa et al., 2009).

Arsenic

Arsenic (As) occurs in both organic and inorganic forms. The organic form is obtained from the diets like fish, fruits, and vegetables. However, inorganic As is found in water, poultry, cereals, and meat. Almost 60%–70% of As gets absorbed in the body. However, As is primarily toxic and could affect every organ of the body. As ranks first among toxicants and is also a well-known co-carcinogen. There is no estimated RDA range established for As. Yet, studies show that even at low concentrations, As could damage the neurological function (Tyler & Allan, 2014). As is found to cause the aggregation of insoluble SYN. The aggregated SYN fails to maintain the DA homeostasis (Cholanians et al., 2016). Hence, excess cytosolic DA forms DA-quinone (DAQ) and associated oxyradicals, leading to DA neuronal death. DA further accelerates and stabilizes the SYN oligomerization/aggregation (Perez & Hastings, 2004). Also, As generates ROS in mitochondria by inhibiting succinic dehydrogenase, in response to Nox enzyme, Endoplasmin Reticulum-DMA III, and various other factors. The ROS results in neuronal apoptosis (Hu et al., 2020).

Cobalt

Cobalt (Co) is required for the biosynthesis of the water-soluble vitamin B12 (González-Montaña et al., 2020). It comprises various forms such as methyl-, deoxyadenosyl-, cyano-, and hydroxy-cobalamin. The cyano form is present in food supplements. Animal tissues provide a rich source for B12, like liver (26–58 μg), beef (1–3 μg), chicken (trace-1 μg), dairy foods (0.3–2.4 μg), and eggs (1–2.5 μg). There are no bioactive forms of B12 from plant sources (O'Leary & Samman, 2010). The RDA for vitamin B12 is 2.4 g/day for adults. It possesses antioxidant properties and maintains neuronal health. The potential antioxidant features of B12 include (1) low homocysteine levels, (2) antioxidant glutathione (GSH) preservation, (3) reducing the activity of advanced glycation end products (AGE), and (4) protective cytokine and growth factor modulation. However, subclinical deficiency of B12 can contribute to ROS-mediated OxS and further neurodegeneration in PD. OxS results through homocysteine (Hcy) accumulation, reduced ROS scavengers, and GSH, AGE formation, and decreased immune response by increased cytokines and growth factors (van de Lagemaat, de Groot, & van den Heuvel, 2019). Evidence reports that Co has also been shown to accelerate SYN aggregation and lead to neurodegeneration (Szybińska & Leśniak, 2017).

Chromosomal genes and PD

ATP13A2/PARK9

ATPase cation-transporting 13A2 (*ATP13A2*) gene is localized in 1p36.13 with 29 exons. ATPase is an exporter of lysosomal polyamine along with greater spermine affinity (van Veen et al., 2020). Furthermore, it protects against toxicity of polyamine along with cellular uptake stimulation. It has a role in maintaining the integrity of the brain and cationic intracellular homeostasis (Ramonet et al., 2012). ATP13A2 is responsible for the proper maintenance of the level of zinc in the body, which is needed for the regulation of lysosome and mitochondria (Estrada-Cuzcano et al., 2017; Kong et al., 2014). In Lewy's bodies, it is noted that the level of ATP13A2 is increased and accumulated in the cytosol of nigral dopaminergic neurons. Mutations in *ATP13A2* disintegrate proteins and form autophagosomes, observed in PD (Dehay et al., 2012).

PARK7/DJ-1

Parkinsonism-associated deglycase (*PARK7/DJ-1*) is located in the 1p36.23 with an exon count of 8. PARK7 is a multipurpose protein with a combative cellular function, of preventing OxS and cell death with the help of proteases. Furthermore, PARK7 is involved in PINK1 and NFE2L2 stabilization, cell growth, NF-kappa-B signaling pathway, and androgen signaling pathway (Shendelman, Jonason, Martinat, Leete, & Abeliovich, 2004). It is also reported that DJ-1 has a role to protect neuronal cells from oxidative stress as well as maintaining function of mitochondria. Mutation in DJ-1 causes tremors, rigidity, dyskinesia, psychotic disturbance, cognitive decline, and anxiety; those symptoms are strongly associated with PD pathogenesis.

DNAJC6

DNAJC6 belongs to HSP40 conserved protein family and is localized in 1p31.3. The gene counts for 21 exons. It mediates chaperone function by provoking the action of ATPase. DNAJ consists of three distinct domains, namely, cysteine-rich domain, phenylalanine/glycine-rich domain, and a preserved 70-amino acid J domain (Ohtsuka & Hata, 2000). DNAJC6 is associated with the clathrin-mediated endocytosis, which acts in some cellular functions, which include regulation of surface protein expression, developmental, plasma membrane homeostasis, synaptic signaling, and synaptic neurotransmission (Kaksonen & Roux, 2018). The association of DNAJC6 with the pathogenesis of PD is still under study. DNAJ6 encodes for auxilin, which is clathrin-associated protein majorly present in nerve terminals that are involved in vesicle trafficking that is associated with pathogenesis of PD (Eisenberg & Greene, 2007).

GBA

The protein β-glucocerebrosidase is synthesized from the *GBA* gene present in 1q22. It plays a crucial role in glycolipid metabolism and has 12 exons. It is proved that mutation in *GBA* has a higher prevalence of developing PD (Lwin, Orvisky, Goker-Alpan, LaMarca, & Sidransky, 2004). Furthermore, families with heterozygous GBA mutation along with Gaucher disease also have high risk of PD (Goker-Alpan et al., 2004).

PINK1

PTEN-induced kinase 1 is located in 1p36.12 with 8 exons. Mutation in *PINK1* gene is caused by nonsense or missense mutation and rarely entire-exon deletion (Cazeneuve et al., 2009). Almost 61 nonsense mutation and missense mutation were found to affect exons of *PINK1*. *PINK1* encodes kinase protein, which is a 581-amino acid protein. PD-associated *PINK1* genes were related to the signs and symptoms of psychiatric diseases. The vital function of the Parkin and *PINK1* is to detect and degrade the damaged mitochondria. About two-thirds of mutation affects the function of kinase domain and inhibits the enzymatic activity, which shows the significance of *PINK1* in PD pathogenesis (Youle & Narendra, 2011).

DCTN1

DCTN1, localized on 2p13.1, contains 66 exons. It encodes for dynactin subunit protein that is responsible for Perry syndrome (PS), which is characterized by autosomal dominant parkinsonism (Konno et al., 2017). Neuron-enriched dynactin is a heteromultimeric motor protein and required for axonal transport (Waterman-Storer et al., 1997). Dynactin complex binds and activates the cytoplasmic dynein and induces retrograde transport (Konno et al., 2017). Any impairment in dynein-facilitated axonal transport leads to neurodegenerative diseases (Guo, StoklundDittlau, & Van Den Bosch, 2020).

MAP4K4

MAP4K4 of MAP4 kinase subfamily is localized on 2q11.2 with 33 exons and identified to promote the JNK and c-Jun pathway (Larhammar, Huntwork-Rodriguez, Rudhard, Sengupta-Ghosh, & Lewcock, 2017). The c-Jun-*N*-terminal kinase (JNK) signaling pathway mediates neurodegeneration in PD (Pan et al., 2009). Dual leucine zipper kinase (DLK) is essential for stress-induced neuronal JNK signaling. After neuronal injury, DLK gets activated and propagates downstream stress signals such as JNK phosphorylation (p-JNK). The p-JNK is translocated to the nucleus, leading to c-Jun phosphorylation by dynein-associated scaffolding protein (JIP3) (Larhammar et al., 2017). Experimental animal studies reveal that blocking the MAP4K4 could possibly attenuate dopaminergic neurodegeneration (Pan et al., 2009).

IP6K2

IP6K2 or inositol hexaphosphate (*IP6*) localized on 3p21.31 with 3 exons exhibits neuroprotective and antiapoptotic effects in dopaminergic cells. Animal studies reveal that *IP6* reverses cytochrome *c* release, loss of mitochondrial membrane potential, Bcl-2 upregulation, and caspases activation in PD (Zhang et al., 2016).

DNAJC13

DNAJC13 gene contains 56 exons, which is localized on 3q22.1. The encoded protein performs cell surface recycling and lysosomal degradation of transferrin (TF). Mutated *DNAJC13* causes TF accumulation in endosomal compartments, signifying disrupted endosomal trafficking (Hasegawa, Yoshida, Sugeno, Kobayashi, & Aoki, 2018; Vilariño-Güell et al., 2014).

SATB1

SATB1 gene is located at 3p24.3 with 17 exons. *SATB1* is a DNA binding protein associated with PD. *SATB1* prevents cellular senescence in postmitotic dopaminergic neurons. Loss of *SATB1* causes the activation of a cellular senescence transcriptional program in dopamine neurons, both in human stem cell-derived dopaminergic neurons and in mice. *SATB1* suppresses the expression of p21, a key regulator of cellular senescence. The senescence phenotype is improved when p21 is inhibited in *SATB1*-mutant cells (Riessland et al., 2019).

FAM47E

FAM47E is located at 4q21.1 and has 10 exons. *FAM47E* (scavenger receptor class B member 2) lies in close proximity to *SCARB2*, and the most frequently associated SNP (rs6812193) is situated between the two genes. However, among the two, SCARB2 seems to be a more promising candidate for a PD risk factor (Ran & Belin, 2014).

BST1

BST1 gene is located at 4p15.32 and has an exon count of 10. *BST1* (bone marrow stromal cell antigen-1), a member of CD38 family, is a glycosylphosphatidylinositol-linked cell surface protein. It has both ADP ribosyl cyclase and cyclic ADP ribose hydrolase activity. Cyclic ADP-ribose, a secondary messenger, causes calcium influx by releasing calcium from intracellular deposits. Calcium regulates neurotransmitter release from presynaptic terminal; any change in calcium homeostasis could affect dopaminergic neurons (Chen, Lin, Lee, & Wu, 2014; Ran & Belin, 2014).

SNCA

SNCA gene is located at 4q22.1 with 6 exons. Synuclein (SNCA) is a small, acidic protein with a molecular weight of 14.5 kDa and 140 amino acids (Rotter et al., 2019). SYN contributes to Lewy bodies, a key characteristic of PD pathogenesis. It has two main physiological roles: (1) control of synaptic membrane processes and (2) neurotransmitter release via interactions with SNARE family (Bellucci et al., 2012). *SNCA* is also known to be involved in the recycling pathway. Misfolded and fibrillar *SNCA* activates microglia via toll-like receptors (TLR2 and TLR4) and activation of NLRP3 inflammasome (Gustot et al., 2015). These processes contribute to dopaminergic neuronal death (Dzamko, Geczy, & Halliday, 2015). In addition, *SNCA* mutations can lead to dysfunctions in several other cellular pathways, such as mitochondrial function, oxidative stress, cytochrome *c* release, ER stress, and apoptosis (Gallegos, Pacheco, Peters, Opazo, & Aguayo, 2015).

GAK

GAK gene is located at 4p16.3 with 31 exons. GAK is a particularly promising candidate for a risk gene, because it is differentially expressed in the SN region of PD (Pankratz et al., 2009). Multiple stages of clathrin-mediated vesicular transport are facilitated by the GAK. Recently, depletion of GAK has been shown to influence cathepsin D, the main lysosomal enzyme involved in SNCA degradation (Latourelle et al., 2009). Knockdown of GAK in primary rat neurons increased SNCA toxicity, which might increase the risk of PD pathogenesis (Perrett, Alexopoulou, & Tofaris, 2015).

PARK2

PARKIN encoded by PARK2 is located at 6q25.2-27 with 12 exons. PARKIN was found to be associated with autosomal recessive PD (Kitada et al., 1998). It is inactivated as a result of nitrosative stress, dopaminergic stress, and OxS, all of them are crucial pathogenic events in PD. Parkin is required for mitochondrial integrity in cells. PARKIN regulates protein degradation and plays an essential function in the ubiquitin-proteasome system. It is translocated to the mitochondrial surface and causes the destruction of a variety of membrane proteins. PARKIN mutations result in misfolded or aggregated proteins as well as mitochondrial degeneration, represent sporadic PD.

GPNMB

GPNMB is located at 7p15.3 with 12 exons. GPNMB (glycoprotein nonmetastatic melanoma protein B) has been linked to a higher chance of PD. It presumably plays an important role in neuronal survival and neuroprotection (Xu et al., 2016). It is also involved in phagocytosis and helps to recruit an autophagy protein LC3-II to the phagosome. It is essential for the fusion of phagosome and lysosome to degrade the phagosome content (Herrero, Estrada, Maatouk, & Vyas, 2015).

POLG

POLG is localized in 15q26.1 and possesses 23 exons. The gene encodes for polymerase γ enzyme (DeBrosse & Bass, 2014). The mitochondrial polymerase (POLG1) plays a key role in synthesizing mitochondrial DNA (mtDNA) and repair. Mutation in the *POLG* may lead to several clinical manifestations, including parkinsonism. *POLG* consists of three domains, namely, exonuclease (*exo*), linker region, and polymerase (*pol*) (Hsieh et al., 2019). The *POLG* mutations clustering in the pol domain raise the dominant inheritance pattern in PD. It is found that these mutations aggravate OxS in the dopaminergic neurons of SN region (Hsieh et al., 2019).

VPS13C

VPS13C (*vacuolar protein sorting 13C)* containing 86 exons belongs to chorein family of proteins, and it is localized on15q22.2 (Lesage et al., 2016). In eukaryotes, four orthologues *of VPS13 (A, B, C,* and D) were found (Park et al., 2016). Mutations in *VPS13C* are associated with early-onset PD, which is characterized by severe and rapid disease progression and early cognitive decline. *VSP13* proteins are primarily involved in mitochondrial maintenance. Experiments on cell models show that VPS13C is localized in the outer membrane and its depletion leads to lower membrane potential, fragmentation, and impaired *PINK1*/Parkin-dependent mitophagy. Neurons produce ATP through oxidative phosphorylation. During VPS13C depletion, they were observed to undergo acute mitochondrial stress and are unable to switch to glycolysis for ATP. Such conditions result in the generation of ROS and irreversible mitochondrial damage. In addition, *PINK1*/Parkin-dependent mitophagy was also upregulated. These findings suggest that *VPS13C* depletion makes mitochondria vulnerable to stress and eventually causes PD (Lesage et al., 2016).

CD19

CD19 antigen, a transmembrane glycoprotein of the immunoglobulin (Ig) superfamily, is encoded by the *CD19* gene localized on 16p11.2. It aids in the stimulation of B cells and the development of an immunological response (Wang, Wei, & Liu, 2012). *LRRK2* expression is not limited to neurons but also extended to innate and adaptive immune systems. Studies show that increased *LRRK2* mRNA and protein are found in circulating CD19+ B cells following pro-inflammatory IFN-γ stimulation. The elevated *LRRK2* suggests that the protein is linked to the PD progression (Cook et al., 2017).

NOD2

NOD2 or nucleotide-binding oligomerization domain-containing protein is encoded by *NOD2* gene localized on 16q12.1. They contain 12 exons and are NOD-like receptors playing a regulatory role in apoptotic pathway. A study on 6-OHDA-induced PD model resulted in a significant increase of *NOD2* and DA neuronal apoptosis. *NOD2* exerts pro-apoptotic effects by regulating cytochrome C, Bax, and Bcl-2, thus promoting the activation of caspase-3. These signal transduction cascades promote pro-apoptotic function of 6-OHDA neurotoxins. Alternatively, NOD2 deficiency detoured the apoptotic pathway induced by 6-OHDA (Cheng et al., 2018).

VPS35

VPS35 (vacuolar protein sorting 35) gene localized on 16q11.2 containing 17 exons recently emerged as a cause of late-onset familial PD (Eleuteri & Albanese, 2019). The protein encoded by the gene confines on endosomes and forms the retromer complex with other peripheral membrane proteins. *VPS35* mutations induce dopaminergic neurodegeneration through several ways. D620N mutation observed in *VPS35* impairs WASH endosomal sorting complex and macroautophagy. Overexpressed mutant D620N promotes defective trafficking or altered synaptic transmission. Also, it can induce mitochondrial fragmentation and neuronal loss. These effects can be released by mitochondrial fission protein dynamin-like protein 1 (DLP1) inhibition. Collectively, these mechanisms eventually lead to neurodegeneration in PD (Williams, Chen, & Moore, 2017).

CRHR1

CRHR1 or corticotropin-releasing hormone receptor 1 gene localized on chr17 *encodes for* corticotropin-releasing factor (CRF) or corticotropin-releasing hormone (CRH). The gene contains 14 exons. Neurogenesis is regulated tightly by a network of growth, transcriptional, and hormonal factors. During stress conditions, high glucocorticoid levels can reduce neurogenesis. Studies performed in PD show that CRH could reverse glucocorticoid effects and offer neuroprotection (Cheng, Zhu, & Zhang, 2020). However, genetic deficiency can lead to apoptosis (Koutmani et al., 2013).

FAM171A2

FAM171A2 (family with sequence similarity 171 member A2) localized on 17q21.31 with 9 exons is involved in regulating Progranulin (PGRN) production (Xu et al., 2020). PGRN is a pleiotropic glycoprotein controlling neuroinflammation (Van Kampen, Baranowski, & Kay, 2014). Genetic deficiency observed in PD causes reduced PGRN expression (Xu et al., 2020).

UBTF

UBTF gene localized on 17q21.31 with 23 exons encodes for upstream binding transcription factor (UBTF), a multifaceted protein. The protein is found to have a role in DNA binding and transcriptional activation of ribosomal RNA (rRNA). UBTF mutation observed in PD causes nucleolar stress, aberrant rRNA expression, and further DNA damage (Toro et al., 2018).

RIT2

RIT2 is localized on 18q12.3 with 6 exons also known as Rin. The gene encodes for Ras superfamily of guanosine triphosphate-binding proteins of plasma membrane. They are preferentially expressed in dopaminergic neurons of SN region. RIT2 is observed to interact directly with the dopamine transporter (Lu, Liu, Tan, et al., 2015). But in case of PD, RIT2 expression is lowered (Daneshmandpour, Darvish, & Emamalizadeh, 2018).

SPPL2B

SPPL2B is identified to be a risk factor of PD, is localized on 19p13.3 with 16 exons. It encodes for aspartic proteases of GXGD family. It catalyzes the cleavage of TNF-alpha and triggers the pro-inflammatory cytokine expression (Chang et al., 2015).

CRLS1

CRLS1 is localized on 20p12.3 with 11 exons, encodes for cardiolipin synthase 1 (CRLS1). The protein is required for Rg3, a ginsenoside exhibiting neuroprotective effects with antiinflammation and antioxidative stress. Any aberrant change in the cardiolipin (CL), a mitochondrial phospholipid, is linked to impaired neuron biogenesis and dysfunction (Falabella, Vernon, Hanna, Claypool, & Pitceathly, 2021).

TMEM230

TMEM230 gene containing 5 exons is localized on 20p13-p12.3. The protein encoded is identified to be localized in the dopaminergic neurons of SN region. The protein has a diverse role in synaptic vesicles and retromer trafficking and secretory autophagy. It shares pathogenic pathways in common with major PD risk genes such as *LRRK2*, *VPS35*, *SNCA*, and *PINK1*. LRRK2 kinase mediates phosphorylation of Rab family of proteins that performs intracellular vesicular trafficking. The process is found to be impaired when TMEM230 loses its function, similar to the lack of LRRK2. TMEM230 protein also co-localizes partially with VPS35 and regulates retromer trafficking. Alpha SYN from mutated *SNCA*, the key constituent of Lewy's bodies, was found to enclose TMEM230 protein. In addition, *TMEM230* variants impair synaptic vesicles movement and α-synuclein level (Deng, Fan, & Jankovic, 2018).

DYRK1A

DYRK1A is localized on 21q22.13 with 13 exons, encodes for dual-specificity tyrosine-(Y)-phosphorylation-regulated kinase 1A, which is a pleiotropic kinase contributing to neural proliferation and apoptosis. Experimental studies show that the kinase hinders dopaminergic neuronal apoptosis by Cas9 phosphorylation (Barallobre et al., 2014). On contrary, DYRK1A phosphorylates α-synuclein and causes aggregation/fibrillization in PD (Cen et al., 2016; Wegiel, Gong, & Hwang, 2011).

SYNJ1

SYNJ1 (Synaptojanin 1) is localized on 21q22.11, with 32 exons. It encodes for a phosphoinositide phosphatase that performs synaptic vesicle endocytosis and recycling. The protein enriched in nerve terminals consists of two functional inositol phosphatase domains: Sac1 domain dephosphorylates phosphatidylinositol monophosphates, while the 5′-phosphatase domain dephosphorylates phosphatidylinositol bis- or trisphosphates. The dual phosphatase activity is crucial for the protein function. In case of PD, the mutated *SYNJ1* impairs Sac1 domain activity and synaptic vesicle recycling (Drouet & Lesage, 2014).

FBXO7

FBXO7 gene is localized on 22q12.3 and contains 9 exons. The F-box only protein 7 is involved in mitochondrial quality control via Parkin-mediated mitophagy. During mitophagy, both FBXO7 and PINK1 recruit Parkin to depolarized mitochondria. The recruitment leads to ubiquitination of various targets, including mitochondrial fusion proteins (Mfns), which in turn promote mitochondrial fragmentation. A mutation in FBXO7 has been identified in severe forms of autosomal recessive early-onset PD. It includes impaired Parkin translocation and Mfn1 ubiquitination (Burchell et al., 2013).

PLA2G6

PLA2G6 gene is localized on 22q13.1 with 19 exons, encodes for iPLA$_2$-VI, a calcium-independent phospholipase A2. *PLA2G6* mutations were found to be associated with atypical autosomal recessive PD and sporadic early-onset PD (Morgan et al., 2006). The mutant cases presented alpha-synuclein accumulation in the substantia nigra (Shen et al., 2019).

Mitochondrial genes and PD

HSPA9 encoding mortalin is a mitochondrial chaperone that belongs to the heat shock protein 70 (HSP70) family. It plays a key role in mitochondrial biogenesis by direct import of nuclear-encoded proteins (De Mena et al., 2009). Mitochondria associated membranes (MAMs) are structures that maintain the physiological process between endoplasmic reticulum (ER) and mitochondria to regulate mitochondrial biogenesis (Rodríguez-Arribas et al., 2017). Mortalin expression connects the mitochondria and ER of MAM through proteins. Lower expression of mortalin or impaired MAM observed in PD (Ferré et al., 2021) can lead to mitochondrial dysfunction, including Parkin-defective functions, DJ-1 downregulation, and α-synuclein mutations (Rodríguez-Arribas et al., 2017). Ubiquitin carboxy-terminal hydrolase-L1 *(UCHL1)* constitutes 1%–2% of brain proteins and functions in the ubiquitin-proteasome system. It has been found that in sporadic PD, wild-type UCH-L1 protein is deposited in Lewy's bodies. Mitochondrial transcription factor A (TFAM) is a nuclear-encoded protein with an essential role in mtDNA metabolism. TFAM controls the transcription of mtDNA and regulates the mtDNA-copy number, thus being important for maintaining ATP production. TFAM dysfunction or mutations may also be involved in PD. *CLN8* gene was co-expressed with a monogenic PD gene already implicated in mitochondrial function such as *PARK7*. Furthermore, with the exception of *CLN8* (*FASN*, *MPI*, *MRPL43*, and *MRPS3*), the novel mitochondrial gene was assigned to a co-expression module enriched for neuronal markers.

Genes and dietary metals in PD

Zinc

Zinc is the biometal involved in the variety of cellular functions, including ATP13A2, which is enriched in brain. Zinc binding proteins interact with a fragment of ATP13A2 peptide. Dysregulation of zinc has been linked to the neurodegenerative diseases, which can cause PD (Medici, Peana, Delogu, & Zoroddu, 2012). Cation pump is encoded by ATP13A2, which acts as a transporter that is involved in the Zn^{2+}-assisted cytoplasmic lysosomal confiscation. MHC class HLA-DRB5 binds to the zinc to form a C-terminal domain residue by streptococcal pyrogenic exotoxin C (SPE C); zinc binding site connects to the β-chain of class II major histocompatibility complex (MHC) molecule. The zinc binds to the residues 167, 201, and 293 on SPE C and residue 81 on the β-chain of HLA-DR2a (DRA*0101, DRB5*0101). Any gene alteration in SPEC C residue will lead to a decrease in the biological activity. Majority of the biological processes of SPE C depend on zinc MHC II binding sites (Tripp, McCormick, Webb, & Schlievert, 2003). The sequence alignment of the residue GCH1-encoded protein is transparently organized with zinc ions particularly in their active sites. GCH1 proteins are conserved in the viral species of *Rickettsia monacensis*. So mutation in GCH1 gene increases the risk of PD (Auerbach et al., 2000). Muscle blind-like 2 (MBNL2) belongs to a zinc finger protein family; it plays a major role in the development of the eye and muscle. MBNL2 has conserved zinc finger, which binds to CUG RNA expansions to persuade subsequent splicing of RNA. So it is involved in synaptic mechanism; any dysfunction of gene may lead to cause neurodegenerative disease like PD and other disease (Miller et al., 2000). FYN is a proto-oncogene that belongs to the family of Src family tyrosine kinase, which is a negative regulator of Nrf2; FYN exports Nrf2 to the cytosol from the nuclei and degrades it. Phosphorylation of Akt and GSK-3β is accordingly stimulated by zinc, which in turn reduces the expression of FYN (Jain & Jaiswal, 2007; Kaspar & Jaiswal, 2011). Parkin interacting substrate (PARIS) is a zinc finger transcriptional repressor, which has five domains, four zinc fingers that mediate the binding of DNA to the specific sequences and the Kruppel-associated Box (KRAB) that inhibits transcription. Improper regulation of parkin gene may lead to PD in humans (Shin et al., 2011).

Calcium

Calcium ion influx initiates and accelerates synaptic vesicles endocytosis (SVE) (Wu et al., 2014). The SVE recycles or regenerates vesicles at nerve terminals and secretory cells. The regeneration of SVs following neurotransmission involves various synaptic proteins. Clathrin-mediated endocytosis, a type of SVE, includes five steps. (1) Invagination: Endophilin A1 recruits dynamin to the neck of clathrin-coated vesicle (CCV) and regulates its curvature; (2) Fission: Dynamin constricts the neck of CCV and mediates fission from the membrane; (3) Dephosphorylation: EndophilinA1 recruits synaptojanin 1 to dephosphorylate SV membrane lipids in order to release adaptor proteins and allow auxilin to bind to CCV; (4) Clathrin uncoating: Auxilin removes the clathrin coat via its ATPase activity; (5) Packaging: Dopamine is packaged into the vesicle and available for neurotransmitter release. Studies show that endocytic protein dephosphorylation can also be performed by calcineurin (CaN), a calcium/calmodulin-activated phosphatase. During neuron stimulation, Ca^{2+} influx activates CaN activity that rapidly dephosphorylates endocytic proteins. Animal knockout studies reflect that *SH3GL2* genes have a link to PD pathogenesis implicating for impaired SVE in neurodegeneration (Nguyen, Wong, Ysselstein, Severino, & Krainc, 2019). LRRK2 mutations represent a cause for familial PD. Neurons of SN region are vulnerable to mitochondrial Ca^{2+} dysregulation. The rapid sequestration of intracellular Ca^{2+} by mitochondria is achieved through mitochondrial calcium uniporter (MCU). It is balanced by the release of Ca^{2+} back to the cytosol via sodium/calcium antiporter (NCLX). Besides, other regulatory proteins such as Mitochondrial Calcium Uptake 1 (MICU1) and Mitochondrial Calcium Uptake 2 (MICU2) regulate the mitochondrial Ca^{2+} uptake through MCU. Mutant LRRK2 (G2019S and R1441C) is found to increase mitochondrial and cytosolic Ca^{2+} levels by upregulating MCU and MICU1 expression through the ERK1/2 (MAPK3/1) pathway (Verma et al., 2017). Also, autophagic degradation of mitochondria and dendrite shortening are caused by LRRK2-G2019S mutation. Dendrites have deficient Ca^{2+} buffering capacity and elevated Ca^{2+} influx, leading to mitochondrial stress (Cherra, Steer, Gusdon, Kiselyov, & Chu, 2013). SNCA gene is responsible for the alpha SYN expression. High levels of Ca^{2+} are the crucial factor for SYN aggregation. Ca^{2+} is found to possess high affinity toward C terminus of SYN, thereby forming oligomers. Also, calpain, a Ca^{2+}-activated protease, cleaves SYN, resulting in the deletion of Ca^{2+} binding domain to produce fragments that cause dopamine toxicity and oxidative stress (Mishizen-Eberz et al., 2003). Ca^{2+} localizes at mitochondria-associated membranes (MAMs) to regulate Ca^{2+} transfer between ER and mitochondria. Mutations in alpha SYN cause a decreased association with MAM followed by mitochondrial fragmentation and autophagy (Guardia-Laguarta et al., 2014).

Molybdenum

Dietary sources of purines such as xanthosine, adenosine, inosine, and guanosine are catabolized by the Mo-dependent enzyme xanthine oxidase–dehydrogenase and finally excreted. Under basal conditions, XOD activity is high and catabolizes dietary purines in the gut and liver without salvage. During Mo-deficient conditions, these purines are not catabolized by XOD and undergo salvage pathway to reach CNS. The purine accumulation in CNS affects the astrocyte and its receptor function. Astrocytes regulate extracellular glutamate levels in the nigrostriatum through the A2a adenosine receptor. Purine loading-induced receptor dysfunction causes either increased or decreased glutamate uptake. In this way, glutamatergic transmission is either chronically increased or reduced. Even though astrocyte dysfunction is a primary consequence of purine loading, neurodegeneration appears as a secondary consequence. Both *DJ-1* and *SNCA* genes have been found to control Mo-induced glutamate uptake by astrocyte (Bourke et al., 2018).

Cobalt

5′-Deoxyadenosylcobalamin (AdoCbl) is one of the physiological forms of cobalamine. The hyperactivity of LRRK2 kinase encoded by *LRRK2* gene is due to the pathogenic mutations of neurotoxicity. AdoCbl obtained from the diet affects the LRRK2 kinase conformation and dimerization via allosteric inhibition. Hence, AdoCbl prevents neurotoxicity and Parkinson-like phenotypes in PD (Schaffner et al., 2019).

Arsenic

Synucleinopathy observed in PD is characterized by alpha SYN accumulation, misfold, and cytotoxic oligomers/aggregates. The expression of synuclein is regulated by *SNCA* gene. Physiological levels of soluble synuclein monomers are not toxic but insoluble oligomers/aggregates exhibit cytotoxicity. Arsenic aggravates the oligomerization of synuclein and promotes proteotoxicity and apoptosis. Also, it interferes with cellular antioxidants, depletion of thiol content, increased ROS, and chaperone inhibition (Cholanians et al., 2016).

Other components of interest

Many epidemiological studies showed that PD is associated with metals such as aluminum, bismuth, copper, iron, lead, manganese, mercury, thallium, zinc, and others. Only dietary metals and minerals are summarized in this book chapter and the others were excluded. The major risk factors of PD are age, gender, smoking, environmental, genetic, and chemical toxicants. The major genetic factors include gene polymorphisms and hereditary. Some polymorphic genes are reported to be associated with PD pathogenesis, which includes PARK7, PINK1, SNCA, and LRRK2. Recent studies aimed to explore the connection between the gene polymorphisms and the metals in the progression of PD.

Key facts

- Irons, calcium, magnesium, potassium, and other metals are essential for maintaining blood formation, regulating blood pressure, bone and teeth formation, brain signaling pathway, gene regulation, etc.
- Iron is one of the abundant metals in human body, which regulates oxidative stress and maintains homeostasis.
- Metal deficiency and metal abundance in human body affects the homeostasis of human body.
- Gene polymorphisms are associated with the PD pathogenesis.
- Metals and gene polymorphism plays an important role in PD.

Summary points

- Parkinson's disease (PD) is a chronic age-related neurodegenerative disease that affects 1% of the population above the age of 60.
- Lewy's bodies and loss of dopaminergic neurons in the substantia nigra region are the two pathological hallmarks of PD.
- PD is influenced by a complex interplay between genetic and environmental factors.
- Recent studies reveal that several dietary metals contribute in PD pathogenesis.
- Iron, magnesium, calcium, copper, cobalt, molybdenum, selenium, vanadium, and zinc reduce the risk of PD.
- Exposure to manganese and arsenic promotes the pathogenesis.

References

Adrogué, H. J., & Madias, N. E. (2000). Hyponatremia. *The New England Journal of Medicine*, *342*(21), 1581–1589.

AfesehNgwa, H., Kanthasamy, A., Anantharam, V., Song, C., Witte, T., Houk, R., et al. (2009). Vanadium induces dopaminergic neurotoxicity via protein kinase Cdelta dependent oxidative signaling mechanisms: Relevance to etiopathogenesis of Parkinson's disease. *Toxicology and Applied Pharmacology*, *240*(2), 273–285.

Anderson, R. A., & Cefalu, W. T. (2010). Chromium. In P. M. Coates, J. M. Betz, M. R. Blackman, et al. (Eds.), *Encyclopedia of dietary supplements* (2nd ed.). New York: NY Informa Healthcare.

Anirudhan, A., Angulo-Bejarano, P. I., Paramasivam, P., Manokaran, K., Kamath, S. M., Murugesan, R., et al. (2021). RPL6: A key molecule regulating zinc- and magnesium-bound metalloproteins of Parkinson's disease. *Frontiers in Neuroscience*, *15*, 631892.

Auerbach, G., Herrmann, A., Bracher, A., Bader, G., Gutlich, M., Fischer, M., et al. (2000). Zinc plays a key role in human and bacterial GTP cyclohydrolase I. *Proceedings of the National Academy of Sciences of the United States of America*, *97*(25), 13567–13572.

Authier, F. J., Cherin, P., Creange, A., Bonnotte, B., Ferrer, X., Abdelmoumni, A., et al. (2001). Central nervous system disease in patients with macrophagic myofasciitis. *Brain: A Journal of Neurology*, *124*(Pt 5), 974–983.

Barallobre, M. J., Perier, C., Bové, J., Laguna, A., Delabar, J. M., Vila, M., et al. (2014). DYRK1A promotes dopaminergic neuron survival in the developing brain and in a mouse model of Parkinson's disease. *Cell Death & Disease*, *5*(6), e1289.

Bellucci, A., Zaltieri, M., Navarria, L., Grigoletto, J., Missale, C., & Spano, P. (2012). From α-synuclein to synaptic dysfunctions: New insights into the pathophysiology of Parkinson's disease. *Brain Research*, *1476*, 183–202.

Bourke, C. A. (2018). Astrocyte dysfunction following molybdenum-associated purine loading could initiate Parkinson's disease with dementia. *NPJ Parkinson's Disease*, *4*, 7.

Brown, R. O., Forloines-Lynn, S., Cross, R. E., & Heizer, W. D. (1986). Chromium deficiency after long-term total parenteral nutrition. *Digestive Diseases and Sciences*, *31*(6), 661–664.

Burchell, V. S., Nelson, D. E., Sanchez-Martinez, A., Delgado-Camprubi, M., Ivatt, R. M., Pogson, J. H., et al. (2013). The Parkinson's disease-linked proteins Fbxo7 and Parkin interact to mediate mitophagy. *Nature Neuroscience*, *16*(9), 1257–1265.

Cardoso, B. R., Roberts, B. R., Bush, A. I., & Hare, D. J. (2015). Selenium, selenoproteins and neurodegenerative diseases. *Metallomics: Integrated Biometal Science*, *7*(8), 1213–1228.

Cazeneuve, C., Sân, C., Ibrahim, S. A., Mukhtar, M. M., Kheir, M. M., Leguern, E., et al. (2009). A new complex homozygous large rearrangement of the PINK1 gene in a Sudanese family with early onset Parkinson's disease. *Neurogenetics*, *10*(3), 265–270.

Cen, L., Xiao, Y., Wei, L., Mo, M., Chen, X., Li, S., et al. (2016). Association of DYRK1A polymorphisms with sporadic Parkinson's disease in Chinese Han population. *Neuroscience Letters*, *632*, 39–43.

Chang, K. H., Wu, Y. R., Chen, Y. C., Fung, H. C., Lee-Chen, G. J., & Chen, C. M. (2015). STK39, but not BST1, HLA-DQB1, and SPPL2B polymorphism, is associated with Han-Chinese Parkinson's disease in Taiwan. *Medicine*, *94*(41), e1690.

Chen, P., Chakraborty, S., Peres, T. V., Bowman, A. B., & Aschner, M. (2015). Manganese-induced neurotoxicity: From *C. elegans* to humans. *Toxicology Research*, *4*(2), 191–202.

Chen, M. L., Lin, C. H., Lee, M. J., & Wu, R. M. (2014). BST1 rs11724635 interacts with environmental factors to increase the risk of Parkinson's disease in a Taiwanese population. *Parkinsonism & Related Disorders*, *20*(3), 280–283.

Chen, P., Parmalee, N., & Aschner, M. (2014). Genetic factors and manganese-induced neurotoxicity. *Frontiers in Genetics*, *5*, 265.

Cheng, L., Chen, L., Wei, X., Wang, Y., Ren, Z., Zeng, S., et al. (2018). NOD2 promotes dopaminergic degeneration regulated by NADPH oxidase 2 in 6-hydroxydopamine model of Parkinson's disease. *Journal of Neuroinflammation*, *15*(1), 243.

Cheng, W. W., Zhu, Q., & Zhang, H. Y. (2020). Identifying risk genes and interpreting pathogenesis for Parkinson's disease by a multiomics analysis. *Genes*, *11*(9), 1100. https://doi.org/10.3390/genes11091100.

Cherra, S. J., 3rd, Steer, E., Gusdon, A. M., Kiselyov, K., & Chu, C. T. (2013). Mutant LRRK2 elicits calcium imbalance and depletion of dendritic mitochondria in neurons. *The American Journal of Pathology*, *182*(2), 474–484.

Cholanians, A. B., Phan, A. V., Ditzel, E. J., Camenisch, T. D., Lau, S. S., & Monks, T. J. (2016). From the cover: Arsenic induces accumulation of α-synuclein: Implications for synucleinopathies and neurodegeneration. *Toxicological Sciences: An Official Journal of the Society of Toxicology*, *153*(2), 271–281.

Cook, D. A., Kannarkat, G. T., Cintron, A. F., Butkovich, L. M., Fraser, K. B., Chang, J., et al. (2017). LRRK2 levels in immune cells are increased in Parkinson's disease. *NPJ Parkinson's Disease*, *3*, 11.

Cousins, R. J. (1994). Metal elements and gene expression. *Annual Review of Nutrition*, *14*, 449–469. https://doi.org/10.1146/annurev.nu.14.070194.002313.

Cracan, V., & Banerjee, R. (2013). Cobalt and corrinoid transport and biochemistry. In *Metallomics and the Cell* (pp. 333–374). Springer.

Daneshmandpour, Y., Darvish, H., & Emamalizadeh, B. (2018). RIT2: Responsible and susceptible gene for neurological and psychiatric disorders. *Molecular and General Genetics MGG*, *293*(4), 785–792.
De Mena, L., Coto, E., Sánchez-Ferrero, E., Ribacoba, R., Guisasola, L. M., Salvador, C., et al. (2009). Mutational screening of the mortalin gene (HSPA9) in Parkinson's disease. *Journal of Neural Transmission (Vienna, Austria: 1996)*, *116*(10), 1289–1293.
Deng, H., Fan, K., & Jankovic, J. (2018). The role of TMEM230 gene in Parkinson's disease. *Journal of Parkinson's Disease*, *8*(4), 469–477.
DeBrosse, S. D., & Bass, N. E. (2014). Alpers' disease (progressive infantile poliodystrophy). In *Encyclopedia of the neurological sciences* (pp. 114–115).
Dehay, B., Ramirez, A., Martinez-Vicente, M., Perier, C., Canron, M. H., Doudnikoff, E., et al. (2012). Loss of P-type ATPase ATP13A2/PARK9 function induces general lysosomal deficiency and leads to Parkinson disease neurodegeneration. *Proceedings of the National Academy of Sciences of the United States of America*, *109*(24), 9611–9616.
Del Valle, H. B., Yaktine, A. L., Taylor, C. L., & Ross, A. C. (Eds.). (2011). *Dietary reference intakes for calcium and vitamin D* National Academies Press.
DeMaagd, G., & Philip, A. (2015). Parkinson's disease and its management: Part 1: Disease entity, risk factors, pathophysiology, clinical presentation, and diagnosis. *P & T: A Peer-Reviewed Journal for Formulary Management*, *40*(8), 504–532.
Doyon, N., Vinay, L., Prescott, S. A., & De Koninck, Y. (2016). Chloride regulation: A dynamic equilibrium crucial for synaptic inhibition. *Neuron*, *89*(6), 1157–1172.
Dragicevic, E., Schiemann, J., & Liss, B. (2015). Dopamine midbrain neurons in health and Parkinson's disease: Emerging roles of voltage gated calcium channels and ATP-sensitive potassium channels. *Neuroscience*, *284*, 798–814.
Drouet, V., & Lesage, S. (2014). Synaptojanin 1 mutation in Parkinson's disease brings further insight into the neuropathological mechanisms. *BioMed Research International*, *2014*, 289728.
Dzamko, N., Geczy, C. L., & Halliday, G. M. (2015). Inflammation is genetically implicated in Parkinson's disease. *Neuroscience*, *302*, 89–102.
Eisenberg, E., & Greene, L. E. (2007). Multiple roles of auxilin and hsc70 in clathrin-mediated endocytosis. *Traffic (Copenhagen, Denmark)*, *8*(6), 640–646.
Eleuteri, S., & Albanese, A. (2019). VPS35-based approach: A potential innovative treatment in Parkinson's disease. *Frontiers in Neurology*, *10*, 1272.
Ellwanger, J. H., Franke, S. I., Bordin, D. L., Prá, D., & Henriques, J. A. (2016). Biological functions of selenium and its potential influence on Parkinson's disease. *Anais da Academia Brasileira de Ciências*, *88*(3 Suppl), 1655–1674.
Erdman, J. W., Jr., Macdonald, I. A., & Zeisel, S. H. (Eds.). (2012). *Present knowledge in nutrition* John Wiley & Sons.
Estevez, A. O., Mueller, C. L., Morgan, K. L., Szewczyk, N. J., Teece, L., Miranda-Vizuete, A., et al. (2012). Selenium induces cholinergic motor neuron degeneration in *Caenorhabditis elegans*. *Neurotoxicology*, *33*(5), 1021–1032.
Estrada-Cuzcano, A., Martin, S., Chamova, T., Synofzik, M., Timmann, D., Holemans, T., et al. (2017). Loss-of-function mutations in the ATP13A2/PARK9 gene cause complicated hereditary spastic paraplegia (SPG78). *Brain: A Journal of Neurology*, *140*(2), 287–305.
Falabella, M., Vernon, H. J., Hanna, M. G., Claypool, S. M., & Pitceathly, R. (2021). Cardiolipin, mitochondria, and neurological disease. *Trends in Endocrinology and Metabolism: TEM*, *32*(4), 224–237.
Ferguson, L. W., Rajput, A. H., & Rajput, A. (2016). Early-onset vs. late-onset Parkinson's disease: A clinical-pathological study. *Canadian Journal of Neurological Sciences*, *43*(1), 113–119.
Ferré, C. A., Thouard, A., Bétourné, A., Le Dorze, A. L., Belenguer, P., Miquel, M. C., et al. (2021). HSPA9/Mortalin mediates axo-protection and modulates mitochondrial dynamics in neurons. *Scientific Reports*, *11*(1), 17705.
Gallegos, S., Pacheco, C., Peters, C., Opazo, C. M., & Aguayo, L. G. (2015). Features of alpha-synuclein that could explain the progression and irreversibility of Parkinson's disease. *Frontiers in Neuroscience*, *9*, 59.
Gankam Kengne, F., & Decaux, G. (2017). Hyponatremia and the brain. *Kidney International Reports*, *3*(1), 24–35.

García-Moreno, J. M., & Chacón, J. (2002). Hipotiroidismo enmascarado por enfermedad de Parkinson [Hypothyroidism concealed by Parkinson's disease]. *Revista de Neurologia*, *35*(8), 741–742.

Goker-Alpan, O., Schiffmann, R., LaMarca, M. E., Nussbaum, R. L., McInerney-Leo, A., & Sidransky, E. (2004). Parkinsonism among Gaucher disease carriers. *Journal of Medical Genetics*, *41*(12), 937–940.

Goldberg, J. A., Guzman, J. N., Estep, C. M., Ilijic, E., Kondapalli, J., Sanchez-Padilla, J., et al. (2012). Calcium entry induces mitochondrial oxidant stress in vagal neurons at risk in Parkinson's disease. *Nature Neuroscience*, *15*(10), 1414–1421.

González-Montaña, J. R., Escalera-Valente, F., Alonso, A. J., Lomillos, J. M., Robles, R., & Alonso, M. E. (2020). Relationship between vitamin B12 and cobalt metabolism in domestic ruminant: An update. *Animals: An Open Access Journal From MDPI*, *10*(10), 1855.

Gruzewska, K., Michno, A., Pawelczyk, T., & Bielarczyk, H. (2014). Essentiality and toxicity of vanadium supplements in health and pathology. *Journal of Physiology and Pharmacology: An Official Journal of the Polish Physiological Society*, *65*(5), 603–611.

Guardia-Laguarta, C., Area-Gomez, E., Rüb, C., Liu, Y., Magrané, J., Becker, D., et al. (2014). α-Synuclein is localized to mitochondria-associated ER membranes. *The Journal of Neuroscience: The Official Journal of the Society for Neuroscience*, *34*(1), 249–259.

Guo, W., StoklundDittlau, K., & Van Den Bosch, L. (2020). Axonal transport defects and neurodegeneration: Molecular mechanisms and therapeutic implications. *Seminars in Cell & Developmental Biology*, *99*, 133–150.

Gustot, A., Gallea, J. I., Sarroukh, R., Celej, M. S., Ruysschaert, J. M., & Raussens, V. (2015). Amyloid fibrils are the molecular trigger of inflammation in Parkinson's disease. *Biochemical Journal*, *471*(3), 323–333.

Hare, D. J., Cardoso, B. R., Raven, E. P., Double, K. L., Finkelstein, D. I., Szymlek-Gay, E. A., et al. (2017). Excessive early-life dietary exposure: A potential source of elevated brain iron and a risk factor for Parkinson's disease. *NPJ Parkinson's Disease*, *3*, 1. https://doi.org/10.1038/s41531-016-0004-y.

Hasegawa, T., Yoshida, S., Sugeno, N., Kobayashi, J., & Aoki, M. (2018). DnaJ/Hsp40 family and Parkinson's disease. *Frontiers in Neuroscience*, *11*, 743.

Herrero, M. T., Estrada, C., Maatouk, L., & Vyas, S. (2015). Inflammation in Parkinson's disease: Role of glucocorticoids. *Frontiers in Neuroanatomy*, *9*, 32.

Hsieh, P. C., Wang, C. C., Tsai, C. L., Yeh, Y. M., Lee, Y. S., & Wu, Y. R. (2019). POLG R964C and GBA L444P mutations in familial Parkinson's disease: Case report and literature review. *Brain and Neuroinflammation*, *9*(5), e01281.

Hu, Y., Li, J., Lou, B., Wu, R., Wang, G., Lu, C., et al. (2020). The role of reactive oxygen species in arsenic toxicity. *Biomolecules*, *10*(2), 240.

Jain, A. K., & Jaiswal, A. K. (2007). GSK-3beta acts upstream of Fyn kinase in regulation of nuclear export and degradation of NF-E2 related factor 2. *The Journal of Biological Chemistry*, *282*(22), 16502–16510.

Kaksonen, M., & Roux, A. (2018). Mechanisms of clathrin-mediated endocytosis. Nature reviews. *Molecular and Cell Biology*, *19*(5), 313–326.

Kaspar, J. W., & Jaiswal, A. K. (2011). Tyrosine phosphorylation controls nuclear export of Fyn, allowing Nrf2 activation of cytoprotective gene expression. *FASEB Journal: Official Publication of the Federation of American Societies for Experimental Biology*, *25*(3), 1076–1087.

Kitada, T., Asakawa, S., Hattori, N., Matsumine, H., Yamamura, Y., Minoshima, S., et al. (1998). Mutations in the parkin gene cause autosomal recessive juvenile parkinsonism. *Nature*, *392*(6676), 605–608.

Koibuchi, N. (2008). Hormonal regulation of cerebellar development and plasticity. *Cerebellum (London, England)*, 7(1), 1–3.

Kong, S. M., Chan, B. K., Park, J. S., Hill, K. J., Aitken, J. B., Cottle, L., et al. (2014). Parkinson's disease-linked human PARK9/ATP13A2 maintains zinc homeostasis and promotes α-Synuclein externalization via exosomes. *Human Molecular Genetics*, *23*(11), 2816–2833.

Konno, T., Ross, O. A., Teive, H., Sławek, J., Dickson, D. W., & Wszolek, Z. K. (2017). DCTN1-related neurodegeneration: Perry syndrome and beyond. *Parkinsonism & Related Disorders*, *41*, 14–24.

Koutmani, Y., Politis, P. K., Elkouris, M., Agrogiannis, G., Kemerli, M., Patsouris, E., et al. (2013). Corticotropin-releasing hormone exerts direct effects on neuronal progenitor cells: Implications for neuroprotection. *Molecular Psychiatry*, *18*(3), 300–307.

Larhammar, M., Huntwork-Rodriguez, S., Rudhard, Y., Sengupta-Ghosh, A., & Lewcock, J. W. (2017). The Ste20 family kinases MAP4K4, MINK1, and TNIK converge to regulate stress-induced JNK signaling in neurons. *The Journal of Neuroscience: The Official Journal of the Society for Neuroscience*, *37*(46), 11074–11084.

Lashuel, H. A., Petre, B. M., Wall, J., Simon, M., Nowak, R. J., Walz, T., et al. (2002). Alpha-synuclein, especially the Parkinson's disease-associated mutants, forms pore-like annular and tubular protofibrils. *Journal of Molecular Biology*, *322*(5), 1089–1102.

Latourelle, J. C., Pankratz, N., Dumitriu, A., Wilk, J. B., Goldwurm, S., Pezzoli, G., et al. (2009). Coordinators and molecular genetic laboratories; GenePD investigators, coordinators and molecular genetic laboratories: Genomewide association study for onset age in Parkinson disease. *BMC Medical Genetics*, *10*, 98.

Lesage, S., Drouet, V., Majounie, E., Deramecourt, V., Jacoupy, M., Nicolas, A., et al., & International Parkinson's Disease Genomics Consortium (IPDGC). (2016). Loss of VPS13C function in autosomal-recessive Parkinsonism causes mitochondrial dysfunction and increases PINK1/Parkin-dependent Mitophagy. *American Journal of Human Genetics*, *98*(3), 500–513.

Lu, Y., Liu, W., Tan, K., et al. (2015). Genetic association of RIT2 rs12456492 polymorphism and Parkinson's disease susceptibility in Asian populations: A meta-analysis. *Scientific Reports*, *5*, 13805.

Lwin, A., Orvisky, E., Goker-Alpan, O., LaMarca, M. E., & Sidransky, E. (2004). Glucocerebrosidase mutations in subjects with parkinsonism. *Molecular Genetics and Metabolism*, *81*(1), 70–73.

Madelin, G., Kline, R., Walvick, R., & Regatte, R. R. (2014). A method for estimating intracellular sodium concentration and extracellular volume fraction in brain in vivo using sodium magnetic resonance imaging. *Scientific Reports*, *4*, 4763. https://doi.org/10.1038/srep04763.

Medici, S., Peana, M., Delogu, L. G., & Zoroddu, M. A. (2012). Mn(II) and Zn(II) interactions with peptide fragments from Parkinson's disease genes. *Dalton Transactions (Cambridge, England: 2003)*, *41*(15), 4378–4388. https://doi.org/10.1039/c2dt12168a.

Mehta, N. J., Asmaro, K., Hermiz, D. J., Njus, M. M., Saleh, A. H., Beningo, K. A., et al. (2016). Hypochlorite converts cysteinyl-dopamine into a cytotoxic product: A possible factor in Parkinson's disease. *Free Radical Biology & Medicine*, *101*, 44–52. https://doi.org/10.1016/j.freeradbiomed.2016.09.023.

Melton, J. E., & Nattie, E. E. (1983). Brain and CSF water and ions during dilutional and isosmotic hyponatremia in the rat. *The American Journal of Physiology*, *244*(5), R724–R732.

Mendel, R. R., & Kruse, T. (2012). Cell biology of molybdenum in plants and humans. *Biochimica et Biophysica Acta*, *1823*(9), 1568–1579.

Miller, J. W., Urbinati, C. R., Teng-Umnuay, P., Stenberg, M. G., Byrne, B. J., Thornton, C. A., et al. (2000). Recruitment of human muscleblind proteins to (CUG)(n) expansions associated with myotonic dystrophy. *The EMBO Journal*, *19*(17), 4439–4448.

Mishizen-Eberz, A. J., Guttmann, R. P., Giasson, B. I., Day, G. A., 3rd, Hodara, R., Ischiropoulos, H., et al. (2003). Distinct cleavage patterns of normal and pathologic forms of alpha-synuclein by calpain I in vitro. *Journal of Neurochemistry*, *86*(4), 836–847.

Mocchegiani, E., Romeo, J., Malavolta, M., Costarelli, L., Giacconi, R., Diaz, L. E., et al. (2013). Zinc: Dietary intake and impact of supplementation on immune function in elderly. *Age (Dordrecht, Netherlands)*, *35*(3), 839–860.

Mochizuki, H., Choong, C. J., & Baba, K. (2020). Parkinson's disease and iron. *Journal of Neural Transmission (Vienna, Austria: 1996)*, *127*(2), 181–187.

Morgan, N. V., Westaway, S. K., Morton, J. E., Gregory, A., Gissen, P., Sonek, S., et al. (2006). PLA2G6, encoding a phospholipase A2, is mutated in neurodegenerative disorders with high brain iron. *Nature Genetics*, *38*(7), 752–754.

Nguyen, M., Wong, Y. C., Ysselstein, D., Severino, A., & Krainc, D. (2019). Synaptic, mitochondrial, and lysosomal dysfunction in Parkinson's disease. *Trends in Neurosciences*, *42*(2), 140–149.

Nielsen, F. H. (2012). Manganese, molybdenum, boron, chromium, and other trace elements. In *Present knowledge in nutrition* (pp. 586–607). Elsevier.

Novotny, J. A., & Peterson, C. A. (2018). Molybdenum. *Advances in Nutrition (Bethesda, MD)*, *9*(3), 272–273.

Nuytemans, K., Theuns, J., Cruts, M., & Van Broeckhoven, C. (2010). Genetic etiology of Parkinson disease associated with mutations in the SNCA, PARK2, PINK1, PARK7, and LRRK2 genes: A mutation update. *Human Mutation*, *31*(7), 763–780.

O'Leary, F., & Samman, S. (2010). Vitamin B12 in health and disease. *Nutrients*, *2*(3), 299–316.

Ohtsuka, K., & Hata, M. (2000). Mammalian HSP40/DNAJ homologs: Cloning of novel cDNAs and a proposal for their classification and nomenclature. *Cell Stress & Chaperones*, *5*(2), 98–112.

Okuda, B., Iwamoto, Y., Tachibana, H., & Sugita, M. (1997). Parkinsonism after acute cadmium poisoning. *Clinical Neurology and Neurosurgery*, *99*(4), 263–265. https://doi.org/10.1016/s0303-8467(97)00090-5.

Pan, J., Xiao, Q., Sheng, C. Y., Hong, Z., Yang, H. Q., Wang, G., et al. (2009). Blockade of the translocation and activation of c-Jun N-terminal kinase 3 (JNK3) attenuates dopaminergic neuronal damage in mouse model of Parkinson's disease. *Neurochemistry International*, *54*(7), 418–425.

Pankratz, N., Wilk, J. B., Latourelle, J. C., DeStefano, A. L., Halter, C., Pugh, E. W., et al. (2009). Genome-wide association study for susceptibility genes contributing to familial Parkinson disease. *Human Genetics*, *124*(6), 593–605.

Park, J. S., Thorsness, M. K., Policastro, R., McGoldrick, L. L., Hollingsworth, N. M., Thorsness, P. E., et al. (2016). Yeast Vps13 promotes mitochondrial function and is localized at membrane contact sites. *Molecular Biology of the Cell*, *27*(15), 2435–2449.

Perez, R. G., & Hastings, T. G. (2004). Could a loss of alpha-synuclein function put dopaminergic neurons at risk? *Journal of Neurochemistry*, *89*(6), 1318–1324.

Perrett, R. M., Alexopoulou, Z., & Tofaris, G. K. (2015). The endosomal pathway in Parkinson's disease. *Molecular and Cellular Neuroscience*, *66*, 21–28.

Ramonet, D., Podhajska, A., Stafa, K., Sonnay, S., Trancikova, A., Tsika, E., et al. (2012). PARK9-associated ATP13A2 localizes to intracellular acidic vesicles and regulates cation homeostasis and neuronal integrity. *Human Molecular Genetics*, *21*(8), 1725–1743.

Ran, C., & Belin, A. C. (2014). The genetics of Parkinson's disease: Review of current and emerging candidates. *Journal of Parkinsonism and Restless Legs Syndrome*, *2014*(4), 63–75.

Rivas, M., & Naranjo, J. R. (2007). Thyroid hormones, learning and memory. *Genes, Brain, and Behavior*, *6*(Suppl. 1), 40–44.

Riessland, M., Kolisnyk, B., Kim, T. W., Cheng, J., Ni, J., Pearson, J. A., et al. (2019). Loss of SATB1 induces p21-dependent cellular senescence in post-mitotic dopaminergic neurons. *Cell Stem Cell*, *25*(4), 514–530.e8.

Rodríguez-Arribas, M., Yakhine-Diop, S., Pedro, J., Gómez-Suaga, P., Gómez-Sánchez, R., Martínez-Chacón, G., et al. (2017). Mitochondria-associated membranes (MAMs): Overview and its role in Parkinson's disease. *Molecular Neurobiology*, *54*(8), 6287–6303.

Rotter, A., Lenz, B., Pitsch, R., Richter-Schmidinger, T., Kornhuber, J., & Rhein, C. (2019). Alpha-synuclein RNA expression is increased in major depression. *International Journal of Molecular Sciences*, *20*(8), 2029.

Saeed, U., Compagnone, J., Aviv, R. I., Strafella, A. P., Black, S. E., Lang, A. E., et al. (2017). Imaging biomarkers in Parkinson's disease and Parkinsonian syndromes: Current and emerging concepts. *Translational Neurodegeneration*, *6*(1), 1–25.

Schaffner, A., Li, X., Gomez-Llorente, Y., Leandrou, E., Memou, A., Clemente, N., et al. (2019). Vitamin B12 modulates Parkinson's disease LRRK2 kinase activity through allosteric regulation and confers neuroprotection. *Cell Research*, *29*(4), 313–329.

Shaw, C. A., & Tomljenovic, L. (2013). Aluminum in the central nervous system (CNS): Toxicity in humans and animals, vaccine adjuvants, and autoimmunity. *Immunologic Research*, *56*(2–3), 304–316.

Shen, T., Hu, J., Jiang, Y., Zhao, S., Lin, C., Yin, X., et al. (2019). Early-onset Parkinson's disease caused by PLA2G6 compound heterozygous mutation, a case report and literature review. *Frontiers in Neurology*, *10*, 915.

Shendelman, S., Jonason, A., Martinat, C., Leete, T., & Abeliovich, A. (2004). DJ-1 is a redox-dependent molecular chaperone that inhibits alpha-synuclein aggregate formation. *PLoS Biology*, *2*(11), e362.

Shin, J. H., Ko, H. S., Kang, H., Lee, Y., Lee, Y. I., Pletinkova, O., et al. (2011). PARIS (ZNF746) repression of PGC-1α contributes to neurodegeneration in Parkinson's disease. *Cell*, *144*(5), 689–702.

Sikora, J., Kieffer, B. L., Paoletti, P., & Ouagazzal, A. M. (2020). Synaptic zinc contributes to motor and cognitive deficits in 6-hydroxydopamine mouse models of Parkinson's disease. *Neurobiology of Disease*, *134*, 104681.

Solovyev, N. D. (2015). Importance of selenium and selenoprotein for brain function: From antioxidant protection to neuronal signalling. *Journal of Inorganic Biochemistry*, *153*, 1–12.

Szybińska, A., & Leśniak, W. (2017). P53 dysfunction in neurodegenerative diseases—the cause or effect of pathological changes? *Aging and Disease*, *8*(4), 506–518.

Tanner, C. M., Ottman, R., Goldman, S. M., Ellenberg, J., Chan, P., Mayeux, R., et al. (1999). Parkinson disease in twins: An etiologic study. *JAMA*, *281*(4), 341–346.

Thor, M. Y., Harnack, L., King, D., Jasthi, B., & Pettit, J. (2011). Evaluation of the comprehensiveness and reliability of the chromium composition of foods in the literature. *Journal of Food Composition and Analysis: An Official Publication of the United Nations University, International Network of Food Data Systems*, *24*(8), 1147–1152.

Toro, C., Hori, R. T., Malicdan, M., Tifft, C. J., Goldstein, A., Gahl, W. A., et al. (2018). A recurrent de novo missense mutation in UBTF causes developmental neuroregression. *Human Molecular Genetics*, *27*(4), 691–705.

Tripp, T. J., McCormick, J. K., Webb, J. M., & Schlievert, P. M. (2003). The zinc-dependent major histocompatibility complex class II binding site of streptococcal pyrogenic exotoxin C is critical for maximal superantigen function and toxic activity. *Infection and Immunity*, *71*(3), 1548–1550. https://doi.org/10.1128/IAI.71.3.1548-1550.2003.

Trumbo, P., Yates, A. A., Schlicker, S., & Poos, M. (2001). Dietary reference intakes: Vitamin A, vitamin K, arsenic, boron, chromium, copper, iodine, iron, manganese, molybdenum, nickel, silicon, vanadium, and zinc. *Journal of the American Dietetic Association*, *101*(3), 294–301.

Tyler, C. R., & Allan, A. M. (2014). The effects of arsenic exposure on neurological and cognitive dysfunction in human and rodent studies: A review. *Current Environmental Health Reports*, *1*(2), 132–147.

Ullah, I., Zhao, L., Hai, Y., Fahim, M., Alwayli, D., Wang, X., et al. (2021). Metal elements and pesticides as risk factors for Parkinson's disease—A review. *Toxicology Reports*, *8*, 607–616.

van de Lagemaat, E. E., de Groot, L., & van den Heuvel, E. (2019). Vitamin B12 in relation to oxidative stress: A systematic review. *Nutrients*, *11*(2), 482.

Van Kampen, J. M., Baranowski, D., & Kay, D. G. (2014). Progranulin gene delivery protects dopaminergic neurons in a mouse model of Parkinson's disease. *PLoS One*, *9*(5), e97032.

van Veen, S., Martin, S., Van den Haute, C., Benoy, V., Lyons, J., Vanhoutte, R., et al. (2020). ATP13A2 deficiency disrupts lysosomal polyamine export. *Nature*, *578*(7795), 419–424.

Verma, M., Callio, J., Otero, P. A., Sekler, I., Wills, Z. P., & Chu, C. T. (2017). Mitochondrial calcium dysregulation contributes to dendrite degeneration mediated by PD/LBD-associated LRRK2 mutants. *The Journal of Neuroscience: The Official Journal of the Society for Neuroscience*, *37*(46), 11151–11165.

Vilariño-Güell, C., Rajput, A., Milnerwood, A. J., Shah, B., Szu-Tu, C., Trinh, J., et al. (2014). DNAJC13 mutations in Parkinson disease. *Human Molecular Genetics*, *23*(7), 1794–1801.

Vinceti, M., Filippini, T., & Wise, L. A. (2018). Environmental selenium and human health: An update. *Current Environmental Health Reports*, *5*(4), 464–485.

Wang, K., Wei, G., & Liu, D. (2012). CD19: A biomarker for B cell development, lymphoma diagnosis and therapy. *Experimental Hematology & Oncology*, *1*(1), 36.

Waterman-Storer, C. M., Karki, S. B., Kuznetsov, S. A., Tabb, J. S., Weiss, D. G., Langford, G. M., et al. (1997). The interaction between cytoplasmic dynein and dynactin is required for fast axonal transport. *Proceedings of the National Academy of Sciences of the United States of America*, *94*(22), 12180–12185.

Wegiel, J., Gong, C. X., & Hwang, Y. W. (2011). The role of DYRK1A in neurodegenerative diseases. *The FEBS Journal*, *278*(2), 236–245.

Williams, E. T., Chen, X., & Moore, D. J. (2017). VPS35, the retromer complex and Parkinson's disease. *Journal of Parkinson's Disease*, 7(2), 219–233.

Wu, X. S., Zhang, Z., Zhao, W. D., Wang, D., Luo, F., & Wu, L. G. (2014). Calcineurin is universally involved in vesicle endocytosis at neuronal and nonneuronal secretory cells. *Cell Reports*, 7(4), 982–988.

Xu, Y., Chen, Y., Ou, R., Wei, Q. Q., Cao, B., Chen, K., et al. (2016). No association of GPNMB rs156429 polymorphism with Parkinson's disease, amyotrophic lateral sclerosis and multiple system atrophy in Chinese population. *Neuroscience Letters*, *622*, 113–117.

Xu, W., Han, S. D., Zhang, C., Li, J. Q., Wang, Y. J., Tan, C. C., et al. (2020). The FAM171A2 gene is a key regulator of progranulin expression and modifies the risk of multiple neurodegenerative diseases. *Science Advances*, *6*(43), eabb3063.

Yamanaka, R., Shindo, Y., & Oka, K. (2019). Magnesium is a key player in neuronal maturation and neuropathology. *International Journal of Molecular Sciences*, *20*(14), 3439.
Youle, R. J., & Narendra, D. P. (2011). Mechanisms of mitophagy. Nature reviews. *Molecular and Cell Biology*, *12*(1), 9–14.
Yunos, N. M., Bellomo, R., Story, D., & Kellum, J. (2010). Bench-to-bedside review: Chloride in critical illness. *Critical care (London, England)*, *14*(4), 226.
Zaichick, S. V., McGrath, K. M., & Caraveo, G. (2017). The role of Ca2+ signaling in Parkinson's disease. *Disease Models & Mechanisms*, *10*(5), 519–535.
Zhang, Z., Hou, L., Li, X., Ju, C., Zhang, J., Li, X., et al. (2016). Neuroprotection of inositol hexaphosphate and changes of mitochondrion mediated apoptotic pathway and α-synuclein aggregation in 6-OHDA induced Parkinson's disease cell model. *Brain Research*, *1633*, 87–95.

PART XI

Peripheral neuropathy

CHAPTER 36

Vitamin D and diabetic peripheral neuropathy

Rizaldy Taslim Pinzon, Angela, Vanessa Veronica, Febrina Eva Susanto, and Andryawan Wahyu Pradana
Department of Neurology, Faculty of Medicine, Duta Wacana Christian University, Yogyakarta, Indonesia

Abbreviations

ALA	α-lipoic acid
BPPV	benign paroxysmal positional vertigo
EBV	Epstein-Barr virus
IL-1	interleukin-1
Mncv	motor nerve conduction velocity
MS	multiple sclerosis
NO	nitric oxide
NADPH	nicotinamide adenine dinucleotide phosphate
Nbf	nerve blood flow
PD	Parkinson's disease
TNF-α	tumor necrosis factor-α
VDR	vitamin D receptor
ω-3 PUFA	omega-3 polyunsaturated fatty acids

Introduction

Vitamin D is a fat soluble and found naturally in a small number of foods, can be incorporated with other foods, and can be purchased as a dietary supplement (Shipton & Shipton, 2015). Vitamin D derived from the supplement, food, and sunlight is biologically inert and must be activated in the body by two hydroxylations. Vitamin D is implicated in the homeostasis of a number of other metabolic pathways, including cell multiplication and differentiation, blood pressure management, production of inflammatory interleukins, and autoimmunity regulation (Jorge, Cordeiro, Rosa, & Bianchi, 2018).

Diabetes, which is defined by a rise in blood glucose levels, is becoming more common. Several of the sequential events that allow the beta cells (β-cells) in the pancreatic islets of Langerhans to produce the insulin required to regulate blood glucose levels appear to be maintained by vitamin D. Insulin levels begin to drop as soon as insulin resistance develops. In order to compensate for the decreased efficacy of insulin activity,

Vitamins and Minerals in Neurological Disorders
https://doi.org/10.1016/B978-0-323-89835-5.00007-7

the β-cells will raise the quantity of insulin produced. However, as glucose levels increase, glucotoxicity enhances hyperactivity, which overpowers the β-cells, causing them to die. Diabetes is characterized by an elevation of blood glucose as a result of the decrease in insulin synthesis (Berridge, 2017).

Previous studies showed that vitamin D deficiency is common in diabetes. Diabetic neuropathy is a series of clinical manifestations that affect the peripheral nervous system as a complication of diabetes mellitus and is a main source of neuropathy worldwide and causes amputation and disability (Nascimento, Pupe, & Cavalcanti, 2016). Current symptomatic therapy such as administration of gabapentin or amitriptyline only affects one-third of patients with painful diabetic neuropathy and only reduces pain by about 50% (McCall, 2018).

Vitamin D deficiency is significantly correlated with an increased risk of developing painful diabetic neuropathy in patients that have type 2 diabetes (Lv et al., 2015). Antioxidants contained in vitamin D can act as neuroprotective substances (Celikbilek et al., 2014). There are two mechanisms regarding the function of vitamin D in the condition of diabetic neuropathy. First is the neurotropic effect of vitamin D on nerve function through the regulation of neurotropin levels and neuronal calcium hemostasis, which has a neuroprotective effect. Second is the role of vitamin D in reducing musculoskeletal pain associated with the improvement of nociceptor function with an increase in pain threshold (Meijer et al., 2002).

Vitamin D can also affect the pain process, which acts as a neuroactive steroid that modulates neuronal excitability of the brain and activation of the signal transduction system (Habib, Nagi, Thillaiappan, Sukumaran, & Akhtar, 2020; Helde-Frankling & Björkhem-Bergman, 2017). Vitamin D affects the action of prostaglandins by inhibiting the release of COX-2 and stimulating the 15-prostaglandin threshold in sensory nerves; inhibits the formation of nitric oxide synthase that produces nitric oxide (NO) in macrophages, thereby activating small Glial cells and astrocytes; increase the synthesis of neurohormones, which will repair and enhance the ability of neurons; increase the transforming growth factor beta 1 or TGF-β1 in astrocytes and microglia, thereby reducing the effects of inflammatory cytokines, reducing tumor necrosis factor alpha (TNF-α) and macrophage colony-stimulating factor in astrocytes and microglia; and inhibiting pain pathways (Alam et al., 2020; Helde-Frankling & Björkhem-Bergman, 2017). This chapter goes on to discuss some clinical studies on vitamin D levels and neurological condition.

Applications to other neurological conditions

Vitamin D has been shown to improve various diseases, one of which is to protect against peripheral neuropathy. Many studies discuss the broad application of vitamin D to other neurological conditions, and some of these studies will be discussed in the section (Holick et al., 2011) (Table 1).

Table 1 Vitamin D status category according to the Endocrine Society (Holick et al., 2011).

Vitamin D status	25(OH)D levels (ng/mL)	25(OH)D levels (nmol/L)
Severe deficiency	<10	<25
Deficiency	<20	<50
Insufficiency	<30	<75
Sufficiency	30–44	75–110
Toxicity	>100	>250

Throughout this chapter, we will explain the wide applicability of the vitamin D for other neurological conditions. In this study, we found that hypovitaminosis D is one of the aspects that contribute to the risk of multiple sclerosis (MS) development (Kimlin, 2008; Orton et al., 2011). An critical scientific affiliation among MS and diet D is that populations placed similarly from the equator and, thus, obtain minimum publicity to UVB radiation are much more likely to stand diet D deficiency and concurrently a better chance of growing MS. Ramagopalan et al. (2009) found that the MS HLA-DRB1 1501 susceptibility gene is governed by a vitamin D-delivering promoter, and a clinically compatible UVB-independent effect of vitamin D on MS risk seems likely. 1,25 (OH) 2 D_3 and vitamin D receptors have also been proven to collaborate with the Epstein-Barr virus (EBV), Epstein-Barr nuclear antigen, which is considered to be a major contributor to the pathogenesis of MS (Salzer et al., 2013). There is a hypothesis that hypovitaminosis D enhances the autoimmune symptoms of EBV infection, raising the chances of MS (Marcucci & Obeidat, 2020).

A study confirmed that the superiority of diet that lacks vitamin D changed into better among sufferers with Parkinson's disease (PD), although that they had regular ambulation and gastrointestinal functions (Evatt et al., 2011). Research has proofed the presence of metabolizing enzymes CYP24A1, CYP27B1, vitamin D receptor (VDR), and vitamin D metabolites inside the brain. This suggests that our brain can manage 1,25 (OH) $2D_3$ levels and that vitamin D has a major function in maintaining the nervous system to operate optimally (Harms, Burne, Eyles, & McGrath, 2011). In addition, CYP27B1 and VDR expression was commonly found in the substantia nigra (a region in brain that has a lot of dopaminergic neurons) based on immunofluorescence (Eyles, Smith, Kinobe, Hewison, & McGrath, 2005). Furthermore, vitamin D serves as a neuroprotective agent to offer a partial safety for dopaminergic neurons. Thus, insufficient nutrition D reputation can play a vital position in PD, ensuing in innovative lack of dopaminergic neurons with inside the human brain (Smith, Fletcher-Turner, Yurek, & Cass, 2006).

Laboratory statistics additionally display a capability affiliation of diet D deficiency as a danger factor for stroke, through direct and oblique stimulation of systemic and vascular infection that reasons atherogenesis (Nargesi et al., 2016). This occurs due to upregulation of interleukin-6 induced ischemia (IL-6) in low vitamin D. Vitamin D is a nutraceutical that has a major role in modulating the inflammatory process in ischemic stroke, through its anti-inflammatory and antioxidant properties, especially in interleukin modulation (Agbalalah, Hughes, Freeborn, & Mushtaq, 2017; Muscogiuri et al., 2017).

A study in China confirmed that 25(OH)D levels within the BPPV organization were significantly reduced than within the manipulator organization. Lower levels of the D diet can invade with the formation of otoconia or otoconial membrane, causing an imbalance of calcium homeostasis in the vestibular endolymph. which can cause otoconia. Epithelial Ca^{2+} channel dispatch device in semicircular canal duct can keep Ca^{2+} attention low in vestibular endolymph; in particular, they discovered that the epithelial Ca^{2+} channel can be upregulated with the help of 1,25-dihydroxyvitamin D (Song et al., 2020; Yamauchi et al., 2005).

Decreased vitamin D levels have an impact on constant low back pain in human body, and decreasing levels can also cause extended pain sensitivity and impaired neurological and muscular functions. Vitamin D deficiency raises the risk of inflammatory activity inside the vertebral endplates, resulting in a lowered ache threshold and, consequently, common discomfort in muscles and bones that leads to weakening (DeLuca, 2004; Holick, 2007; Lodh et al., 2015).

Others components of interest

In this chapter, we have discussed the role of vitamin D for diabetic peripheral neuropathy. Vitamin E (Rajanandh, Kosey, & Prathiksha, 2014) has also shown positive impacts on peripheral diabetic neuropathy. As a natural antioxidant, vitamin E could reduce the pain in diabetic neuropathy patients. A significant decrease in the random blood glucose level and total pain score was reported in the intervention group that obtained vitamin E along with regular medication (pregabalin and oral hypoglycemic agent) after 12 weeks of treatment.

Diabetes mellitus is associated with peripheral neuropathy, which is a common complication for diabetic mellitus. In rats, diabetes reduced their pain threshold significantly. Training (swimming) and vitamin C supplements for 8 weeks significantly reduced diabetes-induced hyperalgesia (Bigdeli & Heidarianpour, 2012). One of the antioxidant vitamins is vitamin C. This vitamin, as other antioxidant supplements, has the ability to scavenge free radicals (Maritim, Sanders, & Watkins III, 2003). Vitamin C normalizes sorbitol levels in the blood, inactivates lipid peroxides, and increases glutathione in diabetic neuropathy (Cunningham, Mearkle, & Brown, 1994; Eriksson & Kohvakka, 1995; Fadupin, Akpoghor, & Okunade, 2007; Je et al., 2002; Mazloom et al., 2011).

A systematic review of clinically controlled trials found that the impact of vitamin B12 on somatic and autonomic symptoms of peripheral diabetic neuropathy was beneficial.

There are three basic types of vitamin B12: cyanocobalamin, hydroxocobalamin, and methylcobalamin. The first one is the least expensive and perhaps most easily accessible, although some experts favor the other two types. Medications for diabetes, like metformin, can cause a deficit of vitamin B12. Thus, treatment with vitamin B12 can modify this state of deficiency and likely change its associated neuropathy. Vitamin B12 plays a key role in the metabolism of fatty acids, which are important for nerve myelin maintenance. Continuous deficiency of B12 may result in nerve degeneration and permanent damage to the nerve (Gilligan, 2002; Scalabrino, Buccellato, Veber, & Mutti, 2003; Sun, Lai, & Lu, 2005).

Supplementation of α-lipoic acid (ALA) could be a recommended treatment for patients with peripheral diabetic neuropathy. When given by parenteral supplementation over a 3-week duration, ALA tends to enhance neuropathic deficits and symptoms. ALA works as an antioxidant, much like vitamin E and vitamin C (Foster, 2007).

The omega-3 polyunsaturated fatty acids (ω-3 PUFA) contained in seal oil have also been shown to slow the development of diabetic neuropathy. Individuals with type 1 diabetes who took seal oil omega-3 polyunsaturated fatty acids (ω-3 PUFA) supplements for a year had longer corneal nerve fibers (Lewis et al., 2017).

In diabetic neuropathy patients, folic acid has been linked to nerve conduction velocity. Supplementing 1 mg folic acid for 16 weeks significantly increased sensory sural amplitude, motor nerve amplitude, and motor nerve velocity (Mottaghi, Khorvash, Maracy, Bellissimo, & Askari, 2018).

Another study showed beneficial effects of zinc (Liu et al., 2014) and resveratrol (Kumar, Kaundal, Iyer, & Sharma, 2007) in animal models of diabetes. The motor nerve conduction velocity (Mncv) at week 6 and week 8 postinduction of diabetes was significantly improved by resveratrol and zinc supplementation, respectively. Moreover, the tactile response threshold in diabetic rats receiving zinc supplementation was higher than in diabetic rats receiving normal saline. Nerve blood flow (Nbf) and increased thermal hyperalgesia were also enhanced by resveratrol. Zinc acts by inducing metallothionein synthesis and inhibiting oxidative stress, whereas the protective effect of resveratrol is modulated by oxidative stress reduction and fragmentation of DNA; thus, zinc and resveratrol have a defensive effect toward peripheral nerve damage caused by diabetes. Human studies (Powell, 2000) showed that 8 weeks of zinc supplementation in healthy subjects demonstrated decreased plasma levels of lipid peroxidation products and DNA adducts and decreased mRNA for tumor necrosis factor-α (TNF-α) and interleukin-1 (IL-1) relative to control groups; hence, zinc supplementation has also antioxidant effects on human studies.

Nano-curcumin supplementation has also been shown to be of benefit in peripheral diabetic neuropathy. A mouse model research has shown that curcumin supplementation for 2 until 4 weeks can relieve neuropathic pain by inhibiting tumor necrosis factor-α (TNF-α), nitric oxide (NO), and nicotinamide adenine dinucleotide phosphate

(NADPH) oxidase (Sharma, Kulkarni, Agrewala, & Chopra, 2006; Zhao et al., 2014). Results in human trials showed reduced foot discomfort, numbness, fatigue, pinprick, light touch, and vibration following nano-curcumin administration for 8 weeks (Asadi et al., 2019).

A study in 1984 examined the effect of oxygen supplementation on diabetic rats (Low, Tuck, Dyck, Schmelzer, & Yao, 1984). The effects of 40% oxygen supplementation on diabetic rats for 4 weeks reported that the time to 50% block of nerve conduction and nerve conduction velocity were no longer statistically different from controls, and elevated endoneurial free sugars were decreased significantly, while plasma glucose levels remained unchanged. Another animal model research analyzed the impact of 2% L-glutamine supplementation (Pereira et al., 2011) on peripheral diabetic neuropathy in male Wistar rats. In diabetic rats supplemented with 2% L-glutamine relative to diabetic rats not supplemented with 2% L-glutamine, L-glutamine (2%) inhibited neuronal death caused by diabetes. After 2% L-glutamine supplementation, the glial density per unit area was reduced significantly. L-Glutamine is an amino acid that acts as a precursor for the glutathione synthesis and provides protection mechanism, the most abundant endogenous cellular antioxidant; as a result, L-glutamine has neuroprotective properties.

The protective role of fish oil against diabetic neuropathy was observed in a study by Gerbi et al. (1999). Fish oil supplementation for 8 weeks improved nerve conduction velocity as well as Na, K-ATPase activity in diabetic rats. These results were linked to reduced anatomical damage and a preventative effect on myelinated fiber diameter bimodal distribution. Fish oil, as hypothesized in this study, prevents membrane modification, which in turn inhibits changes in Na, K-ATPase activity, histological disruption, and partial loss of neuronal conduction velocity. Furthermore, fish oil has a gene expression effect on the Na, K-ATPase gene, which results in functional and histological parameters being prevented.

Key facts of Vitamin D

- Vitamin D is a vitamin that can be soluble in fat and found naturally in a few foods, added to others, or taken as a supplement.
- When ultraviolet (UV) rays from sunlight penetrates the skin, the synthesis of vitamin D is formed endogenously.
- There are two types of vitamin D: D_2 (ergocalciferol) and D_3 (cholecalciferol).
- Vitamin D could regulate immune and neuromuscular functions, glucose metabolism, and cell growth as well as reducing inflammation.
- Vitamin D deficiency is significantly correlated with an increased risk of developing painful diabetic neuropathy in patients with type 2 diabetes.

Mini-dictionary of terms

Diabetic peripheral neuropathy. A common complication of diabetes mellitus, which results in a reduced pain threshold. Not only is the somatic system affected in diabetic peripheral neuropathy, but also the autonomic system.

Multiple Sclerosis. Autoimmune disease, which attacks the central nervous system including axons and myelin.

EBV infection. Infection that occurs due to the spread of the Epstein-Barr virus through an intermediary through saliva.

BPPV. A type of peripheral vertigo, which affects the vestibular system of the inner ear where balance is involved.

Parkinson's disease. Neurodegenerative diseases that interfere with dopaminergic nerve cells.

Summary points

- This chapter showed that vitamin D supplementation is very promising for painful diabetic neuropathy.
- Vitamin D deficiency is significantly correlated with an increment risk of developing painful diabetic neuropathy in patients that have type 2 diabetes.
- Wide application of vitamin D to other neurological conditions has been shown to improve a variety of diseases, such as multiple sclerosis, Parkinson's disease, stroke, BPPV, and low back pain.
- Vitamin D is not the only supplement that has positive impacts on diabetic peripheral neuropathy. Supplementation of vitamin E, vitamin C, vitamin B12, ALA, ω-3 PUFA, folic acid, zinc, resveratrol, nano-curcumin, oxygen, L-glutamine, and fish oil has been shown to improve diabetic peripheral neuropathy.

References

Agbalalah, T., Hughes, S. F., Freeborn, E. J., & Mushtaq, S. (2017). Impact of vitamin D supplementation on endothelial and inflammatory markers in adults: A systematic review. *The Journal of Steroid Biochemistry and Molecular Biology*, *173*, 292–300.

Alam, U., Petropoulos, I. N., Ponirakis, G., Ferdousi, M., Asghar, O., Jeziorska, M., et al. (2020). Vitamin D deficiency is associated with painful diabetic neuropathy. *Diabetes/Metabolism Research and Reviews*. https://doi.org/10.1002/dmrr.3361.

Asadi, S., Gholami, M. S., Siassi, F., Qorbani, M., Khamoshian, K., & Sotoudeh, G. (2019). Nano curcumin supplementation reduced the severity of diabetic sensorimotor polyneuropathy in patients with type 2 diabetes mellitus: A randomized double-blind placebo-controlled clinical trial. *Complementary Therapies in Medicine*, *43*, 253–260.

Berridge, M. J. (2017). Vitamin D deficiency accelerates ageing and age-related diseases: A novel hypothesis. *The Journal of Physiology*, *595*(22), 6825–6836. https://doi.org/10.1113/JP274887.

Bigdeli, Y., & Heidarianpour, A. (2012). Effect of regular exercise and vitamin C on pain threshold in diabetic rats. *Journal of Arak University Medical Sciences*, *15*(4), 10–17.

Celikbilek, A., Gocmen, A. Y., Tanik, N., Borekci, E., Adam, M., Celikbilek, M., et al. (2014). Decreased serum vitamin D levels are associated with diabetic peripheral neuropathy in a rural area of Turkey. *Acta Neurologica Belgica, 115*, 47–52.

Cunningham, J. J., Mearkle, P. L., & Brown, R. G. (1994). Vitamin C: An aldose reductase inhibitor that normalizes erythrocyte sorbitol in insulin-dependent diabetes mellitus. *Journal of the American College of Nutrition, 13*(4), 344–350.

DeLuca, H. F. (2004). Overview of general physiologic features and functions of vitamin D. *The American Journal of Clinical Nutrition, 80*, 1689–1696.

Eriksson, J., & Kohvakka, A. (1995). Magnesium and ascorbic acid supplementation in diabetes mellitus. *Annals of Nutrition and Metabolism, 39*(4), 217–223.

Evatt, M. L., DeLong, M. R., Kumari, M., Auinger, P., McDermott, M. P., Tangpricha, V., et al. (2011). High prevalence of hypovitaminosis D status in patients with early Parkinson disease. *Archives of Neurology, 68*, 314–319.

Eyles, D. W., Smith, S., Kinobe, R., Hewison, M., & McGrath, J. J. (2005). Distribution of the vitamin D receptor and 1 alpha-hydroxylase in human brain. *Journal of Chemical Neuroanatomy, 29*, 21–30.

Fadupin, G. T., Akpoghor, A. U., & Okunade, K. A. (2007). A comparative study of serum ascorbic acid level in people with and without type 2 diabetes in Ibadan, Nigeria. *African Journal of Medicine and Medical Sciences, 36*(4), 335–339.

Foster, T. S. (2007). Efficacy and safety of α-lipoic acid supplementation in the treatment of symptomatic diabetic neuropathy. *The Diabetes Educator, 33*(1), 111–117.

Gerbi, A., Maixent, J., Ansaldi, J., Pierlovisi, M., Coste, T., Pelissier, J., et al. (1999). Fish oil supplementation prevents diabetes-induced nerve conduction velocity and neuroanatomical changes in rats. *Journal of Nutrition, 129*(1), 207–213.

Gilligan, M. A. (2002). Metformin and vitamin B12 deficiency. *Archives of Internal Medicine, 162*(4), 484–485.

Habib, A. M., Nagi, K., Thillaiappan, N. B., Sukumaran, V., & Akhtar, S. (2020). Vitamin D and its potential interplay with pain signaling pathways. *Frontiers in Immunology, 11*, 820. https://doi.org/10.3389/fimmu.2020.00820.

Harms, L. R., Burne, T. H., Eyles, D. W., & McGrath, J. J. (2011). Vitamin D and the brain. *Best Practice & Research. Clinical Endocrinology & Metabolism, 25*, 657–669.

Helde-Frankling, M., & Björkhem-Bergman, L. (2017). Vitamin D in pain management. *International Journal of Molecular Sciences, 18*(10), 2170.

Holick, M. F. (2007). Vitamin D deficiency. *The New England Journal of Medicine, 357*, 260–281.

Holick, M. F., Binkley, N. C., Bischoff-Ferrari, H. A., Gordon, C. M., Hanley, D. A., Heaney, R. P., et al. (2011). Evaluation, treatment, and prevention of vitamin D deficiency: An Endocrine Society clinical practice guideline. *The Journal of Clinical Endocrinology and Metabolism, 96*, 1911–1930.

Je, H. D., Shin, C. Y., Park, S. Y., Yim, S. H., Kum, C., Huh, I. H., et al. (2002). Combination of vitamin C and rutin on neuropathy and lung damage of diabetes mellitus rat. *Archives of Pharmacal Research, 25*(2), 184–190.

Jorge, A. J. L., Cordeiro, J. R., Rosa, M. L. G., & Bianchi, D. B. C. (2018). Vitamin D deficiency and cardiovascular diseases. *International Journal of Cardiovascular Sciences, 31*(4), 422–432.

Kimlin, M. G. (2008). Geographic location and vitamin D synthesis. *Molecular Aspects of Medicine, 29*, 453–461.

Kumar, A., Kaundal, R. K., Iyer, S., & Sharma, S. S. (2007). Effects of resveratrol on nerve functions, oxidative stress and DNA fragmentation in experimental diabetic neuropathy. *Life Sciences, 80*(13), 1236–1244.

Lewis, E. J. H., Perkins, B. A., Lovblom, L. E., Bazinet, R. P., Wolever, T. M. S., & Bril, V. (2017). Effect of omega-3 supplementation on neuropathy in type 1 diabetes: A 12-month pilot trial. *Neurology, 88*(24), 2294–2301.

Liu, F., Ma, F., Kong, G., Wu, K., Deng, Z., & Wang, H. (2014). Zinc supplementation alleviates diabetic peripheral neuropathy by inhibiting oxidative stress and upregulating methallothionein in peripheral nerves of diabetic rats. *Biological Trace Element Research, 158*, 211–218.

Lodh, M., Goswami, B., Mahajan, R. D., Sen, D., Jajodia, N., & Roy, A. (2015). Assessment of vitamin D status in patients of chronic low back pain of unknown etiology. *Indian Journal of Clinical Biochemistry, 30*, 174–179.

Low, P. A., Tuck, R. R., Dyck, P. J., Schmelzer, J. D., & Yao, J. K. (1984). Prevention of some electrophysiologic and biochemical abnormalities with oxygen supplementation in experimental diabetic neuropathy. *PNAS, 81*(21), 6894–6898.

Lv, W. S., Zhao, W. J., Gong, S. L., Fang, D. D., Wang, B., Fu, Z. J., et al. (2015). Serum 25-hydroxyvitamin D levels and peripheral neuropathy in patients with type 2 diabetes: A systematic review and meta-analysis. *Journal of Endocrinological Investigation, 38*(5), 513–518.

Marcucci, S. B., & Obeidat, A. Z. (2020). EBNA1, EBNA2, and EBNA3 link Epstein-Barr virus and hypovitaminosis D in multiple sclerosis pathogenesis. *Journal of Neuroimmunology, 339*, 577116.

Maritim, A. C., Sanders, R. A., & Watkins, J. B., III. (2003). Diabetes, oxidative stress, and antioxidants: A review. *Journal of Biochemical and Molecular Toxicology, 17*(1), 24–38.

Mazloom, Z., Hejazi, N., Dabbaghmanesh, M.-H., Tabatabaei, H.-R., Ahmadi, A., & Ansar, H. (2011). Effect of vitamin C supplementation on postprandial oxidative stress and lipid profile in type 2 diabetic patients. *Pakistan Journal of Biological Sciences, 14*(19), 900–904.

McCall, B. (2018). *Low vitamin D levels in painful diabetic peripheral neuropathy*.

Meijer, J. W., Smit, A. J., van Sonderen, E., Groothoff, J. W., Eisma, W. H., & Links, T. P. (2002). Symptom scoring systems to diagnose distal polyneuropathy in diabetes: The Diabetic Neuropathy Symptom score. *Diabetic Medicine, 19*(11), 962–965.

Mottaghi, T., Khorvash, F., Maracy, M., Bellissimo, N., & Askari, G. (2018). Effect of folic acid supplementation on nerve conduction velocity in diabetic polyneuropathy patients. *Neurological Research, 41*(4), 364–368.

Muscogiuri, G., Annweiler, C., Duval, G., Karras, S., Tirabassi, G., & Salvio, G. (2017). Vitamin D and cardiovascular disease: From atherosclerosis to myocardial infarction and stroke. *International Journal of Cardiology, 230*, 577–584.

Nargesi, A. A., Heidari, B., Esteghamati, S., Hafezi-Nejad, N., Sheikhbahaei, S., Pajouhi, A., et al. (2016). Contribution of vitamin D deficiency to the risk of coronary heart disease in subjects with essential hypertension. *Atherosclerosis, 244*, 165–171.

Nascimento, O., Pupe, C., & Cavalcanti, E. (2016). Diabetic neuropathy. *Revista Dor. São Paulo, 17*(1), 46–51.

Orton, S. M., Wald, L., Confavreux, C., Vukusic, S., Krohn, J. P., Ramagopalan, S. V., et al. (2011). Association of UV radiation with multiple sclerosis prevalence and sex ratio in France. *Neurology, 76*, 425–431.

Pereira, R. V. F., Tronchini, E. A., Tashima, C. M., Alves, E. P. B., Lima, M. M., & Zanoni, J. N. (2011). L-glutamine supplementation prevents myenteric neuron loss and has gliatrophic effects in the ileum of diabetic rats. *Digestive Diseases and Sciences, 56*, 3507–3516.

Powell, S. R. (2000). The antioxidant properties of zinc. *The Journal of Nutrition, 130*(5), 1447S–1454S.

Rajanandh, M. G., Kosey, S., & Prathiksha, G. (2014). Assessment of antioxidant supplementation on the neuropathic pain score and quality of life in diabetic neuropathy patients—A randomized controlled study. *Pharmacological Reports, 66*(1), 44–48.

Ramagopalan, S. V., Maugeri, N. J., Handunnetthi, L., Lincoln, M. R., Orton, S. M., Dyment, D. A., et al. (2009). Expression of the multiple sclerosis-associated MHC class II allele HLA-DRB1*1501 is regulated by vitamin D. *PLoS Genetics, 5*(2), e1000369.

Salzer, J., Nyström, M., Hallmans, G., Stenlund, H., Wadell, G., & Sundström, P. (2013). Epstein-Barr virus antibodies and vitamin D in prospective multiple sclerosis biobank samples. *Multiple Sclerosis, 19*, 1587–1591.

Scalabrino, G., Buccellato, F., Veber, D., & Mutti, E. (2003). New basis of the neurotrophic action of vitamin B12. *Clinical Chemistry and Laboratory Medicine, 41*(11), 1435–1437.

Sharma, S., Kulkarni, S. K., Agrewala, J. N., & Chopra, K. (2006). Curcumin attenuates thermal hyperalgesia in a diabetic mouse model of neuropathic pain. *European Journal of Pharmacology, 563*(3), 256–261.

Shipton, E. A., & Shipton, E. E. (2015). Vitamin D and pain: Vitamin D and its role in the aetiology and maintenance of chronic pain states and associated comorbidities. *Pain Research and Treatment, 2015*, 904967. https://doi.org/10.1155/2015/904967.

Smith, M. P., Fletcher-Turner, A., Yurek, D. M., & Cass, W. A. (2006). Calcitriol protection against dopamine loss induced by intracerebroventricular administration of 6-hydroxydopamine. *Neurochemical Research*, *31*, 533–539.

Song, P., Zhao, X., Xu, Y., Zhao, Z., Wang, L., Liu, Y., et al. (2020). Correlation between benign paroxysmal positional Vertigo and 25-hydroxyvitamin D. *Frontiers in Neurology*, *11*, 576.

Sun, Y., Lai, M.-S., & Lu, C.-J. (2005). Effectiveness of vitamin B12 on diabetic neuropathy: Systematic review of clinical controlled trials. *Acta Neurologica Taiwanica*, *14*(2), 48–54.

Yamauchi, D., Raveendran, N. N., Pondugula, S. R., Kampalli, S. B., Sanneman, J. D., Harbidge, D. G., et al. (2005). Vitamin D upregulates expression of ECaC1 mRNA in semicircular canal. *Biochemical and Biophysical Research Communications*, *331*, 1353–1357.

Zhao, W.-C., Zhang, B., Liao, M.-J., Zhang, W.-X., He, W.-Y., Wang, H.-B., et al. (2014). Curcumin ameliorated diabetic neuropathy partially by inhibition of NADPH oxidase mediating oxidative stress in the spinal cord. *Neuroscience Letters*, *560*, 81–85.

CHAPTER 37

B vitamins on the nervous system: A focus on peripheral neuropathy

Ana María Paez-Hurtado[a], Maria Camila Cortes-Albornoz[a], Diego Alejandro Rodríguez-Gomez[a], Carlos-Alberto Calderón-Ospina[b], and Mauricio Orlando Nava-Mesa[a]

[a]Neuroscience Research Group (NEUROS) – Centro Neurovitae, Neuroscience Unit, School of Medicine and Health Sciences, Universidad del Rosario, Bogotá, Colombia

[b]Center for Research in Genetics and Genomics (CIGGUR), GENIUROS Research Group, Pharmacology Unit, Department of Biomedical Sciences, School of Medicine and Health Sciences, Universidad del Rosario, Bogotá, Colombia

Abbreviations

5-HTP	5-hydroxytryptophan
AADC	L-amino acid decarboxylase
Acetyl CoA	acetyl coenzyme A
AKD/alpha-KGDH	alpha-ketoglutarate
ATP	adenosine triphosphate
BDBF	brain-derived neurotrophic factors
CNS	central nervous system
CoA	coenzyme A
GABA	gamma-aminobutyric acid
Hcy	homocysteine
INH	isonicotinic acid hydrazide
L-DOPA	Levodopa
NADPH	nicotinamide adenine dinucleotide phosphate
NMDA	*N*-methyl-D-aspartate
P2X	purinergic receptors
PDH	pyruvate dehydrogenase complex
PKC	protein kinase
PLP	pyridoxal 5′-phosphate
PN	peripheral neuropathy
PNS	peripheral nervous system
SAM	S-adenosylmethionine
SCD	Subacute combined degeneration
TCSS	Toronto Clinical Neuropathy Scoring System
TDP	thiamine diphosphate
TPP	thiamine pyrophosphate
TSS	total symptom score
VAS	visual analogue scale
WKS	Wernicke-Korsakoff syndrome

Vitamins and Minerals in Neurological Disorders
https://doi.org/10.1016/B978-0-323-89835-5.00013-2

Introduction

The eight B vitamins that are known today—vitamins B1 (thiamine), B2 (riboflavin), B3 (niacin), B5 (pantothenic acid), B6 (pyridoxine), B7 (biotin), B9 (folate), and B12 (cobalamin)—are the result of years of investigation derived from the initial discovery of "factor B" and the clinical observation of vitamin deficiencies (Semba, 2012). Although they are chemically heterogeneous, these molecules form a group of essential nutrients that share two properties: water solubility and their ability to act as a cofactor (or a precursor) for key metabolic cellular processes (Calderón-Ospina & Nava-Mesa, 2020; Kennedy, 2016).

Since mammals are unable to synthetize B vitamins, they must acquire them from dietary or microbial sources (Calderón-Ospina & Nava-Mesa, 2020; Kennedy, 2016; Zempleni, Suttie, Gregory III, & Stover, 2013). Most of these vitamins, except for cyanocobalamin, are plant-based nutrients, though the consumption of some animal products can also provide these vitamins, albeit frequently in different biochemical forms (Kennedy, 2016). On the other hand, vitamin B12 is produced by bacteria in the foregut of ruminants or the colon of humans, and thus, it can only be found in animal derivatives such as meat, eggs, and dairy products. Despite this, humans do not have the ability to assimilate this vitamin produced by the intestinal microbiota, since its absorption takes place in the ileal mucosa through an intrinsic factor-mediated mechanism (Kennedy, 2016; Zempleni et al., 2013). This highlights the importance of an adequate diet that meets all the nutritional requirements necessary for the functioning of the different systems.

The presence of these B vitamins is of great importance in different enzymatic reactions in the human metabolism because they act as cofactors, enabling not only various biological functions, but also specific neurological processes. These so-called neurotropic B vitamins, specifically vitamins B1, B6, and B12, play a special role in the central (CNS) and peripheral (PNS) nervous systems (Wendolowicz, Stefanska, & Ostrowska, 2018). For this reason, supplementation with these vitamins through dietary intake is of utmost importance for health (Geller et al., 2017; Kennedy, 2016). Both cobalamin and pyridoxine are involved in the methionine-cysteine cycle, as well as in the citric acid cycle, together with other B vitamins such as thiamine, suggesting that their interrelation in the biochemical processes of these systems may affect cognitive impairment when these vitamins are in low doses (Kennedy, 2016; Mikkelsen, Stojanovska, Tangalakis, Bosevski, & Apostolopoulos, 2016).

There is a considerable proportion of the population that presents some degree of deficiency of one or more of these B vitamins, which may be evidenced by neurological pathologies such as depression, beriberi, seizures, Wernicke's encephalopathy (WKS), subacute combined degeneration of the spinal cord, or peripheral neuropathy (PN) (Calderón-Ospina & Nava-Mesa, 2020; Kennedy, 2016; Mikkelsen, Stojanovska, Tangalakis, et al., 2016). Despite this, an improvement in certain neurological conditions such as neuropathy, motor control, and nociceptive and neuropathic pain (Hakim et al., 2018; Wang, Gan, Rupert, Zeng, & Song, 2005) has been observed with adequate supplementation of vitamins B1, B6, and B12, even when their deficiency has not been proven (Calderon-Ospina, Nava-Mesa, & Arbeláez Ariza, 2020; Kennedy, 2016). For this reason, knowledge about the properties and biochemical functions of B vitamins is increasingly important due to the myriad of clinical applications in the diagnosis and treatment of several neurologic and nonneurologic conditions. This chapter aims to compile the physiological role that B vitamins play in the nervous system (focusing on the neurotrophic B vitamins B1, B6, and B12), and how their deficiency can lead to the development of certain PN. Finally, we will address the importance of the administration of these vitamins in the management and treatment of these neurological pathologies, as well as their safety profile.

Peripheral neuropathy due to vitamin B deficiency

B vitamins have an important effect on the nervous system due to their neurotrophic effects on cellular metabolism (Wendolowicz et al., 2018). B1, B6, and B12 are involved in critical biochemical pathways for the functioning of neurons and glial cells, such as the citric acid and methionine cycles. Their neural value is such that a deficiency may be related to neurological diseases (Table 1), and recent evidence suggests a potential therapeutic role in neuropathy and pain management (Calderón-Ospina & Nava-Mesa, 2020; Nava-Mesa & Aispuru Lanche, 2021).

PN is the result of nerve injury derived from different comorbidities such as alcoholism, diabetes mellitus, autoimmune diseases, and chronic malabsorption, among others, in which metabolic stress and inflammation may coexist with nutritional deficiencies. Indeed, a common cause of many cases of PN is B vitamin deficiency, inadequate food intake, or history of gastric surgery (Staff & Windebank, 2014). As with neurological manifestations, some experimental studies have shown that lower levels of vitamin B are associated with acute peripheral nerve injury, and its regeneration may be accelerated with adequate supplementation of these vitamins, thereby also improving neural symptoms due to their antiinflammatory and analgesic effects (Geller et al., 2017).

Thiamine deficiency is mainly associated with progressive motor neuropathy, mimicking Guillain-Barré syndrome and cranial nerve palsy (tongue, facial and laryngeal weakness). Its deficiency may cause nervous system dysfunction and neuroinflammation that manifests as dry Beriberi (Spinas, Saggini, Kritas, Cerulli, & Caraffa, 2015). Though this pathology is rare in developed countries, it can be found in patients with gastrectomy or total parenteral feeding (Zak, Burns, Lingenfelser, Steyn, & Marks, 1991). Thiamine level monitoring and supplementation should be considered in patients with symmetric motor-sensory polyneuropathy after gastrectomy. Some genetic causes have also been identified as possible risk factors for individual susceptibility to thiamine deficiency. Some

Table 1 Biochemical pathways of vitamins B1, B6, and B12 involved in the nervous system and the development of peripheral neuropathy due to their deficiency or toxicity.

Vitamin	Biochemical pathway	Cofactor of	Role in nervous system	Implications in peripheral neuropathy	Safety profile
B1	Glycolysis Pentose phosphate pathway Krebs cycle	Pyruvate dehydrogenase Transketolase Alpha-ketoglutarate dehydrogenase	Maintenance of GABA and aspartate levels Synthesis of neurotransmitters (acetylcholine, serotonin), myelin, and glutathione Conservation of nerve membrane Neural excitability, metabolic regulation, and antioxidant effects	Progressive motor neuropathy Cranial nerve palsy Dry beriberi Symmetric sensory polyneuropathy Alcoholic neuropathy	No adverse reactions associated with oral intake. Phlebitis and hypersensitivity reactions related to parenteral administration
B6	One-carbon unit metabolism Hcy metabolism Methionine metabolism Adrenergic, serotonergic, and glutamatergic pathways	Aromatic L-amino acid decarboxylase Serine-hydroxymethyltransferase Cystathionine-beta-synthase/lyase	Myelin synthesis Synthesis of dopamine, serotonin, and GABA Cofactor in myelin formation through sphingolipid synthesis Metabolism of amino acids, neurotransmitters, and DNA/RNA	PN due to deficiency (related to drugs that lower B6 levels or form complexes that reduce bioavailability) PN due to toxicity of B6: their excess affect the soma of the dorsal root ganglion, leading to peripheral and central fiber necrosis or suppression of pyridoxine active form	Reversible neuropathy at higher daily doses over 500 mg/day and/or with treatments longer than 6 months A possible association with hip fractures and lung cancer has not been confirmed
B12	Hcy metabolism Methylmalonyl CoA pathway	Methionine synthase Methylmalonyl CoA mutase	Myelin synthesis and regeneration after neuron injury Immunomodulatory and neurotrophic effects Nerve metabolism, nucleic acid production, cell maturation, and neurotransmitter synthesis Metabolism of fatty acids, amino acids, and DNA/RNA	Small-fiber neuropathy Diabetic neuropathy related to metformin administration Symmetric sensorimotor neuropathy with axonal degeneration	A possible association with hip fractures and lung cancer has not been confirmed

Note: Content based on references: Calderón-Ospina and Nava-Mesa (2020); Calderon-Ospina, Nava-Mesa, and Paez-Hurtado (2020).

of the identified mutations are located in the transketolase-related pathways such as the high-affinity thiamine transporter and thiaminase resistance genes (Koike, Ito, Morozumi, Kawagashira, & Iijima, 2008).

Among the causal mechanisms in which thiamine deficiency may lead to axonal and synaptic abnormalities is its cofactor role in the activation of the pyruvate dehydrogenase complex (PDH) and alpha-ketoglutarate dehydrogenase (alpha-KGDH) in the Krebs cycle and in cellular respiration. Proper PDH function is essential for acetylcholine production and myelin formation. In addition, both alpha-KGDH and the active form of thiamine diphosphate (TDP) participate in the maintenance of myelin and in the neural levels of glutamate, GABA, and aspartate (Hammond, Wang, Dimachkie, & Barohn, 2013). Furthermore, with vitamin B1 deficiency, an alternate pathway converts pyruvate into lactate, which results in a high serum concentration of the latter. All these altered mechanisms lead to a progression of symmetric sensory polyneuropathy with painful components, ataxia and areflexia with involvement of lower limbs (Dhaliwal, Larson, Dhindsa, Bhogal, & Rochling, 2020; Wiley & Gupta, 2021). It has been shown that vitamin B1 supplementation is associated with decreased protein glycation, decreased activity of diacylglycerol protein kinase C (PKC), and in the hexosamine pathway, modulation of neural excitability and Na^+ currents, all of which partially reverse induced nerve injury (Geller et al., 2017; Stein, Geisel, & Obeid, 2021).

Likewise, vitamin B6 imbalances may also contribute to the development of PN. Several drugs, including INH, cycloserine, penicillamine, and L-DOPA, form complexes with B6 that narrows its bioavailability. This lowers B6 serum levels that could lead to PN due to deficiency. Since pyridoxine is involved in the methionine metabolisms essential for neuronal function, suboptimal concentrations of B6 could develop PN-related symptomatology. On the other hand, pyridoxine toxicity cases are derived from supratherapeutic dosing of supplements or iatrogenesis (Hemminger & Wills, 2021). Cell death occurs in a concentration-dependent manner through two main mechanisms: the first one takes place when the inactive form of vitamin B6 competitively suppresses pyridoxal-5′-phosphate (the active form), which causes symptoms similar to those of pyridoxine deficiency (Hemminger & Wills, 2021). In the second, high levels of pyridoxine may directly affect the soma of the dorsal root ganglion, leading to peripheral and central fiber necrosis and neuronal death. With this, necrotic cells stimulate inflammatory responses that worsen the symptomatology related to sensory or sensory-motor neuropathy (Hasannejad, Ansar, Rostampour, Mahdavi Fikijivar, & Khakpour Taleghani, 2019).

Deficiencies of cobalamin generate subacute combined degeneration of the spinal cord and myelin damage of dorsal and lateral columns. These changes are associated with symmetric sensorimotor neuropathy with axonal degeneration, in some cases cognitive impairment, and upper and lower motor neuron dysfunctions (Staff & Windebank,

2014). Vitamin B12 is related to the conversion of L-methylmalonyl CoA into succinyl-CoA. This biochemical process is essential for myelin sheath formation due to its relationship with cellular biosynthesis through the activation of the PDH (Della-Flora Nunes, Mueller, Silvestri, Patel, & Wrabetz, 2017; Hammond et al., 2013). Other molecular mechanisms involved in PN induced by vitamin B12 deficiency are hypomethylation, phosphorus metabolism, and the neurotoxic effect of Hcy (Stein et al., 2021).

Considering that around 6%–30% of patients treated with metformin show vitamin B12 deficiency (Kim, Ahn, Fang, Lee, & Park, 2019), their administration in type 2 diabetes would be related to the progression of diabetic neuropathy in patients without adequate monitoring. Indeed, several reports suggest that long-term treatment and higher doses of metformin may induce cobalamin decrease and exacerbates diabetic peripheral neuropathy symptoms (Hashem, Esmael, Nassar, & El-Sherif, 2021; Roy et al., 2016). Metformin decreased B12 absorption in the gastrointestinal tract affecting calcium-mediated channels and B12-intrinsic factor complex uptake in the ileum (Bauman, Shaw, Jayatilleke, Spungen, & Herbert, 2000). As we mentioned before, cobalamin is necessary as a cofactor for the remethylation pathway in the metabolism of HCy. Deficiency in vitamin B12 (or even B6 and folate) may lead to hyperhomocysteinemia, which can contribute to the production of neurotoxic agents (Roy et al., 2016; Stein et al., 2021). In a cross-sectional study with patients who received metformin, it was found a reduction in serum levels of cobalamin and folic acid with increased levels of HCy, in contrast to patients without metformin therapy (Roy et al., 2016). Thus, metformin administration may lead to a higher risk of sensory neuropathy that may be associated with hyperglycemia-induced damage in the peripheral neurons in diabetic patients without adequate glycemic control (Roy et al., 2016). Some authors suggest the beneficial effects of screening vitamin B12, HCy, and methylmalonic acid as part of the preventive measures for patients at risk of neuropathy (Stein et al., 2021).

Finally, another common cause of PN is derived from alcohol abuse. Alcoholic patients may also have a higher risk of neuropathy due to dietary deficiencies. According to a recent systematic review and meta-analysis, the prevalence of peripheral neuropathy among chronic alcoholic abusers is around 46% (between 35% and 57%) (Julian, Glascow, Syeed, & Zis, 2019). It mainly involves direct nerve injury caused by excessive long-term alcohol drinking as well as thiamine deficiency, which results in hyperalgesia, allodynia, and burning pain (Chopra & Tiwari, 2012). Pathogenesis is still under debate; however, the current described mechanism begins with diminished thiamine absorption in the intestine due to ethanol effects. This leads to impaired hepatic storage and thiamine phosphorylation, which reduces its active form. As is well known, vitamin B1 is involved in key metabolic processes in the neurons, control of inflammation, and oxidative stress resistance (Zhang et al., 2020).

A major pathophysiological mechanism resulting from cobalamin deficiency is hypomethylation in the CNS due to the inhibition of methionine synthase. This leads to a reduced conversion from S-adenosylmethionine (SAM) to S-adenosylhomocysteine, which participates in the methylation of myelin proteins. Furthermore, the pathological pathway ends in axonopathy secondary to small fiber degeneration (Chopra & Tiwari, 2012). Although deficiency of vitamins B2, B6, and folate can contribute to clinical manifestations of alcoholic neuropathy, they are not considered a causative entity, unlike ethanol and deficiency in vitamin B12 (Chopra & Tiwari, 2012). With this in mind, depending on the extent and severity of the injury involved, a nutritionally balanced diet that includes all B vitamins, and alcohol abstinence, could restore the nerve damage to their normal functioning (Chopra & Tiwari, 2012).

B vitamins in the management of peripheral neuropathies

Once a diagnosis of B vitamin deficiency has been suspected or confirmed, vitamin replacement should be implemented until adequate nutrition is restored (confirmed by monitoring plasma levels) and clinical improvement is found. In the case of thiamine deficiency, treatment is usually given by the parenteral route (intravenously or intramuscularly) at an initial dose of 100 mg, followed by 100 mg per day (Hammond et al., 2013). Neurological improvement may take around three or six months with motor impairments responding better than sensory manifestations (Koike et al., 2001). Vitamin B6 doses of 50 mg to 100 mg daily should be provided in cases with pyridoxine-dependent seizures and symptomatic vitamin deficiencies (Hammond et al., 2013; Staff & Windebank, 2014) or in patients treated with medications with B6-antagonistic effects such as INH or hydralazine and chronic hemodialysis patients (Corken & Porter, 2011). In patients with neurological damage after cobalamin deficiency, treatment includes parenteral administration (intramuscularly) of 1000 μg daily for 5–7 days, followed by 1000 μg monthly; or the same dose once a week for 4 weeks, and later one injection monthly. Usually, neurological improvement occurs within the first six months of therapy (Hammond et al., 2013).

In recent years, B vitamins have been used in the management of PN, even in those cases where there are no demonstrated nutritional deficiencies (Geller et al., 2017). According to a recent prospective, open-label, multicenter, observational study ($n=411$), subjects with PN of different etiologies (i.e., diabetic PN, carpal tunnel syndrome, and idiopathic) treated with a combination of thiamine (100 mg), pyridoxine (100 mg), and cyanocobalamin (5 mg) for a period of 3 months (oral route) felt progressive relief of different neuropathic symptoms such as stabbing and burning pain, paresthesia, and numbness (Hakim et al., 2018). In that study, there was no prior determination of B vitamin levels. The same treatment also provides benefits in different etiological subgroups of peripheral neuropathy according to the same authors (Hakim et al., 2019).

There are other clinical reports which showed the efficacy of the combination of thiamine, pyridoxine, and cobalamin in the symptomatic management of diabetic PN (Rizvi, Ahmad, & Rizvi, 2013). In the case of isolated B vitamins treatment, there are some controversial results. For instance, according to randomized clinical trials, benfotiamine (a lipid-soluble vitamin B1 derivative) has shown significant symptomatic benefits in patients with diabetic polyneuropathy (Stracke, Gaus, Achenbach, Federlin, & Bretzel, 2008). However, according to a systematic review, the use of vitamin B12 supplements alone for diabetic neuropathy has shown limited clinical and electrophysiological improvements (Jayabalan & Low, 2016). Therefore, some benefits of B vitamins are likely to be attributed to synergic and complementary mechanisms of the vitamin combination in patients with neuropathy of overlapping etiologies.

Although it is true that in the case of alcoholic neuropathy, replenishment of thiamine produces clinical benefits, vitamin supplementation alone is not enough to alleviate some symptoms in patients who persistently consume alcohol (Chopra & Tiwari, 2012). According to in vitro, animal and human studies, vitamin B complex may have a preventive role in chemotherapy-induced peripheral neuropathy; however, further studies are required to recommend its prophylactic use (Schloss & Colosimo, 2017). Indeed, to date, there are no large-scale, randomized, multicenter, double-blind clinical trials that compare the effect of B vitamins vs. placebo in neuropathy without nutritional deficiency; however, the above studies suggest a clear clinical benefit in patients with PN regardless of the etiology and without prior determination of B vitamin levels. Some authors recommended the use of neurotropic B vitamins in other neurological disorders such as trigeminal and plantar neuralgia, sciatica, lumbago, and shoulder-arm syndrome. In addition, methyl-cobalamin has been studied in the management of trigeminal and glossopharyngeal neuralgia, and subacute herpetic neuralgia (Zhang, Han, Hu, & Xu, 2013). However, there is no consensus about posology, type of B vitamin combination, and administration route because vitamin requirements depend on individual patient case and adherence.

Several studies have shown the role of B vitamins in promoting nerve repair and recovery of nerve function. For instance, in patients with uremic and uremic-diabetic polyneuropathy, treatment with methylcobalamin (IV injection) for 6 months improved motor and sensory nerve conduction test scores. Similarly, oral administration of methylcobalamin for 3 months improved nerve conduction velocity in patients with chronic diabetic polyneuropathy. These functional improvements are associated with alleviation of neuropathic pain and neurological disability scores, according to similar reports (Zhang et al., 2013). In addition, animal models have shown that vitamin B complex and vitamin B12 may increase nerve regeneration after acute peripheral nerve traumatic injury, in experimental acrylamide neuropathy and in streptozotocin-diabetic rats (Sonobe et al., 1988). In a preliminary clinical study in patients with symptomatic diabetic peripheral neuropathy, the combination of L-methylfolate, methylcobalamin, and PLP for 6 months improved epidermal nerve fiber density, which is associated with a reduction of paresthesia and/or dysesthesias (Jacobs & Cheng, 2011).

As we mentioned above, B vitamins are involved in the regeneration of myelin and, therefore, in nerve functional recovery. In fact, methylcobalamin increases the methylation cycle and promotes neurite outgrowth and neuronal survival (Okada et al., 2010). Other mechanisms that explain some of the effects of B vitamins in the alleviation of neuropathic pain symptoms include antiinflammatory effects (Hosseinzadeh, Moallem, Moshiri, Sarnavazi, & Etemad, 2012; Moallem, Hosseinzadeh, & Farahi, 2008), reduction in nerve injury-induced hyperexcitability by sodium currents, inhibition of spontaneous ectopic discharges (Zhang et al., 2013), inhibitory synaptic control, and blockade of purinergic receptors (P2X) in sensory neurons at the dorsal root ganglion level (Thériault et al., 2014).

Conclusions

B vitamins have several specific functions in the CNS and PNS, especially in metabolic, antioxidant, and nerve regeneration processes; myelination; and neurotransmitter synthesis. Taking into account the high energy demand of neurons for the generation of action potentials, the maintenance of ionic pumps, and the metabolism of different excitatory neurotransmission systems, the nervous system is especially susceptible to the effects of glucose and oxygen deprivation, as well as the deficiency of cofactors necessary for its metabolism, such as B vitamins. Many of the degeneration mechanisms of peripheral nerves are explained by these deficiencies. However, there are many types of PN where there is no demonstrated deficiency of B vitamins, and even so, these vitamins can have a role that is both therapeutic (in neuronal regeneration) and symptomatic (in the reduction of neuropathic pain and sensory symptoms), as well as a prophylactic one in pathologies where the said deficiency is expected, or where inflammatory or pharmacological conditions suppose a future deficiency. However, there is a need for further large-scale clinical trials to establish posology, route of administration, type of combination of B vitamins, and doses for the treatment of PN depending on their etiology.

Finally, considering that B vitamins are water soluble and are excreted by the kidney even when used at high doses, they have a good safety profile and their benefits outweigh the potential risks. Likewise, taking into account their multiple mechanisms of action and synergistic effects at metabolic and cellular levels, their use in combination has a potential therapeutic effect in patients with neuropathy of overlapping etiologies.

Applications to other neurological conditions

Neurological conditions induced by B vitamin deficiencies

Several mechanisms are associated with nutritional deficiency of B vitamins. These include malnutrition or inadequate dietary intake, psychiatric conditions (such as selective eating disorder or anorexia nervosa), use of certain medications, and malignancies. In some cases, specific conditions can cause nutritional deficiencies, particularly strict vegan diets, bariatric surgery, and pernicious anemia with vitamin B12 deficiency (Harmandar, Dolu, & Çekin,

2020; Lederer et al., 2019). Regardless of the etiology, vitamin B deficiencies are associated with several neurologic manifestations other than PN.

For instance, low thiamine intake due to nutritional deficiencies, commonly occurring among chronic alcoholic patients, is associated with energy deprivation, increased oxidative stress, microglial activation, and NMDA-mediated excitotoxicity (Thomson & Marshall, 2006). Eventually, these lead to irreversible brain damage and the development of WKS, characterized by an acute confusional state, confabulation, memory loss, and gait abnormalities (Chandrakumar, Bhardwaj, & 't Jong, 2018). Diagnosis of WKS is based on clinical findings, and even when aggressive thiamine supplementation is promptly started, these patients usually require complex long-term medical care (Day, Bentham, Callaghan, Kuruvilla, & George, 2004).

With respect to niacin, its deficiency causes pellagra, a clinical entity that manifests as a triad of dementia or neuropsychiatric symptoms, diarrhea, and dermatitis in sun-exposed areas. Neuropsychiatric manifestations of pellagra include hallucinations, apathy, irritability, confusion, altered mental status, and coma (Hammond et al., 2013).

Subacute combined degeneration (SCD) is a clinical syndrome caused by vitamin B12 deficiency that usually presents with dorsal column dysfunction. It is characterized by subacute onset of symmetrical paresthesia, loss of distal proprioception and vibration sensitivity, and ataxic gait (Lee, Poon, Chuah, & So, 2020). Treatment with supplemental vitamin B12 and physical therapy is useful to relieve the clinical manifestations as well as other associated vitamin B12 deficiency conditions such as optic neuropathy, megaloblastic anemia, and neuropsychiatric manifestations (Issac, Soundarya, Christopher, & Chandra, 2015).

Although rare, vitamin B6 deficiency can also be a cause of neurologic manifestations. Altered mental status and seizures are signs of vitamin B6 deficiency, which most commonly occur in association with other nutritional deficiencies, such as in alcoholic patients, or when consuming certain medications, such as isoniazid (INH) (Gerlach et al., 2011).

Long-term deficiency or inadequate vitamin B intake has been associated with neurodegenerative pathologies and brain atrophy (Sechi, Sechi, Fois, & Kumar, 2016). As a result, patients with dementia and Alzheimer's disease often present with decreased levels of B vitamins. Identification and treatment of these nutritional conditions are associated with a reduced progression of the disease and improvement of cognitive function (Mikkelsen, Stojanovska, Tangalakis, et al., 2016). Lastly, nutritional deficiency of B vitamins is also related to the development of neuropsychiatric disorders and chronic pain. It is also a risk factor for the development of poor mood status and, possibly, depression (Mikkelsen, Stojanovska, & Apostolopoulos, 2016). In fact, studies show that vitamin B supplementation in at-risk populations improves mood status and stress. Due to its analgesic and antiinflammatory properties, the combination of B vitamins has been studied as an adjunctive treatment for pain conditions, including back pain, headaches, and fibromyalgia (Buesing, Costa, Schilling, & Moeller-Bertram, 2019; Nava-Mesa & Aispuru Lanche, 2021). Knowledge of neurological and nonneurological manifestations of nutritional deficiencies is essential in order to maintain a high clinical suspicion and therefore timely diagnose and treat these patients.

Other components of interest

Complementary to the effects of vitamins B1, B6, and B12 in PN, in a recent qualitative analysis, the role of other B vitamins (including niacin), their derivatives (benfotiamine), and metabolic cofactors (myoinositol) in the management of diabetic pain and neuropathy was described (Karaganis & Song, 2021). The same authors suggest that a therapeutic role of B vitamins as monotherapy and in combination with standard treatments (i.e., pregabalin) is plausible; however, the evidence for a benefit remains moderate, and larger, placebo-controlled studies are needed. In a recent systematic review and meta-analysis, it was found that low B12 vitamin plasma concentrations, high plasma methylmalonic acid, and increased plasma total homocysteine were associated with neuropathy, regardless of the etiology (Stein et al., 2021). Curiously, although in the same study a reduction of B1 and B6 vitamin levels was not directly associated with neuropathy, treatment with thiamine improved PN symptoms. In addition, a role of folic acid as a neuroprotective nutrient is plausible, but further well-designed studies are needed.

In a prospective, noncomparative clinical study ($n=497$), the combination of methylcobalamin, alpha-lipoic acid, folic acid, biotin, benfotiamine, and vitamin B6 improved motor and sensory symptoms in patients with PN (Maladkar, Tekchandani, & Dave, 2014). Similarly, a recent pilot study ($n=40$), evaluated the efficacy of methylcobalamin vs. alpha-lipoic acid in patients with diabetic PN. Although both treatments reduced several neuropathic scores (i.e., TCSS, TSS), it was found that the reduction in the VAS score for burning pain was higher in the alpha-lipoic group, while vitamin B12 significantly reduced numbness and paresthesia (Han et al., 2018). The antioxidant mechanism was involved in the effects of alpha-lipoic acid but not in the case of cobalamin.

Key facts of peripheral neuropathy induced by B vitamins deficiency

- Reduction of thiamine, pyridoxine, or cyanocobalamin levels is a common association factor of neuropathy regardless of the etiology.
- Peripheral neuropathy affects around a half of chronic alcoholic abusers, most of them induced by thiamine deficiency, oxidative stress, and neuroinflammation.
- Alcoholic neuropathy is characterized by motor and sensory symptoms such as burning pain, hyperalgesia, and allodynia.
- High-dose and long-term use of some oral hypoglycemic drugs may exacerbate diabetic neuropathy by decreasing vitamin B12 absorption.

- Most of patients with diabetic neuropathy have neuropathic pain and sensory symptoms such as sensory loss and paresthesia.
- Combination of B vitamins (especially at high doses) may have a therapeutic and symptomatic effect in peripheral neuropathies of different etiologies such as diabetic, alcoholic, idiopathic, and toxic.

Mini-dictionary of terms

- **Peripheral neuropathy:** is a damage to one or more of the peripheral nerves. It usually starts with loss of sensation or pain in hands and feet, which later extends to other areas. It also manifests as weakness, numbness, and burning sensation in the affected area.
- **Neuropathic pain:** pain caused by a damage or malfunctioning of the somatosensory system. It is characterized by burning and scalding pain, prickling, tingling, pins-and-needle sensation, among others. Although in some cases it can resolve spontaneously, it is usually a chronic condition.
- **Allodynia:** is a condition in which an unexpected painful response is caused by stimuli that does not normally induce pain.
- **Hyperalgesia:** is a condition that involves an increased sensitivity to stimuli resulting in an extreme pain. It is usually due to a damage to the nerves or biochemical dysfunction in the metabolic pathways of the nervous system responsible for pain processing.
- **Paresthesia:** refers to a painless, numbness, and itching sensation usually presented in the upper or lower limbs. It is described as prickling perception that could also manifest in other parts of the body without any warning.

Summary points

- The present chapter is focus on the role of B vitamins in the pathophysiology of peripheral neuropathy.
- Neurotrophic B vitamins, such as thiamine, pyridoxine, and cyanocobalamin, serve as cofactors in several neuronal metabolic processes.
- B vitamins have a role in myelin regeneration, neurotransmitter synthesis, antioxidative mechanisms, and controlling neuronal excitability.
- Its deficiency is associated with multiple neurological syndromes but also in several types of peripheral neuropathies, such as alcoholic and diabetic.
- Neurotrophic B vitamins have a therapeutic potential in the management of neuropathies, even without a demonstrated nutritional deficiency.

References

Bauman, W. A., Shaw, S., Jayatilleke, E., Spungen, A. M., & Herbert, V. (2000). Increased intake of calcium reverses vitamin B12 malabsorption induced by metformin. *Diabetes Care*, *23*(9), 1227–1231.

Buesing, S., Costa, M., Schilling, J. M., & Moeller-Bertram, T. (2019). Vitamin B12 as a treatment for pain. *Pain Physician*, *22*(1), E45–E52.

Calderon-Ospina, C. A., Nava-Mesa, M. O., & Paez-Hurtado, A. M. (2020). Update on safety profiles of vitamins B1, B6, and B12: A narrative review. *Therapeutics and Clinical Risk Management*, *16*, 1275.

Calderón-Ospina, C. A., & Nava-Mesa, M. O. (2020). B vitamins in the nervous system: Current knowledge of the biochemical modes of action and synergies of thiamine, pyridoxine, and cobalamin. *CNS Neuroscience & Therapeutics*, *26*(1), 5–13.

Calderon-Ospina, C.-A., Nava-Mesa, M. O., & Arbeláez Ariza, C. E. (2020). Effect of combined diclofenac and B vitamins (thiamine, pyridoxine, and cyanocobalamin) for low back pain management: Systematic review and meta-analysis. *Pain Medicine*, *21*(4), 766–781.

Chandrakumar, A., Bhardwaj, A., & 't Jong, G. W. (2018). Review of thiamine deficiency disorders: Wernicke encephalopathy and Korsakoff psychosis. *Journal of Basic and Clinical Physiology and Pharmacology*, *30*(2), 153–162.

Chopra, K., & Tiwari, V. (2012). Alcoholic neuropathy: Possible mechanisms and future treatment possibilities. *British Journal of Clinical Pharmacology*, *73*(3), 348–362.

Corken, M., & Porter, J. (2011). Is vitamin B(6) deficiency an under-recognized risk in patients receiving haemodialysis? A systematic review: 2000–2010. *Nephrology (Carlton, Vic.)*, *16*(7), 619–625.

Day, E., Bentham, P., Callaghan, R., Kuruvilla, T., & George, S. (2004). Thiamine for Wernicke-Korsakoff syndrome in people at risk from alcohol abuse. *The Cochrane Database of Systematic Reviews*, *1*, CD004033.

Della-Flora Nunes, G., Mueller, L., Silvestri, N., Patel, M. S., Wrabetz, L., et al. (2017). Acetyl-CoA production from pyruvate is not necessary for preservation of myelin. *Glia*, *65*(10), 1626–1639.

Dhaliwal, A., Larson, J. L., Dhindsa, B. S., Bhogal, N., & Rochling, F. A. (2020). Dry beriberi manifesting as acute inflammatory demyelinating polyneuropathy in a patient with decompensated alcohol-induced cirrhosis. *Cureus*, *12*(10), e11281.

Geller, M., Oliveira, L., Nigri, R., Mezitis, S., Ribeiro, M., Fonseca, A., et al. (2017). B vitamins for neuropathy and neuropathic pain. *Vitamins and Minerals*, *6*(2), 161.

Gerlach, A. T., Thomas, S., Stawicki, S. P., Whitmill, M. L., Steinberg, S. M., & Cook, C. H. (2011). Vitamin B6 deficiency: A potential cause of refractory seizures in adults. *Journal of Parenteral and Enteral Nutrition*, *35*(2), 272–275.

Hakim, M., Kurniani, N., Pinzon, R., Tugasworo, D., Basuki, M., & Haddani, H. (2019). A review on prevalence and causes of peripheral neuropathy and treatment of different etiologic subgroups with neurotropic B vitamins. *Journal of Clinical and Experimental Pharmacology*, *9*(262). ISSN: 2161-1459.

Hakim, M., Kurniani, N., Pinzon, R. T., Tugasworo, D., Basuki, M., Haddani, H., et al. (2018). Management of peripheral neuropathy symptoms with a fixed dose combination of high-dose vitamin B1, B6 and B12: A 12-week prospective non-interventional study in Indonesia. *Asian Journal of Medical Sciences*, *9*(1), 32–40.

Hammond, N., Wang, Y., Dimachkie, M. M., & Barohn, R. J. (2013). Nutritional neuropathies. *Neurologic Clinics*, *31*(2), 477–489.

Han, Y., Wang, M., Shen, J., Zhang, Z., Zhao, M., Huang, J., et al. (2018). Differential efficacy of methylcobalamin and alpha-lipoic acid treatment on symptoms of diabetic peripheral neuropathy. *Minerva Endocrinologica*, *43*(1), 11–18.

Harmandar, F. A., Dolu, S., & Çekin, A. H. (2020). Role of pernicious Anemia in patients admitted to internal medicine with vitamin B12 deficiency and oral replacement therapy as a treatment option. *Clinical Laboratory*, *66*(3).

Hasannejad, F., Ansar, M. M., Rostampour, M., Mahdavi Fikijivar, E., & Khakpour Taleghani, B. (2019). Improvement of pyridoxine-induced peripheral neuropathy by *Cichorium intybus* hydroalcoholic extract through GABAergic system. *The Journal of Physiological Sciences*, *69*(3), 465–476.

Hashem, M. M., Esmael, A., Nassar, A. K., & El-Sherif, M. (2021). The relationship between exacerbated diabetic peripheral neuropathy and metformin treatment in type 2 diabetes mellitus. *Scientific Reports*, *11*(1), 1940.

Hemminger, A., & Wills, B. K. (2021). Vitamin B6 toxicity. In *StatPearls* StatPearls Publishing. http://www.ncbi.nlm.nih.gov/books/NBK554500/.

Hosseinzadeh, H., Moallem, S. A., Moshiri, M., Sarnavazi, M. S., & Etemad, L. (2012). Anti-nociceptive and anti-inflammatory effects of cyanocobalamin (vitamin B12) against acute and chronic pain and inflammation in mice. *Arzneimittel-Forschung*, *62*(7), 324–329.

Issac, T. G., Soundarya, S., Christopher, R., & Chandra, S. R. (2015). Vitamin B12 deficiency: An important reversible co-morbidity in neuropsychiatric manifestations. *Indian Journal of Psychological Medicine*, *37*(1), 26–29.

Jacobs, A. M., & Cheng, D. (2011). Management of diabetic small-fiber neuropathy with combination L-methylfolate, methylcobalamin, and pyridoxal 5′-phosphate. *Reviews in Neurological Diseases*, *8*(1–2), 39–47.

Jayabalan, B., & Low, L. L. (2016). Vitamin B supplementation for diabetic peripheral neuropathy. *Singapore Medical Journal*, *57*(2), 55–59.

Julian, T., Glascow, N., Syeed, R., & Zis, P. (2019). Alcohol-related peripheral neuropathy: A systematic review and meta-analysis. *Journal of Neurology*, *266*(12), 2907–2919.

Karaganis, S., & Song, X. J. (2021). B vitamins as a treatment for diabetic pain and neuropathy. *Journal of Clinical Pharmacy and Therapeutics*, *46*(5), 1199–1212.

Kennedy, D. O. (2016). B vitamins and the brain: Mechanisms, dose and efficacy—A review. *Nutrients*, *8*(2), 68.

Kim, J., Ahn, C. W., Fang, S., Lee, H. S., & Park, J. S. (2019). Association between metformin dose and vitamin B12 deficiency in patients with type 2 diabetes. *Medicine*, *98*(46), e17918.

Koike, H., Ito, S., Morozumi, S., Kawagashira, Y., Iijima, M., et al. (2008). Rapidly developing weakness mimicking Guillain-Barré syndrome in beriberi neuropathy: Two case reports. *Nutrition*, *27*(7–8), 776–780.

Koike, H., Misu, K., Hattori, N., Ito, S., Ichimura, M., Ito, H., et al. (2001). Postgastrectomy polyneuropathy with thiamine deficiency. *Journal of Neurology, Neurosurgery, and Psychiatry*, *71*(3), 357–362.

Lederer, A.-K., Hannibal, L., Hettich, M., Behringer, S., Spiekerkoetter, U., Steinborn, C., et al. (2019). Vitamin B12 status upon short-term intervention with a vegan diet—A randomized controlled trial in healthy participants. *Nutrients*, *11*(11).

Lee, W. J., Poon, Y. C., Chuah, J. H., & So, S. C.-T. (2020). Subacute combined degeneration of the spinal cord. *The American Journal of Medicine*, *133*(12), 1421–1423.

Maladkar, M., Tekchandani, C., & Dave, U. (2014). Post-marketing surveillance of fixed dose combination of methylcobalamin, alpha lipoic acid, folic acid, biotin, benfotiamine & vitamin B6-nutripathy for the management of peripheral neuropathy. *Journal of Diabetes Mellitus*, *4*(02), 124.

Mikkelsen, K., Stojanovska, L., & Apostolopoulos, V. (2016). The effects of vitamin B in depression. *Current Medicinal Chemistry*, *23*(38), 4317–4337.

Mikkelsen, K., Stojanovska, L., Tangalakis, K., Bosevski, M., & Apostolopoulos, V. (2016). Cognitive decline: A vitamin B perspective. *Maturitas*, *93*, 108–113.

Moallem, S. A., Hosseinzadeh, H., & Farahi, S. (2008). A study of acute and chronic anti-nociceptive and anti-inflammatory effects of thiamine in mice. *Iranian Biomedical Journal*, *12*(3), 173–178.

Nava-Mesa, M. O., & Aispuru Lanche, G. R. (2021). Role of B vitamins, thiamine, pyridoxine, and cyanocobalamin in back pain and other musculoskeletal conditions: A narrative review. *SEMERGEN*.

Okada, K., Tanaka, H., Temporin, K., Okamoto, M., Kuroda, Y., Moritomo, H., et al. (2010). Methylcobalamin increases Erk1/2 and Akt activities through the methylation cycle and promotes nerve regeneration in a rat sciatic nerve injury model. *Experimental Neurology*, *222*(2), 191–203.

Rizvi, A., Ahmad, A., & Rizvi, Z. (2013). Efficacy of combination of vitamin B1, B6 and B12 in management of diabetic peripheral neuropathy. *Pakistan Journal of Medical and Health Sciences*, 7(3), 801–804.

Roy, R. P., Ghosh, K., Ghosh, M., Acharyya, A., Bhattacharya, A., Pal, M., et al. (2016). Study of vitamin B12 deficiency and peripheral neuropathy in metformin-treated early type 2 diabetes mellitus. *Indian Journal of Endocrinology and Metabolism*, *20*(5), 631–637.

Schloss, J., & Colosimo, M. (2017). B vitamin complex and chemotherapy-induced peripheral neuropathy. *Current Oncology Reports*, *19*(12), 76.

Sechi, G., Sechi, E., Fois, C., & Kumar, N. (2016). Advances in clinical determinants and neurological manifestations of B vitamin deficiency in adults. *Nutrition Reviews*, *74*(5), 281–300.
Semba, R. D. (2012). The discovery of the vitamins. *International Journal for Vitamin and Nutrition Research*, *82*(5), 310–315.
Sonobe, M., Yasuda, H., Hatanaka, I., Terada, M., Yamashita, M., Kikkawa, R., et al. (1988). Methylcobalamin improves nerve conduction in streptozotocin-diabetic rats without affecting sorbitol and myoinositol contents of sciatic nerve. *Hormone and Metabolic Research*, *20*(11), 717–718.
Spinas, E., Saggini, A., Kritas, S. K., Cerulli, G., & Caraffa, A. (2015). Crosstalk between vitamin B and immunity. *Journal of Biological Regulators and Homeostatic Agents*, *29*(2), 283–288.
Staff, N. P., & Windebank, A. J. (2014). Peripheral neuropathy due to vitamin deficiency, toxins, and medications. *Continuum (Minneapolis, Minn.)*, *20*(5), 1293–1306. Peripheral Nervous System Disorders.
Stein, J., Geisel, J., & Obeid, R. (2021). Association between neuropathy and B-vitamins: A systematic review and meta-analysis. *European Journal of Neurology*, *28*(6), 2054–2064.
Stracke, H., Gaus, W., Achenbach, U., Federlin, K., & Bretzel, R. G. (2008). Benfotiamine in diabetic polyneuropathy (BENDIP): Results of a randomised, double blind, placebo-controlled clinical study. *Experimental and Clinical Endocrinology & Diabetes: Official Journal, German Society of Endocrinology [and] German Diabetes Association*, *116*(10), 600–605.
Thériault, O., Poulin, H., Thomas, G. R., Friesen, A. D., Al-Shaqha, W. A., & Chahine, M. (2014). Pyridoxal-5′-phosphate (MC-1), a vitamin B6 derivative, inhibits expressed P2X receptors. *Canadian Journal of Physiology and Pharmacology*, *92*(3), 189–196.
Thomson, A. D., & Marshall, E. J. (2006). The natural history and pathophysiology of Wernicke's encephalopathy and Korsakoff's psychosis. *Alcohol and Alcoholism (Oxford, Oxfordshire)*, *41*(2), 151–158.
Wang, Z.-B., Gan, Q., Rupert, R. L., Zeng, Y.-M., & Song, X.-J. (2005). Thiamine, pyridoxine, cyanocobalamin and their combination inhibit thermal, but not mechanical hyperalgesia in rats with primary sensory neuron injury. *Pain*, *114*(1–2), 266–277.
Wendolowicz, A., Stefanska, E., & Ostrowska, L. (2018). Influence of selected dietary components on the functioning of the human nervous system. *Roczniki Państwowego Zakładu Higieny*, *69*(1).
Wiley, K. D., & Gupta, M. (2021). *Vitamin B1 thiamine deficiency*. En *StatPearls*: StatPearls Publishing. http://www.ncbi.nlm.nih.gov/books/NBK537204/.
Zak, J., Burns, D., Lingenfelser, T., Steyn, E., & Marks, I. N. (1991). Dry beriberi: Unusual complication of prolonged parenteral nutrition. *Journal of Parenteral and Enteral Nutrition*, *15*(2), 200–201.
Zempleni, J., Suttie, J. W., Gregory, J. F., III, & Stover, P. J. (2013). *Handbook of vitamins*. CRC Press.
Zhang, M., Han, W., Hu, S., & Xu, H. (2013). Methylcobalamin: A potential vitamin of pain killer. *Neural Plasticity*, *2013*, 424651.
Zhang, H., Peng, A. L., Zhao, F. F., Yu, L. H., Wang, M. Z., Osorio, J. S., et al. (2020). Thiamine ameliorates inflammation of the ruminal epithelium of Saanen goats suffering from subacute ruminal acidosis. *Journal of Dairy Science*, *103*(2), 1931–1943.

Further reading

Paez-Hurtado, A. M., Calderon-Ospina, C. A., & Nava-Mesa, M. O. (2022). Mechanisms of action of vitamin B1 (thiamine), B6 (pyridoxine), and B12 (cobalamin) in pain: A narrative review. *Nutritional Neuroscience*, 1–19.

PART XII

Prenatal effects and neurodevelopment

CHAPTER 38

Dietary B-vitamin deficiencies and maternal oversupplementation on neurodevelopment: An updated narrative

McCoy Clementson[a,b], Jesse Jauhal[a], and Nafisa M. Jadavji[a,c,d]
[a]Biomedical Sciences, College of Graduate Studies, Midwestern University, Glendale, AZ, United States
[b]College of Osteopathic Medicine, Midwestern University, Glendale, AZ, United States
[c]College of Veterinary Medicine, Midwestern University, Glendale, AZ, United States
[d]Department of Neuroscience, Carleton University, Ottawa, ON, Canada

Abbreviations

1C	one-carbon metabolism
BDNF	brain-derived growth factor
MTHFR	methylenetetrahydrofolate reductase
NTD	neural tube defects
SAM	S-adenosylmethionine
SGA	small for gestational age

Introduction

In 1998, the United States and Canada, other countries followed suite by implementing mandatory fortification of folate (vitamin B9) in grains. The U.S. Public Health Service recommended that women should consume 400 μg/day to prevent neural tube defects (NTDs), such as anencephaly and spina bifida (Williams et al., 2015). Since 1998, grains were fortified with 140 μg/g of folate and the incidence of NTDs decreased (Williams et al., 2015), a resounding success of the public health initiative. However, women who are pregnant and seeking to become pregnant are recommended to take prenatal vitamins, which often contain higher than the recommended dosage of folate (Wilson et al., 2015). The current recommendations of the World Health Organization specifically cite that folic acid (400 μg/day) supplementation should begin when a woman starts trying to conceive until 12 weeks of gestation (Boukerdenna, Pena-Rosas, & Rogers, 2019).

Folate, a B-vitamin, is the main component of one-carbon (1C) metabolism (Fig. 1). Folic acid is the chemically synthesized form of folate, which is often taken by women of childbearing age. One-carbon metabolism also includes other B-vitamins, such as

Vitamins and Minerals in Neurological Disorders
https://doi.org/10.1016/B978-0-323-89835-5.00041-7

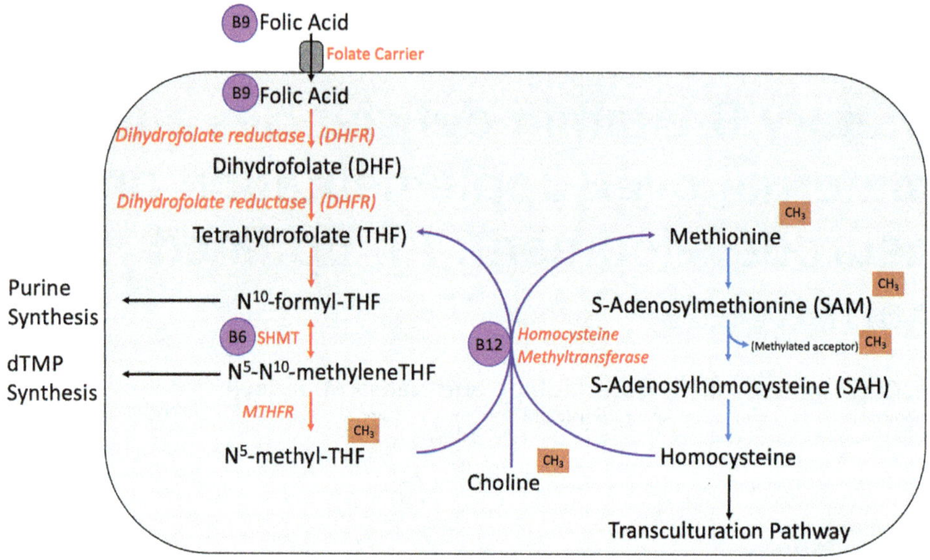

Fig. 1 One-carbon metabolism in the cell-related biological roles (purine synthesis, DNA repair, and methylation). Abbreviations: *BHMT*, betaine homocysteine methyltransferase; *DHFR*, dihydrofolate reductase; *dTMP*, deoxythymidine monophosphate; *dTTP*, deoxythymidine triphosphate; *dUMP*, deoxyuridine monophosphate; *dUTP*, deoxyuridine triphosphate; *MTHFR*, methylenetetrahydrofolate reductase; *SAH*, *S*-adenosylhomocysteine; *SAM*, *S*-adenosylmethionine.

vitamin B12 and riboflavin (vitamin B2). The 1C metabolic pathway integrates nutritional signals with biosynthesis, redox homeostasis, and epigenetics; it also plays essential roles in the regulation of cell proliferation, stress resistance, and embryo development. Dietary maternal contributions of 1C are essential for the developing embryo. The aim of this chapter is to review clinical and preclinical data on how maternal deficiencies and oversupplementation impact neurodevelopment after birth.

Dietary deficiencies in B-vitamins

Clinical studies

Maternal deficiencies in B-vitamins are known to have various detrimental effects on the neurodevelopment of offspring beginning with the closure of the neural tube during in utero development (Tulchinsky, 2010). An Argentinian study of 96 female patients reported that a polymorphism in methylenetetrahydrofolate reductase (*MTHFR*) 677C ➔T (Fig. 1) led to a producing an inefficient and thermolabile enzyme that along with poor maternal nutritional status may increase the risk of irreversible hypoxic-ischemic encephalopathy of newborns (Brown et al., 2007). Furthermore, a metaanalysis with over 8000 mother-child pairs investigated the significance of *MTHFR* polymorphisms on fetal development, and results from the study indicate a significant correlation between the TT *MTHFR* polymorphism and the birth of a Down syndrome-affected child (Kaur & Kaur, 2016; Pilsner et al., 2010). Results of population-based cohort study in Mexico using a lead-exposed experimental population indicate that the maternal *MTHFR* 667T allele is an independent predictor of impaired offspring neurodevelopment (Pilsner et al., 2010). Lead-exposed populations have also shown more exaggerated cognitive decline when offspring are also deficient in folate and iron (Solon et al., 2008).

There has been an association between elevated levels of homocysteine during the third trimester of pregnancy and increased risk of schizophrenia in a population cohort study in which offspring were followed for up to 40 years after birth (Brown et al., 2007). The authors postulate that developmental alterations in brain structure lead to damaged vasculature, which prevented adequate oxygen delivery to the fetus, and thus increased risk of schizophrenia (Brown et al., 2007). Maternal deficiencies in vitamin B12 have also been linked to increased levels of homocysteine as well as neurological abnormalities in Chinese children aged 2 years old (Guerra-Shinohara et al., 2002). A 2019 study conducted in Singapore using 1247 mother-child pairs also noted that vitamin B12-deficient mothers produced offspring that consistently displayed poorer cognitive scores on the Bayless Scales of Infant and Toddler Development compared to offspring from nondeficient mothers (Lai et al., 2019). In an Indian study using new born babies and their mothers, results indicated that impaired maternal 1C metabolism during pregnancy may result in a decreased development of the fetus, which can be combatted by reducing levels of homocysteine (Yajnik et al., 2014). A 2005 study from Birmingham, Alabama (United States), has indicated that folate status in mothers during late pregnancy did not appear to have an effect on offspring neurodevelopment at the age of five (Tamura, 2005).

Long-term effects of prenatal maternal folate deficiency include large-scale alterations in brain morphology and development, as well as increased levels of homocysteine, both of which have been associated with poorer cognitive performance in school-aged children from the Netherlands (Ars et al., 2019). A study including 459 2-year-old children from China indicates that maternal serum folate levels during early pregnancy are inversely associated with fine motor development, with lower folate levels leading to worse performance in the Chinese Gesell Development Scale (Huang et al., 2020). Furthermore, deficiencies in maternal dietary folate levels during offspring early-life development have been preliminarily associated with decreased brain size and increased hyperactivity, inattention, and peer problems in childhood in a 2010 study with a mean child age of 8 years and 9 months (Schlotz et al., 2010). An increased risk of emotional problems in offspring has also been tied to maternal folate deficiencies in a large-scale population-based cohort study that compared MTHFR genotype and the impact of folic acid levels (Steenweg-de Graaff et al., 2012).

A European study using 1988 mother-child pairs measured the levels of genome-wide epigenetic methylation modulated by folate levels. Results indicated that 320 genes are associated with maternal dietary folate levels and that these genes have been linked with neural tube defects and decreases in neural function (Joubert et al., 2016).

Supplementation of methyl group donors such as folate and choline during the periconception period of pregnancy has been shown to influence offspring buccal DNA methylation in genes related to metabolism, appetite regulation, and DNA methylation in a Belgian study using 114 mother-child pairs (Pauwels et al., 2017).

Model system studies

Animal models have been used to study the mechanisms through which maternal dietary deficiencies of these key vitamins and nutrients will lead to impaired neurodevelopment of offspring (Kumar et al., 2019).

Decreased dietary maternal folate has been shown to hinder normal early embryonic growth as well as increase the risk for cancer in old age through epigenetic modification in a rat model (Kim, Friso, & Choi, 2009). Furthermore, reductions in dietary B12 and folic acid in 10-week-old male mice have been shown to decrease neurological development and performance in behavior, such as the Morris water maze, novel object recognition, locomotor activity, and the accelerating rotarod used to test spatial memory and learning, spontaneous locomotor activity, and motor coordination and balance, respectively (Kumar et al., 2019). This is likely due to the decrease in size and weight of the brain, body, and liver that has been associated with maternal dietary folic acid deficiency during pregnancy and lactation, as well as distension of the lateral ventricles and rounding of the cerebral aqueduct (Middaugh et al., 1976; Whitley, Dell, Hogan, & Agri, 1951). Also, these dietary deficiencies are associated with decreases in cerebellum plasticity through impaired regulation of the ER-a pathway, ultimately leading to reduced synapsin activity (Pourie et al., 2015). Altered micronutrient levels during pregnancy can negatively affect neurotrophin expression at birth, which can impair plasticity and the ability to mediate the association between the environment and the phenotype (Sable, Dangat, Kale, & Joshi, 2011).

Furthermore, decreased levels of B-vitamins have been shown to lead to hyperhomocysteinemia due to decreased methyl group availability. High levels of homocysteine have been associated with increased apoptosis in the hippocampus, cerebellum, striatum, and subventricular zone due to DNA fragmentation and p53 activation (Blaise et al., 2007). This is linked with persistent functional disabilities in the rat model (Blaise et al., 2007). The effects of hyperhomocysteinemia have been shown to be long lasting in rats, likely due to alterations in monoamine neurotransmitters in the midbrain and brainstem, which affect molecular and cellular plasticity of nerve tissue (Shcherbitskaya et al., 2017). The formation of short- and long-term memory is also impaired in mice that experience hyperhomocysteinemia (Shcherbitskaya et al., 2017). The use of omega fatty acids as a supplement has been shown to play a role in the neurodevelopment of offspring. More specifically, through the regulation of DNA methylation, omega fatty acids have shown to improve cognition and have neuroprotective properties in a rat model (Sable et al., 2015).

There is a link between early-life nutritional deficiencies, during fetal development, and mood disorders. Prenatal folate deficiencies are correlated with increased anxiety in offspring using a mouse model (Ferguson et al., 2005). Research that has been conducted on the impact of B-vitamins on neurodevelopment indicates that deficiencies in these nutrients can lead to impaired development of the brain and body that can hinder memory, motor function, as well as have novel effects on mood within current animal models. Adequate maternal supplementation during pregnancy is extremely important in ensuring the healthy development of offspring.

Dietary oversupplementation of folates

Clinical studies

Pregnant women who eat grains and take prenatal vitamins might be exposing their fetuses to an extremely high level of folate. At this point, it is unknown what the long-term consequences of this excess in utero folate exposure are. This is an area of emerging research, and the data are unclear.

If excess folate maternal folate consumption is causing adverse effects, what could those effects be? An American study published in 2011 examined data from 1976 to 1997 in the city of Rochester, Minnesota. They discovered that there was a relatively strong association between high maternal folate consumption and the development of autism spectrum disorder (Beard, Panser, & Katusic, 2011). It is important to note that this study only showed correlation; further research must be done to determine if there are any causative effects. Yet, this study does demonstrate a link between excess folate and an adverse effect.

Given folate's necessary role in neurodevelopment including the closure of the neural tube (Williams et al., 2015), one could assume an abnormal amount of folate could affect the development of the brain. A comprehensive study in Spain discovered just that. Researchers surveyed over 2500 pregnant women in 4 different regions of Spain to determine the level of folate consumption based on their diet and prenatal vitamins. Once their children reached 4–5 years of age, they underwent a series of cognitive assessments. Results from these tests showed that children of mothers who consumed $\geq$1000 μg/day of folic acid had significantly lower global memory and verbal memory scores, and they also demonstrated lower cognitive function of the posterior cortex (Valera-Gran et al., 2017). This data was adjusted for several confounding variables, including the mother's verbal reasoning skills, whether or not the baby was breastfed, and excluding preterm and low birthweight babies. Even after adjustment, these scores were still significantly below the children of mothers who consumed the recommended amount of folate (400–999 μg/day) (Valera-Gran et al., 2017). A 2014 study demonstrated that women who consumed more than 1000 μg/day of folic acid had children who had delayed psychomotor development (Valera-Gran et al., 2014). This study demonstrates that there are strong links between impaired cognitive function and high maternal folate supplementation; however, there is also reason to believe that high maternal folic acid supplementation can be protective in other developmental areas.

A mother-child cohort study performed in Crete, Greece, found that mothers who consumed 5 mg/day of supplemental folic acid were associated with a 31% decrease in preterm birth, a 60% decreased risk in delivering a low birthweight neonate, and a 66% decreased risk for delivering a small for gestational age (SGA) baby (Papadopoulou et al., 2013). In fact, a 2016 Iranian study discovered that women who consumed 5 mg/day of folic acid had significantly higher birth weight, compared to women who only consumed 0.5 mg/day (Sayyah-Melli et al., 2016). This randomized clinical trial also showed that levels of homocysteine, an important player in 1C metabolism and a marker for inflammation (El Oudi et al., 2010), were decreased in the high-dose group (Sayyah-Melli et al., 2016). This supports data which showed that high-dose maternal folic acid supplementation was not associated with preeclampsia. In addition, women who were suffering from preeclampsia that received high-dose folic acid (5 mg/day) improved their endothelial dysfunction to the point that mean blood pressure can be reduced (Hashemi, Heshmat-Ghahdarijani, Zarean, Baktash, & Mortazavi, 2016). Overall, results from clinical trials are mixed in terms of the effects of high-dose folic acid supplementation on neurodevelopment in offspring after birth.

Model system studies

A study in C57Bl/6 male offspring fed a high dietary folate throughout pregnancy and lactation led to offspring that had short-term memory impairment. In a novel object recognition task, the pups were familiarized to two identical copies of an object. Then, one object was replaced and the pups were allowed to explore the new toy and the old toy. What researchers found is that the mice born to mothers who were fed 10 times the recommended amount of folate throughout pregnancy and lactation spent significantly less time with the novel object than with the familiar object. This suggests that these mice had short-term memory impairment (Bahous et al., 2017). High maternal folic acid has been shown to decrease the levels of enzymes in the brain. In particular, experiments using a mice model showed decreased levels of a DNA methyltransferase, Dnmt3a (DNA (cytosine-5)-methyltransferase 3A), in both the cortex and hippocampus (Bahous et al., 2017). Mutations in Dnmt3a will affect the methylation patterns in DNA and have epigenetic consequences (Gao et al., 2020). What those consequences are remains to be seen. However, even with mandatory fortification and a mother who is diligently taking her prenatal vitamins, it is unlikely that she would consume

10 times the recommended amount of folate as was done in the study (Navarrete-Muñoz et al., 2015). Thus, it is hard to extrapolate these findings to humans, but it does confirm that there is a tolerable upper limit to maternal folic acid consumption in regard to memory, brain development, and methylation patterns.

Excess dietary folic acid consumption can impair brain function (Bahous et al., 2017; Sable et al., 2011; Valera-Gran et al., 2017). Researchers found that rats whose mothers were supplemented with folic acid at 8 mg/kg displayed reduced levels of brain-derived neurotrophic factor (BDNF), an important protein that promotes growth and survival of neurons, including in the hippocampus and cortex (Binder & Scharfman, 2004; Sable et al., 2011). The hippocampus is an important part of the limbic system involved in learning and memory consolidation (Anand & Dhikav, 2012). It has been shown that reduced BDNF is implicated in neurodegenerative diseases like Alzheimer's, Parkinson's, and Huntington's disease (Binder & Scharfman, 2004). Based on these findings, it is reasonable to suspect that mothers who consume excess folate during pregnancy could produce offspring with memory deficits and possibly predispose them to debilitating neurodegenerative diseases later in life.

High-dose maternal folic acid supplementation also seems to show some protective effects in animal studies as well. High-dose B vitamins improved anxiety and depressive symptoms in mice offspring whose mothers were exposed to PM2.5, a common pollutant which is particulate matter that is less than 2.5 μm. In addition, mice who were subjected to maternal deprivation, an extreme stressor for the animals, developed depressive symptoms. However, these depressive symptoms could be alleviated by folic acid and omega-3 fatty acid supplementation (Réus et al., 2018). In addition, it has been demonstrated that excess maternal folic acid can aid cortical layering patterns (Bahous et al., 2017). Hematoxylin and eosin staining of the cortex of pups sacrificed at embryonic day 17.5 revealed more organized and developed cortical, subventricular zone, and subventricular zone layers (Bahous et al., 2017). It is unknown at this time what effects more efficient layering will have on the offspring at this time, but it is reasonable to assume that this would not harm brain function. Further research should be done to determine if there are any benefits to more organized layering. Based on the results of these experiments, it is not possible to definitively say whether or not high-dose maternal folic acid supplementation is a net positive or negative on the health of the offspring.

Conclusion

In this chapter, we reported that too little or too much maternal dietary B-vitamin intake during pregnancy and lactation results in change in offspring neurological function. These changes range from behavioral alterations to brain tissue. The importance of mandatory fortification to reduce the emphasis of neural tube defects is emphasized in this chapter, especially in populations who do not have access to adequate prenatal medical care. Emphasis on the tolerable upper limit of 1000 μg of folic acid should be communicated to healthcare providers and women of child-bearing age, so that new problems are not created. The data summarized in this chapter outline the importance of adequate levels of B-vitamins throughout pregnancy to facilitate neurological development. More investigation is required to understand the mechanisms through which oversupplementation impairs neurological function in offspring.

Applications to other neurological conditions

Maternal dietary contributions of folates play a vital role in the closure of the neural tube during in utero development; the role of this vitamin after birth remains relatively unknown. In this chapter, we have reviewed the impact of maternal dietary deficiencies and oversupplementation of dietary folates on offspring neurodevelopment after birth. Both clinical and preclinical studies have been incorporated in the review of the literature. A maternal diet deficient or oversupplementation can have a negative impact on offspring neurological function. Too much maternal dietary folate during early development has been linked to neurodevelopmental diseases, such as autism spectrum disorder. More research is required to understand the impact of maternal oversupplementation of folates on offspring brain function after birth.

Other components of interest

In this chapter, we describe the impact of maternal dietary levels of B-vitamins on offspring neurological function after birth. Both clinical and preclinical studies have been incorporated in the review of the literature. A metaanalysis with over 8000 mother-child pairs investigated the significance of *MTHFR* polymorphisms on fetal development, and results from the study indicate a significant correlation between the TT *MTHFR* polymorphism and Down syndrome (Kaur & Kaur, 2016; Pilsner et al., 2010). Results from a population-based cohort study in Mexico using a lead-exposed experimental population indicate that the maternal *MTHFR* 667T allele is an independent predictor of impaired offspring neurodevelopment (Pilsner et al., 2010). Lead-exposed populations have also shown more exaggerated cognitive decline when offspring are also deficient in folate and iron (Solon et al., 2008). Maternal deficiencies in vitamin B12 have also been linked to increased levels of homocysteine as well as neurological abnormalities in Chinese children aged 2 years old (Guerra-Shinohara et al., 2002). A 2019 study conducted in Singapore using 1247 mother-child pairs also noted that vitamin B12-deficient mothers produced offspring that consistently displayed poorer cognitive scores on the Bayless Scales of Infant and Toddler Development compared to offspring from nondeficient mothers (Lai et al., 2019).

Key facts

- Adequate levels of folates (B-vitamins) are important in the closure of the neural tube of the developing fetus.
- Folic acid is the chemically synthesized form of folates (vitamin B9).

Mini-dictionary of terms

- **Homocysteine.** A naturally occurring amino acid found in blood plasma. High levels of homocysteine in the blood are believed to increase the chance of heart disease, stroke, Alzheimer's disease, and osteoporosis.
- **One-carbon metabolism.** Comprises a series of interlinking metabolic pathways that include the methionine and folate cycles that are central to cellular function, providing 1C units (methyl groups) for the synthesis of DNA, polyamines, amino acids, creatine, and phospholipids.
- **Methylenetetrahydrofolate reductase.** A gene on chromosome 1p36.3 that encodes an enzyme that converts 5,10-methylenetetrahydrofolate to 5-methyltetrahydrofolate, a co-substrate for homocysteine remethylation to methionine.
- **Neural Tube Defects.** A congenital defect in closure of the bony encasement of the spinal cord or of the skull.
- **Vitamin B12.** A water-soluble vitamin of animal origin required for DNA synthesis.
- **Folic Acid.** A water-soluble vitamin belonging to the B complex group of vitamins. These vitamins help the body break down complex carbohydrates into simple sugars to be used for energy. Excess B vitamins are excreted from the body rather than stored for later use. This is why, sufficient daily intake of folic acid is necessary.

Summary points

- Maternal dietary contributions of B-vitamins, such as folic acid, are essential to the closure of the neural tube during in utero development.
- In the cell, folic acid (vitamin B9) is involved in nucleotide synthesis, DNA repair, lipid metabolism, methylation, and neurotransmitter synthesis.
- The impact of maternal deficiencies and oversupplementation of B-vitamins on offspring neurological function after birth remains relatively unknown.
- The studies described in the present chapter have reported that maternal deficiencies or oversupplementation in folates during pregnancy result in changes in behavioral function of offspring.

- There are also changes reported in blood and brain tissue in offspring, including altered methylation, reduced levels of *S*-adenosylmethionine (SAM), a global methyl donor, and increased levels of oxidative stress.

References

Anand, K. S., & Dhikav, V. (2012). Hippocampus in health and disease: An overview. *Annals of Indian Academy of Neurology*, *15*(4), 239–246.

Ars, C. L., et al. (2019). Prenatal folate, homocysteine and vitamin B_{12} levels and child brain volumes, cognitive development and psychological functioning: The generation R study. *British Journal of Nutrition*, *122*(s1), S1–S9.

Bahous, R. H., Jadavji, N. M., Deng, L., Cosín-Tomás, M., Lu, J., Malysheva, O., et al. (2017). High dietary folate in pregnant mice leads to pseudo-MTHFR deficiency and altered methyl metabolism, with embryonic growth delay and short-term memory impairment in offspring. *Human Molecular Genetics*, 26(5), 888–900.

Beard, C. M., Panser, L. A., & Katusic, S. K. (2011). Is excess folic acid supplementation a risk factor for autism? *Medical Hypotheses*, 77(1), 15–17.

Binder, D. K., & Scharfman, H. E. (2004). Brain-derived neurotrophic factor. *Growth Factors*, *22*(3), 123–131.

Blaise, S. A., et al. (2007). Gestational vitamin B deficiency leads to homocysteine-associated brain apoptosis and alters neurobehavioral development in rats. *American Journal of Pathology*, *170*(2), 667–679.

Boukerdenna, H., Pena-Rosas, J. P., & Rogers, L. (2019). *Periconceptional daily folic acid (400 μg) supplementation for prevention of neural tube defects*. World Health Organization. https://www.who.int/selection_medicines/committees/expert/20/applications/Folic_acid.pdf.

Brown, A. S., et al. (2007). Elevated prenatal homocysteine levels as a risk factor for schizophrenia. *Archives of General Psychiatry*, *64*(1), 31.

El Oudi, M., Aouni, Z., Mazigh, C., Khochkar, R., Gazoueni, E., Haouela, H., et al. (2010). Homocysteine and markers of inflammation in acute coronary syndrome. *Experimental and Clinical Cardiology*, *15*(2), e25–e28.

Ferguson, S. A., et al. (2005). Behavioral effects of prenatal folate deficiency in mice. *Birth Defects Research. Part A, Clinical and Molecular Teratology*, *73*(4), 249–252.

Gao, L., Emperle, M., Guo, Y., Grimm, S. A., Ren, W., Adam, S., et al. (2020). Comprehensive structure-function characterization of DNMT3B and DNMT3A reveals distinctive de novo DNA methylation mechanisms. *Nature Communications*, *11*(1), 3355.

Guerra-Shinohara, E. M., et al. (2002). Relationship between total homocysteine and folate levels in pregnant women and their newborn babies according to maternal serum levels of vitamin B12. *BJOG: An International Journal of Obstetrics and Gynaecology*, *109*(7), 784–791.

Hashemi, M., Heshmat-Ghahdarijani, K., Zarean, E., Baktash, F., & Mortazavi, Z. S. (2016). Evaluation of the effect of high-dose folic acid on endothelial dysfunction in pre-eclamptic patients: A randomized clinical trial. *Journal of Research in Medical Sciences*, *21*, 114.

Huang, X., et al. (2020). Maternal folate levels during pregnancy and children's neuropsychological development at 2 years of age. *European Journal of Clinical Nutrition*. http://feeds.nature.com/~r/ejcn/rss/current/~3/W8sTuWAhgac/s41430-020-0612-9?utm_source=researcher_app&utm_medium=referral&utm_campaign=RESR_MRKT_Researcher_inbound.

Joubert, B. R., et al. (2016). Maternal plasma folate impacts differential DNA methylation in an epigenome-wide meta-analysis of newborns. *Nature Communications*, 7(May 2015), 10577.

Kaur, A., & Kaur, A. (2016). Maternal MTHFR polymorphism (677 C–T) and risk of Down's syndrome child: Meta-analysis. *Journal of Genetics*, *95*(3), 505–513.

Kim, K.-c., Friso, S., & Choi, S.-W. (2009). DNA methylation, an epigenetic mechanism connecting folate to healthy embryonic development and aging. *The Journal of Nutritional Biochemistry*, *20*(12), 917–926.

Kumar, M., et al. (2019). Effects of altered maternal folate and vitamin B12 on neurobehavioral outcomes in F1 male mice. *Brain Research Bulletin*, *153*, 93–101.

Lai, J. S., et al. (2019). Maternal plasma vitamin B12 concentrations during pregnancy and infant cognitive outcomes at 2 years of age. *British Journal of Nutrition*, 1303–1312.

Middaugh, L. D., et al. (1976). Neurochemical and behavioral effects of diet related perinatal folic acid restriction. *Pharmacology, Biochemistry, and Behavior*, *5*(2), 129–134.

Navarrete-Muñoz, E. M., Valera-Gran, D., García de la Hera, M., Gimenez-Monzo, D., Morales, E., Julvez, J., et al. (2015). Use of high doses of folic acid supplements in pregnant women in Spain: An INMA cohort study. *BMJ Open*, *5*(11), e009202.

Papadopoulou, E., Stratakis, N., Roumeliotaki, T., Sarri, K., Merlo, D. F., Kogevinas, M., et al. (2013). The effect of high doses of folic acid and iron supplementation in early-to-mid pregnancy on prematurity and fetal growth retardation: The mother-child cohort study in Crete, Greece (Rhea study). *European Journal of Nutrition*, *52*(1), 327–336.

Pauwels, S., et al. (2017). Maternal intake of methyl-group donors affects DNA methylation of metabolic genes in infants. *Clinical Epigenetics*, *9*(1), 16.

Pilsner, J. R., et al. (2010). Maternal MTHFR genotype and haplotype predict deficits in early cognitive development in a lead-exposed birth cohort in Mexico City. *The American Journal of Clinical Nutrition*, *92*(1), 226–234.

Pourie, G., et al. (2015). Folate- and vitamin B12-deficient diet during gestation and lactation alters cerebellar synapsin expression via impaired influence of estrogen nuclear receptor. *The FASEB Journal*, *29*(9), 1–13.

Réus, G. Z., Maciel, A. L., Abelaira, H. M., de Moura, A. B., de Souza, T. G., Dos Santos, T. R., et al. (2018). ω-3 and folic acid act against depressive-like behavior and oxidative damage in the brain of rats subjected to early- or late-life stress. *Nutrition*, *53*, 120–133.

Sable, P., Dangat, K., Kale, A., & Joshi, S. (2011). Altered brain neurotrophins at birth: Consequence of imbalance in maternal folic acid and vitamin B_{12} metabolism. *Neuroscience*, *190*, 127–134.

Sable, P., et al. (2015). Maternal micronutrients and brain global methylation patterns in the offspring. *Nutritional Neuroscience*, *18*(1), 30–36.

Sayyah-Melli, M., Ghorbanihaghjo, A., Alizadeh, M., Kazemi-Shishvan, M., Ghojazadeh, M., & Bidadi, S. (2016). The effect of high dose folic acid throughout pregnancy on homocysteine (Hcy) concentration and pre-eclampsia: A randomized clinical trial. *PLoS One*, *11*(5), e0154400.

Schlotz, W., et al. (2010). Lower maternal folate status in early pregnancy is associated with childhood hyperactivity and peer problems in offspring. *Journal of Child Psychology and Psychiatry*, *51*(5), 594–602.

Shcherbitskaya, A. D., et al. (2017). The effects of prenatal hyperhomocysteinemia on the formation of memory and the contents of biogenic amines in the rat hippocampus. *Neurochemical Journal*, *11*(4), 296–301.

Solon, O., et al. (2008). Nutrition among children in the Central Philippines. *The Journal of Pediatrics*, *152*(2), 237–244.

Steenweg-de Graaff, J., et al. (2012). Maternal folate status in early pregnancy and child emotional and behavioral problems: The generation R study. *American Journal of Clinical Nutrition*, *95*(C), 1413–1421.

Tamura, T. (2005). Folate status of mothers during pregnancy and mental and psychomotor development of their children at five years of age. *Pediatrics*, *116*(3), 703–708.

Tulchinsky, T. H. (2010). Micronutrient deficiency conditions: Global Health issues. *Public Health Reviews*, *32*(1), 243–255.

Whitley, J. R., Dell, B. L. O., Hogan, A. G., & Agri, M. (1951). Effect of diet on maze learning. *The Journal of Nutrition*, *45*(1), 153–160.

Valera-Gran, D., García de la Hera, M., Navarrete-Muñoz, E. M., Fernandez-Somoano, A., Tardón, A., Julvez, J., et al. (2014). Folic acid supplements during pregnancy and child psychomotor development after the first year of life. *JAMA Pediatrics*, *168*(11), e142611.

Valera-Gran, D., Navarrete-Muñoz, E. M., Garcia de la Hera, M., Fernández-Somoano, A., Tardón, A., Ibarluzea, J., et al. (2017). Effect of maternal high dosages of folic acid supplements on neurocognitive development in children at 4-5 y of age: The prospective birth cohort Infancia y Medio Ambiente (INMA) study. *The American Journal of Clinical Nutrition*, *106*(3), 878–887.

Williams, J., Mai, C. T., Mulinare, J., Isenburg, J., Flood, T. J., Ethen, M., et al. (2015). Updated estimates of neural tube defects prevented by mandatory folic acid fortification – United States, 1995-2011. *Morbidity and Mortality Weekly Report*, *64*(1), 1–5.

Wilson, R. D., Committee, Genetics, Wilson, R. D., Audibert, F., Brock, J. A., Carroll, J., et al. (2015). Preconception folic acid and multivitamin supplementation for the primary and secondary prevention of neural tube defects and other folic acid-sensitive congenital anomalies. *Journal of Obstetrics and Gynaecology Canada*, *37*(6), 534–552.

Yajnik, C. S., et al. (2014). Maternal homocysteine in pregnancy and offspring birthweight: Epidemiological associations and Mendelian randomization analysis. *International Journal of Epidemiology*, *43*(July), 1–11.

Further reading

De Kremer, R. D., & Grosso, C. (2004). Maternal mutation 677C > T in the methylenetetrahydrofolate reductase gene associated with severe brain injury in offspring: Severe brain damage and maternal 677C > T allele. *Clinical Genetics*, *67*(1), 69–80.

CHAPTER 39

Essential trace elements in neurodevelopment: An updated narrative

Anatoly Skalny[a], Lyubov Chernova[b], and Anastasia Skalnaya[c]
[a]The Center for Bioelementology and Human Ecology, Sechenov University, Moscow, Russian Federation
[b]The Laboratory of Molecular Dietetics, The Center for Bioelementology and Human Ecology, Sechenov University, Moscow, Russian Federation
[c]Laboratory of Invasive Neurointerfaces of Research Institute "Technobiomed", A.I. Yevdokimov Moscow State University of Medicine and Dentistry, Moscow, Russian Federation

List of abbreviations

AD	Alzheimer's disease
ADHD	attention deficit hyperactivity disorder
ASD	autism spectrum disorder
ATP	adenosine triphosphate
CNS	central nervous system
Cu	copper
FASDs	fetal alcohol spectrum disorders
Fe	iron
GABA	γ-aminobutyric acid
I	iodine
ID	iron deficiency
IQ	intelligence quotient
IVF	in vitro fertilization
Mn	manganese
mTOR	mammalian target of rapamycin
MS	multiple sclerosis
NDDs	neurodevelopmental disorders
NMDAR	*N*-methyl-D-aspartate receptor
PD	Parkinson's disease
ProSAP/Shank	family of multidomain proteins of the postsynaptic density
Se	selenium
T3	3,5,3′-triiodothyronine
T4	thyroxine
TE(s)	trace element(s)
Zn	zinc

Vitamins and Minerals in Neurological Disorders
https://doi.org/10.1016/B978-0-323-89835-5.00033-8

Introduction

The study of brain physiology proceeded simultaneously with the investigation of the participation of various biologically active substances in the functioning of the central nervous system (CNS).

Table 1 shows the list of essential and toxic trace elements that are needed for brain metabolism according to World Health Organization classification. An element is considered essential when a tiny quantity of it provides a physiologically important function, or when it's an integral part of an organic structure, performing a vital function in the organism (Mertz, 1998).

It's well known that essential trace elements (TEs) influence basic processes in the nervous system as neurogenesis, neuronal differentiation, myelination, and synaptogenesis (González & Visentin, 2016). The effect of a nutrient deficiency depends on its duration and the sensitivity of certain brain areas to the shortage. The vulnerability of the brain increases during the so-called "critical" and "sensitive" periods. Alteration to neurodevelopment in the critical period results in irreversible consequences to the structure and function of the child's brain, but in the sensitive period, consequences are not such as serious and small errors may be corrected (Krebs, Lozoff, & Georgieff, 2017). Defining and dividing these periods is a difficult task. Therefore, the supply of sufficient levels of TEs to the mother and the fetus is one of the crucial factors in assuring and maintaining mental health and normal behavior in the future.

Key essential TEs with a well-established role in neurodevelopment include iron (Fe), zinc (Zn), iodine (I), selenium (Se), copper (Cu), and manganese (Mn) (Yan, Zhao, Li, He, & Xu, 2018). Below, we discuss the role of nutrients and their imbalance on neurodevelopment.

Table 1 Essential and toxic TEs for brain metabolism.

Essential TEs	Toxic TEs
Zinc	Aluminum
Iron	Arsenic
Copper	Beryllium
Manganese	Cadmium
Selenium	Mercury
Iodine	Nickel
Cobalt	Lead
Chromium	Tin
Silicon	
Lithium	

Iron

Distribution of Fe in the brain seems to be connected to areas of motor function: the highest level is in the globus pallidus, putamen, substantia nigra, caudate nucleus, red nucleus, dentate nucleus, and locus coeruleus. The concentration of Fe in these regions is compatible to the liver, which is Fe-storage tissue (Rouault & Cooperman, 2006). Also, it has been shown that large quantities of Fe are contained in the hippocampus and prefrontal cortex (Bastian, Von Hohenberg, Mickelson, Lanier, & Georgieff, 2016). The possible reason is the presence of Fe-dependent enzymes in the extrapyramidal system and related areas, for example, the tyrosine hydroxylase, crucial in dopamine production (Hare, Cardoso, Szymlek-Gay, & Biggs, 2018).

Oligodendrocytes are responsible for the production of myelin in the CNS and have a high requirement for Fe. The cytochrome oxidase system, glucose-6-phosphate dehydrogenase, and succinic dehydrogenase are Fe-requiring enzymes, so activity and maturation of oligodendrocytes depends, to a large extent, on Fe. Also, Fe is the cofactor for cholesterol and lipid biosynthesis, so it plays an even greater role in myelination. Astrocytes also have strong Fe transport capacity and upregulate Fe efflux and influx in response to ongoing physiological and pathophysiological processes in the CNS (Cheli, Correale, Paez, & Pasquini, 2020).

Fe is a cofactor of the enzymes tryptophan hydroxylase and tyrosine hydroxylase, which are involved in the synthesis of neurotransmitters (such as serotonin, GABA, dopamine), and is also associated with catecholamines. Dopamine is a major neurotransmitter for striatal networks, related to motor function and cognitive and emotional processes. So there is a possible link between Fe deficiency, the dopaminergic system, and anxiety-like behavior. Iron deficiency (ID) is one of the causes of an impaired serotonin pathway, which is important for proper functioning of neural circuits, and may be related to neurodevelopmental disorders, such as autism, anxiety, or depression (Radlowski & Johnson, 2013).

It is believed that there are three critical windows for Fe and neurodevelopment. The first risk factors are in the preconception and in utero period: maternal diabetes mellitus, pregnancy-induced hypertension, and intrauterine growth restriction. Maternal diabetes results in Fe redistribution during erythropoiesis to ensure a rapid somatic growth and a fight against the hypoxia of the fetus. The result is the emptying of fetal hepatic Fe stores, indicated by low ferritin levels, which correlate with insufficient supply of Fe to the brain. The second important period is the first 6 months after birth, when active myelination occurs. The establishment of neurotransmitter pathways in the first 6–24 months of a baby's development is the third critical period. Fetuses and infants require large Fe stores during periods of active development (McCann, Amadó, & Moore, 2020). This applies to healthy full-term infants, who are self-sufficient with regard to Fe because of prenatally emerging adequate Fe stores. Premature babies do not have a sufficient store of Fe, and it is significantly consumed for rapid growth; therefore, it is recommended to include an Fe-fortified formula in their diet from the first days after birth (Wang, Wu, Li, Wang, & Zhu, 2019).

Neurodevelopmental abnormalities occur when the neonatal ferritin concentration decreases below 35 μg/L. Perinatal Fe deficiency can lead to delayed myelination in specific brain regions (Doom & Georgieff, 2014). Subsequently, ID might result in affected motor development (initiation and control of movement), impact the cognitive and behavioral function (including impaired learning and memory, poorer auditory recognition, and a less cooperative, confident, and persistent personality), and in severe cases affect the global maturation of the CNS. This has been demonstrated in the study of outcomes over a 25-year period of a cohort of 33 patients from Costa Rica with chronic ID in infancy. There were direct and indirect associations of chronic ID not only with poorer emotional health, behavioral problems, and feelings of dissociation/detachment, but also with poorer cognitive functioning in early adolescence (Lozoff, Castillo, Clark, Smith, & Sturza, 2014). ID of in-term infants more often resulted in cognitive impairment, whereas preterm babies faced affected movement function expressed in fidgety movements (small amplitude, moderate speed, and variable acceleration occurring in the neck, trunk, and limbs in all directions) (Bruggink et al., 2008). A study of 8–18-month-old children from different countries has shown that Fe deficiency is associated with lower cognitive scores on the Bayley Scales of Infant Development-III test and causes slower speed of cognitive processing (Fuglestad, Kroupina, Johnson, & Georgieff, 2016).

Zinc

The Zn-rich areas of the brain are the hippocampus and the amygdala. The amygdala and olfactory bulb are regions with a high density of Zn-containing neuron terminals (vesicular Zn) (Takeda, 2001). Also, in CNS regions like the cerebellum and the prefrontal cortex have high concentrations of a free form of Zn that is associated with presynaptic vesicles of glutamatergic neurons (Krebs et al., 2017).

Metallothioneins bind the vast majority of Zn; in such a bound form, Zn is found in neurons and glial cells. A small part, about 10%, of Zn is contained in vesicles in ionic form, and likely acts as an endogenous neuromodulator in synaptic neurotransmission, especially excitatory transmission via glutamate (Takeda, 2001). Zn modulates the activity and gene expression of NMDAR, and also the activity of g-aminobutyric acid receptors and voltage-gated calcium channels (Wang et al., 2019).

Data on the possible mechanisms of Zn in epileptogenesis are very ambiguous; it may produce either pro- and anticonvulsant effects, according to the results of many studies. Both extracellular and intracellular pre- and postsynaptic Zn was found to modulate a variety of ion channels or even acts like a neurotransmitter (Doboszewska et al., 2019).

Hippocampal sclerosis, as well as other epilepsy-associated pathologies (for example, focal cortical dysplasia, tuberous sclerosis, Rasmussen's encephalitis) is associated with mTOR pathway activation in dysmorphic neurons and microglia. Zn can be involved not only in the pathogenesis of epilepsy, but also in the development of cognitive impairment and depression associated with it. It has also been noted that a decrease in Zn in the hippocampus leads to the development of memory and learning impairment (Skalny, Aschner, & Tinkov, 2021).

Zn deficiency and/or high toxic metal concentrations are also important factors for the development of ASD. Correspondingly, prenatally Zn-deficient mice also considered on the ASD model are characterized by altered brain lateralization, increased striatal volume, and impaired striatal lateralization of dopamine receptor 1 expression. Zn deficiency alters the plasticity, formation and maturation of synapses, presumably due to the disruption of the ProSAP/Shank mechanism, which is associated with behavior associated with ASD (Grabrucker et al., 2018). One of the Zn-dependent mechanisms in ASD pathogenesis is the modulation of dopaminergic dysfunction. Zn deficiency has been shown to be associated with impaired dopaminergic transmission in the striatum due to a decrease in tyrosine hydroxylase activity and an increase in mTOR (Skalny et al., 2021). Zn deficiency in 8–18-month-old children has also been associated with compromised memory function (Fuglestad et al., 2016).

In metalloproteins, Zn accumulates in fetal liver after the 24th week of gestation in neonates born at term. Based on a review of available data, the critical period for Zn deficiency is assumed from 26th weeks of pregnancy to the 44th week of the postnatal period (Brion, Heyne, & Lair, 2020). Placental transfer of Zn is an active process, and fetal Zn levels are maintained at a predetermined high level (1,5 mg/kg). So, significant fetal Zn deficiency can develop only with a serious deficiency in the mother; however, prophylactic Zn intake by the mother may not have an effect on the fetus development (Donangelo & King, 2012). Limited supply of Zn during gestation and early postnatal development is not teratogenic but affects neurotransmission and cell signaling, reduces stem cell proliferation, increases neuronal precursor apoptosis, and impairs neuronal differentiation (Levenson & Morris, 2011). Excessive umbilical cord Zn is neurotoxic in newborns, because it leads to neuronal damage by excitotoxicity, oxidative stress, and impairment of neuronal energy production (Morris & Levenson, 2012).

Iodine

Optimal availability of I-containing thyroid hormones (T3, T4) is important for normal brain development. Until mid-pregnancy, the mother is the only source of cerebral T3, which is produced in the fetal brain by type II 5-iodothyronine deiodinase from maternal T4. The first stage of thyroid hormone-dependent nervous system development relies on adequate maternal T4 level and begins in the second half of the first trimester. This stage includes the proliferation of neurons and the onset of neuronal migration in the cerebral cortex, the hippocampus, and the medial ganglionic eminence. The fetal thyroid gland begins to produce hormones approximately at the beginning of the second trimester. Full development of the pituitary-portal system in the fetus occurs no earlier than 18–20 weeks of gestation. The second stage includes neuronal migration, axonal growth, dendritic branching and synaptogenesis, differentiation and migration of glial cells, and the onset of myelination. However, the fetal thyroid hormone concentration before birth depends on maternal thyroid hormone levels (Skeaff, 2011).

Animal studies confirm that I deficiency causes disruption in neurogenesis, myelination, synaptogenesis, and neurotransmitter balance. Thus, nerve fiber malfunction and impairments in hippocampal development were noted in rats with developmental I deficiency (Gong et al., 2010). Gilbert et al. (2013) showed that even a 15% decrease in cortical T4 in the fetal brain in I-deficient mice was sufficient to cause an irreversible decrease in synaptic function in adults. Marginal I deficiency in pregnant rats has led to impaired cell migration in the developing brain. In turn, earlier I supplementation was associated with better behavior performance (Zhang et al., 2016).

Now nearly one-third of children worldwide remain at risk for insufficient I intake and impaired mental development (Pearce, 2014). Negative effects include intellectual deficits, attention deficit hyperactivity disorder (ADHD), and psychiatric disorders (Verheesen & Schweitzer, 2008). The most serious consequence of I deficiency is cretinism, characterized by profound mental retardation. Severe I deficiency during pregnancy has also been associated with spontaneous abortion, stillbirth, congenital anomalies, and infant mortality (Zimmermann, 2012).

By the way, the results of cohort studies and meta-analyses of I supplementation efficacy during pregnancy are still under debate (Wassie, Smithers, & Zhou, 2021). Potential randomized controlled trials in this case should begin with supplementation no later than the first trimester (Levie et al., 2019). In addition, studies on the effectiveness of I supplementation in preterm infants also do not always demonstrate positive effects on the development of the nervous system (Walsh, Brown, & McGuire, 2019).

Manganese

Mn plays unique roles in normal development that cannot be substituted by other metals, such as Mn-dependent enzymes such as arginase, agmatinase, glutamine synthetase, and Mn superoxide dismutase. Optimal Mn level is necessary to support the brain development, and the lack of this essential TE can cause cognitive deficits (Balachandran et al., 2020). The liver, the pancreas, bones, kidneys, and the brain are the organs containing the highest Mn levels in humans (Chen, Bornhorst, & Aschner, 2018).

Mn deficiency is uncommon, its toxicity is more significant, and the primary target in instances of excessive Mn is the nervous system. Mn accumulates mainly in the basal ganglia, including the substantia nigra, striatum, and globus pallidum. Molecular mechanisms of Mn toxicity include oxidative stress, mitochondrial dysfunction, protein misfolding, endoplasmic reticulum stress, autophagy dysregulation, apoptosis, and disruption of other metal homeostasis. The accumulation of Mn in the brain has the potential to alter several neurotransmitter systems and their activity. Exposure to Mn, even at levels that are lower than cytotoxic to neurons (e.g., 100 μM), causes abnormalities in GABAergic neurotransmission, resulting in disinhibition. Excessive accumulation of Mn in the brain causes brain abnormalities through the activation of excitotoxic mechanisms. An excessive release of glutamate disrupts NMDAR expression and damages neurons and glia (Balachandran et al., 2020). Mn accumulation in the basal ganglia leads to an impairment of dopaminergic function. The consequences of Mn excess affect both motor and cognitive functions (Martins et al., 2020). Anxiety, hyperactivity, learning, and memory deficits are functional consequences of exposure to excessive Mn during development in rats. Mn neurotoxicity is higher when prenatal and postnatal exposures are combined than when it is limited to the prenatal or postnatal period and is mediated by altered neurotransmission, apoptosis and abnormal neuronal migration, excitotoxicity, and oxidative stress (Coetzee et al., 2016). Excessive exposure to Mn has been associated with decreased academic performance, behavioral problems, decreased IQ, and frontal lobe abnormalities (Modabbernia, Arora, & Reichenberg, 2016). Most studies that evaluated the effects of Mn exposure on neurodevelopment found negative associations between Mn level, IQ, and speech skills. These associations were detected both in children of the first year of life and in children aged 6–13 years (Rodríguez-Barranco et al., 2013). Higher Mn concentrations in hair have consistently been associated with lower IQ scores, while studies on blood biomarkers and IQ scores have yielded conflicting results (Coetzee et al., 2016).

Consumption with food is the primary route of Mn intake for the majority of children. Infant formulas are the major Mn food source that may result in Mn accumulation in infants (Martins et al., 2020). Exposure to Mn in children is a public health concern, and there are no consistent guidelines for safe levels of Mn exposure (Coetzee et al., 2016).

The manifestations of Mn neurotoxicity can be described by term "dysfunction continuum": manganism, neurological disorder in response to high Mn exposures; neuropsychological abnormalities; and idiopathic Parkinson's disease (PD). In recent years, an inverse U-shaped relationship has been reported between Mn exposure and adverse effects on neurodevelopment in infants and children, suggesting adverse effects of both low and high Mn exposure in humans and animals. This pattern is consistent with the fact that Mn acts as an important nutrient and as a toxicant (Balachandran et al., 2020).

Copper

Cu is one of many metal ions that are essential, but neurotoxic in excess (Desai & Kaler, 2008). Cu plays an important role in the development of the CNS. The amount of this element in an adult human body ranges from 50 to 120 mg, and the highest concentrations of Cu are found in the brain and the liver (Grzeszczak, Kwiatkowski, & Kosik-Bogacka, 2020). Cu is present in the brain prominently in the basal ganglia, the hippocampus, the cerebellum, and numerous synaptic membranes. CNS Cu-dependent enzymes include tyrosinase, peptidyl-glycine alpha-amidating monooxygenase, Cu/Zn superoxide dismutase, ceruloplasmin, hephaestin, dopamine hydroxylase, and cytochrome C oxidase (Desai & Kaler, 2008).

The main effects of Cu in the brain include the regulation of dopamine monooxygenase activity and the uptake of reactive oxygen species by Cu-Zn superoxide dismutase. Furthermore, it regulates the transport of Fe across the blood-brain barrier by donating Cu multi-oxidase protons to convert ferrous Fe to its transportable ferric form. Thus, Cu deficiency alters the status of dopamine and cytochrome C in the brain, especially in areas with a high metabolic rate such as the hippocampus and the cerebellum (Georgieff, Brunette, & Tran, 2015). Recent evidence suggests that excess Cu is likely to harm neuropsychological, cognitive functions by impairing neurotransmission and causing mitochondrial dysfunction (Modabbernia et al., 2016).

Little is known about the prevalence and possible health effects of marginal Cu deficiency. Risk factors include low birth weight, inadequate nutritional status in infancy, prolonged diarrhea, and intestinal malabsorption (Domellöf, 2014). In infants, a lack of Cu leads to neutropenia, anemia, impairment of growth, abnormalities in glucose and cholesterol metabolism, and increased rates of infection (Özden et al., 2015). Cu deficiency has also been associated with pathological pregnancy (habitual abortion, imminent abortion, missed abortion, missed labor, spontaneous abortion, premature rupture of membranes) and risk of preterm birth (Vukelić et al., 2012).

Selenium

It is now well known that Se plays a key role in several major metabolic pathways such as antioxidant defense, thyroid hormone metabolism, and immune functions (Vinceti et al., 2014). Brain tissue contains comparatively low concentration of Se (only 2.3% of total body Se), and around 20% of total Se in the brain is incorporated in glutathione peroxidase (Steinbrenner & Sies, 2013). Selenoproteins maintain optimal brain functions via redox regulation (Vinceti et al., 2014). Due to particularities of the regulation of certain essential selenoproteins' expression, especially in the brain, the organism retains these important selenoproteins at a high level even in situations of Se deficiency (Solovyev, Drobyshev, Blume, & Michalke, 2021).

Reduced brain Se bioavailability leads to oxidative stress and neuronal dysfunction (Steinbrenner & Sies, 2013). Despite its well-known essentiality, excess of Se can have neurotoxic effects including induction of neuronal apoptosis and degeneration, changes in dopamine levels, inhibition of glutamate uptake, decrease of total antioxidant status, etc. Some regions of the world such as China, the USA, and Italy have high Se content in soil, which leads to chronic exposure and neurotoxicity at an early stage of development (Vinceti et al., 2014). In addition, the role of Se in the activation of thyroid hormones reported in animal studies may be implicated in Se's effects on cognitive maturation (Polanska et al., 2016).

Pregnant women and babies are at risk for Se deficiency. Low maternal Se levels during pregnancy were found to be associated with adverse outcomes, including miscarriages, preeclampsia, gestational diabetes, and intrauterine growth restriction (Yang, Yu, Fu, Li, & Ren, 2013), fetal malformations such as neural tube defects, an increased risk of infection in infants during the first 6 weeks of life, and lower psychomotor levels after 6 months (Varsi et al., 2017). A negative correlation was shown between maternal blood Se levels and infant's cerebellar length (Močenić et al., 2019). Women's Se status during pregnancy is also associated with cognitive development (Močenić et al., 2019) and children's psychomotor and speech abilities in the first years of life (Polanska et al., 2016). Se imbalance has been reported in ASD and sleep disorders (Anatoly et al., 2018).

Applications to other neurological conditions

Menkes disease is an X-linked recessive disorder caused by mutations of ATP7A, a Cu-transporting P-type ATPase, which regulates the absorption of Cu. Cu deficiency leads to the inhibition of Cu-associated enzymes. In such cases, the synthesis of dopamine is disrupted. It leads to a delay in mental and motor development, and later, neurodegeneration occurs. Another genetically determined disease, known as Wilson's disease, which is caused by mutations in gene ATP7B, encoding the transmembrane Cu-transporting ATPase, leads to Cu overload in the liver, brain, and other organs. In adults, the disease often presents as movement disorders and neurobehavioral changes, depending on the region of Cu accumulation. Cu deficiency may be a complication of Zn supplementation, including the treatment of Wilson's disease. Acquired Cu deficiency can manifest as myelopathy, similar to the subacute combined degeneration, associated with vitamin B12 deficiency (Kumar, 2017).

In conditions of Zn deficiency, the absorption of Cu increases and the excess of Cu becomes neurotoxic. Altered Zn and Cu homeostasis, such as a high serum Cu/Zn ratio, has been demonstrated in AD and PD. Zn and Cu are likely to be involved in the pathogenesis of these diseases and to interact with amyloid precursor protein, β-amyloid, and tau protein. Zn supplementation has been shown to improve cognitive function in patients, but its effect is dependent on the homeostasis of other nutrients, such as a decrease of serum free Cu. It is known that Zn is directly related to the immune system, but its effect on the immune-mediated processes underlying the pathogenesis of multiple sclerosis is ambiguous.

Mn predominantly accumulates in the basal ganglia system. The globus pallidus is thought to become a primary target of Mn-induced degeneration. The cerebellum, red nucleus, pons, the cortex, the thalamus, and anterior horn of the spinal cord may also be affected by Mn toxicity. Manganism is manifested by cognitive deficit, neuropsychological abnormalities, and parkinsonism. In PD, dopaminergic neurons of the substantia nigra pars compacta are affected, while in manganism, dysfunction of GABAergic neurons in the globus pallidus is proposed. So, Mn-induced parkinsonism is an extrapyramidal, neurodegenerative disorder with Parkinson-like motor deficit and cognitive impairment. Furthermore, Fe deficiency has been associated with Mn overaccumulation in the caudate putamen and the globus pallidus. Fe is essential for the proper functioning of the brain, but Fe-induced neurotoxicity is thought to be involved in the pathogenesis of various diseases. In PD patients, Fe may accumulate in the substantia nigra, while in AD there is a diffuse accumulation of Fe that occurs in the cortex and the hippocampus, in multiple sclerosis (MS)—in the thalamus and striatum. Accumulation of Fe in cerebral deep gray matter has been considered to be a global marker for neurodegeneration in MS. Using Fe-sensitive magnetic resonance imaging, a moderate association has been reported between Fe accumulation in the globus pallidus and lower cognitive composite scores in MS patients, despite the absence of the globus pallidus atrophy. It is also assumed that the accumulation of Fe and Cu promotes both the formation and accumulation of β-amyloid in AD plaques, although conflicting findings have also been reported (Mezzaroba, Alfieri, Simão, & Reiche, 2019).

Other components of interest

It is widely accepted that the status of essential TEs in pregnant women can influence the development of the fetal brain and the maturation of the cognitive functions, psychomotor and behavioral abilities of their children. In this regard, the supplementation of essential TE before and during pregnancy can have a significant impact on the development and maturation of the child's nervous system (Martins et al., 2020).

Excess of toxic elements and a deficiency of essential TEs are associated with low human fertility. At the same time, the TE balance is one of the important factors for the successful outcome of in vitro fertilization (IVF). Studies have shown that women with IVF pregnancy have deficiencies of various essential elements compared to controls. The most heavily documented deficiencies are reduced Fe and Cu contents in the hair of these women. Exposure to toxic elements (Cr, As, and Cd) is associated with poor IVF outcomes. Se, as a heavy metal antagonist, has been shown to improve outcomes (Skalny, Tinkov, et al., 2018).

Mothers' exposure to prenatal alcohol alters Zn metabolism and causes not only physical but also neurological developmental disorders in children, with the phenomena known as fetal alcohol spectrum disorders (FASD). Alcohol-induced Zn deficiency may be associated with defective neurotransmission, severe cognitive and behavioral problems in infants. Decreased Zn in the cortex and elevated Zn in the cerebellum have been demonstrated after even short-term consumption of alcohol by female rats before conception (Keen et al., 2010).

Key facts

- An imbalance of essential TEs in critical periods of brain maturation can lead to impaired brain development.
- Areas of the brain with the highest metabolic rate (the hippocampus, the cortex, the cerebellum) are at greatest risk for metabolic disturbances, including TEs imbalance (deficiency or toxicity) (Table 2).
- The most studied TEs involved in neurogenesis are Fe, Zn, I, Se, Cu, and Mn (Table 3).

Table 2 The accumulation of essential TEs in brain regions.

Brain region	TEs
The cortex	Zn, Fe, I
The hippocampus	Zn, Fe, I, Cu
The basal ganglia system and associated structures	Fe, Cu, Mn (the globus pallidus)
The cerebellum	Zn, Cu, Mn
The amygdala	Zn
Olfactory bulb	Zn
The thalamus	Mn, Fe

Table 3 The essential TEs and processes of neurodevelopment.

Process	TEs
Synaptogenesis	Zn, I
Neurotransmission and production of neurotransmitters (especially dopaminergic system)	Zn, Fe, Cu, Mn
Myelination	Fe, I, Cu
Cellular proliferation and differentiation	Zn, I
Antioxidant defense systems	Se, Cu, Zn

- Fe and Mn mainly accumulate in structures associated with motor and cognitive functions—the basal ganglia system. Also, Fe is found in areas functionally associated with the basal ganglia, including the cerebellum, red nucleus, and locus coeruleus. Cu is located prominently in the basal ganglia, the hippocampus, and the cerebellum.
- Zn distribution is mainly related to the structures of the limbic system: the hippocampus, the amygdala, and olfactory bulb.
- I is accumulated in the cortex and the hippocampus. There is a low concentration of Se in the brain, but it performs an important function, protecting against the toxicity of heavy metals.
- I is involved in all stages of brain development and is required from the first trimester, while for other elements, critical periods usually begin closer to the third trimester and infancy.
- Zn, Fe, Cu, and Mn regulate neurotransmission and have a particularly significant effect on the dopaminergic system.
- Despite its essentiality, Fe, Cu, and Mn are toxic to the CNS at excessive levels. Therefore, the maintenance of homeostasis of these metals is critical, as any substantial change can lead to neurotoxic outcomes. Zn, Cu, Mn, and Fe also likely exert critical roles in the pathophysiology of AD, PD, and MS (Table 4).
- Low levels of Zn, as well as high levels of Mn, Cu, and Fe, contribute to the activation of signaling pathways involved in oxidative stress. Se, meanwhile, acts as an antioxidant.

Mini-dictionary of terms

Autism spectrum disorders are a group of disorders of children characterized by repetitive patterns of behavior and difficulties with social communication and interaction.

Fetal alcohol spectrum disorders are caused by the mother's drinking alcohol while pregnant. The consequences include physical, mental, behavioral, and learning disabilities.

Table 4 The essential TEs and some other neurological conditions.

Disorder	Deficiency	Excess
AD	Zn	Cu, Fe
PD	Zn	Cu, Mn, Fe
MS	Zn	Fe
FASD	Zn	–
Menkes disease	Cu	–
Wilson's disease	–	Cu
Myelopathy	Cu	–

Iron deficiency—there are different stages when Fe-dependent activities are downregulated: Fe depletion (ferritin reduction), ID (serum iron and transferrin saturation become low), and Fe deficiency anemia, characterized by reduced hemoglobin concentration.

Myelination is the formation of the myelin sheath around axons. This process provides faster transmission of the electrical signal.

Neurodevelopmental disorders are a group of disorders affecting the development of the CNS, leading to abnormal brain function, i.e., intellectual disability, ASD, ADHD, one form of FASD, conduct disorder, learning disability, and various genetic disorders.

Applications to other areas

The interaction of TE is an important factor for biological effects on an organism. There are many factors that influence TEs absorption and metabolism. Effectiveness of TEs intake depends on the species of TEs. The so-called chelated elements absorb and metabolize better in the human body but there is a lack of data in this field. Factors that improve metabolism of TEs include a protein-rich diet, organic acids, and some vitamins (e.g., vitamin C for Fe absorption, vitamin D for Ca and P absorption). Phytates, tannins, alcohol, simple sugars, and fibers inhibit TEs absorption.

One of the most important factors of TEs metabolisms is interaction between essential (Zn/Cu, Fe/Mn) and other essential/toxic elements. For example, even a moderate increase in As or Hg intake accompanied by a low Se level influences human morbidity (Hatfield, Berry, & Gladyshev, 2011).

Speciation analysis is a new method in analytical chemistry that allows to determine qualitatively and quantitatively the forms of various chemical elements in different biological materials. The distribution of a metal over these different forms is referred to as metal speciation. Speciation analysis will expand our understanding of the influence of chemical elements on numerous biological processes, including neurodevelopment (Michalke, Halbach, & Nischwitz, 2009).

Summary points

- It is difficult to isolate TEs specific function. In the maturation and development of the brain, TEs mainly take part in oxidative and antioxidative processes, synaptogenesis, and neurotransmission, as well as myelination to varying degrees (Table 5). Zn and Fe are the prevalent cerebral metals.
- Perinatal Fe deficiency may result in affected motor development, impact cognitive and behavioral function, and in severe cases, affect the global maturation of the CNS.

Table 5 The essential TEs imbalance and its effects on infant brain function.

TEs	Effects (Def—deficiency, Ex—excess)
Fe	Def—Motor, cognitive and behavioral impairment, autism, anxiety, depression Ex—Risk of infections, impaired growth
Zn	Def—Cognitive impairment, epilepsy, depression, autism, neural tube defects Ex—Neuronal damage
I	Def—Cognitive impairment (even to mental retardation), attention deficit hyperactivity disorder, psychiatric disorders, congenital anomalies
Mn	Def—Cognitive impairment Ex—Cognitive and behavioral problems, motor impairment
Cu	Def—Motor and cognitive, growth impairment Ex—Cognitive and neuropsychological impairment
Se	Def—risk of infection, lower psychomotor development, cognitive impairment, neural tube defects Ex—Neuronal degeneration, impaired neurotransmission and antioxidant processes

- Zn deficiency is one of the causes of disturbances of the higher nervous system, including cognitive deficit, memory impairment, reduced learning ability, and minor psychoneurological deviations.
- The consequences of I deficiency during neurodevelopment can include mental retardation, and the most serious is cretinism.
- Deficiency of Zn, Fe, Cu, Mn, Se, and I in fetuses and infants can result in impaired cognitive, behavioral, and motor development, and later, these children may have ASD, ADHD, FASD, epilepsy, and psychiatric disorders like schizophrenia, depression, and anxiety. In adults, Cu and Mn neurotoxicity is related to Zn and Fe deficiency, respectively, and also leads to motor and cognitive impairment, but other diseases may develop as well, including AD, PD, MS, Huntington's disease, prion disease, etc.
- TEs appear to be good candidates as biomarkers for NDDs. The estimation of the elemental status of children and their mothers (before and during pregnancy and lactation) can be an effective predictive tool for neurodevelopmental disorders.
- An adequate level of micronutrients, including TEs, in pregnant women and children under 5 years is one of the key goals in human nutrition. Fig. 1 shows the impact of essential TEs on neurodevelopment in different stages of life.

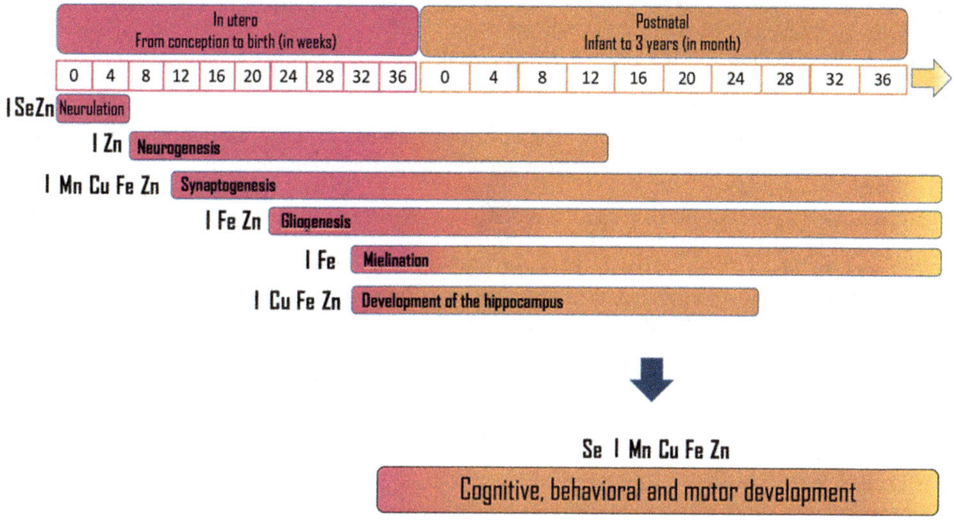

Fig. 1 The timeline showing the influence of essential TEs on neurodevelopment in different stages of life.

References

Anatoly, V., Skalny, M. G., Skalnaya, G. B., Gritsenko, V. A., Aaseth, J., & Tinkov, A. A. (2018). Selenium and autism spectrum disorder. In B. Michalke (Ed.), *Selenium molecular and integrative toxicology* (pp. 193–210). Cham: Springer.

Balachandran, R. C., Mukhopadhyay, S., McBride, D., Veevers, J., Harrison, F. E., Aschner, M., et al. (2020). Brain manganese and the balance between essential roles and neurotoxicity. *Journal of Biological Chemistry*, *295*(19), 6312–6329.

Bastian, T. W., Von Hohenberg, W. C., Mickelson, D. J., Lanier, L. M., & Georgieff, M. K. (2016). Iron deficiency impairs developing hippocampal neuron gene expression, energy metabolism, and dendrite complexity. *Developmental Neuroscience*, *38*(4), 264–276.

Brion, L. P., Heyne, R., & Lair, C. S. (2020). Role of zinc in neonatal growth and brain growth: Review and scoping review. *Pediatric Research*, 1–14. April.

Bruggink, J. L. M., Einspieler, C., Butcher, P. R., Van Braeckel, K. N. J. A., Prechtl, H. F. R., & Bos, A. F. (2008). The quality of the early motor repertoire in preterm infants predicts minor neurologic dysfunction at school age. *Journal of Pediatrics*, *153*(1), 32–39.

Cheli, V. T., Correale, J., Paez, P. M., & Pasquini, J. M. (2020). Iron metabolism in oligodendrocytes and astrocytes, implications for myelination and remyelination. *ASN Neuro*, *12*, 1–15.

Chen, P., Bornhorst, J., & Aschner, M. A. (2018). Manganese metabolism in humans. *Frontiers in Bioscience—Landmark*, *23*(9), 1655–1679.

Coetzee, D. J., McGovern, P. M., Rao, R., Harnack, L. J., Georgieff, M. K., & Stepanov, I. (2016). Measuring the impact of manganese exposure on children's neurodevelopment: Advances and research gaps in biomarker-based approaches. *Environmental Health*, *15*(1), 1–20.

Desai, V., & Kaler, S. G. (2008). Role of copper in human neurological disorders. *The American Journal of Clinical Nutrition*, *88*(3), 855S–858S.

Doboszewska, U., Młyniec, K., Wlaź, A., Poleszak, E., Nowak, G., & Wlaź, P. (2019). Zinc signaling and epilepsy. *Pharmacology & Therapeutics*, *193*, 156–177.

Domellöf, M. (2014). Nutritional care of premature infants: Microminerals. *World Review of Nutrition and Dietetics*, *110*, 121–139. https://doi.org/10.1159/000358462.

Donangelo, C. M., & King, J. C. (2012). Maternal zinc intakes and homeostatic adjustments during pregnancy and lactation. *Nutrients*, *4*(7), 782–798.

Doom, J. R., & Georgieff, M. K. (2014). Striking while the iron is hot: Understanding the biological and neurodevelopmental effects of iron deficiency to optimize intervention in early childhood. *Current Pediatrics Reports*, *2*(4), 291–298.

Fuglestad, A. J., Kroupina, M. G., Johnson, D. E., & Georgieff, M. K. (2016). Micronutrient status and neurodevelopment in internationally adopted children. *Acta Paediatrica, International Journal of Paediatrics*, *105*(2), e67–e76.

Georgieff, M. K., Brunette, K. E., & Tran, P. V. (2015). Early life nutrition and neural plasticity. *Development and Psychopathology*, *27*(2), 411–423.

Gilbert, M. E., Hedge, J. M., Valentín-Blasini, L., Blount, B. C., Kannan, K., Tietge, J., et al. (2013). An animal model of marginal iodine deficiency during development: The thyroid axis and neurodevelopmental outcome. *Toxicological Sciences*, *132*(1), 177–195.

Gong, J., Dong, J., Wang, Y., Xu, H., Wei, W., Zhong, J., et al. (2010). Developmental iodine deficiency and hypothyroidism impair neural development, up-regulate caveolin-1 and down-regulate synaptophysin in rat hippocampus. *Journal of Neuroendocrinology*, *22*(2), 129–139.

González, H. F., & Visentin, S. (2016). Micronutrients and neurodevelopment: An update. *Archivos Argentinos de Pediatría*, *114*(6), 570–575.

Grabrucker, S., Haderspeck, J. C., Sauer, A. K., Kittelberger, N., Asoglu, H., Abaei, A., et al. (2018). Brain lateralization in mice is associated with zinc signaling and altered in prenatal zinc deficient mice that display features of autism spectrum disorder. *Frontiers in Molecular Neuroscience*, *10*, 450.

Grzeszczak, K., Kwiatkowski, S., & Kosik-Bogacka, D. (2020). The role of Fe, Zn, and Cu in pregnancy. *Biomolecules*, *10*(8), 1176.

Hare, D. J., Cardoso, B. R., Szymlek-Gay, E. A., & Biggs, B. A. (2018). Neurological effects of iron supplementation in infancy: Finding the balance between health and harm in iron-replete infants. *The Lancet Child and Adolescent Health*, *2*(2), 144.

Hatfield, D. L., Berry, M. J., & Gladyshev, V. N. (2011). *Selenium: Its molecular biology and role in human health* (3rd ed.). Springer Science & Business Media. Chapter 41.

Keen, C. L., Uriu-Adams, J. Y., Skalny, A., Grabeklis, A., Grabeklis, S., Green, K., et al. (2010). The plausibility of maternal nutritional status being a contributing factor to the risk for fetal alcohol spectrum disorders: The potential influence of zinc status as an example. *BioFactors*, *36*(2), 125–135.

Krebs, N. F., Lozoff, B., & Georgieff, M. K. (2017). Neurodevelopment: The impact of nutrition and inflammation during infancy in low-resource settings. *Pediatrics*, *139*(April), S50–S58.

Kumar, N. (2017). Nutrients and neurology. *Continuum (Minneapolis, Minn.)*, *23*(3, Neurology of Systemic Disease), 822–861.

Levenson, C. W., & Morris, D. (2011). Zinc and neurogenesis: Making new neurons from development to adulthood. *Advances in Nutrition*, *2*(2), 96–100.

Levie, D., Korevaar, T. I. M., Bath, S. C., Murcia, M., Dineva, M., Llop, S., et al. (2019). Association of maternal iodine status with child IQ: A meta-analysis of individual participant data. *The Journal of Clinical Endocrinology & Metabolism*, *104*(12), 5957–5967.

Lozoff, B., Castillo, M., Clark, K. M., Smith, J. B., & Sturza, J. (2014). Iron supplementation in infancy contributes to more adaptive behavior at 10 years of age. *Journal of Nutrition*, *144*(6), 838–845.

Martins, A. C., Krum, B. N., Queirós, L., Tinkov, A. A., Skalny, A. V., Bowman, A. B., et al. (2020). Manganese in the diet: Bioaccessibility, adequate intake, and neurotoxicological effects. *Journal of Agricultural and Food Chemistry*, *68*(46), 12893–12903.

McCann, S., Amadó, M. P., & Moore, S. E. (2020). The role of iron in brain development: A systematic review. *Nutrients*, *12*(7), 1–23.

Mertz, W. (1998). Review of the scientific basis for establishing the essentiality of trace elements. *Biological Trace Element Research*, *66*(1), 185–191.

Mezzaroba, L., Alfieri, D. F., Simão, A. N. C., & Reiche, E. M. V. (2019). The role of zinc, copper, manganese and iron in neurodegenerative diseases. *Neurotoxicology*, *74*, 230–241.

Michalke, B., Halbach, S., & Nischwitz, V. (2009). JEM spotlight: Metal speciation related to neurotoxicity in humans. *Journal of Environmental Monitoring*, *11*(5), 939–954.

Močenić, I., Kolić, I., Nišević, J. R., Belančić, A., Tratnik, J. S., Mazej, D., et al. (2019). Prenatal selenium status, neonatal cerebellum measures and child neurodevelopment at the age of 18 months. *Environmental Research*, *176*, 108529.

Modabbernia, A., Arora, M., & Reichenberg, A. (2016). Environmental exposure to metals, neurodevelopment, and psychosis. *Current Opinion in Pediatrics*, *28*(2), 243–249.

Morris, D. R., & Levenson, C. W. (2012). Ion channels and zinc: Mechanisms of neurotoxicity and neurodegeneration. *Journal of Toxicology*, *2012*(5), 785647.

Özden, T. A., Gökçay, G., Cantez, M. S., Durmaz, Ö., İşsever, H., Ömer, B., et al. (2015). Copper, zinc and iron levels in infants and their mothers during the first year of life: A prospective study. *BMC Pediatrics*, *15*(1), 1–11.

Pearce, E. N. (2014). Iodine deficiency in children. *Paediatric Thyroidology*, *26*, 130–138.

Polanska, K., Krol, A., Sobala, W., Gromadzinska, J., Brodzka, R., Calamandrei, G., et al. (2016). Selenium status during pregnancy and child psychomotor development—Polish mother and child cohort study. *Pediatric Research*, *79*(6), 863–869.

Radlowski, E. C., & Johnson, R. W. (2013). Perinatal iron deficiency and neurocognitive development. *Frontiers in Human Neuroscience*, 7(SEP), 1–11.

Rodríguez-Barranco, M., Lacasaña, M., Aguilar-Garduño, C., Alguacil, J., Gil, F., González-Alzaga, B., et al. (2013). Association of arsenic, cadmium and manganese exposure with neurodevelopment and behavioural disorders in children: A systematic review and meta-analysis. *Science of the Total Environment*, *454*, 562–577.

Rouault, T. A., & Cooperman, S. (2006). Brain iron metabolism. *Seminars in Pediatric Neurology*, *13*(3), 142–148.

Skalny, A. V., Aschner, M., & Tinkov, A. A. (2021). Zinc. In N. A. M. Eskin (Ed.), *Vol. 96. The latest research and development of minerals in human nutrition* (pp. 251–310). Academic Press (Chapter 8).

Skalny, A. V., Tinkov, A. A., Voronina, I., Terekhina, O., Skalnaya, M. G., & Kovas, Y. (2018). Hair trace element and electrolyte content in women with natural and in vitro fertilization-induced pregnancy. *Biological Trace Element Research*, *181*(1), 1–9.

Skeaff, S. A. (2011). Iodine deficiency in pregnancy: The effect on neurodevelopment in the child. *Nutrients*, *3*(2), 265–273.

Solovyev, N., Drobyshev, E., Blume, B., & Michalke, B. (2021). Selenium at the neural barriers: A review. *Frontiers in Neuroscience*, *15*, 88–106.

Steinbrenner, H., & Sies, H. (2013). Selenium homeostasis and antioxidant selenoproteins in brain: Implications for disorders in the central nervous system. *Archives of Biochemistry and Biophysics*, *536*(2), 152–157.

Takeda, A. (2001). Zinc homeostasis and functions of zinc in the brain. *Biometals*, *14*(3–4), 343–351.

Varsi, K., Bolann, B., Torsvik, I., Rosvold Eik, T. C., Høl, P. J., & Bjørke-Monsen, A.-L. (2017). Impact of maternal selenium status on infant outcome during the first 6 months of life. *Nutrients*, *9*(5), 486.

Verheesen, R. H., & Schweitzer, C. M. (2008). Iodine deficiency, more than cretinism and goiter. *Medical Hypotheses*, *71*(5), 645–648.

Vinceti, M., Mandrioli, J., Borella, P., Michalke, B., Tsatsakis, A., & Finkelstein, Y. (2014). Selenium neurotoxicity in humans: Bridging laboratory and epidemiologic studies. *Toxicology Letters*, *230*(2), 295–303.

Vukelić, J., Kapamadžija, A., Petrović, Đ., Grujić, Z., Novakov-Mikić, A., Kopitović, V., et al. (2012). Variations of serum copper values in pregnancy. *Srpski Arhiv za Celokupno Lekarstvo*, *140*(1–2), 42–46.

Walsh, V., Brown, J. V. E., & McGuire, W. (2019). Iodine supplementation for the prevention of mortality and adverse neurodevelopmental outcomes in preterm infants. *Cochrane Database of Systematic Reviews*, *2*, CD005253.

Wang, Y., Wu, Y., Li, T., Wang, X., & Zhu, C. (2019). Iron metabolism and brain development in premature infants. *Frontiers in Physiology*, *10*, 463.

Wassie, M. M., Smithers, L. G., & Zhou, S. J. (2021). Association between newborn thyroid-stimulating-hormone concentration and neurodevelopment and growth: A systematic review. *Biological Trace Element Research*, 1–15.

Yan, X., Zhao, X., Li, J., He, L., & Xu, M. (2018). Effects of early-life malnutrition on neurodevelopment and neuropsychiatric disorders and the potential mechanisms. *Progress in Neuro-Psychopharmacology and Biological Psychiatry*, *83*, 64–75.

Yang, X., Yu, X. D., Fu, H. H., Li, L. L., & Ren, T. H. (2013). Different levels of prenatal zinc and selenium had different effects on neonatal neurobehavioral development. *Neurotoxicology*, *37*, 35–39.

Zhang, L., Zhai, X., Liu, Y., Li, J., Shan, Z., & Teng, W. (2016). Treatment with iodine in pregnant rats with marginal iodine deficiency improves cell migration in the developing brain of the progeny. *Molecular Neurobiology*, *53*(4), 2212–2221.

Zimmermann, M. B. (2012). The effects of iodine deficiency in pregnancy and infancy. *Paediatric and Perinatal Epidemiology*, *26*, 108–117.

Further reading

Herlenius, E., & Lagercrantz, H. (2001). Neurotransmitters and neuromodulators during early human development. *Early Human Development*, *65*(1), 21–37.

Rebagliato, M., Murcia, M., Álvarez-Pedrerol, M., Espada, M., Fernández-Somoano, A., Lertxundi, N., et al. (2013). Iodine supplementation during pregnancy and infant neuropsychological development: INMA mother and child cohort study. *American Journal of Epidemiology*, *177*(9), 944–953.

Skröder, H. M., Hamadani, J. D., Tofail, F., Persson, L.Å., Vahter, M. E., & Kippler, M. J. (2015). Selenium status in pregnancy influences children's cognitive function at 1.5 years of age. *Clinical Nutrition*, *34*(5), 923–930.

Verhagen, N. J. E., Gowachirapant, S., Winichagoon, P., Andersson, M., Melse-Boonstra, A., & Zimmermann, M. B. (2020). Iodine supplementation in mildly iodine-deficient pregnant women does not improve maternal thyroid function or child development: A secondary analysis of a randomized controlled trial. *Frontiers in Endocrinology*, *11*, 572984.

CHAPTER 40

Role of folic acid in neuronal development, neurotransmission and hypothalamic regulation of energy intake

Gerald Harvey Anderson[a,b], Emanuela Pannia[c,d], Rola Hammoud[e], Neil Victor Yang[f], and Ruslan Kubant[a]
[a]Department of Nutritional Sciences, Faculty of Medicine, University of Toronto, Toronto, ON, Canada
[b]Department of Physiology, Faculty of Medicine, University of Toronto, Toronto, ON, Canada
[c]Department of Molecular Genetics, Dr. Jim Dowling's Research Group, Faculty of Medicine, University of Toronto, Toronto, ON, Canada
[d]Program in Genetics and Genome Biology, Dr. Jim Dowling's Research Group, The Hospital for Sick Children, Toronto, ON, Canada
[e]Lunenfeld-Tanenbaum Research Institute, Dr. Daniel J. Drucker Lab, Mount Sinai Hospital, Toronto, ON, Canada
[f]Department of Nutritional Sciences and Toxicology, Comparative Biochemistry Graduate Group, Dr. Ronald Krauss's Research Group, University of California, Berkeley, CA, United States

Abbreviations

5-HTR	5-hydroxytryptamine receptor
AgRP	agouti-related protein
ARC	arcuate nucleus of the hypothalamus
BDNF	brain-derived neurotrophic factor
BH4	tetrahydrobiopterin
CART	cocaine-amphetamine-regulated transcript
DFE	dietary folate equivalent
ERK1/2	extracellular signal-regulated kinase ½
FA	folic acid
GABA	gamma aminobutyric acid
Hes5	Hes family bHLH transcription factor 5
MEK1/2	mitogen-activated protein kinase/extracellular signal-regulated kinase ½
MTHF	[6S]-5-methyltetrahydrofolic acid
MTHFR	methylenetetrahydrofolate reductase
NPCs	neural progenitor cells
NPY	neuropeptide-Y
NSC	neural stem cell
NTDs	neural tube defects
POMC	proopiomelanocortin
SAM	S-adenosylmethionine
TH	tyrosine hydroxylase
THF	tetrahydrofolate
TPH	tryptophan hydroxylase

Vitamins and Minerals in Neurological Disorders
https://doi.org/10.1016/B978-0-323-89835-5.00008-9

Introduction

Folate is an essential water-soluble vitamin and the most well-known methyl donor in the B vitamin family that is naturally occurring in food or added in its synthetic form to dietary supplements and fortified foods. Dietary folates are mainly found in the tetrahydrofolate (THF) form and have additional glutamic acid residues, making them polyglutamated and thus less bioavailable (Wusigale & Liang, 2020). Intakes of naturally occurring dietary folates are inadequate to meet the needs of pregnant women to support maternal and fetal health outcomes. Therefore, the more stable monoglutamate synthetic form of folate, namely, folic acid (FA), is incorporated in prenatal supplements and fortified foods (Crider, Bailey, & Berry, 2011). The National Academy of Medicine recommends a minimum of 400–600 μg of dietary folate equivalent (DFE) prior to and during pregnancy (National Institutes of Health, 2021). Total DFEs include intakes from synthetic and natural folate sources. Additional recommendations have been set for FA as a preventative strategy for neural tube defects (NTDs) (Wilson et al., 2015). The tolerable upper intake level (UL) for FA, defined as the maximum level of the nutrient that poses no adverse risk to the general population, was set at 1000 μg/day. This dose was based on the potential of higher doses of FA, but not other folate forms, to aggravate neurological and hematological consequences of vitamin B12-deficiency.

While folate supplementation during pregnancy has been imperative in mitigating negative pregnancy outcomes, its increased popularity and use has raised concerns (Pannia et al., 2016). On average, North American women under adequate care not only meet FA recommendations but exceed them by 2–7-fold the recommended daily allowance (RDA) (Pannia et al., 2020). While in utero folate deficiency is a known determinant of later-life disease risk, the effects of excess folates in utero are not well known. The potential adverse programming effects of high FA intakes and related unmetabolized FA levels have been recently discussed by a folate expert panel (Lamers, MacFarlane, O'Connor, & Fontaine-Bisson, 2018). These discussions have encouraged clinical and preclinical folate research to confirm the potential adverse effects of folate oversupplementation and delineate the potential mechanisms of action.

The uncertainly surrounding high FA intakes have also encouraged intakes of other forms of folate. More recently, the naturally occurring calcium salt of the bioactive folate form, [6S]-5-methyltetrahydrofolic acid (MTHF), has also been widely incorporated into prenatal supplements (Saldanha, Dwyer, Haggans, Mills, & Potischman, 2020). MTHF has increased bioefficacy compared to FA, though its effects when consumed during pregnancy have only begun to be explored.

Overall, folate serves a myriad of critical biological functions. It is a coenzyme in single-carbon transfers via the interrelated pathway known as the 1-carbon cycle (Ducker & Rabinowitz, 2017) (Fig. 1). The 1-carbon cycle is ubiquitously present and requires an adequate pool of methyl nutrients, including folate, vitamins B2, B6, B12, and choline. The 1-carbon cycle supports three key metabolic functions: (1) de

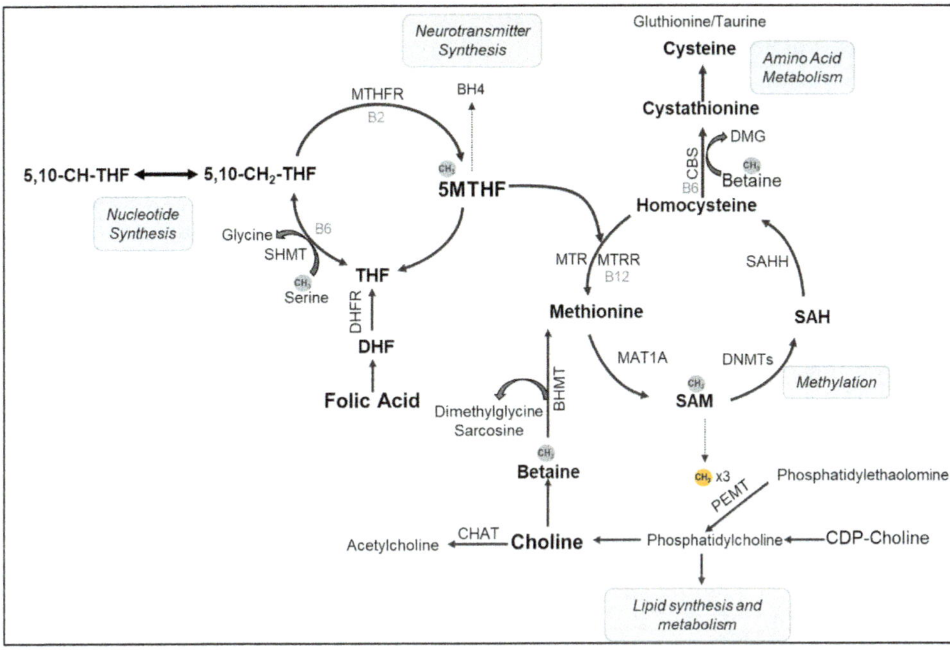

Fig. 1 Role of Methyl-nutrients Folate, Vitamins B2, B6, B12, and Choline in the 1-Carbon Cycle. FA is reduced to DHF and then to THF, via the dihydrofolate reductase enzyme (DHFR). A methyl group is then transferred from serine to THF to form glycine and 5,10-methylenetetrahyrofolate (5,10-MTHF) via the vitamin B6-dependent serine hydroxy methyltransferase (SHMT) enzyme supporting nucleotide synthesis reactions. 5,10-MTHF is then reduced to the bioactive folate form 5-MTHF, via the vitamin B2-dependent enzyme MTHFR. 5-MTHF donates a methyl group to homocysteine, thus forming methionine, a reaction catalyzed by the vitamin B12-dependent methionine synthase and MTRR. 5-MTHF, as the only folate form that can pass the blood–brain-barrier, is involved in the regeneration of BH4 required for neurotransmitter synthesis and neurotransmission. Methionine is adenosylated via the enzyme methionine adenosyltransferase to form s-adenosylmethionine the universal methyl donor regulating phospholipid and neurotransmitters synthesis and epigenetic control of gene expression. Dietary choline provides a complementary source of methyl groups via its oxidation into betaine. SAM is then converted to S-adenosylhomocysteine and then homocysteine, which may enter the trans-sulfuration pathway to form cystathionine, cysteine, and ultimately producing glutathione utilized in redox reactions and amino acid metabolism. Choline may also impact methylation capacity via the phosphatidylethanolamine-*N*-methyltransferase pathway required for several processes including synthesis of several phospholipids required for lipid metabolism and neuronal development. Abbreviations: MTHF, 5-methytetrahydrofolic acid; DHF, dihydrofolate; THF, tetrahydrofolate; SHMT, serine hydroxymethyltransferase; MTHFD, methylenetetrahydrofolate dehydrogenase; MTHFR, methylenetetrahydrofolate reductase; BHMT, betaine-homocysteine methyltransferase; MTR/MTRR, methyltetrahydrofolate-homocysteine methyltransferase/methionine synthase; MAT1a, methionine adenosyltransferase; SAM, S-adenosylmethionine; SAH, S-adenosylhomocysteine; MTase, methyltransferase; DNMT, DNA methyltransferase; phosphatidylethanolamine-*N*-methyltransferase (PEMT), PC, phosphatidylcholine; PE, phosphatidylethanolamine; ChAT, choline acetyltransferase; CBS, cystathionine-β-synthase. BH4, tetrahydrobiopterin; CH3, methyl groups (gray = methyl donors, yellow = methyl transfer).

novo synthesis of nucleotides, (2) re-methylation of homocysteine to methionine, and (3) the production of S-adenosylmethionine (SAM) required for methylation reactions supporting the synthesis of DNA, neurotransmitters, and lipid metabolism. Through these processes, the 1-carbon cycle functions to regulate the expression of genes, amino acid synthesis, redox regulation, and metabolism during phases of organogenesis, rapid cellular growth, and differentiation.

Folate also plays an essential role in neuronal development and function throughout the life phases. Folate deficiency, as well as excess, has been associated with various congenital abnormalities, cognitive dysfunction, autism spectrum disorder, dementia, and Alzheimer's disease (McGarel, Pentieva, Strain, & McNulty, 2015). In addition, folate has been shown to be an in utero modifier of energy regulation and the later-life metabolic phenotype of offspring (Pannia et al., 2016). While the pathomechanisms remain unknown, several preclinical studies have described a role for folate in programming hypothalamic development. The hypothalamus is the central regulator of several important physiological processes, including appetite and energy homeostasis (Schwartz, Woods, Porte Jr., Seeley, & Baskin, 2000).

This review examines the role of folate in neuronal development and programming of physiological functions of the hypothalamus, with a focus on food intake and body weight regulation. As evidence derived from human studies is limited, this chapter will provide a focused overview of support derived from preclinical in vitro and/or in vivo studies.

Effect of deficiency and excess of folate on neuronal development

Folate metabolism plays a key role in regulating early neuronal developmental processes: from neural stem cell (NSC) proliferation and apoptosis, to cerebellar, hippocampal, and hypothalamic development. The generation of neurons occurs over a wide range of processes encompassing neurogenesis, synaptogenesis, and early myelination of the fetal brain (Lee & Ahima, 2012). Proper fetal brain development is dependent on NSCs, which are highly proliferative cells that regenerate into new NSCs or differentiate into neural progenitor cells (NPCs). These cells later become mature neurons, glial cells, or astrocytes. Folate is needed in NSC proliferation and differentiation during the development of the brain. Both high and low folate levels in utero can also affect intracellular signaling pathways involved in the development of the central nervous system from gestation to later life.

Several human observational studies have shown a role for folate in offspring neuronal development, by which adequate maternal folate intake is associated with improved cognitive function and behavioral outcomes in the child (McGarel et al., 2015), but randomized clinical studies are limited. Nevertheless, this evidence is supported by preclinical studies. The following sections provide an overview of folate deficiency and/or oversupplementation on neuronal development derived from in vitro and in vivo studies.

Several in vitro studies have linked FA to NSC proliferation and differentiation. In cultures of primary embryonic cells from rat fetuses, maternal FA deficiency led to reduced cell proliferation impairing mitosis, and increased apoptosis (Zhang, Huang, Liu, Chang, & Wilson, 2012). FA supplementation rescued these effects (Liu, Huang, Zhang, Ren, & Wilson, 2010). Treatment of primary mouse NSC/NPS cells in medium containing low folate (folic acid 0.66 mg/l) resulted in decreased SAM and DNA hypomethylation of genes involved in neuronal differentiation and maturation indicating the importance of folate in neuronal development (Araki et al., 2021). High levels of folate can also affect neurogenesis and excitability in neuronal cells. NSCs from rat cortices exposed to high levels of FA (40 mg/L) for 6 days stimulated cell growth and neurogenesis in the brain (Liu et al., 2010). This may be due to increased activation of *Notch-1* and *Hes* family bHLH transcription factor 5 (Hes5) signaling pathways (Liu et al., 2010). Specifically, rat neuronal cells isolated from the hippocampus supplemented with folate metabolites (10 mM THF and MTHF) resulted in increased synaptic density but disrupted neuronal signaling (Girotto et al., 2013).

In vivo studies also support a role of FA in regulating neuronal development. FA deficiency in utero resulted in a decrease in NSC proliferation and an increase in apoptosis in fetal brain during embryonic days E11–E17 (Craciunescu et al., 2004). Rhesus monkey fed diets depleted in FA had fetuses with fewer embryonic stem cells, and decreased cell proliferation and neuronal differentiation (Chen et al., 2012). Maternal FA deficiency has also been shown to impair NSC differentiation and increase apoptosis of neurons in the fetal brain via its effects on *ERK1/2* transcription factors involved in the *MEK1/2* signaling pathway (Zhang, Huang, Tian, Ren, & Wilson, 2009). *ERK1/2* genes are expressed in all neurons within the brain and are associated with cell growth, proliferation, differentiation, and migration, as well as synaptic remodeling, axonal growth, and neuronal excitability (Medina & Viola, 2018). These deleterious effects of FA deficiency may be due to lowered levels of purines, thymidine, and methionine; raised levels of the neurotoxin homocysteine; or changes in intracellular signaling pathways (Kado et al., 2005).

Similarly, high FA exposure affects neuronal pathways regulating motor and cognitive functions. Male Sprague Dawley rats born from dams fed high gestational FA had altered neural development of the cerebellum, which is required for sensory input and motor functions. Similarly, maternal feeding of moderate- to high-dose FA (i.e., 8 or 40 mg/kg) resulted in >40% of pups with cerebellar morphological defects (Partearroyo et al., 2016). Folates have also been reported to affect the hippocampus required for learning and memory. Long-Evans rats fed 4 mg/mL FA in 10% sucrose solution during gestation developed behavioral seizures by postnatal day P14, compared to control solution (Girotto et al., 2013). Girotto et al., also discovered that hippocampal cells had decreased brain activity, resulting in less stable and hyperexcitable neuronal networks. Another study that fed 20 mg/kg (10-fold recommendations) FA for 6 weeks to mice throughout gestation resulted in 3-wk-old male offspring with impaired visual memory and reduced hippocampal density (Bahous et al., 2017). However, moderate supplementation of FA

(8 mg/kg, 4-fold recommendations) during pregnancy and lactation gave rise to male pups that did not exhibit gross morphological changes in hippocampal development up to 3 months postnatal. Notably, DNA hypermethylation was found within the cortex of the FA-exposed groups at this time. This suggests a role for FA in epigenetic modifications, which may have later-life health implications that were not explored.

In addition to diet, genetic polymorphisms that disrupt the expression of key folate metabolism genes can influence folate status. Methylenetetrahydrofolate reductase (MTHFR) is a critical enzyme in the 1-carbon cycle and is responsible for catalyzing the conversion of FA into 5-methyltetrahydrofolate, the bioactive folate form that is used for cellular functions (Greenberg, Bell, Guan, & Yu, 2011). Mice with *Mthfr* deficiency have low folate and elevated levels of SAH and homocysteine. A high (10-fold)-FA diet fed to *Mthfr*-deficient mice throughout gestation resulted in 3-wk-old offspring with memory impairments, reduced hippocampal volume, and reduced choline concentrations in the brain (Bahous et al., 2017). These results are consistent with another study that identified changes in choline metabolism in the hippocampus and cortex of *Mthfr* +/− mice (Jadavji et al., 2012). Taken together, these studies suggest possible perturbations in 1-carbon metabolism and a greater reliance on the choline pathway for brain development in response to folate inadequacy.

Role of folate in neurotransmission

The process of neurotransmission is critical for neuronal development as it facilitates cellular organization, stability, and execution of all metabolic and physiological functions. These processes require coordination between pre- and post-synaptic neurons to induce membrane depolarization and calcium influx. As neurotransmission is plastic and responds to fluctuations in the 1-carbon cycle (Miller, 2008), variations in folate intake during neurogenic periods may have lasting effects on the development and function of neurotransmission pathways.

Several preclinical studies support a role for folate and neurotransmission. The in vitro study by Wiens et al., using dorsal root ganglia from the chick embryo as a model of neurogenesis, has shown that excess FA can reduce the activity of neurons in a dose-dependent manner by reducing neurite extension, growth cone activity, and synapse formation (Wiens et al., 2016). In addition, a study conducted using a medium of neuronal cells found that FA supplementation led to upregulation of genes and proteins related to GABA receptor subunits (Vasquez, Kuizon, Junaid, & Idrissi, 2013). Furthermore, in vivo studies have shown an effect of maternal folate intake in the programming of metabolism via altered neurotransmission. FA at 2.5 × provided during mating and pregnancy to Sprague–Dawley rats influenced metabolites related to glycine, tryptophan, serine, and GABA in the brain of male offspring along with several proteins involved in the regulation of synaptic plasticity (Liu, Liu, Li, & Sun, 2021). Another study that conducted whole transcriptome analyses on the arcuate nucleus of the hypothalamus (ARC) from dams fed high-folate diets during pregnancy revealed differential expression of > 200 genes related to neurotransmission, including synaptic plasticity, calcium signaling, GABAergic, glutaminergic and retrograde signaling, and neuronal systems (Pannia, Hammoud, et al., 2020).

Human studies also support a potential role for folate in determining neurotransmitter synthesis and synaptic function, of which may predict behavior and metabolic outcomes. For example, patients with clinical depression are associated with low plasma folate, high plasma homocysteine, and significantly lower cerebrospinal fluid levels of SAM and monoamine neurotransmitters (Miller, 2008). The use of high (7.5 ×) MTHF, in replacement of FA as an adjunctive therapy for the treatment for major depressive disorder, highlights the importance of adequate 5-MTHF availability in maintaining brain neurochemical concentrations and signaling capacity (Jain, Manning, & Cutler, 2020).

The mechanisms by which folate may modulate neuronal function remain uncertain, but several direct and indirect mechanisms have been proposed. Disruption in folate metabolism may affect the availability of 1-carbon methyl donors (e.g., SAM and betaine), homocysteine catabolism, and neurotransmitter synthesis, including the main excitatory (i.e., glutamate) and inhibitory (i.e., GABA) neurotransmitters. A simplified overview of these mechanisms is provided in Fig. 2. As 5-MTHF is the only folate form that has been shown to pass the blood–brain barrier (Stover, Durga, & Field, 2017), factors that influence the availability of 5-MTHF can affect the regeneration of the essential cofactor, tetrahydrobiopterin (BH4). BH4 regulates the rate-limiting enzymes tyrosine hydroxylase (TH) and tryptophan hydroxylase (TPH) required for the synthesis of the monoaminergic neurotransmitters dopamine, norepinephrine, and serotonin (Stahl, 2008). Notably,

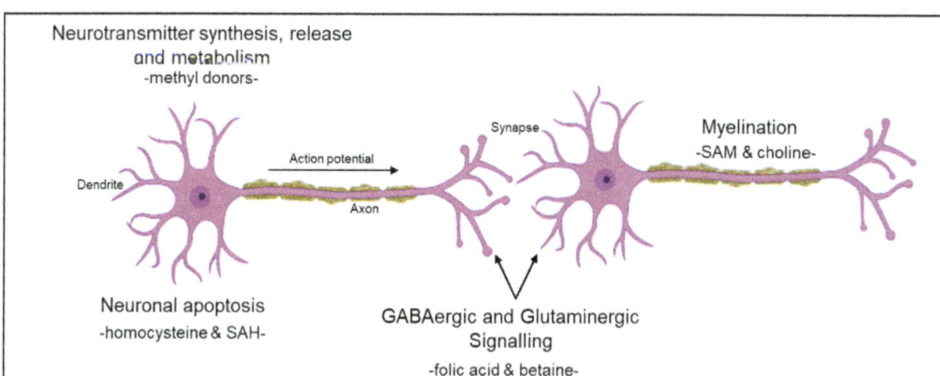

Fig. 2 1-Carbon Metabolism Regulates Neurotransmission. A simplified diagram that identifies the key roles of folate and other components of one-carbon cycle in neurotransmitter synthesis, signaling, neuronal apoptosis, and myelination.

overproduction of neurotransmitters can also have inhibitory actions and contribute to hypofunctioning of neurotransmitter systems (Seo, Patrick, & Kennealy, 2008), where more of the rewarding substance (e.g., food) is required to achieve gratification. Excessive or deficient intakes of folate can also affect the availability of SAM and thus the synthesis and metabolism of monoamines (Mischoulon & Fava, 2002). SAM is also critical for axon myelin sheath production to regulate neuronal impulse conduction (Varela-Rey et al., 2014). Folate inadequacies may also limit the catabolism of homocysteine and consequently induce neuronal apoptosis (Mattson & Shea, 2003). High FA has also been shown to induce dependency on alternative methyl donor pathways, which can dysregulate neuronal activity and signaling. For example, the methyl donor betaine competes with GABA on the same receptor for uptake into the neuron and thus may modulate GABAerigic signaling (Knight, Piibe, Lambie, Perkins, & Yancey, 2017). Moreover, FA has been shown to produce an excitatory effect and modify synaptic plasticity by enhancing presynaptic disinhibition of GABAergic neurons (Otis, Madison, & Nicoll, 1985). Lastly, because of the structural similarity of FA with glutamate, FA can competitively bind to glutamate receptors in the synapse and alter the calcium ion concentration gradient produced by glutamate on its ionotropic *N*-methyl-D-aspartate receptor (Wiens et al., 2016).

Hypothalamic development and food intake regulation

While many studies have explored the effects of inadequate maternal FA intakes on the development of various brain regions, including the hippocampus, cerebellum, and cortex (McGarel et al., 2015), the hypothalamus has received less attention. The hypothalamus is the central regulator of several important physiological processes, including appetite and energy homeostasis, stress responsiveness, and reproductive function (Schwartz et al., 2000).

The hypothalamus consists of many neurocircuits that decipher nutritional and hormonal signals from the environment, translating them into physiological or behavioral responses that favor energy balance. As shown in Fig. 3, central homeostatic food intake regulation begins with the first-order neurons located in the ARC of the hypothalamus. Neurons within the ARC are termed first-phase energy sensors, which translate orexigenic and anorectic signals to second-order neurons within the hypothalamus. Their projections also extend to extra-hypothalamic nuclei controlling hedonic (reward driven) aspects of feeding behavior.

Within the ARC, the most well-defined first-order neurons include the (1) orexigenic neurons co-expressing the inhibitory neurotransmitter GABA and the neuropeptides, neuropeptide-Y (NPY) and agouti-related protein (AgRP); and, (2) anorexigenic neurons co-expressing neuropeptides pro-opiomelanocortin (POMC) and its derivative cocaine-amphetamine-regulated transcript (CART) (Schwartz et al., 2000).

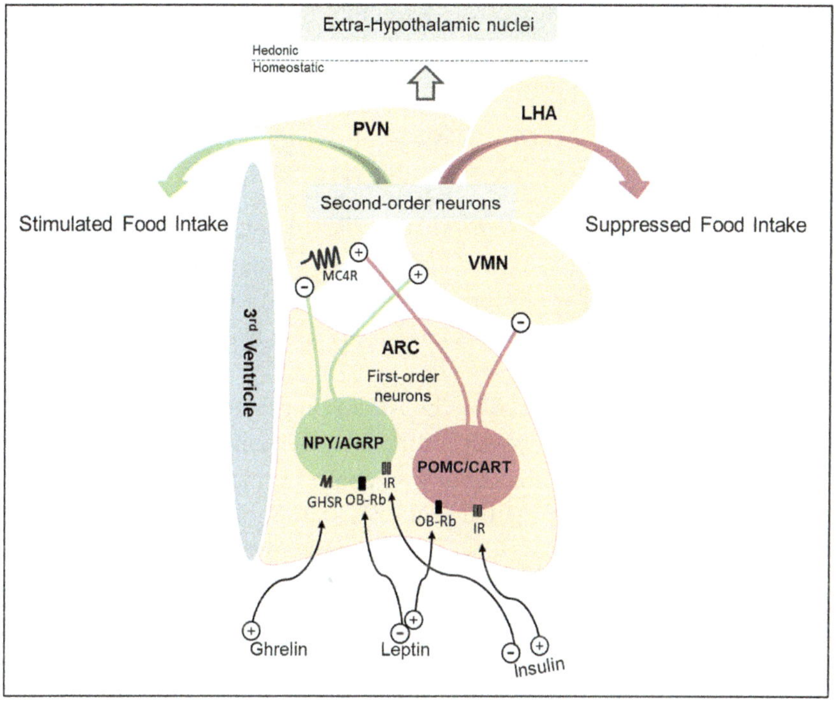

Fig. 3 A simplified diagram of hypothalamic food intake regulator pathway. Abbreviations: AGRP: Agouti-related protein, NPY: Neuropeptide Y, POMC: Pro-opiomelanocortin, CART: Cocaine-and amphetamine-regulated transcript, OB-Rb: The long form of the leptin receptor, IR: Insulin receptor, GHSR: Growth hormone secretagogue receptor, ARC: Arcuate nucleus, VMN: Ventromedial nucleus, LHA: Lateral hypothalamic area, PVN: Paraventricular nucleus, MC4R: melanocortin-4-receptor.

Approximately one-third of POMC/CART neurons also release the excitatory neurotransmitter glutamate and contain cholinergic fibers that release acetylcholine (Meister et al., 2006).

Development of the hypothalamus initiates in the proliferative zone of the third ventricle, containing the neural progenitor pool for all neurons involved in food intake regulatory pathways (Markakis, 2002). Notably, the stages of hypothalamic neuronal development differ between species. In humans and other primates, hypothalamic development predominantly occurs prior to birth (Xie & Dorsky, 2017). However, in rodents, it begins during the second week of gestation and continues up to 2 weeks postbirth.

Hypothalamic neurons differentiate and mature following a fixed schedule based on their cell lineage. In rodents, neuronal progenitor cells within the ARC express POMC at embryonic day 12 and give rise to NPY/AgRP expression neurons by E13.5-E14.5 (Bouret, 2010). Peak expression of both neuronal populations is observed by day 15 postbirth. Neuronal projections from the ARC to neighboring hypothalamic nuclei (e.g., dorsomedial nucleus, lateral hypothalamic area, and paraventricular nucleus) occur in early postnatal life when pups begin to consume solid foods. Therefore, the pre- and early postnatal period is a critical window of susceptibility to environmental factors, such as maternal nutrition, determining the offspring's health and disease risk.

Programming effects of maternal methyl donor vitamins on hypothalamic regulation of food intake and metabolism

The long-term regulatory effects of high-vitamin diets consumed during pregnancy on hypothalamic food intake regulation, metabolism, and the development of obesity have been described in several preclinical studies. They have led to the hypothesis that characteristics of the metabolic syndrome in rat offspring arise from alteration in the hypothalamic regulation of food intake.

The first series of studies pioneered by the Anderson group, using an outbred rodent strain which is relevant to human obesity, showed adverse metabolic programming of high methyl vitamin intakes in utero. Feeding pregnant Wistar rats 10-fold above recommended intakes of multivitamins, methyl group vitamins (i.e., FA, B12, and B6), or FA alone produced male offspring with characteristics of the metabolic syndrome, including increased body weight, visceral adiposity, insulin resistance, and hyperphagia (Cho et al., 2013b; Reza-Lopez, Anderson, Szeto, Taha, & Ma, 2009; Szeto et al., 2008). The adverse effects of gestational FA supplementation on metabolic dysregulation of the offspring have been reproduced by others (Cho et al., 2015; Cho et al., 2013a, 2013b; Hammoud et al., 2021; Huang et al., 2014; Kelly et al., 2016; Mjaaseth et al., 2021; Morakinyo, Samuel, Awobajo, Oludare, & Mofolorunso, 2019; Reza Lopez, Poon, Szeto, Ma, & Anderson, 2013; Xie et al., 2018; Yang et al., 2018). These in utero programming effects are accelerated when males are exposed to an obesogenic post-weaning diet (Szeto, Das, Aziz, & Anderson, 2009), highlighting an important interaction between the gestational and later-life environment. The high multivitamin maternal diets programmed lower hypothalamic gene expression of several anorexigenic neuropeptides, including brain-derived neurotrophic factor *(Bdnf)*, serotonin receptors (*5-Htr1a/2a/2c*), and *Pomc* at weaning and later during adulthood, thus favoring overeating (Cho et al., 2013a, 2013b). Male offspring born to dams fed high methyl group vitamins had later-life disruptions in hypothalamic leptin and insulin receptor signaling, as well as changes in the expression of hypothalamic *Pomc* and insulin receptor *(Ir)* genes acquired through DNA methylation (Cho et al., 2013a, 2015). FA alone was identified as the main methyl vitamin contributing to many of the observed effects. Both low (no added, 0-fold) and high (2.5–5-fold) intakes of FA during pregnancy have also been shown to result in greater body weight gain and food intake that are associated with a greater number of mature NPY-positive neurons in the ARC at birth and lower POMC neuronal activation at 9 weeks post-weaning (Yang et al., 2020).

Notably, the in utero programming effects of inadequate FA on male offspring extend to the female offspring and the mother rat post-birth (Huot et al., 2013, 2016; Pannia et al., 2015, 2020, 2021; Pannia, Hammoud, et al., 2020). However, females may be more resistant to increased FA fluctuations in the maternal diet. Female pups born from dams fed moderately high FA diets (2.5–5-fold) had either no change or delayed weight gain when fed a recommended diet post-weaning, but consistently had altered expression of hypothalamic energy regulatory genes (Huot et al., 2013, 2016; Pannia et al., 2021; Sie et al., 2013)]. However, when exposed to very high levels of gestational FA (10–20-fold), or a highly palatable obesogenic diet post-weaning, female offspring also exhibit similar body weight gain, food intake, and metabolic dysregulation as their male siblings (Keating et al., 2015; Reza-Lopez et al., 2009; Szeto et al., 2009; Tojal et al., 2019). Similarly, high (10 ×) multivitamins and specifically methyl group vitamins (i.e., FA, B12, and B6) during pregnancy programmed greater maternal postpartum weight gain, which is associated with dysregulation of central hypothalamic, hedonic, and peripheral energy regulatory systems (Pannia et al., 2015). Subsequently, four other rodent studies have supported both below and above the recommended content of FA in the gestational diet to program later-life maternal metabolic dysregulation (Henderson et al., 2018; Keating et al., 2015; Pannia, Hammoud, et al., 2020; Pannia, Yang, et al., 2020). A novel role of the form of folate, in addition to the dose, has also been shown to program maternal and female offspring hypothalamic gene expression and later-life metabolism (Pannia et al., 2021; Pannia, Hammoud, et al., 2020).

The effects of inadequate folate exposure in utero have recently been attributed to the imbalance of folate with other methyl nutrients, namely, choline. A gestational diet with imbalanced FA and choline contents contributed to an obesogenic phenotype of male and female offspring post-weaning (Hammoud et al., 2020, 2021; Mjaaseth et al., 2021). Consistently, choline supplementation in a background of high FA during pregnancy mitigated high FA-induced perturbations in 1-carbon metabolism and obesity in the offspring. These effects were also associated with differential gene expression related to brain development, axon guidance, and oxidative stress in the hypothalamus of newborn pups as observed by bulk gene expression analyses (Hammoud et al., 2021).

Together, these observations show that FA in the prenatal diet is essential in the development of neuronal pathways in the brain that regulate food intake and metabolism in rodents. However, its effects are dependent on the FA dose, form, duration of exposure, and nutrient imbalances in the maternal and/or post-weaning diet.

Conclusion

Folate nutriture in utero is essential for neuronal developmental processes: from NSC proliferation, differentiation, and apoptosis to the development of the cerebellar cortex, hippocampus, and hypothalamus. As provided by the example in this chapter on the role of folate in the development of neuronal pathways in the hypothalamus that regulate food intake and metabolism in rodents, it has long-lasting effects on the offspring. Although evidence is mainly derived from preclinical studies, continued investigation to determine translatability to humans is justified. Addressing these knowledge gaps is a timely first step to help define an optimal folate dose and form of folate to support proper neuronal development and long-term metabolic outcomes of mothers and their children.

Applications to other neurological conditions

Folate deficiency, as well as excess, has been associated with various congenital abnormalities, cognitive dysfunction, autism spectrum disorder, dementia, and Alzheimer's disease. This review adds plausibility to these associations by showing the function of folate nutriture in utero on neuronal development of brain systems regulating physiological functions and behaviors.

Other components of interest

The role of folate in in utero development is dependent on other methyl group vitamins in the one-carbon cycle and its balance with choline sets the methylation potential for epigenetic modifications of gene expression.

Mini-dictionary of terms

Folic Acid: A monoglutamate form of folate, is a stable form of folate and used as a supplement in pills and in fortified foods.
Folate-mediated 1-carbon metabolism: A cellular process of transfer of 1C units required for the synthesis of nucleotides, re-methylation of homocysteine, and production of S-adenosylmethionine for DNA methylation reactions.
Neuronal development: A process of formation and reshaping of the neural pathways from the earliest stages of fetal development to adulthood.

Key facts of folate

- Folate is a water-soluble B-group methyl donor vitamin.
- Dietary folates are mainly found in the tetrahydrofolate form and have additional glutamic acid residues, making them polyglutamated.
- Folate deficiency, as well as excess, has been associated with various congenital abnormalities, metabolic and neurological disorders.
- The naturally occurring calcium salt of the bioactive folate form, [6S]-5-methyltetrahydrofolic acid, has now been widely incorporated into prenatal supplements.

Summary points

- In utero exposure of the fetus to folate below requirements leads to congenital disorders.
- Both folate deficiency and excess lead to behavioral and metabolic disorders.
- Both folate deficiency and excess affect neuronal development of regulatory systems in the brain.
- Development of the regulation of homeostatic systems in the hypothalamus is sensitive to dietary folate in gestational diets.
- An optimal folate dose and form of folate to support proper neuronal development and long-term metabolic outcomes of mothers and their children remain to be defined.

References

Araki, R., Nishida, S., Nakajima, Y., Iwakumo, A., Tachioka, H., Kita, A., et al. (2021). Low folate induces abnormal neuronal maturation and DNA hypomethylation of neuronal differentiation-related genes in cultured mouse neural stem and progenitor cells. *Heliyon*, 7(9), e08071. https://doi.org/10.1016/j.heliyon.2021.e08071.

Bahous, R. H., Jadavji, N. M., Deng, L., Cosin-Tomas, M., Lu, J., Malysheva, O., et al. (2017). High dietary folate in pregnant mice leads to pseudo-MTHFR deficiency and altered methyl metabolism, with embryonic growth delay and short-term memory impairment in offspring. *Human Molecular Genetics*, *26*(5), 888–900. https://doi.org/10.1093/hmg/ddx004.

Bouret, S. G. (2010). Development of hypothalamic neural networks controlling appetite. *Forum of Nutrition*, *63*, 84–93. https://doi.org/10.1159/000264396.

Chen, Y., Wang, Z., Xie, Y., Guo, X., Tang, X., Wang, S., et al. (2012). Folic acid deficiency inhibits neural rosette formation and neuronal differentiation from rhesus monkey embryonic stem cells. *Journal of Neuroscience Research*, *90*(7), 1382–1391. https://doi.org/10.1002/jnr.23030.

Cho, C. E., Pannia, E., Huot, P. S., Sanchez-Hernandez, D., Kubant, R., Dodington, D. W., et al. (2015). Methyl vitamins contribute to obesogenic effects of a high multivitamin gestational diet and epigenetic alterations in hypothalamic feeding pathways in Wistar rat offspring. *Molecular Nutrition & Food Research*, *59*(3), 476–489. https://doi.org/10.1002/mnfr.201400663.

Cho, C. E., Sanchez-Hernandez, D., Reza-Lopez, S. A., Huot, P. S., Kim, Y. I., & Anderson, G. H. (2013a). High folate gestational and post-weaning diets alter hypothalamic feeding pathways by DNA methylation in Wistar rat offspring. *Epigenetics*, *8*(7), 710–719. https://doi.org/10.4161/epi.24948.

Cho, C. E., Sanchez-Hernandez, D., Reza-Lopez, S. A., Huot, P. S., Kim, Y. I., & Anderson, G. H. (2013b). Obesogenic phenotype of offspring of dams fed a high multivitamin diet is prevented by a post-weaning high multivitamin or high folate diet. *International Journal of Obesity*, *37*(9), 1177–1182. https://doi.org/10.1038/ijo.2012.210.

Craciunescu, C. N., Brown, E. C., Mar, M. H., Albright, C. D., Nadeau, M. R., & Zeisel, S. H. (2004). Folic acid deficiency during late gestation decreases progenitor cell proliferation and increases apoptosis in fetal mouse brain. *The Journal of Nutrition*, *134*(1), 162–166. https://doi.org/10.1093/jn/134.1.162.

Crider, K. S., Bailey, L. B., & Berry, R. J. (2011). Folic acid food fortification-its history, effect, concerns, and future directions. *Nutrients*, *3*(3), 370–384. https://doi.org/10.3390/nu3030370.

Ducker, G. S., & Rabinowitz, J. D. (2017). One-carbon metabolism in health and disease. *Cell Metabolism*, *25*(1), 27–42. https://doi.org/10.1016/j.cmet.2016.08.009.

Girotto, F., Scott, L., Avchalumov, Y., Harris, J., Iannattone, S., Drummond-Main, C., et al. (2013). High dose folic acid supplementation of rats alters synaptic transmission and seizure susceptibility in offspring. *Scientific Reports*, *3*, 1465. https://doi.org/10.1038/srep01465.

Greenberg, J. A., Bell, S. J., Guan, Y., & Yu, Y. H. (2011). Folic acid supplementation and pregnancy: More than just neural tube defect prevention. *Reviews in Obstetrics and Gynecology*, *4*(2), 52–59. Retrieved from http://www.ncbi.nlm.nih.gov/pubmed/22102928.

Hammoud, R., Pannia, E., Kubant, R., Liao, C. S., Ho, M., Yang, N. V., et al. (2020). Maternal choline intake programs hypothalamic energy regulation and later-life phenotype of male Wistar rat offspring. *Molecular Nutrition & Food Research*, *64*(9), e1901178. https://doi.org/10.1002/mnfr.201901178.

Hammoud, R., Pannia, E., Kubant, R., Wasek, B., Bottiglieri, T., Malysheva, O. V., et al. (2021). Choline and folic acid in diets consumed during pregnancy interact to program food intake and metabolic regulation of male Wistar rat offspring. *The Journal of Nutrition*, *151*(4), 857–865. https://doi.org/10.1093/jn/nxaa419.

Henderson, A. M., Tai, D. C., Aleliunas, R. E., Aljaadi, A. M., Glier, M. B., Xu, E. E., et al. (2018). Maternal folic acid supplementation with vitamin B12 deficiency during pregnancy and lactation affects the metabolic health of adult female offspring but is dependent on offspring diet. *The FASEB Journal*, *32*(9), 5039–5050. https://doi.org/10.1096/fj.201701503RR.

Huang, Y., He, Y., Sun, X., He, Y., Li, Y., & Sun, C. (2014). Maternal high folic acid supplement promotes glucose intolerance and insulin resistance in male mouse offspring fed a high-fat diet. *International Journal of Molecular Sciences*, *15*(4), 6298–6313. https://doi.org/10.3390/ijms15046298.

Huot, P. S., Dodington, D. W., Mollard, R. C., Reza-Lopez, S. A., Sanchez-Hernandez, D., Cho, C. E., et al. (2013). High folic acid intake during pregnancy lowers body weight and reduces femoral area and strength in female rat offspring. *Journal of Osteoporosis*, *2013*, 154109. https://doi.org/10.1155/2013/154109.

Huot, P. S., Ly, A., Szeto, I. M., Reza-Lopez, S. A., Cho, D., Kim, Y. I., et al. (2016). Maternal and post-weaning folic acid supplementation interact to influence body weight, insulin resistance, and food intake regulatory gene expression in rat offspring in a sex-specific manner. *Applied Physiology, Nutrition, and Metabolism*, *41*(4), 411–420. https://doi.org/10.1139/apnm-2015-0503.

Jadavji, N. M., Deng, L., Leclerc, D., Malysheva, O., Bedell, B. J., Caudill, M. A., et al. (2012). Severe methylenetetrahydrofolate reductase deficiency in mice results in behavioral anomalies with morphological and biochemical changes in hippocampus. *Molecular Genetics and Metabolism*, *106*(2), 149–159. https://doi.org/10.1016/j.ymgme.2012.03.020.

Jain, R., Manning, S., & Cutler, A. J. (2020). Good, better, best: Clinical scenarios for the use of L-methylfolate in patients with MDD. *CNS Spectrums*, *25*(6), 750–764. https://doi.org/10.1017/S1092852919001469.

Kado, D. M., Karlamangla, A. S., Huang, M. H., Troen, A., Rowe, J. W., Selhub, J., et al. (2005). Homocysteine versus the vitamins folate, B6, and B12 as predictors of cognitive function and decline in older high-functioning adults: MacArthur studies of successful aging. *The American Journal of Medicine*, *118*(2), 161–167. https://doi.org/10.1016/j.amjmed.2004.08.019.

Keating, E., Correia-Branco, A., Araujo, J. R., Meireles, M., Fernandes, R., Guardao, L., et al. (2015). Excess perigestational folic acid exposure induces metabolic dysfunction in post-natal life. *The Journal of Endocrinology*, *224*(3), 245–259. https://doi.org/10.1530/JOE-14-0448.

Kelly, K. B., Kennelly, J. P., Ordonez, M., Nelson, R., Leonard, K., Stabler, S., et al. (2016). Excess folic acid increases lipid storage, weight gain, and adipose tissue inflammation in high fat diet-fed rats. *Nutrients*, *8*(10). https://doi.org/10.3390/nu8100594.

Knight, L. S., Piibe, Q., Lambie, I., Perkins, C., & Yancey, P. H. (2017). Betaine in the brain: Characterization of betaine uptake, its influence on other osmolytes and its potential role in neuroprotection from osmotic stress. *Neurochemical Research*, *42*(12), 3490–3503. https://doi.org/10.1007/s11064-017-2397-3.

Lamers, Y., MacFarlane, A. J., O'Connor, D. L., & Fontaine-Bisson, B. (2018). Periconceptional intake of folic acid among low-risk women in Canada: Summary of a workshop aiming to align prenatal folic acid supplement composition with current expert guidelines. *The American Journal of Clinical Nutrition*, *108*(6), 1357–1368. https://doi.org/10.1093/ajcn/nqy212.

Lee, E. B., & Ahima, R. S. (2012). Alteration of hypothalamic cellular dynamics in obesity. *The Journal of Clinical Investigation*, *122*(1), 22–25. https://doi.org/10.1172/JCI61562.

Liu, H., Huang, G. W., Zhang, X. M., Ren, D. L., & Wilson, J. X. (2010). Folic acid supplementation stimulates notch signaling and cell proliferation in embryonic neural stem cells. *Journal of Clinical Biochemistry and Nutrition*, *47*(2), 174–180. https://doi.org/10.3164/jcbn.10-47.

Liu, L., Liu, Z., Li, Y., & Sun, C. (2021). Integration of metabolomics and proteomics to highlight altered neural development related pathways in the adult offspring after maternal folic acid supplement. *Clinical Nutrition*, *40*(2), 476–487. https://doi.org/10.1016/j.clnu.2020.05.042.

Markakis, E. A. (2002). Development of the neuroendocrine hypothalamus. *Frontiers in Neuroendocrinology*, *23*(3), 257–291. https://doi.org/10.1016/s0091-3022(02)00003-1.

Mattson, M. P., & Shea, T. B. (2003). Folate and homocysteine metabolism in neural plasticity and neurodegenerative disorders. *Trends in Neurosciences*, *26*(3), 137–146. https://doi.org/10.1016/S0166-2236(03)00032-8.

McGarel, C., Pentieva, K., Strain, J. J., & McNulty, H. (2015). Emerging roles for folate and related B-vitamins in brain health across the lifecycle. *The Proceedings of the Nutrition Society*, *74*(1), 46–55. https://doi.org/10.1017/S0029665114001554.

Medina, J. H., & Viola, H. (2018). ERK1/2: A key cellular component for the formation, retrieval, reconsolidation and persistence of memory. *Frontiers in Molecular Neuroscience*, *11*, 361. https://doi.org/10.3389/fnmol.2018.00361.

Meister, B., Gomuc, B., Suarez, E., Ishii, Y., Durr, K., & Gillberg, L. (2006). Hypothalamic proopiomelanocortin (POMC) neurons have a cholinergic phenotype. *The European Journal of Neuroscience*, *24*(10), 2731–2740. https://doi.org/10.1111/j.1460-9568.2006.05157.x.

Miller, A. L. (2008). The methylation, neurotransmitter, and antioxidant connections between folate and depression. *Alternative Medicine Review*, *13*(3), 216–226. Retrieved from https://www.ncbi.nlm.nih.gov/pubmed/18950248.

Mischoulon, D., & Fava, M. (2002). Role of S-adenosyl-L-methionine in the treatment of depression: A review of the evidence. *The American Journal of Clinical Nutrition*, *76*(5), 1158S–1161S. https://doi.org/10.1093/ajcn/76/5.1158S.

Mjaaseth, U. N., Norris, J. C., Aardema, N. D. J., Bunnell, M. L., Ward, R. E., Hintze, K. J., et al. (2021). Excess vitamins or imbalance of folic acid and choline in the gestational diet alter the gut microbiota and obesogenic effects in Wistar rat offspring. *Nutrients*, *13*(12). https://doi.org/10.3390/nu13124510.

Morakinyo, A. O., Samuel, T. A., Awobajo, F. O., Oludare, G. O., & Mofolorunso, A. (2019). High-dose perinatal folic-acid supplementation alters insulin sensitivity in Sprague Dawley rats and diminishes the expression of adiponectin. *Journal of Dietary Supplements*, *16*(1), 14–26. https://doi.org/10.1080/19390211.2018.1426076.

National Institutes of Health. (2021). *Folate: fact sheet for health professionals*. Retrieved from https://ods.od.nih.gov/factsheets/Folate-HealthProfessional/#en2.

Otis, L. C., Madison, D. V., & Nicoll, R. A. (1985). Folic acid has a disinhibitory action in the rat hippocampal slice preparation. *Brain Research*, *346*(2), 281–286. https://doi.org/10.1016/0006-8993(85)90861-3.

Pannia, E., Cho, C. E., Kubant, R., Sanchez-Hernandez, D., Huot, P. S., Chatterjee, D., et al. (2015). A high multivitamin diet fed to Wistar rat dams during pregnancy increases maternal weight gain later in life and alters homeostatic, hedonic and peripheral regulatory systems of energy balance. *Behavioural Brain Research*, *278*, 1–11. https://doi.org/10.1016/j.bbr.2014.09.019.

Pannia, E., Cho, C. E., Kubant, R., Sanchez-Hernandez, D., Huot, P. S., & Harvey Anderson, G. (2016). Role of maternal vitamins in programming health and chronic disease. *Nutrition Reviews*, *74*(3), 166–180. https://doi.org/10.1093/nutrit/nuv103.

Pannia, E., Hammoud, R., Kubant, R., Sa, J. Y., Simonian, R., Wasek, B., et al. (2021). High intakes of [6S]-5-Methyltetrahydrofolic acid compared with folic acid during pregnancy programs central and peripheral mechanisms favouring increased food intake and body weight of mature female offspring. *Nutrients*, *13*(5). https://doi.org/10.3390/nu13051477.

Pannia, E., Hammoud, R., Simonian, R., Arning, E., Ashcraft, P., Wasek, B., et al. (2020). [6S]-5-Methyltetrahydrofolic acid and folic acid pregnancy diets differentially program metabolic phenotype and hypothalamic gene expression of Wistar rat dams post-birth. *Nutrients*, *13*(1). https://doi.org/10.3390/nu13010048.

Pannia, E., Yang, N. V., Ho, M., Chatterjee, D., Hammoud, R., Kubant, R., et al. (2020). Folic acid content of diet during pregnancy determines post-birth re-set of metabolism in Wistar rat dams. *The Journal of Nutritional Biochemistry*, *83*, 108414. https://doi.org/10.1016/j.jnutbio.2020.108414.

Partearroyo, T., Perez-Miguelsanz, J., Pena-Melian, A., Maestro-de-Las-Casas, C., Ubeda, N., & Varela-Moreiras, G. (2016). Low and high dietary folic acid levels perturb postnatal cerebellar morphology in growing rats. *The British Journal of Nutrition*, *115*(11), 1967–1977. https://doi.org/10.1017/S0007114516001008.

Reza Lopez, S. A., Poon, A. N., Szeto, I. M., Ma, D. W., & Anderson, G. H. (2013). High multivitamin intakes during pregnancy and postweaning obesogenic diets interact to affect the relationship between expression of PPAR genes and glucose regulation in the offspring. *The Journal of Nutritional Biochemistry*, *24*(5), 877–881. https://doi.org/10.1016/j.jnutbio.2012.06.001.

Reza-Lopez, S. A., Anderson, G. H., Szeto, I. M., Taha, A. Y., & Ma, D. W. (2009). High vitamin intake by Wistar rats during pregnancy alters tissue fatty acid concentration in the offspring fed an obesogenic diet. *Metabolism*, *58*(5), 722–730. https://doi.org/10.1016/j.metabol.2009.01.014.

Saldanha, L. G., Dwyer, J. T., Haggans, C. J., Mills, J. L., & Potischman, N. (2020). Perspective: Time to resolve confusion on folate amounts, units, and forms in prenatal supplements. *Advances in Nutrition*. https://doi.org/10.1093/advances/nmaa017.

Schwartz, M. W., Woods, S. C., Porte, D., Jr., Seeley, R. J., & Baskin, D. G. (2000). Central nervous system control of food intake. *Nature*, *404*(6778), 661–671. https://doi.org/10.1038/35007534.

Seo, D., Patrick, C. J., & Kennealy, P. J. (2008). Role of serotonin and dopamine system interactions in the neurobiology of impulsive aggression and its comorbidity with other clinical disorders. *Aggression and Violent Behavior*, *13*(5), 383–395. https://doi.org/10.1016/j.avb.2008.06.003.

Sie, K. K., Li, J., Ly, A., Sohn, K. J., Croxford, R., & Kim, Y. I. (2013). Effect of maternal and postweaning folic acid supplementation on global and gene-specific DNA methylation in the liver of the rat offspring. *Molecular Nutrition & Food Research*, *57*(4), 677–685. https://doi.org/10.1002/mnfr.201200186.

Stahl, S. M. (2008). L-methylfolate: A vitamin for your monoamines. *The Journal of Clinical Psychiatry*, *69*(9), 1352–1353. https://doi.org/10.4088/jcp.v69n0901.

Stover, P. J., Durga, J., & Field, M. S. (2017). Folate nutrition and blood-brain barrier dysfunction. *Current Opinion in Biotechnology*, *44*, 146–152. https://doi.org/10.1016/j.copbio.2017.01.006.

Szeto, I. M., Aziz, A., Das, P. J., Taha, A. Y., Okubo, N., Reza-Lopez, S., et al. (2008). High multivitamin intake by Wistar rats during pregnancy results in increased food intake and components of the metabolic syndrome in male offspring. *American Journal of Physiology. Regulatory, Integrative and Comparative Physiology*, *295*(2), R575–R582. https://doi.org/10.1152/ajpregu.90354.2008.

Szeto, I. M., Das, P. J., Aziz, A., & Anderson, G. H. (2009). Multivitamin supplementation of Wistar rats during pregnancy accelerates the development of obesity in offspring fed an obesogenic diet. *International Journal of Obesity*, *33*(3), 364–372. https://doi.org/10.1038/ijo.2008.281.

Tojal, A., Neves, C., Veiga, H., Ferreira, S., Rodrigues, I., Martel, F., et al. (2019). Perigestational high folic acid: Impact on offspring's peripheral metabolic response. *Food & Function*, *10*(11), 7216–7226. https://doi.org/10.1039/c9fo01807g.

Varela-Rey, M., Iruarrizaga-Lejarreta, M., Lozano, J. J., Aransay, A. M., Fernandez, A. F., Lavin, J. L., et al. (2014). S-adenosylmethionine levels regulate the schwann cell DNA methylome. *Neuron*, *81*(5), 1024–1039. https://doi.org/10.1016/j.neuron.2014.01.037.

Vasquez, K., Kuizon, S., Junaid, M., & Idrissi, A. E. (2013). The effect of folic acid on GABA(a)-B 1 receptor subunit. *Advances in Experimental Medicine and Biology*, *775*, 101–109. https://doi.org/10.1007/978-1-4614-6130-2_8.

Wiens, D., DeWitt, A., Kosar, M., Underriner, C., Finsand, M., & Freese, M. (2016). Influence of folic acid on neural connectivity during dorsal root ganglion neurogenesis. *Cells, Tissues, Organs*, *201*(5), 342–353. https://doi.org/10.1159/000444389.

Wilson, R. D., Genetics, C., Wilson, R. D., Audibert, F., Brock, J. A., Carroll, J., et al. (2015). Pre-conception folic acid and multivitamin supplementation for the primary and secondary prevention of neural tube defects and other folic acid-sensitive congenital anomalies. *Journal of Obstetrics and Gynaecology Canada*, *37*(6), 534–552. https://doi.org/10.1016/s1701-2163(15)30230-9.

Wusigale, & Liang, L. (2020). Folates: Stability and interaction with biological molecules. *Journal of Agriculture and Food Research*, *2*.

Xie, Y., & Dorsky, R. I. (2017). Development of the hypothalamus: Conservation, modification and innovation. *Development*, *144*(9), 1588–1599. https://doi.org/10.1242/dev.139055.

Xie, K., Fu, Z., Li, H., Gu, X., Cai, Z., Xu, P., et al. (2018). High folate intake contributes to the risk of large for gestational age birth and obesity in male offspring. *Journal of Cellular Physiology*, *233*(12), 9383–9389. https://doi.org/10.1002/jcp.26520.

Yang, N. V., Pannia, E., Chatterjee, D., Kubant, R., Ho, M., Hammoud, R., et al. (2018). Gestational folic acid content alters the development and function of hypothalamic food intake regulating neurons in Wistar rat offspring post-weaning. *Nutritional Neuroscience*, 1–12. https://doi.org/10.1080/1028415X.2018.1479628.

Yang, N. V., Pannia, E., Chatterjee, D., Kubant, R., Ho, M., Hammoud, R., et al. (2020). Gestational folic acid content alters the development and function of hypothalamic food intake regulating neurons in Wistar rat offspring post-weaning. *Nutritional Neuroscience*, *23*(2), 149–160. https://doi.org/10.1080/1028415X.2018.1479628.

Zhang, X., Huang, G., Liu, H., Chang, H., & Wilson, J. X. (2012). Folic acid enhances notch signaling, hippocampal neurogenesis, and cognitive function in a rat model of cerebral ischemia. *Nutritional Neuroscience*, *15*(2), 55–61. https://doi.org/10.1179/1476830511Y.0000000025.

Zhang, X. M., Huang, G. W., Tian, Z. H., Ren, D. L., & Wilson, J. X. (2009). Folate stimulates ERK1/2 phosphorylation and cell proliferation in fetal neural stem cells. *Nutritional Neuroscience*, *12*(5), 226–232. https://doi.org/10.1179/147683009X423418.

PART XIII

Stroke

CHAPTER 41

Vitamin B12, hyperhomocysteinemia, and stroke

J. David Spence[a,b,*]
[a]Neurology & Clinical Pharmacology, Western University, London, ON, Canada
[b]Stroke Prevention & Atherosclerosis Research Centre, Robarts Research Institute, London, ON, Canada

Abbreviations

B12	cobalamin
B6	pyridoxine
MTH	methylenetetrahydrofolate
MTHFR	methylenetetrahydrofolate reductase
SAM	S-adenosylmethionine
SAH	S-adenosylhomocysteine
5-Me THF	5-methyl tetrahydrofolate

Introduction

Vitamin B12 deficiency is important in neurology for several reasons: it causes neuropathy, myelopathy (subacute combined degeneration), and dementia, and because it causes hyperhomocysteinemia, increases the risk of stroke. Furthermore, it is far more common than supposed. It is easily treated, with inexpensive vitamins, and the consequences are serious, so this condition should receive more attention than it does. In this review, I focus on hyperhomocysteinemia and stroke.

Forms of B12

B12 isoforms, their transport, and their roles were reviewed in 2015 by Obeid et al. (Obeid, Fedosov, & Nexo, 2015). The first form to be discovered was cyanocobalamin, an inexpensive and stable form that is the usual supplement in use. Other forms include methylcobalamin (MeCbl), the cofactor for methionine synthase, 5′-deoxy-5′-adenosylcobalamin (AdoCbl), the cofactor for mitochondrial methylmalonyl-CoA

* In Absentia Contact Person: Ms. Cher Hickey, Stroke Prevention & Atherosclerosis Research Centre, 1400 Western Road, London, ON, Canada N6G 2V4, Phone 1-519-931-5731, Fax 1-519-931-5737, Email Cher.Hickey@lhsc.on.ca.

Vitamins and Minerals in Neurological Disorders
https://doi.org/10.1016/B978-0-323-89835-5.00037-5

(MM-CoA) mutase, and hydroxycobalamin (HOCbl), produced when MeCbl or AdoCbl is exposed to light.

"Cobalamin (Cbl) is necessary for synthesis of DNA bases, transfer of methyl group [i.e. regeneration of S-adenosylmethionine (SAM)], as well as metabolism of branched chain amino acids and fatty acids with an odd number of carbon atoms" (Obeid et al., 2015).

The physiological forms of B12 are MeCbl, AdoCbl, and HOCbl; cyanocobalamin must be decyanated to become physiologically active. That becomes important when considering B12 supplementation in patients with renal failure, who have high levels of thiocyanate (Koyama et al., 1997).

Homocysteine metabolism

Homocysteine is remethylated to methionine by the action of methylenetetrahydrofolate reductase (MTHFR), the cofactor for which is folic acid, and methionine synthase, the cofactor for which is cobalamin (Fig. 1). It is also removed by the trans-sulfuration pathway, to cystathionine via cystathionine beta synthase (cofactor pyridoxine/vitamin B6 and riboflavin), in turn converted to cysteine via cystathionine γ-lyase (cofactors B6 and riboflavin). Elevated levels of cystathionine and cysteine are associated with increased cardiovascular risk (Dhar et al., 2018; El-Khairy et al., 2001), so shunting of homocysteine from remethylation to trans-sulfuration may be harmful. Methionine synthase is inhibited by nitrous oxide anesthesia. This can result in postoperative increased plasma total homocysteine (Badner, Drader, Freeman, & Spence, 1998) and increased myocardial ischemia (Badner, Beattie, Freeman, & Spence, 2000). It seems likely that this would also contribute to postoperative stroke and venous thrombosis. Preoperative B vitamins can prevent this (Badner, Freeman, & Spence, 2001). Nitrous oxide anesthesia (Patel et al., 2018) and recreational use of nitrous oxide (Lan et al., 2019) have also resulted in subacute combined degeneration of the spinal cord.

When homocysteine is measured in blood, what is actually measured is a combination of homocysteine, its dimer homocystine, and homocysteine-cysteine mixed disulfide; this is called total homocysteine (tHcy) (Mudd et al., 2000).

B12 deficiency: "Biochemical" and "metabolic" deficiency

Deficiency of vitamin B12 is common for several reasons: in the diet, it is derived mostly from meat; so deficiency is common in vegetarians (particularly vegans); furthermore, the absorption of B12 is extremely complex. Haptocorrin from saliva, gastric acid (to strip the B12 from the meat), intrinsic factor, and pancreatic proteases to release cobalamin from haptocorrin are all required for absorption, and transcobalamin 1 and 2 are required for transport (Stabler, 2013). Deficiency of intrinsic factor due to autoimmune gastritis may

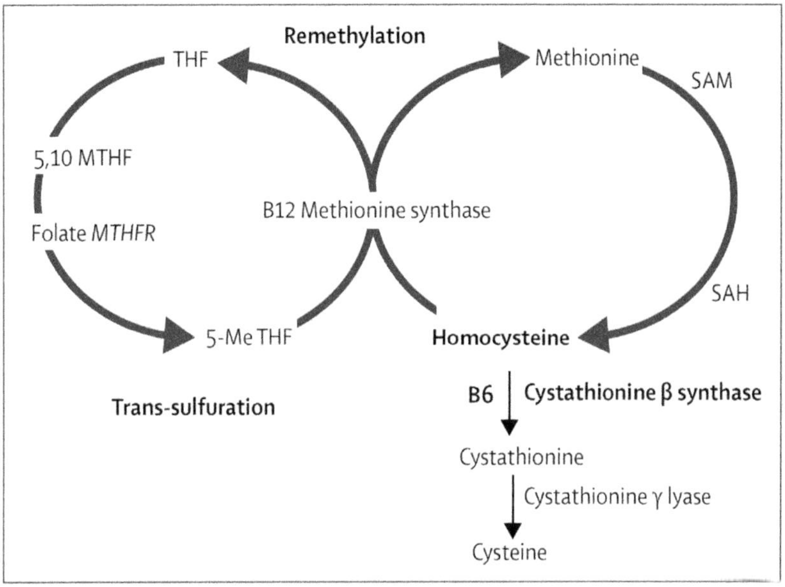

Fig. 1 Homocysteine metabolism. Homocysteine is removed by two pathways: remethylation to methionine by the actions of MTHFR (cofactor folic acid) and methionine synthase (cofactor B12), and trans-sulfuration to cystathionine by cystathionine beta synthase (cofactor B6) and then to cysteine by cystathionine γ lyase (cofactor B6). Riboflavin is required for the activation of B6 to its active form, pyridoxal 5′ phosphate. Other enzymes not shown are betaine/homocysteine methyltransferase (cofactors betaine and zinc), and methionine synthase reductase, required to activate methionine synthase. Variants of the genes encoding all these enzymes may result in hyperhomocysteinemia. B12 = cobalamin, B6 = pyridoxine, MTH = methylenetetrahydrofolate, MTHFR = methylenetetrahydrofolate reductase, SAM = S-adenosylmethionine, SAH = S-adenosylhomocysteine, 5-Me THF = 5-methyl tetrahydrofolate. *(Reproduced by permission of Elsevier from: Spence, J. D., Yi, Q., & Hankey, G. J. (2017). B vitamins in stroke prevention: Time to reconsider. Lancet Neurology, 16(9), 750–760.)*

be the commonest cause of B12 deficiency. (Stabler, 2013) Proton pump inhibitors can reduce B12 absorption by preventing the secretion of gastric acid, and the frequency of achlorhydria increases with age. Some of the protein factors for absorption can be inherited, and antibodies can develop that inactivate them. B12 deficiency increases with age, probably for all those reasons, including diet ("tea and toaster" behavior).

A little-recognized problem is that both folate deficiency and B12 deficiency can interfere with the absorption of oral vitamins, because of atrophy of the intestinal villi (Arvanitakis, 1978; Arvantakis, Lyford, & Folscroft, 1977). Folate and B12 are required for DNA synthesis, so cells that turn over quickly are affected by their deficiency. Red blood cells turn over in about 100 days and platelets turn over in about 9 days; the mucosal cells of the intestinal villi turn over in about 5 days. Megaloblastic anemia, hypersegmentation of neutrophils, and changes in platelets are well-recognized consequences of B12 deficiency, but villous atrophy is seldom considered. Although most patients will respond to high doses of oral vitamins, some may require a series of B12 injections to restore the intestinal villi and absorption from oral sources. (I have seen this with folate deficiency as well.)

Deficiency of cobalamin is under-recognized because many/most physicians assume that a serum B12 level in the "normal" or reference range (the 95% of the population within 2 standard deviations of the mean) represents adequacy of cobalamin. A serum B12 below the reference range (~ 160 pmol/L) might be called "biochemical B12 deficiency." However, a small and variable percentage of total serum B12 is active (~ 6–20%), so in a given individual, a serum B12 in the normal range may not represent sufficient metabolically active B12; i.e., the patient has "metabolic B12 deficiency." To assess metabolic adequacy of B12, it is necessary to measure holotranscobalamin, or one of the metabolites that becomes elevated in B12 deficiency: methylmalonic acid (MMA), or plasma total homocysteine (tHcy) (Spence, 2016; Stabler, 2013).

As shown in Fig. 2, from the US National Health and Nutrition Examination Survey (Bang, Mazumdar, & Spence, 2006), the serum B12 below which plasma tHcy rises steeply is around 400 pmol/L. In the Hordaland study, both tHcy and MMA levels rose steeply with a serum B12 < 400 pmol/L (Vogiatzoglou et al., 2009). Stabler defined metabolic B12 deficiency as a serum B12 below 258 pmol/L (the median in some laboratories), with elevation of MMA, or a plasma total homocysteine >14 μmol/L (Stabler, Lindenbaum, & Allen, 1996). Although the recommended daily intake of B12 is usually quoted as being quite low (4–6 μg/day), patients with B12 deficiency may need high doses, > 500 μg/day to achieve adequate serum B12 levels (Stabler, 2013). I know of patients who require 1000 μg twice daily. In my view, the appropriate dose of B12 is methylcobalamin or oxocobalamin in a sufficient dose to maintain a serum B12 > 400 pmol/L (more readily available than serum MMA or tHcy).

Both biochemical B12 deficiency and metabolic B12 deficiency are very common. Fig. 3 shows the distribution of serum B12 levels among 3025 patients referred to me for secondary stroke prevention. Among 4055 patients referred to University Hospital in London, Canada, for secondary stroke prevention between 2000 and 2012, biochemical B12 deficiency was present in 8.5% of patients referred; it was present in 10.9% before 2009 and 5.4% thereafter (Ahmed et al., 2019). It is likely that this decline resulted from a request that referring doctors routinely measure serum B12 in patients referred for stroke prevention, after I reported in 2009 that hyperhomocysteinemia was very common in patients referred for stroke prevention (~ 10% at the age of 35, 25% at the age of 65, and 40% above the age of 80) (Spence, 2009). It seems likely that B12 deficiency may be more common in other cities where a routine request for serum B12 at the time of referral was not implemented.

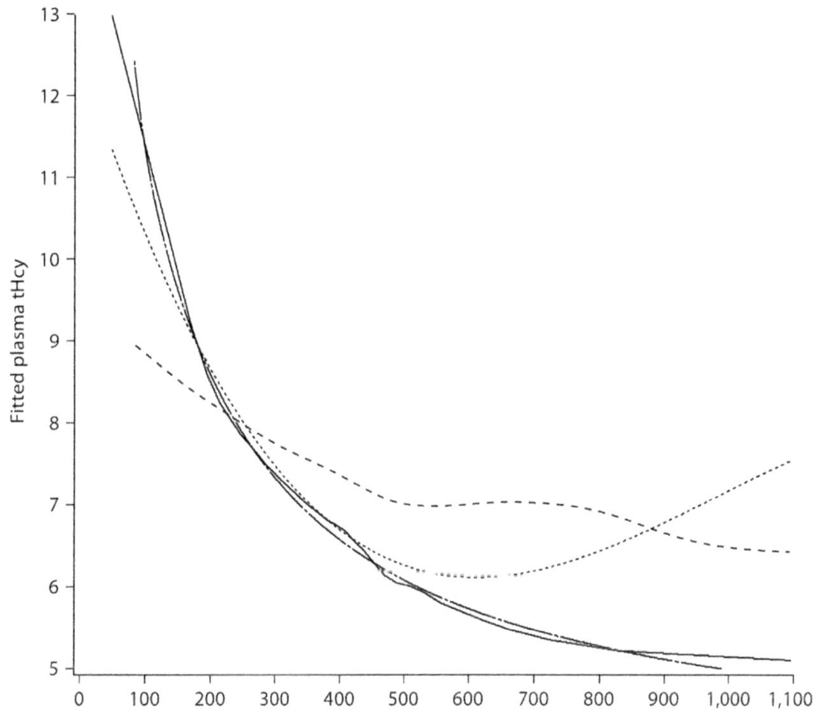

Fig. 2 Polynomial and inverse regression, spline, loess fits for plasma tHcy with B12. Unadjusted analysis. Solid, dotted, long dashed, and short dashed lines represent loess, cubic polynomials, cubic polynomials of the inverse covariate, and smooth splines, respectively. Plasma tHcy levels rise steeply when the serum B12 is below 400 pmol/L. *(Reproduced by permission of Karger from: Bang, H., Mazumdar, M., & Spence, J. D. (2006). Tutorial in biostatistics: Analyzing associations between total plasma homocysteine and B vitamins using optimal categorization and segmented regression.* Neuroepidemiology, 27, *188–200.)*

Over the period 2000–2012, metabolic B12 deficiency was present in 3.9% of patients aged <50 years, increasing to 15% at age ≥80 years. Renal dysfunction raises tHcy to 14.5 μmol/L at an estimated glomerular filtration rate < 60 mL/min/1.73 m^2. When patients with eGFR <60 mL/min/1.73 m^2 were excluded, the percentage of patients with metabolic B12 deficiency was 3.4% below the age of 50 and 12.3% at age ≥80 years. Even using that more conservative estimate, 15% of patients aged 60–69, and > 20% of patients aged ≥80 years, had either biochemical or metabolic B12 deficiency (Ahmed et al., 2019). Because B12 deficiency is so common, has serious consequences, and is easily treated, it should be routine to measure serum B12 and tHcy in all patients with stroke.

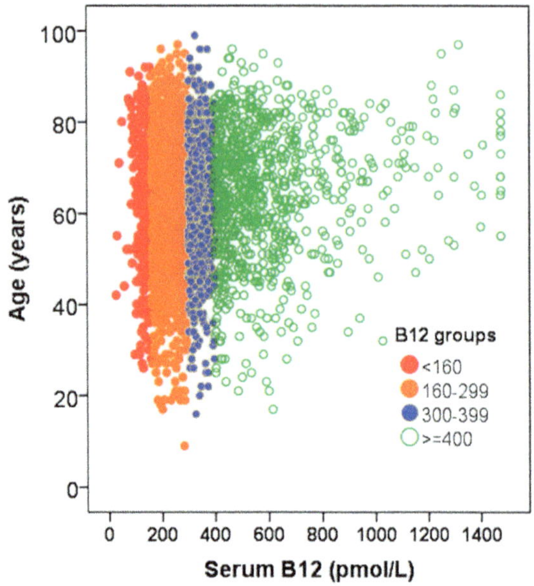

Fig. 3 Serum B12 by age in patients at the Stroke Prevention Clinic at University Hospital, London, Ontario, Canada. Of 3025 patients referred for secondary stroke prevention, 281 (9%) had a serum B12 < 160 pmol/L (the lower limit of the reference range for the hospital laboratory), and 1317 (44%) had a B12 level in the range of 160–299 pmol/L that requires further testing such as MMA or tHcy to assess metabolic adequacy of B12. Only 797 (26%) had clearly adequate levels of serum B12 > 400 pmol/L associated with optimal function of B12. *(Reproduced by permission of Elsevier from: Spence, J. D. (2016). Metabolic vitamin B12 deficiency: A missed opportunity to prevent dementia and stroke. Nutrition Research, 36, 109–116.)*

Risk of stroke with hyperhomocysteinemia

The history of the recognition that hyperhomocysteinemia causes vascular disease is analogous to the discovery that familial hypercholesterolemia with very high levels of low-density lipoprotein cholesterol (LDL-C) caused premature coronary disease, followed by the recognition that even moderate elevation of LDL-C was an important treatable cause of vascular disease. The adverse effects of homocysteine were first recognized in young patients with thrombosis of their arteries and veins, with very high levels of tHcy in classical homocystinuria, due to a defective gene for cystathionine synthase. McCully, who proposed the homocysteine theory of atherosclerosis in 1969 (Mccully, 1969), reviewed the history of this discovery in 2007 (Mccully, 2007), taking the history back to a case report at the Massachusetts General Hospital in 1933, and a 1962 paper from Ireland. In subsequent years, it became apparent that even moderate elevation of tHcy increased vascular risk. In 1985, Boers et al. (Boers et al., 1985) reported that about 30% of patients with premature vascular disease were heterozygous for homocystinuria. In 1997, the

Concerted European Action Project reported that in a case/control study, the relative risk of vascular disease was 3-fold higher with fasting tHcy >13.4 vs. <7, and the relative risk was 4 with a post-methionine tHcy level > 43.5 vs. <21.5 (Graham et al., 1997). They also reported a strong interaction between tHcy and smoking. Also, Nygård et al. (1997) reported on a prospective study in 587 patients with coronary disease followed for a median of 4.6 years. Mortality rose 8-fold in a linear fashion with tHcy increasing from 5 to >20 μmol/L. Over 4.6 years, mortality was 3.8% with tHcy <9, vs. 24.7% with tHcy >15 μmol/L (Nygård et al., 1997).

There are many mechanisms that contribute to vascular harm with homocysteine, including thrombosis, oxidative stress, impairment of endothelial function, inflammation (Mccully, 2009), and homocysteinylation of metallothionein (Colgan & Austin, 2007). Hyperhomocysteinemia quadruples the risk of stroke in atrial fibrillation (Poli et al., 2005).

Some authors define hyperhomocysteinemia (HHcy) by a tHcy >15 μmol/L; however, in a study of serum B12, tHcy and carotid plaque following folate fortification of the grain supply in North America, it seemed that an optimal tHcy was <7 μmol/L, and plaque burden was not higher with tHcy >12 μmol/L. I have defined it as a tHcy ≥14 μmol/L.

The prevalence of HHcy varies substantially around the world. It is particularly high in India and Pakistan, and in China. Among ischemic stroke patients in Pakistan, 58.3% were reported to have tHcy >15 μmol/L (Sadiq, Alam, Kanpurwala, & Khan, 2014). In India, the main reasons may be vegetarianism (~ 60% of the population), causing B12 deficiency, and cooking methods that apply prolonged high temperatures, destroying folate. In China, HHcy is so prevalent that the term "H-type hypertension" is used to refer to hypertension with HHcy. One study reported that 16.7% of hypertensive patients with stroke in China had tHcy >30 μmmol/L (Wang et al., 2014). In China, the reason may be largely genetic; the T allele of MTHFR is more common there. In the population of the China Stroke Primary Prevention Trial, MTHFR genotypes were CC 27.3%, CT49.1%, and TT 23.6% (Huo et al., 2015). In contrast, among patients referred for vascular prevention in London, Canada, the genotypes were CC 40.4%, CT 46.6%, and TT 13.0% (Spence, Cordy, Kortas, & Freeman, 1999).

Prevention of stroke with B vitamins

The widespread and mistaken belief that B vitamins do not prevent stroke is curiously persistent. Because of the roles of folic acid, B12, and B6 in homocysteine metabolism, it seemed likely that a combination of those vitamins should lower tHcy and reduce the risk of stroke. The first major trial of B vitamins to prevent stroke was the Vitamin Intervention in Stroke Prevention (VISP) Trial, published in 2004 (Toole et al., 2004). We found no reduction of stroke or other cardiovascular events, a result that seemed implausible. However, there are important reasons for that absence of benefit. Since then, it has taken many years to understand the complexity of the evidence. B vitamins do reduce the risk of stroke substantially, but it has taken some detective work to sort this out.

In VISP, we randomized patients who had experienced a stroke to either high-dose B vitamins (folic acid 2.5 mg, B6 25 mg, and cyanocobalamin 400 μg) or low-dose vitamins (folic acid 20 μg, B6 200 μg, or cyanocobalamin 6 μg) daily (Toole et al., 2004). It was my fault that we used low-dose vitamins instead of placebo (I wanted to avoid patients taking over the counter vitamins outside of their randomization, to guard against being randomized to placebo), and it was Dr. James Toole's decision to include 6 μg of cyanocobalamin, the recommended daily intake of B12, for what he regarded as ethical reasons. Furthermore, also for ethical reasons, it was decided that participants in either arm of the study who had B12 deficiency would receive monthly B12 injections, thus negating the effect of B12 in the very participants who would have benefited most. Furthermore, folate fortification of the grain supply in North America began at the same time as the trial, largely negating the benefit of folic acid.

In order to address those concerns, we carried out a subgroup analysis (Spence, Bang, Chambless, & Stampfer, 2005) from which we excluded VISP participants who received B12 outside their randomized assignment, and for the wrong reason, we also excluded those with impaired renal function, with an estimated glomerular filtration rate (eGFR) in the lowest 10%; this turned out to be an eGFR $<46\,mL/min/1.73\,m^2$. (The reason was that I had found that higher doses of folic acid did not reduce tHcy in dialysis patients (Spence et al., 1999), so I thought patients with impaired renal function would be less responsive.) As shown in Fig. 4, we found that participants with good renal function who entered the study with a serum B12 above the median of 322 pmol/L (meaning that they could absorb B12 well) and who received high-dose vitamins, had a 34% reduction of stroke/myocardial infarction/vascular death over 750 days of follow-up ($P=0.03$).

I suspect that the widespread belief that B vitamins do not prevent stroke originated with several pronouncements by authorities. The first was the pronouncement that "Homocysteine is dead" in September 2005 at the European Society of Cardiology Congress in Stockholm by Dr. Kaare Harald Bønaa. He was describing the results of the Norwegian Vitamin Trial (NORVIT) (Bønaa et al., 2006) in patients with coronary artery disease, in which a B vitamin combination containing cyanocobalamin was actually slightly harmful, and other B vitamin combinations were not beneficial. That study was published in the *New England Journal of Medicine* in 2006; in the same issue, the Hypertension Optimal Prevention Evaluation 2 (HOPE-2) Trial was reported (Lonn et al., 2006), along with an accompanying editorial by Loscalzo (Loscalzo, 2006), who pronounced that the results were clear; B vitamins did not prevent vascular events, but the reasons were complex: he postulated that the reason was unmetabolized folic acid. The HOPE 2 Trial actually showed a statistically significant reduction of stroke, but the

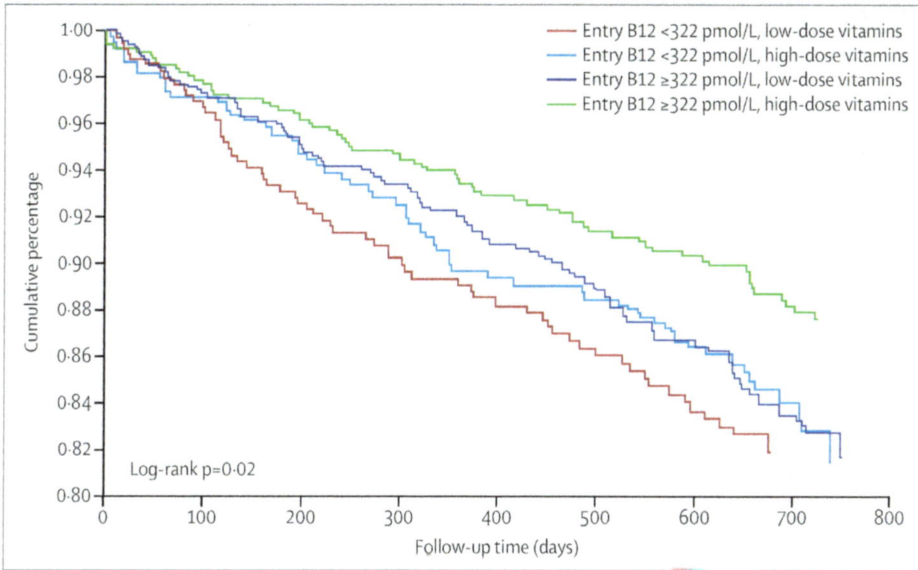

Fig. 4 Benefit of B vitamins in the VISP Trial. In this post hoc analysis of the VISP Trial, the following participants were excluded: participants with serum cyanocobalamin below the 25th percentile (250 pmol/L; to exclude those who received cyanocobalamin injections); participants with serum cyanocobalamin above the 95th percentile (950 pmol/L; to exclude those receiving cyanocobalamin supplements outside the study); and participants with the lowest 10% of eGFRs, as calculated by the Cockcroft-Gault equations (eGFR <46·18 mL/min per 1·73 m^2). Survival free of stroke, or myocardial infarction, and cardiovascular death was stratified by treatment arm (high-dose vs low-dose B vitamins) and by baseline serum B12, cutoff at the median (322 pmol/L) to define groups more able and less able to absorb the vitamin. A 34% reduction in the number of events was observed in the participants with a baseline serum cyanocobalamin above the median who received high-dose B vitamins including cyanocobalamin (400 µg daily) compared with those with a baseline serumB12 below the median who received low-dose study vitamins. *(Reproduced by permission of Elsevier from: Spence, J. D., Yi, Q., & Hankey, G. J. (2017). B vitamins in stroke prevention: Time to reconsider. Lancet Neurology, 16(9), 750–760.)*

authors, being cardiologists and therefore innocent of the cerebral circulation, pronounced that since myocardial infarction was not prevented, the reduction of stroke was a chance finding. In 2006, I reviewed the differences between myocardial infarction and stroke, suggesting that B vitamins actually did prevent stroke.

Then in 2010, the French SuFolOM3 Trial was published, reporting a 43% reduction of stroke with B vitamins. Crucially, that study had the best renal function of all the trials, and used a far lower dose of cyanocobalamin: only 20 µg daily, vs. 400–1000 µg daily in the other trials (Table 1). Also, in 2010, we published the results of the Diabetic Intervention with Vitamins in Diabetic Nephropathy (DIVINe) Trial. Patients with diabetic nephropathy were randomized to high-dose B vitamins vs. placebo; the B vitamin doses were folic acid 2.5 mg, B6 25 mg, and cyanocobalamin 1000 µg daily. There was a

Table 1 Clinical trials of vitamin B therapy for the prevention of stroke.

	Dose of cyanocobalamin (μg)	Serum creatinine (μmol/L)[a]	Stroke hazard ratio (95% CI)	*P* value
DIVINe[45]	1000	141.4 (97.2); control 123.8 (79.6)	6.6 (0.8–54.4)	0.08
VISP[1 b]	400	99.9 (55.7); control 97.2 (47.7)	1.0 (0.8–1.3)	0.80
SEARCH[42]	1000	NA	1.02 (0.86–1.21)	NA
VITATOPS[37 c]	500	92.4 (40.3); control 91.4 (34.6)	0.92 (0.81–1.06)	0.25
NORVIT[2]	400	91 (27); control 91 (24)	0.83 (0.47–1.47)	0.52
HOPE-2[3]	1000	88.4 (26.5); control 88.4 (26.5)	0.75 (0.59–0.97)	0.03
Su.Fol.OM3[43]	20	78.0 (70.0–88.0); control 78.0 (69.0–88.0)	0.57 (0.33–0.97)	0.04
CSPPT[5] with eGFR <60[d]	0	126.6 (72.7); control 130.6 (68.6)	0.88 (0.33–2.36)	0.81
CSPPT[4](all)	0	65.95 (19.0); control 65.95 (19.0)	0.79 (0.68–0.93)	0.003
CSPPT[5] excluding eGFR ≤60[d]	0	64.7 (13.8); control 64.6 (13.6)	0.79 (0.67–0.92)	

Individual data were obtained from both VISP and VITATOPS trials. NA = not applicable. eGFR = estimated glomerular filtration rate.

Reproduced by permission of Elsevier from: Spence, J. D., Yi, Q., & Hankey, G. J. (2017). B vitamins in stroke prevention: Time to reconsider. Lancet Neurology, 16(9), 750–760.

[a]Data are presented as the mean (SD), unless otherwise specified.

[b]Patients with an eGFR of less than 50 mL/min per 1·73 m^2 had a mean eGFR rate (SD) of 38·7 mL/min per 1·73 m^2 (9·2; control 39·3 mL/min per 1·73 m^2 [8·6]; composite outcomes HR [95% CI] 1·01 [0·74–1·37; *P*=0·977]). Patients with an eGFR of 50 mL/min per 1·73 m^2 or more had mean eGFR rate of 89·2 mL/min per 1·73 m^2 (44·6; control 87·3 [31·1]; composite outcomes HR 0·92 (0·77–1·11; *P*=0·382)]. Composite outcomes refer to myocardial infarction, stroke, or death.

[c]Patients with an eGFR of 50 mL/min per 1·73 m^2 or less had a mean eGFR rate (SD) of 38·6 mL/min per 1·73 m^2 (10·0; control 38·3 [9·5]; composite outcomes HR [95% CI] 0·88 [0·59–1·32; *P*=0·54]). Patients with an eGFR of more than 50 mL/min per 1·73 m^2 had a mean eGFR rate of 80·2 (18·1; control 80·9 [18·3]; composite outcomes HR 0·82 [0·68–0·98; *P*=0·03]). Composite outcomes refers to myocardial infarction, stroke, or death.

[d]mL/min per 1·73 m^2.

Reproduced by permission of Elsevier from: Spence, J. D., Yi, Q., & Hankey, G. J. (2017). B vitamins in stroke prevention: Time to reconsider. Lancet Neurology, 16(9), 750–760.

significant reduction of tHcy with the B vitamins, but to our surprise, there was a doubling of cardiovascular events with the high-dose B vitamins (Fig. 5). Following that study, I received emails from all over the world, essentially telling me (like Inspector Clouseau), "yes, you fool, this is what we have been saying all along!" An email from Japan informed me of the work of Koyama, who had shown that dialysis patients had high levels of thiocyanate (Koyama et al., 1997) and that methylcobalamin was better

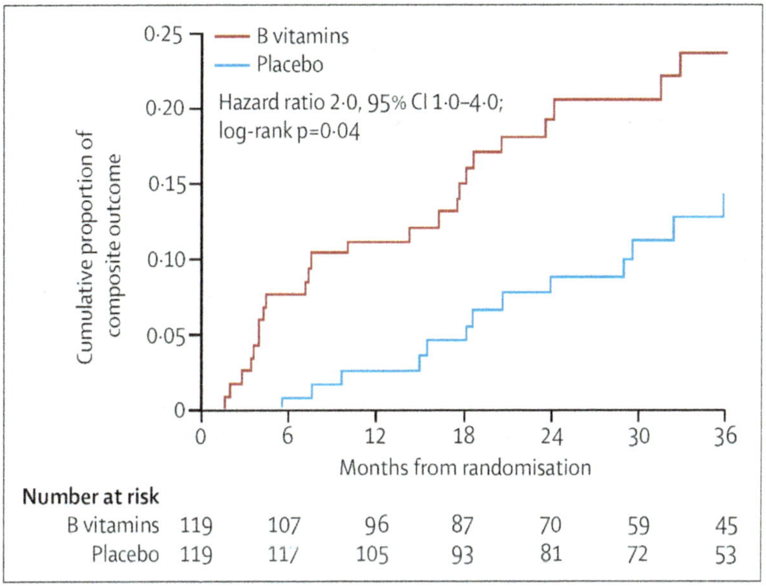

Fig. 5 Effect of B vitamins on the risk of a composite cardiovascular outcome in patients with diabetic nephropathy in the DIVINe Trial. Cumulative proportion of myocardial infarction, stroke, revascularization, and all-cause mortality in the DIVINe Trial.45 Following treatment with B vitamins (2·5 mg folic acid, 25 mg pyridoxine, and 1000 µg cyanocobalamin daily), the 36-month risk of the composite outcome was 23·5% in the treatment group and 14·4% in the placebo group (log-rank $P = 0{\cdot}04$). All the events occurred in participants with a glomerular filtration rate of less than 50 mL/min per 1·73 m^2. *(Reproduced by permission of Elsevier from: Spence, J. D., Yi, Q., & Hankey, G. J. (2017). B vitamins in stroke prevention: Time to reconsider. Lancet Neurology, 16(9), 750–760.)*

for lowering both tHcy and asymmetric dimethylarginine in dialysis patients (Koyama, Usami, Takeuchi, Morozumi, & Kimura, 2002). An email from England informed me about concerns about cyanocobalamin in tobacco amblyopia (Freeman, 1992, 1996). In 2011, Stampfer and I hypothesized that harm from cyanocobalamin among study participants with impaired renal function may have accounted for the null results in the early trials of B vitamin combinations containing cyanocobalamin.

Then in 2015, the China Stroke Primary Prevention Trial (CSPPT) established clearly that B vitamins to lower homocysteine do indeed reduce the risk of stroke. Probably, the absence of folate fortification in China helped make that possible. Patients with hypertension ($n = 20{,}702$) were randomized to folic acid 0.8 mg plus enalapril 10 mg vs. placebo plus enalapril 10 mg daily, and followed for a median of 4.5 years. Ischemic stroke was reduced by 24% with folate (HR, 0.76; 95% CI, 0.64–0.91); a composite of cardiovascular death, MI, and stroke was reduced by 20% (HR, 0.80; 95%CI, 0.69–0.92) (Huo et al., 2015). The risk reduction was greater among participants at higher risk. Among those with a low-density lipoprotein cholesterol >200 mg/dL (5.17 mmol/L), the reduction of stroke was 36% (HR, 0.64; 95% CI 0.52–0.80) (Qin et al., 2016); among those with a platelet count in the lowest quartile and a tHcy $\geq$15 μmol/L, stroke was reduced by with folate by 73% (HR, 0.27; 95% CI, 0.11–0.64) (Kong et al., 2018).

Crucially, folate was also beneficial in participants with impaired renal function, slowing severe decline of renal function (OR, 0.44; 95%CI, 0.26–0.75) and reducing a composite event of severe decline in renal function, all-cause death, rapid decline in renal function, and rate of eGFR decline (OR, 0.62; 95% CI, 0.43–0.90). That finding disproves Loscalzo's hypothesis (Loscalzo, 2006) that unmetabolized folic acid accounted for the null result of the early trials, and supports the hypothesis that the reason was toxicity from cyanocobalamin among participants with impaired renal function. A 2017 meta-analysis stratified by cyanocobalamin dose and by renal function confirmed the hypothesis (Spence, Yi, & Hankey, 2017). Two more recent meta-analyses also confirm that folate and B vitamin combinations containing B12 reduce the risk of stroke Jenkins et al., 2018.

Although folic acid alone reduced stroke in China, it is likely that in countries with folate fortification B12 supplements would be as or more important than folate for stroke prevention. An interaction between MTHFR genotype, serum B12, and response to folate indicates that among persons with the T allele of MTHFR, higher doses of folate are probably required (Qin et al., 2020).

To summarize this issue, B vitamins to lower homocysteine do indeed reduce the risk of stroke, but in addition to folate, we should be using methylcobalamin or hydroxycobalamin, not cyanocobalamin, particularly in patients with impaired renal function, including older patients. Freeman pointed out that in 1995, the World Health Organization list of essential drugs replaced cyanocobalamin with oxocobalamin (Freeman, 1996).

Applications to other neurological conditions

Other conditions in which B12 deficiency and replacement are important include neuropathy and myelopathy (subacute combined degeneration). B12 is also important for tobacco amblyopia, in which cyanide is a key factor, and as pointed out by Freeman (Freeman, 1992, 1996, 1999), it is important not to use cyanocobalamin for that condition.

Other components of interest

Besides B12, folate, betaine, pyridoxine (B6), and riboflavin may be important to lower tHcy. B6 is particularly important for persons with deficiency of cystathionine beta synthase and cystathionine gamma lyase, and riboflavin is needed to convert pyridoxine to its active form, pyridoxal phosphate.

Mini-dictionary of terms

Plasma total homocysteine (tHcy): it includes homocysteine, its dimer homocystine and cysteine/homocysteine mixed disulfide.
Hyperhomocysteinemia: elevated tHcy (variously $\geq$14 or 15 μmol/L).
Biochemical B12 deficiency: a serum B12 below the reference range 9–150–160 pmol/L.
Metabolic B12 deficiency: a serum B12 in the lower half of the reference range (< 258 pmol/L) with hyperhomocysteinemia.
Ischemic stroke: cerebral infarction due to ischemia.

Summary points

- Vitamin B12 deficiency is very common, increasing with age.
- It causes neuropathy, myelopathy, and dementia, and increases the risk of stroke.
- Metabolic B12 deficiency is often missed because the serum B12 is "normal."
- Hyperhomocysteinemia is a powerful risk factor for stroke.
- Lowering homocysteine with B vitamins reduces the risk of stroke.
- Cyanocobalamin should be replaced by methyl- or hydroxycobalamin, especially in persons with impaired renal function (including old persons).

References

Ahmed, S., Bogiatzi, C., Hackam, D. G., Rutledge, A. C., Sposato, L. A., Khaw, A., et al. (2019). Vitamin B 12 deficiency and hyperhomocysteinaemia in outpatients with stroke or transient ischaemic attack: A cohort study at an academic medical Centre. *BMJ Open*, *9*, e026564.

Arvanitakis, C. (1978). Functional and morphological abnormalities of the small intestinal mucosa in pernicious anemia—A prospective study. *Acta Hepato-Gastroenterologica*, *25*, 313–318.

Arvantakis, C., Lyford, C. L., & Folscroft, J. (1977). Intestinal absorptive and digestive function in pernicious anemia. *The American Journal of Medicine*, *63*, 859–864.

Badner, N. H., Beattie, W. S., Freeman, D., & Spence, J. D. (2000). Nitrous oxide-induced increased homocysteine concentrations are associated with increased postoperative myocardial ischemia in patients undergoing carotid endarterectomy. *Anesthesia and Analgesia*, *91*, 1073–1079.

Badner, N. H., Drader, K., Freeman, D., & Spence, J. D. (1998). The use of intraoperative nitrous oxide leads to postoperative increases in plasma homocysteine. *Anesthesia and Analgesia*, *87*, 711–713.

Badner, N. H., Freeman, D., & Spence, J. D. (2001). Preoperative oral B vitamins prevent nitrous oxide-induced postoperative plasma homocysteine increases. *Anesthesia and Analgesia*, *93*, 1507–1510.

Bang, H., Mazumdar, M., & Spence, J. D. (2006). Tutorial in biostatistics: Analyzing associations between total plasma homocysteine and B vitamins using optimal categorization and segmented regression. *Neuroepidemiology*, *27*, 188–200.

Boers, G. H. J., Trijbels, F. J. M., Fowler, B., Bakkeren, J. A. J. M., Schoonderwaldt, H. C., Kleiher, W. J., et al. (1985). Heterozygosity for homocystinuria in premature peripheral and cerebral occlusive arterial disease. *The New England Journal of Medicine*, *313*, 709–714.

Bønaa, K. H., Njolstad, I., Ueland, P. M., Schirmer, H., Tverdal, A., Steigen, T., et al. (2006). Homocysteine lowering and cardiovascular events after acute myocardial infarction. *The New England Journal of Medicine*, *354*, 1578–1588.

Colgan, S. M., & Austin, R. C. (2007). Homocysteinylation of metallothionein impairs intracellular redox homeostasis: The enemy within! *Arteriosclerosis, Thrombosis, and Vascular Biology*, *27*, 8–11.

Dhar, I., Svingen, G. F. T., Ueland, P. M., Lysne, V., Svenningsson, M. M., Tell, G. S., et al. (2018). Plasma cystathionine and risk of incident stroke in patients with suspected stable angina pectoris. *Journal of the American Heart Association*, *7*, e008824.

El-Khairy, L., Ueland, P. M., Refsum, H., Graham, I. M., Vollset, S. E., & European Concerted Action, P. (2001). Plasma total cysteine as a risk factor for vascular disease: The European concerted action project. *Circulation*, *103*, 2544–2549.

Freeman, A. G. (1992). Cyanocobalamin—A case for withdrawal: Discussion paper. *Journal of the Royal Society of Medicine*, *85*(11), 686–687.

Freeman, A. G. (1996). Hydroxocobalamin versus cyanocobalamin. *Journal of the Royal Society of Medicine*, *89*, 659.

Freeman, AG. (1999). Oral or parenteral therapy for vitamin B12 deficiency. *Lancet*, *353*(9150), 410–411.

Graham, I. M., Daly, L. E., Refsum, H. M., Robinson, K., Brattstrom, L. E., Ueland, P. M., et al. (1997). Plasma homocysteine as a risk factor for vascular disease. The European concerted action project. *JAMA*, *277*, 1775–1781.

Huo, Y., Li, J., Qin, X., Huang, Y., Wang, X., Gottesman, R. F., et al. (2015). Efficacy of folic acid therapy in primary prevention of stroke among adults with hypertension in China: The CSPPT randomized clinical trial. *JAMA*, *313*(13), 1325–1335.

Jenkins, D. J. A., Spence, J. D., Giovannucci, E. L., Kim, Y. I., Josse, R., Vieth, R., et al. (2018). Supplemental vitamins and minerals for CVD prevention and treatment. *Journal of the American College of Cardiology*, *71*(22), 2570–2584.

Kong, X, Huang, X, Zhao, M, Xu, B, Xu, R, Song, Y, et al. (2018). Platelet count affects efficacy of folic acid in preventing first stroke. *Journal of the American College of Cardiology*, *71*(19), 2136–2146.

Koyama, K., Usami, T., Takeuchi, O., Morozumi, K., & Kimura, G. (2002). Efficacy of methylcobalamin on lowering total homocysteine plasma concentrations in haemodialysis patients receiving high-dose folic acid supplementation. *Nephrology, Dialysis, Transplantation*, *17*, 916–922.

Koyama, K., Yoshida, A., Takeda, A., Morozumi, K., Fujinami, T., & Tanaka, N. (1997). Abnormal cyanide metabolism in uraemic patients. *Nephrology, Dialysis, Transplantation*, *12*, 1622–1628.

Lan, S. Y., Kuo, C. Y., Chou, C. C., Kong, S. S., Hung, P. C., Tsai, H. Y., et al. (2019). Recreational nitrous oxide abuse related subacute combined degeneration of the spinal cord in adolescents—A case series and literature review. *Brain and Development*, *41*, 428–435.

Lonn, E., Yusuf, S., Arnold, M. J., Sheridan, P., Pogue, J., Micks, M., et al. (2006). Homocysteine lowering with folic acid and B vitamins in vascular disease. *The New England Journal of Medicine*, *354*, 1567–1577.

Loscalzo, J. (2006). Homocysteine trials—Clear outcomes for complex reasons. *The New England Journal of Medicine*, *354*, 1629–1632.

Mccully, K. S. (1969). Vascular pathology of homocysteinemia: Implications for the pathogenesis of atherosclerosis. *The American Journal of Pathology*, *56*, 111–128.

Mccully, K. S. (2007). Homocysteine, vitamins, and vascular disease prevention. *The American Journal of Clinical Nutrition*, *86*, 1563S–1568S.

Mccully, K. S. (2009). Chemical pathology of homocysteine. IV. Excitotoxicity, oxidative stress, endothelial dysfunction, and inflammation. *Annals of Clinical and Laboratory Science*, *39*, 219–232.

Mudd, S. H., Finkelstein, J. D., Refsum, H., Ueland, P. M., Malinow, M. R., Lentz, S. R., et al. (2000). Homocysteine and its disulfide derivatives: A suggested consensus terminology. *Arteriosclerosis, Thrombosis, and Vascular Biology*, *20*, 1704–1706.

Nygård, O., Nordehaug, J. E., Refsum, H., Ueland, P. M., Farstad, M., & Vollset, S. E. (1997). Plasma homocysteine levels and mortality in patients with coronary artery disease. *The New England Journal of Medicine*, *337*, 230–236.

Obeid, R., Fedosov, S. N., & Nexo, E. (2015). Cobalamin coenzyme forms are not likely to be superior to cyano- and hydroxyl-cobalamin in prevention or treatment of cobalamin deficiency. *Molecular Nutrition & Food Research*, *59*, 1364–1372.

Patel, K. K., Mejia Munne, J. C., Gunness, V. R. N., Hersey, D., Alshafai, N., Sciubba, D., et al. (2018). Subacute combined degeneration of the spinal cord following nitrous oxide anesthesia: A systematic review of cases. *Clinical Neurology and Neurosurgery*, *173*, 163–168.

Poli, D., Antonucci, E., Cecchi, E., Marcucci, R., Liotta, A. A., Cellai, A. P., et al. (2005). Culprit factors for the failure of well-conducted warfarin therapy to prevent ischemic events in patients with atrial fibrillation: The role of homocysteine. *Stroke*, *36*, 2159–2163.

Qin, X., Li, J., Spence, J. D., Zhang, Y., Li, Y., Wang, X., et al. (2016). Folic acid therapy reduces the first stroke risk associated with hypercholesterolemia among hypertensive patients. *Stroke*, *47*(11), 2805–2812.

Qin, X, Spence, JD, Li, J, Zhang, Y, Li, Y, Sun, N, et al. (2020). Interaction of serum vitamin B12 and folate with MTHFR genotypes on risk of ischemic stroke. *Neurology*, *94*(11), e1126–e1136.

Sadiq, M., Alam, M. T., Kanpurwala, M. A., & Khan, M. S. (2014). Frequency of hyper-homocysteinaemia in ischaemic stroke patients of Karachi. *Journal of the Pakistan Medical Association*, *64*(9), 1063–1066.

Spence, D. (2009). Mechanisms of thrombogenesis in atrial fibrillation. *Lancet*, *373*, 1006–1007.

Spence, J. D. (2016). Metabolic vitamin B12 deficiency: A missed opportunity to prevent dementia and stroke. *Nutrition Research*, *36*, 109–116.

Spence, J. D., Bang, H., Chambless, L. E., & Stampfer, M. J. (2005). Vitamin intervention for stroke prevention trial: An efficacy analysis. *Stroke*, *36*, 2404–2409.

Spence, J. D., Cordy, P., Kortas, C., & Freeman, D. (1999). Effect of usual doses of folate supplementation on elevated plasma homocyst(e)ine in hemodialysis patients: No difference between 1 and 5 mg daily. *American Journal of Nephrology*, *19*, 405–410.

Spence, J. D., Yi, Q., & Hankey, G. J. (2017). B vitamins in stroke prevention: Time to reconsider. *Lancet Neurology*, *16*(9), 750–760.

Stabler, S. P. (2013). Clinical practice. Vitamin B12 deficiency. *The New England Journal of Medicine*, *368*, 149–160.

Stabler, S. P., Lindenbaum, J., & Allen, R. H. (1996). The use of homocysteine and other metabolites in the specific diagnosis of vitamin B-12 deficiency. *The Journal of Nutrition*, *126*, 1266S–1272S.

Toole, J. F., Malinow, M. R., Chambless, L. E., Spence, J. D., Pettigrew, L. C., Howard, V. J., et al. (2004). Lowering plasma total homocysteine to prevent recurrent stroke, myocardial infarction, and death in ischemic stroke patients: Results of the vitamin intervention for stroke prevention (VISP) randomized trial. *Journal of the American Medical Association*, *291*, 565–575.

Vogiatzoglou, A., Oulhaj, A., Smith, A. D., Nurk, E., Drevon, C. A., Ueland, P. M., et al. (2009). Determinants of plasma methylmalonic acid in a large population: Implications for assessment of vitamin B12 status. *Clinical Chemistry*, *55*, 2198–2206.

Wang, C. Y., Chen, Z. W., Zhang, T., Liu, J., Chen, S. H., Liu, S. Y., et al. (2014). Elevated plasma homocysteine level is associated with ischemic stroke in Chinese hypertensive patients. *European Journal of Internal Medicine*, *25*(6), 538–544.

CHAPTER 42

Dietary sodium and stroke: A new narrative

Ahmad Jayedi[a] and Sakineh Shab-Bidar[b]
[a]Food Safety Research Center (Salt), Semnan University of Medical Sciences, Semnan, Iran
[b]Department of Community Nutrition, School of Nutritional Sciences and Dietetics, Tehran University of Medical Sciences, Tehran, Iran

Abbreviations

CNS central nervous system
CVD cardiovascular disease
DALY disability-adjusted life-years
FFQ food frequency questionnaire
HTN hypertension
ICH intracerebral hemorrhage
SAH subarachnoid hemorrhage
WHO World Health Organization

Introduction

According to the definition presented by the American Heart Association/American Stroke Association, central nervous system (CNS) infarction is defined as the death of the brain, spinal cord, or retinal cells due to ischemia diagnosed by established evidence of neuropathological injury, neuroimaging techniques, or clinical evidence of permanent neurological injury (Sacco et al., 2013).

Stroke includes a broad clinical spectrum, including CNS infarction and hemorrhagic accidents. CNS infarction is defined as the death of the brain, spinal cord, or retinal cells resulting from an ischemic event and includes ischemic stroke and silent CNS infarction (Sacco et al., 2013). Ischemic stroke is characterized by the evident clinical manifestation of neurological dysfunction resulting from a focal infarction in the brain, spinal cord, or retina (Sacco et al., 2013).

Stroke can also occur due to a hemorrhagic accident. Intracerebral hemorrhage (ICH) is defined as an accumulation of blood in the brain ventricular system or parenchyma not caused by a trauma, and may include the accumulation of blood in the brain parenchyma following a CNS infarction (Sacco et al., 2013). ICH is commonly accompanied by rapid onset of clinical manifestation of neurological dysfunction. Hemorrhagic stroke also includes subarachnoid hemorrhage (SAH), which is a focal collection of blood in the

Vitamins and Minerals in Neurological Disorders
https://doi.org/10.1016/B978-0-323-89835-5.00025-9

subarachnoid space of the brain. SAH is also accompanied by rapid onset of clinical manifestation of neurological dysfunction and/or headache (Sacco et al., 2013).

Stroke events can lead to neuronal injuries, necrosis, and death and thus can cause sustained disabilities such as cognitive impairment (Lopez et al., 2003), depression (Robinson & Jorge, 2016), impaired ability to move and mobility (Van De Port, Kwakkel, Van Wijk, & Lindeman, 2006), and death. Although the worldwide age-standardized stroke mortality rate has been accompanied by a sharp decrease over the past 30 years, it is still responsible for about 5.5 million deaths across the globe. Stroke was therefore the second leading cause of death and disability-adjusted life-years (DALY) in the world in 2016 (Gorelick, 2019).

Cardiometabolic risk factors including high blood pressure, hyperglycemia, adiposity, and blood lipid abnormalities are responsible for 72% of stroke DALYs, and lifestyle-related factors including sedentary lifestyle, unhealthy eating pattern, and smoking are responsible for 66% of stroke DALYs (GBD 2016 Lifetime Risk of Stroke Collaborators, 2018). Almost all aforementioned risk factors are modifiable and have therefore been proven to diminish stroke risk (Virani et al., 2020). Of those, hypertension (HTN) is the major cause of stroke death and disability in the world. Among 87 different risk factors, high systolic blood pressure was the most important risk factor of death in 2019, responsible for about 10.8 million deaths, 19% of all deaths, in the world (Murray et al., 2020).

High sodium intake is one of the key dietary factors related to high blood pressure (Karppanen & Mervaala, 2006). Existing evidence suggests an evident linear dose–response association between dietary sodium intake and blood pressure (Filippini et al., 2021). Human interventional studies have indicated that even a modest reduction in dietary sodium intake, independent of blood pressure status, can exert favorable and clinically important effects on levels of blood pressure (Filippini et al., 2021).

In this chapter, we discussed current evidence relating dietary sodium intake to stroke risk, potential mechanisms explaining this association, and current recommendations for reducing sodium intake to decrease the risk of cardiovascular disease (CVD), including stroke.

Epidemiologic studies

The association between dietary sodium intake and the risk of stroke incidence has been evaluated in several prospective cohort studies over the past 30 years. A decade of follow-up among Hawaiian Japanese men indicated a nonsignificant association between dietary sodium intake and the risk of total stroke and its subtypes (Kagan, Popper, Rhoads, & Yano, 1985). A report from the first National Health and Nutrition Examination Survey Epidemiologic Follow-up Study in 1999 indicated that every 100 mmol per day increase in dietary sodium intake, as assessed by a single 24-h dietary recall, was associated with a 32% higher risk of stroke incidence and an 89% higher risk of stroke mortality among overweight adults aged 25–74 years, after 19 years of follow-up (He et al., 1999).

However, high sodium intake was not related to the risk of stroke among normal-weight individuals.

Large-scale prospective cohort studies in Japan also showed positive associations between sodium intake and stroke risk (Nagata, Takatsuka, Shimizu, & Shimizu, 2004; Umesawa et al., 2008). Both studies were conducted in Japanese general populations and used an FFQ for dietary assessments.

However, prospective cohort studies conducted in European countries indicated no associations between sodium intake and stroke risk (Geleijnse et al., 2007; Tuomilehto et al., 2001). Both studies evaluated urinary sodium excretion.

Potential mechanisms

The underlying mechanism explaining the link between high sodium intake and the risk of stroke is the effect of high sodium intake on levels of blood pressure (Karppanen & Mervaala, 2006). High blood pressure is the leading risk factor of stroke incidence and mortality (Guzik & Bushnell, 2017).

High blood pressure can exert extra strain on blood vessels and thus can make blood vessels stiff and narrow, leading to the formation of clots (Lawes, Bennett, Feigin, & Rodgers, 2004). Formation of the clots in brain vessels or traveling them from other organs into the brain vessels can lead to ischemic stroke. Elevated blood pressure can also damage blood vessels and weaken them, leading to a leak of the blood or burst and subsequent bleeding in the brain, called hemorrhagic stroke (Lawes et al., 2004).

Other potential mechanisms linking sodium intake to stroke risk include adverse effects of high sodium intake on renal dysfunction, vascular remolding, endothelial dysfunction, arterial fibrosis, and left ventricular hypertrophy, all of which are associated with an increased risk of CVD events (Appel et al., 2011). High sodium intake may also induce oxidative stress (Bayorh, Ganafa, Socci, Silvestrov, & Abukhalaf, 2004), leading to adverse impacts on cerebral vasculature and ischemic brain injury (Allen & Bayraktutan, 2009). Potential mechanisms linking high sodium intake to stroke risk are summarized in Table 1.

Table 1 Potential mechanisms linking high sodium intake to stroke risk.

High blood pressure
Renal dysfunction
Vascular remodeling
Endothelial dysfunction
Arterial fibrosis
Left ventricular hypertrophy
Oxidative stress

Regional difference in the sodium-stroke association

A recent meta-analysis of prospective cohort studies suggested a potential regional difference in the association between sodium intake and stroke risk (Jayedi, Ghomashi, Zargar, & Shab-Bidar, 2019). The analyses of cohort studies indicated that higher sodium intake was not related to the risk of stroke in the US and especially Europe, but was positively associated with stroke risk in Asian countries.

However, there were no significant subgroup effects (P for subgroup difference = 0.18). In addition, Asian studies were large-scale prospective cohort studies and thus, showed a stronger effect size. The inconsistent findings can also be attributable to the higher rate of stroke incidence and mortality in Asian countries as compared to Western countries (Johnson et al., 2019), as well as higher sodium intake in Asian communities (Powles et al., 2013). A worldwide survey in 2010 indicated that compared with the global mean sodium intake of 3.95 g/day, sodium intake was higher in Asian and Eastern European countries (>4.2 g/d), but was lower in North America, Western Europe, and Australia (ranged between 3.4 and 3.8 g/d) (Powles et al., 2013).

The shape of the association between sodium intake and stroke risk

Results from epidemiologic studies presented strong and convincing evidence that higher sodium intake is positively associated with a higher risk of developing CVD (O'donnell, Mente, Smyth, & Yusuf, 2013). Evidence from clinical trials has consistently indicated that lowering dietary sodium intake can exert favorable effects on levels of blood pressure (Filippini et al., 2021). Therefore, current guidelines recommended that dietary sodium intake should be reduced to 1500–2300 mg/d (equal to 3.75–5.75 g of salt/d) in adults to lower CVD risk (Appel et al., 2011; U.S. Department of Health and Human Services, 2015; World Health Organization, 2012).

However, some observational studies suggested that there might be a potential J- or U shaped association between sodium intake and the risk of CVD (Stolarz-Skrzypek et al., 2011). A meta-analysis of cohort studies suggested that as compared to moderate sodium intake (2.7 to 5 g/d), low sodium intake (<2.7 g/d) may be associated with a higher risk of all-cause mortality and CVD events (Graudal, Jürgens, Baslund, & Alderman, 2014).

It is suggested that the observed J-shaped association between sodium intake and CVD risk may be due to reverse causality bias resulting from a higher prevalence of preexisting comorbidities in individuals with low sodium intake (Cobb et al., 2014). Also, residual confounding and short follow-up duration may distort the association between sodium intake and CVD risk.

Our recent dose–response meta-analysis of prospective cohort studies (Jayedi et al., 2019) also suggested a positive monotonic association between sodium intake and stroke

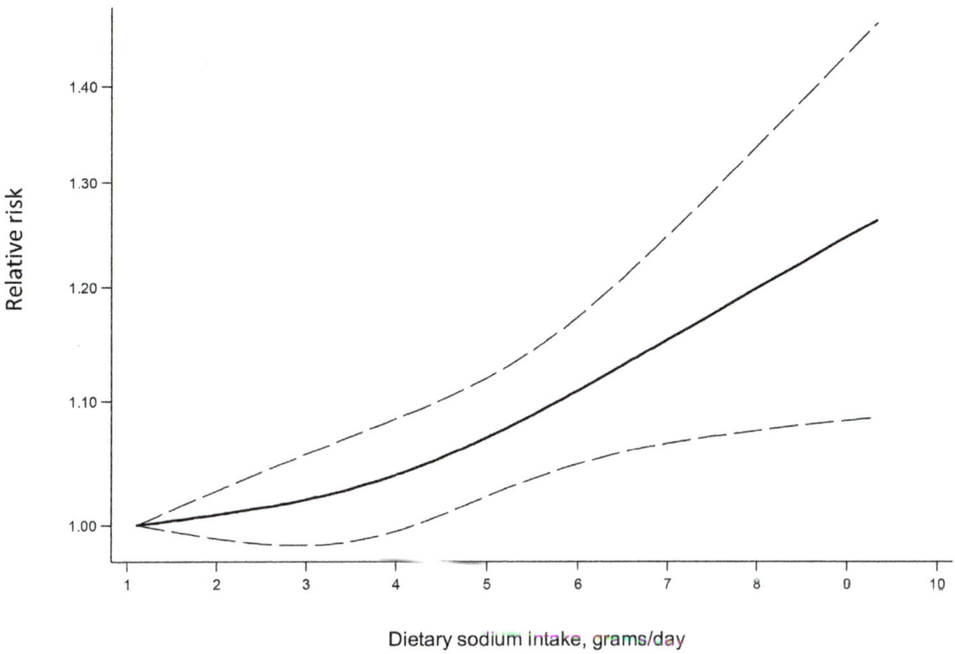

Total stroke, P-nonlinearity = 0.23

Fig. 1 Dose–response association between dietary sodium intake and the risk of stroke. *(Reprinted from "A systematic review and nonlinear dose-response meta-analysis," by Ahmad Jayedi, Farnoosh Ghomashi, Mahdieh Sadat Zargar, Sakineh Shab-Bidar (2019).* Clinical Nutrition, 38(3), *1092–1100. Copyright 2018 by "Elsevier Ltd and European Society for Clinical Nutrition and Metabolism".)*

risk in the general population, with risk increasing significantly at sodium intake of about 5 g/d and with no evidence of an increased risk in individuals with low sodium intake (Fig. 1). The analyses indicated that sodium intake higher than 5 g/d was associated with a sharp increase in the risk of stroke incidence and mortality.

Measuring sodium intake

Results from randomized trials have indicated that lowering sodium intake is an effective dietary intervention for reducing blood pressure levels (Filippini et al., 2021) and subsequent risk of CVD and stroke (Appel et al., 2011). Reduction in mean population sodium or salt intake by 30% is one of the key targets outlined in the World Health Organization (WHO) Global Action Plan for the Prevention and Control of Non-Communicable Disease 2013–2020 (House, 2013). To provide effective recommendations, researchers should use accurate tools for measuring sodium intake.

Dietary tools for the assessment of nutrient intake include FFQ, dietary recall, or weighted food records. Dietary assessment methods may underestimate sodium intake because of difficulties in measuring sodium concentration in foods and recipes and under-reporting of dietary intake (McLean, 2014). Sodium intake is highly correlated with energy intake (Whelton et al., 2012) and thus, under-reporting of energy intake (Bailey, Mitchell, Miller, & Smiciklas-Wright, 2007), which is highly prevalent in the world due to the high prevalence of adiposity, may lead to underestimation of sodium intake.

People may also change their eating behaviors when recording their dietary data. In addition, large variations in the sodium content of both home-made and processed foods, differences in dietary sources of sodium across countries (Anderson et al., 2010), and discretionary use of salt make it difficult to perform an accurate estimation of sodium intake using dietary assessment tools. However, dietary assessment methods can be used in epidemiologic studies, can determine dietary sources of sodium and, thus, are useful to provide more effective recommendations for lowering sodium intake. They also allow to estimate energy-adjusted intake and thus control for confounding factors associated with total energy intake (Willett, Howe, & Kushi, 1997).

About 90% of sodium intake is excreted in the urine, and thus, measuring urinary sodium excretion is a reliable tool for measuring dietary sodium intake. 24-h urine sample is the gold standard method for measuring sodium intake and validates other tools for the assessment of sodium intake (Domnariu, Cucu, & Furtunescu, 2013). However, collection of 24 urine is difficult and, thus, may limit its utilization in epidemiologic research. Due to a lower response rate, using 24-h urine may introduce bias in epidemiologic research (McLean, 2014). Research also uses spot urine samples for estimating 24-h urinary sodium excretion (Tanaka et al., 2002). However, more research is needed to determine whether a spot urine sample is a convenient alternative to 24-h urine in population-based surveys (Ji, Dary, Campbell, & Cappuccio, 2013). Therefore, it may be a good approach to collect 24-h urine samples, when using a spot urine sample in epidemiologic research, to provide valid equations to do accurate estimation from spot urine samples.

Applications to other neurological conditions

Evidence regarding the association of dietary sodium intake with other neurological conditions is sparse. Higher sodium intake may induce a proinflammatory state and affect autoimmunity (Kleinewietfeld et al., 2013) and thus, may be a potential hazard for autoimmune disease such as multiple sclerosis (MS) (Probst, Mowbray, Svensen, & Thompson, 2019).

Only two studies evaluated the association between sodium intake and risk of MS: one a case–control study in pediatrics (McDonald et al., 2016) and another a cohort study of about 80,000 female nurses (Cortese, Yuan, Chitnis, Ascherio, & Munger, 2017), both of which reported null associations. It seems that the adverse effects of sodium on immune cells found in animal studies are not powerful enough to induce MS. Evidence regarding the adverse effects of high sodium intake on disease activity and progression in patients diagnosed with MS is also conflicting (Farez, Fiol, Gaitán, Quintana, & Correale, 2015; Fitzgerald et al., 2017).

There are also some suggestions of the adverse effects of high sodium intake on cognitive function and dementia (Mohan et al., 2020). Higher sodium intake is a key dietary risk factor of high blood pressure, an important risk factor for cognitive decline (Manolio, Olson, & Longstreth, 2003). High sodium intake can also induce a proinflammatory state and, thus, can exert unfavorable impacts on cognitive function (Mohan et al., 2020). Other potential mechanisms such as gut-initiated pathways (Faraco et al., 2018) have also been reported to explain the potential link observed between high sodium intake and cognitive decline. However, existing evidence is heterogeneous and conflicting and is derived from observational studies with poor methodological quality, highlighting the need for more research in the field.

A cross-sectional evaluation within the 1999–2004 National Health and Nutrition Examination Survey suggested an inverse association between sodium intake and odds of having migraine history (Pogoda et al., 2016b). The unexpected inverse relationship between sodium intake and migraine may be due to altered sodium hemostasis in the brain in migraine and the use of antihypertensive medications (Pogoda et al., 2016a). In addition, another small cross-sectional study in Iran suggested that higher sodium intake is positively associated with the frequency and duration of migraine headaches and migraine headache index score (Arab, Khorvash, Heidari, & Askari, 2021).

Other components of interest

Epidemiologic studies have reported associations for other dietary components. In general, higher intake of plant-based foods and food groups has been associated with stroke risk. Higher intake of fruit (Hu, Huang, Wang, Zhang, & Qu, 2014), especially those rich in flavonoids (Gao et al., 2021), is related to a lower risk of developing stroke in prospective cohort studies. Higher intakes of vegetables, especially allium, carrots, cruciferous, and green leafy vegetables, have been associated with a lower risk of incidence of stroke and other CVD (Zurbau et al., 2020). Fruit and vegetables are rich in nutrients with strong antiinflammatory and antioxidant activities. Low-grade systemic inflammation and oxidative stress are the two underlying mechanisms involved in the pathogenesis of stroke, especially those with ischemic background (Tiwari, Mishra, & Rajavashisth, 2019).

Other plant-based foods such as nuts and whole grains are also inversely associated with stroke risk (Aune et al., 2016a, 2016b). Consistently, results from prospective cohort studies illustrated that diets rich in plant-based foods such as dietary approaches to stop hypertension (DASH) diet (Chiavaroli et al., 2019) and the Mediterranean dietary pattern (Rosato et al., 2019) have been associated with a lower risk of developing stroke.

In addition, there have been associations for animal-based foods. Greater fish intake, the main dietary source of long-chain omega-3 fatty acids, has been related to lower stroke risk (Chen et al., 2021). In contrast, higher red meat intake, especially those of processed type, has been positively but slightly associated with the risk of ischemic stroke, but not hemorrhagic stroke (Yang et al., 2016). Consistently, a higher intake of ultra-processed foods, rich in sodium, is associated with a higher risk of developing cerebrovascular disease (Srour et al., 2019).

The associations between dietary carbohydrates with stroke risk have also been investigated. A recent comprehensive review of prospective cohort studies indicated that higher dietary glycemic load, as being representative of both quality and quantity of dietary carbohydrates, was associated with a higher risk of stroke, with the certainty of evidence being rated moderate (Jayedi, Soltani, Jenkins, Sievenpiper, & Shab-Bidar, 2020). However, no association has been reported between dietary glycemic index and stroke risk.

Another important nutrient associated with stroke risk is potassium. Potassium intake is a key dietary determinant of blood pressure and, thus, can be associated with stroke risk (Vinceti et al., 2016). Randomized trials have indicated that potassium supplementation can significantly decrease blood pressure, especially in those with high sodium intake (Filippini, Violi, D'Amico, & Vinceti, 2017). There is evidence that a higher dietary sodium-to-potassium ratio is strongly associated with a higher risk of stroke. A recent meta-analysis of cohort studies illustrated evidence of a positive monotonic association between dietary sodium-to-potassium ratio and stroke risk (Fig. 2) (Jayedi et al., 2019). It is proposed that the inverse associations found between plant-based foods and diets rich in plant-based foods such as DASH dietary pattern is mediated, in part, by their higher potassium content. The association between different dietary components and the risk of stroke is summarized in Table 2.

Current recommendations for sodium intake

In 2012, the World Health Organization confirmed its previous recommendations (World Health Organization, 2007) for lowering sodium intake to <2 g/d (5 g of salt/d) as optimal for adults (World Health Organization, 2012). The World Health Organization Global Action Plan for the Prevention and Control of Non-Communicable Disease 2013–2020 asserted that relative reduction in mean population sodium intake by 30% is one of the key targets for reducing the risk of chronic disease (House, 2013). The US Department of Agriculture (U.S. Department of Health and Human Services, 2015) and the American Heart Associations (Appel et al., 2011) currently recommend daily sodium intake of 2 g and 1.5 g (equal to 5 and 3.75 g of salt/d, respectively) for adults.

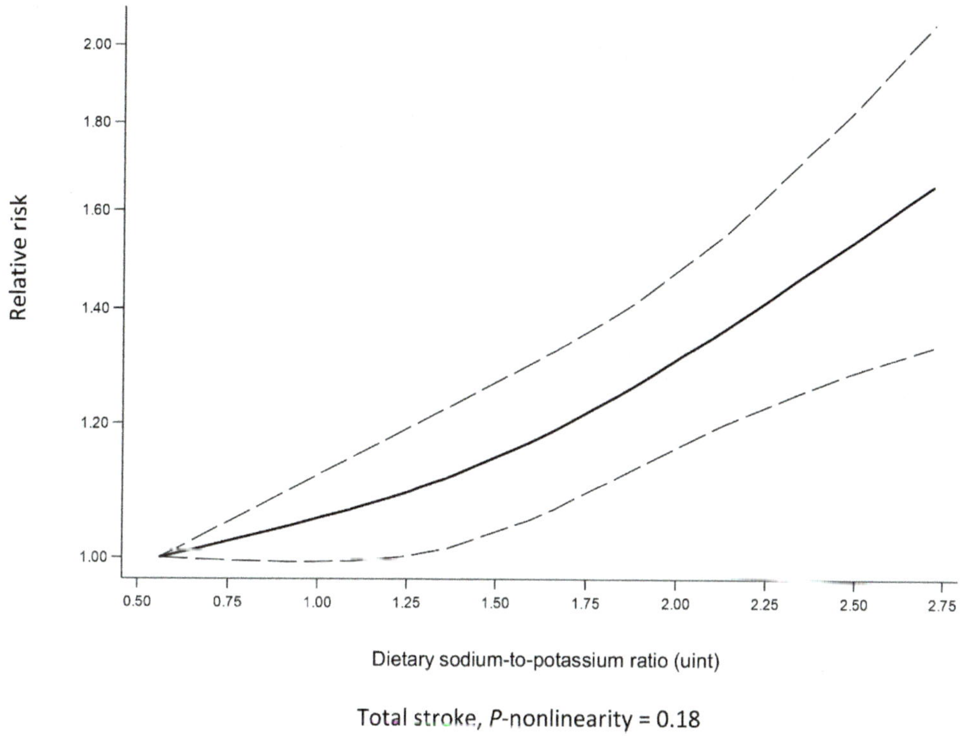

Fig. 2 Dose–response association between dietary sodium-to-potassium ratio and the risk of stroke. *(Reprinted from "A systematic review and nonlinear dose-response meta-analysis," by Ahmad Jayedi, Farnoosh Ghomashi, Mahdieh Sadat Zargar, Sakineh Shab-Bidar (2019).* Clinical Nutrition, 38(3), *1092–1100. Copyright 2018 by "Elsevier Ltd and European Society for Clinical Nutrition and Metabolism".)*

The global mean sodium intake was estimated to be about 3.95 g/d in 2010 (Powles et al., 2013), which was far greater than the current recommendations for daily sodium intake. Human interventional studies have indicated that lowering sodium intake is an effective dietary strategy for reducing high blood pressure (Filippini et al., 2021), a key risk factor for CVD and especially stroke. Thus, the relative reduction in population sodium intake can be a convenient approach to reduce the burden of stroke death and disabilities.

Conclusions

Higher sodium intake is a key dietary factor related to high blood pressure levels, the underlying risk factor for stroke death and disability. Currently, the global mean population sodium intake is far higher than the recommended daily sodium intake.

Table 2 The association between different dietary components and the risk of stroke.

Dietary component	Association with the risk of stroke
Fruit	↓
Vegetables	↓
Whole grains	↓
Nuts	↓
Fish	↓
Mediterranean diet	↓
DASH dietary pattern	↓
Plant-based diet	↓
Vegetarian diet	↔
Red meant	↑
Processed meat	↑
Ultra-processed foods	↑
Saturated fats	↓
Linoleic acid	↓
Omega-3 fatty acids	↓
Trans fats	↔
Vitamin C	↓
Vitamin E	↓
Potassium	↓
Magnesium	↓
Sodium	↑

DASH, dietary approaches to stop hypertension.

The relative reduction in sodium intake, in parallel with adopting a healthy plant-based eating style and lower consumption of ultra-processed foods, may be a convenient dietary suggestion for lowering the large burden of stroke death and disability across the globe.

Mini-dictionary of terms

- **Dose–response association:** The association of different values of dietary sodium intake with the risk of stroke incidence, compared to the lowest intake.
- **Regional difference:** Existing difference in the magnitude and direction of the association between dietary sodium intake and stroke risk across different geographical regions.
- **Plant-based diets:** Diets rich in foods originated from plants such as whole grains, legumes, fruit, vegetables, nuts, and legumes.
- **Ultra-processed foods:** Foods (such as soft drinks, chips, chocolate, candy, ice-cream, sweetened breakfast cereals, packaged soups, chicken nuggets, hotdogs, fries)

that change under influence of different processes such as milling, extrusion, or molding. They are highly manipulated and also may contain added ingredients.
- **Sodium-to-potassium ratio:** The ratio of dietary sodium intake to dietary potassium intake.

Key facts of stroke

- Stroke is defined as the death of the brain, spinal cord, or retinal cells due to ischemia or haemorrhagic accidents.
- Stroke events can lead to neuronal injuries, necrosis and death and thus can cause sustained disabilities.
- Stroke is responsible for about 5.5 million deaths and therefore, is the second leading cause of death and disability-adjusted life-years in the world.
- High systolic blood pressure was the most important risk factor of death in 2019, responsible for about 10.8 million deaths.

Key facts of stroke

- Current recommendations for sodium intake in adults range from 1.5 g/d by the American Heart Association to 2 g/d by the World Health Organization and the US Department of Agriculture.
- The global mean sodium intake was estimated to be about 3.95 g/d, ranging from 3.4 to 3.8 g/d in the North America, Western Europe, and Australia to >4.2 g/d in Asian and Eastern European countries.
- Relative reduction in mean population sodium intake by 30% is one of the key targets for reducing the risk of chronic disease.

Summary points

- Stroke and associated disabilities are one of the leading causes of death and impaired mobility in the world.
- Elevated blood pressure is the leading risk factor for stroke incidence and mortality.
- High salt intake is the key dietary determinant of blood pressure levels.
- Reducing dietary sodium intake can reduce blood pressure levels.
- Relative reduction in sodium intake may be a convenient dietary suggestion for lowering the large burden of stroke death and disability across the globe.
- Adopting a healthy plant-based eating style and lower consumption of ultra-processed foods can help to lower the risk of stroke death and disabilities.

References

Allen, C. L., & Bayraktutan, U. (2009). Oxidative stress and its role in the pathogenesis of ischaemic stroke. *International Journal of Stroke*, *4*(6), 461–470.

Anderson, C. A., Appel, L. J., Okuda, N., Brown, I. J., Chan, Q., Zhao, L., et al. (2010). Dietary sources of sodium in China, Japan, the United Kingdom, and the United States, women and men aged 40 to 59 years: The INTERMAP study. *Journal of the American Dietetic Association*, *110*(5), 736–745.

Appel, L. J., Frohlich, E. D., Hall, J. E., Pearson, T. A., Sacco, R. L., Seals, D. R., et al. (2011). The importance of population-wide sodium reduction as a means to prevent cardiovascular disease and stroke: A call to action from the American Heart Association. *Circulation*, *123*(10), 1138–1143.

Arab, A., Khorvash, F., Heidari, Z., & Askari, G. (2021). Is there a relationship between dietary sodium and potassium intake and clinical findings of a migraine headache? *British Journal of Nutrition*, 1–10.

Aune, D., Keum, N., Giovannucci, E., Fadnes, L. T., Boffetta, P., Greenwood, D. C., et al. (2016a). Nut consumption and risk of cardiovascular disease, total cancer, all-cause and cause-specific mortality: A systematic review and dose-response meta-analysis of prospective studies. *BMC Medicine*, *14*(1), 1–14.

Aune, D., Keum, N., Giovannucci, E., Fadnes, L. T., Boffetta, P., Greenwood, D. C., et al. (2016b). Whole grain consumption and risk of cardiovascular disease, cancer, and all cause and cause specific mortality: Systematic review and dose-response meta-analysis of prospective studies. *BMJ*, *353*.

Bailey, R. L., Mitchell, D. C., Miller, C., & Smiciklas-Wright, H. (2007). Assessing the effect of underreporting energy intake on dietary patterns and weight status. *Journal of the American Dietetic Association*, *107*(1), 64–71.

Bayorh, M. A., Ganafa, A. A., Socci, R. R., Silvestrov, N., & Abukhalaf, I. K. (2004). The role of oxidative stress in salt-induced hypertension. *American Journal of Hypertension*, *17*(1), 31–36.

Chen, C., Huang, H., Dai, Q.-Q., Ren, J., Cai, H.-H., Hu, W.-J., et al. (2021). Fish consumption, long-chain omega-3 fatty acids intake and risk of stroke: An updated systematic review and meta-analysis. *Asia Pacific Clinical Nutrition Society*, *30*(1), 140–152.

Chiavaroli, L., Viguiliouk, E., Nishi, S. K., Blanco Mejia, S., Rahelić, D., Kahleová, H., et al. (2019). DASH dietary pattern and cardiometabolic outcomes: An umbrella review of systematic reviews and meta-analyses. *Nutrients*, *11*(2), 338.

Cobb, L. K., Anderson, C. A., Elliott, P., Hu, F. B., Liu, K., Neaton, J. D., et al. (2014). Methodological issues in cohort studies that relate sodium intake to cardiovascular disease outcomes: A science advisory from the American Heart Association. *Circulation*, *129*(10), 1173–1186.

Cortese, M., Yuan, C., Chitnis, T., Ascherio, A., & Munger, K. L. (2017). No association between dietary sodium intake and the risk of multiple sclerosis. *Neurology*, *89*(13), 1322–1329.

Domnariu, C. D., Cucu, A., & Furtunescu, F. L. (2013). World Health Organization guidelines on salt intake in adults and children. *Acta Medica Transilvanica*, *18*(1), 166.

Faraco, G., Brea, D., Garcia-Bonilla, L., Wang, G., Racchumi, G., Chang, H., et al. (2018). Dietary salt promotes neurovascular and cognitive dysfunction through a gut-initiated TH17 response. *Nature Neuroscience*, *21*(2), 240–249.

Farez, M. F., Fiol, M. P., Gaitán, M. I., Quintana, F. J., & Correale, J. (2015). Sodium intake is associated with increased disease activity in multiple sclerosis. *Journal of Neurology, Neurosurgery & Psychiatry*, *86*(1), 26–31.

Filippini, T., Malavolti, M., Whelton, P. K., Naska, A., Orsini, N., & Vinceti, M. (2021). Blood pressure effects of sodium reduction: Dose-response meta-analysis of experimental studies. *Circulation*, *143*(16), 1542–1567.

Filippini, T., Violi, F., D'Amico, R., & Vinceti, M. (2017). The effect of potassium supplementation on blood pressure in hypertensive subjects: A systematic review and meta-analysis. *International Journal of Cardiology*, *230*, 127–135.

Fitzgerald, K. C., Munger, K. L., Hartung, H. P., Freedman, M. S., Montalbán, X., Edan, G., et al. (2017). Sodium intake and multiple sclerosis activity and progression in BENEFIT. *Annals of Neurology*, *82*(1), 20–29.

Gao, Q., Dong, J.-Y., Cui, R., Muraki, I., Yamagishi, K., Sawada, N., et al. (2021). Consumption of flavonoid-rich fruits, flavonoids from fruits and stroke risk: A prospective cohort study. *British Journal of Nutrition*, 1–8.

GBD 2016 Lifetime Risk of Stroke Collaborators. (2018). Global, regional, and country-specific lifetime risks of stroke, 1990 and 2016. *New England Journal of Medicine, 379*(25), 2429–2437.

Geleijnse, J. M., Witteman, J. C., Stijnen, T., Kloos, M. W., Hofman, A., & Grobbee, D. E. (2007). Sodium and potassium intake and risk of cardiovascular events and all-cause mortality: The Rotterdam study. *European Journal of Epidemiology, 22*(11), 763–770.

Gorelick, P. B. (2019). The global burden of stroke: Persistent and disabling. *The Lancet Neurology, 18*(5), 417–418.

Graudal, N., Jürgens, G., Baslund, B., & Alderman, M. H. (2014). Compared with usual sodium intake, low- and excessive-sodium diets are associated with increased mortality: A meta-analysis. *American Journal of Hypertension, 27*(9), 1129–1137.

Guzik, A., & Bushnell, C. (2017). Stroke epidemiology and risk factor management. *Continuum: Lifelong Learning in Neurology, 23*(1), 15–39.

He, J., Ogden, L. G., Vupputuri, S., Bazzano, L. A., Loria, C., & Whelton, P. K. (1999). Dietary sodium intake and subsequent risk of cardiovascular disease in overweight adults. *JAMA, 282*(21), 2027–2034.

House, W. (2013). *Follow-up to the political declaration of the high-level meeting of the general assembly on the prevention and control of non-communicable diseases.* World Health Organization.

Hu, D., Huang, J., Wang, Y., Zhang, D., & Qu, Y. (2014). Fruits and vegetables consumption and risk of stroke: A meta-analysis of prospective cohort studies. *Stroke, 45*(6), 1613–1619.

Jayedi, A., Ghomashi, F., Zargar, M. S., & Shab-Bidar, S. (2019). Dietary sodium, sodium-to-potassium ratio, and risk of stroke: A systematic review and nonlinear dose-response meta-analysis. *Clinical Nutrition, 38*(3), 1092–1100.

Jayedi, A., Soltani, S., Jenkins, D., Sievenpiper, J., & Shab-Bidar, S. (2020). Dietary glycemic index, glycemic load, and chronic disease: An umbrella review of meta-analyses of prospective cohort studies. *Critical Reviews in Food Science and Nutrition*, 1–10.

Ji, C., Dary, O., Campbell, N. R., & Cappuccio, F. P. (2013). Spot and overnight urine are inappropriate to assess population sodium intake. *Revista Panamericana de Salud Pública, 34*(4), 283.

Johnson, C. O., Nguyen, M., Roth, G. A., Nichols, E., Alam, T., Abate, D., et al. (2019). Global, regional, and national burden of stroke, 1990–2016: A systematic analysis for the global burden of disease study 2016. *The Lancet Neurology, 18*(5), 439–458.

Kagan, A., Popper, J., Rhoads, G., & Yano, K. (1985). Dietary and other risk factors for stroke in Hawaiian Japanese men. *Stroke, 16*(3), 390–396.

Karppanen, H., & Mervaala, E. (2006). Sodium intake and hypertension. *Progress in Cardiovascular Diseases, 49*(2), 59–75.

Kleinewietfeld, M., Manzel, A., Titze, J., Kvakan, H., Yosef, N., Linker, R. A., et al. (2013). Sodium chloride drives autoimmune disease by the induction of pathogenic TH 17 cells. *Nature, 496*(7446), 518–522.

Lawes, C. M., Bennett, D. A., Feigin, V. L., & Rodgers, A. (2004). Blood pressure and stroke: An overview of published reviews. *Stroke, 35*(3), 776–785.

Lopez, O. L., Jagust, W. J., Dulberg, C., Becker, J. T., DeKosky, S. T., Fitzpatrick, A., et al. (2003). Risk factors for mild cognitive impairment in the cardiovascular health study cognition study: Part 2. *Archives of Neurology, 60*(10), 1394–1399.

Manolio, T. A., Olson, J., & Longstreth, W. (2003). Hypertension and cognitive function: Pathophysiologic effects of hypertension on the brain. *Current Hypertension Reports, 5*(3), 255–261.

McDonald, J., Graves, J., Waldman, A., Lotze, T., Schreiner, T., Belman, A., et al. (2016). A case-control study of dietary salt intake in pediatric-onset multiple sclerosis. *Multiple Sclerosis and Related Disorders, 6*, 87–92.

McLean, R. M. (2014). Measuring population sodium intake: A review of methods. *Nutrients, 6*(11), 4651–4662.

Mohan, D., Yap, K. H., Reidpath, D., Soh, Y. C., McGrattan, A., Stephan, B., et al. (2020). Link between dietary sodium intake, cognitive function, and dementia risk in middle-aged and older adults: A systematic review. *Journal of Alzheimer's Disease*, 1–27.

Murray, C. J., Aravkin, A. Y., Zheng, P., Abbafati, C., Abbas, K. M., Abbasi-Kangevari, M., et al. (2020). Global burden of 87 risk factors in 204 countries and territories, 1990–2019: A systematic analysis for the global burden of disease study 2019. *The Lancet, 396*(10258), 1223–1249.

Nagata, C., Takatsuka, N., Shimizu, N., & Shimizu, H. (2004). Sodium intake and risk of death from stroke in Japanese men and women. *Stroke, 35*(7), 1543–1547.

O'donnell, M., Mente, A., Smyth, A., & Yusuf, S. (2013). Salt intake and cardiovascular disease: Why are the data inconsistent? *European Heart Journal, 34*(14), 1034–1040.

Pogoda, J. M., Gross, N. B., Arakaki, X., Fonteh, A. N., Cowan, R. P., & Harrington, M. G. (2016a). Severe headache or migraine history is inversely correlated with dietary sodium intake: NHANES 1999-2004: A response. *Headache, 56*(7), 1216–1218.

Pogoda, J. M., Gross, N. B., Arakaki, X., Fonteh, A. N., Cowan, R. P., & Harrington, M. G. (2016b). Severe headache or migraine history is inversely correlated with dietary sodium intake: NHANES 1999–2004. *Headache: The Journal of Head and Face Pain, 56*(4), 688–698.

Powles, J., Fahimi, S., Micha, R., Khatibzadeh, S., Shi, P., Ezzati, M., et al. (2013). Global, regional and national sodium intakes in 1990 and 2010: A systematic analysis of 24 h urinary sodium excretion and dietary surveys worldwide. *BMJ Open, 3*(12).

Probst, Y., Mowbray, E., Svensen, E., & Thompson, K. (2019). A systematic review of the impact of dietary sodium on autoimmunity and inflammation related to multiple sclerosis. *Advances in Nutrition, 10*(5), 902–910.

Robinson, R. G., & Jorge, R. E. (2016). Post-stroke depression: A review. *American Journal of Psychiatry, 173*(3), 221–231.

Rosato, V., Temple, N. J., La Vecchia, C., Castellan, G., Tavani, A., & Guercio, V. (2019). Mediterranean diet and cardiovascular disease: A systematic review and meta-analysis of observational studies. *European Journal of Nutrition, 58*(1), 173–191.

Sacco, R. L., Kasner, S. E., Broderick, J. P., Caplan, L. R., Connors, J., Culebras, A., et al. (2013). An updated definition of stroke for the 21st century: A statement for healthcare professionals from the American Heart Association/American Stroke Association. *Stroke, 44*(7), 2064–2089.

Srour, B., Fezeu, L. K., Kesse-Guyot, E., Allès, B., Méjean, C., Andrianasolo, R. M., et al. (2019). Ultra-processed food intake and risk of cardiovascular disease: Prospective cohort study (NutriNet-Santé). *BMJ, 365*.

Stolarz-Skrzypek, K., Kuznetsova, T., Thijs, L., Tikhonoff, V., Seidlerová, J., Richart, T., et al. (2011). Fatal and nonfatal outcomes, incidence of hypertension, and blood pressure changes in relation to urinary sodium excretion. *JAMA, 305*(17), 1777–1785.

Tanaka, T., Okamura, T., Miura, K., Kadowaki, T., Ueshima, H., Nakagawa, H., et al. (2002). A simple method to estimate populational 24-h urinary sodium and potassium excretion using a casual urine specimen. *Journal of Human Hypertension, 16*(2), 97–103. https://doi.org/10.1038/sj.jhh.1001307.

Tiwari, S. K., Mishra, P., & Rajavashisth, T. (2019). Inflammation, oxidative stress, and cerebral stroke: Basic principles. In *Advancement in the Pathophysiology of Cerebral Stroke* (pp. 11–21). Springer.

Tuomilehto, J., Jousilahti, P., Rastenyte, D., Moltchanov, V., Tanskanen, A., Pietinen, P., et al. (2001). Urinary sodium excretion and cardiovascular mortality in Finland: A prospective study. *The Lancet, 357*(9259), 848–851.

U.S. Department of Health and Human Services. (2015). *2015–2020 dietary guidelines for Americans*. Washington, DC: US Department of Agriculture.

Umesawa, M., Iso, H., Date, C., Yamamoto, A., Toyoshima, H., Watanabe, Y., et al. (2008). Relations between dietary sodium and potassium intakes and mortality from cardiovascular disease: The Japan collaborative cohort study for evaluation of Cancer risks. *The American Journal of Clinical Nutrition, 88*(1), 195–202.

Van De Port, I. G., Kwakkel, G., Van Wijk, I., & Lindeman, E. (2006). Susceptibility to deterioration of mobility long-term after stroke: A prospective cohort study. *Stroke, 37*(1), 167–171.

Vinceti, M., Filippini, T., Crippa, A., de Sesmaisons, A., Wise, L. A., & Orsini, N. (2016). Meta-analysis of potassium intake and the risk of stroke. *Journal of the American Heart Association, 5*(10), e004210.

Virani, S. S., Alonso, A., Benjamin, E. J., Bittencourt, M. S., Callaway, C. W., Carson, A. P., et al. (2020). Heart disease and stroke statistics—2020 update: A report from the American Heart Association. *Circulation, 141*(9), e139–e596.

Whelton, P. K., Appel, L. J., Sacco, R. L., Anderson, C. A., Antman, E. M., Campbell, N., et al. (2012). Sodium, blood pressure, and cardiovascular disease: Further evidence supporting the American Heart Association sodium reduction recommendations. *Circulation, 126*(24), 2880–2889.

Willett, W. C., Howe, G. R., & Kushi, L. H. (1997). Adjustment for total energy intake in epidemiologic studies. *The American Journal of Clinical Nutrition*, *65*(4), 1220S–1228S.

World Health Organization. (2007). *Reducing salt intake in populations: Report of a WHO forum and technical meeting, 5–7 October 2006, Paris, France.*

World Health Organization. (2012). *Guideline: Sodium intake for adults and children.* World Health Organization.

Yang, C., Pan, L., Sun, C., Xi, Y., Wang, L., & Li, D. (2016). Red meat consumption and the risk of stroke: A dose–response meta-analysis of prospective cohort studies. *Journal of Stroke and Cerebrovascular Diseases*, *25*(5), 1177–1186.

Zurbau, A., Au-Yeung, F., Blanco Mejia, S., Khan, T. A., Vuksan, V., Jovanovski, E., et al. (2020). Relation of different fruit and vegetable sources with incident cardiovascular outcomes: A systematic review and meta-analysis of prospective cohort studies. *Journal of the American Heart Association*, *9*(19), e017728.

CHAPTER 43

Calcium intake and the risk of stroke

Daniel T. Dibaba[a] **and Ka Kahe**[b]
[a]Cancer Research And Biostatistics (CRAB), Seattle, WA, United States
[b]Department of Obstetrics and Gynecology and Department of Epidemiology, Columbia University, New York, NY, United States

Introduction

Stroke is a leading cause of death and disability worldwide (World Health Organization Geneva. WHO, 2022). In the USA, on average, someone develops stroke every 40 min, and someone dies of stroke every 3 min and 35 seconds (Virani et al., 2020). There are two main types of strokes: ischemic stroke (~85%) and hemorrhagic stroke. Ischemic stroke is the result of blockage of blood flow to the brain, while the hemorrhagic stroke happens because of bleeding into the intracerebral or subarachnoid spaces (Center for Disease Prevention and Control, 2020).

The primary prevention of stroke requires a holistic approach, and there is no single magic bullet for this purpose. The efforts for the primary prevention of stroke should be aimed at preventing the established stroke risk factors through behavioral modification. The established risk factors for the ischemic stroke include atherosclerosis, hypertension, diabetes, overweight and obesity, physical inactivity, and smoking. All these risk factors can be prevented through lifestyle or behavioral modifications. Dietary habit can be modified and is an important factor that can either prevent disease or increase the risk of a disease. There is a growing interest on the association between dietary and supplemental calcium intake and the risk of cardiovascular diseases and stroke. Calcium has been suggested to be associated with the risk factors of stroke for decades. The main sources of dietary calcium are dairy foods (such as milk, yogurt, cheese), calcium-fortified foods, salmon, tofu, vegetables, fruits, and grains (Table 1). Calcium is a critical component in muscle contraction, including vascular smooth muscles (Adelstein & Sellers, 1987; Reid, Birstow, & Bolland, 2017). Calcium is also required for the cardiac signaling and function (Fearnley, Roderick, & Bootman, 2011). However, calcium is involved in atherosclerotic plaque formation (Kalampogias et al., 2016), though studies indicate that supplemental calcium, not dietary calcium intake, is the one associated with incident coronary artery calcification (Anderson et al., 2016). There is an increased supplemental calcium intake in the developed countries especially among women past the age of menopause for the prevention of osteoporosis and fracture (Tankeu, Ndip Agbor, & Noubiap, 2017). In fact, more than 50% of elderly men and 70% of elderly women in

Vitamins and Minerals in Neurological Disorders
https://doi.org/10.1016/B978-0-323-89835-5.00022-3

Table 1 Dietary sources of calcium (National Institute of Health Office of Dietary Supplements, 2022).

Yogurt	Nonfat milk
Fortified orange juice	Calcium-fortified soymilk
Mozzarella cheese	Reduced fat milk
Sardines	Tofu
Cheddar cheese	Salmon

Table 2 The RDA for calcium intake by age group and sex.

	RDA for calcium (National Institute of Health Office of Dietary Supplements, 2022)	
Age	**Men**	**Women**
19–50 years	1000 mg/day	1000 mg/day
51–70 years	1000 mg/day	1000 mg/day
71 + years	1200 mg/day	1200 mg/day

RDA, recommended dietary allowances.

the USA are said to take supplemental calcium (Morelli, Santulli, & Gambardella, 2020). The recommended daily allowance of calcium are shown in Table 2. Despite the potential benefit of supplemental calcium on bone health, high supplemental calcium intake has been reported with unfavorable cardiovascular health.

Several original studies and meta-analyses studies have been conducted to examine the associations between calcium intake and the risk of stroke (Table 3). The findings from those studies range from inverse associations to nonlinear associations for dietary or total calcium intake to direct association for supplemental calcium intake. In this chapter, we will discuss the associations between calcium intake and stroke, and the associations between calcium intake and the risk factors of stroke.

Calcium and stroke

Individual cohort studies report an inverse association between dietary calcium intake and the risk of stroke. In a study conducted among 41,526 Japanese participants, the authors found that dietary calcium intake was associated with a lower risk of total stroke and ischemic stroke (Umesawa et al., 2008). In a cohort of 19,553 American participants, investigators found an inverse association between total calcium intake and the risk of ischemic stroke; however, there was no dose–response relationship but a threshold-like nonlinear association (Dibaba et al., 2019). Of note, the association of dietary calcium intake with ischemic stroke was mediated by fasting blood glucose, type 2 diabetes, systolic blood pressure, and hypertension. Supplemental calcium had a nonlinear association

Table 3 Meta-analyses studies on calcium intake and the risk of stroke.

Studies	Exposure	Outcome	Type of study	Number of participants	Population	Findings
Yang et al. (2020)	Dietary calcium	Stroke	Meta-analysis of cohort studies	26 cohort studies	Both men and women	No association with the risk of stroke
	Calcium supplement	Stroke	Meta-analysis	16 clinical trials	Both men and women	Increased risk
Adebamowo, Spiegelman, Willett, and Rexrode (2015)	Dietary calcium	Total stroke	Meta-analysis	2 cohort studies	Women	No association
Bolland, Grey, Avenell, Gamble, and Reid (2011)	Calcium supplement	Stroke	Meta-analysis	3 clinical trials	Women	Increased risk of stroke
Larsson, Orsini, and Wolk (2013)	Dietary calcium	Risk of stroke	Meta-analysis	11 cohort studies	Both men and women	Nonlinear association
Tian et al. (2015)	Dietary calcium	Risk of stroke	Meta-analysis	10 cohort studies	Both men and women	Reduced risk with diary calcium, and long-term follow-up ($\geq$14 years) and nonlinear association

with the risk of ischemic stroke; the risk increased with low supplemental calcium intake ($\sim<300$ mg/day) and high supplemental calcium intake (>1000 mg/day). In another large early study among women in the Nurse's Health Study cohort of 85,764 participants, dietary calcium intake was inversely associated with the risk of ischemic stroke but not with other stroke subtypes. The risk was higher in those with a lower calcium intake, and the magnitude of the inverse association did not increase in groups with intake >600 mg/day (Iso et al., 1999). Another prospective cohort study among 34,468 participants found a reduced risk of stroke and cardiovascular mortality with increased dietary calcium intake; comparing the highest quartile to the lowest, the risk of cardiovascular and stroke mortality was lower by 16% (Khan et al., 2015).

A meta-analysis of eleven longitudinal studies found a nonlinear association between dietary calcium intake and the risk of stroke. As the amount of calcium intake increased, the risk of stroke also increased, while below certain levels of dietary calcium intake (<700 mg/day), the risk was reduced (Larsson et al., 2013). Another meta-analysis that included ten studies with over 371,495 participants found that dietary calcium intake was inversely associated with lower risk of stroke (Tian et al., 2015).

However, not all studies found a statistically significant association between dietary calcium intake and the risk of stroke. Two other meta-analysis of longitudinal studies (26 and 2 studies) did not find a significant association between dietary calcium intake and the risk of stroke. (Adebamowo et al., 2015; Yang et al., 2020) Of the ten studies that reported on stroke risk or mortality, eight of the cohort studies included in the larger meta-analysis (Yang et al., 2020) did not find a significant association between dietary calcium intake and the risk of stroke or stroke-related mortality.

Similarly, a study on the association between calcium or calcium and vitamin D intake with the risk of coronary heart diseases among women aged 52–62 found that the risk of coronary heart diseases was higher by 26% in multivariate model comparing users to nonusers (Pentti et al., 2009). The meta-analysis of 16 clinical trials found that supplemental calcium was associated with 8%–21% increased risk of coronary heart disease and myocardial infraction, respectively (Yang et al., 2020). Another meta-analysis of clinical trials found an increased risk of stroke, myocardial infarction, or both in those with supplemental calcium intake with or without vitamin D intake (Bolland et al., 2011). In general, there is no strong evidence for the benefit of supplemental calcium intake in relation to the risk of stroke or stroke-related mortality but there are multiple evidence suggesting that there might be an increased risk of stroke or stroke-related mortality associated with supplemental calcium intake.

Given that atherosclerosis, hypertension, type 2 diabetes, and obesity are the risk factors associated with the risk of ischemic stroke, we will briefly discuss the association between calcium intake and these risk factors.

Calcium and atherosclerosis

A study after study suggests that calcium supplementation is related to an increased risk of atherosclerosis and high cholesterol levels. This indirectly suggests that calcium supplementation might increase the risk for ischemic stroke. A Multi-Ethnic Study of Atherosclerosis found that calcium supplementation was associated with an increased risk of atherosclerosis, while dietary calcium intake was not (Anderson et al., 2016). In this study, total calcium intake was associated with a 27% reduced risk of incident atherosclerosis comparing the highest quintile of intake to the lowest, but the association disappeared for dietary calcium intake alone and the risk increased for supplemental calcium intake. Moreover, in a recently published study, the authors reported that calcium supplementation use and amount were significantly associated with the progress of abdominal aorta calcification in a study conducted in one center of the Canadian Multi-Centre Osteoporosis Study (Hulbert et al., 2020). A long-term calcium supplementation trial in postmenopausal women in China indicated increased carotid intima-media thickness after 2 years (Li et al., 2013). Similarly, a calcium supplementation clinical trial conducted in Iran among women with overweight or obesity indicated statistically significant increase in very low density lipoprotein cholesterol (VLDL) and triglyceride (Karandish, Shockravi, Jalali, & Haghighizadeh, 2009). A meta-analysis of 15 clinical trials indicated an elevated risk of myocardial infarction or myocardial infarction and stroke in those with calcium supplementation (Bolland et al., 2010). An observational study based in Italy also found a modest direct correlation between serum calcium levels and LDL and total cholesterol (Gallo et al., 2016). However, older studies suggest that calcium supplementation is associated with improved cholesterol profile (Carlson, Olsson, Oro, & Rossner, 1971; Reid et al., 2002).

Calcium and hypertension

Several studies indicate that dietary calcium intake and calcium supplementation are inversely related to hypertension. A meta-analysis of calcium supplementation found that calcium supplementation reduced both systolic and diastolic blood pressure. The reduction was higher in younger (≤35 years) participants and those who received higher doses (>1500 mg/day) (Cormick, Ciapponi, Cafferata, Cormick, & Belizán, 2015). An older meta-analysis of calcium supplementation clinical trial found a similar effect of calcium supplementation on systolic and diastolic blood pressure. The reduction was higher in those with low (≤800 mg/day) calcium intake (van Mierlo et al., 2006). Overall calcium has been known to modestly lower both systolic and diastolic blood pressure, and such effect may reduce the risk of cardiovascular complications such as hypertension (Nakamura et al., 2019). However, the atherogenic effect of calcium makes the benefit of calcium on hypertension questionable. This makes calcium a doubled-edged sword having benefit on hypertension and bone health while also having an atherogenic effect.

Calcium intake and the risk of type 2 diabetes

A meta-analysis of six cohort studies found that calcium intake was associated with a reduced risk of type 2 diabetes but when the meta-analysis was limited to the studies that adjusted for magnesium intake, the association was attenuated (Dong & Qin, 2012). The authors of the study concluded that calcium may not have a beneficial effect on the risk of type 2 diabetes independent of magnesium intake. Moreover, calcium supplementation along with vitamin D was not associated with the risk of incident type 2 diabetes in the Women Health Initiative study (de Boer et al., 2008). The result of another study also supports that calcium supplementation may not have a beneficial effect on fasting blood glucose or glycated Hg A1C, while combined vitamin D and calcium supplementation might have a beneficial effect against the risk of type 2 diabetes (Mitri, Dawson-Hughes, Hu, & Pittas, 2011).

Calcium intake and the risk of obesity

Overweight or obesity is one of the risk factors of ischemic stroke. A substantial number of observational studies have documented a direct association of obesity with the risk of stroke. Every unit increase in BMI has been associated with over 5% increase in the risk of stroke (Kernan, Inzucchi, Sawan, Macko, & Furie, 2013). Moreover, there are several studies that examined the association between calcium intake and the risk of weight gain, body fat accumulation, and obesity. Many of these studies documented an inverse association between calcium supplementation or dietary calcium intake and weight gain, body fat accumulation, or obesity (Villarroel, Villalobos, Reyes, & Cifuentes, 2014). Some of the studies have limitation in that they were cross-sectional study or were not designed to study the association between calcium intake and the risk of obesity. On the other hand, several studies have found no association between calcium supplementation and obesity (Vergnaud et al., 2008; Yanovski et al., 2009). This indicates there is a notable inconsistency in the findings of studies on the effect of calcium intake on body weight and obesity.

Conclusion

In summary, there is a strong evidence against a recommendation for a widespread calcium supplementation due to the high atherosclerosis risk that may outweigh its benefits on osteoporosis, bone fracture, and hypertension. Despite the adverse cardiovascular disease and stroke risk that might be related to high calcium supplementation, dietary calcium intake is associated with a lower stroke risk (Larsson et al., 2013; Tian et al., 2015). The evidence from individual studies and meta-analysis also suggests that the association between calcium intake and the risk of stroke might be nonlinear. This means that the risk is higher at extreme levels and lower with moderate dietary calcium intake. A recent cohort study also suggested that moderate intake of dietary calcium was related to a reduced risk of cardiovascular mortality and risk of stroke but the association was attenuated for the highest intake (Pana et al., 2020). There are also two other meta-analyses that did not find a significant association between dietary calcium intake and the risk of stroke or stroke mortality, (Adebamowo et al., 2015; Yang et al., 2020) which makes the benefit of dietary calcium for cardiovascular health and stroke questionable.

A recent umbrella review of the association between nutritional supplements and health concluded that calcium supplementation alone and along with vitamin D was associated with higher risk of stroke. The authors of the study concluded that calcium and vitamin D supplementation are harmful for stroke (Khan et al., 2019). A more recent narrative review of the role of calcium supplementation on cardiovascular health and stroke concluded that the purported link with cardiovascular outcome is unconvincing (Curtis, Cooper, & Harvey, 2021).

Even though there is no strong evidence against dietary calcium intake, calcium supplementation should be used with caution and within the limits of the recommended dietary allowance. Supplementation should be limited to those deficient of this nutrient. Specially, people at high risk for stroke such as those with uncontrolled hypertension, atherosclerosis, or obesity, and smokers or those with sedentary lifestyle should be cautious about the use of high supplemental calcium.

Summary points

- Dietary calcium intake is inversely associated with the risk of stroke.
- In some developed countries, about 50% men and 70% postmenopausal women take supplemental calcium.
- Some studies and meta-analyses indicate the association between calcium intake and the risk of stroke is nonlinear with increased stroke risk below (<300 mg/day) and above (>700 mg/day) certain levels and lower risk between those ranges.
- While supplemental calcium is recommended for osteoporosis, bone fracture, and hypertension, high supplemental calcium intake above the RDA might be related to an increased risk of cardiovascular diseases, atherosclerosis, and stroke.
- Calcium from dietary sources is unlikely to cause disease risk within the RDA and might be beneficial for the primary prevention of stroke but high supplemental calcium might increase the risk of atherosclerosis and stroke.

Definition of terms and abbreviations

1. BMI: body mass index
2. VLDL: very low-density lipoprotein cholesterol
3. RDA: recommended dietary allowance
4. Nonlinear: an association between calcium intake and the risk of stroke that is variable at different levels of calcium intake. Nonlinear means the association might be direct at lower intake and higher intake levels and inverse in between.
5. Meta-analysis: is the pooling of the results from individual studies into one aggregate value. In meta-analysis, results from multiple studies are pooled into a single study, and thus, meta-analysis has a better power to detect when difference exists. It is also helpful when findings from studies are inconsistent.

References

Adebamowo, S. N., Spiegelman, D., Willett, W. C., & Rexrode, K. M. (2015). Association between intakes of magnesium, potassium, and calcium and risk of stroke: 2 cohorts of US women and updated meta-analyses. *The American Journal of Clinical Nutrition*, *101*(6), 1269–1277. https://doi.org/10.3945/ajcn.114.100354 (published Online First: 2015/05/08).

Adelstein, R. S., & Sellers, J. R. (1987). Effects of calcium on vascular smooth muscle contraction. *The American Journal of Cardiology*, *59*(3), 4b–10b. https://doi.org/10.1016/0002-9149(87)90076-2 [published Online First: 1987/01/30].

Anderson, J. J. B., Kruszka, B., Delaney, J. A. C., He, K., Burke, G. L., Alonso, A., et al. (2016). Calcium intake from diet and supplements and the risk of coronary artery calcification and its progression among older adults: 10-year follow-up of the multi-ethnic study of atherosclerosis (MESA). *Journal of the American Heart Association*, *5*(10), e003815. https://doi.org/10.1161/JAHA.116.003815.

Bolland, M. J., Avenell, A., Baron, J. A., Grey, A., MacLennan, G. S., Gamble, G. D., et al. (2010). Effect of calcium supplements on risk of myocardial infarction and cardiovascular events: meta-analysis. *BMJ*, *341*, c3691. https://doi.org/10.1136/bmj.c3691.

Bolland, M. J., Grey, A., Avenell, A., Gamble, G. D., & Reid, I. R. (2011). Calcium supplements with or without vitamin D and risk of cardiovascular events: Reanalysis of the Women's Health Initiative limited access dataset and meta-analysis. *BMJ*, *342*, d2040. https://doi.org/10.1136/bmj.d2040 [published Online First: 2011/04/21].

Carlson, L. A., Olsson, A. G., Oro, L., & Rossner, S. (1971). Effects of oral calcium upon serum cholesterol and triglycerides in patients with hyperlipidemia. *Atherosclerosis*, *14*(3), 391–400. https://doi.org/10.1016/0021-9150(71)90067-0.

Center for Disease Prevention and Control. (2020). *Types of stroke*. cdc.gov.

Cormick, G., Ciapponi, A., Cafferata, M. L., Cormick, M. S., & Belizán, J. M. (2015). Calcium supplementation for prevention of primary hypertension. *Cochrane Database of Systematic Reviews*, *6*. https://doi.org/10.1002/14651858.CD010037.pub2.

Curtis, E. M., Cooper, C., & Harvey, N. C. (2021). Cardiovascular safety of calcium, magnesium and strontium: What does the evidence say? *Aging Clinical and Experimental Research*, *33*(3), 479–494. https://doi.org/10.1007/s40520-021-01799-x (published Online First: 2021/02/11).

de Boer, I. H., Tinker, L. F., Connelly, S., Curb, J. D., Howard, B. V., Kestenbaum, B., et al. (2008). Calcium plus vitamin D supplementation and the risk of incident diabetes in the Women's Health Initiative. *Diabetes Care*, *31*(4), 701–707. https://doi.org/10.2337/dc07-1829.

Dibaba, D. T., Xun, P., Fly, A. D., Bidulescu, A., Tsinovoi, C. L., Judd, S. E., et al. (2019). Calcium intake and serum calcium level in relation to the risk of ischemic stroke: Findings from the REGARDS study. *Journal of Stroke*, *21*(3), 312–323. https://doi.org/10.5853/jos.2019.00542 [published Online First: 2019/09/30].

Dong, J. Y., & Qin, L. Q. (2012). Dietary calcium intake and risk of type 2 diabetes: Possible confounding by magnesium. *European Journal of Clinical Nutrition*, *66*(3), 408–410. https://doi.org/10.1038/ejcn.2012.5.

Fearnley, C. J., Roderick, H. L., & Bootman, M. D. (2011). Calcium signaling in cardiac myocytes. *Cold Spring Harbor Perspectives in Biology*, *3*(11), a004242. https://doi.org/10.1101/cshperspect.a004242.

Gallo, L., Faniello, M. C., Canino, G., Tripolino, C., Gnasso, A., Cuda, G., et al. (2016). Serum calcium increase correlates with worsening of lipid profile: An observational study on a large cohort from South Italy. *Medicine*, *95*(8), e2774. https://doi.org/10.1097/MD.0000000000002774.

Hulbert, M., Turner, M. E., Hopman, W. M., Anastassiades, T., Adams, M. A., & Holden, R. M. (2020). Changes in vascular calcification and bone mineral density in calcium supplement users from the Canadian multi-center osteoporosis study (CaMOS). *Atherosclerosis*, *296*, 83–90. https://doi.org/10.1016/j.atherosclerosis.2019.12.003 [published Online First: 2019/12/21].

Iso, H., Stampfer, M. J., Manson, J. E., Rexrode, K., Hennekens, C. H., Colditz, G. A., et al. (1999). Prospective study of calcium, potassium, and magnesium intake and risk of stroke in women. *Stroke*, *30*(9), 1772–1779. https://doi.org/10.1161/01.str.30.9.1772 (published Online First: 1999/09/02).

Kalampogias, A., Siasos, G., Oikonomou, E., Tsalamandris, S., Mourouzis, K., Tsigkou, V., et al. (2016). Basic mechanisms in atherosclerosis: The role of calcium. *Medicinal Chemistry*, *12*(2), 103–113. https://doi.org/10.2174/1573406411666150928111446 (published Online First: 2015/09/29).

Karandish, M., Shockravi, S., Jalali, M. T., & Haghighizadeh, M. T. (2009). Effect of calcium supplementation on lipid profile in overweight or obese Iranian women: A double-blind randomized clinical trial. *European Journal of Clinical Nutrition*, *63*(2), 268–272. https://doi.org/10.1038/sj.ejcn.1602921.

Khan, B., Nowson, C. A., Daly, R. M., English, D. R., Hodge, A. M., Giles, G. G., et al. (2015). Higher dietary calcium intakes are associated with reduced risks of fractures, cardiovascular events, and mortality: A prospective cohort study of older men and women. *Journal of Bone and Mineral Research*, *30*(10), 1758–1766. https://doi.org/10.1002/jbmr.2515 (published Online First: 2015/04/02).

Kernan, W. N., Inzucchi, S. E., Sawan, C., Macko, R. M., & Furie, K. L. (2013). Obesity. *Stroke*, *44*(1), 278–286. https://doi.org/10.1161/STROKEAHA.111.639922.

Khan, S. U., Khan, M. U., Riaz, H., Valavoor, S., Zhao, D., Vaughan, L., et al. (2019). Effects of nutritional supplements and dietary interventions on cardiovascular outcomes: An umbrella review and evidence map. *Annals of Internal Medicine*, *171*(3), 190–198. https://doi.org/10.7326/m19-0341 (published Online First: 2019/07/10).

Larsson, S. C., Orsini, N., & Wolk, A. (2013). Dietary calcium intake and risk of stroke: A dose-response meta-analysis. *The American Journal of Clinical Nutrition*, *97*(5), 951–957. https://doi.org/10.3945/ajcn.112.052449.

Li, S., Na, L., Li, Y., Gong, L., Yuan, F., Niu, Y., et al. (2013). Long-term calcium supplementation may have adverse effects on serum cholesterol and carotid intima-media thickness in postmenopausal women: A double-blind, randomized, placebo-controlled trial. *The American Journal of Clinical Nutrition*, *98*(5), 1353–1359. https://doi.org/10.3945/ajcn.113.062844 (published Online First: 2013/09/21).

Mitri, J., Dawson-Hughes, B., Hu, F. B., & Pittas, A. G. (2011). Effects of vitamin D and calcium supplementation on pancreatic β cell function, insulin sensitivity, and glycemia in adults at high risk of diabetes: The calcium and vitamin D for diabetes mellitus (CaDDM) randomized controlled trial. *The American Journal of Clinical Nutrition*, *94*(2), 486–494. https://doi.org/10.3945/ajcn.111.011684 (published Online First: 2011/06/29).

Morelli, M. B., Santulli, G., & Gambardella, J. (2020). Calcium supplements: Good for the bone, bad for the heart? A systematic updated appraisal. *Atherosclerosis*, *296*, 68–73. https://doi.org/10.1016/j.atherosclerosis.2020.01.008.

Nakamura, H., Tsujiguchi, H., Hara, A., Kambayashi, Y., Miyagi, S., Nguyen, T. T. T., et al. (2019). Dietary calcium intake and hypertension: Importance of serum concentrations of 25-Hydroxyvitamin D. *Nutrients*, *11*(4), 911. https://doi.org/10.3390/nu11040911.

National Institute of Health Office of Dietary Supplements. (2022). *Calcium fact sheet for health professionals*. Available from https://ods.od.nih.gov/factsheets/Calcium-HealthProfessional/.

Pana, T. A., Dehghani, M., Baradaran, H. R., Neal, S. R., Wood, A. D., Kwok, C. S., et al. (2020). Calcium intake, calcium supplementation and cardiovascular disease and mortality in the British population: EPIC-norfolk prospective cohort study and meta-analysis. *European Journal of Epidemiology*. https://doi.org/10.1007/s10654-020-00710-8 (published Online First: 2021/01/01).

Pentti, K., Tuppurainen, M. T., Honkanen, R., Sandini, L., Kröger, H., Alhava, E., et al. (2009). Use of calcium supplements and the risk of coronary heart disease in 52-62-year-old women: The Kuopio osteoporosis risk factor and prevention study. *Maturitas*, *63*(1), 73–78. https://doi.org/10.1016/j.maturitas.2009.03.006 (published Online First: 2009/04/28).

Reid, I. R., Birstow, S. M., & Bolland, M. J. (2017). Calcium and cardiovascular disease. *Endocrinology & Metabolism*, *32*(3), 339–349. https://doi.org/10.3803/EnM.2017.32.3.339.

Reid, I. R., Mason, B., Horne, A., Ames, R., Clearwater, J., Bava, U., et al. (2002). Effects of calcium supplementation on serum lipid concentrations in normal older women: A randomized controlled trial. *The American Journal of Medicine*, *112*(5), 343–347. https://doi.org/10.1016/S0002-9343(01)01138-X.

Tankeu, A. T., Ndip Agbor, V., & Noubiap, J. J. (2017). Calcium supplementation and cardiovascular risk: A rising concern. *The Journal of Clinical Hypertension*, *19*(6), 640–646. https://doi.org/10.1111/jch.13010.

Tian, D. Y., Tian, J., Shi, C. H., Song, B., Wu, J., Ji, Y., et al. (2015). Calcium intake and the risk of stroke: An up-dated meta-analysis of prospective studies. *Asia Pacific Journal of Clinical Nutrition, 24*(2), 245–252. https://doi.org/10.6133/apjcn.2015.24.2.22 (published Online First: 2015/06/17).

Umesawa, M., Iso, H., Ishihara, J., Saito, I., Kokubo, Y., Inoue, M., et al. (2008). Dietary calcium intake and risks of stroke, its subtypes, and coronary heart disease in Japanese: The JPHC study cohort I. *Stroke, 39*(9), 2449–2456. https://doi.org/10.1161/strokeaha.107.512236 [published Online First: 2008/07/19].

van Mierlo, L. A. J., Arends, L. R., Streppel, M. T., Zeegers, M. P. A., Kok, F. J., Grobbee, D. E., et al. (2006). Blood pressure response to calcium supplementation: A meta-analysis of randomized controlled trials. *Journal of Human Hypertension, 20*(8), 571–580. https://doi.org/10.1038/sj.jhh.1002038.

Vergnaud, A. C., Péneau, S., Chat-Yung, S., Kesse, E., Czernichow, S., Galan, P., et al. (2008). Dairy consumption and 6-y changes in body weight and waist circumference in middle-aged French adults. *The American Journal of Clinical Nutrition, 88*(5), 1248–1255. https://doi.org/10.3945/ajcn.2007.25151 (published Online First: 2008/11/11).

Villarroel, P., Villalobos, E., Reyes, M., & Cifuentes, M. (2014). Calcium, obesity, and the role of the calcium-sensing receptor. *Nutrition Reviews, 72*(10), 627–637. https://doi.org/10.1111/nure.12135.

Virani, S. S., Alonso, A., Benjamin, E. J., Bittencourt, M. S., Callaway, C. W., Carson, A. P., et al. (2020). Heart disease and stroke statistics—2020 Update: A report from the American Heart Association. *Circulation, 141*(9), e139–e596. https://doi.org/10.1161/CIR.0000000000000757.

World Health Organization Geneva. WHO. (2022). *Disease burden and mortality estimates 2016*. Available from http://www.who.int/healthinfo/global_burden_disease/estimates/en/. (Accessed 23 April 2021).

Yang, C., Shi, X., Xia, H., Yang, X., Liu, H., Pan, D., et al. (2020). The evidence and controversy between dietary calcium intake and calcium supplementation and the risk of cardiovascular disease: A systematic review and meta-analysis of cohort studies and randomized controlled trials. *Journal of the American College of Nutrition, 39*(4), 352–370. https://doi.org/10.1080/07315724.2019.1649219.

Yanovski, J. A., Parikh, S. J., Yanoff, L. B., Denkinger, B. I., Calis, K. A., Reynolds, J. C., et al. (2009). Effects of calcium supplementation on body weight and adiposity in overweight and obese adults: A randomized trial. *Annals of Internal Medicine, 150*(12), 821–829. w145–6 https://doi.org/10.7326/0003-4819-150-12-200906160-00005 (published Online First: 2009/06/17).

CHAPTER 44

Recommended resources for the study and investigation of vitamins and minerals in neurological disorders

Rajkumar Rajendram[a,b], Vinood B. Patel[c], and Victor R. Preedy[d]
[a]College of Medicine, King Saud bin Abdulaziz University for Health Sciences, Riyadh, Saudi Arabia
[b]Department of Medicine, King Abdulaziz Medical City, King Abdullah International Medical Research Center, Ministry of National Guard Health Affairs, Riyadh, Saudi Arabia
[c]School of Life Sciences, University of Westminster, London, United Kingdom
[d]Department of Nutritional Sciences, Faculty of Life Science and Medicine, School of Life Course and Population Sciences, King's College London, London, United Kingdom

Introduction

Vitamins and minerals have several functions within the nervous system that promote its health (Rutjes, Denton, Di Nisio, et al., 2018). The optimal function of the nervous system requires a constant supply of nutrients (Kumar, 2010). Vitamins and minerals that are especially important for the neuronal function include copper, folic acid, B vitamins (e.g., thiamine, niacin, vitamin B12) and tocopherol (i.e., vitamin E) (Kumar, 2010). It is commonly suggested that the use of dietary supplementation with select vitamins and minerals may help to maintain the cognitive function in later life (Rutjes et al., 2018). However, any mineral or vitamin deficiency will have a physiological impact, affecting variety of organ systems, including neurological tissues. Furthermore, nutrition is important in the treatment and prevention of neurodegenerative diseases, especially in the elderly (Mostafavi & Hosseini, 2015). Select vitamins and minerals in foods can prevent or ameliorate the symptoms of chronic neurodegenerative diseases or associated conditions (Farah & Yammine, 2022; Olasehinde, Oyeleye, Ogunsuyi, & Ogunruku, 2017). Thus, the consumption of food supplements (i.e., functional foods) may be beneficial in the treatment of age-related neurological diseases (e.g., Parkinson's disease) (Evatt, 2007).

Malnutrition can cause neurologic complications (Kumar, 2010). In developed countries, risk groups include the elderly, patients on prolonged parenteral nutrition, individuals with eating disorders such as bulimia, and those suffering from a malabsorption syndrome (e.g., celiac disease or inflammatory bowel disease) (Kumar, 2010). Childhood micronutrient deficiencies can have devastating effects. For example, maternal and infant iodine deficiency is reported to be the largest cause of mental retardation and impairment on a global basis (Bougma, Aboud, Harding, & Marquis, 2013). It is also important to note that nutrient deficiencies induced by bariatric surgery may result in neurologic

Vitamins and Minerals in Neurological Disorders
https://doi.org/10.1016/B978-0-323-89835-5.00045-4

complications (Kumar, 2010). The coexistence of multiple nutritional deficiencies is common. The prognosis is improved by rapid diagnosis and the initiation of appropriate supplementation with vitamins and minerals.

Thiamine, the first vital amine (i.e., vitamin), was identified in 1926 (Mozaffarian, Rosenberg, & Uauy, 2018). Studies conducted since then have provided great insight into the effect of nutrition on specific biochemical pathways that are important to the cognitive function (Gómez-Pinilla, 2008). Thus, the evidence base on vitamins and minerals for neurological disorders has grown enormously, and more recently, there have been considerable advances in molecular biology and genomics (for example: Dionisio, Amaral, Rodrigues, Spetz, & Galluzzi, 2020; Okdahl & Brock, 2021).

Keeping up to date can be problematic even for experienced scientists, researchers, and clinicians. We have therefore produced tables containing resources as recommended by active researchers and practitioners. The aim of this compilation is to assist those colleagues who are interested in understanding more about vitamins, minerals, and neurological disorders. Essentially, the sites and resources presented here draw upon the wealth of experiences and "know how" accrued over many years. The list below acknowledges all the experts who helped to prepare these valuable resources.

Resources

Tables 1–5 list the most up-to-date information on the regulatory bodies (Table 1), professional societies (Table 2), books (Table 3), emerging technologies, platforms (Table 4),

Table 1 Regulatory bodies or Organizations dealing with vitamins, minerals, or neurological disorders.

Regulatory body or organization	Web address
Alzheimer's Association	https://www.alz.org/
Alzheimer's Drug Discovery Foundation	https://www.alzdiscovery.org/
Brain and Spine Foundation	https://www.brainandspine.org.uk/
Brain Trauma Foundation	https://www.braintrauma.org
Centers for Disease Control and Prevention (CDC)	https://www.cdc.gov/
Dayton Center of Neurological Disorders	https://dcndinc.com/
Food and Agriculture Organization of the United Nations (FAO)	https://www.fao.org/home/en/
Food and Drug Administration Center for Food Safety and Applied Nutrition	http://www.fda.gov/food/
Global Alliance for Improved Nutrition	https://www.gainhealth.org/
Indian Council of Medical Research (ICMR)-National Institute of Nutrition (NIN), India	https://www.nin.res.in/

Table 1 Regulatory bodies or Organizations dealing with vitamins, minerals, or neurological disorders—cont'd

Regulatory body or organization	Web address
Linus Pauling Institute: Oregon State University	https://lpi.oregonstate.edu
Micronutrient Forum	https://micronutrientforum.org/
Migraine Trust	https://www.migrainetrust.org/
National Academy of Medicine	https://nam.edu/
National Agency for Food and Drug Administration and Control	https://www.nafdac.gov.ng
National Center for Complementary and Integrative Health (NCCIH)	https://www.nccih.nih.gov/
National Headache Foundation	https://headaches.org/
National Health Service (NHS)	https://www.nhs.uk/
National Health Surveillance Agency	https://www.gov.br/anvisa/pt-br
National Institutes of Health	https://www.nih.gov/
National Institute of Neurological Disorders and Stroke	https://www.ninds.nih.gov/
National Institute on Aging	https://www.nia.nih.gov/
National Institute on Aging	https://www.alzheimers.gov/
Nutrition International	https://www.nutritionintl.org/
Nutritionist Resource	https://www.nutritionist-resource.org.uk/
Office of Dietary Supplements: National Institutes of Health	https://ods.od.nih.gov/
Physicians Committee for Responsible Medicine	https://www.pcrm.org/
Standards Organisation of Nigeria	https://son.gov.ng
United Nations Children's Emergency Fund (UNICEF)	https://www.unicef.org/
United States Department of Agriculture (USDA)	https://www.nutrition.gov/
United States Department of Agriculture (USDA) Center for Nutrition Policy and Promotion	https://www.fns.usda.gov/cnpp
World Food Safety Organisation (WFSO)	https://worldfoodsafety.org/
World Health Organization (WHO)	https://www.who.int/
World Iodine Association	https://www.worldiodineassociation.com/

This table lists the regulatory bodies and organizations involved with vitamins, minerals, or neurological disorders. The links were accurate at the time of going to press but may move or alter. In these cases, the use of the "Search" tabs should be explored at the parent address or site. In some cases, links direct the reader to pages related to vitamins, minerals, or new neurological disorders within parent sites. Some societies and organizations have a preference for shortened terms, such as acronyms and abbreviations. See also Table 2.

Table 2 Professional societies relevant to the study of vitamins, minerals, or neurological disorders.

Society name	Web address
African Nutrition Society	https://www.ansnet.org
Alzheimer's Association	https://www.alz.org/
Alzheimer's Disease International	https://www.alzint.org/
Alzheimer's Society	https://www.alzheimers.org.uk/
American Academy of Neurological and Orthopaedic Surgeons	https://aanos.org/
American Academy of Neurology	https://www.aan.com/
American Association of Neurological Surgeons	https://www.aans.org/
American Association of Neuropathologists	http://www.aanp-jnen.com/
American Board of Neurological Surgery	https://abns.org/
American Council on Science and Health	https://www.acsh.org/
American Neurological Association	https://myana.org/
American Nutrition Association	https://theana.org/
American Society for Clinical Nutrition	https://nutrition.org/
American Society for Parenteral and Enteral Nutrition	https://www.nutritioncare.org
American Trauma Society	https://www.amtrauma.org
Australasian Society for Parenteral Nutrition	www.auspen.org.au
Brain Trauma Foundation	https://www.braintrauma.org
British Association for Parenteral and Enteral Nutrition	www.bapen.org.uk
British Dietetic Association	https://www.bda.uk.com/resource/vitamin-d.html
Canadian Nutrition Society	www.cns-scn.ca
Child Neurology Society	https://www.childneurologysociety.org/
European Academy of Neurology	https://www.ean.org/
European Iron Club	https://ironmasterclass.eu/
European Society for Clinical Nutrition and Metabolism	https://www.espen.org/
Federation of European Nutrition Societies	https://fensnutrition.org/
Federation of European Societies on Trace Elements and Minerals (FESTEM)	http://festem.eu/
Hellenic Society of Medical/Clinical Nutrition and Metabolism (GrESPEN)	www.grespen.org
International Society for the Development of Research on Magnesium (SDRM)	https://www.sdrmsociety.org/
International Society for the Study of Iron in Biology and Medicine	https://bioiron.org/
International Society for Trace Element Research in Humans	https://www.isterhgroup.org/
International Society for Zinc Biology	https://www.zinc-net.com/ISZB.html
International Society to Advance Alzheimer's Research and Treatment (ISTAART)	https://action.alz.org/personifyebusiness/default.aspx?tabid=1516

Table 2 Professional societies relevant to the study of vitamins, minerals, or neurological disorders—cont'd

Society name	Web address
Iodine Global Network	https://www.ign.org/
Irish Society for Clinical Nutrition and Metabolism	www.irspen.ie
National Association of Nutrition Professionals	https://nanp.org/
Neurocritical Care Society	https://www.neurocriticalcare.org/home
Neuroscience Society of Nigeria (NSN)	http://www.neurosciencenigeria.org
Nutrition Society	https://www.nutritionsociety.org
Parkinson's Foundation	https://www.parkinson.org
Society for Neuroscience	https://www.sfn.org/
Society of Neuroscientists of Africa (SONA)	https://sonafrica.org
United Council for Neurologic Subspecialties	https://www.ucns.org/
Universal Society of Food and Nutrition	https://www.usfn.net
Vitamins Society of Japan	https://www.vitamin-society.jp/en/
World Federation of Neurology	https://wfneurology.org/

This table lists some societies and organizations devoted to vitamins, Minerals, and neurological disorders. Please note that the location of the websites or web address changes. Some societies and organizations have a preference for shortened terms, such as acronyms and abbreviations. See also Table 1.

Table 3 Books on vitamins, minerals, or neurological disorders.

Book title	Authors or editors	Publisher	Year of publication
Advanced Human Nutrition	Wildman REC, Medeiros DM	Jones and Bartlett Learning	2018
Advances in Food and Nutrition Research	Toldra F	Elsevier Science	2017
Bioactive Nutraceuticals and Dietary Supplements in Neurological and Brain Disease	Watson R, Preedy VR	Academic Press	2014
Cognitive Plasticity in Neurologic Disorders	Tracy JI, Hampstead BM, Sathian K	Oxford University Press	2015
Communicating Nutrition: The Authoritative Guide	Mayfield BJ	Academy of Nutrition and Dietetics	2020
Diet and Exercise in Cognitive Function and Neurological Diseases	Farooqui T, Farooqui AA	John Wiley and Sons, Inc	2015
Diet and Nutrition in Dementia and Cognitive Decline	Martin C, Preedy VR	Academic Press	2014
Electrical Brain Stimulation for the Treatment of Neurological Disorders	Zohuri B, McDaniel PJ	Apple Academic Press	2019

Continued

Table 3 Books on vitamins, minerals, or neurological disorders—cont'd

Book title	Authors or editors	Publisher	Year of publication
Essential and Toxic Trace Elements and Vitamins in Human Health	Brewer G, Prasad A	Elsevier	2020
Frontiers in Clinical Drug Research: CNS and Neurological Disorders	ur Rahman A, Amtul Z	Bentham Science Publishers	2020
Handbook of Nutrition, Diet, and Epigenetics	Patel V, Preedy VR	Springer International Publishing	2019
Health and Disease: Role of Micronutrients and Trace Elements	Nath R	APH Publishing Corp.	2020
Human Nutrition	Geissler C, Powers H	Oxford University Press	2017
Incidence and Economic Burden of Injuries in the United States. New York	Finkelstein E, Corso P, Miller T	Oxford University Press.	2006
Magnesium: Applications in Clinical Medicine	Kupetsky EA	CRC Press/ Taylor and Francis Group.	2019
Magnesium in Human Health and Disease	Watson RR, Preedy VR, Zibadi S	Humana Press	2013
Magnesium in the Central Nervous System [Internet]	Vink R, Nechifor M	University of Adelaide Press	2011
Neurological Consequences of Nutritional Disorders	Misra UK, Kalita J	CRC Press	2021
Neurological Disorders	Parkes JD	Springer International	2012
Neurological Disorders and Imaging Physics, Volume 5: Applications in Dyslexia, Epilepsy And Parkinson's	Suri JS, El-Baz A	Institute of Physics Publishing	2020
Neuroprotective Effects of Phytochemicals in Neurological Disorders	Farooqui T, Farooqui AA	John Wiley and Sons, Inc.	2017
Nutrition	Insel PM, Turner RE, Ross D	Jones and Bartlett	2004
Nutrition Maintaining and Improving Health	Webb GP	CRC Press	2019
Nutrition and Diet Therapy	Roth RA, Wehrle KL	Cengage Learning	2017
Nutrition and Lifestyle in Neurological Autoimmune Diseases	Watson RR, Kilgore WDS	Elsevier	2016

Table 3 Books on vitamins, minerals, or neurological disorders—cont'd

Book title	Authors or editors	Publisher	Year of publication
Nutrition and the Developing Brain	Moran VH, Lowe NM	CRC Press	2017
Nutrition Counseling and Education Skills: A Guide for Professionals	Beto JA, Holli BB	Jones and Bartlett Learning	2020
Nutrition Essentials for Mental Health: A Complete Guide to the Food-Mood Connection Hardcover – Illustrated	Korn L, Lake J	W. W. Norton and Company	2016
Nutrition for the Ailing Brain: Your Guide for Parkinson's Disease and Other Neurological Disorders	Hogan JB	F4L Publishing	2018
Nutrition in Neurologic Disorders A Practical Guide	Arsava E. M.	Springer International Publishing	2017
Nutrizione e DietologiaAspetti Clinici Dell'Alimentazione	Liguri G	Zanichelli	2015
Olives and Olive Oil in Health and Disease Prevention	Preedy VR, Watson RR	Academic Press	2021
Omega-3 Fatty Acids in Brain and Neurological Health	De Meester F, Watson RR	Elsevier Science	2014
Oxidative Stress and Dietary Antioxidants in Neurological Diseases	Martin C, Preedy VR	Academic Press	2020
Personal Nutrition	Boyle MA	Cengage Learning	2018
Role of the Mediterranean Diet in the Brain and Neurodegenerative Diseases	Farooqui T, Farooqui AA	Elsevier Inc.	2018
Textbook of Nutrition in Health and Disease	Chakrabarty K, Chakrabarty AS	Springer International	2019
Trace Elements, Micronutrients, and Free Radicals	Dreosti IE	Springer	1991

This table lists books relevant to vitamins, minerals or neurological disorders.

Table 4 Emerging techniques and platforms related to the study and investigation of vitamins, minerals, or neurological disorders.

Organization or company name	Web address
Alector	https://alector.com/
Denali Therapeutics	https://www.denalitherapeutics.com/
Mass spectrometry	https://www.degruyter.com/document/doi/10.1515/psr-2017-8002/html
Nestle Health Science	https://www.nestlehealthscience.com/health-management/pediatrics/neurological-disorders
Neural Analytics	https://novasignal.com/
NeuroPace	https://www.neuropace.com/
Neuroventis	https://neuroventis.care/our-solution/
Noom	https://www.noom.com/#/
Skyhawk Therapeutics	https://skyhawktx.com/
Virta	https://www.virtahealth.com/
Zoe	https://joinzoe.com/

This table lists technologies or platforms relevant to the study and investigation of vitamins, minerals and neurological disorders. Please note, occasionally the location of the websites or web address changes.

Table 5 Other resources of interest or relevance for health care professionals or patients related to biomedical aspects of vitamins, minerals or neurological disorders.

Name of resource or organization	Web address
Academy of Nutrition and Dietetics	www.eatright.org
Brain and Spine Foundation: Material on Nutrition	https://www.brainandspine.org.uk/information-and-support/health-and-lifestyle-tips/brain-food/
Centers for Disease Control and Prevention: Material on Dementia	https://www.cdc.gov/aging/dementia/index.html
Central College: Nutrition and Neurology	https://central.edu/writing-anthology/2019/04/09/nutrition-and-neurology/
Dementia Research Centre: University College London	https://www.ucl.ac.uk/drc/
European Federation of Pharmaceutical Industries and Associations: Working with Patient Groups	https://www.efpia.eu/relationships-code/patient-organisations/
European Society for Clinical Nutrition and Metabolism (ESPEN): Guideline on Clinical Nutrition in Neurology	https://www.espen.org/files/ESPEN-Guidelines/ESPEN-guideline_clinical_nutrition_in_neurology.pdf
Findhope: Material on Nutrition in Functional Neurological Disorder	https://fndhope.org/living-fnd/healthy-living/nutrition/

Table 5 Other resources of interest or relevance for health care professionals or patients related to biomedical aspects of vitamins, minerals or neurological disorders—cont'd

Name of resource or organization	Web address
FY2021 Referral Guidelines: Office of Dietary Supplements (ODS) Research Scholars Program (National Institute of Health, US)	https://ods.od.nih.gov/Research/Scholars.aspx
Healthline	https://www.healthline.com/nutrition
Iodine Related Resources for Program Managers	https://www.ign.org/rp.htm
Linus Pauling Institute	https://lpi.oregonstate.edu/mic
Mercy: Material on Diet and Neurological Health	https://www.mercy.net/service/neurology-services/eating-for-your-brain-two-diets-that-can-enhance-neurological-health/
National Center for Complementary and Integrative Health (NCCIH): Material on Herbs	https://www.nccih.nih.gov/health/herbsataglance
National Center for Complementary and Integrative Health (NCCIH): Safe Use of Complementary Health Products and Practices	https://www.nccih.nih.gov/health/safety
National Health Service (NHS): Material on Alzheimer's Disease	https://www.nhs.uk/conditions/alzheimers-disease/
National Institute of Neurological Disorders and Stroke: Patient Organisations	https://www.ninds.nih.gov/Disorders/Support-Resources/Patient-Organizations
National Institute of Neurological Disorders and Stroke: Information on Alzheimer's Disease and Related Dementias	https://www.ninds.nih.gov/Current-Research/Focus-Disorders/Alzheimers-Related-Dementias
National Institute on Aging: Material on Alzheimer's Disease?	https://www.nia.nih.gov/health/what-alzheimers-disease
National Institutes of Health: Copper Fact Sheet for Health Professionals	https://ods.od.nih.gov/factsheets/Copper-HealthProfessional/
Nestle Health Science: Material on Nutrition and Neurological Disorders In Children	https://www.nestlehealthscience.com/health-management/pediatrics/neurological-disorders
Neurological Disorders—World Health Organization (WHO)	https://www.who.int/publications/i/item/9789241563369
Neurosymptoms	https://www.neurosymptoms.org/
Nutrition Link	http://www.nutritionlink.org
Nutrition Support for Neurologically Impaired Children	https://naspghan.org/files/documents/pdfs/position-papers/nutriSupportNeuro_Impaired_Child.pdf
Nutritional Neuroscience Lab	https://www.american.edu/cas/health/nutrneurolab/
Nutritional Support in Neurological Diseases	https://lllnutrition.com/mod_lll/TOPIC25/m251.pdf

Continued

Table 5 Other resources of interest or relevance for health care professionals or patients related to biomedical aspects of vitamins, minerals or neurological disorders—cont'd

Name of resource or organization	Web address
Nutritionist Resource	https://www.nutritionist-resource.org.uk/
Oxford Medicine Online: Dietary Therapy for Neurological Disorders	https://oxfordmedicine.com/view/10.1093/med/9780190497996.001.0001/med-9780190497996-chapter-18
Physicians Committee for Responsible Medicine: Material on Nutrition	https://nutritionguide.pcrm.org/nutritionguide/
The Patients Association	https://www.patients-association.org.uk/
World Health Organization (WHO): Material on Dementia	https://www.who.int/news-room/fact-sheets/detail/dementia

This table lists other resources of interest or relevance to the study and biomedical study of vitamins, minerals, or neurological disorders. Please note, occasionally the location of the websites or web address changes.

and other resources of interest (Table 5) that are relevant to an evidence-based approach to the study of micronutrients and neurological disorders. Some organizations are listed in more than one table as they occasionally fulfill multiple roles.

Other resources

The Wellcome Collection (https://wellcomecollection.org/collections) and the British Library (https://www.bl.uk/) also list material on topics related to micronutrients and neurological disorders. Some holdings may be digital and can be accessed online.

Other chapters on resources (recommended by authors and practitioners) may also be relevant to vitamins, minerals, and neurological disorders. These include diet quality (Rajendram, Rajendram, Patel, & Preedy, 2013a), food fortification (Rajendram, Rajendram, Patel, & Preedy, 2013b), famine and starvation (Rajendram, Patel, & Preedy, 2018a, 2018b), branched chain amino acids (Rajendram, Patel, & Preedy, 2015), glutamine (Rajendram, Patel, & Preedy, 2014), nutrition and epigenetics (Rajendram et al., 2018a, 2018b); development (Rajendram, Patel, & Preedy, 2021a), aging (Rajendram, Patel, & Preedy, 2021b), Parkinson's disease (Rajendram, Patel, & Preedy, 2020), diet and nutrition in critical care (Alzaid, Rajendram, Patel, & Preedy, 2015), and the metabolism and physiology of bariatric surgery (Rajendram, Martin, & Preedy, 2016).

This list of material in these tables is included to provide general information only. It does not constitute any recommendation or endorsement of the activities of these sites, facilities, or other resources listed in this chapter, by the authors or editors of this book.

Summary points

- Select vitamins and minerals are important for neuronal function.
- Deficiency of vitamins and minerals can cause adverse neurological symptoms.
- Thiamine, the first vital amine (i.e., vitamin), was identified in 1926.
- There have been significant advances in the understanding of vitamins and minerals in neurological disease in recent years especially with our advances in molecular techniques.
- Without the guidance contained in this resource, it can be problematic to keep up with recent advances relevant to vitamins, minerals, and neurological disorders, and these tables essentially summarize the know how accrued over many years.

Acknowledgments (in alphabetical order)

The authors thank the following authors for their contributions to the development of this resource. The authors apologize if some of the suggested material was not included in this chapter or has been moved to another chapter. Agarwal, Renu; Akhtar, Ansab; Borges, Julita; Borjac, Jamilah; Choudhary, Aakansha; Choudhary, Ajay; Harrison, Fiona; Kaddoumi, Amal; Kaushik, Kaviraj; Kumar, Anoop; Kumar, Ashok; Leichtle, Stefan; Lezhitsa, Igor; Liampas, Ioannis; Marino, Angela; Matar, Alaa; Onaolapo, Adejoke; Onaolapo, Olakunle; Pilkhwal Sah, Sangeeta; Rizzo, Gianluca; Saadat, Payam; Saleki, Kiarash; Santos, Gesivaldo; Squitti, Rosanna; Toscano, Ana Elisa.

References

Alzaid, F., Rajendram, R., Patel, V. B., & Preedy, V. R. (2015). Expanding the knowledge base in diet, nutrition and critical care. Electronic and published resources. In R. Rajendram, V. R. Preedy, & V. B. Patel (Eds.), *Diet and nutrition in critical care*. Germany: Springer.

Bougma, K., Aboud, F. E., Harding, K. B., & Marquis, G. S. (2013). Iodine and mental development of children 5 years old and under: A systematic review and meta-analysis. *Nutrients*, *5*(4), 1387–1416.

Dionisio, P. A., Amaral, J. D., Rodrigues, C. M. P., Spetz, J. K. E., & Galluzzi, L. (2020). Molecular mechanisms of necroptosis and relevance for neurodegenerative diseases. *International Review of Cell and Molecular Biology*, *353*, 31–82.

Evatt, M. L. (2007). Nutritional therapies in Parkinson's disease. *Current Treatment Options in Neurology*, *9*, 198–204. https://doi.org/10.1007/BF02938409.

Farah, S., & Yammine, K. (2022). A systematic review on the efficacy of vitamin B supplementation on diabetic peripheral neuropathy. *Nutrition Reviews*, *80*(5), 1340–1355.

Gómez-Pinilla, F. (2008). Brain foods: The effects of nutrients on brain function. *Nature Reviews. Neuroscience*, *9*, 568–578.

Kumar, N. (2010). Neurologic presentations of nutritional deficiencies. *Neurologic Clinics*, *28*(1), 107–170.

Mostafavi, S.-A., & Hosseini, S. (2015). Foods and dietary supplements in the prevention and treatment of neurodegenerative diseases in older adults. In R. Watson (Ed.), *Foods and dietary supplements in the prevention and treatment of disease in older adults* (pp. 63–67). Amsterdam, The Netherlands: Elsevier.

Mozaffarian, D., Rosenberg, I., & Uauy, R. (2018). History of modern nutrition science—Implications for current research, dietary guidelines, and food policy. *BMJ*, *361*, k2392.

Okdahl, T., & Brock, C. (2021). Molecular aspects in the potential of vitamins and supplements for treating diabetic neuropathy. *Current Diabetes Reports*, *21*(9) (no pagination).

Olasehinde, T., Oyeleye, S. I., Ogunsuyi, O. B., & Ogunruku, O. (2017). Functional foods in the management of neurodegenerative diseases. In G. Oboh (Ed.), *Functional foods: Unlocking the medicine in foods* (pp. 72–81). Memphis, TN: Graceland Prints.

Rajendram, R., Martin, C. R., & Preedy, V. R. (2016). Recommended resources on metabolism and physiology of bariatric surgery. In R. Rajendram, C. R. Martin, & V. R. Preedy (Eds.), *Metabolism and physiology of bariatric surgery* Elsevier, UK.

Rajendram, R., Patel, V. B., & Preedy, V. R. (2014). Web based resources and suggested reading. In R. Rajendram, V. B. Patel, & V. R. Preedy (Eds.), *Glutamine in health and disease* Springer, USA.

Rajendram, R., Patel, V. B., & Preedy, V. R. (2015). Web based resources and suggested reading. In R. Rajendram, V. B. Patel, & V. R. Preedy (Eds.), *Branched chain amino acids in health and disease* Springer, USA.

Rajendram, R., Patel, V. B., & Preedy, V. R. (2018a). Resources in famine, starvation, and nutrient deprivation. In V. B. Patel, & V. R. Preedy (Eds.), *Famine, starvation, and nutrient deprivation* Springer, USA.

Rajendram, R., Patel, V. B., & Preedy, V. R. (2018b). Resources in diet, nutrition and epigenetics. In V. B. Patel, & V. R. Preedy (Eds.), *Nutrition and epigenetics* Springer, USA.

Rajendram, R., Patel, V. B., & Preedy, V. R. (2020). Recommended research and resources on the neuroscience of Parkinson's disease. In V. B. Patel, & V. R. Preedy (Eds.), *Neuroscience of Parkinson's disease* Elsevier, UK.

Rajendram, R., Patel, V. B., & Preedy, V. R. (2021a). Recommended resources on the neuroscience of development. In C. R. Martin, & V. R. Preedy (Eds.), *Neuroscience of development disease* Elsevier, UK.

Rajendram, R., Patel, V. B., & Preedy, V. R. (2021b). Recommended resources on the neuroscience of aging. In V. B. Patel, & V. R. Preedy (Eds.), *Neuroscience of aging* Elsevier, UK.

Rajendram, R., Rajendram, R., Patel, V. B., & Preedy, V. R. (2013a). Diet quality: What more is there to know? In V. R. Preedy, & V. B. Patel (Eds.), *Diet quality: An evidence-based approach.* Germany: Springer.

Rajendram, R., Rajendram, R., Patel, V. B., & Preedy, V. R. (2013b). Food fortification: What more is there to know? In V. R. Preedy, R. Srirajaskanthan, & V. B. Patel (Eds.), *Handbook of food fortification and health: From concepts to public health.* Germany: Springer.

Rutjes, A. W., Denton, D. A., Di Nisio, M., et al. (2018). Vitamin and mineral supplementation for maintaining cognitive function in cognitively healthy people in mid and late life. *Cochrane Database of Systematic Reviews, 12*(12), CD011906.

Index

Note: Page numbers followed by *f* indicate figures, and *t* indicate tables.

D

E

F

O

P

W

Z